1989 Recommended Dietary Allowances (RDA)

Age (yr)	Energy (kcal)	Protein (g)	Vitamin A (μg RE)	Vitamin E (mg α-TE)	Vitamin K (μg)	Vitamin C (mg)	Iron (mg)	Zinc (mg)	Iodine (μg)	Selenium (μg)
Infants										
0.0–0.5	650	13	375	3	5	30	6	5	40	10
0.5–1.0	850	14	375	4	10	35	10	5	50	15
Children										
1–3	1300	16	400	6	15	40	10	10	70	20
4–6	1800	24	500	7	20	45	10	10	90	20
7–10	2000	28	700	7	30	45	10	10	120	30
Males										
11–14	2500	45	1000	10	45	50	12	15	150	40
15–18	3000	59	1000	10	65	60	12	15	150	50
19–24	2900	58	1000	10	70	60	10	15	150	70
25–50	2900	63	1000	10	80	60	10	15	150	70
51+	2300	63	1000	10	80	60	10	15	150	70
Females										
11–14	2200	46	800	8	45	50	15	12	150	45
15–18	2200	44	800	8	55	60	15	12	150	50
19–24	2200	46	800	8	60	60	15	12	150	55
25–50	2200	50	800	8	65	60	15	12	150	55
51+	1900	50	800	8	65	60	10	12	150	55
Pregnancy	+300	60	800	10	65	70	30	15	175	65
Lactation										
1st 6 mo.	+500	65	1300	12	65	95	15	19	200	75
2nd 6 mo.	+500	62	1200	11	65	90	15	16	200	75

In addition to the values that serve as goals for nutrient intakes (presented in the adjacent tables), the Dietary Reference Intakes include a set of values called Tolerable Upper Intake Levels—the maximum amount of a nutrient that appears safe for most healthy people to consume on a regular basis.

Tolerable Upper Intake Levels for Selected Nutrients (per day)

Age (yr)	Vitamin D (μg)[a]	Niacin (mg)	Vitamin B_6 (mg)	Folate (μg)	Choline (mg)	Calcium (mg)	Phosphorus (mg)	Magnesium (mg)	Fluoride (mg)
Infants									
0.0–0.5	25	—[b]	—[b]	—[b]	—[b]	—[b]	—[b]	—[b]	0.7
0.5–1.0	25	—[b]	—[b]	—[b]	—[b]	—[b]	—[b]	—[b]	0.9
Children									
1–3	50	10	30	300	1000	2500	3000	65	1.3
4–8	50	15	40	400	1000	2500	3000	110	2.2
9–13	50	20	60	600	2000	2500	4000	350	10.0
14–18	50	30	80	800	3000	2500	4000	350	10.0
Adults									
19–70	50	35	100	1000	3500	2500	4000	350	10.0
>70	50	35	100	1000	3500	2500	3000	350	10.0
Pregnancy	50	35	100	1000	3500	2500	3500	350	10.0
Lactation	50	35	100	1000	3500	2500	4000	350	10.0

NOTE: An Upper Level was not established for thiamin, riboflavin, vitamin B_{12}, panthothenic acid, and biotin because of a lack of data, not because these nutrients are safe to consume at any level of intake; all nutrients can have adverse effects when intakes are excessive.

[a] To convert μg to IU, multiply by 40. For example, 50 μg × 40 = 2000 IU.

[b] Upper Levels were not established for many nutrients in the infant category because of a lack of data.

Source: Adapted from the first two of the *Dietary Reference Intakes* series, National Academy Press. Copyright 1997 and 1998, by the National Academy of Sciences. Courtesy of the National Academy Press, Washington, D.C.

Nutrition and Diet Therapy

PRINCIPLES AND PRACTICE

Fifth Edition

Corinne Balog Cataldo
Linda Kelly DeBruyne
Eleanor Noss Whitney

Wadsworth Publishing Company
I(T)P® An International Thomson Publishing Company

Belmont, CA • Albany, NY • Boston • Cincinnati • Johannesburg • London • Madrid • Melbourne • Mexico City • New York • Pacific Grove, CA • Scottsdale, AZ • Singapore • Tokyo • Toronto

Nutrition Publisher: Peter Marshall
Developmental Editor: Laura Graham
Editorial Assistant: Tangelique Williams
Marketing Manager: Becky Tollerson
Project Editor: Sandra Craig
Print Buyer: Barbara Britton
Permissions Editor: Robert Kauser
Production: The Book Company
Text and Cover Design: Carolyn Deacy
Copyediting: Patricia Lewis
Photo Research: Stephen Forsling
Illustrations: Impact Publications, McMahon Medical Art
Cover Image: Jennie Oppenheimer
Index: Barbara Farabaugh
Composition: Parkwood Composition Service, Inc.
Printer: Von Hoffmann Press, Inc.

Printed in the United States of America

1 2 3 4 5 6 7 8 9 10

For more information, contact Wadsworth Publishing Company, 10 Davis Drive, Belmont, CA 94002, or electronically at http://www.wadsworth.com

International Thomson Publishing Europe
Berkshire House
168-173 High Holborn
London, WC1V 7AA, United Kingdom

Nelson ITP, Australia
102 Dodds Street
South Melbourne
Victoria 3205 Australia

Nelson Canada
1120 Birchmount Road
Scarborough, Ontario
Canada M1K 5G4

International Thomson Publishing
Southern Africa
Building 18, Constantia Square
138 Sixteenth Road, P.O. Box 2459
Halfway House, 1685 South Africa

International Thomson Editores
Seneca, 53
Colonia Polanco
11560 México D.F. México

International Thomson Publishing Asia
60 Albert Street
#15-01 Albert Complex
Singapore 189969

International Thomson Publishing
Japan
Hirakawa-cho Kyowa Building, 3F
2-2-1 Hirakawa-cho, Chiyoda-ku
Tokyo 102 Japan

Library of Congress Cataloging-in-Publication Data
Cataldo, Corinne Balog.
Nutrition & diet therapy : principles and practice / Corinne Balog Cataldo, Linda Kelly DeBruyne, Eleanor Noss Whitney. -- 5th ed.
p. cm.
Includes bibliographical references and index.
ISBN 0-534-54594-7
1. Diet therapy. 2. Nutrition. I. DeBruyne, Linda K. II. Whitney, Eleanor Noss
RM216.C36 1998 98-42478
615.8'54--dc21

See photo and art credits on page I–26 following the index, which is a continuation of this copyright page.

About the Authors

Corinne Balog Cataldo, M.M.Sc., R.D., C.N.S.D., received her B.S. in community health nutrition from Georgia State University in 1976 and her M.M.Sc. in clinical dietetics from Emory University in 1979. She has worked in private practice in Atlanta, as a clinical dietitian and metabolic support nutritionist at Georgia Baptist Medical Center in Atlanta, as a faculty member and dietetic internship coordinator at Emory University, and as a nutritionist for the Infant Formula Council. She has made numerous presentations, and in addition to this book, she has written a manual on tube feedings and the books *Understanding Normal and Clinical Nutrition, Nutrition for Health and Health Care,* and *Understanding Clinical Nutrition.* She maintains professional memberships in the American Dietetic Association, the American Society for Parenteral and Enteral Nutrition, the American Diabetes Association, and Dietitians in Nutrition Support.

Linda Kelly DeBruyne, M.S., R.D., received her B.S. in 1980 and her M.S. in 1982 in nutrition and food science at Florida State University. She is a founding member of Nutrition and Health Associates, an information resource center in Tallahassee, Florida, where her specialty areas are life cycle nutrition and fitness. Her other publications include the textbooks *Nutrition for Health and Health Care, Life Span Nutrition: Conception through Life, Health: Making Life Choices,* and *The Fitness Triad* and a multimedia CD-ROM called *Nutrition Interactive.* As a consultant for a group of Tallahassee pediatricians, she teaches infant nutrition classes to parents. She maintains a professional membership in the American Dietetic Association.

Eleanor Noss Whitney, Ph.D., received her B.A. in biology from Radcliffe College in 1960 and her Ph.D. in biology from Washington University in St. Louis in 1970. She is a registered dietitian and a member of the American Dietetic Association. Formerly on the faculty at Florida State University, she now devotes full time to research, writing, and consulting. Her earlier publications include articles in *Science, Genetics,* and other journals. Her textbooks include *Understanding Nutrition, Understanding Normal and Clinical Nutrition, Nutrition Concepts and Controversies, Life Span Nutrition: Conception through Life, Nutrition for Health and Health Care,* and *Essential Life Choices* for college students and *Making Life Choices* for high school students. Her most intense interests currently include energy conservation, solar energy uses, alternatively fueled vehicles, and ecosystem restoration.

Dedication

To my mother and father, Katherine Nagy Balog and George Balog, whose examples have inspired me to cherish the independence earned through knowledge, dedication, and hard work.
Corkie

To my mother, Dorothy Garfunkel Kelly, whose enthusiasm for life is both contagious and inspiring, from the luckiest daughter in the world.
Linda

To the memory of my parents, Edith Tyler Noss and Henry H. B. Noss, who supported me with love, discipline, and pride.
Ellie

Contents in Brief

PART ONE

Basic Nutrition

PART TWO

Nutrition Throughout Life

PART THREE

Nutrition in Health Care

PART FOUR

Diet Therapy

Contents

PART ONE

Basic Nutrition

How To Features Appear on the Following Pages

Case Studies Appear in Chapters of Parts Two, Three, and Four

Preface

In today's health care environment, all members of a health care team, regardless of their chosen profession, increasingly share responsibility for meeting clients' nutrition needs. This fifth edition of *Nutrition and Diet Therapy* seeks to provide both the basic facts and a wealth of practical information readers need to assume their responsibilities for nutrition care effectively and confidently.

Every chapter has been substantially revised to reflect the many changes that have occurred in the field of nutrition since the last edition. Nutrition is a young and rapidly expanding science, however, with many questions remaining to be answered and new "facts" surfacing virtually every day. One of the missions of this text is to show readers how to ascertain and cautiously interpret new information—a skill vital to developing sound clinical judgment.

Within each chapter, definitions and notes in the margins clarify nutrition information, remind readers of previously defined terms, provide cross-references, and assist in reviewing chapter material. New to this edition are Internet site references (see margin) that may provide additional information on the subject matter. "How to" skill boxes help readers work through mathematical calculations or give practical suggestions for applying nutrition concepts. The Self Check at the end of each chapter provides questions to help review chapter information. Self Studies in the early chapters and Case Studies in the later chapters guide readers in applying information to everyday situations. Later chapters also include Clinical Applications and Nutrition Assessment Checklists. Clinical Applications help the reader practice mathematical calculations, synthesize information from previous chapters, or understand the impact of nutrition care on health care professionals or their clients. New to this edition, Nutrition Assessment Checklists remind readers of the most relevant assessment findings specific to a group of disorders and provide cautions regarding the interpretation of assessment parameters.

Two logos identify Internet sites. The first,

WWW

shows that you can access the site without typing "www" before the address.

The second,

WWW.

shows that you must type "www." before the address.

At the end of each chapter, a Nutrition in Practice section explores current topics, advanced subjects, or specialty areas such as nutrition and dental health. This edition includes new Nutrition in Practice sections exploring the Dietary Reference Intakes, childhood obesity and early development of chronic diseases, food and foodservice in the hospital, alternative therapies, client-centered diabetes care, antioxidants and free radicals, and dialysis and nutrition.

Part One of this book introduces the basics of nutrition and shows how nutrition supports health. Part Two examines how nutrient needs change throughout the life cycle. In Part Three, readers learn to apply what they have learned about nutrition to the health care setting. This part describes the nutrition care process, with an emphasis on nutrition assessment. Part Four examines diet therapy and its role in supporting health and treating diseases and symptoms. Significant revisions have been made in the presentation of GI tract disorders and in the discussion of severe stress. The chapter on diabetes incorporates the new classification and diagnosis criteria for diabetes, as well as the revised exchange lists. The discussion of cardiovascular diseases includes the Dietary Approaches to Stop Hypertension

(DASH) diet, and the chapter on liver diseases emphasizes a highly individualized approach to treating cirrhosis.

The appendixes support the book with a wealth of information on nutrient contents of foods and enteral formulas, Canadian nutrient recommendations and choices, U.S. nutrient intake recommendations and the revised exchange system, additional information regarding nutrition assessments, aids to calculations, and nutrition resources, including many Internet addresses.

We hope that as you discover the many fascinating aspects of nutrition science, you will enthusiastically apply the concepts in both your professional and your personal life. For nutrition updates and other resources, we invite you to visit our Website: http://www.wadsworth.com/nutrition

Acknowledgments

Among the most difficult words to write are those that express the depth of our gratitude to the many dedicated people whose efforts have made this book possible. A special note of appreciation to Sharon Rolfes for her numerous contributions to the chapters and Nutrition in Practice sections as well as the Dietary Reference Intakes table on the inside front cover and the Body Mass Index table on the inside back cover. We thank Pam Schmidt for her valuable work on Chapter 11. Special thanks to Sally Mayo for her expert word processing. Thanks also to Tangelique Williams for her help in securing permissions, Pat Lewis for copyediting our manuscript, and Elizabeth Hand and Bob Geltz for creating an extremely accurate and practical food composition appendix. We are indebted to our editorial team—Peter Marshall, and Laura Graham, and to our production team—Sandra Craig and Dusty Friedman, for seeing this project through from start to finish. We would also like to acknowledge Becky Tollerson for her marketing efforts. To the many others involved in designing, indexing, typesetting, dummying, and marketing, we offer our thanks. We are especially grateful to our associates, family, and friends for their continued encouragement and support and to our reviewers who consistently offer excellent suggestions for improving the text.

Reviewers of Nutrition and Diet Therapy

Anita Dacpano Daus
Charles Stewart Mott Community College

Kelley DeVane-Hart
Troy State University

Susan Ettinger
New York Institute of Technology

Denise Garner
York College of Pennsylvania

Catherine Graziano
Salve Regina University

Corliss Hendrix
Milwaukee Technical College

Debra Hollingsworth
McNeese State University

Ann Hunter
Wichita State University

Louise M. Johnston
East Stroudsburg University

Patricia Lillis
Medical College of Georgia

Myrtle McCulloch
Georgetown University

Nelda Malm
Seminole Community College

Marcia Nahikian-Nelms
Southeast Missouri State University

Jillann Neely
Onondaga Community College

Linda Picklesimer
Greenville Technical College

Mary Ellen Posthauer
University of Southern Indiana

Carol Powell
Indiana Wesleyan University

Padmini Shankar
Georgia Southern University

Nancy B. Shaw
Southeastern Louisiana University

Janet K. Tanner
Midlands Technical College

Janice Young
Los Angeles City College

Jody Yates-Taylor
Portland Community College

1 *Perspectives on Nutrition*

Ethnic meals and family gatherings nourish the spirit as well as the body.

You choose to eat a meal about 1000 times a year. If you live 65 years or longer, you will have eaten more than 65,000 meals in your lifetime. Each day's intakes of food and nutrients may benefit or harm health only slightly, but repeated over years and decades, they accumulate to have major effects on body organs and their functions. The science of nutrition is the study of the nutrients in food and the body's handling of these nutrients.

Sound nutrition throughout life does not ensure good health and long life, but close attention to nutrition each day can certainly help to weight things in their favor. Nevertheless, most people choose foods for reasons other than their nutrient contributions. Food choices are personal and not always sensible, and to a great extent, people resist change when it comes to their food choices. Before undertaking diet planning, the planner must understand the dynamics of food choices, because people will alter their eating habits only if their preferences are honored.

science of nutrition: the study of nutrients in foods and of their ingestion, digestion, absorption, transport, metabolism, interaction, storage, and excretion. A broader definition includes the study of the environment and of human behavior as it relates to these processes.

Food Choices

Why do people choose particular foods? Several reasons come to mind, but nutrition would be only one of them and would not necessarily be at the top of many people's lists. Even people who claim to choose foods primarily for the sake of nutrition will admit that other factors also influence their food choices. They may *know how* to prepare nutritious meals, but that doesn't mean they actually *eat* such meals all the time. Instead, people's food choices tend to be powerfully influenced by a variety of personal factors.

Preference Why do people like certain foods? One reason, of course, is their preference for certain tastes. Some tastes are widely liked, such as the sweetness of sugar and the zest of salt. Research suggests that genetics may influence people's taste preferences, a finding that may eventually have implications for clinical nutrition.[1] For example, the role that genetics may play in food selection is gaining importance in cancer research.

Associations People also like foods with happy associations—foods eaten in the midst of warm family gatherings on traditional holidays or given to them as children by someone who loved them. By the same token, people can attach intense and unalterable dislikes to foods that they ate when they were sick or that were forced on them when they weren't hungry.

Habit Sometimes habit dictates people's food choices. People eat a sandwich for lunch or drink orange juice at breakfast simply because they have always done so.

Ethnic Heritage and Tradition Every country, and every region of a country, has its own typical foods and ways of combining them into meals. The foodways of North America reflect the many different cultural and ethnic backgrounds of its inhabitants. Many foods with ethnic origins are familiar features on North American menus: tacos, egg rolls, lasagna, and gyros, to name a few. Still others, such as pizza, spaghetti, and croissants, are integral in the "American diet." North American regional cuisines like Cajun and TexMex blend the traditions of several cultures. Table 1.1 (pp. 4–6) presents profiles of selected ethnic diets, together with comments on their nutritional merits and drawbacks.

ethnic diets: foodways and cuisines typical of national origins, races, cultural heritages, or geographic locations.

Values People's values, environmental ethics, religious beliefs, and political views also influence their food choices. By choosing to eat some foods or avoid others, people make statements about themselves that reflect their values. For example, people may select only foods that come in containers that can be reused or recycled. Some people choose only brands of canned tuna fish that state "dolphin safe" on the label, meaning that dolphins were not killed when the tuna were netted. (Nutrition in Practice 11 elaborates further on ways the environment relates to food choices.) Religion also influences many people's food choices. Jewish law sets forth an extensive set of dietary rules. Many Christians forgo meat during Lent, the period prior to Easter. Other faiths prohibit some dietary practices and promote others. Diet planners can promote people's use of sound nutrition practices only if they respect and honor such values.

When people choose foods, they are selecting nutrients.

Social Pressure Social pressure is another powerful influence on people's food choices. Such pressure operates in all circles and across all cultural lines. It is often considered rude to refuse food or drink being shared by a group or offered by a host. Often you become accepted as a member of a social gathering only when you "break bread" with the other members.

Emotional Comfort Some people eat in response to emotional stimuli—for example, to relieve boredom or depression or to calm anxiety. A lonely person may choose to eat rather than to call a friend and risk rejection. A person who has returned home from an exciting evening out may unwind with a late-night snack. Eating in response to emotions can easily lead to overeating and obesity, but may be appropriate at times. For example, sharing food at times of bereavement serves both the giver's need to provide comfort and the receiver's need to be cared for and to interact with others.

Availability, Convenience, and Economy The influence of these factors on people's food selections is clear. You cannot eat foods if they are not available, if you do not have the time or skill to prepare them, or if you cannot afford them. Convenience plays a major role in many people's food selections today, but along with convenience, consumers want great taste and quality—foods that are fast, delicious, and nutritious.[2] The demand for foods that are ready to eat or can be easily prepared in a microwave oven demonstrates this influence.

Image Sometimes people select foods that they associate with ideals of body image. The fashion and movie industries, not the medical community, have defined what people believe to be the ideal body—sometimes an excessively thin body for women or an excessively muscular body for men. Both men and women seek "beautiful bodies," and in doing so, they select or avoid foods that they believe will improve or impair their physical appearance. Such intentions are rational when based on sound nutrition and fitness knowledge, but when based on faddism or carried to extremes, they undermine good health.

Medical Conditions Sometimes medical conditions and the medications used to treat the conditions limit the foods a person can select. The second half of this text discusses diets modified for different medical conditions.

Nutrition No matter what dietary tradition people follow, they have no guarantee of diet adequacy. Since sound nutrition promotes health and longevity, it is in every person's own best interest to know how to obtain optimal nourishment. Consumers today cite nutrition as a primary concern in making food choices, yet the foods they choose do not always reflect this concern; they need to know more about the nutrients in food.

People choose the foods they eat for many different personal reasons. Whatever those reasons may be, over the years, people's food choices can either benefit or harm health.

Table 1.1 Characteristics of Selected Ethnic Diets

Staple Foods	Strengths of the Diet	Weaknesses of the Diet
Hispanic Americans from Cuba, Haiti, Puerto Rico		
Include: ■ Steamed white rice; wheat breads. ■ Starchy vegetables (beans, cassavas, yuccas); plantains; green peppers; tomatoes; garlic. ■ Dried, salted fish; chicken; pork. ■ Lard; olive oil; sugar; jams and jellies; sweet pastries; sugared fruit juices; coffee. Exclude: ■ Green, leafy vegetables. ■ Milk as a beverage for adults. ■ Fish other than dried and salted.	Provides adequate protein, many other nutrients, and fiber	May provide too much fat, especially animal fat; may lack calcium
Hispanic Americans from Mexico, Central America		
Include: ■ Steamed rice, corn products such as tortillas. ■ Many varieties of beans; chili peppers; tomatoes; mangoes; prickly pear fruit; potatoes. ■ Meat and sausages; fish; poultry; eggs. ■ Lard; chocolate and coffee drinks; cakes; pastries. Exclude: ■ Green, leafy vegetables; yellow vegetables. ■ Milk as a beverage for adults.	Most nutrients can be obtained	Is high in kcalories and fat, especially saturated fat, and high in sugar
Black Americans from West Indies, Central or South America and Recent African Immigrants		
Include: ■ Millet, corn, wheat, rice, or barley. ■ Starchy roots such as cassavas, yams; plantains; bananas; coconuts; peanuts; fresh fruits; hot peppers; tomatoes; onions; okra. ■ Palm oil; fruit wine; tea; coffee; honey; molasses. Exclude: ■ Milk and milk products (meat and fish limited use).	Is low in fat and salt; is high in fiber	Is low in calcium, iron, and vitamin B_{12}; is potentially low in protein, depending on availability of foods
Southern Black Americans from West Africa (Many Generations in United States)		
Include: ■ Rice; hominy grits; biscuits; cornmeal and cornbread. ■ Legumes; potatoes; onions; tomatoes; hot peppers; green, leafy vegetables; okra; sweet potatoes; squashes; corn; cabbage; melons; peaches. ■ Smoked pork; meats and poultry; fish; thick stews. ■ Pecans; butter, shortening, and lard; sugar; bread puddings, pies, and sweets. Exclude: ■ Milk and milk products. ■ Yeast breads.	Provides ample nutrients of meat	Provides excess protein; is high in kcalories; provides excess fat, especially saturated fat; is high in salt; is low in calcium

Staple Foods	Strengths of the Diet	Weaknesses of the Diet
Chinese Americans from China (Diets Sometimes Vary With Region)		
Include: ■ Rice and rice gruel; wheat noodles; soy bean noodles. ■ Corn; vegetables from the cabbage family; squashes; cucumbers; eggplant; leafy vegetables; various shoots (bamboo, mung, and soy); sweet potatoes; radishes; onions; peas and pods; mushrooms; roots; local vegetables; pickled vegetables; sea vegetables; plums; peaches; tangerines; kumquats; other citrus fruits; litchis; longans; mangoes; papayas; pomegranates. ■ Soybean products (tofu and soy milk); meat; fish with bones; poultry; seafood. ■ Soup or tea as beverage; soy sauce; sugar. Exclude: ■ Milk and most milk products.	Is low in fat; is high in fiber and many nutrients	Depending on availability of protein-rich foods, protein and iron may be low; high in salt
Japanese Americans from Japan		
Include: ■ Rice. ■ Vegetables (including pickled and sea); fruits; salads. ■ Soy (miso, tofu, bean paste); fish with bones; seafood. ■ Sugars as seasoning; ginseng; soy sauce. Exclude: ■ Milk and milk products.	Provides abundant nutrients with little fat	Is high in salt
Korean Americans from South Korea		
Include: ■ Rice; noodles. ■ Leafy vegetables; kimchi (hot pickled cabbage); sea vegetables; hot peppers; seasonal fruits; mushrooms. ■ Small fish with bones; grilled beef; chicken; squid, octopus, and lobster; mussels; eggs. ■ Lard and vegetable fat for frying; sesame oil; nuts and seeds; ginger; sugar as seasoning. Exclude: ■ Milk and milk products.	Provides adequate protein	Is high in fat; monotonous in winter (kimchi is served at each meal, to the exclusion of other vegetables); without the traditional small fish with bones, calcium can be lacking.
Vietnamese Americans from Vietnam		
Include: ■ Rice, rice noodles; french bread and croissants. ■ Hot peppers; curries of asparagus and potatoes; salads; tropical fruits and vegetables; lemons and limes. ■ Small portions of poultry; eggs; fish pâtés; nuoc nam (a strong, fermented fish sauce). ■ Sweets, candies, sweetened drinks; coffee; tea; butter. Exclude: ■ Milk and milk products.	Provides adequate protein and vitamins	Can be low in iron or calcium

continued

Table 1.1 *(continued)*

Staple Foods	Strengths of the Diet	Weaknesses of the Diet
Native Americans		
Include: ■ *Southeast:* corn; cornmeal; coontie (flour from a palmlike plant); fried breads; pumpkins; squashes; papayas; alligator, snake, wild hog, duck, fish, and shellfish. ■ *Northeast:* blueberries; cranberries; beans; corn; pumpkins; fish; lobster; wild game; maple syrup. ■ *Midwest:* bison; beans; corn; melons; squashes; tomatoes. ■ *Southwest:* corn (many varieties); beans; squash; pumpkins; chili peppers; melons; pinenuts; cactus. ■ *Northwest:* salmon; caviar; other fish; otter; seal; elk; whale; bear; other game; wild fruits, nuts, and greens. Exclude: ■ Milk and milk products.	Varies with region; may provide adequate protein and fiber	Varies with region; may be low in calcium
Italian Americans from Italy		
Include: ■ *Northern Italy:* egg-based, ribbon-shaped pastas; cheese; cream; butter; meat; eggs. ■ *Southern Italy:* wheat pastas; artichokes, eggplants, peppers, and tomatoes; beans; olive oil.	Provides adequate protein and most nutrients	Varies with region; can be high in fat

The Nutrients

nutrients: substances obtained from food and used in the body to provide energy and structural materials and to promote growth, maintenance, and repair.

essential nutrients: nutrients a person must obtain from food because the body cannot make them for itself in sufficient quantity to meet physiological needs.

The six classes of nutrients are water, carbohydrate, fat, protein, vitamins, and minerals.

organic: carbon containing. The four organic nutrients are carbohydrate, fat, protein, and vitamins.

Metabolism, the set of processes by which nutrients are rearranged into body structures or broken down to yield energy, is described in Chapter 6.

Almost any food you eat is composed of dozens or even hundreds of different kinds of materials—atoms and molecules too small to see with even the most powerful microscope. A food such as spinach, for example, is composed mostly of water (95 percent), and most of its solid materials are the compounds carbohydrate, fat (properly called *lipid*), and protein. If you could remove these materials, you would find a tiny residue of minerals, vitamins, and other compounds. The nutrients in spinach are of six types: water, carbohydrate, fat, protein, vitamins, and minerals. Some of the other materials in spinach, such as the pigments and some minerals, are not nutrients. The body can make some nutrients for itself, at least in limited quantities, but it cannot make them all, and it makes some in insufficient quantities to meet its needs. Therefore the body must obtain many nutrients from foods. The nutrients that foods must supply are called *essential nutrients.*

Three Energy Nutrients Four of the six classes of nutrients (carbohydrate, fat, protein, and vitamins) contain carbon, which is found in all living things. They are therefore organic (meaning, literally, "alive"). During metabolism, three of these four (carbohydrate, fat, and protein) provide energy the body can use. These energy-yielding nutrients continually replenish the energy you spend daily. Without them you would soon die. Carbohydrate and fat are the major energy-yielding nutrients. Protein becomes a major fuel only when other fuels are unavailable.

energy-yielding nutrients: the fuel nutrients, those that yield energy the body can use.

Vitamins, Minerals, and Water Vitamins are organic but do not provide energy to the body. They facilitate the release of energy from the other three organic nutrients. In contrast, minerals and water are inorganic nutrients. Minerals yield no energy in the human body, but like vitamins, they help to regulate the release of energy. As for water, it is the medium in which all of the body's processes take place.

kCalories: Measures of Energy The amount of energy the energy-yielding nutrients release can be measured in calories (or more properly, kilocalories, or kcalories). kCalories are not constituents of foods; they are a measure of the energy in foods. It is as incorrect to refer to the kcalories in a food as it is to refer to the inches in a person. It is correct to refer to the *energy* in a food and to the *height* of a person. The energy in a food depends on how much carbohydrate, fat, and protein the food contains.

Carbohydrate yields 4 kcalories of energy from each gram, and so does protein. Fat yields 9 kcalories per gram. If you know how many grams of each nutrient a food contains, you can derive the number of kcalories potentially available from the food. Simply multiply the carbohydrate grams times 4, the protein grams times 4, and the fat grams times 9, and add the results together (see the box on the next page).

kcalories: units by which energy is measured. Most people speak of these units simply as calories, but on paper the word *calorie* is prefaced by a k for kilocalorie. We use kcalories and kcal throughout this book.

Food energy can also be measured in kilojoules (kJ). One kcalorie equals 4.2 kJ. The kilojoule is the international unit of energy.

Energy Nutrients in Foods Practically all foods contain mixtures of the energy-yielding nutrients, although foods are sometimes classified by their predominant nutrient. Thus to speak of meat as "a protein" or of bread as "a carbohydrate" is inaccurate. Each is rich in a particular nutrient, but a protein-rich food such as beef contains a lot of fat along with protein, and a carbohydrate-rich food such as cornbread also contains fat (corn oil) and protein. Only a few foods are exceptions to this rule, the common ones being sugar (which is pure carbohydrate) and oil (which is pure fat).

Energy Storage in the Body If your body doesn't use the energy-yielding nutrients to fuel metabolic and physical activities, it rearranges them into storage compounds, primarily body fat, and puts them away for later use. Thus, if you take in more energy than you expend, whether from carbohydrate, fat, or protein, the result is usually a gain of body fat. Too much meat (a protein-rich food) is just as fattening as too many potatoes (a carbohydrate-rich food), depending on the total energy in each.

Alcohol, Not a Nutrient When taken in excess of energy need, alcohol, too, is converted to body fat and stored. The body derives energy from alcohol at the rate of 7 kcalories per gram. Alcohol is not a nutrient, however, because it contributes nothing to the body's growth, maintenance, or repair. When alcohol contributes too much energy to a person's diet, the harm it does extends far beyond the problems of excess body fat. Nutrition in Practice 8 discusses alcohol's effects on nutrition.

Foods provide energy and nutrients—substances the body uses to promote growth, maintenance, and repair. Carbohydrate, fat, and protein yield energy the body uses or stores. Vitamins, minerals, and water facilitate many body processes.

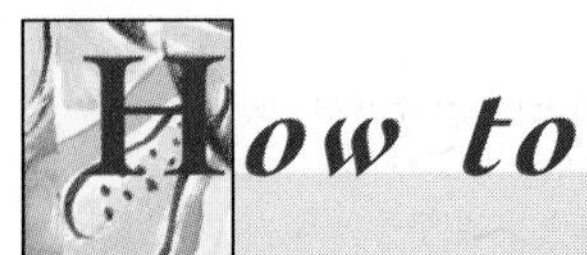

How to

Calculate the Energy in a Food

The following example shows how to calculate the energy available from 1 slice of bread that has 1 teaspoon of butter on it. From food tables such as Appendix A in this book, you can determine that this food contains 15 grams carbohydrate, 2 grams protein, and 5 grams fat:

$$\begin{aligned} 15 \text{ g carbohydrate} \times 4 \text{ kcal/g} &= 60 \text{ kcal.} \\ 2 \text{ g protein} \times 4 \text{ kcal/g} &= 8 \text{ kcal.} \\ 5 \text{ g fat} \times 9 \text{ kcal/g} &= 45 \text{ kcal.} \\ \text{Total} &= 113 \text{ kcal.} \end{aligned}$$

From this information, you can calculate the percentage of kcalories each of the energy nutrients contributes to the total. To determine the percentage of kcalories from fat, for example, divide the 45 fat kcalories by the total 113 kcalories: $45 \div 113 = 0.398$ (round to 0.40). Then multiply by 100 to get the percentage: $0.40 \times 100 = 40\%$.

Health recommendations that urge people to limit fat intake to 30 percent of kcalories refer to the day's total energy intake, not to individual foods. Still, if the proportion of fat in each food choice throughout a day exceeds 30 percent of kcalories, then the day's total surely will, too. Knowing that this snack provides 40 percent of its kcalories from fat alerts a person to the need to make lower-fat selections at other times that day.

Nutrition Standards and Guidelines

Appendix C presents nutrient recommendations developed by two international groups, the Food and Agricultural Organization and the World Health Organization (the FAO/WHO recommendations). The Canadian recommendations are in Appendix B.

To nourish yourself optimally, you need to eat foods that provide adequate amounts of essential nutrients and energy. It is impossible, however, to know exactly how much of each nutrient an individual needs, because each person's body has its own unique set of requirements. Nutrient intake recommendations offer a rough guideline for dietary adequacy.

Recommended Nutrient Intakes

Recommended Dietary Allowances (RDA): the average daily amounts of nutrients considered adequate to meet the known nutrient needs of practically all healthy people; a goal for dietary intake by individuals.

Defining the amounts of energy, nutrients, and other dietary components that best support health is a huge task. For more than 50 years, nutrition experts produced a set of energy and nutrient standards known as the Recommended Dietary Allowances (RDA) for people in the United States. The Canadian equivalent of the RDA is the RNI (Recommended Nutrient Intakes). Over the years, these standards were revised periodically as new evidence became available, but each revision maintained the original goal of protecting against nutrient deficiencies. Given the abundance of research now linking diet and health, that goal has been broadened to include supporting optimal activities within the body and preventing chronic diseases as well. Previous editions also focused narrowly on nutrients known to be essential. With recent research revealing the health benefits of other dietary components such as fiber, their recommended intakes need to be addressed as well.

Dietary Reference Intakes (DRI): a set of values for the dietary nutrient intakes of healthy people in the United States and Canada. These values are used for planning and assessing diets.

Dietary Reference Intakes A major revision of dietary recommendations is currently under way. The revised recommendations are called Dietary Reference Intakes (DRI) and reflect the collaborative efforts of experts from both the United States and Canada.[3] The first in a series of reports was published in 1997 and presents both the framework for developing the DRI and the revised recommendations for the five nutrients that play key roles in bone health—calcium, phosphorus, magnesium, vitamin D, and fluoride. A second report came out in 1998 and features revised recommendations for the eight B vitamins and choline. The recommendations for these 14 nutrients appear on the inside front cover under the heading 1997–1998 Dietary Reference Intakes. For the remaining nutrients and energy, the 1989 RDA will continue to serve health professionals until DRI can be established. The values for these nutrients also appear on the inside front cover under the heading 1989 Recommended Dietary Allowances. The nutrition in practice that follows this chapter offers many more details about the DRI.

For more information on Dietary Reference Intakes, go to this site on the World Wide Web—the portion of the Internet that contains sites hosted by government agencies, academic institutions, and other groups and individuals. Throughout this text you will see addresses for other sites that provide additional information on the topic at hand.

WWW. nas.edu
Dietary Reference Intakes

requirement: the lowest continuing intake of a nutrient that will maintain a specified criterion of adequacy.

Nutrient Recommendations in Perspective Nutrient recommendations have been much misunderstood. Nutrient recommendations are just that—recommendations, not laws. The following facts will help put the recommendations in perspective:

- Nutrient recommendations are not requirements; they are allowances, and except for energy, they are generous. Even so, they do not necessarily cover every individual for every nutrient. Figure 1.1 presents an accurate view of how a person's nutrient needs fall within a range, with marginal and danger zones both below and above it.
- Nutrient recommendations are estimates of the needs of healthy persons only. Medical problems alter nutrient needs as later chapters describe.
- Nutrient recommendations take into account the differences among individuals and define a range within which most healthy persons' intakes of nutrients probably should fall. Individuals whose needs are higher than the average are included within this range.
- Separate recommendations are made for different sets of people: men, women, pregnant women, lactating women, children, and infants.

In contrast to the vitamin and mineral recommendations, the energy recommendation is not generous, but rather is set at the mean of the population's estimated requirement (see Figure 1.2 on p. 10). Although not enough energy may cause undernutrition, too much energy is as bad for health as too little because excess energy leads to obesity. In contrast, in the case of vitamins and minerals, small amounts above the daily requirement do no harm, whereas amounts below the requirement lead to health problems. When people's intakes are consistently deficient (less than the requirement), their nutrient stores decline, and over time this decline leads to poor health and deficiency symptoms. Therefore, to ensure

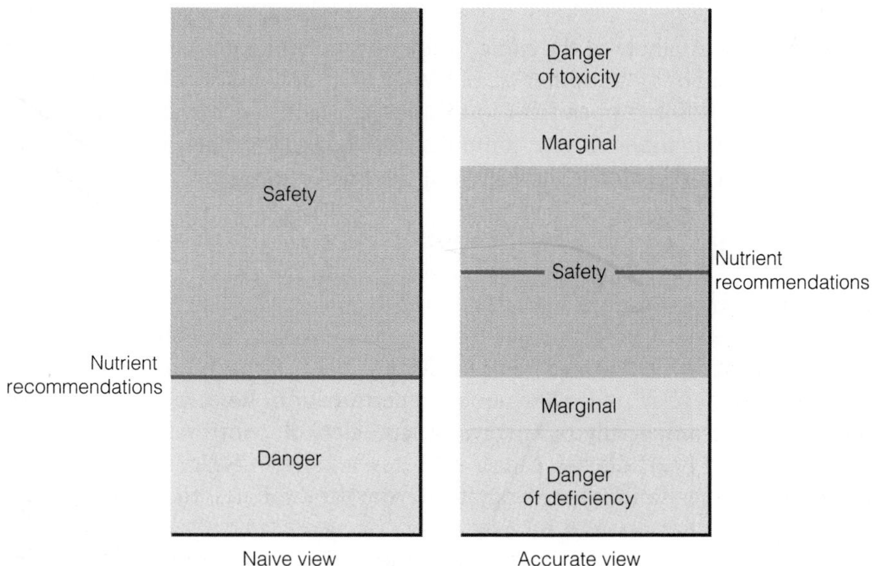

Figure 1.1
Naive versus Accurate View of Nutrient Recommendations
The recommendation for a given nutrient represents a point within a range of appropriate and reasonable intakes that lies between toxicity and deficiency. The recommendation is high enough to provide reserves in times of short-term dietary inadequacies, but not so high as to approach toxicity. Nutrient intakes above or below this range might be equally harmful.

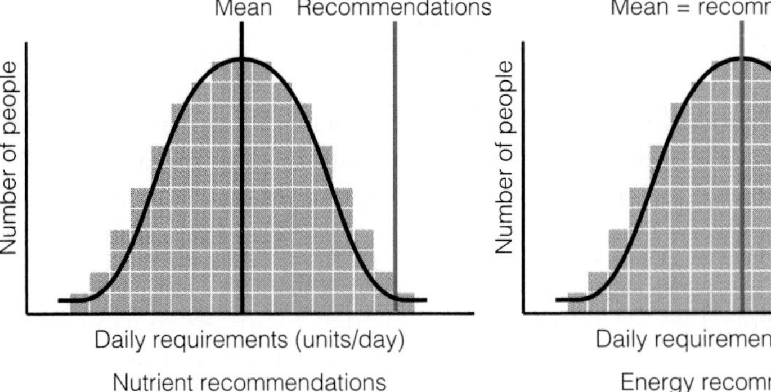

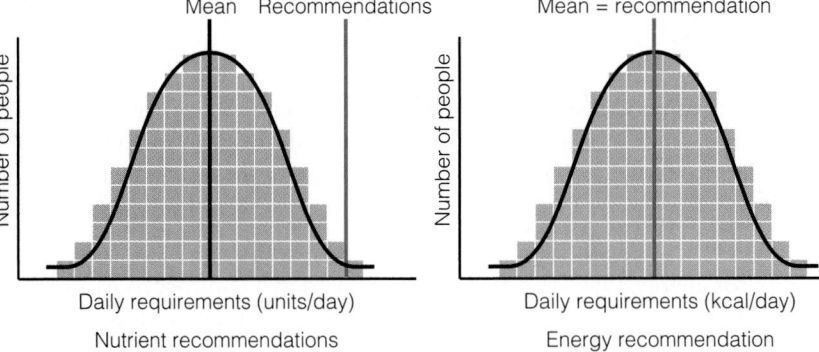

Figure 1.2
Nutrient Recommendations and the Energy Recommendation Compared

that the vitamin and mineral recommendations meet the needs of as many people as possible, the recommendations are set near the top end of the range of the population's estimated requirements (see Figure 1.2).

Uses of Nutrient Recommendations With the understanding that nutrient recommendations are approximate, flexible, and generous, they can be used as a yardstick to measure the adequacy of diets in entire populations, such as those of a certain school or community. In addition, diets of individuals are often assessed by comparing them to nutrient recommendations because, as mentioned earlier, it is impractical, if not impossible, to determine any one person's precise needs. Such comparisons provide a rough estimate of adequacy and are usually sufficient.

Tools Used with Nutrient Recommendations: Diet History To evaluate a diet, your own or someone else's, you need not only the recommendations but also an accurate listing of the foods eaten over a period of time. An assessor uses a food intake form to record the types of foods, with portion sizes, an individual eats (Chapter 15 provides the details). Then the assessor calculates the nutrients and energy the person consumed and compares the totals with recommendations such as the RDA.

Diet Analysis Analyses of the energy and nutrients someone has consumed can be done by hand or by computer (see Chapter 15). For hand calculation, the table of food composition, Appendix A in this book, lists the energy and nutrients in several thousand foods. Many computer diet analysis programs also contain these data and can perform and print out the calculations.

Dietary Guidelines

Cholesterol is a member of the lipid family and receives attention in Chapter 3.

overnutrition: overconsumption of food energy or nutrients sufficient to cause disease or increased susceptibility to disease; a form of malnutrition.

undernutrition: underconsumption of food energy or nutrients severe enough to cause disease or increased susceptibility to disease; a form of malnutrition.

Nutrient recommendations were developed to ensure *adequate* nutrient intakes, but they do little to protect people from *excess* intakes of fat, cholesterol, sugar, salt, and alcohol. Government authorities are now as much concerned about overnutrition as they once were about undernutrition. Research confirms that dietary excesses, especially of energy, fat, and alcohol, contribute to many diseases, including heart disease, cancer, diabetes, and liver disease.[4] Several sets of dietary recommendations have originated from the awareness that overnutrition contributes to disease.

Dietary Recommendations Many sets of dietary recommendations have been published in the United States with titles such as *dietary goals* or *dietary guide-*

Table 1.2 *Diet and Health* Recommendations

Nutrient and Energy Recommendations	Suggested Food Choices
Reduce total *fat* intake to 30% or less of kcalories. Reduce saturated fatty acid intake to less than 10% of kcalories and intake of cholesterol to less than 300 mg daily.	Reduce fat and cholesterol intake by substituting fish, poultry without skin, lean meats, and low-fat or nonfat dairy products for fatty meats and whole-milk products; by choosing more vegetables, fruits, cereals, and legumes; and by limiting fats, oils, egg yolks, and fried and other fatty foods.
Increase intake of starches and other *complex carbohydrates.*	Every day eat 5 or more servings of a combination of vegetables and fruits, especially green and yellow vegetables and citrus fruits, and six or more daily servings of a combination of breads, cereals, and legumes. The committee does not recommend increasing intake of added sugars, because their consumption is strongly associated with dental caries.
Maintain *protein* intake at moderate levels.	Meet at least the RDA for protein; do not exceed twice the RDA.
Balance food intake and physical activity to maintain appropriate *body weight.*	
For those who drink *alcoholic beverages,* the committee recommends limiting consumption to the equivalent of less than 1 oz pure alcohol in a single day. Pregnant women should avoid alcoholic beverages.	The committee does not recommend alcohol consumption. One ounce of pure alcohol is the equivalent of two cans of beer, two small glasses of wine, or two average cocktails.
Limit total daily intake of *salt* (sodium chloride) to 6 g or less.	Limit the use of salt in cooking, and avoid adding it to food at the table. Salty, highly processed salty, salt-preserved, and salt-pickled foods should be consumed sparingly.
Maintain adequate *calcium* intake.	
Avoid taking dietary *supplements* in excess of the RDA in any one day.	
Maintain an optimal intake of *fluoride,* particularly during the years of primary and secondary tooth formation and growth.	

Source: Adapted from the National Academy of Sciences report *Diet and Health: Implications for Reducing Chronic Disease Risk* (Washington, D.C.: National Academy Press, 1989).

lines. These sets of recommendations differ only a little from one another, and all emphasize prevention of overnutrition and disease. Table 1.2 summarizes the dietary recommendations from a National Academy of Sciences report titled *Diet and Health: Implications for Reducing Chronic Disease Risk,* and Table 1.3 on p. 12 lists the 1995 *Dietary Guidelines for Americans.*

WWW **usda.gov/fcs/cnpp/guide.htm**
1995 *Dietary Guidelines for Americans*

Weight Maintenance and Physical Activity Dietary recommendations state not only what people should eat but also what they should avoid. In addition, the recommendations refer to weight maintenance and physical activity. Some people's diets are close to these recommendations, but the typical North American diet, which emphasizes meat, falls far short of the recommendations. So does the sedentary lifestyle of many North Americans who fail to meet even minimal guidelines for physical activity.[5]

Diet-Planning Principles

How can people juggle the foods available to them to create a diet that supplies all the needed nutrients in the appropriate amounts for good health? The principle is simple enough: select a variety of foods that supply the nutrients your body needs. In practice, how do you do this? It helps to keep in mind six basic diet-planning principles, which are listed in the margin in alphabetical order for ease in remembering them.

This cola and bunch of grapes illustrate nutrient density. Each provides about 150 kcalories, but the grapes offer a trace of protein, some vitamins, minerals, and fiber along with the energy; the cola beverage offers only "empty" kcalories. Grapes, or any fruit for that matter, are more nutrient dense than cola beverages.

Diet-planning principles:
- *Adequacy.*
- *Balance.*
- *kCalorie control.*
- *Nutrient Density.*
- *Moderation.*
- *Variety.*

dietary adequacy: the characteristic of a diet that provides all the essential nutrients, fiber, and energy necessary to maintain health and body weight.

dietary balance: providing foods of a number of types in balance with one another such that foods rich in one nutrient do not crowd out foods that are rich in another nutrient.

kcalorie control: management of food energy intake.

nutrient density: a measure of the nutrients a food provides relative to the energy it provides. The more nutrients and the fewer kcalories, the higher the nutrient density.

moderation: providing enough, but not too much of a dietary constituent.

Table 1.3 ***Dietary Guidelines for Americans***

- Eat a variety of foods.
- Balance the food you eat with physical activity—maintain or improve your weight.
- Choose a diet with plenty of grain products, vegetables, and fruits.
- Choose a diet low in fat, saturated fat, and cholesterol.
- Choose a diet moderate in sugars.
- Choose a diet moderate in salt and sodium.
- If you drink alcoholic beverages, do so in moderation.

Source: Nutrition and Your Health: Dietary Guidelines for Americans, U.S. Department of Agriculture, Home and Garden Bulletin Number 232 (Washington, D.C.: Government Printing Office 1995).

Adequacy The ideal of dietary adequacy has already been touched on. A diet that provides enough energy and enough of every nutrient to meet the needs of healthy people is adequate.

Balance As for dietary balance, the essential minerals calcium and iron illustrate its importance. Meats, fish, poultry, and legumes are rich in iron but poor sources of calcium. Similarly, milk and milk products are rich in calcium but poor sources of iron. In fact, milk (except breast milk) and milk products are so low in iron that overuse of these foods can actually lead to iron-deficiency anemia by displacing iron-rich foods from the diet. Yet milk is the single most nutritious food for infants and can be an important calcium source for people of all ages.

Use some meat and meat alternates for iron; use some milk and milk products for calcium. Save some space, too, for other foods, for a diet consisting only of milk and meat would be far from adequate. To obtain the other needed nutrients, you have to eat vegetables, fruits, grains, and other foods. In short, balance in the diet helps to ensure adequacy.

kCalorie Control While it takes thought and skill to design an adequate, balanced diet, incorporating kcalorie control presents an added challenge—to eat an adequate, balanced diet without overeating. (Energy balance and weight control are discussed in Chapter 9.)

Nutrient Density To this end, the concept of nutrient density helps: seek out foods that deliver the highest nutrient values per kcalorie. For example, among foods containing calcium, a 1½-ounce portion of cheddar cheese and 1 cup of nonfat milk both provide about the same amount of calcium; but the cheese contributes twice as much food energy as the nonfat milk (see Appendix A). The nonfat milk, then, is more calcium dense: it offers the same amount of calcium for half the kcalories.

Moderate Sugar and Fat Intakes Moderation in diet planning refers to control of food constituents that are undesirable in excess—fat, sugar, and salt. Since fat and sugar add kcalories to foods, moderation in their use automatically helps with kcalorie control, too. Consider the example of the cheese and the milk once again. The nonfat milk not only offers the same amount of calcium at a lower food energy cost, but it also contains far less fat than the cheese. The cola and grapes shown in the margin are another example: the cola contains only sugar and water, whereas the grapes deliver many nutrients. People who exercise moderation in their sugar intakes will choose grapes over cola and will meet their daily nutrient needs using fewer total kcalories. It all ties together: the most nutrient-dense foods are the best choices for the sake of both kcalorie control and moderation.

Moderate Salt Intakes Controlling salt intake might seem to require another approach: salt provides no kcalories, so added salt does not reduce the nutrient density of foods. Still, choosing nutrient-dense, low-fat, low-sugar foods does help keep your salt intake down. Why? The reason is that manufacturers usually add all three—salt, sugar, and fat—to enhance the taste of foods. Thus pursuing nutrient density helps keep your salt intake moderate. This recommendation may seem to be suggesting that you seek out mostly *unprocessed* foods. That is true in the sense that additions of sugar, salt, and fat are forms of processing. But other forms of food processing are beneficial, so wait until you've learned more about processing from Chapter 11 before attempting to generalize.

Variety helps to ensure an adequate and balanced diet.

Variety Your diet can have all of the characteristics just described but still lack variety if you eat the same foods day after day. Vary your choices within each class of foods from day to day, for at least two reasons. First, different foods in the same group contain different arrays of nutrients. Among the fruits, for example, strawberries are especially rich in vitamin C, while peaches are rich in vitamin A. Thus variety helps ensure adequacy. Second, no food is guaranteed entirely free of substances that in excess could harm you. (Contamination of foods is discussed in Chapter 11.) By choosing strawberries today, peaches tomorrow, and watermelon the day after, you ensure that your total diet will have diluted concentrations of any contaminants that may be present in the foods available to you.

variety (dietary): using different foods to obtain the same nutrients on different occasions.

These diet-planning principles offer a framework of excellence to strive for in planning diets; they are dietary ideals. To plan a diet that achieves these ideals, the planner needs knowledge and skill. It helps to know which kinds of foods offer which nutrients and how many daily servings of the different foods are recommended.

Food Group Plans

About 40 vitamins and minerals are needed altogether. Each nutrient has its own unique pattern of distribution in foods. Although working all the nutrients into the meals you eat might seem quite a challenge, people all over the world obtain fine nutrition from an astonishing variety of diets.

food group plans: diet-planning tools that sort foods of similar origin and nutrient content into groups and then specify that people should eat certain numbers of servings from each group.

Food Groups For many people, food group plans, such as the Four Food Group Plan of past decades or the Daily Food Guide used today, serve as the basis for planning adequate, balanced diets. Food group plans sort foods into clusters that make the same key nutrient contributions. For example, cheese and yogurt fall within the milk group because each is notable for calcium, riboflavin, and protein, as well as small amounts of other nutrients. Legumes are included in the meat, fish, and poultry group because these foods are notable for their protein, phosphorus, vitamin B_6, iron, and zinc.

Daily Food Guide: a food group plan for ensuring dietary adequacy that offers five categories of foods to choose from. (It treats vegetables and fruits as two separate groups.)

Miscellaneous Foods Food group plans exclude some foods from the main clusters because they make no notable nutrient contributions to people's intakes. These foods, called "miscellaneous" foods, are allowed in diet planning only as extras in small quantities once basic nutrient needs have been met by nutritious foods. Examples of miscellaneous foods are soft drinks, jams, salad dressings, and butter.

The Daily Food Guide Figure 1.3 presents the Daily Food Guide and includes the most notable nutrients within each food group, the number of servings recommended, the serving sizes, and the foods within each group categorized by nutrient density. Figure 1.3 also includes the Food Guide Pyramid, which presents the Daily Food Guide in pictorial form. The pyramid shape, with grains at the base, is designed to convey the idea that people should eat more grain foods

WWW **nalusda.gov/fnic/Fpyr/pyramid.html**
Food Guide Pyramid

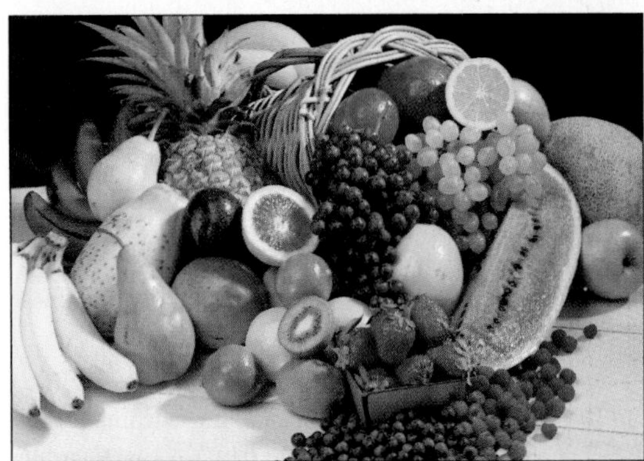

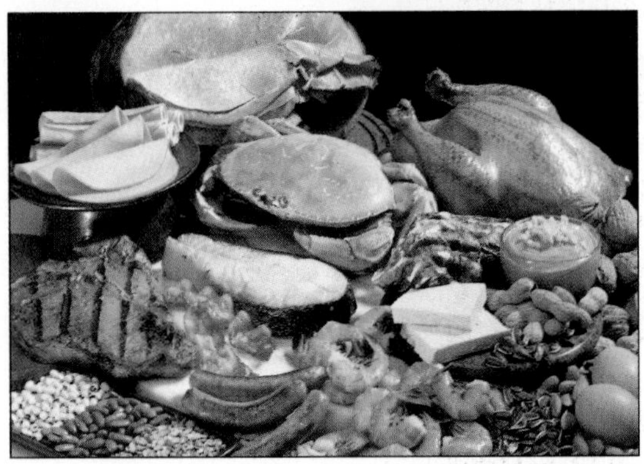

Figure 1.3

The Daily Food Guide and The Food Guide Pyramid

Breads, Cereals, and Other Grain Products

These foods are notable for their contributions of complex carbohydrates, riboflavin, thiamin, niacin, iron, protein, magnesium, and fiber.

6 to 11 servings per day.

Serving = 1 slice bread; ½ c cooked cereal, rice, or pasta; 1 oz ready-to-eat cereal; ½ bun, bagel, or English muffin; 1 small roll, biscuit, or muffin; 3 to 4 small or 2 large crackers.

- ■ Whole grains (wheat, oats, barley, millet, rye, bulgur), enriched breads, rolls, tortillas, cereals, bagels, rice, pastas (macaroni, spaghetti), air-popped corn.
- ▪ Pancakes, muffins, cornbread, crackers, low-fat cookies, biscuits, presweetened cereals.
- ▫ Croissants, pastries, biscuits, fried rice, granola.

Vegetables

These foods are notable for their contributions of vitamin A, vitamin C, folate, potassium, magnesium, and fiber, and for their lack of fat and cholesterol.

3 to 5 servings per day (use dark green, leafy vegetables and legumes several times a week).

Serving = ½ c cooked or raw vegetables; 1 c leafy raw vegetables; ½ c cooked legumes; ¾ c vegetable juice.

- ■ Bean sprouts, broccoli, brussels sprouts, cabbage, carrots, cauliflower, corn, cucumbers, eggplant, green beans, green peas, leafy greens (spinach, mustard, and collard greens), legumes, lettuce, mushrooms, potatoes, tomatoes, water chestnuts, winter squash.
- ▪ Candied sweet potatoes.
- ▫ French fries, fried okra, olives, tempura vegetables.

Fruits

These foods are notable for their contributions of vitamin A, vitamin C, potassium, and fiber, and for their lack of sodium, fat, and cholesterol.

2 to 4 servings per day.

Serving = typical portion (such as 1 medium apple, banana, or orange, ½ grapefruit, 1 melon wedge); ¾ c juice; ½ c berries; ½ c diced, cooked, or canned fruit; ¼ c dried fruit.

- ■ Apricots, cantaloupe, grapefruit, oranges, orange juice, peaches, strawberries, apples, bananas, pears.
- ▪ Canned or frozen fruit in syrup.
- ▫ Avocados, dried fruit, coconut.

Meat, Poultry, Fish, and Alternates

These foods are notable for their contributions of protein, phosphorus, vitamin B_6, vitamin B_{12}, zinc, magnesium, iron, niacin, and thiamin.

2 to 3 servings per day.

Servings = 2 to 3 oz lean, cooked meat, poultry, or fish (total 5 to 7 oz per day); count 1 egg, ½ c cooked legumes, or 2 tbs peanut butter as 1 oz meat (or about ⅓ serving).

- ■ Poultry, fish, lean meat (beef, lamb, pork, veal), legumes, egg whites.
- ▪ Fat-trimmed beef, lamb, pork; refried beans; egg yolks, tofu, tempeh.
- ▫ Hot dogs, luncheon meats, peanut butter, nuts, sausage, bacon, fried fish or poultry, duck.

Key:

- ■ Foods generally highest in nutrient density (good first choice).
- ▪ Foods moderate in nutrient density (reasonable second choice).
- ▫ Foods lowest in nutrition density (limit selections).

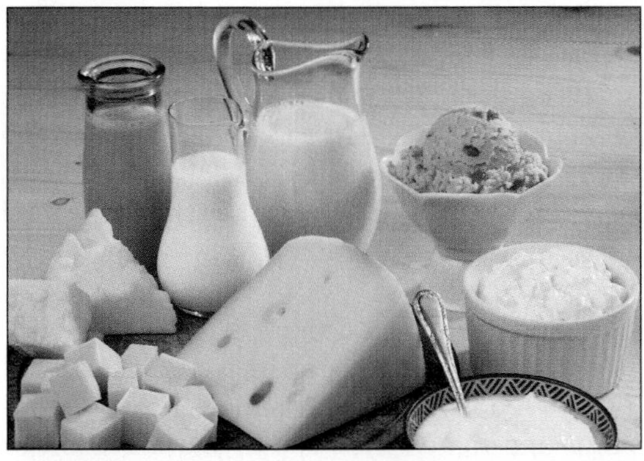

Milk, Cheese, and Yogurt

These foods are notable for their contributions of calcium, riboflavin, protein, vitamin B_{12}, and, when fortified, vitamin D and vitamin A.

2 servings per day.

3 servings per day for teenagers and young adults, pregnant/lactating women, women past menopause.

4 servings per day for pregnant/lactating teenagers.

Serving = 1 c milk or yogurt; 2 oz processed cheese food; 1½ oz cheese.

- Nonfat and 1% low-fat milk (and nonfat products such as buttermilk, cottage cheese, cheese, yogurt); fortified soy milk.
- 2% reduced-fat milk (and reduced-fat products such as yogurt, cheese, cottage cheese); sherbet; ice milk.
- Whole milk (and whole-milk products such as cheese, yogurt, cottage cheese); custard; milk shakes; pudding; ice cream.

Miscellaneous Group

These foods are notable for their contributions of sugar, fat, alcohol, and food energy. These foods are not in the pattern because they provide few nutrients. Note that some of the following items could be placed in more than one group or in a combination group. For example, doughnuts are high in both sugar and fat.

- Miscellaneous foods, not high in kcalories, include spices, herbs, coffee, tea, and diet soft drinks.
- Foods high in fat include margarine, salad dressings, oils, mayonnaise, cream, cream cheese, sour cream, butter, gravy, sauces, potato chips, doughnuts, and chocolate bars.
- Foods high in sugar include cake, pie, cookies, doughnuts, sweet rolls, candy, soft drinks, fruit drinks, jelly, syrup, gelatin, desserts, sugar, and honey.
- Alcoholic beverages include wine, beer, and liquor.

Note: Serve children at least the lower number of servings from each group, but in smaller amounts (for example, ¼ to ⅓ cup rice). Children should receive the equivalent of 2 cups of milk each day, but again in smaller quantities per serving (for example: 4 half-cup portions). Pregnant women may require additional servings of fruits, vegetables, meats, and breads to meet their higher needs for energy, vitamins, and minerals.

Food Guide Pyramid

A Guide to Daily Food Choices

The breadth of the base shows that grains (breads, cereals, rice, and pasta) deserve most emphasis in the diet. The tip is smallest: use fats, oils, and sweets sparingly.

KEY

● Fat (naturally occurring and added)

▼ Sugars (added)

These symbols show fats, oils and added sugars in foods.

Fats, Oils & Sweets
Use sparingly

Milk, Yogurt & Cheese Group
2–3 servings

Meat, Poultry, Fish, Dry Beans, Eggs & Nuts Group
2–3 servings

Vegetable Group
3–5 servings

Fruit Group
2–4 servings

Bread, Cereal, Rice & Pasta Group
6–11 servings

than anything else—grains form the foundation of a sound diet. Fruits and vegetables share the next level of the pyramid, indicating that they, too, are prominent components of a sound diet. Meats and milk are dense in protein and other nutrients, but can also contribute fat and kcalories. Their location just below the pyramid's apex implies that servings must be limited. Fats, oils, and sweets occupy only a tiny triangle at the pyramid's apex, indicating they should be used sparingly.

Different Energy Intakes Possible The Daily Food Guide offers a strong foundation for a healthy diet, but it fails to specify food energy intakes. Large fat and energy differences exist within a single food group—for example, between nonfat milk and ice cream, green beans and french fries, apples and avocados, or bread and croissants. Yet, according to the Daily Food Guide, any of these choices would be acceptable. Figure 1.3 provides a nutrient density key for foods *within each group*. People can therefore control their food energy intakes while using the Daily Food Guide. The foods color-coded with green boxes are the highest in nutrient density; they contribute the most nutrients for the fewest kcalories. Those with yellow boxes are moderate in nutrient density; and those with red boxes are lowest in nutrient density. People who choose the smallest recommended number of servings, choose them all from the green-coded foods, and strictly limit foods from the miscellaneous group will obtain the needed nutrients at an energy intake of about 1600 kcalories. People who choose more servings or less nutrient-dense foods can easily boost their energy intakes. Depending on their energy needs, such choices might add to their body weight.

vegetarian diets: a general term used to describe diets that exclude meat, poultry, fish, or other animal-derived foods. The two main categories are *lacto-ovo vegetarian* and *vegan* diets.

lacto-ovo vegetarian diets: diets that include milk, cheese, and eggs (animal products) but exclude meat, fish, and poultry (animal flesh).

vegan diets: diets that exclude all animal products and include only plant foods; also known as **strict vegetarian diets.**

WWW. **vrg.org**
The Vegetarian Resource Group

Flexible Food Choices The beauty of the Daily Food Guide is that it is simple and easy to learn. It may appear rigid, but it actually offers great flexibility once its intent is understood. For example, cheese can be substituted for milk because both supply the key nutrients for the milk group (protein, calcium, and riboflavin) in about the same amounts. To limit kcalories, choose nonfat milk; to add kcalories, choose cheese. Legumes are alternative choices for meats, so vegetarians can adapt the pattern by using legumes in place of meat selections (see Table 1.4).

Planning Vegetarian Diets Making vegetarian diets adequate and balanced is not difficult, but does require skill. People adopt vegetarian diets for a variety of religious, ethical, social, or economic reasons, but to the diet planner all vegetarian diets present similar challenges. A few nutrients require careful attention if a vegetarian diet is to offer nutrition and health benefits to adults.[6] Nutrition in Practice 4 shows how to make vegetarian diets nutritious.

Nutrition Surveys

Researchers use nutrition surveys to determine which foods people are eating, to assess people's nutritional health, and to measure people's knowledge, attitude, and behaviors about nutrition and how these relate to health.[7] The resulting wealth of information can be used for a variety of purposes. For example, Congress uses this information to establish public policy on nutrition education, assess food assistance programs, and regulate the food supply. Scientists use the information to establish research priorities. One of the first nutrition surveys, taken before World War II, suggested that up to a third of the U.S. population might be eating poorly. Programs to correct malnutrition have been evolving ever since.

malnutrition: any condition caused by deficient or excess energy or nutrient intake or by an imbalance of nutrients.

food consumption survey: a survey that measures the amounts and kinds of food people consume (using diet histories), estimates the nutrient intakes, and compares them with a standard such as the RDA.

Food Consumption Surveys Food consumption surveys determine the kinds and amounts of foods people eat. Researchers calculate the energy and nutrients in the foods and compare the amounts consumed with a standard such as the RDA. An example of this type of survey is the Nationwide Food Consumption

Table 1.4 Daily Food Guide for Vegetarians		
Food Group	**Suggested Daily Servings**	**Serving Sizes**
Breads, cereals, and other grain products	6 or more	1 slice bread ½ bun, bagel, or English muffin ½ c cooked cereal, rice, or pasta 1 oz dry cereal
Vegetables	4 or more[a]	½ c cooked or 1 c raw
Fruits	3 or more	1 piece fresh fruit ¾ c fruit juice ½ c canned or cooked fruit
Legumes and other meat alternates	2 to 3	½ c cooked beans 4 oz tofu or tempeh 8 oz soy milk 2 tbs nuts or seeds (these tend to be high in fat, so use sparingly) 1 egg or 2 egg whites
Milk and milk products	2 to 3 servings[b]	1 c low-fat or nonfat milk 1 c low-fat or nonfat yogurt 1½ oz low-fat cheese

[a]Include 1 cup of dark green vegetables daily to help meet iron requirements.

[b]If not using milk or milk products: use soy milk fortified with calcium and vitamin B_{12}.

Source: Adapted with permission from Position of The American Dietetic Association: Vegetarian diets, *Journal of the American Dietetic Association* 93 (1993): 1318.

Survey (NFCS). For the third NFCS (1994–1996), researchers gathered information from 15,000 people using food intake records for two nonconsecutive days.

Nutrition Status Surveys Nutrition status surveys examine the people themselves, using nutrition assessment methods. The National Health and Nutrition Examination Survey (NHANES) is an example of a nutrition status survey. The third NHANES (1988–1996) gathered information from between 40,000 and 70,000 people using food intake data, anthropometric measurements, physical examinations, and laboratory tests. The data provide information on several nutrition-related conditions, such as growth retardation, heart disease, and nutrient deficiencies. Both the NFCS and the NHANES oversample high-risk groups (low-income families, infants and children, and the elderly) in order to glean an accurate estimate of their health and nutrition status.

nutrition status survey: a survey that evaluates people's nutrition status using food intake data, anthropometric measures, physical examinations, and laboratory tests.

Coordinated Effort Until 1990, findings from the nation's many nutrition surveys were almost impossible to compare and synthesize into a single cohesive report. Then the National Nutrition Monitoring and Related Research Act was enacted to coordinate the many nutrition-related activities that had been under way within 22 different federal agencies. The law mandated that the U.S. Department of Agriculture (USDA) and the Department of Health and Human Services (DHHS) establish and implement a Ten-Year Comprehensive Plan for nutrition monitoring and related research.[8] All major reports that examine the contribution of diet and nutrition status to the health of the people of the United States depend on information collected and coordinated by this national program. These data provided the basis for the mid-1990s report on Healthy People 2000.[9] This report shows we are not meeting many of our health goals; in fact, we are not even heading in the right direction for some goals, such as reducing the prevalence of overweight in the United States.[10] Table 1.5 on the next page lists the 21 nutrition-related priorities of *Healthy People 2000.*

WWW.

usda.gov
U.S. Department of Agriculture

os.dhhs.gov
Department of Health and Human Services

The *Healthy People 2000* report sets national objectives in health promotion and disease prevention for the year 2000.

WWW

cdc.gov/nchswww/about/otheract/hp2000/hp2000.htm
Healthy People 2000

Table 1.5 *Healthy People 2000* Nutrition Objectives

Health-related objectives

- Reduce coronary heart disease deaths to no more than 100 per 100,000 people.
- Reverse the rise in cancer deaths to achieve a rate of no more than 130 per 100,000 people.
- Reduce overweight to a prevalence of no more than 20% among people aged 20 years and older and maintain prevalence at no more than 15% among adolescents aged 12 through 19 years.
- Reduce growth retardation among low-income children aged 5 years and younger to less than 10%.

Nutrient intake objectives

- Reduce dietary fat intake to an average of 30% of energy or less and average saturated fat intake to less than 10% of energy among people aged 2 years and older.
- Increase complex carbohydrate and fiber-containing foods in the diets of adults to five or more daily servings for vegetables (including legumes) and fruits and to six or more daily servings for grain products.
- Increase to at least 50% the proportion of overweight people aged 12 years and older who have adopted sound dietary practices combined with regular physical activity to attain an appropriate body weight.
- Increase calcium intake, so that at least 50% of youth aged 12 through 24 years and 50% of pregnant and lactating women consume three or more servings of calcium-rich foods daily and at least 50% of people aged 25 years and older consume two or more servings of calcium-rich foods daily.
- Decrease salt and sodium intake so at least 65% of home meal preparers prepare foods without adding salt, at least 80% of people avoid using salt at the table, and at least 40% of adults regularly purchase foods modified or lower in sodium.
- Reduce iron deficiency to less than 3% among children aged 1 to 4 and women of childbearing age.
- Increase to at least 75% the proportion of mothers who breastfeed their babies in the early weeks and to at least 50% the proportion who continue breastfeeding until their babies are 5 to 6 months old.
- Increase to at least 75% the proportion of parents and caregivers who use feeding practices that prevent nursing bottle tooth decay.
- Increase to at least 85% the proportion of people aged 18 and older who use food labels to make nutritious food selections.

Services and information objectives

- Achieve useful and informative nutrition labeling for virtually all processed foods and at least 40% of fresh meats, poultry, fish, fruits, vegetables, baked goods, and ready-to-eat carry-away foods.
- Increase to at least 5000 brand items the number of processed food products that are reduced in fat and saturated fat.
- Increase to at least 90% the proportion of restaurants and institutional foodservice operations that offer identifiable low-fat, low-kcalorie food choices, consistent with the *Dietary Guidelines for Americans.*
- Increase to at least 90% the proportion of school lunch and breakfast services and increase to at least 50% the proportion of child care foodservices with menus that are consistent with the nutrition principles in the *Dietary Guidelines for Americans.*
- Increase to at least 80% the receipt of home foodservices by people aged 65 and older who have difficulty in preparing their own meals or are otherwise in need of home-delivered meals.
- Increase to at least 75% the proportion of the nation's schools that provide nutrition education from preschool through grade 12, preferably as part of quality school health education.
- Increase to at least 50% the proportion of worksites with 50 or more employees that offer nutrition education and/or weight management programs for employees.
- Increase to at least 75% the proportion of primary care providers who provide nutrition assessment and counseling and/or referral to qualified nutritionists or dietitians.

Source: Healthy People 2000: National Health Promotion and Disease Prevention Objectives (Washington, D.C.: U.S. Department of Health and Human Services, 1990).

Nutrient intake recommendations such as the RDA are recommendations for daily intakes of selected nutrients considered adequate to meet the needs of healthy people. The RDA are used to assess the adequacy of diets. Dietary guidelines and recommendations such as the *Diet and Health* recommendations and the *Dietary Guidelines for Americans* emphasize disease prevention. Diet-planning principles and food group plans help people choose from different groups of similar foods to provide adequacy, balance, and variety.

Nutrition Labeling

Today consumers know more about the links between diet and disease than they did in the past, and they are demanding still more information on disease prevention. Surveys show that many people rely on food labels to tell them which substances to avoid for health reasons and to provide them with information about nutrition.[11] The Nutrition Labeling and Education Act of 1990 made extensive changes in the requirements for label information to better accommodate consumers' nutrition and health needs. Most food labels must conform to all these requirements.[12] Exceptions include plain coffee, tea, spices, and other foods contributing few nutrients; foods produced by small businesses; and those prepared and sold in the same establishment as long as the foods do not make nutrient or health claims.[13]*

WWW **vm.cfsan.fda.gov/~lrd/newlabel.html**
Background on Food Labels

Information on Food Labels According to law, every food label must state the following:

- The common or usual name of the product.
- The name and address of the manufacturer, packer, or distributor.
- The net contents in terms of weight, measure, or count.

Then, the label must list in ordinary language the following:

- The ingredients, in descending order of predominance by weight.

The Ingredient List All labels must state at least this much. Even if they say no more, you can learn a lot about the nutritional value of a product from its ingredient list. Consider the following products:

- An orange-flavored drink that contains "water, sugar, citric acid, orange flavor . . . ."
- A juice that contains "water, tomato concentrate, concentrated juices of carrots, celery . . . ."

Knowing that the first ingredient named predominates by weight, you can tell that the first product is mostly sugar and that the second is mostly vegetable juice.

Information on Fresh Foods Voluntary labeling or point-of-purchase nutrition information is encouraged for the 20 most frequently eaten fresh fruits, vegetables, and seafoods.[14] Grocers often post nutrition-information placards or pamphlets near displays of these foods.

For raw meat and poultry products, producers who choose to provide nutrition information on labels must adhere to the requirements of the mandatory program. In other words, if a package of raw chicken breast carries a food label, it must provide the information required on all other food labels.

Nutrition Facts The "Nutrition Facts" panel on a food label must provide:

- Standard serving size expressed in both common household and metric measures to allow comparison with other foods in the same category.
- Number of servings or portions per container.
- Total food energy (kcalories) per serving.

*For example, as of 1997, restaurants making "heart healthy" claims for menu items are required to provide nutrient information for consumers.

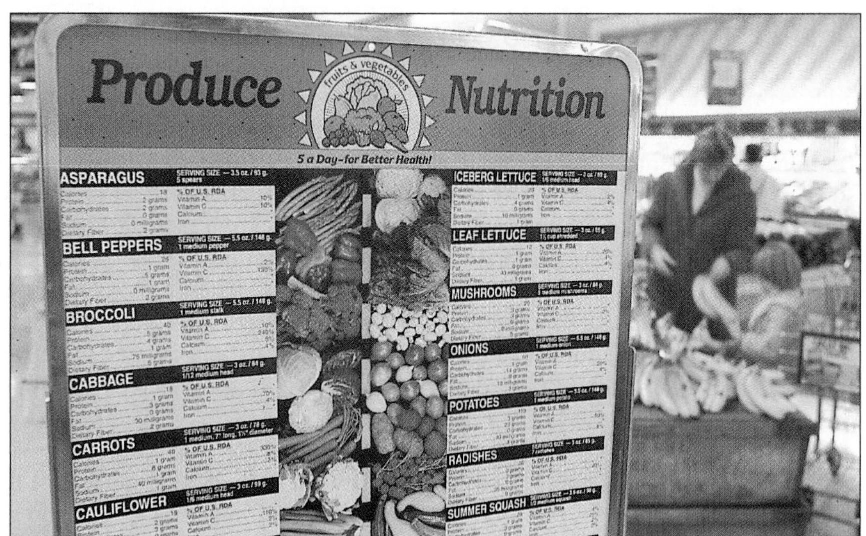

Posters in the produce department present nutrition information for nonpackaged items such as raw fruits and vegetables.

- Fat (grams) per serving with a breakdown showing saturated fat (grams) and cholesterol (milligrams).
- Sodium (milligrams) per serving.
- Total carbohydrate (grams) per serving.
- Fiber and sugars (grams) per serving.
- Total protein (grams) per serving.

In addition, for each of the nutrients just named plus vitamin A, vitamin C, calcium, and iron, the amount in a serving of the food must be expressed as a percentage of the Daily Value for a person who requires 2000 kcalories per day.[15]*

Daily Values: reference values developed by the FDA specifically for food labels.

The Daily Values on Labels The Daily Values (inside back cover) are a set of nutrient standards designed strictly for use on food labels. The Daily Values are based partly on the RDA and partly on a set of standards created by the Food and Drug Administration (FDA) especially for food labels. The Daily Values do two things: they set *adequacy* standards for nutrients that are desirable in the diet such as protein, vitamins, minerals, and fiber, and they also set *moderation* standards for other nutrients that must be limited, such as fat, cholesterol, and sodium, according to the *Dietary Guidelines.*

Labels on Different-Sized Packages A package's size helps determine how much information it must deliver. Large labels of 40 or more square inches must include all the information listed earlier and must also list the Daily Values for two people: one who consumes 2000 kcalories per day, and one who consumes 2500 kcalories per day. The label may also provide a reminder that a gram of carbohydrate or protein supplies 4 kcalories and that a gram of fat supplies 9 kcalories. The side panel of the box of cereal in Figure 1.4 provides all this information. Packages with labels smaller than 40 square inches, such as tuna fish cans, may omit the second set of Daily Values, but must deliver the other information listed earlier. Figure 1.4 shows a can of chicken with a condensed label format. Tiny packages, such as a roll of mints, may provide only a phone number for consumers to call and obtain information.

*The percent Daily Value for protein is not mandatory on all labels, but is required whenever a food makes a protein claim or is intended for consumption by children under 4 years old.

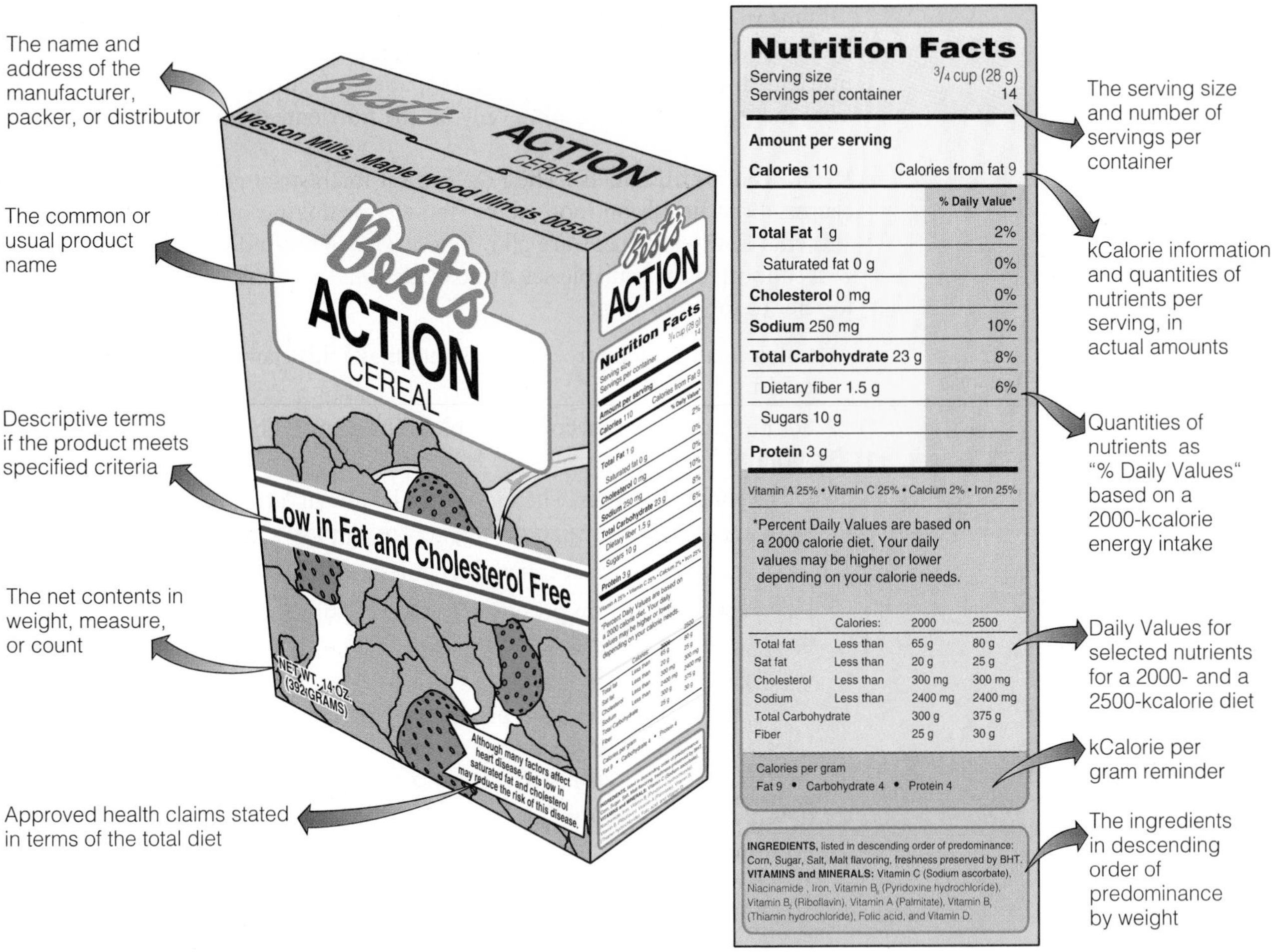

The name and address of the manufacturer, packer, or distributor

The common or usual product name

Descriptive terms if the product meets specified criteria

The net contents in weight, measure, or count

Approved health claims stated in terms of the total diet

Weston Mills, Maple Wood Illinois 00550

Best's ACTION CEREAL

Low in Fat and Cholesterol Free

NET WT. 14 OZ. (392 GRAMS)

Although many factors affect heart disease, diets low in saturated fat and cholesterol may reduce the risk of this disease.

Nutrition Facts

Serving size 3/4 cup (28 g)
Servings per container 14

Amount per serving	
Calories 110	Calories from fat 9
	% Daily Value*
Total Fat 1 g	2%
Saturated fat 0 g	0%
Cholesterol 0 mg	0%
Sodium 250 mg	10%
Total Carbohydrate 23 g	8%
Dietary fiber 1.5 g	6%
Sugars 10 g	
Protein 3 g	

Vitamin A 25% • Vitamin C 25% • Calcium 2% • Iron 25%

*Percent Daily Values are based on a 2000 calorie diet. Your daily values may be higher or lower depending on your calorie needs.

	Calories:	2000	2500
Total fat	Less than	65 g	80 g
Sat fat	Less than	20 g	25 g
Cholesterol	Less than	300 mg	300 mg
Sodium	Less than	2400 mg	2400 mg
Total Carbohydrate		300 g	375 g
Fiber		25 g	30 g

Calories per gram
Fat 9 • Carbohydrate 4 • Protein 4

INGREDIENTS, listed in descending order of predominance: Corn, Sugar, Salt, Malt flavoring, freshness preserved by BHT. **VITAMINS and MINERALS:** Vitamin C (Sodium ascorbate), Niacinamide, Iron, Vitamin B_6 (Pyridoxine hydrochloride), Vitamin B_2 (Riboflavin), Vitamin A (Palmitate), Vitamin B_1 (Thiamin hydrochloride), Folic acid, and Vitamin D.

The serving size and number of servings per container

kCalorie information and quantities of nutrients per serving, in actual amounts

Quantities of nutrients as "% Daily Values" based on a 2000-kcalorie energy intake

Daily Values for selected nutrients for a 2000- and a 2500-kcalorie diet

kCalorie per gram reminder

The ingredients in descending order of predominance by weight

A container with fewer than 40 square inches of surface area can present fewer facts in this format.

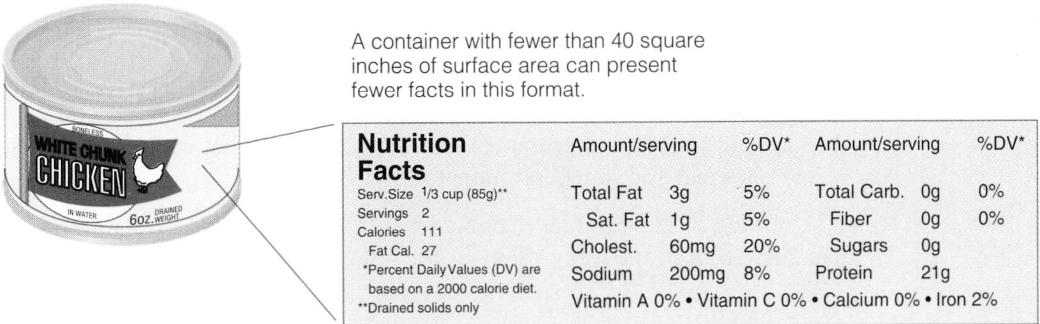

BONELESS
WHITE CHUNK
CHICKEN
IN WATER
6oz. DRAINED WEIGHT

Nutrition Facts
Serv.Size 1/3 cup (85g)**
Servings 2
Calories 111
Fat Cal. 27
*Percent Daily Values (DV) are based on a 2000 calorie diet.
**Drained solids only

Amount/serving		%DV*	Amount/serving		%DV*
Total Fat	3g	5%	Total Carb.	0g	0%
Sat. Fat	1g	5%	Fiber	0g	0%
Cholest.	60mg	20%	Sugars	0g	
Sodium	200mg	8%	Protein	21g	

Vitamin A 0% • Vitamin C 0% • Calcium 0% • Iron 2%

Packages with fewer than 12 square inches of surface area need not carry nutrition information, but they must provide an address or telephone number for obtaining information.

Figure 1.4
Examples of Food Labels

One more factor helps to determine requirements for a label—the nutrients in the food. A food that contains insignificant amounts of more than half of the nutrients listed in the "Nutrition Facts" panel in Figure 1.4 may bear a simplified label that provides information only on the nutrients the food does contain.

Health Claims on Labels The FDA has set forth strict guidelines pertaining to claims about health on food labels. Ten claims linking nutrients and food constituents to disease states are allowable under FDA guidelines because they are well supported by available scientific evidence. A statement on a label is allowed to refer to the following:

1. Calcium and osteoporosis. Foods that make this claim must be high in calcium.
2. Sodium and hypertension (high blood pressure). The food must be low in sodium.
3. Dietary fat and cancer. The food must be low in fat.
4. Dietary saturated fat, dietary cholesterol, and coronary heart disease. The food must be low in saturated fat, cholesterol, and fat.
5. Fiber-containing grain products, fruits, vegetables, and cancer. The food must be low in fat, and without added fiber, it must be a good source of dietary fiber.
6. Fruits, vegetables, and grain products that contain fiber, particularly soluble fiber, and risk of coronary heart disease. The food must be low in saturated fat, fat, and cholesterol. It must also contain at least 0.6 grams of soluble fiber (explained in Chapter 2) per serving.
7. Fruits and vegetables and cancer. The food must be low in fat, and without added nutrients, it must be a good source of fiber, vitamin A, or vitamin C.
8. Oatmeal or oat bran and risk of coronary heart disease. The food must be low in saturated fat, cholesterol, and fat; contain at least 13 grams of oat bran or 20 grams of oatmeal; and, without fortification, provide at least 1 gram of soluble fiber per serving.[16]
9. The vitamin folate and birth defects of the brain and spinal cord (neural tube defects). The food must provide 10 percent or more of the Daily Value for folate per serving and not contain more than 100 percent of the Daily Values for vitamin A and D.[17]
10. Sugar alcohols do not promote tooth decay. Frequent between-meal snacks, high in sugar and starch, promote tooth decay.

A serving of any product making a health claim may also contain no more than 20 percent of the Daily Value for the following:

- Total fat.
- Saturated fat.
- Cholesterol.
- Sodium.

Thus the fats and sodium in foods that make health claims are controlled by law. Therefore, whole milk, even though it is high in calcium, may not make a claim about osteoporosis because it contains too much saturated fat to qualify. Low-fat and nonfat milk, however, do qualify to bear the calcium and osteoporosis claim. Table 1.6 (pp. 23–24) defines the related terms consumers may see on food labels.

Those who design food labels must proceed carefully. For example, use of the word "healthy" in a name such as "Healthy Start" or the use of a heart-shaped logo may imply that a food is health promoting. Foods bearing such words or logos must not exceed limits set for fat, saturated fat, cholesterol, and sodium contents.

Table 1.6 Terms Used on Food Labels

General Terms

Free, without, no, zero: Contains no amount or a trivial amount. *kCalorie-free* means containing fewer than 5 kcal per serving; *sugar-free* or *fat-free* means containing less than 0.5 g per serving.

Fresh: Raw, unprocessed, or minimally processed (blanched or irradiated) with no added preservatives.

Good source: Provides 10 to 19% of the Daily Value of a given nutrient per serving.

Healthy: A food that is low in fat, saturated fat, cholesterol, and sodium and that contains at least 10% of the Daily Value for vitamin A, vitamin C, iron, calcium, protein, or fiber.

High: Provides 20% or more of the Daily Value per serving.

Less, fewer, reduced: Provides 25% less of a nutrient or kcalories than a reference food. This may occur naturally or as a result of altering the food. For example, pretzels, which are usually low in fat, can claim to provide less fat than potato chips, a comparable food.

Light: This descriptor has three meanings on labels:

- A serving provides one-third fewer kcalories or half the fat of the regular product.
- A serving of a low-kcalorie, low-fat food provides half the sodium normally present.
- The product is light in color and texture; the label must make this intent clear, as in "light brown sugar."

More: Contains at least 10% more of the Daily Value for a given nutrient than a comparable food. The nutrient may be added or may occur naturally.

Carbohydrates: Fiber and Sugar Terms

High fiber: Provides 5 g or more fiber per serving; a high-fiber claim made on a food that contains more than 3 g fat per serving and per 100 g of food must also declare total fat.

Sugar-free: Provides less than 0.5 g sugar per serving.

Energy Terms

kCalorie-free: Contains fewer than 5 kcal per serving.

Low kcalorie: Contains 40 kcal or less per serving.

Reduced kcalorie: Contains at least 25% fewer kcalories per serving than the comparison product.

Fat Terms (Meat and Poultry Products)

Extra lean: A product contains:

- Less than 5 g fat.
- Less than 2 g saturated fat.
- Less than 95 mg cholesterol per serving.

Lean:[a] A product contains:

- Less than 10 g fat.
- Less than 4.5 g saturated fat.
- Less than 95 mg cholesterol per serving.

Fat and Cholesterol Terms (All Products)

Cholesterol-free: Contains less than 2 mg cholesterol per serving and 2 g or less saturated fat per serving.

Fat-free: Contains less than 0.5 g fat per serving.

Low cholesterol: Contains 20 mg or less cholesterol per serving and 2 g or less saturated fat per serving.

Low fat: Contains 3 g or less fat per serving.

Low saturated fat: Contains 1 g or less saturated fat per serving.

Percent fat free: May be used only if the product meets the definition of low fat or fat-free. Requires disclosure of grams fat per 100 g food.

Reduced or **less cholesterol:** Contains 25% or less cholesterol than the comparison food and 2 g or less saturated fat per serving.

Reduced or **less fat:** Contains 25% or less fat than the comparison food.

(continued)

Table 1.6 *(continued)*

Fat and Cholesterol Terms (All Products) *(continued)*
Reduced or **less saturated fat:** Contains 25% or less saturated fat than the comparison food and reduced by more than 1 g per serving.
Saturated fat–free: Contains less than 0.5 g saturated fat and less than 0.5 g *trans*-fatty acids.
Sodium Terms
Low sodium: Contains 140 mg or less sodium per serving.
Sodium-free: Contains less than 5 mg sodium per serving.
Very low sodium: Contains 35 mg or less sodium per serving.

[a]The word *lean* as part of the brand name (as in "Lean Supreme") indicates that the product contains fewer than 10 grams of fat per serving. *Lean* ground beef can contain up to 22.5 percent fat by weight.

Sources: The new food label, *FDA Backgrounder,* December 10, 1992; Nutrition labeling of meat and poultry products, *FSIS Backgrounder,* January 1993.

Consumers may wonder whether they are reducing their health risks if they eat only "healthy" foods. They may be, and that is all that a label is allowed to say: that a substance "may" or "might" reduce disease risks. This wording is tentative because scientists are still accumulating evidence concerning the roles of diet in disease. A claim must also state that the development of a disease rests on many factors. A permissible health claim reads like this: "Development of heart disease depends on many factors. A healthful diet low in saturated fat and cholesterol may lower blood cholesterol levels and may reduce the risk of heart disease." Health claims on labels are so carefully controlled that they can be an asset to consumers who would rather not worry about grams, percentages, and other mathematical stumbling blocks as they shop for foods.

The Nutrition Labeling and Education Act of 1990 significantly changed requirements for food label information to help consumers choose foods to meet their nutrition and health goals. Interested consumers will find food labels fascinating reading.

Self Check

1. People choose particular foods because:
   a. the foods taste good to them.
   b. the foods remind them of pleasant times.
   c. the foods are foods they always eat.
   d. all of the above.

2. Minerals:
   a. yield energy in the body.
   b. are inorganic nutrients.
   c. are organic nutrients.
   d. are both organic and inorganic nutrients.

3. Energy-yielding nutrients include all of the following **except:**
   a. vitamins. b. carbohydrates.
   c. protein. d. fat.

4. Health recommendations urge people to limit fat to ______ of total energy intake:
   a. 40 percent. b. 30 percent.
   c. 50 percent. d. 60 percent.

5. Alcohol is not a nutrient because:
   a. the body derives no energy from it.
   b. the body derives too much energy from it.
   c. it is converted to body fat.
   d. it does not contribute to the body's growth or repair.

6. Nutrient intake recommendations offer:
   a. a rough guideline for dietary adequacy.
   b. exact guidelines for each individual's nutrient needs.
   c. no guidelines for children.
   d. none of the above.

7. One of the characteristics of a nutritious diet is that it provides no constituent in excess. This principle of diet planning is called:
   a. adequacy. b. balance.
   c. moderation. d. variety.

8. A slice of apple pie supplies 350 calories with 3 grams of fiber; an apple provides 80 calories and the same 3 grams of fiber. This is an example of:
   a. calorie control. b. nutrient density.
   c. variety. d. essential nutrients.

9. According to the Food Guide Pyramid, which foods should form the foundation of a healthy diet?
   a. vegetables b. breads, cereals, rice, and pasta
   c. fruits d. milk, yogurt, and cheese

10. "Miscellaneous" foods are excluded from the main cluster of food group plans because they are:
    a. usually too high in fat, sugar, or both and provide few nutrients.
    b. foods that people do not usually eat.
    c. foods that provide too much sodium and too many nutrients.
    d. foods that do not belong in any other group.

Answers to these questions appear in Appendix H.

Self Study

How Balanced Is Your Diet?

Our purpose in providing these exercises is to encourage you to study your own diet. Your reaction to them may be mixed. They will slow you down, and filling out all the forms can be tedious. Like your checkbook, to be accurate and meaningful, they have to be done carefully, with frequent checking of arithmetic and clear handwriting.

The benefits, however, may well outweigh the drawbacks. Most students who do these activities with thoughtful attention report that, unlike a checkbook, they are intriguing, informative, and often reassuring. They are also rewarding—in direct proportion to your accuracy. They not only teach you about yourself but enable you to assess clients' diets once you are working in a health profession.

In this first Self Study, make a record of your typical food intake, and analyze it for the nutrients it contains. You will use the results over and over again in succeeding Self Studies, so invest time and effort now to achieve maximum accuracy. You can undertake this analysis before you have learned about the nutrients; having the results in front of you as you work will help make the reading meaningful.

1. Using three copies of Form 1 or the Diet Analysis Plus software,[a] record all the foods you eat during a three-day period. If, like most people, you eat differently on weekdays than on weekends, then you should probably record for two weekdays and one weekend day to get a true average, or record your food intake for a week. You will learn the most from these Self Studies if you select days that are truly typical of your food intake and physical activity.

As you record each food, make careful note of the serving size.[b] Estimate the amount to the nearest ounce, quarter cup, tablespoon, or other common measure. In guessing at the sizes of meat portions, it helps to know that a piece of meat the size of the palm of your hand weighs about 3 or 4 ounces. If you are unable to estimate serving sizes in cups, tablespoons, or teaspoons, try measuring out servings in those proportions to see how they look. It also helps to know that a slice of cheese (like sliced American cheese) or a 1½-inch cube of cheese weighs about 1 ounce.

You may have to break down mixed dishes to their ingredients. Appendix A includes sections on both mixed dishes and soups. Other mixtures are simple to analyze. A ham and cheese sandwich, for example, can be listed as 2 slices of bread, 1 tablespoon of mayonnaise, 2 ounces of ham, 1 ounce of cheese, and so on. If you can't identify all the ingredients, just estimate the amounts of the major ones, like the beef, tomatoes, and potatoes in a beef-vegetable soup.

You will, of course, make errors in estimating amounts. In calculations of this kind, errors of up to 20 percent are expected and tolerated. Still, your rough approximation will enable you to compare your nutrient intakes with the recommended ones.

Do not record any nutrient supplements you take. It is important for you to discover whether your food choices alone deliver the nutrients you need. If they don't, you'll have the first clues to a solution, perhaps one that involves better food choices rather than supplements.

2. Calculate for each day your total intakes of kcalories, protein, fat, carbohydrate, calcium, iron, magnesium, phosphorus, sodium, zinc, vitamin A, thiamin, vitamin E, riboflavin, niacin, vitamin B_6, folate, and vitamin C.[c]

If a packaged food you have eaten does not appear in Appendix A, read the label on the package. If you eat a packaged food in which the nutrient amounts are listed on the label as "percent Daily Values," use the table on the inside back cover of the text to convert to grams, milligrams, micrograms, or retinol equivalents. Suppose a food label states that a serving contains 25 percent of the Daily Value for iron, for example. The table shows that the Daily Value for iron is 18 milligrams. The food portion, therefore, contributes 25 percent of 18 milligrams, or 4.5 milligrams of iron.

For foods you eat in restaurants, at friends' houses, or elsewhere, ask the composition or use your ingenuity and make a guess. Use the most similar food you can find as a guide. For example, if you ate smoked cod (which is not listed in Appendix A), you would not be far off in using the values for smoked halibut. If you ate cream of broccoli soup, you might substitute the values for cream of mushroom soup.

Be careful in recording the nutrient amounts in odd-sized portions. For example, if you use ¼ cup of milk, you will have to look up the nutrient values for a cup of milk and divide each one by 4. Also note the units in which the nutrients are measured:

- Energy is measured in kcalories (kcal).
- Protein, fat, carbohydrate, and fiber are measured in grams (g).

[a]When the program has been installed on your hard drive, click the DA plus icon to open it. Go to file and select "new." The program will ask you to complete a personal profile. When you have filled in the required information, click "OK." The program will calculate your nutrition profile for you. To input your daily food intakes, select Profile, Daily Intake 1, 2, etc. Type the name of the food or the code number for the food you would like (see Appendix A). If you type the name of the food, such as "cereal," the program will provide a list of more specific choices. Select the food choice or the closest match.

[b]If you are using the software, the program will require you to estimate the amount and type of measure consumed (i.e., nonfat milk, 1 cup).

[c]If you are using the forms provided, refer to Appendix A. If you are using Diet Analysis Plus, the program will provide these calculations. Go to Foodlist and select Analyses/Reports. View them in whatever form you choose by selecting the bar graph, single nutrient, or spreadsheet report.

Form 1 Nutrient Intakes (Use One Form for Each Day)

Food	Approximate Measure or Weight	Energy[a] (kcal)	Prot[a] (g)	Carb[a] (g)	Fiber[a] (g)	Fat[a] (g)	Fat Breakdown Sat (g)	Mono (g)	Poly (g)	Chol[a] (mg)	Calcium[a] (mg)	Iron[b] (mg)	Magn[b] (mg)	Pota[a] (mg)	Sodium[b] (mg)	Zinc[b] (mg)	Vit A[a] (RE)	Thia (mg)	Vit E[b] (α-TE)	Ribo[b] (mg)	Niac[b] (mg)	B_6[b] (mg)	Fol[a] (µg)	Vit C[b] (mg)
Totals																								

[a] Compute these values to the nearest whole number..
[b] Compute these values to two decimal places.

(continued)

- Cholesterol, calcium, iron, magnesium, potassium, sodium, zinc, thiamin, riboflavin, niacin, vitamin B_6, and vitamin C are measured in milligrams (mg)—thousandths of a gram (0.001 g).
- Folate is measured in micrograms (μg)—thousandths of a milligram or millionths of a gram (0.001 mg or 0.000001 g). Thus, 800 mg calcium is the same as 0.8 g calcium, and 400 μg folate is the same as 0.4 mg folate. Be sure to convert all calcium amounts to milligrams and all folate amounts to micrograms before adding them together and comparing them with standards.
- Vitamin A can be measured in international units (IU) or retinol equivalents (RE). Appendix A lists vitamin A in RE to ease comparison with the recommended intake, which is also in RE. If you eat a packaged food that lists vitamin A in IU on the label, be sure to convert to RE before calculating. (For more details, see Chapter 7.)
- Appendix A lists vitamin E in tocopherol equivalents (TE) to ease comparison with the recommended intake. One TE equals 1 milligram of active vitamin E.

3. Now total the amount of each nutrient you've consumed each day, and transfer your totals to Form 2. Form 2 provides a convenient means of deriving an average intake for each nutrient.[d]
4. As a final step, transfer your average intakes to Form 3 for future reference. For comparison, enter the intakes recommended for a person of your age and sex, using either the RDA (on the inside front cover of the text) or the Recommended Nutrient Intakes for Canadians (in Appendix B), whichever you prefer. For fat, cholesterol, carbohydrate, and fiber, use the Daily Values (inside back cover).[e]

Suspend judgment about the adequacy of your intakes for the moment. You have much to learn about your individuality, the nutrients, and the recommendations before you can reach any reasonable conclusions.

5. Now, check your overall food intake for balance. You can get an indication of whether you are choosing a balanced selection of foods by using the Daily Food Selection Scorecard (Form 4—one copy for each day). How does your diet score by these criteria?
6. Another way to check for balance is to evaluate your diet using the guideline that about 58 percent of your kcalories should come from carbohydrate, about 12 percent from protein, and not more than 30 percent from fat. Use Form 5 to calculate these percentages.[f] What percentage of the food energy you consume comes from protein? ________ percent. Fat? ________ percent. Carbohydrate? ________ percent. Is your diet balanced using these criteria?

As soon as you have perfected these skills on yourself, you can use them on clients to their benefit. In fact, dietitians use a similar approach in assessing a client's nutrient intake (see Chapter 15).

[d]If you are using the software, you can print a daily average report for all of the days you have input. Go to your profile. Under File, select Print, and select Average All.

[e]If you are using the software, your recommended intakes will be calculated for you in your nutrition profile. You can list these in the Standard box.

[f]You can view the number of grams/day of these nutrients in the Analysis/Reports section of your Diet Analysis Plus program.

Form 2 Average Daily Energy and Nutrient Intakes

Day	Energy (kcal)	Prot (g)	Carb (g)	Fiber (g)	Fat (g)	Fat Breakdown: Sat (g)	Mono (g)	Poly (g)	Chol (mg)	Calcium (mg)	Iron (mg)	Magn (mg)	Pota[a] (mg)	Sodium (mg)	Zinc (mg)	Vit A (RE)	Thia (mg)	Vit E (α-TE)	Ribo (mg)	Niac (mg)	B_6 (mg)	Fol (µg)	Vit C (mg)
1																							
2																							
3																							
Total																							
Average daily intake (divide total by 3)																							

Form 3 Comparison with a Standard Intake

Day	Energy (kcal)	Prot (g)	Carb (g)	Fiber (g)	Fat (g)	Fat Breakdown: Sat (g)	Mono (g)	Poly (g)	Chol (mg)	Calcium (mg)	Iron (mg)	Magn (mg)	Pota[a] (mg)	Sodium (mg)	Zinc (mg)	Vit A (RE)	Thia (mg)	Vit E (α-TE)	Ribo (mg)	Niac (mg)	B_6 (mg)	Fol (µg)	Vit C (mg)
Average daily intake (from Form 2)																							
Standard[a]																							
Intake as percentage of standard[b]																							

[a] Use the DRI and RDA tables (inside front cover, and Appendix C) or the Canadian standards in Appendix B, or in the case of fat, fiber, and cholesterol, the Daily Values (inside back cover). Alternatively for energy, use your calculation from Self Study 6. If you are using Diet Analysis Plus, refer to your profile for your recommended amounts.

[b] For example, if your intake was 50 g and the standard for a person your age and sex was 46 g, you consumed $(50 \div 46) \times 100$, or 109% of the standard.

(continued)

Form 4 Food Selection Scorecard

Record your day's food intake on a separate piece of paper (or review your record from Form 1 or your Daily Intake for Day 1 if you used Diet Analysis Plus). Assign each food you ate to its appropriate group and estimate the number of servings you ate (Figure 1.3 presents the foods found within each food group and serving sizes). Record the number of servings for each group in column 2. Then multiply that number by the points per serving to determine your score. You can earn up to 20 points for each of the five food groups.

Food Group and Recommended Intake	Number of Servings Eaten	Your Score
Breads and cereals: 6 or more servings		
1 serving = 3.5 points	× 3.5 =	
Subtotal (no more than 20 points allowed)		
Vegetables: 3 or more servings		
1 serving vitamin A–rich dark green or deep yellow-orange vegetable (such as spinach, broccoli, carrots, squash) = 10 points (no more than 10 points allowed—count additional servings as other vegetables below)		
1 serving other vegetables = 5 points	× 10 =	
Subtotal (no more than 20 points allowed)	× 5 =	
Fruits: 2 or more servings		
1 serving vitamin C–rich fruits (such as oranges, strawberries, watermelon) = 10 points (no more than 10 points allowed—count additional servings as other fruits below)		
1 serving other fruits = 10 points	× 10 =	
Subtotal (no more than 20 points allowed)	× 10 =	
Meats and meat alternates: 2 to 3 servings		
1 serving = 10 points	× 10 =	
Subtotal (no more than 20 points allowed)		
Milk and milk products: 2 servings		
1 serving = 10 points	× 10 =	
Subtotal (no more than 20 points allowed)		
GRAND TOTAL (no more than 100 points allowed)		

The above foods are foundation foods. Additional foods are those that do not fit into any of the five food groups, but add flavor, interest, variety, and often kcalories. List additional foods:

Form 5 Percentage of kCalories from Protein, Fat, and Carbohydrate

Average daily intakes from Form 2:

Protein: g/day × 4 kcal/g = (*P*) ________ kcal/day.
Fat: g/day × 9 kcal/g = (*F*) ________ kcal/day.
Carbohydrate: g/day × 4 kcal/g = (*C*) ________ kcal/day.
Total kcal/day =(*T*) ________ kcal/day.

Percentage of kcalories from protein: $\frac{(P)}{(T)} \times 100 =$ ________________ % of total kcalories.

Percentage of kcalories from fat: $\frac{(F)}{(T)} \times 100 =$ ________________ % of total kcalories.

Percentage of kcalories from carbohydrate: $\frac{(C)}{(T)} \times 100 =$ ________________ % of total kcalories.

Note: The three percentages can total 99, 100, or 101, depending on the way figures were rounded off earlier.

If you used an alcoholic beverage, you have to add a line for kcalories from alcohol. To find out how many kcalories in the beverage were from alcohol, look the beverage up in Appendix A. Figure out how many kcalories were from carbohydrate (multiply carbohydrate grams times 4), fat (fat grams times 9), and protein (protein grams times 4). The remaining kcalories were from alcohol.

Notes

1. A. Drewnoski and C. L. Rock, The influence of genetic taste markers on food acceptance, *American Journal of Clinical Nutrition* 62 (1995): 506–511.

2. A. E. Sloan, America's appetite '96: The top ten trends to watch and work on, *Food Technology* 50 (1996): 55–71.

3. E. R. Monsen, New Dietary Reference Intakes proposed to replace the Recommended Dietary Allowances, *Journal of the American Dietetic Association* 96 (1996): 754–755.

4. N. S. Scrimshaw, Nutrition and health from womb to tomb, *Nutrition Today* 31 (1996): 55–67. J. M. McGinnis and W. H. Foege, Actual causes of death in the United States, *Journal of the American Medical Association* 270 (1993): 2207–2212.

5. R. R. Pate and coauthors, Physical activity and public health: A recommendation from the Centers for Disease Control and Prevention and the American College of Sports Medicine, *Journal of the American Medical Association* 273 (1995): 402–407.

6. Position of The American Dietetic Association: Vegetarian diets, *Journal of the American Dietetic Association* 93 (1993): 1317–1319.

7. R. R. Briefel, Nutrition monitoring in the United States, in *Present Knowledge in Nutrition,* 7th ed., eds. E. E. Ziegler and L. J. Filer (Washington, D.C.: International Life Sciences Institute Press, 1996), pp. 517–529.

8. Ten-year comprehensive plan for the national nutrition monitoring and related research program, *Federal Register,* June 11, 1993.

9. *Healthy People 2000: National Health Promotion and Disease Prevention Objectives* (Washington, D.C.: U.S. Department of Health and Human Services, 1990).

10. J. M. McGinnis and P. R. Lee, *Healthy People 2000* at mid decade, *Journal of the American Medical Association* 273 (1995): 1123–1129.

11. Food label importance, *FDA Consumer,* May 1990, p. 3.

12. J. E. Foulke, Cooking up the new food label, *FDA Consumer,* May 1993, pp. 33–38.

13. P. Kurtzweil, Today's special: Nutrition information, *FDA Consumer,* May/June 1997, pp. 21–25.

14. U.S. Department of Agriculture, Food Safety and Inspection Service, *FSIS Backgrounder,* January 1993, pp. 1–6.

15. E. C. Henley, Food and Drug Administration's proposed labeling rules for protein, *Journal of the American Dietetic Association* 92 (1992): 293–296.

16. Federal update, Health claim for oat products, *Journal of the American Dietetic Association* 96 (1996): 332.

17. P. Kurtzweil, How folate can help prevent birth defects, *FDA Consumer,* September 1997, pp. 7–10.

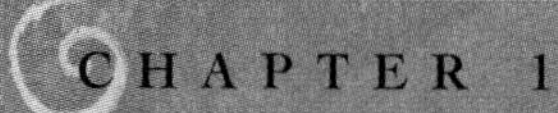

Nutrition in Practice

▪ The New Dietary Reference Intakes (DRI) ▪

As discussed in Chapter 1, nutrient recommendations are sets of standards used to measure the adequacy of healthy people's diets. Since the last editions of the RDA and RNI, scientific research has revealed new understandings about the influence of nutrients and other food constituents on the development of chronic diseases such as heart disease and cancer. As a result, the Food and Nutrition Board of the National Institute of Medicine together with Health Canada is revising the RDA and RNI and calling the new standards Dietary Reference Intakes (DRI).

WWW.

nas.edu
Dietary Reference Intakes

hc-sc.gc.ca/datahpb/datafood/english/pub/bns/dietrie.htm
Nutrition and Dietary Guidelines (from Health Canada)

The idea of revised nutrient recommendations sounds good, but why change the name?

Until now, the RDA have been used for many different purposes. Scientists used them to assess the diets of groups; individuals used them to judge the adequacy of their own diets; agencies used them to set dietary goals for communities and the nation. In short, the RDA were used to fill every conceivable need for nutrient intake standards. No one standard, however, can perfectly perform every task to which the RDA have been applied. The new DRI comprise four sets of values: the Recommended Dietary Allowances, Adequate Intakes, Tolerable Upper Intake Levels, and Estimated Average Requirements. Each set of values was designed with a specific purpose in mind (see Table NP1.1).[1]

Having four sets of values sounds complicated.

Yes, it does, and students of nutrition may well wonder how to keep all four sets of values and their uses straight. Fortunately, they really are not too complicated when you consider one set at a time. Let's begin with the familiar RDA.

Good. Tell me about the RDA.

The new RDA values are similar to the 1989 RDA in that they are reliable standards based on scientific evidence. As Chapter 1 explained, the RDA for a vitamin or mineral is set near the top end of the range of the population's requirements so that almost everybody's needs are met. Even people whose needs are higher than average will be covered if they achieve this dietary goal.

I understand the RDA. Please explain Adequate Intakes.

For some nutrients, evidence is insufficient to calculate an RDA. In these cases, an Adequate Intake (AI) value is established. Like the RDA, AI may be used as a dietary goal for individuals.

While both RDA and AI serve as nutrient intake goals for individuals, their differences are noteworthy. An RDA for a given nutrient is based on enough scientific evidence to expect that the needs of almost all healthy people will be met. An AI, on the other hand, must rely more heavily on scientific judgments because sufficient evidence is lacking. The percentage of people covered by an AI is unknown; it is expected that an AI exceeds average requirements, but it may or may not cover as many people as an RDA would (if an RDA could be determined). For these reasons, AI values are more tentative than RDA. Later chapters will present the RDA and AI values for the vitamins and minerals.

What about the other parts of the DRI?

As Table NP1.1 shows, in addition to RDA and AI, the DRI include Tolerable Upper Intake Levels (UL) and Estimated Average Requirements (EAR). The recommended intakes for nutrients are generous, and although they do not necessarily cover every individual

Table NP1.1 Four Dietary Reference Intake Values and Their Uses

Reference Value	Description
Recommended Dietary Allowances (RDA)	Nutrient intake goals for individuals. RDA values are based on scientific data and aim for prevention of deficiencies.
Adequate Intakes (AI)	Suggested nutrient intake targets for individuals. AI values are to be used wherever RDA values are lacking.
Tolerable Upper Intake Levels (UL)	A guide to the risks of higher nutrient intakes.
Estimated Average Requirements (EAR)	Values scientists use to evaluate intakes of populations and to generate new RDA values.

Source: Adapted from Food and Nutrition Board, Institute of Medicine, National Academy of Sciences, Dietary Reference Intakes (DRIs) for calcium, phosphorus, magnesium, vitamin D and fluoride, *Nutrition Today* 32 (1997): 182–190.

for every nutrient, they probably should not be exceeded by much. People's tolerances for high doses of nutrients vary, and somewhere above the recommended level is a *Tolerable Upper Intake Level* beyond which a nutrient is likely to become toxic.[2] The Tolerable Upper Intake Levels specify safe upper limits on nutrient intakes to help consumers, manufacturers, and planners of food fortification programs judge the safety of nutrients in supplements and fortified foods. The EAR are useful only to scientists, who need them to assess the nutrient intakes of groups and to develop new RDA.

Notes

1. Dietary Reference Intakes, *Nutrition Reviews* 55 (1997): 319–326.

2. Standing Committee on the Scientific Evaluation of Dietary Reference Intakes, Food and Nutrition Board, Institute of Medicine, *Dietary Reference Intakes for Calcium, Phosphorus, Magnesium, Vitamin D, and Fluoride* (Washington, D.C.: National Academy Press, 1997), p. S-5.

2 *Carbohydrates*

Carbohydrates are found in plant foods and in milk.

carbohydrates: energy nutrients composed of monosaccharides.
carbo = carbon
hydrate = water

simple carbohydrates (sugars): the monosaccharides (glucose, fructose, and galactose) and the disaccharides (sucrose, lactose, and maltose).

complex carbohydrates: long chains of sugars arranged as starch or fiber; also called **polysaccharides.**

monosaccharide: a single sugar unit.
mono = one
saccharide = sugar

disaccharide: a pair of sugar units bonded together.
di = two

glucose: a monosaccharide, the sugar common to all disaccharides and polysaccharides; also called blood sugar or dextrose.

insulin: a hormone secreted by the pancreas in response to high blood glucose; it promotes cellular glucose uptake and use or storage.

homeostasis: the maintenance of constant internal conditions (such as chemistry, temperature, and blood pressure) by the body's control systems.
homeo = the same
stasis = staying

fructose: a monosaccharide; sometimes known as fruit sugar. It is abundant in fruits, honey, and saps.
fruct = fruit

Most people would like to feel good all the time. No matter what each day may bring, the potential enjoyment available can be tremendous when a person's body and mind are tuned for it. The feeling of well-being that comes with energy, alertness, clear thinking, and confidence is so rewarding that if you know how to produce it, you will probably make the necessary effort. Part of the secret of feeling well is keeping your energy supply going with food. That means choosing foods that contain the energy nutrients—carbohydrate and fat, primarily. But which to choose?

Carbohydrate is the preferred energy source for most of the body's functions. As long as carbohydrate is available, the human brain depends exclusively on it as an energy source. Athletes eat a "high-carb" diet to store as much muscle fuel as possible, and dietary recommendations urge people to eat carbohydrate-rich foods for better health. And where can you find carbohydrate-rich foods? Almost exclusively among the plants; milk is the only animal-derived food that contains significant amounts of carbohydrate.

Carbohydrate is the body's fuel of choice; fat is next. Fat, however, has disadvantages: it normally is not used as fuel by the brain and central nervous system, and diets high in fat are associated with many diseases. The other energy sources available to the body—protein and alcohol—offer no advantage as fuels. Protein is best left to serve its own diverse functions, as you will see in Chapter 4. Alcohol, of course, has well-known undesirable side effects when used in excess. Alcohol abuse and its relationships with nutrition are the subject of Nutrition in Practice 8.

The Chemist's View of Carbohydrates

Chemists divide the carbohydrates into two categories: simple and complex. The simple carbohydrates (sugars) are:

- Monosaccharides (single sugars).
- Disaccharides (double sugars).

The complex carbohydrates (polysaccharides) are:

- Starch.
- Glycogen.
- Some fibers.

All of these carbohydrates are composed of the simple sugar glucose and other compounds that are much like glucose in composition and structure. Figure 2.1 shows the chemical structure of glucose.

Monosaccharides (Single Sugars)

Three monosaccharides are important in nutrition: glucose, fructose, and galactose. Almost all the body's cells use glucose as their chief energy source. The body can obtain this glucose from all plant carbohydrates. Plants capture the sun's radiant energy and through the process of photosynthesis trap this energy in glucose. In the human body, the hormone insulin, among others, enables cells to take up glucose from the blood and so helps to keep blood glucose constant (homeostasis).

Plants also make fructose, which is the sweetest of the sugars. It is abundant in fruits, honey, and saps. The body can convert fructose to glucose, or it can break fructose down to fragments and make fat from them. Glucose and fructose are the most common monosaccharides in nature. The third single sugar, galactose, appears in nature only as part of lactose, a disaccharide also known as milk sugar. Only during digestion is galactose freed as a single sugar.

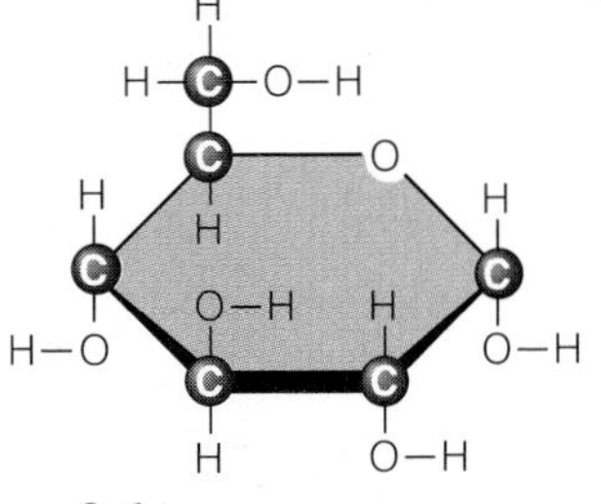

Figure 2.1
Chemical Structure of Glucose
On paper, the structure of glucose has to be drawn flat, but in nature the five carbons and oxygen are roughly in a plane, with the H, OH, and CH_2OH extending out above and below it.

Disaccharides (Double Sugars)

In disaccharides, pairs of single sugars are linked together. Lactose is an example of a disaccharide, and there are two others. All three have glucose as one of their single sugars. As Table 2.1 shows, the other monosaccharide is either another glucose (in maltose), or galactose (in lactose), or fructose (in sucrose). The shapes of the sugars in Table 2.1 reflect their chemical structures as drawn on paper.

Sucrose: Table Sugar Sucrose (table, or white, sugar) is the most familiar of the three disaccharides and is what people mean when they speak of "sugar." This sugar is usually obtained by refining the juice from sugar beets or sugarcane to provide the brown, white, and powdered sugars available in the supermarket, but it occurs naturally in many fruits and vegetables.

When you eat a food containing sucrose, enzymes in your digestive tract split the sucrose into its glucose and fructose components. Because the body can convert fructose to glucose, one molecule of sucrose can ultimately yield two molecules of glucose.

galactose: a monosaccharide; part of the disaccharide lactose.

sucrose: a disaccharide composed of glucose and fructose; commonly known as table sugar, beet sugar, or cane sugar.
sucro = sugar

Honey versus Sugar People often ask: What is the difference between honey and white sugar? Is honey more nutritious? Honey, like white sugar, contains glucose and fructose. The difference is that in white sugar, the glucose and fructose are bonded together in pairs, whereas in honey some of them are paired and some are free single sugars. When you eat either white sugar or honey, though, your body breaks all of the sugars apart into single sugars. It ultimately makes no difference, then, whether you eat single sugars linked together, as in white sugar, or the same sugars unlinked, as in honey; they will end up as single sugars in your body. True, honey contains trace amounts of a few vitamins and minerals, but to say that honey is nutritious is misleading.

Digestion and absorption are the topics of Chapter 5.

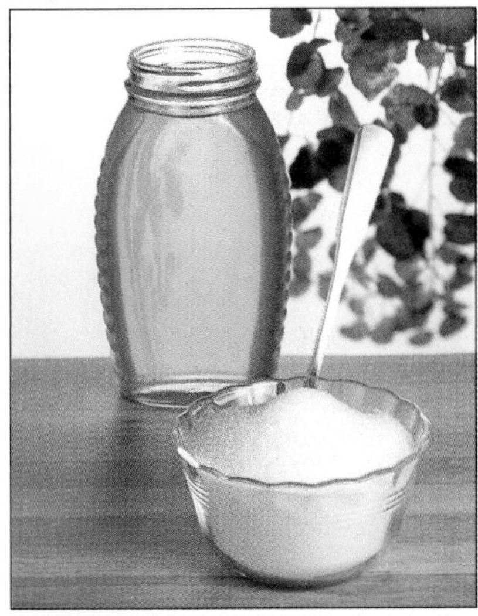

Honey, like sugar, contains glucose and fructose.

Fruits versus Sugar Some sugar sources are more nutritious than others, though. Consider a fruit such as an orange. You receive the same sugars and about the same energy from an orange as from a tablespoon of sugar or honey, but the packaging makes a big difference in nutrient density. The sugars of the orange are diluted in a large volume of fluid that contains valuable vitamins and minerals, and the flesh and skin of the orange are supported by fibers that also offer health benefits. A tablespoon of honey offers no such bonuses.

Cola Beverages and Sweets A cola beverage, containing many teaspoons of sugar, offers no advantages either. Table 2.2 shows sample nutrients supplied by

Table 2.1 The Major Sugars

Monosaccharides	Disaccharides
Glucose	Sucrose (glucose + fructose)
Fructose	Lactose (glucose + galactose)
Galactose (found only as part of lactose)	Maltose (glucose + glucose)

An orange provides the same sugars as honey or sugar, but the packaging makes a big nutrition difference.

Table 2.2 Sample Nutrients in Sugars and Other Foods

The indicated portion of any of these foods provides approximately 100 kcalories. Notice that for a similar number of kcalories, milk, legumes, fruits, and grains offer more of the other nutrients than do the sugars.

Food	Protein (g)	Calcium (mg)	Vitamin A (RE)
Sugar, white (2 tbs)	0	Trace	0
Cola beverage (1 c)	0	6	0
Honey, strained or extracted (1½ tbs)	Trace	2	0
Milk, 1% low-fat (1 c)	8	300	145
Kidney beans (½ c)	7	35	0
Apricots (6)	2	30	554
Bread, whole wheat (1½ slices)	3	30	Trace

some sugar sources; note the "0s" and "traces" by honey, sugar, and the cola beverage and the substantial numbers by the others. Sucrose is often the principal ingredient of carbonated beverages, candy, cakes, frostings, cookies, and other concentrated sweets, so they are often not nutritious.

lactose: a disaccharide composed of glucose and galactose; commonly known as milk sugar.
lact = milk

Lactose: Milk Sugar Lactose is the principal carbohydrate of milk. Most human babies are born with the digestive enzymes necessary to split lactose into its two monosaccharide parts, glucose and galactose, so as to absorb it. Breast milk thus provides a simple, easily digested carbohydrate that meets a baby's energy needs; most formulas do, too, because they are made from milk.

Lactose Intolerance Many people lose the ability to digest lactose after infancy. This condition, known as lactose intolerance, occurs in 70 to 90 percent of Native American, Asian, African, Mediterranean, and Middle Eastern people.[1] It is less common among northern Europeans. Lactose intolerance is not the same as milk allergy, which is caused by an immune reaction to the protein in milk. Lactose intolerance is discussed further in Chapter 20, and milk allergy in Chapter 13.

maltose: a disaccharide composed of two glucose units; sometimes known as malt sugar.

Maltose The third disaccharide, maltose, is a plant sugar that consists of two glucose units. Maltose appears at only one stage in the life of a plant—when the plant is digesting its stored starch for energy and starting to sprout.

zestec.nl/crf
Carbohydrate Research Foundation

In summary, then, the major simple carbohydrates, or sugars, are the three single sugars and the three double sugars shown earlier in Table 2.1. Glucose, fructose, maltose, and sucrose come from plants; lactose and galactose from milk and milk products.

Starch and Glycogen (Energy-Yielding Polysaccharides)

Unlike the sugars, which contain the three monosaccharides—glucose, fructose, and galactose—in different combinations, the polysaccharides, starch and glycogen, are composed almost entirely of glucose. They differ from each other only in the nature of the bonds that link the glucose units together.

starch: a plant polysaccharide composed of glucose and digestible by human beings.

Starch Starch is a long, straight or branched chain of hundreds of glucose units linked together. These giant molecules are packed side by side in a rice grain or potato root—as many as a million per cubic inch of food. When you eat the plant, your body splits the starch into glucose units and uses the glucose for energy.

The short chains of glucose units that result from the breakdown of starch are known as **dextrins.** The word sometimes appears on food labels because dextrins can be used as thickening agents in foods.

Starchy Foods All starchy foods are plant foods. Grains are the richest food source of starch. In most human societies, people depend on a staple grain for much of their food energy: rice in Asia; wheat in Canada, the United States, and Europe; corn in much of Central and South America; and millet, rye, barley, and oats elsewhere. A second important source of starch is the legume (bean and pea) family. Legumes include peanuts and "dry" beans such as butter beans, kidney beans, "baked" beans, black-eyed peas (cowpeas), chickpeas (garbanzo beans), and soybeans. Root vegetables (tubers) such as potatoes and yams are a third major source of starch, and in many non-Western societies, they are the primary starch sources.

Grains, legumes, and tubers not only are rich in starch, but also contain abundant dietary fiber, protein, and other nutrients. When nutrition experts advise you to seek out carbohydrate-rich foods to meet most of your energy needs, these are the foods they are recommending. Grains, legumes, and tubers are excellent energy foods and sources of many nutrients.

Glycogen Glycogen molecules, which are also made of chains of glucose, are more highly branched than starch molecules. As starch stores energy for plants, glycogen stores energy for human beings and animals. Because glycogen does not occur in plants and is found in meats only to a limited extent, it is not important as a nutrient. Nevertheless, glycogen plays an important role in the body as a readily available source of glucose, especially during exercise.

glycogen (GLY-co-gen): a polysaccharide composed of glucose, made and stored by liver and muscle tissues of human beings and animals as a storage form of glucose. Glycogen is not a significant food source of carbohydrate and is not counted as one of the complex carbohydrates in foods.

The energy-yielding complex carbohydrates or polysaccharides are starch, the storage form of glucose in plants, and glycogen, the storage form of glucose in animals and human beings.

The Fibers

The fibers of plants are constituents of plant cell walls. Most fibers are polysaccharides—chains of sugars—just as starch is, but in fibers the sugar units are held together by bonds that human digestive enzymes cannot break. Figure 2.2 shows the difference between starch and the plant fiber cellulose. In addition to cellulose, fibers include the polysaccharides hemicellulose, pectins, gums, and mucilages, as well as the nonpolysaccharide lignins.

fibers: a general term denoting in plant foods the polysaccharides cellulose, hemicellulose, pectins, gums, and mucilages, as well as the nonpolysaccharide lignins, that are not attacked by human digestive enzymes.

Fibers in Foods Cellulose is the main constituent of plant cell walls, so it is found in all vegetables, fruits, and legumes. Hemicellulose is the main constituent of cereal fibers. Pectins are abundant in vegetables and fruits, especially citrus fruits and apples. The food industry uses pectins to thicken jelly and keep salad dressing from separating. Gums and mucilages have similar structures.

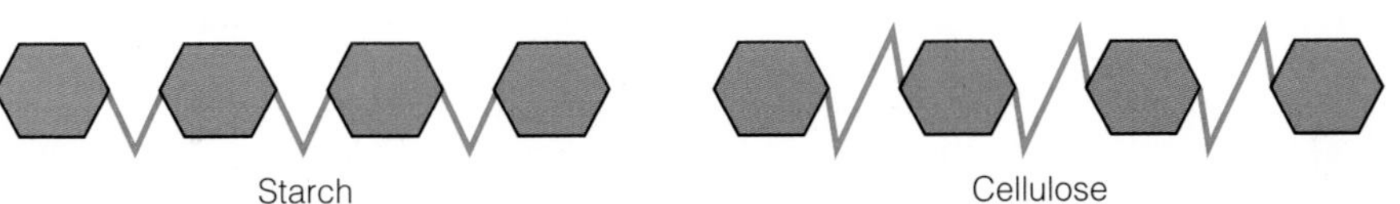

Figure 2.2
Starch and Cellulose Molecules Compared (Small Segments)
The bonds that link the glucose units together in cellulose are different from the bonds in starch (and glycogen). Human enzymes cannot digest cellulose.

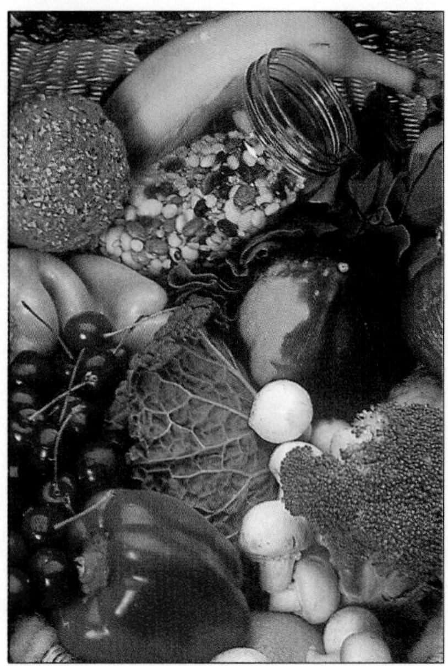

Starch- and fiber-rich foods are the foods to emphasize.

insoluble fibers: the tough, fibrous structures of fruits, vegetables, and grains; indigestible food components that do not dissolve in water.

soluble fibers: indigestible food components that readily dissolve in water and often impart gummy or gel-like characteristics to foods. An example is pectin from fruit, which is used to thicken jellies.

They are used as additives or stabilizers by the food industry. Lignins are the tough, woody parts of plants. Few foods people eat contain much lignin.

Fibers in the Body Although cellulose and other fibers are not attacked by human enzymes, some fibers can be digested by bacteria in the human digestive tract and can generate some absorbable products which can yield energy when metabolized. Food fibers, therefore, can contribute some energy, depending on the extent to which they break down in the body.

Soluble versus Insoluble Fibers Fibers are classified according to several characteristics, including their chemical structure, their digestibility by bacterial enzymes, and their solubility in water. Some fibers are insoluble, meaning that they do not dissolve readily in water; other fibers are soluble—they do dissolve in water. These distinctions influence the health effects of fibers, which are discussed in a later section.

Most fibers are polysaccharides, but their bonds cannot be broken by human digestive enzymes. The fibers therefore yield little energy.

Health Effects of Sugars and Alternative Sweeteners

Starch-rich and fiber-rich foods such as vegetables, grains, legumes, and fruits should predominate in people's diets; concentrated sweets should account for only 10 percent or less of total kcalories. For most people this means that total carbohydrate intake should increase while sugar intake should decline. The sugars in vegetables, fruits, grains, and milk are acceptable because they are accompanied by many nutrients. In contrast, concentrated sweets contribute many kcalories, but relatively few nutrients, and so should be limited. People who want to limit their use of sugar may choose from two sets of alternative sweeteners: sugar alcohols and artificial sweeteners.

Sugars

In the United States, it is estimated that each man, woman, and child consumes about 45 pounds of sugar per year, or a little less than a pound per week.[2] As a percentage of daily energy intake, this amount is roughly equivalent to current recommendations that sugar contribute no more than 10 percent of energy intake. Recommendations that people reduce their consumption of concentrated sugars to 10 percent or less of total kcalories stem from widely published reports of research performed during the 1970s.[3] These reports implicated sugars (mainly sucrose) as a possible contributing factor in several diseases. Since then, many accusations have been made against the sugar sucrose, but the Food and Drug Administration (FDA) and the National Academy of Sciences have concluded that in the amounts people currently consume, sugar poses no major health risk. Still, some controversy continues. A brief description of accusations pertaining to sugar's effects on health may help clarify the issues. Most commonly, reports accuse sugar of causing (1) obesity, (2) diabetes, (3) heart disease, (4) hyperactivity and aggressive behavior in children, and (5) dental caries.

Sugar and Obesity On the first accusation—that sugar causes obesity—the facts are these. Population studies show that obesity rises as sugar consumption increases, but that sugar is not the sole cause. Whenever sugar intakes increase, usually fat and total energy intakes do, too. Simultaneously, physical activity declines. On the other hand, obesity also occurs where sugar intakes are low. In some instances, obese people eat less sugar than thin people.[4] Fat is more kcalorie dense than sugar and can easily contribute to obesity. Thus sugar can contribute to obesity, but does not, by itself, cause obesity.

WWW diabetes.org/ada/c40c.html
Sugars and Artificial Sweeteners
(from the American Diabetes Association)

Sugar and Diabetes On sugar's relation to diabetes, the evidence is conflicting and interesting. In many parts of the world, as sugar consumption has increased, a profound increase in the incidence of one type of diabetes (type 2) has occurred. Yet, in other populations, no relationship has been found between diabetes and sugar consumption. Body fatness is one known risk factor: the majority of people with type 2 diabetes are overweight or obese, and evidence shows that weight reduction helps to prevent diabetes or relieve its symptoms. Genetics, age, and physical inactivity are also considered risk factors for type 2 diabetes.[5] Thus sugar may be a causative factor only when it contributes to obesity. Chapter 25 discusses diabetes further.

type 2 diabetes: the more common type of diabetes in which the fat cells resist insulin.

Sugar and Behavior The accusation that sugar causes hyperactive or aggressive behavior in children stands unproven.[6] One group of researchers tested the effects of sugar on children already diagnosed as hyperactive.[7] The hyperactive children did become more distracted (paid less attention) than usual after eating a sugary breakfast, but did not behave more aggressively. Many other studies have failed to demonstrate any consistent effect of sugar on behavior in either normal or hyperactive children.[8] If sugar is related to behavior problems in children, it may be because the sugary foods replace nutrient-dense foods in children's diets, making nutrient deficiencies likely. Many different nutrient deficiencies adversely affect behavior. A lack of nutrients in children's diets, not sugar itself, can in some cases contribute to undesirable behavior.

Chapter 13 offers further discussion of children's nutrition.

Sugar and Dental Caries As to whether sugar contributes to dental caries, the evidence says yes. Any carbohydrate-containing food, including bread, bananas, or milk, as well as sugar, can support the bacterial growth in the mouth that produces the acid that eats away tooth enamel. Total sugar intake still plays a major role in caries incidence; populations with diets of more than 10 percent of kcalories from sugar have an unacceptably high incidence of dental caries.[9] Nutrition in Practice 2 discusses nutrition and dental health.

dental caries: the gradual decay and disintegration of a tooth.

WWW. ada.org
American Dental Association

Recommended Sugar Intakes Moderate sugar intakes (5 to 10 percent of total kcalories)—enough for pleasure, but not enough to displace more nutritious foods—are not harmful. Sugar is a delicious, concentrated source of food energy, but it contains no protein, vitamins, or minerals. Eaten in the place of nutrient- and fiber-rich foods, it makes malnutrition likely.

Recognizing Sugars People often fail to recognize sugar in all its forms and so fail to realize how much they consume. To estimate how much sugar you consume, treat all of the following concentrated sweets as equivalent to 1 teaspoon of white sugar:

- 1 teaspoon brown sugar, candy, jam, jelly, any corn sweetener, syrup, honey, molasses, or maple sugar.
- 1 tablespoon catsup.
- 1½ ounces carbonated soft drink (that's 8 teaspoons of sugar per 12-ounce can).

These portions of sugar all provide about the same number of kcalories. Some are closer to 10 kcalories (for example, 14 kcalories for molasses), while some are over 20 (22 kcalories for honey), so an average figure of 20 kcalories is an acceptable approximation. The glossary on the next page presents the multitude of names that denote sugar on food labels. The next section discusses sugar substitutes.

Sugar Glossary

brown sugar: refined white sugar crystals to which manufacturers have added molasses syrup with natural flavor and color; 91 to 96% pure sucrose.

concentrated fruit juice sweetener: a concentrated sugar syrup made from dehydrated, deflavored fruit juice, commonly grape juice; used to sweeten products that can then claim to be "all fruit."

confectioners' sugar: finely powdered sucrose; 99.9% pure.

corn sweeteners: corn syrup and sugars derived from corn.

corn syrup: a syrup produced by the action of enzymes on cornstarch, containing mostly glucose. See also *high-fructose corn syrup (HFCS).*

dextrose: an older name for glucose and the name used for glucose in intravenous solutions.

fructose, galactose, glucose: already defined (pp. 36 and 37).

granulated sugar: common table sugar, crystalline sucrose; 99.9% pure.

high-fructose corn syrup (HFCS): the predominant sweetener used in processed foods today. HFCS is mostly fructose; glucose makes up the balance.

honey: sugar (mostly sucrose) formed from nectar gathered by bees. An enzyme splits the sucrose into glucose and fructose. Composition and flavor vary, but honey always contains a mixture of sucrose, fructose, and glucose.

invert sugar: a mixture of glucose and fructose formed by splitting sucrose in a chemical process; sold only in liquid form, sweeter than sucrose. Invert sugar is used as an additive to help preserve food freshness and prevent shrinkage.

lactose: already defined (p. 38).

levulose: an older name for fructose.

maltose: already defined (p. 38).

maple sugar: a sugar (mostly sucrose) purified from concentrated sap of the sugar maple tree. Maple sugar is expensive compared with other sweeteners.

molasses: a thick brown syrup, left over from sugarcane juice during sugar refining. Blackstrap molasses contains iron, which comes from the machinery used to process it. This iron is not as well absorbed as the iron in meats and other foods.

raw sugar: the first crop of crystals harvested during sugar processing. Raw sugar cannot be sold in the United States because it contains too much filth (dirt, insect fragments, and the like). Sugar sold domestically as raw sugar has actually gone through about half of the refining steps.

sucrose: already defined (p. 37).

turbinado (ter-bih-NOD-oh) **sugar:** raw (brown) sugar from which the filth has been washed; legal to sell in the United States.

white sugar: pure sucrose, produced by dissolving, concentrating, and recrystallizing raw sugar.

In amounts typically consumed, sugars pose no major health risk, except for an increased risk of dental caries.

Alternative Sweeteners: Sugar Alcohols

sugar alcohols: sugarlike compounds; like sugars, they are sweet to taste but yield 2 to 3 kcal per gram, slightly less than sucrose. Examples are maltitol, mannitol, sorbitol, and xylitol.

nutritive sweeteners: sweeteners that yield energy, including both the sugars and the sugar alcohols.

The sugar alcohols are carbohydrates, but they yield slightly less energy (2 to 3 kcalories per gram) than sucrose (4 kcalories per gram) because they are not absorbed completely.[10] The sugar alcohols are sometimes called nutritive sweeteners.

The sugar alcohols occur naturally in fruits. Mannitol, maltitol, and sorbitol are less sweet than sucrose, while xylitol is equivalent to sucrose in sweetness.[11] Unlike sucrose, significant amounts of the sugar alcohols a person consumes reach the large intestine and are fermented to gases by intestinal bacteria. Therefore, side effects such as gas, abdominal discomfort, and diarrhea make the sugar alcohols less attractive than the artificial sweeteners.

The sugar alcohols offer a distinct advantage over sucrose when it comes to dental caries. Bacteria in the mouth metabolize sugar alcohols much more slowly than sucrose, thereby inhibiting the production of acids that promote caries formation. Sugar alcohols are therefore less caries promoting than sugar is. They are valuable in chewing gums, breath mints, and other products that people keep in their mouths a while. The FDA allows food labels to carry a health claim (see p. 22 in Chapter 1) about the relationship between sugar alcohols and the non-promotion of dental caries as long as the FDA criteria for sugar-free status and other criteria are met.[12]

Alternative Sweeteners: Artificial Sweeteners

The artificial sweeteners are not carbohydrates. They yield virtually no energy in the amounts typically used and are sometimes called nonnutritive sweeteners. Like the sugar alcohols, artificial sweeteners make foods taste sweet without promoting tooth decay. The glossary on the next page offers details on six artificial sweeteners of interest.

These chewing gums contain sugar alcohols, which are better than sugar for the teeth, but are not kcalorie-free.

artificial sweeteners: noncarbohydrate, nonkcaloric synthetic sweetening agents; sometimes called **nonnutritive sweeteners.**

Saccharin Saccharin has been used for more than 100 years in the United States and is currently used by more than 50 million people, mainly in soft drinks and as a tabletop sweetener. Questions about the safety of saccharin arose in 1977 when experiments suggested that it caused bladder tumors in second-generation rats fed high doses. The FDA proposed banning saccharin as a result. Public outcry in favor of saccharin was so loud that Congress declared a moratorium on the ban, a moratorium that was repeatedly extended. In 1991, the FDA withdrew its proposal to ban saccharin, but labels must still carry a consumer warning: "Use of this product may be hazardous to your health. This product contains saccharin, which has been determined to cause cancer in laboratory animals."[13]

Does saccharin cause cancer? The largest population study to date, involving 9000 men and women, showed overall that saccharin use did not raise the risk of cancer. Among certain small groups of the population, however, such as those who both smoked heavily and used saccharin, the risk of bladder cancer was slightly greater. Other studies involving more than 5000 people with bladder cancer showed no association between bladder cancer and saccharin use.[14] Common sense dictates that consuming large amounts of saccharin is probably not safe, but at moderate intake levels, saccharin is currently assumed to be safe for most people. The FDA has set Acceptable Daily Intake (ADI) levels for some of the artificial sweeteners used in the United States.

Acceptable Daily Intake (ADI): the amount of a sweetener that individuals can safely consume each day over the course of a lifetime without adverse effect. It includes a 100-fold safety factor.

Aspartame Aspartame is the active ingredient in NutraSweet, which is used in many commercially prepared foods, and in Equal, a tabletop sweetener. Aspartame was approved by the FDA in 1981 and currently dominates the world market for artificial sweeteners. Aspartame is one of the most studied of all food additives: extensive animal and human studies document its safety. Long-term consumption of aspartame is safe and is not associated with any adverse health effects.[15] Aspartame is approved for use in more than 90 countries.

The FDA's approval of aspartame is based on the assumption that no one will consume more than the ADI of 50 milligrams per kilogram of body weight in a day. This daily intake is indeed a lot: for a 132-pound person, it adds up to 80 packets of Equal or 15 soft drinks sweetened only with aspartame. Most people consume between 2 and 10 milligrams per kilogram of body weight per day. Still, a child who drinks a quart of Kool-Aid sweetened with aspartame on a hot day, and who also has pudding, chewing gum, cereal, and other products sweetened with aspartame, takes in more than the ADI. Although this presents no proven hazard, it seems wise to offer children other foods so as not to exceed the limit.

WWW **wolfenet.com/~kronmal**
National PKU News

Aspartame and PKU Although aspartame is considered safe for most people, individuals with the metabolic disorder phenylketonuria (PKU) are an exception. The labels of products that contain aspartame must include information for individuals with PKU. Aspartame contains the amino acid phenylalanine, and people with PKU cannot dispose of it efficiently. Adults with PKU can use some pure aspartame, but children with PKU need to get all their phenylalanine from nutrient-rich foods such as milk and meat (see Nutrition in Practice 28).

Acesulfame Potassium The FDA approved acesulfame potassium (acesulfame-K) in 1988 after reviewing more than 90 safety studies, conducted over 15 years. Some consumer groups believe that acesulfame-K causes tumors in rats and

Artificial Sweeteners Glossary

acesulfame potassium: a low-kcalorie sweetener approved by the FDA and Health Canada; also known as acesulfame-K because K is the chemical symbol for potassium.

alitame: a compound of two amino acids (alanine and aspartic acid) that is 2000 times sweeter than sucrose; FDA approval pending.

aspartame: a compound of two amino acids (phenylalanine and aspartic acid) that tastes like the sugar sucrose but is much sweeter. It provides about 4 kcalories per gram, as does protein, but because so little is used, it is virtually kcalorie-free. In powdered form it is sometimes mixed with lactose, however, so a 1-gram packet may contain 4 kcalories. It is used in both the United States and Canada.

cyclamate: a 0-kcalorie sweetener; FDA approval pending in the United States; available in Canada on grocery-store shelves but only as a tabletop sweetener, not as an additive.

saccharin: a 0-kcalorie sweetener used in the United States, but available in Canada only in pharmacies and only as a sweetener, not as an additive.

sucralose: a 0-kcalorie sweetener that is 600 times sweeter than sucrose; FDA approval pending in the United States; approved in Canada.

should not have been approved. The FDA counters that the tumors were not caused by the sweetener, but were typical of those routinely found in rats.

Marketed under the trade names Sunette and Sweet One, acesulfame-K is about as sweet as aspartame. It is used in chewing gum, beverages, instant coffee and tea, gelatins, and puddings, as well as for table use. Unlike aspartame, acesulfame-K holds up well during cooking.

Other Artificial Sweeteners FDA approval for three other sweeteners—cyclamate, alitame, and sucralose—is still pending. To date, no safety issues have been raised for alitame or sucralose. Cyclamate, on the other hand, has been battling safety issues for 50 years. Approved by the FDA in 1949, cyclamate was banned in 1970 because of evidence indicating that it caused bladder cancer in rats. In 1985, the National Academy of Sciences concluded that evidence to date indicated that cyclamate did not cause cancer in human beings, but warranted further studies. In Canada, cyclamate is restricted to use as a tabletop sweetener on the advice of a physician. In the United States, the FDA is currently reviewing a petition to reapprove the use of cyclamate.

Artificial Sweeteners and Weight Control Many people eat and drink products sweetened with artificial sweeteners to help them control weight. Does this work? Ironically, a few studies have reported that after consuming such products, people experience heightened feelings of hunger. Despite these reports, most studies find that artificial sweeteners do not heighten feelings of hunger, enhance food intake, or cause weight gain in people.[16] Some studies find, however, that when people reduce their energy intakes by replacing sugar in their diets with artificial sweeteners, they compensate for the reduced energy at later meals.[17] Using artificial sweeteners will not automatically lower energy intake; to successfully control energy intake, a person needs to make informed diet and activity decisions throughout the day.

Recommendations For people who choose to use artificial sweeteners, the most important consideration is to use them in moderation. After all, almost every substance, even water, can be harmful in excess. The American Dietetic Association recommends that artificial sweeteners be used in moderation and only as part of a well-balanced diet.[18]

Alternative sweeteners include the sugar alcohols and artificial sweeteners. The sugar alcohols yield slightly less energy than sucrose and do not promote dental caries. Artificial sweeteners such as aspartame and saccharin yield no energy and do not promote dental caries.

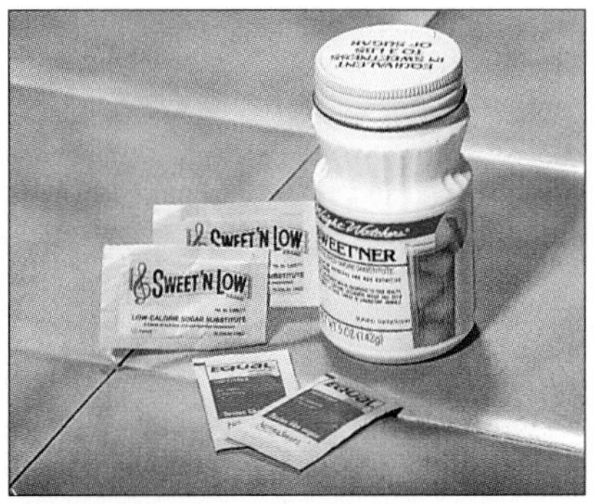
Artificial sweeteners offer the sweet taste of sugar without the kcalories.

Health Effects of Complex Carbohydrates

Despite dietary recommendations that people should eat generous servings of complex carbohydrate–rich foods for their health, many people still believe that carbohydrate is the "fattening" component of foods. Gram for gram, carbohydrates contribute fewer kcalories to the body than dietary fat, so a diet of high-carbohydrate foods is likely to be *lower* in kcalories than a diet of high-fat foods.

The health benefits a person can expect from a diet high in complex carbohydrate foods and low in concentrated sugar are many and wonderful. Such a diet is almost invariably low in fat, low in food energy, and high in fiber, vitamins, and minerals. All these factors working together can help reduce the risks of obesity, cancer, cardiovascular disease, diabetes, dental caries, gastrointestinal disorders, and malnutrition.

It is difficult to sort out which complex carbohydrates contribute to which health benefits. Starch and fibers almost always occur together in foods (except refined foods), so it is hard to distinguish their effects. Some health effects appear to be especially closely associated with fibers, however, and these are discussed next.

Fibers

Fiber-rich foods benefit health in many ways. Fibers in foods are thought to play a beneficial role in the prevention or management of:

- *Weight control* Fiber-rich foods contribute little energy and take longer to eat than low-fiber foods, thereby enhancing satiety.[19] High-fiber foods also promote a feeling of fullness as they absorb water. In addition, soluble fibers in a meal slow the movement of food through the upper digestive tract, so a person feels fuller longer. In a study of more than 200 men, researchers assessed body fat and diet composition (energy, carbohydrate, fat, protein, and fiber). Dietary fiber had the strongest correlation with body fat: the men with the highest percentage of body fat consumed significantly less fiber than leaner men.[20] In a different study, researchers found that lean men and women consumed significantly more fiber than obese men and women.[21] A diet high in fiber-rich foods can promote weight loss if those foods displace concentrated fats and sweets.
- *Constipation, hemorrhoids, and diarrhea* Fibers that both enlarge and soften stools (such as insoluble wheat bran) ease elimination for the rectal muscles, which alleviates or prevents constipation and hemorrhoids.[22] Other fibers help to solidify watery stools.
- *Appendicitis* Some fibers (again, such as wheat bran) help keep the contents of the intestinal tract moving easily. This action helps prevent compaction of the intestinal contents, which could obstruct the appendix and permit bacteria to invade and infect it.
- *Diverticulosis* Fibers stimulate the muscles of the digestive tract so that they retain their health and tone. This prevents the muscles from becoming weak and the lining of the digestive tract from bulging out in places, as occurs in diverticulosis. In a study of more than 40,000 men, those who ate high-fiber diets had lower risks of diverticulosis than men who ate low-fiber diets.[23]

Chapter 19 describes diverticulosis.

- *Colon cancer* Evidence strongly suggests that high-fiber diets protect against colon cancer. Populations who consume high-fiber diets have lower rates of colon cancer than similar populations consuming low-fiber diets.[24] Fiber may help prevent colon cancer by diluting, binding, and rapidly removing potentially cancer-causing agents from the colon.
- *Heart disease* Diets high in fiber, especially cereal fiber, significantly reduce the risk of heart disease.[25] Soluble fibers bind cholesterol compounds and carry them out of the body with the feces, thus lowering the body's cholesterol concentration.[26] High-fiber foods may also lower blood cholesterol indirectly by displacing fatty, cholesterol-raising foods from the diet.[27] Even when dietary fat intake is low, research shows that high intakes of soluble fiber (such as that in apples and oat bran) exert a separate and significant blood cholesterol–lowering effect.[28]
- *Diabetes* Some fibers delay the passage of nutrients from the stomach to the small intestine. This delay slows glucose absorption, thus eliciting a moderate insulin response and a moderate rise in blood glucose. This slow, sustained rise in blood glucose is considered desirable; fast glucose absorption and a surge in blood glucose are undesirable. The term *glycemic effect* describes the effect of food on blood glucose. An extensive long-term study of more than 65,000 women showed that women whose diets had the highest glycemic effect and the lowest cereal fiber content were most vulnerable to type 2 diabetes independent of other dietary constituents or known risk factors.[29]

glycemic effect: a measure of the extent to which a food raises the blood glucose concentration and elicits an insulin response, as compared with pure glucose.

Fibers and Health Claims The FDA authorizes three health claims on food labels concerning fiber. One is for "fiber-containing grain products, fruits, and vegetables and cancer." Another is for "fruits, vegetables, and grain products that contain fiber, and the risk of coronary heart disease," and a third is for "foods that contain oat bran or oatmeal and the risk of coronary heart disease." Chapter 1 describes the criteria foods must meet to bear these health claims.

Even with all these advantages, carbohydrate in the form of raw fiber is not a wonder cure. In some cases it can be detrimental. When too much fiber is consumed, some essential vitamins and minerals may bind to it and be excreted with it without becoming available for the body to use. Also, consuming purified fiber such as cellulose may not confer the same health benefits as consuming cellulose from a food source such as whole grains.

Fibers in Foods Meals selected according to the Food Guide Pyramid deliver ample fiber. Most people in the United States do not eat this way, though, and their average fiber intakes are about one-half of the recommended 20 to 35 grams daily.[30] Research shows that only when people meet the grain recommendation *and* one or both of the vegetable and fruit recommendations (based on the Food Guide Pyramid), do they achieve fiber intakes higher than 20 grams per day.[31] To obtain enough fiber, eat ample servings of whole foods, using Figure 2.3 as a guide.

Miscellaneous whole foods also add some fiber: you can get about 1 gram from ½ ounce of nuts, a tablespoon of peanut butter, a large pickle, or 5 olives. A day's meals based on the Daily Food Guide, such as those shown in Figure 2.4 (on p. 50), not only meet carbohydrate recommendations, but provide abundant fiber, too.

Fiber-rich foods benefit health in many ways. Fibers help maintain the health of the digestive tract and may protect against or control certain chronic diseases such as heart disease, cancer, and diabetes.

Figure 2.3
Fiber in Selected Foods

Bread, Cereal, Rice, and Pasta Group

Whole-grain products provide 1 to 2 g of fiber or more per serving:

- 1 slice whole-wheat or rye bread (1 g).
- 1 slice pumpernickel bread (2 g).
- ½ c ready-to-eat 100% bran cereal (10 g).
- ½ c cooked barley, bulgur, grits, oatmeal (2 to 3 g).

Vegetable Group

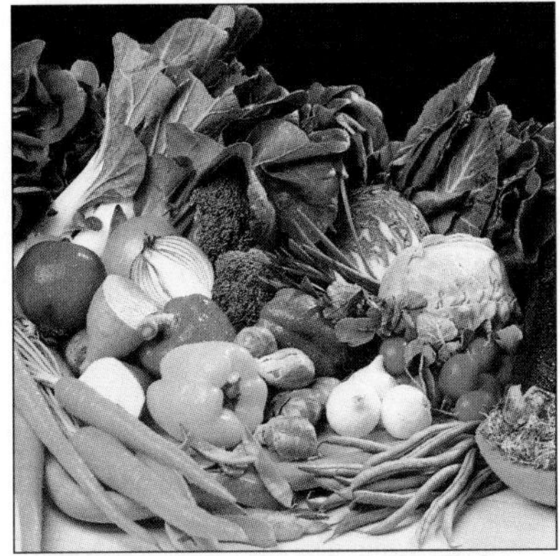

Most vegetables contain 2 to 3 g of fiber per serving:

- 1 c raw bean sprouts.
- ½ c cooked broccoli, brussels sprouts, cabbage, carrots, cauliflower, collards, corn, eggplant, green beans, green peas, kale, mushrooms, okra, parsnips, potatoes, pumpkin, spinach, sweet potatoes, swiss chard, winter squash.
- ½ c chopped raw carrots, peppers.

Fruit Group

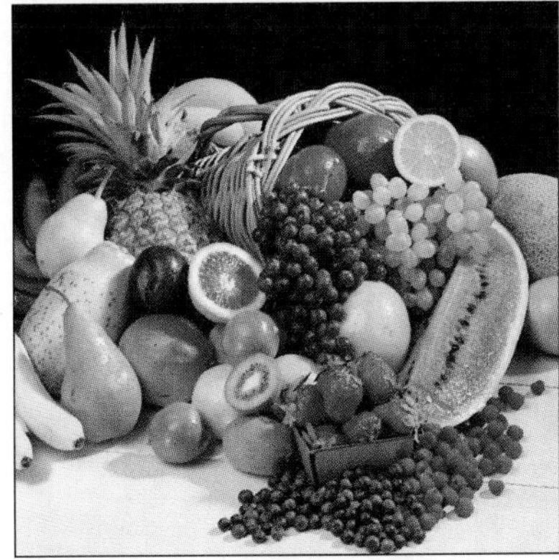

Fresh, frozen, and dried fruits have about 2 g of fiber per serving:

- 1 medium apple, banana, kiwi, nectarine, orange, pear.
- ½ c applesauce, blackberries, blueberries, raspberries, strawberries.
- Fruit juices contain very little fiber.

Legumes

Many legumes provide about 8 g of fiber per serving:

- ½ c cooked baked beans, black beans, black-eyed peas, kidney beans, navy beans, pinto beans.

Some legumes provide about 5 g of fiber per serving:

- ½ c cooked garbanzo beans, great northern beans, lentils, lima beans, split peas.

Note: Appendix A provides fiber grams for over 2000 foods.

Carbohydrates: Food Sources and Recommendations

Grains, vegetables, fruits, and milk are noted for the valuable energy-yielding carbohydrates they contain: starches and dilute sugars. Each class of foods makes its own typical carbohydrate contribution. The Daily Food Guide and the Food Guide Pyramid in Chapter 1 can help you obtain the carbohydrate-rich foods you need.

Breads, Cereal, Rice, and Pasta A serving of most foods in this group—a slice of whole-wheat bread, half an English muffin or bagel, a 6-inch tortilla, or ½ cup of rice, pasta or cooked cereal—provides abundant carbohydrate as starch.* Some foods in this group, especially baked goods such as biscuits, croissants, muffins, and snack crackers, also contain sugar and/or fat.

Vegetables Some vegetables are major contributors of starch in the diet. Just a small white or sweet potato or ½ cup of cooked dry beans, corn, peas, plantain, or winter squash provides 15 grams of carbohydrate, as much as in a slice of bread, though as a mixture of sugars and starch. A ½-cup portion of carrots, okra, onions, tomatoes, cooked greens, or most other nonstarchy vegetables or a cup of salad greens provides about 5 grams as a mixture of starch and sugars. Each of these foods also contributes a little protein and no fat.

Fruits The size of a typical serving of fruit varies depending on the form of the fruit: ¾ cup of juice; a small banana, apple, or orange; ½-cup of most canned or fresh fruit; or ¼ cup of dried fruit. A typical fruit serving contains an average of about 15 grams of carbohydrate, mostly as sugars, including the fruit sugar fructose. Fruits vary greatly in their water and fiber contents, and therefore their sugar concentrations vary also. With the exception of avocado, which is high in fat, the fruits contain insignificant amounts of fat and protein.

Milk, Cheese, and Yogurt One cup of milk or yogurt or the equivalent (1 cup of buttermilk, ⅓ cup of dry milk powder, or ½ cup of evaporated milk) provides a generous 12 grams of carbohydrate. Among cheeses, cottage cheese provides about 6 grams carbohydrate per cup, while most other types contain little if any carbohydrate. All milk products vary in fat content, an important consideration in choosing among them; Chapter 3 provides the details.

Cream and butter, although dairy products, are not equivalent to milk because they contain little or no carbohydrate and insignificant amounts of the other nutrients important in milk. They are appropriately placed with the fats at the top of the Food Guide Pyramid.

Meat, Poultry, Fish, Dry Beans, Eggs, and Nuts With two exceptions, foods of this group provide almost no carbohydrate to the diet. The exceptions are nuts, which provide a little starch and fiber along with their abundant fat, and dry beans, which are low-fat sources of both starch and fiber. Just a ½-cup serving of beans provides 15 grams of carbohydrate, an amount equal to the richest sources in the Food Guide Pyramid.

Recommendations Carbohydrate has no RDA, but there is a recommendation that 55 to 60 percent of total kcalories should come from carbohydrate. The FDA used this guideline in establishing a Daily Value for carbohydrate of 300 grams per day, or 60 percent of kcalories based on 2000 kcalories per day. For people who eat about 1700 kcalories a day, the recommended intake translates to some 900 to 1000 kcalories from carbohydrate per day or, at 4 kcalories per gram, some

*Gram values in this section are adapted from the 1995 Exchange System.

225 to 250 grams. Most of this carbohydrate should be starch, not sugars, thus many servings of starchy foods are needed to meet this recommendation.

Figure 2.4 on the next page offers an example of meals that provide about 1700 kcalories and 225 grams of carbohydrate from nutrient-dense, fiber-rich foods. The carbohydrate content of a diet can be determined by using a nutrient composition table such as that found in Appendix A or by using the exchange list system described in Chapter 25.

Carbohydrates on Food Labels Food labels list the amount, in grams, of total carbohydrate—including starch, fibers, and sugars—per serving. Fiber grams are also listed separately, as are the grams of sugars. (With this information, you could calculate starch grams by subtracting the grams of fibers and sugars from the total carbohydrate.) Sugars reflect both added sugars and those that occur naturally in foods. Total carbohydrate and dietary fiber are also expressed as "% Daily Values" for a person consuming 2000 kcalories; there is no Daily Value for sugars.

Grains, vegetables, fruits, and milk, as well as legumes, are carbohydrate-rich. Recommendations advise that 55 to 60 percent of daily energy intake should come from carbohydrates.

Energy Nutrients in Perspective

An uninterrupted flow of energy is so vital to life that other functions are sometimes sacrificed to maintain it. For example, when a child is fed too little food, the food the child does consume will be used for energy to keep the heart and lungs going, while growth comes to a standstill. To go totally without an energy supply, even for a few minutes, is to die. The urgency of the need for energy has ensured that all creatures have developed ways of accumulating built-in reserves to protect themselves from being deprived of energy. One major provision against this sort of emergency is glycogen, the storage form of glucose. (The other is body fat, about which the next chapter says more.)

Glycogen Used for Energy When you do not eat carbohydrate, your body rapidly devours its glycogen stores. Stored glycogen can return glucose to the blood whenever the supply runs short, but the liver cells can store only limited amounts of glycogen. Once this supply is depleted, the body must turn to the other energy nutrients—fat and protein—to meet its energy needs.

Fat Used for Energy Unlike the liver cells, which can store only a limited amount of glycogen, the body's fat cells can store virtually unlimited quantities of fat, so supplies almost never run out. Fat normally is used to meet two-thirds of the body's energy needs, and most tissues can use it as is. The brain and nerves, however, need their energy as glucose, and the body cannot convert fat to glucose. After a long period of glucose deprivation, brain and nerve cells develop the ability to derive about half of their energy from a special form of fat known as ketones, but they still require glucose as well. This means that people wanting to lose weight need to eat a certain minimum amount of carbohydrate to meet their energy needs, even when they are limiting their food intakes.

Chapter 9 offers guidelines for weight loss.

ketones (KEY-tones): acidic, fat-related compounds formed from the incomplete breakdown of fat when carbohydrate is not available; technically known as *ketone bodies.*

Protein Used for Energy During a fast, when the available glycogen is gone and no carbohydrate is coming in from food, brain and nerve cells demand the glucose they need from the only available source—protein. The body begins to dismantle its own muscles and other lean tissues to generate glucose. Only

Figure 2.4

A Day's Meals that Offer 1700 kCalories and Meet the Carbohydrate Recommendation

Breakfast

Before heading off to classes, a student eats breakfast:
- 2 shredded wheat biscuits.
- 1 c 1% low-fat milk.
- ½ banana (sliced).

Lunch

Then goes home for a quick lunch:
- 1 turkey sandwich on whole-wheat bread with mayonnaise and mustard.
- 1 c vegetable juice (canned).

Afternoon Snack

While studying that afternoon, the student eats a snack:
- 4 whole-wheat crackers.
- 1 oz cheddar cheese.
- 1 apple.

Dinner

That night, the student makes dinner:

A salad made with:
- 1 c raw spinach leaves, shredded carrots, and sliced mushrooms.
- ⅓ c garbanzo beans.
- 5 lg olives.
- 1 tbs ranch salad dressing.

A dinner of:
- 1 c spaghetti with meat sauce.
- ½ c green beans.
- 2 tsp butter.

And, for dessert:
- 1¼ c strawberries (fresh).

Evening Snack

Later that evening, the student enjoys a bedtime snack:
- 3 graham crackers.
- 1 c 1% low-fat milk.

Total kcal: 1725
53% kcal from carbohydrate
29% kcal from fat
18% kcal from protein

Self Study

How's Your Carbohydrate Intake?

Refer to the forms you filled out for the Self Study in Chapter 1, or your Diet Analysis Plus profile to answer the following questions and complete Form 6.

1. How many grams of carbohydrate do you consume in an average day (from Form 2 or your Daily Average Report)? ________ grams
2. How many kcalories does this represent? ________ kcalories (Remember, 1 gram of carbohydrate contributes 4 kcalories.)
3. It is estimated that you should have at least 125 grams of carbohydrate in a day. How does your intake compare with this minimum? ________
4. What percentage of your total kcalories is contributed by carbohydrate (carbohydrate kcalories divided by total kcalories times 100—or use the answer you obtained on Form 5)? ________ percent
5. How does this figure compare with the recommendation that about 55 to 60 percent of the kcalories in your diet should come from carbohydrate? ________
6. Another dietary goal is that no more than 10 percent of total kcalories should come from refined and other processed sugars and foods high in such sugars. To assess your intake against this standard, sort the carbohydrate-containing foods you ate into three groups:
   - *Group A* Foods containing complex carbohydrate (foods found among the bread, cereal, rice, and pasta group and the vegetable group) contributed ________ kcalories.
   - *Group B* Nutritious foods containing simple carbohydrate (foods in the milk and fruit groups) contributed ________ kcalories.
   - *Group C* Foods containing mostly concentrated simple carbohydrate (sugar, honey, molasses, syrup, jam, jelly, candy, cakes, doughnuts, sweet rolls, cola beverages, and so on) contributed ________ kcalories.

   Does your concentrated sugar intake (Group C) fall within the recommended maximum of 10 percent of total daily kcalories? ________ If not, what food choices account for the excess sugar? ________________
7. Estimate the number of pounds of sugar (concentrated simple carbohydrate) you eat in a year (1 pound = 454 grams):

   ________ kcalories from Group C ÷ 4 = ________ grams/day.

   ________ grams/day × 365 days/year = ________ grams/year.

   ________ grams/year ÷ 454 grams/pound = ________ pounds/year.

   How does your yearly sugar intake compare with the estimated U.S. and Canadian average of about 45 pounds per person per year? Comment on this. ________________

When dietitians analyze a client's diet with respect to carbohydrate intake, they check for total carbohydrate, complex carbohydrate, sugar, and fiber. From this information, they can reinforce a client's current habits or make recommendations for changes.

adequate dietary carbohydrate can prevent this use of protein for energy, and its action in doing so is known as the protein-sparing effect of carbohydrate.

protein-sparing effect: the effect of carbohydrate in providing energy that allows protein to be used for other purposes.

Ultimately, after half of the body's protein is used, death occurs. Death from loss of lean body tissues will occur even in an obese person who fasts too long. It should be clear, then, that although carbohydrate is an ideal energy source, fat and sometimes protein are also extremely important in meeting energy demands.

Form 6 Carbohydrates and Sugar kCalories

Average daily carbohydrate intake: ________ grams
Total kcalories from carbohydrate: ________ kcalories (grams × 4)
Total kcalories from all sources: ________ kcalories
Total kcalories from carbohydrates should equal 55 to 60% of total kcalories from all sources. Percentage of kcalories from carbohydrate: ________%
Breakdown of carbohydrate kcalories:
A. ________ complex
B. ________ nutritious simple
C. ________ concentrated simple
kCalories from concentrated simple sugars should equal 10% or less of total kcalories from all sources. Percentage of kcalories from concentrated simple sugars: ________%

Self Check

1. Carbohydrates are found in virtually all foods **except:**
   a. milks. b. meats.
   c. breads. d. vegetables.
2. Complex carbohydrates include:
   a. galactose, starch, and glycogen.
   b. starch, glycogen, and fiber.
   c. lactose, maltose, and glycogen.
   d. sucrose, fructose, and glucose.
3. The chief energy source of the body is:
   a. sucrose. b. glycogen.
   c. glucose. d. fructose.
4. The primary form of stored glucose in animals is:
   a. glycogen. b. starch.
   c. cellulose. d. maltose.
5. The polysaccharide that helps form the cell walls of plants is:
   a. cellulose. b. starch.
   c. glycogen. d. lactose.
6. The two types of alternative sweeteners are:
   a. saccharin and cyclamate.
   b. sugar alcohols and artificial sweeteners.
   c. sorbitol and xylitol.
   d. sucrose and fructose.
7. Which of the following items may denote sugar on food labels?
   a. corn syrup b. molasses
   c. honey d. all of the above
8. A diet high in complex carbohydrates is:
   a. most likely low in fat.
   b. most likely low in fiber.
   c. most likely poor in vitamins and minerals.
   d. most likely disease promoting.
9. A fiber-rich diet may help to prevent or control:
   a. some types of cancer. b. heart disease.
   c. constipation. d. all of the above.
10. The recommended fiber intake is:
   a. 10 to 15 grams per day. b. 15 to 25 grams per day.
   c. 20 to 35 grams per day. d. 40 to 55 grams per day.

Answers to these questions appear in Appendix H.

Notes

1. B. Levine, Most frequently asked questions about lactose intolerance, *Nutrition Today* 31 (1996): 78–79; E. Gudmand-Hoyer, The clinical significance of disaccharide maldigestion, *American Journal of Clinical Nutrition* 59 (1994): 735S–741S.

2. W. H. Glinsmann and Y. K. Park, Perspective on the 1986 Food and Drug Administration assessment of the safety of carbohydrate sweeteners: Uniform definitions and recommendations for future assessments, *American Journal of Clinical Nutrition* 62 (1995): 161S–169S.

3. B. Szepesi, Carbohydrates, in *Present Knowledge in Nutrition,* 7th ed., eds. E. E. Ziegler and L. J. Filer (Washington, D.C.: International Life Sciences Institute Press, 1996), pp. 33–43.

4. J. O. Hill and A. M. Prentice, Sugar and body weight regulation, *American Journal of Clinical Nutrition* 62 (1995): 264S–274S; C. J. Lewis and coauthors, Nutrient intakes and body weights of persons consuming high and moderate levels of added sugars, *Journal of the American Dietetic Association* 92 (1992): 708–713.

5. A. R. Folsom and coauthors, Increase in fasting insulin and glucose over seven years with increasing weight and inactivity of young adults: The CARDIA study, *American Journal of Epidemiology* 144 (1996): 235–246; G. A. Colditz and coauthors, Weight gain as risk factor for clinical diabetes mellitus in women, *Annals of Internal Medicine* 122 (1995): 481–486.

6. M. Kinsbourne, Sugar and the hyperactive child, *New England Journal of Medicine* 330 (1994): 355–356; M. L. Wolraich and coauthors, Effects of diets high in sucrose or aspartame on the behavior and cognitive performance of children, *New England Journal of Medicine* 330 (1994): 301–307.

7. E. H. Wender and M. V. Solanto, Effects of sugar on aggressive and inattentive behavior in children with attention deficit disorder

with hyperactivity and normal children, *Pediatrics* 88 (1991): 960–966.

8. M. L. Wolraich, D. B. Wilson, and J. W. White, The effect of sugar on behavior or cognition in children: A meta-analysis, *Journal of the American Medical Association* 274 (1995): 1617–1621.

9. K. G. Konig and J. M. Navia, Nutritional role of sugars in oral health, *American Journal of Clinical Nutrition* 62 (1995): 275S–283S; A. Sheilham, Why free sugar consumption should be below 15 kg per person per year in industrial countries: The dental evidence, *British Dental Journal* 171 (1991): 63–65.

10. S. S. Natah and coauthors, Metabolic response to lactitol and xylitol in healthy men, *American Journal of Clinical Nutrition* 65 (1997): 942–950; J. W. Finley and G. A. Leveille, Macronutrient substitutes, in *Present Knowledge in Nutrition,* 7th ed., eds. E. E. Ziegler and L. J. Filer (Washington, D.C.: International Life Sciences Institute Press, 1996), pp. 581–595.

11. Finley and Leveille, 1996; W. L. Dills, Sugar alcohols as bulk sweeteners, *Annual Review of Nutrition* 9 (1989): 161–186.

12. Food and Drug Administration, Food labeling: Health claims—Sugar alcohols and dental caries, *Federal Register* 61 (August 23, 1996): 43433–43447.

13. Withdrawal of certain pre-1986 proposed rules: Final action, *Federal Register* 56 (1991): 67442, as cited in Position of The American Dietetic Association: Use of nutritive and nonnutritive sweeteners, *Journal of the American Dietetic Association* 93 (1993): 816–821.

14. Position of The American Dietetic Association, 1993.

15. Position of The American Dietetic Association, 1993.

16. Position of The American Dietetic Association, 1993; B. J. Rolls, Effects of intense sweeteners on hunger, food intake, and body weight: A review, *American Journal of Clinical Nutrition* 53 (1991): 872–877.

17. R. M. Black and G. H. Anderson, Sweeteners, food intake, and selection, in *Appetite and Body Weight Regulation: Sugar, Fat and Macronutrient Substitutes,* eds. J. D. Fernstrom and G. D. Miller (Boca Raton, Fla.: CRC Press, 1994), pp. 125–136; R. Mattes, Effects of aspartame and sucrose on energy intake and hunger in humans, *Physiology and Behavior* 47 (1990): 1037–1044.

18. Position of The American Dietetic Association, 1993.

19. J. W. Anderson, B. M. Smith, and N. J. Gustafson, Health benefits and practical aspects of high-fiber diets, *American Journal of Clinical Nutrition* 59 (1994): 1242S–1247S.

20. L. H. Nelson and L. A. Tucker, Diet composition related to body fat in a multivariate study of 203 men, *Journal of the American Dietetic Association* 96 (1996): 771–777.

21. W. C. Miller and coauthors, Dietary fat, sugar, and fiber predict body fat content, *Journal of the American Dietetic Association* 94 (1994): 612–615.

22. J. H. Cummings and H. N. Englyst, Gastrointestinal effects of food carbohydrate, *American Journal of Clinical Nutrition* 61 (1995): 938S–945S.

23. W. H. Aldoori and coauthors, A prospective study of diet and the risk of symptomatic diverticular disease in men, *American Journal of Clinical Nutrition* 60 (1994): 757–764.

24. Y. Kim and J. B. Mason, Nutrition chemoprevention of gastrointestinal cancers: A critical review, *Nutrition Reviews* 54 (1996): 259–279.

25. E. B. Rimm and coauthors, Vegetable, fruit, and cereal fiber intake and risk of coronary heart disease among men, *Journal of the American Medical Association* 275 (1996): 447–451.

26. C. M. Ripsin and coauthors, Oat products and lipid lowering: A meta-analysis, *Journal of the American Medical Association* 267 (1992): 3317–3325.

27. J. F. Swain and coauthors, Comparison of the effects of oat bran and low-fiber wheat on serum lipoprotein levels and blood pressure, *New England Journal of Medicine* 322 (1990): 147–152.

28. D. J. A. Jenkins and coauthors, Effect on blood lipids of very high intakes of fiber in diets low in saturated fat and cholesterol, *New England Journal of Medicine* 329 (1993): 21–26; J. W. Anderson and coauthors, Prospective, randomized controlled comparison of the effects of low-fat plus high-fiber diets on serum lipid concentrations, *American Journal of Clinical Nutrition* 56 (1992): 887–894.

29. J. Salmeron and coauthors, Dietary fiber, glycemic load, and risk of noninsulin-dependent diabetes mellitus in women, *Journal of the American Medical Association* 277 (1997): 472–477.

30. Position of The American Dietetic Association: Health implications of dieteary fiber, *Journal of the American Dietetic Association* 97 (1997): 1157–1159.

31. S. M. Krebs-Smith and coauthors, Characterizing food intake patterns of American adults, *American Journal of Clinical Nutrition* 65 (1997): 1264S–1268S.

CHAPTER 2

Nutrition in Practice

▪ Nutrition and Dental Health ▪

The health value of eating complex carbohydrate–rich foods was emphasized throughout Chapter 2. Eating complex carbohydrates is a sound nutrition practice that promotes overall health, but unfortunately, it does not necessarily promote dental health. Nutrients in general support dental health in the same ways they support whole-body health, but carbohydrates in particular have to be handled in special ways to maximize dental health.

Nutrition plays a major role in facilitating the development of healthy teeth.[1] Table NP2.1 shows the effects of nutrient deficiencies on tooth development. The foods you eat and the times you eat them also play a major role in promoting or preventing dental caries—a pervasive health problem throughout the world.

Table NP2.1 Nutrient Deficiencies Affecting Tooth Development

Nutrient Deficiency	Effect on Tooth Development
Protein	Small, irregularly shaped teeth; delayed eruption; high caries susceptibility
Vitamin C	Disturbance of dentin formation
Vitamin A	Disturbance of enamel formation, delayed eruption
Vitamin D	Poor mineralization, pitting, striations
Calcium	Poor mineralization
Phosphorus	Poor mineralization
Magnesium	Enamel underdeveloped
Iron	High caries susceptibility
Zinc	High caries susceptibility
Fluoride	High caries susceptibility

What is dental caries?

Dental caries is an infectious oral disease that develops in the tooth enamel (see Figure NP2.1). Caries develops when bacteria that reside in the plaque of teeth consume and metabolize carbohydrates, producing acids that attack the tooth enamel. Thus at least two main ingredients are required to make dental caries: bacteria and carbohydrates. In addition, factors such as heredity, nutrition status during early tooth development, dental hygiene practices, and fluoride intake influence a person's susceptibility to caries.

How do sugar and other carbohydrate-rich foods promote caries development?

The bacteria that promote dental caries thrive on food particles that contain carbohydrate. Both sugar and starch can support bacterial growth. Equally important is the length of time the food stays in the mouth, and this depends on how soon you brush your teeth after eating and how sticky the food is. The damage a food does relates to both its carbohydrate content and its stickiness. For example, raisins and granola, which adhere to the teeth, cause more caries than a food that is easily rinsed off such as a sugary beverage.

Sugar can be eaten without inviting tooth decay if it is removed from tooth surfaces promptly. Bacterial action is maximal in the first 20 minutes after the first contact. If immediate brushing is not possible, water or other beverages swished in the mouth after a meal can effectively rinse the teeth. Once-a-day flossing may also effectively control formation of caries, regardless of the carbohydrate content of the diet. Some people may never get caries because they have inherited resistance to them.

Can you rinse your mouth with your own saliva?

Yes, and some foods stimulate more saliva flow than others. Saliva protects against caries formation in several ways. It not only rinses the mouth, but also dilutes the caries-causing acid produced by bacteria, exerts antibacterial action, and provides protective minerals.[2] Foods that elicit saliva flow may therefore defend

WWW.
ada.org
American Dental Association

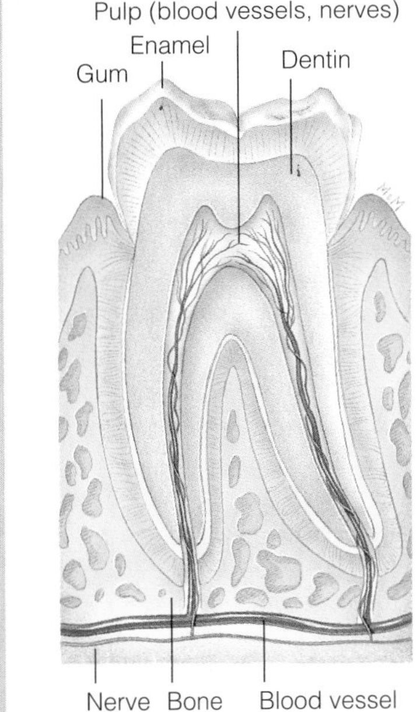

Figure NP2.1
A Tooth
The inner layer of dentin is bonelike material. The outer layer is enamel, which is harder than bone. Caries begins when acid dissolves the tooth's enamel.

against caries formation, but not all of them are protective; foods that also liberate sugar may promote acid formation. Apples are an example: they stimulate saliva flow, but they also liberate sugar, so they have both caries-preventing and caries-promoting effects. As you can see, many different factors influence caries development, making it difficult to predict exactly which foods are cariogenic.

WWW.
adha.org
American Dental Hygienists' Association

Are any foods strictly caries preventing?

Yes. Some foods stimulate saliva flow and do not contribute to acid formation in the mouth: cheese is an example. Such foods are good choices to eat at the end of a meal. Cheese is a powerful saliva stimulant and does not promote acid formation, so it reduces the cariogenicity of a meal. Furthermore, the high calcium and phosphorus content of cheese contributes to dental health.

High-fiber foods are, in general, anticariogenic, especially if their carbohydrate content is low. For example, raw vegetables do not stick to the teeth, and because they require vigorous chewing, they stimulate saliva flow. Cocoa products (including chocolate), coffee, tea, and beer all contain tannin, an acid that prevents caries formation. Table NP2.2 (p. 56) lists some dietary recommendations for controlling dental caries.

Besides foods, what other factors protect against dental caries development?

Research shows that when fluoride is added to the water supply, the children in the community have fewer dental caries than children who drink nonfluoridated water. In fact, water fluoridation is the most effective, least expensive way to provide dental care to everyone. The following recommendations will maximize protection against dental caries:

- Watch for hidden sugars in foods; use low-sugar or sugar-free products whenever possible.
- Restrict sweets to mealtimes.
- After you eat a meal or a between-meal snack, brush and floss or, at least, rinse with water.
- Limit the time that teeth are exposed to sticky foods.
- In any case, brush at least twice daily, and floss at least once daily.
- Visit a dentist regularly.
- Drink fluoridated water; provide infants and children with fluoride supplements when such water is not available.
- Eat a balanced diet composed of a variety of foods that will maintain adequate nutrition status.
- Eat foods that are rich in calcium and phosphorus.
- Eat a variety of firm, fibrous foods that will stimulate saliva flow.

In summary, learning and practicing sound dental hygiene habits, as well as developing eating habits that are consistent with both dental health and nutritional health, will serve a person throughout life.

Nutrition in Practice

Table NP2.2 Dietary Recommendations for Controlling Dental Caries

Food Group	Low Cariogenicity (Use When Teeth Cannot Be Brushed Immediately)	High Cariogenicity (Do Not Use unless Followed by Prompt and Thorough Dental Hygiene)
Dairy	Milk, cheese, plain yogurt	Ice cream, ice milk, milk shakes, fruited yogurts, eggnog
Meats/meat alternates	Meat, fish, poultry, eggs, legumes	Peanut butter with added sugar, luncheon meats with added sugar, meats with sugared glazes
Fruits[a]	Fresh, packed in water	Dried (raisins, figs, dates), packed in syrup or juice, jams, jellies, preserves, fruit juices and drinks
Vegetables	Most vegetables	Candied sweet potatoes, glazed carrots
Breads/cereals[b]	Popcorn, toast, hard rolls, pretzels, pizza, bagels	Cookies, sweet rolls, pies, cakes, dry sugared cereals as between-meal snacks, doughnuts, potato chips, granola bars
Other	Sugarless gum, coffee or tea without sugar, nuts	Sugared soft drinks, candy, fudge, caramels, honey, sugars, syrups

[a]Tiny particles of bananas can get lodged between teeth and decompose, increasing risk of caries.

[b]Tiny particles of breads, crackers, and chips can also become lodged in teeth, promoting caries formation.

Notes

1. D. Fitzsimons and coauthors, Nutritional and oral health guidelines for pregnant women, infants, and children, *Journal of the American Dietetic Association* 98 (1998): 182–189.

2. K. G. Konig and J. M. Navia, Nutritional role of sugars in oral health, *American Journal of Clinical Nutrition* 62 (1995): 275S–283S.

3 *Lipids*

lipids: a family of compounds that includes triglycerides (fats and oils), phospholipids, and sterols.

adipose tissue: the body's fat, which consists of masses of fat-storing cells called *adipose cells.*

You probably know that too much fat in the diet imposes health risks, but you may be surprised to learn that too little does, too. It is true, though, that people in the United States are more likely to eat too much fat than too little.

Fat is a member of the class of compounds called lipids. The lipids in foods and in the human body include triglycerides (fats and oils), phospholipids, and sterols.

Energy from Fat Lipids perform many tasks in the body, but most importantly, they provide energy. A constant flow of energy is so vital to life that, in a pinch, any other function is sacrificed to maintain it. Chapter 2 described one safeguard against such an emergency—the stores of glycogen in the liver that provide glucose to the blood whenever the supply runs short. The body's stores of glycogen are limited, however. In contrast, the body's capacity to store fat for energy is virtually unlimited due to the fat-storing cells of the adipose tissue. Unlike most body cells, which can store only limited amounts of fat, the adipose, or fat, cells seem able to expand almost indefinitely. The more fat they store, the larger they grow. Figure 3.1 shows a fat cell. The fat stored in these cells supplies 60 percent of the body's ongoing energy needs during rest. During physical activity or prolonged periods of food deprivation, fat stores may make an even greater energy contribution.

Roles of Body Fat In addition to supplying energy, fat serves other roles in the body. Natural oils in the skin provide a radiant complexion; in the scalp, they help nourish the hair and make it glossy. The layer of fat beneath the skin insulates the body from extremes of temperature. A pad of hard fat beneath each kidney protects it from being jarred and damaged, even during a motorcycle ride on a bumpy road. The soft fat in a woman's breasts protects her mammary glands from heat and cold and cushions them against shock. The fat embedded in muscle tissue shares with muscle glycogen the task of providing energy when the muscles are active. The phospholipids and the sterol cholesterol are cell membrane constituents that help maintain the structure and health of all cells. Table 3.1 on the next page summarizes the major functions of fats in the body.

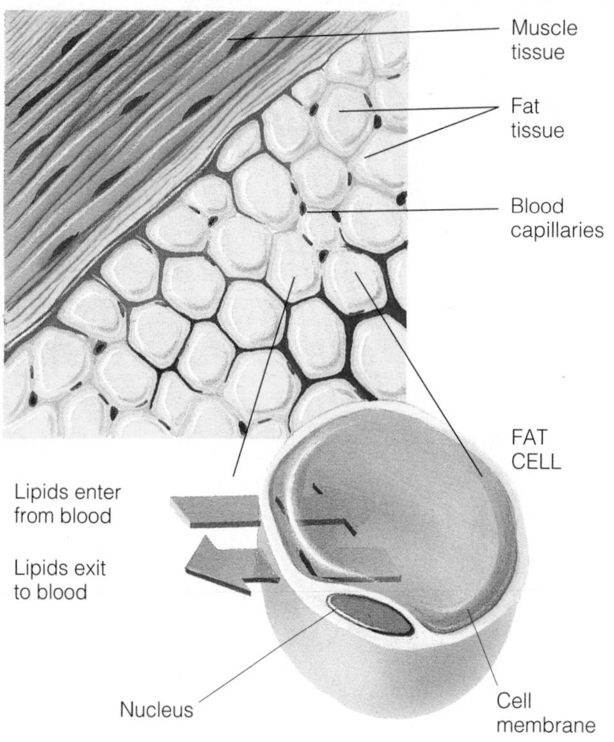

Figure 3.1
A Fat Cell
Within the fat cell, lipid is stored in a droplet. This droplet can greatly enlarge, and the fat cell membrane will grow to accommodate its swollen contents.

The Chemist's View of Lipids

The diverse and vital functions that lipids play in the body reveal why eating too little fat can be harmful. As mentioned earlier, though, too much fat in the diet seems to be the bigger problem for most people. To understand both the beneficial and harmful effects that fats exert on the body, a closer look at the structure and function of members of the lipid family is in order.

Table 3.1 The Functions of Fats in the Body
Fats in the body: ■ Provide energy. ■ Insulate the body against temperature extremes. ■ Protect the body's vital organs from shock. ■ Form the major material of cell membranes.

Triglycerides

When people talk about fat—for example, "I'm too fat" or "That meat is fatty"—they are usually referring to triglycerides. Among lipids, triglycerides predominate—both in the diet and in the body. The name *triglyceride* almost explains itself: three fatty acids (*tri*) attached to a glycerol "backbone." Figure 3.2 (p. 60) shows how three fatty acids combine with glycerol to make a triglyceride.

triglycerides: the chief form of fat in the diet and the major storage form of fat in the body; composed of glycerol with three fatty acids attached.
tri = three
glyceride = a compound of glycerol

Fatty Acids

When energy from any energy-yielding nutrient is to be stored as fat, the nutrient is first broken into small fragments. Then the fragments are linked together into chains known as fatty acids. The fatty acids are then packaged, three at a time, with glycerol to make triglycerides.

fatty acids: organic compounds composed of a chain of carbon atoms with hydrogens attached and an acid group at one end.

glycerol (GLISS-er-ol): an organic compound, three carbons long, that can form the backbone of triglycerides and phospholipids.

Chain Length and Saturation Fatty acids may differ from one another in two ways—in chain length and in degree of saturation. The chain length of a fatty acid affects the way it is absorbed (see Chapter 5). Saturation refers to its chemical structure—specifically, to the number of hydrogens the fatty acid chain is holding. If every available bond from the carbons is holding a hydrogen, the chain is called a saturated fatty acid, meaning that the chain is filled to capacity with hydrogen. The first zigzag structure in Figure 3.3 (see p. 61) represents a saturated fatty acid.

saturated fatty acid: a fatty acid carrying the maximum possible number of hydrogen atoms (having no points of unsaturation). A saturated fat is a triglyceride that contains three saturated fatty acids.

Unsaturated Fatty Acids In some fatty acids, especially those of plants and fish, hydrogens are missing in the fatty acid chains. The points where the hydrogens are missing are called points of unsaturation, and a chain containing such points is called an unsaturated fatty acid. If there is one point of unsaturation (as in oleic acid), the chain is monounsaturated. The second structure in Figure 3.3 is an example.

If there are *two* or more points of unsaturation, then the fatty acid is polyunsaturated (see the third structure in Figure 3.3). You sometimes see polyunsaturated fatty acids abbreviated on food labels as PUFA.

unsaturated fatty acid: a fatty acid with one or more points of unsaturation where hydrogens are missing (includes monounsaturated and polyunsaturated fatty acids).

monounsaturated fatty acid: a fatty acid that has one point of unsaturation; for example, oleic acid.

polyunsaturated fatty acid (PUFA): a fatty acid with two or more points of unsaturation. For example, linoleic acid has two such points, and linolenic acid has three. Thus a polyunsaturated *fat* is composed of triglycerides containing a high percentage of PUFA.

Hard and Soft Fat A triglyceride can contain any combination of fatty acids—long chain or short chain; saturated, monounsaturated, or polyunsaturated. Whether a fat is soft or hard depends on which fatty acids it contains. Fats that contain the shorter-chain or the more unsaturated fatty acids are softer and melt more readily (discussed later in this chapter). Each animal species (including human beings) obeys its own genetic program and makes its own characteristic kinds of triglycerides; but fats in the diet can, within limits, affect the types of triglycerides made. For example, animals raised for food can be fed diets containing softer or harder fats to give their meat or other products softer or harder fat, whichever consumers demand. People, too, incorporate the fatty acids they eat into their own tissues.

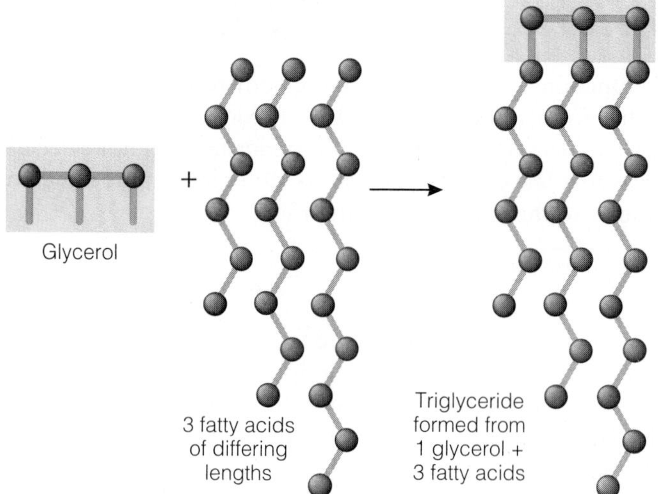

Figure 3.2
Triglyceride Formation
Glycerol, a small, water-soluble compound, plus three fatty acids, equals a triglyceride.

linoleic acid, linolenic acid: polyunsaturated fatty acids, essential for human beings.

essential fatty acids: fatty acids that the body requires but cannot make in amounts sufficient to meet its physiological needs.

omega: the last letter of the Greek alphabet (ω, sometimes replaced by the letter *m* or *n*), used by chemists to refer to the position of the last double bond in a fatty acid.

omega-6 fatty acid: a polyunsaturated fatty acid with its endmost double bond six carbons back from the end of its carbon chain; long recognized as important in nutrition. Linoleic acid is an example.

omega-3 fatty acid: a polyunsaturated fatty acid with its endmost double bond three carbons back from the end of its carbon chain; relatively newly recognized as important in nutrition. Linolenic acid is an example.

EPA, DHA: omega-3 fatty acids made from linolenic acid. The full name for EPA is eicosapentaenoic (EYE-cosa-PENTA-ee-NO-ick) acid. The full name for DHA is docosahexaenoic (DOE-cosa-HEXA-ee-NO-ick) acid.

Essential Fatty Acids The human body can synthesize all the fatty acids it needs from carbohydrate, fat, or protein except for two—linoleic and linolenic acid. Both linoleic and linolenic acid are polyunsaturated fatty acids and cannot be made from other substances in the body. They must be obtained from food and are therefore called *essential* fatty acids. Linoleic and linolenic acid are found in small amounts in plant and fish oils, and the body readily stores them, making deficiencies unlikely. From both of these essential fatty acids, the body makes important hormonelike substances that help regulate a wide range of body functions: blood pressure, clot formation, blood lipid concentration, the immune response, the inflammatory response to injury, and many others.[1] These two essential nutrients also serve as structural components of cell membranes.

Linoleic Acid: An Omega-6 Fatty Acid Linoleic acid is an omega-6 fatty acid found in the seeds of plants and in the oils produced from the seeds. Any diet that contains vegetable oils, seeds, nuts, and whole-grain foods provides enough linoleic acid to meet the body's needs. Researchers have long known and appreciated the importance of the omega-6 fatty acid family.

Linolenic Acid and Other Omega-3 Fatty Acids Linolenic acid belongs to a family of polyunsaturated fatty acids known as omega-3 fatty acids, a family that also includes EPA and DHA, seen on supplement labels. EPA and DHA are found primarily in fish oils. As mentioned, the human body cannot make linolenic acid, but given dietary linolenic acid, it can make EPA and DHA, although the process is slow.

The importance of omega-3 fatty acids was recognized during the 1980s when extensive worldwide research efforts began to unveil impressive roles for EPA and DHA in metabolism and disease prevention. DHA is one of the most abundant structural lipids in the brain, and both EPA and DHA are needed for normal brain development.[2] EPA and DHA are also especially active in the rods and cones of the retina of the eye.[3] Today researchers know that these omega-3 fatty acids are essential for normal growth and development and that they may play an important role in the prevention and treatment of heart disease, hypertension, arthritis, and cancer.[4]

Omega-3 Supplements Omega-3 fatty acid supplements are being aggressively marketed as a cure-all for many different diseases without regard for consumer safety. Although some claims are based on research, confirmation that fish oil can prevent or treat heart disease, cancer, or other diseases in individuals or populations is lacking. Experts do agree that adding *fish* to the diet two or three times

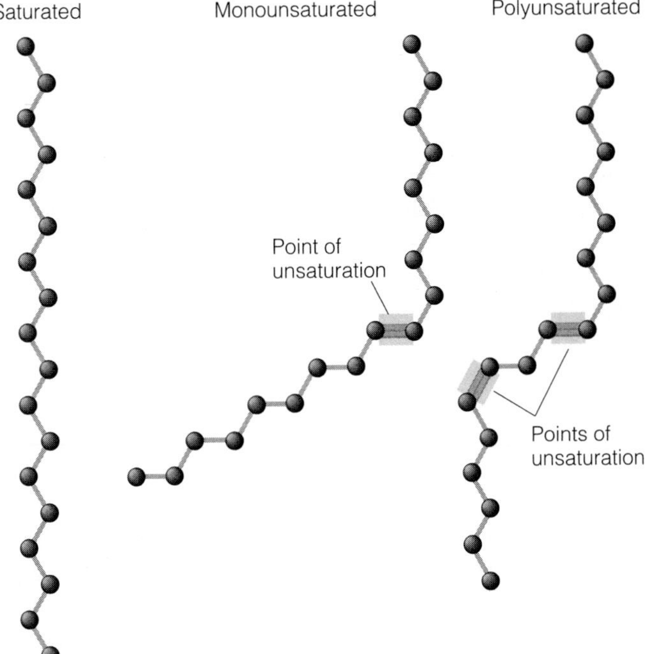

Figure 3.3
Three Fatty Acids
The more carbon atoms in a fatty acid, the longer its chain. The more hydrogen atoms attached to those carbons, the more saturated the fatty acid.

per week may be of benefit in preventing heart disease.[5] The idea that fish oil *supplements* are beneficial and safe in any amount is erroneous, however.

In the first place, the supplements themselves may carry hazards. They may contain toxic amounts of fat-soluble vitamins or of pesticide residues. Even when the supplements are not toxic, they may have harmful effects on the body. One potential problem is that the supplement form may not function the same way as the form available directly from foods. Furthermore, omega-3 and omega-6 fatty acids compete for the same slots in the body. Consequently, taking supplements of one can easily induce a deficiency of the other. Still another drawback is that fish oil can cause or aggravate illness by altering blood lipids or blood clotting. Finally, the quantities of omega-3 fatty acids in supplements vary widely from what the labels say they contain.

These fish provide at least 1 g of omega-3 fatty acids, including EPA and DHA, in 100 g of fish (about 3.5 oz):

- *Anchovy, European.*
- *Bluefish.*
- *Capelin conch.*
- *Herring, Atlantic or Pacific.*
- *Mackerel, Atlantic, chub, Japanese horse, or king.*
- *Mullet.*
- *Sablefish.*
- *Salmon, all varieties.*
- *Sturgeon, Atlantic or common.*
- *Trout, lake.*
- *Tuna, white albacore or bluefin (not canned light tuna).*
- *Whitefish, lake*

Eat Dark-Meat Fish Questions also remain about the appropriate amount of omega-3 fatty acids for a day's intake. While these questions await answers, the best way to increase your intake of omega-3 fatty acids is to eat several portions of fish each week, particularly the kinds listed in the margin. The darker flesh of fish has the highest fat content, and this is where most omega-3 fatty acids are found.

Triglycerides (glycerol with three fatty acids attached) are the major type of fat in the diet and in the body. Fatty acids can be saturated (filled with hydrogens) or monounsaturated (with one point of unsaturation) or polyunsaturated (with more than one point of unsaturation). The degree of saturation of the fatty acids in a fat determines the softness or hardness of the fat. Two polyunsaturated fatty acids, linoleic acid and linolenic acid, are essential nutrients used to make hormonelike substances that regulate a wide range of body functions.

Phospholipids

phospholipids: one of the three main classes of lipids; these compounds are similar to triglycerides but have choline (or another compound) and a phosphorus-containing acid in place of one of the fatty acids.

Up to now, this discussion has focused on one of the three classes of lipids, the triglycerides (fats and oils), and their component parts, the fatty acids (see Table

Table 3.2 The Lipid Family

Triglycerides (fats and oils)
- Glycerol (1 per triglyceride)
- Fatty acids (3 per triglyceride)
  - Saturated
  - Monounsaturated
  - Polyunsaturated
    - Omega-6
    - Omega-3

Phospholipids (such as lecithin)

Sterols (such as cholesterol)

lecithins: one type of phospholipids.

choline: a nonessential nutrient that can be made in the body from an amino acid.

emulsifiers: substances that mix with both fat and water and that permanently disperse the fat in the water, forming an emulsion.

3.2. The other two classes of lipids, the phospholipids and sterols, make up only 5 percent of the lipids in the diet, but they are nevertheless worthy of attention. Among the phospholipids, the lecithins are of particular interest.

Structure of Phospholipids Like the triglycerides, the lecithins and some other phospholipids have a backbone of glycerol; they differ from the triglycerides in having only two fatty acids attached to them. In place of the third fatty acid is a molecule of choline or a similar compound.

Roles of Phospholipids Lecithins and other phospholipids are important constituents of cell membranes. They also act as emulsifiers, helping to keep other fats in solution in the blood and body fluids.

Phospholipids: Not Essential Lecithins periodically receive noisy attention in the popular press and are credited with great deeds. You may have heard that they are a major constituent of cell membranes (true), that the functioning of all cells depends on the integrity of the cell membranes (true), and that you must therefore purchase bottles of lecithin and give yourself daily doses (false). The body digests lecithins before it absorbs them, so the lecithins you eat do not reach the body tissues intact. Instead, the lecithins used for building cell membranes are made from scratch by the liver. In other words, the lecithins are not essential nutrients.

Before buying bottles of lecithin or any other supplement products, ask yourself, "Do I really need this? What is the evidence that my body is likely to be deficient?"

Phospholipids, including the lecithins, are important constituents of cell membranes and act as emulsifiers in the blood and other body fluids.

Sterols

sterols: one of the three main classes of lipids; sterols include cholesterol, vitamin D, and the sex hormones (such as testosterone).

Sterols are large, complex molecules consisting of interconnected rings of carbon. Cholesterol is the most familiar sterol, but others, such as vitamin D and the sex hormones (for example, testosterone), are important, too.

Cholesterol Synthesis Like the lecithins, cholesterol can be made by the body, so it is not an essential nutrient. Your liver is manufacturing it now, as you read, at the rate of perhaps 50,000,000,000,000,000 molecules per second. The raw materials that the liver uses to make cholesterol can all be taken from glucose or saturated fatty acids. In other words, cholesterol can be made from either carbohydrate or fat. More than 90 percent of all the body's cholesterol ends up in the cells, where it performs vital structural and metabolic functions.

Cholesterol's Two Routes in the Body After being made, cholesterol either leaves the liver or is transformed there into related compounds such as vitamin D. Cholesterol leaves the liver by two routes:

1. It may be made into bile and pass into the intestine.
2. It may travel, via the bloodstream, to all the body's cells.

bile: a compound made by the liver from cholesterol and stored in the gallbladder. Bile prepares fat for digestion.

Cholesterol Recycled The bile that is made from cholesterol in the liver is released into the intestine to aid in the digestion and absorption of fat (see Chapter 5). After it does its job, some of the bile is reabsorbed and recycled and some is excreted in the feces.

Cholesterol Excreted While bile is in the intestine, some of it may be trapped by certain kinds of dietary fibers or by some medications, which carry it out of the body in feces. The excretion of bile reduces the total amount of cholesterol remaining in the body.

Cholesterol Transport Some cholesterol, packaged with other lipids and protein, leaves the liver via the arteries and is transported to the body tissues by the blood. The packages are called lipoproteins. As the lipoproteins travel through the body, tissues can extract lipids from them.

Lipoproteins are made by both the intestine and the liver. Chapter 5 tells the story of lipid transport.

Sterols are large complex molecules that include cholesterol, vitamin D, and the sex hormones.

Fats and Health

Of all the dietary factors related to chronic diseases prevalent in developed countries, fat is by far the most significant. Excessive intakes of dietary fat contribute to obesity, diabetes, cancer, coronary heart disease (CHD), and probably other diseases and disorders as well. Heart disease is the number one killer of adults in the United States. In the interest of good health and disease prevention, the one change that most people should make in their diets is to limit their intakes of total fat. It is especially important to limit saturated fat.

coronary heart disease (CHD): disease of the arteries that supply blood and oxygen to the heart (see Chapter 26).

Fats and Fatty Acids

The cholesterol that accumulates in arteries is manufactured largely from fragments derived from saturated fat. Most people realize that elevated blood cholesterol is an important risk factor for CHD. The higher the blood cholesterol, the greater the risk of heart disease. Most people may not realize, though, that cholesterol in *food* is not the main influential factor in raising *blood* cholesterol. It is total fat, especially saturated fat, that raises blood cholesterol. An extensive review of well-controlled studies of the effects of dietary fats on blood cholesterol reached the following conclusions:

- Saturated fatty acids elevate blood cholesterol and are the main dietary determinants of blood cholesterol levels.[6*]

Less influential, but still significant, were the following factors:

- Polyunsaturated fatty acids lower blood cholesterol.
- Monounsaturated fatty acids may lower blood cholesterol slightly or have a neutral effect on blood cholesterol.

Saturated Fat and Blood Cholesterol In support of these conclusions, a recent study found that diets low in saturated fatty acids helped lower blood cholesterol in men.[7] Adding polyunsaturated fatty acids to the diet lowered blood cholesterol further. Probably, the researchers speculated, reducing dietary cholesterol would have reduced blood cholesterol still further. These results are consistent with population studies that show that people who eat diets high in saturated fat are more likely to die from heart disease than those who eat the same diet, but also eat fish (rich in polyunsaturated fatty acids) a couple of times a week.[8]

WWW.

amhrt.org
American Heart Association

cancer.org
American Cancer Society

nci.nih.gov
National Cancer Institute

*It should be noted that not all saturated fatty acids have the same cholesterol-raising effect. For example, stearic acid, an 18-carbon fatty acid, does not raise blood cholesterol.

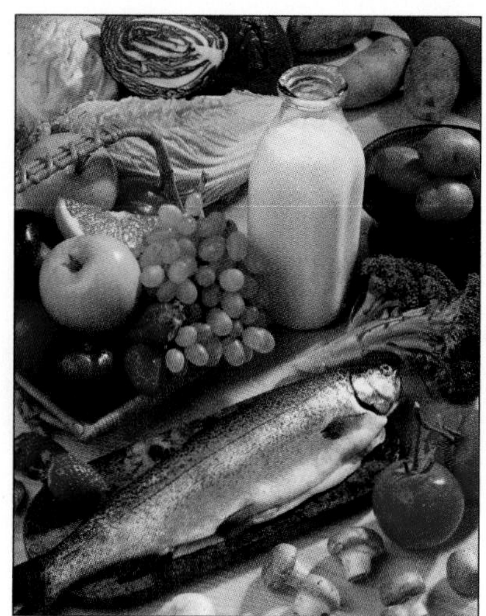

Enjoy low-fat foods for good health.

Omega-3 Fatty Acids and Blood Cholesterol Of the polyunsaturated fatty acids, the omega-3 fatty acids appear to reduce blood cholesterol the most. This effect was first noticed when researchers learned that the Inuit peoples of Alaska and Greenland, despite high-energy, high-fat, high-cholesterol diets, enjoyed relative freedom from heart diseases—especially atherosclerosis. Analysis of the foods common in Inuit diets, which derive primarily from marine animals, revealed that they were rich in omega-3 fatty acids, particularly EPA and DHA. Research findings such as these have led the American Heart Association to recommend two to three fish meals a week.

Omega-3 Fatty Acids and Cancer Omega-3 fatty acids may also help to prevent cancer. High *fat* intakes appear to *promote* cancer, and animal studies suggest that the fatty acids most likely to promote cancer are polyunsaturated fatty acids from vegetable oils (mostly omega-6 fatty acids). In comparison, saturated fats have little or no effect, and polyunsaturated fats from fish oils (mostly omega-3 fatty acids) may delay cancer development, slow tumor growth rates, and reduce the size and number of tumors.[9]

Monounsaturated Fatty Acids and Blood Cholesterol Whether monounsaturated fatty acids exert an independent blood cholesterol–lowering effect or have a neutral effect is unclear. Some research suggests that the cholesterol-lowering effects of monounsaturated fatty acids depend on the amount of cholesterol-elevating saturated fatty acids in the diet.[10] When dietary saturated fatty acids are low, monounsaturated fatty acids seem to have a cholesterol-lowering effect. When dietary saturated fatty acids are high, the cholesterol-lowering effects of monounsaturated fatty acids become less apparent or neutral. In general, compared with saturated fatty acids, monounsaturated fatty acids are cholesterol lowering, although less so than polyunsaturated fatty acids.[11] A benefit to health is seen when monounsaturated fat (such as olive oil or canola oil) replaces saturated fat (such as beef fat) in the diet.[12]

Recommendations Dietary guidelines recommend that *total* fat intake should not exceed 30 percent of the day's total energy intake; 20 percent might be ideal. Saturated fats should contribute less than 10 percent; polyunsaturated fats should not exceed 10 percent; monounsaturates should provide the remaining 10 percent or so. Limiting dietary fat intake to 30 percent or less of total energy requires careful planning at every meal. According to research on people's food choices, people eat about 34 percent of their energy intakes as fat.[13] Even though this is less than in the recent past, it is more than at the turn of the century, when foods were less highly processed, and it is more than people need. Guidelines for cholesterol intake recommend less than 300 milligrams daily. A later "How to" box offers specific suggestions for putting these guidelines into practice.

Excess dietary fat contributes to many diseases, including heart disease. Elevated blood cholesterol is a risk factor for heart disease. It is not cholesterol intake, however, but dietary fat intake, especially saturated fat, that is the major dietary factor that raises blood cholesterol.

WWW

ificinfo.health.org/review/ir-fat.htm
International Food Information Council Review: Sorting Out the Facts on Fat

fda.gov:80/fdac/features/696_fat.html
Taking the Fat Out of Food (from *FDA Consumer*)

Fat Substitutes

As people learn more and more about the health consequences of high-fat diets, fat substitutes, or fat replacements, offer hope for the prevention and treatment of heart disease and obesity. Skeptics say that people will use fat replacers the same way they use artificial sweeteners: in *addition* to fats, rather than *instead* of fats. Some research shows that when people use fat substitutes, they eat less fat but more sugar and starch, so their energy intakes remain constant.[14]

Types of Fat Replacers Food chemists have been working for decades on ways to reduce the fat in foods. Because the functions of fat in food are both diverse and desirable, it is not an easy ingredient to replace. In general, fat replacers fall into three categories: carbohydrate based (Oatrim and Z-Trim), protein based (Simplesse), or fat based (Salatrim and olestra).[15] Fat-based replacers can be further categorized into modified and synthetic. Modified fat-based replacers are triglycerides that have been altered to contain specific mixtures of fatty acids or specific arrangements of fatty acids. Salatrim is an example of a modified fat-based replacer. Olestra is the only synthetic fat-based replacer approved so far. Olestra is synthetic because its chemical configuration does not occur in nature.

Today shoppers are overwhelmed by the number of new fat-reduced products on grocery shelves. Fat-free bakery products, cheeses, frozen desserts, and many other products are available that taste rich but offer less than half a gram of fat in a serving. Products using traditional ingredients such as sugar, starch, nonfat milk, soluble fiber, or egg whites in place of fat are selling well.[16]

Oatrim and Z-Trim Researchers at the U.S. Department of Agriculture (USDA) developed Oatrim, which is derived from oat fiber. Oatrim is stable when heated and can be used as a fat substitute to reduce the energy content of foods such as frozen desserts, salad dressings, soups, and high-fiber baked goods. Oatrim and other carbohydrate-based replacers cannot be used for frying, however. Researchers say Oatrim reduces the energy content of frozen desserts like ice cream by half. Unlike Simplesse or other fat substitutes, Oatrim retains the qualities of its fiber, so it lowers cholesterol not only by replacing saturated fat but also by providing fiber.

Z-Trim is made from the seed hulls of oats, peas, soybeans, or rice or the bran from corn or wheat. Z-Trim provides no calories and imparts no flavor to foods. It can be used in baked goods, cheese, and meat products.[17]

Simplesse The Food and Drug Administration (FDA) declared Simplesse safe for use in ice cream and frozen desserts in 1990. Simplesse is made from protein—either egg white or milk—which is processed into mistlike particles similar in consistency to fat. Because the components of Simplesse have long been used in foods, safety studies are not required. This fat substitute mimics the rich taste and texture of fat but cuts the kcalorie content by up to 80 percent. Because proteins coagulate at high temperatures, Simplesse cannot be used for frying or cooking, but can be used in ice creams, yogurts, salad dressings, mayonnaise, and butter.

Olestra Olestra, which was formerly known as sucrose polyester (SPE), is a synthetic combination of sucrose and fatty acids that looks, feels, and tastes like food fat. Unlike sucrose or fatty acids, though, olestra is indigestible; the body has no way to take it apart. Olestra can therefore be substituted for fats without adding kcalories or raising a person's blood lipids.

From some points of view, Olestra is the most successful of the fat replacers, for its properties are identical to those of fats and oils when used in frying, cooking, and baking. It can be heated to frying temperatures without breaking down; it performs all of the functions of fat in cakes, pie crusts, and other baked goods; and most remarkably—aside from a slight aftertaste—it tastes like fat. However, the attributes of olestra must be weighed against evidence concerning its safety.

The Safety of Olestra The company that invented olestra has been studying its safety for more than two decades. The results of the studies revealed that olestra causes digestive distress and nutrient losses. The company has addressed each of these problems to the satisfaction of an FDA panel, which approved olestra for use in snack foods in 1996. Some nutrition experts disagree with the panel, however, and have come out strongly against olestra's approval. These issues deserve a moment's attention.

Olestra's side effects are rooted in two aspects of its nature. For one thing, because olestra replaces a major food constituent, fat, the substance is consumed in large amounts, measured in many grams per serving. Nonfat potato chips, for example, derive about a third of their weight from olestra. Also, by design, olestra is indigestible. All of the oily olestra eaten in a food passes through the digestive tract and is excreted from the body.

Digestive Problems The presence of olestra in the large intestine causes diarrhea, gas, cramping, and an urgent need for defecation in some people. Further, the oil can creep through the feces and leak uncontrollably from the anus. No one yet knows who is most likely to encounter these effects, but some of these symptoms almost always occur when olestra is consumed in large quantities. The FDA panel that approved olestra decided that these digestive problems were unpleasant, but did not constitute a safety problem.

The FDA requires that olestra-containing foods bear this warning:

"This Product Contains Olestra. Olestra may cause abdominal cramping and loose stools. Olestra inhibits the absorption of some vitamins and other nutrients. Vitamins A, D, E, and K have been added."

Nutrient Losses Whenever olestra is present in the digestive tract, it dissolves fat-soluble substances in the foods being digested. For example, some vitamins dissolve in fat, and all of these (vitamins A, D, E, and K) become unavailable for absorption when olestra is present in a meal. Thus, when olestra-containing chips are eaten with other foods, the olestra traps the vitamins those foods contain, robbing the eater of those vitamins. To compensate for this effect, olestra is fortified with vitamins A, D, E, and K. The FDA ruled that fortification removes the threat of harm from malnutrition that olestra could otherwise cause.

Do Fat Substitutes Work? Research has not yet shown that fat substitutes promote weight loss or lower blood lipid concentrations. If people use fat replacers literally as replacers for fat in their diets, then perhaps potential health benefits will be realized. Some experts are concerned, however, that people may feel at liberty to eat *more* high-fat foods by rationalizing that they obtain fat "credit" when they eat foods containing fat substitutes. Indeed, this seems to be the case with artificial sweeteners. Although consumption of sugar substitutes has risen since their introduction, sugar consumption has risen as well. Another concern is that people may become so carried away with eating foods containing fat substitutes that they will neglect to eat more nutrient-dense foods such as fresh fruits and vegetables. It seems that with fat substitutes, as with most things in life, moderation is the key to appropriate use.

WWW **fda.gov:80/opacom/backgrounders/olestra.html**
FDA Backgrounder on Olestra and Other Fat Substitutes

Fat replacers such as Oatrim, Z-Trim, Simplesse, and Olestra have been developed to replace fat in many foods. Olestra is a zero-kcalorie, synthetic fat-based fat replacer approved for use in snack foods. Olestra can cause unpleasant side effects such as diarrhea and cramping.

Fats in Foods

Fats are important in foods as well as in the body. Many of the compounds that give foods their flavor and aroma are found in fats and oils. The delicious aromas associated with bacon, ham, and other meats, as well as with onions being fried, come from fats. Fats also influence the texture of many foods, enhancing smoothness, creaminess, moistness, or crispness.[18] Four vitamins—A, D, E, and K—are soluble in fat. When the fat is removed from a food, many fat-soluble compounds, including these vitamins, are also removed. Table 3.3 summarizes the roles of fats in foods.

Fats are also an important part of most people's ethnic or national cuisines. Each culture has its own favorite food sources of fats and oils. In Canada, canola oil (also known as rapeseed oil) is widely used. In the Mediterranean area,

Greeks, Italians, and Spaniards rely heavily on olive oil. Both canola oil and olive oil are rich sources of monounsaturated fatty acids. Asians use the polyunsaturated oil of soybeans. Jewish people traditionally employ chicken fat. Everywhere in North America, butter and margarine are widely used.

These cultural associations along with the very real pleasures fats contribute to meals help explain why fat consumption is so high. Nevertheless, reducing dietary fat is the single most important thing you can do for health; it is the top recommendation of almost every nutrition authority. The first step to reducing dietary fat is finding out where the fat is.

Table 3.3 The Functions of Fats in Foods

Fats in foods:

- Contribute flavor and aroma.
- Influence the texture, adding creaminess, smoothness, moistness, or crispness.
- Help make foods tender.
- Carry fat-soluble vitamins.

Finding the Fats in Foods

Of the groups in the Food Guide Pyramid, the fats and the meats (and nuts) always contain fat, but two other food groups—the milk, cheese, and yogurt group and the breads—sometimes contain fat as well. Vegetables and fruits, if unprocessed, are virtually fat-free, but two exceptions are rich in monounsaturated fat—avocados and olives. Grains, too, in their natural state contain little or no fat.

Remember, fat is a more concentrated energy source than the other energy nutrients: 1 g carbohydrate or protein = 4 kcal, but 1 gram fat = 9 kcal.

Added Fats in Foods A dollop of dessert topping, a spread of butter on bread, oil or shortening in a recipe, dressing on a salad—all of these are examples of *added* fats. Indeed, all sorts of fats can be added to foods during commercial or home preparation or at the table. The following amounts of these fats contain about 5 grams of pure fat, providing 45 kcalories and negligible protein and carbohydrate:

- 1 teaspoon of oil or shortening.
- 1½ teaspoons of mayonnaise, butter, or margarine.
- 1 tablespoon of regular salad dressing, cream cheese, or heavy cream.
- 1½ tablespoons of sour cream.

These foods provide the majority of added fats to the diet. They are the fats of fried foods or baked goods, sauces and mixed dishes, and dips and spreads.

WWW **nalusda.gov/fnic/Fpyr/pyramid.html**
Food Guide Pyramid

Hidden Fats in Foods Other foods that are significant contributors of fat include convenience foods, lunch meats, and other prepared meats. The fat in these foods is sometimes referred to as invisible fat because it does not have the obvious appearance of fat. An ounce of lean meat or low-fat cheese supplies about half its kcalories from fat (28 kcalories from protein and 27 kcalories from fat). An ounce of a high-fat meat (such as bologna) or most cheeses supplies 72 percent of its energy from fat (28 kcalories from protein and 72 kcalories from fat). Two tablespoons of peanut butter supply 72 percent of their energy as fat (32 kcalories from protein, 24 kcalories from carbohydrate, and 140 kcalories from fat)! Thus foods that are usually thought of as protein-rich foods may actually contain more fat energy than protein energy. Note that the values for meat given here are for 1-ounce portions. An average serving of hamburger is usually 3 or 4 ounces. An average dinner steak may be 8 ounces or larger.

Remember that an ounce of meat is not an ounce of protein. An ounce (30 g) of lean meat contains 7 g protein and 3 g fat. The other 20 g are largely water with associated vitamins and minerals.

Fats in the Milk, Yogurt, and Cheese Group The fat in milk is about 63 percent saturated fat; the cholesterol content is 33 milligrams per cup for whole milk or 4 milligrams for nonfat milk. Thus choosing nonfat in place of whole milk reduces your intake of cholesterol as well as of saturated fat.

Note that cream and butter do not appear in the milk group. Milk and yogurt are rich in calcium and protein, but cream and butter are not. Cream and butter are fats, as are whipped cream, sour cream, and cream cheese. That is why the food group that includes milk is carefully labeled the "milk, yogurt, and cheese group," not the "dairy group."

1 c whole milk:
8 g fat
5 g saturated fat
33 mg cholesterol

1 c 2% reduced-fat milk:
5 g fat
3 g saturated fat
18 mg cholesterol

1 c nonfat milk:
<1 g fat
<1 g saturated fat
4 mg cholesterol

At room temperature, unsaturated fats (such as those found in oil), are usually liquid, whereas saturated fats (such as those found in butter) are solid.

Fats in the Meat, Poultry, Fish, Dry Beans and Nuts, and Eggs Group The fats in meats and eggs are about half saturated. The fats in poultry, fish, and nuts are more unsaturated than saturated, a healthier balance. Eating fish instead of meat two or three times a week supports heart health. Fish is not only leaner than most other animal-protein sources, but is a source of omega-3 fatty acids as well. As noted earlier, research shows that a diet rich in fish oils can lower blood cholesterol, just as a diet low in fat and saturated fat can.[19]

Fats in the Bread, Cereal, Rice, and Pasta Group Breads and cereals in their natural state are very low in fat, but fat may be added during processing or cooking. The fat in these foods can be particularly hard to detect, so people must remember which foods stand out as being high in fat. Notable are granola, croissants, biscuits, cornbread, dinner rolls, quick breads, snack and party crackers, muffins, pancakes, and waffles. Packaged breakfast bars often resemble candy bars in their fat and sugar contents.

Many different fats are added to foods during home or commercial preparation. Examples of added fats include oils, butter, and cream. Some fat in foods is not readily apparent and is known as invisible or hidden fat. Convenience foods, meats, and fried foods are major contributors of fat in people's diets. In the Food Guide Pyramid, the fats and meats always contain fat. In addition, foods in the milk group and the bread group sometimes contain fat.

Cutting Fat Intake and Choosing Unsaturated Fats

Knowing which foods contain the most fat is the first step toward meeting the recommendation to reduce dietary fat in general and saturated fat in particular. As a general rule, a person who eats meat and wishes to reduce both saturated fat and cholesterol intake can accomplish these objectives by eating fewer high-fat meats and dairy foods, fewer eggs, and more poultry (without the skin), fish, and nonfat dairy products. A vegetarian who eats dairy products and eggs can shift to nonfat milk and low-fat cheeses and limit butter and egg intake. Vegetarians who omit animal-derived foods already eat a diet low in fat and consume no cholesterol because plant foods do not contain cholesterol. The accompanying box offers strategies for lowering fat, food group by food group.

Fats and kCalories Removing fat from food also removes energy as Figure 3.4 (see p. 70) shows. A small pork chop with the fat trimmed to within a half-inch of the lean provides 275 kcalories; with the fat trimmed off completely, it supplies 165 kcalories. A baked potato with butter and sour cream (1 tablespoon each) has 350 kcalories; a plain baked potato has 220 kcalories. The single most effective step you can take to reduce the energy value of a food is to eat it with less fat.

Choosing Unsaturated Fats To the extent that a person does eat fats, those to choose are the unsaturated ones. The hardness of fats at room temperature is an indicator: the softer a fat is, the more unsaturated it is. Chicken fat is softer than pork fat, which is softer than beef tallow. Of the three, chicken fat is the most unsaturated, and beef tallow is the most saturated. Unsaturated fats melt more readily. Generally speaking, vegetable and fish oils are rich in polyunsaturates, olive oil and canola oil are rich in monounsaturates, and the harder fats—animal fats—are more saturated (see Figure 3.5 on p. 71).

Don't Overdo Fat Restriction Although it is very difficult to do, some people actually manage to eat too *little* fat—to their detriment. Among them are young women and men with eating disorders, described in Nutrition in Practice 9. As a practical guideline, it is wise to include the equivalent of at least a teaspoon of fat in every meal.

How to

Lower Fat Intake—by Food Group

Fats can sneak into the diet in every food group. Inspect each group to discover where the fats are, and control them as follows.

Meat, Fish, and Poultry

- Choose fish, poultry, or lean cuts of pork or beef; look for cuts named *round* or *loin* (eye of round, top round, round tip, tenderloin, sirloin, and top loin).
- Trim the fat from pork and beef; remove the skin from poultry.
- Grill, roast, broil, bake, stir-fry, stew, or braise meats. (Don't fry.) When possible, place meat on a rack while cooking so that fat can drain.
- Use lean ground turkey instead of hamburger in recipes.
- Brown ground meats without added fat; then drain off fat.
- Refrigerate meat pan drippings and broth; when the broth solidifies, remove the fat and use the defatted broth in recipes.
- Select tuna packed in water; rinse oil-packed tuna with hot water to remove much of the fat.
- Fill kabob skewers with lots of vegetables and slivers of meat; create main dishes and casseroles by combining a little meat, fish, or poultry with a lot of pasta, rice, or vegetables.
- Make meatless spaghetti sauces and casseroles.
- Eat a meatless meal or two daily (use the milk and other food groups with care as suggested next).

Milk and Cheeses

- Drink nonfat, low-fat, and reduced-fat milk instead of whole milk.
- Use nonfat, low-fat, and reduced-fat cheeses (such as part-skim ricotta and low-fat mozzarella) instead of regular cheeses.
- Use nonfat yogurt or sour cream instead of regular sour cream.
- Use evaporated nonfat milk instead of cream.
- Enjoy nonfat frozen yogurt, sherbet, or ice milk instead of ice cream.

Fruits and Vegetables

- Use butter-flavored granules on vegetables instead of butter or margarine.
- Use nonfat yogurt or nonfat salad dressing instead of sour cream, cheese, mayonnaise, or other sauces on vegetables and in casseroles.
- Select nonfat or low-fat mayonnaise or salad dressings, or use herbs, lemon juice, and spices instead of regular salad dressing.
- Add a little water to thick, bottled salad dressings to dilute the amount of fat each serving provides.
- Eat at least two vegetables (in addition to a salad) with dinner.
- Snack on raw vegetables or fruits instead of high-fat foods like potato chips.
- Enjoy fruit for dessert.

Breads and Cereals

- Use fruit butters or jellies on bread instead of butter or margarine.
- Select breads, cereals, and crackers that are low in fat (for example, bagels instead of croissants).

Other Foods and Cooking Tips

- Use a nonstick pan or coat the pan lightly with cooking oil.
- Use egg substitutes in recipes instead of whole eggs or use 2 egg whites in place of each whole egg.
- Use half the margarine, butter, or oil called for in a recipe. (The minimum amount of fat for muffins, quick breads, and biscuits is 1 to 2 tablespoons per cup of flour; for cakes and cookies, 2 tablespoons per cup.)
- Select whipped types of butter, margarine, or cream cheese for use at the table; they contain half the kcalories of the regular types.
- Use wine, lemon juice, or broth instead of butter or margarine when cooking.
- Stir-fry in a small amount of oil; add moisture and flavor with broth, tomato juice, or wine.
- Use variety to enhance enjoyment of the meal: vary colors, textures, and temperatures—hot cooked versus cool raw foods—and use garnishes to complement the food.

Pork chop with a half-inch of fat (275 kcal and 19 g fat).

Potato with 1 tbs butter and 1 tbs sour cream (350 kcal and 14 g fat).

Whole milk, 1 c (150 kcal and 8 g fat).

Pork chop with fat trimmed off (165 kcal and 8 g fat).

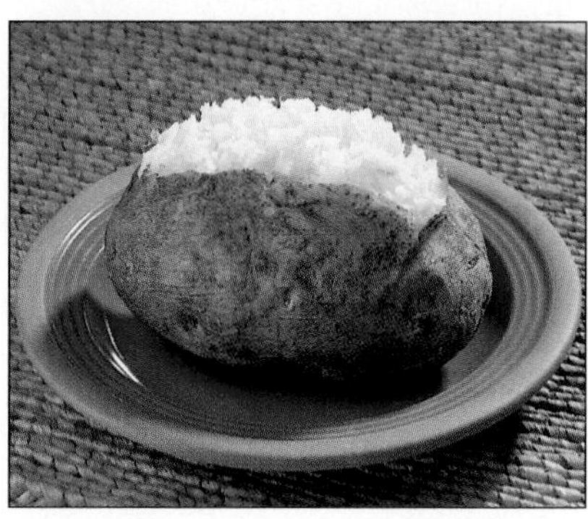

Plain potato (220 kcal and <1 g fat).

Nonfat milk, 1 c (90 kcal and 1 g fat).

Figure 3.4
Food Fat and kCalories

Cautions If you wish to make choices consistent with current recommendations, you should learn how to read food labels, limit fat in general, and seek out the polyunsaturated and monounsaturated fats in preference to the saturated ones. But beware: vegetable fat or vegetable oil doesn't always mean unsaturated fat. Both coconut oil and palm oil, for example, which are often used in nondairy creamers, are saturated fats, and both raise blood cholesterol.

hydrogenation: the process of adding hydrogen to unsaturated fat to make it more solid and resistant to chemical change.

rancid: the term used to describe fats when they have deteriorated, usually by oxidation; rancid fats often have an "off" odor.

Hydrogenated Fats Vegetable oils that are hydrogenated have lost their polyunsaturated character and the health benefits that go along with it. Food producers hydrogenate unsaturated fatty acids to prevent spoilage and make them harder—as when corn oil is hydrogenated to make spreadable margarine. The points of unsaturation in fatty acids are vulnerable to attack by oxygen, which makes them rancid.

***Trans*-Fatty Acids** When polyunsaturated oils are hydrogenated, the fatty acids not only become more saturated, but the hydrogenation process changes the shape of the molecules. In nature, most unsaturated fatty acids are *cis*-fatty acids—meaning that the hydrogens next to the double bonds are on the same side of the carbon chain. When hydrogenated, some of the fatty acids become *trans*-fatty acids—meaning that some of their hydrogens jump to the opposite side of the chain (see Figure 3.6).

Trans-fatty acids have implications for the body's health. In terms of the health of the heart and arteries, they carry a risk similar to saturated fats. *Trans*-fatty acids elevate blood cholesterol and thus raise the risk of heart disease and heart attack.[20] A high total fat consumption is also associated with cancer susceptibility, and *trans*-fatty acids contribute to the total fat intake. No evidence suggests that *trans*-fatty acids by *themselves* play any specific role in promoting or causing cancer, however.[21]

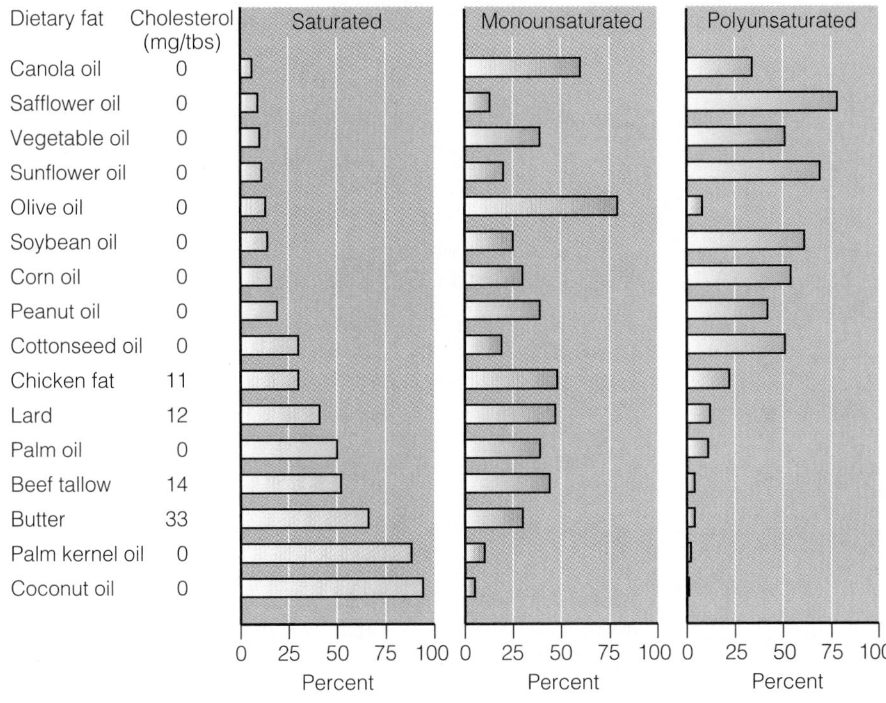

Figure 3.5
Comparison of Dietary Fats
Most fats are mixtures of saturated, monounsaturated, and polyunsaturated fatty acids.

When news of *trans*-fatty acids' effects on heart health was first emerging, some people hastily switched from using margarine back to butter, believing oversimplified reports that margarine provided no heart health advantage over butter. It is true that most margarines and virtually all shortenings are made from mostly hydrogenated fats and therefore contain substantial *trans*-fatty acids—up to 40 percent. However, some margarines, especially the soft or liquid varieties, are made from unhydrogenated oils. These have long proved to be less likely to elevate serum cholesterol than the saturated fats of butter.

In regard to serum cholesterol, margarine may be just a small contributor. Foods other than margarine contribute far more *trans*-fatty acids to the diet—and more total fat, too.[22] Fast foods, chips, baked goods, and other commercially prepared foods are high in fats containing up to 50 percent *trans*-fatty acids. Overall, consumers are eating more fats containing *trans*-fatty acids than ever before, because manufacturers are adding more hydrogenated fats to processed foods.

Trans-fatty acids contribute about 2 to 4 percent of energy intake, whereas total fat accounts for about 34 percent of total energy intake. By limiting total fat, one can limit the health risks posed by all types of fat.

These foods are major contributors of *trans*-fatty acids.

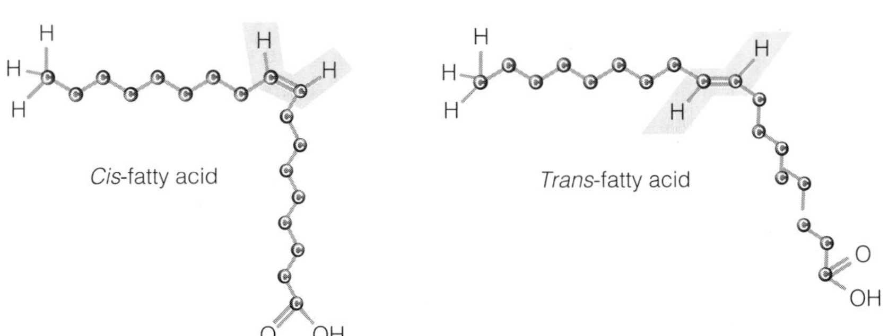

Figure 3.6
***Cis*- and *Trans*-Fatty Acids Compared**

Self Study

How's Your Fat Intake?

Refer to the forms you filled out for the Self Study in Chapter 1, or your Diet Analysis Plus profile to answer the following questions:

1. How many grams of fat do you consume in an average day (from Form 2 or your Daily Average report)? ___ grams. How many grams of saturated fat do you consume in an average day? ___ grams
2. How many kcalories does this represent? (Remember, 1 gram of fat contributes 9 kcalories.) ___ kcalories from total fat and ___ kcalories from saturated fat
3. What percentage of your total kcalories is contributed by fat (fat kcalories divided by total kcalories times 100—or use the answer you obtained on Form 5)? ___ percent from fat
4. What percentage of your total kcalories is from saturated fat? ___ percent from saturated fat
5. A dietary guideline says that fat should contribute no more than 30 percent of total kcalories and saturated fat no more than 10 percent. How does your fat intake compare with these recommendations? ___ If it is higher, look over your food records. What specific foods could you cut down on or eliminate, and what foods could you add to your diet to bring your total fat intake into line? ___
6. You may not be aware of how much fat you are eating when you eat meat. Weigh your meat portions for a day or so; then calculate how much fat you derive from meat in a day (use Appendix A). To visualize this amount of fat, weigh out an equal amount from a bottle of oil, a can of cooking fat, or a tub of butter or margarine. Try the same demonstration with a fast-food meal that includes fried foods. How much fat do you eat in a day? ___

The connections between the overconsumption of fats and chronic diseases (obesity, diabetes, cancer, and cardiovascular disease) alert dietitians to carefully consider a client's fat intake. Using food intake data, dietitians determine the total amount of fat, the types of fat, and the total cholesterol intake. If fat intakes are excessive, dietitians recommend the same fat-cutting strategies recommended in this chapter.

In conducting the preceding Self Study, imagine yourself to be a person with a high risk of one of the chronic diseases related to a high fat and saturated fat intake. Critique your diet on this basis and make realistic recommendations for change.

Food labels can be misleading in this regard. On the "Nutrition Facts" panels, *trans*-fatty acids are counted among the polyunsaturated fats from which they arose, and not with the saturated fats whose health effects they mimic. Further, fast-food chains advertise foods fried in "vegetable oil" when that oil is a hydrogenated type containing abundant *trans*-fatty acids. Some experts are calling for a separate statement of *trans*-fatty acids on food labels and an end to deceptive advertising.[23]

To reduce dietary fat, eliminate added fats in food preparation, and eat fewer high-fat red meats and dairy foods and more lean meats such as poultry and fish and low-fat or nonfat dairy products. *Trans*-fatty acids carry a risk of heart disease similar to saturated fats.

Chapters 2 and 3 have looked briefly at the two major energy fuels in the body—carbohydrate and fat. When used for energy, each has desirable characteristics. The glucose derived from carbohydrate is needed by the brain and nerve tissues and is easily used for energy in other cells. Fat is a particularly useful fuel because the body stores it efficiently and in generous amounts and can use it for energy if carbohydrate is not available. Chapter 4 looks at protein, a nutrient that can be used as fuel, but whose primary role is to provide machinery for getting things done.

Self Check

1. Fat in the body:
   a. provides energy.
   b. insulates the body against extreme temperatures.
   c. is a part of cell membranes.
   d. all of the above.
2. Three classes of lipids in the body are:
   a. triglycerides, fatty acids, and cholesterol.
   b. triglycerides, phospholipids, and sterols.
   c. fatty acids, phospholipids, and cholesterol.
3. In the body, essential fatty acids are used to make substances that:
   a. regulate blood pressure.
   b. help with blood clot formation.
   c. participate in the immune response.
   d. all of the above.
4. To include omega-3 fatty acids in the diet, the American Heart Association recommends eating:
   a. cholesterol-free margarine.
   b. fish oil supplements.
   c. hydrogenated margarine.
   d. two to three fish meals per week.
5. Excess dietary fat contributes to:
   a. heart disease, obesity, diabetes, cancer.
   b. liver disease, sickle-cell anemia, fatty liver, ulcerative colitis.
   c. iron-deficiency anemia, Wilson's disease, fatty liver.
   d. food allergies, hyperglycemia, hepatitis, ulcerative colitis.
6. Some examples of hidden fats in foods are:
   a. cheese, lettuce, and fruit juices.
   b. peanut butter, cheese, and lunch meats.
   c. fish, rice, and potatoes.
   d. baked potatoes, vegetables, and fruits.
7. A fatty acid that has the maximum possible number of hydrogen atoms is known as a:
   a. saturated fatty acid. b. monounsaturated fatty acid.
   c. PUFA. d. none of the above.
8. Generally speaking, vegetable and fish oils are rich in:
   a. polyunsaturated fat. b. saturated fat.
   c. cholesterol. d. *trans*-fatty acids.
9. Lecithins and other phospholipids in the body function as:
   a. emulsifiers. b. enzymes.
   c. constituents of cell membranes. d. a and c.
10. Two ways to lower fat intake are to:
   a. eat no red meat or vegetables.
   b. fry meat for a shorter time so it absorbs less fat and substitute sour cream for butter.
   c. look for cuts of meat named round or loin and remove the skin from poultry.
   d. all of the above.

Answers to these questions appear in Appendix H.

Notes

1. E. J. Schaefer, Effects of dietary fatty acids on lipoproteins and cardiovascular disease risk: Summary, *American Journal of Clinical Nutrition* 65 (1997): 1655S–1656S; C. A. Drevon, Marine oils and their effects, *Nutrition Reviews* 50 (1992): 38–45.

2. M. A. Crawford, The role of essential fatty acids in neural development: Implications of perinatal nutrition, *American Journal of Clinical Nutrition* 57 (1993): 703S–710S.

3. J. Jumpsen and M. T. Clandinin, Lipids and essential fatty acids in brain development, in *Brain Development: Relationship to Dietary Lipid and Lipid Metabolism* (Champaign, Ill.: AOCS Press, 1995), pp. 20–36; W. E. Connor, M. Neuringer, and S. Reisbick, Essential fatty acids: The importance of *n*-3 fatty acids in the retina and brain, *Nutrition Reviews* 50 (1992): 21–29.

4. A. P. Simopoulos, Epidemiological aspects of omega-3 fatty acids in disease states, in *Handbook of Lipids in Human Nutrition,* ed. G. A. Spiller (New York: CRC Press, 1996), pp. 75–89; A. P. Simopoulos and coauthors, The First Congress of the International Society for the Study of Fatty Acids and Lipids (ISSFAL): Fatty acids and lipids from cell biology to human disease, *Nutrition Today* 29 (1994): 24–27.

5. D. S. Siscovick and coauthors, Dietary intake and cell membrane levels of long-chain n-3 polyunsaturated fatty acids and the risk of primary cardiac arrest, *Journal of the American Medical Association* 274 (1995): 1363–1367.

6. P. M. Kris-Etherton and S. Yu, Individual fatty acid effects on plasma lipids and lipoproteins: Human Studies, *American Journal of Clinical Nutrition* 65 (1997): 1628S–1644S; R. McPherson and G. A. Spiller, Effects of dietary fatty acids and cholesterol on cardiovascular disease risk factors in man, in *Handbook of Lipids in Human Nutrition,* ed. G. A. Spiller (New York: CRC Press, 1996), pp. 41–49.

7. A. Nordöy and coauthors, Individual effects of dietary saturated fatty acids and fish oil on plasma lipids and lipoproteins in normal men, *American Journal of Clinical Nutrition* 57 (1993): 634–639.

8. Siscovick and coauthors, 1995; M. L. Burr and coauthors, Effects of changes in fats, fish and fibre intakes on death and myocardial infarction: Diet and reinfarction trial (DART), *Lancet* 2 (1989): 757–761, as cited in Nordöy and coauthors, 1993.

9. Simopoulos, 1996; Y. Kim and J. B. Mason, Nutrition chemoprevention of gastrointestinal cancers: A critical review, *Nutrition Reviews* 54 (1996): 259–279.

10. S. Yu and coauthors, Plasma cholesterol predictive equations demonstrate that stearic acid is neutral and monounsaturated fatty acids are hyprocholesteremic, *American Journal of Clinical Nutrition* 61 (1995): 1129–1139.

11. Kris-Etherton and Yu, 1997.

12. M. B. Katan, P. L. Zock, and R. P. Mensink, Effects of fats and fatty acids on blood lipids in humans: An overview, *American Journal of Clinical Nutrition* 60 (1994): 1017S–1022S.

13. Federation of American Societies for Experimental Biology, Executive summary from the third report on nutrition monitoring in the United States, *Journal of Nutrition* 126 (1996): 1907S–1936S.

14. J. R. Cotton, J. A. Westrate, and J. E. Blundell, Replacements of dietary fat with sucrose polyester: Effects on energy intake and appetite control in nonobese males, *American Journal of Clinical Nutrition* 63 (1996): 891–896; S. N. Gershoff, Nutrition evaluation of dietary fat substitutes, *Nutrition Reviews* 53 (1995): 305–313; B. J. Rolls and coauthors, Effects of olestra, a noncaloric fat substitute, on daily energy and fat intakes in lean men, *American Journal of Clinical Nutrition* 56 (1992): 84–92.

15. N. I. Hahn, Replacing fat with food technology, *Journal of the American Dietetic Association* 97 (1997): 15–16.

16. Position of The American Dietetic Association: Fat replacements, *Journal of the American Dietetic Association* 91 (1991): 1285–1288.

17. Hahn, 1997.

18. A. Drewnowski, Why do we like fat? *Journal of the American Dietetic Association* 97 (1997): S58–S62.

19. Nordöy and coauthors, 1993.

20. J. T. Judd and coauthors, Dietary *trans* fatty acids: Effects of plasma lipids and lipoproteins of healthy men and women, *American Journal of Clinical Nutrition* 59 (1994): 861–868; A. Ascherio and coauthors, *Trans*-fatty acids intake and risk of myocardial infarction, *Circulation* 89 (1994): 94–101; M. B. Katan, P. L. Zock, and R. P. Mensink, *Trans* fatty acids and their effects on lipoproteins in humans, *Annual Review of Nutrition* 15 (1995): 473–493; M. B. Katan, Commentary on the supplement *trans* fatty acids and coronary heart disease risk, *American Journal of Clinical Nutrition* 62 (1995): 518–519.

21. M. B. Katan and R. P. Mensink, Isomeric fatty acids and serum lipoproteins, *Nutrition Reviews* 50 (1992): 46–48.

22. ASCN/AIN Task Force on *Trans*-fatty acids, Position paper on *trans* fatty acids, *American Journal of Clinical Nutrition* 63 (1996): 663–670.

23. W. C. Willet and A. Ascherio, Response to the International Life Sciences Institute report on *trans* fatty acids, *American Journal of Clinical Nutrition* 62 (1995): 524–526; M. B. Katan, European researcher calls for reconsideration of *trans* fatty acid, *Journal of the American Dietetic Association* 94 (1994): 1097–1098.

CHAPTER 3

Nutrition in Practice

▪ Are Fat kCalories More Fattening? ▪

Dietary recommendations to eat a diet low in fat and rich in complex carbohydrates originated almost two decades ago from the awareness that overnutrition (excess dietary fat and cholesterol) contributed to degenerative diseases such as heart disease and cancer. When following the recommendations resulted in weight loss, it was attributed to the effect of eating a greater bulk of fruits, vegetables, and whole grains, which satisfied the appetite sooner than foods high in fat, so that total energy intake decreased.

Today accumulating evidence offers new support for these not-so-new recommendations, but the long-held premise that a kcalorie is a kcalorie, regardless of its source, is the subject of debate. People seem to gain more body fat when they eat excess energy from fat than when they eat excess energy from carbohydrate.[1]

Where did this idea—that fat kcalories are more fattening—come from?

As often happens in science, researchers were testing a different question, whether a fat-rich diet could lead rats to overeat, when they stumbled onto this idea. The researchers found that rats became fatter on high-fat diets than on low-fat diets, even though the total *kcalories* in the diets were the same. This surprise finding opened a whole new field of research on human beings. Researchers have been investigating whether fat in the diet may influence body fatness more than the diet's total kcalorie content.[2]

Conducting research to answer such questions is harder with human beings as subjects than with rats. Scientists can control perfectly which foods, and how much of those foods, rats eat over a lifetime, but with human beings, they can usually only observe and estimate. Also, researchers cannot use genetically identical human beings, and different people carry different genes that may make them more or less predisposed to obesity. Some people tend toward excessive body fatness whatever the diet composition.[3] Given these handicaps, conclusions must be tentative, but most studies seem to show that fat in the diet does add more fat to the human body than does an energy-equivalent amount of carbohydrate.[4]

Well, that's an easy concept to grasp: if people eat a lot of fat, they gain a lot of fat, right?

So it seems, and population studies seem to bear this out. For example, obesity is less prevalent in countries such as China and Japan where the diet is low in fat than in countries such as the United States where the diet is high in fat. When exposed to high-fat western diets, immigrants from these countries develop obesity. To conclude from population data, however, that it is specifically the high fat intakes that promote obesity would fail to recognize other significant factors such as total energy intakes and physical activity. Population data do not show that a higher dietary fat intake causes obesity. Still other factors—smoking, alcohol consumption, eating patterns, metabolic differences, and genetics—further complicate the relationship between dietary fat and body fat.

Can't researchers study individuals, rather than whole populations, to answer this question?

Yes, and research results implicate fat's high energy density in promoting body fatness. Men ate freely from three different diet plans: a low-fat diet (about 20 percent of total energy from fat), a medium-fat diet (about 40 percent), and a high-fat diet (about 60 percent).[5] The foods in each diet plan were similar in taste and appearance; they differed only in the percentages of total energy contributed by fat. Unlike rats, the men failed to adjust perfectly. The more fat in the food, the more food energy the men consumed. The men ate the same total bulk of food whether the diet was high in fat or not. The researchers confirmed these results with a follow-up study.[6]

One reason the men may have failed to compensate is that fat occupies so little bulk. For example, only 2 teaspoons of fat in an 8-ounce glass of milk nearly doubles the energy content of the milk (a cup of nonfat milk has 90 kcalories; a cup of whole milk, 150). Therefore, to keep kcalories constant while using whole milk in place of nonfat milk, one would have to reduce one's milk portion by almost half. The men did not cut down at all on total bulk of high-fat foods, so they consumed considerably more food energy. Other

Nutrition in Practice

experiments designed the same way have confirmed that people do tend to overconsume food energy when given high-fat diets.[7]

Does the research offer any clues as to why fat kcalories seem to be more fattening than carbohydrate kcalories?

Research has begun to focus on the biochemical reasons for body fat buildup. What does the body do with fat and carbohydrate from a meal? Researchers gave men mixed meals of moderate energy value and found that their bodies used much of the carbohydrate for energy and converted much of the fat into body fat.[8]

Why do you suppose the body prefers to store fat over carbohydrate?

Evidence is accumulating that seems to indicate that the body regulates carbohydrate, fat, and protein separately and that overall body energy regulation relates to all three.[9] The details of how this might occur are only now under study, but of the three energy-yielding nutrients, fat seems to be the most efficiently stored, even with moderate food energy intakes.

The picture that emerges is this. Immediately after a meal, when food fat has enriched the blood with lipids, fat-storage cells avidly take them up and store them. Fat storage is a highly efficient process, requiring little energy. Once stored in adipose tissue, fat becomes less readily available to body tissues as an energy source than fuels still traveling in the bloodstream.

The body's handling of carbohydrate from food starts out the same as for fat. Glucose from food is readily taken up by tissues that store it by converting it to glycogen. However, unlike adipose tissue, which stores virtually unlimited quantities of fat, glycogen stores are extremely limited. The body can store many months' worth of extra energy as fat, but can only accommodate about a day's worth of energy as glycogen. Once glycogen stores are full, excess glucose cannot remain in the blood. Any glucose beyond the amount needed to fill glycogen stores must be disposed of by the body.

The body has two options for dealing with excess energy from glucose: it may either use the glucose up immediately for fuel or convert it to fat by way of a long, energy-intensive series of reactions. Finally, this fat can be stored in fat cells. Normally, however, when presented with excesses of both fat and carbohydrate from a meal, the body apparently opts for the most energy-efficient process. It tends to store the fat, while oxidizing the glucose for energy.

So the bottom line is: the less fat you eat, the less fat you store—right?

Yes, more and more research shows that the amount of fat in the diet is an important factor in the body's fat storage.[10]

If the body doesn't convert much carbohydrate to fat, does this mean that people can indulge freely in carbohydrate and not get fat?

Most people today seem to believe that it does. Since 1970, Americans have reduced their fat intakes somewhat, but at the same time they have consumed more energy from carbohydrate. During the same period, they have become fatter. They seem to have gone overboard eating "low-fat" foods while ignoring the carbohydrate energy those foods contribute to the diet.

Does this mean that* carbohydrate, *not fat, has caused the average adult's recent weight gain?

Actually, no. People did eat a smaller *percentage* of fat in 1990 than in 1970, but because it was a percentage of more total energy, the total *grams* of fat they ate each day stayed about the same. They increased their total food energy intakes, especially from carbohydrate, and gained weight proportionately.

This real-life "experiment," conducted spontaneously by the whole U.S. population, showed that vastly overconsuming food energy from a mixed diet causes weight gain, whatever the diet's composition. The weight gain may also reflect an extraordinarily sedentary lifestyle, typical for the people of this nation.[11]

While academically interesting, these insights into the body's regulation of energy metabolism do not change the old truth: people who eat too much and exercise too little get fat. Fat may be *more* fattening than carbohydrate, but it seems that people who overeat either one must pay the price by gaining weight.

Clearly, from the research reported here, a diet's total fat content remains an important predictor of how much fat the body will store, but total food energy can also contribute to weight gain. People must control both their fat *and* their energy intakes *and* engage in frequent and regular physical activity.

Nutrition in Practice

Notes

1. C. Proserpi and coauthors, Ad libitum intake of a high carbohydrate or high-fat diet in young men: Effects on nutrient balances, *American Journal of Clinical Nutrition* 66 (1997): 539–545; E. Ravussin and P. A. Tataranni, Dietary fat and human obesity, *Journal of the American Dietetic Association* 97 (1997): S42–S46; L. H. Nelson and L. A. Tucker, Diet composition related to body fat in a multivariate study of 203 men, *Journal of the American Dietetic Association* 96 (1996): 771–777; T. J. Horton and coauthors, Fat and carbohydrate overfeeding in humans: Different effects on energy storage, *American Journal of Clinical Nutrition* 62 (1995): 19–29.

2. K. R. Westerterp, Food quotient, respiratory quotient, and energy balance, *American Journal of Clinical Nutrition* 57 (1993): 759S–765S; B. J. Rolls and V. A. Hammer, Fat, carbohydrate, and the regulation of energy intake, *American Journal of Clinical Nutrition* 62 (1995): 1086S–1095S.

3. Ravussin and Tataranni, 1997.

4. Proserpi and coauthors, 1997; Ravussin and Tataranni, 1997; Nelson and Tucker, 1996; Horton and coauthors, 1995.

5. R. J. Stubbs and coauthors, Covert manipulation of dietary fat and energy density effect on substrate flux and food intake in men eating ad libitum, *American Journal of Clinical Nutrition* 62 (1995): 316–329.

6. R. J. Stubbs and coauthors, Covert manipulation of the ratio of dietary fat to carbohydrate and energy density: Effect on food intakes and energy balance in free-living men eating ad libitum, *American Journal of Clinical Nutrition* 62 (1995): 330–337.

7. Prosperi and coauthors, 1997; J. E. Blundell and coauthors, Control of human appetite: Implications for the intake of dietary fat, *Annual Review of Nutrition* 16 (1996): 285–319; A. Tremblay and coauthors, Impact of dietary fat content and fat oxidation on energy intake in humans, *American Journal of Clinical Nutrition* 49 (1989): 799–805.

8. O. E. Owen and coauthors, Oxidative and nonoxidative macronutrient disposal in lean and obese men after mixed meals, *American Journal of Clinical Nutrition* 55 (1992): 630–636.

9. J. P. Flatt, Use and storage of carbohydrate and fat, *American Journal of Clinical Nutrition* 61 (1995): 952S–959S.

10. C. N. Bouzer, A. Brasseur, and R. L. Atkinson, Dietary fat affects weight loss and adiposity during energy restriction in rats, *American Journal of Clinical Nutrition* 58 (1993): 846–852; W. C. Miller and coauthors, Diet composition, energy intake, and exercise in relation to body fat in men and women, *American Journal of Clinical Nutrition* 52 (1990): 426–430.

11. Interagency Board for Nutrition Monitoring and Related Research, *Third Report on Nutrition Monitoring in the United States* (Washington, D.C.: Government Printing Office, 1995), pp. 194–195.

4 *Proteins and Amino Acids*

People think of proteins as body-building nutrients, the material of strong muscles, and rightly so. No new living tissue can be built without them, for proteins are part of every cell, every bone, the blood, and every other tissue. Proteins constitute the cells' machinery—they do the cells' work. The energy to fuel that work comes from carbohydrate and fat.

The Chemist's View of Proteins

proteins: compounds composed of carbon, hydrogen, oxygen, *and* nitrogen atoms arranged into strands of amino acids. Some amino acids also contain sulfur atoms.

Proteins are chemical compounds that contain the same atoms as carbohydrate and lipid—carbon (C), hydrogen (H), and oxygen (O)—but proteins are different in that they also contain nitrogen (N) atoms. These nitrogen atoms give the name *amino* (nitrogen containing) to the amino acids that are the links in the chains we call proteins.

The Structure of Proteins

amino (a-MEEN-oh) **acids:** building blocks of protein; each has an amino group and an acid group attached to a central carbon, which also carries a distinctive side chain.
amino = containing nitrogen

About 20 different amino acids may appear in proteins.* All amino acids share a common chemical "backbone," and it is these backbones that are linked together to form proteins. Each amino acid also carries a side chain, which varies from one amino acid to another (see Figure 4.1). The side chains make the amino acids differ in size, shape, and electrical charge. The side chains on amino acids are what make proteins so varied in comparison with either carbohydrates or lipids.

dipeptide: two amino acids bonded together.
di = two
peptide = amino acid

tripeptide: three amino acids bonded together.
tri = three

polypeptide: many amino acids bonded together. *Many* refers to ten or more. An intermediate strand of between four and ten amino acids is an **oligopeptide.**
poly = many
oligo = few

Protein Chains The 20 amino acids can be linked end-to-end in a virtually infinite variety of sequences to form proteins. When two amino acids bond together, the resulting structure is known as a dipeptide. Three amino acids bonded together form a tripeptide. As additional amino acids join the chain, the structure becomes a polypeptide. Most proteins are polypeptides that are 100 to 300 amino acids long.

Protein Shapes These polypeptide chains twist into complex shapes. Each amino acid has special characteristics that attract it to, or repel it from, the surrounding fluids and other amino acids. Because of these interactions, polypeptide chains fold and intertwine into intricate coils (see Figure 4.2). The amino acid sequence of a protein determines the specific way the chain will fold.

Protein Functions The dramatically different shapes of proteins enable them to perform different tasks in the body. Some, such as hemoglobin in the blood (see

Figure 4.1
Amino Acid Structure and Examples of Amino Acids

Side chain
Amino group
Acid group
Backbone

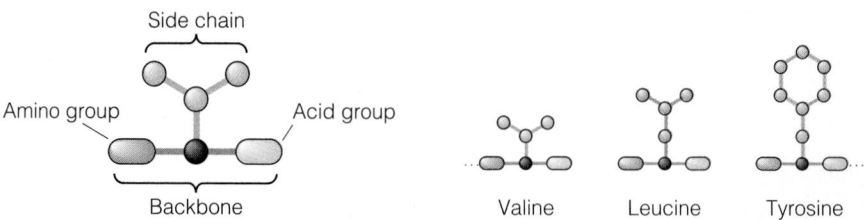

The "backbone" is the same for all amino acids. The nitrogen is in the amino group.

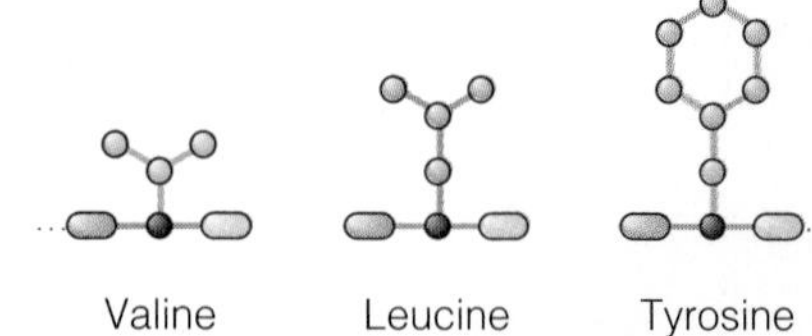

The side chain differs from one amino acid to the next.

*Besides the 20 common amino acids, which can all be components of proteins, others occur individually (for example, ornithine). Chemists can make still others.

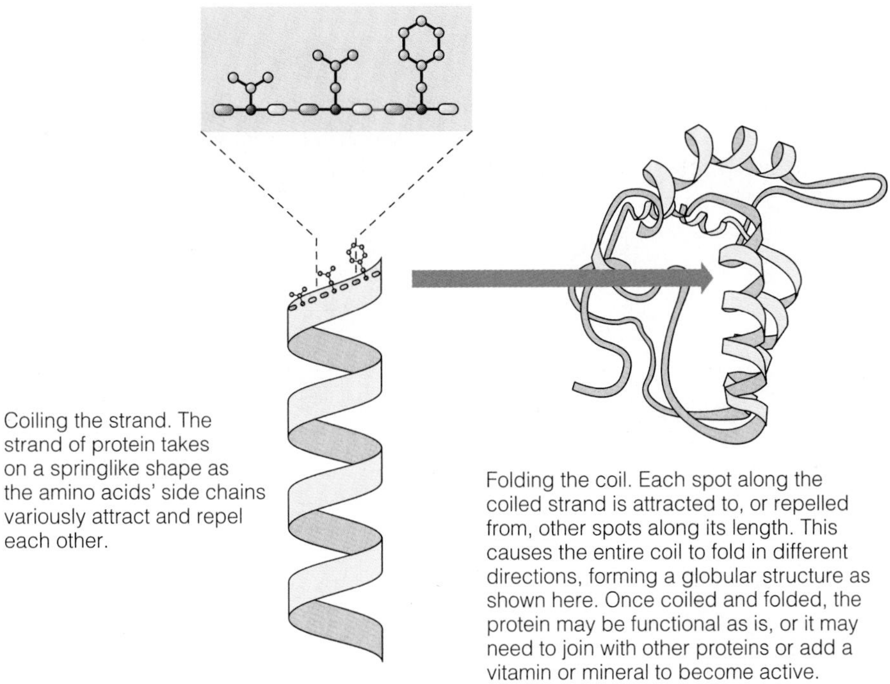

Coiling the strand. The strand of protein takes on a springlike shape as the amino acids' side chains variously attract and repel each other.
Folding the coil. Each spot along the coiled strand is attracted to, or repelled from, other spots along its length. This causes the entire coil to fold in different directions, forming a globular structure as shown here. Once coiled and folded, the protein may be functional as is, or it may need to join with other proteins or add a vitamin or mineral to become active.

Figure 4.2
The Coiling and Folding of a Protein

Figure 4.3 on p. 82), are globular in shape, some are hollow balls that can carry and store materials within them, and some, such as those that form tendons, are more than ten times as long as they are wide, forming stiff, sturdy, rodlike structures.

Essential Amino Acids

Proteins in foods do not provide body proteins directly, but rather supply the amino acids from which the body makes its own proteins. The body can make over half of the amino acids for itself; the protein in food does not need to supply these. But there are other amino acids that the body cannot make at all, and some that it cannot make fast enough to meet its needs. The proteins in foods must supply these amino acids to the body; they are therefore called the *essential* amino acids. Nine amino acids are essential.

Sometimes a nonessential amino acid can become essential. During illness or conditions of trauma, or in other special circumstances, the need for an amino acid that is normally nonessential may become greater than the body's ability to produce it. In such circumstances, that amino acid becomes essential for the ill person. Amino acids that behave this way are referred to as "conditionally essential" amino acids. Research suggests that glutamine, normally a nonessential amino acid, may be a conditionally essential amino acid for critically ill people.[1]

Proteins are unique among the energy nutrients because proteins contain nitrogen atoms. The 20 different nitrogen-containing amino acids link together to form proteins. Nine of the amino acids are essential—the proteins in foods must supply these amino acids to the body.

essential amino acids: amino acids that the body cannot synthesize in amounts sufficient to meet physiological need. Also called *indispensable amino acids.* Nine amino acids are known to be essential for human adults:

- ***histidine*** *(HISS-tuh-deen).*
- ***isoleucine*** *(eye-so-LOO-seen).*
- ***leucine*** *(LOO-seen).*
- ***lysine*** *(LYE-seen).*
- ***methionine*** *(meh-THIGH-oh-neen).*
- ***phenylalanine*** *(fen-il-AL-uh-neen).*
- ***threonine*** *(THREE-oh-neen).*
- ***tryptophan*** *(TRIP-toe-fane).*
- ***valine*** *(VAY-leen)* .

conditionally essential amino acid: an amino acid that is normally nonessential but must be supplied by the diet in special circumstances when the need for it becomes greater than the body's ability to produce it.

Proteins in the Body

What distinguishes you chemically from any other human being are minute differences in your particular body proteins (enzymes, antibodies, and others). These differences are determined by your proteins' amino acid sequences, which are written into the genes you inherited from your parents and ancestors. The genes direct the making of all the body's proteins.

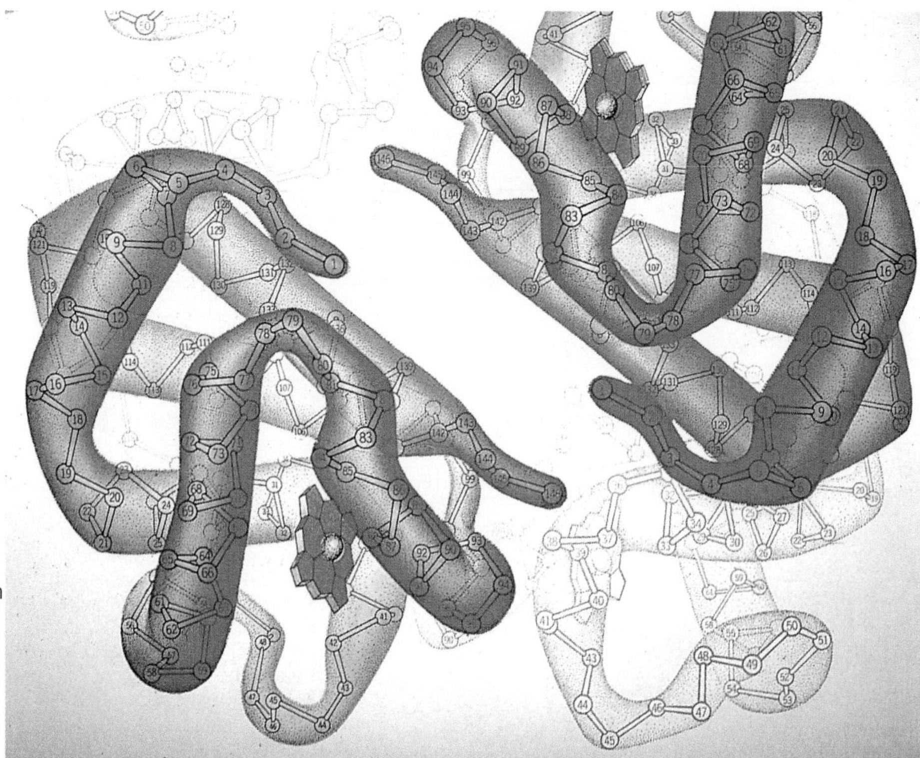

Figure 4.3
A Portion of a Hemoglobin Molecule
One molecule of the protein hemoglobin consists of four highly folded polypeptide chains. Two of the chains are shown here; the jagged-edge structure in each chain is an iron-containing heme group (heme iron is discussed in Chapter 8). This model represents a portion of a hemoglobin molecule magnified millions of times.

The human body contains an estimated 10,000 to 50,000 different kinds of proteins. The roles of more than 1000 of these proteins are now known. Only a few of the many roles proteins play are described here, but these should serve to illustrate proteins' versatility, uniqueness, and importance.

enzymes: protein catalysts. A catalyst is a compound that facilitates chemical reactions without itself being changed in the process.

Enzymes Enzymes are catalysts that are essential to all life processes. Enzymes in the cells of plants or animals put together the pairs of sugars that make disaccharides and the long strands of sugars that make starch, cellulose, or glycogen. Enzymes also dismantle these compounds to free their constituent parts and release energy. Enzymes also assemble and disassemble lipids, assemble all other compounds that the body makes, and disassemble all compounds that the body can use for building tissue and other metabolic work. As Figure 4.4 shows, enzymes themselves are not altered by the reactions they facilitate. All enzymes are proteins, and when amino acids have to be put together to make proteins, it is enzymes that put them together, too. In other words, these proteins can even make other proteins.

The protein story moves in a circle. To follow the circle in nutrition, start with a person eating food proteins. The proteins are broken down by proteins (digestive enzymes) into amino acids. The amino acids enter the cells of the body, where proteins (enzymes) put the amino acids together in long chains whose sequences are specified by the genes. The chains fold and twist back on themselves to form proteins, and some of these proteins become enzymes themselves. Some of these enzymes break apart compounds; others put compounds together. Day by day, in billions of reactions, these processes repeat themselves, and life goes on. Only living systems can achieve such self-renewal. A toaster cannot produce another toaster; a car cannot fix a broken-down car. Only living creatures and the parts they are composed of—the cells—can duplicate and repair themselves.

Fluid and Electrolyte Balance Proteins help maintain the body's fluid and electrolyte balance. As Figure 4.5 shows, the body's fluids are contained in three major

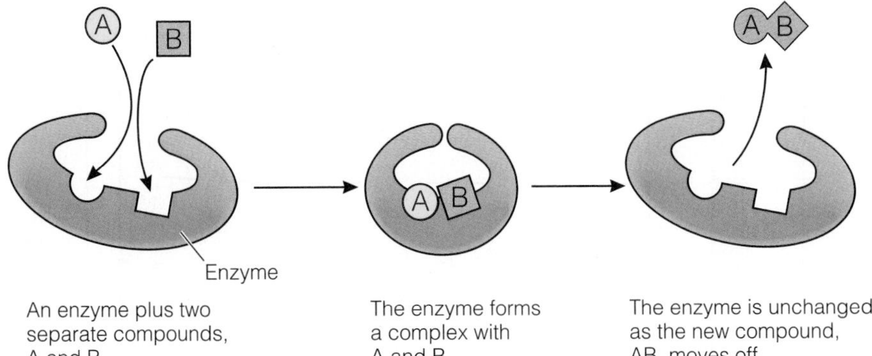

Figure 4.4
Enzyme Action
Each enzyme facilitates a specific chemical reaction that would happen anyway, but much more slowly. The enzyme in this diagram promotes the linking of two compounds to make a more complex compound.

body compartments: (1) the spaces inside the blood vessels; (2) the spaces within the cells; and (3) the spaces between the cells (outside the blood vessels). Fluids flow back and forth between these compartments, and proteins in the fluids, together with minerals, help to maintain the needed distribution of these fluids.

fluid and electrolyte balance: maintenance of the necessary amounts and types of fluid and minerals in each compartment of the body fluids.

Minerals are helper nutrients. The attraction of protein and mineral particles to water is due to osmotic pressure (see Chapter 8).

Proteins are able to help determine the distribution of fluids in living systems for two reasons: first, proteins cannot pass freely across the membranes that separate body compartments, and second, they are attracted to water. A cell that "wants" a certain amount of water in its interior space cannot move the water around directly, but it can manufacture proteins, and these proteins will hold water. Thus the cell can use proteins to help regulate the distribution of water indirectly. Similarly, the body makes proteins for the blood and the interstitial (intercellular) spaces. These proteins help maintain the fluid volume in those spaces. When too much fluid collects in the interstitial spaces, edema results.

edema (eh-DEEM-uh): the swelling of body tissue caused by leakage of fluid from the blood vessels and accumulation of the fluid in the interstitial spaces.

Not only is the quantity of the body fluids vital to life, but so is their composition. Special transport proteins in the membranes of cells continuously transfer substances into and out of cells to maintain balance. For example, sodium is concentrated outside the cells, and potassium is concentrated inside. The balance of these two electrolytes is critical to nerve transmission and muscle contraction. Any disturbance in this balance triggers a major medical emergency. Such imbalances can cause irregular heartbeats, kidney failure, muscular weakness, and even death.

acids: compounds that release hydrogen ions in a solution.

bases: compounds that accept hydrogen ions in a solution.

acid-base balance: the balance maintained between acid and base concentrations in the blood and body fluids.

Acid-Base Balance Proteins also help maintain the balance between acids and bases within the body's fluids. Normal body processes continually produce acids and bases, which must be carried by the blood to the kidneys and lungs for excretion. The blood must do this without upsetting its own acid-base balance. Blood

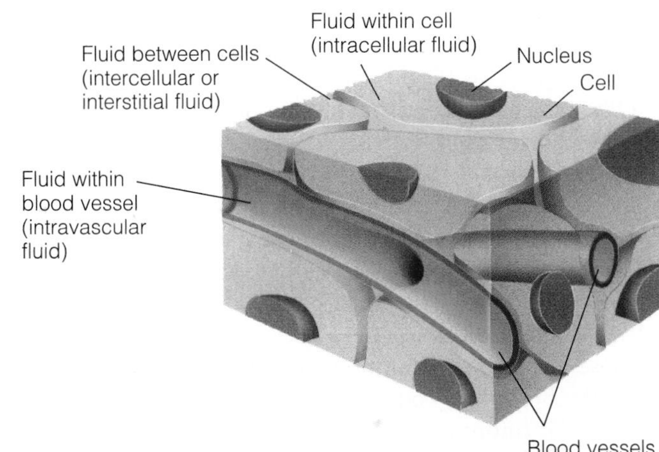

Figure 4.5
One Cell and Its Associated Fluids

Figure 4.6
The pH Scale
A substance's acidity or alkalinity is measured in pH units. Each step down the scale indicates a tenfold increase in the concentration of hydrogen ions. Notice how small the range of normal blood pH is.

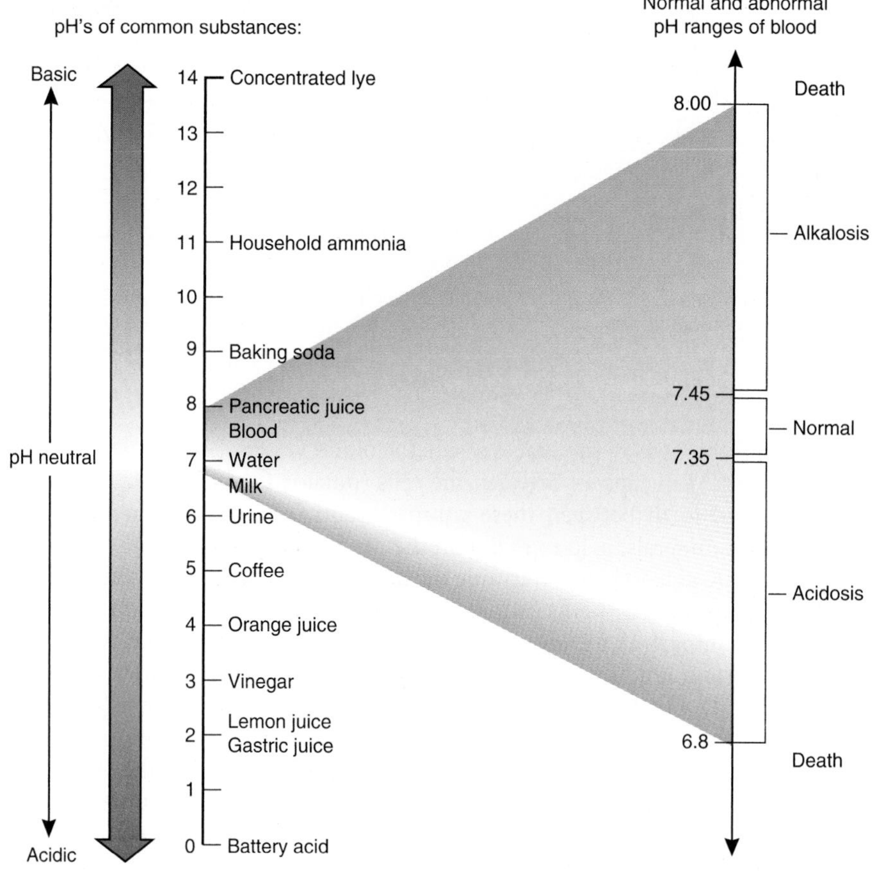

pH: the concentration of hydrogen ions. The lower the pH, the stronger the acid. Thus pH 2 is a strong acid; pH 6 is a weak acid; pH 7 is neutral; and a pH above 7 is alkaline.

denaturation (dee-nay-cher-AY-shun): the change in a protein's shape brought about by heat, acid, or other agents. Past a certain point, denaturation is irreversible.

acidosis: too much acid in the blood and body fluids.

alkalosis: too much base in the blood and body fluids.

buffers: compounds that can reversibly combine with hydrogen ions to help keep a solution's acidity or alkalinity constant.

antibodies: large proteins of the blood and body fluids, produced in response to invasion of the body by unfamiliar molecules (mostly proteins); antibodies inactivate the invaders and so protect the body. The invaders are called **antigens.**
anti = against

pH is one of the most tightly controlled conditions in the body. If the blood becomes too acidic, vital proteins may undergo denaturation, losing their shape and ability to function. A similar situation arises when the balance tips too far toward base. These imbalances are known as acidosis and alkalosis, respectively, and either can be fatal. Figure 4.6 shows the normal and abnormal pH ranges of body fluids, as well as the pH's of some common substances.

Proteins such as albumin in plasma help to prevent these imbalances from arising. In a sense, the proteins protect one another by gathering up extra acid (hydrogen) ions when there are too many in the surrounding medium and by releasing them when there are too few. By accepting and releasing hydrogen ions, proteins act as buffers, maintaining the acid-base balance of the blood and body fluids.

Antibodies Other major proteins in the blood—the antibodies—act against viruses, bacteria, and other disease agents. The antibodies work so efficiently that if a million bacterial cells are injected into the skin of a healthy person, fewer than ten are likely to survive for five hours. This is why in a normal, healthy, individual, most diseases never have a chance to get started. Without sufficient protein, the body cannot maintain its resistance to disease.

Hormones The blood also carries messenger molecules known as hormones, and some of these are made of amino acids. (Recall that other hormones are sterols.) Among the hormones composed of amino acids are the thyroid hormones and insulin. Hormones have many profound effects, which will become evident in subsequent chapters.

Transport Proteins Some transport proteins are not attached to membranes, but move about in the body fluids, carrying nutrients and other molecules from one organ to another. The protein hemoglobin, which carries oxygen from the lungs to the body's cells, is a prime example. The lipoproteins transport lipids around the body. In addition, special proteins also carry vitamins and minerals.

hormones: chemical messengers. Hormones are secreted by a variety of glands in the body in response to altered conditions. Each travels to one or more target tissues or organs and elicits specific responses to restore normal conditions.

Growth, Maintenance, and Repair The body uses amino acids to build the proteins of all its new tissues. The new tissues may be in an embryo, in a growing child, or in new hair and nails. Proteins also help replace worn-out cells in everyone's body all the time. For example, the millions of cells that line the intestinal tract live for three days; they are constantly being shed and must be replaced. The cells of the skin die and rub off, and new ones grow from underneath. The body uses amino acids to repair damaged tissues too. The protein collagen serves as the mending material of torn tissue, forming scars to hold the separated parts together.

Both inside and outside the body, then, cells constantly make and break down their proteins. When proteins break down, their component amino acids are liberated to join the general circulation. Some of these amino acids may be promptly recycled into other proteins; others may be stripped of their nitrogen and used for energy. By reusing amino acids to build proteins, however, the body conserves and recycles a valuable commodity.

People need to eat protein-rich foods every day to replace the protein they continuously lose. If the body is growing, it needs more protein than is necessary just for maintenance. Children end each day with more blood cells, more muscle cells, and more skin cells than they had at the beginning of the day. So protein is needed both for routine maintenance (replacement) and growth (addition).

Nitrogen Balance If the body maintains the same amount of protein in its tissues from day to day, it is in nitrogen balance. If the body adds protein, it is in positive nitrogen balance; if it loses protein, it is in negative nitrogen balance.

Normally, healthy adults are in nitrogen balance; that is, their nitrogen intakes equal their nitrogen outputs. Growing children and pregnant women are in positive nitrogen balance, because they are adding new blood, bone, and muscle cells to their bodies. People who are fasting or starving, such as those with anorexia nervosa, and people who suffer from traumas, such as burns (see Chapter 23), are in negative nitrogen balance because their bodies are forced to use protein for energy.

nitrogen balance: the amount of nitrogen consumed (N in) as compared with the amount of nitrogen excreted (N out) in a given period of time. The laboratory scientist can estimate the protein in a sample of food, body tissue, or excreta by measuring the nitrogen in it.

Nitrogen equilibrium (zero nitrogen balance): N in = N out.
Positive nitrogen balance: N in > N out.
Negative nitrogen balance: N in < N out.

Energy Even though amino acids are needed to do the work that only they can perform—build vital proteins—they will be sacrificed to provide energy and glucose if need be. Keeping energy and glucose available are among the body's highest priorities; without energy, cells die; without glucose, the brain and nervous system falter. When glucose or fatty acids are limited, then, cells are forced to use amino acids for energy and glucose. The body does not make a specialized storage form of protein as it does for carbohydrate and fat. Glucose is stored as glycogen, fat as triglycerides, but body protein is available only as the working and structural components of the tissues. When the need arises, the body dismantles its tissue proteins and uses them for energy. Thus, over time, energy deprivation (starvation) always incurs wasting of lean body tissue as well as fat loss.

The list of protein functions discussed here and summarized in Table 4.1 is by no means exhaustive. Nevertheless, it does give you some sense of the immense variety of proteins and their importance in the body. Proteins perform diverse and important roles in the body as enzymes, as regulators of fluid and electrolyte bal-

Table 4.1 Summary of Functions of Proteins	
■ *Enzymes.* Proteins facilitate needed chemical reactions. ■ *Fluid and electrolyte balance.* Proteins help to maintain the fluid and mineral composition of various body fluids. ■ *Acid-base balance.* Proteins help maintain the acid-base balance of various body fluids by acting as buffers. ■ *Antibodies.* Proteins act against disease agents to fight diseases. ■ *Hormones.* Proteins regulate body processes. Some, but not all, hormones are made of protein.	■ *Transportation.* Proteins help transport needed substances, such as lipids, minerals, and oxygen, around the body. ■ *Structural components.* Proteins form integral parts of most body structures such as skin, tendons, ligaments, membranes, muscles, organs, and bones. ■ *Growth and maintenance.* Proteins serve as building materials for growth and repair of body tissues. ■ *Energy.* Proteins provide some fuel for the body's energy needs.

ance, as acid-base buffers, as antibodies and hormones, as transporters, as building blocks for new or damaged tissues, and at times for energy.

Protein and Health

In the short time that scientists have been studying nutrition, no nutrient has been more intensely scrutinized than protein. As you know by now, it is indispensable to life. And it should come as no surprise that protein deficiency can have devastating effects on people's health. But as with the other nutrients, protein in excess can also be harmful; the end of this section discusses the consequences of protein excess.

Protein-Energy Malnutrition

protein-energy malnutrition (PEM): a deficiency of protein and food energy; the world's most widespread malnutrition problem, including both marasmus and kwashiorkor.
mal = bad, poor

WWW
secondharvest.org
Second Harvest

wfp.org
World Food Programme

When people are deprived of food and suffer an energy deficit, they degrade their own body protein for energy and indirectly suffer a protein deficiency, as well as an energy deficiency. Because protein and energy deprivation go hand in hand, public health officials have adopted an abbreviation for the overlapping pair: protein-energy malnutrition (PEM). PEM often strikes early in childhood, but it endangers many adults as well. PEM is the most widespread form of malnutrition in the world today. Most of the 33,000 children who die each day are malnourished and suffer from infectious diseases.[2] PEM is prevalent in Africa, Central America, South America, the Middle East, and East Asia, but developed countries including the United States are not immune to it. PEM is common among some population groups in the United States: impoverished people living on U.S. Indian reservations, in inner cities, and in rural areas; many elderly people; homeless children; and those suffering from the eating disorder anorexia nervosa.[3] PEM has been recognized in people with many chronic diseases such as cancer and AIDS and in those who have severe stresses such as burns or extensive infections (see Chapter 23). The consequences of PEM as a world malnutrition problem are considered here; the problems associated with PEM and illness are described throughout later chapters.

Of all population groups, children are most seriously affected by malnutrition. Children who are thin for their heights may have recently developed PEM whereas children who are short for their ages may have experienced PEM for longer periods of time. Stunted growth due to PEM is easy to overlook because a small child may look quite normal, but it may be the most common sign of malnutrition in the developing countries.

Marasmus and Kwashiorkor—One Disease or Two? The long-held idea that the two forms of PEM—marasmus and kwashiorkor—are different diseases caused by different nutrient deficiencies has been challenged in recent years. Marasmus was thought to be caused by energy deficiency and kwashiorkor by protein deficiency. In reality, though, marasmus reflects inadequate food intake and therefore inadequate energy, vitamins, and minerals as well as too little protein. Some researchers now maintain that marasmus and kwashiorkor are two stages of the same disease.[4] Kwashiorkor is usually precipitated by an illness such as measles or other infection. The stress of the illness rapidly depletes energy stores and body protein so that too little protein remains to support optimal body function.[5] Researchers continue to question the exact causes of kwashiorkor, but clearly, an absolute or relative protein deficiency is involved.

marasmus (ma-RAZZ-mus): a disease related to PEM; marasmus results from severe deprivation, or impaired absorption, of protein, energy, vitamins, and minerals.

kwashiorkor (kwash-ee-OR-core or kwash-ee-or-CORE): a disease related to PEM.

Marasmus Marasmus commonly occurs in children from 6 to 18 months of age in all the overpopulated urban slums of the world. Children in impoverished nations subsist on a weak cereal drink that supplies scant energy and protein of low quality: such food can barely sustain life, much less support growth. Consequently, marasmic children look like little old people—just skin and bones.

Without adequate nutrition, muscles, including the heart muscle, waste and weaken. Because the brain normally grows to almost its full adult size within the first two years of life, marasmus impairs brain development and learning ability. Reduced synthesis of key hormones leads to a metabolism so slow that the body temperature is subnormal. There is little or no fat under the skin to insulate against cold. Hospital workers find that the primary need of marasmic victims is to be wrapped up and kept warm.

The starving child faces this threat to life by engaging in as little activity as possible—not even crying for food. The body gathers all its forces to meet the crisis, so it cuts down on any expenditure of protein not needed for the heart, lungs, and brain to function. Growth ceases; the child is no larger at age four than at age two. Digestive enzymes are in short supply, the digestive tract lining deteriorates, and absorption fails. The child cannot assimilate what little food is eaten.

Blood proteins, including hemoglobin, are no longer synthesized. Antibodies to fight off invading bacteria are degraded to provide amino acids for other uses, rendering the child vulnerable to infection. Then dysentery, an infection of the digestive tract, causes diarrhea, further depleting the body of nutrients. In the marasmic child, once infection has set in, kwashiorkor often follows.[6] The infection that occurs with malnutrition is responsible for two-thirds of the deaths of young children in developing countries.[7]

dysentery (DIS-en-terry): an infection of the gastrointestinal tract caused by an amoeba or bacterium that gives rise to severe diarrhea.
dys = bad
entery = intestine

When two variables interact so that each increases the other, **synergism** (SIN-er-jiz-um) is said to be occurring. Malnutrition and infection are a deadly combination because they work in this way.
syn = with, together
ergism = work

If caught in time, a child's starvation may be reversed by careful nutrition therapy. The fluid balances are most critical. Diarrhea will have depleted the body's potassium and disturbed other electrolyte balances. Careful correction of fluid and electrolyte imbalances usually raises the blood pressure and strengthens the heart. After the first 24 to 48 hours, protein and energy may be given in small quantities, with intakes gradually increased as tolerated.[8]

Kwashiorkor Kwashiorkor was originally a Ghanaian word meaning an "evil spirit that infects the first child when the second child is born." If you consider how kwashiorkor often develops, you can easily see how the Ghanaians arrived at this name for the disease. When a mother who has been nursing her first child bears a second child, she weans the first child and puts the second one on the breast. The first child, suddenly switched from nutrient-dense, protein-rich breast milk to a starchy, protein-poor gruel, soon begins to sicken and die. Kwashiorkor typically sets in at about the age of two. As mentioned earlier, some researchers believe that kwashiorkor and marasmus are two stages of the same

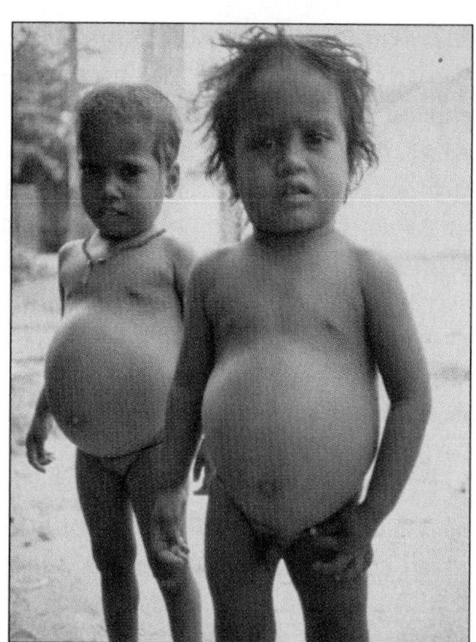

In the photo on the left, the extreme loss of muscle and fat, characteristic of marasmus is apparent in the child's matchstick arms and legs. In contrast, the edema and enlarged liver characteristic of kwashiorkor is apparent in the swollen bellies of the children in the photo on the right.

disease. Some research indicates that marasmus represents the body's attempt to adapt to starvation and that kwashiorkor develops when stresses tax nutrient status further and adaptation fails.

Some symptoms of kwashiorkor resemble those of marasmus, but without severe wasting of body fat. Proteins and hormones that previously maintained fluid balance diminish, and fluid leaks out of the blood. The child's limbs and face become swollen with edema, a distinguishing feature of kwashiorkor; the belly bulges with a fatty liver caused by lack of the protein carriers that transport fat out of the liver. The child's hair loses its color; the skin becomes patchy and scaly, sometimes with ulcers and sores that fail to heal.

fatty liver: an accumulation of fat in the liver. In PEM, fat accumulates in the liver because no protein is available to form the lipoproteins that normally escort fat molecules in the blood (see Chapter 28).

Experts assure us that we possess the knowledge, technology, and resources to end hunger. Programs that have involved the local people in the process of identifying the problem and devising its solution have met with some success. But until those who have the food, technology, and resources make fighting hunger a priority, the war on hunger will not be won.

Protein Excess

While many of the world's people struggle to obtain enough food and enough protein to survive, in the developed nations protein is so abundant that problems of protein excess are seen. Overconsumption of protein offers no benefits and may pose health risks. For example, protein-rich foods are often high-fat foods that contribute to obesity with its accompanying health risks. Some studies suggest a link between high-meat diets and colon cancer.[9] As in studies of heart disease, however, the effects of protein and fats cannot easily be separated. One recent study reported an increase in the risk of a type of cancer known as non-Hodgkin lymphoma with a high intake of animal protein and red meat.[10] The higher a person's intake of protein-rich foods such as meat and milk, the more likely that fruits, vegetables, and grains will be crowded out, making the diet inadequate in other nutrients.

Some research suggests that diets high in protein promote calcium excretion, depleting the bones of their chief mineral. When high protein intakes are accompanied by low calcium intakes, as is typical of many women's diets in the United States, calcium losses incurred by excess protein may compromise bone health.[11] One study of the relationship between protein intake and bone fractures in women reported a higher risk of arm fractures in women with high animal-protein intakes.[12] There are evidently no benefits to be gained by consuming a diet that derives more than 15 percent of its energy from protein. The *Diet and Health* recommendations (see Table 1.2 in Chapter 1) advise a moderate protein intake—one that falls between the RDA and twice the RDA.

Amino Acid Supplements Unnecessary In view of the high protein intakes of people in the United States and other developed nations, it is absurd that many people feel compelled to take protein and amino acid supplements. Why do people take protein or amino acid supplements? Athletes take them to build muscle. Dieters take them to spare their bodies' protein while losing weight. Popular reports that the amino acid tryptophan could relieve pain and cure depression and insomnia led to widespread public use. More than 1500 people who elected to take tryptophan developed an illness (EMS, short for *eosinophilia-myalgia syndrome*). EMS is characterized by severe muscle and joint pain, limb swelling, an elevated white blood cell count, extremely high fever, and, in at least 38 cases, death.[13] Some studies suggested that changes in procedures at a major Japanese tryptophan processing plant may have introduced contaminants that caused the disease. Later research suggests multiple factors are involved, and the exact causes of EMS remain uncertain. The Food and Drug Administration (FDA) issued a recall of all products containing tryptophan except for specific medical formulas. If you own a bottle of tryptophan, throw it away. It is safer to derive your amino acids from protein-rich foods.

Muscle work builds muscle; protein supplements do not, and athletes do not need them. Chapter 10 describes how muscles are built and the diet that best supports them.

WWW.
nemsn.org
National Eosinophilia-Myalgia Syndrome Network

Recently, the FDA asked a panel of scientists from a well-known scientific research group to review the safety of amino acid supplements.[14] When the scientists began to search the literature for well-controlled studies on the supplements, they found next to none. The panel did find evidence of adverse health effects from amino acids, however, and therefore concluded that, without appropriate scientific research, no level of intake of these supplements can be considered safe. The scientists also warned that any use of amino acids as dietary supplements is inappropriate and that some (serine and proline) present a high risk of toxicity.

The panel also singled out some groups of people whose growth or altered metabolism makes them especially likely to suffer harm from amino acid supplements:

- All women of childbearing age.
- Pregnant or lactating women.
- Infants, children, and adolescents.
- Elderly people.
- People with inborn errors of metabolism that affect their bodies' handling of amino acids.
- Smokers.
- People on low-protein diets.
- People with chronic or acute mental or physical illnesses who take amino acids without medical supervision.

Also, because they may take frequent, massive amino acid doses, weight lifters and bodybuilders may suffer harm from the supplements while believing false promises of benefits. Anyone considering taking amino acid supplements should check with a physician first.

Protein Recommendations The committee that established the RDA states that a generous daily protein allowance for a healthy adult is 0.8 gram per kilogram (2.2 pounds) of appropriate or average body weight for height. The protein RDA is adjusted to cover additional needs for building new tissue and so is higher for infants, children, and pregnant and lactating women. (The protein needs of athletes are discussed in Chapter 10.) Protein RDA for people of average height at all ages are presented in the RDA table (inside front cover). If your height is not average, you can compute your own individualized RDA for protein (see the Self Study at the end of this chapter).

In setting the RDA, the committee assumes that the protein eaten will be of high quality, that it will be consumed together with adequate energy from carbohydrate and fat, and that other nutrients in the diet will be adequate. The committee also assumes that the RDA will be applied only to healthy individuals with no unusual alteration of protein metabolism. Most people in this country already eat this much protein, and more.

Protein deficiencies arise from both energy-poor and protein-poor diets and lead to the devastating diseases marasmus and kwashiorkor. Together these diseases are known as PEM (protein-energy malnutrition). Excess protein can also be harmful.

Protein in Foods

To make body protein, a cell must have all the needed amino acids available simultaneously. Therefore, the first important requirement for protein in the diet of a healthy adult is that it should supply at least the 9 essential amino acids and enough nitrogen and energy for the synthesis of the other 11.

complete protein: a protein containing all the amino acids essential in human nutrition in amounts adequate for human use.

incomplete protein: a protein lacking or low in one or more of the essential amino acids.

Complete Protein A protein that fits this description is called a complete protein. It contains all the essential amino acids in amounts adequate for human use; it may or may not contain all the others. Generally, proteins derived from animal foods (meats, fish, poultry, eggs, and milk) are complete, although gelatin is an exception. Proteins derived from plant foods (legumes, grains, and vegetables) vary more. Some plant proteins are notoriously incomplete—for example, corn protein. Others are complete—for example, soy protein. Using plant foods alone, the educated vegetarian can design a diet that is adequate in protein by choosing a variety of legumes, grains, and vegetables. (Vegetarian diets are discussed further in Nutrition in Practice 4.) Table 4.2 lists the protein contents of foods based on the food groups of the Daily Food Guide in Chapter 1.

high-quality protein: an easily digestible, complete protein.

Protein Digestibility and Quality Ideally, a protein should be not only complete, but easily digestible as well so that sufficient numbers of amino acids reach the body's cells to permit them to make the proteins they need. Such a protein is called a high-quality protein. One of the finest proteins available by these standards is egg protein. Eggs are a highly valued protein source in developing nations where protein-rich foods are scarce.

limiting amino acid: the essential amino acid found in the shortest supply relative to the amounts needed for protein synthesis in the body. It limits the amount of protein the body can make.

Limiting Amino Acids Ideally, dietary protein supplies each amino acid in the amount needed for protein synthesis in the body. If one amino acid is supplied in an amount smaller than is needed, then the total amount of protein that can be synthesized will be limited. The body makes only complete proteins. (By analogy, suppose that a sign maker plans to make 100 identical signs saying "LEFT TURN ONLY." The sign maker needs 200 *Ls*, 200 *Ns*, 200 *Ts*, and 100 of each of the other letters. If only 20 *Ls* are available, only 10 signs can be made, even if all the other letters are available in unlimited quantities. Suppose further that the sign maker has no place to keep leftover letters—just as the body has no storage place for extra amino acids. If the sign maker doesn't get some additional *Ls* right away, he will have to throw away all the other letters.)

Table 4.2 Protein-Containing Foods

Milk, Cheese, and Yogurt
Each of the following provides about 8 grams of protein:[a] ■ 1 cup of milk, buttermilk, or yogurt (choose reduced-fat, low-fat, or nonfat). ■ 1 ounce of regular cheese (for example, cheddar or swiss).[b] ■ ¼ cup of cottage cheese (choose reduced-fat, low-fat, or nonfat).
Meat, Poultry, Fish, and Alternates
Each of the following provides about 7 grams of protein: ■ 1 ounce of meat, poultry, or fish (choose lean meats to limit fat intake). ■ ½ cup of legumes (navy beans, pinto beans, black beans, lentils, soybeans, and other dried beans and peas). ■ 1 egg.[b] ■ ½ cup tofu (soybean curd).[b] ■ 2 tablespoons of peanut butter.[b] ■ 1 to 2 ounces of nuts or seeds.[b]
Breads, Cereals, and Other Grain Products
Each of the following provides about 3 grams of protein: ■ 1 slice of bread. ■ ½ cup of cooked rice, pasta, cereals, or other grain foods.
Vegetables
Each of the following provides about 2 grams of protein: ■ ½ cup of cooked vegetables. ■ 1 cup of raw vegetables.

[a]For reference, an adult might need 40 to 100 grams of protein in a day.
[b]These are medium-fat or high-fat choices.

Protein Sparing Dietary protein—no matter how high the quality—will not be used efficiently and will not support growth when energy from carbohydrate and fat is lacking. The body assigns top priority to meeting its energy need and, if necessary, will break down protein to meet this need. After stripping off and excreting the nitrogen from the amino acids, the body will use their carbon skeletons in much the same way it uses those from glucose or fat. A major reason why people must have ample carbohydrate and fat in the diet is to prevent this wasting of protein.

Reminder: Carbohydrate and fat allow amino acids to be used to build body proteins. This is known as the *protein-sparing effect* of carbohydrate and fat.

Dietary protein should supply at least the essential amino acids and enough nitrogen and energy for synthesis of the others. A complete protein has all the essential amino acids; an incomplete protein lacks one or more essential amino acids.

How's Your Protein Intake?

These exercises use the information recorded on Forms 1 to 5 or input in the Daily Food Intakes section of Diet Analysis Plus.

1. How many grams of protein do you consume in a day? ___ grams
2. How many kcalories does this represent? (Remember, 1 gram of protein contributes 4 kcalories.) ___ kcalories
3. What percentage of your total kcalories is contributed by protein? ___ percent
4. Dietary guidelines suggest that protein should contribute about 10 to 15 percent of total kcalories. How does your protein intake compare with this recommendation? (Note: If you are on a kcalorie-restricted diet, a higher percentage of your kcalories should come from protein.) If your protein intake is out of line, what foods could you consume more or less of to bring it into line? ______________________ ______________________
5. Compare your protein intake (from item 1) with the recommendation for an "average" person of your age and sex as shown in the RDA tables (inside front cover) or in the Canadian recommendations (Appendix B). If you are using Diet Analysis Plus software, the program will calculate your recommended protein intake. ___ If you are not of average height or weight, calculate your protein RDA:
   a. Look up the midpoint weight for a person your height (Table 9.1). Assume this is an appropriate weight for you.
   b. Change pounds to kilograms: 2.2 pounds equal 1 kilogram.
   c. Multiply kilograms by 0.8 gram/kilogram.

   Example (for a medium-frame male 5 feet 10 inches tall):

   a. "Appropriate" weight: about 153 pounds.
   b. 153 pounds ÷ 2.2 pounds/kilogram = 70 kilograms (rounded off).
   c. 70 kilograms × 0.8 gram/kilogram = 56 grams protein (rounded off).
6. Compare your average daily protein intake with your RDA. About what percentage of that intake are you consuming each day? ___ percent. If you are "average" and healthy, the recommendation is probably generous for you, yet you may be eating more protein than that. If so, you may be spending protein prices for an energy nutrient. What substitutions could you make in your day's food choices that would enable you to derive the kcalories you need for energy from carbohydrate rather than from protein? ______________________ ______________________
7. How many of your protein grams are from animal foods? ___ How many are from plant foods? ___ Assuming that the animal protein is of high quality, no more than 20 percent of your total protein intake need come from this source. Should you alter the ratio of plant to animal protein in your diet? ___ If you were to do so, what effect would this have on the total fat content of your diet? ______________________ ______________________

PEM is a common finding among people in the hospital. The many negative effects of protein deficiency described in this chapter can help you to see why such a deficiency can be particularly dangerous at a time of severe illness (see Chapter 23). Dietitians assess protein status not only by checking protein intake but also by using anthropometric and biochemical measures. Chapter 16 describes these tests in detail.

Self Check

1. Proteins are chemically different from carbohydrate and fat because they also contain:
   a. phosphorus. b. nitrogen.
   c. sodium. d. iron.

2. The basic building blocks for protein are:
   a. glucose units. b. amino acids.
   c. side chains. d. saturated bonds.

3. Enzymes are proteins that:
   a. are essential to all life processes.
   b. help assemble disaccharides into starch, cellulose, or glycogen.
   c. facilitate chemical reactions without themselves being changed in the process.
   d. all of the above.

4. Functions of proteins in the body include:
   a. growth and maintenance, supplying hormones to regulate body processes, and maintaining fluid and electrolyte balance.
   b. supplying omega-3 fatty acids for growth, helping lower serum cholesterol, and helping with weight control.
   c. supplying fiber to aid digestion and being made into cellulose, the main fuel source for muscles.
   d. all of the above.

5. The swelling of body tissue caused by the leakage of fluid from the blood vessels and accumulation of the fluid in the interstitial spaces is called:
   a. anemia. b. edema.
   c. sickle-cell anemia. d. acidosis.

6. Major proteins in the blood that protect against bacteria and other disease agents are called:
   a. antigens. b. acids.
   c. buffers. d. antibodies.

7. Marasmus can be distinguished from kwashiorkor because in marasmus:
   a. the limbs and face swell with edema, and the belly bulges with a fatty liver.
   b. severe wasting of body fat and muscle occurs.
   c. only adults are victims.
   d. all of the above.

8. The RDA for protein for a healthy adult is ____ gram(s) per kilogram of appropriate or average body weight for height.
   a. 0.5 b. 0.8
   c. 1.1 d. 1.4

9. Generally speaking, from which of the following foods are complete proteins derived?
   a. vegetables, grains, and fruits
   b. meats, fish, poultry, eggs, and milk
   c. corn, rice, potatoes, and barley
   d. none of the above

10. An incomplete protein lacks one or more:
    a. essential fatty acids.
    b. essential amino acids.
    c. hydrogen bonds.
    d. saturated fatty acids.

Answers to these questions appear in Appendix H.

Notes

1. J. C. Teran, K. D. Mullen, and A. J. McCullough, Glutamine—A conditionally essential amino acid in cirrhosis? *American Journal of Clinical Nutrition* 62 (1995): 897–900; S. Mobrahan, Glutamine: A conditionally essential nutrient or another nutritional puzzle, *Nutrition Reviews* 50 (1992): 331–333.

2. D. G. Schroeder and R. Martorell, Enhancing child survival by preventing malnutrition, *American Journal of Clinical Nutrition* 65 (1997): 1080–1081.

3. J. E. Miller and S. Korenman, Nutritional status of poor children in the United States, *Journal of the American Dietetic Association* 95 (1995): 248–250; M. Mowe, T. Bohmer, and E. Kindt, Reduced nutritional status in an elderly population (>70 yrs) is probable before disease and possibly contributes to the development of disease, *American Journal of Clinical Nutrition* 59 (1994): 317–324; J. Dye, Malnourished children in the United States: Caught in the cycle of poverty, *Journal of the American Dietetic Association* 94 (1994): 218; M. L. Taylor and S. A. Koblinsky, Dietary intake and growth status of young homeless children, *Journal of the American Dietetic Association* 93 (1993): 464–466; J. C. Wolgemuth and coauthors, Wasting malnutrition and inadequate nutrient intakes identified in a multiethnic homeless population, *Journal of the American Dietetic Association* 92 (1992): 834–839; M. Nestle and S. Guttmacher, Hunger in the United States: Rationale, methods, and policy implications of state hunger surveys, *Journal of Nutrition Education* 24 (1992): 18S–22S.

4. C. Goplan, The contribution of nutrition research to the control of undernutrition: The Indian experience, *Annual Review of Nutrition* 12 (1992): 1–17.

5. J. C. Waterlow, Childhood malnutrition in developing nations: Looking back and forward, *Annual Review of Nutrition* 14 (1994): 1–19.

6. L. Lewinter-Suskind and coauthors, The malnourished child, in *Textbook of Pediatric Nutrition,* 2nd ed., eds. R. M. Suskind and L. Lewinter-Suskind (New York: Raven Press, 1993), pp. 127–140.

7. P. W. Yoon and coauthors, The effect of malnutrition on the risk of diarrhea and respiratory mortality in children <2 y of age in Cebu, Philippines, *American Journal of Clinical Nutrition* 65 (1997): 1070–1077.

8. Lewinter-Suskind and coauthors, 1993.

9. L. H. Kushi, E. B. Lenart, and W. C. Willet, Health implications of Mediterranean diets in light of contemporary knowledge: Meat, wine, fats, and oils, *American Journal of Clinical Nutrition* 61 (1995): 1416S–1427S.

10. B. C. H. Chiu and coauthors, Diet and risk of non-Hodgkin lymphoma in older women, *Journal of the American Medical Association* 275 (1996): 1315–1321.

11. R. P. Heaney, Protein intake and the calcium economy, *Journal of the American Dietetic Association* 93 (1993): 1259–1260.

12. D. Feskanich and coauthors, Protein consumption and bone fractures in women, *American Journal of Epidemiology* 143 (1996): 472–479.

13. D. Farley, Making sure hype doesn't overwhelm science, *FDA Consumer,* November 1993, pp. 9–13.

14. S. A. Anderson and D. J. Raiten, eds., *Safety of Amino Acids Used as Supplements* (Bethesda, Md.: Federation of American Societies for Experimental Biology, 1992).

CHAPTER 4

Nutrition in Practice

▪ VEGETARIAN DIETS ▪

Eating patterns all along the continuum of dietary choices—from one end, where people eat no foods of animal origin, to the other end, where they eat generous quantities of meat every day—can support or compromise nutritional health. The quality of the diet depends not on whether it consists of all plant foods or centers on meat, but on whether the eater's food choices are based on sound nutrition principles: adequacy of nutrient intakes, balance and variety of foods chosen, appropriate energy intake, and moderation in intakes of substances such as fat, sodium, alcohol, and caffeine that are harmful in excess.

People choose to exclude meat and other animal-derived foods from their diets for various reasons—philosophies, health attitudes, or convenience. Some believe that vegetarianism is better for the environment; some, that it is healthier; and some, that it is less costly than the meat-eating alternative. Some just like it better. Whatever the reasons, vegetarians and health professionals who work with them should be aware of the nutrition and health implications of vegetarian diets.

Because vegetarian diets vary in both the types and amounts of animal-derived foods they include, these differences must be considered when evaluating the health status of vegetarians. The glossary on the next page defines the various kinds of vegetarian diets.

WWW.

vrg.org
The Vegetarian Resource Group

veg.org/veg
Vegetarian Pages

I've been thinking about becoming a vegetarian, but I'm not sure vegetarian diets are nutritionally sound. Are they?

The American Dietetic Association takes the position that well-planned vegetarian diets offer nutrition and health benefits to adults in general.[1] Research suggests that meat-eating adults who switch to vegetarian diets reduce their risks of heart disease, hypertension, diabetes, some types of cancer, and obesity.[2]

What should be my main concerns when planning a nutritionally sound vegetarian diet?

A vegetarian diet planner faces the same task as other diet planners—obtaining a variety of foods that provide all the needed nutrients within an energy allowance that maintains a healthy body weight. The challenge is to do so using at least one less food group. Since all vegetarians omit meat, and some omit other animal-derived foods, protein, the nutrient that meat is famous for, merits some discussion here.

Yes, I've heard that vegetarian diets are low in protein. Is this true?

No, protein is not the problem it was once thought to be in vegetarian diets. People who include animal-derived foods such as milk and eggs in their diets need not worry at all about protein deficiency. Even for those who eat only plant-derived foods, protein intakes are usually satisfactory as long as energy intakes are adequate and protein sources are varied.[3] A mixture of proteins from whole grains, legumes, seeds, nuts, and vegetables can provide adequate amounts of all the amino acids.

The idea persists that vegans must carefully combine their plant-protein foods in order to obtain the full array of essential amino acids, but new knowledge indicates that this is not necessary. Plant foods can provide more than enough of all the essential amino acids and can sustain people in good health, as long as the diet supplies sufficient energy and does not include too many empty-kcalorie foods. It is true, however, that a meal delivers higher-quality protein when it combines two or more different individual plant-protein sources, each of which supplies amino acids limited in, or missing from, the others. This strategy is called *mutual supplementation.* The protein foods that mutually supplement each other are called *complementary proteins.* Figure NP4.1 (p. 97) shows combinations of plant proteins that provide higher-quality protein than the individual foods alone could supply.

CHAPTER 4

Nutrition in Practice

Glossary of Vegetarian Terms

complementary proteins: two or more proteins whose amino acid assortments complement each other in such a way that the essential amino acids missing from one are supplied by the other.

lacto-vegetarians: people who include milk or milk products, but exclude meat, poultry, fish, shellfish, and eggs from their diets.

lacto-ovo vegetarians: people who include milk or milk products and eggs, but omit meat, fish, shellfish, and poultry from their diets.

mutual supplementation: the strategy of combining two protein foods in a meal so that each food provides the essential amino acid(s) lacking in the other.

semivegetarians: people who include some, but not all, groups of animal-derived foods in their diets; they usually exclude meat and may occasionally include poultry, fish, and shellfish; also called *partial vegetarians.*

vegans: people who exclude all animal-derived foods (including meat, poultry, fish, shellfish, eggs, cheese, and milk) from their diets; also called *strict vegetarians* or *total vegetarians.*

What sorts of food energy intakes do vegetarian diets provide?

Researchers find that vegetarians as a group are closer to a healthy body weight than nonvegetarians. Because obesity impairs health in a number of ways, vegetarians therefore have a health advantage. Vegetarian diets tend to be high in complex carbohydrates and low in fat, characteristics that are consistent with current dietary recommendations aimed at reducing the incidence of obesity and other degenerative diseases in this country.

Not all vegetarians fit the average pattern, though. Obesity does threaten vegetarians who include milk, eggs, and cheese in their diets. They can easily consume both a high-fat diet and excess food energy and so must be careful to select nonfat and low-fat dairy foods and to avoid relying too heavily on these foods in general.

In contrast, people who exclude all animal-derived foods (vegans) may have trouble obtaining *enough* food energy. This is especially true for children and pregnant and lactating women.[4] Vegan diets can fail to provide food energy sufficient to support the growth of a child within a bulk of food small enough for the child to eat.[5] Plant foods that are best suited to meeting energy needs in a small volume are cereals, legumes, and nuts; these foods should be emphasized in a vegan child's diet. Table 1.4 in Chapter 1 offers a suggested daily food guide for vegetarians.

Tell me about vitamins and minerals. If I choose to eat a vegetarian diet, will I need to take vitamin supplements?

That depends on the kind of vegetarian diet you follow. The lacto-ovo vegetarian diet can be complete in all vitamins, but for the vegan, several vitamins may be a problem. One such vitamin is B_{12}. Because vitamin B_{12} occurs only in animal-derived foods, supplements are necessary to prevent deficiency. Women who have adhered to all-plant diets for many years are especially likely to have low vitamin B_{12} stores. Pregnant vegan women, whose needs for vitamin B_{12} are especially high, find it virtually impossible to maintain adequate vitamin B_{12} status without taking supplements or including a reliable food source of the nutrient.

A vitamin B_{12} deficiency can take a long time to develop in adults because up to four years' worth of the vitamin can be stored in the body. But when the deficiency sets in, it does severe damage to the nervous system (see Chapter 7). In infants, deficiencies set in more rapidly and so threaten their nervous systems early. All vegan mothers must be sure to take the appropriate supplements or to use vitamin B_{12}–fortified products.

What other vitamins do vegans need?

Another vitamin of concern is vitamin D. The milk drinker is protected, provided the milk is fortified with vitamin D, but there is no practical source of vitamin D in plant foods. Regular exposure to the sun will prevent a deficiency, but vegans who are homebound or live in a northern climate or smoggy city probably should take vitamin D supplements. Excesses of vitamin D are toxic, and one should not exceed the recommended daily amount of 5 micrograms.

Figure NP4.1
Nonmeat Mixtures That Provide High-Quality Protein

Vegetarians who eat no foods from animal sources select foods from two or more of these columns to create high-quality protein combinations:

Grains	Legumes	Seeds and Nuts	Vegetables
Barley	Dried beans (pintos, kidney, navy, etc.)	Almonds	Broccoli
Bulgur	Dried peas	Cashews	Cabbage
Oats	Dried lentils	Nut butters	Peppers
Pasta	Peanuts	Sesame seeds	Spinach
Rice	Soy products	Sunflower seeds	Squash
Whole-grain breads		Walnuts	

Black beans and rice, a favorite Hispanic combination.

Tofu and stir-fried vegetables with rice, an Asian dish.

Riboflavin, another vitamin often obtained from milk, is not a problem for the vegan who eats dark greens frequently in ample servings. The vegan who doesn't consume a lot of greens, however, may not meet riboflavin needs. Nutritional yeast is a rich source of riboflavin for the vegetarian.

So, on a vegan diet, vitamin B_{12}, vitamin D, and riboflavin can be problems if I'm not careful. What about minerals?

For *all* vegetarians, not just the vegan, two minerals may be of concern—iron and zinc. Legumes are an important source of iron in the vegetarian diet. The iron in legumes, however, is not as absorbable as that in meat. In fact, people absorb three times as much iron from a meal that includes meat as from one that does not. Because vitamin C in fruits and vegetables can triple iron absorption from other foods eaten at the same meal, vegetarian meals should be rich in foods offering vitamin C.

Zinc may also be a problem nutrient for vegetarians. It is widespread in plant foods, but its availability may be hindered by the fibers and other binders found in fruits and vegetables. The zinc needs of vegetarians and the effects of mineral binders are subjects of intensive study at the present time. While research continues, vegetarians are advised to eat varied diets that include whole-grain breads well leavened with yeast, which improves the availability of their minerals.

What about calcium for the vegan?

Good thinking. Yes, of course, calcium is of concern. The milk-drinking vegetarian is protected from deficiency, but the vegan must find other sources of calcium. Some good calcium sources are regular and ample servings of stone-ground meal, self-rising flour and meal, legumes, calcium-fortified soy milk, calcium-fortified orange juice, some nuts such as almonds, and certain seeds such as sesame seeds. The choices should be varied because binders in some of these foods may hinder calcium absorption. The vegetarian is urged to use calcium-fortified soy milk in ample quantities regularly. This is especially important for children. Infant formula based on soy is fortified with calcium and can easily be used in cooking foods, even for adults.

It sounds as though I can do well with a vegetarian diet as long as I do a little planning. I could use some help with weight control. Are there any other health advantages to the vegetarian diet?

Yes. Vegetarian protein foods are often higher in fiber, richer in certain vitamins and minerals, and lower in fat than meats. Vegetarians can enjoy a nutritious diet very low in fat provided that they limit other high-fat foods such as butter, cream cheese, sour cream, and nuts. If vegetarians follow the guidelines presented here and plan carefully, they can support their health as well as, or perhaps better than, nonvegetarians.

Abundant evidence supports the idea that vegetarians may actually be healthier than meat eaters. Informed vegetarians are not only more likely to be at the desired weights for their heights, but to have lower blood cholesterol levels, lower rates of certain kinds of cancer, better digestive function, and more. Even among people who are health conscious, generally vegetarians experience fewer deaths from cardiovascular disease than meat eaters do. Since vegetarians also often abstain from smoking and the consumption of alcohol and emphasize a supportive family life, dietary practices alone probably do not account for all the aspects of improved health. Clearly, however, they contribute significantly to it.

Notes

1. Position of The American Dietetic Association: Vegetarian diets, *Journal of the American Dietetic Association* 97 (1997): 1317–1321.

2. P. Walter, Effects of vegetarian diets on aging and longevity, *Nutrition Reviews* 55 (1997): S61–S68; T. C. Campbell and C. Junshi, Diet and chronic degenerative diseases: Perspectives from China, *American Journal of Clinical Nutrition* 59 (1994): 1153S–1161S.

3. J. T. Dwyer, Vegetarian eating patterns: Science, values, and food choices—Where do we go from here? *American Journal of Clinical Nutrition* 59 (1994): 1255S–1262S.

4. T. Sanders and S. Reddy, Vegetarian diets and children, *American Journal of Clinical Nutrition* 59 (1994): 1176S–1181S; L. H. Allen and coauthors, The interactive effects of dietary quality on the growth and attained size of young Mexican children, *American Journal of Clinical Nutrition* 56 (1992): 353–364.

5. Sanders and Reddy, 1994.

5

Digestion and Absorption

Your body's ability to transform the foods you eat into the nutrients that fuel your body's work is quite remarkable. Yet most people probably give little, if any, thought to all the body does with food once it is eaten. This chapter offers you the opportunity to learn how the body digests, absorbs, and transports the nutrients and how it excretes the unwanted substances in foods. The next chapter shows you how the body assimilates the nutrients once they have been absorbed and are traveling in the blood and lymph.

One of the beauties of the digestive tract is that it is selective. Materials that are nutritive for the body are broken down into particles that can be absorbed into the bloodstream. Most of the nonnutritive materials are left undigested and pass out the other end of the digestive tract.

Anatomy of the Digestive Tract

GI tract: the gastrointestinal tract or digestive tract; the principal organs are the stomach and intestines.
gastro = stomach

The gastrointestinal (GI) tract is a flexible muscular tube measuring about 15 feet in length from the mouth to the anus.[1] Figure 5.1 (pp. 102–103) traces the path followed by food from one end to the other. The accompanying glossary defines GI anatomy terms. In a sense, the human body surrounds the GI tract. Only when a nutrient or other substance passes through the cells of the digestive tract wall does it actually enter the body.

The Digestive Organs

digestion: the process by which complex food particles are broken down to smaller absorbable particles.

The process of digestion begins in the mouth as you begin to chew. The teeth crush and soften foods, while saliva mixes with the food mass and moistens it for comfortable swallowing. Saliva also helps dissolve the food so that you can taste it; only particles in solution can react with taste buds.

bolus (BOH-lus): the portion of food swallowed at one time.

Mouth to the Esophagus Once a mouthful of food has been swallowed, it is called a bolus. Each bolus first slides across your epiglottis, bypassing the entrance to your lungs. During each swallow, the epiglottis closes off your air passages so that you do not choke.

sphincter (SFINK-ter): a circular muscle surrounding, and able to close, a body opening.
sphincter = band (binder)

chyme (KIME): the semiliquid mass of partly digested food expelled by the stomach into the duodenum (the top portion of the small intestine).

Esophagus to the Stomach Next, the bolus slides down the esophagus, which conducts it through the diaphragm to the stomach. The cardiac sphincter, a band of muscle surrounding the esophagus where it enters the stomach, closes behind the bolus so that it cannot slip back. The stomach retains the bolus for a while, adds juices to it (gastric juices are discussed on p. 106), and grinds it into a semiliquid mass called chyme. Then bit by bit, the stomach releases the chyme through another sphincter, the pyloric sphincter, which opens into the small intestine and then closes behind the chyme.

The Small Intestine At the top of the small intestine, the chyme passes by an opening from the common bile duct. Through this opening, the gallbladder secretes fluids into the small intestine from outside the GI tract. The chyme travels on down the small intestine through its three segments—the duodenum, the jejunum, and the ileum—a total of about 10 feet of tubing coiled within the abdomen.[2]

American College of Gastroenterology

The Colon (Large Intestine) Having traveled the length of the small intestine, the chyme passes through another sphincter, the ileocecal valve, into the beginning of the colon (large intestine) in the lower right-hand side of the abdomen.

In the colon, the chyme travels up the right-hand side of the abdomen, across the front to the left-hand side, down to the lower left-hand side, and finally below the other folds of the intestines to the back side of the body above the rectum.

The Rectum During chyme's passage to the rectum, the colon withdraws water from it, leaving semisolid waste. The strong muscles of the rectum hold back this waste until it is time to defecate. Then the rectal muscles relax, and the last sphincter in the system, the anus, opens to allow the wastes to pass.

In summary, food follows the path shown in the margin. Considering all that happens on the way, the route is remarkably simple.

The Path of Food through the Digestive Tract

- *Mouth.*
- *Esophagus.*
- *Cardiac sphincter (or lower esophageal sphincter).*
- *Stomach.*
- *Pyloric sphincter.*
- *Duodenum (common bile duct enters here), jejunum, ileum.*
- *Ileocecal valve.*
- *Colon.*
- *Rectum.*
- *Anus.*

The Involuntary Muscles and the Glands

You are usually unaware of all the activity that goes on between the time you swallow and the time you defecate. As is the case with so much else that happens in the body, the muscles and glands of the digestive tract meet internal needs without your having to exert any conscious effort to get the work done.

People consciously chew and swallow, but even in the mouth there are some processes over which you have no control. The salivary glands secrete just enough saliva to moisten each mouthful of food so that it can pass easily down your esophagus.

gland: a cell or group of cells that secretes materials for special uses in the body. Glands may be *exocrine glands,* secreting their materials "out" (into the digestive tract or onto the surface of the skin), or *endocrine glands,* secreting their materials "in" (into the blood).
exo = outside
endo = inside
krine = to separate

Glossary of GI Terms

These terms are listed in order from the beginning of the digestive tract to the end.

epiglottis (epp-ee-GLOT-tiss): a cartilage structure in the throat that prevents fluid or food from entering the trachea when a person swallows.
epi = upon (over)
glottis = back of tongue

trachea (TRAKE-ee-uh): the windpipe; the passageway from the mouth and nose to the lungs.

esophagus (e-SOFF-uh-gus): the food pipe; the conduit from the mouth to the stomach.

cardiac sphincter (CARD-ee-ack SFINK-ter): the sphincter muscle at the junction between the esophagus and the stomach.
cardiac = heart

pyloric (pie-LORE-ic) **sphincter:** the sphincter muscle separating the stomach from the small intestine (also called *pylorus* or *pyloric valve*).
pylorus = gatekeeper

gallbladder: the organ that stores and concentrates bile. When it receives the signal that fat is present in the duodenum, the gallbladder contracts and squirts bile through the bile duct into the duodenum.

pancreas: a gland that secretes enzymes and digestive juices into the duodenum. (This is its exocrine function; it also has the endocrine function of secreting insulin and other hormones into the blood.)

small intestine: a 10-foot length of small-diameter (1-inch) intestine that is the major site of digestion of food and absorption of nutrients.

duodenum (doo-oh-DEEN-um or doo-ODD-num): the top portion of the small intestine (about "12 fingers' breadth" long, in ancient terminology).
duodecim = twelve

jejunum (je-JOON-um): the first two-fifths of the small intestine beyond the duodenum.

ileum (ILL-ee-um): the last segment of the small intestine.

ileocecal (ill-ee-oh-SEEK-ul) **valve:** the sphincter muscle separating the small and large intestines.

colon or **large intestine:** the last portion of the intestine, which absorbs water. Its main segments are the ascending colon, the transverse colon, the descending colon, and the sigmoid colon.
sigmoid = shaped like the letter S (*sigma* in Greek)

appendix: a narrow blind sac extending from the beginning of the colon; the appendix stores lymphocytes.

rectum: the muscular terminal part of the GI tract extending from the sigmoid colon to the anus; the rectum stores waste prior to elimination.

anus (AY-nus): the terminal sphincter muscle of the GI tract.

Figure 5.1
The Gastrointestinal Tract

FIBER	CARBOHYDRATE
Mouth The mechanical action of the mouth and teeth crushes and tears fiber in food and mixes it with saliva to moisten it for swallowing.	The salivary glands secrete a watery fluid into the mouth to moisten the food. The salivary enzyme amylase begins digestion: $\text{Starch} \xrightarrow{\text{amylase}} \text{small polysaccharides, maltose.}$
Esophagus Fiber is unchanged.	Digestion of starch continues as swallowed food moves down the esophagus.
Stomach Fiber is unchanged.	Stomach acid and enzymes start to digest salivary enzymes, halting starch digestion. To a small extent, stomach acid hydrolyzes maltose and sucrose.
Small intestine Fiber is unchanged.	The pancreas produces enzymes and releases them through the pancreatic duct into the small intestine: $\text{Polysaccharides} \xrightarrow[\text{amylase}]{\text{pancreatic}} \text{disaccharides.}$ Then enzymes on the surfaces of the small intestinal cells break disaccharides into monosaccharides, and the cells absorb them: $\text{Maltose} \xrightarrow{\text{maltase}} \text{glucose} + \text{glucose.}$ $\text{Sucrose} \xrightarrow{\text{sucrase}} \text{fructose} + \text{glucose.}$ $\text{Lactose} \xrightarrow{\text{lactase}} \text{galactose} + \text{glucose.}$
Colon (large intestine) Most fiber passes intact through the digestive tract to the colon. Here, bacterial enzymes digest some fiber: $\text{Some fiber} \xrightarrow[\text{enzymes}]{\text{bacterial}} \text{fatty acids, gas.}$ Fiber holds water; regulates bowel activity; and binds cholesterol and some minerals, carrying them out of the body as it is excreted with feces.	

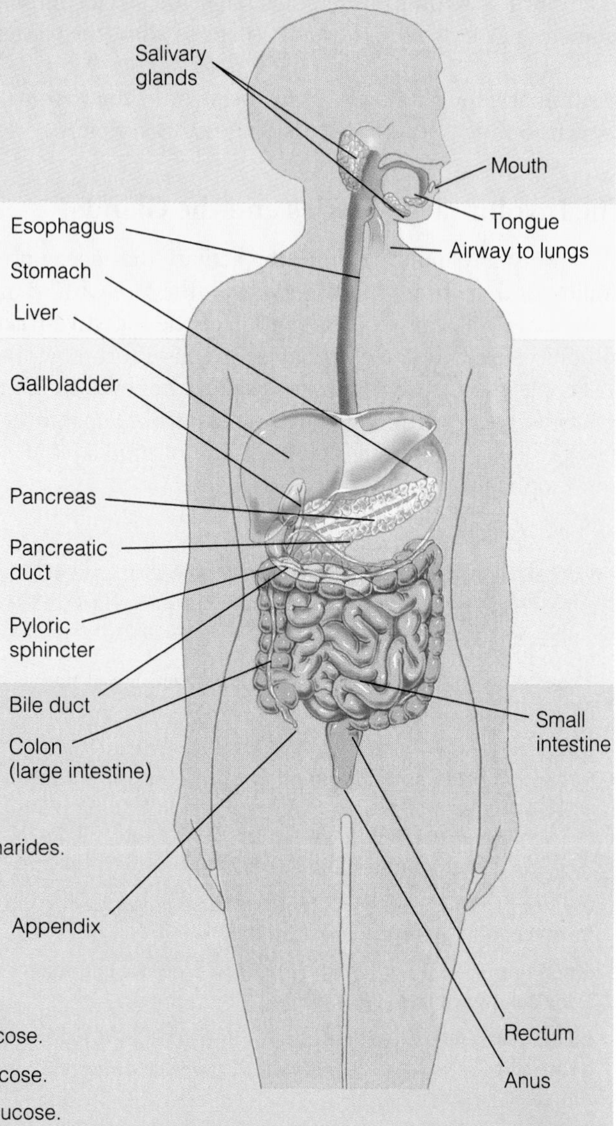

Salivary glands
Mouth
Tongue
Airway to lungs
Esophagus
Stomach
Liver
Gallbladder
Pancreas
Pancreatic duct
Pyloric sphincter
Bile duct
Small intestine
Colon (large intestine)
Appendix
Rectum
Anus

(continued)

FAT	PROTEIN	VITAMINS	MINERALS AND WATER
Mouth Glands in the base of the tongue secrete a fat-digesting enzyme known as lingual lipase. Some hard fats begin to melt as they reach body temperature.	Chewing and crushing moisten protein-rich foods and mix them with saliva to be swallowed.	No action.	The salivary glands add water to disperse and carry food.
Esophagus Fat is unchanged.	No action.	No action.	No action.
Stomach The acid-stable lingual lipase splits one bond of triglycerides to produce diglycerides and fatty acids. The degree of hydrolysis is slight for most fats but may be appreciable for milk fats. The stomach's churning action mixes fat with water and acid. A gastric lipase accesses and hydrolyzes a very small amount of fat.	Stomach acid uncoils protein strands and activates stomach enzymes: $\text{Protein} \xrightarrow[\text{HCl}]{\text{pepsin}} \text{smaller polypeptides.}$	Intrinsic factor (see Chapter 7) attaches to vitamin B_{12}.	Stomach acid (HCl) acts on iron to reduce it, making it more absorbable (see Chapter 8). The stomach secretes enough watery fluid to turn a moist, chewed mass of solid food into liquid chyme.
Small intestine Bile flows in from the liver and gallbladder (via the common bile duct): $\text{Fat} \xrightarrow{\text{bile}} \text{emulsified fat.}$ Pancreatic lipase flows in from the pancreas (via the pancreatic duct): $\text{Emulsified fat} \xrightarrow[\text{lipase}]{\text{pancreatic}}$ monoglycerides, glycerol, fatty acids (absorbed).	Pancreatic and small intestinal enzymes split polypeptides further: $\text{Polypeptides} \xrightarrow[\text{and intestinal proteases}]{\text{pancreatic}}$ dipeptides, tripeptides, and amino acids. Then enzymes on the surface of the small intestinal cells hydrolyze these peptides, and the cells absorb them: $\text{Peptides} \xrightarrow[\text{dipeptidases and tripeptidases}]{\text{intestinal}}$ amino acids (absorbed)	Bile emulsifies fat-soluble vitamins and aids in their absorption with other fats. Water-soluble vitamins are absorbed.	The small intestine, pancreas, and liver add enough fluid so that approximately 2 gallons are secreted into the intestine in a day. Many minerals are absorbed. Vitamin D aids in the absorption of calcium.
Colon Some fat and cholesterol, trapped in fiber, exit in feces.		Bacteria produce vitamin K, which is absorbed.	More minerals and most of the water are absorbed.

Gastric Motility Once you have swallowed, materials are moved through the rest of the GI tract by involuntary muscular contractions. This motion, known as gastric motility, consists of two types of movement, peristalsis and segmentation (see Figure 5.2). Peristalsis propels, or pushes; segmentation mixes, with more gradual pushing.

gastric motility: spontaneous motion in the digestive tract accomplished by involuntary muscular contractions.

Peristalsis Peristalsis begins when the bolus enters the esophagus. The entire GI tract is ringed with circular muscles that can squeeze it tightly. Surrounding these rings of muscle are longitudinal muscles. When the rings tighten and the long muscles relax, the tube is constricted. When the rings relax and the long muscles tighten, the tube bulges. These actions follow each other continuously and push the intestinal contents along. If you have ever watched a bolus of food pass along the body of a snake, you have a good picture of how these muscles work. The waves of contraction ripple through the GI tract at varying rates and intensities depending on the part of the GI tract and on whether food is present. Peristalsis, aided by the sphincter muscles that surround the tract at key places, keeps things moving along.

peristalsis (peri-STALL-sis): successive waves of involuntary muscular contractions passing along the walls of the GI tract that push the contents along.
peri = around
stellein = wrap

Segmentation The intestines not only push but also periodically squeeze their contents as if a string tied around the intestines were being pulled tight. This motion, called segmentation, forces the contents back a few inches, mixing them and promoting close contact with the digestive juices and the absorbing cells of the intestinal walls before letting them slowly move along again.

segmentation: a periodic squeezing or partitioning of the intestine by its circular muscles that both mixes and slowly pushes the contents along.

Liquefying Process Besides forcing the intestinal contents along, the muscles of the GI tract help to liquefy them to chyme so that the digestive enzymes will have access to all their nutrients. The mouth initiates this liquefying process by chewing, adding saliva, and stirring with the tongue to reduce the food to a coarse mash suitable for swallowing. The stomach then further mixes and kneads the food.

Stomach Action The stomach has the thickest walls and strongest muscles of all the GI tract organs. In addition to the circular and longitudinal muscles, the stomach has a third layer of diagonal muscles that also alternately contract and relax (see Figure 5.3 on p. 106). These three sets of muscles work to force the chyme downward, but the pyloric sphincter usually remains tightly closed so that the stomach's contents are thoroughly mixed and squeezed. Meanwhile, the gastric glands are adding juices. When the chyme is thoroughly liquefied, the pyloric sphincter opens briefly, about three times a minute, to allow small portions through. At this point, the intestinal contents no longer resemble food in the least.

The Process of Digestion

One person eats nothing but vegetables, fruits, and nuts; another, nothing but meat, milk, and potatoes. How is it that both people wind up with essentially the same body composition? It all comes down to the fact that the body renders food—whatever it is to start with—into the basic units that make up carbohydrate, fat, and protein. The body absorbs these units and builds its tissues from them.

To digest food, five different body organs secrete digestive juices: the salivary glands, the stomach, the small intestine, the liver (via the gallbladder), and the pancreas. Each of the juices has a turn to mix with the food and promotes its breakdown to small units that can be absorbed into the body. The glossary (p. 107) defines some of the digestive glands and their juices.

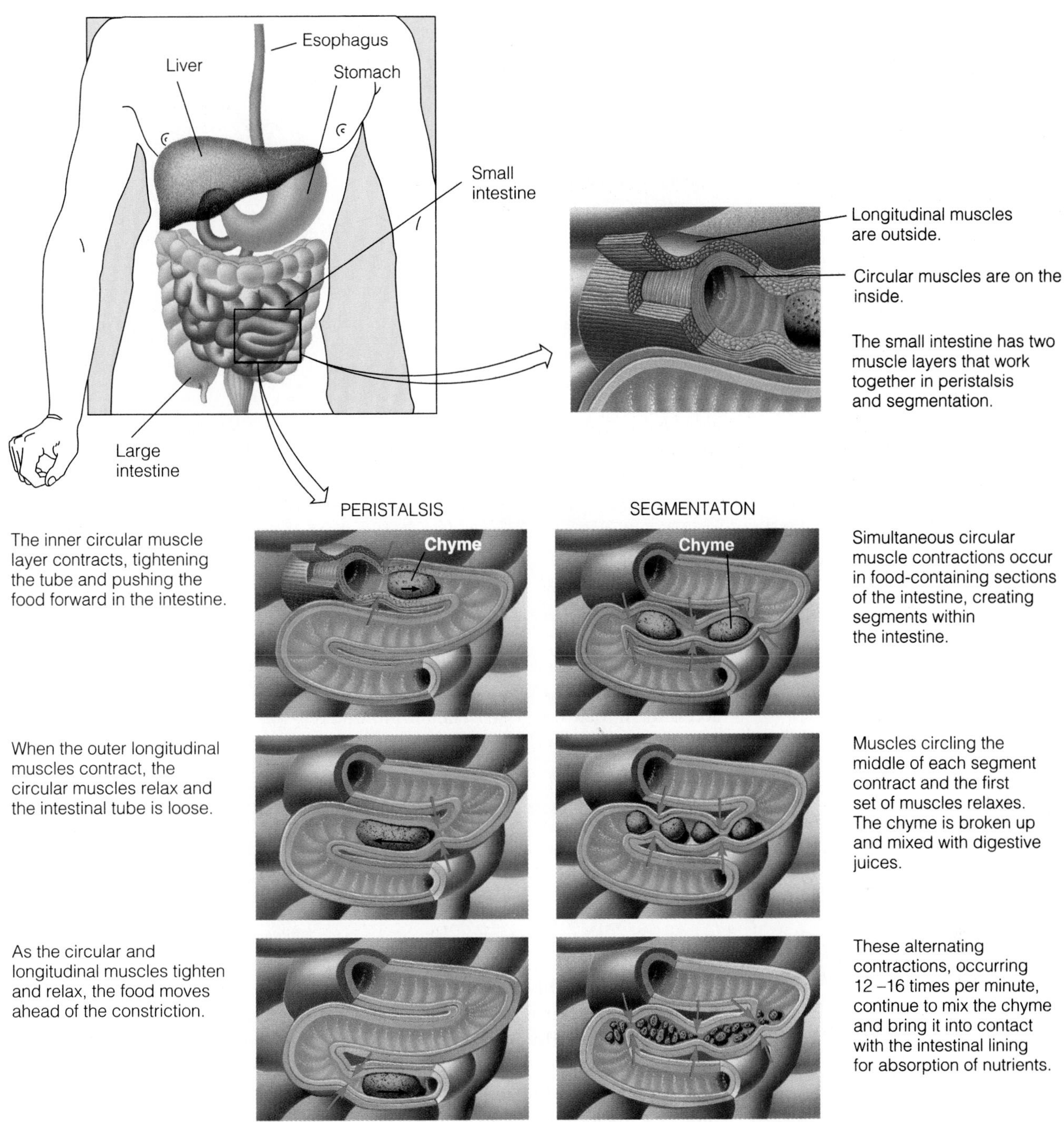

Esophagus
Liver
Stomach
Small
intestine
Large
intestine
Longitudinal muscles
are outside.
Circular muscles are on the
inside.
The small intestine has two
muscle layers that work
together in peristalsis
and segmentation.
PERISTALSIS
SEGMENTATON
Chyme
Chyme
The inner circular muscle
layer contracts, tightening
the tube and pushing the
food forward in the intestine.
Simultaneous circular
muscle contractions occur
in food-containing sections
of the intestine, creating
segments within
the intestine.
When the outer longitudinal
muscles contract, the
circular muscles relax and
the intestinal tube is loose.
Muscles circling the
middle of each segment
contract and the first
set of muscles relaxes.
The chyme is broken up
and mixed with digestive
juices.
As the circular and
longitudinal muscles tighten
and relax, the food moves
ahead of the constriction.
These alternating
contractions, occurring
12–16 times per minute,
continue to mix the chyme
and bring it into contact
with the intestinal lining
for absorption of nutrients.

Figure 5.2
Peristalsis and Segmentation

Digestion in the Mouth

Digestion of carbohydrate begins in the mouth, where the salivary glands secrete saliva, which contains water, salts, and enzymes (including salivary amylase) that break the bonds in the chains of starch. Saliva also protects the tooth surfaces and linings of the mouth, esophagus, and stomach from attack by molecules that might harm them. The enzymes in the mouth do not, for the most part, affect the fats, proteins, vitamins, minerals, and fiber that are present in the foods people eat.

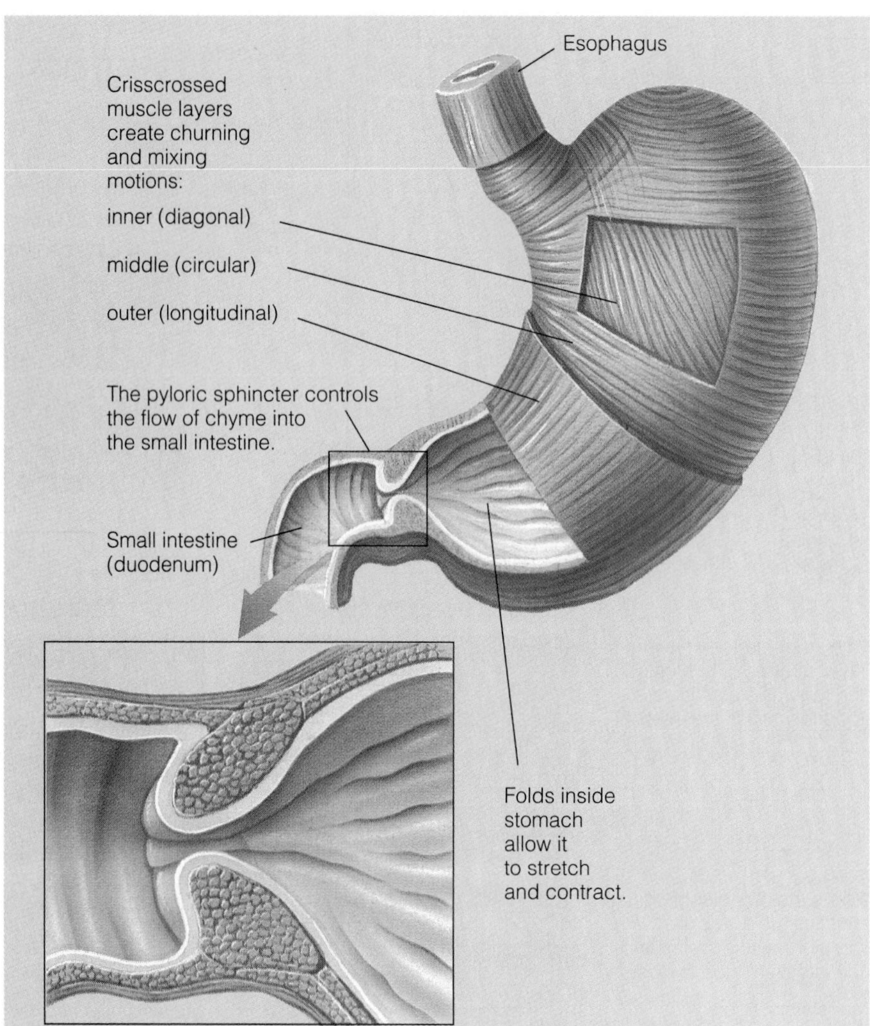
Esophagus
Crisscrossed muscle layers create churning and mixing motions:
inner (diagonal)
middle (circular)
outer (longitudinal)
The pyloric sphincter controls the flow of chyme into the small intestine.
Small intestine (duodenum)
Folds inside stomach allow it to stretch and contract.

Figure 5.3
Stomach Muscles
The stomach has three layers of muscles.

Digestion in the Stomach

Gastric juice is composed of water, enzymes, and hydrochloric acid. The acid is so strong that it burns the throat if it chances to reflux into the upper esophagus and mouth. The strong acidity of the stomach prevents bacterial growth and kills most bacteria that enter the body with food. You might expect that the stomach's acid would attack the stomach itself, but the cells of the stomach wall secrete mucus, a thick, slimy, white polysaccharide that coats the stomach's lining.

Digestive Activities The major digestive event in the stomach is the initial breakdown of proteins. Other than being crushed and mixed with saliva in the mouth, nothing happens to protein until it comes in contact with the gastric juices in the stomach. There, the acid helps to uncoil (denature) the protein's tangled strands so that the stomach enzymes can attack the bonds. Both the enzyme pepsin and the stomach acid itself act as catalysts in the process. Minor events are the digestion of some fat by a gastric lipase, the digestion of sucrose (to a very small extent) by the stomach acid, and the attachment of a protein carrier to vitamin B_{12}.

The stomach enzymes work most efficiently in the stomach's strong acid, but salivary amylase, which is swallowed with food, does not work in acid this strong. Consequently, the digestion of starch gradually ceases as the acid penetrates the bolus. In fact, salivary amylase becomes just another protein to be

Glossary of Digestive Glands and Their Secretions

These terms are listed in order from the beginning of the digestive tract to the end.

salivary glands: exocrine glands that secrete saliva into the mouth.

saliva: the secretion of the salivary glands; the principal enzyme is salivary amylase.

amylase (AM-uh-lace): an enzyme that splits amylose (a form of starch). Amylase is a carbohydrase. The ending -ase indicates an enzyme; the root tells what it digests. Other examples: protease, lipase.

gastric glands: exocrine glands in the stomach wall that secrete gastric juice into the stomach.
gastro = stomach

gastric juice: the digestive secretion of the gastric glands containing a mixture of water, hydrochloric acid, and enzymes. The principal enzymes are pepsin (acts on proteins) and lipase (acts on emulsified fats).

hydrochloric acid (HCl): an acid composed of hydrogen and chloride atoms. The gastric glands normally produce this acid.

mucus (MYOO-cuss): a mucopolysaccharide (a relative of carbohydrate) secreted by cells of the stomach wall that protects the cells from exposure to digestive juices (and other destructive agents). The cellular lining of the stomach wall with its coat of mucus is known as the mucous membrane. (The noun is mucus; the adjective is mucous.)

pepsin: a protein-digesting enzyme (gastric protease) in the stomach. It circulates as a precursor, pepsinogen, and is converted to pepsin by the action of stomach acid.

intestinal juice: the secretion of the intestinal glands; contains enzymes for the digestion of carbohydrate and protein and a minor enzyme for fat digestion.

bile: an emulsifier that prepares fats and oils for digestion; made by the liver, stored in the gallbladder, and released into the small intestine when needed.

pancreatic (pank-ree-AT-ic) **juice:** the exocrine secretion of the pancreas, containing enzymes for the digestion of carbohydrate, fat, and protein. Juice flows from the pancreas into the small intestine through the pancreatic duct. The pancreas also has an endocrine function, the secretion of insulin and other hormones.

bicarbonate: an alkaline secretion of the pancreas; part of the pancreatic juice. (Bicarbonate also occurs widely in all cell fluids.)

digested. The amino acids in amylase end up being absorbed and recycled into other body proteins.

Antacids: Their Use and Misuse Note that the strong acidity of the stomach is a desirable condition, television commercials for antacids notwithstanding. In a person who overeats or who "inhales" food, the stomach is likely to react with such violence as to cause regurgitation. When this happens, the stomach acid makes a bad taste in the mouth, which the person may interpret as "acid indigestion." Responding to television commercials, an overeater may take antacids to neutralize the stomach acid. But then the stomach will have to secrete more acid to enable the digestive enzymes to do their work. The consumer's stomach ends up with the same amount of acid but has had to work against the antacid to produce it.

Antacids are not designed to relieve the digestive discomfort of the hasty eater. Their proper use is to correct an abnormal condition, such as that of the person with ulcers, whose stomach or duodenal lining has been attacked by acid. To avoid falling into the same trap as our misguided consumer, hasty eaters should remember to chew food more thoroughly, eat it more slowly, and possibly eat less at a sitting.

Digestion in the Small and Large Intestines

By the time food leaves the stomach, digestion of all three energy-yielding nutrients has begun, but the process gains momentum in the small intestine. There, the pancreas, the liver, and intestine contribute additional digestive juices through the duct leading into the duodenum. These juices contain enzymes, bicarbonate, and bile.

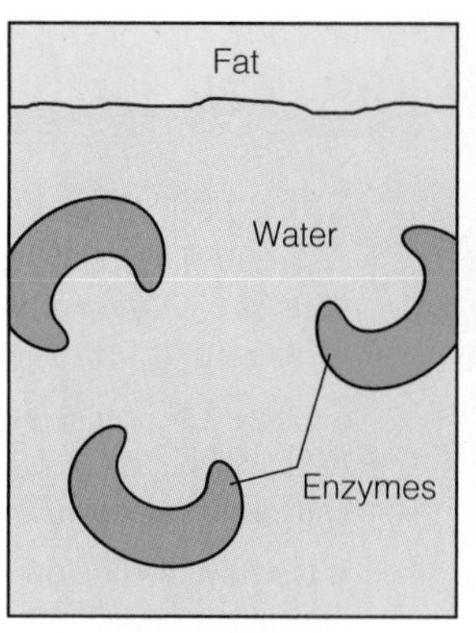

In the stomach, the fat and water are separate. The enzymes are in the water and can't get at the fat.

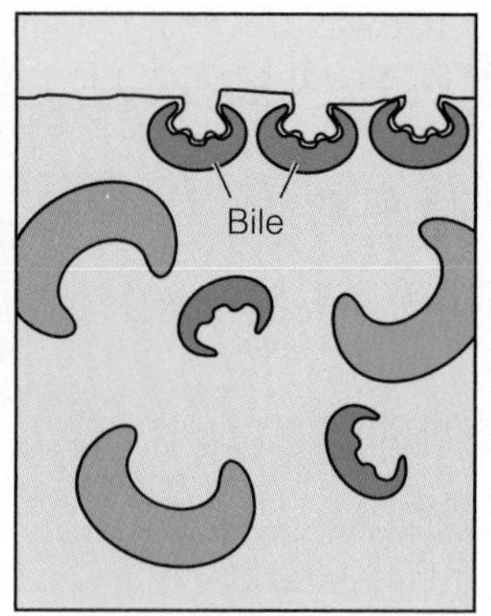

In the small intestine, bile (an emulsifier) arrives. Bile has an affinity for both fat and water and can therefore bring the fat into solution in the water.

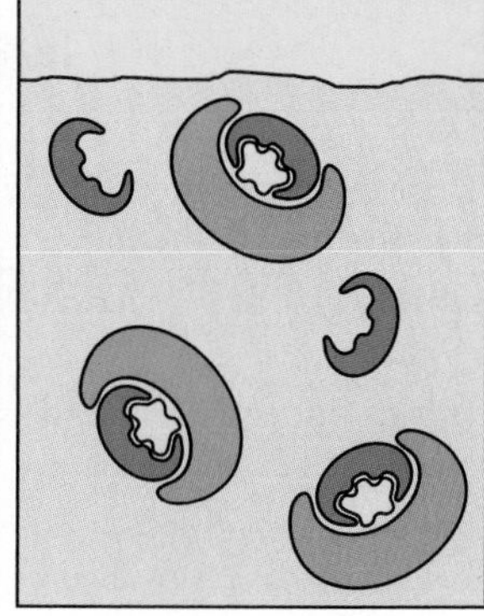
After emulsification, the fat is mixed in the water solution, so the enzymes have access to it.

Figure 5.4
The Action of Bile in Fat Digestion

Digestive Juice Pancreatic juice contributes enzymes that act on fats, proteins, and carbohydrates. Glands in the intestinal wall also secrete digestive enzymes. The pancreatic juice also contains sodium bicarbonate, which neutralizes the acidic chyme as it enters the small intestine. From this point on, the contents of the digestive tract are neutral or slightly alkaline. The enzymes of both the intestine and the pancreas work best in this environment.

Bile Bile is secreted by the liver continuously and is concentrated and stored in the gallbladder. The gallbladder squirts bile into the duodenum whenever fat arrives there. Bile is not an enzyme but an emulsifier that brings fats into suspension in water (see Figure 5.4). After the fats are emulsified, enzymes can work on them, and they can be absorbed. Thanks to all these secretions, all three energy-yielding nutrients are digested in the small intestine.

Reminder: An emulsifier is a substance that mixes with both fat and water and permanently disperses the fat in the water, forming an *emulsion.*

Mayonnaise, made from vinegar and oil, would separate as other vinegar-and-oil salad dressings do if food chemists did not blend the vinegar and oil with a third ingredient—an emulsifier. The emulsifier mixes well with the fatty oil and the watery vinegar. In the case of mayonnaise, the emulsifier is lecithin from egg yolks.

The Rate of Digestion The rate of digestion of the energy nutrients depends upon the contents of the meal. If the meal is high in simple sugars, digestion proceeds fairly rapidly. On the other hand, if the meal is rich in fat, digestion is slower.

intestinal flora: the bacterial inhabitants of the GI tract.
flora = plant growth

Protective Factors The intestine also contains bacteria that produce a variety of vitamins, including biotin and vitamin K (although bacteria alone cannot meet the need for these vitamins). The GI bacteria also protect people from infections. Provided that the normal intestinal flora are thriving, infectious bacteria have a hard time getting established and launching an attack on the system. In addition, the small intestine and the entire GI tract manufacture and maintain a strong arsenal of defenses against foreign invaders. Several different types of defending cells are present there and confer specific immunity against intestinal diseases.

The Final Stage The story of how food is broken down into nutrients that can be absorbed is now nearly complete. The three energy-yielding nutrients—carbohydrate, fat, and protein—are disassembled to basic building blocks before they are absorbed. Most of the other nutrients—vitamins, minerals, and water—are absorbed as they are. Undigested residues, such as some fibers, are not absorbed but continue through the digestive tract, providing a semisolid mass that helps stimulate the muscles of the GI tract so that they will remain strong and perform peristalsis efficiently. Fiber also retains water, keeping the stools soft, and carries bile acids, sterols, and fat with it out of the body.

The process of absorbing the nutrients into the body is discussed in the next section. For the moment, let us assume that the digested nutrients simply disappear from the GI tract as they are ready. Virtually all are gone by the time the contents of the GI tract reach the end of the small intestine. Little remains but water, a few undissolved salts and body secretions, and undigested materials such as fiber. These enter the large intestine (colon).

In the colon, intestinal bacteria degrade some of the fiber to simpler compounds. The colon itself retrieves from its contents the materials that the body is designed to recycle—water and dissolved salts. The waste that is finally excreted has little or nothing of value left in it. The body has extracted all that it can use from the food.

The Absorptive System

Within three or four hours after you have eaten a meal, your body must find a way to absorb some two hundred thousand, million amino acid molecules one by one and a comparable number of monosaccharide, monoglyceride, glycerol, fatty acid, vitamin, and mineral molecules as well. The absorptive system is ingeniously designed to accomplish this task.

The Small Intestine

The small intestine is a tube about 10 feet long and about an inch across, yet it provides a surface comparable in area to a tennis court. Nutrient molecules make contact with this surface and are absorbed. To remove these molecules rapidly and provide room for more to be absorbed, a rush of circulation continuously bathes the underside of the surface, washing away the absorbed nutrients and carrying them to the liver and other parts of the body.

Villi and Microvilli How does the intestine manage to provide such a large surface area? Its inner surface looks smooth, but viewed through a microscope, it turns out to be wrinkled into hundreds of folds. Each fold is covered with thousands of fingerlike projections called villi. The villi are as numerous as the hairs on velvet fabric. A single villus, magnified still more, turns out to be composed of several hundred cells, each covered with microscopic hairs called microvilli (see Figure 5.5 on the next page).

The villi are in constant motion. Each villus is lined by a thin sheet of muscle so that it can wave, squirm, and wiggle like the tentacles of a sea anemone. Any nutrient molecule small enough to be absorbed is trapped among the microvilli and drawn into a cell beneath them. Some partially digested nutrients are caught in the microvilli, digested further by enzymes there, and then absorbed into the cells.

villi (VILL-ee or VILL-eye): fingerlike projections from the folds of the small intestine. The singular form is **villus.**
villus = shaggy hair

microvilli (MY-cro-VILL-ee or MY-cro-VILL-eye): tiny, hairlike projections on each cell of every villus that can trap nutrient particles and transport them into the cells. The singular form is **microvillus.**

Specialization in the Intestinal Tract As you can see, the intestinal tract is beautifully designed to perform its functions. A further refinement of the system is that the cells of successive portions of the tract are specialized to absorb different nutrients. The nutrients that are ready for absorption early on are absorbed near the top of the tract; those that take longer to be digested are absorbed further down. The rate at which the nutrients travel through the GI tract is finely adjusted to maximize their availability to the appropriate absorptive segment of the tract when they are ready. The lowly "gut" turns out to be one of the most elegantly designed organ systems in the body.

Figure 5.5
The Small Intestinal Villi

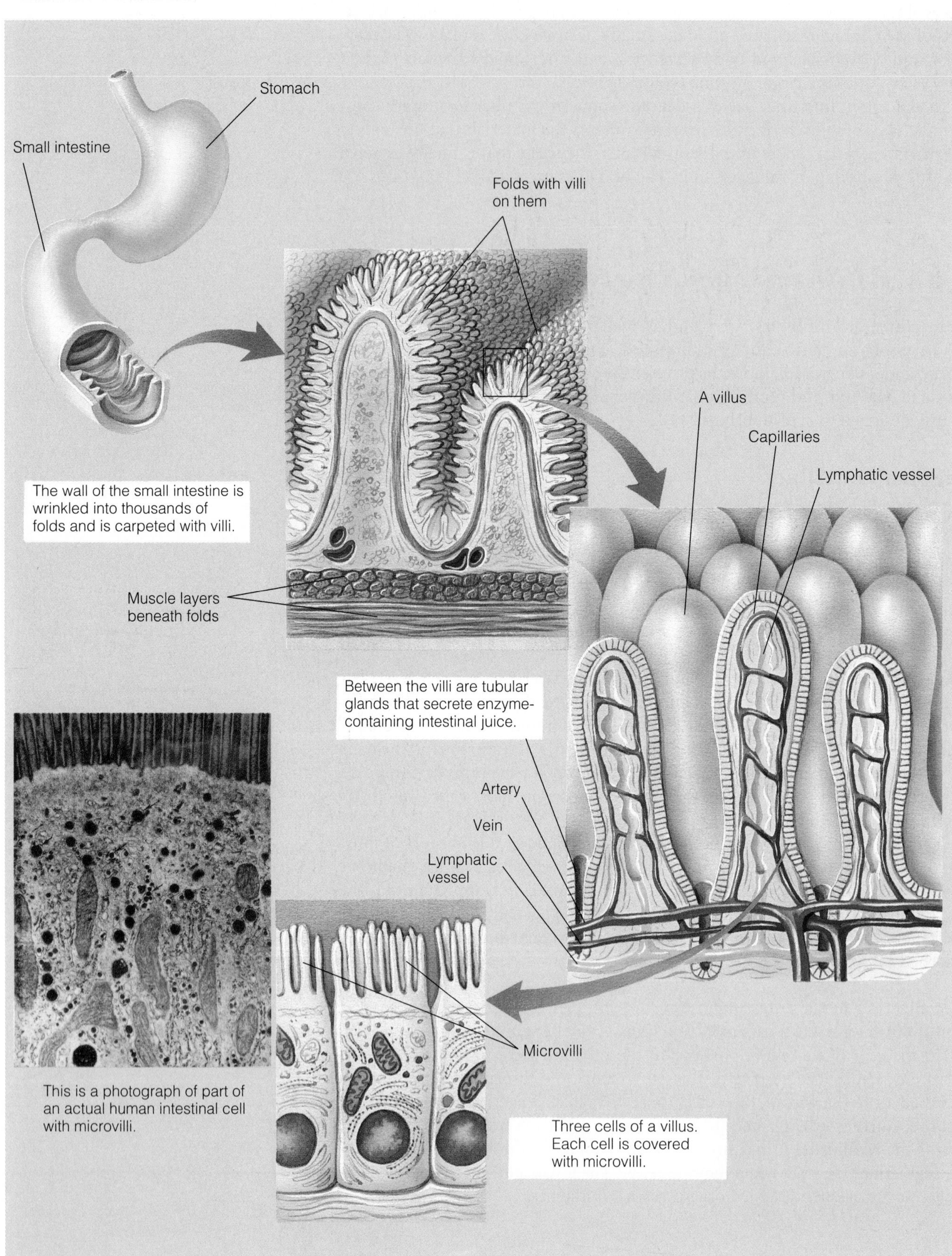

Stomach
Small intestine
Folds with villi on them
The wall of the small intestine is wrinkled into thousands of folds and is carpeted with villi.
Muscle layers beneath folds
A villus
Capillaries
Lymphatic vessel
Between the villi are tubular glands that secrete enzyme-containing intestinal juice.
Artery
Vein
Lymphatic vessel
Microvilli
This is a photograph of part of an actual human intestinal cell with microvilli.
Three cells of a villus. Each cell is covered with microvilli.

Release of Absorbed Nutrients

Once a molecule has entered a cell in a villus, the next step is to transmit it to a destination elsewhere in the body by way of the body's two transport systems—the bloodstream and the lymphatic system. As Figure 5.5 shows, both systems supply vessels to each villus. Through these vessels, the nutrients leave the cell and enter either the lymph or the blood. In either case, the nutrients end up in the blood, at least for a while. The water-soluble nutrients (including the smaller products of fat digestion) are released directly into the bloodstream by way of the capillaries, but the larger fats and the fat-soluble vitamins find access directly into the capillaries impossible because these nutrients are insoluble in water (and blood is mostly water). They require some packaging before they are released.

The intestinal cells assemble the monoglycerides and long-chain fatty acids into larger triglyceride molecules. These triglycerides, fat-soluble vitamins (when present), and other large lipids (cholesterol and the phospholipids) are then packaged for transport. They cluster together with special proteins to form chylomicrons, one kind of lipoprotein (lipoproteins are described beginning on p. 112). Finally, the cells release the chylomicrons into the lymphatic system. They can then glide through the lymph spaces until they arrive at a point of entry into the bloodstream near the heart.

lymphatic system: a loosely organized system of vessels and ducts that conveys the products of digestion toward the heart.

lymph (LIMF): the body fluid found in lymphatic vessels; lymph consists of all the constituents of blood except red blood cells.

lipoproteins: clusters of lipids associated with proteins that serve as transport vehicles for lipids in the lymph and blood.

chylomicrons (kye-lo-MY-crons): the lipoproteins that transport lipids from the intestinal cells into the body. The cells of the body remove the lipids they need from the chylomicrons, leaving chylomicron remnants to be picked up by the liver cells.

Transport of Nutrients

Once a nutrient has entered the bloodstream or the lymphatic system, it may be transported to any part of the body and thus becomes available to any of the cells, from the tips of the toes to the roots of the hair. The circulatory systems are arranged to deliver nutrients wherever they are needed.

The Vascular System

The vascular or blood circulatory system is a closed system of vessels through which blood flows continuously in a figure eight, with the heart serving as a pump at the crossover point. On each loop of the figure eight, blood travels a simple route: heart to arteries to capillaries to veins to heart.

The routing of the blood through the digestive system is different, however. The blood is carried to the digestive system (as it is to all organs) by way of an artery, which (as in all organs) branches into capillaries to reach every cell. Blood leaving the digestive system, however, goes by way of a vein, not back to the heart, but to the liver. This vein *again* branches into capillaries so that every cell of the liver has access to the newly absorbed nutrients that the blood is carrying. Blood leaving the liver then returns to the heart by way of a vein. The route is heart to arteries to capillaries (in intestines) to vein to capillaries (in liver) to vein to heart.

An anatomist studying this system knows there must be a reason for this special arrangement. The liver is placed in the circulation at this point so that it will have the first chance at the materials absorbed from the GI tract. In fact, the liver is the body's major metabolic organ (see Figure 5.6 on the next page). It must prepare the absorbed nutrients for use by the body, and it has many jobs to perform in this process. Furthermore, the liver stands as gatekeeper to waylay intruders that might otherwise harm the heart or brain. Chapter 28 offers more information about this noble organ.

artery: a vessel that carries blood away from the heart.

capillary (CAP-ill-ary): a small vessel that branches from an artery. Capillaries connect arteries to veins. Oxygen, nutrients, and waste materials are exchanged across capillary walls.

vein: a vessel that carries blood back to the heart.

The blood arriving at the intestines flows through the *mesentery* (MEZ-en-terry), a strong, flexible membrane that surrounds and supports the abdominal organs.
mes = middle

The vein that collects blood from the mesentery and conducts it to capillaries in the liver is the *portal vein.*
portal = gateway

The vein that collects blood from the liver capillaries and returns it to the heart is the *hepatic vein.*
hepat = liver

The artery that delivers oxygen-rich blood from the heart to the liver is the *hepatic artery.*

The Lymphatic System

The lymphatic system is a one-way route that fluids can travel to get from tissue spaces to the blood. The lymphatic system has no pump; instead, lymph is

Figure 5.6
The Liver and Its Circulatory System

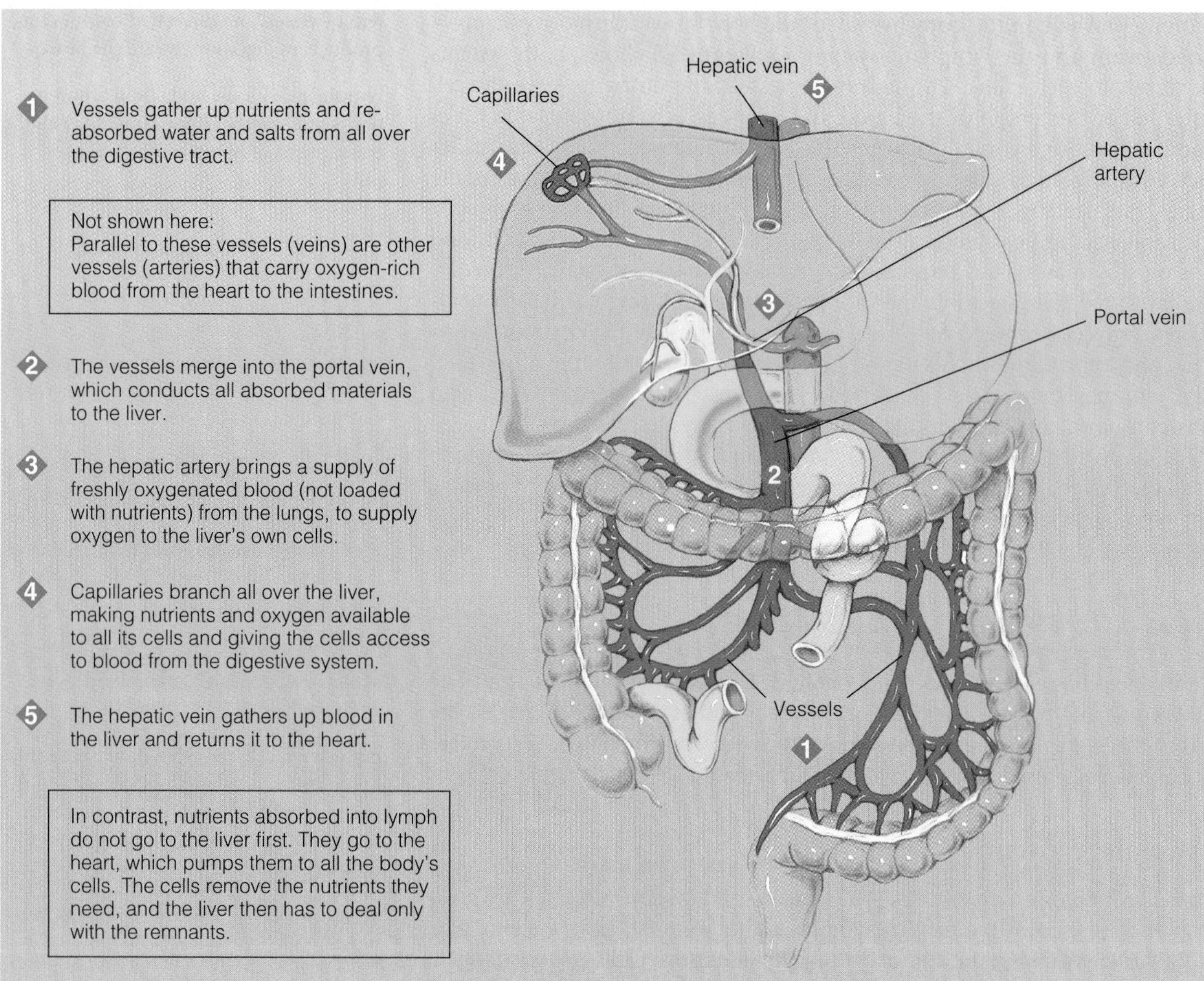

1 Vessels gather up nutrients and reabsorbed water and salts from all over the digestive tract.
Not shown here:
Parallel to these vessels (veins) are other vessels (arteries) that carry oxygen-rich blood from the heart to the intestines.
2 The vessels merge into the portal vein, which conducts all absorbed materials to the liver.
3 The hepatic artery brings a supply of freshly oxygenated blood (not loaded with nutrients) from the lungs, to supply oxygen to the liver's own cells.
4 Capillaries branch all over the liver, making nutrients and oxygen available to all its cells and giving the cells access to blood from the digestive system.
5 The hepatic vein gathers up blood in the liver and returns it to the heart.
In contrast, nutrients absorbed into lymph do not go to the liver first. They go to the heart, which pumps them to all the body's cells. The cells remove the nutrients they need, and the liver then has to deal only with the remnants.
Hepatic vein
Capillaries
Hepatic artery
Portal vein
Vessels

The duct that conveys lymph toward the heart is the *thoracic* (thor-ASS-ic) *duct.* The *subclavian vein* connects this duct with the right upper chamber of the heart, providing a passageway by which lymph can be returned to the vascular system.

squeezed from one portion of the body to another like water in a sponge, as muscles contract and create pressure here and there. Ultimately, the lymph collects in a large duct behind the heart. This duct terminates in a vein that conducts the lymph into the heart. Thus some materials from the GI tract enter the lymphatic system at first, then later enter the bloodstream.

Transport of Lipids: Lipoproteins

Within the circulatory system, lipids always travel from place to place bundled with protein, that is, as lipoproteins. When physicians measure a person's blood lipid profile, they are interested not only in the types of fat present (such as triglycerides and cholesterol) but also in the types of lipoproteins that carry them.

VLDL (very-low-density lipoprotein): the type of lipoprotein made primarily by liver cells to transport lipids to various tissues in the body; composed primarily of triglycerides.

VLDL, LDL, and HDL As mentioned earlier, chylomicrons transport newly absorbed (*diet-derived*) lipids from the intestinal cells to the rest of the body. As chylomicrons circulate through the body, cells remove their lipid contents, so the chylomicrons get smaller and smaller. The liver picks up and dismantles the chylomicron remnants and assembles new lipoproteins, which are known as very-low-density lipoproteins (VLDL). As the body's cells remove triglycerides from the VLDL, the VLDL gather cholesterol from other lipoproteins in the blood and become low-density lipoproteins (LDL). This exchange explains why LDL contain few triglycerides but are loaded with cholesterol. The LDL circulate through the body making their contents available to all the body's cells. Thus LDL carry cholesterol and triglycerides *from* the liver to the rest of the body. Lipids returning *to* the liver from other parts of the body are packaged in lipoproteins known as high-density lipoproteins (HDL).

The more lipid in the lipoprotein molecule, the lower the density; the more protein, the higher the density. Both LDL and HDL carry lipids around in the blood, but LDL are larger, lighter, and more lipid filled; HDL are smaller, denser, and packaged with more protein. LDL deliver cholesterol and triglycerides from the liver to the tissues; HDL scavenge excess cholesterol and phospholipids from the tissues and return them to the liver for disposal. Figure 5.7 (p. 114) shows the relative sizes and composition of the lipoproteins.

LDL (low-density lipoprotein): the type of lipoprotein derived from VLDL as cells remove triglycerides from them. LDL carry cholesterol and triglycerides from the liver to the cells of the body and are composed primarily of cholesterol.

HDL (high-density lipoprotein): the type of lipoprotein that transports cholesterol back to the liver from peripheral cells; composed primarily of protein.

Health Implications of LDL and HDL The distinction between LDL and HDL has implications for the health of the heart and blood vessels. Elevated LDL concentrations in the blood are associated with a high risk of heart attack, and elevated HDL concentrations are associated with a low risk.[3] This is why some people refer to LDL as "bad" cholesterol and HDL as "good" cholesterol. Keep in mind, though, that there is only *one* kind of cholesterol; the differences between LDL and HDL reflect *proportions* of lipids and proteins within them—not the type of cholesterol. Factors that improve the LDL-to-HDL ratio include the following:

- Weight control.
- Polyunsaturated or monounsaturated, instead of saturated, fatty acids in the diet.
- Soluble fibers (see Chapter 2).
- Antioxidants (see Chapter 7).
- Physical activity.

Lipoproteins and heart disease are discussed in Chapter 26.

The System at Its Best

The GI tract is the first organ in the body to deal with the nutrients that will ultimately maintain the health and nutrition status of the whole body. The intricate architecture of the GI tract makes it sensitive and responsive to conditions in its environment. One condition indispensable to its performance is its own good health. Such lifestyle factors as sleep, physical activity, state of mind, and nutrition affect GI tract health. For example, adequate sleep allows for repair and maintenance of tissue. Physical activity promotes healthy muscle tone and may protect against cancer of the colon.[4] Mental state profoundly affects digestion and absorption through the activity of nerves and hormones that help regulate these processes. A relaxed, peaceful attitude during a meal enhances digestion and absorption.

Among the characteristics of meals that promote optimal absorption of nutrients are adequacy, balance, moderation, and variety, because every nutrient depends on every other. The nutrients all work together, and all are present in the cells of a healthy digestive tract.

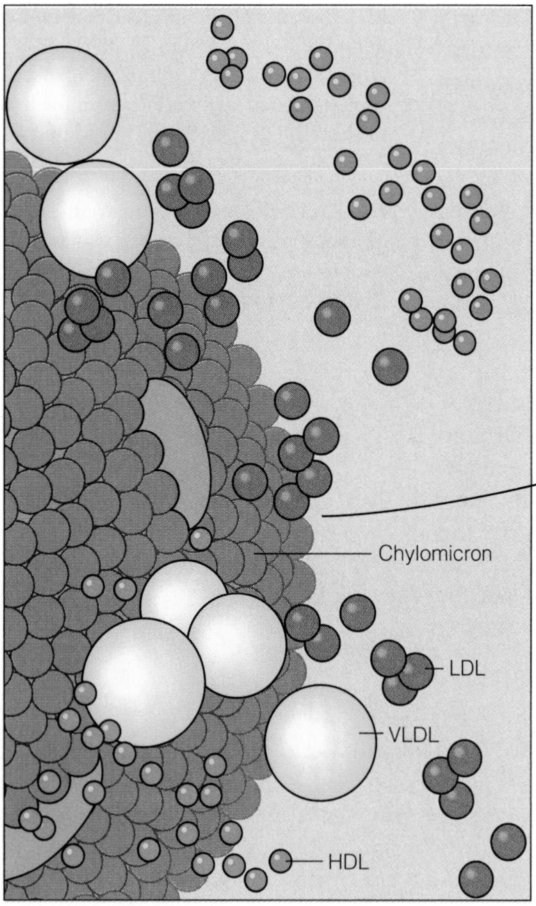

Chylomicron
LDL
VLDL
HDL

This solar system of lipoproteins shows their relative sizes. Notice how large the fat-filled chylomicron is compared with the others and how the others get progressively smaller as their proportion of fat declines and protein increases.

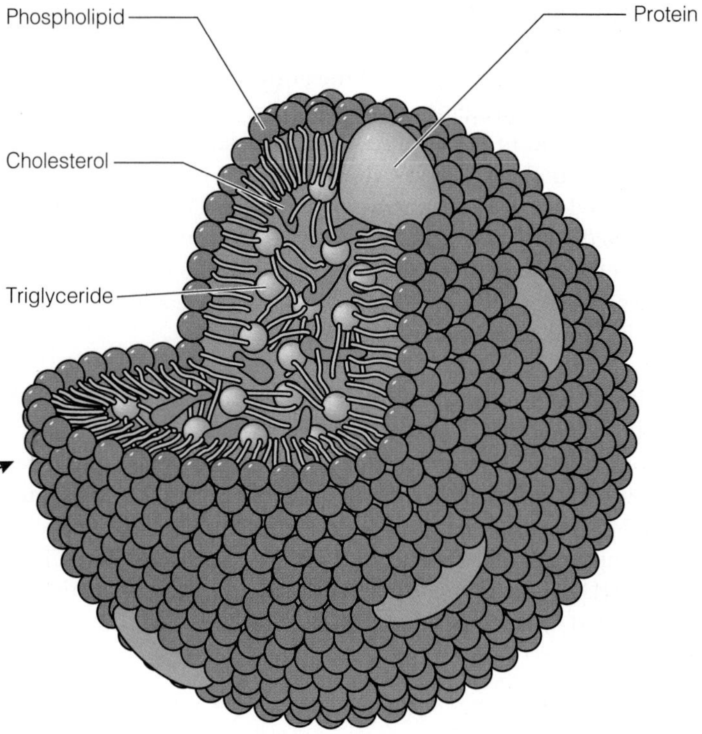

Phospholipid
Protein
Cholesterol
Triglyceride

A typical lipoprotein contains an interior of triglycerides and cholesterol surrounded by phospholipids. The phospholipids' fatty acid "tails" point toward the interior, where the lipids are. Proteins near the outer ends of the phospholipids cover the structure. This arrangement of hydrophobic molecules on the inside and hydrophilic molecules on the outside allows lipids to travel through the watery fluids of the blood.

Chylomicrons contain so little protein and so much triglyceride that they are the lowest in density.

Very-low-density lipoproteins (VLDL) are half triglycerides, accounting for their low density.

Low-density lipoproteins (LDL) are half cholesterol, accounting for their implication in heart disease.

High-density lipoproteins (HDL) are half protein, accounting for their high density.

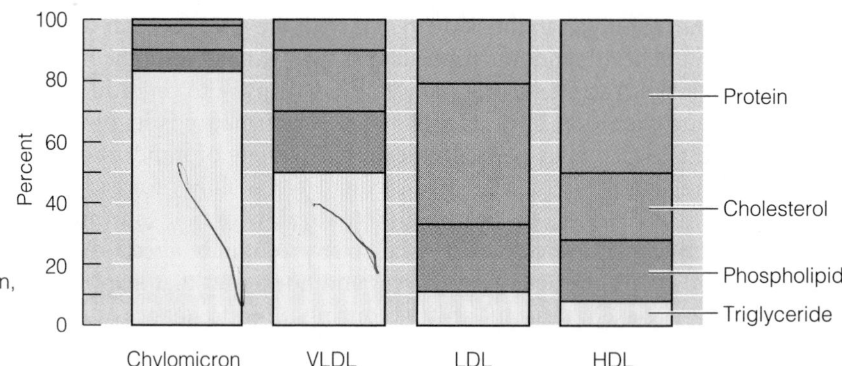

Figure 5.7
The Lipoproteins

Self Check

1. Once food is swallowed, it travels through the digestive tract in this order:
   a. esophagus, stomach, colon, liver.
   b. small intestine, large intestine, stomach, colon.
   c. esophagus, stomach, small intestine, colon.
   d. small intestine, stomach, esophagus, colon.
2. Once chyme travels the length of the small intestine, it passes through the ileocecal valve at the beginning of the:
   a. colon. b. stomach.
   c. esophagus. d. jejunum.
3. The periodic squeezing or partitioning of the intestine by its circular muscles that both mixes and slowly pushes the contents along is known as:
   a. persistalsis. b. absorption.
   c. segmentation. d. secretion.
4. Match the following:
   ______a. bile ______b. amylase
   ______c. gastric juice ______d. pepsin
   ______e. bicarbonate
   1. alkaline secretion of the pancreas, part of the pancreatic juice
   2. stomach secretion containing water, hydrochloric acid, and the enzymes pepsin and lipase
   3. emulsifier made by the liver that prepares fats and oils for digestion
   4. enzyme that splits starch
   5. protein-digesting enzyme in the stomach
5. An enzyme in saliva begins digestion of which nutrient in the mouth?
   a. starch b. vitamins
   c. protein d. all of the above
6. Which nutrient passes through the large intestine mostly unabsorbed?
   a. starch b. vitamins
   c. minerals d. fiber
7. The two major nutrient transport systems in the body are:
   a. lipoproteins and chylomicrons.
   b. vascular and lymphatic systems.
   c. LDL and HDL.
   d. digestion and absorption.
8. Within the circulatory system, lipids always travel from place to place bundled with proteins called:
   a. lipoproteins. b. chylomicrons.
   c. microvilli. d. phospholipids
9. Elevated LDL concentrations in the blood are associated with:
   a. a high risk of heart attack.
   b. a high-protein diet.
   c. a low risk of heart attack.
   d. none of the above.
10. Three factors that improve the LDL-to-HDL ratio include:
   a. saturated fat, antioxidants, and insoluble fibers.
   b. antioxidants, insoluble fibers, and low HDL.
   c. weight control, soluble fibers, and physical activity.
   d. polyunsaturated fat, rest, and low HDL.

Answers to these questions appear in Appendix H.

Notes

1. W. F. Ganong, *Review of Medical Physiology* (Norwalk, Conn.: Appleton & Lang, 1993), pp. 438–465.

2. The small intestine in living adults is almost two and a half times shorter than at death, when muscles are relaxed and elongated. Ganong, 1993.

3. J. M. Dietschy, Theoretical considerations of what regulates low-density lipoprotein and high-density-lipoprotein cholesterol, *American Journal of Clinical Nutrition* 65 (1997): 1581S–1589S; NIH Consensus Conference, Triglyceride, high-density lipoprotein, and coronary heart diseae, *Journal of the American Medical Association* 269 (1993): 505–510.

4. The American Cancer Society 1996 Dietary Guidelines Advisory Committee, *Guidelines on Diet, Nutrition, and Cancer Prevention: Reducing the Risk of Cancer with Healthy Food Choices and Physical Activity,* a booklet.

CHAPTER 5

Nutrition in Practice

▪ Common Digestive Problems ▪

The facts of anatomy and physiology presented in Chapter 5 permit easy understanding of some common situations. Everyone, at one time or another, has to deal with various kinds of digestive distress. Short bouts of vomiting and diarrhea that do not seriously impair nutrition status are discussed here, together with ordinary constipation, belching, and gas. Later chapters deal with more serious causes and consequences of vomiting and diarrhea.

Please explain what causes vomiting and when it is serious.

Vomiting can be a symptom of many different diseases or may arise in any situation that upsets the body's equilibrium, such as air or sea travel. For whatever reason, the diaphragm and abdominal muscles contract, the intra-abdominal pressure increases, and the contents of the stomach are propelled up through the esophagus to the mouth and expelled.

If vomiting continues long enough or is severe enough, the increased pressure forces not only the stomach contents but also the contents of the duodenum, with their green bile salts, into the stomach and then up the esophagus. Although unpleasant and wearying for the nauseated person, vomiting such as this is no cause for alarm. Vomiting is one of the body's adaptive mechanisms to rid itself of something irritating. The best advice is to rest and drink fluids (small amounts as tolerated) until the nausea subsides.

Vomiting can be serious, however, when large quantities of fluid are lost from the GI tract, causing dehydration. As Chapter 22 describes, prompt emergency care may be needed in those cases.

What are some cases in which vomiting requires medical attention?

In an infant, vomiting is likely to become serious early in its course, and a physician should be contacted soon after onset. Infants have more fluid between their body cells than adults do, so more fluid can move into the digestive tract and be lost from the body. Consequently, the body water of infants becomes depleted and fluid balance becomes upset earlier than in adults.

Self-induced vomiting, such as occurs in bulimia, also has serious consequences. (Bulimia is discussed in Nutrition in Practice 9.) Besides fluid and salt imbalances, repeated vomiting can irritate and infect the pharynx, esophagus, and salivary glands; erode the teeth; and cause dental caries. The esophagus or stomach may rupture or tear. Sometimes the eyes become red from pressure during vomiting. Bulimic behavior reflects underlying problems that require attention.

Projectile vomiting is also serious. The contents of the stomach are expelled with such force that they leave the mouth in a wide arc like a bullet leaving a gun. This type of vomiting requires immediate medical attention.

What about diarrhea? What should be done about it, and when is medical help needed?

The frequent, loose, watery stools of diarrhea indicate that the intestinal contents have moved too quickly through the intestines for fluid absorption to take place, or that water from the GI tract lining has been added to the food residue. Like vomiting, diarrhea can lead to considerable fluid and salt losses, but the acid-base composition of the fluids is different. Stomach fluids lost in vomiting are highly acidic; intestinal fluids lost in diarrhea are nearly neutral. When fluid losses require medical attention, correct replacement is crucial.

For short bouts of diarrhea, rest and drink fluids to replace losses. If diarrhea continues, call for help. For an infant, get help promptly, for diarrhea may quickly lead to dehydration so severe as to require emergency medical treatment.

WWW.

acg.gi.ord
American College of Gastroenterology

gihealth.com
Center for Digestive Health and Nutrition

CHAPTER 5

Nutrition in Practice

What is constipation, and when does it require medical attention?

Constipation is generally not a cause for concern. Each person's GI tract responds to food in its own way, with its own rhythm. Food is digested and its residue becomes ready for excretion in a predictable number of hours. Each GI tract thus has its own cycle, which depends on its owner's physical makeup, the type of food eaten, when it was eaten, and when the person makes time to defecate.

Even if several days pass between movements, so long as these movements take place without discomfort and the time that has passed since the last bowel movement is typical for that person, the person is not constipated. But, if a movement is passed with difficulty, discomfort, or pain, the person is constipated.

Often a person's own lifestyle may cause constipation. If a person receives the signal to defecate and ignores it, the signal may not return for several hours. In the meantime, water continues to be withdrawn from the fecal matter, so when the person does defecate, the bowel movement is dry and hard.

Careful review of daily habits may reveal such a cause of constipation. Being too busy to respond to the defecation signal is a common complaint. A person's daily regimen may need to be revised, instituting regular eating and sleeping times that will allow time in the day's schedule to have a bowel movement when the body sends its signal. This may mean going to bed earlier in order to rise earlier, allowing ample time for a leisurely breakfast and a movement.

Another cause of constipation is lack of physical activity.[1] In today's society many people drive cars or ride buses to work, stand at assembly lines or sit behind desks, and then sit in front of television sets in the evening. Increasing physical activity may require some rearrangement of one's lifestyle. People can work out in spas or health clubs, or more simply, they can park their cars a few blocks from the office and walk, or walk up several flights of stairs a day rather than taking the elevator. Such planning enables people to engage in activity that will improve the muscle tone, not just of the outer body, but also of the digestive tract.

Although constipation usually reflects lifestyle habits, in some cases it may be a side effect of medication or may reflect a medical problem such as tumors that are obstructing the passage of waste. If discomfort or distress is associated with passing fecal matter, a physician's help should be sought to rule out disease. Once this has been done, dietary or other measures for correction can be considered.

Does fiber, water, or prune juice help relieve ordinary constipation?

Yes to all three. The fibers in cereal products are especially helpful. In the GI tract, these fibers attract water, thus creating soft, bulky stools that stimulate bowel contractions to push the contents along. These contractions strengthen the intestinal muscles; the improved muscle tone and the water content of the stools not only aid elimination but also reduce pressure on the rectal veins, helping to prevent hemorrhoids.

Drinking plenty of water in conjunction with eating foods high in fiber supplies fluid for the fiber to take up. The resulting increased bulk physically stimulates the upper GI tract, promoting peristalsis throughout.

Prunes are not only high in fiber but also contain a laxative substance. If a morning defecation is desired, a person can drink prune juice at bedtime; if the evening is preferred, the person can drink prune juice with breakfast.

Another simple tactic that can help with some constipation is to add some fat to the diet. Fat summons bile into the duodenum, bile's salts draw water from the intestinal wall, and the water stimulates peristalsis and softens the fecal matter.

Is it ever necessary to take laxatives, enemas, or mineral oil for constipation?

The changes in diet or lifestyle just suggested should correct chronic constipation without the use of laxatives, enemas, or mineral oil, although television commercials often try to persuade people otherwise. One of the fallacies often perpetrated by television commercials is that one person's successful use of a product is a good recommendation for others to use that product.

As a matter of fact, even diet changes that relieve constipation for one person may increase the constipation of another. For instance, increasing fiber intake stimulates peristalsis and helps the person with a sluggish colon. Some people, though, have a spastic type of constipation, in which peristalsis promotes strong contractions that close off a segment of the colon and prevent passage. For these people, increasing fiber intake would be exactly the wrong thing to do.

A person who seems to need products such as laxatives should therefore seek a physician's opinion.

Nutrition in Practice

Advice from friends may do more harm than good. Frequent use of laxatives and enemas can lead to dependency; can upset the body's fluid, salt, and mineral balances; and, in the case of mineral oil, can interfere with the absorption of fat-soluble vitamins. (Mineral oil dissolves the vitamins, but is not itself absorbed; instead, it leaves the body, carrying the vitamins with it.)

You promised to discuss belching and gas, too.

Many people complain of problems that they attribute to excessive gas. For some, belching is the complaint. Others blame intestinal gas for abdominal discomforts and embarrassment. Most people believe that the problems occur after they eat certain foods. This may be the case with intestinal gas, but belching results from swallowing air.

Everyone swallows a little air with every mouthful of food, but people who eat too fast may swallow too much air and then have to belch. Ill-fitting dentures, carbonated beverages, and chewing gum can also contribute to air swallowing and belching. Occasionally, belching can be a sign of a more serious disorder, such as gallbladder pain, colonic distress, or impending obstruction of a coronary blood vessel.

Little is known about intestinal gas, but techniques to collect gas directly from the abdomen have allowed researchers to gain some knowledge. While expelling gas can be a humiliating experience, it is quite normal. (People experiencing painful bloating from malabsorption diseases, however, require medical treatment.) Healthy people expel several hundred milliliters of gas several times a day. Almost all (99 percent) of the gases expelled—nitrogen, oxygen, hydrogen, methane, and carbon dioxide—are odorless. The remaining "volatile" gases are the infamous ones.

Foods that produce gas usually must be determined individually. Table 19.4, later in the book, lists some of the likely candidates. The most common offenders are foods rich in the carbohydrates—sugars, starches, and fibers. When partially digested carbohydrates reach the large intestine, bacteria digest them, giving off gas as a by-product. People can test foods suspected of forming gas by omitting them individually for a trial period and seeing if there is any improvement. The best advice seems to be to eat slowly, chew thoroughly, and relax while eating.

What about more serious digestive problems?

Some illnesses seriously affect the digestion and absorption of nutrients. Individuals with these disorders risk severe malnutrition, and the diet must be altered to ensure health. The details of these disorders and their diet therapies are described in Chapters 18 through 28.

Notes

1. R. S. Sandler, M. C. Jordan, and B. J. Shelton, Demographic and dietary determinants of constipation in the US population, *American Journal of Public Health* 80 (1990): 185–189.

6 Metabolism and Energy Balance

metabolism: the sum total of all the chemical reactions that go on in living cells.
meta = among
bole = change

Every organ, every tissue, and every cell of the body engages in metabolism, the chemical reactions involved in breaking down compounds, making new compounds, and transporting compounds from place to place. Viewed from this perspective, the body is a giant chemical factory that works with astounding efficiency to produce a myriad of products and dispose of a myriad of wastes. All these processes are regulated by hormonal signals that coordinate supply and demand in much the same way as a superb communication system coordinates a smoothly functioning economy.

In disease, metabolic processes always become disturbed, and some diseases are caused by metabolic abnormalities. This chapter provides the normal background against which the disruptions caused by diseases can best be understood.

The Organs and Their Metabolic Roles

When everything in the body is functioning normally, every organ plays a metabolic role that serves the others. Metabolic reactions also consume or release energy and therefore affect body weight, with consequences for health.

The Principal Organs

Of particular concern to metabolism are the digestive organs, the liver, the pancreas, the circulatory system, and the kidneys. Together, they perform much of the work of breaking down compounds, making new ones, transporting nutrients and oxygen throughout the body, and removing the waste products generated by metabolic processes.

The Digestive Organs The digestive system has just received attention in Chapter 5. Notable among the digestive system's activities are the physical processes that allow foods to be transported throughout the GI tract; the production of digestive juices and enzymes; the absorption of nutrients; the making of transport proteins to carry lipids and vitamins to other sites in the body; and in the lower digestive tract, the reabsorption of salts and fluids. The digestive system also possesses the body's most rapidly multiplying cells: when healthy, they replace themselves every few days. Disorders affecting the GI tract lead to failures to ingest, digest, absorb, and metabolize nutrients, as described in Chapters 18, 19, and 20.

The Liver Nutrients absorbed into the bloodstream are conducted to the liver, as described in Chapter 5. The liver is one of the body's most active metabolic factories. It receives nutrients and metabolizes, packages, stores, or ships them out for uses by other organs. It metabolizes and stores most vitamins and many minerals. It manufactures bile, which the body uses in emulsifying fat for digestion and absorption. It metabolizes and detoxifies drugs, prepares waste products for excretion, and participates in iron recycling and blood cell manufacture. It also makes many proteins necessary for health, including immune factors, transport proteins and clotting factors. When liver disorders disrupt metabolism, they profoundly affect both nutrition and general health status, as described in Chapter 28.

The Pancreas The pancreas contributes digestive juices to the GI tract, but also has another metabolic function: it produces insulin and other hormones that regulate the body's use of glucose. It is insulin that prompts cells to take glucose up and use it as fuel; insulin also prompts liver cells in particular to store glucose as glycogen. The liver cells can later release glucose back into the blood as needed. Glucose is an indispensable fuel for brain and nerve cells and for red blood cells as well. Its availability is therefore crucial to normal nervous system activity and normal blood chemistry. Abnormalities associated with the digestive functions of the pancreas are described in Chapter 20 and those associated with its hormonal functions, diabetes and hypoglycemia, are described in Chapter 25.

The Heart and Blood Vessels The heart and blood vessels conduct blood with its cargo of nutrients and oxygen to all other body cells and carry wastes from them. Diseases of the heart and arteries therefore affect the health of the whole body. Metabolic reactions that affect the heart and blood vessels include, most importantly, the making and transport of lipoproteins, the carriers of cholesterol and other lipids from the liver to the tissues and back again. When lipoproteins deposit cholesterol in the artery walls, coronary heart disease and high blood pressure can result and increase the risk of disability or death from heart attacks and strokes. Chapter 26 is devoted to these conditions.

W.W.W.
amhrt.org
American Heart Association

The Kidneys The kidneys are also active metabolic organs. Unceasingly, for 24 hours of every day, they filter waste products from the blood to be excreted in the urine and reabsorb needed nutrients thereby maintaining the blood's delicate chemical balances. The kidneys' cells also produce compounds that help to regulate blood pressure and convert a precursor compound to active vitamin D, thereby helping to maintain the bones. Disorders of the kidneys nearly always involve the heart and the skeleton; kidney disorders are the subject of Chapter 27.

W.W.W.
kidney.org
National Kidney Foundation

asn-online.com
American Society of Nephrology

Energy for Metabolic Work

The metabolic work that the body's cells do, like all work, requires energy, and food supplies that energy. Food in turn gets its energy from the sun, either directly (in the case of photosynthesizing plants) or indirectly (in the case of animals that eat plants). When chemical reactions in cells release stored energy from energy-yielding nutrients, that energy becomes available to do the cells' work.

Heat Energy and Body Temperature The cells of each organ conduct metabolic activities specific to that organ, but in addition, all cells must maintain themselves, and many must reproduce. To do this, they must have all the essential nutrients available to them: the energy nutrients, the vitamins, and the minerals, as well as water. As cells do their metabolic work, the chemical reactions involved release heat, and this heat keeps the body warm. By regulating the rates at which these metabolic reactions release heat energy, the body maintains its constant normal temperature of about 98.6°F.

Fast Metabolism, Fever, and Wasting During stress, metabolism speeds up. Fever sometimes develops. An accelerated metabolism signifies that fuels are being burned at a rate more rapid than normal; this may lead to wasting of body organs and loss of weight including loss of vital lean tissue. Chapter 23 describes the metabolic consequences of severe stress and, in particular, of severe infections, major surgery, and burns. Chapter 24 describes the metabolic consequences of the so-called wasting diseases, cancer and AIDS.

As this brief discussion has shown, metabolism occurs all through the body, all the time, and supports normal health. The remainder of this chapter delves into one aspect of metabolism—the metabolism of the energy nutrients and the resulting consequences of underweight and overweight.

The Body's Energy Metabolism

The body manages its energy supply with amazing precision. Consider, for example, that a consistent 1 percent error in energy intake can cause a person to become more than 200 pounds overweight in a lifetime. Yet most people maintain their weight within about a 10- to 20-pound range throughout their lives. How do they do this? How does the body manage excess energy? And how does it manage to do without food for prolonged periods—as when someone is starving or fasting? The answers to questions like these lie in an understanding of metabolism.

energy metabolism: all the reactions by which the body obtains and spends the energy from food or body stores.

Energy metabolism is defined as the sum total of all the chemical reactions that manage energy nutrients in the body. Earlier chapters introduced the energy-yielding nutrients—carbohydrate, fat, and protein—and showed how they are broken down into basic units that are absorbed into the blood. Picking up from there, what becomes of these nutrients? The question is important because it provides insight into proper and improper ways to aid the body in maintaining or losing weight.

Energy metabolism centers on four basic units:

- From carbohydrates: glucose.
- From lipids: glycerol and fatty acids.
- From proteins: amino acids.

Building Body Compounds

anabolism (an-ABB-o-lism): reactions in which small molecules are put together to build larger ones. Anabolic reactions consume energy.
ana = up

catabolism (ca-TAB-o-lism): reactions in which large molecules are broken down to smaller ones. Catabolic reactions usually release energy.
kata = down

When not needed by the cells for energy, the basic units of energy-yielding nutrients can be used to build body compounds. The building up of body compounds is known as anabolism; this book represents anabolic reactions, wherever possible, by "up" arrows in chemical diagrams (such as those shown in Figure 6.1). Glucose units can be strung together to make glycogen chains. Glycerol and fatty acids can be assembled into triglycerides. Amino acids can be linked together to make proteins. These anabolic reactions, in which simple compounds are put together to form larger, more complex structures, involve doing work and so require energy.

Breaking Down Nutrients for Energy

glycolysis (gligh-COLL-uh-sis): the metabolic breakdown of glucose to pyruvate.
glyco = glucose
lysis = breakdown

pyruvate (PIE-roo-vate): pyruvic acid, a 3-carbon compound derived from glucose, glycerol, and certain amino acids in metabolism.

CoA (coh-AY): a nickname for a small molecule (coenzyme A) that participates in metabolism. As pyruvate breaks down to the smaller compound acetic acid, a molecule of CoA is attached to it, making acetyl CoA (ASS-uh-teel or uh-SEET-ul co-AY).

The breaking down of body compounds is known as catabolism; catabolic reactions usually release energy and are represented, wherever possible, by "down" arrows in chemical diagrams (see Figure 6.1). Glycogen can be broken down to glucose, triglycerides to fatty acids and glycerol, and protein to amino acids. When the body needs energy, it breaks any or all of the four basic units—glucose, fatty acids, glycerol, and amino acids—into even smaller units.

Glucose Breakdown Glucose breakdown occurs via a pathway known as glycolysis. In glycolysis, glucose is broken down to pyruvate.* Most often, pyruvate is then converted to a smaller compound, acetyl CoA. In a series of metabolic reactions called the tricarboxylic acid (TCA) cycle, acetyl CoA splits, and its energy is donated to storage compounds, used to do the body's work, or used to produce heat.

The following sequence is central to an understanding of metabolism:

Glucose ↔ pyruvate → acetyl CoA → energy.

*The term *pyruvate* means a salt of pyruvic acid. Throughout this book, the ending *-ate* is used interchangeably with *-ic acid;* for our purposes they mean the same thing.

Figure 6.1
Anabolic and Catabolic Reactions Compared

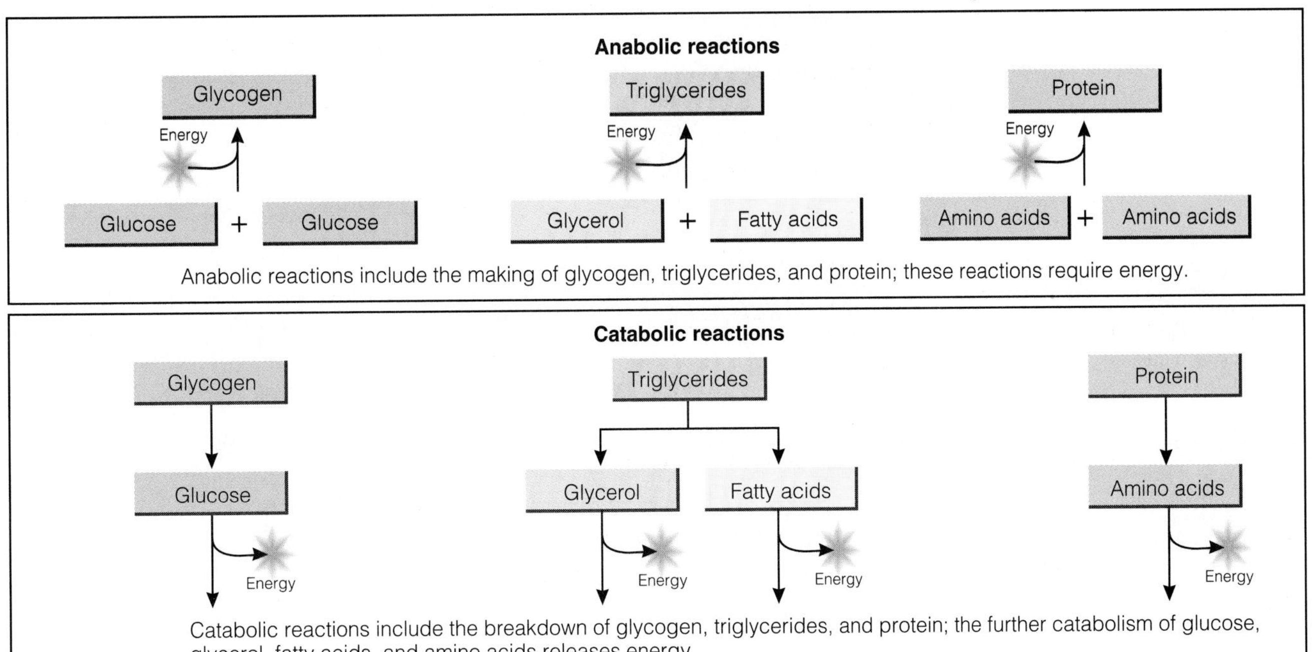

Notice the two-way arrow between glucose and pyruvate and the one-way arrows after pyruvate. They show that pyruvate can be reconverted to glucose but that acetyl CoA cannot. Any compound that can be converted to pyruvate can be used to make glucose. Any compound that is converted to acetyl CoA cannot be used to make glucose.

The reactions by which the complete oxidation of acetyl CoA is accomplished are those of the TCA cycle, or Krebs cycle, and oxidative phosphorylation. The net result is that acetyl CoA splits, and some of its energy is made available for the body's use.

Fat Breakdown Triglycerides (the primary form of fat in the body) cannot be converted to glucose, for the most part, because they consist mostly of fatty acids. Fatty acids are broken down into 2-carbon fragments that combine with CoA to form acetyl CoA. Because fatty acids are broken down to acetyl CoA, they cannot be used to make glucose. Glycerol is interconvertible with pyruvate and can yield glucose, but glycerol represents only about 5 percent of the weight of a triglyceride molecule. Thus fat is an inefficient source of glucose. About 95 percent of it cannot be converted to glucose at all; therefore fat, for the most part, normally cannot provide energy for the organs (brain and nervous system) that require glucose as fuel. This leaves the task of fueling the nervous system's activities mainly to protein and carbohydrate.

Amino Acid Breakdown Ideally, amino acids are used to maintain supplies of needed body proteins and will not be used for energy. If they are needed for energy, or if they are consumed in excess, they first undergo deamination, a reaction in which they are stripped of their nitrogen. The nitrogen can be used to make other compounds, including the nonessential amino acids, or it can be excreted. With nitrogen removed, about half of the amino acids can be converted to pyruvate and can therefore provide glucose. The other amino acids are converted to acetyl CoA directly or enter the TCA cycle at another point. Thus protein, unlike fat, is a fairly efficient source of glucose when carbohydrate is not available.

deamination: removal of the amino (NH_2) group from a compound such as an amino acid.

The principal nitrogen-excretion product of metabolism is **urea** (you-REE-uh).

The making of glucose from protein or fat is **gluconeogenesis** (gloo-co-nee-o-JEN-uh-sis). About 5% of fat (the glycerol portion of triglycerides) and about 50% of protein (the glycogenic amino acids) can be converted to glucose.
gluco, glyco = glucose
neo = new
genesis = making

Figure 6.2 (p. 124) depicts the major metabolic pathways involving carbohydrates, fats, and amino acids. Note the central pathway from glucose to pyruvate

Figure 6.2
The Central Pathways of Energy Metabolism

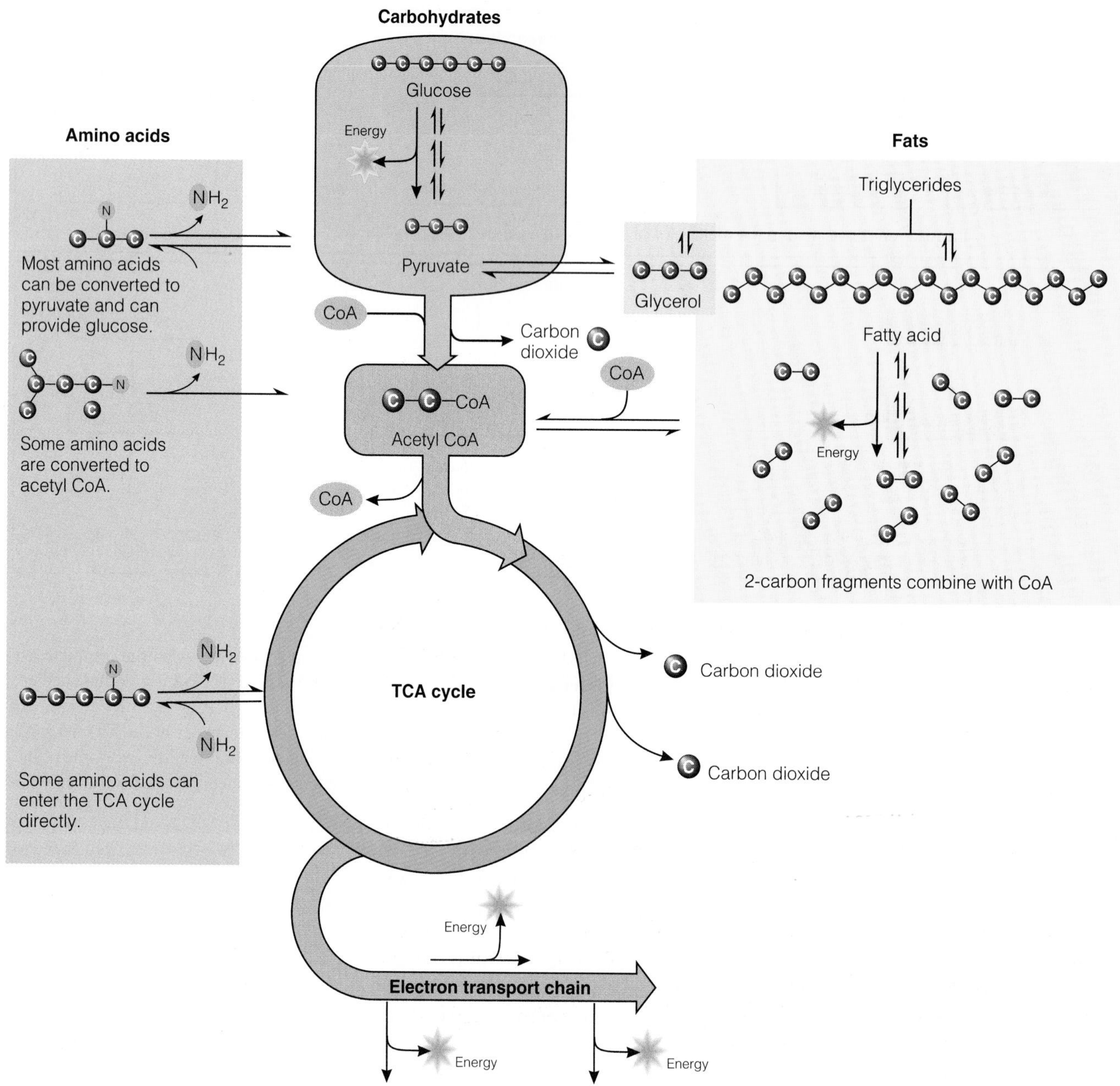

to acetyl CoA to energy. Also note that all carbohydrates, some amino acids, and the glycerol from fat can be converted to pyruvate and then to glucose. Finally, note that the vast majority of fragments from fat can be used only for energy and not to make glucose. With these understandings, you can follow the events that lead to weight gain and weight loss.

The Body's Energy Budget

The average person takes in close to a million kcalories a year and expends more than 99 percent of them, maintaining a stable weight for years on end. In other words, the body's energy budget is balanced. Some people, however, eat too much and get fat; others eat too little and get thin. This section examines metabolism from the perspective of the energy budget, looking first at the two forms of unbalanced budget, feasting and fasting, and then at a balanced budget.

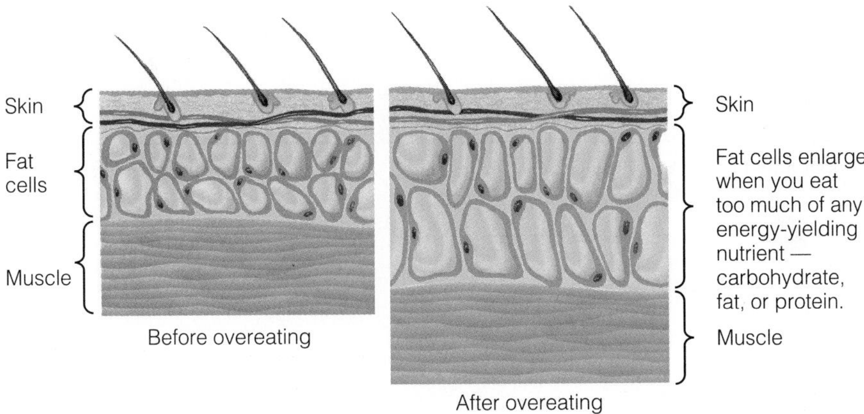
Skin
Fat cells
Muscle
Before overeating
After overeating
Skin
Fat cells enlarge when you eat too much of any energy-yielding nutrient — carbohydrate, fat, or protein.
Muscle

Figure 6.3
Fat Cell Enlargement

The Economics of Feasting

Everyone knows that when you consume more energy than you expend, much of the excess is stored as body fat. Fat can be made from an excess of any energy-yielding nutrient that you eat. In addition, excess energy from alcohol is also stored as fat.[1] Fat cells enlarge as they fill with fat, and the body's fat-storing capacity seems to be able to expand indefinitely, as Figure 6.3 shows.

Excess Carbohydrate Surplus carbohydrate (glucose) is first stored as glycogen, but the glycogen-storing cells have limited capacity. Once glycogen stores are filled, any additional carbohydrate is routed to fat. Thus excess carbohydrate can contribute to obesity.

Excess Fat Surplus dietary fat contributes easily to the body's fat stores. During digestion and metabolism, fat may break down into fragments, such as acetyl CoA, but if the flow of these fragments is rapid enough to meet the body's need for energy, any excess fragments that are available will be stored as triglycerides in the fat cells.

Excess Protein Surplus protein may encounter the same fate. If not needed to build body protein or to meet energy needs, amino acids will lose their nitrogens and will be converted through the intermediates, pyruvate and acetyl CoA, to triglycerides. These, too, swell the fat cells and add to body weight. Figure 6.4 on the next page, shows the metabolic events of feasting.

In summary, the following points deserve repeating:

- Excess energy from carbohydrate, fat, protein, and alcohol will be stored in the body as fat.
- Fat from food, as compared with carbohydrate and protein, is especially easy for the body to store in fat tissue.

The Economics of Fasting

The body spends energy all the time. Even when a person is asleep and totally relaxed, the cells of many organs are hard at work. In fact, this cellular work, which maintains all life processes without a person's awareness, represents about two-thirds of the total energy a person spends in a day. (The other one-third is the work that a person's muscles do voluntarily during waking hours.)

Figure 6.4
Feasting
When people overeat, they store energy.

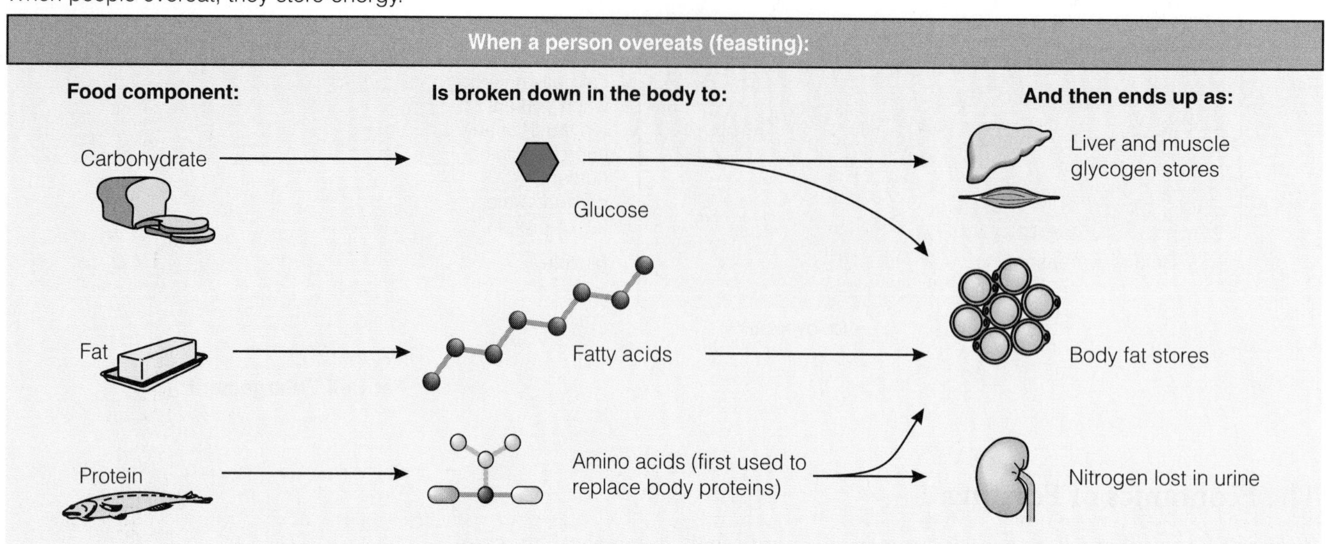

In *fasting,* a person voluntarily stops eating food, whereas in *starvation* the failure to eat is involuntary. The body, however, makes no distinction between the two—metabolically, fasting and starvation are identical.

Energy Deficit The body's top priority is to meet the energy needs for this ongoing activity. Its normal way of doing so is by periodic refueling, that is, by eating several times a day. When food is not available, the body uses other fuel sources from its own tissues. If people choose not to eat, we say they are fasting; if they have no choice (as in a famine), we say they are starving. In the body, no metabolic difference exists between fasting and starving. In either case, the body is forced to switch to a wasting metabolism, drawing on its reserves of carbohydrate and fat and, within a day or so, on its vital protein tissues as well.

Glycogen Used First As a fast or period of starvation begins, glucose from the liver's stored glycogen and fatty acids from the body's stored fat both flow into cells to fuel their work. Several hours later, however, most of the glucose is used up—liver glycogen is exhausted. Low blood glucose concentrations serve as a signal to promote further fat breakdown.

Reminder: The liver releases glucose, and the fat cells release fat to fuel the body's cells, but the brain can use only glucose.

Glucose Needed for Brain At this point, a few hours into a fast, most of the cells are depending on fatty acids to continue providing fuel. But the nervous system (brain and nerves) cannot use fatty acids; it still needs glucose. Even if other energy fuel is available, glucose has to be present to permit the brain's energy-metabolizing machinery to work. Normally, the nervous system consumes about two-thirds of the total glucose used each day—about 400 to 600 kcalories' worth.

Fasting = living on the body's fat and protein.

In fasting, muscle and lean tissues give up protein to supply amino acids for conversion to glucose. This glucose, with ketones produced from fat, fuels the brain's activities.

Protein Breakdown and Ketosis Because fat stores cannot provide the glucose needed by the brain and nerves, body protein tissues (such as liver and muscle) always break down to some extent during fasting. In the first few days of a fast, body protein provides about 90 percent of the needed glucose, and glycerol provides about 10 percent. If body protein loss were to continue at this rate, death would ensue within about three weeks. As the fast continues, however, the body finds a way to use its fat to fuel the brain. It adapts by condensing together acetyl CoA fragments derived from fatty acids to produce ketones, which can serve as fuel for some brain cells. Ketone production rises until, after several weeks of fasting, it is meeting much of the nervous system's energy needs. Still, many areas of the brain rely exclusively on glucose, and body protein continues to be sacrificed to produce it. Figure 6.5 shows the metabolic events that occur during fasting.

Figure 6.5
Fasting
When people are fasting, they draw on stored energy.

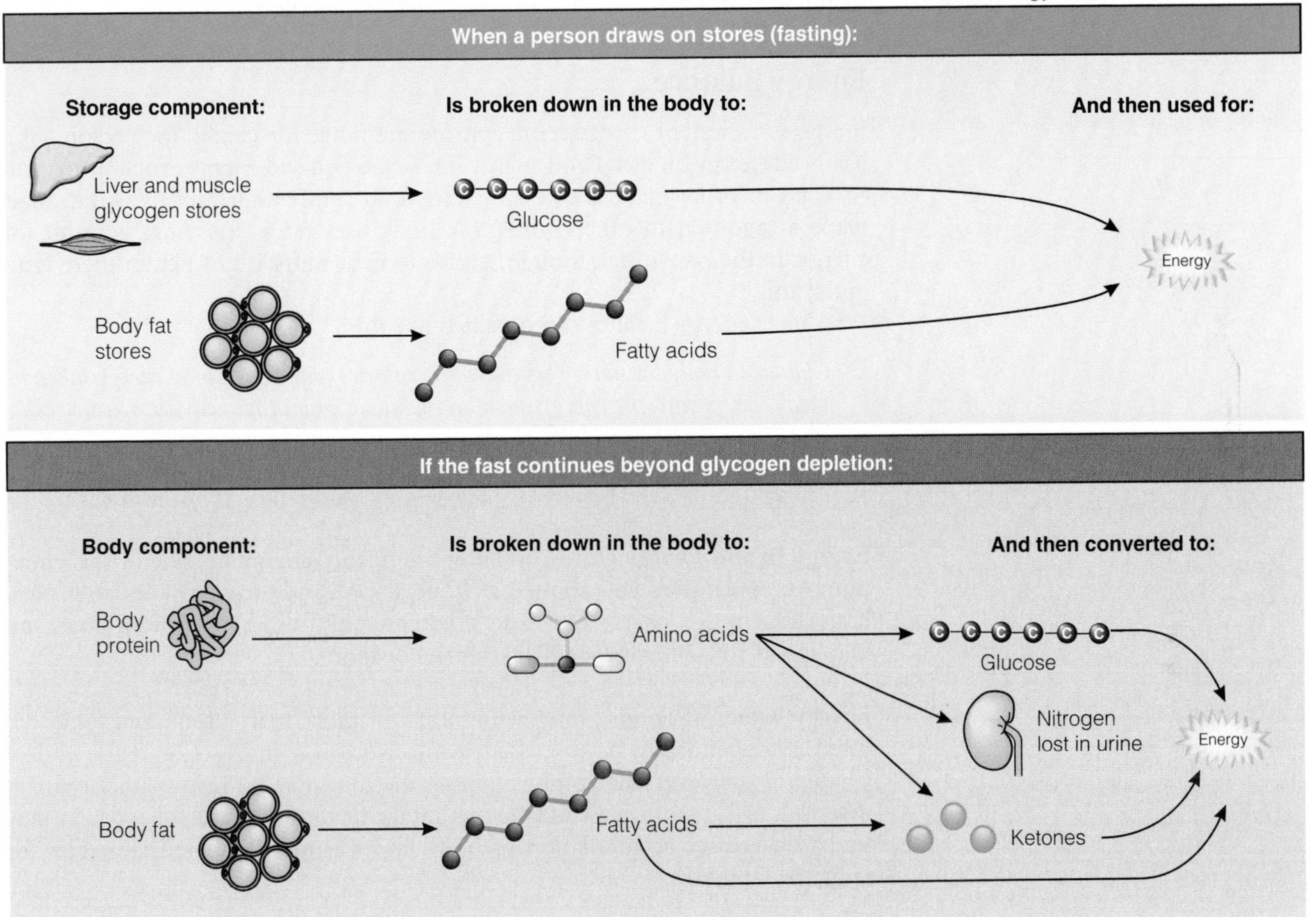

Slowed Metabolism As fasting continues and the body is shifting to partial dependence on ketones for energy, the body simultaneously reduces its energy output (metabolic rate) and conserves both fat and lean tissue. Because of the slowed metabolism, the loss of fat falls to a bare minimum. Thus, although *weight* loss during fasting may be quite dramatic, *fat* loss may actually be less than when at least some food is supplied.

Reminder: *Ketones* are acidic, fat-related compounds formed from the incomplete breakdown of fat when carbohydrate is not available. Small amounts of ketones are normally produced during energy metabolism but when their blood concentration rises, they spill into the urine. The combination of high blood ketones (*ketonemia*) and ketones in the urine (*ketonuria*) is termed *ketosis*.

Hazards of Fasting The body's adaptations to fasting are sufficient to maintain life for a long period. Mental alertness need not be diminished. Even physical energy may remain unimpaired for a surprisingly long time. Still, fasting is not without its hazards. Among the many changes that take place in the body are:

- Wasting of lean tissues.
- Impairment of disease resistance.
- Lowering of body temperature.
- Disturbances of the body's salt and water balance.

For the person who wants to lose weight, fasting is a dangerous way to go. The body's lean tissue continues to be degraded, sometimes amounting to as much as 50 percent of the weight lost. Over the long term, a diet only moderately restricted in energy can actually promote a greater rate of *weight* loss, a faster rate of *fat* loss, and the retention of more lean tissue than a severely restricted fast.[2]

Alterations similar to those in fasting are seen in low-carbohydrate dieting. Renewed food intake, especially of carbohydrate, results in dramatic changes in the body's salt and water balance, accounting for most of the wide swings in body weight seen in people on fasts or low-carbohydrate diets.

Energy Balance

If a person's weight is within the appropriate range for height, the person has a balanced energy budget. Food energy intake has equaled energy expenditure, and so, deposits of fat made at one time have been compensated for by withdrawals made at another. In other words, the body uses fat as a savings account for energy. In the case of fat, though, unlike money, more is not better; there is an optimum.

A day's energy balance can be stated like this:

Change in body fat stores (expressed in kcalories) equals the food energy taken in (kcalories) minus the energy spent on metabolic and other activities (kcalories).

More simply:

Change in fat stores (kcalories) = energy in (kcalories) − energy out (kcalories).

Energy In and Energy Out You know about the "energy in" side of this equation. An apple gives you about 125 kcalories; a candy bar provides about 300 kcalories. On the "energy out" side, if you are physically active for an hour, you may spend 100, 300, or even 500 kcalories or more.

Energy Values of Foods Energy amounts for more than a thousand foods are listed in Appendix A. As discussed in the Nutrition in Practice at the end of Chapter 3, however, the composition of the diet may be just as important in determining fat storage as total energy intake or output. Excess dietary fat may lead to a greater accumulation of body fat.[3] Chapter 3 offered strategies for reducing fat intake.

Energy Expenditures The body spends energy in two major ways: to fuel its basal metabolism and to fuel its voluntary activities. People can change their voluntary activities to spend more or less energy in a day, and over time they can also change their basal metabolism by building up the body's metabolically active lean tissue, as explained in Chapter 9.

basal metabolism: the energy needed to maintain life when a person is at complete rest after a 12-hour fast. Basal metabolism, sometimes called *basal metabolic rate (BMR),* is normally the largest part of a person's daily energy expenditure.

voluntary activities: the component of a person's daily energy expenditure that involves conscious and deliberate muscular work—walking, lifting, climbing, and other physical activities. Voluntary activities normally require less energy in a day than basal metabolism does.

Basal Metabolism The basal metabolism supports the work that goes on all the time without conscious awareness. The beating of the heart, the inhaling and exhaling of air, the maintenance of body temperature, and the transmission of nerve and hormonal messages to direct these activities are the basal processes that maintain life.

Basal Metabolic Rate The basal metabolic rate (BMR) is the rate at which the body spends energy for these maintenance activities. This rate varies from person to person and may vary for a single individual with a change in circumstance or physical condition. For example, an infant's metabolic rate relative to body weight is much faster than an adult's to support the infant's extraordinary growth rate. In general, BMR is fast in people with considerable lean body mass (growing children, physically active people, pregnant women, and males). One way to increase the BMR then is to maximize lean body tissue by participating regularly in endurance and strength-building activities.[4] BMR is also fast in people who are tall and so have a large surface area for their weight, in people with fever or under stress, in people taking certain medications, and in people with highly active thyroid glands. BMR is slowed down by loss of lean tissue and depression of thyroid hormone activity due to disease, inactivity, fasting, or malnutrition. Table 6.1 summarizes the factors that speed up and slow down the BMR.

Table 6.1 Factors That Affect BMR

Factor	Effect on BMR
Age	In youth, the BMR is higher; lean body mass diminishes with age, slowing the BMR.[a]
Height	In tall, thin people, the BMR is higher.[b]
Growth	In children and pregnant women, the BMR is higher.
Body composition	The more lean tissue, the higher the BMR. The more fat tissue, the lower the BMR.[c]
Fever	Fever raises the BMR.[d]
Stresses	Some stresses and certain medications raise the BMR.
Environmental temperature	Both heat and cold raise the BMR.
Fasting/starvation	Fasting/starvation lowers the BMR.[e]
Malnutrition	Malnutrition lowers the BMR.
Hormones	The thyroid hormone thyroxine, for example, is a key BMR regulator; the more thyroxine produced, the higher the BMR.[f]

[a]The BMR begins to decrease in early adulthood (after growth and development cease) at a rate of about 2 percent/decade. A reduction in voluntary activity as well brings the total decline in energy expenditure to 5 percent/decade.
[b]If two people weigh the same, the taller, thinner person will have the faster metabolic rate, reflecting the greater skin surface, through which heat is lost by radiation, in proportion to the body's volume.
[c]In general, males tend to have a higher BMR than females due to their greater lean body mass.
[d]Fever raises BMR by 7 percent for each degree Fahrenheit.
[e]Prolonged starvation reduces the total amount of metabolically active lean tissue in the body, although the decline occurs sooner and to a greater extent than body losses alone can explain. More likely, the neural and hormonal changes that accompany fasting are responsible for changes in BMR.
[f]The thyroid gland releases hormones that travel to the cells and influence cellular metabolism. Thyroid hormone activity can speed up or slow down the rate of metabolism by as much as 50 percent.

Basal Metabolic Needs Basal metabolic needs are surprisingly large. A person whose total energy needs are 2000 kcalories a day spends 1200 to 1400 of them to support basal metabolism. The "How to" box on the next page shows how to estimate energy expenditure for basal metabolism.

Energy for Activities The number of kcalories spent on voluntary activities depends on three factors: muscle mass, body weight, and activity. The larger the muscle mass required for the activity and the heavier the weight of the body part being moved, the more kcalories are spent. The activity's duration, frequency, and intensity also influence energy costs: the longer, the more frequent, and the more intense the activity, the more kcalories spent per minute. Table 6.2 on the next page shows the energy expended on various types of activities. The energy spent on activities, added to the energy required for basal metabolism, equals the total energy you spend in a day (see the "How to" box).

Energy to Manage Food One component of energy expenditure is not taken into account in the calculations in the box: the energy required for the body to process food. When food is taken into the body, many cells that have been dormant begin to be active. The muscles that move the food through the intestinal tract speed up their rhythmic contractions; the cells that manufacture and secrete digestive juices begin their tasks. All these and other cells need extra energy as they come alive to participate in the digestion, absorption, and metabolism of food. This stimulation of cellular activity produces heat and is known as the thermic effect of food. The thermic effect of food is generally thought to represent about 10 percent of the total food energy taken in. For purposes of rough

thermic effect of food: an estimation of the energy required to process food (digest, absorb, transport, metabolize, and store ingested nutrients).

How to

Estimate a Day's Energy Output

The calculation shown here exemplifies one way of calculating energy needs. Another way is often used in clinical practice: the BEE (basal energy expenditure) method, explained in Chapter 23, Table 23.2.

- *Basal metabolism* Convert your body weight from pounds to kilograms. Then multiply by the factor 1.0 kcalorie per kilogram of body weight per hour for men (or 0.9 for women).[a] Then multiply by the 24 hours in a day. For example, for a 160-pound man:

1. Change pounds to kilograms:

$$160 \text{ lb} \div 2.2 \text{ lb/kg} = 72.7 \text{ kg}.$$

2. Multiply weight in kilograms by the BMR factor:

$$72.7 \text{ kg} \times 1 \text{ kcal/kg/hr} = 72.7 \text{ kcal/hr}.$$

3. Multiply kcalories used in one hour by hours in a day:

$$72.7 \text{ kcal/hr} \times 24 \text{ hr/day} = 1744.8 \text{ kcal/day}.$$

Energy for BMR equals 1745 (rounded) kcalories per day.

- *Voluntary muscular activity* To estimate the energy for your activities, determine from Table 6.2 (p. 130) the level of intensity that typifies your average daily activity. Then multiply your BMR by the corresponding activity factor. For example, if our 160-pound man engages in mostly light activity, his activity factor would be 1.6. Multiply this factor by his BMR kcalories:

$$1.6 \times 1745 \text{ kcal/day} = 2792 \text{ kcal/day}.$$

- *Total energy needs* The result, 2792 kcalories/day, expresses his total daily energy needs.

Alternatively, total energy expenditure can be estimated in one step based on body weight as shown in the last column of Table 6.2. As an example, for a 160-pound man engaged in mostly light activity:

$$38 \text{ kcal/kg/day} \times 72.7 \text{ kg} = 2763 \text{ kcal/day}.$$

The difference between 2792 and 2763 is insignificant and acceptable. Either way, the man's total energy needs are about 2800 kcalories per day.

[a]Men's metabolic energy needs are assumed to be higher than women's because their hormones induce them to develop more lean tissue than do most women, and lean tissue burns more energy per hour.

Table 6.2 Estimating Daily Energy RDA at Various Levels of Physical Activity

Level of Intensity	Type of Activity	Activity Factor (x BMR)	Energy Expenditure (kcal/kg/day)
Very light	Seated and standing activities, painting trades, driving, laboratory work, typing, sewing, ironing, cooking, playing cards, playing a musical instrument	1.3 Men 1.3 Women	31 30
Light	Walking on a level surface at 2.5 to 3 mph, garage work, electrical trades, carpentry, restaurant trades, housecleaning, child care, golf, sailing, table tennis	1.6 Men 1.5 Women	38 35
Moderate	Walking 3.5 to 4 mph, weeding and hoeing, carrying a load, cycling, skiing, tennis, dancing	1.7 Men 1.6 Women	41 37
Heavy	Walking with a load uphill, tree felling, heavy manual digging, basketball, climbing, football, soccer	2.1 Men 1.9 Women	50 44
Exceptional	Athletes training in professional or world-class events	2.4 Men 2.2 Women	58 51

Source: Reprinted with permission from *Recommended Dietary Allowances: 10th edition.* Copyright 1989 by the National Academy of Sciences. Courtesy of the National Academy Press, Washington, D.C.

Self Study

How Much Energy Do You Spend in a Day?

Use the box on p. 130 (How to Estimate a Day's Energy Output) to estimate the energy you spend in a day. Form 7 will help you record your calculations.

Form 7 Estimating Energy Output

Basal Metabolism

Step 1: My weight in pounds (___ lb) divided by 2.2 lb/kg equals my weight in kilograms: ___ kg.

Step 2: My weight in kilograms (___ kg) times 1.0 kcal/kg/hr for men or 0.9 kcal/kg/hr for women equals the number of kcalories I spend on basal metabolism in an hour: ___ kcal/hr.

Step 3: My energy expenditure per hour (___ kcal/hr) times the hours in a day (24) equals the number of kcalories I spend on basal metabolism in a day: ___ kcal/day.

Activities

MEN (do *one* of the following five calculations):

___ I am very lightly active, so my activity factor (see Table 6.2) is 1.3. Multiply this factor by your basal metabolic energy (step 3): 1.3 × ___ (BMR) = ___ kcal/day.

Your result, ___ kcal/day, expresses your total daily needs.

___ I am lightly active, so my activity factor (see Table 6.2) is 1.6. Multiply this factor by your basal metabolic energy (step 3): 1.6 × ___ (BMR) = kcal/day.

Your result, ___ kcal/day, expresses your total daily needs.

___ I am moderately active, so my activity factor (see Table 6.2) is 1.7. Multiply this factor by your basal metabolic energy (step 3): 1.7 × ___ (BMR) = ___ kcal/day.

Your result, ___ kcal/day, expresses your total daily needs.

___ I am very active (heavy activity) so my activity factor (see Table 6.2) is 2.1. Multiply this factor by your basal metabolic energy (step 3): 2.1 × ___ (BMR) = ___ kcal/day.

Your result, ___ kcal/day, expresses your total daily needs.

___ I am exceptionally active, so my activity factor (see Table 6.2) is 2.4. Multiply this factor by your basal metabolic energy (step 3): 2.4 × ___ (BMR) = ___ kcal/day.

Your result, ___ kcal/day, expresses your total daily needs.

WOMEN (do *one* of the following five calculations):

___ I am very lightly active, so my activity factor (see Table 6.2) is 1.3. Multiply this factor by your basal metabolic energy (step 3): 1.3 × ___ (BMR) = ___ kcal/day.

Your result, ___ kcal/day, expresses your total daily needs.

___ I am lightly active, so my activity factor (see Table 6.2) is 1.5. Multiply this factor by your basal metabolic energy (step 3): 1.5 × ___ (BMR) = ___ kcal/day.

Your result, ___ kcal/day, expresses your total daily needs.

___ I am moderately active, so my activity factor (see Table 6.2) is 1.6. Multiply this factor by your basal metabolic energy (step 3): 1.6 × ___ (BMR) = ___ kcal/day.

Your result, ___ kcal/day, expresses your total daily needs.

___ I am very active (heavy activity) so my activity factor (see Table 6.2) is 1.9. Multiply this factor by your basal metabolic energy (step 3): 1.9 × ___ (BMR) = ___ kcal/day.

Your result, ___ kcal/day, expresses your total daily needs.

___ I am exceptionally active, so my activity factor (see Table 6.2) is 2.2. Multiply this factor by your basal metabolic energy (step 3): 2.2 × ___ (BMR) = ___ kcal/day.

Your result, ___ kcal/day, expresses your total daily needs.

estimates, though, the thermic effect of food can be ignored; the 10 percent it might contribute to total energy output is smaller than the probable errors involved in estimating energy input from food or output for activities.

In summary, the energy you spend in a day is about equal to the sum of two components—your basal metabolic energy and your activity energy. The Self Study and accompanying Form 7 take you through the steps for computing the range of total energy you spend in a day. This expenditure represents your approximate energy need, which you meet by eating food. If you eat more than this, you will store the excess energy in your body, mostly as fat. If you eat less than this, you will use up body tissue as fuel to make up the deficit.

Self Check

1. The principal organs of metabolism are:
   a. the digestive organs, liver, pancreas, circulatory system, and kidneys.
   b. the brain, digestive system, heart, stomach, and lungs.
   c. the lungs, bloodstream, pancreas, liver, and heart.
   d. the stomach, heart, lungs, liver, and brain.
2. Anabolism is defined as:
   a. reactions in which small molecules are put together to build larger ones and energy is consumed.
   b. reactions in which large molecules are broken down to smaller ones and energy is released.
   c. removal of the amino group from a compound such as an amino acid.
   d. the metabolic breakdown of glucose to pyruvate.
3. During glycolysis and the tricarboxylic acid (TCA) cycle, glucose is first broken down into _____, then _____, and finally _____ for use in the body.
   a. pyruvate, acetyl CoA, energy
   b. acetyl CoA, pyruvate, energy
   c. amino acids, acetyl CoA, energy
   d. glycerol, pyruvate, energy
4. Fats are catabolized to _____ and _____ for use in the body.
   a. glycerol, fatty acids b. glycogen, fatty acids
   c. amino acids, energy d. glucose, energy
5. Two functions of protein in the body are:
   a. maintenance of the body's supply of amino acids and conversion of protein to glucose for energy.
   b. catabolism to glycerol and fatty acids for storage and synthesis of new proteins.
   c. catabolism to glycogen and synthesis of triglycerides for storage.
   d. anabolism of glucose to glycogen for storage.
6. As carbohydrate and fat stores are depleted during fasting or starvation, the body then uses _____ as its fuel source.
   a. triglycerides b. protein
   c. alcohol d. glucose
7. When carbohydrate is not available to provide energy for the brain, as in starvation, incomplete catabolism of _____ produces ketones.
   a. carbohydrate b. protein
   c. fat d. amino acids
8. Three hazards of fasting are:
   a. wasting of lean tissue, impairment of disease resistance, and disturbances of the body's salt and water balance.
   b. water weight gain, decrease in mental alertness, and impairment of disease resistance.
   c. water weight gain, impairment of disease resistance, and lowering of body temperature.
   d. water weight loss, decrease in mental alertness, and wasting of lean tissue.
9. Two activities that contribute to the basal metabolic rate are:
   a. maintenance of heartbeat and body temperature.
   b. maintenance of body temperature and walking.
   c. maintenance of heartbeat and running.
   d. walking and running.
10. Three factors that affect the body's basal metabolic rate are:
   a. weight, fever, and environmental temperature.
   b. fever, body composition, and altitude.
   c. height, weight, and energy intake.
   d. age, body composition, and height.

Answers to these questions appear in Appendix H.

Notes

1. P. M. Suter, E. Häsler, and W. Vetter, Effects of alcohol on energy metabolism and body weight regulation: Is alcohol a risk factor for obesity? *Nutrition Reviews* 55 (1997): 157–171; P. M. Suter, Y. Schutz, and E. Jequier, The effect of ethanol on fat storage in healthy subjects, *New England Journal of Medicine* 326 (1992): 983–987.

2. M. E. Sweeney and coauthors, Severe vs. moderate energy restriction with and without exercise in the treatment of obesity: Efficiency of weight loss, *American Journal of Clinical Nutrition* 57 (1993): 127–134.

3. T. J. Horton and coauthors, Fat and carbohydrate overfeeding in humans: Different effects on energy storage, *American Journal of Clinical Nutrition* 62 (1995): 19–29.

4. T. J. Horton and C. A. Geissler, Effect of habitual exercise on daily energy expenditure and metabolic rate during standardized activity, *American Journal of Clinical Nutrition* 59 (1994): 13–19.

CHAPTER 6

Nutrition in Practice

Finding the Truth about Nutrition

Nutrition and health receive so much attention on television, in the popular press, and on the Internet that it is easy to be overwhelmed with conflicting, confusing information. In fact, determining whether nutrition information is accurate can be a challenging task. It is also an important task, because nutrition affects both your professional and your personal life.

I see a nutrition report today on television and then read a conflicting report the next day in the newspaper. Why do nutrition news reports seem to contradict each other so often?

The problem of conflicting messages arises for several reasons:

- Popular media, often faced with tight deadlines and limited time or space to report new information, rush to present the latest "breakthrough" in a headline or a 60-second spot. They can hardly help omitting important facts about the study or studies that the "breakthrough" is based on.
- Despite tremendous advances in the last few decades, scientists still have much to learn about the human body and nutrition. Scientists themselves often disagree on their first tentative interpretations of new research findings, and these are the very findings that the public hears most about.
- The popular media often broadcast preliminary findings in hopes of grabbing attention and boosting readership or television ratings.
- Commercial promoters turn preliminary findings into advertisements for products or supplements long before the findings have been validated—or disproved. The scientific process requires many experiments or trials to confirm a new finding. Seldom do promoters wait as long as they should to make their claims.
- Consumers like to try new products or treatments even though they probably will not withstand the tests of time and scientific scrutiny.

So how can I tell what claims to believe?

The Food and Nutrition Science Alliance (FANSA), whose partners include the American Dietetic Association (ADA), the American Society for Nutritional Sciences (ASNS), and the Institute of Food Technologists (IFT), attempts to help consumers distinguish valid from misleading nutrition information. FANSA has created a list of ten red flags for detecting "junk science" (see Table NP6.1).[1]

The ADA asserts that nutrition misinformation harms the health and economic status of consumers. The ADA recognizes its responsibility to work with health care professionals and educators to present sound nutrition information to the public and to actively confront nutrition misinformation.[2] Table NP6.2 offers a list of credible sources of nutrition information.

WWW.

eatright.org
American Dietetic Association

faseb.org/asns
American Society for Nutritional Sciences

ift.org
Institute of Food Technologists

ncahf.org
National Council Against Health Fraud

Everyone seems to be giving advice on nutrition. How can I tell whom to listen to?

Registered dietitians (R.D.'s) and nutrition professionals with advanced degrees (M.S., Ph.D.) are experts (see the glossary on p. 135). These professionals are probably in the best position to answer your nutrition questions. On the other hand, "nutritionists" may be experts or quacks, depending on the state where they practice. Some states require people who use this title to meet strict standards. In other states, a "nutritionist" may be any individual who claims a career connection with the nutrition field.

Nutrition in Practice

Table NP6.1 FANSA's Red Flags of Junk Science

1. Recommendations that promise a quick fix.
2. Dire warnings of danger from a single product or regimen.
3. Claims that sound too good to be true.
4. Simplistic conclusions drawn from a complex study.
5. Recommendations based on a single study.
6. Dramatic statements that are refuted by reputable scientific organizations.
7. Lists of "good" and "bad" foods.
8. Recommendations made to help sell a product.
9. Recommendations based on studies published without peer review.
10. Recommendations from studies that ignore differences among individuals or groups.

Source: Reprinted with permission from B. Hansen, President's address, 1996: A virtual organization for nutrition in the 21st century, *American Journal of Clinical Nutrition* 64 (1996): 796–799. © American Society for Clinical Nutrition.

Table NP6.2 Credible Sources of Nutrition Information

Professional health organizations, government health agencies, volunteer health agencies, and consumer groups provide consumers with reliable health and nutrition information. Credible sources of nutrition information include:

- Professional health organizations, especially the American Dietetic Association's National Center for Nutrition and Dietetics (NCND); also the Society for Nutrition Education and the American Medical Association.
- Government health agencies such as the Federal Trade Commission (FTC), the U.S. Department of Health and Human Services (U.S. DHHS), the Food and Drug Administration (FDA), and the U.S. Department of Agriculture (USDA).
- Volunteer health agencies such as the American Cancer Society, the American Diabetes Association, and the American Heart Association.
- Reputable consumer groups such as the Better Business Bureau, the Consumers Union, the American Council on Science and Health, and the National Council on Science and Health, and the National Council Against Health Fraud.

Appendix D provides addresses for these and other organizations.

Source: Data from J. M. Ashley and W. T. Jarvis, Position of The American Dietetic Association: Food and nutrition misinformation, *Journal of the American Dietetic Association* 95 (1995): 705–707.

Other purveyors of nutrition information may also lack credentials. A health food store owner may be in the nutrition business simply because it is a lucrative market. Such a person may have a background in business or sales and no education in nutrition at all. Such a person is not qualified to provide nutrition information to customers. For accurate nutrition information, seek out a trained professional with a knowledge of nutrition—an expert in the field of dietetics.

What about other health care professionals?

All members of the health care team share responsibility for helping each client to achieve optimal health, but the registered dietitian (R.D.) is usually the primary nutrition expert. Each of the other team members has a related specialty. Some physicians are specialists in clinical nutrition and are also experts in the field. Other physicians, nurses, and dietetic technicians (D.T.R.'s) often assist dietitians in providing nutrition information and may help to administer direct nutrition care. Nurses play central roles in client care management and client relationships. Visiting nurses and home health care nurses may become intimately involved in clients' nutrition care at home, teaching them both theory and cooking techniques. Physical therapists can provide individualized exercise programs related to nutrition—for example, to help control obesity. Social workers may provide practical and emotional support.

What roles might these health professionals play in nutrition care?

Some of the responsibilities of the health care professional might be:

- Helping people understand why nutrition is important to them.
- Answering questions about food and diet.
- Explaining to clients how modified diets work.
- Collecting information about clients that may influence their nutrition health.
- Identifying clients at risk for poor nutrition status (see Chapters 15 and 16) and recommending or taking appropriate action.
- Recognizing when clients need extra help with nutrition problems (in such cases, the problems should be referred to a dietitian or physician).

Health care professionals might routinely perform these nutrition-related tasks:

Glossary of Nutrition Experts

dietetic technician registered (D.T.R.): a professional who has earned an associate degree or higher; has completed a dietetic technician program approved by the American Dietetic Association (ADA); has passed a national registration exam; and assists in planning, implementing, and evaluating nutritional care.

dietetics: the practical application of nutrition, including the assessment of nutrition status, recommendation of appropriate diets, nutrition education, and the planning and serving of meals.

nutritionist: a person who specializes in the study of nutrition. Some nutritionists are registered dietitians, but others are self-described experts whose training may be minimal or nonexistent. Some states make the term meaningful by allowing it to apply only to people who have master's (M.S.) or doctoral (Ph.D.) degrees from institutions accredited to offer such degrees in nutrition or related fields.

registered dietitian (R.D.): a dietitian who has graduated from a university or college after completing a program of dietetics that has been accredited by the American Dietetic Association (or Dietitians of Canada). The dietitian must serve in an approved internship or coordinated program to practice the necessary skills, pass the association registration examination, and maintain competency through continuing education. Many states require licensing for practicing dietitians. Licensed dietitians (L.D.'s) have met all *state* requirements to offer nutrition advice.

- Obtaining diet histories.
- Taking weight and height measurements.
- Feeding clients who cannot feed themselves.
- Recording what clients eat or drink.
- Observing clients' responses and reactions to foods.
- Helping clients mark menus.
- Monitoring weight changes.
- Monitoring food and drug interactions.
- Encouraging clients to eat.
- Assisting clients at home in planning their diets and managing their kitchen chores.
- Alerting the physician or dietitian when nutrition problems are identified.
- Charting actions taken and communicating on these matters with other professionals as needed.

As you can see, although the dietitian assumes the primary role as the nutrition expert on a health care team, other health care professionals play important roles in administering nutrition care.

Notes

1. B. Hansen, President's address, 1996: A virtual organization for nutrition in the 21st century, *American Journal of Clinical Nutrition* 64 (1996): 796–799.

2. J. M. Ashley and W. T. Jarvis, Position of The American Dietetic Association: Food and nutrition misinformation, *Journal of the American Dietetic Association* 95 (1995): 705–707.

7 *The Vitamins*

Table 7.1 Vitamin Names
Fat-Soluble Vitamins
Vitamin A
Vitamin D
Vitamin E
Vitamin K
Water-Soluble Vitamins
B vitamins
Thiamin
Riboflavin
Niacin
Pantothenic acid
Biotin
Vitamin B_6
Folate
Vitamin B_{12}
Vitamin C

vitamins: essential, noncaloric, organic nutrients needed in tiny amounts in the diet.

vita-men.com
Vita-Men

home.hyperlink.net.au/~bookman/index.html
Vitamin Update

Earlier chapters focused primarily on the energy-yielding nutrients—carbohydrate, fat, and protein. This chapter and the next one discuss the nutrients everyone thinks of when nutrition is mentioned—the vitamins and minerals.

The vitamins occur in foods in much smaller quantities than do the energy-yielding nutrients, and they themselves contribute no energy to the body. Instead, they serve mostly as facilitators of body processes. They are a powerful group of substances, as their absence attests. Vitamin A deficiency can cause blindness; a lack of niacin can cause mental illness; and a lack of vitamin D can retard growth. The consequences of deficiencies are so dire and the effects of restoring the needed nutrients so dramatic that people spend billions of dollars each year on vitamin supplements to cure many different ailments. Vitamins certainly contribute to sound nutritional health, but they do not cure all ills. Actually, a vitamin can cure only the disease caused by a deficiency of that vitamin. The vitamins' roles in supporting optimal health extend far beyond preventing deficiency diseases, however. Emerging evidence points to relationships between low intakes of vitamins and chronic diseases such as cancer and heart disease.

A child once defined vitamins as "what, if you don't eat, you get sick." The description is both insightful and accurate. A more prosaic definition is that vitamins are potent, essential, noncaloric, organic nutrients needed from foods in trace amounts to perform specific functions that promote growth or reproduction or maintain health and life. Two characteristics distinguish vitamins from energy nutrients:

1. Vitamins do not yield energy when broken down, but assist the enzymes that release energy from carbohydrate, fat, and protein.
2. Vitamins are needed in much smaller amounts than the energy nutrients.

As the individual vitamins were discovered, they were named or given letters, numbers, or both. This led to the confusion that still exists today. This chapter uses the names shown in Table 7.1; alternative names are given in Tables 7.3 and 7.4, which appear later in the chapter.

Vitamins fall naturally into two classes—fat soluble and water soluble. The solubility of a vitamin confers on it many characteristic behaviors and determines how it is absorbed and transported, whether it can be stored, and how easily it is lost from the body. This discussion of vitamins begins with the fat-soluble vitamins.

The Fat-Soluble Vitamins

The fat-soluble vitamins—A, D, E, and K—usually occur together in the fats and oils of foods, and the body absorbs them in the same way it absorbs lipids. Therefore, any condition that interferes with fat absorption can precipitate a deficiency of the fat-soluble vitamins. Once absorbed, fat-soluble vitamins are stored in the liver and fatty tissues until the body needs them. They are not readily excreted, and unlike most of the water-soluble vitamins, they can build up to toxic concentrations.

The capacity to store fat-soluble vitamins affords a person some flexibility as to dietary intakes. When blood concentrations begin to decline, the body can retrieve the vitamins from storage. Thus a person need not eat a day's allowance of each fat-soluble vitamin every day, but need only make sure that over time, average daily intakes approximate recommended intakes. In contrast, water-soluble vitamins must be consumed more regularly because the body does not store them to any great extent.

Vitamin A and Beta-Carotene

Vitamin A has the distinction of being the first fat-soluble vitamin to be recognized. Today, after more than 75 years of research and revelations, vitamin A and its plant-derived precursor, beta-carotene, are the focus of attention and interest for researchers around the world. Much of this intensive research effort is based on accumulating evidence that both the active vitamin and beta-carotene may protect against certain types of cancer.

vitamin A: a fat-soluble vitamin. Its three chemical forms are *retinol* (the alcohol form), *retinal* (the aldehyde form, which is active in the pigments of the eye), and *retinoic acid* (the acid form).

precursor: a compound that can be converted into another compound; with regard to vitamins, compounds that can be converted into active vitamins; also known as **provitamins.**

beta-carotene: a vitamin A precursor made by plants and stored in human fat tissue; an orange pigment.

retinol-binding protein (RBP): the specific protein responsible for transporting retinol.

Measurement of the blood concentration of RBP is a sensitive test of vitamin A status.

Metabolic Roles of Vitamin A Vitamin A is a versatile vitamin, playing diverse roles in vision, protein synthesis and cell differentiation (and thereby maintaining the health of body linings and skin), and reproduction and growth. Three different forms of vitamin A are active in the body: retinol, retinal, and retinoic acid. Each form of vitamin A performs specific tasks. Retinol is the major transport and storage form of the vitamin; the cells convert retinol to its other active forms as needed. A special transport protein, retinol-binding protein (RBP), picks up retinol from the liver where it is stored and carries it in the blood.

Vitamin A in Vision Vitamin A plays two indispensable roles in the eye. It helps maintain a healthy, crystal-clear outer window, the cornea; and it participates in the events of light detection at the retina. Figure 7.1 shows vitamin A's site of action inside the eye.

When vitamin A is lacking, the eye has difficulty adapting to changing light levels. At night, after the eye has adapted to darkness, a lag occurs before the eye can see again after a flash of bright light. This lag in the recovery of night vision is known as night blindness. Because night blindness is easy to test, it aids in the diagnosis of vitamin A deficiency. Night blindness is only a symptom, however, and may indicate a condition other than vitamin A deficiency.

cornea (KOR-nee-uh): the hard, transparent membrane covering the outside of the eye.

retina (RET-in-uh): the layer of light-sensitive nerve cells lining the back of the inside of the eye; consists of rods and cones.

night blindness: the slow recovery of vision after exposure to flashes of bright light at night; an early symptom of vitamin A deficiency.

Vitamin A in Protein Synthesis and Cell Differentiation The role that vitamin A plays in vision is undeniably important, but only one-thousandth of the body's vitamin A is in the retina. Much more is in the skin and the linings of organs, where it participates in protein synthesis and cell differentiation.

All body surfaces, both inside and out, are covered by layers of cells known as epithelial cells. The epithelial tissue on the outside of the body is, of course, the skin. The epithelial tissues inside the body include the linings of the mouth, stomach, and intestines; the linings of the lungs and the passages leading to them; the lining of the bladder; the linings of the uterus and vagina; and the linings of the eyelids and sinus passageways. The epithelial tissues on the inside of the body must be kept smooth. To ensure that they are, the epithelial cells on their surfaces secrete a smooth, slippery substance (mucus) that coats and protects the tissues

differentiation: development of specific functions different from those of the original.

epithelial (ep-i-THEE-lee-ul) **cells:** cells on the surface of the skin and mucous membranes.

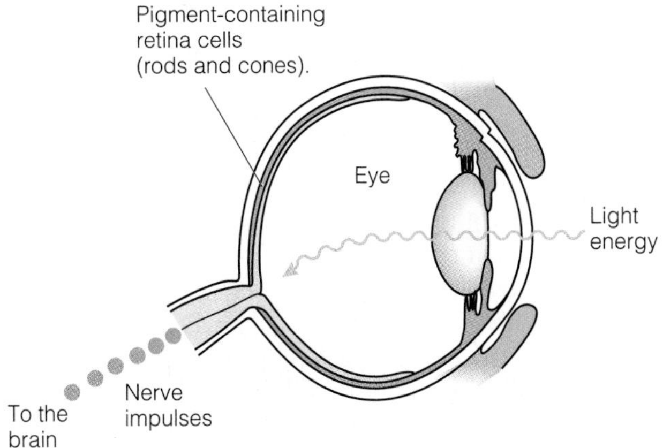

Figure 7.1
Vitamin A's Role in Vision
As light enters the eye, pigments within the cells of the retina absorb the light and generate nerve impulses that travel to the brain. Each pigment contains retinal, the active form of vitamin A.

from invasive microorganisms and other harmful particles. The mucous membrane that lines the stomach also shields its cells from digestion by gastric juices. Vitamin A helps to maintain the integrity of the epithelial cells.

epithelial tissues: the layers of the body that serve as selective barriers between the body's interior and the environment (examples are the cornea, the skin, the respiratory lining, and the lining of the digestive tract).

mucous membrane: the membranes composed of mucus-secreting cells that line the surfaces of body tissues. (Reminder: *Mucus* is the smooth, slippery substance secreted by these cells.)

Vitamin A in Reproduction and Growth Vitamin A also supports reproduction and growth. In men vitamin A participates in sperm development, and in women vitamin A promotes normal fetal growth and development.[1]

Metabolic Roles of Beta-Carotene For many years scientists believed beta-carotene to be of interest solely as a vitamin A precursor. Eventually, researchers began to recognize beta-carotene as an extremely effective antioxidant in the body. Antioxidants are compounds that protect other compounds (such as lipids in cell membranes) from attack by oxygen. Oxygen triggers the formation of compounds known as free radicals that can start chain reactions in cell membranes. If left uncontrolled, these chain reactions can damage cell structures and impair cell functions. Oxidative and free-radical damage to cells is suspected of instigating some early stages of cancer and heart disease.[2]

antioxidant: a compound that protects other compounds from oxygen by itself reacting with oxygen.

free radical: a highly reactive chemical form that can cause destructive changes in nearby compounds, sometimes setting up a chain reaction.

Nutrition in Practice 26 offers information about free radicals and heart disease.

WWW. cancer.org
American Cancer Society

Beta-Carotene and Cancer Studies of populations suggest that people whose diets are low in beta-carotene have higher incidences of certain types of cancer than those whose diets contain generous amounts of foods rich in beta-carotene.[3] Based on findings that beta-carotene in foods may protect against cancer, researchers designed a study to determine the effects of beta-carotene *supplements* on the incidence of lung cancer among smokers. The researchers expected to see a beneficial effect, but instead found that smokers taking the beta-carotene supplements suffered a *greater* incidence of lung cancer than those taking placebos.[4] Beta-carotene in foods is just one among many antioxidant nutrients and other nutrients present in foods. Furthermore, as Table 7.2 shows, many other compounds known as phytochemicals are also present in foods and may be at least partially responsible for the effect. Until more is known, eating beta-carotene–rich foods, not supplements, is in the best interest of health.

phytochemicals: nonnutrient compounds found in plant-derived foods that have biological activity in the body.
phyto = plant

Vitamin A Deficiency In vitamin A deficiency, the epithelial cells flatten and begin to produce keratin—the hard, inflexible protein of hair and nails. In the eye, this process leads to drying and hardening of the cornea, which may progress to permanent blindness.

keratin (KERR-uh-tin): a water-insoluble protein; the normal protein of hair and nails. Keratin-producing cells may replace mucus-producing cells in vitamin A deficiency.

Blindness Vitamin A deficiency is the major cause of childhood blindness in the world, causing more than half a million children to lose their sight every year. More than 100 million children worldwide endure less severe forms of vitamin A deficiency, making them vulnerable to infectious diseases.

xerophthalmia (zer-off-THAL-mee-uh): progressive blindness caused by vitamin A deficiency.
xero = dry
ophthalm = eye

An early sign is **xerosis** (drying of the cornea); the last and most severe stage is **keratomalacia** (kerr-uh-to-mal-AY-shuh), or total blindness.
malacia = softening, weakening

Infection Proneness All body surfaces, both inside and out, maintain their integrity with the help of vitamin A. When vitamin A is lacking, cells of the skin harden and flatten, making it dry, rough, scaly, and hard. An accumulation of keratin makes a lump around each hair follicle (keratinization).

follicle (FOLL-i-cul): a group of cells in the skin from which a hair grows.

The accumulation of the hard material keratin around each hair follicle is **follicular hyperkeratosis.**

In the mouth, a vitamin A deficiency results in drying and hardening of the salivary glands, making them susceptible to infection. Secretions of mucus in the stomach and intestines are reduced, hindering normal digestion and absorption of nutrients. Infections of other mucous membranes also become likely.

Vitamin A's role in maintaining the body's defensive barriers may partially explain the relationship between vitamin A deficiency and susceptibility to infection.[5] When children with measles complicated by diarrhea or infections such as pneumonia, or both were given vitamin A supplements, their recovery times and hospital stays were shorter, and their overall survival rates significantly greater, than those of similar children who did not receive vitamin A.[6]

Table 7.2 Phytochemicals—Their Food Sources and Actions		
Food Source	**Name**	**Action in the Body**
Deeply pigmented fruits and vegetables (carrots, sweet potatoes, tomatoes, spinach, broccoli, cantaloupe, pumpkin, apricots)	Carotenoids[a] (including beta-carotene and lycopene)	Act as antioxidants, reducing the risk of cancer.
Citrus fruits	Limonene	Triggers enzyme production to facilitate carcinogen excretion.
	Phenols	Inhibit lipid oxidation; block formation of carcinogenic nitrosamines in the body.
Garlic, onions, leeks, chives	Allyl sulfides	Trigger enzyme production to facilitate carcinogen excretion.
Broccoli, broccoli sprouts, and other cruciferous vegetables (cauliflower, cabbage, kale, brussels sprouts)	Sulforaphane	Protects against cancer.
	Dithiolthiones	Trigger enzyme production to block carcinogen damage to cells' DNA.
	Indoles	Trigger enzymes to inhibit estrogen action, reducing the risk of breast cancer.
	Isothiocyanates	Trigger enzyme production to block carcinogen damage of cells' DNA.
Grapes	Ellagic acid	Scavenges carcinogens.
Soy/legumes	Protease inhibitors	Suppress enzyme production in cancer cells, slowing tumor growth.
	Phytosterols	Inhibit cell reproduction in GI tract, preventing colon cancer.
	Isoflavones[b]	Block estrogen activity in cells, reducing the risk of breast and ovarian cancer.
	Saponins	Interfere with DNA reproduction, preventing cancer cell multiplication.
Flaxseed	Lignans[b]	Block estrogen activity in cells, reducing the risk of breast and ovarian cancer.
Fruits (blueberries, prunes, grapes), oats, soybeans	Caffeic acid	Triggers enzyme production to make carcinogens water soluble, facilitating excretion.
	Ferulic acid	Binds to nitrates in stomach, preventing the conversion to nitrosamines.
Grains	Phytic acid	Binds to minerals, preventing cancer-causing free-radical formation.
Fruits, vegetables, tea, wine, oregano	Flavonoids (including quercetin)	Act as antioxidants, reducing the risk of cancer and heart disease.

[a]In addition to beta-carotene, other carotenoids include alpha-carotene, beta-cryptoxanthin, lutein, zeaxanthin, and lycopene.
[b]Isoflavones and lignans are types of phytoestrogens—compounds that bind to estrogen receptors and reduce estrogen activity.

The evidence that vitamin A reduces the severity of measles and measles-related infections and diarrhea has prompted the World Health Organization (WHO) and UNICEF (the United Nations International Children's Emergency Fund) to make control of vitamin A deficiency a major goal in their quest to improve child survival throughout the developing world. The American Academy of Pediatrics has issued a formal statement recommending vitamin A supplementation for certain groups of measles-infected infants and children in the United States.[7]

who.ch
World Health Organization

unicef.org
UNICEF

aap.org
American Academy of Pediatrics

Timing of Deficiency Up to a year's supply of vitamin A can be stored in the body, 90 percent of it in the liver. If a healthy adult were to stop eating vitamin A–rich foods, deficiency symptoms would not begin to appear until after stores were depleted, which would take one to two years. Then, however, the consequences would be profound and severe. Table 7.3, later in this chapter, itemizes some of them.

Vitamin A Toxicity When the body stores excess vitamin A, toxicity is possible. Normally, toxicity symptoms are likely only when animal-derived foods or supplements are the source of the excess vitamin, for in these sources the vitamin is already active; it is called *preformed* vitamin A. Plant foods contain the vitamin only as beta-carotene, its inactive, precursor form. The precursor does not convert to active vitamin A rapidly enough to cause toxicity.

preformed vitamin A: vitamin A in its active form.

Overdoses of vitamin A damage the same body systems that exhibit symptoms in vitamin A deficiency (see Table 7.3 later in the chapter). Children are most vulnerable to vitamin A toxicity because, being smaller, they need less than adults, and it is easy to give them too much in pill form. The availability of breakfast cereals, instant meals, fortified milk, and chewable candylike vitamins, each containing 100 percent or more of the recommended daily intake of vitamin A, makes it possible for a well-meaning parent to provide several times the daily allowance of the vitamin to a child within a few hours. Serious toxicity is seen in infants and young children when they are given more than ten times the recommended amount every day for weeks at a time.

Certain vitamin A relatives are available by prescription as acne treatments. When applied directly to the skin surface, these preparations help relieve the symptoms of acne. Taking massive doses of vitamin A internally will *not* cure acne, however, and may cause the miseries itemized in Table 7.3. Foods are always a better choice than supplements for needed nutrients. The best way to ensure a safe vitamin A intake is to eat generous servings of vitamin A–rich foods. Well-nourished, healthy people need no supplements.

Beta-Carotene Conversion and Toxicity When beta-carotene is converted to retinol in the body, losses occur. This is why, rather than expressing the amounts of beta-carotene in foods, nutrition scientists use the RE (retinol equivalent), which expresses the amount of retinol the body actually derives from a plant food after conversion. The body can make one unit of retinol from about three molecules, or six units, of beta-carotene.

The units in which vitamin A amounts in foods are expressed are **RE (retinol equivalents).** A unit used earlier was the **IU (international unit).**

Beta-carotene from plant foods is not converted to the active form of vitamin A rapidly enough to be hazardous. It has, however, been known to turn people bright yellow if they eat too much. Beta-carotene builds up in the fat just beneath the skin and imparts a yellow cast.

Vitamin A in Foods In the United States, about half of the vitamin A activity consumed in foods comes from fruits and vegetables that supply the vitamin as its precursor, beta-carotene. The other half of vitamin A activity consumed in foods comes from preformed vitamin A supplied in milk, cheese, butter, and other dairy products; eggs; and meats. Liver is a rich source of preformed vitamin A. Liver offers many nutrients, so eating it periodically may benefit nutritional health, but once every week or so is enough.

1 RE = 3.33 IU from animal foods or 10 IU from plant foods. (On the average, 1 RE = about 5 IU.)

Because vitamin A is fat soluble, it is lost when milk is skimmed. Nonfat milk is thus often fortified with vitamin A to compensate. Margarine is usually fortified so as to provide the same amount of vitamin A as butter. Snapshot 7.1 shows a sampling of the richest food sources of both preformed vitamin A and beta-carotene.

Fast-food meals often lack vitamin A. When fast-food restaurants offer salads with cheese, carrots, and other vitamin A–rich foods, the nutritional quality of their meals greatly improves.

Beta-Carotene in Foods Many foods from plants contain beta-carotene. Carrots, sweet potatoes, pumpkins, cantaloupe, and apricots are all rich sources, and their bright orange color enhances the eye appeal of the plate. Another colorful group, *dark* green vegetables, such as spinach, other greens, and broccoli, owe their color to both chlorophyll and beta-carotene. The orange and green pigments together give a deep, murky, dark green color to the vegetables. Other col-

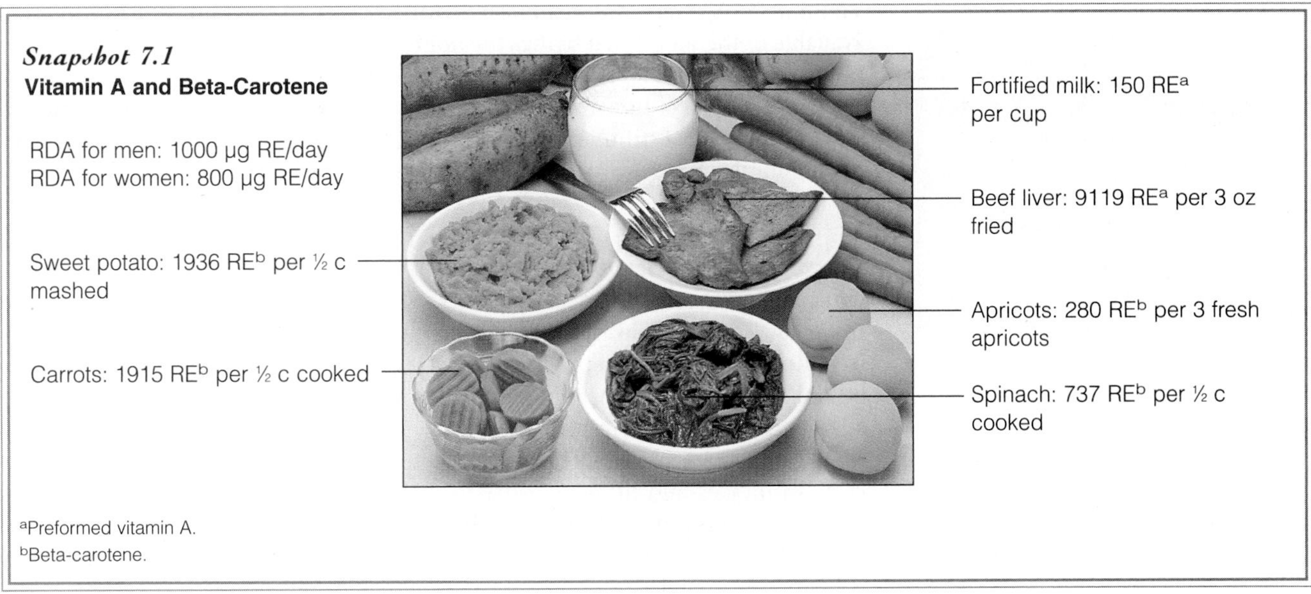

orful vegetables, such as iceberg lettuce, beets, and sweet corn, can fool you into thinking they contain beta-carotene, but these foods derive their color from other pigments and are poor sources of beta-carotene. As for "white" plant foods such as grains and potatoes, they have none. Recommendations to eat *dark* green or *deep* orange vegetables and fruits at least every other day help people to meet their vitamin A needs.

Vitamin D

Vitamin D is different from all the other nutrients in that the body can synthesize it with the help of sunlight. Therefore, in a sense, vitamin D is not an essential nutrient. Given enough sun, people need no vitamin D from foods.

Sunlight promotes vitamin D formation in the skin.

Vitamin D's Metabolic Conversions The liver manufactures a vitamin D precursor, which migrates to the skin where it is converted to a second precursor with the help of the sun's ultraviolet rays. Next, the liver and then the kidneys alter the second precursor to produce the active vitamin. Vitamin D precursors from plants require the same two conversions by the liver and kidneys to become active. The biologic activity of the active vitamin is 500- to 1000-fold higher than that of its precursor. Diseases that affect either the liver or the kidneys may impair the transformations of precursor vitamin D to active vitamin D and therefore produce symptoms of vitamin D deficiency.

The precursor of vitamin D made in the liver is 7-dehydrocholesterol, which is made from cholesterol. This is one of the body's many "good" uses for cholesterol.

The final, active vitamin is 1-25 dihydroxycholecalciferol, or, more simply, dihydroxy vitamin D.

Vitamin D's Actions Although known as a vitamin, vitamin D is actually a hormone—a compound manufactured by one organ of the body that has effects on another. The best-known vitamin D target organs are the intestine, the kidneys, and the bones, but scientists have discovered many other vitamin D target tissues, including the brain, the pancreas, the skin, the reproductive organs, and many cancer cells.[8] The abundance of these discoveries suggests that numerous additional functions for vitamin D may surface, including the regulation of the immune system.[9]

Vitamin D's Roles in Bone Vitamin D is a member of a large, cooperative bone-making and maintenance team composed of nutrients and other compounds, including vitamins A, C, and K; the hormones parathormone and calcitonin; the protein collagen; and the minerals calcium, phosphorus, magnesium, and fluoride.

Sunlight promotes vitamin D synthesis in the skin. Exposure to the sun should be moderate, however; excessive exposure may cause skin cancer.

rickets: the vitamin D–deficiency disease in children.

osteomalacia (os-tee-o-mal-AY-shuh): a bone disease characterized by softening of the bones; symptoms include bending of the spine and bowing of the legs. The disease occurs most often in adult women.
osteo = bone
mal = bad (soft)

osteoporosis: literally, porous bones; reduced density of the bones, also known as *adult bone loss.*

WWW. **nof.org**
National Osteoporosis Foundation

WWW. **fda.gov**
Food and Drug Administration

Vitamin D's special role in bone growth is to make calcium and phosphorus available in the blood that bathes the bones. The bones grow denser and stronger as the minerals are deposited from the blood.

Vitamin D acts in three ways to maintain blood concentrations of calcium and phosphorus: it stimulates their absorption from the GI tract; it mobilizes calcium and phosphorus from bones into the blood; and it stimulates their retention by the kidneys.

Vitamin D Deficiency The symptoms of vitamin D deficiency are those of calcium deficiency, as shown in Table 7.3. The bones fail to calcify normally and may grow so weak that they become bent when they have to support the body's weight. A child with rickets who is old enough to walk characteristically develops bowed legs, often the most obvious sign of the disease. Worldwide, rickets afflicts a large number of children.

Adult rickets, or osteomalacia, occurs most often in women who have low calcium intakes and little exposure to sun and who go through repeated pregnancies and periods of lactation. The bones of the legs may soften to such an extent that a young woman who is tall and straight at 20 may be condemned by repeated pregnancies to become bent, bowlegged, and stooped before she is 30.

Inadequate vitamin D is recognized as a risk factor in osteoporosis (reduced bone density). Without sufficient vitamin D, calcium is mobilized from the bones, and bone remodeling is impaired. This combination leads to a loss of bone mass.

Vitamin D Toxicity Whereas vitamin D deficiency depresses calcium absorption, blood calcium, and bone mineralization, an excess of vitamin D does the opposite, as shown in Table 7.3. It enhances calcium absorption, produces high blood calcium, and promotes return of bone calcium into the blood. The excess calcium then tends to precipitate in the soft tissues, forming stones, including kidney stones. Calcification may also harden the blood vessels and is especially dangerous in the major arteries of the heart and lungs, where it can cause death.

Vitamin D in excess is the most toxic of all the vitamins. The amounts of vitamin D in foods available in the United States and Canada are well within safe limits, but supplements containing the vitamin in concentrated form are not. Adults should use caution when taking vitamin D supplements and keep them out of the reach of children.

Vitamin D from the Sun Most of the world's population relies on natural exposure to sunlight to maintain adequate vitamin D nutrition. The sun imposes no risk of vitamin D toxicity. Prolonged exposure to sunlight degrades the vitamin D precursor in the skin, preventing its conversion to the active vitamin. Even lifeguards on southern beaches are safe from vitamin D toxicity from the sun.

Effects of Sunscreens Prolonged exposure to sunlight has other undesirable consequences such as premature wrinkling of the skin and risk of skin cancer. These risks may be reduced by using sunscreens. Unfortunately, sunscreens with sun protection factors (SPF) of 8 and above also retard vitamin D synthesis. A strategy to avoid this dilemma is to apply sunscreen after enough time has elapsed to provide sufficient vitamin D. For most people, exposing hands, face, and arms on a clear summer day for 10 minutes, a few times a week, should be sufficient to maintain vitamin D nutrition. Dark-skinned people require longer exposure than light-skinned people, but by three hours, vitamin D synthesis in heavily pigmented skin arrives at the same plateau as in fair skin after 30 minutes.

Tanning Lamp Effects The ultraviolet rays from tanning lamps and tanning booths may also stimulate vitamin D synthesis, but the hazards outweigh any possible benefits. The Food and Drug Administration (FDA) warns that if the lamps are not properly filtered, people using tanning booths risk burns, damage to blood vessels, skin cancer, and damage to the eyes.[10]

Smog Effects The ultraviolet rays of the sun that promote vitamin D synthesis may be filtered out by heavy clouds, smoke, or smog. Together with skin pigmentation, smog probably accounts for the fact that dark-skinned people in northern, smoggy cities are prone to rickets. For these people, and for those who are unable to go outdoors frequently, dietary vitamin D is important.

Smog filters out ultraviolet rays of the sun.

Vitamin D in Foods Only a few animal foods, notably, eggs, liver, butter, some fish, and fortified milk supply significant amounts of vitamin D. For those who use margarine in place of butter, fortified margarine is a significant source. Infant formulas are fortified with vitamin D in amounts adequate for daily intake. Breast milk is low in vitamin D, so vitamin D supplements are prescribed for dark-skinned, breastfed infants who do not have adequate exposure to the sun.[11] These sources, plus any exposure to the sun, provide babies with more than enough of this vitamin.

Vitamin D activity was previously expressed in international units (IU), but as of 1980, it is expressed in micrograms of cholecalciferol, the active form of vitamin D. To convert, use the following factor:

100 IU = 2.5 μg
400 IU = 10 μg

Importance of Milk The fortification of milk with vitamin D is the best guarantee that children will meet their vitamin D needs and underscores the importance of milk in children's diets. Unlike milk, cheese and yogurt are not fortified with vitamin D. Strict vegetarians, and especially their children, may have low vitamin D intakes because no fortified plant source except margarine exists. In the United States, breakfast cereals may be fortified with vitamin D, as their labels indicate.[12]

Reminder: Milk is also an excellent source of calcium and other bone-building nutrients.

Vitamin D for Adults Most adults, especially in sunny regions, need not make special efforts to obtain vitamin D in food. People who are not outdoors much or who live in northern or predominantly cloudy or smoggy areas, however, are advised to make sure their milk is fortified with vitamin D and to drink at least 2 cups a day.

Vitamin E

Like beta-carotene, vitamin E is a fat-soluble antioxidant; it protects other substances from oxidation by being oxidized itself. It is one of the body's primary defenders against oxidation.

Reminder: *Oxidation* is a type of chemical reaction, so named because oxygen is one of the agents that often brings it about.

Cell Membrane Antioxidant If there is plenty of vitamin E in the membranes of cells exposed to an oxidant, chances are this vitamin will take the brunt of any oxidative attack, protecting the lipids and other vulnerable components of the membranes. Vitamin E is especially effective in preventing the oxidation of the polyunsaturated fatty acids (PUFA), but it protects all other lipids (for example, vitamin A) as well. Table 7.3 summarizes important information about vitamin E.

Lung Antioxidant Vitamin E exerts an especially important antioxidant effect in the lungs, where the cells are exposed to high concentrations of oxygen. Vitamin E also protects the lungs from air pollutants that are strong oxidants.

Vitamin E and Heart Disease Prevention Accumulating evidence suggests vitamin E offers protection against heart disease by protecting LDL (low-density lipoproteins) from oxidation.[13] The oxidation of LDL encourages development of atherosclerosis. Both vitamin E from foods and vitamin E from supplements seem to be protective against heart disease.[14]

Vitamin E Myths While research continues to reveal possible roles for vitamin E, it has also clearly discredited claims that vitamin E improves athletic skill, enhances sexual performance, or cures sexual dysfunction in males. Vitamin E also does not prevent or cure hereditary muscular dystrophy, nor does it slow or prevent processes of aging, such as graying of the hair, wrinkling of the skin, or reduced activity of body organs.

muscular dystrophy (DIS-tro-fee): a hereditary disease in which the muscles gradually weaken; its most debilitating effects arise in the lungs. This disease should not be confused with *nutritional* muscular dystrophy, a vitamin E–deficiency disease of animals characterized by gradual paralysis of the muscles.

Vitamin E Deficiency When the blood concentration of vitamin E falls below a certain critical level, the red blood cells tend to break open and spill their contents, probably because the PUFA in their membranes oxidize. This classic vitamin E–deficiency symptom, known as erythrocyte hemolysis, is seen in premature infants, born before the transfer of vitamin E from the mother to the fetus that takes place in the last weeks of pregnancy. Vitamin E treatment corrects erythrocyte hemolysis.

Two other conditions appear to respond to vitamin E therapy. One is a non-malignant breast disease (fibrocystic breast disease), and the other (intermittent claudication) is an abnormality of blood flow that causes cramping in the legs.

erythrocyte (er-REETH-ro-cite) **hemolysis** (he-MOLL-uh-sis): rupture of the red blood cells, caused by vitamin E deficiency.
erythro = red
cyte = cell
hemo = blood
lysis = breaking

fibrocystic breast disease: a harmless condition in which the breasts develop lumps, sometimes associated with caffeine consumption. In some, it responds to treatment by abstinence from caffeine; in others, it can be treated with vitamin E.
fibr = fibrous lumps
cystic = in sacs

intermittent claudication: severe calf pain caused by inadequate blood supply; it occurs when walking and subsides during rest.
intermittent = at intervals
claudicare = to limp

Caution: Other serious conditions can cause lumps in the breast and pain in the legs. Don't self-diagnose; see a physician.

Causes of Vitamin E Deficiency In human beings, vitamin E deficiency is usually associated with diseases, notably those that cause malabsorption of fat. These include diseases of the liver, gallbladder, and pancreas, as well as various hereditary diseases involving digestion and use of nutrients.

It may be, however, that rare vitamin E deficiencies are seen in people without diseases. Most likely, such deficiencies occur in people who for years eat diets extremely low in fat; or use fat substitutes, such as diet margarines and salad dressings, as their only sources of fat; or consume diets composed of highly processed or "convenience" foods. Vitamin E is destroyed by extensive heating in the processing of foods.

Vitamin E Toxicity Vitamin E supplement use has risen in recent years as its antioxidant action against disease has been recognized. As a result, signs of toxicity are now known or suspected, although vitamin E toxicity is not nearly as common, and its effects are not as serious, as vitamin A or vitamin D toxicity. High doses of vitamin E interfere with the blood-clotting action of vitamin K and enhance the action of anticoagulant medications, leading to hemorrhage. For most individuals, however, daily doses below 300 milligrams seem to be harmless.[15]

On vitamin bottles, vitamin E activity is often expressed in IU. One IU is the same as 1 mg of the active form of vitamin E.

The RDA gives values for vitamin E in units known as alpha-TE (alpha-tocopherol equivalents). One of these units, 1 alpha-TE, equals 1 mg of active vitamin E.

Vitamin E in Foods Vitamin E is widespread in foods. About 20 percent of the vitamin E in the diet comes directly or indirectly from vegetable oils and the products made from them, such as margarine, salad dressings, and shortenings (see Snapshot 7.2). Another 20 percent comes from fruits and vegetables. Soybean oil and wheat germ oil have especially high concentrations of vitamin E. Animal fats, such as butter and milk fat, contain little or no vitamin E. Because vitamin E is readily destroyed by heat processing and oxidation, fresh or lightly processed foods are the most desirable sources of this vitamin.

Vitamin K

K stands for the Danish word *koagulation* (coagulation or "clotting").

Vitamin K has long been known for its role in blood clotting, where its presence can make the difference between life and death. The vitamin also participates in the synthesis of several bone proteins.[16] Research is underway to determine the specific roles of these vitamin K–dependent proteins in bone metabolism and the risk of osteoporosis.[17]

hemorrhagic (hem-o-RAJ-ik) **disease:** the vitamin K–deficiency disease in which blood fails to clot.

Blood Clotting At least 13 different proteins and the mineral calcium are involved in making a blood clot. Vitamin K is essential for the activation of seven of these proteins, among them prothrombin, the precursor of the protein thrombin (see Figure 7.2 on p. 148).[18] When any of the blood-clotting factors is lacking, hem-

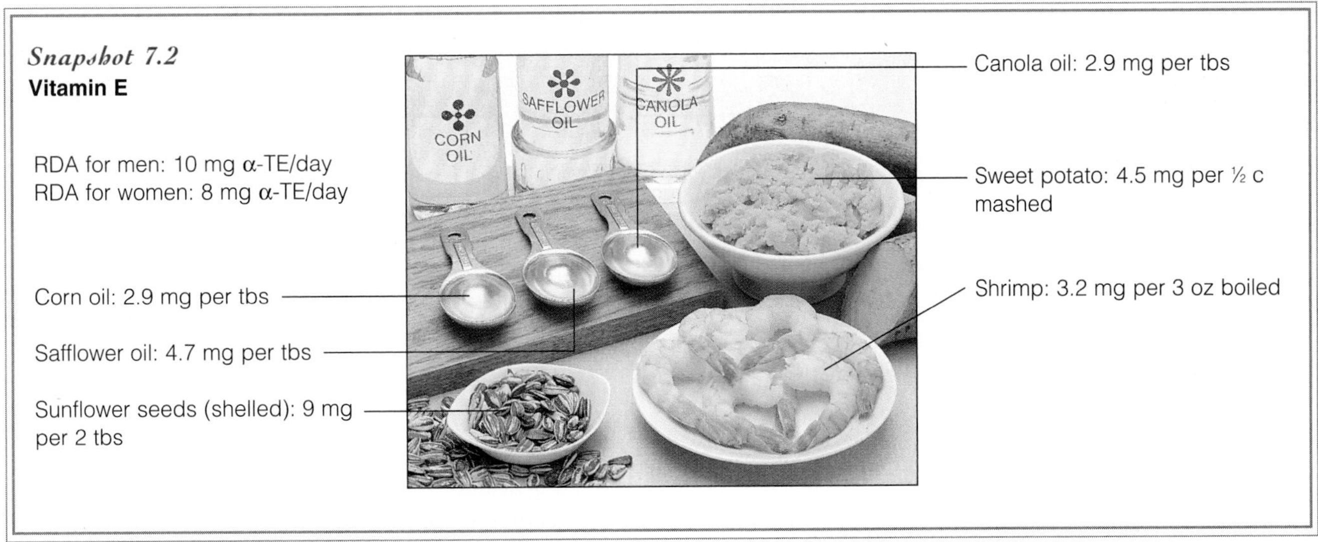

orrhagic disease results. If an artery or vein is cut or broken, bleeding goes unchecked. Of course, this is not to say that the cause of hemorrhaging is always a vitamin K deficiency.

Intestinal Synthesis Like vitamin D, vitamin K can be obtained from a nonfood source. Bacteria in the intestinal tract synthesize vitamin K that the body can absorb, but people cannot depend on this source alone for their vitamin K.

Reminder: The bacterial inhabitants of the digestive tract are known as the *intestinal flora.*

flora = plant inhabitants

Vitamin K Deficiency Vitamin K deficiency is rare, but may occur in two circumstances. First, it may arise in conditions of fat malabsorption. Second, some medications interfere with vitamin K's synthesis and action in the body: antibiotics kill the vitamin K–producing bacteria in the intestine, and anticoagulant medications interfere with vitamin K metabolism and activity. When vitamin K deficiency does occur, it can be fatal.

sterile: free of microorganisms, such as bacteria.

jaundice: yellowing of the skin due to spillover of bile pigments from the liver into the general circulation.

Vitamin K for Newborns Newborn infants present a unique case of vitamin K nutrition. An infant is born with a sterile digestive tract, and some weeks pass before the vitamin K–producing bacteria become fully established in the infant's intestines. At the same time, plasma prothrombin concentrations are low (this helps prevent blood clotting during the stress of birth, which might otherwise be fatal). A modest dose of vitamin K, usually in a water-soluble form, may therefore be given at birth to prevent hemorrhagic disease in the newborn.

Vitamin K Toxicity A high intake of vitamin K can reduce the effectiveness of anticoagulant medications used to prevent the blood from clotting. People taking these medications should eat consistent amounts of vitamin K–rich foods from day to day. Vitamin K–toxicity symptoms include red cell hemolysis, jaundice, and brain damage (see Table 7.3 on pp. 150–151).

Vitamin K in Foods Many foods contain ample amounts of vitamin K, notably green leafy vegetables, members of the cabbage family, and liver. Milk, meats, eggs, cereal, fruits, and vegetables provide smaller, but still significant, amounts.

Notable food sources of vitamin K include milk, eggs, brussels sprouts, cabbage, and spinach.

The four fat-soluble vitamins play many specific roles in the growth and maintenance of the body. Their presence affects the health and function of the eyes,

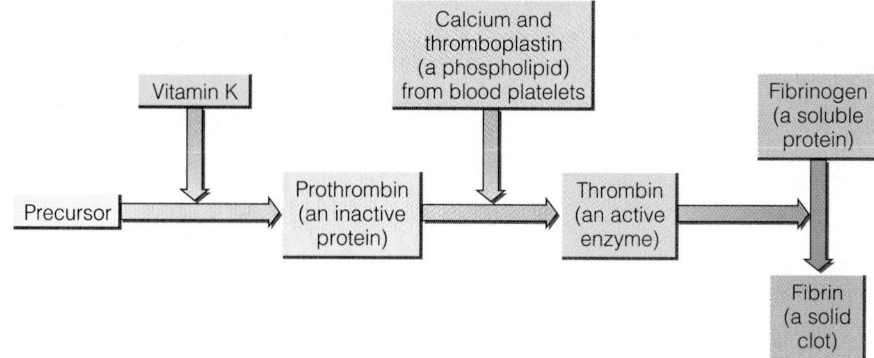

Figure 7.2
Blood-Clotting Process
When blood is exposed to air, foreign substances, or secretions from injured tissues, platelets (small, cell-like structures in the blood) release a phospholipid known as thromboplastin. Thromboplastin catalyzes the conversion of the inactive protein prothrombin to the active enzyme thrombin. Thrombin then catalyzes the conversion of the precursor protein fibrinogen to the active protein fibrin that forms the clot.

skin, GI tract, lungs, bones, teeth, nervous system, and blood; their deficiencies become apparent in these same areas. Toxicities of the fat-soluble vitamins are possible, especially when people use supplements, because the body stores excesses.

The Water-Soluble Vitamins

The B vitamins and vitamin C are the water-soluble vitamins. These vitamins, found in the watery compartments of foods, are distributed into water-filled compartments of the body. They are easily absorbed into the bloodstream and are just as easily excreted if their blood concentration rises too high. Thus the water-soluble vitamins are less likely to reach toxic concentrations in the body than are the fat-soluble vitamins. Foods never deliver toxic doses of the water-soluble vitamins, but the large doses concentrated in vitamin supplements can reach toxic levels.

The B Vitamins

Despite advertisements that claim otherwise, the B vitamins do not give people energy. Carbohydrate, fat, and protein—the *energy-yielding* nutrients—are used for fuel. The B vitamins help to burn that fuel, but do not serve as fuel themselves.

Figure 7.3
Coenzyme Action

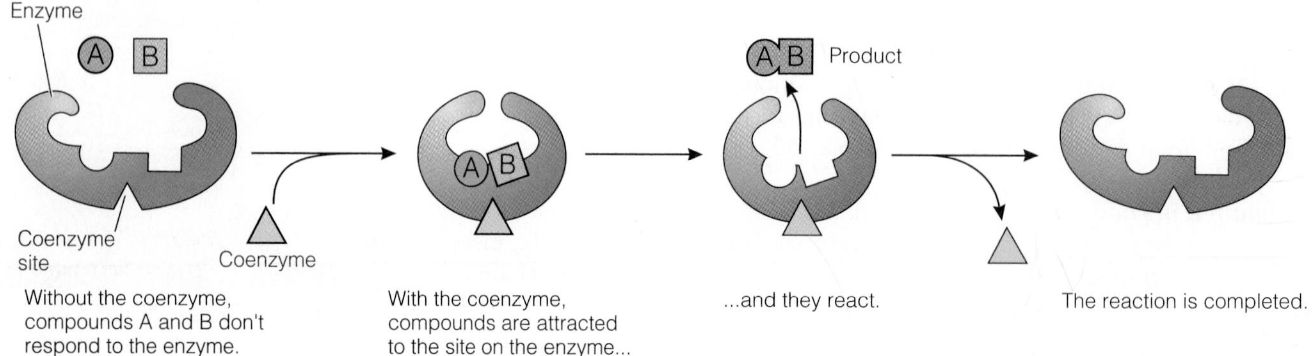

Coenzymes The eight B vitamins were listed in Table 7.1. Each is part of an enzyme helper known as a coenzyme. Each B vitamin has other important functions in the body as well, but the roles these vitamins play as parts of coenzymes are the best understood. A coenzyme is a small molecule that combines with an enzyme to make it active. With the coenzyme in place, the substance to be worked on is attracted to the enzyme, and the reaction proceeds instantaneously. Figure 7.3 illustrates coenzyme action.

Thiamin, riboflavin, niacin, pantothenic acid, and biotin are each part of a distinct coenzyme necessary for the production of energy from glucose, amino acids, and fats. A coenzyme containing vitamin B_6 assists enzymes that metabolize amino acids. The making of new cells depends on a folate coenzyme, and the making of this coenzyme depends on vitamin B_{12}. Folate and vitamin B_{12} together are, among other things, involved in duplicating genetic material when cells divide. A coenzyme already mentioned in the last chapter was coenzyme A, or CoA, made from the vitamin pantothenic acid.

These eight B vitamins play many specific roles in helping the enzymes to perform thousands of different molecular conversions in the body. They must be present in every cell continuously for the cells to function as they should. As for vitamin C, its primary role, discussed later, is as an antioxidant.

coenzyme (co-EN-zime): a small molecule that works with an enzyme to promote the enzyme's activity. Many coenzymes have B vitamins as part of their structure.
co = with

B Vitamin Deficiencies In academic and clinical discussions of the vitamins, different sets of deficiency symptoms are ascribed to each individual vitamin. Such clear-cut symptoms are found only in laboratory animals that have been fed contrived diets that lack just one nutrient. In reality, a deficiency of any single B vitamin seldom shows up in isolation because people do not eat nutrients one by one; they eat foods containing mixtures of nutrients. If a major class of foods is missing from the diet, all of the nutrients delivered by those foods will be lacking to various extents.

In only two cases have dietary deficiencies associated with single B vitamins been observed on a large scale in human populations. Diseases have been named for these deficiency states. One of them, beriberi, was first observed in Southeast Asia when the custom of polishing rice became widespread. Rice contributed 80 percent of the energy intake of the people in these areas, and rice hulls were their principal source of thiamin. When the hulls were removed to make the rice whiter, beriberi spread like wildfire.

The niacin-deficiency disease, pellagra, became widespread in the southern United States in the early part of the twentieth century among people who subsisted on a low-protein diet with a staple grain of corn. This diet was unusual in that it supplied neither enough niacin nor enough of its amino acid precursor tryptophan to make the niacin intake adequate.

Even in the cases of beriberi and pellagra, the deficiencies were probably not pure. When foods were provided containing the one vitamin known to be needed, other vitamins that may have been in short supply came as part of the package.

beriberi: the thiamin-deficiency disease; characterized by loss of sensation in the hands and feet, muscular weakness, advancing paralysis, and abnormal heart action.

pellagra (pell-AY-gra): the niacin-deficiency disease. Symptoms include the "4 Ds": diarrhea, dermatitis, dementia, and, ultimately, death.
pellis = skin
agra = seizure

Interdependent Systems Table 7.4, at the end of this chapter, sums up a few of the better-established facts about B vitamin deficiencies. A look at the table will make another generalization possible. Different body systems depend to different extents on these vitamins. Processes in nerves and in their responding tissues, the muscles, depend heavily on glucose metabolism and hence on thiamin, so paralysis sets in when this vitamin is lacking, but thiamin is important in all cells, not just in nerves and muscles. Similarly, because the red blood cells and GI tract cells divide the most rapidly, two of the first symptoms of a deficiency of folate are a type of anemia and GI deterioration—but again, all systems depend on folate, not just these. The list of symptoms in Table 7.4 is far from complete.

refined grain: a product from which the bran, germ, and husk have been removed, leaving only the endosperm.

Table 7.3 The Fat-Soluble Vitamins—A Summary

Vitamin A		
Other Names	**Deficiency Symptoms**	**Toxicity Symptoms**
Retinol, retinal, retinoic acid; main precursor is beta-carotene		
	Blood/Circulatory System	
Chief Functions in the Body	Anemia (small-cell type)[a]	Red blood cell breakage, nosebleeds
	Bones/Teeth	
Vision: health of cornea, epithelial cells, mucous membranes; skin health; bone and tooth growth; reproduction; hormone synthesis and regulation; immunity; cancer protection	Cessation of bone growth, painful joints; impaired enamel formation, cracks in teeth, tendency to decay	Bone pain; growth retardation; increase of pressure inside skull mimicking brain tumor; headaches
	Digestive System	
Deficiency Disease Name	Diarrhea, changes in lining	Abdominal cramps and pain, nausea, vomiting, diarrhea, weight loss
	Immune System	
Hypovitaminosis A	Depression; frequent respiratory, digestive, bladder, vaginal, and other infections	Overreactivity
Significant Sources		
	Nervous/Muscular Systems	
Retinol: fortified milk, cheese, cream, butter, fortified margarine, eggs, liver	Night blindness (retinal)	Blurred vision, pain in calves, fatigue, irritability, loss of appetite
Beta-carotene: spinach and other dark leafy greens; broccoli; deep orange fruits (apricots, cantaloupe) and vegetables (squash, carrots, sweet potatoes, pumpkin)	Skin and Cornea	
	Keratinization, corneal degeneration leading to blindness,[b] rashes	Dry skin, rashes, loss of hair
	Other	
	Kidney stones, impaired growth	Cessation of menstruation, liver and spleen enlargement

Vitamin D		
Other Names	**Deficiency Symptoms**	**Toxicity Symptoms**
Calciferol, cholecalciferol, dihydroxy vitamin D; precursor is cholesterol	Blood/Circulatory System	
		Raised blood calcium
Chief Functions in the Body	Bones/Teeth	
Mineralization of bones (raises blood calcium and phosphorus via absorption from digestive tract, and by withdrawing calcium from bones and stimulating retention by kidneys)	Abnormal growth, misshapen bones (bowing of legs), soft bones, joint pain, malformed teeth	Increased calcium withdrawal
	Nervous/Muscular Systems	
	Muscle spasms	Excessive thirst, headaches, irritability, loss of appetite, weakness, nausea
Deficiency Disease Name	Other	
Rickets, osteomalacia		Kidney stones, stones in arteries, death
Significant Sources		
Self-synthesis with sunlight; fortified milk, margarine, butter, and cereals; eggs, liver, small fish (sardines)		

[a]Small-cell anemia is termed *microcytic anemia;* large-cell type is *macrocytic* or *megaloblastic anemia.*
[b]Corneal degeneration progresses from *keratinization* (hardening) to *xerosis* (drying) to *xerophthalmia* (thickening, opacity, and irreversible blindness).

Table 7.3 *(continued)*

Vitamin E

Other Names	Deficiency Symptoms	Toxicity Symptoms
Alpha-tocopherol, tocopherol	**Blood/Circulatory System**	
Chief Functions in the Body	Red blood cell damage, anemia	Augments the effects of anticlotting medication
Antioxidant (detoxification of strong oxidants), stabilization of cell membranes, regulation of oxidation reactions, protection of PUFA and vitamin A	**Digestive System**	
		General discomfort
	Nervous/Muscular System	
Deficiency Disease Name	Degeneration, weakness, difficulty walking, leg cramps	(No symptoms reported)
(No name)	**Other**	
	Fibrocystic breast disease	(No symptoms reported)
Significant Sources		
Polyunsaturated plant oils (margarine, salad dressings, shortenings), green and leafy vegetables, wheat germ, whole-grain products, nuts, seeds		

Vitamin K

Other Names	Deficiency Symptoms	Toxicity Symptoms
Phylloquinone, menaquinone, naphthoquinone	**Blood/Circulatory System**	
	Hemorrhaging	Interference with anticlotting medication; vitamin K analogues may cause jaundice
Chief Functions in the Body		
Synthesis of blood-clotting proteins and a protein that binds calcium in the bones		
Deficiency Disease Name		
(No name)		
Significant Sources		
Bacterial synthesis in the digestive tract; liver, green leafy vegetables, cabbage-type vegetables, milk		

Subtle Deficiencies Major deficiency diseases such as pellagra and beriberi no longer occur in the United States and Canada, but more subtle deficiencies of nutrients, including the B vitamins, sometimes are observed. When they do occur, it is usually in people whose food choices are poor because of poverty, ignorance, illness, or poor health habits such as alcohol abuse. If the staple grain food is made from refined grain, vitamin B deficiencies are especially likely.

B Vitamin Enrichment of Foods One way to protect people from deficiencies is to add nutrients to their staple food, a process known as fortification or enrichment. The enrichment of refined breads and cereals has drastically reduced the incidence of iron and B vitamin deficiencies.

fortification: the addition to a food of nutrients that were either not originally present or present in insignificant amounts. Fortification can be used to correct or prevent a widespread nutrient deficiency, to balance the total nutrient profile of a food, or to restore nutrients lost in processing.

enrichment: the addition of nutrients to a food to meet a specified standard. In the case of refined bread or cereal, five nutrients have been added: thiamin, riboflavin, niacin, and folate in amounts approximately equivalent to, or higher than, those originally present and iron in amounts to alleviate the prevalence of iron-deficiency anemia.

The preceding discussion has shown both the great importance of the B vitamins in promoting normal, healthy functioning of all body systems and the severe consequences of deficiency. Now you may want to know how to be sure you are getting enough of these vital nutrients. The next sections present information on each B vitamin. While reading further, keep in mind that *foods* can provide all the needed nutrients and that supplements are a poor second choice. Some supplements are absurdly costly; but even if they are inexpensive, most people don't need them. Supplementing with fat-soluble vitamins can be dangerous, and if you are eating an adequate diet, supplementing with water-soluble vitamins offers no benefit other than to increase the dollar value of your urine. Nutrition in Practice 7 discusses uses and choices of supplements in more detail.

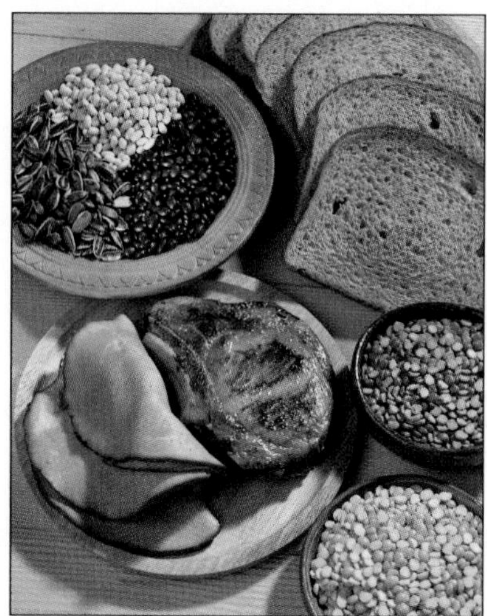

Nutritious foods such as pork, legumes, sunflower seeds, and whole-grain breads are valuable sources of thiamin.

Thiamin

All cells use thiamin, which plays a critical role in their energy metabolism. Thiamin also occupies a special site on nerve cell membranes. Consequently, processes in nerves and in their responding tissues, the muscles, depend heavily on thiamin.

Thiamin Need As long as people consume enough food energy to meet their needs—and obtain that energy from thiamin-containing foods—thiamin needs will be met. People who derive a large proportion of their energy from empty-kcalorie items, like sugar or alcohol, however, risk thiamin deficiency, a condition that seems to be reappearing as the population of malnourished and homeless people rises. A person who is fasting or who has adopted a very-low-kcalorie diet needs as much thiamin as when eating enough to meet energy needs.

Thiamin in Foods Thiamin occurs in small quantities in virtually all nutritious foods, but it is concentrated in only a few foods, of which pork and ham are the most commonly eaten. A useful guideline for meeting thiamin needs is to keep empty-kcalorie foods to a minimum in the diet and to include ten or more different servings of nutritious foods each day, assuming that each serving will contribute, on the average, about 10 percent of needs. Foods chosen from the bread and cereal group should be either whole grain or enriched. Thiamin is not stored in the body to any great extent, so daily intake is important.

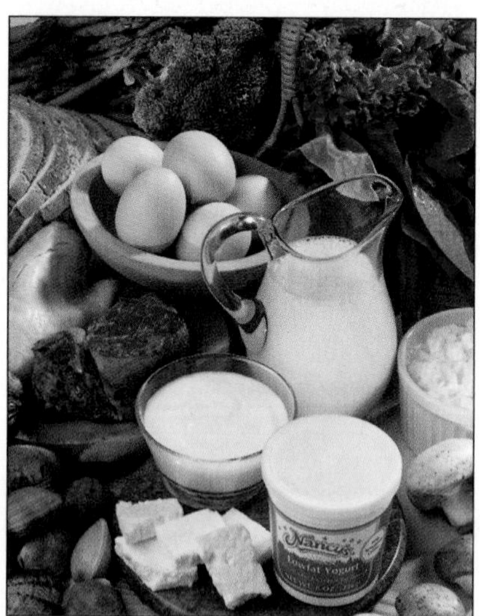

Milk and milk products supply much (about 50%) of the riboflavin in people's diets, but meats, eggs, green vegetables, and whole-wheat bread are good sources, too.

Riboflavin

Like thiamin, riboflavin facilitates energy production in the body. Riboflavin recommendations are stated in terms of milligrams per 1000 kcalories. Differences in the riboflavin RDA for different age-sex groups primarily reflect differences in energy intakes. The needs of infants, children, and pregnant women rise rapidly during periods of active growth.

Riboflavin in Foods Unlike thiamin, riboflavin is not evenly distributed among the food groups. The major contributors of riboflavin to people's diets are milk, milk products, meats, and green vegetables (broccoli, turnip greens, asparagus, and spinach). The riboflavin richness of milk and milk products is a good reason to include these foods in every day's meals. No other commonly eaten food can make such a substantial contribution. People who omit milk and milk products from their diets can substitute generous servings of dark green, leafy vegetables. Among the meats, liver and heart are the richest sources, but all lean meats, as well as eggs, offer some riboflavin.

Effects of Light Riboflavin is light sensitive; it can be destroyed by the ultraviolet rays of the sun or of fluorescent lamps. For this reason, milk is sold in cardboard or opaque plastic containers to protect the riboflavin in the milk from light. In contrast, riboflavin is heat stable, so ordinary cooking does not destroy it.

Niacin

Like thiamin and riboflavin, niacin participates in the energy metabolism of every body cell. Niacin is unique among the B vitamins in that the body can make it from protein. The amino acid tryptophan can be converted to niacin in the body: 60 milligrams of tryptophan yield 1 milligram of niacin. Recommended intakes are therefore stated in "niacin equivalents," reflecting the body's ability to convert tryptophan to niacin.

A food containing 1 mg of niacin and 60 mg tryptophan contains the niacin equivalent of 2 mg, or 2 mg NE.

niacin equivalents: the amount of niacin present in food, including the niacin that can theoretically be made from its precursor tryptophan present in the food.

Niacin Used as a Medication Certain forms of niacin supplements in amounts ten times or more the RDA cause "niacin flush," a dilation of the capillaries of the skin with perceptible tingling that, if intense, can be painful. Physicians sometimes use diet and large doses of niacin to lower blood cholesterol, to treat atherosclerosis. When used this way, niacin leaves the realm of nutrition to become a pharmacological agent, a drug. As with any medication, self-dosing with niacin is ill-advised; large doses may injure the liver, cause ulcers, and produce some symptoms of diabetes.[19]

When a normal dose of a nutrient clears up a deficiency condition, the effect is a physiological one. When a megadose (100 times larger) overwhelms some system and acts like a drug, the effect is a pharmacological one.

Niacin in Foods Meat, poultry, and fish contribute about half the niacin equivalents most people consume; enriched breads and cereals contribute about a fourth. Among the vegetables, mushrooms, asparagus, and green leafy vegetables are the richest niacin sources. Niacin is less vulnerable to losses during food preparation and storage than other water-soluble vitamins. Being fairly heat-resistant, niacin can withstand reasonable cooking times, but like other water-soluble vitamins, it will leach into cooking water.

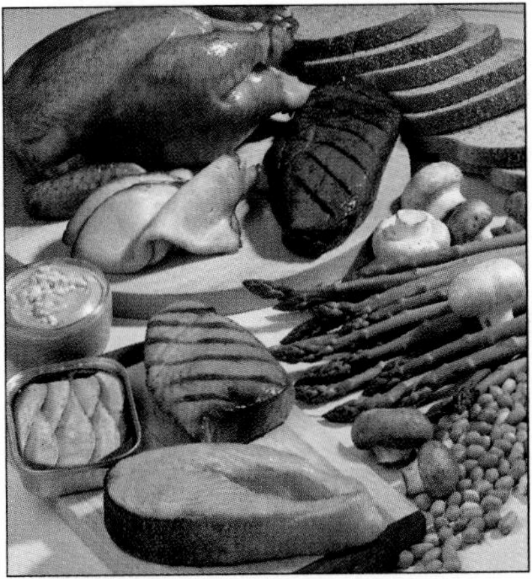

Niacin-rich foods include meat, fish, poultry, and peanut butter, as well as enriched breads and cereals and a few vegetables.

Pantothenic Acid and Biotin

Two other B vitamins—pantothenic acid and biotin—are also important in energy metabolism. Pantothenic acid was first recognized as a substance that stimulates growth. It is a component of a key enzyme that makes possible the release of energy from the energy nutrients. Pantothenic acid is involved in more than 100 different steps in the synthesis of lipids, neurotransmitters, steroid hormones, and hemoglobin. Biotin plays an important role in metabolism as a coenzyme that serves as a carrier of carbon dioxide. This role is critical in the TCA cycle.

Pantothenic Acid and Biotin in Foods Both pantothenic acid and biotin are more widespread in foods than the other vitamins discussed so far. There seems to be no danger that people who consume a variety of foods will suffer deficiencies. Claims that pantothenic acid and biotin are needed in pill form to prevent or cure disease conditions are at best unfounded and at worst intentionally misleading.

Biotin Deficiency in the Hospital Biotin deficiencies are rare, but have been reported in adults fed artificially by vein without biotin supplementation. Researchers can induce biotin deficiency in animals or human beings by feeding them raw egg whites, which contain a protein that binds biotin and prevents its absorption.

The protein **avidin** in egg whites binds biotin.

Vitamin B_6

Vitamin B_6 has been called the "sleeping giant" of vitamins. A surge of research interest in the last decade has not only revealed new knowledge about it but has

Meat, chicken, and fish, as well as some fruits and vegetables, are good sources of vitamin B_6.

also raised new questions. For example, unlike other water-soluble vitamins, vitamin B_6 is stored extensively in muscle tissue.

Metabolic Roles of Vitamin B_6 Vitamin B_6 has long been known to play roles in protein and amino acid metabolism. In the cells, vitamin B_6 helps to convert one kind of amino acid, which the cells have in abundance, to another, which they need in larger amounts. It also aids in the conversion of tryptophan to niacin and plays important roles in the synthesis of hemoglobin. Vitamin B_6 also assists in releasing stored glucose from glycogen and thus contributes to the regulation of blood glucose. Research suggests new roles for vitamin B_6 in immune function and hormone response.[20] The association between vitamin B_6 and immune function is related to the critical role the vitamin plays in protein metabolism. Vitamin B_6 deficiency can significantly impair the immune response, perhaps by way of impaired antibody production.[21]

A possible role for vitamin B_6 in the origins and treatments of some diseases is also being investigated. For example, elevated blood levels of the amino acid homocysteine may be a risk factor for heart disease.[22] Evidence suggests that low blood concentrations of vitamin B_6 and folate are associated with elevated homocysteine concentrations.[23] Researchers conducted a study of more than 80,000 women to determine the relationship between vitamin B_6 and folate intake and the risk of heart disease.[24] Women with the highest intakes of vitamin B_6 from food and supplements combined had lower risks of heart disease than women with low intakes. This was true of folate as well, and the risk was lower still in women with the highest intakes of both nutrients.

Vitamin B_6 Deficiency Besides a weakening immune response, vitamin B_6 deficiency is expressed in general symptoms, such as weakness, irritability, and insomnia. Other symptoms include a greasy, flaky dermatitis, anemia, and, in advanced cases, convulsions.

Vitamin B_6 Toxicity For years it was believed that vitamin B_6, like other water-soluble vitamins, could not reach toxic concentrations in the body. Toxic effects of vitamin B_6 became known when a physician reported them in women who had been taking more than 2 grams of vitamin B_6 daily for two months or more. Most of these women had been attempting to cure symptoms of premenstrual syndrome (PMS), the cluster of physical, emotional, and psychological symptoms that some women experience prior to menstruation. The first symptom of toxicity was numb feet; then the women lost sensation in their hands; then they became unable to walk. The women recovered after they discontinued the supplements.

The specific cause or causes of PMS remain undefined, although researchers agree that the hormonal changes of the menstrual cycle must be responsible. Despite a lack of conclusive evidence that vitamin B_6 is an effective treatment for PMS, the vitamin and many other unproven remedies remain popular among PMS sufferers.

Vitamin B_6 and Protein Intake Vitamin B_6's many roles in amino acid metabolism are reflected in dietary needs that are roughly proportional to protein intakes. The RDA for vitamin B_6 is more than adequate to handle average protein intakes of 100 grams per day for men and 60 grams per day for women.

Vitamin B_6 in Foods The richest food sources of vitamin B_6 are protein-rich meat, fish, and poultry. Potatoes, a few other vegetables, and some fruits are good sources, too. Foods lose vitamin B_6 when heated.

Folate

The B vitamin folate is active in cell division. During periods of rapid growth and cell division, such as pregnancy and adolescence, folate needs increase, and deficiency is especially likely. When a deficiency occurs, the rapidly dividing cells of the blood and the GI tract stop dividing and begin to deteriorate. Not surprisingly, then, two of the first symptoms of a folate deficiency are a type of anemia and GI tract deterioration (see Table 7.4).

Folate, Alcohol, and Drugs Of all the vitamins, folate appears to be most vulnerable to interactions with alcohol and other drugs. As discussed in Nutrition in Practice 8, alcohol-addicted people are at risk of folate deficiency because alcohol impairs folate's absorption and increases its excretion. Furthermore, as people's alcohol intakes rise, their folate intakes decline. Many medications, including aspirin, oral contraceptives, and anticonvulsants, impair folate status. Smoking also exerts a negative effect on folate status.

Folate and Neural Tube Defects Research studies confirm the importance of folate in preventing neural tube defects.[25] The brain and spinal cord develop from the neural tube, and defects in its orderly formation during the early weeks of pregnancy may result in various central nervous system disorders and death. Folate supplements taken one month before conception and continued throughout the first trimester of pregnancy can prevent neural tube defects.[26] For this reason, the Public Health Services recommends that all women of childbearing age who are capable of becoming pregnant take 0.4 milligram (400 micrograms) of folate daily.[27]

neural tube defects: malformations of the brain, spinal cord, or both during embryonic development.

The two main types of neural tube defects are **spina bifida** (literally "split spine") and **anencephaly** ("no brain").

Although this amount can be met through diet by eating the recommended five servings of fruits and vegetables daily, most women's dietary folate intakes fall short. Furthermore, some research indicates that folate status improves more with supplementation or fortification than with a dietary intake that meets recommendations.[28]

Because most pregnancies are unplanned and neural tube defects arise early in pregnancy before most women realize they're pregnant, the FDA has mandated that refined grain products (flour, cornmeal, pasta, and rice) be fortified to increase average folate intakes. Fortification is expected to prevent half of the 4000 neural tube defects that occur each year. Folate fortification, however, raises safety concerns as well. High doses of folate can complicate the diagnosis of vitamin B_{12} deficiency, as discussed below. Thus folate intakes should not exceed 1 milligram per day.[29]

Bread products, flour, corn grits, and pasta must be fortified with 1.4 mg per 100 g of food (about ½ c cooked food or 1 slice of bread.

March of Dimes

Folate in Foods The best food sources (see Snapshot 7.3 on p. 156) of folate are liver, legumes, green leafy vegetables (the name of the vitamin is related to the word *foliage*), and beets. Among the fruits, oranges, orange juice, and cantaloupe are the best sources. With fortification, grain products are good sources of folate, too. The bioavailability of added folate is good, and fortification is expected to increase women's average folate intakes by 100 micrograms per day.[30] Heat and oxidation during cooking and storage can destroy up to half of the folate in foods.

Vitamin B_{12}

Vitamin B_{12} and folate share a special relationship: vitamin B_{12} assists folate in its role in cell division. Their roles intertwine, but each performs a specific task that the other cannot perform.

Vitamin B_{12}, Folate, and Cell Division Vitamin B_{12} (in coenzyme form) stands by to accept carbon groups from folate as folate removes them from other compounds. The passing of these carbon groups from folate to vitamin B_{12} regenerates the active form of folate so that it can continue its dismantling tasks. In the absence of vitamin B_{12}, folate is trapped in its inactive, metabolically useless form, unable to do its job. When folate is either trapped due to a vitamin B_{12}

Snapshot 7.3
Folate

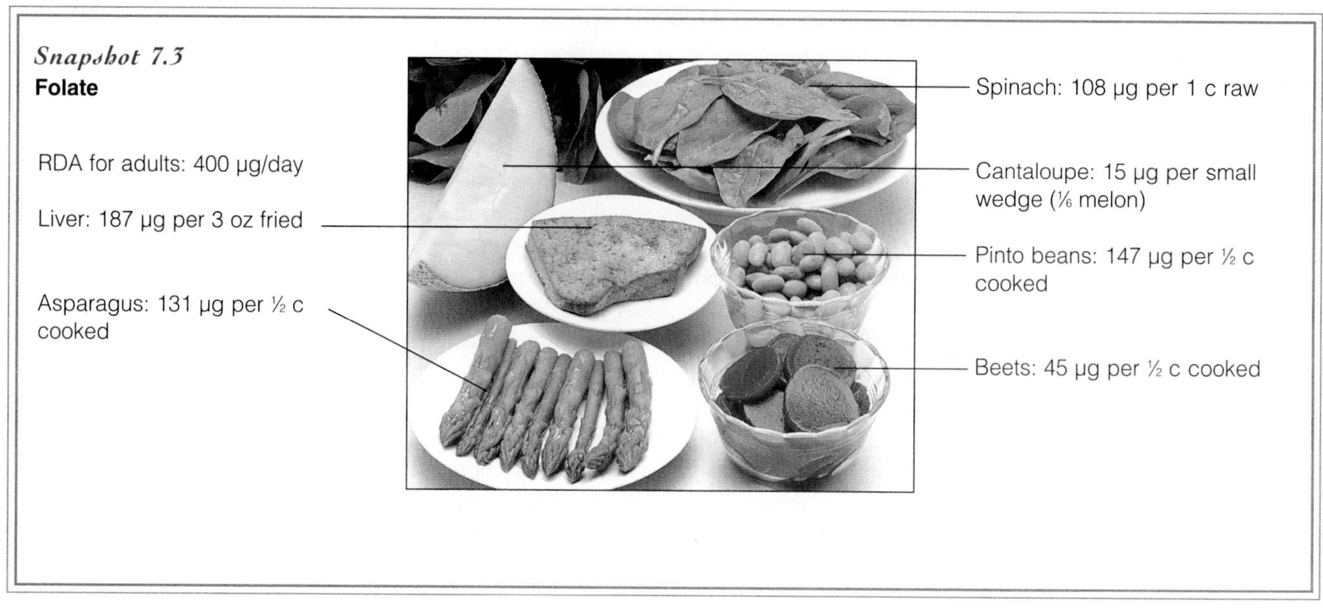

deficiency or unavailable due to a deficiency of folate itself, cells that are growing most rapidly, notably, the blood cells, are the first to be affected. Thus a deficiency of either nutrient—vitamin B_{12} or folate—impairs maturation of the blood cells and produces anemia. The anemia is identifiable by microscopic examination of the blood, which reveals many large, immature red blood cells. Either vitamin B_{12} or folate will clear up the anemia.

The large-cell anemia of folate deficiency is known as **macrocytic** or **megaloblastic** anemia.
macro = large
cyte = cell
mega = large

Vitamin B_{12} and the Nervous System Although either vitamin will clear up the anemia caused by vitamin B_{12} deficiency, if folate is given when vitamin B_{12} is needed, the result is disastrous, not to the blood but to the nervous system. The reason: vitamin B_{12} also helps maintain nerve fibers. Devastating neurological symptoms, undetectable by a blood test, can ultimately result from an undetected vitamin B_{12} deficiency. A deceptive folate "cure" of the blood symptoms in vitamin B_{12} deficiency allows the nerve symptoms to progress, leading to paralysis and permanent nerve damage. This interaction between folate and vitamin B_{12} raises safety concerns about the use of folate supplements and fortification of foods.

The way folate masks vitamin B_{12} deficiency underlines a point already made several times: it takes a skilled diagnostician to make a correct diagnosis. The risk you take when you diagnose yourself on the basis of a single observed symptom is clearly intolerable.

intrinsic: inside the system. The **intrinsic factor** necessary to prevent pernicious anemia is now known to be made in the stomach and to aid in the absorption of vitamin B_{12}. The extrinsic factor necessary to prevent pernicious anemia is vitamin B_{12} itself, which must be obtained outside the system from food.

Vitamin B_{12} Absorption Vitamin B_{12} requires an "intrinsic factor"—a compound made inside the body—for absorption from the intestinal tract into the bloodstream. The intrinsic factor is made in the stomach, where it attaches to the vitamin; the complex then passes to the small intestine and is gradually absorbed.

Loss of Intrinsic Factor In some cases, intrinsic factor production becomes inadequate or ceases altogether—for example, after surgical removal of the stomach. Some people inherit a defective gene for intrinsic factor. Because vitamin B_{12} deficiency in the body may be caused either by a lack of the vitamin in the diet or by the body's inability to absorb the vitamin, a change in diet alone may not correct the deficiency. When absorption failure is the problem, vitamin B_{12} must be supplied by injection to prevent vitamin B_{12}–deficiency symptoms from developing. The vitamin B_{12} deficiency caused by lack of intrinsic factor is known as pernicious anemia.

Vitamin B_{12} in Foods A unique characteristic of vitamin B_{12} is that it is found almost exclusively in foods derived from animals. People who eat meat are guaranteed an adequate intake, and lacto-ovo vegetarians (who use milk, cheese, and eggs) are also protected from deficiency. It is a myth, however, that fermented soy products, such as miso (a soybean paste), or sea algae, such as spirulina, provide vitamin B_{12} in its active form. Extensive research shows that the amounts of vitamin B_{12} listed on the labels of these plant products are inaccurate and misleading because the vitamin B_{12} in these products occurs in an inactive, unavailable form. Strict vegetarians must take vitamin B_{12} supplements or find other sources of active vitamin B_{12}. Some loss of vitamin B_{12} occurs when foods are heated in microwave ovens.[31]

Vitamin B_{12} is found exclusively in foods derived from animals.

Vitamin B_{12} Deficiency in Vegans Strict vegetarians are at special risk for undetected vitamin B_{12} deficiency for two reasons: first, they receive none in their diets, and second, they consume large amounts of folate in the vegetables they eat. Because the body can store 1000 times the amount of vitamin B_{12} used each day, a deficiency may take years to develop in a new vegetarian. When a deficiency does develop, though, it may progress to a dangerous extreme because the deficiency of vitamin B_{12} may be masked by the high folate intake.

Non-B Vitamins

Other compounds are sometimes inappropriately called B vitamins because, like the true B vitamins, they serve as coenzymes in metabolism. Even if they were essential, however, supplements would be unnecessary because these compounds are abundant in foods.

Inositol, Choline, and Carnitine Among the non-B vitamins are a trio of coenzymes known as inositol, choline, and carnitine. Researchers are exploring the possibility that these substances may be essential. As this text goes to press, an Adequate Intake (AI) has been established for choline.

Other Non-B Vitamins Other substances have also been mistaken for essential nutrients. They include para-aminobenzoic acid (PABA), bioflavonoids (vitamin P or hesperidin), and ubiquinone. Other names you may hear are "vitamin B_{15}" (a hoax) and "vitamin B_{17}" (laetrile, a fake cancer-curing drug and not a vitamin by any stretch of the imagination). There is, however, one other water-soluble vitamin of great interest and importance—vitamin C.

Vitamin C

Two hundred years ago, any man who joined the crew of a seagoing ship knew he had only half a chance of returning alive—not because he might be slain by pirates or die in a storm, but because he might contract the dread disease scurvy. Then a physician with the British navy found that citrus fruits could cure the disease, and thereafter, all ships were required to carry lime juice for every sailor. (This is why British sailors are still called "limeys" today.) Nearly 200 years later, the antiscurvy factor in citrus fruits was isolated from lemon juice and named ascorbic acid. Today, hundreds of millions of vitamin C pills are produced in pharmaceutical laboratories.

scurvy: the vitamin C–deficiency disease.

ascorbic acid: one of the two active forms of vitamin C. Many people refer to vitamin C by this name.
a = without
scorbic = having scurvy

Metabolic Roles of Vitamin C Vitamin C's action defies a simple, tidy description. It plays many important roles in the body, and its modes of action differ in different situations.

Collagen Formation The best-understood action of vitamin C is its role in helping to form collagen, the single most important protein of connective tissue.

collagen: the characteristic protein of connective tissue.
kolla = glue
gennan = produce

Collagen serves as the matrix on which bone is formed, the material of scars, and an important part of the "glue" that attaches one cell to another. This latter function is especially important in the artery walls, which must expand and contract with each beat of the heart, and in the walls of the capillaries, which are thin and fragile.

Antioxidant Vitamin C is also an important antioxidant. Recall that the antioxidants beta-carotene and vitamin E protect fat-soluble substances from oxidizing agents; vitamin C protects water-soluble substances the same way. Vitamin C's antioxidant action is twofold. First, by being oxidized itself, vitamin C regenerates already-oxidized substances such as iron and copper to their original, active form. Second, in the process, the vitamin removes the damaging oxidizing agent. In the intestines, it protects iron from oxidation and so enhances iron absorption. In the cells and body fluids, it helps to protect other molecules, including the fat-soluble compounds vitamin A, vitamin E, and the polyunsaturated fatty acids.

Amino Acid Metabolism Vitamin C is also involved in the metabolism of several amino acids. Some of these amino acids may end up being converted to hormones of great importance in body functioning, among them norepinephrine and thyroxine.

Role of Stress During stress, the adrenal glands release large quantities of vitamin C together with the stress hormones epinephrine and norepinephrine. What the vitamin has to do with the stress reaction is unclear, but it is known that stress increases vitamin C needs somewhat.

Possible Antihistamine Newspaper headlines touting vitamin C as a cure for colds and cancer have appeared frequently over the years. Some research suggests that vitamin C (2 grams per day for two weeks) may reduce the severity and duration of cold and allergy symptoms by reducing blood histamine concentrations.[32] In other words, vitamin C acts as an antihistamine. If further research confirms vitamin C's antihistamine effect, its use may permit people to rely less heavily on antihistamine drugs when suffering from cold and allergy symptoms.

Role in Cancer Prevention and Treatment The role of vitamin C in the prevention of, or therapy for, cancer is still being studied. A large-scale population study of over 11,000 U.S. adults reported an inverse relationship between all causes of death and vitamin C intake up to a few hundred milligrams.[33] The relationship was stronger for men than for women. In a dozen or so different well-controlled studies, researchers identified individuals with and without cancer and assessed their dietary intakes of vitamin C. They found that people with high vitamin C intakes had lower risks of these cancers than did people with low intakes. The correlation may reflect not just an association with vitamin C, but the broader benefits of a diet rich in fruits and vegetables and low in fat. It does not support the taking of vitamin C supplements to treat or prevent cancer.

latent: the period in the course of a disease when the conditions are present but the symptoms have not begun to appear.
latens = lying hidden

overt: out in the open, full-blown.
ouvrire = to open

Vitamin C Deficiency When intake of vitamin C is inadequate, the body's vitamin C pool dwindles in size, and latent scurvy appears. The blood vessels show the first deficiency signs. The gums around the teeth begin to bleed easily, and capillaries under the skin break spontaneously, producing pinpoint hemorrhages. Then the symptoms of overt scurvy appear. Muscles, including the heart muscle, may degenerate. The skin becomes rough, brown, scaly, and dry. Wounds fail to heal because scar tissue will not form. Bone rebuilding falters; the ends of the long bones become softened, malformed, and painful; and fractures occur. The teeth may become loose in the jawbone, and fillings may loosen and fall out. Anemia and infections are common. Sudden death is likely, perhaps because of massive bleeding into the joints and body cavities.

It takes only 10 or so milligrams of vitamin C a day to prevent overt scurvy, and not much more than that to cure it. Once diagnosed, scurvy is readily reversible with moderate doses, in the neighborhood of 100 milligrams per day. Such an intake is easily achieved by including vitamin C–rich foods in the diet.

Vitamin C Toxicity The easy availability of vitamin C in pill form and the publication of books recommending vitamin C to prevent colds and cancer have led thousands of people to take megadoses of vitamin C. Not surprisingly, instances of vitamin C causing harm have surfaced.

Doses of 10 to 30 or more times the recommended intake of a nutrient are termed megadoses. In the case of vitamin C, any amount over 1 g (1000 mg) is considered a megadose.

Some of the suspected toxic effects of vitamin C megadoses have not been confirmed, but others have been seen often enough to warrant concern. Nausea, abdominal cramps, and diarrhea are often reported. Several instances of interference with medical regimens are known. The large amounts of vitamin C excreted in the urine obscure the results of tests used to detect diabetes. People taking anticoagulants may unwittingly counteract the effect of these medicines if they also take massive doses of vitamin C. Vitamin C megadoses can enhance iron absorption too much, resulting in iron overload (see Chapter 8).

The anticoagulants with which vitamin C interferes are warfarin and dicumarol.

People with sickle-cell anemia may be especially vulnerable to megadoses of vitamin C. Those who have a tendency toward gout, as well as those who have a genetic abnormality that alters the way they metabolize vitamin C, are more prone to forming stones if they take megadoses of vitamin C.

gout (GOWT): a metabolic disease in which crystals of uric acid precipitate in the joints.

Withdrawal Reaction A person who has taken large doses of vitamin C for a long time may adapt by limiting absorption and destroying and excreting more of the vitamin than usual. If the person then suddenly reduces intake to normal, the accelerated disposal system can't put on its brakes fast enough to avoid destroying too much of the vitamin. Some case histories have shown that adults who discontinue megadosing develop scurvy on intakes that would protect a normal adult.

The temporary condition manifested by withdrawal symptoms and experienced by the person who stops overdosing is vitamin C dependency. The body has adjusted to a high intake and so "needs" a high intake until it can readjust.

Recommended Intakes of Vitamin C The RDA for vitamin C is 60 milligrams for adults, with an extra 10 milligrams recommended for pregnant women and an additional 35 milligrams for lactating women. This amount is midway between two extremes. At one extreme is the requirement to prevent overt scurvy, or 10 milligrams per day; at the other extreme is the amount at which the body's pool of vitamin C would be full to overflowing: about 100 milligrams per day.

Special Needs for Vitamin C As is true of all nutrients, unusual circumstances may raise vitamin C needs. Among the stresses known to do so are infections; burns; surgery; extremely high or low temperatures; toxic doses of heavy metals, such as lead, mercury, and cadmium; and the chronic use of certain medications, including aspirin, barbiturates, and oral contraceptives. Smoking, too, has adverse effects on vitamin C status. Cigarette smoke contains oxidants, which deplete this potent antioxidant.[34] Accordingly, the vitamin C recommendation for people who smoke is 100 milligrams per day.

Safe Limits Few instances warrant the taking of more than 100 to 300 milligrams a day. Adults may not be exposing themselves to severe risks if they choose to dose themselves with 1 to 2 grams a day, but those taking more than 2 grams, and especially those taking above 8 grams per day, should be aware of the distinct possibility of harm.

Vitamin C in Foods The inclusion of intelligently selected fruits and vegetables in the daily diet guarantees a generous intake of vitamin C. Even those who wish to ingest amounts well above the recommended 60 milligrams can easily meet their goals by eating certain foods (see Snapshot 7.4 on p. 160). Citrus fruits are rightly famous for their vitamin C contents. Some other fruits and certain

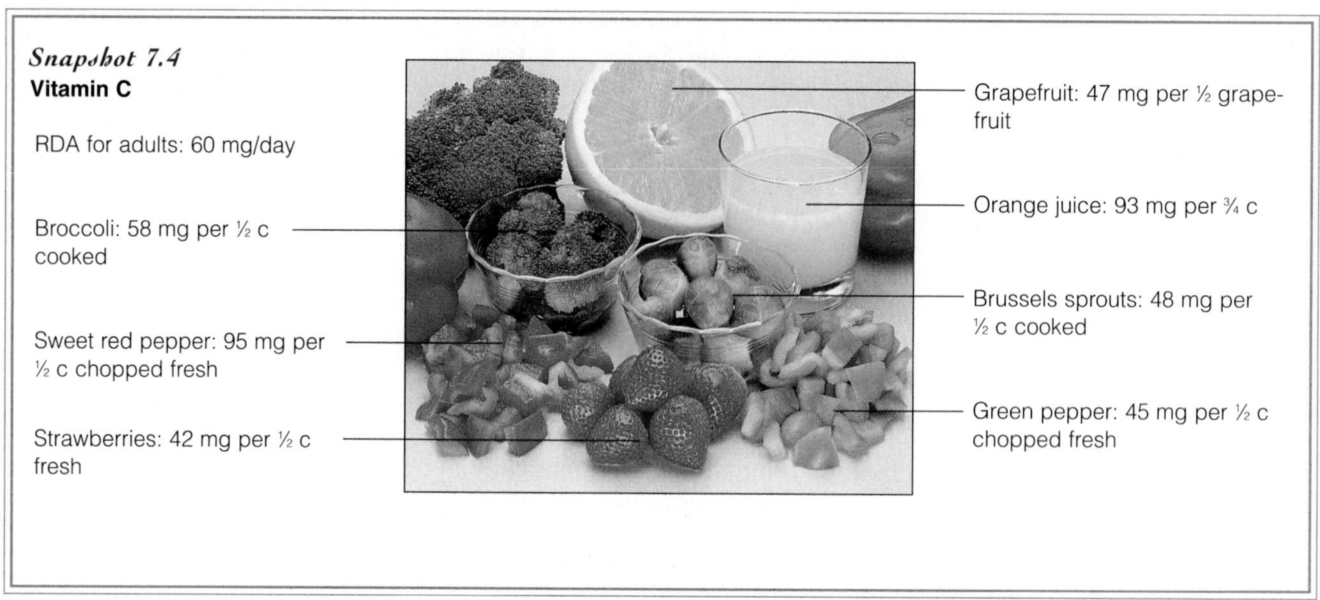

vegetables are also rich sources: cantaloupe, strawberries, broccoli, and brussels sprouts. No animal foods other than organ meats, such as liver and kidneys, contain vitamin C. The humble potato is an important source of vitamin C in Western countries, where potatoes are eaten so frequently that they make substantial vitamin C contributions overall. They provide about 20 percent of all the vitamin C in the diet. Vitamin C in foods is easily oxidized, so store cut produce in airtight wrappers and juices in closed containers.

Vitamin C and Iron Absorption Eating foods containing vitamin C at the same meal with foods containing iron can double or triple the absorption of iron from those foods. For women and children, whose energy intakes are not large enough to guarantee that they will get enough iron from the foods they eat, this strategy is highly recommended. Table 7.4 (pp. 161–162) summarizes functions, deficiency and toxicity symptoms, and food sources of vitamin C and the other water-soluble vitamins.

In conclusion, a summary of the key points about vitamins—both fat soluble and water soluble—seems in order:

- Vitamins are essential, noncaloric, organic nutrients needed in tiny amounts in the diet.
- Vitamins A, D, E, and K are the fat-soluble vitamins. The B vitamins and vitamin C are the water-soluble vitamins.
- The only disease a vitamin will cure is the one caused by a deficiency of that vitamin.
- Vitamins yield no energy for the body when broken down, but they facilitate the release of energy from carbohydrate, fat, and protein.
- Nutritious foods, not supplements, are the best sources of vitamins.

Table 7.4 The Water-Soluble Vitamins—A Summary

Thiamin

Other Names	Deficiency Symptoms	Toxicity Symptoms
Vitamin B_1	**Blood/Circulatory System**	
	Edema, enlarged heart, abnormal heart rhythms, heart failure	(No symptoms reported)
Chief Functions in the Body	**Nervous/Muscular Systems**	
Part of a coenzyme used in energy metabolism, supports normal appetite and nervous system function	Degeneration, wasting, weakness, pain, low morale, difficulty walking, loss of reflexes, mental confusion, paralysis	(No symptoms reported)
Deficiency Disease Name		
Beriberi		
Significant Sources		
Occurs in all nutritious foods in moderate amounts; pork, ham, bacon, liver, whole grains, legumes, nuts		

Riboflavin

Other Names	Deficiency Symptoms	Toxicity Symptoms
Vitamin B_2	**Mouth, Gums, Tongue**	
	Cracks at corners of mouth,[a] magenta tongue	(No symptoms reported)
Chief Functions in the Body	**Nervous System and Eyes**	
Part of a coenzyme used in energy metabolism, supports normal vision and skin health	Hypersensitivity to light,[b] reddening of cornea	(No symptoms reported)
	Other	
Deficiency Disease Name	Skin rash	(No symptoms reported)
Ariboflavinosis		
Significant Sources		
Milk, yogurt, cottage cheese, meat, leafy green vegetables, whole-grain or enriched breads and cereals		

Niacin

Other Names	Deficiency Symptoms	Toxicity Symptoms
Nicotinic acid, nicotinamide, niacinamide, vitamin B_3; precursor is dietary tryptophan	**Digestive System**	
	Diarrhea	Diarrhea, heartburn, nausea, ulcer irritation, vomiting
Chief Functions in the Body	**Mouth, Gums, Tongue**	
Part of a coenzyme used in energy metabolism; supports health of skin, nervous system, and digestive system	Black, smooth tongue[c]	(No symptoms reported)

(continued)

[a]Cracks at the corners of the mouth are termed *cheilosis* (kee-LOH-sis).
[b]Hypersensitivity to light is *photophobia*.
[c]Smoothness of the tongue is caused by loss of its surface structures and is termed *glossitis* (gloss-EYE-tis).

Table 7.4 *(continued)*

Niacin *(continued)*

	Deficiency Symptoms	Toxicity Symptoms
Deficiency Disease Name	*Nervous System*	
Pellagra	Irritability, loss of appetite, weakness, dizziness, mental confusion progressing to psychosis or delirium	Fainting, dizziness
Significant Sources	*Skin*	
Milk, eggs, meat, poultry, fish, whole-grain and enriched breads and cereals, nuts, and all protein-containing food	Flaky skin rash on areas exposed to sun	Painful flush and rash ("niacin flush"), sweating
	Other	
		Abnormal liver function, low blood pressure

Pantothenic Acid

Other Names	Deficiency Symptoms	Toxicity Symptoms
(None)	*Digestive System*	
	Vomiting, intestinal distress	(No symptoms reported)
Chief Functions in the Body	*Nervous System*	
Part of a coenzyme used in energy metabolism	Insomnia, fatigue	(No symptoms reported)
	Other	
		Water retention (infrequent)
Deficiency Disease Name		
(No name)		
Significant Sources		
Widespread in foods		

Biotin

Other Names	Deficiency Symptoms	Toxicity Symptoms
(None)	*Blood/Circulatory System*	
	Abnormal heart action	(No symptoms reported)
Chief Functions in the Body	*Digestive System*	
	Loss of appetite, nausea	(No symptoms reported)
Part of a coenzyme used in energy metabolism, fat synthesis, amino acid metabolism, and glycogen synthesis	*Nervous/Muscular Systems*	
	Depression, muscle pain, weakness, fatigue	(No symptoms reported)
Deficiency Disease	*Skin*	
(No name)	Drying, rash, loss of hair	(No symptoms reported)
Significant Sources		
Widespread in foods		

Table 7.4 *(continued)*

Vitamin B_6

Other Names	Deficiency Symptoms	Toxicity Symptoms
Pyridoxine, pyridoxal, pyridoxamine	Blood/Circulatory System	
	Anemia (small-cell type)[d]	Bloating
Chief Functions in the Body	Mouth, Gums, Tongue	
Part of a coenzyme used in amino acid and fatty acid metabolism, helps convert tryptophan to niacin, helps make red blood cells	Smooth tongue[c]	(No symptoms reported)
	Nervous/Muscular Systems	
	Abnormal brain wave pattern, irritability, muscle twitching, convulsions	Depression, fatigue, impaired memory, irritability, headaches, numbness, damage to nerves, difficulty walking, loss of reflexes, weakness, restlessness
Deficiency Disease Name		
(No name)	Skin	
	Irritation of sweat glands, rashes, greasy dermatitis	(No symptoms reported)
Significant Sources	Other	
Green and leafy vegetables, meats, fish, poultry, shellfish, legumes, fruits, whole grains		(No symptoms reported)

Folate

Other Names	Deficiency Symptoms	Toxicity Symptoms
Folic acid, folacin, pteroylglutamic acid	Blood/Circulatory System	
	Anemia (large-cell type)[d]	(No symptoms reported)
Chief Functions in the Body	Digestive System	
Part of a coenzyme used in new cell synthesis	Heartburn, diarrhea, constipation	(No symptoms reported)
	Immune System	
	Suppression, frequent infections	(No symptoms reported)
Deficiency Disease	Mouth, Gums, Tongue	
(No name)	Smooth red tongue[c]	(No symptoms reported)
Significant Sources	Nervous System	
Leafy green vegetables, legumes, seeds, liver	Depression, mental confusion, fainting	(No symptoms reported)
	Other	
		Masks vitamin B_{12} deficiency

Vitaming B_{12}

Other Names	Deficiency Symptoms	Toxicity Symptoms
Cyanocobalamin	Blood/Circulatory System	
	Anemia (large-cell type)[d]	(No symptoms reported)
Chief Functions in the Body	Mouth, Gums, Tongue	
Part of a coenzyme used in new cell synthesis, helps maintain nerve cells	Smooth tongue[c]	(No symptoms reported)

(continued)

[c]Smoothness of the tongue is caused by loss of its surface structures and is termed *glossitis* (gloss-EYE-tis).
[d]Small-cell anemia is termed *microcytic anemia;* large-cell type is *macrocytic* or *megaloblastic anemia.*

Table 7.4 *(continued)*

Vitamin B_{12} *(continued)*

	Deficiency Symptoms	Toxicity Symptoms
Deficiency Disease		
(No name[e])		
Significant Sources		
Animal products (meat, fish, poultry, milk, cheese, eggs)		
	Nervous/Muscular Systems	
	Fatigue, degeneration progressing to paralysis	(No symptoms reported)
	Skin	
	Hypersensitivity	(No symptoms reported)

Vitamin C

	Deficiency Symptoms	Toxicity Symptoms
Other Names		
Ascorbic acid		
Chief Functions in the Body		
Helps in collagen synthesis (strengthens blood vessel walls, forms scar tissue, provides matrix for bone growth), thyroxine synthesis, and amino acid metabolism; serves as an antioxidant; strengthens resistance to infection; helps in absorption of iron		
Deficiency Disease Name		
Scurvy		
Significant Sources		
Citrus fruits, cabbage-type vegetables, dark green vegetables, cantaloupe, strawberries, peppers, lettuce, tomatoes, potatoes, papayas, mangoes		
	Blood/Circulatory System	
	Anemia (small-cell type),[d] atherosclerotic plaques, pinpoint hemorrhages	(No symptoms reported)
	Digestive System	
		Nausea, abdominal cramps, diarrhea, excessive urination
	Immune System	
	Suppression, frequent infections	(No symptoms reported)
	Mouth, Gums, Tongue	
	Bleeding gums, loosened teeth	(No symptoms reported)
	Muscular/Nervous Systems	
	Muscle degeneration and pain, hysteria, depression	Headache, fatigue, insomnia
	Skeletal System	
	Bone fragility, joint pain	(No symptoms reported)
	Skin	
	Rough skin, blotchy bruises	Rashes
	Other	
	Failure of wounds to heal	Interference with medical tests, aggravation of gout symptoms, deficiency symptoms may appear at first on withdrawal of high doses.

[d]Small-cell anemia is termed *microcytic anemia;* large-cell type is *macrocytic* or *megaloblastic anemia.*
[e]The name *pernicious anemia* refers to the vitamin B_{12} deficiency caused by lack of intrinsic factor, but not to that caused by inadequate dietary intake.

Self Study

How Are Your Vitamin Intakes?

Compare your average daily intake of vitamins with the RDA (inside front cover) or the Recommended Nutrient Intakes for Canadians (Appendix B) or refer to the intake values calculated in your Diet Analysis Plus profile. Express each intake as a percentage of the recommended intake. For example, suppose you ingested 0.9 milligram thiamin, and your RDA is 1.1 milligrams. You ingested $(0.9 \div 1.1) \times 100$, or 82 percent of your RDA. If you had ingested 1.4 milligrams of thiamin, you would have ingested $(1.4 \div 1.1) \times 100$, or 127 percent of your RDA. Use Form 8 to record your findings.

Comment on your intakes. Look closely at any vitamins for which your intakes fell below 80 percent of the recommendations. What are your best food sources of those vitamins? Could you eat more of these foods to bring your intake up to the recommended level? If not, what food or foods could you eat to improve your intake?

Health care professionals often see clients whose poor nutrition hampers their recovery; those who are alert to this possibility can often provide significant help and advice. Describe the differences between vitamin deficiency diseases and diseases that increase the likelihood of vitamin deficiency. Discuss ways a vitamin deficiency might weaken the body's resistance to disease.

Pull together information from Chapter 1 about the vitamins in different food groups and the significant sources of vitamins shown in the snapshots throughout this chapter. Consider which vitamins might be lacking in the diet of a client who reports the following:

- Dislikes green leafy vegetables.
- Never uses milk, milk products, or cheese.
- Follows a very-low-fat diet.
- Eats a fruit or vegetable once a day.

What additional information would help you pinpoint a problem with vitamin intake?

Form 8 Vitamin Intakes Compared with Recommended Intakes

	Vitamin A	Vitamin C	Thiamin	Riboflavin	Niacin	Folate	Vitamin B_6
My intake							
Recommended intake[a]							
My intake as a percentage of the recommended intake							

[a]RDA or RNI (Appendix B).

Self Check

1. Which of the following vitamins are fat soluble?
   a. vitamins B, C, and E
   b. vitamins B, C, D, and E
   c. vitamins A, C, E, and K
   d. vitamins A, D, E, and K
2. Which of the following describes fat-soluble vitamins?
   a. They include thiamin, vitamin A, and vitamin K.
   b. They cannot be stored to any great extent and so must be consumed daily.
   c. Toxic levels can be reached by consuming citrus fruits and vegetables.
   d. They can be stored in the liver and fatty tissues and can build up toxic concentrations.
3. Night blindness and susceptibility to infection are the result of a deficiency of which vitamin?
   a. niacin
   b. vitamin C
   c. vitamin A
   d. vitamin K
4. Good sources of vitamin D are:
   a. eggs, fortified milk, and sunlight.
   b. citrus fruits, sweet potatoes, and spinach.
   c. green leafy vegetables, cabbage, and liver.
   d. breast milk, polyunsaturated plant oils, and citrus fruits.
5. Which of the following describes water-soluble vitamins?
   a. They include vitamins D and E.
   b. They are frequently toxic.

c. They are stored extensively in tissues.
d. They are easily absorbed and excreted.

6. A coenzyme is:
   a. a fat-soluble vitamin.
   b. an energy-yielding nutrient.
   c. a source of vitamin K.
   d. a molecule that combines with an enzyme to make it active.

7. Good food sources of folate are:
   a. citrus fruits, dairy products, and eggs.
   b. liver, legumes, and leafy green vegetables.
   c. dark green vegetables, corn, and cabbage.
   d. potatoes, broccoli, and whole-wheat bread.

8. Which vitamin is present only in foods of animal origin?
   a. riboflavin b. pantothenic acid
   c. vitamin B_{12} d. the inactive form of vitamin A

9. Which of the following nutrients is an antioxidant that protects water-soluble substances from oxidizing agents?
   a. beta-carotene b. thiamin
   c. vitamin C d. vitamin D

10. Eating foods containing vitamin C at the same meal can increase the absorption of which mineral?
   a. iron b. calcium
   c. magnesium d. folate

Answers to these questions appear in Appendix H.

Notes

1. J. A. Olson, Vitamin A, in *Present Knowledge in Nutrition,* 7th ed., eds. E. E. Ziegler and L. J. Filer (Washington, D.C.: International Life Sciences Institute Press, 1996), pp. 109–119.

2. B. Halliwell, Antioxidants and human disease: A general introduction, *Nutrition Reviews* 55 (1997): 544–549; C. L. Rock, R. A. Jacob, and P. E. Bowen, Update on the biological characteristics of the antioxidant micro-nutrients: Vitamin C, vitamin E, and the carotenoids, *Journal of the American Dietetic Association* 96 (1996): 693–702.

3. Rock, Jacob, and Bowen, 1996.

4. K. Smigel, Beta-carotene fails to prevent cancer in two major studies: CARET intervention stopped, *Journal of the National Cancer Institute* 88 (1996): 145; C. S. Omenn and coauthors. Effects of a combination of beta-carotene and vitamin A on lung cancer and cardiovascular disease, *New England Journal of Medicine* 334 (1996): 1150–1155.

5. N. S. Scrimshaw and J. P. SanGiovanni, Synergism of nutrition, infection, and immunity, *American Journal of Clinical Nutrition* 66 (1997): 464S–477S.

6. W. W. Fawzi and coauthors, Dietary vitamin A intake and the risk of mortality among children, *American Journal of Clinical Nutrition* 59 (1994): 401–408; W. W. Fawzi and coauthors, Vitamin A supplementation and child mortality: A meta-analysis, *Journal of the American Medical Association* 269 (1993): 898–903.

7. American Academy of Pediatrics, Committee on Infectious Diseases, Vitamin A treatment of measles, *Pediatrics* 91 (1993): 1014–1015.

8. A. W. Norman, Vitamin D, in *Present Knowledge in Nutrition,* 7th ed., eds. E. E. Ziegler and L. J. Filer (Washington, D.C.: International Life Sciences Institute Press, 1996), pp. 120–129.

9. Norman, 1996; H. F. DeLuca, New concepts of vitamin D functions, *Annals of the New York Academy of Sciences* 669 (1992): 59–68.

10. A. Greely, Dodging the rays, *FDA Consumer,* July/August 1993, pp. 30–33.

11. American Academy of Pediatrics, *Pediatric Nutrition Handbook,* 3rd ed., ed. L. A. Barness (Elk Grove Village, Ill.: American Academy of Pediatrics, 1993), pp. 34–42.

12. C. Lamberg-Allardt and coauthors, Low serum 25-hydroxy vitamin D concentrations and secondary hyperthyroidism in middle-aged white, strict vegetarians, *American Journal of Clinical Nutrition* 58 (1993): 684–689.

13. J. Regnström and coauthors, Inverse relation between the concentration of low-density lipoprotein vitamin E and severity of coronary artery disease, *American Journal of Clinical Nutrition* 63 (1996): 377–385; K. G. Losonczy, T. B. Harris, and R. J. Havlik, Vitamin E and vitamin C supplements use and risk of all-cause and coronary heart disease mortality in older persons: The Established Populations (or Epidemiologic Studies of the Elderly), *American Journal of Clinical Nutrition* 64 (1996): 190–196; L. H. Kushi and coauthors, Dietary antioxidant vitamins and death from coronary heart disease in postmenopausal women, *New England Journal of Medicine* 334 (1996): 1156–1162.

14. Kushi and coauthors, 1996; M. J. Stampfer and coauthors, Vitamin E consumption and the risk of coronary disease in women, *New England Journal of Medicine* 328 (1993): 1444–1449; E. B. Rimm and coauthors, Vitamin E consumption and the risk of coronary disease in men, *New England Journal of Medicine* 328 (1993): 1450–1456.

15. H. Kappus and A. T. Diplock, Tolerance and safety of vitamin E: A toxicological position report, *Free Radical Biology and Medicine* 13 (1992): 55–74.

16. P. Dowd and coauthors, The mechanism of action of vitamin K, *Annual Review of Nutrition* 15 (1995): 419–440.

17. L. J. Sokoll and coauthors, Changes in serum osteocalcin, plasma phylloquinone and urinary (γ-carboxyglutamic acid in response to altered intakes of dietary phylloquinone in human subjects, *American Journal of Clinical Nutrition* 65 (1997): 779–784; N. C. Brinkley and J. W. Suttie, Vitamin K nutrition and osteoporosis, *Journal of Nutrition* 125 (1995): 1812–1821.

18. Dowd and coauthors, 1995.

19. M. L. Schwartz, Severe reversible hyperglycemia as a consequence of niacin therapy, *Archives of Internal Medicine* 153 (1993): 2050–2052.

20. J. E. Leklem, Vitamin B-6, in *Present Knowledge in Nutrition,* 7th ed., eds. E. E. Ziegler and L. J. Filer (Washington, D.C.: International Life Sciences Institute Press, 1996), pp. 174–183.

21. L. C. Rall and S. N. Meydani, Vitamin B_6 and immune competence, *Nutrition Reviews* 51 (1993): 217–225.

22. J. Selhub and coauthors, Association between plasma homocysteine concentrations and extracranial carotid-artery stenosis, *New England Journal of Medicine* 332 (1995): 286–291; M. J. Stampfer and M. R. Malinow, Can lowering homocysteine levels reduce cardiovascular risk? *New England Journal of Medicine* 332 (1995): 328–329; J. B. Ubbink, Vitamin nutrition status and homocysteine: An atherogenic risk factor, *Nutrition Reviews* 52 (1994): 383–393.

23. J. V. Woodside and coauthors, Effect of B-group vitamins and antioxidant vitamins on hyperhomocysteinemia: A double-blind, randomized, factorial-design, controlled trial, *American Journal of Clinical Nutrition* 67 (1998): 858–866; Selhub and coauthors, 1995.

24. E. B. Rimm and coauthors, Folate and vitamin B_6 from diet and supplements in relation to risk of coronary heart disease among women, *Journal of the American Medical Association* 279 (1998): 359–364.

25. C. E. Butterworth, Jr., and A. Bendich, Folic acid and the prevention of birth defects, *Annual Review of Nutrition* 16 (1996): 73–97.

26. Committee on Genetics, Folic acid for the prevention of neural tube defects, *Pediatrics* 92 (1993): 493–494.

27. Centers for Disease Control and Prevention, Recommendations for use of folic acid to reduce number of spina bifida cases and other neural tube defects, *Journal of the American Medical Association* 269 (1993): 1233, 1236, 1238.

28. C. J. Cuskelly, H. McNulty, and J. M. Scott, Effect of increasing dietary folate on red-cell folate: Implications for prevention of neural tube defects, *The Lancet* 347 (1996): 657–659.

29. Folate and neural tube defects: U.S. policy evolves, *Nutrition Reviews* 51 (1993): 358–361.

30. C. M. Pfeiffer and coauthors, Absorption of folate from fortified cereal-grain products and of supplemental folate consumed with or without food determined by using dual-label-stable isotope protocol, *American Journal of Clinical Nutrition* 66 (1997): 1388–1397.

31. F. Watanabe and coauthors, Effects of microwave heating on the loss of vitamin B_{12} in foods, *Journal of Agricultural and Food Chemistry* 46 (1998): 206–210.

32. C. S. Johnston, K. R. Retrum, and J. C. Srilaskshmi, Antihistamine effects and complications of supplemental vitamin C, *Journal of the American Dietetic Association* 92 (1992): 988–989.

33. J. E. Enstrom, L. E. Kanim, and M. A. Klein, Vitamin C intake and mortality among a sample of the United States population, *Epidemiology* 43 (1992): 194–202.

34. J. Lykkesfeldt and coauthors, Ascorbic acid and dehydroascorbic acid as biomarkers of oxidative stress caused by smoking, *American Journal of Clinical Nutrition* 66 (1997): 1388–1397.

CHAPTER 7

Nutrition in Practice

▪ VITAMIN SUPPLEMENTS ▪

About 50 percent of U.S. adults collectively spend close to $4 billion a year on vitamin supplements.[1] This trend is accelerating as scientists discover more and more links between nutrition and disease prevention. As discussed in the chapter, women of childbearing age may need folate supplements to reduce the risk of neural tube defects. Many recent reports also indicate that the antioxidant nutrients beta-carotene, vitamin C, and vitamin E may be potent protectors against both cancer and heart disease. Before you race out to buy bottles of antioxidant supplements, it is important to know that the role of antioxidants in preventing disease remains to be confirmed by hundreds of scientific studies that are currently under way. Some findings so far are promising, but nevertheless tentative. The main message of this nutrition in practice is that most healthy people can get the nutrients they need from foods. Supplements cannot substitute for a healthy diet.

WWW
eatright.org/asupple.html
Position of the American Dietetic Association: Vitamin and Mineral Supplementation

Do foods really contain enough vitamins and minerals to supply all that most people need?

Emphatically, yes, for both healthy adults and children who choose a variety of foods in moderation. The Daily Food Guide and the Food Guide Pyramid described in Chapter 1 are the guides to follow to achieve adequate intakes. People who meet their nutrient needs from foods, rather than supplements, have low risks of toxicity as well as of deficiency.

You said most healthy people do not need supplements. Do some people need them?

Yes, some people may suffer marginal nutrient deficiencies due to illness, alcohol or drug addiction, or other conditions that limit food intake. People who may benefit from nutrient supplements in amounts consistent with the RDA include:

- People with nutrient deficiencies.
- People with low energy intakes (less than 1200 kcalories per day), such as habitual dieters.
- People with illnesses that take away the appetite.
- People with illnesses that impair nutrient absorption, such as diseases of the liver, gallbladder, pancreas, and digestive system.
- People taking medications that interfere with nutrient metabolism.
- Women who bleed excessively during menstruation (iron supplements).
- Women who are pregnant or lactating (iron, calcium, and folate).
- Strict vegetarians (calcium, iron, zinc, vitamin B_{12}, and vitamin D).
- Newborn infants (a single dose of vitamin K at birth under the direction of a physician).

A few other special cases exist:

- People who have infections or injuries or who have undergone major surgery. The increased metabolic needs associated with these severe stresses are discussed in Chapter 23.
- Infants, depending on whether they are receiving breast milk or formula and on whether their water contains fluoride (see Chapter 12).

These people are in the minority. For the majority, supplements are not recommended. Nutrients are potentially toxic when taken in large doses, and individual tolerances vary depending on health and age. Whenever a health care professional finds a person's diet inadequate, the right corrective step is to improve the person's food choices and eating patterns, not to begin supplementation.

If only a few situations merit the use of supplements, why do so many people take them?

People frequently take supplements for the wrong reasons, such as "They give me energy" or "They make me strong." Other invalid reasons why people may take supplements include:

- Their feeling of insecurity about the nutrient content of the food supply.
- Their belief that extra vitamins and minerals will help them cope with stress.
- Their belief that supplements can enhance athletic performance or build lean body tissue without physical work.
- Their desire to prevent, treat, or cure symptoms or diseases ranging from the common cold to cancer.

Ironically, at least one study showed that supplement users eat more nutrient-dense diets than nonusers and therefore need supplements less. In addition, little relationship exists between the nutrients people need and the ones they take in supplements. In fact, an argument against supplements is that they may lull people into a false sense of security. A person might eat irresponsibly, thinking, "My supplement will cover my needs."

Okay, so people can eat well enough to meet normal needs. But what about taking antioxidant supplements to prevent cancer and heart disease?

Again, it is better advice to eat a very nutritious diet. Evidence from population studies shows a correlation between low intakes of antioxidant nutrients and a high incidence of disease. Many studies show that low intakes of vegetables and fruits are consistently linked with an increased incidence of cancer.[2]

More than 200 population studies have examined the effects of fruits and vegetables on cancer risk, and the vast majority show that people who eat more of these foods are less likely to develop cancer.[3] Many experts agree that the antioxidant vitamins in these foods are probably the most important protective factors, but they also note that other constituents of fruits and vegetables (see Table 7.2, p. 141) have not been ruled out as contributors to the effect.

The way to apply this information is to eat nutritious foods. Before supplementation is recommended as a strategy to prevent cancer or other diseases, researchers must find out the optimal doses to reduce risk and the potential adverse effects of long-term supplementation.

When I do need a vitamin-mineral supplement, what kind should I use?

Take your health care professional's advice, if it is offered. If you are selecting a supplement yourself, a single, balanced vitamin-mineral supplement should suffice. Choose the kind that provides all the RDA nutrients in amounts equal to, or very close to, the RDA (remember, you get some nutrients from foods). Avoid individual nutrients, unless prescribed by a health care professional. Avoid preparations that are in excess of the RDA. Avoid preparations that contain items not needed in human nutrition such as inositol.

Can I trust what I read on supplement labels?

Yes. To enable consumers to make more informed choices about nutrient supplements, the Food and Drug Administration (FDA), with the encouragement of the American Dietetic Association (ADA), has published new labeling regulations for supplements.[4] The new regulations subject supplements to the same general labeling requirements that apply to foods. Specifically:

- Nutrition labeling for dietary supplements is now required. The nutrition panel on supplements is called "Supplement Facts" (see Figure NP7.1 on p. 170). The Supplement Facts panel lists the quantity and the percentage of the Daily Value for each nutrient in the supplement. Ingredients that have no Daily Value—for example, sugars and gelatin—appear in a list below the Supplement Facts panel.
- Labels may describe nutrient contents (as "high" or "low") according to specific criteria.
- The FDA authorizes health claims on supplement labels about the relationship between folate and the risk of neural tube defects, calcium and osteoporosis, soluble fiber from whole oats and psylium husks and heart disease, and sugar alcohols and dental caries.
- Supplement labels are not allowed to include health claims on a number of other nutrient-disease relationships that have been approved for foods.
- Products may not bear claims to diagnose, treat, cure, or relieve a specific disease.
- Labels may describe the role a nutrient plays in the body, explain how the nutrient performs its function, and indicate that consuming the nutrient is associated with general well-being.

Health claims made on supplement labels must be based on significant scientific evidence. The $4 billion-a-year supplement industry is expected to continue fighting against the regulations, however, by trying to get overruling legislation.

CHAPTER 7

Nutrition in Practice

Figure NP7.1

An Example of a Supplement Label

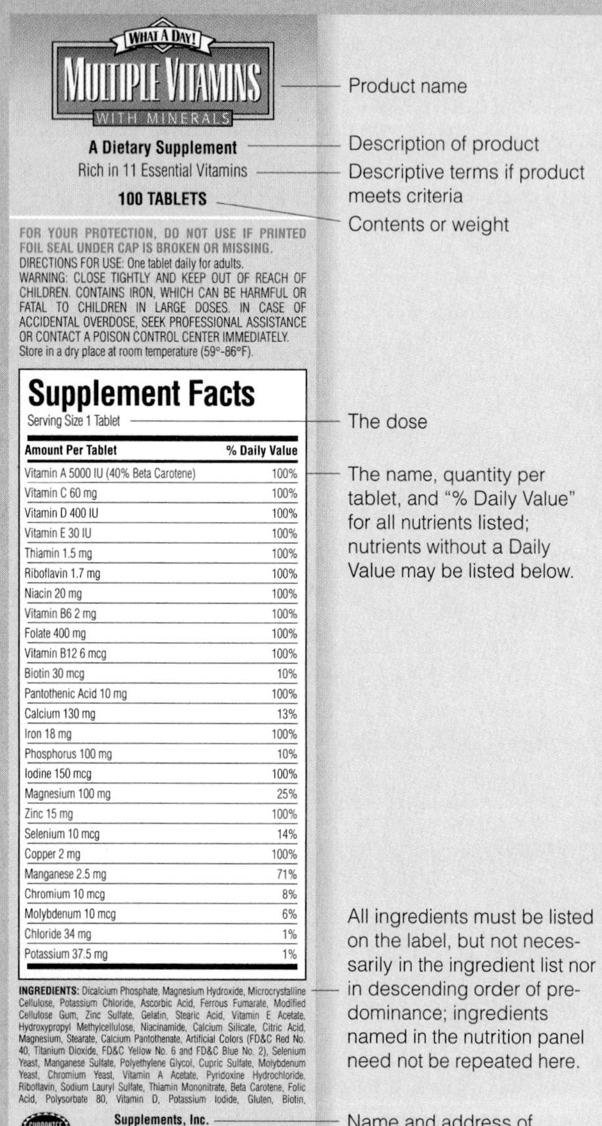

WHAT A DAY!
MULTIPLE VITAMINS
WITH MINERALS
A Dietary Supplement
Rich in 11 Essential Vitamins
100 TABLETS
FOR YOUR PROTECTION, DO NOT USE IF PRINTED FOIL SEAL UNDER CAP IS BROKEN OR MISSING.
DIRECTIONS FOR USE: One tablet daily for adults.
WARNING: CLOSE TIGHTLY AND KEEP OUT OF REACH OF CHILDREN. CONTAINS IRON, WHICH CAN BE HARMFUL OR FATAL TO CHILDREN IN LARGE DOSES. IN CASE OF ACCIDENTAL OVERDOSE, SEEK PROFESSIONAL ASSISTANCE OR CONTACT A POISON CONTROL CENTER IMMEDIATELY.
Store in a dry place at room temperature (59°-86°F).
Supplement Facts
Serving Size 1 Tablet
Amount Per Tablet % Daily Value
Vitamin A 5000 IU (40% Beta Carotene) 100%
Vitamin C 60 mg 100%
Vitamin D 400 IU 100%
Vitamin E 30 IU 100%
Thiamin 1.5 mg 100%
Riboflavin 1.7 mg 100%
Niacin 20 mg 100%
Vitamin B6 2 mg 100%
Folate 400 mg 100%
Vitamin B12 6 mcg 100%
Biotin 30 mcg 10%
Pantothenic Acid 10 mg 100%
Calcium 130 mg 13%
Iron 18 mg 100%
Phosphorus 100 mg 10%
Iodine 150 mcg 100%
Magnesium 100 mg 25%
Zinc 15 mg 100%
Selenium 10 mcg 14%
Copper 2 mg 100%
Manganese 2.5 mg 71%
Chromium 10 mcg 8%
Molybdenum 10 mcg 6%
Chloride 34 mg 1%
Potassium 37.5 mg 1%
INGREDIENTS: Dicalcium Phosphate, Magnesium Hydroxide, Microcrystalline Cellulose, Potassium Chloride, Ascorbic Acid, Ferrous Fumarate, Modified Cellulose Gum, Zinc Sulfate, Gelatin, Stearic Acid, Vitamin E Acetate, Hydroxypropyl Methylcellulose, Niacinamide, Calcium Silicate, Citric Acid, Magnesium, Stearate, Calcium Pantothenate, Artificial Colors (FD&C Red No. 40, Titanium Dioxide, FD&C Yellow No. 6 and FD&C Blue No. 2), Selenium Yeast, Manganese Sulfate, Polyethylene Glycol, Cupric Sulfate, Molybdenum Yeast, Chromium Yeast, Vitamin A Acetate, Pyridoxine Hydrochloride, Riboflavin, Sodium Lauryl Sulfate, Thiamin Mononitrate, Beta Carotene, Folic Acid, Polysorbate 80, Vitamin D, Potassium Iodide, Gluten, Biotin,
GUARANTEE
Complete Satisfaction or Your Money Back
Supplements, Inc.
1234 Fifth Avenue
Anywhere, USA
Product name
Description of product
Descriptive terms if product meets criteria
Contents or weight
The dose
The name, quantity per tablet, and "% Daily Value" for all nutrients listed; nutrients without a Daily Value may be listed below.
All ingredients must be listed on the label, but not necessarily in the ingredient list nor in descending order of predominance; ingredients named in the nutrition panel need not be repeated here.
Name and address of manufacturer

CHAPTER 7

Nutrition in Practice

Notes

1. M. C. Nesheim, Regulation of dietary supplements, *Nutrition Today* 33 (1998): 62–67.

2. K. A. Steinmetz and J. D. Potter, Vegetables, fruit, and cancer prevention: A review, *Journal of the American Dietetic Association* 96 (1996): 1027–1039.

3. Steinmetz and Potter, 1996; K. Meister, Fruits, vegetables, and cancer, *Priorities* (a publication of the American Council on Science and Health), Fall/Winter 1993, pp. 27–30.

4. Commission on Dietary Supplement Labels issues final report, *Journal of the American Dietetic Association* 98 (1998): 270.

Water and Minerals

The body's water cannot be considered separately from the minerals dissolved in it. A person can drink pure water, but in the body it mingles with minerals to become fluids in which all life processes take place. This chapter begins by discussing the body's fluids and their chief minerals. The focus then shifts to other functions of the minerals.

Chapter 11 addresses consumer concerns about the safety of the water people drink.

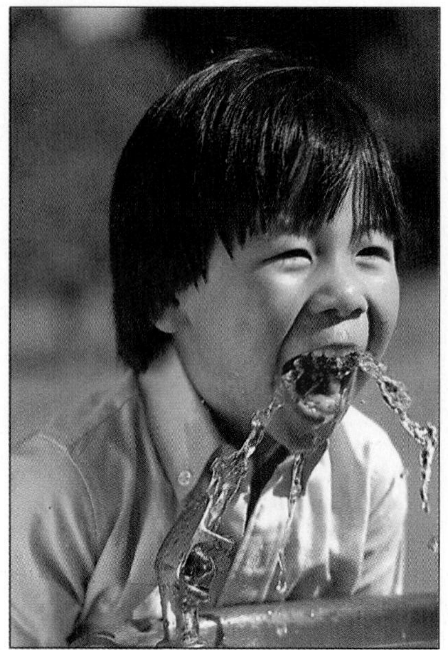

Water is the most vital nutrient of all.

Water and Body Fluids

Water constitutes about 60 percent of an adult's body weight and a higher percentage of a child's. Every cell in the body is bathed in a fluid of the exact composition that is best for that cell. The body fluids bring to each cell the ingredients it requires and carry away the end products of the life-sustaining reactions that take place within the cell's boundaries. The water in the fluids:

- Carries nutrients and waste products throughout the body.
- Participates in chemical reactions.
- Serves as the solvent for minerals, vitamins, amino acids, glucose, and many other small molecules.
- Aids in maintaining the body's blood pressure and temperature.
- Acts as a lubricant and cushion around joints.
- Serves as a shock absorber inside the eyes, spinal cord, and amniotic sac surrounding a fetus in the womb.

Because water is so vital to these and other functions, the body accurately regulates its influx to match outflow.

Water Balance

water balance: the balance between water intake and water excretion, which keeps the body's water content constant.

dehydration: the loss of water from the body that occurs when water output exceeds water input. The symptoms progress rapidly from thirst, to weakness, to exhaustion and delirium and end in death if not corrected.

water intoxication: the rare condition in which body water contents are too high. The symptoms may include confusion, convulsion, coma, and even death in extreme cases.

hypothalamus (high-poh-THALL-uh-mus): a part of the brain that helps regulate many body balances, including fluid balance.

Chapter 10 addresses the fluid needs of athletes.

The cells themselves regulate the composition and amounts of fluids within and surrounding them. The entire system of cells and fluids remains in a delicate but firmly maintained state of dynamic equilibrium. Imbalances such as dehydration and water intoxication can occur, but the body quickly restores the balance to normal if it can. The body controls both water intake and water excretion.

Water Intake Regulation The body can survive for only a few days without water. In healthy people, thirst and satiety govern water intake. Thirst is finely adjusted to ensure a water intake that meets the body's needs. When the blood becomes too concentrated (having lost water but not salt and other dissolved substances), the mouth becomes dry, and the brain center known as the hypothalamus initiates drinking behavior.

Thirst lags behind water lack. A water deficiency that develops slowly can switch on drinking behavior in time to prevent serious dehydration, but a deficiency that develops quickly may not. Also, thirst itself does not remedy a water deficiency; a person must pay attention to the thirst signal and take the time to get a drink. The long-distance runner, the gardener in hot weather, and the busy child at play can experience serious dehydration if they fail to drink promptly in response to their needs for water. With aging, thirst sensations may diminish. Dehydration can threaten elderly people who do not develop the habit of drinking water regularly.

Water Excretion Regulation Water excretion involves the brain and the kidneys. The cells of the brain's hypothalamus, which monitor blood salts, stimulate the pituitary gland to release antidiuretic hormone (ADH) whenever the salts are too concentrated, or the blood volume or blood pressure is too low. ADH stimulates the kidneys to reabsorb water rather than excrete it. Thus the more water you need, the less you excrete.

If too much water is lost from the body, blood volume and blood pressure fall. Cells in the kidneys respond to the low blood pressure by releasing an enzyme. Through a complex series of events, involving the hormone aldosterone, this enzyme also causes the kidneys to retain more water. Again, the effect is that when more water is needed, less is excreted.

Minimum Water Needed These mechanisms can maintain water balance, but only if a person drinks enough water. The body must excrete a minimum of about 500 milliliters each day as urine—enough to carry away the waste products generated by a day's metabolic activities. Above this amount, excretion adjusts to balance intake, so the more you drink, the more dilute your urine becomes. In addition to urine, some water is lost from the lungs as vapor, some is excreted in feces, and some evaporates from the skin. A person's water losses from all of these routes total about 2½ liters (about 2½ quarts) a day on the average. Table 8.1 shows how fluid intake and output naturally balance out.

Water Recommendations and Sources Water needs vary greatly depending on the foods a person eats, the environmental temperature and humidity, the person's activity level, and other factors. Accordingly, a general water requirement is difficult to establish. Recommendations for adults are expressed in proportion to the amount of energy expended under normal environmental conditions.[1] For the person who expends about 2000 kcalories a day, this works out to 2 to 3 liters, or about 8 to 12 cups. You can tell from the color of your urine whether you need more water. Pale yellow urine reflects appropriate dilution.

Plain water best meets people's fluid needs, but milk and juice can also contribute to the day's recommended intake. Alcoholic beverages and those containing caffeine such as coffee, tea, and some sodas are not good water substitutes. Both alcohol and caffeine act as diuretics, causing the body to lose fluids. Foods provide water, too. Most fruits and vegetables contain up to 95 percent water; many meats and cheeses contain at least 50 percent. The energy nutrients in foods also give up water during metabolism.

In summary, water makes up about 60 percent of the body's weight. Water participates in chemical reactions, acts as a solvent, lubricates and cushions the joints, and aids in the maintenance of blood pressure and body temperature. To maintain water balance, intake from liquids and foods must equal losses.

Fluid and Electrolyte Balance

When mineral salts dissolve in water, they separate (dissociate) into charged particles known as ions, which can conduct electricity. For this reason, a salt that partly dissociates in water is known as an electrolyte. The body fluids, which contain water and partly dissociated salts, are electrolyte solutions.

The body's electrolytes are vital to the life of the cells and therefore must be closely regulated to help maintain the appropriate distribution of body fluids. The major minerals form salts that dissolve in the body fluids; the cells direct where these salts go; and the movement of the salts determines where the fluids flow because water follows salt. Cells use this force to move fluids back and forth across their membranes. Thanks to the electrolytes, water can be held in compartments where it is needed.

pituitary (pit-TOO-ih-tary) **gland:** in the brain, the "king gland" that regulates the operation of many other glands.

ADH (antidiuretic hormone): a hormone released by the pituitary gland in response to high salt concentrations in the blood. The kidneys respond by reabsorbing water.

This is the enzyme **renin** (REN-in), released by the kidneys in response to low blood pressure. Renin aids the kidneys in retaining water through the **renin-angiotensin mechanism.**

aldosterone (al-DOS-ter-own): a hormone secreted by the adrenal glands that stimulates the reabsorption of sodium by the kidneys: aldosterone also regulates chloride and potassium concentrations.

500 ml = about ½ qt.

Table 8.1 Water Balance

Water Source	Amount (ml)
Liquids	550 to 1500
Foods	700 to 1000
Metabolic water	200 to 300
	1450 to 2800
Water Output	
Kidneys	500 to 1400
Skin	450 to 900
Lungs	350
Feces	150
	1450 to 2800

salts: compounds composed of charged particles (ions). An example is potassium chloride (K^+Cl^-).

Exceptions: A compound in which the positive ions are hydrogen ions (H^+) is an acid (example: hydrochloric acid, or H^+Cl^-); a compound in which the negative ions are hydroxyl ions (OH^-) is a base (example: potassium hydroxide, or K^+OH^-).

electrolytes: salts that dissolve in water and dissociate into charged particles called **ions.**

electrolyte solutions: solutions that can conduct electricity.

The simple statement that water follows salt describes the force that chemists call *osmosis.*

Proteins in the cell membranes move ions in or out of the cells. These protein pumps tend to concentrate sodium and chloride outside cells and potassium and other ions inside. By maintaining certain amounts of sodium outside and potassium inside, cells can regulate the exact amounts of water inside and outside their boundaries. Physicians apply the same principle when treating kidney failure; they use an electrolyte solution to draw excess fluid out of the blood (see the discussion of *dialysis* in Nutrition in Practice 27).

Healthy kidneys regulate the body's sodium, as well as its water, with remarkable precision. The intestinal tract absorbs sodium readily, and it travels freely in the blood, but the kidneys excrete unneeded amounts. The kidneys actually filter all of the sodium out of the blood; then, with great precision, they return to the bloodstream the exact amount the body needs to retain. Thus the body's total electrolytes remain constant, while the urinary electrolytes fluctuate according to what is eaten.

In some cases the body's mechanisms for maintaining fluid and electrolyte balances may falter. This is the case when large amounts of fluid and electrolytes are suddenly lost and the thirst instinct, cell membranes, and kidneys cannot compensate. Vomiting, diarrhea, heavy sweating, fever, burns, wounds, and the like may incur great fluid and electrolyte losses, precipitating an emergency that demands medical management.

In summary, the body fluids contain water and partly dissociated salts, called electrolytes. Because water follows salt, electrolytes in the fluids help distribute the fluids inside and outside of cells, thus ensuring the appropriate balance to support all life processes.

Acid-Base Balance

The body uses its ions not only to help maintain water balance but also to regulate the acidity (pH) of its fluids. Electrolyte mixtures in the body fluids, as well as proteins, protect the body against changes in acidity by acting as buffers—substances that can accommodate excess acids or bases.

Reminder: *Buffers* are compounds that help keep a solution's acidity or alkalinity constant; buffers are capable of neutralizing both acids and bases and thereby maintaining the original concentration of hydrogen ions (pH) in the solution.

The body's buffer systems serve as a first line of defense against changes in the fluids' acid-base balance. The lungs, skin, GI tract, and kidneys provide other defenses. Of these organ systems, the kidneys play the primary role in maintaining acid-base balance.

Disorders of the kidneys, therefore, impair the body's ability to regulate its acid-base balance, as well as its fluid and electrolyte balances. For a person with renal disease, the physician may order, in addition to many medical procedures, adjustment of the fluid and electrolyte intake from food. Chapter 27 gives more information about renal disease.

Not only do all of the major minerals help maintain fluid and electrolyte balance in the body, but they have other important functions as well, as the next sections describe. Table 8.6, at the end of this chapter, offers a summary of the minerals and their functions.

The Major Minerals

Table 8.2 lists the major and trace minerals in the body, and Figure 8.1 shows the amounts found in the body. As you can see, the most prevalent minerals are calcium and phosphorus, the chief minerals of bone. The distinction between the major and the trace minerals does not mean that one group is more important than the other. A deficiency of the few micrograms of iodine needed daily is just as serious as a deficiency of the several hundred milligrams of calcium. The major minerals are so named because they are present, and needed, in larger amounts in the body.

While all the major minerals influence the body's fluid balance, sodium, chloride, and potassium are most noted for that role. For this reason, these three minerals are discussed first. Each major mineral also plays other specific roles in the body. Sodium, potassium, calcium, and magnesium are critical to nerve transmission and muscle contractions. Phosphorus and magnesium are involved in energy metabolism. Calcium, phosphorus, and magnesium contribute to the structure of the bones. Sulfur helps determine the contour of proteins.

Table 8.2 The Major and Trace Minerals

Major Minerals	Trace Minerals
Calcium	Arsenic
Chloride	Boron
Magnesium	Chromium
Phosphorus	Cobalt
Potassium	Copper
Sodium	Fluoride
Sulfur	Iodine
	Iron
	Manganese
	Molybdenum
	Nickel
	Selenium
	Silicon
	Zinc

Sodium

Sodium is the principal electrolyte in the extracellular fluid and the primary regulator of the extracellular fluid volume. When the blood concentration of sodium rises, as when a person eats salted foods, thirst ensures that the person will drink water until the appropriate sodium-to-water ratio is restored. Sodium also helps maintain acid-base balance and is essential to muscle contraction and nerve transmission.

Sodium Recommendations and Food Sources Diets rarely lack sodium. For this reason, recommended intakes have not been set; instead the estimated *minimum* sodium requirement was established. Health recommendations advise a *maximum* intake of *salt,* primarily to help prevent high blood pressure.[2]

Estimated minimum requirement for sodium: 500 mg/day. Recommended maximum intake of salt: 6 g/day (2400 mg sodium).

5 g salt is about 2 g sodium.
1 g salt = ⅕ tsp salt.

Cultures vary in their use of salt. In the United States, men consume an average of 3300 milligrams of sodium (equivalent to about 8 grams of salt) a day. Asian people, whose staple sauces and flavorings are based on soy sauce and monosodium glutamate (MSG), consume the equivalent of about 30 to 40 grams of salt per day. In China, Japan, and Korea, high blood pressure is as prevalent, or more so, as in the United States.[3]

Sodium intakes vary widely. People who eat mostly processed foods have the highest sodium intakes, while those who eat mostly whole, unprocessed foods, such as fresh fruits and vegetables, have the lowest intakes. In fact, about three-fourths of the sodium in people's diets comes from salt added to foods by manufacturers. Figure 8.2 on the next page shows that processed foods contain not only more sodium but also less potassium than their less-processed counterparts.

Sodium and Blood Pressure Sodium, or the salt that delivers it, contributes to high blood pressure in susceptible people.[4] Susceptibility may increase with

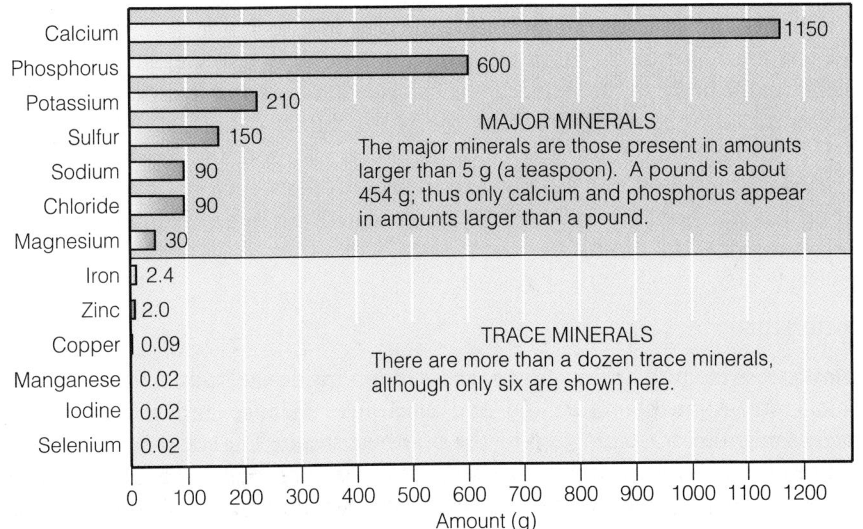

Figure 8.1
The Amounts of Minerals in a 60-kilogram (132-pound) Human Body

Figure 8.2

What Processing Does to the Sodium and Potassium Contents of Foods

Note how potassium is lost and sodium is gained as foods become more processed.

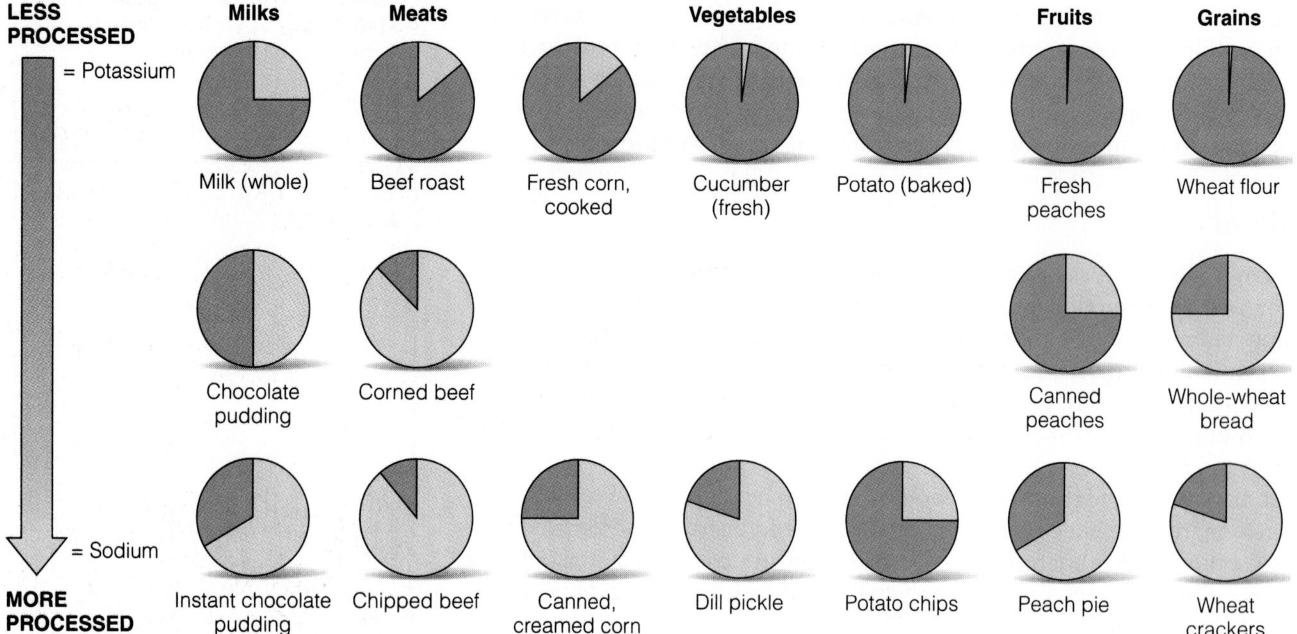

potassium, calcium, and magesium deficiencies. Chapter 26 offers suggestions for avoiding excessive salt intakes, and describes the relationship of dietary factors to blood pressure.

WWW. nof.org National Osteoporosis Foundation

Sodium and Osteoporosis A high sodium intake has also been associated with calcium losses and bone losses.[5] Dietary advice to prevent osteoporosis might suggest both eating more calcium-rich foods and restricting foods high in sodium.

Chloride

The chloride ion is the major negative ion of the fluids outside the cells, where it occurs primarily in association with sodium. Chloride can move freely across cell membranes and so is also found inside the cells in association with potassium. Like sodium, chloride is critical to maintaining fluid, electrolyte, and acid-base balance in the body. In the stomach, the chloride ion is part of hydrochloric acid, which maintains the strong acidity of the gastric fluids.

Estimated minimum requirement for chloride: 750 mg/day.

Salt is a major food source of chloride, and as with sodium, processed foods are a major contributor of this nutrient to people's diets. A chloride recommendation has not been established, but an estimated minimum requirement has been determined for adults.

Potassium

Potassium is the principal positively charged ion inside the body cells. It plays a major role in maintaining fluid and electrolyte balance and cell integrity. Potassium is also critical to keeping the heartbeat steady. The sudden deaths that occur in severe diarrhea and in children with kwashiorkor are likely due to heart failure caused by potassium loss. Potassium also assists in carbohydrate and protein metabolism.

Potassium Deficiency and Toxicity Potassium deficiency results more often from excessive losses than from deficient intakes. Deficiency arises in abnormal conditions such as diabetic acidosis, dehydration, or prolonged vomiting or diarrhea; potassium deficiency can also result from the regular use of certain medications, including diuretics, steroids, and cathartics. One of the earliest symptoms of deficiency is muscle weakness. Low potassium intakes are possible with diets low in fresh fruits and vegetables, but out-and-out deficiencies of potassium are unlikely in healthy people.

In healthy people, potassium toxicity from foods is not a problem because the kidneys excrete excess potassium. Intakes from potassium supplements can reach toxic levels, however, and can cause death.

diuretic (dye-yoo-RET-ic): a medication that promotes the excretion of water through the kidneys. Only some diuretics increase the urinary loss of potassium. Others, called potassium-sparing diuretics, are less likely to result in a potassium deficiency (see Chapter 26).

steroid (STARE-oid): a medication used to reduce tissue inflammation, to suppress the immune response, or to replace certain steroid hormones in people who cannot synthesize them.

cathartic (ca-THART-ic): a strong laxative.

Potassium Recommendations and Food Sources As with sodium, an estimated minimum requirement for potassium has been determined. Surveys show wide variations in potassium intakes in the United States; people who emphasize fresh fruits and vegetables in their diets have high intakes. Potassium is abundant inside all living cells, both plant and animal, and because cells remain intact unless foods are processed, the richest sources of potassium are *fresh* foods of all kinds—especially fruits and vegetables.

Estimated minimum requirement for potassium: 2000 mg/day.

Potassium and Blood Pressure Diets low in potassium seem to play an important role in the development of high blood pressure. Research suggests that increasing potassium intakes may both prevent and help to correct hypertension.[6]

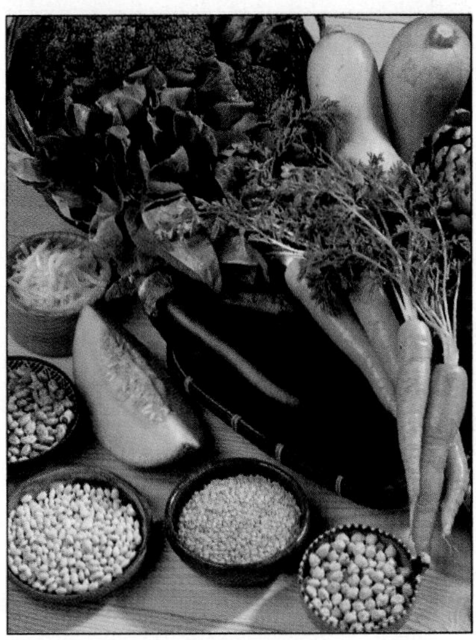

Fresh fruits and vegetables provide potassium in abundance.

Potassium, the major positive ion inside cells, is important in maintaining fluid and electrolyte balance. Fresh foods are the best sources of potassium.

Calcium

Calcium occupies more space in this chapter than does any other major mineral. Other minerals are revisited later in this book, where they play key roles in heart disease and kidney disease. Calcium, though, deserves emphasis here in the normal nutrition part of the book because an adequate intake of calcium early in life helps grow a healthy skeleton and prevent bone disease in later life.

Calcium Roles in the Body Calcium owns the distinction of being the most abundant mineral in the body. Ninety-nine percent of the body's calcium is stored in the bones, where it plays two important roles. First, it is an integral part of bone structure. Second, it serves as a calcium bank available to the body fluids should a drop in blood calcium occur.

Calcium in Bone As bones begin to form, calcium salts form crystals on a matrix of the protein collagen. As the crystals become denser, they give strength and rigidity to the maturing bones. As a result, the long leg bones of children can support their weight by the time they have learned to walk. Many people have the idea that bones are inert, like rocks. Not so. Bones continuously gain and lose minerals in an ongoing process of remodeling. Growing children gain more bone than they lose, and healthy adults maintain a reasonable balance. When withdrawals substantially exceed deposits, however, problems such as osteoporosis develop. Figure 8.3 (p. 180) shows the lacy network of calcium-containing crystals in the bone.

Calcium in Body Fluids The 1 percent of the body's calcium that circulates in the fluids as ionized calcium is vital to life. It helps regulate muscle contractions, transmit nerve impulses, clot blood, and secrete hormones, digestive enzymes, and neurotransmitters. Calcium also helps convey signals received at the cell surface to the inside of the cell. Calcium is a cofactor for several enzymes as well.

cofactor: a mineral element that, like a coenzyme, works with an enzyme to facilitate a chemical reaction.

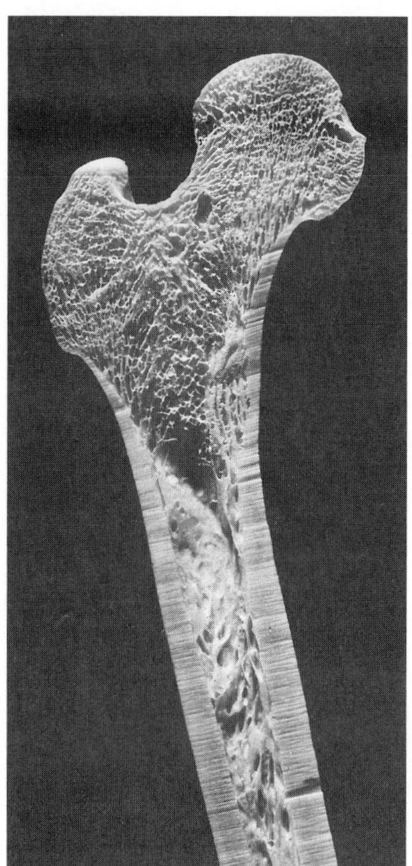

Figure 8.3
Cross Section of Bone
The lacy structural elements are **trabeculae** (tra-BECK-you-lee), which can be drawn on to replenish blood calcium.

The regulators are hormones from the thyroid and parathyroid glands, as well as vitamin D. One hormone, parathormone, raises blood calcium. Others, calcitonin and thyrocalcitonin, lower blood calcium by inhibiting release of calcium from bone. The hormonelike vitamin D raises blood calcium by acting at the three sites listed.

calcium rigor: hardness or stiffness of the muscles caused by high blood calcium.

calcium tetany: intermittent spasms of the extremities due to nervous and muscular excitability, which is caused by low blood calcium.

WWW. nof.org
National Osteoporosis Foundation

osteoporosis (oss-tee-oh-pore-OH-sis): literally, porous bones; reduced density of the bones. Also known as *adult bone loss,* it is a condition in which the bones become porous and fragile. The causes of osteoporosis are multiple.
osteo = bone

Calcium Balance Blood calcium concentration is tightly controlled. Whenever blood calcium rises too high, a system of hormones and vitamin D promotes its deposit into bone. Whenever blood calcium falls too low, the regulatory system acts in three locations to raise it:

1. The intestine absorbs more calcium.
2. The bones release more calcium.
3. The kidneys excrete less calcium.

Thus blood calcium rises to normal.

The calcium stored in bone provides a nearly inexhaustible source of calcium for the blood. Even in a calcium deficiency, blood calcium remains normal. Blood calcium changes only in response to abnormal regulatory control, not to diet. Blood calcium above normal causes calcium rigor: the muscles contract and cannot relax. Blood calcium below normal causes calcium tetany—also characterized by uncontrolled muscle contraction. These conditions are caused by a lack of vitamin D or by abnormal concentrations of the hormones that regulate calcium homeostasis.

Although a chronic *dietary* deficiency of calcium or a chronic deficiency due to poor absorption does not change blood calcium, it does deplete the savings account in the bones. Because this is an important concept, we repeat: it is the bones, not the blood, that are robbed by calcium deficiency.

Calcium and Osteoporosis Bone mass peaks at the time of skeletal maturity (about age 30), and a high peak bone mass is the best protection against later age-related bone loss and fracture. Adequate calcium nutrition during the growing years is essential to achieving optimal peak bone mass.[7] Following menopause, women lose about 15 percent of their bone mass, as do middle-aged and older men. When bone loss has reached such an extreme that bones fracture under even common, everyday stresses, the condition is known as osteoporosis. Osteoporosis afflicts as many as 25 million people, mostly women 45 years of age or older.

Both genetic and environmental factors contribute to osteoporosis. Table 8.3 summarizes risk and protective factors for osteoporosis. Osteoporosis is more prevalent in women than men for several reasons. First, women consume only about half as much dietary calcium as men do. Second, at all ages, women's bone mass is lower than men's because women generally have a smaller body size. Finally, bone loss begins earlier in women than in men, and it accelerates after menopause.

Bone Loss Prevention Many minerals and vitamins, including phosphorus, magnesium, fluoride, and vitamin D, help to form and stabilize the structure of bones. Any or all of these elements are needed to prevent bone loss. The first, most obvious lines of defense, however, are to maintain a lifelong adequate intake of calcium and to "exercise it into place." Active bones are denser than sedentary bones.[8] Weight-bearing physical activity, such as walking, running or dancing, prompts the bones to lay down minerals. It has long been known that when people are confined to bed, both their muscles and their bones lose strength. Muscle strength and bone strength go together: when muscles work, they pull on the bones, and both are stimulated to grow stronger.

Calcium and Hypertension Emerging evidence suggests that calcium helps prevent hypertension.[9] Studies of populations prone to developing hypertension show that low dietary calcium correlates with a high blood pressure.[10] Reports indicate that calcium supplements can sometimes lower blood pressure.[11] Some researchers speculate that calcium's effect on blood pressure is related to its action on the smooth muscle surrounding blood vessels.

Table 8.3 Risk and Protective Factors That Correlate with Osteoporosis

Risk Factors	Protective Factors
High Correlation	
Advanced age; postmenopausal	African American
Alcohol abuse	Estrogens, long-term use
Anorexia nervosa	
Caucasian	
Chronic steroid use	
Female sex	
Rheumatoid arthritis	
Surgical removal of ovaries	
Thinness	
Moderate Correlation	
Chronic thyroid hormone use	Having given birth
Cigarette smoking	High body weight
Diabetes (type 1)	
Early menopause	
Excessive antacid use	
Low-calcium diet	
Sedentary lifestyle or immobility	
Vitamin D deficiency	
Probably Important But Not Yet Proved	
Caffeine use	High-calcium diet
Family history of osteoporosis	Regular physical activity
High-fiber diet	
High-protein diet	
High-sodium diet	

Source: Adapted from C. D. Arnaud and S. D. Sanchez, The role of calcium in osteoporosis, *Annual Review of Nutrition* 10 (1990): 397–414.

Calcium Recommendations As mentioned earlier, blood calcium concentration does not reflect calcium status. Calcium recommendations are therefore based on balance studies, which measure daily intake and excretion. An optimal calcium intake reflects the amount needed to retain the most calcium. The more calcium retained, the greater the bone density (within genetic limits) and, potentially, the lower the risk of osteoporosis. Calcium recommendations during adolescence are set high (1300 milligrams) to help ensure that the skeleton will be strong and dense. Between the ages of 19 and 50, recommendations are lowered to 1000 milligrams a day. For those over 50, recommendations are raised again, to 1200 milligrams, to minimize bone loss. Some authorities advocate calcium recommendations as high as 1500 milligrams per day for women over 50. Many women have intakes well below recommendations.

Calcium AI:
- Adults (19–50 yr): 1000 mg/day.
- Adults (51 and older): 1200 mg/day.

Calcium in Foods Calcium is found almost exclusively in a single class of foods—milk and milk products. For this reason, dietary recommendations advise daily consumption of reduced-fat, low-fat, or nonfat milk products. A cup of nonfat milk offers about 300 milligrams of calcium, so an adult who drinks 2 to 3 cups of milk a day is well on the way to meeting daily calcium needs. Pregnant and lactating teenagers need more (see Table 8.4 on p. 182). The other dairy food that contains comparable amounts of calcium is cheese.

Table 8.4 Recommended Fluid Milk Intakes	
Age	Recommended Intake
Children	2 c
Teenagers and young adults	3 c
Adults	2 c
Pregnant and lactating women	3 c
Pregnant and lactating teens	4 c
Women past menopause	3 c

Apparently, all fibers in plant foods—cellulose, hemicellulose, pectin, and others—bind calcium to some extent, as do phytate and oxalate. Phytate and oxalate are *binders* that combine with minerals to form complexes that the body cannot absorb.

One slice of cheese (1 ounce) contains about two-thirds as much calcium as a cup of milk. Cottage cheese, however, is a poor source of calcium. Snapshot 8.1 shows foods that are rich in calcium, and the accompanying box suggests ways of adding calcium to meals.

Some foods offer large amounts of calcium because of fortification. Calcium-fortified juice, high-calcium milk (milk with extra calcium added), and calcium-fortified cereal are examples.

Among the vegetables, mustard greens, kale, parsley, watercress, and broccoli are good sources of available calcium. Some dark green, leafy vegetables—notably, spinach and swiss chard—appear to be calcium-rich but actually provide very little, if any, calcium to the body. These foods contain binders that prevent calcium absorption.

People may think that taking a calcium supplement is preferable to getting calcium from food, but foods offer important fringe benefits. For example, drinking 2 cups of milk fortified with vitamins A and D will supply substantial amounts of other nutrients. Furthermore, calcium absorption is enhanced by the vitamin D, lactose, fat, and possibly other nutrients in the milk. A calcium supplement supplies only calcium, and in a less absorbable form.

The body is able to regulate its absorption of calcium by altering its production of the calcium-binding protein aided by vitamin D. More of this protein is made if more calcium is needed. Infants and children absorb up to 75 percent of ingested calcium, and pregnant women, about 50 percent. Other adults, who are not growing, absorb about 30 percent. Also, calcium seems to be better absorbed if accompanied by an approximately equal amount of phosphorus.

Calcium, the most abundant mineral in the body, is vital to life. Calcium in bone is integral to bone structure and serves as a calcium bank for the body. Calcium in the body's fluids helps regulate muscle contractions, nerve impulses, blood clotting, and other processes. Adequate calcium intake during the growing years is essential to achieving optimal bone mass and protection from osteoporosis in later life. Milk and milk products are calcium-rich.

Phosphorus

Phosphorus is the second most abundant mineral in the body. About 85 percent of it is found combined with calcium in the crystals of the bones and teeth. As

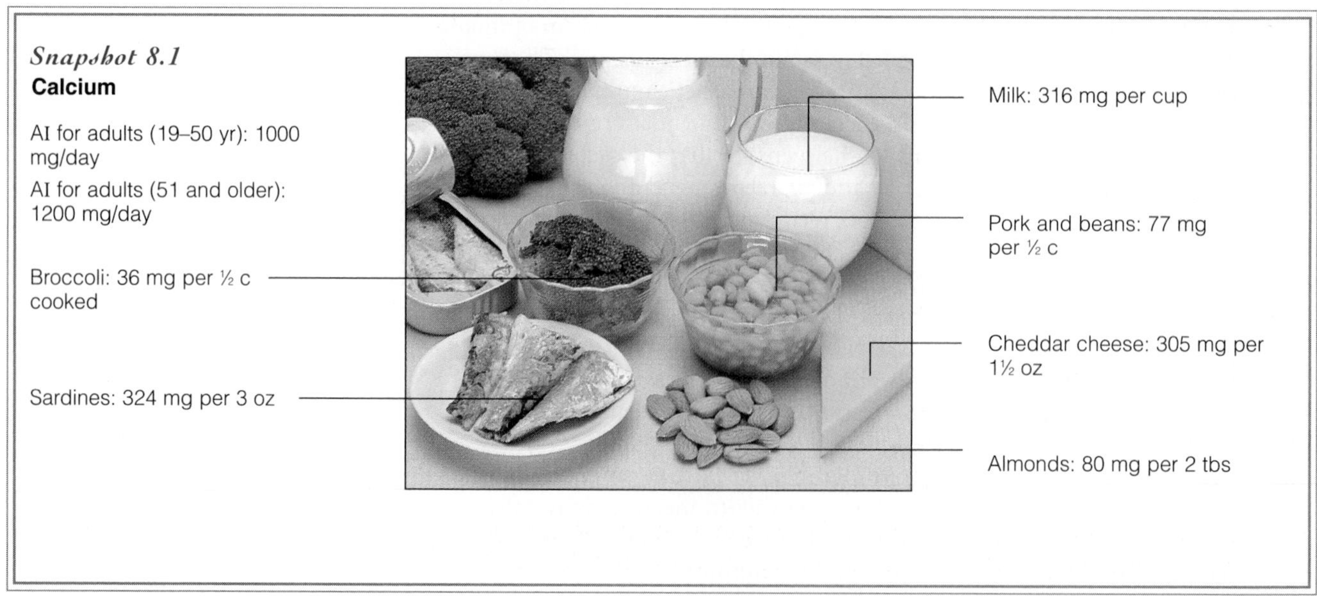

How to Add Calcium to Daily Meals

Many people cannot or will not drink enough milk to meet recommendations. To help deliver calcium, try the following suggestions:

1. Powdered nonfat milk is an excellent and inexpensive source of calcium and can be added to many foods (such as baked products and meat loaf) during preparation.
2. Yogurt and kefir (fermented dairy products) are acceptable substitutes for regular milk.
3. Puddings, custards, and baked goods can be prepared using appreciable amounts of milk.
4. Strict vegetarians and people who are either allergic to milk or lactose intolerant can use calcium-rich milk and cheese substitutes such as calcium-fortified soy milk or tofu (bean curd).
5. Small fish, such as canned sardines, and other canned fish prepared with their bones, such as canned salmon, are also rich in calcium.

foodallergy.org
Food Allergy Network

part of one of the body's major buffers (phosphoric acid), phosphorus is also found in all body cells. Phosphorus is a part of DNA and RNA, the genetic code material present in every cell. Thus phosphorus is necessary for all growth. Phosphorus also plays many key roles in energy transfers occurring during cellular metabolism.

Phosphorus-containing lipids (phospholipids) help transport other lipids in the blood. Phospholipids are also principal components of cell membranes. Animal protein is the best source of phosphorus because the mineral is so abundant in the cells of animals. Recommended intakes for phosphorus were recently revised and now differ from those for calcium. Deficiencies are unknown. A summary of facts about phosphorus appears in Table 8.6.

Phosphorus RDA for adults: 700 mg/day.

Most of the phosphorus in the body is in the bones and teeth. Phosphorus helps maintain acid-base balance in the blood, is part of the genetic materials in cells, assists in energy metabolism, and is part of cell membranes. Deficiencies of phosphorus are unknown.

Magnesium

Magnesium barely qualifies as a major mineral. Only about 1¾ ounces of magnesium are present in the body of a 130-pound person, over half of it in the bones. Most of the rest is in the muscles, heart, liver, and other soft tissues, with only 1 percent in the body fluids. Bone magnesium seems to be a reservoir to ensure that some will be on hand for vital reactions regardless of recent dietary intake.

Magnesium is critical to the operation of hundreds of enzymes. Magnesium acts in all the cells of the soft tissues, where it forms part of the protein-making machinery and is necessary for the release of energy. Magnesium helps relax muscles after contraction and promotes resistance to tooth decay by holding calcium in tooth enamel.

Magnesium Deficiency Magnesium deficiency can result from vomiting, diarrhea, alcohol abuse, or protein malnutrition; after surgery in people who have been fed incomplete fluids intravenously for too long; or in people using diuretics. A severe deficiency causes tetany, an extreme and prolonged contraction of the muscles much like the reaction of the muscles when calcium levels fall.

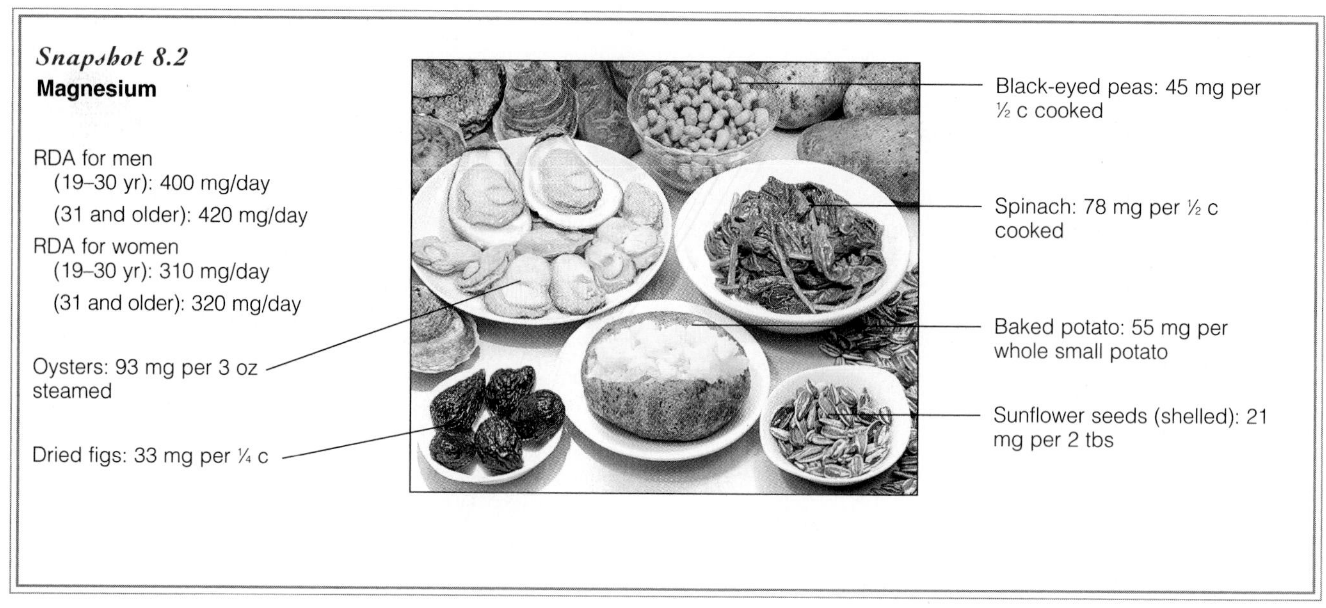

Snapshot 8.2
Magnesium
RDA for men
(19–30 yr): 400 mg/day
(31 and older): 420 mg/day
RDA for women
(19–30 yr): 310 mg/day
(31 and older): 320 mg/day
Oysters: 93 mg per 3 oz steamed
Dried figs: 33 mg per ¼ c
Black-eyed peas: 45 mg per ½ c cooked
Spinach: 78 mg per ½ c cooked
Baked potato: 55 mg per whole small potato
Sunflower seeds (shelled): 21 mg per 2 tbs

Magnesium deficiencies are also thought to cause the hallucinations experienced during withdrawal from alcohol intoxication.

Magnesium Toxicity Magnesium toxicity is most often reported in older adults who abuse magnesium-containing laxatives, antacids, and other medications. The consequences can be severe: lack of coordination, confusion, coma, and, in extreme cases, death.

Magnesium RDA:
Men (19–30 yr): 400 mg/day.
(31 and older): 420 mg/day.
Women (19–30 yr): 310 mg/day.
(31 and older): 320 mg/day.

Magnesium Intakes and Food Sources Dietary intakes of magnesium average about three-quarters of the recommended intake for both men and women in the United States. Dietary intake data, however, do not assess the nutrient contribution of water. In various parts of the country, the water contains both calcium and magnesium and is known as "hard" water. Hard water can contribute significantly to magnesium intakes.

Magnesium-rich food sources (Snapshot 8.2) include dark green, leafy vegetables, nuts, legumes, whole-grain breads and cereals, seafood, chocolate, and cocoa. Magnesium is easily lost from foods during processing, so unprocessed foods are the best choices.

Most of the body's magnesium is in the bones and can be drawn out for all the cells to use in building protein and releasing energy. Most people's intakes of magnesium fall short of recommendations.

Sulfur

The body does not use sulfur by itself as a nutrient. Sulfur is included here because it occurs in essential nutrients that the body does use, such as thiamin and certain amino acids. Sulfur is present in all proteins and plays its most important role in helping strands of protein to assume a particular shape and hold it. Thus sulfur helps the proteins to do their specific jobs, such as enzyme work. Skin, hair, and nails contain some of the body's more rigid proteins, and they have a high sulfur content.

Amino acids containing sulfur are methionine and cysteine. Cysteine in one part of a protein chain can bind to cysteine in another part of the chain by way of a sulfur-sulfur bridge, thus helping to stabilize the protein structure.

There is no recommended intake for sulfur, and no deficiencies are known. Only a person who lacks protein to the point of severe deficiency will lack the sulfur-containing amino acids.

The Trace Minerals

Figure 8.1, earlier in this chapter, shows how tiny the quantities of trace minerals in the human body are. If you could remove all of them from your body, you would have only a bit of dust, hardly enough to fill a teaspoon. Yet each of the trace minerals performs some vital role for which no substitute will do. A deficiency of any of them can be fatal, and an excess of many can be equally deadly.

Recommendations have been established for the best-known trace minerals—iron, zinc, iodine, selenium, and fluoride. Tentative ranges for safe and adequate daily intakes of others are also published. Still others are recognized as essential nutrients for some animals, but have not been proved to be required for human beings (see Table 8.5). Still others are under study to determine whether they, too, perform indispensable roles in the body.

Table 8.5 Trace Minerals
RDA
Iron Zinc Iodine Selenium
Adequate Intake (AI)
Fluoride
Safe and Adequate Daily Dietary Intakes
Copper Manganese Chromium Molybdenum
Known Essential for Animals; Human Requirements under Study
Arsenic Nickel Silicon Boron
Known Essential for Some Animals; No Evidence That Intake by Humans Is Ever Limiting; No Recommendation Necessary
Cobalt

Note: The evidence for requirements and essentiality is weak for the trace minerals cadmium, lead, lithium, tin, and vanadium.

Iron

Every living cell—both plant and animal—contains iron. Most of the iron in the body is a component of the proteins hemoglobin in red blood cells and myoglobin in muscle cells. Hemoglobin in the blood carries oxygen from the lungs to tissues throughout the body. Myoglobin holds oxygen for the muscles to use when they contract. Both the hemoglobin and myoglobin molecules contain iron, which helps them carry and hold oxygen and then release it. As part of many enzymes, iron is vital to the processes by which cells generate energy. Iron is also needed to make new cells, amino acids, hormones, and neurotransmitters.

The special provisions the body makes for iron's handling show that it is the body's gold, a precious mineral to be tightly hoarded. For example, when a red blood cell dies, the liver saves the iron and returns it to the bone marrow, which uses it to build new red blood cells. Thus only tiny amounts of iron are lost, principally in urine, sweat, shed skin, and blood (if bleeding occurs).

Normally, only about 10 to 15 percent of dietary iron is absorbed; but if the body's supply is diminished or if the need increases for any reason (such as pregnancy), absorption increases. The body makes several provisions for absorbing iron. A special protein in the intestinal cells captures iron and holds it in reserve for release into the body as needed; another protein transfers the iron to a special iron-carrier in the blood. The blood protein (transferrin) carries the iron to tissues throughout the body. When more iron is needed, more of these special proteins are produced so that more than the usual amount of iron can be absorbed and carried. If there is a surplus of iron, special storage proteins in the liver, bone marrow, and other organs store it.

hemoglobin: the oxygen-carrying protein of the red blood cells.
hemo = blood
globin = globular protein

myoglobin: the oxygen-carrying protein of the muscle cells.
myo = muscle

transferrin (trans-FERR-in): the body's iron-carrying protein.

The storage proteins are **ferritin** (FERR-i-tin) and **hemosiderin** (heem-oh-SID-er-in).

Iron Deficiency Worldwide, iron deficiency is the most common nutrient deficiency, affecting more than one billion people. In developing countries, one-third of the children and women of childbearing age suffer from iron-deficiency anemia.[12] In the United States, iron deficiency is less prevalent, but still affects about 10 percent of toddlers, adolescent girls, and women of childbearing age.[13]

Vulnerability to Iron Deficiency Women are especially prone to iron deficiency during their reproductive years because of blood losses during menstruation. Pregnancy places further iron demands on women. Iron is needed to support the added blood volume, the growth of the fetus, and blood loss during childbirth. Infants (six months or older) and young children receive little iron from their high-milk diets, yet need extra iron to support growth. The rapid growth of adolescence and, for females, the blood losses of menstruation also demand extra iron that a typical teen diet may not provide.

One common test for anemia measures the **hemoglobin concentration** of blood.

- *Norms for adults:*
  Men: ≥13.5 g/100 ml.
  Women: ≥12 g/100 ml.
- *Norms for children:*
  Ages 2–5: ≥11 g/100 ml.
  Ages 6–12: ≥11.5 g/100 ml.

Note that hemoglobin is measured in grams per 100 ml, but often just the number of grams alone is used in speaking of it: "hemoglobin, 14."

Another common test, the **hematocrit,** represents the percentage of red blood cells in a whole blood sample.

- *Norms for adults:*
  Men: ≥41%.
  Women: ≥36%.
- *Norms for children:*
  Ages 2–5: ≥34%.
  Ages 6–12: ≥35%.

Transferrin can be measured directly or estimated by measuring the **total iron-binding capacity (TIBC)** and the **transferrin saturation.**

iron deficiency: having depleted iron stores.

iron-deficiency anemia: a blood iron deficiency characterized by small, pale red blood cells; also called **microcytic hypochromic anemia.**
micro = small
cytic = cells
hypo = too little
chrom = color

In all people including those who are dark skinned, a sign of iron deficiency can be observed by looking in the corner of the eye. The eye lining, normally pink, will be very pale, even white. The skin of a fair person who is anemic may be noticeably pale.

There is more about the effects of iron deficiency on children's behavior in Chapter 13.

pica (PIE-ka): a craving for nonfood substances; also known as *geophagia* (jee-oh-FAY-jee-uh) when referring to clay-eating behavior.
picus = woodpecker or magpie
geo = earth
phagein = to eat

Causes of Iron Deficiency The cause of iron deficiency is usually inadequate intake from ignorance of what foods to choose, from sheer lack of food altogether, or from high consumption of iron-poor foods. In the Western world, high sugar and fat intakes are often responsible for low iron intakes. Blood loss is the primary nonnutritional cause, especially in poor regions of the world where parasitic infections of the GI tract may lead to blood loss.

Tests for Iron Deficiency The most common tests for iron deficiency anemia measure the number and size of the red blood cells and the cells' hemoglobin content. Before these levels fall, at the very beginning of an iron deficiency, the transferrin concentration *rises.* A sensitive test that will detect a developing iron deficiency before it is full-blown measures the amount of transferrin in the blood and the amount of iron it is carrying. Other tests measure iron stores.

Anemia and Iron Deficiency The distinction between anemia and iron deficiency is important. They often go hand in hand, but people can be anemic without being iron deficient and iron deficient without being anemic. Anemia is a symptom of a wide variety of disorders, some unrelated to nutrition, and some related to nutrients other than iron, such as folate and vitamin B_{12}. (Appendix E lists tests useful in identifying anemia and distinguishing between the major types of nutritional anemias.) In iron-deficiency anemia, new red blood cells are smaller and lighter red than normal (see Figure 8.4). The depleted cells cannot carry enough oxygen from the lungs to the tissues, so energy release in all the cells is hindered. The entire body feels the effect.

Anemia is a clinical sign of severe iron deficiency. Other classic symptoms include fatigue, weakness, headaches, apathy, and pallor. A more recently recognized symptom is poor tolerance to cold. One way the body accelerates heat production when the environmental temperature falls involves the neurotransmitter norepinephrine and the thyroid hormones, which speed up the metabolic rate. Iron deficiency impairs temperature regulation in both animals and human beings, probably by interfering with the normal production of these compounds.

Less severe iron deficiency produces symptoms, too. Long before the mass of the red blood cells is affected and anemia is diagnosed, a developing iron deficiency affects behavior. Even at slightly lowered iron levels, the complete oxidation of pyruvate is impaired, reducing physical work capacity and productivity. Children deprived of iron become irritable, restless, and unable to pay attention. These symptoms are among the first to appear when the body's iron begins to fall and among the first to disappear when iron intake is increased again.

Several mechanisms by which iron deficiency may affect behavior have been proposed. The one most often discussed and researched proposes that even in the earliest stages of iron deficiency, a deficit of iron-dependent neurotransmitter receptors in the brain alters behavior.[14]

Iron Deficiency and Pica A curious symptom seen in some iron-deficient individuals is an appetite for ice, clay, paste, and other nonnutritious substances. Such people have been known to eat as many as eight trays of ice in a day, for example. This behavior, which has been named *pica,* has been observed for years, especially in women and children of low-income groups who are deficient in either iron or zinc. After iron is given, pica clears up dramatically within days, long before the red blood cells respond.

Caution on Self-Diagnosis Low hemoglobin may reflect an inadequate iron intake, and if it does, the physician may prescribe iron supplements. However, any nutrient deficiency or disease or agent that interferes with hemoglobin synthesis, disrupts hemoglobin function, or causes a loss of red blood cells can precipitate anemia.

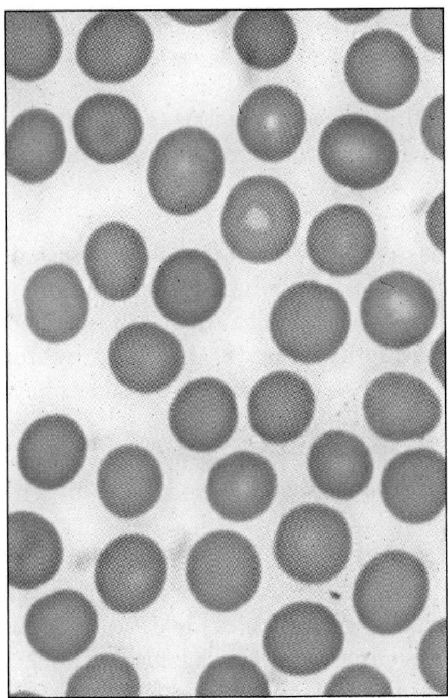

Normal blood cells. Both size and color are normal. The one large, purple cell is a normal white blood cell, stained purple.

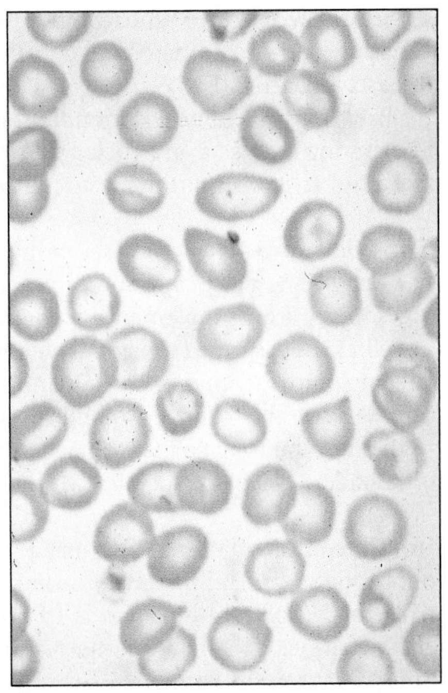

Blood cells in microcytic hypochromic anemia such as that caused by iron deficiency. These cells are small and pale because they contain less hemoglobin.

Figure 8.4
Normal and Anemic Blood Cells

Feeling fatigued, weak, and apathetic is thus a sign that something is wrong, but does not indicate that a person should take iron supplements; it means that the person should consult a physician. In fact, taking iron supplements may be the worst possible thing a person can do, because such supplements can mask a serious medical condition, such as hidden bleeding from cancer or an ulcer. Furthermore, a person can waste precious time in not seeking treatment. Remember, don't self-diagnose.

Iron Overload Normally, the body protects itself against absorbing too much iron by setting up a block in the intestinal cells. The system can be overwhelmed, however, and iron overload is the result. Once considered rare, iron overload has emerged as an important disorder of iron metabolism and regulation.[15]

Iron overload is known as hemochromatosis and is usually caused by a genetic disorder that enhances iron absorption. Hereditary hemochromatosis is the most common genetic disorder in the United States, affecting about 1.5 million people.[16] Other causes of iron overload include repeated blood transfusions, massive doses of supplementary iron, and other rare metabolic disorders. Long-term overconsumption of iron may cause hemosiderosis, a condition characterized by large deposits of the iron-storage protein hemosiderin in the liver and other tissues.

Some of the signs and symptoms of iron overload are similar to those of iron deficiency: apathy, lethargy, and fatigue. Therefore, taking iron supplements before assessing iron status is clearly unwise; hemoglobin tests alone would fail to make the distinction.

Iron overload is characterized by tissue damage, especially in iron-storing organs such as the liver. Infections are likely because bacteria thrive on iron-rich blood. Symptoms are most severe in alcohol abusers because alcohol damages

Binding proteins in the intestinal cells (mucosal ferritin and mucosal transferrin) capture and hold unneeded iron to be shed with the cells, thereby forming a **mucosal block** to iron absorption.

iron overload: toxicity from excess iron.

hemochromatosis (heem-oh-crome-a-TOE-sis): iron overload characterized by deposits of iron-containing pigment in many tissues, with tissue damage. Hemochromatosis is a hereditary defect in iron metabolism.

hemosiderosis (heem-oh-sid-er-OH-sis): a condition characterized by the deposition of the iron-storage protein hemosiderin in the liver and other tissues.

the intestine, further impairing its defenses against absorbing excess iron. Untreated hemochromatosis aggravates the risk of diabetes, liver cancer, heart disease, and arthritis.

Iron overload is more common in men than in women and is twice as prevalent among men as iron deficiency. The fortification of foods with iron (to protect women) makes it difficult to follow a low-iron diet.

WWW.
aap.org
American Academy of Pediatrics

fda.gov
U.S. Food and Drug Administration

Iron Poisoning The rapid ingestion of massive amounts of iron can cause sudden death. The most common cause of accidental poisoning in small children is ingestion of iron supplements or vitamins with iron.[17] The American Academy of Pediatrics has urged the Food and Drug Administration (FDA) to improve the labeling of iron-containing drugs and supplements. As few as 6 to 12 tablets have caused death in a child.[18] A child suspected of iron poisoning should be rushed to the hospital to have the stomach pumped. Thirty minutes can make a crucial difference.

Iron RDA:
Men (19 and older): 10 mg/day.
Women (19–50 yr): 15 mg/day.
(>50 yr): 10 mg/day.

Iron Recommendations The usual mixed diet in the United States provides only about 6 to 7 milligrams of iron in every 1000 kcalories. The recommended daily intake for an adult man is 10 milligrams; most men easily eat more than 2000 kcalories, so a man can meet his iron needs without special effort. The recommendation for women during childbearing years, however, is 15 milligrams. Because women have higher iron needs and typically consume fewer than 2000 kcalories per day, they have trouble achieving appropriate iron intakes. On the average, women receive only 10 to 11 milligrams of iron per day. A woman who wants to meet her iron needs from foods must emphasize the most iron-rich foods in every food group.

About 40% of the iron in meat, fish, and poultry is bound into molecules of **heme** (HEEM), the iron-holding part of the hemoglobin and myoglobin proteins. This heme iron is much more absorbable than nonheme iron.

Iron in Foods Iron occurs in two forms in foods, one of which is up to ten times more absorbable than the other. The absorbable form is heme iron, which is bound into the iron-carrying proteins hemoglobin and myoglobin in meats, poultry, and fish. The less absorbable form is nonheme iron, found in meats and also in plant foods. Heme iron contributes a smaller portion of the iron consumed by most people, but healthy people absorb it at a fairly constant rate of about 23 percent. People absorb nonheme iron at a lower rate (2 to 20 percent); its absorption depends on dietary factors and iron stores. Most of the iron people consume is nonheme iron from vegetables, grains, eggs, meat, fish, and poultry. Snapshot 8.3 shows the iron found in usual serving sizes of different foods.

To absorb a maximum of iron from the foods you eat, you need to know what enhances iron absorption: MFP factor and vitamin C. Meat, fish, and poultry contain a factor (MFP factor) other than heme that promotes the absorption of iron. MFP factor even enhances the absorption of nonheme iron from other foods eaten at the same time. Vitamin C eaten in the same meal also doubles or triples nonheme iron absorption. Additionally, cooking with iron skillets can contribute iron to the diet. Tea and coffee interfere with iron absorption. The accompanying box offers suggestions on obtaining adequate iron.

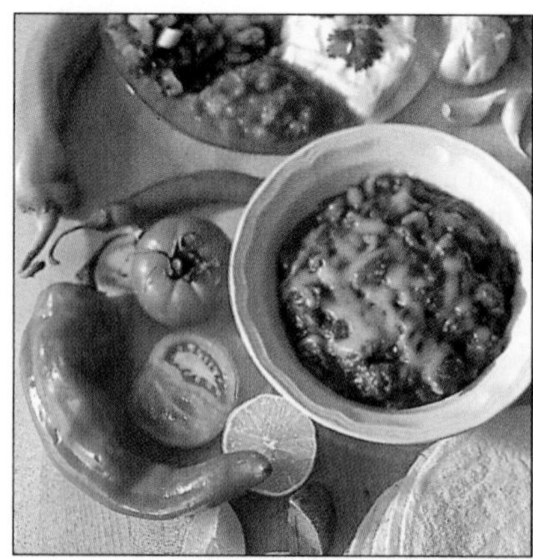

This chili dinner provides iron and MFP factor from meat, iron from legumes, and vitamin C from tomatoes. The combination of heme iron, nonheme iron, MFP factor, and vitamin C helps to achieve maximum iron absorption.

In summary, most of the body's iron is in hemoglobin and myoglobin. Special proteins help with iron absorption, transport, and storage. Iron deficiency is a worldwide problem and is most common among infants, children, adolescents, and women of childbearing age. Iron overload is more common among men than women. Heme iron in foods is better absorbed than nonheme iron. Nonheme iron absorption is improved by eating iron-containing foods with foods containing the MFP factor and vitamin C.

How to

Add Iron to Daily Meals

The following set of guidelines can be used for planning an iron-rich diet:

- *Breads and cereals* Use only whole-grain, enriched, and fortified products (iron is one of the enrichment nutrients).
- *Vegetables* The dark green, leafy vegetables are rich in vitamin C and iron. Eat vitamin C–rich vegetables often to enhance absorption of the iron from foods eaten with them.
- *Fruits* Dried fruits, such as raisins, apricots, peaches, and prunes, are high in iron. Eat vitamin C–rich fruits often with iron-containing foods.
- *Milk and cheese* Don't overdo foods from the milk group; they are poor sources of iron. But don't omit them either, because they are rich in calcium. Drink nonfat milk to free kcalories to be invested in iron-rich foods.
- *Meats* Meat, fish, and poultry are excellent iron sources.
- *Meat alternates* Include legumes frequently. A cup of peas or beans can supply up to 7 milligrams of iron.

Zinc

Zinc is a versatile trace mineral required as a cofactor by more than 100 enzymes. These zinc-requiring enzymes perform tasks in the eyes, liver, kidneys, muscles, skin, bones, and male reproductive organs. Zinc works with the enzymes that make genetic material; manufacture heme; digest food; metabolize carbohydrate, protein, and fat; liberate vitamin A from storage in the liver; and dispose of damaging free radicals. Zinc also interacts with platelets in blood clotting, affects thyroid hormone function, assists in immune function, and affects behavior and learning performance. Zinc is needed to produce the active form of vitamin A in visual pigments and is essential to wound healing, taste perception, the making of sperm, and fetal development. When zinc deficiency occurs, it impairs all these and other functions.

The body's handling of zinc differs from that of iron, but with some interesting similarities. For example, like iron, extra zinc that enters the body is held within the intestinal cells, and only the amount needed is released into the

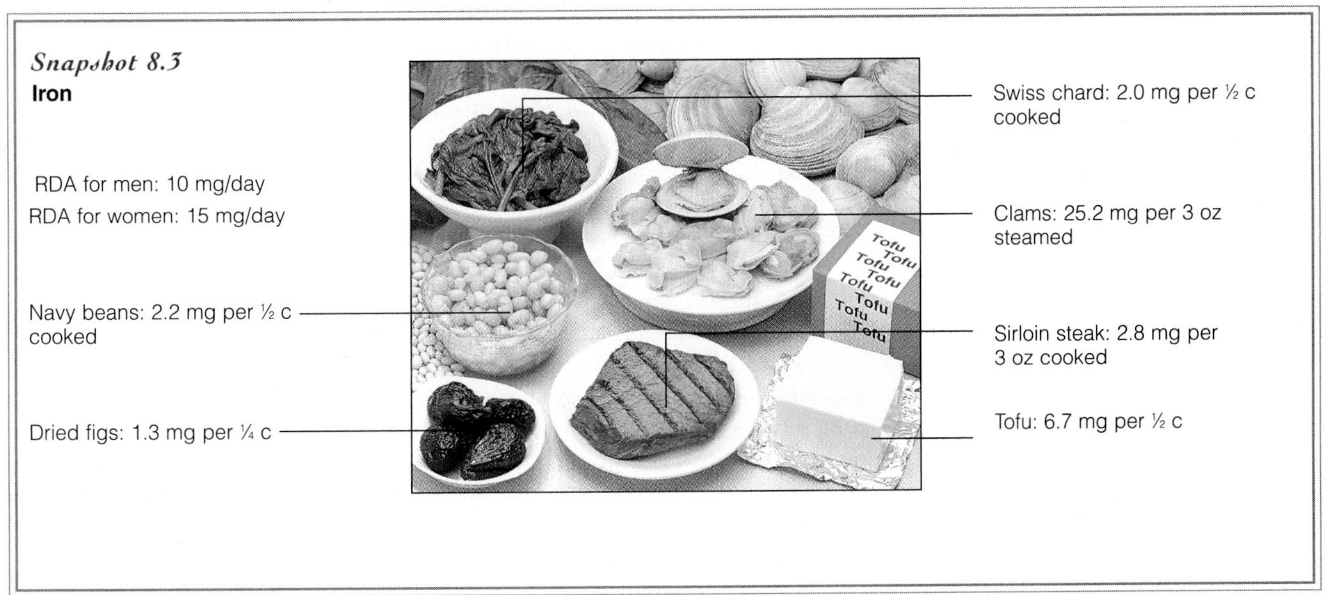

bloodstream. As with iron, a person's zinc status influences the percentage of zinc the person absorbs from the diet; if more is needed, more is absorbed.

Zinc's main transport vehicle in the blood is the protein albumin. Research suggests that circulating albumin is a main determinant of zinc absorption. This may account for observations that zinc absorption declines in conditions that lower plasma albumin concentrations—for example, pregnancy and malnutrition.

phytates: nonnutrient components of grains, legumes, and seeds; phytates can bind minerals such as zinc, iron, calcium, and magnesium in insoluble complexes in the intestine, which the body excretes unused.

Zinc Deficiency Zinc deficiency in human beings was first reported in the 1960s from studies of growing children and male adolescents in Egypt, Iran, and Turkey. Their diets were typically low in zinc and high in fiber and phytates (which impair zinc absorption). The zinc deficiency was marked by dwarfism or severe growth retardation, as well as arrested sexual maturation—symptoms that were responsive to zinc supplementation.

Since that time, zinc deficiency has been recognized elsewhere and is known to affect more than growth. It drastically impairs immune function, causes loss of appetite, and, during pregnancy, may lead to developmental disorders. A detailed list of symptoms of zinc deficiency is presented later in Table 8.6. Conditions other than poor diet that contribute to the development of zinc deficiency include loss of blood due to parasitic infections, climates that increase sweat losses, and the practice of clay eating.

Clay eating: see *pica*, p. 186.

Pronounced zinc deficiency is not widespread in developed countries, but deficiencies do occur in the most vulnerable groups of the U.S. population—pregnant women, young children, the elderly, and the poor. Even mild zinc deficiency can result in metabolic changes such as impaired immune response, abnormal taste, and abnormal dark adaptation (zinc is required to produce the active form of vitamin A, retinal, in visual pigments).

Pregnant teenagers are particularly vulnerable because they need zinc for their own ongoing growth, as well as for the developing fetus. Vegetarians, especially pregnant vegetarians, who consume large amounts of fiber, phytate, and dairy foods or low levels of protein need to scrutinize their diets for possible zinc deficiency. Research shows that both zinc intake and zinc absorption are reduced in people eating vegetarian diets that include milk and eggs compared with those whose diets include meat, fish, and poultry.[19] The researchers note, however, that despite a greater risk of zinc deficiency in people consuming vegetarian diets, zinc balance can be maintained with the inclusion of zinc-rich whole-grain breads and cereals and legumes.

Zinc Toxicity Zinc can be toxic if consumed in large enough quantities. A high zinc intake is known to produce copper-deficiency anemia by inducing the intestinal cells to synthesize large amounts of a protein that captures copper in a nonabsorbable form. Accidental consumption of high levels of zinc can cause vomiting, diarrhea, fever, exhaustion, and a host of other symptoms (see Table 8.6, later in the chapter). Large doses can even be fatal.

Zinc RDA:
Men: 15 mg/day.
Women: 12 mg/day.

Zinc Recommendations and Food Sources The zinc recommendation for men is 15 milligrams per day; for women, 12 milligrams. Zinc intakes of adults in the United States fall short of recommendations.

Zinc is most abundant in foods high in protein, such as shellfish (especially oysters), meats, and liver. In general, two ordinary servings a day of animal protein provide most of the zinc a healthy person needs. Milk, eggs, and whole-grain products are good sources of zinc if eaten in large quantities. For infants, breast milk is a good source of zinc, which is more efficiently absorbed from human milk than from cow's milk. Commercial infant formulas are fortified with zinc, of course. Snapshot 8.4 shows zinc-rich foods.

Zinc supplements are not recommended except for an accurately diagnosed zinc deficiency or when needed for use as a medication to displace other ions in

Snapshot 8.4
Zinc

RDA for men: 15 mg/day
RDA for women: 12 mg/day

Black beans: 1.0 mg per ½ c cooked

Crabmeat: 3.6 mg per 3 oz steamed

Yogurt: 2.2 mg per cup

Green peas: 1.0 mg per ½ c

Sirloin steak: 5.5 mg per 3 oz cooked

Oysters: 154.0 mg per 3 oz steamed

unusual medical circumstances. Normally, it should be possible to obtain enough zinc from the diet.

Zinc assists more than 100 enzymes in reactions affecting growth, vitamin A activity, digestion, and sperm and fetal development. Like iron, zinc absorption is influenced by zinc status; if more is needed, more is absorbed. Zinc deficiency retards growth and delays sexual maturation. Protein-rich foods are the best sources of zinc. Fiber and phytates in cereals can bind zinc and limit absorption.

Selenium

Selenium is an essential trace mineral that functions as part of an antioxidant enzyme called glutathione peroxidase. Glutathione peroxidase prevents free-radical formation, thus blocking the damaging chain reaction before it begins. Glutathione peroxidase and vitamin E work in concert. If free radicals do form, and a chain reaction starts, vitamin E halts it. Selenium also plays a role in converting thyroid hormone to its active form.[20]

selenium (se-LEEN-ee-um): a trace element.

Selenium and Cancer The question of whether selenium protects against the development of some cancers is under investigation. Some research suggests that selenium supplements may reduce the incidence of some types of cancers, but given the potential for harm and the lack of additional evidence, recommendations to take selenium supplements would be premature.[21]

Selenium Deficiency Selenium deficiency is associated with heart disease in children in regions of China where the soil and foods lack selenium. The heart disease is named *Keshan disease* for one of the provinces of China where it was studied.

Selenium Toxicity High doses of selenium are toxic. Selenium toxicity causes vomiting, diarrhea, loss of hair and nails, and lesions of the skin and nervous system.

Selenium RDA:
Men: 70 µg/day.
Women: 55 µg/day.

Selenium Recommendations and Intakes Anyone who eats a normal diet composed mostly of unprocessed foods need not worry about meeting the selenium recommendations. Selenium is widely distributed in foods such as meats and shellfish and in vegetables and grains grown on selenium-rich soil. Some regions in the United States and Canada produce crops on selenium-poor soil, but people are protected from deficiency because they eat selenium-rich meat and supermarket foods transported from other regions.

Iodine

Iodine occurs in the body in minuscule amounts, but its principal role in human nutrition is well known, and the amount needed is well established. Iodine is an integral part of the thyroid hormones, which regulate body temperature, metabolic rate, reproduction, growth, the making of blood cells, nerve and muscle function, and more.

goiter (GOY-ter): an enlargement of the thyroid gland due to an iodine deficiency, malfunction of the gland, or overconsumption of a thyroid antagonist. Goiter caused by iodine deficiency is *simple goiter.*

A thyroid antagonist found in food, which causes *toxic goiter,* is called a **goitrogen.**

cretinism (CREE-tin-ism): an iodine-deficiency disease characterized by mental and physical retardation.

Iodine Deficiency When the iodine concentration in the blood is low, the cells of the thyroid gland enlarge in an attempt to trap as many particles of iodine as possible. If the gland enlarges until it is visible, the swelling is called a simple goiter. As many as 800 million people are at the borderline of iodine deficiency, and 200 million people worldwide have goiter. In all but 4 percent of these cases, the cause is iodine deficiency. As for the 4 percent (8 million), those people have goiter because they overconsume plants of the cabbage family and others that contain an antithyroid substance whose effect is not counteracted by dietary iodine.

In addition to causing sluggishness and weight gain, an iodine deficiency may have serious effects on fetal development. Severe thyroid undersecretion during pregnancy causes the extreme and irreversible mental and physical retardation known as cretinism. A cretin has an IQ as low as 20 (100 is normal) and a face and body with many abnormalities. Iodine deficiency is one of the world's most common preventable causes of mental retardation and can be averted if the pregnant woman's deficiency is detected and treated in time.[22]

WWW. thyroid.org
American Thyroid Association

Iodine Toxicity Excessive intakes of iodine can enlarge the thyroid gland, just as deficiencies can. In infants, the goiterlike condition can be so severe as to block the airways and cause suffocation.

Iodine Sources and Intakes The ocean is the world's major source of iodine. In coastal areas, seafood, water, and even iodine-containing sea mist are important iodine sources. Further inland, the amount of iodine in the diet is variable and generally reflects the amount present in the soil in which plants are grown or on which animals graze. In the United States and Canada, the use of iodized salt has largely wiped out the iodine deficiency that once was widespread.

Iodine RDA: 150 μg/day.

The need for iodine is easy to meet by consuming seafood, vegetables grown in iodine-rich soil, and iodized salt. In the United States, you have to read the label to find out whether salt is iodized; in Canada, all table salt is iodized.

Copper

The body contains about 100 milligrams of copper. About one-fourth is in the muscles, one-fourth is in the liver, brain, and blood, and the rest is in the bones, kidneys, and other tissues. The primary function of copper in the body is to serve as a constituent of enzymes.[23] The copper-containing enzymes have diverse metabolic roles: they catalyze the formation of hemoglobin, help manufacture the protein collagen, assist in the healing of wounds, and help maintain the sheaths around nerve fibers. One of copper's most vital roles is to help cells use iron. Like iron, copper is needed in many reactions related to respiration and energy release.

Copper Deficiency Copper deficiency is rare but not unknown. It has been seen in malnourished children. High intakes of vitamin C and iron interfere with copper absorption and can lead to deficiency.[24]

Copper Toxicity Some genetic disorders create a copper toxicity. Copper toxicity from foods, however, is unlikely.

Copper Recommendations and Food Sources An estimated safe and adequate daily dietary intake has been established for copper. The best food sources of copper include legumes, whole grains, seafood, nuts, and seeds.

Estimated safe and adequate dietary intake for copper: 1.5 to 3.0 mg/day.

Manganese

The human body contains a tiny 20 milligrams of manganese, mostly in the bones and glands. Manganese is a cofactor for many enzymes, helping to facilitate dozens of different metabolic processes. Deficiencies of manganese have not been noted in people, but toxicity may be severe. Miners who inhale large quantities of manganese dust on the job over prolonged periods show many symptoms of a brain disease, along with abnormalities in appearance and behavior.

Manganese requirements are low, and plant foods such as nuts, whole grains, and green, leafy vegetables contain significant amounts of this trace mineral. Deficiencies are therefore unlikely.

Estimated safe and adequate dietary intake for manganese: 2.5 to 5.0 mg/day.

Fluoride

Only a trace of fluoride occurs in the human body, but research demonstrates that where diets are high in fluoride during the growing years, crystalline deposits in bones and teeth are larger and more perfectly formed. When bones and teeth become mineralized, first a crystal called hydroxyapatite forms from calcium and phosphorus. Then fluoride replaces the hydroxy portion of hydroxyapatite, forming fluorapatite, which makes the bones and teeth more resistant to decay. Once the teeth have erupted, the topical application of fluoride by way of toothpaste or mouth rinse continues to exert a caries-reducing effect.

fluorapatite (floor-APP-uh-tite): the stabilized form of bone and tooth crystal, in which fluoride has replaced the hydroxy portion of hydroxyapatite.

Fluoride Deficiency Where fluoride is lacking in the water supply, the incidence of dental decay is high. Fluoridation of water to raise its fluoride concentration to 1 part per million is recommended as an important public health measure. Those fortunate enough to have had sufficient fluoride during the tooth-forming years of infancy and childhood are protected throughout life from dental decay. Dental problems are of great concern because they can lead to a multitude of other health problems affecting the whole body. Despite fluoride's value, violent disagreement often surrounds the introduction of fluoride to a community. Figure 8.5 on the next page shows the extent of fluoridation nationwide.

Fluoride AI:
Men: 3.8 mg/day.
Women: 3.1 mg/day.

fluorosis (floor-OH-sis): mottling of the tooth enamel from ingestion of too much fluoride during tooth development.

Fluoride Sources All normal diets include some fluoride, but drinking water is usually the most significant source. Fish and tea may supply substantial amounts.

In some areas, the natural fluoride concentration in water is high, and children's teeth develop with mottled enamel. Although this condition, called fluorosis, may not be harmful, it violates some people's prejudice that teeth should be white. Fluorosis occurs only during tooth development and cannot be reversed, making its prevention a high priority.

To prevent fluorosis:
- *Monitor the fluoride content of the local water supply.*
- *Supervise toddlers when they brush their teeth and use only a pea-size amount of toothpaste.*
- *Use fluoride supplements only as prescribed by a physician.*

Chromium

Chromium is an essential mineral that participates in carbohydrate and lipid metabolism. Chromium enhances the activity of the hormone insulin.[25] Consequently, less insulin is needed to control blood glucose. When chromium is lacking, a diabetes-like condition may develop.

Small organic compounds that enhance insulin's activity are called glucose tolerance factors (GTF). Some glucose tolerance factors contain chromium.

Chromium deficiency is unlikely, given the small amount required and its presence in a variety of foods. The more refined foods people eat, however, the

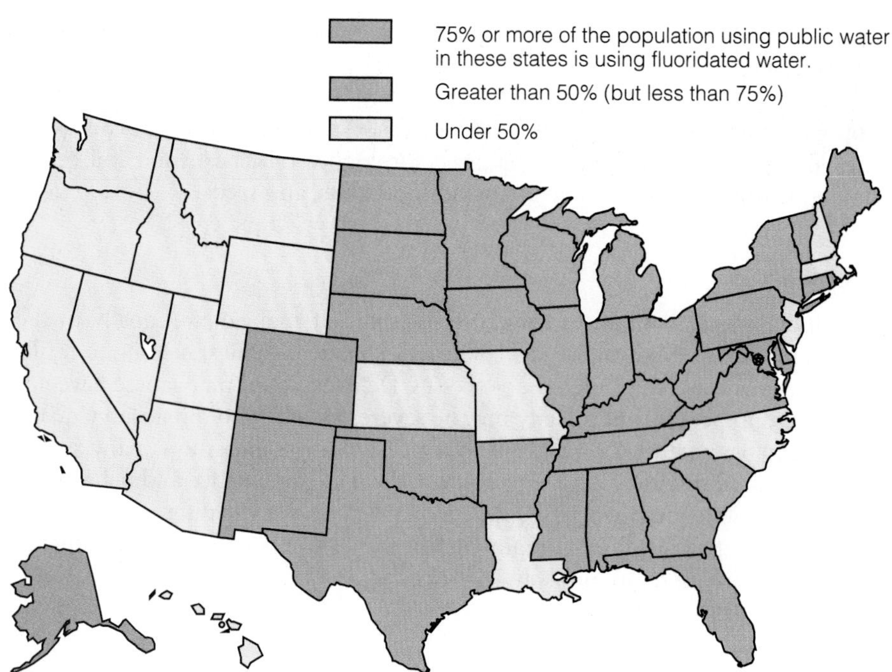
75% or more of the population using public water in these states is using fluoridated water.
Greater than 50% (but less than 75%)
Under 50%

Figure 8.5
Fluoridation in the United States
Source: Fluoridation Census 1989 Summary, U.S. Department of Health and Human Services, Public Health Service, Centers for Disease Control and Prevention, National Center for Prevention Services, Division of Oral Health. Atlanta, Ga., April 1993.

Estimated safe and adequate dietary intake for chromium: 50 to 200 µg/day.

less chromium they obtain from their diets. Unrefined foods such as liver, brewer's yeast, whole grains, nuts, and cheeses are the best sources.

Other Trace Minerals

molybdenum (mo-LIB-duh-num): a trace element.

Estimated safe and adequate dietary intake for molybdenum: 75 to 250 µg/day.

The trace minerals have been known for decades, but their roles as nutrients are a recent surprise. Molybdenum functions as a working part of several metal-containing enzymes, some of which are giant proteins. Deficiencies or toxicities of molybdenum are unknown.

Nickel is recognized as important for the health of many body tissues. Nickel deficiencies harm the liver and other organs. Silicon is known to be involved in bone calcification, at least in animals. Tin is necessary for growth in animals and probably in people also. Cobalt is recognized as the mineral in the large vitamin B_{12} molecule. The future may reveal that other trace minerals also play key roles. Even arsenic—famous as the death potion in many murder mysteries and known to be a carcinogen—may turn out to be an essential nutrient in tiny quantities.

In summary, the body requires trace minerals in tiny amounts, and they function in similar ways—assisting enzymes all over the body. Eating a diet that consists of a variety of foods is the best way to ensure an adequate intake of these important nutrients. Many dietary factors, including the trace minerals themselves, affect the absorption and availability of these nutrients.

Like the vitamins, the minerals perform a multitude of functions throughout the body. Table 8.6 offers a summary of facts about minerals in the body.

Table 8.6 The Minerals—A Summary

Mineral Name	Chief Functions In the Body	Deficiency Symptoms	Toxicity Symptoms	Significant Sources
Major Minerals				
Sodium	With chloride and potassium (electrolytes), maintains cells' normal fluid balance and acid-base balance in the body. Also critical to nerve impulse transmission.	Muscle cramps, mental apathy, loss of appetite.	Hypertension.	Salt, soy sauce, processed foods.
Chloride	Part of the hydrochloric acid found in the stomach and necessary for proper digestion.	Growth failure in children; muscle cramps, mental apathy, loss of appetite; can cause death (uncommon).	Normally harmless (the gas chlorine is a poison but evaporates from water); can cause vomiting.	Salt, soy sauce; moderate quantities in whole, unprocessed foods, large amounts in processed foods.
Potassium	Facilitates reactions, including the making of protein; the maintenance of fluid and electrolyte balance; the support of cell integrity; the transmission of nerve impulses; and the contraction of muscles, including the heart.	Deficiency accompanies dehydration; causes muscular weakness, paralysis, and confusion; can cause death.	Causes muscular weakness; triggers vomiting; if given into a vein, can stop the heart.	All whole foods: meats, milk, fruits, vegetables, grains, legumes.
Calcium	The principal mineral of bones and teeth. Also acts in normal muscle contraction and relaxation, nerve functioning, blood clotting, blood pressure, and immune defenses.	Stunted growth in children; adult bone loss (osteoporosis).	Excess calcium is excreted except in hormonal imbalance states (not caused by nutritional deficiency).	Milk and milk products, oysters, small fish (with bones), tofu (bean curd), greens, legumes.
Phosphorus	Important in cells' genetic material, in cell membranes as phospholipids, in energy transfer, and in buffering systems.	Phosphorus deficiency unknown.	Excess phosphorus may cause calcium excretion.	All animal tissues.
Magnesium	Another factor involved in bone mineralization, the building of protein, enzyme action, normal muscular contraction, transmission of nerve impulses, and maintenance of teeth.	Weakness; confusion; depressed pancreatic hormone secretion; if extreme, convulsions, bizarre movements (especially of eyes and face), hallucinations, and difficulty in swallowing. In children, growth failure.[a]	Not known; large doses have been taken in the form of the laxative Epsom salts, without ill effects except diarrhea.	Nuts, legumes, whole grains, dark green vegetables, seafoods, chocolate, cocoa.
Sulfur	A component of certain amino acids; part of the vitamins biotin and thiamin and the hormone insulin; combines with toxic substances to form harmless compounds; stabilizes protein shape by forming sulfur-sulfur bridges.	None known; protein deficiency would occur first.	Would occur only if sulfur amino acids were eaten in excess; this (in animals) depresses growth.	All protein-containing foods.

continued

[a]A still more severe deficiency causes tetany, an extreme, prolonged contraction of the muscles similar to that caused by low blood calcium.

Table 8.6 *(continued)*

Mineral Name	Chief Functions In the Body	Deficiency Symptoms	Toxicity Symptoms	Significant Sources
Trace Minerals				
Iron	Part of the protein hemoglobin, which carries oxygen in the blood; part of the protein myoglobin in muscles, which makes oxygen available for muscle contraction; necessary for the utilization of energy.	Anemia: weakness, pallor, headaches, reduced resistance to infection, inability to concentrate, lowered cold tolerance.	Iron overload: infections, liver injury, possible increased risk of heart attack, acidosis, bloody stools, shock.	Red meats, fish, poultry, shellfish, eggs, legumes, dried fruits.
Zinc	Part of the hormone insulin and many enzymes; involved in making genetic material and proteins, immune reactions, transport of vitamin A, taste perception, wound healing, the making of sperm, and normal fetal development.	Growth failure in children, sexual retardation, loss of taste, poor wound healing.	Fever, nausea, vomiting, diarrhea, muscle incoordination, dizziness, anemia, accelerated atherosclerosis, kidney failure.	Protein-containing foods: meats, fish, shellfish, poultry, grains, vegetables.
Selenium	Part of an enzyme that breaks down reactive chemicals that harm cells; works with vitamin E.	Muscle discomfort, weakness, predisposition to heart disease characterized by cardiac tissue becoming fibrous.	Nausea, abdominal pain, nail and hair changes, nerve damage.	Seafoods, organ meats, other meats, grains and vegetables depending on soil conditions.
Iodine	A component of two thyroid hormones, which help to regulate growth, development, and metabolic rate.	Goiter, cretinism.	Depressed thyroid activity; goiterlike thyroid enlargement.	Iodized salt; seafood; bread; plants grown in most parts of the country and animals fed those plants.
Copper	Necessary for the absorption and use of iron in the formation of hemoglobin; part of several enzymes.	Anemia, bone abnormalities (rare in human beings).	Vomiting, diarrhea.	Meat, drinking water.
Manganese	Facilitator, with enzymes, of many cell processes.	(In experimental animals): poor growth, nervous system disorders, reproductive abnormalities.	Nervous system disorders.	Widely distributed in foods.
Fluoride	An element involved in the formation of bones and teeth; helps to make teeth resistant to decay.	Susceptibility to tooth decay.	Fluorosis (discoloration of teeth), nausea, diarrhea, chest pain, itching, vomiting.	Drinking water (if fluoride containing or fluoridated), tea, seafood.
Chromium	Associated with insulin and required for the release of energy from glucose.	Diabetes-like condition marked by an inability to use glucose normally.	Unknown as a nutrition disorder; occupational exposures damage skin and kidneys.	Meat, unrefined foods, fats, vegetable oils.

How Are Your Mineral Intakes?

1. Compare your intakes of minerals with the RDA or AI (inside front cover) or RNI (Appendix B) or refer to the intake values calculated in your Diet Analysis Plus profile. Express each intake as a percentage of the recommended intake. For example, suppose you ingested 640 milligrams of calcium and the AI is 1000 milligrams. You ingested 64 percent (640 ÷ 1000 × 100) of your AI. If you had ingested 1400 milligrams of calcium, you would have ingested 140 percent (1400 ÷ 1000 × 100) of your AI. Use Form 9 to record your findings.

   Comment on your mineral intakes. For any mineral for which your intake fell below 80 percent of the recommendation, what were your best food sources? Could you eat more of them to bring your intake up to the recommended level? If not, what food or foods could you eat to increase your intake?

2. Compute your iron absorption from a meal of your choosing. Three factors go into the calculation. First, how much of the iron in the meal was heme iron and how much was nonheme iron? Second, how much vitamin C was in the meal? Third, how much total meat, fish, and poultry (MFP) was consumed? Here's how it works. Begin by answering these six questions:

   a. How much iron was from animal tissues (MFP)? ________ milligrams.

   b. Forty percent of this is heme iron. ________ milligrams heme iron.

   c. How much iron was from other sources? ________ milligrams.

   d. This, plus 60 percent of the iron from animal tissues (MFP), is nonheme iron. ________ milligrams nonheme iron.

   e. How much vitamin C was in the meal? Less than 25 milligrams is low; 25 to 75 milligrams is medium; more than 75 milligrams is high.

   f. How much MFP was in the meal? Less than 1 ounce lean MFP is low; 1 to 3 ounces is medium; more than 3 ounces is high.

   Now you're ready to calculate your iron absorption. You absorbed 23 percent of the heme iron (see step b) or ____ milligrams heme iron. Now, for nonheme iron, take your best response from step e or f. If either vitamin C or MFP was high, the availability of your nonheme iron was high. If neither was high but either was average, the availability of your nonheme iron was medium. If both were low, your nonheme iron had poor availability. You absorbed:

   - High availability: 8 percent of the nonheme iron.
   - Medium availability: 5 percent of the nonheme iron.
   - Poor availability: 3 percent of the nonheme iron.
   - Your absorption: ________ milligrams nonheme iron absorbed

   Now compute your iron absorption by adding the two together:

   ________ milligrams heme iron absorbed.

   ________ milligrams nonheme iron absorbed.

   Total = ________ milligrams iron absorbed.

Form 9 Mineral Intakes Compared with Recommended Intakes

	Calcium	Iron	Zinc	Magnesium	Phosphorus	Potassium	Sodium
My intake							
Recommended intake[a]							
My intake as a percentage of the recommended intake							

[a]RDA, AI, or RNI (Appendix B).

Self Check

1. Water excretion is governed by the:
   a. liver.
   b. kidneys.
   c. brain.
   d. b and c.
2. People's fluid needs are best met by:
   a. any clear liquid.
   b. plain water, milk, and juice.
   c. plain water, coffee, and tea.
   d. plain water, coffee, and alcohol.
3. Two situations in which a person may experience fluid and electrolyte imbalances are:
   a. vomiting and burns.
   b. diarrhea and cuts.
   c. broken bones and fever.
   d. heavy sweating and excessive carbohydrate intake.
4. Three-fourths of the sodium in people's diets comes from:
   a. fresh meats.
   b. home-cooked foods.
   c. frozen vegetables and meats.
   d. salt added to food by manufacturers.
5. Which mineral is critical to keeping the heartbeat steady and plays a major role in maintaining fluid and electrolyte balance?
   a. sodium
   b. calcium
   c. potassium
   d. magnesium
6. Three good food sources of calcium are:
   a. milk, sardines, and broccoli.
   b. spinach, yogurt, and sardines.
   c. cottage cheese, spinach, and tofu.
   d. swiss chard, mustard greens, and broccoli.
7. The two best ways to prevent age-related bone loss and fracture are to:
   a. take calcium supplements and estrogen.
   b. participate in aerobic activity and drink 8 glasses of milk daily.
   c. eat a diet low in fat and salt and refrain from smoking.
   d. maintain a lifelong adequate calcium intake and engage in weight-bearing physical activity.
8. Foods high in iron that help prevent or treat anemia include:
   a. green peas and cheese.
   b. dairy foods and fresh fruits.
   c. homemade breads and most fresh vegetables.
   d. meat and tomato chili and dark green, leafy vegetables.
9. Two groups of people who are especially at risk for zinc deficiency are:
   a. Asians and children.
   b. infants and the elderly.
   c. smokers and athletes.
   d. pregnant teenagers and vegetarians.
10. A deficiency of ______ is one of the world's most common preventable causes of mental retardation.
   a. zinc
   b. iodine
   c. selenium
   d. magnesium

Answers to these questions appear in Appendix H.

Notes

1. Food and Nutrition Board, *Recommended Dietary Allowances,* 10th ed. (Washington, D.C.: National Academy Press, 1989), pp. 247–261.

2. USDA Center for Nutrition Policy and Promotion, Dietary guidance on sodium: Should we take it with a grain of salt? *Nutrition Today* 32 (1997): 250.

3. Committee on Diet and Health, Food and Nutrition Board, *Diet and Health: Implications for Reducing Chronic Disease Risk* (Washington, D.C.: National Academy Press, 1989), pp. 99–135.

4. A. W. Cowley, Jr., Genetic and nongenetic determinants of salt sensitivity and blood pressure, *American Journal of Clinical Nutrition* 65 (1997): 587S–593S.

5. R. Itoh and Y. Suyama, Sodium excretion in relation to calcium and hydroxyproline excretion in a healthy Japanese population, *American Journal of Clinical Nutrition* 63 (1996): 735–740; A. Devine and coauthors, A longitudinal study of the effect of sodium and calcium intakes on regional bone density in postmenopausal women, *American Journal of Clinical Nutrition* 62 (1995): 740–745.

6. P. K. Whelton and coauthors, Effects of oral potassium on blood pressure: Meta-analysis of randomized controlled clinical trial, *Journal of the American Medical Association* 277 (1997): 1624–1632.

7. R. P. Heaney, Nutrition factors in osteoporosis, *Annual Review of Nutrition* 13 (1993): 287–316.

8. I. Vuori, Peak bone mass and physical activity: A short review, *Nutrition Reviews* 54 (1996): S11–S14; L. Alekel and coauthors, Contributions of exercise, body composition, and age to bone mineral density in premenopausal women, *Medicine and Science in Sports and Exercise* 27 (1995): 1477–1485.

9. D. A. McCarron, Role of adequate dietary calcium intake in the prevention and management of salt sensitive hypertension, *American Journal of Clinical Nutrition* 65 (1997): 712S–716S; Joint National Committee on Detection, Evaluation, and Treatment of High Blood Pressure, The Fifth Report of the Joint National Committee on Detection, Evaluation, and Treatment of High Blood Pressure, *Archives of Internal Medicine* 153 (1993): 154–183.

10. C. G. Osborne and coauthors, Evidence for the relationship of calcium to blood pressure, *Nutrition Reviews* 54 (1996): 365–381;

D. A. McCarron and coauthors, Dietary calcium and blood pressure: Modifying factors in specific populations, *American Journal of Clinical Nutrition* (supplement) 54 (1991): 215–219.

11. H. C. Bucher and coauthors, Effects of dietary calcium supplementation on blood pressure: A meta-analysis of randomized controlled trials, *Journal of the American Medical Association* 275 (1996): 1016–1022.

12. C. E. West, Strategies to control nutritional anemia, *American Journal of Clinical Nutrition* 64 (1996): 789–790.

13. A. C. Looker and coauthors, Prevalence of iron deficiency in the United States, *Journal of the American Medical Association* 277 (1997): 973–976.

14. E. Pollitt, Iron deficiency and cognitive function, *Annual Review of Nutrition* 13 (1993): 521–537.

15. J. C. Fleet, Discovery of the hemochromatosis gene will require rethinking the regulation of iron metabolism, *Nutrition Reviews* 54 (1996): 285–292.

16. Iron overload disorders among Hispanics—San Diego, California, 1995, *Morbidity and Mortality Weekly Report* 45 (1996): 991–993; D. H. G. Crawford and coauthors, Factors influencing disease expression in hemochromatosis, *Annual Review of Nutrition* 16 (1996): 139–160; C. E. McLaren and coauthors, Prevalence of heterozygotes for hemochromatosis in the white population of the United States, *Blood* 84 (1995): 2121–2127.

17. Pediatricians seek FDA's help in preventing poisoning deaths, *Journal of the American Dietetic Association* 93 (1993): 529.

18. Keep iron tablets away from children, *FDA Consumer,* May 1993, p. 2.

19. J. R. Hunt, L. A. Matthys, and L. K. Johnson, Zinc absorption, mineral balance, and blood lipids in women consuming controlled lactovegetarian and omnivorous diets for 8 wk, *American Journal of Clinical Nutrition* 67 (1998): 421–430.

20. J. R. Arthur, F. Nicol, and G. J. Beckett, Selenium deficiency, thyroid hormone metabolism, and thyroid hormone deiodinases, *American Journal of Clinical Nutrition* 57 (1993): 236S–239S.

21. L. C. Clark and coauthors, Effects of selenium supplementation for cancer prevention in patients with carcinoma of the skin—A randomized controlled trial, *Journal of the American Medical Association* 276 (1996): 1984–1985; Letters from V. Herbert, L. H. Kuller, J. S. Parker, and L. C. Clark, Selenium supplementation and cancer rates, *Journal of the American Medical Association* 277 (1997): 880–881.

22. G. R. Delong, Effects of nutrition on brain development in humans, *American Journal of Clinical Nutrition* 57 (1993): 286S–290S.

23. R. Uauy, M. Olivares, and M. Gonzalez, Essentiality of copper in humans, *American Journal of Clinical Nutrition* 67 (1998): 952S–959S.

24. B. Lonnerdal, Copper nutrition during infancy and childhood, *American Journal of Clinical Nutrition* 67 (1998): 1046S–1053S.

25. S. Fairweather-Tait and R. F. Hurrell, Bioavailability of minerals and trace elements, *Nutrition Research Reviews* 9 (1996): 295–324.

CHAPTER 8

Nutrition in Practice

▪ Nutrition and the Alcohol Abuser ▪

Chapter 8 has discussed the last of the nutrients—water and the minerals. Next to the nutrients, probably the most influential substance people normally ingest is alcohol. Its impacts on nutrition are so profound that they deserve attention here.

Like all drugs, alcohol—properly termed ethanol, the active ingredient of alcoholic beverages—offers both benefits and hazards. Wine, beer, and other fermented beverages have been associated with pleasure and relaxation for more than 5000 years. People have always known that these beverages affected their moods, sensations, and behavior. Taken in moderation, alcohol can relax people, reduce their inhibitions, and encourage social interactions. Taken in excess, alcohol can be devastatingly destructive. This discussion focuses on the nutrition implications of alcohol abuse and alcohol addiction. Chapter 12 addresses the dangers of alcohol use during pregnancy.

How many drinks constitute moderate use? And how much is a "drink"?

A drink is any alcoholic beverage that delivers ½ ounce of pure ethanol:

- 4 to 5 ounces of wine.
- 10 ounces of wine cooler.
- 12 ounces of beer.
- 1½ ounce of hard liquor (80 proof whiskey, scotch, brandy, rum, gin, or vodka).

Because people's tolerances to alcohol differ, it is impossible to name an exact amount of alcohol per day that is appropriate for everyone, but authorities have attempted to set limits that are acceptable for most healthy adults. An accepted definition of moderation is not more than two drinks a day for the average-sized man and not more than one drink a day for the average-sized woman. This amount is supposed to be enough to elevate mood without causing any long-term harm to health. Doubtless, some people could consume slightly more; others could not handle nearly so much without significant risk. The amount a person can drink safely is highly individual, depending on genetics, health condition, sex, body composition, age, and family history.

WWW

niaaa.nih.gov
National Institute on Alcohol Abuse and Alcoholism

ncadd.org
National Council on Alcoholism and Drug Dependence

health.org/aboutn.htm
National Clearinghouse for Alcohol and Drug Information

Many people drink much more than one or two drinks a day, don't they?

Yes, and they may incur long-term harm to health as a result. Alcohol is the most widely abused drug in the world. Most people who choose to drink alcohol do so with few, if any, adverse consequences. Some people, however, encounter problems related to alcohol consumption. *Alcohol abuse* refers to patterns of drinking that result in health problems, social problems, or both. Alcohol abusers can often change their drinking behavior in response to simple warnings or explanations, thereby alleviating their alcohol-related problems. *Alcohol addiction,* often called alcoholism, refers to a disease that is characterized by abnormal alcohol-seeking behavior that leads to impaired control over drinking.[1*] Alcohol abusers and alcohol-addicted individuals experience many of the same harmful effects of alcohol consumption; the distinguishing characteristics of alcohol addiction are physical dependence on alcohol and an impaired ability to control alcohol intake. Alcohol abuse and addiction exert a heavy toll on the health of the 18 million people in the United States who meet the criteria for alcohol abuse, alcohol addiction, or both. The effects of alcohol on nutrition and

*Alcohol addiction means the same thing as alcohol dependence. This book uses the term *addiction* because it is more self-explanatory.

metabolism—both directly and as a consequence of alcohol-related diseases—are significant. Every alcohol abuser and alcohol-addicted person should be considered at risk for poor nutrition status.

Would you please clarify exactly what alcoholism is?

In the *Eighth Special Report to the U.S. Congress on Alcohol and Health,* the U.S. Department of Health and Human Services defines alcoholism as a disease. The term is essentially synonymous with alcohol addiction. Alcoholism has four main clinical features:

- Tolerance—more and more alcohol is needed to produce the desired effects.
- Physical dependence—when alcohol consumption is interrupted, a characteristic withdrawal syndrome appears that is relieved by more alcohol.
- Impaired ability to regulate alcohol intake—impairment can occur at any time, once drinking has begun.
- Discomfort with abstinence—the person experiences a "craving" for alcohol that can lead to relapse.

The alcohol-addicted person's craving for alcohol becomes marked by several features. The person thinks about alcohol a lot (*obsession*), drinks in spite of resolving not to (*broken promises*), and then suffers *remorse.* Such strong feelings about any substance signify addiction, but note that these feelings do not reflect personal inferiority. The person is simply someone whose internal makeup reacts in a special way to alcohol.

Some misconceptions about alcoholism can be dangerous and demand correction. For example, some people believe that alcoholism is related to the type of alcohol-containing beverage a person drinks. This is not true. People who drink only beer and wine can become alcohol addicts just as readily as people who drink hard liquors. It is not what people drink, but how much, that makes the difference. Another common misconception is that only morally degenerate people become addicted. Alcohol addiction does not single out people of low moral character or people of any particular age, race, education, social class, or income. It is true, however, that people of certain races and cultures tend to be more susceptible to alcohol addiction than others. Environment and heredity can contribute to such differences.

Alcohol addiction sets in much more quickly in young people than in adults.[2] Those who start drinking at an early age are more likely to become alcohol addicted than those who start later in life.

Table NP8.1 kCalories in Alcoholic Beverages and Mixers

Beverage	Amount (oz)	Energy (kcal)
Beer		
Regular	12	150
Light	12	78–131
Nonalcoholic	12	32–82
Distilled liquor (gin, rum, vodka, whiskey)		
80 proof	1½	100
86 proof	1½	105
90 proof	1½	110
Liqueurs		
Coffee liqueur	1½	175
Coffee and cream liqueur	1½	155
Crème de menthe	1½	185
Mixers		
Club soda	12	0
Cola	12	150
Cranberry juice cocktail	8	145
Diet drinks	12	2
Ginger ale	12	125
Grapefruit juice	8	95
Orange juice	8	110
Tomato or vegetable juice	8	45
Tonic	12	124
Wine		
Dessert	3½	160
Nonalcoholic	8	14
Red	3½	75
Rosé	3½	75
White	3½	70
Wine cooler	12	150

Why do you say that all alcohol abusers and alcohol-addicted people are at risk for poor nutrition status?

Alcohol produces euphoria, which depresses appetite, so heavy drinkers tend to eat poorly. Alcohol is rich in energy (7 kcalories per gram), but like pure fat or sugar kcalories, the kcalories from alcohol are empty kcalories. The more alcohol a person drinks, the less likely that he or she will eat enough food to obtain adequate nutrients. Table NP8.1 shows the kcalories in typical alcoholic beverages. Nutrient deficiencies are an almost inevitable result of alcohol abuse, not only because the person who drinks obtains fewer nutrients from food but also because alcohol interferes with the body's ingestion, digestion, absorption, metabolism, and excretion of nutrients.

Alcohol is directly toxic to the liver. Studies of human beings and animals show that even when the diet is adequate, alcohol damages the liver.[3] Since alcohol can affect virtually every organ, other complications frequently develop that also change nutrient requirements. Some of these include anemia (Chapter 8), gastritis and ulcers (Chapter 18), pancreatitis (Chapter 20), and liver disease (Chapter 28).

Just how does alcohol alter nutrient metabolism?

With alcohol in the system, major changes occur in the way the body metabolizes many nutrients. Dietary glucose and dietary fat are diverted to making fat, which may accumulate in the liver. In addition, alcohol metabolism in the liver results in structural changes in liver cells that can permanently alter the liver's ability to metabolize fatty acids, causing further fat accumulation.[4] Fatty liver, the first stage of liver deterioration in the heavy drinker (which can be reversed by abstinence from alcohol), can progress to cirrhosis (which is irreversible).

Does alcohol affect vitamin and mineral metabolism, too?

Yes. Alcohol interferes with the availability and activation of virtually every vitamin and many minerals. For example, as Figure NP8.1 shows, alcohol impairs absorption of thiamin, folate, and other nutrients. Alcohol-induced liver injury impairs the activity of the liver enzyme that activates thiamin to its active form.

Alcohol's effect on folate is dramatic. When alcohol is present, the body behaves as if it were actively trying to expel folate from all its sites of action and storage. The liver, which normally contains enough folate to meet all needs, leaks folate into the blood. As the blood folate concentration rises, the body appears to have an excess of folate, and the kidneys are deceived into excreting it. The intestine normally releases and retrieves folate continuously, but it becomes damaged by folate deficiency and alcohol toxicity, so it fails to retrieve its own folate and misses out on any that may trickle in from food as well. Alcohol also interferes with the action of what little folate is left, inhibiting the production of new cells, especially the rapidly dividing cells of the intestine and the blood. The combination of poor folate status and alcohol consumption has been implicated in promoting colorectal cancer.[5] Alcohol abuse causes a folate deficiency that devastates digestive system function.

Alcohol abuse alters metabolism of many other vitamins, including vitamins B_6, B_{12}, and A. One of the products of alcohol metabolism dislodges vitamin B_6 from its protective binding protein so that it is destroyed. Alcohol inhibits vitamin B_{12} absorption, both directly and indirectly, by suppressing the secretion of the factor that facilitates the vitamin's absorption from the intestine to the bloodstream. Alcohol does not impair absorption of vitamin A, but even moderate alcohol use has been shown to deplete liver stores of the vitamin.

Alcohol also promotes water excretion by the kidneys. Important minerals, such as zinc, magnesium, and potassium, are lost with the water. In short, alcohol profoundly affects nutrition status.

I see. What about alcohol's effect on body weight? I thought alcohol had a lot of kcalories, but I had an uncle who abused alcohol and became thinner and thinner.

It is true that the energy contribution of alcohol is relatively high. Alcohol yields about 7 kcalories per gram—more than protein and carbohydrate (4 kcalories per gram) and only slightly less than fat (9 kcalories per gram). In some cases, as mentioned in Chapter 6, excess energy from alcohol can contribute to obesity.[6] That is, when alcohol is consumed as *added* energy, it seems to be a risk factor for obesity. Chronic alcohol ingestion seems to have the opposite effect, however. Some researchers attribute the incomplete energy utilization of large doses of alcohol to an alternate metabolic pathway for alcohol that consumes energy rather than generating it.

What are the long-term nutrition effects of alcohol abuse?

By far, the longest-term effect of alcohol is the damage done to a child whose mother has abused alcohol during pregnancy. The devastating effects of alcohol on the unborn, and the messages pregnant women should hear, are presented in Chapter 12. For nonpregnant adults, long-term alcohol abuse damages different organs in different individuals, but always affects all organs and organ systems to various extents.

The most common damage is to the liver, where the function of affected cells may be lost forever unless

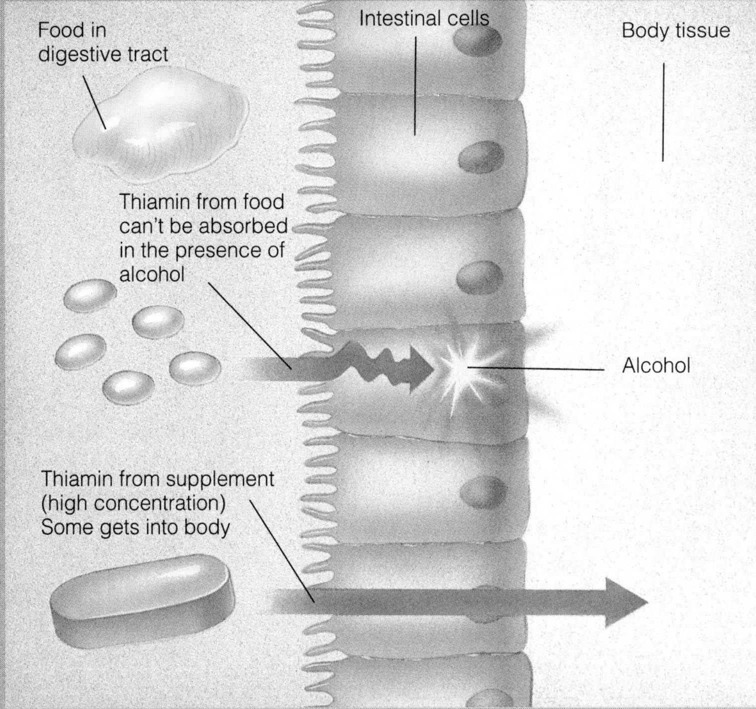

Figure NP8.1
Alcohol's Effect on Vitamin Absorption (Example)
In the presence of alcohol, intestinal cells fail to absorb thiamin, except at very high concentrations.

abstinence from alcohol and sound nutrition intervene in time to reverse the damage. The injury to the liver results in many side effects; one is high blood pressure, which may lead to heart damage and stroke. As the synthesis of fat speeds up, fat is deposited in the heart, arteries, and liver.

Alcohol consumption also increases the risks of cancer of the mouth, throat, esophagus, and lungs.[7] Women who drink alcohol have a greater risk of breast cancer than women who don't drink.[8] In one study, women who drank at least one alcoholic beverage per day had a 30% higher risk of breast cancer than women who did not drink. Alcohol interferes with the body's ability to detoxify and excrete cancer-causing compounds.

The central nervous system is particularly sensitive to alcohol. The brain shrinks even in people who drink only moderately, but the extent of damage from alcohol is proportional to the amount consumed. The following are a few of the other effects of alcohol:

- Inflammation of the intestines; ulcers of the stomach and intestines.
- Deterioration of the muscles, including the heart muscle.
- Reduced capacity for physical activity; heart discomfort sooner during physical activity.
- Kidney, bladder, and prostate gland damage.
- Reduced resistance to disease.
- Loss of function of the testicles and damage to the adrenal glands, leading to feminization and sexual impotence in men.
- Failure of the ovaries and early menopause in women.
- Increased susceptibility to lung infections.

If you make a point of eating well, can you drink alcohol and escape harm to health?

No. Eating well and even taking supplements of protein, vitamins, and minerals do not protect the drinker. There is no set level of safe drinking where no adverse effects take place. Even just a couple of drinks set in motion the destructive processes described, but if the drinking has been moderate, the next day's abstinence can repair the damage. Someplace between total abstinence and the extreme of alcoholism, there may be alcohol intakes moderate enough not to harm health, but the more a person drinks, the closer to a dangerous extreme that person comes.

Notes

1. Secretary of Health and Human Services, *Eighth Special Report to the U.S. Congress on Alcohol and Health* (Rockville, Md.: U.S. Department of Health and Human Services, 1993), pp. 1–15.

2. Committee on Substance Abuse, Alcohol use and abuse: A pediatric concern, *Pediatrics* 95 (1995): 439–442.

3. C. S. Lieber, Herman Award Lecture, 1993: A personal perspective on alcohol, nutrition, and the liver, *American Journal of Clinical Nutrition* 58 (1993): 430–442.

4. Lieber, 1993.

5. A. E. Rogers, Methyl donors in the diet and responses to chemical carcinogens, *American Journal of Clinical Nutrition* 61 (1995): 659S–665S; Folate, alcohol, methionine, and colon cancer risk: Is there a unifying theme? *Nutrition Reviews* 52 (1994): 18–20.

6. P. M. Suter, E. Häsler, and W. Vetter, Effects of alcohol on energy metabolism and body weight regulation: Is alcohol a risk factor for obesity? *Nutrition Reviews* 55 (1997): 157–171; P. M. Suter, Y. Schutz, and E. Jequier, The effect of ethanol on fat storage in healthy subjects, *New England Journal of Medicine* 326 (1992): 983–987.

7. M. J. Thun and coauthors, Alcohol consumption and mortality among middle-aged and elderly U.S. adults, *New England Journal of Medicine* 337 (1997): 1705–1714; D. M. Winn, Diet and nutrition in the etiology of oral cancer, *American Journal of Clinical Nutrition* 61 (1995): 437S–445S; Lieber, 1993.

8. S. A. Smith-Warner and coauthors, Alcohol and breast cancer in women: A pooled analysis of cohort studies, *Journal of the American Medical Association* 279 (1998): 535–540.

9 *Overweight, Underweight, and Weight Control*

Shape Up America!

niddk.nih.gov/health/nutrit/win.htm
Weight-control Information Network

Are you pleased with your body weight? If you answered yes, you are a rare individual. Nearly all people in our society think they should weigh more or less (mostly less) than they do. Usually, their primary reason is appearance, but they often perceive, correctly, that physical health is also somehow related to weight. At the extremes, both overweight and underweight present definite health risks.

Overweight and underweight both result from unbalanced energy budgets. The simple picture is as follows. Overweight people have consumed more food energy than they have spent and have banked the surplus in their body fat. To reduce body fat, overweight people need to spend more energy than they take in from food. In contrast, underweight people have consumed too little food energy to support their activities and so have depleted their bodies' fat stores and possibly some of their lean tissues as well. To gain weight, they need to take in more food energy than they expend. As you will see, though, the details of the body's weight regulation are quite complex.

This chapter's missions are to examine the problems associated with excessive and deficient body fatness; to present strategies toward solving these problems; and to point out how appropriate body composition, once achieved, can be maintained. The chapter emphasizes overweight because it has been more intensively studied and is a more widespread health problem in the developed countries.

Body Weight and Body Composition

The body's weight reflects its composition—the proportions of its bone, muscle, fat, fluid, and other tissue. All of these body components can vary in quantity and quality—the bones can be dense or porous; the muscles can be well developed or underdeveloped; fat can be abundant or scarce; and so on. By far the most variable tissue, though, is body fat. More than any other component, fat responds to changes in food intake or physical activity, so it is fat that is usually the target of efforts at weight control.

Defining Healthy Body Weight

How much should a person weigh? How can a person know if her weight is appropriate for her height and age? How can a person know if his weight is jeopardizing his health? Such questions seem so simple, yet even the experts cannot agree on the answers.[1] Most often, they try to identify the weights associated with lowest mortality. With this in mind, healthy body weight is defined by three criteria:

- A weight within the suggested range for height, as shown in Table 9.1.
- A fat distribution pattern that is associated with a low risk of illness or death.
- Freedom from all medical conditions that would suggest a need for weight loss.

People who meet all of these criteria may not gain any health advantage by changing their weights. Those who mistakenly think of themselves as overweight even though they meet these criteria for healthy weight may need to revise their self-image. Such people may still want to improve their eating and exercise habits, but they should do so to reap the rewards of being physically fit, not for the sake of weight loss. Anyone who does not meet all of the above criteria may want to consult with a health care professional, who should carefully consider each criterion in relation to the others. The rest of the chapter examines these three criteria in more detail.

Weight for Height Scale weight fails to reflect body fatness accurately. Still, health care providers typically compare people's weights with weight-for-height tables. Normally, the assessor uses the midpoint of the weight range for a person of a given height and medium build as a standard. If the person's actual weight is 10 to 20 percent above that, then the person is considered overweight; if 20 percent or more above the standard, the person is obese; and if 10 percent below the standard, the person is underweight. Note, however, that weight tables such as Table 9.1 present ranges rather than pinpointing one ideal weight, a good reminder that there is no one perfect weight that suits everyone. The traditional Metropolitan Height and Weight tables (see Appendix E) are not reliable for identifying the weights most closely associated with minimal health risks.[2]

Standards for desirable weights have steadily increased over the past 35 years. Authorities argue over which weight standards are most appropriate. Some approve, and some disapprove, of the current weight table for not specifying recommendations by sex; the table simply states that higher weights in the ranges generally apply to men and lower weights more often apply to women. Similar controversy arose when the current guidelines revised the standards so that weight standards for all adults are now the same as those issued earlier for younger adults; the reason for the change, as the guidelines explained, is that "health risks due to excess weight appear to be the same for older as for younger adults."

Body Mass Index A single standard, derived mathematically from the height and weight measures, is the body mass index (BMI):

$$\text{BMI} = \frac{\text{weight (kg)}}{\text{height (m)}^2}.$$

A person who takes measurements in pounds and inches can convert them to metric units or can use this modified equation.[3]

$$\text{BMI} = \frac{\text{weight (lb)}}{\text{height (in)}^2} \times 705.$$

The upper ends of the suggested weight ranges in Table 9.1 were calculated using a BMI of 25, which represents a healthy target—either for overweight people to achieve or for others not to exceed.[4] Obesity-related diseases become evident beyond this upper limit. Figure 9.1 on the next page presents visual images associated with various BMI values.

BMI values correlate with disease risks as Figure 9.2 (p. 208) shows.[5] Most people with a BMI between 18.5 and 25 have few of the health risks typically associated with too-low or too-high body weight. Risks increase as BMI falls below 18.5 or rises above 25, reflecting the reality that both underweight and overweight impair health status. Factors such as adverse family medical history, high blood pressure, or tobacco use raise risks independently of BMI.

Weight measures are inexpensive, easy to take, and highly accurate, but they fail to reveal two valuable pieces of information in assessing disease risk. They don't reveal how much of the weight is fat, and they don't show where the fat is located. For this knowledge, measures of body composition are needed.

Body Composition

For many people, being overweight compared with the standard means that they are overfat. This is not the case, though, for athletes with dense bones and well-developed muscles; they may carry little body fat. Conversely, inactive people may seem to have acceptable weights, but may carry too much body fat. In addition, the distribution of fat on the body, as discussed later, may be even more critical than overfatness alone. Chapter 16 and Appendix E describe clinical techniques for estimating body fat and its distribution, including fatfold tests, waist-to-hip ratio, bioelectrical impedance, and underwater weighing.

Table 9.1 Suggested Weights for Adults, 1995 Guidelines

Height[a]	Weight (lb)[a]	
	Midpoint	Range
4'10"	105	91–119
4'11"	109	94–124
5'0"	112	97–128
5'1"	116	101–132
5'2"	120	104–137
5'3"	124	107–141
5'4"	128	111–146
5'5"	132	114–150
5'6"	136	118–155
5'7"	140	121–160
5'8"	144	125–164
5'9"	149	129–169
5'10"	153	132–174
5'11"	157	136–179
6'0"	162	140–184
6'1"	166	144–189
6'2"	171	148–195
6'3"	176	152–200
6'4"	180	156–205
6'5"	185	160–211
6'6"	190	164–216

Note: The higher weights in the ranges generally apply to men, who tend to have more muscle and bone; the lower weights more often apply to women, who have less muscle and bone.

[a]Without shoes or clothes.

Source: Report of the Dietary Guidelines Advisory Committee on the Dietary Guidelines for Americans (Washington, D.C.: Government Printing Office, 1995).

body mass index (BMI): an index of a person's weight in relation to height, determined by dividing the weight (in kilograms) by the square of the height (in meters).

To convert pounds to kilograms, divide by 2.2.

To convert inches to meters, divide by 39.37.

The inside back covers show weights for various heights using the BMI to define underweight, healthy weight, overweight, and obesity.

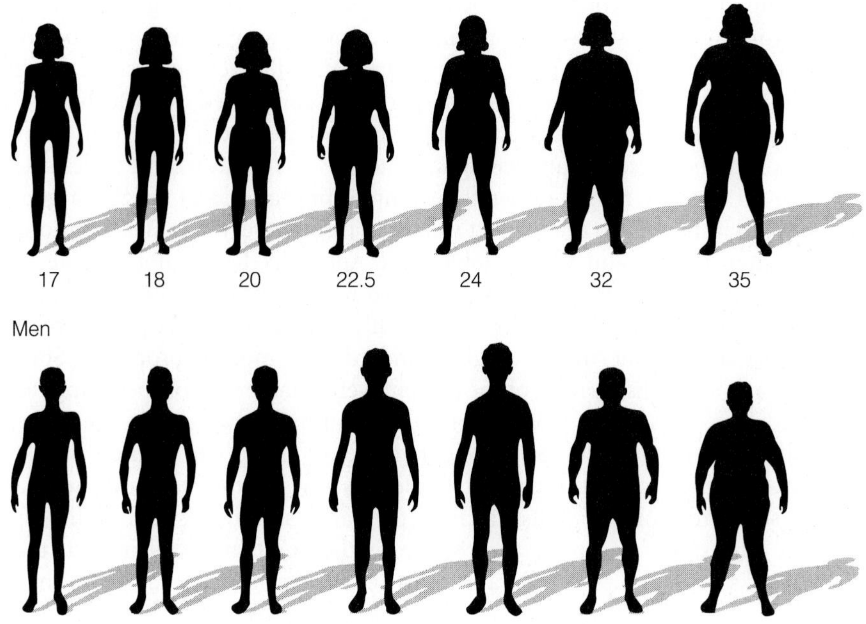

Figure 9.1

Silhouettes and BMI (Actual BMI Shown)

Source: Reprinted from material of the Canadian Dietetic Association.

- *BMI <18.5 = underweight.*
- *BMI 18.5 to 24.9 = normal.*
- *BMI 25 to 29.9 = overweight.*
- *BMI 30 or above = obese.*

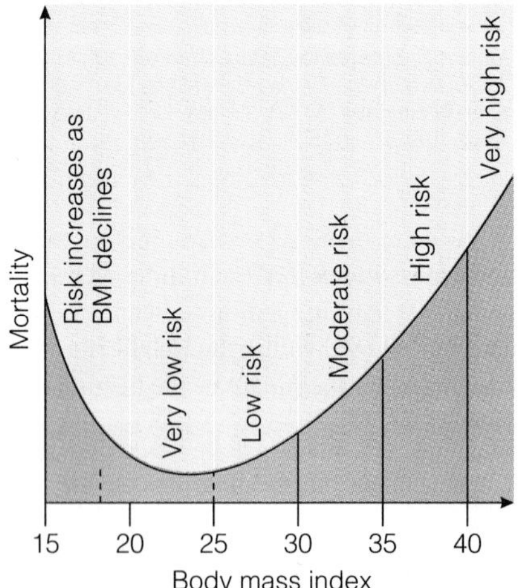

Figure 9.2

Body Mass Index and Mortality

Both underweight and overweight present risks of a premature death. This J-shaped curve describes the relationship between body mass index (BMI) and mortality and shows that optimal BMI is between 18.5 and 25.

How Much Is Too Much Body Fat?

The ideal amount of body fat depends partly on the person. A man with a BMI within the recommended range may have from 10 to 25 percent body fat; a woman, because of her greater quantity of indispensable fat, 18 to 32 percent. For many athletes, a lower-than-average percentage of body fat may be ideal—just enough fat to provide fuel, insulate and protect the body, assist in nerve impulse transmissions, and support normal hormone activity, but not so much as to contribute excess weight for the muscles to carry. For athletes, then, ideal body fat might be 5 to 10 percent for men and 15 to 20 percent for women.

For an Alaskan fisherman, a higher-than-average percentage of body fat is probably beneficial because fat helps prevent heat loss in cold weather. A woman starting a pregnancy needs sufficient body fat to support conception and fetal growth. Below a certain threshold for body fat, individuals may become infertile, develop depression, experience abnormal hunger regulation, or become unable to keep warm. These thresholds differ for each function and for each individual; much remains to be learned about them.

Clearly, the most important criterion of appropriate fatness is health. Researchers find health problems develop when body fat exceeds 22 percent in young men, 25 percent in older men, 32 percent in younger women, and 35 percent in older women; these are the values used to define obesity and age 40 is the dividing line.

In summary, no single body weight is best for everyone; different people have different needs and goals. Health care providers usually compare people's weights with weight-for-height tables, but the tables fail to indicate how much of the weight is fat and where the fat is located. The body mass index correlates heights and weights with health risks. The ideal amount of body fat depends partly on the person. The most important criterion of appropriate body fatness is health.

Causes of Obesity

Henceforth, this chapter will use the term *obesity* to refer to excess body fat. Excess body fat accumulates when people take in more food energy than they spend. Why do they do this? Is it genetic? Metabolic? Psychological? Behavioral? All of these? Most likely, obesity has many interrelated causes: some experts in the field speak of several different *obesities.*

obesity: a chronic disease characterized by excessively high body fat in relation to lean body tissue. See p. 208 for body fat percentages that define obesity.

Genetics and Weight When both parents are obese, the chances that their children will be obese are quite high (up to 80 percent), whereas when neither parent is obese, the chances are relatively small (less than 10 percent). Adoption studies find a similarity in obesity between biological parents and their natural children, but not between adoptive parents and their adopted children.[6]

To determine the relative contributions of genetic and environmental factors to body weight, one group of researchers studied identical and fraternal twins, some of whom were reared together and some apart. Like previous studies, this study found that identical twins were twice as likely to have similar weights as fraternal twins—even when reared apart. These findings suggest an important role for genetics in determining a person's susceptibility to obesity.[7]

Genetics may influence the way energy is *stored,* for example. When identical twins are given an extra 1000 kcalories a day for 100 days, some pairs may gain less than 10 pounds while others may gain up to 30 pounds. Within each pair, the amount of weight gain, percentage of body fat, and distribution of fat may be similar.

Genetics and Energy Expenditure Genetics may also influence how much energy the body spends. For example, the differences in basal metabolic rate (BMR) between individuals are greater than can be explained by age, sex, and body composition alone. Similarities within families suggest a genetic influence on BMR. A low metabolic rate is a major risk factor for weight gain.

Lipoprotein Lipase Some of the research investigating genetic influence on obesity focuses on the enzyme lipoprotein lipase (LPL), which promotes fat storage in fat cells and muscle cells. People with high LPL activity are especially efficient at storing fat. As you might expect, obese people have much more LPL activity than lean people.

lipoprotein lipase (LPL): an enzyme mounted on the surface of fat cells (and other cells). It hydrolyzes triglycerides in the blood into fatty acids and glycerol for absorption into the cells. There they are metabolized or reassembled for storage.

Leptin Researchers have discovered a gene in humans called the obesity (*ob*) gene. The obesity gene codes for the protein leptin in fat cells.[8] Leptin appears to act on the hypothalamus, influencing appetite and energy balance. As the photo on the next page shows, mice with a defective obesity gene do not produce leptin and can weigh up to three times as much as normal mice. When injected with a synthetic form of leptin, the mice lose weight. (Because leptin is a protein, it would be destroyed during digestion if given orally; consequently, it must be given by injection.)

Genes instruct cells to make proteins, and each protein performs a unique function.

leptin: a protein produced by fat cells under the direction of the obesity gene that increases satiety and energy expenditure.
leptos = thin

Researchers have identified a genetic deficiency of leptin in human beings as well.[9] An error in the gene that codes for leptin was discovered in two extremely obese children whose blood levels of leptin are barely detectable. Without leptin, their satiety is impaired; the children are constantly hungry and eat considerably more than their siblings or peers.

Most obese people do not have leptin deficiency, however, In fact, in obese people, the more body fat, the more leptin.[10] Researchers speculate that leptin rises in an effort to suppress appetite and inhibit fat storage when fat cells are ample. Obese people with elevated leptin concentrations may be resistant to its satiating effect.[11] The absence of or resistance to leptin in obesity parallels the scenario of insulin in diabetes: some people have an insulin deficiency (type 1), whereas many others have elevated insulin, but are resistant to its glucose-storing effect (type 2).

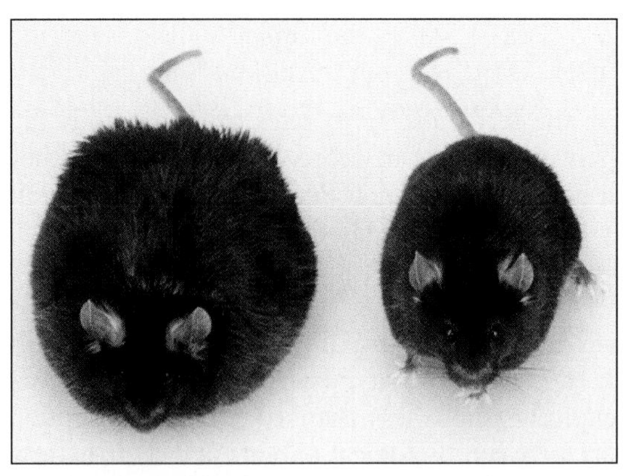

The mouse on the left is genetically obese—it lacks the gene for producing leptin. The mouse on the right is *also* genetically obese, but because it receives leptin, it eats less, expends more energy, and is less obese than it would be had it not received the leptin.

Fat Cell Development Another cause of obesity may be the development of excess fat cells during childhood. The amount of fat on a person's body reflects both fat cell *number* and *size.* The number of fat cells increases most rapidly during the growing years of late childhood and early puberty. Fat cell number increases more rapidly in obese children than in lean children, and obese children entering their teen years may already have as many fat cells as do adults of normal weight.

Fat cells can also expand in size. Upon reaching their maximum size, the cells may divide. Thus obesity develops when a person's fat cells increase in number, in size, or quite often both. With fat loss, the size of the fat cells shrinks, but not their number. For this reason, people with extra fat cells may tend to regain lost weight rapidly. Prevention of obesity, then, is most critical during the growing years when fat cell number is increasing.

set-point theory: the theory that proposes that the body tends to maintain a certain weight by means of its own internal controls.

Set-Point Theory One popular theory of why the obese person's body may store too much fat is the set-point theory. The set-point theory proposes that body weight, like body temperature, is physiologically regulated. Researchers have noted that most people who lose weight on reducing diets later quickly regain all the lost weight. This suggests that somehow the body chooses a weight that it wants to be and defends that weight by regulating eating behaviors and hormonal actions. Research confirms that the body adjusts its metabolism whenever it gains or loses weight—in the direction that returns to the initial body weight: energy expenditure increases with weight gain and decreases with weight loss.[12] These changes in energy expenditure are greater than those predicted based on body composition and help to explain why it is so difficult for an obese person to maintain weight losses. Researchers speculate that an individual's set point for body weight is adjustable, shifting over the life span in response to physiological changes and influenced by genetic, dietary, and other factors.[13]

Environmental Stimuli To a degree, obesity may be environmentally determined. People may overeat as a response to stimuli in their surroundings—primarily, the availability of many delectable foods. One food constituent is perceived as especially palatable—fat. Not only does fat deliver twice the kcalories, gram for gram, as protein and carbohydrate, but it also seems to be stored preferentially by the body, and with great efficiency. Of the three energy nutrients, fat stimulates the least energy expenditure after a meal and is least powerful in signaling satiety.[14]

Reminder: The energy required to process food in the body is known as the thermic effect of food.

Learned Behavior Psychological stimuli also trigger inappropriate eating behavior in some people. Appropriate eating behavior is a response to hunger. Hunger is a drive programmed into people by their heredity. Appetite, in contrast, is learned and can lead people to ignore hunger or to overrespond to it. Hunger is physiological, whereas appetite is psychological, and the two do not always coincide.

hunger: the physiological need to eat, experienced as a drive for obtaining food; an unpleasant sensation that demands relief.

appetite: the psychological desire to eat; a learned motivation that is experienced as a pleasant sensation that accompanies the sight, smell, or thought of appealing foods.

Food behavior is also intimately connected to deep emotional needs such as the primitive fear of starvation. Yearnings, cravings, and addictions with profound psychological significance can express themselves in people's eating behavior. An emotionally insecure person might eat rather than call a friend and risk rejection. Another person might use eating to relieve boredom or to ward off depression.

Physical Inactivity The possible causes of obesity mentioned so far all relate to the input side of the energy equation. What about output? People may be obese, not because they eat too much, but because they spend too little energy. More than one-third of the overweight population report no physical activity during their leisure time.[15] Obese people observed closely are often seen to eat less than lean people, but they are sometimes so extraordinarily inactive that they still manage to accumulate an energy surplus. Reducing their food intake further would jeopardize health and incur nutrient deficiencies.

Physical activity, then, is a necessary component of nutritional health. People must be physically active if they are to eat enough food to deliver all the nutrients needed without unhealthy weight gain.

One hundred years ago, 30 percent of the energy used in farm and factory work came from muscle power; today only 1 percent does.[16] Modern technology has replaced physical activity at home, at work, and in transportation. Underactivity is probably the single most important contributor to obesity. In turn, television watching may contribute most to physical inactivity.[17]

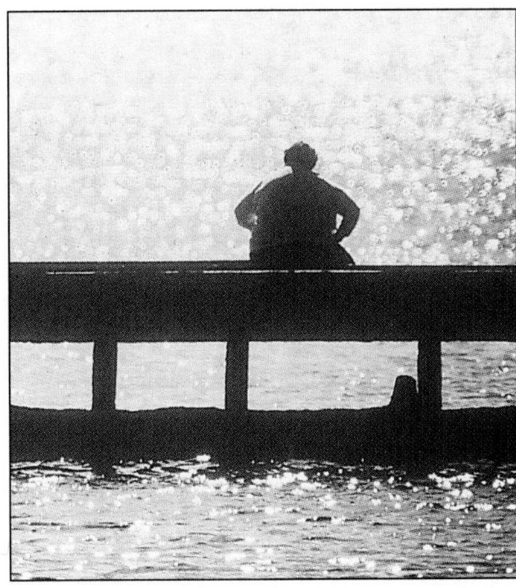

Lack of physical activity fosters obesity.

Watching television contributes to obesity in several ways. First, television viewing requires little energy beyond the resting metabolic rate. Second, it replaces time spent in more vigorous activities. Third, watching television correlates with between-meal snacking, eating the high-kcalorie, high-fat foods most heavily advertised on programs, and influencing family food purchases.

Like all the other "causes" of obesity, inactivity alone fails to explain it fully. Genetics, fat cell development, set point, and overeating all offer possible, but still incomplete, explanations. Most likely, obesity has not one cause, but different causes and combinations of causes in different people. After all, no two people are alike either physically or psychologically. Some causes may be within a person's control and some may be beyond it. In recent years, the view has been gaining ground that obesity is no one's "fault"—it is not a matter of undisciplined gluttony. Philosophies of weight control and treatment have been evolving to square with this view.

WWW

nhlbi.nih.gov/nhlbi/cardio/obes/prof/guidelns/ob_home.htm
Clinical Guidelines on the Identification, Evaluation, and Treatment of Overweight and Obesity in Adults

naaso.org
North American Association for the Study of Obesity

Health Risks of Overweight and Obesity

Overweight is associated with disease risks. For example, it can precipitate hypertension and bring on strokes.[18] Often weight loss alone can normalize the blood pressure of an overfat person; some people with hypertension can tell you exactly at what weight their blood pressure begins to rise. Weight gain can also precipitate diabetes in genetically susceptible people. If hypertension or diabetes runs in your family, you urgently need to attend to weight control.

overweight: body weight above some standard of acceptable weight that is usually defined in relation to height (such as the weight-for-height tables).

Despite our nation's preoccupation with body image and weight loss, the incidence of overweight continues to rise dramatically (see Figure 9.3). Approximately one out of three adults, one out of nine teenagers, and one out of seven children in the United States are now overweight.[19] The prevalence of overweight has been increasing and is especially high among women, the poor, and some ethnic groups. If this trend continues, some obesity experts predict that by the year 2230, every adult in the United States will be overweight. Such a dramatic statement sounds preposterous, but it speaks to the reality: obesity is a major public health problem without a solution.[20]

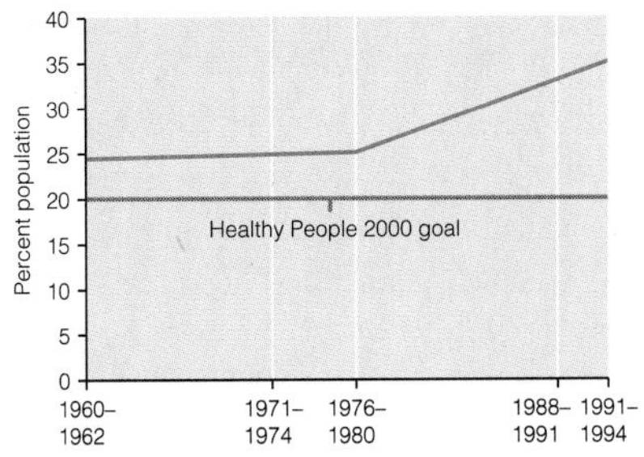

Figure 9.3
Prevalence of Overweight among Adults in the United States

Health Risks of Obesity The health risks of overfatness are so many that it has been declared a disease: obesity. In the United States, obesity is second only to tobacco use as the most significant cause of preventable death.[21] Besides diabetes and hypertension already mentioned, other risks threaten obese adults. Among

A healthy body contains enough lean tissue to support health and the right amount of fat to meet body needs.

them are high blood lipids, cardiovascular disease, sleep apnea (abnormal ceasing of breathing during sleep), osteoarthritis, abdominal hernias, some cancers, varicose veins, gout, gallbladder disease, respiratory problems (including Pickwickian syndrome, a breathing blockage linked with sudden death), liver malfunction, complications in pregnancy and surgery, flat feet, and even a high accident rate. Each year these obesity-related illnesses cost our nation billions of dollars.[22] The cost in terms of lives is also great. People with lifelong obesity are twice as likely to die prematurely as others.

People want to know exactly how much fat is too fat for health. Ideally, a person has enough fat to meet basic needs but not so much as to incur health risks. Some evidence indicates that being even mildly or moderately overweight aggravates the risk of heart disease.[23] The degree of overweight is related to the rate of development of cardiovascular disease.[24] Overweight may even have effects years after the excess weight is lost. Even if lean as adults, people who were 20 pounds or more overweight as teenagers may be more likely than others to die of heart disease.[25] On the other hand, some obese people seem to remain healthy and live long despite their body fatness. It may be that genetics determines who among the overweight will be susceptible to diseases and who will stay well. Still, the majority of obese people do develop associated health problems.

central obesity: excess fat on the abdomen and around the trunk of the body.

Central Obesity Even more than total fatness, fat that collects in the central abdominal area of the body may be especially likely to lead to diabetes, stroke, hypertension, and coronary artery disease.[26] The risk of death from all causes may be higher for those with central obesity than for those whose fat accumulates elsewhere in the body.[27] Unlike the fat layers lying just beneath the skin of the abdomen and elsewhere, intra-abdominal fat, when mobilized, goes directly to the liver where it is made into cholesterol-carrying low-density lipoprotein (LDL).[28] Fat from elsewhere may arrive in the liver eventually, but it takes a circuitous route that first allows other tissues the chance to pull it from the circulation and metabolize it.

intra-abdominal fat: fat stored within the abdominal cavity in association with the internal abdominal organs, as opposed to the fat stored directly under the abdominal skin (subcutaneous fat).

waist-to-hip ratio: a valuable, commonly used indicator of fat distribution; waist-to-hip ratio = waist circumference ÷ hip circumference. A ratio of 0.8 or greater for a woman or 0.95 or greater for a man suggests a risk to health.

Intra-abdominal fat creates the "apple" profile of central obesity. Fat around the hips and thighs creates more of a "pear" profile. Men of all ages carry more intra-abdominal fat than either premenopausal or postmenopausal women.[29] For those women with abdominal fat, the risks of cardiovascular disease and mortality are increased, just as they are for men.[30] Smokers, too, may carry more of their body fat centrally. A smoker may weigh less than the average nonsmoker, but the smoker's waist-to-hip ratio may be greater, leading researchers to think that smoking may directly affect body fat distribution.[31] Two other factors that may affect body fat distribution are intakes of alcohol and physical activity. Alcohol consumption, particularly beer drinking, may favor central obesity.[32] In contrast, regular physical activity seems to prevent abdominal fat accumulation.[33]

Other Risks of Obesity While some obese people seem to escape health problems, no one who is fat in our society quite escapes the social and economic handicaps. People who have been overweight as adolescents are still, seven years later, less likely to be married and more likely to have low household incomes than those who were not overweight earlier.[34] This is especially true for women. In contrast, people with other chronic conditions such as asthma, diabetes, and epilepsy do not differ in these characteristics from nonoverweight people.

Our society places enormous value on thinness. Obese people pay more for insurance and for clothing. Psychologically, too, fat people are made to feel rejected and embarrassed, and this hurts self-esteem.

Traditional medical advice urges all obese people to reduce their weight to reduce associated health risks. Lately, though, experts have been debating whether this advice applies equally to all obese people. Some people may risk more losing weight than remaining obese.

Risks of Obesity Treatment An estimated 30 to 40 percent of all U.S. women (and 20 to 25 percent of all U.S. men) are trying to lose weight at any given time, spending up to $40 billion each year to do so. Some of these people do not even need to lose weight. Others need to lose weight, but are not successful; few succeed, and even fewer succeed permanently. In fact, only 5 percent of people who try to lose weight achieve long-term success.

Many people assume that every overweight person can achieve slenderness and should pursue that goal. Consider, however, that most overweight people cannot become slender. People vary in their weight tendencies just as they vary in their potentials for height and degrees of health. The question of whether a person should lose weight depends on many factors: the extent of overweight, age, health, and genetics, to name a few. Just as there are unhealthy, normal-weight people, there are healthy, overweight people. Weight-loss advice, then, does not apply equally to all overweight people. Some people may risk more in the process of losing weight than in remaining overweight. Others may reap significant health benefits with just modest weight loss.

The risks people incur in attempting to lose weight often depend on how they go about it. Weight-loss plans and obesity treatments flourish—some are adequate; many are ineffective and possibly dangerous. Inappropriate ways of treating obesity are listed in the glossary on the next page. The next section addresses aggressive approaches to obesity for those obese people with high risks of medical problems who must lose weight rapidly. The section after that discusses reasonable approaches to overweight for those seeking gradual weight loss.

At 6 feet 3 inches tall and 245 pounds, Mike O'Hearn has a BMI greater than 30 and would be considered overweight by most weight-for-height standards. Yet he is clearly not overfat. In fact, his body fat is only 8 percent.

In summary, excessive body fatness presents health risks as well as social and economic problems. Central obesity may be more hazardous to health than other forms of obesity.

Aggressive Treatments of Obesity

For some obese people, the medical problems caused by their obesity demand treatment approaches that may, themselves, incur some risks. The health benefits of weight loss, however, may in some cases outweigh these risks.

Obesity Drugs

Several prescription medications for weight loss have been tried over the years. When used as part of a long-term comprehensive weight-loss program, medications can help obese people to lose approximately 10 percent of their weight and maintain that loss for at least a year.[35] Because weight regain commonly occurs with the discontinuation of drug therapy, treatment is long term. And the long-term use of medications poses risks. Medical experts do not yet know whether a person would benefit more from maintaining a 20-pound excess or from taking a medication for a decade to keep the 20 pounds off.

The challenge, then, is to develop an effective medication that can be used over time without adverse side effects or the potential for abuse. No such medication currently exists. Several medications, however, are under study.

Very-Low kCalorie Diets

Very-low-kcalorie diet (VLCD) plans provide 800 kcalories, at least 1 gram of high-quality protein per kilogram of body weight, little or no fat, and a minimum of 50 grams of carbohydrate (not enough to spare protein). Clients receive

Glossary of Poor Treatment Choices for Obesity

diet pills: pills that depress the appetite temporarily; often, physician-prescribed amphetamines (speed). It is generally agreed that these drugs are of little value for weight loss and that their use can cause a dangerous dependency.

diuretic abuse: use of diuretics to promote water excretion by dieters who believe their weight excesses are due to water accumulation. Diuretics promote water loss, not fat loss, and their use can cause dehydration and mineral imbalances.

fad diets: diets based on exaggerated or false theories of weight loss; such diets are usually inadequate in energy and nutrients. Most fad diets, including the currently popular Zone Diet, advocate essentially the same high-protein, low-carbohydrate diet. Such diets may offer short-term weight-loss success to some who try them, but they fail to produce long-lasting results for most people. Furthermore, high-protein, low-carbohydrate diets are often high in fat and low in fiber, vitamins, and some minerals. Long-term use of such diets may produce adverse side effects such as nausea, fatigue, constipation, and low blood pressure. Some fad diets are more dangerous to health than obesity itself.

herbal laxatives: laxatives containing senna, aloe, rhubarb root, castor oil, or buckthorn are commonly sold as "dieter's tea." Such products cause nausea, vomiting, diarrhea, fainting, and, in some users, possibly death.[a]

herbal products: substances extracted from plants to mimic drugs that suppress appetite. Such substances may have dangerous side effects. St. John's wort, for example, is often prepared in combination with the herbal stimulant ephedrine, extracted from the Chinese plant ma huang. Ephedrine has been implicated in several cases of heart attacks and seizures. (See Nutrition in Practice 24.)

low-carbohydrate diets: diets designed to bring about metabolic responses similar to those of fasting (see Chapter 6, pp. 126–127). Without sufficient carbohydrate, the body cannot use its fat in the normal way, and ketosis results. Many physiological hazards accompany low-carbohydrate diets: high blood cholesterol, mineral imbalances, hypoglycemia, and more.

[a]P. Kurtzweil, Dieter's brews make tea time a dangerous affair, *FDA Consumer*, July/August 1997, pp. 6–11.

an assortment of vitamins and minerals from supplements. Meals consist of a limited number of foods (primarily lean meats, fish, and poultry) each day, a powdered formula available by prescription, or a combination of the two.

VLCD formulas are designed to be nutritionally adequate, but the body responds to this severe energy restriction as if the person were starving—conserving energy and preparing to regain weight at the first opportunity. Several changes occur in hormone concentrations, metabolic activities, fluid and electrolyte balances, and organ functions in the effort to meet the challenge of living on a much-less-than-adequate energy intake. For these reasons, a VLCD is appropriate only for short-term use (four months) and under close medical supervision. Common side effects of VLCD include headache, fatigue, dry skin, nausea, and hair loss, as well as others.

Surgery

Surgery as an approach to weight loss is justified in some specific cases of clinically severe obesity. Two gastric partitioning procedures have gained wide acceptance. Both procedures limit food intake by effectively reducing the size of the stomach. They reduce the size of the outlet as well, so they delay the passage of food from the stomach into the intestine for digestion and absorption.

clinically severe obesity: a BMI of 40 or greater or 100 lb or more overweight for an average adult. A less preferred term used to describe the same condition is *morbid obesity.*

gastric partitioning: a surgical procedure used to treat clinically severe obesity. The operation limits food intake by effectively reducing the size of the stomach and delays gastric emptying by restricting the outlet.

The long-term safety and effectiveness of gastric surgery depend, in large part, on compliance with dietary instructions. Common immediate postsurgical complications include infections, nausea, vomiting, and dehydration; in the long term, vitamin and mineral deficiencies and psychiatric disorders are common. Lifelong medical supervision is necessary for those who choose the surgical route, but in suitable candidates the benefits of weight loss prove worth the risks.

Obese people with high risks of medical problems may need aggressive treatment, including drugs, VLCD, or surgery.

Reasonable Strategies for Weight Loss

The 1995 *Dietary Guidelines* (Table 1.3 on p. 13) suggest that for good health, a person should "maintain or improve your weight." The focus is not so much on weight loss as on health gains. In fact, the *Guidelines* go on to say, "if you're overweight and cannot lose weight, try not to gain weight." Modest weight loss, even when a person is still overweight, can improve control of diabetes and reduce the risks of heart disease by lowering blood pressure and blood cholesterol, especially for those with abdominal fat.

1 lb body fat = 3500 kcal.
To lose a pound a week, cut 500 kcal/day.

Of course, the same eating and activity habits that improve health often lead to a healthier body weight and composition as well.[36] A loss of 10 to 15 pounds can improve a person's BMI by 2 units, which can significantly improve health—even if the person is still overweight. Successful weight loss, then, is not defined by weight-for-height tables, but by reductions in disease risks.[37] People less concerned with disease risks may prefer to set goals for personal fitness, such as being able to play with children or climb stairs without becoming short of breath.

Whether the goal is health or fitness, weight-loss expectations need to be reasonable. Unreachable targets ensure frustration and failure. If goals are achieved and exceeded, there will be rewards instead of disappointments. The box on p. 216 offers a way of judging weight-loss diets and programs based on sound nutrition principles.

Weight-loss pointer:
- *Adopt reasonable expectations about health and weight goals and about how long it will take to achieve them.*

Diet

No particular eating plan is magical, and no particular food must either be included or avoided. You are the one who will have to live with the plan, so you had better be involved in its planning. Don't think of it as a diet you are going "on"—because then you may be tempted to go "off." The diet is successful only if the pounds do not return. Think of it as an eating plan that you will adopt for life. It must consist of foods that you like, that are available to you, and that are within your means.

- *Be involved in planning.*
- *Keep in mind that you will want to maintain your lost weight. Practice needed behaviors as you go.*

A Realistic Energy Intake Choose an energy intake you can live with. You need at least 10 kcalories per pound of current weight each day to lose fat while retaining lean tissue. Nutritional adequacy is difficult for most people to achieve on fewer than 1200 kcalories a day, and most healthy adults should not consume any less than that. You will experience a healthier, more successful weight loss with a small energy deficit that provides an adequate intake than with a large energy deficit that creates feelings of starvation and deprivation, which can lead to an irresistible urge to binge.

- *Adopt a realistic plan.*

Nutritional Adequacy Nutritional adequacy should be a high priority. Take a look at the 1200-kcalorie food plan in Table 9.2 on p. 217. Notice that this pattern offers the minimum number of servings suggested in the Daily Food Guide (introduced in Chapter 1) and allows a teaspoon of fat at each of three meals. Such an intake would allow most people to lose weight at a satisfactory rate and still meet their nutrient needs with careful food selections. (Women might need an iron supplement.) The other patterns provide for higher energy intakes.

- *Make the diet adequate by emphasizing nutrient-dense foods.*

Small Portions Overweight people usually need to learn to eat less food at each meal—one piece of chicken for dinner instead of two, a teaspoon of butter on the vegetables instead of a tablespoon, and one cookie for dessert instead of six. The

- *Eat small portions of foods at each meal.*

How to

Rate Sound and Unsound Weight-Loss Schemes and Diets

Start by giving each diet or program 160 points. Subtract points as instructed, whenever a diet falls short of ideals.

Scoring: 160 = fine
140–150 = possibly safe with some disadvantage
120–130 = needs improvement
110 or below = dangerous to use

Does the diet or program:

1. Provide a reasonable number of kcalories (not fewer than 1200 kcalories for an average-size person)? If not, give it a minus 10.
2. Provide enough, but not too much, protein (at least the recommended intake or RDA, but not more than twice that much)? If no, minus 10.
3. Provide enough fat for balance but not so much fat as to go against current recommendations (between 20 and 30 percent of kcalories from fat)? If no, minus 10.
4. Provide enough carbohydrate to spare protein and prevent ketosis (100 grams of carbohydrate for the average-size person)? Is it mostly complex carbohydrate (not more than 10 percent of the kcalories as concentrated sugar)? If no to either or both, minus 10.
5. Offer a balanced assortment of vitamins and minerals—that is, foods from all food groups? If it omits a food group (for example, meats), does it provide a suitable substitute? Count five food groups in all: milk/milk products, meat/fish/poultry/eggs/ legumes, fruits, vegetables, and starches/grains. For *each* food group omitted and not adequately substituted for, subtract 10 points.
6. Offer variety, in the sense that different foods can be selected each day? If you'd class it as boring or monotonous, give it a minus 10.
7. Consist of ordinary foods that are available locally (for example, in the main grocery stores) at the prices people normally pay? Or does the dieter have to buy special, expensive, or unusual foods to adhere to the diet? If you would class it as "bizarre" or "requiring special foods," minus 10.
8. Promise dramatic, rapid weight loss (substantially more than 1 percent of total body weight per week)? If yes, minus 10.
9. Encourage permanent, realistic lifestyle changes, including regular physical activity and the behavioral changes needed for weight maintenance? If not, minus 10.
10. Misrepresent salespeople as "counselors" supposedly qualified to give guidance in nutrition and/or general health without a profit motive, or collect large sums of money at the start, or require that clients sign contracts for expensive long-term programs? If so, minus 10.
11. Fail to inform clients about the risks associated with weight loss in general or the specific program being promoted? If so, minus 10.
12. Promote unproven or spurious weight-loss aids such as starch blockers, diuretics, sauna belts, body wraps, passive exercise, ear stapling, acupuncture, electric muscle stimulating (EMS) devices, spirulina, amino acid supplements (e.g., arginine, ornithine), glucomannan, appetite suppressants, "unique" ingredients, and so forth? If so, minus 10.

goal is to eat enough food for energy, nutrients, and pleasure, but not more. This amount should leave a person feeling satisfied—not necessarily full. Keep in mind that even low-fat foods can deliver a lot of kcalories when a person eats large quantities.

▪ *Make grains, legumes, vegetables, and fruits central to your diet plan.*

Carbohydrates, Not Fats Center meals and snacks on complex carbohydrate foods. Fresh fruits, vegetables, legumes, and whole grains offer abundant vitamins and minerals. They also offer more fiber (which provides bulk and satiety) and far less fat and food energy than smooth, refined foods. Researchers compared a diet that restricted fat, but allowed complex carbohydrates to be eaten freely with a more conventional energy-restricted diet, used over six months by obese women.[38] Women in both diet groups lost substantial weight, but those who ate the fat-restricted, complex carbohydrate–rich diet rated it higher in

Table 9.2 Diet Patterns for Different Energy Intakes

Food Group	Energy Level (kcal)						
	1200	1500	1800	2000	2200	2600	3000
Bread, cereal, rice, and pasta	6	7	8	9	11	13	15
Meat and meat alternates (ounces)	4	5	6	6	6	7	8
Vegetable	3	4	5	5	5	6	6
Fruit	2	3	4	4	4	5	6
Milk and milk products (nonfat)	2	2	2	3	3	3	3
Fat	3	5	6	7	8	10	12

Note: These patterns follow the Daily Food Guide plan and supply less than 30 percent of kcalories as fat.

Delicious, low-fat, carbohydrate-rich foods such as fresh fruits, vegetables, whole grains, and legumes offer abundant vitamins, minerals, and fiber.

terms of satiety and taste. A person who makes low-fat food selections habitually—even without purposely limiting energy intake—eats less food, satisfies hunger, and diminishes the desire to eat.[39] How much weight is lost on a low-fat diet depends on the extent of fat reduction and the degree of obesity. A person limiting energy intake would also lose weight, but maintaining that weight loss would be easier following a low-fat diet.[40]

■ *Select low-fat foods regularly.*

Sugar and Alcohol A person trying to achieve or maintain a healthy weight needs to pay attention not only to fat, but to sugar and alcohol, too. Using them for pleasure on occasion is compatible with health as long as most daily choices are of nutrient-dense foods.

■ *Limit concentrated sweets and alcoholic beverages.*

Adequate Water Learn to satisfy thirst with water. Water fills the stomach between meals and dilutes the metabolic wastes generated from the breakdown of fat, easing their excretion. It meets the water need that was formerly met by eating extra food (remember that food provides water).

■ *Drink plenty of water (8 glasses or more a day).*

In summary, adopt an "eating plan for good health" rather than a "diet for weight loss." That way, you will be able to keep the lost weight off.

■ *Learn, practice, and follow a healthful eating plan for the rest of your life.*

Physical Activity

Either dieting or physical activity alone can produce some weight loss. Clearly, however, the combination is most effective.[41]

weight cycling: repeated cycles of weight loss and subsequent regain that affect body composition and metabolism. With intermittent dieting, a person rebounds to a higher weight (and a higher body fat content) after each round. The weight-cycling pattern is popularly called the *ratchet effect* or *yo-yo effect* of dieting.

Weight Cycling Those who endeavor to lose weight without physical activity often become trapped in weight cycling, the endless repeating rounds of weight loss and regain from "yo-yo" dieting (see Figure 9.4 on the next page). Nearly a third of all women interviewed in one poll reported themselves to be perpetual dieters who dieted at least once a month.

Harm from Weight Cycling Such fluctuations in body weight appear to increase the risks of chronic diseases and even premature death, independently of obesity itself.[42] Maintaining a stable weight, even if it is overweight, may be

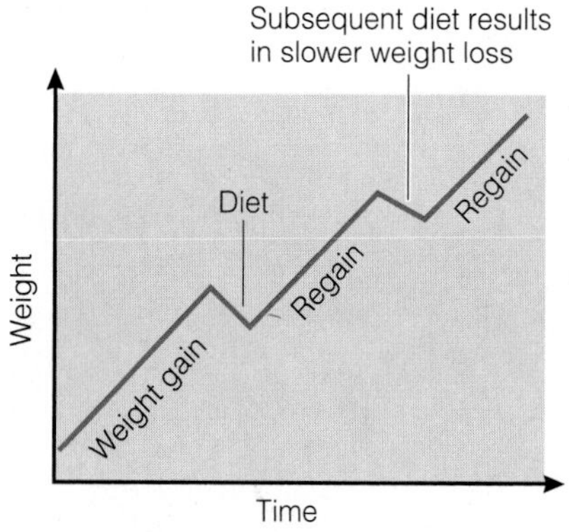

Figure 9.4

The Weight-Cycling Effect of Repeated Dieting

Each round of dieting is followed by a rebound of weight to a higher level than before.

less harmful to health than repeated bouts of weight gain and loss. Such concerns should not deter obese people who want to lose weight from trying, but rather should encourage them to commit to lifelong changes (including physical activity) that will maintain weight losses.[43] People who combine diet and physical activity are more likely to lose more fat, retain more muscle, and regain less weight than those who only diet.[44]

Physical Activity and Energy Expenditure Physical activity makes many contributions to weight loss and maintenance. For one thing, it directly increases energy output by the muscles and cardiovascular system. A 150-pound person walking a brisk 4 miles per hour for 30 minutes spends an extra 185 kcalories on that activity. A football player in training may spend several thousand extra kcalories on a day of heavy training.

Activity and BMR Activity also contributes to energy output in an indirect way—by speeding up basal metabolism.[45] It does this both immediately and over the long term. On any given day, after intense and prolonged exercise, basal metabolism remains elevated for several hours. Over the long term, daily vigorous activity for many weeks gradually shifts body composition toward more lean tissue, which is more active metabolically than fat tissue. Then the ongoing metabolic rate rises accordingly, and this makes a contribution toward continued weight loss or maintenance.

The raised metabolic rate continues for as long as the person is physically active on a regular basis. The more energy expended in metabolic activities, the greater the energy requirement. This means that a person can eat more without gaining weight, and this, in turn, brings both pleasure and nutrients.

Benefits of physical activity in a weight-control program:

- *Short-term increase in energy expenditure (from exercise and from a slight rise in BMR).*
- *Long-term increase (slight) in BMR.*
- *Appetite control.*
- *Stress reduction and control of stress eating.*
- *Physical, and therefore psychological, well-being.*
- *High self-esteem.*

Activity and Appetite Control Physical activity also helps to control appetite. People think that exercising will make them hungry, but this is not entirely true.[46] Yes, active people do have healthy appetites, but immediately after a good workout, most people do not feel like eating. They want to shower and may be thirsty, but they are not hungry. The reason is that the body has responded to the stress of activity by mobilizing fuels from storage: glucose and fatty acids are abundant in the blood. At the same time, the body has suppressed its digestive functions. Hard physical work and eating are not compatible.

Physical activity helps especially to curb the inappropriate appetite that prompts a person to eat when bored, anxious, or depressed. Weight-control programs encourage people to go out and be active when they're tempted to eat but not really hungry.

Activity and Psychological Benefits Activity also helps to reduce stress. Since stress itself is a cue to inappropriate eating behavior for many people, activity can help here, too.

Activity offers still more psychological advantages. The fit person looks and feels healthy, and high self-esteem accompanies these benefits. High self-esteem tends to support a person's resolve to persist in a weight-control effort, rounding out a beneficial cycle.

Choosing Activities What kind of physical activity is best? People seeking to lose weight should choose activities that they enjoy and are willing to do regularly. In addition to activities such as walking or aerobic dance, there are hundreds of ways to incorporate energy-spending activities into daily routines: take the stairs instead of the elevator, walk to the neighbor's apartment instead of making a phone call, and rake the grass clippings instead of using a bagger. These activities burn only a few kcalories each, but over a year's time they become significant.

Spot Reducing People sometimes ask about "spot reducing." Unfortunately, no one part of the body gives up fat in preference to another. Fat cells all over the body release fat in response to demand, and the fat is then used by whatever muscles are active. No exercise can remove the fat from any one particular area—and, incidentally, neither can a massage machine that claims to break up fat on trouble spots.

Physical activity can help with trouble spots in another way, though. Strengthening muscles in a trouble area can help to improve their tone; stretching to gain flexibility can help with associated posture problems. Thus aerobic, strength, and flexibility workouts all have a place in fitness programs (see Chapter 10, which is devoted entirely to physical activity).

In summary, those who attempt to lose weight without physical activity may find themselves trapped in weight cycling, which can sometimes be harmful. People who combine diet and physical activity are more likely to lose fat, gain muscle, and regain less weight than those who only diet.

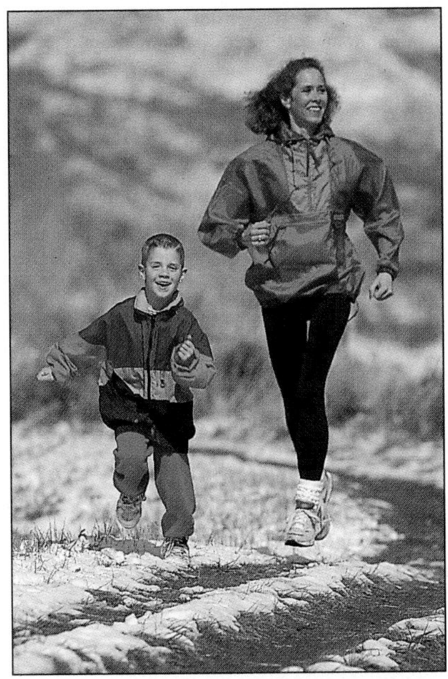

Regular physical activity helps people achieve and maintain healthy weights.

Behavior and Attitude

Behavior modification once held a key position in weight-loss programs, but its status has diminished.[47] Still, behavior and attitude are important supporting factors in achieving and maintaining appropriate body weight and composition. Changing the behaviors of overeating and underexercising that lead to, and perpetuate, obesity requires time and effort.

behavior modification: the changing of behavior by the manipulation of *antecedents* (cues or environmental factors that trigger behavior), the behavior itself, and *consequences* (the penalties or rewards attached to behavior).

Becoming Aware of Behaviors A person who is aware of all the behaviors that create a problem has a head start toward solving the problem. First, establish a baseline (a record of present eating and exercise behaviors) against which to measure future progress. It is best to keep a diary (see Figure 9.5 on p. 220) that includes the time and place of meals and snacks, the type and amount of foods eaten, the persons present when food is eaten, and a description of the individual's feelings when eating. The diary should also record physical activities: the kind, the intensity level, the duration, and the person's feelings about them. These entries will help the individual identify possible behaviors to change.

Making Small Changes The box on p. 221 describes strategies to support weight control. A particularly attractive feature of these strategies is that they do not involve blaming oneself or putting oneself down—an important element in fostering self-esteem.

Maintaining Weight Finally, be aware that it can be hard to maintain weight loss. On arriving at the goal weight after months of self-discipline and new habit formation, the victorious weight loser must not "celebrate" by resuming old eating habits. Membership in an ongoing weight-control organization and regular, continued physical activity can provide indispensable support for the formerly overweight person who wants to remain trim.

Personal Attitude For many people, overeating and being overweight may have become an integral part of their identity. Changing diet and activity behaviors without attention to a person's self-concept invites failure.

Many people overeat to cope with the stresses of life. To break out of that pattern, they must first identify the particular stressors that trigger their urges to overeat. Then, when faced with these situations, they must learn to practice problem-solving skills. When the problems that trigger the urge to overeat are dealt with in alternative ways, people may find that they eat less. The message is that sound emotional health supports your ability to take care of your health in all ways—including nutrition, weight control, and fitness.

Time	Place	Activity or food eaten	People present	Mood
10:30	School vending machine	6 peanut butter crackers and 12 oz. cola	by myself	starved
12:15	Restaurant	Sub sandwich and 12 oz. cola	friends	relaxed & friendly
3:00	Gym	45 min weight training	work out partner	tired
4:00	Snack bar	Small frozen yogurt	by myself	OK

Figure 9.5

Food and Activity Diary

A record of diet and physical activity habits reveals problem areas, the first step toward improving behaviors.

Underweight

Underweight is far less prevalent than overweight, affecting no more than 10 percent of U.S. adults. The health risks associated with underweight are fewer than those that accompany overweight. Both underweight women and those who have lost a significant amount of weight, however, are more susceptible to osteoporosis. Underweight women may become infertile or may give birth to unhealthy infants. An underweight woman can improve her chances of bearing a healthy infant by gaining weight prior to conception, during pregnancy, or both.

Underweight becomes more hazardous when accompanied by undernutrition. An inadequate supply of nutrients and energy leaves the body underprepared to handle its many metabolic and physical tasks. A person without reserves has a particularly tough battle against medical stresses such as surgery or the wasting diseases of cancer and AIDS. Thus underweight people are urged to gain lean tissue and body fat (as an energy reserve) and to acquire protective amounts of all the nutrients that can be stored.

Strategies for Weight Gain

Weight gain, like weight loss, is an individual matter. People who are healthy at their present weights may stay there; those who are at risk for illness should try to gain.

Some people are unalterably thin by reasons of heredity or early physical influences. Those who wish to gain weight for appearance's sake or to improve athletic performance should be aware that a healthful weight can be achieved only through physical activity, particularly strength training, combined with a high energy intake. Eating many high-kcalorie foods can bring about weight gain, but it will be mostly fat, and this can be as detrimental to health as being slightly underweight. In an athlete, such a weight gain can impair performance. Therefore, in weight gain, as in weight loss, physical activity is an essential component of a sound plan.

How to

Change Behaviors

Start simply and don't try to master all the behavior changes at once. Attempting too many changes at one time is never successful; a person must set priorities. A person may begin by limiting meals to three a day. When between-meal snacking is no longer a problem, the person may want to change another behavior.

1. To eliminate inappropriate eating cues:
   - Don't buy problem foods (such as ready-to-eat foods).
   - Don't shop when you are hungry.
   - Don't serve rich sauces and toppings.
   - Let other family members buy, store, and serve their own sweets.
   - Change channels or look away when the television shows food commercials.
   - Shop only from a list and stay away from convenience stores.
   - Carry appropriate snacks from home and avoid vending machines.
2. To suppress the cues you cannot eliminate:
   - Eat only in one place and in one room.
   - Clear plates directly into the garbage.
   - Create obstacles to the eating of problem foods (for example, make it necessary to unwrap, cook, and serve each one separately).
   - Minimize contact with excessive food (serve individual plates, don't put serving dishes on the table, and leave the table when you have finished eating).
   - Make small portions of food look large (spread food out, serve on small plates).
   - Control states of deprivation (eat regular meals, don't skip meals, avoid getting tired, avoid boredom by keeping cues to fun activities in sight).
3. To strengthen the cues to appropriate eating and physical activity:
   - Eat only at planned times; plan not to eat after a specified time (say, 7:00 or 8:00 P.M.).
   - Encourage others to eat appropriate foods with you.
   - Keep your favorite appropriate foods in the front of the refrigerator.
   - Learn appropriate portion sizes and prepare one portion at a time.
   - Save permitted foods from meals for snacks (and make these your only snacks).
   - Prepare permitted foods attractively.
   - Keep your hiking boots (ski poles, tennis racket) by the door.
4. To engage in desired eating or physical activity behaviors:
   - Slow down (pause several times during a meal, put down utensils between mouthfuls, chew thoroughly before swallowing, swallow before reloading the fork, always use utensils).
   - Leave some food on the plate.
   - Engage in no other activities while eating (such as reading or watching television).
   - Move more (shake a leg, pace, fidget, flex your muscles).
   - Join in and exercise with a group of active people.
5. To arrange or emphasize negative consequences of inappropriate eating:
   - Eat your meals with other people.
   - Ask that others respond neutrally when you deviate from your plan (make no comment). This is a negative consequence because it withholds attention.
6. To arrange or emphasize positive consequences of appropriate behaviors:
   - Update records of food intake, physical activity, and weight change regularly.
   - Arrange for rewards for each unit of behavior change or weight loss.
   - Ask family and friends for reinforcement (praise and encouragement).

Physical Activity to Build Muscles The person who wants to gain weight should use weight training primarily. As activity is increased, energy intake must be increased to support that activity. Eating extra food will then support a gain of both muscle and fat. About 700 to 1000 kcalories a day above normal energy needs is enough to support both the activity and the building of muscle.

Weight-gain pointer:
- *Be physically active and eat to build muscles.*

Energy-Dense Foods Energy-dense foods (the very ones eliminated from a successful weight-loss diet) hold the key to weight gain. Pick the highest-kcalorie items from each food group—that is, milk shakes instead of nonfat milk, peanut butter instead of lean meat, avocados instead of cucumbers, and whole-wheat

- *Eat energy-dense foods regularly.*

muffins instead of whole-wheat bread. Because fat contains more than twice as many kcalories per teaspoon as sugar does, fat adds kcalories without adding much bulk.

Be aware that health experts recommend a low-fat diet for the general U.S. population because the general population is overweight and at risk for heart disease. Consumption of high-fat foods is not healthy for most people, of course, but may be essential for an underweight individual who needs to gain weight. An underweight person who is physically active and eating a nutritionally adequate diet can afford a few extra kcalories from fat.

- *Eat at least three hearty meals a day.*

Three Meals Daily People wanting to gain weight should eat at least three hearty meals a day. Many people who are underweight have simply been too busy (sometimes for months) to eat enough to gain or maintain weight. Therefore, they need to make meals a priority and plan them in advance. Taking time to prepare and eat each meal can help, as can learning to eat more food within the first 20 minutes of a meal. Another suggestion is to eat meaty appetizers or the main course first and leave the soup or salad until later.

- *Eat large portions of foods and expect to feel full.*

Large Portions It is also important to learn to eat more food at each meal. Have two sandwiches for lunch instead of one, drink milk from a larger glass, and eat cereal from a larger bowl.

The person should expect to feel full. Most underweight individuals are accustomed to small quantities of food. When they begin eating significantly more, they feel uncomfortable. This is normal and passes over time.

- *Eat snacks between meals.*

Extra Snacks Since a substantially higher energy intake is needed each day, in addition to eating more food at each meal, it is necessary to eat more frequently. Between-meal snacking offers a solution. For example, a student might make three sandwiches in the morning and eat them between classes in addition to the day's three regular meals.

- *Drink plenty of juice and milk.*

Juice and Milk Beverages provide an easy way to increase energy intake. Consider that 6 cups of cranberry juice add almost 1000 kcalories to the day's intake. kCalories can be added to milk by mixing in powdered milk or packets of instant breakfast.

For people who are underweight due to illness, concentrated liquid formulas are often recommended because a weak person can swallow them easily. A physician or registered dietitian can recommend high-protein, high-kcalorie formulas to help the underweight person maintain or gain weight. Used in addition to regular meals, these can help considerably.

An extreme underweight condition known as anorexia nervosa is sometimes seen in young people who exercise unreasonable self-denial in order to control their weight. They go to such extremes that they become severely undernourished and underweight. The distinguishing feature of a person with anorexia nervosa, as opposed to other thin people, is that the starvation is intentional.

Anorexia nervosa is a major eating disorder seen in our society today. Another is bulimia—compulsive overeating, usually with purging. Eating disorders are the subject of the Nutrition in Practice that follows this chapter.

To gain weight, a person must train physically and increase energy intake by selecting energy-dense foods, eating regular meals, taking larger portions, and consuming extra snacks and beverages.

Self Study

Choose a Goal Weight and Develop a Weight-Control Plan

What weight is appropriate for you? When physical health alone is considered, a wide range of weights is acceptable for a person of a given height. Within the safe range, the choice of a weight is up to the individual.

Choose a Goal Weight

1. Determine whether your current weight is appropriate for your height.
   - Record your height: ___ in (or ___ cm).
   - Record your weight: ___ lb (or ___ kg).
   Look up the weight range for a person of your height in Table 9.1 on p. 207.
   - Record the entire range: ___ to ___ lb.

   Does your weight fall within the suggested range? Now calculate your BMI using the equations on pp. 207–208.
   - Record your BMI: ___

   Look up the disease risk for a person with your BMI value in Figure 9.2 on p. 208.
   - Record your risk of disease based on your BMI:

   If this level of risk is unacceptable, calculate the weight needed for a desired BMI value (divide the desired BMI by the appropriate height factor in the table on the next page). For example, a 165-pound person who is 5 feet 5 inches tall has a BMI of 27.5. To obtain a BMI of 22, the person would need to weigh about 133 pounds (22 ÷ 0.166).
   - Record your desired weight based on your height and desired BMI: ________________________.

   If your weight is below the range in Table 9.1 and your BMI is below 18.5, you may need to gain weight for your health's sake. If your weight is over the suggested weight range and your BMI value is associated with an unacceptable risk of disease, you may want to examine your body's fat distribution.
2. Determine whether your fat distribution is associated with health risks.
   - Record your waist measurement: ___.
   - Record your hip measurement: ___.

   Calculate the waist-to-hip ratio by dividing the number of inches (or centimeters) around your waistline by the number of inches (or centimeters) around your hips.
   - Record your waist-to-hip ratio: ___.

   Women with a ratio of 0.8 or greater and men with a ratio of 0.95 or greater are at high risk of obesity-related health problems.
3. Check your health history. A family or personal medical history of diabetes (type 2), hypertension, or high blood cholesterol signals the need to pay attention to diet and physical activity habits.

Based on these three considerations, how does your current weight compare with standards that are compatible with health? If your current weight compares favorably, you probably want to maintain your weight. If you want to gain weight or lose weight, indicate a sensible goal weight here: ___ lb goal weight.

Choose weight loss or weight gain as a goal for yourself. (If you are at the perfect weight, pretend that you are not, for purposes of this exercise, and develop a plan to change your weight by 10 pounds.)

Develop a Weight-Control Plan

To *lose* weight, you would need to adjust your energy balance by reducing your energy intake, increasing your energy output, or both. To *gain* weight, you would need to increase your energy intake. (It is hardly ever desirable to reduce energy expenditure.)

1. Review your energy intake. Return to Self Study 1, Form 2, and record from it your average daily energy intake:
   - My energy intake is ___ kcal/day.
2. Review your energy output. The Self Study in Chapter 6 helped you estimate your energy expenditure for a day. Record it here:
   - My energy output is ___ kcal/day.
3. Estimate your rate of weight gain or loss. Compare your energy intake with your output. Recall that a difference of 3500 kcalories between energy intake and output will make a difference of 1 pound, and estimate the rate at which you must be gaining or losing weight.

   Example: If your intake is 2000 kcalories per day and your output is 1500 kcalories per day, then you are acquiring an excess of 500 kcalories per day beyond your need, or 3500 kcalories each week. That means you should be gaining 1 pound per week:
   - With an intake of ___ kcal/day and an output of ___ kcal/day, I must be gaining/losing (circle one) 1 pound every ___ days.
4. Now choose a goal and a balance that will achieve it. Do not plan to gain or lose weight at a rate greater than 2 pounds per week, and do not plan to eat less than 10 kcalories per pound of your current weight each day.

(continued)

- I wish to gain/lose (circle one) 1 pound every ___ days. That means my energy intake should be ___ kcal per day and my energy output should be (or remain at) ___ kcal per day.

This exercise assumes you will adjust your energy output appropriately and focuses on the input side, the diet.

5. Choose an appropriate diet. Table 9.2 on p. 217 offers diet patterns for different energy intakes, starting with a 1200-kcalorie plan using the minimum number of servings suggested in the Daily Food Guide. This pattern would allow most people to lose weight at a satisfactory rate and still meet all nutrient needs. The other plans offer patterns for higher energy intakes. All of the patterns in the table supply less than 30 percent of kcalories from fat:

- I choose the ___-kcalorie diet.

6. Now plan a day's menus, following the pattern. What will you have for breakfast, lunch, dinner, and snacks (if any)?

Height	Height Factor	Height	Height Factor	Height	Height Factor
4'7"	0.232	5'3"	0.177	5'11"	0.139
4'8"	0.224	5'4"	0.172	6'0"	0.136
4'9"	0.216	5'5"	0.166	6'1"	0.132
4'10"	0.209	5'6"	0.161	6'2"	0.128
4'11"	0.202	5'7"	0.157	6'3"	0.125
5'0"	0.195	5'8"	0.152	6'4"	0.122
5'1"	0.189	5'9"	0.148	6'5"	0.119
5'2"	0.183	5'10"	0.143	6'6"	0.116

Note: To obtain the weight needed for a certain BMI, divide the desired BMI by the height factor appropriate for your height.

Source: R. P. Abernathy, Body mass index: Determination and use, Copyright the American Dietetic Association, Reprinted by permission from *Journal of the American Dietetic Association* 91 (1991): 843.

Self Check

1. The BMI range that correlates with few health risks is:
   a. 18 to 20.5. b. 18.5 to 25.
   c. 25 to 30. d. 30 to 34.9.
2. Two causes of obesity in humans are:
   a. genetics and low-carbohydrate diets.
   b. mineral imbalances and fat cell imbalance.
   c. set-point theory and BMI.
   d. genetics and physical inactivity.
3. In relation to obesity, genetics may influence the following:
   a. energy storage. b. energy expenditure.
   c. fat cell development. d. all of the above.
4. The obesity theory that suggests the body chooses to be at a specific weight is the:
   a. set-point theory b. enzyme theory.
   c. fat cell theory. d. external cue theory.
5. Which of the following are health risks associated with being overweight?
   a. hypertension and stroke
   b. diabetes and heart disease
   c. gallbladder disease and some cancers
   d. all of the above
6. Even more than total fatness, central obesity may be especially likely to lead to:
   a. diabetes and stroke.
   b. hypertension and heart disease.
   c. a and b.
   d. a "pear" profile.
7. Which of the following are reasonable treatments for obesity?
   a. realistic energy intake and nutrient adequacy

b. small portions and carbohydrates
c. physical activity
d. all of the above

8. Benefits of physical activity in a weight-control program include:
   a. short-term increase in energy expenditure.
   b. appetite control.
   c. stress reduction and control of stress eating.
   d. all of the above.

9. Suggestions to change behaviors for successful weight control include:
   a. let other family members buy, store, and serve their own sweets.
   b. eat in front of the television for distraction.
   c. learn appropriate portion sizes.
   d. a and c.

10. Some strategies for weight gain include:
    a. weight training and eating to build muscles.
    b. eating energy-dense foods regularly.
    c. eating three hearty meals daily with large portions.
    d. all of the above.

Answers to these questions appear in Appendix H.

Notes

1. S. M. Garn, Fractionating healthy weight, *American Journal of Clinical Nutrition* 63 (1996): 412S–414S.

2. Addresses presented at the North American Association for the Study of Obesity and Emory University School of Medicine Conference on Obesity Update: Pathophysiology, Clinical Consequences, and Therapeutic Options, Atlanta, Georgia. August 31–September 2, 1992.

3. S. H. Stensland and S. Margolis, Simplifying the calculation of body mass index for quick reference, *Journal of the American Dietetic Association* 90 (1990): 856.

4. J. G. Meisler and S. St. Jeor, Summary and recommendations for the American Health Foundation's Expert Panel on Healthy Weight, *American Journal of Clinical Nutrition* 63 (1996): 474S–477S.

5. M. Stern, Epidemiology of obesity and its link to heart disease, *Metabolism: Clinical and Experimental* 44 (1995): 1–3; T. B. VanItallie, Body weight, morbidity, and longevity, in *Obesity,* eds. P. Björntorp and B. N. Brodoff (Philadelphia: J. B. Lippincott, 1992), pp. 361–369.

6. C. Bouchard and L. Perusse, Genetics of obesity, *Annual Review of Nutrition* 13 (1993): 337–354.

7. J. P. Foreyt and W. S. C. Poston II, Diet, genetics, and obesity, *Food Technology* 51 (1997): 70–73; C. Bouchard, Human variation in body mass: Evidence for a role of the genes, *Nutrition Reviews* 55 (1997): S21–S30.

8. Y. Zhang and coauthors, Positional cloning of the mouse *obese* gene and its human homologue, *Nature* 372 (1994): 425–431.

9. C. T. Montague and coauthors, Congenital leptin deficiency is associated with severe early-onset obesity in humans, *Nature* 387 (1997): 903–908.

10. R. V. Considine, Serum immunoreactive-leptin concentrations in normal-weight and obese humans, *New England Journal of Medicine* 334 (1996): 292–295; S. G. Hassink and coauthors, Serum leptin in children with obesity: Relationship to gender and development, *Pediatrics* 98 (1996): 201–203.

11. J. Albu and coauthors, Obesity solutions: Report of a meeting, *Nutrition Reviews* 55 (1997): 150–156.

12. R. L. Leibel, M. Rosenbaum, and J. Hirsch, Changes in energy expenditure resulting from altered body weight, *New England Journal of Medicine* 332 (1995): 621–628; J. M. Kinney, Influence of altered body weight on energy expenditure, *Nutrition Reviews* 53 (1995): 265–268; P. Pasquet and M. Apfelbaum, Recovery of initial body weight and composition after long-term massive overfeeding in men, *American Journal of Clinical Nutrition* 60 (1994): 861–863.

13. R. E. Keesey and M. D. Hirvonen, Body weight set-points: Determination and adjustment, *Journal of Nutrition* 127 (1997): 1875S–1883S.

14. J. E. Blundell and coauthors, Control of human appetite: Implications for the intake of dietary fat, *Annual Review of Nutrition* 16 (1996): 285–319.

15. Prevalence of physical inactivity during leisure time among overweight persons—1994, *Morbidity and Mortality Weekly Report* 45 (1996): 185–188.

16. A. P. Simopoulos, Characteristics of obesity, in *Obesity,* eds. P. Björntorp and B. N. Brodoff (Philadelphia: J. B. Lippincott, 1992), pp. 309–319.

17. E. Obartzanek and coauthors, Energy intake and physical activity in relation to indexes of body fat: The National Heart, Lung, and Blood Institute Growth and Health Study, *American Journal of Clinical Nutrition* 60 (1994): 15–22; S. L. Gortmaker, W. H. Dietz, Jr., and L. W. Cheung, Inactivity, diet, and the fattening of America, *Journal of the American Dietetic Association* 90 (1990): 1247–1255.

18. D. A. McCarron and M. E. Reusser, Body weight and blood pressure regulation, *American Journal of Clinical Nutrition* 63 (1996): 423S–425S.

19. Update: Prevalence of overweight among children, adolescents, and adults—United States, 1988–1994, *Morbidity and Mortality Weekly Report* 46 (1997): 199–202.

20. J. Foreyt and K. Goodrick, The ultimate triumph of obesity, *The Lancet* 346 (1995): 134–135.

21. National Institutes of Health, National Heart, Lung, and Blood Institute, *Clinical Guidelines on the Identification, Evaluation, and Treatment of Overweight and Obesity in Adults: The Evidence Report* (Washington, D.C.: U.S. Department of Health and Human Services, 1998), pp. vii–xxvi; Albu and coauthors, 1997.

22. National Task Force on the Prevention and Treatment of

Obesity, Long-term pharmocotherapy in the management of obesity, *Journal of the American Medical Association* 276 (1996): 1907–1915.

23. J. E. Manson and coauthors, Body weight and mortality among women, *New England Journal of Medicine* 333 (1995): 677–685; W. C. Willett and coauthors, Weight, weight change, and coronary heart disease in women, *Journal of the American Medical Association* 273 (1995): 461–465.

24. W. B. Kannel, R. B. D'Agostino, and J. L. Cobb, Effect of weight on cardiovascular disease, *American Journal of Clinical Nutrition* 63 (1996): 419S–422S.

25. A. Must and coauthors, Long-term morbidity and mortality of overweight adolescents, *New England Journal of Medicine* 327 (1992): 1350–1355; G. A. Bray, Adolescent overweight may be tempting fate, *New England Journal of Medicine* 327 (1992): 1378–1380.

26. E. M. Emery and coauthors, A review of the association between abdominal fat distribution, health outcome measures, and modifiable risk factors, *American Journal of Health Promotion,* May/June 1993, pp. 342–353.

27. Emery and coauthors, 1993; A. R. Folsom and coauthors, Body fat distribution and 5-year risk of death in older women, *Journal of the American Medical Association* 269 (1993): 483–487.

28. P. Björntorp, Regional obesity, in *Obesity,* eds., P. Björntorp and B. N. Brodoff (Philadelphia: J. B. Lippincott, 1992), pp. 579–586.

29. S. Lemieux and coauthors, Sex differences in the relation of visceral adipose tissue accumulation to total body fatness, *American Journal of Clinical Nutrition* 58 (1993): 463–467; C. Ley, B. Lees, and J. C. Stevenson, Sex- and menopause-associated changes in body fat distribution, *American Journal of Clinical Nutrition* 55 (1992): 950–954.

30. M. J. Williams and coauthors, Regional fat distribution in women and risk of cardiovascular disease, *American Journal of Clinical Nutrition* 65 (1997): 855–860.

31. Emery and coauthors, 1993; R. J. Troisi, Cigarette smoking, dietary intake, and physical activity: Effects on body fat distribution—The Normative Aging Study, *American Journal of Clinical Nutrition* 53 (1991): 1104–1111.

32. B. B. Duncan and coauthors, Association of the waist-to-hip ratio is different with wine than with beer or hard liquor consumption, *American Journal of Epidemiology* 142 (1995): 1034–1038.

33. G. R. Hunter and coauthors, Fat distribution, physical activity, and cardiovascular risk factors, *Medicine and Science in Sports and Exercise* 29 (1997): 362–369.

34. S. L. Gortmaker and coauthors, Social and economic consequences of overweight in adolescence and young adulthood, *New England Journal of Medicine* 329 (1993): 1008–1012.

35. National Task Force on the Prevention and Treatment of Obesity, 1996.

36. R. P. Abernathy and D. R. Black, Healthy body weights: An alternative perspective, *American Journal of Clinical Nutrition* 63 (1996): 448S–451S.

37. Meisler and St. Jeor, 1996.

38. M. Shah and coauthors, Comparison of a low-fat ad libitum complex-carbohydrate diet with a low-energy diet in moderately obese women, *American Journal of Clinical Nutrition* 59 (1994): 980–984.

39. A. Astrup and coauthors, The role of low-fat diets and fat substitutes in body weight management: What have we learned from clinical studies? *Journal of the American Dietetic Association* 97 (1997): S82–S87; Shah and coauthors, 1994.

40. S. Toubro and A. Astrup, *Ad libitum* low-fat, high-carbohydrate diet *versus* calorie-counting for weight maintenance after major weight loss in obese patients, *British Medical Journal* 314 (1997): 29–34.

41. S. N. Blair, Diet and activity: The synergistic merger, *Nutrition Today* 30 (1996): 108–112; J. H. Wilmore, Increasing physical activity: Alterations in body mass and composition, *American Journal of Clinical Nutrition* 63 (1996): 254–260.

42. R. W. Jeffrey, Does weight cycling present a health risk? *American Journal of Clinical Nutrition* 63 (1996): 452S–455S.

43. National Task Force on the Prevention and Treatment of Obesity, Weight cycling, *Journal of the American Medical Association* 272 (1994): 1196–1202.

44. A. Geliebter and coauthors, Effects of strength or aerobic training on body composition, resting metabolic rate, and peak oxygen consumption in obese dieting subjects, *American Journal of Clinical Nutrition* 66 (1997): 557–563; B. L. Marks and coauthors, Fat-free mass is maintained in women following a moderate diet and exercise program, *Medicine and Science in Sports and Exercise* 27 (1995): 1243–1251; K. P. G. Kempen, W. H. M. Saris, and K. R. Westerterp, Energy balance during an 8-wk energy-restricted diet with and without exercise in obese women, *American Journal of Clinical Nutrition* 62 (1995): 722–729; R. Ross, H. Pedwell, and J. Rissanen, Effects of energy restriction and exercise on skeletal muscle and adipose tissue in women as measured by magnetic resonance imaging, *American Journal of Clinical Nutrition* 61 (1995): 1179–1185; S. B. Racette and coauthors, Effects of aerobic exercise and dietary carbohydrate on energy expenditure and body composition during weight reduction in obese women, *American Journal of Clinical Nutrition* 61 (1995): 486–494; D. D. Hensrud and coauthors, A prospective study of weight maintenance in obese subjects reduced to normal body weight without weight-loss training, *American Journal of Clinical Nutrition* 60 (1994): 688–694.

45. H. M. Sjödin and coauthors, The influence of physical activity on BMR, *Medicine and Science in Sports and Exercise* 28 (1996): 85–91.

46. N. A. King, A. Tremblay, and J. E. Blundell, Effects of exercise on appetite control: Implications for energy balance, *Medicine and Science in Sports and Exercise* 29 (1997): 1076–1089.

47. Albu and coauthors, 1997.

CHAPTER 9

Nutrition in Practice

▪ Eating Disorders ▪

An estimated 2 million people in the United States, primarily girls and young women, suffer from the eating disorders anorexia nervosa and bulimia nervosa. Many more suffer from unspecified eating disorders—conditions that do not meet the strict criteria for anorexia nervosa or bulimia nervosa (see Tables NP9.2 and NP9.3 on p. 231), but still imperil a person's well-being. Characteristics of disordered eating such as restrained eating, binge eating, purging, fear of fatness, and distortion of body image are common, especially among young middle-class girls.[1] In most other societies, these behaviors and attitudes are much less prevalent.

WWW **http://members.aol.com/amanbu**
American Anorexia/Bulimia Association

Why do so many young people in our society suffer from eating disorders?

Excessive pressure to be thin is at least partly to blame. Two-thirds of adolescent girls and one-third of adolescent boys are dissatisfied with their body weight.[2] By making thinness the ideal, society pushes people to view a healthy body of normal weight as too fat. Healthy people then take unhealthy actions to lose weight. Severe restriction of food intake may create intense hunger that leads to binges. Research confirms this theory, showing that unhealthy or dangerous diets often precede binge eating in adolescent girls.[3] Energy restriction followed by bingeing can set in motion a pattern of weight cycling, which may make weight loss and maintenance more difficult over time.[4]

People who attempt extreme weight loss are dissatisfied with their bodies to begin with; they may also be depressed or suffer social anxiety. As weight loss becomes more and more difficult, psychological problems worsen, and the likelihood of developing full-blown eating disorders intensifies.

People with anorexia nervosa suffer from an extreme preoccupation with weight loss that seriously endangers their health and even their lives. People with bulimia engage in episodes of binge eating alternating with periods of severe dieting or self-starvation. Some bulimics also follow binge eating with self-induced vomiting, laxative abuse, or diuretic abuse to "undo the damage." The glossary on the next page defines the relevant terms.

You said girls and young women are most vulnerable to anorexia nervosa and bulimia nervosa. Are there other vulnerable groups?

Yes. Athletes are particularly likely to develop eating disorders.[5] Athletes must often meet stringent weight requirements to compete in their sport. Many athletes report that they engage in behaviors that are typical of people with eating disorders. Female competitors often report being terrified of becoming fat, being obsessed with food, and using laxatives in attempting to control weight. Dancers, jockeys, wrestlers, distance runners, bodybuilders, divers, figure skaters, gymnasts, and others whose body weight and appearance are frequently judged in comparison with an "ideal" are especially prone to develop problems.[6] These athletes may engage in extreme weight-loss practices such as overtraining, prolonged fasting, vomiting, taking diet pills, and using steam baths and saunas to induce sweating.[7]

Men account for about 1 in 20 cases in the general population, but among male athletes and dancers, eating disorders are much more common. Male teenagers normally average about 15 percent of body weight as fat, but some high school athletes strive to carry only 5 percent or so of their body weight as fat.

Even among athletes, however, women are most vulnerable to developing eating disorders. Many female athletes appear healthy, but in fact may easily develop the three interrelated components of the female athlete triad: disordered eating, amenorrhea (the absence of three or more consecutive menstrual cycles), and osteoporosis.[8]

How does the female athlete triad develop?

Many athletic women engage in self-destructive eating behaviors (disordered eating) because they and their

Glossary

anorexia nervosa: an eating disorder characterized by a refusal to maintain a minimally normal body weight, self-starvation to the extreme, and a disturbed perception of body weight and shape, seen (usually) in teenage girls and young women.
anorexia = without appetite
nervosa = of nervous origin

bulimia (byoo-LEEM-ee-uh) **nervosa:** recurring episodes of binge eating combined with a morbid fear of becoming fat, usually followed by self-induced vomiting or purging.

cathartic: a strong laxative.

eating disorder: a disturbance in eating behavior that jeopardizes a person's physical and psychological health.

emetic (em-ETT-ic): an agent that causes vomiting.

female athlete triad: a potentially fatal triad of medical problems: disordered eating, amenorrhea, and osteoporosis.

unspecified eating disorders: eating disorders that do not meet the criteria for specific eating disorders previously defined.

coaches have adopted unsuitable weight standards. An athlete's body must be heavier for height than a non-athlete's body because the athlete's bones and muscles are denser. Weight standards that may be appropriate for others are inappropriate for athletes. Measures such as fatfold measures yield more useful information about body composition.

Many young female athletes severely restrict energy intakes to improve performance, enhance the aesthetic appeal of their performance, or meet the weight guidelines of their specific sports.[9] They fail to realize that the loss of lean tissue that accompanies energy restriction actually impairs their physical performance. Risk factors for the female athlete triad include the following:

- Young age (adolescence).
- Pressure to excel at a chosen physical activity.
- Focus on achieving or maintaining an "ideal" body weight or body fat percentage.
- Participation in endurance sports or competitions that judge performance on aesthetic appeal such as gymnastics, figure skating, or dance.
- Dieting at an early age.
- Unsupervised dieting.

As for amenorrhea, its prevalence among premenopausal women in the United States is about 2 to 5 percent overall, but among female athletes it may be as high as 66 percent.[10] Contrary to previous notions, amenorrhea is *not* a normal adaptation to strenuous physical training: it is a symptom of something going wrong.[11] Amenorrhea is characterized by low blood estrogen, infertility, and often bone mineral losses.

In general, weight-bearing physical activity, dietary calcium, and the hormone estrogen protect against the bone loss of osteoporosis, but in women with disordered eating and amenorrhea, strenuous activity may impair bone health. Vigorous training combined with low food energy intakes and other life stresses seems to trigger amenorrhea and promote bone loss. Low estrogen leads to diminished bone mass and increased bone fragility. Many amenorrheic athletes have decreased bone density, similar to that of 50- to 60-year-old women, when they should have dense, strong bones. Amenorrheic athletes should be encouraged to consume at least 1500 milligrams of calcium each day, to eat nutrient-dense foods, and to obtain enough food energy to cover the energy expended in physical activity. Future research will focus on the question of hormone replacement therapy for these women.

What can be done to prevent eating disorders in athletes and dancers?

To prevent eating disorders in athletes and dancers, both the performers and their coaches must be educated about links between inappropriate body weight ideals, improper weight-loss techniques, eating disorder development, adequate nutrition, and safe weight-control methods. Coaches and dance instructors should never encourage unhealthy weight loss to qualify for competition or to conform with distorted artistic ideals. Frequent weighings can push young people who are striving to lose weight into a cycle of starving to confront the scale, then bingeing uncontrollably afterward. The erosion of self-esteem that accompanies these events can interfere with the normal psychological development of the teen years and set the stage for serious problems later on.

Table NP9.1 provides some suggestions to help athletes and dancers protect themselves against developing eating disorders. The next sections describe eating disorders that anyone, athlete or nonathlete, may experience.

Table NP9.1 Tips for Combating Eating Disorders

General Guidelines

- Never restrict food servings to below the numbers suggested for adequacy by the Food Guide Pyramid.
- Eat frequently. People often do not eat frequent meals because of time constraints, but eating can be incorporated into other activities, such as snacking while studying or commuting. The person who eats frequently never gets so hungry as to allow hunger to dictate food choices.
- If not at a healthy weight, establish a reasonable weight goal based on a healthy body composition.
- Allow a reasonable time to achieve the goal. A reasonable loss of excess fat can be achieved at the rate of about 1 percent of body weight per week.
- Establish a weight-maintenance support group with people who share interests.

Specific Guidelines for Athletes and Dancers

- Remember that eating disorders impair physical performance. Seek confidential help in obtaining treatment if needed.
- Restrict weight-loss activities to the off-season.
- Focus on proper nutrition as an important facet of your training, as important as proper technique.

I remember a high school friend who had anorexia nervosa. She was very bright and seemed to have it all. Is that uncommon for a girl with anorexia nervosa?

Not at all. Most anorexia nervosa victims are females who come from middle- or upper-class families. Family patterns often include parents who oppose one another's authority and who vacillate between defending and condemning the anorexic child's behavior, confusing the child and disrupting normal parental control.[12] The family values achievement and outward appearances more than an inner sense of self-worth and self-actualization.

The person with anorexia nervosa is often a perfectionist who works hard to please her parents. She may identify so strongly with her parents' ideals and goals for her that she sometimes feels she has no identity of her own. She earnestly desires to control her own destiny, but she feels controlled by others. When she does not eat, she gains control.

Many people diet. How do you know when dieting is going too far?

Many young women diet to lose weight. However, when a person loses weight to well below the average for her height and is no longer slim, but too slim, and still doesn't stop, she has gone too far. Regardless of how thin she is, she looks in the mirror and sees herself as fat. Central to the diagnosis of anorexia nervosa is a distorted body image that overestimates body fatness. Table NP9.2 (p. 231) shows the criteria that professionals use to diagnose anorexia nervosa. Anorexia nervosa resembles an addiction. The characteristic behavior is obsessive and compulsive. Before drawing conclusions about someone who is extremely thin or who eats very little, remember that diagnosis of anorexia nervosa requires professional assessment.

What is the harm in being very thin?

Anorexia nervosa damages the body much as starvation does. In young people, growth ceases and normal development falters. They lose so much lean tissue that basal metabolic rate slows, an effect that may remain even after treatment and regain of weight.[13] Additionally, the heart pumps inefficiently and irregularly, the heart muscle becomes weak and thin, the heart chambers diminish in size, and the blood pressure falls. Electrolytes that help to regulate heartbeat become unbalanced. Many deaths from heart failure occur in people with anorexia.

Starvation brings other physical consequences as well: impaired immune response, anemia, and a loss of digestive function that worsens malnutrition. Digestive functioning becomes sluggish, the stomach empties slowly, and the lining of the intestinal tract shrinks. The ailing digestive tract fails to sufficiently digest any food the victim may eat. The pancreas slows its production of digestive enzymes. The person may suffer from diarrhea, further worsening malnutrition.

What kind of treatment helps people with anorexia nervosa?

Treatment of anorexia nervosa requires a multidisciplinary approach that addresses two sets of issues and

Nutrition in Practice

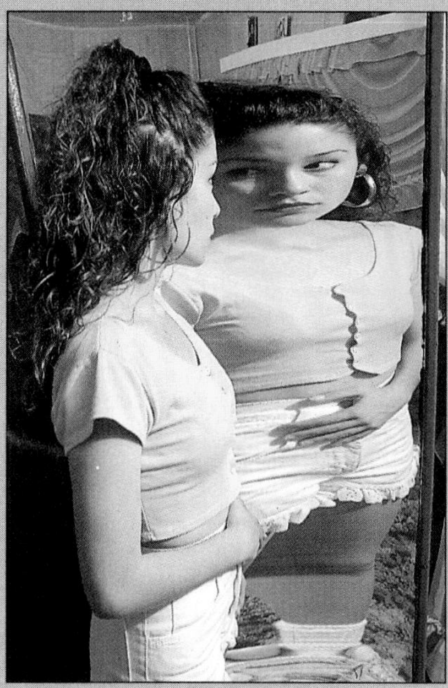

Women with anorexia nervosa see themselves as fat, even when they are dangerously underweight.

For many people with bulimia, guilt, depression, and self-condemnation follow a binge-eating episode.

behaviors: those relating to food and weight, and those involving relationships with oneself and others.[14] Teams of physicians, nurses, psychiatrists, family therapists, and dietitians work together to treat people with anorexia nervosa. Appropriate diet is crucial and must be tailored individually to each client's needs. Seldom are clients willing to eat for themselves, but if they are, chances are they can recover without other interventions.

High-risk clients may require hospitalization and may need to be force-fed by tube at first to forestall death. This step causes psychological trauma. Medications are commonly prescribed, but to date, they play a limited role in treatment.

Denial runs high among those with anorexia nervosa. Few seek treatment on their own. Almost half of the women who are treated can maintain their body weight within 15 percent of healthy weight; at that weight, many of them begin menstruating again. The other half have poor or fair treatment outcomes, and two-thirds of those treated fight an ongoing mental battle with recurring morbid thoughts about food and body weight.[15] Many relapse into abnormal eating behaviors to some extent. About 5 percent die during treatment, 1 percent by suicide.

How does bulimia nervosa differ from anorexia nervosa?

Bulimia nervosa is distinct from anorexia nervosa and is more prevalent. More men suffer from bulimia nervosa than from anorexia, but bulimia is still more common in women. The secretive nature of bulimic behaviors makes recognition of the problem difficult, but once it is recognized, diagnosis is based on the criteria listed in Table NP9.3.

The typical person with bulimia is well educated, in her early twenties, and close to ideal body weight. She is a high achiever, with a strong feeling of dependence on her parents. She experiences considerable social anxiety and has difficulty establishing personal relationships. She is sometimes depressed and often exhibits impulsive behavior.

Like the person with anorexia nervosa, the person with bulimia spends much time thinking about her body weight and food. Her preoccupation with food manifests itself in secretive binge-eating episodes followed by self-induced vomiting, fasting, or the use of laxatives or diuretics. Such behaviors typically begin in late adolescence after a long series of various unsuccessful weight-reduction diets. People with bulimia commonly follow a pattern of restrictive dieting interspersed with bulimic behaviors and experience weight fluctuations of more than 10 pounds up and down over short periods of time.

Unlike the person with anorexia nervosa, the person with bulimia is aware of the consequences of her behavior, feels that it is abnormal, and is deeply ashamed of it. She feels inadequate and unable to control her eating, so she tends to be passive and to look to men for confirmation of her sense of self-worth. When she is rejected, either in reality or in her imagination, her bulimia becomes worse. If her depression deepens, she may seek solace in drug or alcohol abuse or other addictive behaviors. As many as 50 percent of women with bulimia nervosa are alcohol dependent.[16]

Table NP9.2 Criteria for Diagnosis of Anorexia Nervosa

A person with anorexia nervosa demonstrates the following:

A. Refusal to maintain body weight at or above a minimal normal weight for age and height, e.g., weight loss leading to maintenance of body weight less than 85 percent of that expected; or failure to make expected weight gain during period of growth, leading to body weight less than 85 percent of that expected.

B. Intense fear of gaining weight or becoming fat, even though underweight.

C. Disturbance in the way in which one's body weight or shape is experienced; undue influence of body weight or shape on self-evaluation, or denial of the seriousness of the current low body weight.

D. In females past puberty, amenorrhea, i.e., the absence of at least three consecutive menstrual cycles. (A woman is considered to have amenorrhea if her periods occur only following hormone, e.g., estrogen, administration.)

Two types:

- **Restricting type:** During the episode of anorexia nervosa, the person does not regularly engage in binge eating or purging behavior (i.e., self-induced vomiting or the misuse of laxatives, diuretics, or enemas).
- **Binge eating/purging type:** During the episode of anorexia nervosa, the person regularly engages in binge eating or purging behavior (i.e., self-induced vomiting or the misuse of laxatives, diuretics, or enemas).

Source: Reprinted with permission from the *Diagnostic and Statistical Manual of Mental Disorders,* 4th ed. (Washington, D.C.: American Psychiatric Association, 1994). Copyright 1994 American Psychiatric Association.

Table NP9.3 Criteria for Diagnosis of Bulimia Nervosa

A person with bulimia nervosa demonstrates the following:

A. Recurrent episodes of binge eating. An episode of binge eating is characterized by both of the following:

   1. eating, in a discrete period of time (e.g., within any two-hour period), an amount of food that is definitely larger than most people would eat during a similar period of time and under similar circumstances, and,
   2. a sense of lack of control over eating during the episode (e.g., a feeling that one cannot stop eating or control what or how much one is eating).

B. Recurrent inappropriate compensatory behavior in order to prevent weight gain, such as self-induced vomiting; misuse of laxatives, diuretics, enemas, or other medications; fasting; or excessive exercise.

C. Binge eating and inappropriate compensatory behaviors that both occur, on average, at least twice a week for three months.

D. Self-evaluation unduly influenced by body shape and weight.

E. The disturbance does not occur exclusively during episodes of anorexia nervosa.

Two types:

- **Purging type:** The person regularly engages in self-induced vomiting or the misuse of laxatives, diuretics, or enemas.
- **Nonpurging type:** The person uses other inappropriate compensatory behaviors, such as fasting or excessive exercise, but does not regularly engage in self-induced vomiting or the misuse of laxatives, diuretics, or enemas.

Source: Reprinted with permission from the *Diagnostic and Statistical Manual of Mental Disorders,* 4th ed. (Washington, D.C.: American Psychiatric Association, 1994). Copyright 1994 American Psychiatric Association.

What exactly is binge eating?

Binge eating is unlike normal eating, and the food is not consumed for its nutritional value. The binge eater has a compulsion to eat. A typical binge occurs periodically, is done in secret, usually at night, and lasts an hour or more. A binge frequently follows a period of rigid dieting, so the binge eating is accelerated by hunger. During a binge, the person with bulimia may consume from 1000 to many thousands of kcalories of food. The food typically contains little fiber or water, has a smooth texture, and is high in sugar and fat, so it is easy to consume vast amounts rapidly with little chewing.

What are the consequences of this behavior?

After a binge, the person may use a cathartic—a strong laxative that can injure the lower intestinal tract. Or the person may induce vomiting, using an emetic—a drug intended as first aid for poisoning.

On first glance, purging seems to offer a quick and easy solution to the problems of unwanted kcalories and body weight. Many people perceive such behavior as neutral or even positive, when, in fact, bingeing and purging have serious physical consequences.[17] Fluid and electrolyte imbalances caused by vomiting or diarrhea can lead to metabolic alkalosis, a condition characterized by apathy, confusion, and muscle spasms. Vomiting causes irritation and infection of the pharynx, esophagus, and salivary glands; erosion of the teeth; and dental caries. The esophagus may rupture or tear, as may the stomach. Overuse of emetics depletes potassium concentrations and can lead to death by heart failure.[18]

What is the treatment for bulimia?

As for people with anorexia nervosa, a team approach provides the most effective treatment for people with bulimia. Bulimia is easier to treat than anorexia nervosa in many respects because it seems to be more of a chosen behavior. People with bulimia know that their behavior is abnormal, and many are willing to try to cooperate.

The goal of the dietary plan to treat bulimia is to help clients gain control, establish regular eating patterns, and restore nutritional health. Energy intake should not be severely restricted. The person needs to learn to eat a quantity of nutritious food sufficient to nourish her body and to satisfy hunger (at least 1600 kcalories a day). The accompanying box offers some ways to begin correcting bulimia nervosa.

Anorexia nervosa and bulimia nervosa are distinct eating disorders, yet they sometimes overlap. Anorexia victims may purge, and victims of both conditions share an overconcern with body weight and the tendency to drastically undereat. The two disorders can also appear in the same person, or one can lead to the other.

At so tender an age as 12 years, beautifully growing, normal-weight female youngsters are already worried that they are too fat. Most are "on diets." Magazines, newspapers, and television all present the message that to be thin is to be beautiful and happy. Anorexia nervosa and bulimia are not a form of rebellion against these unreasonable expectations, but rather the exaggerated acceptance of them. Perhaps a person's best defense against these disorders is to learn to appreciate his or her own uniqueness.

How to

Combat Bulimia Nervosa

The following advice has proved useful for people fighting bulimia nervosa:

- Avoid finger foods; eat foods that require the use of utensils.
- Enhance satiety by eating warm foods.
- Include vegetables, salad, and/or fruit at meals to prolong eating time.
- Choose whole-grain and high-fiber breads and cereals to maximize bulk.
- Eat a well-balanced diet and meals consisting of a variety of foods.
- Use foods that are naturally divided into portions, such as potatoes (rather than rice or pasta); 4- and 8-ounce containers of yogurt, ice cream, or cottage cheese; precut steak or chicken parts; and frozen entrées.
- Include foods containing ample complex carbohydrates (for satiety) and some fat (to slow gastric emptying).
- Eat meals and snacks sitting down.
- Plan meals and snacks, and record plans in a food diary prior to eating.

Source: Reprinted by permission of Lippincott, Williams & Wilkins.

CHAPTER 9

Nutrition in Practice

Notes

1. G. B. Schreiber and coauthors, Weight modification efforts reported by black and white preadolescent girls: National Heart, Lung, and Blood Institute Growth and Health Study, *Pediatrics* 98 (1996): 63–70; L. M. Mellin, C. E. Irwin, and S. Scully, Prevalence of disordered eating in girls: A survey of middle-class children, *Journal of the American Dietetic Association* 92 (1992): 851–853.

2. D. C. Moore, Body image and eating behavior in adolescents, *Journal of the American College of Nutrition* 12 (1993): 975–980.

3. D. Neumark-Sztainer, R. Butler, and H. Palti, Dieting and binge eating: Which dieters are at risk? *Journal of the American Dietetic Association* 95 (1995): 586–588.

4. D. Neumark-Sztainer, Excessive weight preoccupation, *Nutrition Today,* March/April 1995, pp. 68–74.

5. K. K. Yeager and coauthors, The female athlete triad: Disordered eating, amenorrhea, osteoporosis, *Medicine and Science in Sports and Exercise* 25 (1993): 775–777.

6. A. Yates, Eating disorders in women athletes, *Eating Disorders Review,* July/August 1996, pp. 1–4.

7. Position stand from the Committee on Sports Medicine and Fitness on promotion of healthy weight-control practices in young athletes, *Pediatrics* 97 (1996): 752–753.

8. American College of Sports Medicine, Position stand, The female athlete triad, *Medicine and Science in Sports and Exercise* 29 (1997): i–ix.

9. J. H. Wilson, Nutrition, physical activity, and bone health in women, *Nutrition Research Reviews* 7 (1994): 67–91.

10. Yeager and coauthors, 1993.

11. C. L. Otis, American College of Sports Medicine's Ad Hoc Task Force on Women's Issues in Sports Medicine, as quoted in A. A. Skolnick, "Female athlete triad" risk for women, *Journal of the American Medical Association* 270 (1993): 921–923.

12. G. Szmukler and C. Dare, Family therapy of early-onset, short-history anorexia nervosa, in *Family Approaches in Treatment of Eating Disorders,* eds. D. B. Woodside and L. Shekter-Wolfson (Washington, D.C. : American Psychiatric Press, 1991), pp. 25–47.

13. L. Scalfi and coauthors, Bioimpedance analysis and resting energy expenditure in undernourished and refed anorectic patients, *European Journal of Clinical Nutrition* 47 (1993): 61–67; R. C. Casper and coauthors, Total daily energy expenditure and activity level in anorexia nervosa, *American Journal of Clinical Nutrition* 53 (1991): 1143–1150.

14. Position of The American Dietetic Association: Nutrition intervention in the treatment of anorexia nervosa, bulimia nervosa, and binge eating, *Journal of the American Dietetic Association* 94 (1994): 902–907.

15. American Psychiatric Association Workgroup on Eating Disorders, Practice guidelines for eating disorders, I. Disease definition, epidemiology, and natural history, *American Journal of Psychiatry* 150 (1993): 212–228.

16. C. M. Bulik and coauthors, Drug use in women with anorexia and bulimia nervosa, *International Journal of Eating Disorders* 11 (1992): 213–225.

17. P. W. Meilman, F. A. von Hippel, and M. S. Gaylor, Self-induced vomiting in college women: Its relation to eating, alcohol use, and Greek life, *College Health* 40 (1991): 39–41.

18. J. E. Mitchell, Medical complications of bulimia nervosa, in *Eating Disorders and Obesity: A Comprehensive Handbook,* eds. K. D. Brownell and C. G. Fairburn (New York: Guilford Press, 1995), pp. 271–275.

10 *Fitness and Nutrition*

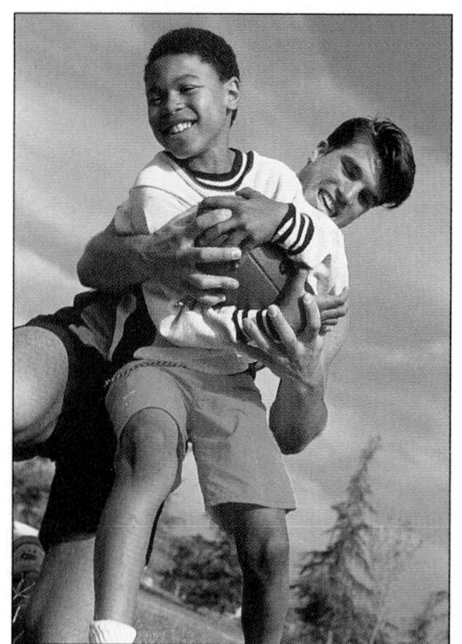

Physical activity helps you look good, feel good, and have fun, and it brings many long-term health benefits as well.

WWW.

shapeup.org
Shape Up America!

primusweb.com/fitnesspartner
Fitness Partner Connection Jumpsite!

dmi-www.mc.duke.edu/dfc/home.html
Duke University Diet & Fitness Center

cooperinst.org
Cooper Institute for Aerobics Research

Perhaps you are already physically fit. If so, the following description applies to you. You are graceful and move with ease. You are strong and meet physical challenges without strain. You have endurance, and you have plenty of energy to handle physical challenges and emergencies. What is more, you are likely to be well able to meet mental and emotional challenges, too—for physical fitness undergirds mental and emotional well-being, as well as physical energy and resilience.

If these statements do not describe you as you are today, then you can gain fitness through practice. Activities that promote fitness are themselves enjoyable, and they quickly lead to rewards in terms of physical improvements. Feeling fit can build your confidence in other areas of life, too: social, academic, and professional. Physical activity and fitness are so closely connected that the rest of this chapter makes no distinction between them.

Fitness

In 1998, the American College of Sports Medicine (ACSM) updated its position statement on the quantity and quality of physical activity recommended for developing and maintaining fitness in healthy adults.[1] The main objective of these guidelines was to outline the types and amounts of physical activity needed for improving *physical fitness.* These familiar guidelines have helped adults develop programs to improve their cardiorespiratory endurance and body composition. However, the types and amounts of physical activity needed to promote *fitness* may differ from those needed to obtain *health* benefits (see Table 10.1). For health's sake, the ACSM specifies that people should spend an accumulated minimum of 30 minutes in some sort of physical activity on most days of each week.[2] The activity need not be sports. A few minutes spent climbing stairs, another few spent pulling weeds, and several more spent walking the dog all contribute to the day's total. The guidelines for developing fitness are still optimal, though, because improving fitness provides additional health benefits (further reduction of cardiovascular disease risk and improved body composition, for example).[3]

This chapter begins by defining fitness and presenting its benefits. Then the chapter goes on to show how nutrition supports fitness.

Three Definitions of Fitness

fitness: the characteristics of the body that enable it to perform physical activity; more broadly, the ability to meet routine physical demands with enough reserve energy to rise to a sudden challenge; or the body's ability to withstand stress of all kinds.

Narrowly defined, the term fitness describes *the characteristics of the body that enable it to perform physical activity.* These characteristics include flexibility of the joints; strength and endurance of the muscles, including the heart muscle; and a healthy body composition. A broader definition of fitness is *the ability to meet routine physical demands with enough reserve energy to rise to a sudden challenge.* This definition shows how fitness relates to everyday life: ordinary tasks such as carrying heavy suitcases, opening a stuck window, or climbing four flights of stairs, which might strain an unfit person, can be well within the capacity of the fit person. Still another definition is *the body's ability to withstand stress,* meaning stress of all kinds, including psychological stress. These three definitions do not contradict each other; all describe the same wonderful condition of the body.

sedentary: physically inactive (literally, "sitting down a lot").

The opposite of a physically active life is a sedentary life, which means, literally, "sitting down a lot." Today's world permits many people to lead inactive lives and even rewards them for it. It provides elevators, escalators, cars, and golf carts so that people can exert a minimum of physical effort. Unfortunately, people are attracted to labor-saving devices, but the more they use the devices, the more

Table 10.1 Physical Activity Guidelines

Guidelines for developing and maintaining *physical fitness:*

- Frequency of activity: three to five days per week.
- Intensity of activity: 50 to 90% of maximum heart rate.
- Duration of activity: 20 to 60 minutes of continuous activity.
- Mode of activity: any activity that uses large muscle groups.
- Resistance activity: strength training of moderate intensity at least two times per week.

Guidelines for obtaining *health* benefits:

- Frequency of activity: every day.
- Intensity of activity: moderate.
- Duration of activity: at least 30 minutes total of activity (can be intermittent).
- Mode of activity: any activity.

Note: Duration and intensity are inversely related. To obtain similar fitness benefits, a person may exercise either at a low intensity for a long duration or at a high intensity for a short duration. For example, a person may choose to walk briskly for 40 to 50 minutes (lower intensity and longer duration) or to jog for 20 to 30 minutes (higher intensity and shorter duration).

weak and unfit they become, and the less well they feel. The body responds to inactivity by losing muscle and skill, just as it responds to activity by gaining them.

Benefits of Fitness

Extensive evidence confirms that regular physical activity promotes health and prevents disease.[4] Still, despite an increasing awareness of the health benefits that physical activity confers, as many as 60 percent of adults in the United States are either irregularly active or completely inactive.[5]

Physical inactivity is linked to the major degenerative diseases—heart disease, cancer, stroke, and hypertension—that are the primary killers of adults in developed countries.[6] Therefore, one of the most important challenges health care professionals face is to motivate more people to become physically active. To motivate others, health care professionals must first become more physically active themselves, thereby enhancing their own health. Second, they can include regular physical activity as a component of therapy for their clients.

People don't have to run marathons to reap the health rewards of physical activity. In fact, anyone who is extremely inactive stands to gain the greatest health benefits by engaging in a regular program of moderate-intensity, endurance-type activity.[7] Researchers who conducted an extensive study on physical fitness and mortality concluded that "moderate levels of physical fitness that are attainable by most adults appear to be protective against early mortality."[8] It makes sense, then, to promote activities that can readily be performed by the least active people, since they can benefit most.

The findings of an extensive study of healthy men from Harvard University support those of previous studies: regular physical activity confers health benefits that can prolong life.[9] This same study also raises new questions, however, because the researchers concluded that only *vigorous* activity prolonged life. Men who expended more than 1500 kcalories per week in vigorous activities such as running, swimming, and cycling had as much as a 25 percent reduced risk of dying compared with others. Any benefits of moderate activity were not apparent, perhaps because this study lumped moderate and minimal exercisers together.

One of the coauthors of this study conducted an earlier study of the Harvard men that showed that moderate physical activity can prolong life.[10] This

For perspective, to spend 1500 kcal/week—

A 175 lb man might:
- *Run 30 min/day, 5 days.*
- *Walk 45 min/day, 5 days.*

A 125 lb woman might:
- *Cycle 1 hr/day, 4 days.*
- *Swim 45 min/day, 5 days.*

researcher, as well as many others, remains confident that moderate activity confers health benefits, but has also concluded that the more vigorous the activity, the greater the benefit—to a point. No one yet knows whether too much physical activity can compromise health. For the vast majority of Americans, however, too much activity is not the problem. The ACSM advises that public health efforts focus on "getting more people more active more of the time" rather than dictating a specific activity level for people to attain.[11]

As a person becomes physically fit, the health of the entire body improves. Table 10.2 summarizes the benefits fitness confers.

In summary, physical activity brings positive rewards: good health, long life, and freedom from disease.

Components of Fitness

Physical fitness expresses itself in body characteristics. Some are health related, some are skill related. The health-related components of fitness include flexibility, muscle endurance and strength, cardiorespiratory endurance, and body composition. Flexibility allows the joints to move without injury. Muscle endurance and strength enable muscles to work without fatigue. Cardiorespiratory endurance supports the ongoing action of the heart and lungs. Fitness also expresses itself in body composition—the proportions of muscle, fat, bone, and other tissue that make up a person's total body weight. Physical activity augments desirable lean body tissue and eliminates excess body fat. Thus, as a person becomes physically fit, the health of the entire body improves.

The person who wants to go beyond general health and enhance athletic performance in specific sports will also value skill-related components of fitness such as agility, balance, coordination, power, reaction time, and speed. The importance of each characteristic varies widely with individual sports, and athletes practice endless hours to develop them. The accompanying glossary describes these characteristics.

flexibility: the capacity of the joints to move through a full range of motion; the ability to bend and recover without injury.

muscle endurance: the ability of a muscle to contract repeatedly within a given time without becoming exhausted.

muscle strength: the ability of muscles to work against resistance.

cardiorespiratory endurance: the ability to perform large-muscle, dynamic exercise of moderate-to-high intensity for prolonged periods.

body composition: the proportions of muscle, bone, fat, and other tissue that make up a person's total body weight.

Principles of Conditioning

Conditioning, or practice sessions, develops fitness and skills. The way to achieve conditioning is by training, primarily by applying overload—that is, by asking a little more of the body in each practice session. During conditioning, the body adapts microscopically to perform the work asked of it. Whatever component of fitness a person seeks to develop, whether flexibility, strength, or endurance, the principles of conditioning apply.

conditioning: the physical effect of *training;* improved flexibility, strength, and endurance.

training: practicing an activity, which leads to conditioning. Training is what you do; conditioning is what you get.

overload: an extra physical demand placed on the body; an increase in the *frequency, duration,* or *intensity* of exercise.

The Overload Principle You can apply the progressive overload principle in several different ways. You can perform the activity more often, that is, increase its frequency; you can perform the activity more strenuously, that is, increase its

Glossary of Skill-Related Fitness Components

agility: the ability to move the entire body quickly.

balance: the ability to maintain equilibrium in a fixed position or in motion.

coordination: the harmonious functioning of the senses and the muscles to accurately perform complex movements, such as hitting a baseball or juggling two or more objects.

power: the combination of strength and speed that allows a person to move quickly and forcefully, as in jumping, shot-putting, or spiking a ball.

reaction time: the amount of time between a stimulus and a response to the stimulus, as when starting a race.

speed: the ability to move fast, as in running or swimming.

Table 10.2 Benefits of Fitness*

- *Restful sleep* Rest and sleep occur naturally after periods of physical activity. During rest, the body repairs injuries, disposes of wastes generated during activity, and builds new physical structures.
- *Nutritional health* Physical activity spends energy and thus allows people to eat more food. If they choose wisely, active people will consume more nutrients and be less likely to develop nutrient deficiencies.
- *Optimal body composition* A balanced program of physical activity limits body fat and maintains lean tissue. Physically active people have relatively less body fat than sedentary people at the same body weight.[a]
- *Optimal bone density* Weight-bearing physical activity builds bone strength and protects against osteoporosis.[b]
- *Resistance to colds and other infectious diseases* Fitness enhances immunity.[†c]
- *Low risks of some types of cancers* Lifelong physical activity may help to protect against colon cancer, breast cancer, and some other types of cancer.[d]
- *Strong circulation and lung function* Physical activity that challenges the heart and lungs slows the aging of the circulatory system.
- *Low risk of cardiovascular disease* Physical activity lowers blood pressure, slows resting pulse rate, and lowers blood cholesterol, thus reducing the risks of heart attack and strokes.[e] Some research suggests that physical activity may reduce the risk of cardiovascular disease in another way as well—by lowering intra-abdominal fat stores.[f]
- *Low risk of diabetes* Physical activity normalizes glucose tolerance, especially via the secretion of insulin.[g]
- *Low incidence and severity of anxiety and depression* Compared with sedentary people, physically active people deal better with stress.
- *Strong self-image* The sense of achievement that comes from meeting physical challenges promotes self-confidence.
- *Long life and high quality of life in the later years* Active people have a lower mortality rate than sedentary people. Even a 2-mile walk daily can add years to a person's life.[h] In addition to extending longevity, physical activity supports independence and mobility in later life by reducing the risks of falls and minimizing the risk of injury should a fall occur.[i]

*The reference notes for this table are on page 260.
†Moderate physical activity can stimulate immune function. Intense, vigorous, prolonged activity such as marathon running, however, may compromise immune function. J. A. Smith, Guidelines, standards, and perspectives in exercise immunology, *Medicine and Science in Sports and Exercise* 27 (1995): 497–506.

intensity; or you can do it for longer times, that is, increase its duration. All three strategies work well, and you can pick one or a combination, depending on your preferences. For example, if you really love your workout, do it more often. If you do not have much time, increase intensity. If you hate hard work, take it easy and go longer. If you desire continuous improvements, remember to overload progressively as you gain higher levels of fitness.

Applying Overload When you are increasing the frequency, intensity, or duration of your workout, exercise to a point that only *slightly* exceeds your comfortable capacity to work. It is better to progress too slowly than to risk serious injury by overexertion. Here are other pointers about applying overload:

- Be active all week. Don't be a weekend athlete.
- Train hard only once or twice a week, not every time you work out. Between times, do moderate workouts and include at least one day of rest each week.
- Pay attention to body signals. Symptoms such as the following demand immediate medical attention: abnormal heartbeats; pain or pressure in the middle of the chest, teeth, jaw, neck, or arm; dizziness; lightheadedness; cold sweat; or confusion.
- Use proper equipment and attire.
- Perform approved exercises using proper form.

progressive overload principle: the training principle that a body system, in order to improve, must be worked at frequencies, durations, or intensities that gradually increase physical demand.

frequency: the number of occurrences per unit of time (for example, the number of exercise sessions per week).

intensity: the degree of exertion while exercising (for example, the amount of weight lifted or the speed of running).

duration: length of time (for example, the length of time spent in each exercise session).

Table 10.3 Major Coronary Risk Factors

Age (men >45 years; women >55 years).
Cigarette smoking.
Diabetes mellitus.[a]
Family history of heart disease.
Hypertension.
Sedentary lifestyle.
Serum cholesterol >200 mg/dL.

[a]Persons with insulin-dependent diabetes mellitus (type 1) who are over 30 years of age or have had type 1 diabetes for more than 15 years and persons with non-insulin-dependent diabetes mellitus (type 2) who are over 35 years of age should be classified as patients with disease and treated according to the guidelines specific for those individuals. Chapter 25 describes coordination of diet and exercise for people with diabetes.

Cautions on Starting Before you begin any fitness program, make sure it is safe for you to do so. The ACSM classifies individuals into three groups based on coronary risk factors (see Table 10.3): "apparently healthy" applies to those individuals who have no more than one of the major coronary risk factors listed in Table 10.3, "individuals at higher risk" pertains to those who have two or more of the risk factors in Table 10.3 and/or symptoms suggestive of disease, and "individuals with disease" applies to those individuals with known cardiac, pulmonary, or metabolic disease.

Most "apparently healthy" people can begin moderate exercise programs such as walking or increasing daily activities without the need for medical examination, but people in either of the other two classifications need medical advice.[12] The ACSM describes *moderate exercise* as activity that can be sustained comfortably for 60 minutes or so. It does not make sense to start with activities so demanding that pain stops you within two days. Learn to enjoy small steps toward improvement. Fitness builds slowly.

moderate exercise: exercise that can be sustained comfortably for 60 minutes or so.

warm-up: five to ten minutes of light activity, such as easy jogging or cycling, to warm up the body in preparation for vigorous activity.

cool-down: five to ten minutes of light activity following a vigorous workout to gradually cool the body's core to near-normal temperature.

Warm-Up and Cool-Down In training sessions, gradualness is a key to success. Sudden intense activity can cause injury, and abrupt discontinuance can hamper recovery, so it is best to ease into and out of activity sessions. All strenuous workouts should therefore be fitted inside a frame composed of warm-up and cool-down activities.

A warm-up facilitates gradual warming of the body and helps prepare muscles, ligaments, and tendons for activity to come. A warm-up also mobilizes fuels to support strength and endurance activities.

Cool-down activity eases the transition from exercising to normal functioning. A few minutes of light activity facilitate the relaxation of tight muscles and enhance the circulation of blood through them. The circulation, in turn, brings accumulated heat from the body's core to the surface, where it can radiate away. As you approach the end of your workout, gradually ease up on the intensity of the activity (for example, if you are running, begin to slow to a light jog), reaching a minimum intensity over five to ten minutes. Stretching the muscles to promote flexibility is particularly well suited to the end of the cool-down.

Cool-down activities can also help to prevent symptoms—dizziness, for example—that you may experience if you abruptly stop exercising. A cool-down facilitates a gradual drop in blood pressure; an abrupt drop would stress the heart. A cool-down can also help to prevent muscle soreness that might otherwise occur.

hypertrophy (high-PER-tro-fee): of muscles, growing larger; an increase in size in response to use.

atrophy (AT-ro-fee): of muscles, becoming smaller; a decrease in size because of disuse, undernutrition, or wasting diseases.

The Body's Response to Physical Activity Fitness develops in response to demand and wanes when demand ceases. Muscles gain size and mass after being made to work repeatedly, a response called hypertrophy. The increases in size and mass, in turn, help improve muscle strength and endurance. Conversely, without activity, muscles diminish in size, a response called atrophy. As muscles atrophy, they lose strength and endurance.

Hypertrophy and atrophy are adaptive responses to the muscles' greater and lesser work demands, respectively. Thus cyclists often have well-developed, strong legs but less arm or chest strength; a tennis player may have one superbly strong arm, while the other is just average. For balanced muscular development, people should work different muscle groups from day to day. This strategy provides a day or two of rest for different muscle groups, giving them time to replenish nutrients and to repair any slight damage incurred by the activity.

weight training (also called **resistance training**): the use of free weights or weight machines to provide resistance for developing muscle strength and endurance. A person's own body weight may also be used to provide resistance as when a person does push-ups, pull-ups, or abdominal crunches.

Weight Training Weight training, also called resistance training, develops muscle strength and endurance and benefits health and overall fitness. Weight training builds lean body mass. Strong muscles in the back and abdomen improve posture and reduce the risk of back injury. Weight training can also help prevent the decline in physical mobility that often accompanies aging.[13] Older adults who participate in weight training programs not only gain muscle strength, but they improve their muscle endurance, which enables them to walk significantly longer before exhaustion. Research has shown that leg strength and walking endurance are powerful predictors of an older adult's physical abilities.[14]

Weight training enhances performance in other sports, too. Swimmers can develop a more efficient stroke and tennis players, a more powerful serve, when they train with weights.

The ACSM advises that weight training to improve muscle strength and endurance may also help maximize and maintain bone mass.[15] Research supports this advice. Young women enrolled in a weight training program increased bone mass density of the spine compared with inactive controls.[16]

Depending on the technique, weight training can emphasize either muscle strength or muscle endurance. To emphasize muscle strength, combine high resistance (heavy weight) with a low number of repetitions. To emphasize muscle endurance, combine less resistance (lighter weight) with more repetitions.

Unlike a poorly maintained car, which will break down when consistently overloaded, the body responds to overload in a positive way—it gets itself into better shape to meet the demand next time. As the next section shows, the overload principle applies to the heart muscle in the same way that it does to the other muscles of the body: the heart becomes stronger.

Cardiorespiratory Endurance

As you know, the heart beats faster during physical activity than during rest. The length of time a person can remain with an elevated heart rate—that is, the ability of the heart, lungs, and blood to sustain a given demand—defines the person's cardiorespiratory endurance. Training can improve ability to sustain a vigorous activity such as running, brisk walking, or swimming. Cardiorespiratory endurance training enhances the ability of the heart, lungs, and blood to deliver oxygen to, and remove waste from, the body's cells during such activity. The benefits of this training are not only physical, though, because all of the body's cells, not just the muscle cells, require oxygen to function. When the cells receive more oxygen more readily, both the body and the mind benefit.

Working muscles need especially large amounts of oxygen to produce energy. Cardiorespiratory endurance training requires the heart and lungs to work extra hard for a sustained period to deliver oxygen to the muscle cells. Cardiorespiratory endurance training, therefore, is *aerobic* (oxygen requiring). As the cardiorespiratory system gradually adapts to the demands of aerobic exercise, the body delivers oxygen more efficiently.

Benefits of Cardiorespiratory Conditioning The changes brought about by endurance activities are called cardiorespiratory conditioning. Among its benefits, cardiac output increases, so the blood can carry more oxygen. The heart becomes larger and stronger, and each beat pumps more blood. As the heart pumps more blood with each beat, fewer beats are necessary, and the pulse rate slows down. The average resting pulse rate for adults is around 70 beats per minute, but people who have cultivated cardiorespiratory conditioning may have resting pulse rates of 50 or even lower. The muscles that work the lungs become stronger, too, so breathing becomes more efficient. Circulation through the arteries and veins improves. Blood moves easily, and blood pressure falls.

Cardiorespiratory endurance is the physical achievement that many people appropriately prize the most highly, because it reflects the health of the heart and circulatory system, on which all other body systems depend. Figure 10.1 (p. 242) shows the major relationships among the heart, circulatory system, and lungs.

To improve cardiorespiratory endurance, the activity must be sustained for 20 minutes or longer and use the large-muscle groups such as the legs and buttocks.

aerobic (air-ROE-bic): refers to energy-producing processes involving the immediate use of oxygen.
aero = air

cardiorespiratory conditioning: improvements in heart and lung function and increased blood volume, brought about by aerobic training.

cardiac output: the volume of blood discharged by the heart each minute. The amount of oxygenated blood the heart ejects toward the tissues at each beat is called the **stroke volume.**

Training for cardiorespiratory conditioning:

- *Increases cardiac output and oxygen delivery.*
- *Increases heart strength and stroke volume.*
- *Slows resting pulse.*
- *Increases breathing efficiency.*
- *Improves circulation.*
- *Reduces blood pressure.*

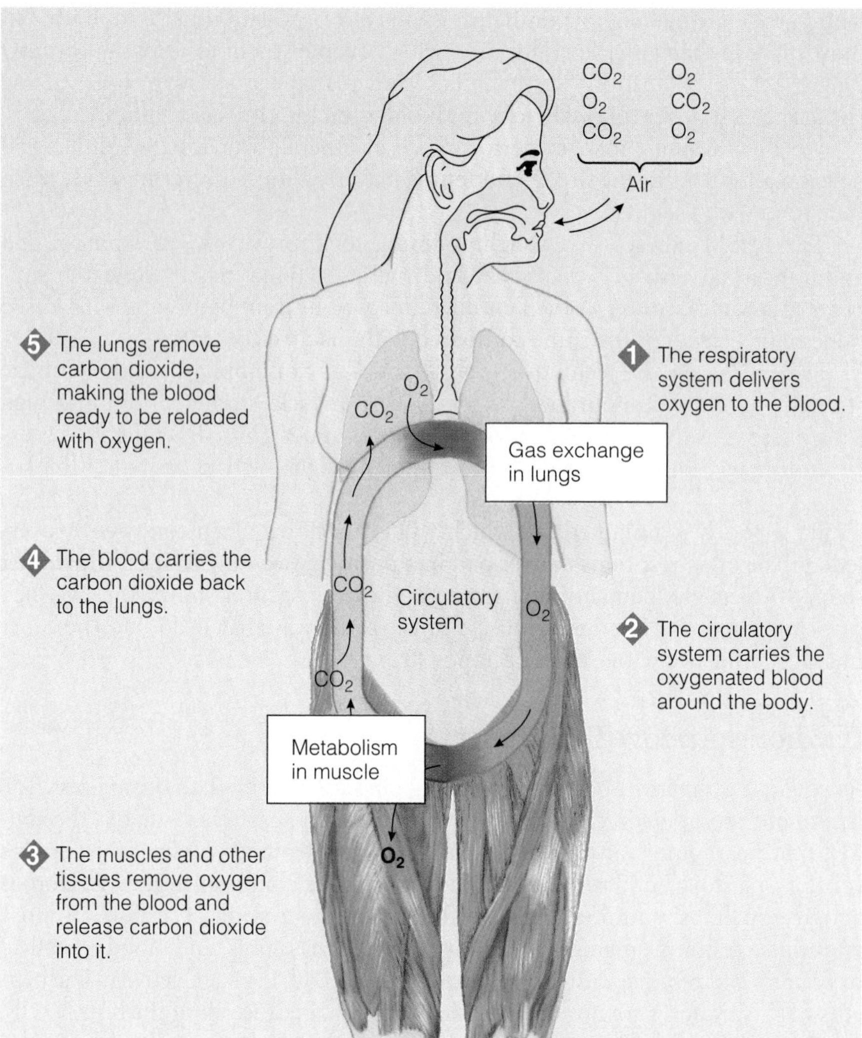
CO_2 O_2
O_2 CO_2
CO_2 O_2
Air
5 The lungs remove carbon dioxide, making the blood ready to be reloaded with oxygen.
1 The respiratory system delivers oxygen to the blood.
O_2
CO_2
Gas exchange in lungs
4 The blood carries the carbon dioxide back to the lungs.
CO_2
Circulatory system
O_2
2 The circulatory system carries the oxygenated blood around the body.
CO_2
Metabolism in muscle
3 The muscles and other tissues remove oxygen from the blood and release carbon dioxide into it.
O_2

Figure 10.1
Delivery of Oxygen by the Heart and Lungs to the Muscles
The cardiorespiratory system responds to the muscles' demand for oxygen by building up its capacity to deliver oxygen. Researchers can measure cardiorespiratory fitness by measuring the amount of oxygen a person consumes per minute while working out, a measured called **VO_2 max.**

A person must also train at an intensity that elevates the heart rate a certain amount above its resting rate.

Although formulas based on maximal oxygen uptake (VO_2 max) or maximal heart rate are available, a person's own perceived effort is usually a reliable indicator of activity intensity. In general, when you're working out, do so at an intensity that raises your breathing and heart rate, but still leaves you able to talk comfortably with a friend. If you are more competitive and want to work to your limit on some days, a treadmill test can reveal your maximal heart rate. You can work out safely at up to 90 percent of that rate. The ACSM guidelines for developing and maintaining cardiorespiratory fitness are listed in Table 10.1 (p. 237).[17]

Cardiorespiratory Training Benefits Muscles A fringe benefit of cardiorespiratory training is its effect on muscles. The more fit a muscle is, the more oxygen it draws from the blood. That oxygen is drawn from the lungs, so the person with more fit muscles extracts more oxygen from the inhaled air than a person with less fit muscles. This improves the efficiency of the cardiorespiratory system still further, reducing the heart's workload. An added bonus is that muscles that can use more oxygen can burn fat longer—an advantage for body composition and weight control.

In a balanced fitness program, aerobic activity improves cardiorespiratory fitness, stretching enhances flexibility, and weight training develops muscle strength and endurance. Table 10.4 shows an example of a balanced fitness program.

The Active Body's Use of Fuels

The body uses different mixtures of fuels at different times, depending on the intensity and duration of its activities and also depending on its own prior conditioning. For athletes who seek the highest levels of achievement in sports, some knowledge of the interplay of fuels and nutrients permits selection of a diet that will best support the chosen activity.

Carbohydrate and Physical Activity

The fuels that support activity are glucose (from carbohydrate), fatty acids (from fat), and, to a small extent, amino acids (from protein). Glucose and its stored form, glycogen, are particularly crucial to athletes who compete in endurance events.

During rest, the body derives a little more than half of its energy from fatty acids and most of the rest from glucose, along with a small percentage from amino acids. During exertion, the liver releases its glucose into the bloodstream. The muscles pick up this glucose and use it in addition to glucose from their own glycogen stores. Although glycogen supplies are ample to support everyday activity, they are less abundant than body fat stores; in other words, glycogen is limited.

Glycogen Storage Depends on Diet The body constantly uses and replenishes its glycogen. How much glycogen a body stores depends partially on the amount of carbohydrate in the diet. How much carbohydrate a person eats affects how much glycogen is stored, which, in turn, influences how much will be used during activity. Thus diet bears on performance because the more glycogen you store, the longer the stores will last as you work.

In a classic study, researchers compared fuel use during physical activity among three groups of runners, each on a different diet. For several days before testing, one of the groups consumed a normal mixed diet (55 percent of kcalories from carbohydrate), the second group consumed a high-carbohydrate diet (83 percent of kcalories from carbohydrate), and the third group consumed a high-fat diet (94 percent of kcalories from fat). Figure 10.2 (p. 245) shows that the high-carbohydrate diet allowed the athletes to keep going longer before they became exhausted. This study and many others that followed have confirmed that a high-carbohydrate diet improves endurance by enhancing glycogen stores. A later section of this chapter describes how to choose a performance diet, paying special attention to carbohydrate.

Intensity of Activity Affects Glycogen Use How long an exercising person's glycogen lasts depends not only on diet, but also on the intensity of the activity. The most intense activities—the kind that make it difficult to "catch your breath" such as a quarter-mile race—use glycogen quickly. Other, less-intense activities, such as jogging, during which breathing is steady and easy use glycogen more slowly. But joggers still use it, and if they run long enough, eventually they run out of it. Glycogen depletion usually occurs within about two hours from the onset of moderately intense physical activity.

During *moderate* activity, the lungs and circulatory system have no trouble keeping up with the muscles' need for oxygen. The individual breathes easily and the heart beats steadily—the activity is aerobic. During aerobic activity, energy derives from both glucose and fatty acids.

During *intense* activity, the demand for oxygen becomes too high to permit much use of fat as fuel. When a person exercises faster than the heart and lungs can supply oxygen to the muscles, aerobic metabolism cannot meet energy needs. The muscles must instead draw more heavily on glucose for energy,

Table 10.4 A Sample Balanced Fitness Program (45 Minutes a Day)

Monday, Wednesday, Friday:

- 10 minutes of warm-up activity.
- 25 minutes of cardiorespiratory activity.
- 10 minutes of cool-down activity and stretching.

Tuesday, Thursday:

- 10 minutes of warm-up activity.
- 25 minutes of weight training.
- 10 minutes of cool-down activity and stretching.

Saturday or Sunday:

- Softball, walking, hiking, biking, or swimming.

During short bursts of intense, anaerobic activity, glucose fuels the muscles without sufficient oxygen to "burn" it completely.

because glucose can serve as a fuel even when oxygen is in short supply, as when a person is "out of breath." Glucose can "burn" without oxygen; it can serve as an anaerobic fuel.

Lactic Acid When muscles are using glucose as a fuel without sufficient oxygen to "burn" it completely (such as during intense activity that lasts for only one to three minutes), they break it down only partway, to a compound known as lactic acid. Lactic acid builds up in the muscles and can lead to muscle exhaustion if it is not cleared away. The blood can, however, carry lactic acid to the liver, which can convert it back into glucose.*

Duration of Activity Affects Glycogen Use Glycogen use during activity depends not only on the *intensity* of the activity, but also on its *duration.* Within the first 20 minutes or so of moderate activity, a person uses mostly glycogen for fuel. A person who continues exercising moderately for longer than 20 minutes begins to use less and less glycogen and more and more fat. Still, glycogen use continues, and if the activity is long and hard enough, glycogen stores run out almost completely. Physical activity can continue for a short time thereafter only because the liver scrambles to produce from the available lactic acid and amino acids a small amount of glucose that can briefly forestall total depletion.

anaerobic: energy-producing processes that do not involve the immediate use of oxygen.

lactic acid: a compound produced in muscles when they break down glucose anaerobically.

Factors that affect glycogen use during physical activity:

- *Dietary and stored carbohydrate.*
- *Intensity of the activity.*
- *Duration of the activity.*
- *Degree of training.*

Glucose Depletion After a couple of hours of strenuous activity, glucose stores are depleted. When depletion hits, it brings nervous system function to a near halt, making continued exertion almost impossible. Marathon runners refer to this point of glucose exhaustion as "hitting the wall."

To avoid such debilitation, endurance athletes try to maintain their blood glucose for as long as they can. To maximize glucose supply, endurance athletes:

- Eat a high-carbohydrate diet (approximately 8 grams of carbohydrate per kilogram of body weight or about 70 percent of energy intake) regularly.†
- Take glucose (usually in diluted fruit juice or other sweet beverages) periodically during activity that lasts for an hour or more.
- Eat carbohydrate-rich foods following activity.
- Train the muscles to store as much glycogen as possible.

The last section of this chapter discusses how to design a high-carbohydrate diet for performance.

Glucose during Activity Muscles can obtain the carbohydrate they need not only from glycogen stores but also from sugar taken during activity, which elevates blood glucose and enhances endurance. Normally, insulin stimulates all tissues of the body to drain glucose from the blood and stow it away—exactly the opposite of what is needed for performance. During physical activity, the body's release of the hormone epinephrine keeps insulin from rising in response to glucose entering the blood. Physical activity also enhances muscle sensitivity to insulin so that the muscles become the primary recipient of blood glucose.

Reminder: *Epinephrine* is one of the stress hormones that is secreted whenever emergency action is called for; it readies body systems for fast action and mobilizes fuel to support that action.

*The recycling process that regenerates glucose from lactic acid is known as the *Cori cycle.*

†Percentage of energy intake is meaningful only when total energy intake is known. Consider that at high energy intakes (say, 5000 kcalories/day), even a moderate carbohydrate diet (40 percent of energy intake) supplies 500 grams of carbohydrate—enough for a 137-pound athlete in heavy training. By comparison, at a moderate energy intake (2000 kcalories/day), a high carbohydrate intake (70 percent of energy intake) supplies 350 grams—plenty of carbohydrate for most people, but not enough for athletes in heavy training.

Figure 10.2
The Effect of Diet on Physical Endurance
A high-carbohydrate diet can triple an athlete's endurance.

Consuming sugar is especially useful during exhausting endurance activity (lasting more than an hour). Endurance athletes often run short of glucose by the end of competitive events, and they are wise to take light carbohydrate snacks or drinks (under 200 kcalories) periodically during activity.[18] During the last stages of an endurance competition, when glycogen is running low, glucose consumed during the event can make its way slowly from the digestive tract to the muscles and augment the body's supply of glucose enough to forestall exhaustion.

Before concluding that sugar might be good for your own performance, though, consider whether you engage in *endurance* activity. Do you run, swim, bike, or ski nonstop at a steady pace for more than an hour at a time? If not, the sugar picture changes. For an everyday jog, swim, or bicycle ride, sugar probably will not help performance because such activity is not limited by carbohydrate availability. The body's glycogen stores are usually sufficient.

Glucose after Activity Research indicates that eating high-carbohydrate foods *after* physical activity also enlarges glycogen stores. A high-carbohydrate meal eaten within 15 minutes after physical activity accelerates the rate of glycogen storage by 300 percent. After two hours, the rate of glycogen storage declines by almost half. Despite this slower glycogen restoration, research shows that as long as athletes eat carbohydrate-rich foods within two hours following activity, high muscle glycogen concentrations will be achieved.[19] This is particularly important to athletes who train extensively more than once a day. A practical tip: after your next workout, enjoy a bagel or a glass of juice for your glycogen's sake.

Degree of Training Affects Glycogen Use Training also affects how much glycogen muscles store—muscles that deplete their glycogen through work adapt to store greater amounts of glycogen to support that work. The more glycogen the muscles store, the longer the stores last during physical activity. Muscles make still another adaptation to training that affects glycogen use during activity—conditioned muscles rely less on glycogen and more on fat for energy, so the rate of glycogen breakdown in trained individuals is lower than in untrained individuals at the same intensity of work.

Fat Use during Physical Activity

An active person who eats a fat-rich diet with little carbohydrate will burn more fat during activity, but will sacrifice athletic performance, as Figure 10.2 showed.

Historically, researchers have emphasized the importance of a high-carbohydrate diet for endurance performance. Recently, though, a few researchers have begun to question this long-held premise. These researchers studied the effects of a high-fat diet (38 percent of total kcalories) versus a high-carbohydrate diet (73 percent of total kcalories) on endurance performance of six conditioned athletes.[20] In this study, the high-fat diet improved endurance. The researchers suggest that high-fat diets benefit endurance performance and that severe dietary fat restriction may be detrimental for some athletes.[21]

Of course, high-fat diets carry risks of heart disease and cannot be recommended without careful consideration. When researchers examined heart disease risk factors such as blood lipids and cholesterol in conditioned athletes eating a high-fat (42 percent of kcalories) diet for a month, they found no adverse effects.[22] An earlier study, however, found that athletes' blood cholesterol concentrations rose when they consumed high-fat diets.[23]

Physical activity offers some protection against cardiovascular disease, but even for athletes, eating a high-fat diet for a prolonged time warrants medical supervision of blood lipids and other risk factors. Overwhelmingly, nutrition experts agree that the potential for adverse health effects from prolonged high-fat diets continues to outweigh any possible benefit to performance. Many more studies are needed before any conclusions about high-fat diets and endurance can be reached. One researcher points out that during endurance activity, the lower oxygen requirement for carbohydrate metabolism produces less cardiovascular stress than does the oxidation of fat.[24] Since even physically active people can suffer heart attacks and strokes, every reliable source speaks out against high-fat diets.

In contrast to *dietary* fat, *body* fat stores are of tremendous importance during physical activity, as long as the activity is not too intense. Unlike glycogen stores, fat stores can fuel hours of activity without running out.

The fat used in physical activity is liberated as fatty acids from the internal fat stores and from the fat under the skin. Areas that have the most fat to spare donate the greatest amounts of fatty acids to the blood (although they may not be the areas that appear most fatty). This is why "spot reducing" doesn't work—muscles do not own the fat that surrounds them. Fat cells release fatty acids into the blood, not into the underlying muscles. Then the blood gives to each muscle the amount of fat that it needs. Proof of this is found in a tennis player's arms—the fatfolds measure the same in both arms, even though the muscles of one arm work much harder and may be larger than those of the other. A balanced fitness program that includes strength training, however, will tighten muscles underneath the fat, improving the overall appearance. Keep in mind that some body fat is essential to good health.

Duration of Activity Affects Fat Use Body fat stores are a virtually unlimited source of energy. Early in an activity, the muscles draw on and use the fatty acids already available to them from the blood. If the activity continues for more than a few minutes, the fat cells get the message that more fat is needed for energy, and they begin rapidly breaking down their stored fat to keep the supply going. After about 20 minutes of sustained, moderate activity, the fat cells are significantly shrinking in size as they empty out their lipid stores.

Intensity of Activity Affects Fat Use The intensity of activity also affects fat use. As intensity increases, fat makes less and less of a contribution to the mix of fuels used. Fat can be broken down for energy in only one way—aerobically. Thus, for fat to fuel activity, oxygen is indispensable. (Remember, if you are breathing easily during physical activity, your muscles are getting all the oxygen they need and are able to burn fat.)

Degree of Training Affects Fat Use The body adapts in response to aerobic activity. For one thing, the trained person's heart and lungs become stronger and better able to deliver oxygen at high activity intensities. For another, as already mentioned, the muscle cells develop greater capacity to use fat as fuel. For still another, the trained person's hormones slow glucose release from the liver and encourage fat use instead. The person who wishes to burn fat can conclude that patient, persistent training is worthwhile.

Recommended Intensities and Durations Health care professionals frequently advise people who want to control their body weight and lose fat to engage in activities of low-to-moderate intensity for a long duration, such as an hour-long fast-paced walk. The reasoning behind such advice is that people exercising at low-to-moderate intensity are likely to stick with their activity for longer times and are less likely to injure themselves. In addition, some research suggests that the longer the duration of activity, the greater the contribution fat will make to the fuel mixture, and consequently, the more body fat will be lost—but this conclusion is controversial.[25] Some research refutes the notion that a person "burns more fat during low-intensity activity" and suggests that weight-control benefits are the same from either low- or high-intensity activity.[26] Regardless of the contribution fat makes to the fuel mix, people who engage in regular, vigorous physical activities have less body fat than those who engage in moderately intense activities.[27] Fat use may continue at an accelerated rate for some time after vigorous physical activity has ceased. The conditioned body that is adapted to strenuous and prolonged aerobic activity uses more fat all day long, not just during activity.[28] The bottom line on physical activity and weight and/or fat loss seems to be that total energy expenditure is the main factor, regardless of how you do it.[29]

Choosing an Activity The intensity and type of physical activities that are best for one person may not be good for another. The intensity to choose depends on your present fitness: work so as to breathe fast, but not so fast as to incur an oxygen debt. A practical rule is that you should be breathing easily enough to talk but not sing. If you can sing, pick up the pace; if you have to huff and puff to talk, slow down. If you have been sedentary for the past few years, the activity intensity that will initially make you breathe slightly fast will differ dramatically from the intensity at which a fit person will breathe slightly fast.

The key to regular physical activity is finding an activity that you enjoy.

The type of physical activity that is best for you depends, too, on what you want to achieve and what you enjoy doing. If you are looking for health benefits, such as reducing your disease risks and lowering your blood cholesterol, then you might want to spend at least 30 minutes each day doing some kind of physical activity. If you are looking to lose weight and improve body composition, then choose an activity that you can sustain for 45 minutes or more at least three days a week. Choose an activity you enjoy: some people love walking, others prefer to dance or ride a bike. If you want to be stronger and firmer, lift weights and do push-ups, sit-ups, and pull-ups. And remember, muscle is more metabolically active than body fat, so the more muscle you have, the more energy you'll burn.

Table 10.5 (p. 248) summarizes fuel use during physical activity as discussed so far. You may wonder why the third energy-yielding nutrient, protein, is not listed in the table. The reason is that protein donates only a little energy to physical activity. It does however, provide the structural material of muscle tissue, so it is important to active people.

Protein and Physical Activity

The body handles protein differently during activity than at rest. Synthesis of body proteins is suppressed during activity and for several hours afterward. In the hours following this period, though, protein synthesis rebounds beyond normal resting levels. The body must adapt and build the tissues it needs for the next

Table 10.5 Carbohydrate and Fat Use during Activity

Fuel Used	Performance Time	Oxygen Needed?	Exercise Intensity	Activity Examples
Carbohydrate	30 seconds to 3 minutes	No	Very high	¼-mile sprint, a football play
Mostly carbohydrate (and some fat)	3 to 20 minutes	Yes	High	Distance swimming or running
Mostly fat (and some carbohydrate)	More than 20 minutes	Yes	Moderate	Distance running or jogging, cross-country skiing

Sources: Adapted in part from M. H. Williams, Human energy, in *Nutritional Aspects of Human Physical Performance,* 2nd ed. (Springfield, Ill.: Charles C. Thomas, 1985), pp. 21–57; E. L. Fox, Sports activities and the energy continuum, in *Sports Physiology,* 2nd ed. (New York: Saunders, 1984), pp. 26–39.

period of activity. Whenever the body remodels a part of itself, it must tear down old structures to make way for new ones. Repeated activity, with just a slight overload, triggers the equipment of each muscle cell to do so—that is, the muscle cells adapt.

Activity Triggers Protein Synthesis The physical work of each muscle cell acts as a signal to its protein-building systems to begin producing the kinds of proteins that best support that work. Take jogging, for example. In the first difficult sessions, the body is not yet well prepared to perform. The muscle fibers have not adapted to producing the energy needed for aerobic work. But with each session, the cells get the message that an overhaul is needed. In the hours that follow the session, muscle cells busily break down any unneeded protein structures and begin producing the needed new structures. Over a few weeks' time, remodeling occurs, and jogging becomes easier.

Diet Affects Protein Use during Activity The factors that affect how much protein is used during activity seem to be the same three that influence the use of fat and carbohydrate—for one, diet. People who consume diets adequate in energy and rich in *carbohydrate* use less protein than those who eat protein- and fat-rich diets. Recall that carbohydrates spare proteins from being broken down to make glucose when needed. Since physical activity requires glucose, a diet lacking in carbohydrate necessitates the conversion of amino acids to glucose. So does a diet high in fat, because fatty acids can never provide glucose.

Intensity and Duration of Activity Affect Protein Use during Activity A second factor, the intensity and duration of activity, also modifies protein use.[30] Endurance athletes who train for over an hour a day, engaging in aerobic activity of moderate intensity and long duration, may deplete their glycogen stores by the end of their workouts and become somewhat more dependent on body protein for energy. The protein needs of bodybuilders and weight lifters are higher than those of sedentary people, but certainly not as high as the protein intakes many bodybuilders consume.

Training Affects Protein Use A third factor that influences a person's use of protein during physical activity is the extent of training. Predictably, the higher the degree of training, the less protein a person uses during an activity.

Protein Recommendations for Active People As mentioned earlier, all active people, and especially those who work like athletes, probably need more protein than do sedentary people. Endurance athletes use more protein for fuel than power athletes do, and they retain some, especially in the muscles used for their sport. Power athletes use less protein for fuel but still use some, and retain much more. Therefore, *all* athletes in training should attend to protein needs, but should back up the protein with ample carbohydrate. Otherwise, they will burn off as fuel the very protein that they wish to retain in muscle.

How much protein, then, should an active person consume? A joint position paper from the American Dietetic Association (ADA) and the Canadian Dietetic Association (CDA) recommends 1.0 to 1.5 grams of protein per kilogram of body weight each day, an amount somewhat higher than the RDA for the general population.[31] Another authority suggests different protein intakes for athletes pursuing different activities.[32] Table 10.6 on the next page lists these recommendations and translates them into daily intakes for active people. Athletes who want to build muscle mass should first meet their energy needs with adequate carbohydrate intakes and then check that they have met protein needs as well. A later section translates protein recommendations into a diet plan and shows that no one needs protein supplements, or even large servings of meat, to obtain the highest recommended protein intakes.

The mixture of fuels the body uses during physical activity depends on diet, the intensity and duration of the activity, and training. During intense activity, the muscles use glucose primarily; during less-intense, moderate activity, fat makes a greater energy contribution, and glycogen use is slower.

Vitamins, Minerals, and Water

Popular belief has it that vitamin supplements can lead to both health benefits and improved performance for those who are physically active. It goes without saying that active people need adequate vitamins and minerals to do what they do. But research confirms that nutrient supplements do not enhance the performance of well-nourished people.[33] Active people do not need supplements—they can get the nutrients they need from food.

Like the vitamins, all of the nutrient minerals are essential to physical activity. In general, the minerals are also probably like the vitamins, in that active people do not need them in supplement form. For the most part, active people who choose foods with care can be sure of meeting their vitamin and mineral needs without supplements. Nutrition in Practice 7 focuses on vitamin and mineral supplements, and Nutrition in Practice 10 offers a discussion of products athletes use to enhance performance.

Iron Deficiency in Women Athletes Iron is an exception to the rule stated above. Physically active young women, especially those who engage in endurance activities such as distance running, are prone to iron deficiency. Iron status may be affected by physical activity in any of several ways. One possibility is that iron lost in sweat contributes to the deficiency.[34] Another possible route to iron loss is red blood cell destruction: blood cells are squashed when body tissues (such as the soles of the feet) make high-impact contact with an unyielding surface such as the ground. In addition, physical activity may cause small blood losses through the digestive tract, at least in some athletes. Habitually low intakes of iron-rich foods, combined with iron losses aggravated by physical activity, may cause iron deficiency in physically active young women.

Iron deficiency impairs physical performance because iron is crucial to the body's handling of oxygen. One consequence of iron-deficiency anemia is impaired oxygen transport. This reduces aerobic work capacity, so the person tires easily. Whether marginal deficiency without anemia impairs physical performance is a point of debate among researchers.[35]

Vitamins and minerals in abundance are best obtained from foods, not from supplements.

Table 10.6 Recommended Protein Intakes for Athletes

	Recommendations (g/kg/day)	Protein Intakes (g/day)	
		Males	Females
RDA for adults	0.8	56	44
ACA/CDA recommended intake	1.0–1.5	70–105	55–83
Recommended intake for power (strength-speed) athletes	1.2–1.7	84–119	66–94
Recommended intake for endurance athletes	1.2–1.4	84–98	66–77
U.S. average intake		95	65

Note: Daily protein intakes are based on a 70-kilogram (154-pound) man and 55-kilogram (121-pound) woman.

Sources: Committee on Dietary Allowances, *Recommended Dietary Allowances,* 10th ed. (Washington, D.C.: National Academy Press, 1989); Position of The American Dietetic Association and The Canadian Dietetic Association: Nutrition for physical fitness and athletic performance for adults, *Journal of the American Dietetic Association* 93 (1993): 691–695; P. W. R. Lemon, Effect of exercise on protein requirements, in *Foods, Nutrition, and Sports Performances: An International Scientific Consensus,* eds. C. Williams and J. T. Devlin (London: E & FN Spon, 1992), pp. 65–86.

Sports Anemia Early in training, athletes may develop low blood hemoglobin for a while. This condition, sometimes called "sports anemia," probably reflects a normal adaptation to physical training. Aerobic training enlarges the blood volume, and with the added fluid, the red blood cell count per unit of blood drops. True iron deficiency requires treatment with prescribed iron supplements, but the temporary reduced red blood cell count seen early in training corrects itself after a while.

Iron Supplements May Be Needed Because true iron deficiency is a real possibility for all people, and especially for active people and athletes, it is important to keep track of your own iron status. (All routine physical examinations that include blood work check you for the extreme deficiency state, anemia, but you should also be aware that such tests will not tell you if your iron stores are low.) Consider your individual needs. Many young menstruating women probably border on iron deficiency even without the additional iron losses incurred through exercise. Active teens of both sexes, because they are growing, have high iron needs, too. Especially for women and teens, then, prescribed supplements may be needed to maintain iron stores or to correct a deficiency of iron, but medical testing should guide decisions on supplementation.

Electrolyte Losses and Replacement Electrolytes—the charged minerals sodium, potassium, chloride, and magnesium—are lost from the body in sweat. Beginners lose electrolytes to a greater extent than do trained athletes; as the body adapts to physical activity, it becomes better at conserving most electrolytes. People normally need make no special effort to replenish lost electrolytes. A regular diet that meets energy and nutrient needs also replenishes all the needed electrolytes.

Moderate Electrolyte Replacement Electrolyte replacement is also not necessary during physical activity, unless a person works up a drenching sweat amounting to the loss of 5 to 10 pounds or more each day (3 percent of body weight) for several consecutive days. In that case, drinking plain water and relying on food to replace lost electrolytes may not suffice, and a commercial sports drink may be drunk for fluid and electrolyte replacement (see Table 10.8 on

p. 253). A homemade mixture of ⅓ teaspoon of table salt and 1 cup of fruit juice added to each quart of water will also serve the purpose. Avoid electrolyte or salt tablets; they can irritate the stomach and cause vomiting. As for potassium, avoid potassium supplements unless prescribed by a physician; although they help some conditions, they make others worse.

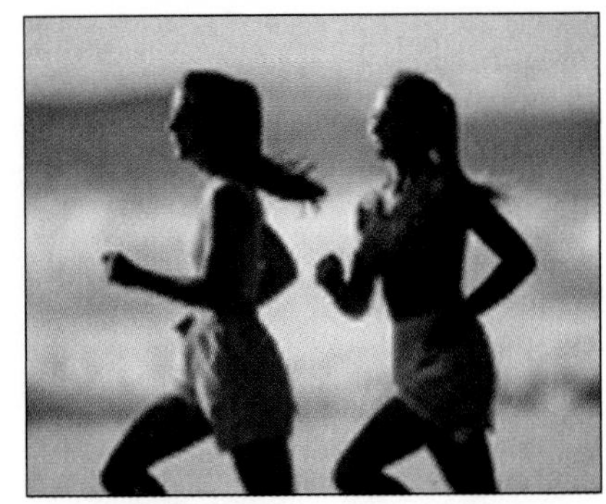
Women athletes may be at special risk of iron deficiency.

Water Is Key to Performance Water is a crucial nutrient for everyone, especially those engaged in physical activity, because the body loses water via sweat. Breathing uses water, too, exhaled as vapor. During physical activity, both routes are significant, and dehydration becomes a threat. Dehydration's first symptom is fatigue: a water loss of even 1 to 2 percent of body weight can reduce a person's capacity to do muscular work.[36] With a water loss of about 7 percent, a person is likely to collapse.[37]

Fluid Losses via Sweat Working muscles produce heat. During intense activity, muscle heat production can be 15 to 20 times greater than at rest.[38] The body cools itself by sweating. Each liter of sweat dissipates almost 600 kcalories of heat, preventing a rise in body temperature of almost 10°C. The body routes its blood supply through the capillaries just under the skin, and the skin secretes sweat to evaporate and cool the skin and the underlying blood. The blood then flows back to cool the deeper body chambers.

Hyperthermia In hot, humid weather, sweat doesn't evaporate well because the surrounding air is already laden with water. Body heat builds up and triggers maximum sweating, but without sweat evaporation, little cooling takes place. In such conditions, active people must take precautions to prevent heat stroke. To reduce the risk of heat stroke, drink enough fluid before and during the activity, rest in the shade when tired, and wear lightweight clothing that allows evaporation.[39] (Hence the danger of rubber or heavy suits that supposedly promote weight loss during physical activity—they promote profuse sweating, prevent sweat evaporation, and invite heat stroke.) If you ever experience any of the symptoms of heat stroke listed in the margin, stop your activity, sip fluids, seek shade, and ask for help. Heat stroke can be fatal, young people often die of it, and these symptoms demand attention.

hyperthermia: an above-normal body temperature.

heat stroke: the dangerous accumulation of body heat with accompanying loss of body fluid.

Symptoms of heat stroke:

- *Headache.*
- *Nausea.*
- *Dizziness.*
- *Clumsiness.*
- *Stumbling.*
- *Excessive or insufficient sweating.*
- *Confusion or other mental changes.*

hypothermia: a below-normal body temperature.

Hypothermia In cold weather, *hypo*thermia, or low body temperature, can pose as serious a threat as heat stroke does in hot weather. Inexperienced, slow runners participating in long races on cold or wet, chilly days are especially vulnerable to hypothermia. Slow runners who produce little heat can become too cold if clothing is inadequate. Early symptoms of hypothermia include shivering and euphoria. As body temperature continues to fall, shivering may stop, and weakness, disorientation, and apathy may occur. Each of these symptoms can impair a person's ability to act against a further drop in body temperature. Even in cold weather, however, the active body still sweats and still needs fluids. The fluids should be warm or at room temperature to help protect against hypothermia.

Fluid Replacement via Hydration Endurance athletes can easily lose 1.5 liters or more of fluid during *each hour* of activity. To prepare for fluid losses, a person must hydrate before activity. To replace fluid losses, the person must rehydrate during and after activity. (Table 10.7 on p. 253 presents one schedule of hydration for physical activity.) Even then, in hot weather, the GI tract may not be able to absorb enough water fast enough to keep up with sweat losses, and some degree of dehydration may be inevitable.

What is the best fluid to support physical activity? For noncompetitive, everyday active people, plain, cool water is recommended, especially in warm weather, for two reasons: it rapidly leaves the digestive tract to enter the tissues where it is needed, and it cools the body from the inside out. For endurance athletes, other

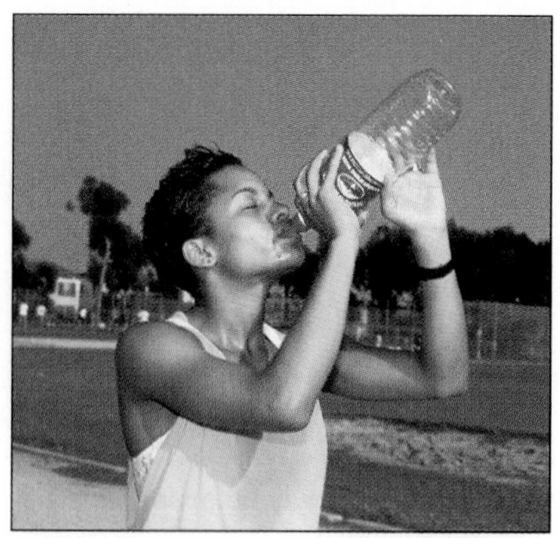

To prevent dehydration and the fatigue that accompanies it, drink plenty of liquids before, during, and after physical activity.

Water recommendation: 1.0 to 1.5 ml/kcal expended.
Note: 1 ml = 0.03 fluid oz.
Easy estimation: ≈ ½ c/100 kcal.

WWW
runnersworld.com/nutrition
Runner's World

eatright.org
American Dietetic Association

beverages may be appropriate. Fluid ingestion during the event has the dual purposes of replenishing water lost through sweating and providing a source of carbohydrate to supplement the body's limited glycogen stores. Carbohydrate depletion brings on fatigue in the athlete, but as already mentioned, fluid loss and the accompanying buildup of body heat can be life-threatening. Thus the first priority for endurance athletes should be to replace fluids. Many good-tasting sports drinks are marketed for active people; Table 10.8 compares them with water.

To recap, with the possible exception of iron, well-nourished active people do not need nutrient supplements. Active people need to drink plenty of water, especially during training or competition.

Food for Fitness

No particular diet best supports physical performance perfectly. Many different diets can be excellent for active people. Food choices must be made within the framework of the diet-planning principles presented in Chapter 1, however.

Choosing a Diet to Support Physical Activity

The active person needs a diet composed mostly of nutrient-dense foods, the kind that supply a maximum of vitamins and minerals for the energy they provide. When active people eat mostly refined, processed foods that have suffered nutrient losses and that contain added sugar and fat, nutrient status suffers.

Balance Active people need to eat both for nutrient density and for energy. Active people are not immune to heart disease and cancer and so must limit fats. A diet that is high in carbohydrate (60 percent of total kcalories or more), low in fat (25 percent or less), and adequate in protein (12 to 15 percent) ensures full glycogen and other nutrient stores. Such a diet helps to control weight (thus reducing risks of diabetes and other diseases) and provides adequate fiber while supplying abundant nutrients. Table 10.9 (p. 254) shows some sample diet plans for people who wish to increase their energy and carbohydrate intakes to allow for a physically active lifestyle, and Figure 10.3 (p. 255) displays a day's meals to fit the recommended pattern. Notice how abundant carbohydrate is in the athlete's meals.

Carbohydrate On two occasions, the active person's high-carbohydrate, fiber-rich diet may require temporary adjustment. Both of these exceptions involve training for competition rather than fitness. During intensive training, energy needs may outstrip the person's capacity to eat enough to meet them. At that point, added sugar and fat may be needed. The person can add concentrated carbohydrate foods such as dried fruits and even high-fat foods such as avocados, nuts, and ice cream. Still, a nutrient-rich diet remains central; energy alone is not enough. The other special occasion is the pregame meal, when fiber-rich, bulky foods are best avoided. The pregame meal is discussed in a later section.

Protein In addition to carbohydrate and some fat (and the energy they provide), physically active people need protein. How much of what kinds of foods supply enough protein to meet their needs? Meats and milk products are rich protein sources, but to recommend that active people emphasize these foods would be narrow advice for many reasons. For one thing, all people must protect themselves from heart disease, and even lean meats and reduced- or low-fat milk products contain fat, much of it saturated fat. For another, as emphasized repeatedly, active people need diets high in carbohydrate, and of course, meats have

Table 10.7 Hydration Schedule for Physical Activity	
When to Drink	**Approximate Amount of Fluid**
2 hr before exercise	3 c
10 to 15 min before exercise	2 c
Every 15 min during exercise	1 c
After exercise	2 c

Source: D. C. Nieman, *Fitness and Sports Medicine: An Introduction* (Palo Alto, Calif.: Bull Publishing, 1990), p. 234.

Those who compete in endurance activities require fluid and carbohydrate fuel.

none to offer. Legumes, grains, and vegetables provide protein with abundant carbohydrate and little fat. Table 10.6 showed some possible protein intakes for active people.

Meals before and after Competition

No single food improves speed, strength, or skill in competitive events, although some *kinds* of foods do support performance better than others as already explained. Still, a competitor may eat a particular food before or after an event

Table 10.8 Sports Drinks

Water best meets the fluid needs of most active people, yet manufacturers market many good-tasting sports drinks. More than 20 "power beverages" compete for their share of the $1 billion market. Sports drinks offer the following:

- *Fluid* Sports drinks offer fluids to help offset the loss of fluids during activity, but plain water can do this, too. Alternatively, fruit juices can be diluted (by one-half to one-third), if preferred to plain water.
- *Glucose* Sports drinks offer simple sugars or glucose polymers that help maintain hydration and blood glucose and enhance performance as effectively as, or maybe even better than, water. Such measures are especially beneficial for strenuous endurance activities lasting longer than an hour. Most sports drinks contain about 7 percent glucose (about half the sugar of ordinary soft drinks, or about 5 teaspoons in each 12 ounces). Less than 6 percent may not enhance performance, and more than 10 percent may cause abdominal cramps, nausea, and diarrhea. Fluid transport to the tissues from beverages containing up to 10 percent glucose is rapid.[a]

  While glucose does enhance endurance performance in strenuous competitive events, for the moderate exerciser, it can be counterproductive if weight loss is the goal. Glucose is sugar, and like candy, it provides only empty kcalories—no vitamins or minerals. Most sports drinks provide between 50 and 100 kcalories per cup.
- *Sodium and other electrolytes* Sports drinks offer sodium and other electrolytes to help replace those lost during physical activity. Sodium in sports drinks also helps to increase the rate of fluid absorption from the GI tract and maintain plasma volume during exercise and recovery.

  Most physically active people do not need to replace the minerals lost in sweat immediately; a meal eaten within hours of competition replaces these minerals soon enough. Most sports drinks are relatively low in sodium, however, so those who choose to use these beverages run little risk of excessive intake.

  In strenuous, world-class competitions lasting 6 hours or more, heavy sweating coupled with consumption of large amounts of plain water dangerously dilutes blood sodium. In these few cases, intravenous fluid and electrolyte repletion is needed.[b]
- *Good taste* Manufacturers reason that if a drink tastes good, people will drink more, thereby ensuring adequate hydration. For athletes who prefer the flavors of sports drinks over water, it may be worth paying for good taste to replace lost fluids.
- *Psychological edge* Sports drinks provide a psychological edge for some people who associate the drinks with athletes and sports. The need to belong is valid. If the drinks boost morale and are used with care, they may do no harm.

For trained endurance athletes who exercise for an hour or more, sports drinks may provide a slight advantage over water. For most physically active people, though, water is the best fluid to replenish lost fluids. The most important thing to do is drink—even if you don't feel thirsty.

glucose polymers: compounds that supply glucose, not as single molecules, but linked in chains somewhat like starch. The objective is to attract less water from the body into the digestive tract (osmotic attraction depends on the number, not the size of particles).

[a] C. V. Gisolfi, Fluid balance for optimal performance, *Nutrition Reviews* 54 (1996): S159–S168.
[b] N. Clark, J. Tobin, and C. Ellis, Feeding the ultraendurance athlete: Practical tips and a case study, *Journal of the American Dietetic Association* 92 (1992): 1258–1262.

Table 10.9 High-Carbohydrate Food Patterns for Various Energy Levels

Food Group	1500 kcal	2000 kcal	2500 kcal	3000 kcal	3500 kcal	4000[a] kcal
	Number of Servings					
Milk	3	3	4	4	4	4
Fruit	5	6	7	9	10	12
Vegetable	3	3	3	5	6	7
Grain	7	11	16	18	20	24
Fat	2	3	5	6	8	10
Meat (ounces)	5	5	5	5	6	6
Percent carbohydrate:	58%	58%	63%	64%	60%	62%[a]

[a]A way to add more energy to the diet without adding much bulk is to include snacks of milkshakes or "complete meal" liquid supplements.

High-carbohydrate, liquid pregame meal ideas:

- *Apple juice, frozen banana, and cinnamon.*
- *Papaya juice, frozen strawberries, and mint.*
- *Nonfat milk, frozen banana, and vanilla.*

for psychological reasons. One eats a steak the night before wrestling, another takes some honey five minutes after diving. As long as these practices remain harmless, they should be respected.

Pregame Meals Science indicates that the pregame meal or snack should include plenty of fluids and be light and easy to digest. It should provide between 300 and 800 kcalories, primarily from carbohydrate-rich foods that are familiar and well tolerated by the athlete. The meal should end three to five hours before competition to allow plenty of time for the stomach to empty before exertion.

Breads, potatoes, pasta, and fruit juices—that is, carbohydrate-rich foods low in fat and fiber—form the basis of the best pregame meal. Bulky, fiber-rich foods such as raw vegetables or high-bran cereals, although usually desirable, are best avoided just before competition. Fiber in the digestive tract attracts water out of the blood and can cause stomach discomfort during performance. Liquid meals are easy to digest, and many such meals are commercially available. Alternatively, athletes can mix nonfat milk or juice, frozen fruits, and flavorings in a blender.

Postgame Meals As mentioned earlier, eating high-carbohydrate foods *after* physical activity enhances glycogen storage. Since people are usually not hungry immediately following physical activity, carbohydrate-containing beverages such as the sports drinks discussed earlier or fruit juice may be preferred. If an active person does feel hungry after an event, then foods high in carbohydrate and low in protein, fat, and fiber are the ones to choose—the same ones recommended prior to competition. Foods high in protein and fat should be avoided during the first few hours after activity as these foods may suppress hunger and thus limit carbohydrate intake.[40]

Nutrient-dense foods best support physical activity. Carbohydrate-rich foods that are light and easy to digest are recommended for both the pregame and postgame meals.

Breakfast:
1 c shredded wheat.
1 c 1% low-fat milk.
1 small banana.
2 slices whole-wheat toast.
4 tsp jelly.
1½ c orange juice.

Snack:
3 c plain popcorn.
A smoothie made from:
1½ c apple juice.
1½ frozen banana.

Lunch:
2 turkey sandwiches.
1½ c 1% low-fat milk.
Large bunch of grapes.

Dinner:
Salad: 1 c spinach, carrots, and mushrooms.
½ c garbanzo beans.
1 tbs sunflower seeds.
1 tbs ranch salad dressing.
1 c spaghetti with meat sauce.
1 c green beans.
1 corn on the cob.
2 slices Italian bread.
4 tsp butter.
1 piece angel food cake.
1¼ c fresh strawberries.
1 tbs whipping cream.
1 c 1% low-fat milk.

Total kcal: 3300
63% kcal from carbohydrate
22% kcal from fat
15% kcal from protein

Figure 10.3
An Athlete's Meal Selections

The person who chooses to live a physically active life can expect to enjoy the rewards of fitness and good overall health as well as the pleasures of the chosen activities themselves. Another benefit accompanies these: when you spend more kcalories, you can eat more food, which can bring you added pleasure and improved nutrition status. The next chapter focuses on foods themselves, and specifically on the food choices that best support health.

Self Study

How Physically Active Are You?

"The road to fitness is physical activity." To find out how physically active you are, answer the following questions and record your answers on Form 10. For each question answered yes, give yourself the number of points indicated. Then total your points to determine your score. This quiz is intended only to help you become aware of opportunities for improving your fitness.

A. Formal, Vigorous Exercise Routines

1. I participate in active recreational sports such as tennis or handball for an hour or more:
   a. about once a week. *(2 points)*
   b. about twice a week. *(4 points)*
   c. three times a week. *(6 points)*
   d. four times a week. *(8 points)*

   (Adjust your point score if your answer is less or more than these. For example, if you play sports six days a week, give yourself more points—say, 12 points. If you play once a week for only 30 minutes, give yourself 1 point. Note: Any session of less than 20 minutes counts as zero.)
2. I participate in vigorous fitness activities like aerobic dancing, jogging, or swimming (at least 20 continuous minutes each session):
   a. about once a week. *(3 points)*
   b. about twice a week. *(6 points)*
   c. three times a week. *(9 points)*
   d. four times a week. *(12 points)*

   (Adjust your point score upward or downward if your answer is slightly different from these. For example, if you work out *six* days a week, give yourself more points—say, 18 points.)

B. Other Formal Exercise Routines

3. I perform floor workouts (sit-ups, push-ups) for at least ten minutes:
   a. two sessions a week. *(2 points)*
   b. three sessions a week. *(3 points)*
   c. four or more sessions a week. *(4 points)*

   (No points for a session of less than ten minutes; no points for only one session a week; maximum is 4 points.)
4. I participate in yoga or perform stretching exercises for at least ten minutes:
   a. two sessions a week. *(2 points)*
   b. three sessions a week. *(3 points)*
   c. four or more sessions a week. *(4 points)*

   (No points for a session of less than ten minutes; no points for only one session a week; maximum is 4 points.)
5. I work out with weights for at least ten minutes:
   a. two sessions a week. *(2 points)*
   b. three sessions a week. *(3 points)*
   c. four or more sessions a week. *(4 points)*

   (No points for a session of less than ten minutes; no points for only one session a week; maximum is 4 points.)

C. Occupational and Daily Activities

6. I walk to and from school, work, and shopping (½ mile or more each way), two or three times a week or more. *(1 point)*
7. I climb stairs rather than using elevators or escalators every other day or more. *(1 point)*
8. My school, job, or household routine involves physical activity that fits the following description:
   a. It is mostly desk work or light physical activity. *(0 points)*
   b. It is mostly farm activities, moderate physical activity, brisk walking, or the like. *(4 points)*
   c. Many of my typical days include several hours of heavy physical activity such as shoveling or lifting. (Don't include sports practice here. See Part A.) *(2 points per day)*

(Part C maximum: 12 points.)

D. Leisure Activities

9. I do several hours of gardening, lawn work, or equally active hobby work each week. *(1 point)*
10. I fish or hunt once a week or more on the average. (This must involve active work such as rowing a boat or tracking game. Dock and truck sitting don't count.) *(1 point)*
11. At least once a week I dance vigorously (folk or square dance) for an hour or more. *(1 point)*
12. In season, I play 9 to 18 holes of golf at least once a week, and I do not use a power cart. *(2 points)*
13. I walk for exercise or recreation.
    a. one to two hours a week. *(1 point per hour)*
    b. three to four hours a week. *(2 points)*
    c. five hours or more a week. *(3 points)*
14. In *addition* to the above, I choose to engage in other forms of physical activity:
    a. one to two hours a week. *(1 point)*
    b. three to four hours a week. *(2 points)*
    c. five hours or more a week. *(3 points)*

(For Part D, don't count sports practice. See Part A.)

(Part D maximum: 11 points.)

Form 10 Physical Activity Scorecard

Record your point scores here for the Self Study:

Category	Score
A. Formal, Vigorous Exercise Routines	___ (A high score would be 20.)
B. Other Formal Exercise Routines	___ (A high score would be 12.)
C. Occupational and Daily Activities	___ (A high score would be 12.)
D. Leisure Activities	___ (A high score would be 11.)
	Total: ___(A high score would be 50.)

Evaluation of total score (circle one):

- Inactive (0 to 5 points).
- Moderately active (6 to 11 points).
- Active (12 to 20 points).
- Very active (21 points or over).

If your score categorized you as inactive or only moderately active, return to Parts A through D, reread the questions, and choose some activities that you would like and could realistically undertake to raise your score to "active" (12 points or more). List these activities below. You are not committing yourself to doing these things, just acknowledging that you could.

A. Formal, Vigorous Exercise Routines

I could: ______________________________

State for how long and how many times a week: ______________________________

B. Other Formal Exercise Routines

I could: ______________________________

State for how long and how many times a week: ______________________________

C. Occupational and Daily Activities

I could: ______________________________

State for how long and how many times a week: ______________________________

D. Leisure Activities

I could: ______________________________

State for how long and how many times a week: ______________________________

Self Check

1. Regular physical activity helps protect against:
   a. backaches, cancer, and emphysema.
   b. cancer, diabetes, and heart disease.
   c. obesity, kidney disease, and anemia.
   d. high blood pressure, cancer, and osteopenia.
2. Fitness benefits health by:
   a. increasing lean body tissue and enhancing resistance to colds.
   b. lowering the risk of heart disease, decreasing muscle mass, and improving nutritional health.
   c. building bone strength, lowering the risk of some cancers, and increasing anxiety.
   d. reducing diabetes risk, compromising lung function, and promoting a strong self-image.
3. Which of the following characteristics is not a health-related component of fitness?
   a. muscle endurance   b. coordination
   c. flexibility   d. muscle strength
4. The progressive overload principle can be applied by performing:
   a. an activity less often.
   b. an activity for a shorter time.
   c. an activity with more intensity.
   d. a different activity each day of the week.
5. Some of the benefits of cardiorespiratory conditioning include:
   a. the blood carries less oxygen, the pulse rate slows down, and blood pressure increases.
   b. the blood carries less oxygen, the pulse rate increases, and blood pressure increases.
   c. the blood carries more oxygen, the blood moves more easily, and blood pressure increases.
   d. the blood carries more oxygen, the pulse rate slows down, and blood pressure falls.
6. Cool-down activities after physical activity:
   a. facilitate the relaxation of tight muscles.
   b. enhance blood circulation through the muscles.
   c. permit a gradual drop in blood pressure.
   d. all of the above.
7. Conditioned muscles rely less on ___ and more on ___ for energy.
   a. protein; fat   b. fat; protein
   c. glycogen; fat   d. fat; glycogen
8. Physically active young women, especially those who are endurance athletes, are prone to:
   a. energy excess.   b. iron deficiency.
   c. protein overload.   d. iodine deficiency.
9. Plain, cool water is the best fluid for everyday active people because it:
   a. rapidly leaves the digestive tract to enter the tissues and cool the body.
   b. tastes good.
   c. provides carbohydrate.
   d. leaves the digestive tract slowly.
10. A recommended pregame meal includes plenty of fluids and provides between:
    a. 300 and 800 kcalories, mostly from fat-rich foods.
    b. 50 and 100 kcalories, mostly from fiber-rich foods.
    c. 1000 and 2000 kcalories, mostly from protein-rich foods.
    d. 300 and 800 kcalories, mostly from carbohydrate-rich foods.

Answers to these questions appear in Appendix H.

Notes

1. American College of Sports Medicine, The recommended quality and quantity of exercise for developing and maintaining cardiorespiratory and muscular fitness and flexibility in healthy adults, *Medicine and Science in Sports and Exercise* 30 (1998): 975–991.

2. U.S. Centers for Disease Control and Prevention and American College of Sports Medicine, Summary statement: Workshop on physical activity and public health, *Sports Medicine Bulletin* 28 (1993): 7.

3. D. A. Leaf, D. L. Parker, and D. Schaad, Changes in VO_{V2max}, physical activity, and body fat with chronic exercise: Effects on plasma lipids, *Medicine and Science in Sports and Exercise* 29 (1997): 1152–1159; P. T. Williams, Relationship of distance run per week to coronary heart disease risk factors in 8283 male runners, *Archives of Internal Medicine* 157 (1997): 191–198.

4. U. M. Kujala and coauthors, Relationship of leisure time, physical activity and mortality: The Finnish twin cohort, *Journal of the American Medical Association* 279 (1998): 440–444; R. R. Pate and coauthors, Physical activity and public health: A recommendation from the Centers for Disease Control and Prevention and the American College of Sports Medicine, *Journal of the American Medical Association* 273 (1995): 402–407; Summary of the Surgeon General's Report addressing physical activity and health, *Nutrition Reviews* 54 (1996): 280–284; R. S. Paffenbarger and coauthors, The association of changes in physical-activity level and other lifestyle characteristics with mortality among men, *New England Journal of Medicine* 328 (1993): 538–545; L. Sandvik and coauthors, Physical fitness as a predictor of mortality among healthy, middle-aged Norwegian men, *New England Journal of Medicine* 328 (1993): 533–537.

5. U.S. Department of Health and Human Services, *Physical Activity and Health: A Report of the Surgeon General Executive Summary* (Washington, D.C.: Government Printing Office, 1996).

6. S. N. Blair, Physical inactivity and cardiovascular disease risk in women, *Medicine and Science in Sports and Exercise* 28 (1997): 9–10; I. Thune and coauthors, Physical activity and the risk of breast cancer, *New England Journal of Medicine* 336 (1997): 1269–1275; NIH Consensus Development Panel on Physical Activity and Cardiovascular Health, Physical activity and cardiovascular health, *Journal of the American Medical Association* 276 (1996): 241–246; Pate and coauthors, 1995; American Heart Association, Position statement on exercise: Benefits and recommendations for physical activity programs for all Americans, *Circulation* 86 (1992): 340–344.

7. NIH Consensus Development Panel on Physical Activity and Cardiovascular Health, 1996; W. L. Haskell, Health consequences of physical activity: Understanding and challenges regarding dose-response, *Medicine and Science in Sports and Exercise* 26 (1994): 649–660.

8. S. N. Blair and coauthors, Changes in physical fitness and all-cause mortality, *Journal of the American Medical Association* 273 (1995): 1093–1098.

9. I. M. Lee, C. Hsieh, and R. S. Paffenbarger, Exercise intensity and longevity in men: The Harvard alumni study, *Journal of the American Medical Association* 272 (1995): 1179–1184.

10. Paffenbarger and coauthors, 1993.

11. American College of Sports Medicine, *ACSM's Guidelines for Exercise Testing and Prescription,* 5th ed. (Philadelphia: Williams & Wilkins, 1995), pp. 3–11.

12. American College of Sports Medicine, 1995.

13. P. A. Ades and coauthors, Weight training improves walking endurance in healthy elderly persons, *Annals of Internal Medicine* 124 (1996): 568–572.

14. J. M. Guralnik and coauthors, Lower-extremity function in persons over the age of 70 years as a predictor of subsequent disability, *New England Journal of Medicine* 332 (1995): 556–561.

15. American College of Sports Medicine, Position stand: Osteoporosis and exercise, *Medicine and Science in Sports and Exercise* 27 (1995): i–vii.

16. C. M. L. Snow-Harter, Effects of resistance and endurance exercise on bone mineral status of young women: A randomized exercise intervention trial, *Journal of Bone Mineral Research* 7 (1992): 761–769.

17. American College of Sports Medicine, 1990.

18. American College of Sports Medicine, Position stand: Exercise and fluid replacement, *Medicine and Science in Sports and Exercise* 28 (1996): i–vii; G. McConell, K. Kloot, and M. Hargreaves, Effect of timing of carbohydrate ingestion on endurance exercise performance, *Medicine and Science in Sports and Exercise* 28 (1996): 1300–1304; P. R. Below and coauthors, Fluid and carbohydrate ingestion independently improve performance during 1 h of intense exercise, *Medicine and Science in Sports and Exercise* 27 (1995): 200–210.

19. J. A. M. Parkin and coauthors, Muscle glycogen storage following prolonged exercise: Effect of timing of ingestion of high glycemic index food, *Medicine and Science in Sports and Exercise* 29 (1997): 220–224.

20. D. M. Muoio and coauthors, Effect of dietary fat on metabolic adjustments to maximal VO_2 and endurance in runners, *Medicine and Science in Sports and Exercise* 26 (1994): 81–88.

21. Muoio, 1994.

22. J. Leddy and coauthors, Effect of a high or a low fat diet on cardiovascular risk factors in male and female runners, *Medicine and Science in Sports and Exercise* 29 (1997): 17–25.

23. S. D. Phinney and coauthors, The human metabolic response to chronic ketosis without caloric restriction: Preservation of submaximal exercise capacity with reduced carbohydrate oxidation, *Metabolism* 32 (1983): 769–776.

24. W. M. Sherman and N. Leenders, Fat loading: The next magic bullet? *International Journal of Sports Nutrition* 5 (1995): S1–S12.

25. P. Arnos, F. Andres, and K. Drowatzky, Fat oxidation and RPE at varied exercise intensities, *Medicine and Science in Sports and Exercise* (supplement) 25 (1993): S9; F. A. Kulling and coauthors, Identification and evaluation of the exercise intensity which maximizes fat oxidation in young women, *Medicine and Science in Sports and Exercise* (supplement) 25 (1993): S179.

26. M. A. Grediagin and coauthors, Exercise intensity does not effect body composition change in untrained, moderately overfat women, *Journal of the American Dietetic Association* 95 (1995): 661–665.

27. A. Tremblay and coauthors, Effect of intensity of physical activity on body fatness and distribution, *American Journal of Clinical Nutrition* 51 (1990): 153–157.

28. T. J. Horton and C. A. Geissler, Effect of habitual exercise on daily energy expenditure and metabolic rate during standardized activity, *American Journal of Clinical Nutrition* 59 (1994): 13–19.

29. Grediagin and coauthors, 1995.

30. P. R. Lemon, Is increased dietary protein necessary or beneficial for individuals with a physically active lifestyle? *Nutrition Reviews* 54 (1996): S169–S175.

31. Position of The American Dietetic Association and The Canadian Dietetic Association: Nutrition for physical fitness and athletic performance for adults, *Journal of the American Dietetic Association* 93 (1993): 691–695.

32. Lemon, 1996.

33. A. Singh, F. M. Moses, and P. A. Deuster, Chronic multivitamin-mineral supplementation does not enhance physical performance, *Medicine and Science in Sports and Exercise* 24 (1992): 726–732.

34. M. F. Walter and E. M. Haymes, The effects of heat and exercise on sweat iron loss, *Medicine and Science in Sports and Exercise* 28 (1996): 197–203.

35. Y. I. Zhu and J. D. Haas, Iron depletion without anemia and physical performance in young women, *American Journal of Clinical Nutrition* 66 (1997): 334–341; J. J. LaManca and E. M. Haymes, Effects of iron repletion on VO_2max, endurance, and blood lactate in women, *Medicine and Science in Sports and Exercise* 25 (1993): 1386–1392.

36. C. V. Gisolfi, Fluid balance for optimal performance, *Nutrition Reviews* 54 (1996): S159–S168.

37. J. E. Greenleaf, Problem: Thirst, drinking behavior, and involuntary dehydration, *Medicine and Science in Sports and Exercise* 24 (1992): 645–656.

38. Gisolfi, 1996.

39. American College of Sports Medicine, Position stand: Heat and cold illnesses during distance running, *Medicine and Science in Sports and Exercise* 28 (1996): i–x.

40. E. F. Coyle, Timing and method of increased carbohydrate intake to cope with heavy training, competition, and recovery, in *Foods, Nutrition, and Sports Performance: An International Scientific Consensus*, eds. C. Williams and J. T. Devlin (London: E & FN Spon, 1992), pp. 37–61.

Table 10.2 Notes

[a]J. H. Wilmore, Increasing physical activity: Alterations in body mass and composition, *American Journal of Clinical Nutrition* 63 (1996): 456S–460S.

[b]D. Teegarden and coauthors, Previous physical activity relates to bone mineral measures in young women, *Medicine and Science in Sports and Exercise* 27 (1995): i–iv.

[c]D. C. Nieman, Exercise, upper respiratory tract infection, and the immune system, *Medicine and Science in Sports and Exercise* 26 (1994): 128–139.

[d]I. Thune and coauthors, Physical activity and the risk of breast cancer, *New England Journal of Medicine* 336 (1997): 1269–1275; M. M. Kramer and C. L. Wells, Does physical activity reduce risk of estrogen-dependent cancer in women? *Medicine and Science in Sports and Exercise* 28 (1996): 322–334; J. A. Woods and J. M. Davis, Exercise, monocyte/macrophage function, and cancer, *Medicine and Science in Sports and Exercise* 26 (1994): 147–157.

[e]S. N. Blair, Physical inactivity and cardiovascular disease risk in women, *Medicine and Science in Sports and Exercise* 28 (1997): 9–10; G. B. M. Mensink and coauthors, Intensity, duration, and frequency of physical activity and coronary risk factors, *Medicine and Science in Sports and Exercise* 29 (1997): 1192–1198; NIH Consensus Development Panel on Physical Activity and Cardiovascular Health, Physical activity and cardiovascular health, *Journal of the American Medical Association* 276 (1996): 241–246; P. T. Williams, High-density lipoprotein cholesterol and other risk factors for coronary heart disease in female runners, *New England Journal of Medicine* 334 (1996): 1298–1303.

[f]G. R. Hunter and coauthors, Fat distribution, physical activity, and cardiovascular risk factors, *Medicine and Science in Sports and Exercise* 29 (1997): 362–369; A. Goulding and coauthors, More exercise, less central fat distribution in women, *Journal of the American Medical Association* 276 (1996): 193–194; A. Tremblay and coauthors, Effect of intensity of physical activity on body fatness and fat distribution, *American Journal of Clinical Nutrition* 51 (1990): 153–157.

[g]G. Perseghin and coauthors, Increased glucose transport-phosphorylation and muscle glycogen synthesis after exercise training in insulin-resistant subjects, *New England Journal of Medicine* 335 (1996): 1357–1362; S. N. Blair and coauthors, Physical activity, nutrition, and chronic disease, *Medicine and Science in Sports and Exercise* 28 (1996): 335–349.

[h]A. A. Hakim and coauthors, Effects of walking on mortality among nonsmoking retired men, *New England Journal of Medicine* 338 (1998): 94–99.

[i]L. DiPietro, The epidemiology of physical activity and physical function in older people, *Medicine and Science in Sports and Exercise* 28 (1996): 596–600; L. E. Voorrips and coauthors, The physical conditions of elderly women differing in habitual physical activity, *Medicine and Science in Sports and Exercise* 25 (1993): 1152–1157.

CHAPTER 10

Nutrition in Practice

▪ Supplements and Ergogenic Aids Athletes Use ▪

In a world where body condition and skill are hard won, athletes gravitate to promises that they can easily improve their performance by taking pills, powders, or potions. Athletes often hear well-intended, but unsubstantiated, advice from their coaches and peers recommending that they use special nutrients, drugs, or procedures to enhance performance. The wish to win is strong, but no amount of wishing can change the fact that an overwhelming majority of supplements sold for athletes are frauds. If the products that are tried have no effect and are harmless, they are only a waste of money; but some products are harmful or actually impair performance, and these are a waste of athletic potential as well. This Nutrition in Practice looks at some of the so-called magical potions that promise to improve physical performance.

WWW
eatright.org
American Dietetic Association

What does ergogenic mean?

Ergogenic means work enhancing or work producing. In connection with athletic performance, ergogenic aids are substances or treatments that purportedly improve athletic performance above and beyond what is possible through training alone. Research findings do not, for the most part, support the claims made for ergogenic aids. When you hear a claim that a product is ergogenic, remember to consider the source of the claim and ask who may gain from the sale.

My coach told me to take protein supplements. Should I take them?

Protein powders and amino acid supplements are among the most common supplements athletes use.[1] The supplements are advocated to improve both strength and endurance, but as discussed in Chapter 10, well-nourished active people and athletes do not need them. Extra protein cannot be forced into the muscles to make them grow. Muscle cells accept nutrients only when they are needed. The cells "decide" what they need, based on the messages they receive from the hormones that regulate them and from the demands put upon them. The way to make muscle cells grow, therefore, is to make them work. The only role for diet in this process is to make protein available, and good diets always do. Although the protein needs of some endurance and strength athletes are higher than those of sedentary people, the additional protein is already present in a well-chosen diet as Chapter 10 described.

Most healthy athletes eating well-balanced diets do not need amino acid supplements either. Advertisers point to research that identifies the branched-chain amino acids as the main ones used as fuel by exercising muscles. What the ads leave out is that compared to glucose and fatty acids, branched-chain amino acids provide almost no fuel and that ordinary foods supply them in abundance anyway.

Furthermore, large doses of branched-chain amino acids can raise plasma ammonia concentrations, which can be toxic to the brain.[2] Branched-chain amino acid supplements are neither effective nor safe and are not recommended.

I know that some athletes, especially endurance athletes, are taking carnitine supplements. What is carnitine?

Carnitine is a nonprotein amino acid. Endurance athletes believe carnitine will help them burn more fat, thereby sparing glycogen during endurance events. Carnitine is also promoted to bodybuilders as a "fat burner."

In the body, carnitine facilitates the transfer of fatty acids across the mitochondrial membrane. Supplement manufacturers suggest that with more carnitine available, fat oxidation will be enhanced, but this does not seem to be the case. Carnitine supplementation for 7 to 14 days neither raised muscle carnitine concentrations nor influenced fat or carbohydrate oxidation.[3] It did, however, produce diarrhea in half of the men tested. Milk and meat products are good sources of carnitine, and supplements are not needed.

What about vitamin or mineral supplements for athletes? I have a friend who is a bodybuilder. She takes several vitamin pills right before each competition.

Tell your friend that this practice is pointless, though probably harmless. For one thing, vitamins taken right before competition have no effect; during the event, they are still waiting in the blood and have not yet been assembled into working molecules. Besides, research shows that most bodybuilders' diets provide adequate amounts of vitamins and minerals, and when a nutrient is lacking, the supplements chosen are seldom the ones needed to remedy the deficiencies. If a diet lacks nutrients, it should be modified to provide the needed nutrients. Only if a health professional identifies a clinical nutrient deficiency should nutrient supplements be prescribed—and then only if dietary modification alone cannot remedy the deficiency.

I've heard that vitamin E supplements may benefit active people and athletes even if they are not vitamin E deficient. Is this true?

Research shows a relationship between physical activity and oxidative stress.[4] Some research suggests that prolonged, high-intensity activity enhances production of damaging free radicals in the body.[5] Another study shows that prolonged, intense physical activity reduces susceptibility to oxidative stress.[6] Meanwhile, many athletes and active people are taking megadoses of vitamin E in hopes of preventing oxidative damage to muscles. The results of some studies lend support to such efforts. Research suggests vitamin E supplements may offer some protection against exercise-induced oxidative stress and damage.[7] In contrast, other research has failed to show a protective effect of vitamin E against oxidative stress during physical activity.[8]

Studies examining the long-term effects of vitamin E supplementation are lacking, however, and others report inconsistent findings. Clearly, more research is needed before conclusions can be made about vitamin E and oxidative stress during exercise. In the meantime, active people can benefit by eating generous servings of antioxidant-rich fruits and vegetables.

My friend who is a bodybuilder also takes a supplement called chromium picolinate. Advertisements in magazines and health-food stores make all kinds of impressive claims for it. Are any of the claims true?

Chapter 8 introduced chromium as an essential trace mineral involved in carbohydrate and lipid metabolism. Advertisements in bodybuilding magazines claim that chromium picolinate, which is supposed to be more easily absorbed than chromium alone, builds muscle, enhances energy, and burns fat. Such claims derive from one or two initial studies on this mineral. Most studies of chromium picolinate and strength training that have followed, however, show no effects of chromium picolinate supplementation on strength, lean body mass, or body fat.[9] Furthermore, large doses of chromium picolinate may lead to iron deficiency.[10]

The latest new supplement I've been reading and hearing about is creatine. A lot of my friends are taking it. Why is it so popular?

Interest in—and use of—creatine monohydrate supplements to enhance energy production during intense activity has grown dramatically in the last few years.[11] Power athletes such as weight lifters use creatine supplements to enhance stores of the high-energy compound creatine phosphate (or phosphocreatine) in muscles. Theoretically, the more creatine phosphate in muscles, the higher the intensity at which an athlete can train. High-intensity training stimulates the muscles to adapt, which, in turn, improves performance.

The results of some studies suggest creatine supplementation enhances performance of high-density strength activity such as weight lifting.[12] Other research findings conflict with reports that creatine supplements improve strength performance.[13]

Some medical and fitness experts voice concern that, like many performance enhancement supplements before it, creatine is being taken in huge doses (5 to 30 grams per day) before evidence of its value has been ascertained.[14] Even people who eat red meat, which is a creatine-rich food, do not consume near the amount athletes are taking. Athletes who take megadoses of creatine risk possible long-term side effects such as organ and muscle damage. Despite the uncertainties, creatine supplements are not illegal in international competition.

I have heard of a technique for improving endurance called glycogen loading. What is glycogen loading?

As you know from reading the chapter, the fuel for intense muscular activity is carbohydrate, stored in the muscle as glycogen. Athletes who compete in long-distance endurance events naturally want to have as much energy stored in their muscles as they can. Various techniques called glycogen loading were used in the past to trick muscles into storing more glycogen than normal. These techniques involved sudden, drastic changes in diet that caused nausea or cramping in some athletes. Other athletes experienced more dangerous effects such as abnormal heart and kidney function.

Exercise physiologists now recommend a modified plan of glycogen loading that confers benefits without such side effects. First, about two or three weeks before competition, the athlete increases activity intensity while eating a normal, high-carbohydrate diet. Then, during the last week before competition, the athlete modifies both physical activity and diet. With respect to activity, the athlete gradually cuts back, resting completely on the day before the event. Meanwhile, with respect to foods, the athlete eats carbohydrate as usual until three days before the competition and then eats a very-high-carbohydrate diet.[15] Endurance athletes who follow this plan can keep going longer than their competitors without ill effects. In a hot climate, extra glycogen confers an additional advantage: as glycogen breaks down, it releases water, which helps to meet the athlete's fluid needs.

Extra glycogen benefits those who must keep going for 90 minutes or longer. The regular, everyday exerciser will not benefit from having larger stores. What that person does need, though, is *adequate* glycogen from eating a diet high in complex carbohydrates.

What about caffeine? I've heard that it can improve endurance performance.

Although some research findings support this notion, other findings suggest that caffeine has no effect on endurance. If caffeine does enhance endurance, the effect probably occurs because caffeine stimulates fatty acid release, thereby slowing glycogen use. Caffeine is a drug that stimulates the nervous system. The possible benefits must be weighed against caffeine's adverse effects—stomach upset, nervousness, irritability, headache, and diarrhea. Caffeine induces fluid losses that can be potentially hazardous if caffeine-containing fluids are used in place of other fluids by athletes competing in hot environments. The use of caffeine is banned by the International Olympic Committee when it exceeds a dosage equivalent to 5 or 6 cups of coffee in a 2-hour period prior to competition.

I have heard that steroids are dangerous, but I have a friend who takes them. His mother is a doctor, and she constantly monitors his blood pressure when he is taking steroids. Are they safe in his case?

Steroids are not safe in your friend's case or in any case; they have dangerous side effects and are illegal. Technically called androgenic-anabolic steroid drugs, they are derivatives of the male sex hormone testosterone. Testosterone promotes the development of male characteristics (androgenic) and lean body mass (anabolic). Athletes take steroids to stimulate muscle bulking. The American College of Sports Medicine and the American Academy of Pediatrics condemn the use of steroids by athletes, and the International Olympic Committee has banned their use.[16] In support of its position, the committee cites the known toxic side effects and maintains that steroid use is a form of cheating. Competitors who use the drugs put other athletes in the difficult position of either conceding an unfair advantage to abusing competitors or taking steroids and accepting the risk of untoward side effects.

The list of hazards and adverse reactions from steroids continues to grow (see Table NP10.1 on p. 264) amid only a slight decline in use of the drugs. Among the side effects and adverse reactions that steroids produce are cancerous liver tumors that cause the liver to rupture and hemorrhage, impairing its function; testicular shrinkage in men and masculinization of women; cardiovascular problems (including high blood pressure); and sterility. Your friend is sure to develop side effects no matter how closely a trainer or doctor monitors him.

The dangers of steroid use cannot be overemphasized. Health care professionals are obligated to warn athletes of these dangers. Speak simply and emphatically: the price for the potential competitive edge that steroids confer is damaged health and sometimes life itself. The safest effective way to build muscle has always been through hard training—and always will be.

What is DHEA and why do some athletes use it?

Some athletes use DHEA as an alternative to anabolic steroids. DHEA is a hormone (dehydroepiandrosterone), made in the adrenal glands, that serves as a precursor to the male hormone testosterone. Because DHEA and testosterone concentrations decline with age, proponents claim DHEA supplements enhance testosterone concentration, thereby slowing aging and building muscle. Advertisements claim it "burns fat,"

Table NP10.1 Anabolic Steroids: Side Effects and Adverse Reactions

Established Side Effects and Adverse Reactions	
Acne	Liver disease
Cancer	Liver tumors
Cholesterol increase	Male pattern baldness (in women—irreversible)
Clitoris enlargement	Oily skin (females only)
Death	Peliosis hepatitis (a liver disease)
Edema (water retention in tissue)	Penis enlargement (young boys)
Fetal damage	Priapism (painful, prolonged erections)
Frequent or continuing erections	Prostate enlargement
HDL (which help reduce cholesterol) decrease	Sterility (reversible)
Heart disease	Stunted growth
Hirsutism (hairiness in women—irreversible)	Swelling of feet or lower legs
Increased risk of coronary artery disease (heart attack, stroke)	Testicular atrophy
Jaundice	Yellowing of the eyes or skin
Other Possible Side Effects and Adverse Reactions	
Abdominal or stomach pains	Insomnia
Aggressive, combative behavior ("roid rage")	Kidney disease
Anaphylactic shock (from injections)	Kidney stones (from hypercalcemia)
Black, tarry, or light-colored stool	Listlessness
Bone pain	Menstrual irregularities
Breast development (sore or swelling—in men)	Muscle cramps
Chills	Nausea or vomiting
Dark-colored urine	Purple- or red-colored spots on body, inside of mouth or nose
Depression	Rash
Diarrhea	Septic shock (blood poisoning from injections)
Fatigue	Sexual problems
Feeling of abdominal or stomach fullness	Sore tongue
Feeling of discomfort	Unexplained darkening of skin
Fever	Unexplained weight loss
Frequent urge to urinate (mature men)	Unnatural hair growth
Gallstones	Unpleasant breath odor
Headache	Unusual bleeding
High blood pressure	Unusual weight gain
Hives	Urination problems
Hypercalcemia (too much calcium)	Vomiting blood
Impotence	
Increased chance of injury to muscles, tendons, and ligaments, plus longer recovery period from injuries	

Sources: K. L. Ropp, No-win situation for athletes, *FDA Consumer,* December 1992, pp. 8–12; National Academy of Sports Medicine policy statement and position paper: Anabolic androgenic steroids, growth hormones, stimulants, ergogenics, and drug use in sports, in B. Goldman and R. Klats, *Death in the Locker Room II: Drugs and Sports* (Chicago: Elite Sports Medicine Publications, 1992) , pp. 328–373.

"builds muscle," and "slows aging," but evidence to support such claims is lacking.

DHEA's short-term side effects include oily skin, acne, body hair growth, liver enlargement, and aggressive behavior.[17] Long-term effects of DHEA use remain to be seen and may take years to become evident. The potential for harm from DHEA supplements is great, and athletes, as well as others, should avoid it. DHEA is banned by the International Olympic Committee and the National Collegiate Athletic Association.

OK, protein supplements and most vitamin supplements are ineffective performance enhancers except when used to treat a true deficiency. More research needs to be done on vitamin E and oxidative stress. Results of studies on chromium picolinate are inconsistent, and experts are concerned about long-term effects of creatine. Glycogen loading works, but is unnecessary unless a person works out hard for longer than 90 minutes at a time. Caffeine may or may not be effective, but can have adverse side effects and is illegal. Steroids pose serious health risks and are illegal. Do any of the substances athletes use to boost performance work?

For the most part, no. Many of these substances have been studied and found to be worthless. The glossary on p. 266 describes commonly used ergogenic aids.

Health professionals can positively influence athletes and others interested in boosting athletic performance by stressing the measures that do help to enhance performance. They are, of course, regular training and sound nutrition.

Notes

1. E. A. Applegate and L. E. Grivetti, Search for the competitive edge: A history of dietary fads and supplements, *Journal of Nutrition* 127 (1997): 869S–873S.

2. E. Coleman, Branched-chain amino acids and fatigue, *Sports Medicine Digest* 18 (1996): 44.

3. M. Vukovich, D. L. Costill, and W. J. Fink, Carnitine supplementation: Effect on muscle carnitine and glycogen content during exercise, *Medicine and Science in Sports and Exercise* 26 (1994): 1122–1129.

4. E. W. Askew, Environmental and physical stress and nutrient requirements, *American Journal of Clinical Nutrition* 61 (1995): 631S–637S; H. M. Alessio, Exercise induced oxidative stress, *Medicine and Science in Sports and Exercise* 25 (1993): 218–224.

5. J. M. McBride and coauthors, Effect of resistance exercise on free radical production, *Medicine and Science in Sports and Exercise* 30 (1998): 67–72; D. A. Leaf and coauthors, The effect of exercise intensity on lipid peroxidation, *Medicine and Science in Sports and Exercise* 29 (1997): 1036–1039; M. Kanter, Free radicals and exercise: Effects of nutritional antioxidant supplementation, *Exercise and Sports Science Review* 23 (1995): 375–397.

6. G. S. Ginsburg and coauthors, Effects of a single bout of ultra endurance exercise on lipid levels and susceptibility of lipids to peroxidation in triathletes, *Journal of the American Medical Association* 276 (1996): 221–225.

7. J. M. McBride and coauthors, 1998; M. Meydani and coauthors, Protective effect of vitamin E on exercise-induced oxidative damage in young and older adults, *American Journal of Physiology* 264 (1993): R992–R998.

8. J. A. Warren and coauthors, Elevated muscle vitamin E does not attenuate eccentric exercise-induced muscle injury, *Journal of Applied Physiology* 72 (1992): 2168–2175; E. H. Witt and coauthors, Exercise, oxidative damage and effects of antioxidant manipulation, *Journal of Nutrition* 122 (1992): 766–773.

9. H. C. Lukaski and coauthors, Chromium supplementation and resistance training: Effects on body composition, strength, and trace element status of men, *American Journal of Clinical Nutrition* 63 (1996): 954–965; M. A. Hallmark and coauthors, Effects of chromium and resistance training on muscle strength and body composition, *Medicine and Science in Sports and Exercise* 28 (1995): 139–144.

10. Lukaski and coauthors, 1996.

11. Applegate and Grivetti, 1997.

12. R. B. Kreider and coauthors, Effects of creatine supplementation on body composition, strength, and sprint performance, *Medicine and Science in Sports and Exercise* 30 (1998): 73–82; J. S. Volek and coauthors, Creatine supplementation enhances muscular performance during high-intensity resistance exercise, *Journal of the American Dietetic Association* 97 (1997): 765–770; S. M. Tolar, Creatine is an ergogen for anaerobic exercise, *Nutrition Reviews* 55 (1997): 21–23; C. P. Earnest and coauthors, The effect of creatine monohydrate ingestion on anaerobic power indices, muscular strength, and body composition, *Acta Physiologica Scandinavica* 153 (1995): 207–709.

13. L. M. Odland and coauthors, Effect of oral creatine supplementation on muscle (PCr) and short-term maximum power output, *Medicine and Science in Sports and Exercise* 29 (1997): 216–219.

14. T. Noakes, as quoted in M. Gaie, Olympic athletes face heat, other health hurdles, *Journal of the American Medical Association* 276 (1996): 178–180.

15. M. H. Williams, *Nutrition for Fitness and Sport*, 4th ed. (New York: McGraw-Hill, 1995), pp. 83–120.

16. D. H. Catlin and T. H. Murray, Performance-enhancing drugs, fair competition and Olympic sport, *Journal of the American Medical Association* 276 (1996): 231–237.

17. E. Coleman, DHEA—An anabolic aid? *Sports Medicine Digest* 18 (1996): 140–141.

CHAPTER 10

Nutrition in Practice

Glossary of Ineffective Ergogenic Aids

bee pollen: a product consisting of bee saliva, plant nectar, and pollen that confers no benefit on athletes and may cause an allergic reaction in individuals sensitive to it.

blood doping: the process of injecting red blood cells to enhance the blood's oxygen-carrying ability. Risks include dangerous blood clotting, especially in athletes who become dehydrated; infections from nonsterile equipment; transfusion reactions; and dangers of improperly transferred blood. Blood doping is banned in Olympic competitions.

branched-chain amino acids: the amino acids leucine, isoleucine, and valine, which are present in large amounts in skeletal muscle tissue; packaged and promoted for athletes as ergogenic aids.

caffeine: a stimulant that in small amounts may produce alertness and reduced reaction time in some people, but that also creates fluid losses. Overdoses cause headaches, trembling, an abnormally fast heart rate, and other undesirable effects.

calcium pangamate: a compound once thought to enhance aerobic metabolism, now known to have no such effect.

carnitine: a nonprotein amino acid found in most body cells that serves as a carrier of unoxidized fatty acids. It is advertised as a "fat burner" to bodybuilders and as a means of sparing glycogen to endurance athletes, but studies show it does neither.

cell salts: a mineral preparation supposedly prepared from living cells.

coenzyme Q10: a lipid found in cells (mitochondria) shown to improve exercise performance in heart disease patients, but not effective in improving performance of healthy athletes.

creatine: a nitrogen-containing compound that combines with phosphate to form the high-energy compound creatine phosphate (or phosphocreatine) in muscles. Claims that creatine enhances energy production and muscle strength need further confirmation.

DHEA (dehydroepiandrosterone): a hormone made in the adrenal glands that serves as a precursor to the male hormone testosterone. Side effects include acne, aggressiveness, and liver enlargement.

DNA and **RNA (deoxyribonucleic acid** and **ribonucleic acid):** the genetic materials of cells necessary in protein synthesis; falsely promoted as ergogenic aids.

epoetin (eh-poy-EE-tin): a drug derived from the human hormone erythropoietin and marketed under the trade name Epogen; illegally used to increase oxygen capacity.

ginseng: a plant whose extract supposedly boosts energy; side effects to chronic use include nervousness, confusion, and depression.

glycine: a nonessential amino acid; promoted as an ergogenic aid because it is a precursor of the high-energy compound creatine phosphate.

growth hormone releasers: herbs or pills falsely promoted as enhancing athletic performance.

guarana: a reddish berry found in Brazil's Amazon valley that is used as an ingredient in carbonated sodas and taken in powder or tablet form. Guarana is marketed as an ergogenic aid to enhance speed and endurance, an aphrodisiac, a "cardiac tonic," an "intestinal disinfectant," and a smart drug that supposedly improves memory and concentration and wards off senility. Because guarana contains seven times as much caffeine as its relative the coffee bean, there are concerns that high doses can stress the heart and cause panic attacks.

herbal steroids or **plant sterols:** lipid extracts of plants, called *ferulic acid, oryzanol, phytosterols,* or *adaptogens;* marketed with false claims that they contain hormones or balance hormonal activity.

inosine: an organic chemical that is falsely said to "activate cells, produce energy, and facilitate exercise," but has been shown actually to reduce the endurance of runners.

octacosanol: an alcohol extracted from wheat germ; often falsely promoted to enhance athletic performance.

phosphate salt: a salt that has been demonstrated to raise the concentration of a metabolically important compound (diphosphoglycerate) in red blood cells and enhance the cells' potential to deliver oxygen to muscle cells; the salts may cause calcium losses from the bones if taken in excess.

royal jelly: a substance produced by worker bees and fed to the queen bees; often falsely promoted as enhancing athletic performance.

sodium bicarbonate: baking soda; an alkaline salt believed to neutralize blood lactic acid and thereby reduce pain and enhance possible workload. Some studies show that sodium bicarbonate in recommended doses can enhance performance of high-intensity exercise (in 1- to 5-minute exercise sessions), but its effects on endurance exercise are unknown; "soda loading" may cause intestinal bloating and diarrhea.

11 *Consumer Concerns about Foods*

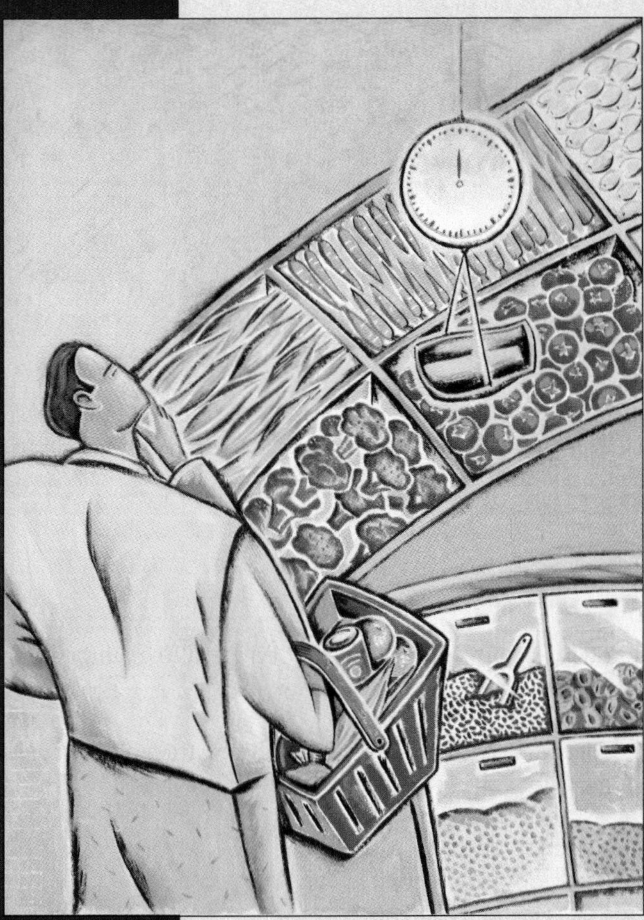

Which foods are most nutritious?

The preceding chapters have presented a basic course in nutrition. This chapter wraps up the course by tying nutrition to the world of foods. After all, nutrition is delivered in the form of foods, and to use foods correctly, consumers need to know how best to select and prepare them. The first half of this chapter is devoted to the nutritional value of grocery-store foods. The second half is devoted to food safety.

Nutritional Value of Grocery-Store Foods

People want to know how to choose the most nutritious foods. Are canned foods OK? Frozen foods? Are fresh foods better? What about fast foods?

Processed Foods

Most of the foods consumed today, whether eaten in restaurants or at home, have been processed in some way by industry. People often ask what processing does to foods and which kinds of foods are most and least nutritious.

Goals of Food Processing Many forms of processing aim to extend the usable life of a food—that is, they preserve the food. To preserve food, a process must prevent three detrimental changes. It must:

- Prevent microbial growth.
- Prevent oxidative changes.
- Prevent enzymatic destruction of food molecules.

Canning, freezing, modified atmosphere packaging, drying, salting, and adding preservatives are all ways of achieving these objectives.

Costs of Processing In general, food processing involves a tradeoff. It makes foods safer, it gives them a longer usable lifetime than fresh foods, or it cuts preparation time—but at the cost of some vitamin and mineral losses. A process such as pasteurization is clearly worth the cost because the safety gains are great and the nutrient losses small. Some processed foods gain a nutritional edge over their unprocessed counterparts; the removal of fat from milk or other foods by processing is an example. The next few sections describe the most common processing techniques and explain their effects on nutrients.

pasteurization: the treatment of food with heat sufficient to kill certain pathogens (disease-causing microbes) commonly transmitted through milk, juices, or foods. It is not a sterilization process; bacteria that cause spoilage are still present.

Canned Foods Canning is one of the better methods for protecting food against the growth of microbes (bacteria, fungi, and yeasts) that might otherwise spoil it, but canning, unfortunately, does incur nutrient losses. Like other heat treatments, the canning process is based on time and temperature. Each small increase in temperature has a major killing effect on microbes with only a minor effect on nutrients, so industry chooses canning treatments that employ high temperatures for short times.

To determine how much of a food's nutritional value is lost in canning, food scientists have performed many experiments. They have paid particular attention to three vulnerable water-soluble vitamins—thiamin, riboflavin, and vitamin C.

Acid stabilizes thiamin, but heat rapidly destroys it, so low-acid canned foods such as lima beans, corn, and meat lose up to half, or even more, of their thiamin. Unlike thiamin, riboflavin is stable to heat but sensitive to light; so glass-packed, not canned, foods are more likely to lose riboflavin. Vitamin C is vulnerable to

To see the effect of canning on thiamin in foods, look at Appendix A, items 890 and 891—½ cup canned green peas versus ½ cup frozen green peas. While you are looking, what other effects of canning on thiamin do you see?

an enzyme present in fruits and vegetables and in microorganisms. By destroying this enzyme, canning actually helps preserve some vitamin C, although some is destroyed by heat in the process. As for the fat-soluble vitamins, they are relatively stable and are not affected much by canning.

Minerals are unaffected by heat processing because they cannot be destroyed as vitamins can be. Some minerals are added when foods are canned. Important in this regard is sodium chloride, table salt, added for flavoring. Because salt tends to raise some people's blood pressure, many food companies have begun making low-salt versions of their products. Unfortunately, though, these may cost more than the higher-salt versions.

Frozen Foods Frozen foods' nutrient contents are similar to those of fresh foods; losses are minimal. The freezing process itself does not destroy any nutrients, but some losses may occur during the steps taken in preparation for freezing, such as the quick immersing in boiling water (blanching), washing, trimming, or grinding. Vitamin C losses are especially likely because they occur whenever tissues are broken and exposed to air (oxygen destroys vitamin C). Uncut fruits do not lose their vitamin C. Strawberries, for example, may be kept frozen for over a year without losing any vitamin C. Mineral contents of frozen foods are much the same as those of fresh foods.[1]

WWW. **nffa.org**
National Frozen Food Association

Fresh foods are often shipped long distances. To ensure that they make the trip without bruising or spoiling, fresh foods are often harvested unripe. Frozen foods are shipped frozen, so the foods are allowed to vine-ripen, enabling nutrients to develop to their fullest potential. Foods frozen and stored under proper conditions will often contain more nutrients when served at the table than fresh fruits and vegetables that have stayed in the produce department of the grocery store for even a day.

Frozen foods have to be kept frozen to retain their nutrients. To be solidly frozen, a food has to be colder than 32°F or 0°C. Vitamin C converts rapidly to its inactive forms at warmer temperatures. Food may seem frozen at 36°F or 2°C, but enzymes can work at these temperatures. Under these conditions, the vitamin C in a frozen food can be completely lost in as short a time as two months. If you want to maximize the nutritive value of the foods you store at home, invest in a freezer thermometer, monitor the temperature of your frozen-food storage place, and keep it at or below 32°F.

Vacuum and Modified Atmosphere Packaging The demand for fresh or nearly fresh foods has aided in the acceptance of vacuum packaging and modified atmosphere packaging (MAP) by both consumers and the food industry. Food manufacturers using vacuum packaging or MAP first package the foods in plastic film or some other wrap that oxygen cannot penetrate. Then, they either remove the air inside the package, creating a vacuum, or replace the air with a mixture of oxygen-free gases, such as carbon dioxide and nitrogen.

vacuum packaging: preservation of a perishable food by packaging it in a gas-impermeable container from which air has been removed.

modified atmosphere packaging (MAP): preservation of a perishable food by packaging it in a gas-impermeable container to which a gas mixture other than air has been added.

One technique combines refrigeration and modified atmosphere to preserve freshness and increase the shelf life of foods such as produce, fresh and cured meats, plated dinners, pasta and pasta sauces, and bakery items.[2] Once exposed to the air, MAP foods retain their vitamins much longer than the same foods processed or packaged using other methods. MAP foods also taste fresh, making them especially popular with consumers. One concern about the MAP method is that it may permit growth of the *Clostridium botulinum* bacterium in moist, low-acid foods, such as lunch meats or cooked dishes, when they are kept for long periods at too warm temperatures.[3] With proper refrigerated storage, however, the foods are as safe and nutritious as fresh foods.

Dried Foods Dried or dehydrated foods have their own special characteristics. Drying offers several advantages. It eliminates microbial spoilage (because

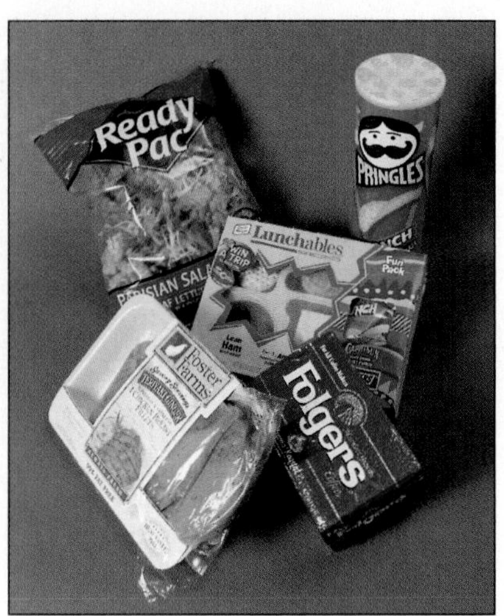

Food manufacturers use vacuum packaging, MAP, or a combination of refrigeration and MAP to preserve the freshness of some foods.

microbes need water to grow), and it greatly reduces the weight and volume of foods (because foods are mostly water). Commercial drying does not destroy many nutrients, but foods dried in heated ovens at home may sustain dramatic nutrient losses. Vacuum puff drying and freeze drying, which take place in cold temperatures, conserve nutrients especially well.[4]

Sulfite additives are added during the drying of fruits such as peaches, grapes (raisins), and plums (prunes) to prevent browning. Sulfur dioxide, a commonly used sulfite, helps to preserve vitamin C as well, but it is highly destructive of thiamin. This is of small concern, however, because most dehydrated products with added sulfur dioxide are not major sources of thiamin anyway.

Some dried foods are heated, ground, and pushed through various kinds of screens to yield different shapes, such as pieces of breakfast cereal or the "bits" sprinkled on salad—so-called food novelties. Considerable nutrient losses occur during these processes, and nutrients are usually added to compensate. Foods this far removed from the original fresh state, however, are still lacking significant nutrients (notably, vitamin E), and consumers should not rely on them as staple foods. Enjoy them, but only as occasional snacks and as additions to enhance the appearance, taste, and variety of meals.

Making Wise Choices

In general, the more heavily processed foods are, the less nutritious they become. Does that mean, then, that everyone should avoid all processed foods? It depends on the food and on the process.

Some Processed Foods Are Fine Consider the case of orange juice and vitamin C. Orange juice is available in several forms, each processed a different way. Fresh juice is simply squeezed from the orange, a process that extracts the fluid juice from the fibrous structures that contain it. The fresh-squeezed juice contains 111 milligrams of vitamin C per 100 kcalories. When this juice is condensed by heat, frozen, and then reconstituted, as is the juice from the freezer case of the grocery store, 100 kcalories of the reconstituted juice contain just 88 milligrams of vitamin C, because vitamin C is destroyed in the condensing process. Canning is even harder on vitamin C: 100 kcalories of canned orange juice have 82 milligrams of vitamin C.

These figures may seem to indicate that fresh juice is the superior food, and so it may be. But consider this: most people's vitamin C RDA (60 milligrams) is easily met by a cup of orange juice in any of these forms. In this case, at least for vitamin C, the losses due to processing are not a problem.

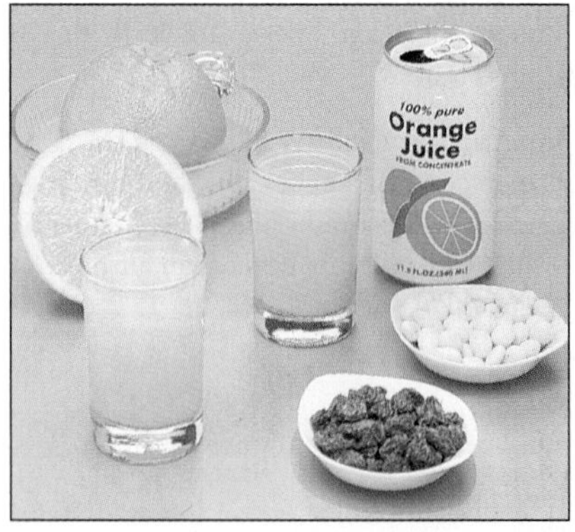

In terms of nutrient density, canned juice is almost as nutritious as fresh, but yogurt-covered raisins are not as nutritious as plain raisins.

On the other hand, refusing to process orange juice would be a big mistake. Fresh orange juice spoils. Shipping fresh juice to distant places in refrigerated trucks would cost much more than shipping frozen juice (which takes up less space) or canned juice (which requires no refrigeration). Fresh juice still contains active enzymes that continue to destroy its nutrients (including vitamin C), but frozen and canned juices remain virtually unchanged for long periods. Without canned or frozen juice, people with limited incomes or those with no access to fresh juice would be deprived of this excellent food.

Some Processed Foods Are Inferior Some processing methods are less desirable. Figure 8.2 in Chapter 8, for example, illustrated how processed foods are often loaded with sodium as their potassium is leached away. A related method of processing is the addition of sugar and fat—palatable, high-kcalorie additives that reduce nutrient density. Nuts and raisins covered with "natural yogurt" are an example. This may sound as though one healthy food is being added to another, but a look at the ingredient panel reveals that generous amounts of fat and sugar accompany the yogurt. About 75 percent of the weight of the product is fat and sugar; only about 8 percent is yogurt.

To pick just one nutrient for an example, here is what happens to the iron density of raisins as they become yogurt-coated raisins: 100 kcalories of raisins contain 0.71 milligram of iron; 100 kcalories of "yogurt" raisins contain 0.26 milligram of iron. These foods taste so good that wishful thinking can easily take hold, but the reality is that fat-coated, sugar-coated food is candy. The word *yogurt* on the label means only that one of the ingredients of the candy coating is some small amount of yogurt. A more nutrient-dense choice would be plain yogurt topped with fresh fruit.

With the privilege of abundance comes the responsibility to choose wisely.

Virtues of Whole Foods A good general rule for making food choices is to select whole foods to the greatest extent possible and, among processed foods, to choose only those that processing has improved nutritionally. The nutrient contents of processed foods exist on a continuum:

Whole-grain bread > refined white bread > sugared doughnuts.

Milk > fruit-flavored yogurt > canned chocolate pudding.

Corn on the cob > canned creamed corn > caramel popcorn.

Oranges > orange juice > orange-flavored drink.

Baked ham > deviled ham > fried bacon.

Another continuum parallels the nutrient continuum—the nutrition status of the consumer.

Chosen by these guidelines, even fast foods can contribute to a nutritious diet. Just choose the tomatoes rather than the catsup, the low-fat milk rather than the milk shake, and the broiled chicken rather than the fried chicken sandwich.

When selecting foods for home preparation, be realistic; don't dismiss all processed foods as inferior. Few people have the time to bake all their own bread from scratch, to shop every few days for fresh meats, or to wash, peel, chop, and cook fresh fruits and vegetables at every meal. This is where food processing comes in. Commercially prepared whole-grain breads, frozen cuts of meats, bags of frozen vegetables, and canned or frozen fruit juices do little disservice to nutrition and enable the consumer to eat a wide variety of foods at great savings in time and human energy.

In modern commercial processing, losses of vitamins seldom exceed 25 percent. In contrast, losses in food preparation at home can be close to 100 percent, and losses in the 60 to 75 percent range are not unusual. These facts put the matter of food processing into perspective and reveal that while the kinds of foods you buy certainly make a difference, what you do with them in your kitchen can make an even greater difference.

Preserving Nutrients

Once you have selected nutritious foods at the market and brought them home, you have the task of storing and preparing them so that they deliver their nutritional benefits to you. This requires some understanding of how cooking and storing foods affect nutrients.

Keep Fresh Produce Cold Vitamins are organic compounds synthesized and broken down by enzymes found in the same foods that contain the vitamins. The enzymes that break down nutrients in fruits and vegetables work best at the temperatures at which the plants grow, which are near room temperature. Chilling fresh produce slows down enzymatic destruction of nutrients. To protect the vitamin content, most fruits and vegetables should be vine-ripened (if possible), chilled immediately after picking, and kept cold until used.

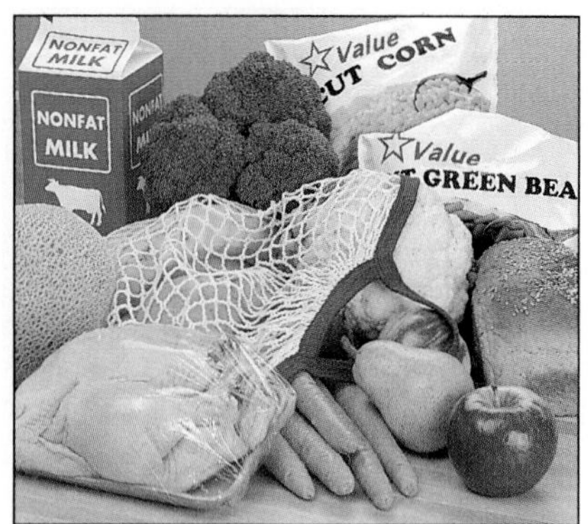

1. Purchase mostly fresh foods or those that processing has benefited nutritionally, for example, nonfat milk.

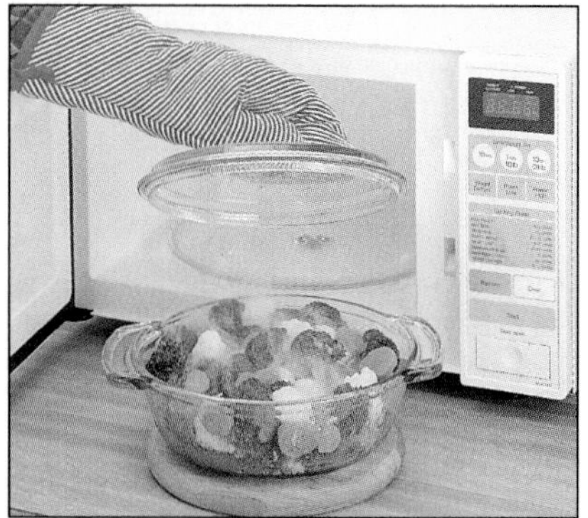
2. Steam vegetables or cook them in a microwave oven to minimize nutrient losses.

3. Wrap foods tightly and refrigerate them. Space foods to allow chilled air to circulate around them.

Keep Meats Cold For meats purchased in bulk, divide the package into portions and immediately refrigerate those to be used within two days. Wrap the portions of meat to be used later in coverings that exclude air, and freeze them.

Keep Milk Products in the Dark Riboflavin is light sensitive; it can be destroyed by the ultraviolet rays of the sun or by fluorescent light. For this reason milk is not sold (and should not be stored) in transparent glass containers. Cardboard or opaque plastic containers screen out light, protecting the riboflavin.

Do Not Cut before Use Some vitamins are acids or antioxidants and so are most stable in an acid solution, away from air. Citrus fruits, tomatoes, and many juices are acid. As long as the skin is uncut or the can is unopened, their vitamins are protected from air. If you store a cut vegetable or fruit, cover it with an airtight wrapper and refrigerate; close an opened carton of juice tightly and store it in the refrigerator.

Save Nutrient-Rich Water Water-soluble vitamins and minerals in cut vegetables readily dissolve into the water in which they are canned, washed, or boiled. If the water is discarded, as much as half of the vitamins and minerals in foods go down the drain with the water. A bit of southern folk wisdom is to serve the liquid with the vegetable rather than throwing it away; this liquid is known as the "pot liquor." The liquid may also be used to moisten cornbread or to make gravies, soups, or stews.

Conserve Water-Soluble Nutrients Other ways to minimize losses of nutrients during cooking are to steam vegetables over water rather than in it, stir-fry them in small amounts of oil, or microwave them. Wash intact foods vigorously and briefly; don't soak them. Cut vegetables after washing, except for those such as broccoli that you have to cut to wash adequately. Add peeled vegetables such as potatoes to water that is vigorously boiling, not to cold water. Microwave cooking is excellent for conserving nutrients and does not require the addition of fats or excess liquid.

Other Tactics During other types of cooking, minimize the destruction of vitamins by avoiding high temperatures and long cooking times. Although iron destroys vitamin C, the increased iron content of foods achieved by cooking in iron utensils may outweigh this disadvantage. Each of these tactics is small by itself, but saving a small percentage of the nutrients in foods daily can add up to significant amounts in a year's time.

Meanwhile, however, a law of diminishing returns operates. Most vitamin losses under reasonable conditions are not catastrophic. Be assured that if you start with fresh, whole foods containing ample amounts of vitamins and are reasonably careful in preparing those foods, you will receive a bounty of the nutrients they contain.

Food Safety

A vitally important aspect of food preparation is safety. Episodes of food poisoning cause illness in at least one-third of the U.S. population each year. Some 20 to 80 million cases of diarrhea, and possibly even more, are caused yearly by food poisoning. Many cases of "the flu" may actually be episodes of food poisoning. Some 6.5 million cases of food poisoning are reported in the United States each year; up to five times as many go unreported. Each year 9000 people die of food poisoning.[5]

food poisoning: illness transmitted to human beings through food and water, caused by infectious agents or by toxins that they produce.

Other aspects of food safety include avoiding the toxins that occur naturally in some foods as part of their normal composition and avoiding the contaminants (including pesticides) that can get into foods before they are harvested. Consumers also want to know whether food additives are safe or should be avoided. The next sections take up these issues in the order just mentioned, which is roughly the order of concern. Food poisoning is far and away the most important issue; food additives are the matter of least concern.

fsis.usda.gov
Food Safety and Inspection Service

foodsafety.org
National Food Safety Database

vm.cfsan.fda.gov/list.hmtl
FDA Center for Food Safety & Applied Nutrition

Preventing Food Poisoning

The term *food poisoning* refers to either food-borne infection or food intoxication. A food-borne infection is an illness caused by microorganisms, such as *Salmonella* varieties, that infect people whereas food intoxication is caused by toxins produced by microorganisms in food or within the digestive tract. In most food-borne illnesses, the symptoms are mild, but for people who are otherwise ill or malnourished, or very old or young, these relatively mild disturbances can be fatal. If abdominal cramps, headache, vomiting, and diarrhea are the major or only symptoms of your next bout of "flu," chances are excellent that what you really have is food poisoning.

toxins: poisons. Toxins produced by bacteria come in two varieties: *enterotoxins,* which act in the GI tract, and *neurotoxins,* which act on the nervous system.

The symptoms of one toxin stand alone as severe and commonly fatal—those of botulism, caused by the toxin of a microbe that grows inside improperly canned, home-canned, or vacuum-packed foods or in homemade garlic- or herb-flavored oils stored at room temperature.[6] Botulism danger signs constitute a true medical emergency (see the margin). Even with medical assistance, survivors can suffer the effects for months, years, or a lifetime. So potent is the botulin toxin that an amount as tiny as a single grain of salt can kill several people within an hour. The botulin toxin is destroyed by heat, so canned foods that have been boiled for ten minutes are generally safe from this threat. Home-canned foods are safe if prepared by following proper canning techniques to the letter.* To prepare herb-flavored oils safely, wash and dry the herbs before adding to the oil. Use the oil that day, and throw out leftovers at the end of the day.[7]

botulism: an often-fatal food poisoning caused by botulin toxin, a toxin produced by bacteria that grow without oxygen.

Warning signs of botulism:

- *Double vision.*
- *Weakening muscles.*
- *Difficulty swallowing.*
- *Difficulty breathing.*
- *Slurred speech.*

Food Safety in the Marketplace Transmission of food-borne illness is changing as the food supply changes. In the past, food-borne illness was caused by one person's error in a small setting, such as improperly refrigerated potato salad at a family picnic, and affected only a few victims. Today, people are eating more foods prepared and packaged by others. Consequently, when a food manufacturer or restaurant chef makes an error, food-borne illness can be epidemic.[8] An estimated 80 percent of reported food-borne illnesses are caused by errors in a commercial setting, such as the improper pasteurization of milk at a large dairy.[9]

In the mid-1990s, when a fast-food restaurant served undercooked burgers tainted with food toxin, hundreds of people became ill and at least three people died. This incident and others focused the national spotlight on two important safety issues: disease-causing organisms are commonly found in raw foods, and thorough cooking kills most pathogens found in foods. These episodes sparked a much needed overhaul of national food safety programs.

pathogens: disease-causing microorganisms.

Industry Controls To make the food supply safer for consumers, government agencies and the food processing industries have developed and implemented new programs to study and control food-borne illness.[10†] For example, the Hazard Analysis Critical Control Points (HACCP) system ensures that food

Hazard Analysis Critical Control Points (HACCP): a systematic plan to identify and correct potential microbial hazards in the manufacturing, distribution, and commercial use of food products.

*Complete, up-to-date, safe home canning instructions are included in the USDA's 172-page *Complete Guide to Home Canning* available for purchase from the Superintendent of Documents, Government Printing Office, Washington, DC 20402.
†These programs include Hazard Analysis Critical Control Points (HACCP), Emerging Infections Program (EIP), Foodborne Diseases Active Surveillance Network (FoodNet), and the Food Safety Inspection Service (FSIS).

manufacturers identify points of contamination and implement controls to prevent food-borne disease.[11] These safety regulations are expected to prevent hundreds of thousands of cases of food-borne illness each year.

Changes in the produce industry provide an example of the effectiveness of HACCP. After tracing two large outbreaks of salmonellosis to imported cantaloupe, producers began using chlorinated water to wash the melons and to make ice for packing and shipping. Since implementing this HACCP plan, no new melon-related cases of salmonellosis have been reported.

Salmonellosis is the food-borne infection caused by the *Salmonella* bacterium.

Consumer Awareness Canned and packaged foods sold in the grocery stores are almost invariably safe, but rare accidents do happen. Batch numbering makes it possible to recall contaminated foods through public announcements via newspapers, television, and radio. In the grocery store, these guidelines can help consumers avoid buying foods that are contaminated:

- Avoid packages with defective seals and wrappers.
- Reject leaking or bulging cans.
- Check safety "buttons" on jars for intact seal.
- Avoid partially frozen foods; those in chest-type freezers should be stored below the frost line.
- Choose packages that have not been damaged, soiled, or punctured.

The "2-40-140" rule will help you to remember the time and temperature danger zone for foods—allow them to stay for no more than 2 hr between 40° and 140°.

Improper handling of foods can occur anywhere along the line, from commercial manufacturers to large supermarkets to small restaurants to private homes. Maintaining a safe food supply requires everyone's efforts.

Food Safety in the Kitchen Whether bacteria multiply and cause illness depends, in part, on what happens in the kitchen—whether the kitchen is in your home, a school cafeteria, a gourmet restaurant, or a canning plant. Foods can provide ideal conditions for bacteria to thrive and produce their toxins. Disease-causing bacteria require:

- Warmth (40° to 140°F).
- Moisture.
- Nutrients.

To prevent bacterial growth, people who prepare foods can do these things: cook foods thoroughly, keep hot foods hot, keep cold foods cold, keep dry foods dry, and keep hands, utensils, and the kitchen clean.

Keep Hot Foods Hot Cook foods long enough to reach an internal temperature that will kill microbes (see Figure 11.1). To prevent bacterial growth when holding cooked food, keep it at 140°F or higher until it is served. Refrigerate leftover food immediately after serving.

Keep Cold Foods Cold Keeping cold foods cold starts when you leave the grocery store. If you are running errands, shop last, so the groceries will not stay in the car too long. (If the ice cream has begun to melt, it has been too long.) Upon arrival home, load foods into the refrigerator or freezer immediately. When serving food cold, let it stay at cool room temperature (about 68°F) for no more than two hours. If the room is warm (about 80°F), refrigerate the food after just one hour. Table 11.1 lists some safe keeping times for foods stored in the refrigerator at 40°F.

Table 11.1 Safe Refrigerator Storage Times (40°F)

Storage time	Foods
1 to 2 Days	Raw ground meats, breakfast or other raw sausages, raw fish or poultry; gravies
3 to 5 Days	Raw steaks, roasts, or chops; cooked meats, vegetables, and mixed dishes; ham slices; mayonnaise salads (chicken, egg, pasta, tuna)
1 Week	Hard-cooked eggs, bacon or hot dogs (opened packages); smoked sausages
2 to 4 Weeks	Raw eggs (in shells); bacon or hot dogs (packages unopened); dry sausages (pepperoni, hard salami); most aged and processed cheeses (swiss, brick)
2 Months	Mayonnaise (opened jar); most dry cheeses (parmesan, romano)

Keeping foods cold applies to defrosting foods before use, too. Bacterial growth begins on thawed portions of food even while the inner core is solidly frozen, so thaw meats or poultry in the refrigerator, not at room temperature. If you must hasten thawing, use cool running water or a microwave oven set to defrost.

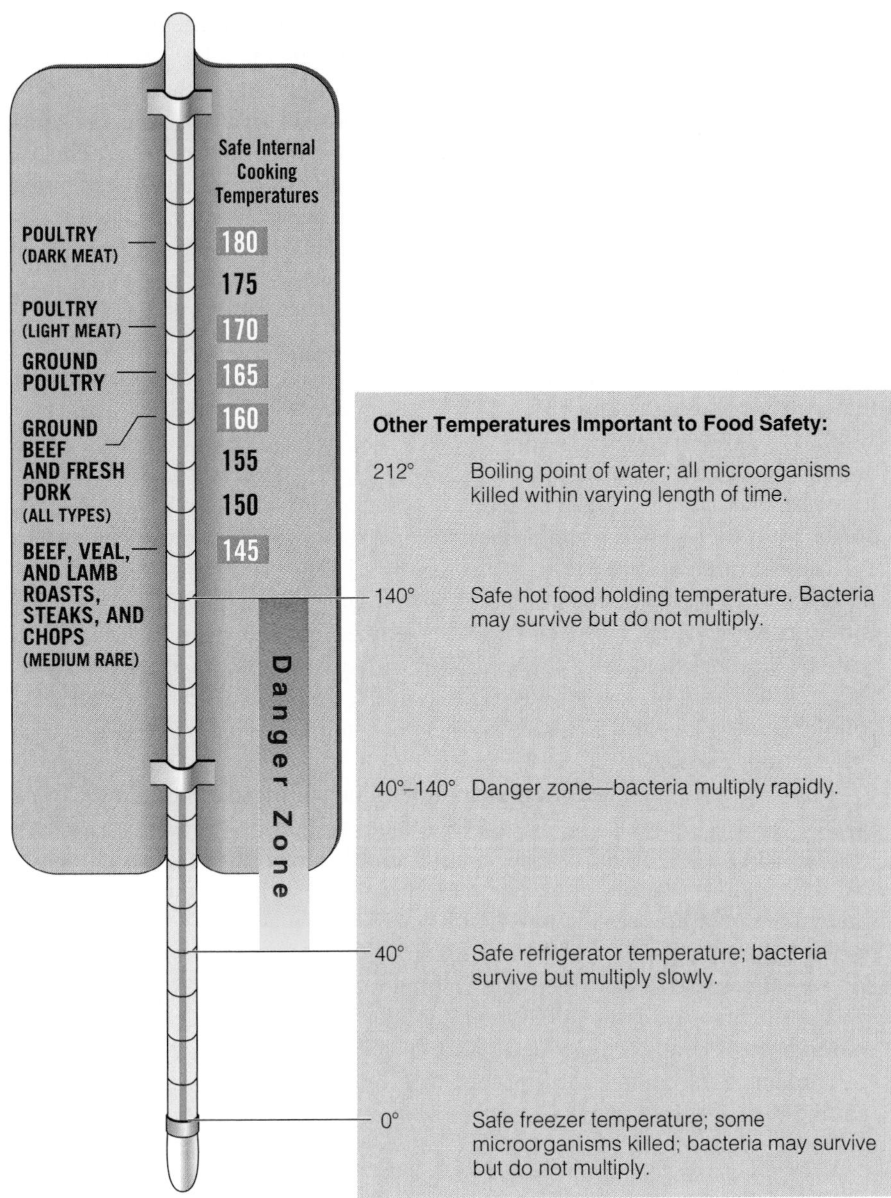
Safe Internal
Cooking
Temperatures
POULTRY
(DARK MEAT)
180
175
POULTRY
(LIGHT MEAT)
170
GROUND
POULTRY
165
GROUND
BEEF
AND FRESH
PORK
(ALL TYPES)
160
155
150
BEEF, VEAL,
AND LAMB
ROASTS,
STEAKS, AND
CHOPS
(MEDIUM RARE)
145
Danger Zone
Other Temperatures Important to Food Safety:
212° Boiling point of water; all microorganisms killed within varying length of time.
140° Safe hot food holding temperature. Bacteria may survive but do not multiply.
40°–140° Danger zone—bacteria multiply rapidly.
40° Safe refrigerator temperature; bacteria survive but multiply slowly.
0° Safe freezer temperature; some microorganisms killed; bacteria may survive but do not multiply.

Figure 11.1
Food Safety Temperatures (Fahrenheit)
Bacteria multiply at temperatures between 40° and 140°F. Cook foods to the temperatures on this thermometer and hold them at 140°F or higher.

Source: U.S. Department of Agriculture, 1993.

Keep Dry Foods Dry Cereals, breads, powdered mixes, and the like will keep for a long time if you don't let moisture get to them. Store them in cool, dry places, and keep them well sealed against humid air.

Keep the Kitchen Clean Keeping the kitchen clean includes using freshly washed utensils and laundered towels, and washing your hands with hot, soapy water before and after each step of food preparation. If you are ill or have open sores, stay away from food so as not to contaminate it.

To eliminate microbes, you have three choices, each with benefits and drawbacks. One is to destroy the microbes where they reside by washing countertops, cutting boards, sponges, and the like with toxic chemicals such as bleach (one capful per gallon of water). The benefit here is that chlorine can kill almost all organisms. The obvious drawback is that chlorine that washes down household drains into the water supply forms chemicals that can harm waterways and fish.

A second option is to use heat. Soapy water heated to 140°F kills most harmful organisms and washes most others away. This method takes effort, though,

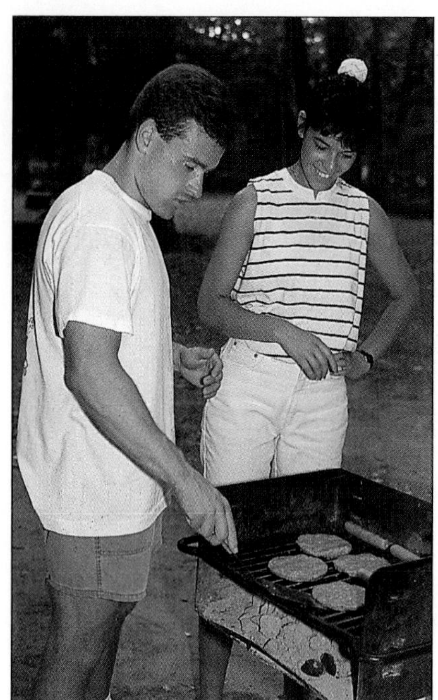

Cook hamburgers until they are brown (not pink) throughout, the juices run clear, and the insides are steaming hot.

for you have to use truly scalding water heated well beyond the temperature of the tap. (Most tap water called "hot" is not hotter than 130°F.)

The third option is to use clean boards for cutting and washable dishcloths that can be laundered often for wiping. (Save sponges for car washing and other heavy cleaning chores, and keep them away from surfaces that come in contact with raw foods.) For a small initial investment, you can create a truly safe environment in which to prepare your food. These measures all help eliminate microbes and decrease the possibility of cross-contamination, the transfer of microbes from one surface to another.

Troublesome Foods Some foods are more hospitable to microbial growth than are others. Especially vulnerable are moist foods, nutrient-rich foods, and foods that are chopped or ground.

Meats and Poultry Because raw meats and poultry require special handling, they now bear labels to instruct consumers on meat safety (see Figure 11.2). Meats and poultry may contain bacteria, and they provide a moist, nutritious environment that is ideal for microbial growth. Do not leave raw meat or poultry at room temperature for more than a very short time. Chill it thoroughly or cook it promptly, preferably the latter. Remember that every surface and utensil that raw meat has touched is contaminated with microbes; wash all such surfaces and utensils with hot, soapy water right away. Never put cooked meat back on a countertop or plate on which the raw meat has stood. The microbes will resume growing in it immediately.*

Ground meat is handled more than other kinds of meat and has much more surface exposed to the air for bacteria to land on, so it poses special risks. It is best to cook hamburger until it is brown (not pink) throughout, the juices run clear, and the inside is steaming hot. For a meat loaf, use a thermometer to test the internal temperature (see Figure 11.1 on p. 275).

Do not use or even taste a food with an "off" appearance or odor. However, don't trust your senses of smell and sight alone to tell you that foods are safe. Most contamination is not detectable by odor, taste, or appearance. Even hot cooked food, if handled improperly prior to cooking, can cause illness.

Though contaminated meat poses a real threat, consumers need to be aware that the media often exaggerate a story beyond its facts. Reports from England on the slaughter of cattle with mad cow disease and stories from Hong Kong on poultry with avian influenza sparked consumer fears, even though U.S. beef and poultry were not infected. Once again, smart consumers consider the source whenever evaluating nutrition information.

mad cow disease: a fatal virus that affects the nervous system of cattle. Formally known as bovine spongiform encephalopathy (BSE), this progressive neurological disorder resulted in the killing of 155,600 cattle in the United Kingdom as a precautionary measure to prevent transmission. Creutzfeldt-Jakob disease (CJD) is the human strain seen in the United Kingdom.

avian influenza: a newly discovered virus transmitted from birds to humans in Hong Kong; also known as "bird flu" or influenza A H5N1. As a precautionary measure to prevent transmission, thousands of birds were slaughtered.

The cardinal rule to protect yourself is to remember that food poisoning is always a possibility. For example, the meatballs in a warming tray at a lovely buffet may be warm but not hot. Despite the beautiful setting, their low temperature is a warning flag. Food at 140°F feels hot, not just warm. The likelihood of illness is strong when food is not hot enough, and the pleasure of eating meatballs isn't worth the risk.

Seafood For adults and children alike, eating raw or lightly steamed seafood is a risky proposition even when it is prepared as sushi by a master Japanese chef. Eating raw oysters can be dangerous for anyone, but people with liver disease, alcoholics, and people with suppressed immune systems are most vulnerable.[12] The microorganisms that lurk in seafood are undetectable even to an expert.†

*The USDA's meat and poultry hotline answers questions about meat and poultry safety: 1-800-535-4555.

†To speak with an expert about seafood safety, call the FDA seafood hotline: 1-800-FDA-4010.

People who like Japanese sushi know that not all varieties are made from raw fish. Many types are made with cooked crabmeat and vegetables, avocado, or other delicacies and are perfectly safe to enjoy. Also, rumor has it that freezing fish will make it safe to eat raw, but this is only partly true. Freezing fish will kill mature parasitic worms, but only cooking can kill all worm eggs and other microorganisms that can cause illness.

As population density increases along the shores of seafood-harvesting waters, pollution of those waters inevitably invades the seafood living there. Watchdog agencies monitor commercial fishing waters and try to keep harvesters out of unsafe waters. To help ensure safe seafood products, the Food and Drug Administration (FDA) requires seafood marketers to adopt food safety practices based on the HACCP system mentioned earlier. Some seafood processors are now pasteurizing raw oysters in the shell to kill pathogens and prevent food-borne illness. According to consumers, this process does not change the texture or the raw flavor. Still, unwholesome foods can reach the market. In one season alone, black-market dealers may sell millions of dollars worth of clams and oysters taken illegally from polluted harvesting areas. Experts are unanimous in saying that the risks of eating raw or lightly cooked seafood have become unacceptably high due to environmental contamination.[13] Chemical pollution and microbial contamination can originate in the water or in the boats and warehouses where seafood is cleaned, prepared, and refrigerated. To keep seafood as fresh as possible, people in the industry "keep it cold and keep it clean." Wise consumers eat it cooked.

At some times of the year, seafood may become contaminated with the so-called red tide toxin that occurs during algae blooms. Consumption of seafood contaminated with red tide causes a paralyzing form of food poisoning. The FDA monitors fishing waters for red tide algae and closes waters to fishing whenever it appears.

Safe Handling Instructions

THIS PRODUCT WAS PREPARED FROM INSPECTED AND PASSED MEAT AND/OR POULTRY. SOME FOOD PRODUCTS MAY CONTAIN BACTERIA THAT CAN CAUSE ILLNESS IF THE PRODUCT IS MISHANDLED OR COOKED IMPROPERLY. FOR YOUR PROTECTION, FOLLOW THESE SAFE HANDLING INSTRUCTIONS.

KEEP REFRIGERATED OR FROZEN.
THAW IN REFRIGERATOR OR MICROWAVE.

KEEP RAW MEAT AND POULTRY SEPARATE FROM OTHER FOODS.
WASH WORKING SURFACES (INCLUDING CUTTING BOARDS), UTENSILS, AND HANDS AFTER TOUCHING RAW MEAT OR POULTRY.

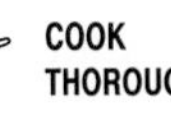

COOK THOROUGHLY.

KEEP HOT FOODS HOT. REFRIGERATE LEFTOVERS IMMEDIATELY OR DISCARD.

Figure 11.2
Safe Handling Instructions for Meat and Poultry

In Florida, containers of raw oysters must bear a warning that a risk is associated with consuming raw oysters or any raw animal protein. The risk is greatest for people who are ill or elderly or have weakened immune systems.

Frequently unsafe:

- *Raw milk and milk products.*
- *Raw or undercooked seafood, meat, poultry, or eggs.*

Occasionally unsafe:

- *Soft cheeses (Mexican style, feta, brie, camembert, blue-veined).*
- *Salad bar items.*
- *Unwashed berries and grapes.*
- *Sandwiches.*
- *Hamburgers.*
- *Airline food.*

Rarely unsafe:

- *Peeled fruit.*
- *High-sugar foods.*
- *Steaming-hot foods.*

Picnics Picnics are fun and can be safe, too. Choose foods that last without refrigeration, such as fresh fruits and vegetables, breads and crackers, and canned spreads and cheeses that you can open and use on the spot. Aged cheeses, such as cheddar and swiss, do well for an hour or two, but for longer periods, carry them in an ice chest. Mayonnaise resists spoilage because of its acid content, but when mixed with chopped ingredients, such as pasta, meat, or vegetable salads, it spoils quickly. The chopped ingredients offer an extensive surface area for bacteria to invade, and the foods have been in contact with cutting boards, hands, and kitchen utensils that have transmitted bacteria to them. Chill chopped salads well before, during, and after the picnic. Keep mayonnaise itself cold.

Honey Honey has been found to contain dormant bacterial spores that can awaken in the human body to produce the deadly botulin toxin mentioned earlier. Adults are big and strong enough to withstand the doses usually encountered, but infants under one year of age should never be fed honey. (It can also be contaminated with environmental pollutants picked up by the bees.) Honey has been implicated in several cases of sudden infant death.

If Illness Occurs Local health departments and the USDA extension service can provide further information about food safety. Should efforts fail and mild food-borne illness develop, drink clear liquids to replace fluids lost through vomiting and diarrhea. If serious food-borne illness is suspected, first call a physician. Then, wrap the remainder of the suspected food and label its container so that it cannot be mistakenly eaten, place it in the refrigerator, and hold it for possible inspection by health authorities.

WWW **foodsafety.org/sea.htm**
Handling Seafood Safely (from the National Food Safety Database)

WWW. **usda.gov**
U.S. Department of Agriculture

traveler's diarrhea: nausea, vomiting, and diarrhea caused by consuming food or water contaminated by any of several organisms, most commonly, *E. coli, Shigella, Campylobacter jejuni,* and *Salmonella.*

Travel Special food safety concerns arise when people travel. In many parts of the world, food-borne illness is likely to strike tourists even while the local people, eating exactly the same foods prepared the same way, remain healthy. That is because the locals have developed immunity to local disease-causing organisms, while tourists have no such protection. The accompanying box offers tips to travelers on avoiding food-borne infection.

Water

Foods are not alone in transmitting food-borne diseases; water is guilty, too. A glass of "water" is more than just water. Some diseases found on fresh fruits and vegetables and in raw oysters are transmitted through contaminated water.[14] In addition to microorganisms, water may contain many of the same impurities that foods do: environmental contaminants, pesticides, and additives such as chlorine used to kill pathogenic microorganisms. These concerns are discussed later in relation to foods.

WWW. **epa.gov**
Environmental Protection Agency

Contamination Contamination can occur as water travels from the main water supply to homes. Lead or asbestos from old, corroded pipes can contaminate drinking water, as can bacteria and dirt from leaking pipes. People who suspect contamination of their water should have it tested where it flows out, at the tap.

Public water systems treat water to remove contaminants that have been detected above acceptable levels. Private well water is usually not treated or cleansed, so people who consume water from private wells are responsible for its safety and should test the water periodically.

Bottled Water Some people turn to bottled water as an alternative to tap water. Bottled water is classified as a food, so it is regulated by the FDA and must meet safety standards similar to those set for public water systems. Bottled water must also be processed and labeled according to FDA regulations. Some bottled waters may have minerals or carbonation added. "Carbonated," "seltzer," and "tonic" waters are not considered waters, however, but soft drinks. The FDA requires labels to disclose the sources of bottled waters and to use legally defined descriptive terms.[15]

WWW **bottledwater.org**
International Bottled Water Association

Safe drinking water is a concern for everyone and must be protected to ensure continued health. To learn more about the water supply in your area, call the local public health agency.*

Advances in Food Technology and Safety

New advances in technology offer promise for the future purity of foods. Someday their use may dramatically improve the safety of foods for sale on the market.

irradiation: sterilizing a food by exposure to energy waves, similar to ultraviolet light and microwaves.[16]

WWW **food-irradiation.com**
Foundation for Food Irradiation Education

ultrahigh temperature (UHT) treatment: sterilizing a food by short-time exposure to temperatures above those normally used in processing.

Irradiation The FDA has approved the use of irradiation on certain foods to improve food safety. Irradiation kills microorganisms and insects on wheat; spices; teas; fresh and frozen beef, lamb, pork, and poultry; and fresh fruits and vegetables. In addition, irradiation inhibits growth of sprouts on potatoes and onions and delays ripening in some fruits such as strawberries and mangoes. Milk products change flavor when irradiated and so are not candidates for the treatment. (Incidentally, the milk in those boxes kept at room temperature on grocery-store shelves is not irradiated, but processed with an ultrahigh temperature treatment for just long enough to sterilize it.)

*For information on safe drinking water in general, call the Environmental Protection Agency's hotline: 1-800-426-4791.

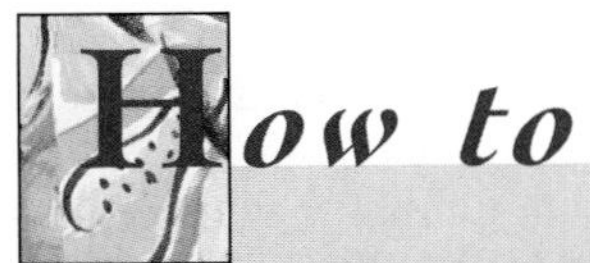

Achieve Food Safety while Traveling

Food-borne illnesses contracted while traveling are colloquially known as traveler's diarrhea. A bout of this ailment can ruin the most enthusiastic tourist's trip. To avoid food-borne illness while traveling:

- Wash your hands often with soap and hot water, especially before handling food or eating.
- Eat only cooked food and canned foods. Eat raw fruits or vegetables only if you have washed them in boiled water and peeled them yourself. Skip salads, raw fish, and shellfish.
- Be aware that water, and ice made from it, may be unsafe, too. Take along disinfecting tablets or an element that boils water in a cup.
- Drink no beverages made with tap water. Drink only treated, boiled, canned, or bottled beverages, and drink them without ice (ice may be made from contaminated water), even if they are not chilled to your liking. Refuse dairy products unless they have been properly pasteurized and refrigerated.
- Do not use the local water, even to brush your teeth, unless you boil or disinfect it first.
- Before you leave on the trip, ask your physician to recommend medicines to take with you in case your efforts to avoid illness fail.

One journalist succinctly sums up these recommendations, "Boil it, cook it, peel it, or forget it."[a] Chances are excellent that if you follow these rules, you will remain well.

[a]R. D. Williams, Boil it, cook it, peel it, or forget it, *FDA Consumer,* September 1991, p. 17.

The use of irradiation on food has been extensively evaluated and is supported by the World Health Organization, the American Medical Association, and other health agencies.[17] The process substantially reduces food-borne pathogens associated with fresh fruits and vegetables and is also effective in eliminating the *Salmonella* bacterium from poultry.[18]

This international symbol identifies retail foods that have been irradiated. The words "Treated by irradiation" or "Treated with irradiation" must accompany the symbol. The irradiation label is not required on commercially prepared foods that contain irradiated ingredients, such as spices.

Consumer Concerns Some consumers, associating radiation with cancer, birth defects, and mutations, have negative emotions about the use of irradiation on foods. Despite consumers' concerns, irradiation does not noticeably change the taste, texture, or appearance of food; nor does it make the food radioactive. Vitamin loss is minimal and comparable to amounts lost in other food processing methods. Irradiation cuts down on food spoilage and can replace some costly pesticides, thus reducing pesticide residues in food.

Regulation of Irradiation The FDA has established regulations governing the uses of irradiation and allowed doses. Each food that has been treated with irradiation must say so on its label.

High-Intensity Pulsed Light A new technology called high-intensity pulsed light has been approved by the FDA to enhance food safety. High-intensity pulsed light uses an intense flash of light to kill microorganisms on the surface of foods, packaging materials, and water. It extends the shelf life of foods without changing their nutritional properties.

These new technologies show promise in the battle against food-borne pathogens. In combination with safe food handling by consumers, these processes will decrease the number of food-borne illnesses contracted each year and increase the safety of our food supply.

Natural Toxins in Foods

Consumers concerned about food contamination may think that they can eliminate all poisons from their diets by eating only "natural" foods. On the contrary, nature has provided natural foods with the natural poisons they need to fend off diseases, insects, and other predators. Nevertheless, although the potential for harm exists, actual harm rarely occurs.

Most people would recognize the names *belladonna* and *hemlock*—both classic deadly poisons in the form of natural herbs. Few people know, however, that the herb *sassafras* contains a cancer-causing agent and is banned from use as an additive in commercially produced foods and beverages. Equally surprising is that cabbage, turnips, mustard greens, and radishes all contain small quantities of harmful goitrogens—compounds that can enlarge the thyroid gland and aggravate thyroid problems.

Cabbages and their relatives are celebrated for containing phytochemicals associated with low cancer incidence. An unexpected twist to the cabbage-family story is that some of the phytochemicals celebrated as protective against cancer are themselves carcinogenic. The protection they confer on the body seems to result because these mild toxins prompt the body to build up defenses, in somewhat the same way as it builds immunity. Then, when a potent carcinogen arrives, the prepared body deals with it swiftly, detoxifies it, and excretes its remnants before cancer can begin.

The cyanogens are another natural poison. These precursors to the deadly poison cyanide are found in lima beans and fruit seeds such as apricot pits. Many countries allow commercial growers to grow only those varieties of lima beans with the lowest cyanogen contents. As for fruit seeds, they are seldom deliberately eaten. An occasional swallowed seed or two presents no danger, but a couple of dozen seeds could be fatal to a small child.

Potatoes contain many natural poisons. One is solanine, a bitter, powerful, narcotic-like substance. The small amounts of solanine normally found in potatoes are harmless, but if potatoes are stored in the light, the solanine in them can build up to toxic levels. Cooking does not destroy solanine, but because most of a potato's solanine is in the green layer that develops just beneath the skin, it can be peeled off, making the potato safe to eat. If the potato tastes bitter, however, discard it.

Environmental Contaminants in Foods

A justifiably high-ranking concern about our food supply is contamination of foods by environmental pollutants. As populations increase worldwide and nations become more industrialized, this problem looms ever larger.

The potential harmfulness of a contaminant depends in part on the extent to which it lingers in the environment or in the human body—how persistent it is. Some contaminants are short-lived, because microorganisms or agents such as sunlight or oxygen can break them down. Some contaminants linger in the body for only a short time, because the body can rapidly excrete them or metabolize them to harmless compounds. These contaminants present little cause for concern. Some contaminants, however, resist breakdown and interact with the body's systems without being metabolized or excreted. These can pass unchanged from food to the eater, and if the same food is eaten every day, larger and larger quantities of the contaminant may accumulate in the eater's body. Figure 11.3 shows how toxins accumulate at higher concentrations at each level of the food chain (bioaccumulation).

persistent: of a stubborn or enduring nature; with respect to food contaminants, the quality of persisting, rather than breaking down, in the bodies of animals and human beings.

bioaccumulation: the accumulation of toxins in living tissues at concentrations that increase at higher levels of the food chain.

Lead A contaminant of great concern in foods today is lead, a heavy metal that appears everywhere in the environment due to human industrial processes and easily finds its way into foods and water. Children are especially vulnerable to lead poisoning, as discussed in Chapter 13.

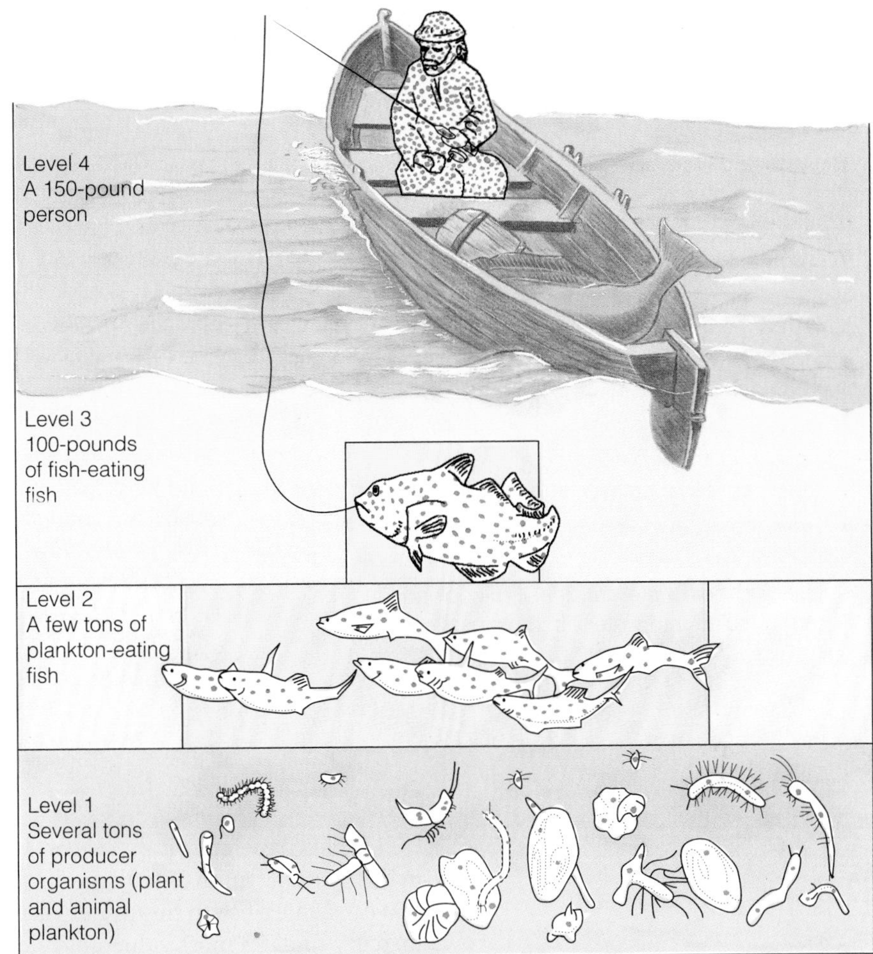
If none of the chemicals are lost along the way, people ultimately receive all of the toxic chemicals that were present in the original plants and plankton.
Contaminants become further concentrated in larger fish that eat the small fish from the lower part of the food chain.
Contaminants become more concentrated in small fish that eat the plants and plankton.
Plants and plankton at the bottom of the food chain become contaminated with toxic chemicals, such as methylmercury.
Level 4
A 150-pound person
Level 3
100-pounds of fish-eating fish
Level 2
A few tons of plankton-eating fish
Level 1
Several tons of producer organisms (plant and animal plankton)
Toxic chemicals

Figure 11.3
Bioaccumulation of Toxins in the Food Chain

People are exposed to lead in some types of gasoline, paint, newspaper ink, batteries, shotgun ammunition, hair dyes, and pesticides as well as in the air and water. Lead works its way through rainfall and soil into plants and animals that people use for food. Lead also enters food from food containers such as tin cans sealed with lead solder and old or imported pottery decorated with lead glazes. Even though the FDA has banned imported foods in cans sealed with lead solder, they still appear on the shelves of small ethnic grocery stores.[19] Similarly, herbal remedies imported from other countries may contain lead.[20] Pipelines soldered with lead also release it into drinking water, especially in public water systems of older communities. Lead is the nation's most significant contaminant. Exposures are highest in urban and industrial areas, near highways, and in slums where old leaded paint peels from the buildings. Consumers would be wise to take ultraconservative measures to protect themselves, and especially their small and unborn children, from lead poisoning; the box on the next page offers suggestions.

WWW **nsc.org/ehc/lead.htm**
Environmental Health Center Lead Program

cdc.gov/nceh/programs/lead/lead.htm
CDC Childhood Lead Poisoning Prevention Program

WWW **epa.gov/opptintr/lead**
Environmental Protection Agency Office of Pollution Prevention and Toxic Lead Programs

Mercury Mercury is another heavy metal that industry releases into the environment. Major contributors are industrial processes such as the burning of coal and the incineration of wastes. Mercury is readily metabolized by living organisms such as fish into the organic form, methylmercury, which is highly toxic, especially to growing infants and children. Displacing other metal ions in the body, mercury can cause blindness, deafness, loss of coordination, and other forms of impaired nervous system function. Higher doses of mercury can cause death.

How to

Avoid Lead Poisoning

- Once you have opened canned food, immediately move it to a lead-free storage container to prevent lead migration into the food.
- Do not store acidic foods or beverages (such as vinegar or orange juice) in ceramic dishware.
- Do not heat coffee in ceramic cups.
- Do not store alcoholic beverages in pewter or crystal decanters.
- Confirm with the publisher that your newspaper uses no lead in its ink before using the paper to wrap food, mulch garden plants, or add to your compost.
- Have the water in your home tested by a competent laboratory.
- Use only cold water for drinking, cooking, and making formula (cold water absorbs less lead).
- When water has been standing in lead-soldered pipes, flush the cold-water pipes until it is as cold as it can get (this may take as long as two minutes). If possible, replumb your home with plastic (PVC) piping, at least for drinking-water lines.
- If lead contamination of your water supply seems probable, obtain additional information and advice from the Environmental Protection Agency and your local public health agency.[a]

By taking these steps, parents can protect themselves and their children from this preventable danger. The National Lead Information Center provides two hotlines: call 1-800-LEAD-FYI (532-3394) for general information or 1-800-424-LEAD (424-5323) for specific questions.

[a]A cure that's worse than the ailment, *Science News* 135 (1989): 135.

Like lead, mercury is persistent and so bioaccumulates to the greatest extent in the tissues of animals high on the food chain, and particularly in the large fish that eat smaller fish in freshwater lakes and rivers and in the ocean. Also like lead, mercury finds its most vulnerable victims among the young. To protect yourself and your family against mercury poisoning, obey public health advisories about the fish caught in your local area, and do not eat any one kind of fish from one source too often.

Lead and mercury have been used here to show the potential seriousness of food and water contamination by persistent contaminants and the best defenses against them. Thousands of other contaminants exist, but in all cases two principles apply. First, remain alert to the possibility of contamination of foods, and keep an ear open for public health announcements and advice. Second, do not eat any one food too often; vary your diet. Switching from food to food is an effective defensive strategy against the accumulation of toxins in your body. This is the principle of dilution: each food eaten dilutes contaminants that may be present in other components of the diet.

Pesticide Residues in Foods

Pesticides are a special category of contaminants, different from those just discussed in that they are applied to foods on purpose and in a manner that is regulated and controlled. Their use is controversial. Pesticides do help to ensure the survival of some crops, but the damage they do to the environment is considerable and increasing. There is also some question about whether the widespread use of pesticides has really improved the overall yield of food. Pesticides and their use are monitored by government agencies.

Hazards of Pesticides Many pesticides are poisons that can damage all living cells, not just those of pests. Their use, therefore, is hazardous to those who work with them: manufacturers, field workers, truck drivers, and anyone else who is exposed to them. The danger of misuses or spills is ever present, and serious accidents may not always be prevented despite safety regulations and precautions regarding pesticide use.

Consumers of produce in the marketplace have reason to be concerned about pesticides, too, because they may still linger in the foods to which they were applied in the field. Risks to health from pesticide exposure are probably small for healthy adults, but children, elderly people, and people with weakened immune systems may be vulnerable to some types of pesticide poisoning.[21]

In some small gardens, handwork can take the place of pesticides.

Regulation of Pesticides The FDA and the Environmental Protection Agency (EPA) set legal limits on the types and amounts of pesticides permitted in foods. Pesticide tolerance levels are based on children. The government agencies set tolerance levels by first identifying foods that children commonly eat in large amounts and then consider the effects of pesticide exposure during each developmental stage.[22] Once tolerances are set, foods and livestock feeds are monitored for pesticides.

WWW.

fda.gov
Food and Drug Administration

epa.gov
Environmental Protection Agency

Pesticides from Other Countries Today, approximately 70 percent of the fruits and vegetables consumed in the United States are imported from other countries, which do not have the same pesticide regulations as the United States and Canada.[23] Indeed, a loophole in federal law allows U.S. companies to produce pesticides that are banned here and sell them in other countries. Those countries then use the banned pesticides on their foods and ship the foods back to U.S. consumers—the so-called circle of poison. Federal inspectors monitor incoming foods, however, and refuse to let them enter the country if they are found to contain illegal residues. The United States, Mexico, and Canada are currently working to establish a pesticide policy for all of North America.[24]* In addition, plans are being developed to allow the FDA to inspect foreign farms and ban any produce that does not meet U.S. food safety standards.[25]

Testing Procedures The FDA analyzes foods using methods that can detect residues well below tolerances. If the FDA finds violative levels, it can seize the products or order them destroyed. Four times a year, FDA surveyors buy over 200 foods in U.S. grocery stores in several cities, prepare the foods table ready, and then analyze them, not only for pesticides but for essential minerals, industrial chemicals, heavy metals, and radioactive materials. Food preparation often reduces levels of contaminants in foods, so the inspectors look for levels at least five times lower than permitted limits. Findings confirm that the bulk of the U.S. food supply is safe from excessive pesticide residues.[26]

A problem, though, is that budget restraints limit the FDA's testing capacity. The FDA does not sample *all* food shipments or test for *all* pesticides. Fewer than 700 inspectors and scientists test food samples from the multitude of farms, groves, docks, airports, warehouses, and processing plants the agency oversees. The FDA cannot (nor can it be expected to) guarantee 100 percent safety in the food supply. Instead, it sets conditions so that substances do not become a hazard and acts promptly when problems or suspicions arise.

Foods imported from other countries may contain residues of pesticides that are banned from use here.

Avoiding Pesticides Consumers, therefore, have some responsibility for their own health and safety with respect to pesticides. They can learn about the potential benefits and dangers of pesticide use, discuss regulations and alternatives with others, advise their government representatives about their findings, and apply pressure wherever it will help change inappropriate procedures. Meanwhile, people can minimize their risks by following the guidelines offered

*These pesticide agreements are under the auspices of the North American Free Trade Agreement (NAFTA).

Pesticide-free produce is a healthy choice.

organically grown crops: crops grown and processed according to USDA regulations defining the use of fertilizers, herbicides, insecticides, fungicides, preservatives, and other chemical ingredients.

WWW iquest.net/ofma
Organic Farmers Marketing Association

incidental food additives: substances that can get into food not through intentional introduction but as a result of contact with the food during growing, processing, packaging, storing, or some other stage before the food is consumed. The terms *accidental additives* and *indirect additives* mean the same thing.

dioxins: toxic organic compounds containing chlorine, arising in industry as (among other things) by-products of the bleaching process.

in the accompanying box. In addition to the suggestions in the box, consumers can buy fresh foods grown locally, especially when they can confirm that the produce has been grown using responsible methods.

Organically Grown Crops Some farmers are turning to organic farming methods as an alternative to heavy pesticide use.[27] These methods are especially useful for farmers who want to produce and market organically grown crops.[28] USDA regulations for organically grown crops are currently being developed and are expected to be in place by spring of 1999. To market products as organic and to be certified, farmers must use methods that meet specific criteria. Most states have organic certification agencies, although exact guidelines vary from state to state. Agricultural products claiming to have been grown organically must meet USDA standards and bear the USDA seal on their labels.

In addition to benefiting from reduced costs of farming, increased soil quality, and decreased chemical impact on the environment, farmers of organic crops stand to increase their share of the market. Many consumers are willing to pay more for organic foods, and sales are increasing in record numbers.

Implied in the definition of organic is that organic products are healthier for consumers than those grown using other methods, which may not be the case. Using unprocessed animal manure as an organic fertilizer, for example, may transmit bacteria, such as *E. coli,* to human beings. Organic and conventional methods both have advantages and disadvantages, and consumers must remain informed.

In short, pesticides can safely improve crop yields when used according to regulations, but can also be hazardous when used inappropriately. The FDA tests both domestic and imported foods for pesticide residues in the fields and in market basket surveys of foods prepared table ready. Consumers can minimize their ingestion of pesticide residues on foods by following the guidelines in the accompanying box. Alternative farming methods may allow farmers to grow crops with few or no pesticides.

Incidental Food Additives

Indirect or incidental additives are really contaminants that find their way into food as a result of some phase of production, processing, storage, or packaging. Examples of incidental additives include tiny bits of plastic, glass, paper, tin, and other substances from packages, as well as chemicals from processing, such as the solvent used to decaffeinate coffee.

Some microwave products are sold in "active packaging" that participates in cooking the food. Pizza, for example, may rest on a cardboard pan coated with a thin film of metal that absorbs microwave energy and may heat up to 500°F. When exposed to the intense heat, some particles of the packaging components migrate into the food.[29] Regular microwave packages heat up less, but particles still migrate, and the materials from both kinds of packaging are under study to determine their safety for consumption. Until more is known, a wise choice is to use only glass or ceramic containers designed for microwaving and to avoid reusing disposable containers, such as margarine tubs, for heating foods.

Coffee filters, milk cartons, paper plates, and frozen food packages can all be made of bleached paper and so can contaminate foods with trace amounts of compounds known as dioxins. Dioxins form during the chlorination step in making bleached paper. Dioxins can migrate into foods that come in contact with bleached paper, but the amounts entering food are infinitesimally small—one part per trillion, or the equivalent of one second in 32,000 years. Such amounts do not appear to present a health risk to people, and drinking milk from bleached cartons appears to be safe. Dioxins are persistent, however, and they leach into the environment by way of both paper-mill effluent and

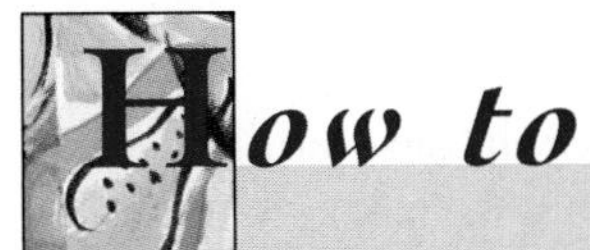

How to

Rid Foods of Pesticide Residues

No matter where your food comes from, it is wise to remove parts that might contain pesticide residues:

- Trim the fat from meat and remove the skin from poultry and fish; discard fats and oils in broths and pan drippings. Avoid fish oil capsules. (Pesticide residues concentrate in the animal's fat.)
- Wash fresh produce in warm water. Use a scrub brush, and rinse thoroughly.
- Use a knife to peel an orange or grapefruit; do not bite into the peel.
- Discard the outer leaves of leafy vegetables such as cabbage and lettuce.
- Peel waxed fruit and vegetables; waxes don't wash off and can seal in pesticide residues.
- Peel vegetables such as carrots and fruits such as apples when appropriate. (Peeling removes pesticides that remain in or on the peel, but also removes fibers, vitamins, and minerals.)

Information about pesticides is available from the EPA's national pesticide hotline (1-800-858-PEST) anytime day or night, 365 days a year.

discarded paper products in landfills. They therefore bioaccumulate as do other contaminants described earlier (see Figure 11.3), becoming more and more concentrated in land, water, and animals until they build up to hazardous levels.

Incidental additives sometimes find their way into foods, but adverse effects are rare. These additives are well regulated. All food packagers are required to perform specific tests to discover whether materials from packages are migrating into foods; if they are, their safety must be confirmed by strict procedures similar to those governing intentional additives, discussed next.

Here's a way to tell if glass or other containers are made of microwave-safe materials. Microwave the empty container for one minute and carefully touch it.

- *Warm = unsafe for microwave.*
- *Lukewarm = safe for short reheating use.*
- *Cool = safe for long microwave cooking times.*

Regulating Intentional Food Additives

Of all consumer concerns about food safety, preventing food poisoning is the most important. Next in importance are natural toxins, environmental contaminants, and pesticides. Last are additives. Manufacturers use food additives to give foods desirable characteristics: color, flavor, texture, stability, higher nutrient content, or resistance to spoilage. Consumers see additives mentioned on labels and wonder or worry about them, but additives are of very little concern compared to the other aspects of foods already covered. Nevertheless, for the sake of completeness, this section will describe the regulations and procedures governing food additives, and the next section will look at some individual additives.

additives: substances that are not normally consumed as foods by themselves, but are added to foods.

Regulations Governing Additives To get permission to use new additives in food products, manufacturers have to go through special procedures that can take many years. The manufacturer must test each new additive to satisfy the FDA of the following:

- The additive is effective (it does what it is supposed to do).
- It can be detected and measured in the final food product.

Then the manufacturer must study the effects of feeding the additive in large doses to animals under strictly controlled conditions to prove that:

- It is safe (it does not cause cancer, birth defects, or other injury).

Finally, the manufacturer must submit all test results to the FDA. Public hearings follow, where consumers are invited to participate and experts present testimony for and against granting permission to use the additive. Thus consumers' rights and responsibilities are written into the provisions for deeming additives safe.

When the FDA approves an additive, it writes a regulation stating in what amounts, for what purposes, and in what foods the additive may be used. No additives are permanently approved; all are periodically reviewed.

GRAS (generally recognized as safe) list: a list of food additives, established by the FDA in 1958, that had long been in use and were believed safe.

GRAS List Additives Many substances such as salt, sugar, caffeine, and herbs were exempted from complying with this procedure at the time it was first instituted because they had been used a long time and their use entailed no known hazards. Some 700 substances in all were put on the generally recognized as safe (GRAS) list. However, when substantial scientific evidence or public outcry has questioned the safety of a substance on the GRAS list, its safety has been reevaluated. All substances about which any legitimate question was raised have been removed or reclassified.

toxicity: the ability of a substance to harm living organisms. All substances are toxic if used in high enough concentrations.

hazard: the ability of a substance to produce injury under the conditions of its use.

Toxicity versus Hazard An important distinction governs decisions about an additive's safety—the distinction between toxicity as a property of substances and hazard associated with substances. Toxicity is a general property of all substances; hazard is the capacity of a chemical to produce injury *under conditions of its use.* All substances can be toxic at some level of consumption, but they are called hazardous only if they are actually consumed in sufficiently large quantities to cause harm. An additive is not considered to be a hazard if some immense amount that people never consume is toxic. The additive is a hazard only if it is toxic under the conditions of its actual use. A food additive is supposed to have a wide margin of safety.

margin of safety: as used when speaking of food additives, a zone between the concentration normally used and that at which a hazard exists. For common table salt, for example, the margin of safety is ⅕ (five times the concentration normally used would be hazardous).

Testing Procedures Most additives that involve risk are allowed in foods only at levels 100 times below those at which the risk is still known to be zero. Experiments to determine the extent of risk involve feeding test animals the substance at different concentrations throughout their lifetimes. The additive is then permitted in foods at 1/100 the level that causes no harmful effect whatever in the animals. In many foods, *naturally* occurring toxins appear at levels that bring their margins of safety closer to ⅒. Even nutrients, as you have seen, involve risks at high dosage levels.

The margin-of-safety concept also applies to nutrients when they are used as additives. Iodine has been added to salt to prevent iodine deficiency, but it has to be added with care because it is a deadly poison in excess. Similarly, iron has been added to refined bread and other grains (enrichment) and has doubtless helped prevent many cases of iron-deficiency anemia in women and children who are prone to that disease. But the addition of too much iron could put men (who usually have enough iron in their bodies) at risk for iron overload. The Tolerable Upper Intake Level has to be remembered.

Benefits versus Risks Most additives used in foods are there because they offer benefits that outweigh their risks or that make the risks worth taking. In the case of color additives that only enhance the appearance of foods and do not improve their health value or safety, no amount of risk may be deemed worth taking. Only 10 of an original 80 synthetic color additives are still approved by the FDA for use in foods, and screening of these continues.[30]

Furthermore, manufacturers must use only the amounts of additives necessary to get the needed effects, not more. Additives must also *not* be used:

- To disguise faulty or inferior products.
- To deceive the consumer.
- Where they significantly destroy nutrients.
- Where their effects can be achieved by economical, sound manufacturing processes.

The regulations in force governing the management of intentional additives are well conceived and have been effective on the whole. Funding shortages limit the capabilities of watchdog agencies such as the FDA, however, and some mistakes and cases of false reporting are bound to slip by.

Two long-used preservatives.

Some Intentional Additives

The next few paragraphs focus on a few individual food additives—notably, those that have received the most negative publicity because people ask questions about them most often. The order is alphabetical; it does not imply an order of importance.

Antimicrobial Agents Foods can spoil in two ways: one dangerous, one not. The dangerous way is by becoming hazardous to health; the other way is by losing their flavor and attractiveness. An example of the dangerous way: bacteria, yeasts, and molds and other fungi growing in foods can cause food poisoning. Preservatives known as antimicrobial agents protect foods from these microbes.

The best-known, most widely used antimicrobial agents are two common substances—salt and sugar. Salt preserves meat and fish; sugar preserves canned and frozen fruits, jams, and jellies. Both salt and sugar work by withdrawing water from the food; microbes cannot grow without water. Today, other additives such as potassium sorbate and sodium propionate are also used to extend the shelf life of baked goods, cheese, beverages, mayonnaise, margarine, and many other products.

Another group of antimicrobial agents, the nitrites, are added to foods for three main purposes: to preserve their color (especially the pink color of hot dogs and other cured meats); to enhance their flavor by inhibiting rancidity (especially in cured meats); and to protect against bacterial growth. In particular, nitrites prevent the growth of the botulinum bacterium that produces the deadly toxin described earlier.

Nitrites clearly perform important jobs, but they have been the object of controversy because in the human body they can be converted to nitrosamines, which cause cancer in animals. Some cured meats are available without nitrites, but reducing nitrites consumed in meats would hardly make a difference in a person's overall exposure to nitrosamine-related compounds. For example, the average cigarette smoker inhales 100 times the nitrosamines that the average bacon eater ingests. Likewise, a beer drinker imbibes up to roughly five times the amount that the bacon eater receives. Cosmetics deliver via absorption through the skin about twice the amount delivered from bacon. Even the air inside automobiles delivers measurable amounts of nitrosamines.

preservatives: antimicrobial agents, antioxidants, chelating agents, radiation, and other additives that retard spoilage or preserve desired qualities, such as softness in baked goods.

antimicrobial agents: substances used as food additives that prevent growth of illness-causing microorganisms in foods.

nitrites: salts added to food to prevent botulism. An example is sodium nitrite.

nitrosamines (nigh-TROHS-uh-meens): derivatives of nitrites that may form when nitrites combine with amines.

Antioxidants The other way foods can go bad is by undergoing changes in color and flavor caused by exposure to oxygen in the air (oxidation). Often these changes involve little hazard to health, but they damage the food's appearance, taste, and nutritional quality. Familiar examples of these changes are sliced apples or potatoes turning brown and oils going rancid. Antioxidant preservatives protect foods from this kind of spoilage. A total of 27 antioxidants are approved for use in foods. Vitamin C (ascorbate) and vitamin E (tocopherol) are among them.

The sulfites are another group of antioxidants. They are used to prevent oxidation in many processed foods, alcoholic beverages (especially wine), and drugs. They used to be popular with restaurant owners for use on salad bars because they keep raw fruits and vegetables looking fresh, but some people experience allergic reactions to the sulfites—reactions that are sometimes dangerous and, for a few, deadly. The FDA now prohibits sulfite use on foods intended to be consumed raw, with the exception of grapes, and it requires sulfite-containing foods and drugs to include a warning on their labels. Restaurants, however, are

antioxidants: defined in Chapter 7 as compounds that protect other compounds from oxygen by reacting with oxygen themselves. Antioxidants are used to prevent rancidity of fats in foods and other damage to food caused by oxygen. Examples are vitamins E and C, BHA, BHT, propyl gallate, and sulfites.

sulfites: salts containing sulfur that are added to fresh and frozen fruits and vegetables to prevent changes in color and texture due to oxidation. Sulfites appear on food labels as:

- *Sulfur dioxide.*
- *Sodium sulfite.*
- *Sodium bisulfite.*
- *Potassium bisulfite.*
- *Sodium metabisulfite.*
- *Potassium metabisulfite.*

Raw grapes may legally be treated with sulfites. Wash grapes thoroughly before eating.

artificial colors: certified food colors, added to enhance appearance (*certified* means approved by the FDA). Common examples of color additives:

- *Carotenoids.*
- *Blue #1 and #2 (brilliant blue and ingotine).*
- *Green #3 (fast green).*
- *Red #40 and #3 (allura red and erythrosine).*
- *Yellow #5 and #6 (tartrazine and sunset yellow).*

Foods containing tartrazine:

- *Orange drinks (Tang, Daybreak, Awake).*
- *Gatorade (lime flavored).*
- *Gelatin desserts (Jell-O, Royal).*
- *Golden Blend Italian dressing (Kraft).*
- *Some cake mixes and icings (Duncan Hines, Pillsbury, Cake Mate).*
- *Imitation banana or pineapple extract (McCormick).*
- *Seasoning salt (French's).*
- *Macaroni and cheese dinner (Kraft).*
- *"Cheez" curls and balls (Planter's).*
- *Fruit chews (Skittles).*
- *Butterscotch squares and candy corn (Brach's).*

artificial flavors, flavor enhancers: chemicals that mimic natural flavors and those that enhance flavor.

MSG symptom complex: an acute and temporary intolerance reaction that may occur after eating MSG (monosodium glutamate). Symptoms include burning sensations, chest and facial flushing and pain, and throbbing headaches.

not required to disclose whether they have used sulfites in food preparation. Therefore, concerned consumers must ask if sulfites have been used.[31] For most people, sulfites do not pose a hazard in the amounts used in products.

Artificial Colors As mentioned, only about ten artificial colors are still on the GRAS list, a highly select group that has survived considerable screening. They are among the most intensively investigated of all additives. In fact, they are much better known than the *natural* pigments of plants, and the limits on the safety of their use can be stated with greater certainty.

Still, the food colors have been more heavily criticized than almost any other group of additives. The reason, simply stated, is that they only make foods attractive, whereas other additives, such as preservatives, make foods safe. Hence, with food colors, we can afford to require that their use entail no risk, whereas with other additives we may have to compromise between the risks of using them and the risks of *not* using them.

The food color tartrazine (yellow dye number 5) causes an allergic reaction in susceptible people. Symptoms include hives, itching, and nasal congestion, sometimes severe enough to require medical treatment. It is not a common problem; only 1 or 2 in 10,000 individuals may have the reaction. Still, that is more than 20,000 individuals in the nation as a whole. These people rightly demand to know what foods contain the dye so that they can avoid it. It is not enough to avoid yellow-colored foods because tartrazine is used to confer turquoise, green, and maroon colors in foods and drugs as well. Legislation is now in force requiring that tartrazine be listed on all labels of foods that contain it.

Artificial Flavors and Flavor Enhancers While only a few artificial colors are currently permitted in foods, close to 2000 artificial flavors and flavor enhancers are approved, making them the largest single group of food additives. One of the best-known members of this group is monosodium glutamate, or MSG (trade name, Accent)—the monosodium salt of the amino acid glutamic acid. MSG is used widely in restaurants, especially Asian restaurants, as a flavor enhancer. Research indicates that in addition to enhancing other flavors, MSG may itself possess a basic taste independent of the well-known sweet, salty, bitter, and sour tastes.[32]*

MSG has received publicity because it may produce an adverse reaction called the MSG symptom complex in 1 to 2 percent of the population.[33] Symptoms include burning sensations, chest and facial flushing or pain, and throbbing headaches. MSG has been investigated extensively enough to be deemed safe for adults to use (except people who react adversely to it, of course), but it is kept out of foods for infants because very large doses have been shown to destroy brain cells in developing mice. Infants have not yet developed the capacity to fully exclude such substances from their brains and so are more sensitive to them. Food labels require ingredient lists to itemize all additives, including MSG.

Texture and Stability Ingredients may be added to foods during processing to maintain emulsions, foams, or suspensions or to lend a desirable thick consistency to foods. Dextrins (short chains of glucose formed as a breakdown product of starch), starch, and pectin are examples. Gums, such as carrageenan, guar, locust bean, agar, and gum arabic, are also added for thickening and stabilizing.

Nutrient Additives Another class of additives includes nutrients added to improve or to maintain the nutritional value of foods.[34] Included among nutrient additives are the nutrients added to refined grains to enrich them; the iodine added to salt; vitamins A and D added to dairy products; and the nutrients added

*The taste produced by MSG is termed *umami.*

to fortified breakfast cereals. Nutrients are sometimes also added for other purposes. The use of vitamins C and E as antioxidants has already been mentioned. Beta-carotene may be added as a selling point because consumers, who have heard media reports of studies linking beta-carotene to reduced risks of diseases, are buying more products that contain it.

Common examples of nutrient additives:

- *Thiamin, niacin, riboflavin, folate, and iron in grain products.*
- *Iodine in salt.*
- *Vitamins A and D in milk.*
- *Vitamin C in fruit drinks.*

As this section has shown, no two additives are alike, and therefore generalizations about them are meaningless. No valid statement can be made that applies to the 3000-odd different substances commonly added to foods. Questions about which additives are safe, under what conditions of use, have to be asked and answered on an item-by-item basis.

The U.S. food supply is well monitored and well protected against hazards that might threaten people's health. Provided that consumers apply common sense in selecting and preparing their foods, they can enjoy the great blessing of an abundant and safe food supply.

Food Biotechnology

Hope for the future purity of foods comes on the crest of new advances in biotechnology. Biotechnology promises to produce greater crop yields, leaner meats, longer shelf lives, better nutrient composition, and fewer pesticides. Overall, biotechnology offers opportunities to enhance the quality, nutritional value, and variety of foods.[35]

biotechnology: the use of biological systems or organisms to create or modify products; also called **biogenetic engineering.**

Genetic Engineering

foodbiotech.org
Food Biotechnology Communications Network

For centuries farmers have manipulated the genetics of plants and animals to shape the characteristics of their crops and livestock. Consider corn, for example. Wild, native corn bears only two or three kernels on a cob, but many years of patient selective breeding have produced the large, full, sweet ears people enjoy today, and many types of wild corn are now all but extinct. Half of the increases in U.S. crop yields in the twentieth century are due to such genetic improvements; the use of irrigation, fertilizers, and pesticides has also contributed. Farmers still use selective breeding to provide consumers with low-fat meats, high-yield grains, and a seemingly endless variety of fruits and vegetables.

Scientists can now speed up the process of genetic change through biotechnology. Farmers need no longer wait patiently for breeding to yield improved crops and animals, nor must they even respect natural lines of reproduction among species. Laboratory scientists can now select desirable traits from any of a number of species and insert those traits into the genetic material of crops and animals.

Today's large, full, sweet ears bear little resemblance to the original wild, native corn with its sparse two or three kernels to a cob.

Among the new products of biotechnology are tomatoes that stay fresh much longer than others and so promise less waste and higher profits. Normally, tomatoes produce a protein that softens them after they have been picked. Scientists introduce into a tomato plant a gene that is a mirror image of the one that codes for the "softening" enzyme. This gene fastens itself to the RNA of the native gene and blocks its action. A vine-ripe tomato with this special gene rots more slowly than a normal tomato, allowing growers to harvest at the most flavorful and nutritious red stage. The tomatoes will still last much longer during shipping and marketing than regular tomatoes harvested when green.

Similarly, soybeans may be implanted with a gene that will upgrade soy protein to a quality approaching that of milk. Corn may be modified to contain lysine and tryptophan, its two limiting amino acids.[36] Fats and oils with a predetermined fatty acid composition may be possible within the decade.[37] Crops that produce their own insecticides upon receiving genes from bacteria may render pesticides unnecessary. Shrimp may soon fight diseases with genetic

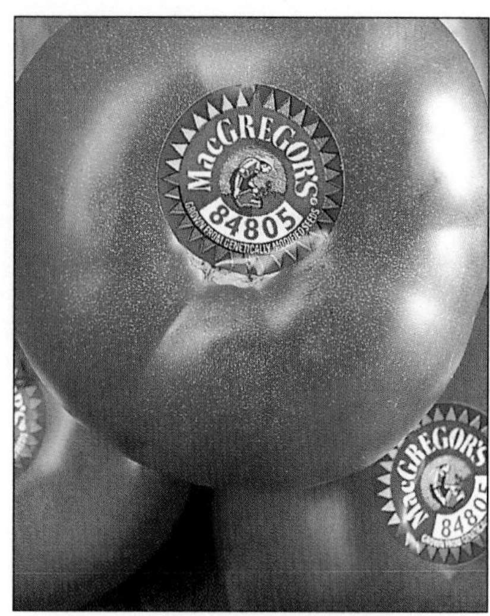

Tomatoes grown from genetically modified seeds stay fresh longer.

ammunition borrowed from sea urchins. Livestock may receive growth-promoting hormones from bacteria. The possibilities seem unlimited, and though they sound fantastic, many are waiting on laboratory shelves for the time when they will be fully employed in agriculture.

While food industrialists hail biotechnology as a miracle, some other people fear that tampering with genetics may change organisms in ways not yet fully understood, even by the scientists who developed the techniques. They wonder what unknown changes take place when the genes of living things are manipulated and what the long-term consequences might be.

Regulations and Labeling

The FDA has taken the position that foods produced through biotechnology are not substantially different from others and require no special safety testing or labeling.[38] A product, such as the tomato described earlier, need not be tested because its new genes *prevent* synthesis of a protein and add nothing but a tiny fragment of genetic material. On the other hand, any substances introduced into a food by way of bioengineering must meet the same safety standards applied to all additives.[39] Some people object to genetic tampering and want labels to help them identify "old-fashioned" tomatoes. They may not realize that most foods available today have already been altered genetically by selective breeding. The new vegetable broccoflower, a product of sophisticated cross-breeding of broccoli with cauliflower, met no testing or approval barriers on its way to the dinner plate. Only after the vegetable became popular with consumers did scientists study its nutrient contents (see Appendix A for their findings).

Scientists are continuing to study the effects not only of biotechnology but of all sorts of new food processing techniques. Their efforts to enhance food production will help meet the challenge of feeding an ever increasing world population.[40]

Self Check

1. Two examples of food processing used to prevent microbial growth are:
   a. canning and air drying.
   b. sun drying and chilling.
   c. adding sulfites and freezing.
   d. irradiation and pasteurization.
2. A method the food industry uses to identify points of contamination and implement controls to ensure food safety is called:
   a. margin of safety.
   b. generally recognized as safe.
   c. North American Free Trade Agreement.
   d. Hazard Analysis Critical Control Points.
3. The temperature danger zone for foods ranges from:
   a. −20°F to 120°F.
   b. 0°F to 100°F.
   c. 20°F to 120°F.
   d. 40°F to 140°F.
4. Examples of foods that frequently cause food-borne illness are:
   a. canned foods.
   b. steaming-hot foods.
   c. fresh fruits and vegetables.
   d. raw milk, seafood, meat, and eggs.
5. Irradiation can help improve the food supply by:
   a. minimizing the use of preservatives.
   b. improving the nutrient content of foods.
   c. cooking foods quickly.
   d. killing microorganisms.
6. A natural toxin present in foods that can cause illness when consumed in excess is:
   a. alar.
   b. solanine.
   c. botulin toxin.
   d. *Salmonella.*
7. Additives approved for use must be:
   a. intentional and undetectable in the final product.
   b. indirect and quality enhancing.
   c. intentional and on the GRAS list.
   d. detectable and measurable in the final product.
8. Common antimicrobial additives include:
   a. salt and nitrites.
   b. carrageenan and MSG.
   c. dioxins and sulfites.
   d. vitamin C and vitamin E.

9. Common antioxidants include:
   a. BHA and BHT.
   b. tartrazine and MSG.
   c. sugar and vitamin E.
   d. nitrosamines and salt.
10. Biotechnological advances that have improved the food supply include:
    a. potatoes with solanine and cabbage without goitrogens.
    b. seafood without methylmercury and onions without an odor.
    c. larger cattle with a higher fat content and rennin produced by sheep.
    d. tomatoes with a longer shelf life and soybeans with a higher-quality protein.

Answers to these questions appear in Appendix H.

Notes

1. M. V. Polo, M. J. Lagarda, and R. Farré, The effect of freezing on mineral element content of vegetables, *Journal of Food Composition and Analysis* 5 (1992): 77–78.

2. J. M. Jones, *Food Safety* (St. Paul, Minn.: Eagan Press, 1993), pp. 182–184.

3. B. Ooraikul and M. E. Stiles, Introduction: Review of the development of modified atmosphere packaging, in *Modified Atmosphere Packaging of Food* (New York: Ellis Horwood, 1991), pp. 1–17.

4. H. Aschkenasy, Working knowledge, *Scientific American,* September 1996, p. 184.

5. S. L. Nightingale, From the Food and Drug Administration: National Food Safety Initiative, *Journal of the American Medical Association* 277 (1997): 1664.

6. C. J. Lackey, Oil, herb, and garlic flavored, http://www.foodsafety.org, site visited on February 5, 1998.

7. Lackey, 1998.

8. R. V. Tauxe and J. M. Hughes, International investigations of outbreaks of foodborne disease: Public health responds to the globalization of food, *British Medical Journal* 313 (1996): 1093–1094; T. W. Hennessey and coauthors, A national outbreak of *Salmonella enteritidis* infections from ice cream, *New England Journal of Medicine* 334 (1996); 1281–1286.

9. Centers for Disease Control and Prevention, Surveillance for foodborne disease outbreaks, United States, 1988–1992, *Morbidity and Mortality Weekly Report CDC Surveillance Survey* 45 (1996): 1–66.

10. Nightingale, 1997; FDA announces a strategy to increase safety of fresh juices, *Nutrition Today* 32 (1997): 190; 1997 food code available, *FDA Consumer,* September/October 1997, pp. 6–9.

11. FSIS Pathogen Reduction/HACCP, *Federal Register,* July 6, 1996.

12. Raw oyster risk for alcoholics, *FDA Consumer,* May 1996, pp. 23–25.

13. Seafood safety: Highlights of the Executive Summary of the 1991 Report by the Committee on Evaluation of the Safety of Fishery Products of the Food and Nutrition Board, Institute of Medicine, National Academy of Sciences, *Nutrition Reviews* 49 (1991): 357–363.

14. Outbreaks of Cyclosporiasis—United States, 1997, *Morbidity and Mortality Weekly Report* 46 (1997): 461–462; *Vibrio vulnificus* infections associated with eating raw oysters—Los Angeles, 1996, *Journal of the American Medical Association* 276 (1996): 937–938; J. W. Besser-Wiek and coauthors, Foodborne outbreak of diarrheal illness associated with *Cryptosporidium parvum*—Minnesota, 1995, *Morbidity and Mortality Weekly Report* 45 (1996): 783–784.

15. New bottled water standards, *FDA Consumer,* April 1996, p. 2; V. Lambert, Bottled water: New Trends, new rules, *FDA Consumer,* June 1993, pp. 8–11.

16. Position of The American Dietetic Association: Food irradiation, *Journal of the American Dietetic Association* 96 (1996): 69–72.

17. S. L. Nightingale, Irradiation of meat approved for pathogen control, *Journal of the American Medical Association* 279 (1998): 9; M. T. Olsterholm, Cyclosporiasis and raspberries—Lessons for the future, *New England Journal of Medicine* 336 (1997): 1597–1598.

18. Olsterholm, 1997; Position of The American Dietetic Association: Food and water safety, *Journal of the American Dietetic Association* 97 (1997): 184–189.

19. Sixth-grader opens lid for FDA investigation, *FDA Consumer,* September/October 1997, pp. 34–35.

20. S. B. Markowitz and coauthors, Lead poisoning due to *Hai Ge Fen:* The porphyrin content of individual erythrocytes, *Journal of the American Medical Association* 271 (1994): 932–934.

21. National Academy of Sciences Committee, as quoted by J. Raloff and D. Pendick, Pesticides in produce may threaten kids, *Science News* 144 (1993): 4–5.

22. Food Quality Protection Act of 1996, Public Law No. 104-170, 110 Statute 1489; C. Marwick, New focus on children's environmental health, *Journal of the American Medical Association* 277 (1997): 871–872.

23. Olsterholm, 1997.

24. Website http://vm.cfsan.fda.gov/~dms/pes96rep.html, visited on March 5, 1998.

25. C. Marwick, "Fresh Produce Initiative" for imports, *Journal of the American Medical Association* 278 (1997): 1481.

26. FDA/CFSAN Pesticide Program, Residue Monitor, http://www.cfsan.gov/list.html, site visited on March 5, 1998; Position of The American Dietetic Association, 1997.

27. P. Kurtzweil, Can your kitchen pass the food safety test? *FDA Consumer,* October 1995, pp. 14–18.

28. Website http://www.ams.usda.gov/nop, visited on March 5, 1998.

29. Kurtzweil, 1995.

30. I. D. Wolf, Critical issues in food safety, 1991–2000, *Food Technology,* January 1992, pp. 64–70.

31. R. Papazian, Sulfites: Safe for most, dangerous for some, *FDA Consumer,* December 1996, pp. 11–14.

32. M. Naim and coauthors, Interaction of MSG taste with nutrition: Perspectives in consummatory behavior and digestion, *Physiology and Behavior* 49 (1991): 1019–1024.

33. D. J. Raiten, J. M. Talbot, and K. D. Fisher, Executive summary from the report analysis of adverse reactions to monosodium glutamate (MSG), *Journal of Nutrition* 125 (1995): 2892S–2906S.

34. W. Mertz, Food fortification in the United States, *Nutrition Reviews* 55 (1997): 44–49.

35. Position of The American Dietetic Association: Biotechnology and the future of food, *Journal of the American Dietetic Association* 95 (1995): 1429–1432.

36. B. A. Larkins, C. R. Lending, and J. C. Wallace, Modification of maize-seed-protein quality, *American Journal of Clinical Nutrition* 58 (1993): 264S–269S.

37. C. R. Somerville, Future prospects for genetic modification of the composition of edible oils from higher plants, *American Journal of Clinical Nutrition* 58 (1993): 270S–275S.

38. J. Henkel, Genetic engineering: Fast forwarding to future foods, *FDA Consumer,* April 1997, pp. 6–11.

39. Biotechnology of food: Background information from the FDA, *Nutrition Today,* July/August 1994, pp. 19–20.

40. T. D. Etherton, The impact of biotechnology on animal agriculture and the consumer, *Nutrition Today,* July/August 1994, pp. 12–18.

CHAPTER 11

Nutrition in Practice

▪ Environmentally Conscious Foodways ▪

The preceding chapter viewed foods from many angles and suggested ways of achieving many dietary goals: nutritional adequacy, protection from food poisoning, avoidance of contaminants, and other elements of safety. People who follow the advice given can be satisfied that they have good answers to the questions "How can I get the best health benefits from my foods?" and "How can I keep my foods safe?"

Some people want to achieve another goal when they shop for foods and cook them. They recognize that they will spend thousands of dollars on foods and that they will cook thousands of meals in a lifetime, and they perceive that their money and actions exert effects in the world outside their own personal lives. Increasingly, people today are asking, "What are the environmental impacts of my food choices? How do my actions in the kitchen affect the environment?" They want to make environmentally responsible choices.

WWW.

unep.org
United Nations Environment Programme

cnie.org/nle
National Library for the Environment

magic.iclei.org
International Council for Local Environmental Initiatives

What kinds of environmental impacts do people's food choices have?

Among the global resources involved in producing food are irrigation water, fertilizers, pesticides, fuel, and land and fisheries. And in the U.S. market, tons of packaging materials and a massive transportation network burning immense quantities of fossil fuel are used to convey foods to consumers (see the glossary on p. 297 for fossil fuels). Each truckload of food produced in this country travels, on the average, 1300 miles to reach the market. It costs 800 kcalories in fuel to make a can of diet soda that contains 1 kcalorie of food energy, and more water is used to make the can than to make the soda.[1] An appetizer of shrimp cocktail may contain 4 ounces of shrimp, but to net those shrimp, the shrimpers had to kill 2½ pounds of young fish that otherwise could have grown up to provide food.[2]

More environmentally benign choices are available. In place of vegetables shipped in from far away, people might choose to eat vegetables grown in their own home states, at least during the growing seasons. In place of several sodas in aluminum cans, a soda drinker might use one large recyclable bottle. In place of shrimp cocktail, the diner might choose a crab salad or a few oysters harvested without killing other sea creatures. Environmentally responsible food choices can also be made in the realms of food shopping, cooking, and cleanup.

What are some examples of environmentally responsible food shopping?

Food shopping involves going to the store, selecting foods once there, choosing among the packages in which those foods are sold, and choosing bags in which to carry the foods home. All of these actions exert impacts on the environment, and consumers can choose to minimize those impacts. Consider the shopping trips first. The environmentally conscious shopper knows that motor vehicles are the world's single largest source of air pollution and so tries to minimize car mileage spent on trips to and from the store. Strategies are to shop nearby and to shop only once a week. To make it possible to shop for a week's meals at a time, a shopper can plan to buy foods with various shelf lives and to eat the most perishable ones first. For example, buy lettuce, cabbage, squash, and carrots. Use up the lettuce first, then the squash. The carrots and cabbage keep longer, so eat these later. Buy fruits of differing ripeness—for example, six bananas: two ripe, two nearly ripe, and two green. Use the ripe ones right

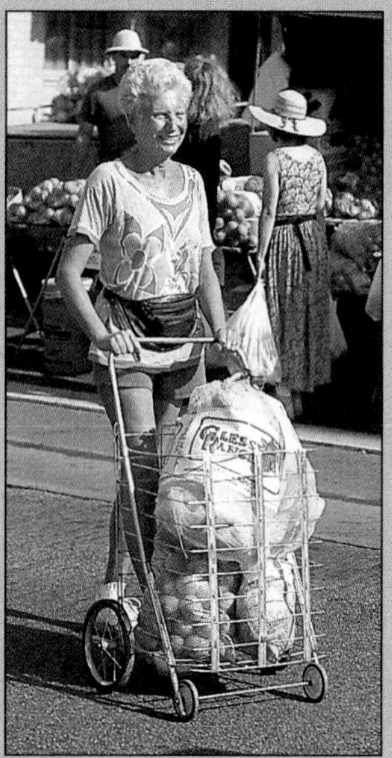

Shopping without a car can be a pleasure, if you can afford the time.

away and the others as they become ripe. On first arriving home, cook the meats for the early meals; portion out the rest, and freeze the bread and meat that won't be needed until midweek. These strategies save time and money as well as fossil fuels.

Give some examples of environmentally responsible food choices.

Perhaps the advice most frequently given to save fuel and resources is to "eat low on the food chain." It takes much less land and fuel, and costs much less in pollution, to produce most plant foods than to produce most meats (see Figure NP11.1). Following this advice benefits nutritional health, too: recall from Chapter 1 that the Daily Food Guide recommends that adults eat 11 or more servings of plant foods (especially vegetables and grains) daily and only 4 or 5 servings of milk products and meats combined. This is also a good strategy for avoiding food contaminants.

Another guideline is to eat foods that are processed as little as possible. Figure NP11.2 on p. 296 shows how much more energy it takes to produce canned or frozen corn than fresh corn. Energy costs mean fuel costs, and fuel costs mean pollution.

Other environmentally aware food-shopping practices include buying products whose production benefits the land, or at least harms it minimally, and boycotting products whose production damages the land. Buying food produced locally by farmers known to use a minimum of fertilizers and pesticides and not to waste water is often a positive choice. It also makes sense to avoid buying canned beef products of any kind, including soups, chili, stews, corned beef, and even beef-flavored pet food. Some of these foods come at the expense of cleared rain forest land. About 200 square feet of rain forest are lost *permanently* for each *pound* of beef produced from cattle raised on the cleared land. Rain forest beef is not labeled, so the only way consumers can be sure they are not buying it is to buy no canned beef at all.

Reusable bags require the fewest resources.

The task for consumers is to become aware of their impacts and to find sustainable ways of doing things. These are ways of living that use up resources at a rate that nature, forestry, or agriculture can replace. They are ways of living that pollute the earth at a rate that nature or human cleanup efforts can keep up with.

Do the packages foods come in exert impacts on the environment?

Yes. It costs energy and resources to make these packages, and it may cost land or pollution to dispose of them. In general, what is best for the environment is no packages; next best are minimal, reusable, or recyclable ones. Even grocery bags represent a huge drain on energy and resources, and many consumers are demanding alternatives to throwaway bags. Many shoppers prefer to carry reusable shopping bags to the store, or they ask for recyclable plastic bags—and then take care to recycle them. The third choice would be paper bags, and last would be nonrecyclable plastic.

What are the best ways to cook food from the environmental standpoint?

Fast cooking saves fuel and so pollutes less. Asian meals exemplify this principle: they are made of precut, bite-sized pieces of food, stir-fried fast in small amounts of oil. This cooking style both saves energy and preserves nutrients.

The pressure cooker or the microwave can also cook food quickly. The pressure cooker can do the occasional big piece of meat the cook wants to serve whole; the microwave can cook vegetables, casseroles, or leftovers. Both may save using several burners on the stove, and both preserve nutrients better than most stove-top methods.

The oven, in contrast, can be a fuel waster. Efficient oven use is possible if the cook bakes or roasts a lot of food at one time. To use the stove top efficiently, a cook can use flat-bottomed pots with close-fitting lids that

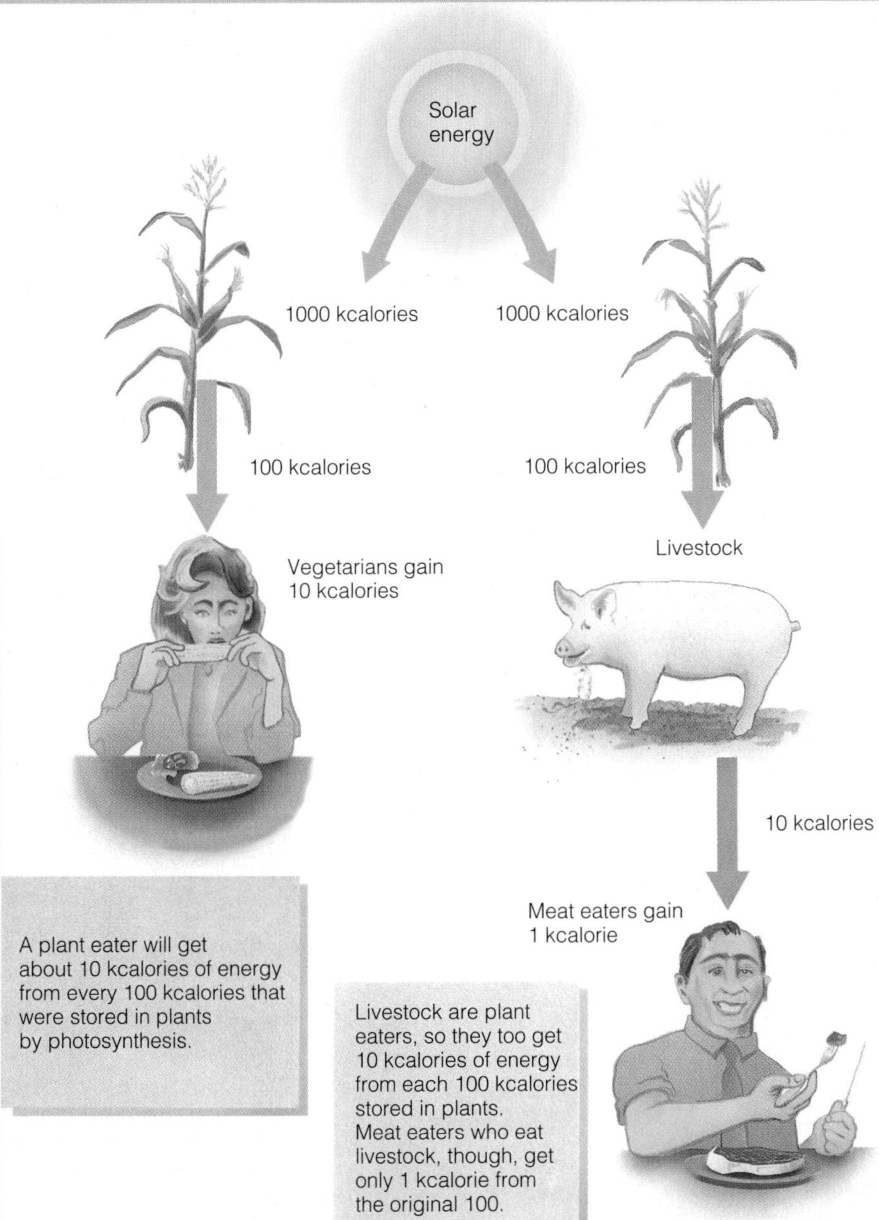
Solar energy
1000 kcalories
1000 kcalories
100 kcalories
100 kcalories
Vegetarians gain 10 kcalories
Livestock
10 kcalories
Meat eaters gain 1 kcalorie
A plant eater will get about 10 kcalories of energy from every 100 kcalories that were stored in plants by photosynthesis.
Livestock are plant eaters, so they too get 10 kcalories of energy from each 100 kcalories stored in plants. Meat eaters who eat livestock, though, get only 1 kcalorie from the original 100.

Figure NP11.1
Eating Low on the Food Chain Saves Resources
It takes ten times as much land and fuel to feed people meat as to feed them plants. (Reminder: The *food chain* is the sequence in which living things depend on other living things for food as shown in Figure 11.3.)

completely cover the burners. That way, each burner will donate all its heat to cooking something, not just heating the kitchen (and the planet). One can also turn electric burners and ovens off before the food is fully cooked and let the cooking finish as the stove cools.

What about cleanup after a meal?

A big energy user associated with food preparation and cleanup is the water heater. The less hot water used, the less fuel must be burned to heat the water. People can save water-heating energy in many ways. They can set the water heater at 140°F, not hotter. They can put it on a timer, so that each day it heats just enough water to meet that day's needs. They can wrap the water heater in insulation to keep it from losing heat to the surroundings, and they can wrap the hot-water pipes all the way to the points of use. Water-saving faucets and showerheads can also save energy.

When replacing a water heater or installing a new one, consumers can purchase a small, instantaneous-type water heater that heats the water only at the point of use, and only when needed. A consumer can choose a gas water heater, rather than an electric one; natural gas is a cleaner fossil fuel than the coal or oil usually burned to make electricity. Solar water heaters work well in sunny regions.

Someone who washes many dishes at a time should consider using a dishwasher, if it is affordable. People may think that the cost of the water, heat, and soap

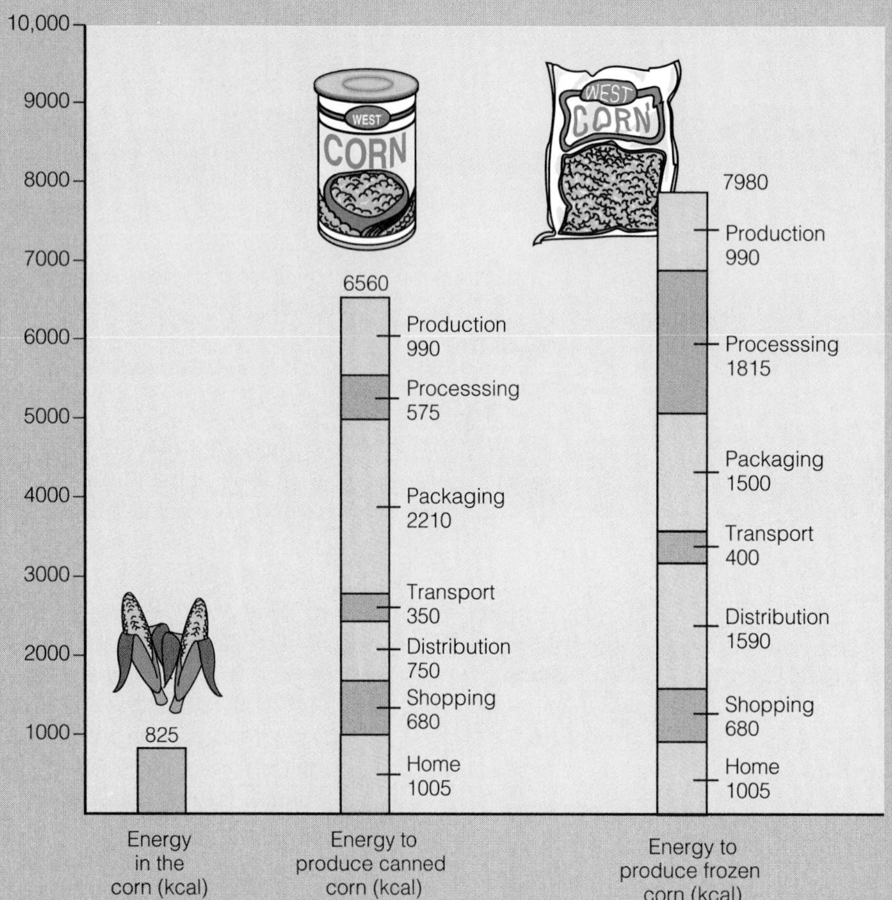

Figure NP11.2

Energy Costs of Canned and Frozen Corn

The corn contains only 825 kcalories per kilogram, but look how much energy goes to produce it.

would be higher than the cost of washing by hand, but this is not the case. One school found that a normal machine cycle used on full loads consumed less than two-thirds the water used in hand washing. Using less hot water also means using less electricity to heat it. The savings are greatest if the dishes are not prerinsed and are allowed to air dry.[3]

Are there "right" and "wrong" ways to throw garbage away, too?

An average American household of four people produces about 100 pounds of trash a week, much of it from the kitchen. National concern has focused on this issue because the nation is running out of landfill space in which to dispose of all the trash. Landfill space is only one of many problems associated with trash, however. Every item thrown away is a resource lost: an aluminum can could be used to make a new aluminum can; a cereal box, but for its clay coating, could become recycled paper; a plastic bottle could become part of a beautiful carpet. Trash need not become an undesirable mess; recycled trash could be viewed as a usable resource. Yet as it is now, about 70 percent of all the metal mined in the

Compost nourishes plants as food nourishes people.

Glossary

fossil fuels: coal, oil, and natural gas, which all come from the fossilized remains of plant life of earlier times.

sustainable: a term used to describe the use of resources at such a rate that the earth can keep on replacing them—for example, a rate of cutting trees no faster than new ones grow.

United States is used only once and then discarded. The aluminum thrown away every three months could rebuild the entire U.S. air fleet.

Garbage is a special case of a resource generated in the kitchen. Vegetable scraps, fruit peelings, and left-over plant foods are organic and biodegradable, like the leaves and grass cuttings people rake up in their yards. All of these materials can be piled up together with some soil and allowed to decompose naturally, forming compost, a rich, crumbly material that can be used as in nature to fertilize growing things. Some communities, recognizing this, conduct composting programs to recycle people's organic debris; some homeowners maintain their own composting piles. College campuses can use composted kitchen waste, mixed with grass cuttings and weeds, to mulch, fertil-ize, and enrich the soil they use in landscaping. Composting can even be done indoors in small odor-free bins and the resulting material used to pot plants.[4]

As you can see, environmental awareness can per-meate every aspect of food management from start to finish. And each choice made to save resources, cut energy use, or refrain from polluting helps to preserve and protect the environment.

Dietitians and foodservice managers have a special role to play, and their efforts can make an impressive difference.[5] The American Dietetic Association (ADA) urges members to conserve resources and minimize waste in both their professional and their personal lives.[6]

Notes

1. J. E. Young, Aluminum's real tab, *World Watch,* March/April 1992, pp. 26–33; Earth Works Group, *50 Simple Things That You Can Do to Save the Earth* (Berkeley, Calif.: Earthworks Press, 1989), pp. 64–65.

2. Regional perspectives: Gulf of Mexico shrimp fishery, *Marine Conservation News,* Winter 1990, p. 11.

3. Ask Garbage, *Garbage,* Spring 1994, p. 63.

4. R. Kourik, As the worm turns, *Garbage,* January/February 1992, pp. 48–51.

5. N. I. Hahn, The greening of a school district: How school foodservice led a recycling revolution, *Journal of the American Dietetic Association* 97 (1997): 371.

6. Position of The American Dietetic Association: Natural resource conservation and waste management, *Journal of the American Dietetic Association* 97 (1997): 425–428.

12 Nutrition during Pregnancy and Infancy

The effects of nutrition extend over years. A woman's nutrition prior to and throughout pregnancy and lactation affects not only her own health but also the growth, development, and health of her child, even long after the child has been born. Similarly, sound nutrition is vital to healthy infant development.

WWW.

acog.com
American College of Obstetricians and Gynecologists

childbirth.org
Childbirth.org

modimes.org
March of Dimes

Pregnancy: The Impact of Nutrition on the Future

The woman who enters pregnancy with full nutrient stores, sound eating habits, and a healthy body weight has done much to ensure an optimal pregnancy outcome. Then, during the pregnancy itself, if she eats a variety of nutrient-dense foods, her own and her infant's health will benefit further.

Preparing for Pregnancy

Full nutrient stores *before* pregnancy are essential both to conception and to healthy infant development during pregnancy. In the early weeks of pregnancy, before many women are even aware that they are pregnant, significant developmental changes occur that depend on a woman's nutrient stores.

Prepregnancy Weight Appropriate weight prior to pregnancy also benefits pregnancy outcome. Being either underweight or overweight (see p. 305) presents medical risks of pregnancy and childbirth. Underweight women are therefore advised to gain weight before becoming pregnant; and overweight women to lose excess weight. Guidelines for weight gain and loss were offered in Chapter 9.

low birthweight (LBW): a birthweight less than 5½ lb (2500 g); indicates probable poor health in the newborn and poor nutrition status of the mother during pregnancy. Normal birthweight for a full-term baby is 6½ to 8¾ lb (about 3000 to 4000 g).

Low-birthweight infants are of two different types. Some are **premature;** they are born early and are of a weight **appropriate for gestational age (AGA).** Others have suffered growth failure in the uterus; they may or may not be born early, but they are **small for gestational age (SGA).**

Infant Birthweight Infant birthweight strongly correlates with prepregnancy weight and is the most potent single predictor of the infant's future health and survival. Low-birthweight babies are statistically more likely than normal-weight babies to contract diseases, and low-birthweight babies are nearly 40 times more likely to die in the first month of life than are normal-weight babies.[1] Low socioeconomic status may limit a mother's access to medical care and nutritious foods and makes low birthweight more likely.[2]

Healthy Support Tissues A major reason why the mother's prepregnancy nutrition is so crucial to a healthy pregnancy is that it determines whether she will be able to grow healthy support tissues: the placenta, the amniotic sac, the umbilical cord, and the expanding uterus (see Figure 12.1). Malnutrition prior to and around conception keeps these tissues from developing fully.[3]

fetus (FEET-us): the developing infant from eight weeks after conception until its birth.

placenta (pla-SEN-tuh): an organ that develops inside the uterus early in pregnancy, in which maternal and fetal blood circulate in close proximity and exchange materials. The fetus receives nutrients and oxygen across the placenta; the mother's blood picks up carbon dioxide and other waste materials to be excreted via her lungs and kidneys.

amniotic (am-nee-OTT-ic) **sac:** the "bag of waters" in the uterus, in which the fetus floats.

umbilical (um-BIL-ih-cul) **cord:** the rope-like structure through which the fetus's veins and arteries reach the placenta; the route of nourishment and oxygen into the fetus and the route of waste disposal from the fetus.

uterus (YOO-ter-us): the womb, the muscular organ within which the infant develops before birth.

A woman's nutrition before and during pregnancy affects both the mother's health and the infant's growth. Appropriate weight prior to pregnancy benefits pregnancy outcome.

Nutrient Needs during Pregnancy

Between the moment of conception and the moment of birth, innumerable events determine the course and outcome of fetal development and, ultimately, the health of the newborn infant. With respect to nutrition, each organ needs nutrients most during its own intensive growth period. A nutrient deficiency during one stage of development might affect the heart and, during another stage, the developing limbs.

A woman's nutrient needs during pregnancy and lactation are higher than at any other time in her adult life and are greater for certain nutrients than for others. Figure 12.2 (p. 302) compares the nutrient needs of nonpregnant, pregnant, and lactating women. A study of the figure reveals some of the key needs.

Pregnancy is often divided into thirds called *trimesters*.

Energy A pregnant woman needs extra food energy, but only a little extra—300 kcalories above the allowance for nonpregnant women—and only during the second and third trimesters. A woman can easily obtain 300 kcalories from just one extra serving from each of the five food groups—a slice of bread, a serving of vegetables, an ounce of lean meat, a piece of fruit, and a cup of nonfat milk. Pregnant teenagers, underweight women, or physically active women may require more.

Energy RDA during pregnancy (2nd and 3rd trimesters):
+300 kcal/day.

Protein RDA during pregnancy:
+10 g/day.

Protein The RDA suggests an added 10 grams of protein per day throughout pregnancy. Many women in the United States exceed the recommended protein intake each day, and so they already receive the 10 grams of additional daily protein recommended for pregnancy. In fact, pregnant women in the United States—even those with low incomes who are not participating in food assistance programs—generally receive between 75 and 110 grams of protein a day.

Obtaining enough protein need not pose a problem, even if the diet excludes all foods of animal origin. Pregnant vegetarian women who meet their energy needs by eating ample servings of protein-containing plant foods such as legumes, whole grains, nuts, and seeds meet their protein needs as well. Use of high-protein supplements during pregnancy can be harmful and is discouraged.

Recommended carbohydrate intake: about 50% of energy intake. In a 2000 kcal/day intake, this represents 1000 kcal of carbohydrate, or about 250 g. Four cups of milk will contribute about 50 g carbohydrate. An apple provides 12 g carbohydrate, and a slice of bread provides 12 g, so generous intakes of fruit and bread are clearly beneficial.

Carbohydrate Pregnant women need generous amounts of carbohydrate to spare the protein they eat. If added energy is needed, it is best obtained from carbohydrate.

Vitamins The vitamins required for rapid cell proliferation—folate and vitamin B_{12}—are needed in large amounts during pregnancy. New cells are laid down at a tremendous pace as the fetus grows and develops. At the same time,

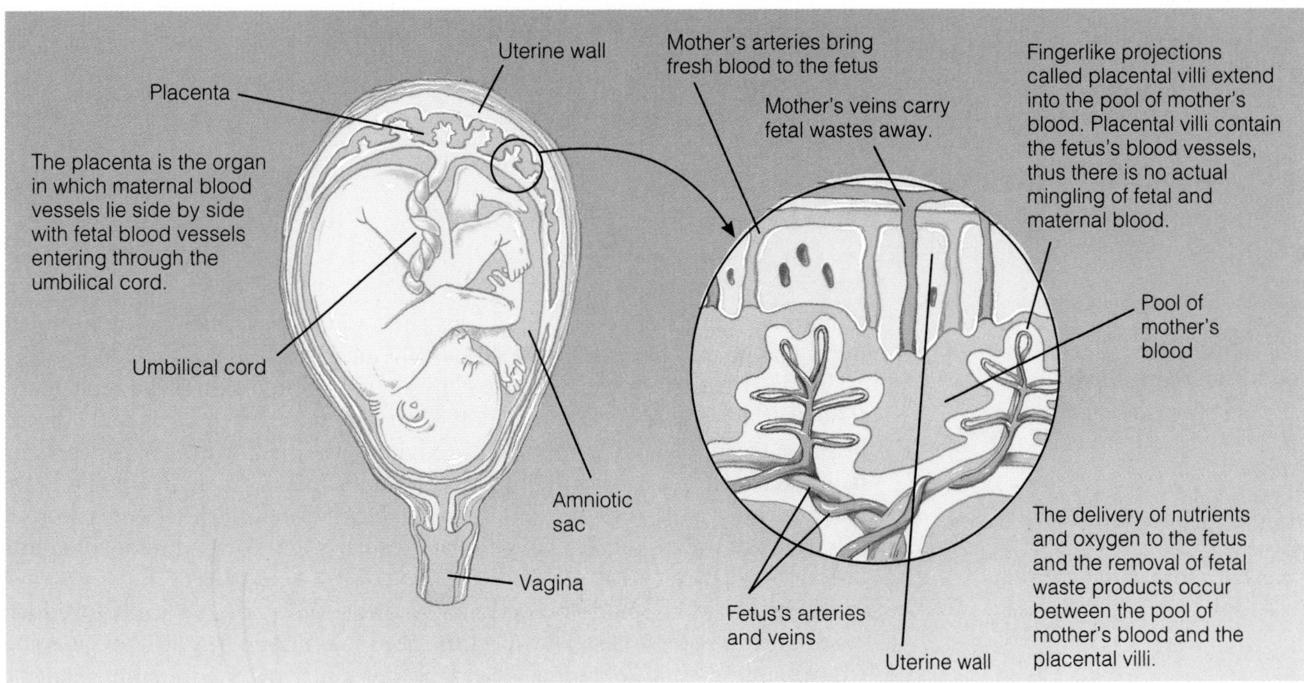

Figure 12.1
The Placenta
The placenta is composed of spongy tissue in which fetal blood and maternal blood flow side by side, each in its own vessels. The maternal blood transfers oxygen and nutrients to the fetus's blood and picks up fetal wastes to be excreted by the mother. Thus the placenta facilitates the nutritive, respiratory, and excretory functions that the fetus's digestive system, lungs, and kidneys will provide after birth.

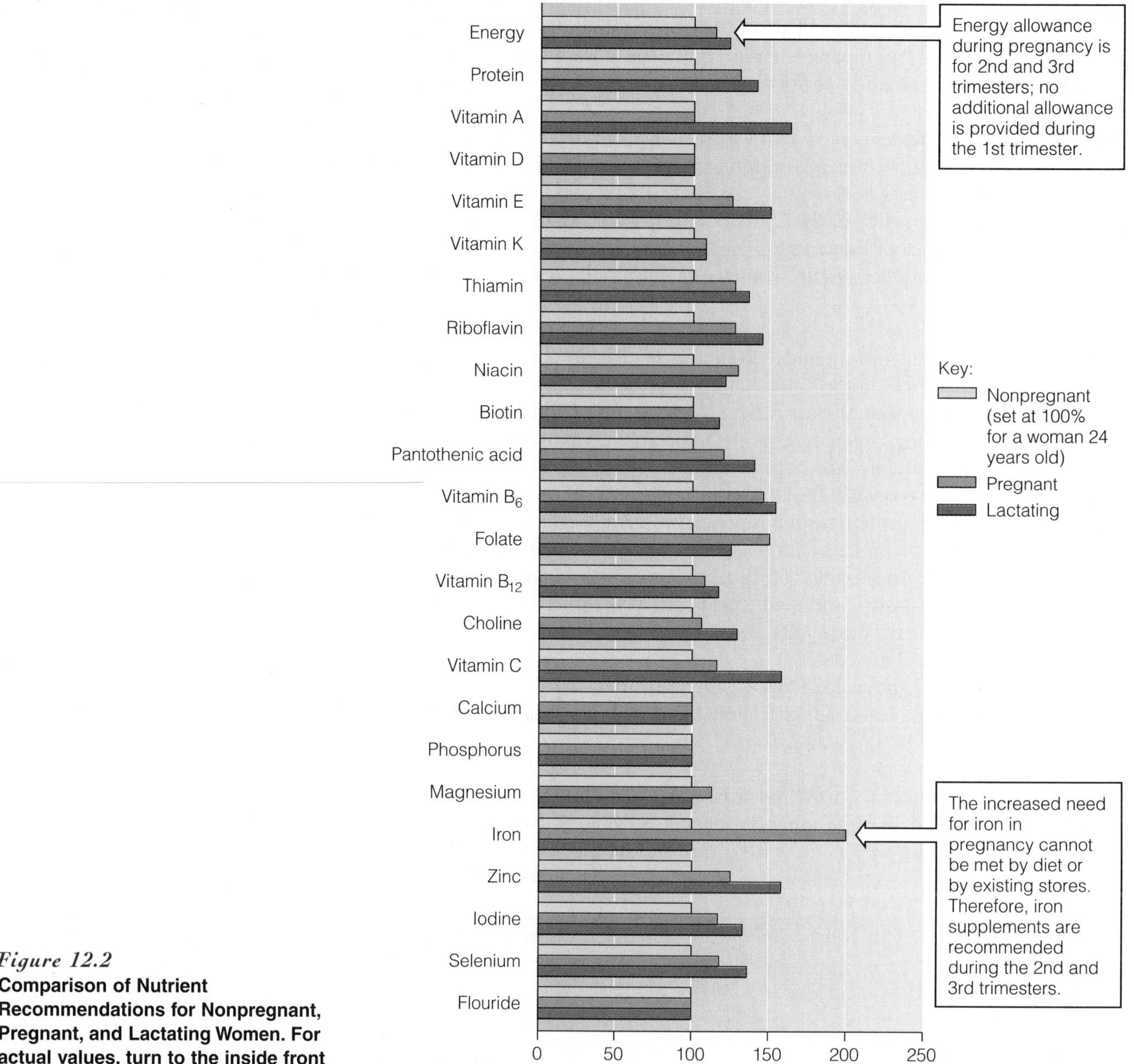

Figure 12.2
Comparison of Nutrient Recommendations for Nonpregnant, Pregnant, and Lactating Women. For actual values, turn to the inside front cover.

Folate RDA during pregnancy: 600 μg/day.

Foods containing folate:

- *Green leafy vegetables.*
- *Legumes.*
- *Liver.*
- *Orange juice and cantaloupe.*
- *Whole-wheat products.*

Vitamin B_{12} RDA during pregnancy: 2.6 μg/day.

the mother's number of red blood cells rises, so the recommendation for folate during pregnancy increases from 400 to 600 micrograms a day.

As described in Chapter 7, folate plays an important role in preventing neural tube defects. Folate supplements (400 micrograms daily) taken one month before conception and continued throughout the first trimester can prevent neural tube defects. Although this amount can be met through diet by eating the recommended 5 servings of fruits and vegetables daily, supplements offer women a convenient way to ingest enough folate regularly and continuously enough to benefit pregnancy.

Generally, even modest amounts of meat, fish, eggs, or milk products together with body stores easily meet the need for vitamin B_{12}. Strict vegetarians who exclude all foods of animal origin, however, may need daily supplements to prevent deficiency.

Dietary improvements are always preferred as the means of correcting nutrient inadequacies, but sometimes nutrient supplementation is necessary. The National Academy of Sciences subcommittee for the publication *Nutrition during Pregnancy* identifies certain women as being nutritionally at risk during pregnancy (see Table 12.1).[4] The subcommittee recommends a multivitamin-mineral supplement containing the nutrient amounts shown in Table 12.2 (p. 304) for these women.

As noted in Chapters 7 and 8, vitamins and minerals can be toxic in excess. During pregnancy, nutrient excesses (especially of vitamin A) can be potentially harmful. Seek advice from a health care provider before taking supplements.

Vitamin D and Calcium for Bones Vitamin D and the minerals involved in building the skeleton—calcium, phosphorus, and magnesium—are in great demand during pregnancy. Insufficient intakes may result in abnormal fetal bone development.

Intestinal absorption of calcium doubles early in pregnancy, when the mother's bones store the mineral. Later, as the fetal bones begin to calcify, there is a dramatic shift of calcium across the placenta. Whether the calcium added to the mother's bones early in pregnancy is withdrawn to build the fetus's bones later is unclear. In the final weeks of pregnancy, more than 300 milligrams a day are transferred to the fetus. Recommendations to ensure an adequate calcium intake during pregnancy are aimed at conserving the mother's bone mass while supplying fetal needs.

For women whose prepregnancy calcium intakes are below recommendations, as most are, increased calcium intakes may be especially important. Milk products offer many advantages over supplements, as emphasized in earlier chapters. Pregnant women under age 25 who consume less than 600 milligrams of calcium a day need to increase their intakes of milk, cheese, yogurt, and other calcium-rich foods. Alternatively, and less preferably, they may need a daily supplement of 600 milligrams of calcium.

Vegetarians who exclude milk products need calcium-fortified foods such as soy milk. It is worth noting that not all soy milk is fortified with calcium; the type fortified with calcium and vitamin D is recommended.

Calcium AI during pregnancy:
1300 mg/day (14 to 18 yr).
1000 mg/day (19 to 50 yr).

Phosphorus RDA during pregnancy:
1250 mg/day (14 to 18 yr).
700 mg/day (19 to 30 yr).

Magnesium RDA during pregnancy:
400 mg/day (14 to 18 yr) .
350 mg/day (19 to 30 yr).
360 mg/day (31 to 50 yr).

Four cups of milk a day will supply 1200 mg calcium. For other food sources, see Chapter 8.

Fluoride Mineralization of the fetus's teeth begins in the fifth month after conception. For this and for bone development, fluoride may be needed. Fluoride does cross the placenta, but whether the placenta can defend against excess intakes is questionable. Therefore, fluoride supplements are not recommended for pregnant women who drink fluoridated water. For women who live in communities without fluoridated water, a fluoride supplement may protect fetal teeth.

Fluoride AI during pregnancy:
2.9 mg/day (14 to 18 yr).
3.1 mg/day (19 to 50 yr).

Iron The body conserves iron especially well during pregnancy: menstruation ceases, and absorption of iron increases up to threefold due to a rise in the

Iron RDA during pregnancy:
30 mg/day.

In pregnancy, hemoglobin values of 12 g are not unusual, and 11 g is where the line defining "too low" is often drawn. Appendix E discusses more sensitive measures of iron status.

Food sources of iron:

- *Liver, oysters.*
- *Red meat, fish, other meat.*
- *Dried fruits (raisins, prunes).*
- *Legumes (dried beans, peas, lima beans).*
- *Dark green vegetables.*

Vitamin C–rich foods enhance iron absorption from foods.

Zinc RDA during pregnancy:
12 mg/day.

Table 12.1 Women at Nutritional Risk during Pregnancy

- Women who ordinarily consume an inadequate diet (for example, those who avoid consumption of all animal-derived foods)
- Women who are lactose intolerant
- Women who are carrying multiple fetuses
- Women who smoke cigarettes or use alcohol or illicit drugs
- Women who are underweight or overweight at conception
- Women who gain insufficient or excessive weight during pregnancy
- Women who lack nutrition knowledge or who have insufficient financial resources to purchase adequate food
- Adolescents

Source: Compiled from information in National Academy of Sciences, Food and Nutrition Board, *Nutrition during Pregnancy* (Washington, D.C.: National Academy Press, 1990).

Table 12.2 Nutrient Supplements during Pregnancy[a]	
Nutrient	**Amount**
Folate	300 µg
Vitamin B_6	2 mg
Vitamin C	50 mg
Vitamin D	5 µg
Calcium	250 mg
Copper	2 mg
Iron	30 mg
Zinc	15 mg

[a]For pregnant women at nutritional risk (see Table 12.1).

Source: Reprinted with permission from *Nutrition during Pregnancy* © 1990 by the National Academy of Sciences. Courtesy of the National Academy Press, Washington, D.C.

blood's iron-absorbing and iron-carrying protein, transferrin. Still, iron needs are so high that stores dwindle during pregnancy.

The developing fetus draws on the mother's iron stores to create stores of its own to last through the first four to six months of life. Research shows that women who enter pregnancy with iron-deficiency anemia have two to three times the normal risk of delivering low-birthweight or preterm infants.[5] Iron losses also occur with the bleeding that is inevitable at birth.

Few women enter pregnancy with adequate iron stores. For all women not taking supplements containing iron, a daily iron supplement containing 30 milligrams is recommended during the second and third trimesters of pregnancy.[6]

Zinc Zinc is another nutrient of vital importance in pregnancy. It is required for DNA and RNA synthesis and thus for protein synthesis, and low blood zinc predicts low birthweight.[7] Zinc is most abundant in foods of high-protein content, such as shellfish, meat, and nuts, but the presence of other trace elements and fiber in foods may adversely affect zinc absorption. Iron interferes with the body's absorption and use of zinc, so women taking iron supplements (more than 30 milligrams per day) may also need zinc supplements.

Food Choices Because food energy needs increase less than nutrient needs, the pregnant woman must select foods of high nutrient density. For most women, appropriate choices include foods like nonfat milk, nonfat plain yogurt, lean meats, eggs, dark green vegetables, vitamin C–rich fruits, legumes, and whole-grain breads and cereals. Table 12.3 provides a suggested food pattern.

WIC, pronounced *WICK,* is the acronym for the federal Special Supplemental Food Program for Women, Infants, and Children. The U.S. Department of Agriculture (USDA) funds WIC, and state health departments administer the program.

WWW **ceo-cap.org/wic.htm**
Women, Infants and Children

WIC A woman of limited financial means may need help in obtaining needed food and information. At the federal level, the WIC program provides nutrition education and nutritious foods to low-income pregnant women and their children. WIC provides eggs, milk, cereal, juice, cheese, legumes, and peanut butter to infants, children up to age five, and pregnant and breastfeeding women who qualify financially and are at medical or nutritional risk. For infants given formula, WIC also provides iron-fortified formulas. Studies of the nutrition and health effects of WIC have found that participation in the program benefits both the iron status and the growth and development of infants and children. WIC participation during pregnancy has been shown to reduce the risks of delivering preterm or low-birthweight infants.[8]

food craving: a deep longing for a particular food.

food aversion: a strong desire to avoid a particular food.

Cravings for nonfood items such as clay, ice, laundry starch, and cornstarch are known as *pica.*

Food Cravings and Aversions Some women develop cravings for, or aversions to, certain foods and beverages during pregnancy. Individual food cravings during pregnancy do not seem to reflect real physiological needs. In other words, a woman who craves pickles does not necessarily need salt, nor does a chocolate craving indicate a need for caffeine or fat. Similarly, cravings for ice cream are common during pregnancy, but do not signify a calcium deficiency. Food aversions and cravings that arise during pregnancy are probably due to hormone-induced changes in taste and sensitivities to smells.

In summary, energy and nutrient needs are high during pregnancy. A balanced diet that includes an extra serving from each of the five food groups can usually meet those needs, with the exception of iron.

Weight Gain

All pregnant women must gain weight: fetal growth and maternal health depend on it. A pregnancy weight gain of 25 to 35 pounds is recommended for women who begin pregnancy at a healthy weight and are carrying a single fetus; for others, see the margin.[9] Some women should strive for gains at the upper ends of the

Table 12.3 Daily Food Choices for Pregnant and Lactating Women		
Food Group	**Number of Servings**	
	Adults	**Pregnant or Lactating Women**
Breads/cereals	6 to 11	7 to 11
Vegetables	3 to 5	4 to 5
Fruits	2 to 4	3 to 4
Meat/meat alternates	2 to 3	3
Milk/milk products	2	3 to 4

Note: Figure 1.3 in Chapter 1 provides a detailed summary of foods in each group with serving sizes.

target ranges, notably, adolescents, who are still growing themselves. Short women (5 feet 2 inches and under) should strive for lower gains. Figure 12.3 (p. 306) shows the components of a weight gain of 30 pounds.

The ideal pattern of weight gain during pregnancy is thought to be about 3½ pounds during the first three months and a pound per week thereafter. Women lose some of the weight gained during pregnancy at delivery and most of the remainder within the following few weeks or months, as blood volume returns to normal and accumulated fluids are lost.

If a woman has gained more than the expected amount of weight early in pregnancy, she should not try to diet in the last weeks. A sudden large weight gain, however, is a danger signal that may indicate the onset of hypertension (see p. 306).

Weight gain during pregnancy, like prepregnancy weight, directly relates to infant birthweight. If the mother does not gain all of the weight recommended, she may give birth to an underweight infant.

Weight-gain recommendations:

- *Underweight women: 28 to 40 lb (12.5 to 18.0 kg).*
- *Normal-weight women: 25 to 35 lb (11.5 to 16.0 kg).*
- *Overweight women: 15 to 25 lb (7.0 to 11.5 kg).*
- *Obese women: 15 lb minimum (6.8 kg minimum).*
- *Twin birth: 35 to 45 lb (16.0 to 20.5 kg).*

Weight-for-height categories for prepregnant women are based on body mass index (BMI) measures (see Appendix E):

- *Underweight BMI: <19.8.*
- *Normal BMI: 19.8 to 26.0.*
- *Overweight BMI: 26.0 to 29.0.*
- *Obese BMI: >29.0.*

A prenatal weight-gain grid (see Appendix E) plots the rate of weight gain during pregnancy.

Physical Activity

Chapter 10 described another lifestyle component that promotes health and well-being: physical activity. The active, physically fit woman experiencing a normal pregnancy can continue her regular activities throughout pregnancy, adjusting the duration and intensity as needed. Her rewards may include reduced stress, better weight control, an easier labor, better posture, less back pain, prevention of gestational diabetes (see the next section), an improved mood and body image, and a faster recovery from childbirth.[10] Potential risks of physical activity during pregnancy include acute conditions such as altered fetal heart rate and hyperthermia and other possible hazards such as premature labor or reduced birthweight, but these seldom outweigh the benefits.

As is true for everyone, the frequency, duration, and intensity of the activity affect the likelihood of the benefits or risks.[11] Physically fit, low-risk pregnant women can probably tolerate longer activity sessions and more exertion than is currently recommended.[12] Table 12.4 (p. 307) provides some guidelines for physical activity during pregnancy, but every pregnant woman should consult with her own health care provider regarding safe activity limits.

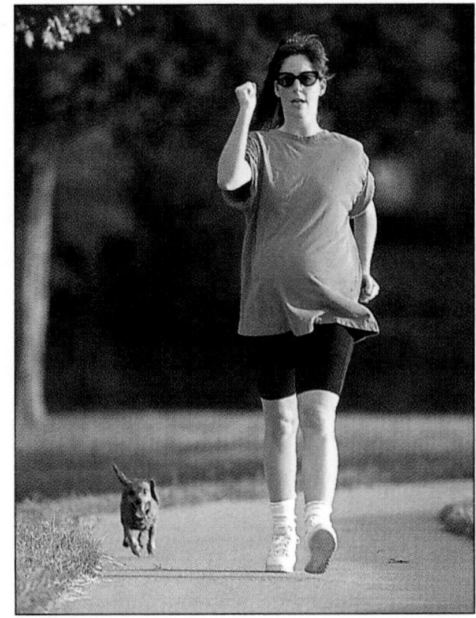

Pregnant women can enjoy the benefits of physical activity.

Problems in Pregnancy

Just as adequate nutrition and normal weight gain support the health of the mother and growth of the fetus, maternal diseases detract from health and growth. If discovered early, many diseases can be controlled—another reason

Figure 12.3
Components of Weight Gain During Pregnancy

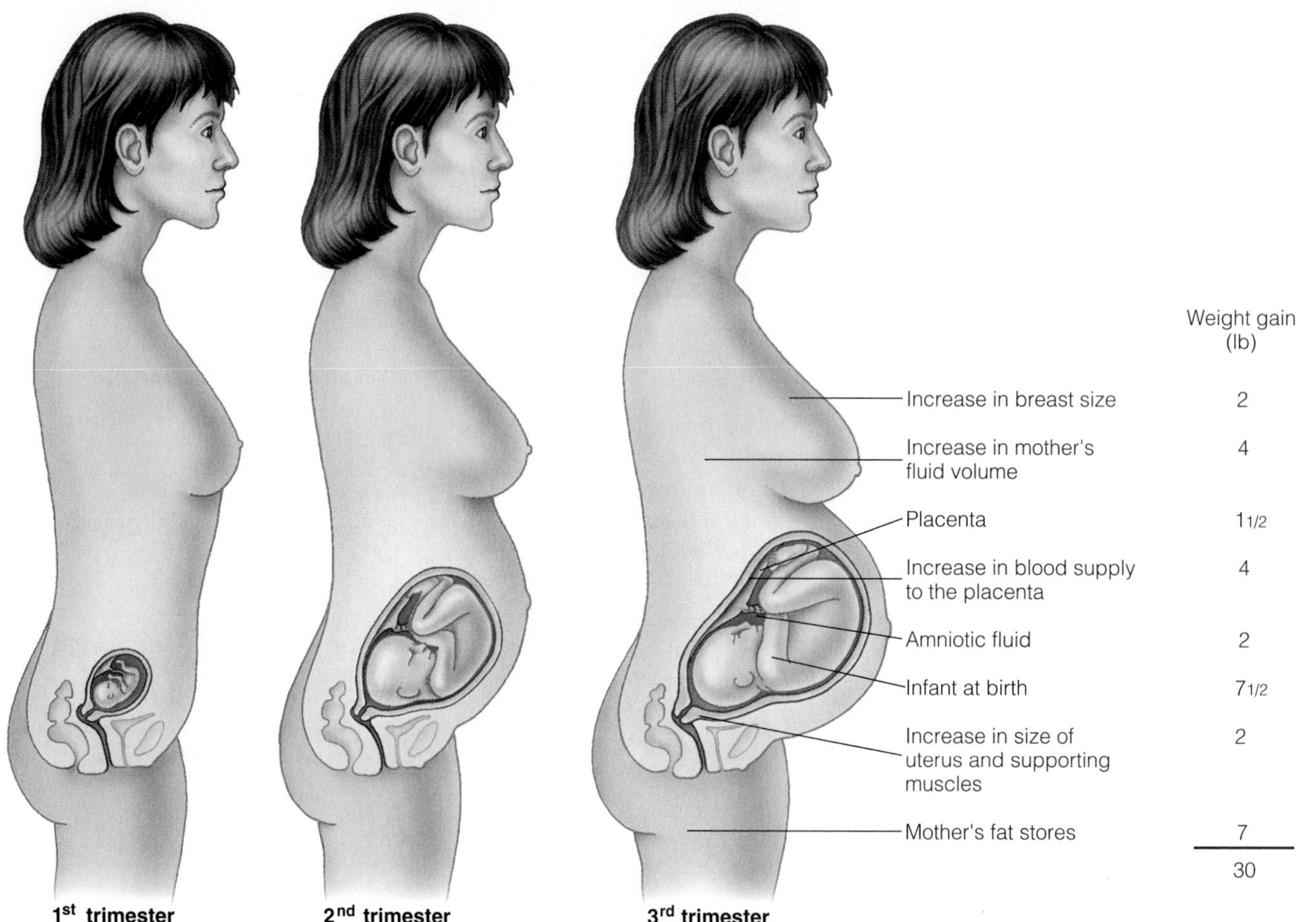

why early prenatal care is recommended. Some nutrition measures can help alleviate the most common problems encountered during pregnancy.

gestational diabetes: the detection of abnormal glucose tolerance during pregnancy, which usually returns to normal postpartum.

Gestational Diabetes Pregnancy precipitates the onset of diabetes in some women because placental hormones alter the way insulin works. This condition is known as gestational diabetes. Blood glucose becomes abnormal during pregnancy but usually returns to normal after the infant is born. In about one-third of such cases, however, diabetes becomes permanent. To ensure that the problems of diabetes are dealt with promptly, most women (see Chapter 25) receive glucose tolerance tests during the sixth month of pregnancy.[13]

WWW. **diabetes.org**
American Diabetes Association

cesarean section: surgical childbirth, in which the infant is delivered through an incision in the woman's abdomen.

Without proper management, gestational diabetes can lead to complications such as a high-birthweight infant and difficult delivery; rates of cesarean section delivery are also exceptionally high (almost 30 percent).[14] An important aspect of nutrition management in gestational diabetes is prevention of excessive weight gain. Women with gestational diabetes need to select foods rich in complex carbohydrates, such as legumes, vegetables, and whole-grain products, and should limit their intakes of concentrated sweets. Optimal protein intakes are also important. Insulin therapy may be required if blood glucose fails to normalize.

Hypertension Hypertension complicates pregnancy and affects its outcome in different ways, depending on when the hypertension first develops and on how severe it becomes.[15] Hypertension can be a preexisting chronic condition that

Table 12.4 Physical Activity Guidelines for Pregnancy

- Be physically active on a regular basis (at least three times a week), not intermittently.
- Stop exercising if you feel overheated.
- Drink plenty of fluids before, during, and after physical activity.
- Avoid exerting yourself in hot, humid weather; avoid overheating.
- Avoid jarring or jerky motions.
- Avoid any activity that has the potential to cause even mild abdominal trauma.
- Discontinue any activity that causes discomfort.
- Do not exercise while lying on your back after the fourth month.
- Do not allow your heart rate to exceed 140 beats per minute.
- Eat enough to support the energy needs of pregnancy and physical activity.

develops before a woman becomes pregnant or a transient condition that develops during the pregnancy and subsides after childbirth. In some cases, hypertension that develops during pregnancy warns of the ominous disorder preeclampsia.

Preexisting Chronic Hypertension In addition to the health risks normally imposed by hypertension (heart attack and stroke), high blood pressure increases the risks of a low-birthweight infant or the separation of the placenta from the wall of the uterus before the birth, resulting in stillbirth. Ideally, before a woman with hypertension becomes pregnant, her blood pressure will be under control.

Transient Hypertension of Pregnancy Some women first develop hypertension during the second half of pregnancy. Most often, the rise in blood pressure is mild and does not affect the pregnancy adversely. Blood pressure usually returns to normal during the first few weeks after childbirth. This transient hypertension of pregnancy differs from the pregnancy-induced hypertension that accompanies preeclampsia.*

Preeclampsia Hypertension may signal the onset of preeclampsia, a condition characterized not only by high blood pressure but by protein in the urine and fluid retention (edema). Preeclampsia, which affects less than 10 percent of pregnant women, usually occurs with first pregnancies and almost always after 20 weeks' gestation. Symptoms typically regress within 48 hours of delivery. The edema of preeclampsia is a whole-body edema, distinct from the localized fluid retention women normally experience late in pregnancy.

Preeclampsia affects almost all of the mother's organs—the circulatory system, liver, kidneys, and brain. If it progresses, she may experience convulsions; when this occurs, the condition is called eclampsia. Maternal mortality during pregnancy is rare in developed countries, but eclampsia is the most common cause.

Preeclampsia demands prompt medical attention. Treatment focuses on regulating blood pressure and preventing convulsions.

preeclampsia: a condition characterized by hypertension, fluid retention, and protein in the urine.

The normal edema of pregnancy responds to gravity: blood pools in the ankles. The edema of preeclampsia is a generalized edema. The distinction helps with diagnosis.

eclampsia: a severe stage that follows preeclampsia in which convulsions occur.

*The Working Group on High Blood Pressure in Pregnancy, convened by the National High Blood Pressure Education Program of the National Heart, Lung, and Blood Institute, has suggested abandoning the term *pregnancy-induced hypertension* because it fails to differentiate between the mild, transient hypertension of pregnancy and the life-threatening hypertension of preeclampsia.

To alleviate the nausea of pregnancy:

- *On waking, arise slowly.*
- *Eat dry toast or crackers before getting out of bed.*
- *Chew gum or suck hard candies.*
- *Eat frequent small meals.*
- *Avoid foods with offensive odors.*
- *When nauseated, do not drink citrus juice, water, milk, coffee, or tea.*
- *Take prenatal vitamin and iron supplements on a full stomach or at a time of day when you feel well.*
- *Avoid shopping for or preparing food when feeling nauseous.*

To prevent or relieve heartburn:

- *Eat frequent small meals.*
- *Drink liquids between meals.*
- *Avoid spicy or greasy foods.*
- *Sit up while eating.*
- *Wait an hour after eating before lying down.*
- *Wait 2 hours after eating before exercising.*

Morning Sickness Unlike the conditions just discussed, the nausea of "morning" (actually, anytime) sickness is usually benign, although distressing to some women. It arises from the hormonal changes taking place early in pregnancy, ranges from mild queasiness to debilitating nausea, and afflicts more than half of all pregnant women. One expert who has worked with women hospitalized with the condition notes that a trigger for morning sickness is what she calls the "radar nose of pregnancy."[16] Many women complain that smells, especially cooking smells, make them sick. Thus minimizing odors is a key to alleviating morning sickness.

Traditional strategies for alleviating nausea are listed in the margin, but many women benefit most from simply eating the foods they want when they feel like eating.

Heartburn Heartburn, a burning sensation in the lower esophagus near the heart, is common during pregnancy and is also benign. As the growing fetus puts increasing pressure on a woman's stomach, acid may back up and create a burning sensation in her throat. Tips to relieve heartburn are listed in the margin.

Constipation As the hormones of pregnancy alter muscle tone and the thriving infant crowds intestinal organs, an expectant mother may complain of constipation, another harmless but annoying condition. A high-fiber diet, physical activity, and a plentiful fluid intake will help relieve this condition. Also, responding promptly to the urge to defecate can help. Laxatives should be used only as prescribed by the physician. Mineral oil should not be used, because it robs the body of fat-soluble vitamins.

Maternal diseases such as diabetes and hypertension can compromise pregnancy outcome. Early prenatal care can help detect and control these conditions to ensure healthy pregnancy outcomes.

Practices to Avoid

A general guideline for the pregnant woman is to eat a normal, healthy diet and practice moderation. A woman's daily choices during pregnancy take on enormous importance. Forewarned, pregnant women can choose to abstain from or avoid potentially harmful practices.

Cigarette Smoking Smoking adversely affects the pregnant woman's nutrition status, which in turn impairs fetal nutrition. Cigarette smoking increases iron needs and reduces the availability of vitamin B_{12}, vitamin C, folate, and zinc.[17] Smoking also restricts the blood supply to the growing fetus and so limits the delivery of oxygen and nutrients and the removal of wastes. Smoking also causes these adverse effects: premature births, spontaneous abortions (fetal deaths), and increased risks of infants dying early in life.[18] In addition, sudden infant death syndrome (SIDS), the sudden, unexplained death of an infant, has been positively linked to the mother's cigarette smoking during pregnancy and even to postnatal exposure to smoke in the household.[19] Research also shows that children born to women who smoke during pregnancy may be intellectually impaired.[20] Cigarette smoking is by far the single most important modifiable risk factor responsible for fetal growth retardation in developed countries.[21] The more a mother smokes, the smaller her baby will be. In short, the scientific and medical literature confirms over and over again that smoking during pregnancy is dangerous.

Caffeine Caffeine crosses the placenta, and the fetus has a limited ability to metabolize it. So far, no convincing evidence indicates that caffeine causes birth defects in human beings (as it does in animals), but limited evidence suggests that moderate-to-heavy use may lower infant birthweight.[22] One well-designed, carefully controlled study of caffeine's effects on pregnancy outcome showed that moderate caffeine consumption (three cups of coffee daily) is not a risk factor for spontaneous abortion or growth retardation.[23] All things considered, it seems most sensible to limit caffeine consumption to the equivalent of one cup of coffee or two 12-ounce cola beverages a day.

A table of the caffeine amounts in beverages and medications is provided in Appendix A.

Medications Medications taken during pregnancy can cause complications during pregnancy and serious birth defects. For these reasons, women are advised to take medications only if their physicians deem it necessary to protect their life and health. Even aspirin can do harm, as revealed by research on more than 3000 pregnant women.[24] Drug labels warn: "As with any drug, if you are pregnant or nursing a baby, seek the advice of a health professional before using this product." For aspirin and ibuprofen, an additional warning immediately follows: "It is especially important not to use aspirin (or ibuprofen) during the last three months of pregnancy unless specifically directed to do so by a doctor, because it may cause problems in the unborn child or excessive bleeding during delivery."

Illicit Drugs The recommendation to avoid drugs during pregnancy includes illicit drugs, of course. Unfortunately, use of illicit drugs is common among pregnant women. One study of about 30,000 women in California found that 11 percent tested positive for alcohol and illicit drugs.[25] Marijuana or cocaine use during pregnancy adversely affects fetal growth and development.[26] Such drugs of abuse pass easily through the placenta and impair fetal development.[27] Moreover, infants born to drug users face low birthweight, cardiovascular problems, and increased risk of death. If they survive, their cries and behavior at birth are abnormal.[28] They may be hypersensitive or underaroused; those who test positive for drugs suffer the greatest effects of toxicity and withdrawal.

Dieting Dieting, even for short periods, is hazardous during pregnancy. Low-carbohydrate diets or fasts that cause ketosis deprive the growing brain of needed glucose and may impair its development. Energy restriction during pregnancy is dangerous for all women, regardless of their prepregnancy weights.

Alcohol Drinking alcohol during pregnancy endangers the fetus. Alcohol can cause irreversible brain damage and mental and physical retardation in the fetus—the abnormalities that define fetal alcohol syndrome (FAS).[29] The potential for fetal damage arises when the mother's liver receives more alcohol than it can detoxify. Alcohol-laden blood then circulates to all parts of the mother's body and freely crosses the placenta to impair fetal development. Alcohol also interferes with placental transport of nutrients to the fetus.

fetal alcohol syndrome (FAS): the cluster of symptoms seen in a person whose mother consumed excess alcohol during her pregnancy; includes mental and physical retardation with facial and other body deformities.

Of the leading causes of mental retardation, FAS is the only one that is totally preventable. The surgeon general has issued a statement that pregnant women should drink absolutely no alcohol. All containers of beer, wine, and distilled liquor now must carry a warning to this effect.

WWW. **nofas.org**
National Organization on Fetal Alcohol Syndrome

Abstinence from smoking, alcohol, and other drugs, as well as moderation in caffeine consumption, improves pregnancy outcome. Over-the-counter and prescription medications should be taken only at the direction of a physician.

Adolescent Pregnancy

Each year in the United States, about one million adolescent girls between the ages of 12 and 19 become pregnant. Of these, about half choose to continue their

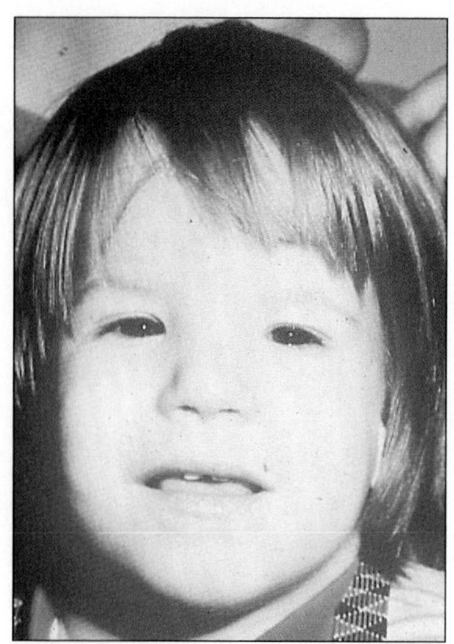

These facial traits are typical of fetal alcohol syndrome, caused by drinking during pregnancy—low nasal bridge, short eyelid opening, underdeveloped groove in center of the upper lip, small midface, short nose, and small head circumference.

pregnancies.[30] Many teenage women, especially the youngest ones, have not had time to store the nutrients needed to support their own rapid growth and development, much less nutrients needed to support pregnancy and the developing fetus. Nutrient shortages place both mother and infant at risk. Pregnant teenagers have more miscarriages, premature births, stillbirths, and low-birthweight infants than do pregnant adult women.[31] Their greatest risk, though, is death of the infant: mothers under 16 bear more babies who die within the first year than do women in any other age group. Clearly, teenage pregnancy is a major public health problem.

Pregnant teenagers suffer many illnesses. The rates of preeclampsia are 50 percent higher in teens than in older women. Other common problems of teen pregnancies are iron-deficiency anemia and prolonged labor.[32]

To support their own and fetal needs, young teenagers (13 to 16 years old) are encouraged to strive for pregnancy weight gains at the upper ends of the ranges recommended for pregnant women.[33] Those who gain between 30 and 35 pounds during pregnancy have lower risks of delivering low-birthweight infants.[34] Adequate nutrition can substantially improve the health of the mother and infant; it is an indispensable component of prenatal care.[35] Pregnant and lactating teenagers can use the Daily Food Guide presented in Table 12.3 (on p. 305), making sure to select at least 4 servings of milk or milk products daily.

High-risk pregnancies, especially for teenagers, threaten the life and health of both mother and infant. Adequate nutrition and prenatal care can improve the health of mother and infant.

WWW.

aap.org
American Academy of Pediatrics

lalecheleague.org
LaLeche League International

eatright.org
American Dietetic Association

Breastfeeding

The American Academy of Pediatrics (AAP) recommends that infants receive breast milk for the first 12 months of life.[36] The American Dietetic Association (ADA) advocates breastfeeding for the nutritional health it confers on the infant as well as for the physiological, social, economic, and other benefits it offers the mother.[37] Breast milk's unique nutrient composition and protective factors promote optimal infant health and development. The only acceptable alternative to breast milk is iron-fortified formula. Adequate nutrition of the mother supports successful lactation, and without it, lactation is likely to falter or fail.

The Mother's Nutrient Needs

By continuing to eat nutrient-dense foods, not restricting weight gain unduly, and enjoying ample food and fluid at frequent intervals throughout lactation, the mother who chooses to breastfeed her infant will be nutritionally prepared to do so. An inadequate diet does not support the stamina, patience, and self-confidence that nursing an infant demands. Figure 12.2 (on p. 302) shows the differences between a lactating woman's nutrient needs and those of a nonpregnant woman, and Table 12.3 (on p. 305) shows a food pattern that meets those needs.

Food Energy A nursing mother produces about 25 ounces of milk a day, more or less, depending primarily on the infant's demand for milk.[38] To produce milk, a woman needs extra energy—almost 650 kcalories a day above her regular need during the first six months of lactation. To meet this energy need, the woman is advised to eat an extra 500 kcalories of foods each day and let the fat reserves she accumulated during pregnancy provide the rest. Some research suggests that many women may need less energy for milk production than current recommendations.[39] Severe energy restriction (less than 1200 kcalories per day), however, hinders milk production and can compromise the mother's health.[40]

Maternal Weight Loss The period of lactation is the natural time for a woman to lose the extra body fat she accumulated during pregnancy. If she chooses nutrient-dense foods, she will gradually lose weight, even though her energy intake may be greater than normal. In one study, researchers found that breastfeeding or a combination of breastfeeding and formula feeding led to faster loss of body weight during the first month after women delivered their babies than formula feeding alone.[41] After the first month, however, no significant effect of breastfeeding on weight loss was observed. Results of other studies examining the relationship between feeding method and loss of body weight and body fat are inconsistent.[42] In most studies where breastfeeding duration was three months or longer, researchers found that lactation had a positive influence on weight loss.[43] Most women lose 1 to 2 pounds a month during the first four to six months of lactation; some may lose more, and others may maintain or even gain weight.

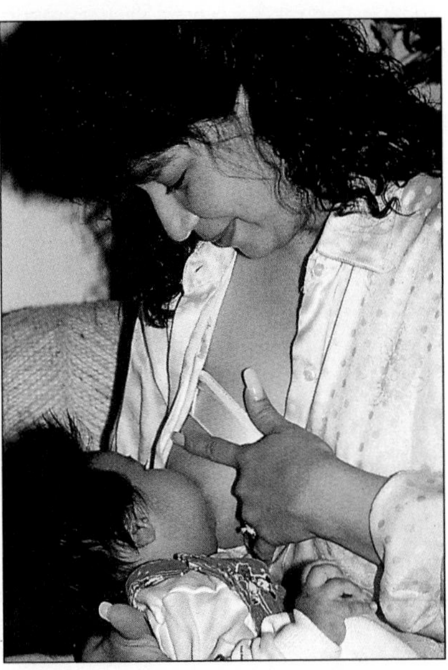

Breastfeeding is a natural extension of pregnancy—the mother's body continues to nourish the infant.

Supplements Most lactating women can obtain all the nutrients they need from a well-balanced diet without taking vitamin-mineral supplements. If a woman's diet is found to be inadequate in one or more nutrients, consumption of foods rich in those nutrients or appropriate nutrient supplementation can be recommended. Women can produce milk that contains adequate protein, carbohydrate, fat, folate, and most minerals, even when their own supplies are limited.[44] The milk's quality is maintained at the expense of maternal stores. The nutrients in breast milk that are most likely to decline in response to prolonged inadequate maternal intakes are vitamins B_6, B_{12}, A, and D.[45]

Water Despite previous misconceptions, drinking more fluid does not enable a mother to produce more breast milk.[46] Nevertheless, a lactating woman needs to drink at least 2 quarts of liquids each day to protect herself from dehydration. A convenient way to ensure adequate fluid consumption is to drink a glass of milk, juice, or water each time the infant nurses, as well as at each meal.

Particular Foods Foods with strong or spicy flavors (such as garlic or onion) may alter the taste of breast milk. A sudden change in the taste of the milk may annoy some infants. Infants who are sensitive to particular foods such as cow's milk protein may become uncomfortable when the mother's diet includes these foods. Most nursing mothers should drink cow's milk, however; the nutrients it provides make a significant contribution to both the infant's and the mother's health.

Contraindications to Breastfeeding

Some substances impair maternal milk production or enter the breast milk and interfere with infant development. Some medical conditions prohibit breastfeeding.

Alcohol Alcohol easily enters breast milk. One study showed that the alcohol concentration of breast milk peaks within one hour after ingestion of even small amounts of alcohol (½ ounce).[47] In this study, alcohol consumption by lactating women significantly reduced the breast milk intakes of their infants. In the past, alcohol has been recommended to lactating mothers to ease milk production, despite a lack of scientific support for such recommendations. The research described here suggests that alcohol consumption by lactating women should be discouraged.

Caffeine Excessive caffeine consumption during lactation may cause irritability and wakefulness in the breastfed infant. As during pregnancy, caffeine consumption should be moderate.

Smoking Health care professionals should actively discourage smoking by lactating women. Research shows that lactating women who smoke produce less milk, and milk with a lower fat content, than mothers who do not smoke.[48] Smoking also exerts numerous harmful effects on both mother and child.[49]

Maternal Illness If a woman has an ordinary cold, she can go on nursing without worry. If susceptible, the infant will catch it from her anyway, and thanks to immunological protection, a breastfed baby may be less susceptible than a formula-fed baby would be. If a woman has a communicable disease such as tuberculosis or hepatitis, which could threaten the infant's health, then mother and baby must be separated. Breastfeeding would be possible only by pumping the mother's breasts several times a day. The Centers for Disease Control recommend that where safe alternatives are available, women who test positive for human immunodeficiency virus (HIV), the virus that causes acquired immune deficiency syndrome (AIDS), should not breastfeed their infants.

Maternal Medication and Drug Use Similarly, if a nursing mother must take medication that is secreted in breast milk and that is known to affect the infant, then breastfeeding is contraindicated.[50] Many prescription medications do not reach nursing infants in sufficient quantities to affect them adversely. Some, however, do. As a precaution, a nursing mother should consult with the prescribing physician prior to ingesting any medication. Drug addicts, including alcohol abusers, are capable of taking such high doses that their infants can become addicts by way of breast milk; in these cases, too, breastfeeding is contraindicated.

A lactating woman is wise to avoid oral contraceptives until after she has weaned her infant and to use another method of contraception in the meantime. Standard oral contraceptives contain estrogen, which reduces both the volume and the protein content of breast milk.[51]

Lactating women need ample fluid and enough energy and nutrients to produce about 25 ounces of milk a day. Alcohol, excessive caffeine consumption, smoking, and medications and drugs may impair milk production or enter breast milk and impair infant development.

Nutrition of the Infant

Early nutrition affects later development, and early feeding sets the stage for eating habits that will influence nutrition status for a lifetime. Trends change, and experts argue about the fine points, but properly nourishing an infant is relatively simple, overall. Common sense in the selection of infant foods and a nurturing, relaxed environment go far to promote an infant's health and well-being.

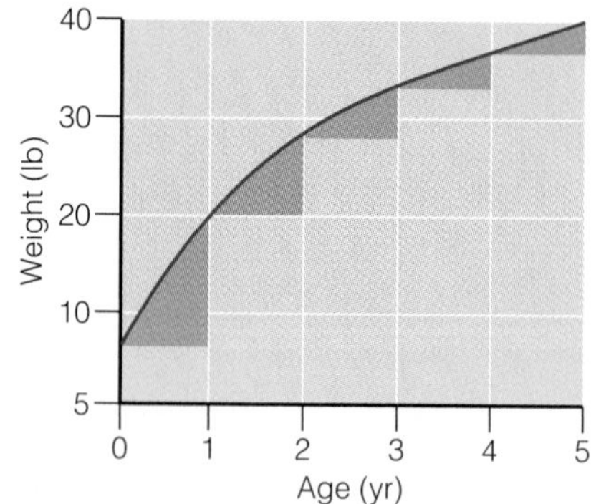

Figure 12.4
Weight Gain of Human Infants in Their First Five Years of Life
In the first year, an infant's birthweight may triple, but over the following several years, the rate of weight gain gradually diminishes.

Nutrient Needs

An infant grows faster during the first year than ever again, as Figure 12.4 shows. The growth of infants and children directly reflects their nutritional well-being and is an important parameter in assessing their nutrition status. Health care professionals use growth charts to evaluate the growth and development of children from birth to 18 years of age (see Appendix E).

Nutrients to Support Growth An infant's birthweight doubles by about four to five months of age, and it triples by the age of one year. (Consider that if an adult, starting at 120 pounds, were to do this, the person's weight would increase to 360 pounds in a single year.) By the end of the first year, the growth rate slows considerably. Between the first and second birthdays, the weight gained amounts to less than 10 pounds.

A newborn infant requires only about 650 kcalories per day, whereas most adults require about 2000 kcalories per day. In comparison to body weight, however, the difference is remarkable. Infants require about 100 kcalories per kilogram of body weight per day; most adults require fewer than 40. A 170-pound adult who tried to eat like an infant would have to ingest over 7000 kcalories a day! Figure 12.5 compares a five-month-old infant's needs per kilogram of body weight with those of an adult male; as you can see, some of the differences are extraordinary. After six months, energy needs increase less rapidly as the growth rate begins to slow, but some of the energy saved by slower growth is spent in increased activity.

Water The most important nutrient of all, for infants as for everyone, is the one easiest to forget: water. Conditions that cause fluid loss, such as vomiting, diarrhea,

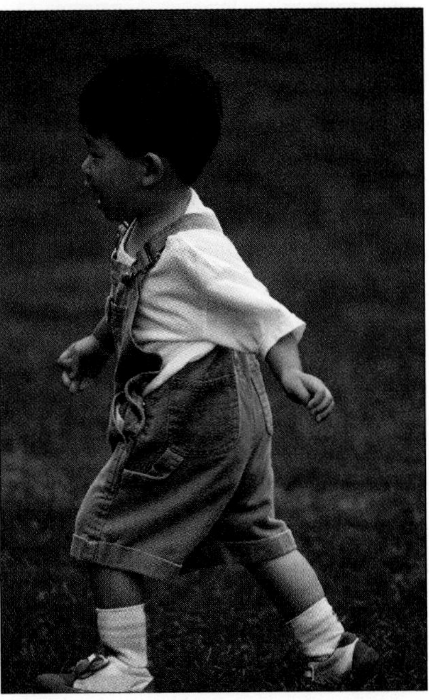

After six months, the energy saved by slower growth is spent on increased activity.

	Infants	Adults
Heart rate (beats/minute)	120 to 140	70 to 80
Respiration rate (breaths/minute)	20 to 40	15 to 20
Energy needs (kcal/body weight)	45/lb (100/kg)	<18/lb (<40/kg)

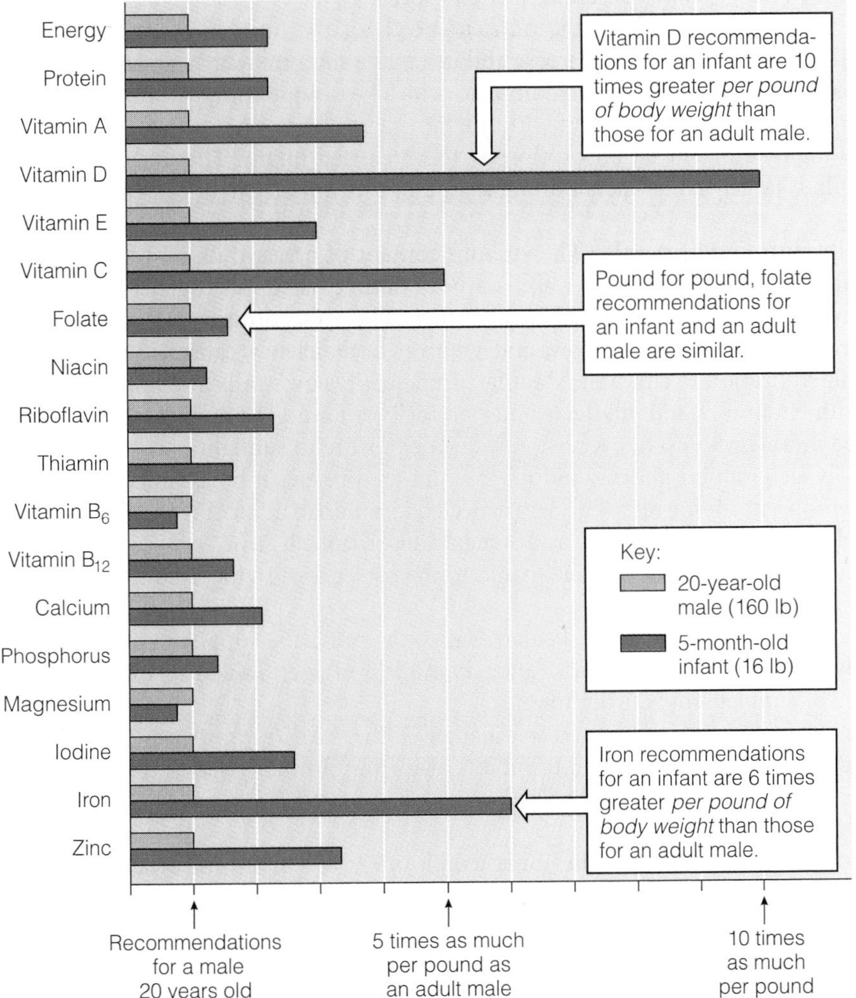

Figure 12.5

Nutrient Needs of a Five-Month-Old Infant and an Adult Male Compared on the Basis of Body Weight

Infants may be relatively small and inactive, but they use large amounts of energy and nutrients, in proportion to their body size, to keep all their metabolic processes going.

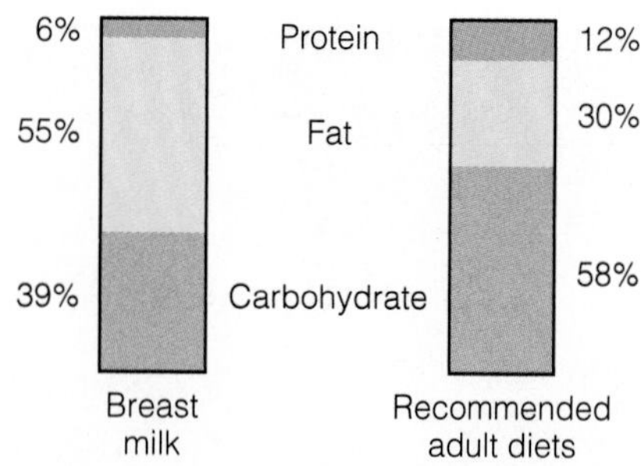

Figure 12.6
Percentages of Energy-Yielding Nutrients in Breast Milk and in Recommended Adult Diets
The balance of energy-yielding nutrients in human breast milk is ideal for infants and does not resemble the balance recommended for adults.

sweating, or obligatory urinary loss without replacement, can rapidly propel an infant into life-threatening dehydration. In early infancy, breast milk or formula normally provides enough water for a healthy infant to replace water losses from the skin, lungs, feces, and urine.[52] An infant who is exposed to hot weather, has diarrhea, or vomits repeatedly, however, needs supplemental water to prevent dehydration. Infants cannot tell you what they are crying for; remember that they may need water, and let them drink it until they quench their thirst.

In developed countries with well-nourished populations, such as the United States and Canada, the dietary practices that influence infants' nutrition status the most center upon which type of milk the infant receives and the age at which solid foods are introduced. The remainder of this discussion is devoted to feeding the infant and identifying the nutrients most often deficient in infant diets.

Breast Milk

Breast milk excels as a source of nutrients for the young infant. With the possible exception of vitamin D, breast milk provides all the nutrients a healthy infant needs for the first four to six months of life, and disease-preventive benefits as well.

Energy Nutrients The energy nutrient balance of breast milk differs dramatically from the balance recommended for adults (see Figure 12.6). Yet for infants, breast milk is the most nearly perfect food, proving that people at different stages of life really do have different nutrient needs.

Tailor-made to meet the nutrient needs of the human infant, breast milk offers its carbohydrate as lactose and its fat as a mixture with a generous proportion of the essential fatty acid linoleic acid. The unique composition of the fat in breast milk, in combination with the fat-digesting enzymes present, contributes to highly efficient fat absorption by the breastfed infant. The protein in breast milk is largely alpha-lactalbumin, a protein the human infant can easily digest.

alpha-lactalbumin (lackt-AL-byoo-min): the chief protein in human breast milk, as **casein** (CAY-seen) is the chief protein in cow's milk.

Vitamins and Minerals The vitamin content of breast milk is ample. Even vitamin C, for which cow's milk is a poor source, is supplied generously by the breast milk of a well-nourished mother. The concentration of vitamin D in breast milk, however, is low, and vitamin D deficiency causes impaired bone mineralization in children. Manufacturers fortify cow's milk and infant formulas with vitamin D, and physicians may prescribe vitamin D supplements for breastfed infants who do not receive sufficient exposure to sunlight.

Calcium, phosphorus, and magnesium are present in breast milk in amounts appropriate for the rate of growth expected in a human infant. Breast milk is healthfully low in sodium, and its iron is highly absorbable. Its zinc, too, is absorbed better than zinc from cow's milk, thanks to the presence of a zinc-binding protein.

Supplements for Infants Pediatricians may prescribe supplements containing vitamin D, iron, and fluoride (after six months of age). Table 12.5 offers a schedule of supplements during infancy.

In addition, as discussed in Chapter 7, the AAP recommends that a single dose of vitamin K be given to infants at birth.[53] In many states, this preventive dose of vitamin K is required by law.

colostrum (co-LAHS-trum): a milklike secretion from the breast that is rich in protective factors; it is present during the first day or so after delivery, before milk appears.

Immunological Protection Breast milk offers the infant unsurpassed protection against infection. Protective factors include antiviral agents, antibacterial agents, and other infection inhibitors.

During the first two or three days of lactation, the breasts produce colostrum, a premilk substance containing antibodies and white cells from the mother's blood. Colostrum is relatively sterile as it leaves the breast, and the infant cannot contract a bacterial infection from it even if the mother has one. Colostrum con-

Table 12.5 Supplements for Full-Term Infants

	Vitamin D[a]	Iron[b]	Fluoride[c]
Breastfed infants:			
Birth to six months of age	√		
Six months to one year	√	√	√
Formula-fed infants:			
Birth to six months of age			
Six months to one year		√	√

[a]Vitamin D supplements are recommended for infants whose mothers are vitamin D deficient and for those who do not receive adequate exposure to sunlight.
[b]Infants four to six months of age need additional iron, preferably in the form of iron-fortified cereal for both breastfed and formula-fed infants and iron-fortified infant formula for formula-fed infants.
[c]The Committee on Nutrition of the American Academy of Pediatrics recommends initiating fluoride supplements at six months of age for breastfed infants, formula-fed infants who receive ready-to-use formulas (these are prepared with water low in fluoride), and those who receive formula mixed with water that contains little or no fluoride (less than 0.3 ppm).

Sources: Adapted from American Academy of Pediatrics, Committee on Nutrition, Vitamin and mineral supplement needs of normal children in the United States, in *Pediatric Nutrition Handbook,* 3rd ed., ed. L. A. Barness (Elk Grove Village, Ill.: American Academy of Pediatrics, 1993), pp. 34–42; American Academy of Pediatrics, Committee on Nutrition, Fluoride supplementation for children: Interim policy recommendations, *Pediatrics* 95 (1995): 777.

tains maternal immune factors that inactivate harmful bacteria within the digestive tract. Later, breast milk also delivers immune factors, although not as many as colostrum. Among them are bifidus factors and lactoferrin.

Breast milk also contains several enzymes, several hormones (including thyroid hormone and prostaglandins), and lipids, all of which protect the infant against infection. Research suggests that breastfeeding offers better protection against wheezing during the first few months of life than formula feeding does.[54] It seems, too, that breastfed babies are less prone to develop stomach and intestinal disorders during the first few months of life and so experience less vomiting and diarrhea than formula-fed infants do. Research shows that exclusive breastfeeding for at least four months protects babies against middle ear infection, one of the most common illnesses of infancy.[55] Researchers speculate that the upright position in which breastfed babies nurse keeps fluid from entering the tubes that drain the middle ear. Furthermore, the antibodies in breast milk keep bacteria from clinging to the tubes. Much remains to be learned about the composition and characteristics of human milk, but clearly it is a very special substance. Nutrition in Practice 12 offers suggestions for successful breastfeeding.

The case study on p. 318 presents a woman who is four months pregnant. Answering the questions offers practice in thinking through some of the issues related to pregnancy and breastfeeding.

bifidus (BIFF-id-us, by-FEED-us) **factors:** factors in colostrum and breast milk that favor the growth of the "friendly" bacteria *Lactobacillus* (lack-toe-ba-SILL-us) *bifidus* in the infant's intestinal tract; these bacteria prevent other, less-desirable intestinal inhabitants from flourishing.

lactoferrin (lack-toe-FERR-in): a factor in breast milk that binds iron and keeps it from supporting the growth of the infant's intestinal bacteria.

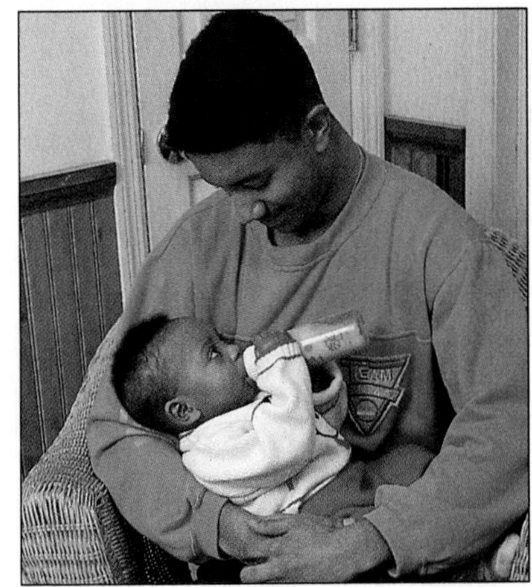

The infant thrives on infant formula offered with affection.

Infant Formula

Breastfeeding offers many benefits to both mother and infant, and it should be encouraged whenever possible. The mother who has decided to use formula, however, should be supported in her choice just as the breastfeeding mother should be. She can offer the same closeness, warmth, and stimulation during feedings as the breastfeeding mother can.

Many mothers choose to breastfeed at first but wean their children within the first 1 to 12 months. Before infants reach a year of age, mothers must wean them onto *infant formula,* not onto plain cow's milk of any kind—whole, reduced fat, low fat, or nonfat.

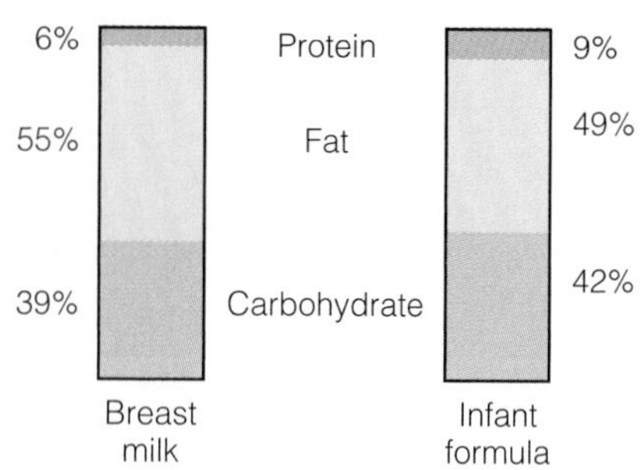

Figure 12.7

Percentages of Energy-Yielding Nutrients in Breast Milk and in Infant Formula

The proportions of energy-yielding nutrients in human breast milk and formula differ slightly.

Infant Formula Composition Formulas can be prepared from cow's milk in such a way that they do not differ significantly from human milk in nutrient content. Figure 12.7 illustrates the energy nutrient balance of both. Formulas contain no protective antibodies for human babies, but preventive medical care (vaccinations) and reliable public health measures help minimize this disadvantage. The educated mother whose water supply is reliable can prepare safe, sanitary formulas. Lead-contaminated water, however, is a major source of lead poisoning in infants.[56]

Infant Formula Standards National and international standards have been set for the nutrient contents of infant formulas. U.S. standards are based on AAP recommendations, and the Food and Drug Administration (FDA) mandates quality control procedures to ensure that the standards are met. All standard formulas are therefore nutritionally similar. Small differences in nutrient content are sometimes confusing but usually not important.

Special Formulas Standard formulas are inappropriate for some infants (see Figure 12.8). For example, premature babies require special formulas. Infants allergic to milk protein can drink special formulas based on soy protein. Soy formulas are usually lactose-free and so can be used for infants with lactose intolerance as well. For infants with other special needs, many other variations have been formulated.

Formula preparation:

- *Liquid concentrate (moderately expensive, relatively easy)—mix with equal part water.*
- *Powdered formula (least expensive, lightest for travel)—read label directions.*
- *Ready-to-feed (easiest, most expensive)—pour directly into clean bottles.*

Risks of Formula Feeding In developing countries and in poor areas of the United States, formula may be unavailable, overdiluted in an attempt to save money, or prepared with contaminated water. Overdilution of formula can cause malnutrition and growth failure. Contaminated formula often causes infections leading to diarrhea, dehydration, and failure to absorb nutrients. Wherever sanitation is poor, breastfeeding should take priority over feeding formula. Breast milk is sterile, and its antibodies enhance an infant's resistance to disease.

Iron in Formula The AAP recommends iron-fortified formula for all formula-fed infants. Low-iron infant formulas have no role in infant feeding. Use of iron-fortified formulas has risen in recent decades and is credited with the decline of iron-deficiency anemia in U.S. infants. Only iron-fortified formula can support normal development in an infant's first year.

nursing bottle tooth decay: extensive tooth decay due to prolonged tooth contact with formula, milk, fruit juice, or other carbohydrate-rich liquid offered to an infant in a bottle.

Nursing Bottle Tooth Decay Dentists advise against putting an infant to bed with a bottle. Salivary flow, which normally cleanses the mouth, diminishes as the baby falls asleep. Sucking for long times pushes the jawline out of shape and causes a bucktoothed profile (protruding upper and receding lower teeth). Furthermore, prolonged sucking on a bottle of formula, milk, or juice bathes the upper teeth in a carbohydrate-rich fluid that nourishes decay-producing bacteria. (The tongue covers and protects most of the lower teeth, but they, too, may be affected.) The result is extensive and rapid tooth decay. To prevent nursing bottle tooth decay, no child should be put to bed with a bottle as a pacifier.

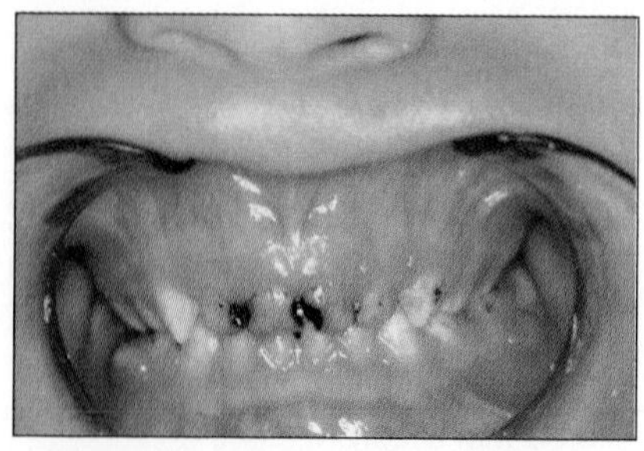

Nursing bottle syndrome, an extreme example.

The Transition to Cow's Milk

During an infant's first six months, formula must supply the nutrients of human milk in similar forms and proportions. Ordinary milk is an inappropriate replacement—primarily because cow's milk provides too little vitamin C and iron and too much sodium and protein. After the first year, the exact formulation of the milk selected is less critical, but milk or a suitable substitute still occupies a place in the diet that no other type of food can fill. Children one to two years of age should not drink reduced-fat, low-fat, or nonfat milk routinely; they need the fat of whole milk. If powdered milk is used, the fat-containing varieties should be chosen.

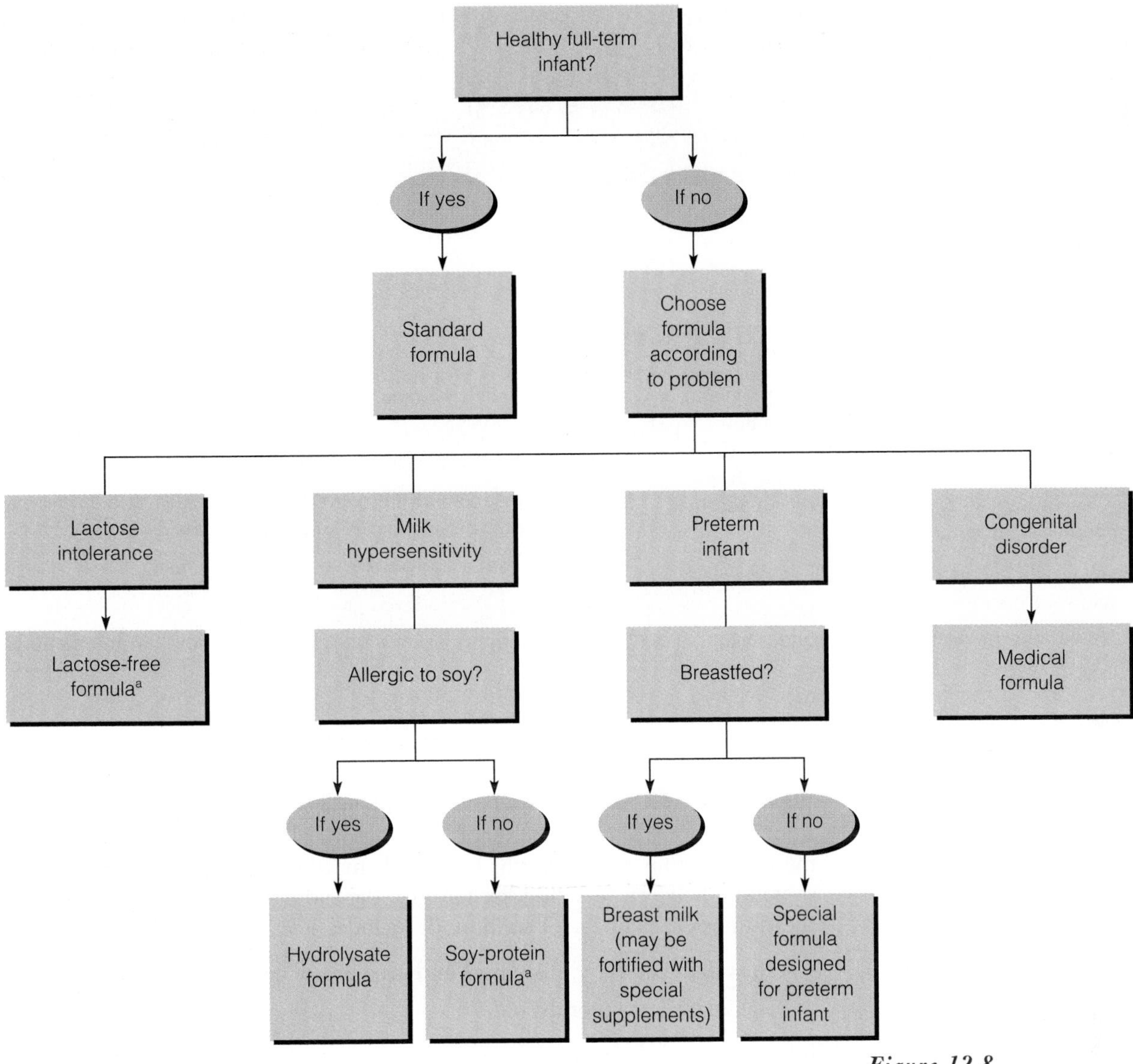

Figure 12.8
Choosing a Formula

[a]Manufacturers design soy-based formulas for infants with milk sensitivities—whether lactose intolerance or milk allergy. These formulas use corn syrup or sucrose in place of lactose.

Introducing First Foods

Changes in the body organs during the first year affect the infant's readiness to accept solid foods. The immature stomach and intestines can digest milk sugar (lactose), but not starch until they are several months old. This is one of the many reasons why breast milk and formula are such good foods for a baby; they provide simple, easily digested carbohydrate that supplies energy for the baby's growth and activity.

The Need for Water The infant's kidneys are unable to concentrate waste efficiently, so an infant must excrete relatively more water than an adult to carry off a comparable amount of waste. The risk of dehydration is ever present, and it becomes greater once solid foods are introduced. Foods high in protein or electrolytes such as meat and eggs can cause dehydration if offered without water. Water should be offered to infants regularly once they are eating solid food.[57] Water also provides fluid without additional food energy. Many adults would no doubt be healthier had they learned early to quench their thirst with water.

Case Study

PREGNANT WOMAN WITH WEIGHT PROBLEM

Ellen is a 24-year-old woman who is four months pregnant. This is her first pregnancy, and she is eager to learn how to feed herself during pregnancy as well as her infant after birth. She is 5 feet 3 inches tall and currently weighs 150 pounds. Her prepregnancy weight was 148 pounds. Ellen is very concerned about her 2-pound weight gain.

- Assess Ellen's weight and determine what her desirable weight would be if she were not pregnant.
- Should Ellen be concerned about her 2-pound weight gain? Why?
- What advice should you give Ellen about her weight gain during pregnancy?
- What other dietary advice would you give her?
- Discuss methods of infant feeding with Ellen and describe some of the advantages breastfeeding would offer her.
- What are the advantages of formula feeding?
- What advice will you give Ellen if she decides to breastfeed?
- What information should Ellen have about formula feeding?

When to Introduce Solid Food For an infant receiving formula or breast milk from a healthy, well-nourished mother, additions to the diet are not needed until the infant is four to six months old. Foods may be started gradually beginning sometime between four and six months, depending on the infant's readiness. Indications of readiness for solid foods include:

- The infant can sit with support and can control head movements.
- The infant is six months old.

Infants vary; and the program of additions depends on the individual baby's developmental readiness, not on any rigid schedule. Table 12.6 presents a suggested sequence for introducing new foods.

The addition of foods to an infant's diet should be governed by three considerations: the infant's nutrient needs, the infant's physical readiness to handle different forms of foods, and the need to detect and control allergic reactions. With respect to nutrient needs, the nutrient needed earliest is iron, then vitamin C.

Foods to Provide Iron and Vitamin C Iron deficiency is common in young children throughout the world, especially between the ages of six months and three years when they are growing fast and milk, which is a poor source of iron, has a large place in their diets. The iron an infant has stored from before birth typically runs out after the birthweight doubles, long before the end of the first year. This is why cow's milk should not be offered during the first year: it not only displaces iron-fortified formula but also causes GI blood loss in many infants. Infants can derive adequate iron first from breast milk or formula with iron, then from iron-fortified cereals, and later from meat or meat alternates such as legumes. Once infants are consuming iron-fortified cereals, parents or caregivers should begin selecting vitamin C–rich foods to go with meals to enhance iron absorption. The best sources of vitamin C are fruits and vegetables (see p. 160).

Physical Readiness for Solid Foods The ability to swallow solid food develops at around four to six months, and food offered by spoon helps to develop swal-

Table 12.6 First Foods for the Infant

Breast milk or iron-fortified formula is the only source of nourishment for the first 4 to 6 months. Throughout the first year, the infant's intake of breast milk or iron-fortified formula will gradually decline as solid food intake increases.

Age (mo)	Addition
4 to 6	Iron-fortified rice cereal, followed by other single-grain cereals, mixed with breast milk, formula, or water
	Pureed vegetables and fruits, one by one (perhaps vegetables before fruits, so the baby will learn to like their less sweet flavors)
6 to 8	Infant breads and crackers
	Mashed vegetables and fruits, and their juices[a]
8 to 10	Breads and cereals from the table
	Soft, cooked vegetables and fruit from the table
	Finely cut meats, fish, chicken, casseroles, cheeses, yogurts, tofu, eggs, and legumes
10 to 12	Continue to introduce a variety of nutritious foods

[a]All baby juices are fortified with vitamin C. Orange juice may cause allergies; apple juice may be a better choice at first. Dilute juices with water and offer in a cup to prevent nursing bottle tooth decay.

Source: Adapted in part from American Academy of Pediatrics, Committee on Nutrition, *Pediatric Nutrition Handbook*, 3rd ed., ed. L. A. Barness (Elk Grove Village, Ill.: American Academy of Pediatrics, 1993), pp. 23–33.

lowing ability. Six months is also a good time to introduce the cup. At eight months to a year, a baby can sit up, can handle finger foods, and begins to teethe. At that time, hard crackers and other hard finger foods may be introduced to promote the development of manual dexterity and control of the jaw muscles. These feedings must occur under the watchful eye of an adult because the infant can also choke on such foods.

Some parents want to feed solids at an earlier age, on the theory that "stuffing the baby" at bedtime promotes sleeping through the night. There is no proof for this theory. On the average, infants start to sleep through the night at about the same age (three to four months) regardless of when solid foods are introduced.

Allergy-Causing Foods New foods should be introduced singly and at intervals spaced to permit detection of allergies. For example, when cereals are introduced, rice cereal is offered first for several days; it causes allergy least often. Wheat cereal is offered last; it is the most common offender. If a cereal causes an allergic reaction (irritability due to skin rash, digestive upset, or respiratory discomfort), its use should be discontinued before going on to the next food.

Choice of Infant Foods Baby foods commercially prepared in the United States and Canada are safe, and except for mixed dinners and heavily sweetened desserts, they generally have high nutrient density. An alternative for the parent who wants the baby to have family foods is to "blenderize" a small portion of the table food (cooked without salt) at each meal.

Foods to Omit Sweets of any kind (including baby food "desserts") have no place in a baby's diet. The added food energy they contribute can promote obesity, and they convey few or no nutrients to support growth. Canned vegetables are inappropriate for infants; they often contain too much sodium. Honey should never be fed to infants because of the risk of botulism. Babies and even young children have difficulty swallowing foods such as popcorn, whole grapes, whole beans, hot dog slices, and nuts, and they can easily choke on these foods.

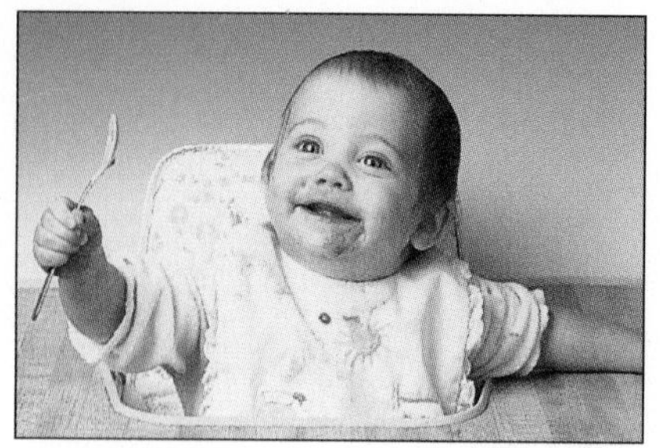

Ideally, a one-year-old eats many of the same foods as the rest of the family.

Reminder: Milk anemia develops when an excessive milk intake displaces iron-rich foods from the diet.

Also, an infant's caregiver must be on guard against food poisoning and take the precautions against it described in Chapter 11.

Foods at One Year Whole milk is the best food to supply most of the nutrients the infant needs at a year of age; 2 to 3½ cups a day meet those needs sufficiently. More milk than this displaces food necessary to provide iron and can cause the iron-deficiency anemia known as milk anemia. Other foods—meat and meat alternates, iron-fortified cereal, enriched or whole-grain bread, fruits, and vegetables—should be supplied in variety and in amounts sufficient to round out total energy needs. Ideally, a one-year-old will sit at the table, eat many of the same foods everyone else eats, and drink liquids from a cup—not a bottle. The sample menu shows a meal plan that meets a one-year-old's requirements.

Looking Ahead

Probably the most important single measure to undertake during the first year is to encourage eating habits that will support continued normal weight as the child grows. This means introducing a variety of nutritious foods in an inviting way, not forcing the baby to finish the bottle or the baby food jar, avoiding concentrated sweets and empty-kcalorie foods, and encouraging physical activity. Parents should avoid teaching infants to seek food as a reward, to expect food as comfort for unhappiness, or to associate food deprivation with punishment. If infants cry from thirst, give them water, not milk or juice. Infants seem to have no internal "kcalorie counter," and they stop eating when their stomachs feel full. Nutrient-dense, low-kcalorie foods will satisfy as long as they provide bulk.

Normal dental development is also promoted by supplying nutritious foods, avoiding sweets, and discouraging the association of food with reward or comfort. Dental health is the subject of Nutrition in Practice 2.

The AAP recommends against a fat-modified diet during infancy. The available evidence does not warrant dietary manipulation to lower serum cholesterol in infants.

Mealtimes

The wise parent of a one-year-old offers nutrition and love together. Both promote growth. Children "fed with love" grow more in both weight and height than children fed the same food in an emotionally negative climate.

Sample Menu for a One-Year Old

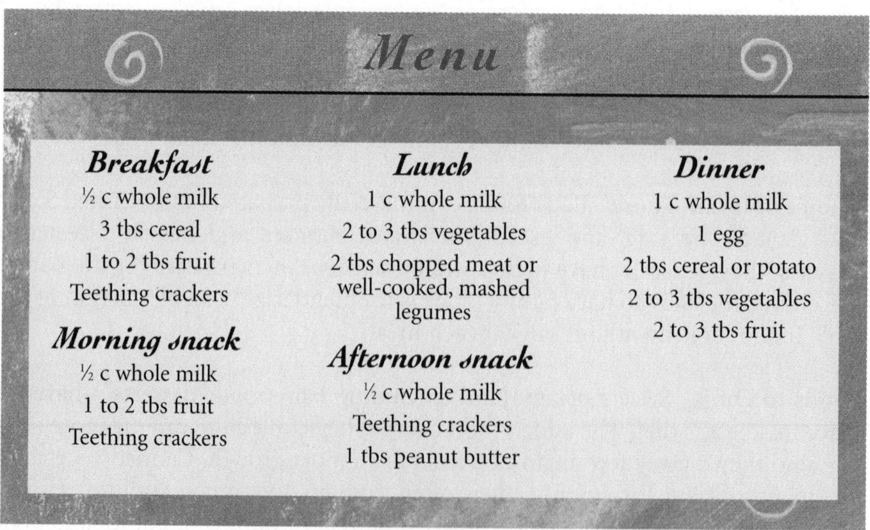

Menu
Breakfast
½ c whole milk
3 tbs cereal
1 to 2 tbs fruit
Teething crackers
Morning snack
½ c whole milk
1 to 2 tbs fruit
Teething crackers
Lunch
1 c whole milk
2 to 3 tbs vegetables
2 tbs chopped meat or well-cooked, mashed legumes
Afternoon snack
½ c whole milk
Teething crackers
1 tbs peanut butter
Dinner
1 c whole milk
1 egg
2 tbs cereal or potato
2 to 3 tbs vegetables
2 to 3 tbs fruit

Note: Fruit choices should include citrus fruits, melons, and berries, and vegetable choices should include dark green, leafy and deep yellow vegetables.

The person feeding a one-year-old should be aware that exploring and experimenting are normal and desirable behaviors at this time in a child's life. The child is developing a sense of autonomy that, if allowed to develop, will lay the foundation for later confidence and effectiveness as an individual. The child's impulses, if consistently denied, can turn to shame and self-doubt. In light of the developmental and nutrient needs of one-year-olds, and in the face of their often contrary and willful behavior, a few feeding guidelines may be helpful:

- *Discourage unacceptable behavior (such as standing at the table or throwing food) by removing the child from the table to wait until later to eat.* Be consistent and firm, not punitive. The child will soon learn to sit and eat.
- *Let the child explore and enjoy food.* This may mean the child eats with fingers for a while. Use of the spoon will come in time.
- *Don't force food on children.* Provide children with nutritious foods, and let them choose which ones and how much they will eat. Gradually, they will acquire a taste for different foods. If children refuse milk, provide cheese, cream soups, and yogurt.
- *Limit sweets strictly.* Infants have little room in their 1000-kcalorie daily energy allowance for empty-kcalorie sweets, except occasionally.

These recommendations reflect a spirit of tolerance that serves the best interests of the child emotionally as well as physically.

In summary, the primary food for infants during the first 6 to 12 months of life is either breast milk or iron-fortified infant formula. In addition to nutrients, breast milk offers immunological protection. At about 4 to 6 months of age, infants should gradually begin eating solid foods so that by one year, they are drinking from a cup and eating many of the same foods as the rest of the family.

Nutrition Assessment

Assessing the nutrition status of pregnant women and of infants offers an opportunity to detect and correct potential problems in their future lives. Accurate histories and measurements of growth and development are especially useful.

Chapter 15 offers details about health, drug, and diet histories; Chapter 16 discusses measures of growth and development.

- The diet history should uncover normal eating habits in all settings. Have the woman's food choices changed since she became pregnant? How did she eat before? Does the same person always feed the infant? Establish baseline eating habits so that realistic improvements can be recommended. Watch for the use of monotonous or bizarre diets.
- For the pregnant woman, does she eat folate-rich foods daily? Are her calcium and iron intakes ample? Does she use alcohol or drugs (prescription or other)? Are her physical activity habits appropriate and sustainable for the future?
- The history should pick up on low socioeconomic status in pregnant women and new mothers. Referral to food assistance programs such as WIC may be appropriate. Also be alert for the risk factors for gestational diabetes.
- Accurately measuring a pregnant woman's weight and plotting it against previous measures are useful for keeping tabs on development over time.
- In the lactating mother, a careful history can determine not only eating and physical activity habits but also alcohol and caffeine use, smoking, and medication and drug use as well as maternal illness that might preclude breastfeeding. Is the breastfeeding mother receiving appropriate supplementation?

- In assessing the infant, a careful history should determine whether the infant is or has been breastfed, what formula is in use, and what solid foods are being fed. Are iron and vitamin C needs covered? Are the timing and style of feeding safe for infant tooth development? Are all foods nutrient-dense? Are they prepared safely?
- In the infant, accurately measuring weight, length, and head circumference and plotting measurements against previous measures will reveal any deviations from normal growth.

Other parameters become important when pregnant women or infants have diseases or disorders that require medical intervention. Later chapters cover these conditions.

Self Check

1. The most important single predictor of an infant's future health and survival is:
   a. the infant's birthweight.
   b. the infant's iron status at birth.
   c. the mother's weight at delivery.
   d. the mother's prepregnancy weight.
2. A mother's prepregnancy nutrition is important to a healthy pregnancy because it determines the development of:
   a. adequate maternal iron stores.
   b. the largest baby possible.
   c. an adequate fat supply for the mother.
   d. healthy support tissues—the placenta, amniotic sac, umbilical cord, and uterus.
3. Two vitamins needed in large amounts during pregnancy for rapid cell proliferation are:
   a. vitamin B_{12} and iron. b. calcium and vitamin B_6.
   c. folate and vitamin B_{12}. d. copper and zinc.
4. For a woman who is at the appropriate weight for height and is carrying a single fetus, the recommended weight gain during pregnancy is:
   a. 40 to 60 pounds. b. 25 to 35 pounds.
   c. 10 to 20 pounds. d. 20 to 40 pounds.
5. Rewards of physical activity during pregnancy may include:
   a. weight loss.
   b. reduced stress, easier labor, and faster recovery from childbirth.
   c. relief from morning sickness.
   d. decreased incidence of pica.
6. A woman's need for _____ is greater during lactation than during pregnancy.
   a. vitamin D b. energy
   c. calcium d. folate
7. Breast milk is recommended for the first 12 months of life because it offers complete nutrition and _____ to the infant.
   a. fluoride b. fructose
   c. immunological protection d. pica
8. An acceptable substitute for breast milk during the first year is:
   a. low-fat cow's milk. b. apple juice.
   c. water. d. iron-fortified infant formula.
9. Indications of readiness for solid foods include:
   a. the infant cries a lot.
   b. the infant is not sleeping through the night.
   c. the infant can sit with support and can control head movements.
   d. the infant is two months old.
10. During the first year of life, the most important step to undertake to encourage healthy eating is to:
    a. give food as a reward or for comfort.
    b. give sweets as a reward for eating vegetables.
    c. introduce a variety of nutritious foods in an inviting way.
    d. all of the above.

Answers to these questions appear in Appendix H.

Notes

1. National Academy of Sciences, Food and Nutrition Board, *Nutrition during Pregnancy* (Washington, D.C.: National Academy Press, 1990), pp. 176–211.

2. A. L. Owen and G. M. Owen, Twenty years of WIC: A review of some effects of the program, *Journal of the American Dietetic Association* 97 (1997): 777–782; C. M. Olson, Promoting positive nutritional practices during pregnancy and lactation, *American Journal of Clinical Nutrition* 59 (1994): 525S–531S.

3. W. W. Hay and coauthors, Workshop summary: Fetal growth: Its regulation and disorders, *Pediatrics* 99 (1997): 585–591; Transplacental nutrient transfer and intrauterine growth retardation, *Nutrition Reviews* 50 (1992): 56–57.

4. National Academy of Sciences, Food and Nutrition Board, 1990, pp. 1–23.

5. T. O. Scholl and M. L. Hediger, Anemia and iron-deficiency anemia: Compilation of data on pregnancy outcome, *American Journal of Clinical Nutrition* 59 (1994): 492S–501S.

6. National Academy of Sciences, Food and Nutrition Board, 1990.

7. Y. H. Neggers and coauthors, A positive association between maternal serum zinc concentration and birth weight, *American Journal of Clinical Nutrition* 51 (1990): 678–684.

8. Owen and Owen, 1997.

9. National Academy of Sciences, Food and Nutrition Board, 1990, pp. 63–96.

10. K. G. Dewey and M. A. McCrory, Effects of dieting and physical activity on pregnancy and lactation, *American Journal of Clinical Nutrition* 59 (1994): 446S–453S.

11. J. M. Pivarnik, Potential effects of maternal physical activity on birth weight: Brief review, *Medicine and Science in Sports and Exercise* 30 (1998): 400–406.

12. M. C. Hatch and coauthors, Maternal exercise during pregnancy, physical fitness, and fetal growth, *American Journal of Epidemiology* 137 (1993): 1105–1114.

13. *ACOG Guide to Planning for Pregnancy, Birth, and Beyond* (Washington, D.C.: The American College of Obstetricians and Gynecologists, 1990), pp. 128–140.

14. C. D. Naylor and coauthors, Cesarean delivery in relation to birth weight and gestational glucose tolerance: Pathophysiology or practice style? *Journal of the American Medical Association* 275 (1996): 1165–1170.

15. B. M. Sibai, Treatment of hypertension in pregnant women, *New England Journal of Medicine* 335 (1996): 257–265.

16. N. I. Hahn and M. Erick, Battling morning (noon and night) sickness: New approaches for treating an age-old problem, *Journal of the American Dietetic Association* 94 (1994): 147–148.

17. National Academy of Sciences, Food and Nutrition Board, 1990.

18. U.S. Department of Health and Human Services, Public Health Service, *The Health Benefits of Smoking Cessation: A Report of the Surgeon General, 1990* (Washington, D.C.: Government Printing Office, 1990).

19. American Academy of Pediatrics, Committee on Environmental Health, Environmental tobacco smoke: A hazard to children, *Pediatrics* 99 (1997): 639–642; E. Cutz and coauthors, Maternal smoking and pulmonary neuroendocrine cells in sudden infant death syndrome, *Pediatrics* 98 (1996): 668–672; H. S. Klonoff-Cohen and coauthors, The effect of passive smoking and tobacco exposure through breast milk on sudden infant death syndrome, *Journal of the American Medical Association* 273 (1995): 795–798; E. A. Mitchell and coauthors, Smoking and sudden infant death syndrome, *Pediatrics* 91 (1993): 893–896; K. C. Schoendorf, Relationship of sudden infant death syndrome to maternal smoking during and after pregnancy, *Pediatrics* 90 (1992): 905–908; B. Haglund and S. Cnattingus, Cigarette smoking as a risk factor for sudden infant death syndrome, *American Journal of Public Health* 80 (1990): 29–32.

20. D. L. Olds, C. R. Henderson, and R. Tatelbaum, Intellectual impairment in children of women who smoke cigarettes during pregnancy, *Pediatrics* 93 (1994): 221–227.

21. National Academy of Sciences, Food and Nutrition Board, 1990, pp. 390–411.

22. T. S. Hinds and coauthors, The effect of caffeine on pregnancy outcome variables, *Nutrition Reviews* 54 (1996): 203–207; National Academy of Sciences, Food and Nutrition Board, 1990.

23. J. L. Mills and coauthors, Moderate caffeine use and the risk of spontaneous abortion and intrauterine growth retardation, *Journal of the American Medical Association* 269 (1993): 593–597.

24. B. M. Sibai and coauthors, Prevention of preeclampsia with low-dose aspirin in healthy, nulliparous, pregnant women, *New England Journal of Medicine* 329 (1993): 1213–1218.

25. W. A. Vega and coauthors, Prevalence and magnitude of perinatal substance exposures in California, *New England Journal of Medicine* 329 (1993): 850–854.

26. F. D. Eyler and coauthors, Birth outcome from a prospective, matched study of prenatal crack/cocaine use I: Interactive and dose effects on health and growth, *Pediatrics* 101 (1998): 229–237; F. D. Eyler and coauthors, Birth outcome from a prospective, matched study of prenatal crack/cocaine use II: Interactive and dose effects on neurobehavioral assessment, *Pediatrics* 101 (1998): 237–241; I. J. Chasnoff and coauthors, Cocaine/polydrug use in pregnancy: Two-year follow-up, *Pediatrics* 89 (1992): 284–289.

27. D. B. Petitti and C. Coleman, Cocaine and the risk of low birth weight, *American Journal of Public Health* 80 (1990): 25–28; J. J. Volpe, Effect of cocaine use on the fetus, *New England Journal of Medicine* 327 (1992): 399–407.

28. L. C. Mayes and coauthors, Neurobehavioral profiles of neonates exposed to cocaine prenatally, *Pediatrics* 91 (1993): 778–783; M. J. Corwin and coauthors, Effects of in utero cocaine exposure on newborn acoustical cry characteristics, *Pediatrics* 89 (1992): 1199–1203.

29. American Academy of Pediatrics, Committee on Substance Abuse and Committee on Children and Disabilities, Fetal alcohol syndrome and fetal alcohol effects, *Pediatrics* 91 (1993): 1004–1006.

30. Position of The American Dietetic Association: Nutrition care for pregnant adolescents, *Journal of the American Dietetic Association* 94 (1994): 449–450.

31. M. Story and I. A. Alton, Nutrition issues and adolescent pregnancy, *Nutrition Today* 30 (1995): 142–151.

32. Position of The American Dietetic Association, 1994; J. L. Beard, Iron deficiency: Assessment during pregnancy and its importance in pregnant adolescents, *American Journal of Clinical Nutrition* 59 (1994): 502S–510S.

33. National Academy of Sciences, Food and Nutrition Board, 1990, pp. 1–23.

34. M. L. Hediger and coauthors, Rate and amount of weight gain during adolescent pregnancy: Associations with maternal weight-for-height and birth weight, *American Journal of Clinical Nutrition* 52 (1990): 793–799.

35. Position of The American Dietetic Association, 1994.

36. American Academy of Pediatrics, Work Group on Breastfeeding, Breastfeeding and the use of human milk, *Pediatrics* 100 (1997): 1035–1039.

37. Position of The American Dietetic Association: Promotion of breastfeeding, *Journal of the American Dietetic Association* 97 (1997): 662–666.

38. K. G. Dewey and coauthors, Maternal versus infant factors related to breast milk and residual milk volume: The DARLING Study, *Pediatrics* 87 (1991): 829–837; National Academy of Sciences, Food and Nutrition Board, *Nutrition during Lactation* (Washington, D.C.: National Academy Press, 1991), pp. 1–19.

39. J. M. A. van Raaij and coauthors, Energy cost of lactation, and energy balances of well-nourished Dutch lactating women: Reappraisal of the extra energy requirements of lactation, *American Journal of Clinical Nutrition* 53 (1991): 612–619.

40. Position of The American Dietetic Association, 1997.

41. F. M. Kramer and coauthors, Breast-feeding reduces maternal lower-body fat, *Journal of the American Dietetic Association* 93 (1993): 429–433.

42. National Academy of Sciences, Food and Nutrition Board, 1991, pp. 197–212.

43. M. J. Heinig and K. G. Dewey, Health effects of breast feeding for mothers: A critical review, *Nutrition Research Reviews* 10 (1997): 35–56.

44. National Academy of Sciences, Food and Nutrition Board, 1991, p. 140.

45. National Academy of Sciences, Food and Nutrition Board, 1991, p. 140.

46. L. B. Dusdieker and coauthors, Prolonged maternal fluid supplementation in breast-feeding, *Pediatrics* 86 (1990): 737–740.

47. J. A. Mennella, and G. K. Beauchamp, The transfer of alcohol to human milk: Effects on flavor and the infant's behavior, *New England Journal of Medicine* 325 (1991): 981–985.

48. Highlights from U.S.D.A. research, *Nutrition Today,* January/February 1993, pp. 4–5.

49. American Academy of Pediatrics, Committee on Drugs, The transfer of drugs and other chemicals into human milk, *Pediatrics* 93 (1994): 137–150.

50. American Academy of Pediatrics, Committee on Drugs, 1994.

51. American Academy of Pediatrics, Committee on Drugs, 1994.

52. American Academy of Pediatrics, Committee on Nutrition, *Pediatric Nutrition Handbook,* 3rd ed., ed. L. A. Barness (Elk Grove Village, Ill.: American Academy of Pediatrics, 1993), pp. 23–33.

53. American Academy of Pediatrics, Committee on Nutrition, 1993, pp. 24–42.

54. American Academy of Pediatrics, Work Group on Breastfeeding, 1997.

55. B. Duncan and coauthors, Exclusive breast-feeding for at least 4 months protects against otitis media, *Pediatrics* 91 (1993): 867–872.

56. M. W. Shannon and J. W. Graef, Lead intoxication in infancy, *Pediatrics* 89 (1992): 87–90.

57. American Academy of Pediatrics, Committee on Nutrition, 1993, pp. 23–33.

CHAPTER 12

Nutrition in Practice

▪ Encouraging Successful Breastfeeding ▪

Breastfeeding offers benefits to both mother and infant. The AAP, the ADA, and the U.S. Department of Health and Human Services all advocate breastfeeding as the preferred means of infant feeding.[1] Promotion of breastfeeding is an integral part of the WIC program's nutrition education component.[2] During the late 1980s, breastfeeding was on the decline after reaching a high of about 60 percent in 1984. Once again, an encourgaging trend of increasing breastfeeding rates is emerging, with almost 60 percent of women initiating breastfeeding in 1995.[3] Nevertheless, only about one in five infants is still being breastfed at five to six months of age. Many of the benefits of breastfeeding are enhanced by breastfeeding for at least four months. The AAP recommends that breastfeeding continue for at least a year and thereafter for as long as mutually desired.[4] Increasing the rates of breastfeeding initiation and duration is one of the goals of *Healthy People 2000:*

> *Increase to at least 75% the proportion of mothers who breastfeed their babies in the early postpartum period, and to at least 50% the proportion who continue breastfeeding until their babies are 5 or 6 months of age.*[5]

Despite the trend toward increasing breastfeeding, the percentage of women choosing to breastfeed their babies and continuing to do so is far below the *Healthy People 2000* goal.

WWW.

eatright.org
American Dietetic Association

aap.org
American Academy of Pediatrics

ceo-cap.org/wic.htm
Women, Infants and Children

who.org
World Health Organization

os.dhhs.gov
U.S. Department of Health and Human Services

Why don't more women choose to breastfeed their babies?

Many experts cite two major deterrents: public advertising of infant formula, and the medical community's failure to encourage breastfeeding. As an example of the medical lack of encouragement, some hospitals routinely separate mother and child soon after birth. The child's first feeding then comes from the bottle rather than the breast. Furthermore, many hospitals send new mothers home with free samples of infant formula. The World Health Organization opposes this practice because it sends a misleading message that medical authorities favor infant formula over breast milk for infants. Even in hospitals where women are encouraged to breastfeed and supported in doing so, little if any assistance is available after hospital discharge, and many breastfeeding women still need assistance. Up to half of mothers who initially breastfeed their infants stop within a month—seemingly due to lack of knowledge. Women who receive early and repeated breastfeeding information and support breastfeed their infants longer than other women do.[6] Information and instruction are especially important during the prenatal period when most women decide whether to breastfeed or to feed formula. Health professionals can play a vital role in encouraging successful breastfeeding by offering women adequate, accurate information about breastfeeding that permits them to make informed choices.

I thought that breastfeeding was a natural process that didn't require any learning.

Although *lactation* is an automatic physiological process, *breastfeeding* requires some learning. This learning is most successful in a supportive environment. It begins with preparatory steps taken before the baby is born.

What are these preparatory steps?

Toward the end of pregnancy and throughout lactation, a woman who intends to breastfeed should stop

Nutrition in Practice

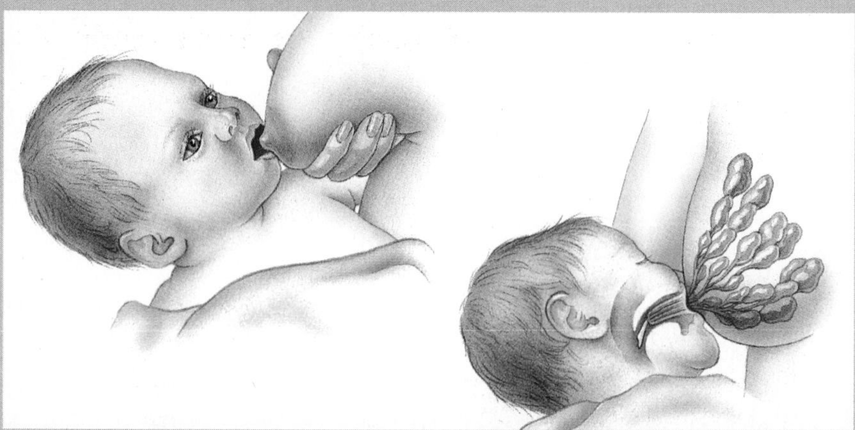

Figure 12.1
Infant's Grasp on Mother's Breast
The mother squeezes the areola, slipping enough of it into the infant's mouth to promote good pumping action. The infant's lips and gums pump the areola, releasing milk from the mammary glands into the milk ducts that lie beneath the areola.

using soap and lotions on her breasts. The natural secretions of the breasts themselves lubricate the nipple area best. A few weeks before the baby is due, the woman should allow her breasts to rub against her outer clothing for a little while each day to toughen the nipples somewhat in preparation for the baby's sucking. Also, she should occasionally go without clothing at home to expose her breasts to air and light.

A woman who plans to breastfeed should acquire at least two nursing bras before her baby is born. The bras should provide good support and have drop-flaps so that either breast can be freed for nursing.

How soon after birth should breastfeeding start?

As soon as possible. Immediately after the delivery, for a short period, the baby is intensely alert and intent on suckling. This is the ideal time for the first breastfeeding and facilitates successful lactation.

What does the new mother need to know in order to breastfeed her infant successfully?

She needs to learn how to relax and position herself so that she and the infant will be comfortable and the infant can breathe freely while nursing. She also needs to understand that infants have a rooting reflex that makes them turn toward any touch on the face. (The accompanying glossary defines this and other relevant terms.) Consequently, she should touch the infant's cheek to her nipple so that the infant will turn the right way and start to nurse. The mother can then squeeze her areola, the colored ring around the nipple, between two fingers and slip enough of it into the infant's mouth to permit a good hold and strong pumping action (see Figure NP12.1). The nipple must rest well back on the infant's tongue so that the infant's gums will squeeze on the glands that release the milk and swallowing will be effortless. To break the suction, if necessary, the mother can slip a finger between the infant's mouth and her breast.

Doesn't it hurt to have the infant sucking so hard on the breast?

No, because the mother has a letdown reflex that forces milk to the front of her breast when the infant begins to nurse, virtually propelling the milk into the infant's mouth. Letdown is necessary for the infant to obtain milk easily, and the mother needs to relax for letdown to occur. The mother who assumes a comfortable position in an environment without interruptions will find it easiest to relax.

How long should the baby be allowed to nurse at each feeding?

Although the infant sucks half the milk from the breast within the first 2 minutes, and 80 to 90 percent of it within 4 minutes, sucking on each breast for 10 to 15 minutes is encouraged. The sucking itself, as well as the complete removal of milk from the breast, stimulates the mammary glands to produce milk for the next nursing session. Successive sessions should start on alternate breasts to ensure that each breast is emptied regularly. This pattern maintains the same supply and demand for each breast and thus prevents either breast from overfilling.

Infants should be fed "on demand" and not held to a rigid schedule. The breastfed baby may average 8 to 12 feedings per 24-hour period during the first month or so. Once the mother's milk supply is well established

Glossary of Breastfeeding Terms

engorgement: overfilling of the breasts with milk.

letdown reflex: the reflex that forces milk to the front of the breast when the infant begins to nurse.

mastitis: infection of a breast.

rooting reflex: a reflex that causes an infant to turn toward whichever cheek is touched, in search of a nipple.

and the infant's capacity has increased, the intervals between feedings will become longer.

What if a mother wants to skip one or two feedings daily—for example, because she works outside the home?

The mother can express breast milk into a bottle ahead of time, freeze the breast milk, and, when needed, substitute the expressed breast milk for a nursing session. Breast milk can be kept refrigerated for 48 hours or frozen (at a freezer temperature below 0°F) for several months.

The mother can hand express her breast milk or use one of several different breast pumps available. The bicycle-horn type of manual breast pump is not recommended, however. These pumps are difficult to keep clean. Cylinder-type manual pumps or electric breast pumps are safer and are also more efficient. Alternatively, a mother can substitute formula for those feedings she will miss and continue to breastfeed at other feedings.

What about problems associated with breastfeeding such as sore nipples or infection of the breast?

Most problems associated with breastfeeding can be resolved. Many mothers experience sore nipples during the initial days of breastfeeding. Sore nipples need to be treated kindly, but nursing can continue. Improper feeding position is a frequent cause of sore nipples: the mother should make sure the infant is taking the entire nipple and part of the areola onto the tongue. She should nurse on the less-sore breast first to get letdown going while the infant is sucking hardest; then she can switch to the sore breast. Between times, she should expose her nipples to light and air to heal them.

Before lactation is well established, when the schedule changes, or when a feeding is missed, the breasts may become full and hard—an uncomfortable condition known as engorgement. The infant cannot grasp an engorged nipple and so cannot provide relief by nursing. A gentle massage or warming the breasts with a heating pad or in a shower helps to initiate letdown and to release some of the accumulated milk; then the mother can pump out some of her milk and allow the infant to nurse.

Infection of the breast, known as mastitis, is best managed by *continuing to breastfeed.* By drawing off the milk, the infant helps to relieve pressure in the infected area. The infant is safe because the infection is between the milk-producing glands, not inside them.

Even if everything is going smoothly, the nursing mother should ideally have enough help and support so that she can rest in bed a few hours each day for the first week or so. Successful breastfeeding requires the support of all those who care. This, plus adequate nutrition, ample fluids, fresh air, and exercise, will do much to enhance the well-being of mother and infant.

Notes

1. American Academy of Pediatrics, Work Group on Breastfeeding, Breastfeeding and the use of human milk, *Pediatrics* 100 (1997): 1035–1039; Position of The American Dietetic Association: Promotion of breastfeeding, *Journal of the American Dietetic Association* 97 (1997): 662–666.

2. A. L. Owen and G. M. Owen, Twenty years of WIC: A review of some effects of the program, *Journal of the American Dietetic Association* 97 (1997): 777–782.

3. Position of The American Dietetic Association, 1997.

4. American Academy of Pediatrics, 1997.

5. *Healthy People 2000: National Health Promotion/Disease Prevention Objectives* (Washington, D.C.: Government Printing Office, 1990).

6. C. I. Dungy and coauthors, Effect of discharge samples on duration of breastfeeding, *Pediatrics* 90 (1992): 233–237.

13 *Nutrition for Children, Teenagers, and Young Adults*

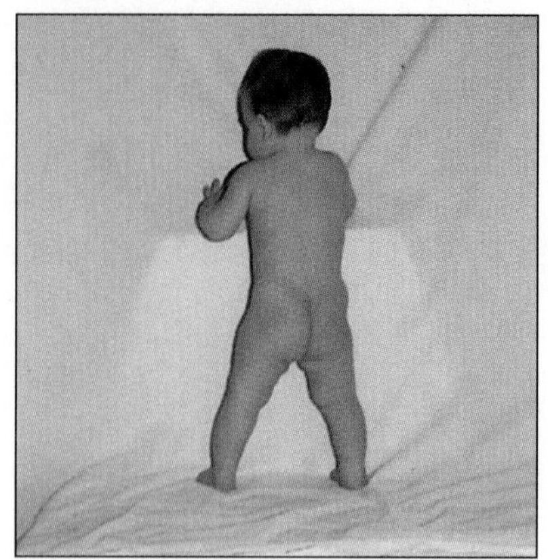

The body shape of a one-year-old (above) changes dramatically by age two (below). The two-year-old has lost much of the baby fat; the muscles (especially in the back, buttocks, and legs) have firmed and strengthened; and the leg bones have lengthened.

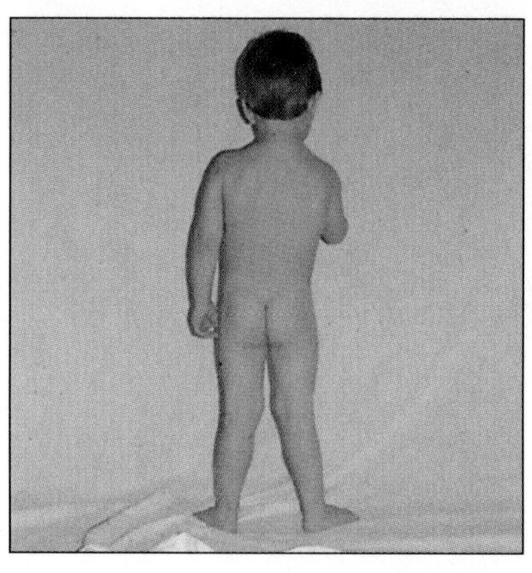

WWW.
kidshealth.org
Kids Health

kidsfood.org
Kids Food Cyber Club

Nutrient needs change throughout life, depending on genetics, rates of growth, activity, and many other factors. Nutrient needs also vary from individual to individual, but generalizations are possible and useful. Sound nutrition throughout childhood promotes normal growth and development; facilitates academic and physical performance; and helps prevent obesity, heart disease, cancer, and other degenerative diseases in adulthood. As children enter the teen years, a foundation built by years of eating nutritious foods best prepares them to meet the upcoming demands of rapid growth.

Early and Middle Childhood

After the age of one, growth rate slows, but the body continues to change dramatically. At one, infants have just learned to stand and toddle; by two, they walk confidently and are learning to run, jump, and climb. Nutrition and physical activity have helped them prepare for these new accomplishments by adding to the mass and density of their bone and muscle tissue. Thereafter, their bones continue to grow longer and their muscles to gain size and strength, though unevenly and more slowly, until adolescence.

Energy and Nutrient Needs

An infant's appetite declines markedly around the first birthday, consistent with the slowed growth rate. Thereafter, the appetite fluctuates. At times children seem to be insatiable, and at other times they seem to live on air and water. Parents need not worry about this—a child will need and demand much more food during periods of rapid growth than during slow periods. The perfect regulation of appetite in children of normal weight guarantees that their food energy intakes will be right for each stage of growth.[1]

Children's Appetites Many people mistakenly believe that they must "make" their children eat the right amounts of food, and children's erratic appetites often reinforce this belief. However, research proves otherwise. Researchers studied preschool children's food intakes for six days. Each child's food energy intake was highly variable from meal to meal, but the total daily energy intake was remarkably constant.[2] The children adjusted their energy intakes at successive meals: if they ate less at one meal, they ate more at the next, and vice versa.

Parents do, however, need to help children choose the right foods, and with overweight children they may need to help more, as described later. Overweight children may not adjust their energy intakes appropriately, but may eat in response to external cues, such as television commercials, disregarding appetite-regulation signals.

Energy Individual children's energy needs vary widely, depending on their growth and physical activity. A one-year-old child needs approximately 1000 kcalories a day; a three-year-old needs slightly more, about 1300 kcalories. By age ten, a child needs about 2000 kcalories a day. Total energy needs increase gradually with age, but energy needs per kilogram of body weight actually decline.

Nutrients Steady growth during childhood necessitates a gradual increase in intakes of most nutrients. Nutrient recommendations cluster children into age groupings that reflect similarities in growth rate, biological changes, and hormone status (see inside front cover).

Ideally, children accumulate stores of nutrients before adolescence. Then, when they take off on the adolescent growth spurt and their nutrient intakes cannot keep pace with the demands of rapid growth, they can draw on the nutrient stores accumulated earlier. This is especially true of calcium; the denser the bones are in childhood, the better prepared they will be to support teen growth and still withstand the inevitable bone losses of later life.[3] The way preteen children eat therefore influences their nutritional health during childhood, during their teen years, and for the rest of their lives.

Food Patterns for Children To provide all the needed nutrients, a child's meals and snacks should include a variety of foods from each food group—in amounts suited to the child's appetite and needs. Table 13.1 offers a guide. Serving sizes increase with age. A portion of grains, meat, fruits, and vegetables for children from age two to the teen years is loosely defined as 1 tablespoon per year. Thus, at four years of age, a serving is about 4 tablespoons (¼ cup).

Children's Food Choices Parents and other caregivers can do much to foster the development of healthy eating habits in a child. The challenge is to deliver nutrients in the form of meals and snacks that are nutritious and delicious to children so that they will learn to enjoy a variety of nutritious foods.

Candy, cola, and other concentrated sweets must be limited in children's diets. If such foods are permitted in large quantities, the only possible outcomes are nutrient deficiencies, obesity, or both. Children can't be trusted to choose nutritious foods on the basis of taste alone; the preference for sweets is innate, and children naturally gravitate to them. In one study, when children were allowed to create meals freely from a variety of foods, they selected foods that provided 25 percent of the kcalories from sugar.[4] When their parents were watching, or even when they were told that their parents would be watching, the children improved their selections. Overweight children, especially, need help in sticking to nutrient-dense foods that will meet their nutrient needs within their energy intake allowances. Underweight children or active, normal-weight children can enjoy higher-kcalorie foods, but these should still be nutritious. Examples are ice cream or pudding in the milk group and whole-grain or enriched pancakes or crackers in the bread group.

Table 13.1 Children's Daily Food Patterns for Good Nutrition

		Average Size of Serving		
Food Group	**Servings per Day**	**1 to 3 Years**	**4 to 6 Years**	**7 to 12 Years**
Bread and cereals (whole grain or enriched)[a]	6 or more	½ slice	1 slice	1 to 2 slices
Vegetables[b]	3 or more	2–4 tbs or ½ c juice	¼–½ c or ½ c juice	½–¾ c or ½ c juice
Fruits[b]	2 or more	2–4 tbs or ½ c juice	¼–½ c or ½ c juice	½–¾ c or ½ c juice
Meat and meat alternates[c]	2 or more	1–2 oz	1–2 oz	2–3 oz
Milk and milk products[d]	3 to 4	½–¾ c	¾ c	¾–1 c

[a]1 slice bread = ¾ c dry cereal, ½ c cooked cereal, ½ c potato, rice, or noodles.
[b]Vitamin C source (citrus fruits, berries, tomatoes, broccoli, cabbage, cantaloupe) daily; vitamin A source (spinach, carrots, squash, tomato, cantaloupe) 3 to 4 times weekly.
[c]1 oz meat, fish, poultry = 1 egg, 1 frankfurter, 2 tbs peanut butter, ½ c legumes.
[d]½ c milk = ½ c cottage cheese, pudding, yogurt; ¾ oz cheese; 2 tbs dried milk.

Source: Adapted from P. M. Queen and R. R. Henry, Growth and nutrient requirements of children, in *Pediatric Nutrition,* eds. R. J. Grand, J. L. Sutphen, and W. H. Dietz, Jr. (Boston: Butterworths, 1987), p. 347.

In summary, children's appetites and nutrient needs reflect their stage of growth. Children's meals and snacks should include a variety of foods from each food group and only a limited amount of concentrated sweets such as candy and cola.

Malnutrition in Children

Most children in the United States and Canada are well nourished. Their average energy intakes are sufficient to support normal growth, and their average nutrient intakes, except for iron, meet or exceed recommendations. Hunger and malnutrition are prevalent among some groups, however. More than 11 million children under 12 years of age in the United States are hungry or are at risk of hunger.

A root cause of this hunger is poverty. Many more homeless people are hungry than people who are housed, and children are the fastest-growing segment of the homeless population. Homeless children from large families and from families with single mothers are especially vulnerable to hunger and its adverse consequences.

WWW **usda.gov/fcs/team.htm**
Team Nutrition

ceo-cap.org/wic.htm
Women, Infants and Children

usda.gov/fcs/cnp.htm
Child Nutrition Programs

Effects of Hunger Both short-term and long-term hunger exert negative effects on behavior and health. Short-term hunger, such as when a child misses a meal, impairs the child's ability to pay attention and to be productive. Hungry children are irritable, apathetic, and uninterested in their environment. Long-term hunger impairs growth and immune defenses. Food assistance programs such as the WIC program and the School Breakfast and National School Lunch Programs are designed to improve the health of children.

Hunger and School Performance Children who eat nutritious breakfasts function better than their peers who do not. Young children who participate in the School Breakfast Program improve their scores on achievement tests and are tardy or absent significantly less often than children who qualify for the program but do not participate. Common sense dictates that it is unreasonable to expect anyone to learn and perform work when no fuel has been provided. By the late morning, discomfort from hunger may become distracting even if a child has eaten breakfast.

The problem children face when attempting morning schoolwork on an empty stomach appears to be at least partly due to low blood glucose. The average child up to the age of ten or so needs to eat every four to six hours to maintain a blood glucose concentration high enough to support the activity of the brain and nervous system. The brain is the body's chief glucose consumer, and a child's brain is as big as an adult's. A child's liver is considerably smaller than an adult's, however—and the liver is the organ responsible for storing glucose (as glycogen) and releasing it into the blood as needed. A child's liver cannot store more than about four hours' worth of glycogen; hence the need to eat fairly often. Teachers aware of the late-morning slump in their classrooms wisely request that a midmorning snack be provided; it improves classroom performance all the way to lunchtime.

WWW **ironpanel.org.au**
Australian Iron Status Advisory Panel

Iron Deficiency and Behavior In U.S. children and adolescents, as in infants after six months, iron-deficiency anemia is the most prevalent nutrient deficiency. The best-known and most widespread effects of iron deficiency are its impacts on behavior. Most people are familiar with the role of iron in carrying oxygen in the blood. Another important function of iron is transporting oxygen within cells, where it is used to help produce energy. A lack of iron not only causes an energy crisis but also directly affects behavior, mood, attention span, and learning ability. Iron is also involved in the function of many molecules in the brain and nervous system. Much of the research on iron and behavior has focused on the proposal that even in the early stages of iron deficiency an iron-dependent neurotransmitter is altered, and this change, in turn, impairs learning ability and behavior.[5]

Iron deficiency is usually diagnosed by a deficit of iron in the *blood*, after the deficiency has progressed all the way to anemia. A child's *brain*, however, is sensitive to slightly lowered iron concentrations long before the blood effects appear. Iron's effects are hard to distinguish from the effects of other factors in children's lives, but it is likely that iron deficiency manifests itself in a lowering of "motivation to persist in intellectually challenging tasks," a shortening of the attention span, and a reduction of overall intellectual performance. Anemic children perform less well on tests and have more conduct disturbances than their nonanemic classmates. The effects of iron-deficiency anemia are especially detrimental to learning when combined with other nutrient deficiencies.[6]

Preventing Iron Deficiency To avert iron-deficiency anemia, children's foods must deliver 10 milligrams of iron or more per day. To achieve this goal, milk intakes must be limited after infancy, because milk is a poor source of iron. Children should receive enough milk products to ensure adequate calcium and riboflavin intakes, but no more. That means 2 to 3 cups of milk per day up to age 3, grading on up to 3 to 4 cups per day from age 6 to 12 (see Table 13.1). After age two, if reduced- or low-fat milk is used instead of whole milk, the saved kcalories can be invested in iron-rich foods such as lean meats, fish, poultry, eggs, and legumes. Whole-grain or enriched breads and cereals also contribute iron. Table 13.2 lists iron-rich foods children like.

The iron status of children in the United States is changing for the better. Infant feeding practices such as breastfeeding and use of iron-fortified formulas may be improving iron status in later childhood.[7] Among children from low-income families, the WIC program, which provides supplemental iron-rich foods during infancy and early childhood, appears to be playing a role in improving iron status.

World Focus on Iron Deficiency The prevalence of iron deficiency among children throughout the world has become the focus of major public health organizations. The U.S. Public Health Service lists the reduction of iron deficiency among young children as one of the nation's foremost health priorities. The World Health Organization is collaborating with a United Nations subcommittee on nutrition to develop a ten-year plan to eliminate iron deficiency.

Other Nutrient Deficiencies Iron is only one of several dozen nutrients that can be displaced in a diet high in nutrient-poor foods. Any of the other nutrients may be lacking as well, and the deficiencies of those nutrients may also cause both behavioral and physical symptoms (see Table 13.3 on the next page). A child with behavioral symptoms of nutrient deficiencies may be irritable, aggressive, disagreeable, or sad and withdrawn. One might label such a child "hyperactive," "depressed," or "unlikable," but in fact these traits might arise from simple, albeit marginal, malnutrition. Should suspicion of dietary inadequacies be raised, *no matter what other causes may be implicated,* the people responsible for feeding the child should take steps to correct those inadequacies promptly.

Children who are chronically hungry and malnourished suffer growth retardation; when hunger is temporary and nutrient deficiencies are mild, the problems are usually more subtle—such as poor academic performance. Iron deficiency is widespread and has both physical and behavioral consequences.

Lead Poisoning in Children

The health impairment caused by malnutrition may be compounded by environmental factors such as lead poisoning. A two-way interaction is typical: lead

Table 13.2 Iron-Rich Foods Children Like[a]

Breads, cereals, and grains
Canned macaroni (½ c)
Canned spaghetti (½ c)
Cream of wheat (¼ c)
Fortified dry cereals (1 oz)[b]
Noodles, rice, or barley (½ c)
Tortillas (1 flour, 2 corn)
Whole-wheat, enriched, or fortified bread (1 slice)
Vegetables
Baked flavored potato skins (½ skin)
Cooked mushrooms (½ c)
Cooked mung bean sprouts or snow peas (½ c)
Green peas (½ c)
Mixed vegetable juice (1 c)
Fruits
Apple juice (1 c)
Canned plums (3 plums)
Cooked dried apricots (½ c)
Dried peaches (4 halves)
Raisins (1 tbs)
Meats and legumes
Bean dip (¼ c)
Canned pork and beans (⅓ c)
Mild chili or other bean/meat dishes (¼ c)
Meat casseroles (½ c)
Peanut butter and jelly sandwich (½ sandwich)
Lean roast beef or cooked ground beef (1 oz)
Sloppy joes (½ sandwich)

[a]Each serving provides at least 1 milligram iron, or one-tenth of a child's RDA for iron. Vitamin C–rich foods included with these snacks increase iron absorption.
[b]Some fortified breakfast cereals contain more than 10 milligrams iron per half-cup serving (read the labels).

WWW.
who.org
World Health Organization

phs.os.dhhs.gov/phs/phs.html
Public Health Service

WWW.
nsc.org/ehc/lead.htm
Environmental Health Center Lead Program

Table 13.3 Signs of Health and Malnutrition in Children

	Healthy	Malnourished
Hair:	Shiny, firm in the scalp	Dull, brittle, dry, loose; falls out
Eyes:	Bright, clear pink membranes; adjust easily to darkness	Pale membranes; spots; redness; adjust slowly to darkness
Teeth and gums:	No pain or cavities, gums firm, teeth bright	Missing, discolored, decayed teeth; gums bleed easily and are swollen and spongy
Face:	Good complexion	Off-color, scaly, flaky, cracked skin
Glands:	No lumps	Swollen at front of neck and cheeks
Tongue:	Red, bumpy, rough	Sore, smooth, purplish, swollen
Skin:	Smooth, firm, good color	Dry, rough, spotty; "sandpaper" feel or sores; lack of fat under skin
Nails:	Firm, pink	Spoon-shaped, brittle, ridged
Behavior:	Alert, attentive, cheerful	Irritable, apathetic, inattentive, hyperactive
Internal systems:	Heart rate, heart rhythm, and blood pressure normal; normal digestive function; reflexes and psychological development normal	Heart rate, heart or blood pressure abnormal; liver and spleen enlarged; abnormal digestion; mental irritability, confusion; burning, tingling of hands and feet; poor balance and coordination
Muscles and bones:	Good muscle tone and posture; long bones straight	"Wasted" appearance of muscles; swollen bumps on skull or ends of bones; small bumps on ribs; bowed legs or knock-knees

Note: The signs here are consistent with malnutrition but not diagnostic of it.

poisoning can cause an iron deficiency, and an iron deficiency can impair the body's defenses against lead absorption.[8] In fact, the interactions between lead poisoning and iron deficiency are so strong that some researchers suggest it may be appropriate to consider lead poisoning an adverse consequence of iron deficiency.[9] Like iron deficiency, mild lead toxicity has nonspecific effects, including diarrhea, irritability, reduced ability of the blood to carry oxygen, and fatigue. The symptoms may be reversible if exposure stops soon enough. With higher levels of lead, the signs become more pronounced, yet pinpointing a cause may still be difficult. Children lose their general cognitive, verbal, and perceptual abilities and develop learning disabilities and behavior problems. Still more severe lead toxicity can cause irreversible nerve damage, paralysis, mental retardation, and death.

Research shows that the ill effects of lead poisoning occur with lower doses than was thought in the past.[10] The lead poisoning threshold—the amount of lead in the blood recognized to cause harm—is now known to be 10 micrograms per 100 milliliters of blood, a considerable drop from the previous 25 micrograms per 100 milliliters. Health agencies label lead poisoning "the most common and societally devastating environmental disease of young children." The Food and Drug Administration (FDA) has proposed reducing the acceptable level of lead in foods tenfold—from its 1958 limit of 10 parts per million to 0.5 to 1.0 parts per million.[11]

Lead toxicity is most prevalent among children under age six—as many as 1.7 million children (10 to 15 percent of all preschoolers) may have blood concentrations high enough to cause mental, behavioral, and other health problems.[12] Table 13.4 lists the symptoms of lead toxicity. Lead aggressively attacks fetuses, infants, and children because the body absorbs lead most efficiently during times of rapid growth. Blood concentrations of lead generally reach a peak in two-year-old children; age two is the typical time of exploring surroundings "hand to mouth" and ingesting lead-tainted dirt and debris. Children's behaviors and activities—putting their hands in their mouths, playing in dirt, and eating non-food items—favor their chances of exposure to lead. Of all the sources of lead for children, paint remains the most important.[13] In the homes of 3 million young children, leaded surfaces are peeling and deteriorating. Each child in these homes is already poisoned or at immediate risk of lead poisoning.

Paint is the primary source of lead in children's lives.

Reductions in the use of leaded products (such as leaded gasoline, lead-soldered food cans, and leaded house paint) mandated by federal law in recent years have helped to limit the amount of lead in the environment, but the problem of exposure to lead still pervades children's lives (see Figure 13.1 on p. 336). Chapter 11 offered a "How to" box on preventing lead exposure, and the accompanying box suggests strategies to protect children from lead poisoning.

Lead toxicity is most prevalent in children under six years of age. Lead poisoning causes mental, behavioral, and other health problems.

Table 13.4 Symptoms of Lead Toxicity

- Learning disabilities
- Low IQ
- Behavior problems
- Slow growth
- Iron-deficiency anemia
- Nervous system disorders
- Impaired concentration
- Reduced short-term memory
- Slow reaction time
- Seizures
- Impaired hearing
- Poor coordination

Food Allergies

Food allergies are frequently blamed for physical and behavioral abnormalities in children. Food allergies are most common during the first few years of life, but then children typically outgrow them.[14] A true food allergy occurs when a whole food protein or other large molecule enters the body and elicits an immunologic response. (Recall that large molecules of food are normally dismantled in the digestive tract to smaller ones before absorption.) The body's immune system reacts to

How to

Protect Children from Lead Exposure

Defensive strategies to protect infants and children from lead exposure include:

- Test children for lead poisoning; effective screening with an appropriate questionnaire is essential to identifying high-risk children and thus treating the devastating effects.[a] About half the pediatricians surveyed report universal screening.[b]
- In contaminated environments, keep small children from putting dirty or old painted objects in their mouths, and make sure children wash their hands before eating. Similarly, keep small children from eating any nonfood items. Lead poisoning has been reported in young children who have eaten pool cue chalk.[c]
- Be aware that other countries do not have the same regulations protecting consumers against lead. Children have been poisoned by eating crayons made in China and drinking fruit juice canned in Mexico.
- Make infant formula from lead-free ingredients. Do not use lead-contaminated water.
- Feed children nutritious meals regularly.

[a]S. J. Schaffer and coauthors, Lead poisoning risk determination in a rural setting, *Pediatrics* 97 (1996): 84–90; M. N. Haan, M. Gerson, and B. A. Zishka, Identification of children at risk for lead poisoning: An evaluation of routine pediatric blood lead screening in an HMO-insured population, *Pediatrics* 97 (1996): 79–83; D. C. Snyder and coauthors, Development of a population-specific risk assessment to predict elevated blood lead levels in Santa Clara County, California, *Pediatrics* 96 (1995): 643–648; American Academy of Pediatrics, Committee on Environmental Health, Lead Poisoning: From screening to primary prevention, *Pediatrics* 92 (1993): 176–183.
[b]J. R. Campbell and coauthors, Blood lead screening practices among U.S. pediatricians, *Pediatrics* 97 (1996): 372–377.
[c]Pool cue chalk: A source of environmental lead, *Pediatrics* 97 (1996): 916–917.

Figure 13.1
Sources of Lead Exposure
Lead finds its way into the bodies of children when they ingest lead-containing foods, water, dust, or paint chips, or when they breathe lead-laden air.

food allergies: adverse reactions to foods that involve an immune response; also called *food-hypersensitivity reactions.*

Reminder: *Antigens* are substances foreign to the body that elicit the formation of antibodies or an inflammation reaction from immune system cells. Food antigens are usually glycoproteins (large proteins with glucose molecules attached).

Reminder: *Antibodies* are large proteins that are produced in response to antigens and then inactivate the antigens.

a food protein or other large molecule as it does to an antigen—by producing antibodies or other defensive agents. A problem that results from exposure to food substances, but does not involve the immune system, is known as a food intolerance.

Asymptomatic and Symptomatic Allergies Allergies may have one or two components. They always involve antibodies; they may or may not involve symptoms. A person may produce antibodies without having any symptoms (known as asymptomatic allergy) or may produce antibodies and have symptoms (known as symptomatic allergy). A person who experiences symptoms without producing antibodies, however, does not have an allergy. This means that allergies have to be diagnosed by testing for antibodies.

Allergy Symptoms A symptomatic allergy will exhibit different symptoms depending on the location of the reaction. In the digestive tract, the allergy may cause nausea or vomiting; in the skin, it may cause rashes; and in the nasal passages and lungs, it may cause inflammation or asthma. A generalized, all-systems shock reaction can also occur.

Parents often mistakenly ascribe children's symptoms to food allergies, especially if the symptoms arise after eating. However, stomachaches, headaches, pain, rapid pulse rate, nausea, wheezing, hives, bronchial irritation, coughs, and the like usually have other causes. Only proper testing can distinguish the many possibilities, and such testing is seldom done.

food intolerance: an adverse response to a food or food additive that does not involve the immune system.

Immediate and Delayed Reactions Allergic reactions to food can occur with different timings, simply classified as immediate and delayed. In both, the antigen interacts immediately with the immune system, but symptoms may appear within minutes or after several (up to 24) hours. Identifying the food that causes an immediate allergic reaction is easy because symptoms correlate closely with the time of eating the food. Identifying the food that caused a delayed reaction is more difficult because the symptoms may not appear until a day after the offending food was eaten; by this time, many other foods will have been eaten, too, complicating the picture.

The foods that most often cause immediate allergic reactions are listed in Table 13.5. Almost 75 percent of reactions are caused by three major foods—eggs, peanuts, or milk.[15] Allergic reactions to single foods are common. Reactions to multiple foods are the exception, not the rule.

WWW

foodallergy.org
Food Allergy Network

aaaai.org
American Academy of Allergy, Asthma and Immunology

Table 13.5 Foods That Most Often Cause Allergies

Eggs	Shellfish
Fish	Soybeans
Milk	Wheat
Peanuts	

Source: Adapted from R. U. Sorensen, M. C. Porch, and L. C. Tu, Food allergy in children, *Textbook of Pediatric Nutrition,* 2nd ed., eds. R. M. Suskind and L. Lewinter-Suskind (New York: Raven Press, 1993), pp. 457–469.

Other Adverse Reaction to Foods Adverse reactions to foods that are not true food allergies include:

- A reaction specific to the flavor enhancer monosodium glutamate, or MSG.
- Reactions to chemicals in foods, such as the natural laxative in prunes.
- Symptoms of digestive diseases such as hernias and ulcers, aggravated by eating any food.
- Enzyme deficiencies, such as lactose intolerance, that cause symptoms superficially indistinguishable from those of food allergy.
- Psychological reactions based on the belief that certain foods cause certain symptoms.

The simple dislike of a food may be a clue to allergy or to any of these reactions.

Food Dislikes Parents are advised to watch for signs of food dislikes and take them seriously. Children's food aversions may be the result of nature's efforts to protect them from allergic or other adverse reactions. Test for allergies, and then apply nutrition knowledge conscientiously in deciding how to alter the diet. Don't risk feeding the child an unbalanced diet, which could lead to nutrient deficiencies. Whenever a food is excluded from the diet, care must be taken to include other foods that offer the same nutrients as the omitted food contains. Remember that children who must avoid certain foods need all their nutrients, just as other children do.

Food allergies always involve an immune response; they may or may not involve symptoms. Most food allergies are caused by three foods: eggs, peanuts, and milk. Food intolerances are adverse reactions to foods that do not involve the immune system.

Hyperactivity

Hyperactivity affects behavior and learning in about 5 percent of young school-aged children. Left untreated, it can interfere with a child's social development and ability to learn. Treatment focuses on relieving the symptoms and controlling the associated problems; there is no cure. Physicians often manage hyperactivity through behavior modification, special educational techniques, psychological counseling, and, in some cases, drug therapy.

hyperactivity: inattentive and impulsive behavior that is more frequent and severe than is typical of others a similar age; professionally called **attention deficit hyperactivity disorder (ADHD).**

WWW. chadd.org
Children and Adults with Attentional Disorders

Parents of hyperactive children sometimes seek help from alternative therapies, including special diets. They mistakenly believe a solution may lie in manipulating the diet—most commonly, by excluding sugar or food additives. Adding carrots or eliminating candy is such a simple solution that many parents eagerly give such diet advice a try. These dietary changes will not solve the problem of true hyperactivity. Studies have consistently found no convincing evidence that sugar causes hyperactivity or worsens behavior.[16] Recommendations to restrict sugar in children's diets to prevent or treat behavior problems are groundless. The accompanying case study offers an opportunity to think about these issues in relation to a specific child.

Children can become excitable, rambunctious, and unruly out of a desire for attention, lack of sleep, overstimulation, watching too much television or playing too many video games, or a lack of physical activity. Such behaviors may suggest that more consistent care is needed. It helps to insist on regular hours of sleep, regular mealtimes, and regular outdoor activity.

Food Choices and Eating Habits of Children

gatekeeper: with respect to nutrition, a key person who controls other people's access to foods and thereby exerts a profound impact on their nutrition. Examples are the spouse who buys and cooks the food, the parent who feeds the children, and the caregiver in a day-care center.

The childhood years are the parents' last chance to influence their children's food choices. Parents are gatekeepers, controlling the availability of foods in their children's environments. Gatekeepers who want to promote nutritious choices and healthful habits provide access to nutrient-dense, delicious foods and opportunities for active play at home. Food choices and regular physical activity not only can promote healthy growth but, as mentioned earlier, also can help prevent the degenerative diseases of later life. Many experts agree that early childhood is the time to put into effect practices that, until recently, were recommended only for adults. Childhood obesity and the early development of chronic diseases are the topic of the nutrition in practice that follows this chapter.

Mealtimes at Home

Feeding children requires not only providing a variety of nutritious foods but also nurturing the children's self-esteem and well-being. Parents face a number of challenges in preparing meals that appeal to their children's tastes as well as provide needed nutrients. Because the interactions between parents and children regarding food intake can set the stage for lifelong attitudes and habits, a child's preferences should be treated with respect, even when nutrient needs must take precedence.

Little children like little tables and little portions.

Honoring Children's Preferences Researchers attempting to explain children's food preferences encounter many contradictions. Children say they like colorful foods, yet most often reject green and yellow vegetables while favoring brown peanut butter and white potatoes, apple wedges, and bread. They do like raw vegetables better than cooked ones, though, so it is wise to offer vegetables that are raw or slightly undercooked and crunchy and bright in color. They should be warm, not hot, because a child's mouth is much more sensitive than an adult's. The flavor should be mild (a child has more taste buds), and smooth foods such as mashed potatoes or pea soup should have no lumps (a child wonders, with some disgust, what the lumps might be). Vegetables should be served separately and be easy to eat.

Young children like to eat at little tables and to be served little portions of food. They also love to eat with other children and have been observed to stay at the table longer and eat much more when in the company of their peers. Parents

Boy with Disruptive Behavior

Freddie is a six-year-old boy who seldom sits still, often misbehaves, and is frequently sick. Freddie's eating habits are erratic and poor, as is his appetite. He often misses breakfast because he is too tired to get up in time to eat before school. By midmorning, Freddie is irritable and disruptive in the classroom. At lunchtime he trades the fruit his mother packed in his lunchbox for a piece of cake. After school he hurries home to watch television while he eats his favorite snack—root beer and potato chips. At dinnertime Freddie picks at his food because he isn't very hungry. Later on, when it's time for bed, Freddie complains that he's hungry. His parents let him stay up to have a bowl of cereal (the kind with marshmallows) before he finally falls asleep.

- What factors in Freddie's daily routine might be contributing to his restless behavior?
- Discuss some changes in diet that might improve Freddie's health and disposition.

who serve food in a relaxed and casual manner, without anxiety, provide the emotional climate in which a child's negative emotions will be minimized.

Avoiding Power Struggles It is not surprising that problems over food often arise during the second or third year, when children begin asserting their independence. Many of these problems stem from the conflict between children's developmental stages and capabilities and parents who, in attempting to do what they think is best for their children, try to control every aspect of eating. Such conflicts can disrupt children's abilities to regulate their own food intakes or to determine their own likes and dislikes. For example, many people share the misconception that children must be persuaded or coerced to try new foods. In fact, the opposite is true. When children are forced to try new foods, even by way of rewards, they are less likely to try those foods again than are children who are left to decide for themselves. The parent is responsible for *what* the child is offered to eat, but the child is responsible for *how much* and even *whether* to eat.[17]

When introducing new foods at the table, parents are advised to offer them one at a time and only in small amounts at first. The more often a food is presented to a young child, the more likely the child will like that food. Whenever possible, the new food should be presented at the beginning of the meal, when the child is hungry, but the child should make the decision to accept or reject it. Parents have their own inclinations and dislikes; so do children. It is best never to make an issue of food acceptance. A power struggle almost invariably sets a firm pattern of resistance and permanently closes the child's mind.

Television's Influence Watching television adversely affects children's nutritional health. As Chapter 9 mentioned, watching television contributes to obesity. Children who watch a lot of television are likely to become obese. Not only are they inactive, but they often snack on the fattening foods that are advertised.

The average child sees about 10,000 commercials a year, and almost all of them urge viewers to purchase sugar-coated breakfast cereals, candy bars, chips, fast foods, and carbonated beverages. Those foods add sugar, fat, and salt to the diet and displace foods that provide needed nutrients. Many parents and pediatricians believe that food ads aimed at children should be banned because they support corporate profits rather than children's health. Alternatively, parents can teach their children how to evaluate food ads and make healthful choices.

Preventing Choking When feeding children, parents must always be alert to the dangers of choking. A choking child is a silent child—an adult should be present whenever a child is eating. Make sure the child sits when eating; choking is more likely when a child is running or falling. Round foods such as grapes, nuts, hard candies, and hot dog pieces are hard to control in a mouth with few teeth, and they can easily become lodged in the small opening of a child's trachea. Other potentially dangerous foods include tough meat, popcorn, and chips.

Play First Ideally, each meal is preceded, not followed, by the activity the child looks forward to the most. A number of schools have discovered that children eat a much better lunch if it is served after, rather than before, recess. Otherwise children "hurry up and eat" so that they can go play.

Child Participation Allowing children to help plan and prepare the family's meals provides enjoyable learning experiences and encourages children to eat the foods they have prepared. Vegetables are pretty, especially when fresh, and provide opportunities for children to learn about color, growing things and their seeds, and shapes and textures—all of which are fascinating to young children. Measuring, stirring, decorating, and arranging foods are skills that even a very small child can practice with enjoyment and pride.

Snacks Parents may find that their children often snack so much that they aren't hungry at mealtimes. Instead of teaching children *not* to snack, teach them *how* to snack. Provide snacks that are as nutritious as the foods served at mealtime. Snacks can even be mealtime foods that are served individually over time, instead of all at once on one plate. When providing snacks to children, think of the food groups and offer such snacks as pieces of cheese, tangerine slices, carrot sticks, and peanut butter on whole-wheat crackers (see Table 13.6). Snacks that are easy to prepare should be readily available to children, especially if they arrive home after school before their parents.

Preventing Dental Caries Children frequently snack on sticky, sugary foods that stay on the teeth and provide an ideal environment for the growth of bacteria that cause dental caries. Teach children to brush and floss after meals, to brush or rinse after eating snacks, to avoid sticky foods, and to select crisp or fibrous foods frequently.

Serving as Role Models In an effort to practice these many tips, parents may overlook perhaps the single most important influence on their children's food habits—themselves. Parents who do not eat carrots should not be surprised when their children refuse to eat carrots. Likewise, parents who dislike the smell of brussels sprouts may not be able to persuade children to try them. Children learn much through imitation. Parents, older siblings, and other caregivers set an irresistible example by sitting with younger children, eating the same foods, and having pleasant conversations during mealtime.[18]

While serving and enjoying food, caregivers can promote both physical and emotional growth at every stage of a child's life. They can help their children to develop both a positive self-concept and a positive attitude toward food. If the beginnings are right, children will grow without the conflicts and confusions over food that can lead to nutrition and health problems.

Nutrition at School

While parents are doing what they can to establish good eating habits in their children at home, others are preparing and serving foods to their children at day-care centers and schools. In addition, children begin learning about food and nutrition in the classroom. Meeting the nutrition and education needs of children is critical to supporting their healthy growth and development.[19]

Table 13.6 Healthful Snack Ideas—Think Food Groups, Alone and in Combination

Selecting two or more foods from different food groups adds variety and nutrient balance to snacks. The combinations are endless, so be creative.

Grain Products

Grain products are filling snacks, especially when combined with other foods:

- Cereal with fruit and milk.
- Crackers and cheese.
- Wheat toast with peanut butter.
- Popcorn with grated cheese.
- Oatmeal raisin cookies with milk.

Vegetables

Cut-up fresh, raw vegetables make great snacks alone or in combination with foods from other food groups:

- Celery with peanut butter.
- Broccoli, cauliflower, and carrot sticks with a flavored cottage cheese dip.

Fruits

Fruits are delicious snacks and can be eaten alone—fresh, dried, or juiced—or combined with other foods:

- Apples and cheese.
- Bananas and peanut butter.
- Peaches with yogurt.
- Raisins mixed with sunflower seeds or nuts.

Meats and Meat Alternates

Meat and meat alternates add protein to snacks:

- Refried beans with nachos and cheese.
- Tuna on crackers.
- Luncheon meat on wheat bread.

Milk and Milk Products

Milk can be used as a beverage with any snack, and many other milk products, such as yogurt and cheese, can be eaten alone or with other foods as listed above.

The U.S. government funds several programs to provide nutritious, high-quality meals for children at school. Both the School Breakfast Program and the National School Lunch Program provide meals at a reasonable cost to children from families with the financial means to pay. Meals are available free or at reduced cost to children from low-income families.

usda.gov/fcs/cnp.htm
Child Nutrition Programs

School Breakfast The School Breakfast Program is available in slightly more than half of the nation's schools, and about 5 million children participate in it. Surveys show that the majority of children who eat school breakfasts are from low-income families. As research results continue to emphasize the positive impact breakfast has on school performance and health, campaigns to expand school breakfast programs are under way.

The school breakfast must contain at a minimum:

- *One serving of fluid milk.*
- *One serving of fruit or vegetable or full-strength juice.*
- *Two servings of bread or bread alternates; or two servings of meat or meat alternates; or one of each.*

School Lunch More than 25 million children receive lunches through the National School Lunch Program—half of them at a free or reduced price. School lunches are designed to provide at least a third of the recommendation for energy, protein, vitamin A, vitamin C, iron, and calcium. They must also include specified numbers of servings of milk, protein-rich food (meat, poultry, fish, cheese, eggs, legumes, or peanut butter), vegetables, fruits, and breads or other grain foods. Table 13.7 shows school lunch patterns for children of different ages.

The Teen Years

adolescence: the period of growth from the beginning of puberty until full maturity. Timing of adolescence varies from person to person.

As children pass through adolescence on their way to becoming adults, they change in many ways. Their physical changes make their nutrient needs high, and their emotional, intellectual, and social changes make meeting those needs a challenge.

Teenagers make many more choices for themselves than they did as children. They are not fed, they eat; they are not sent out to play, they choose to go. At the same time, social pressures thrust choices at them: whether to drink alcoholic beverages and whether to develop their bodies to meet extreme ideals of slimness or athletic prowess. Their interest in nutrition derives from personal, immediate experiences. They are concerned with how diet can improve their lives now—they engage in crash dieting in order to fit into a new bathing suit, avoid greasy foods in an effort to clear acne, or eat a plate of pasta to prepare for a big sporting event. In presenting information on the nutrition and health of adolescents, this chapter includes these many topics of interest to teens.

Growth and Development during Adolescence

With the onset of adolescence, the steady growth of childhood speeds up abruptly and dramatically, and the growth patterns of females and males become distinct. Hormones direct the intensity and duration of the adolescent growth spurt, profoundly affecting every organ of the body, including the brain. After two to three years of intense growth and a few more at a slower pace, physically mature adults emerge.

puberty: the period in life in which a person becomes physically capable of reproduction.

In general, a female's adolescent growth spurt begins at age 10 or 11 and a male's, at 12 or 13. The spurt's duration is about two and a half years. Before puberty, the differences between male and female body composition are minimal. During the adolescent spurt, gender differences become apparent in the skeletal system, lean body mass, and fat stores. In males, the lean body mass—muscle and bone—becomes much greater, and in females, fat becomes a larger percentage of the total body weight. On average, males grow 8 inches taller, and females, 6 inches taller. Males gain approximately 45 pounds, and females, about 35 pounds.

Growth charts used for children must be abandoned when the signs of puberty begin to appear. Age in years indicates little about development. One way to monitor teen growth is to compare height and weight with previous measures taken at intervals. Rating scales based on stages of adolescent development are available and widely used to record developmental changes during puberty.

Energy and Nutrient Needs

The energy needs of teenagers vary greatly, depending on the current rate of growth, body size, and physical activity. Boys' energy needs may be especially high; they experience a more intense growth spurt and, as mentioned, develop more lean body mass than girls do. An active teenage boy of 15 may need 4000 kcalories or more a day just to maintain his weight. In general, because girls enter their growth spurts earlier and grow less than boys, their energy needs peak

Table 13.7 School Lunch Patterns for Different Ages

Food Group	Preschool (Age)		Grade School through High School (Grade)		
	1 to 2	3 to 4	K to 3	4 to 6	7 to 12
Meat or meat alternate					
1 serving:					
Lean meat, poultry, or fish	1 oz	1½ oz	1½ oz	2 oz	3 oz
Cheese	1 oz	1½ oz	1½ oz	2 oz	3 oz
Large egg(s)	½	¾	¾	1	1½
Cooked dry beans or peas	¼ c	⅜ c	⅜ c	½ c	¾ c
Peanut butter	2 tbs	3 tbs	3 tbs	4 tbs	6 tbs
Vegetable and/or fruit					
2 or more servings, both to total	½ c	½ c	½ c	¾ c	¾ c
Bread or bread alternate					
Servings[a]	5 per week	8 per week	8 per week	8 per week	10 per week
Milk					
1 serving of fluid milk	¾ c	¾ c	1 c	1 c	1 c

[a]A serving is 1 slice bread; 1 biscuit, roll, or muffin; ½ c cooked rice, pasta, or cereal grain.
Source: U.S. Department of Agriculture, *Food Program Facts—National School Lunch Program,* 1992.

sooner and decline earlier than those of their male peers. An inactive girl of 15 whose growth is nearly at a standstill may need fewer than 2000 kcalories a day if she is to avoid excessive weight gain. Thus teenage girls need to pay special attention to being physically active and selecting foods of high nutrient density in order to meet their nutrient needs without exceeding their energy needs.

Obesity The insidious problem of obesity becomes ever more apparent in adolescence and often continues into adulthood. One in every five teens is overweight.[20]* The problem is most evident in females, especially those of African American descent. Without intervention, overweight teens will face numerous physical and socioeconomic consequences for years to come. The consequences of obesity are so dramatic and our society's attitude toward obese people is so negative that even teens of normal weight perceive a need to control their weight. When taken to the extremes, restrictive diets bring dramatic physical consequences of their own, as Nutrition in Practice 9 explains.

Vitamins Recommendations for most vitamins increase during the teen years (see the tables on the inside front cover). Several of the vitamin recommendations for adolescents are similar to those for adults, including the new recommendation for vitamin D. During puberty, both the activation of vitamin D and the absorption of calcium are enhanced, thus supporting the intense skeletal growth of the adolescent years without additional vitamin D.

Iron The need for iron during adolescence differs for males and females. Iron needs increase in females as they start to menstruate and in males as their lean

*For boys, obesity is defined as BMI:
- ≥23.0 for 12 to 14 years.
- ≥25.8 for 15 to 17 years.
- ≥26.8 for 18 to 19 years.

For girls, obesity is defined as BMI:
- ≥23.4 for 12 to 14 years.
- ≥24.8 for 15 to 17 years.
- ≥25.7 for 18 to 19 years.

Nutritious snacks play an important role in an active teen's diet.

body mass develops. Iron intakes often fail to keep pace with increasing needs, especially for females, who typically consume less iron-rich foods such as meat and fewer total kcalories than males. For females, the RDA rises at adolescence and remains high into middle age. For males, the RDA returns to preadolescent values in early adulthood.

Calcium Adolescence is a crucial time for bone development, and the requirement for calcium reaches its peak during these years.[21] The calcium recommendation for both males and females is 1300 milligrams per day through age 18. Unfortunately, many adolescents have calcium intakes below current recommendations.[22] Low calcium intakes during the adolescent growth spurt, especially if paired with physical inactivity, may compromise the development of peak bone mass. In contrast, increasing milk products in the diet to meet calcium recommendations greatly increases bone density.[23] The attainment of maximal bone mass is considered the best protection against age-related bone loss and fractures. Once again, teenage girls are most vulnerable, for their milk—and therefore calcium—intakes begin to decline at the time when their calcium needs are greatest. Furthermore, women experience much greater bone losses than men in later life. In addition to dietary calcium, sports activities during adolescence build strong bones.[24]

Food Choices and Health Habits

Teenagers like the freedom to come and go as they choose and eat what they want when they have time. With a multitude of afterschool, social, and job activities, they almost inevitably fall into irregular eating habits. At any given time on any given day, a teenager may be skipping a meal, eating a snack, preparing a meal, or consuming food prepared by a parent or restaurant.

Snacks Snacks typically provide at least a fourth of the average teenager's daily food energy intake. Most often, favorite snacks are high in fat and low in calcium, iron, vitamin A, vitamin C, and folate.[25] Most adolescents need to eat a greater variety of foods to obtain these nutrients. Table 13.6 on p. 341 shows how to combine foods from different food groups to create healthy snacks. Unfortunately, vending machines rarely offer nutrient-dense options, and nutrition information alone does not convince people to make healthy choices.[26]

Beverages Teenagers frequently drink soft drinks with lunch, supper, and snacks. About the only time they select fruit juices is at breakfast. When they drink milk, they are more likely to consume it with a meal (especially breakfast) than as a snack. Because of their greater food intakes, boys are more likely to drink enough milk to meet their calcium needs, whereas girls typically fall short of calcium recommendations.

Appendix A provides a table of the caffeine contents of beverages, foods, and medications.

For teenagers who can afford the kcalories and are meeting their calcium needs, soft drinks are an acceptable part of the diet. Soft drinks may present a problem, however, when caffeine intake becomes excessive. Caffeine is a stimulant added during the manufacture of many soft drinks; on the average, caffeine-containing soft drinks deliver between 30 and 55 milligrams of caffeine per 12-ounce can.[27] Caffeine increases the respiration rate, heart rate, blood pressure, and secretion of stress and other hormones. Caffeine seems to be relatively harmless, however, when used in moderate doses (the equivalent of fewer than, say, four 12-ounce cola beverages a day). In greater amounts, it can cause the symptoms associated with anxiety—sweating, tenseness, and inability to concentrate.

Eating Away from Home Adolescents eat about one-third of their meals away from home, and their nutritional welfare is enhanced or hindered by the choices they make.[28] A lunch of a hamburger, a chocolate shake, and french fries supplies substantial quantities of many nutrients at a kcalorie cost of about 800, an energy intake many adolescents can afford (see Table 13.8). When they eat this sort of lunch, teens can adjust their breakfast and dinner choices to include fruits and vegetables for vitamin A, vitamin C, folate, and fiber, and lean meats for iron and zinc.

Peer Influence Teenagers are intensely engaged in day-to-day life with their peers and preparing for their future lives as adults. Adults need to remember that teenagers have the right to make their own decisions—even if those decisions are not in line with the adults' own views. Gatekeepers can set up the environment so that nutritious foods are available and can stand by with reliable nutrition information and advice, but the rest is up to the teenagers. Ultimately, they make the choices.

Problems Adolescents Face

Physical maturity and growing independence present adolescents with new choices to make. The consequences of those choices will influence their nutritional health both today and throughout life. Some teenagers begin using drugs, alcohol, and tobacco; others wisely refrain. Information about the use of these substances is presented here because most people are first exposed to them during adolescence, but it actually applies to people of all ages.

The dangers of steroid use are presented in Nutrition in Practice 10.

Marijuana One out of every three high school students reports having at least tried marijuana.[29] When inhaled by smoking, the active chemicals are rapidly and almost completely absorbed from the lungs.* They then travel in the blood to the various body tissues that metabolize them. The active ingredients from a single marijuana cigarette can linger in the body's fat a month or more before being excreted in urine.

Marijuana is unique among drugs in that it seems to enhance the enjoyment of eating, especially of sweets, a phenomenon commonly known as "the

*The active ingredient of marijuana, which is primarily responsible for its intoxicating effects, is delta-9-tetrahydrocannabinol, or THC.

Table 13.8 Selected Nutrients in a Hamburger, Low-Fat Chocolate Shake, and Small Serving of French Fries

Nutrient	Percentage of RDA for a Male[a]	Percentage of RDA for a Female[a]
Energy	27	37
Protein	47	63
Fat[b]	24	33
Calcium[c]	39	39
Iron	36	29
Zinc	17	22
Vitamin A	0	0
Folate[d]	12	12
Vitamin C	22	22
Sodium[b]	38	38

[a]RDA for a 15- to 18-year-old, moderately active person of average height and weight.
[b]Daily Values used for fat and sodium.
[c]1997 calcium AI value used.
[d]1998 folate RDA.

munchies." Why or how this effect occurs is not known; it may be a social effect induced by suggestibility, or perhaps the drug stimulates appetite. Whatever the reason, prolonged use of the drug does not seem to bring about a weight gain.

Cocaine One in 20 high school seniors reports having used cocaine at least once.[30] Cocaine stimulates the nervous system and elicits the stress response—constricted blood vessels, raised blood pressure, widened pupils of the eyes, and increased body temperature. It also drives away feelings of fatigue. Cocaine occasionally causes immediate death—usually by heart attack, stroke, or seizure in an already damaged body system.

Nutrition problems of drug abusers:
- *They buy drugs with money that could be spent on food.*
- *They lose interest in food during "highs."*
- *They use drugs that depress appetite.*
- *Their lifestyle fails to promote good eating habits.*
- *They use intravenous (IV) drugs. They may contract AIDS, hepatitis, or other infectious diseases, which increase their nutrient needs. Hepatitis also causes taste changes and loss of appetite.*
- *Medicines used to treat drug abuse may alter nutrition status.*

Weight loss is common, and cocaine abusers often develop eating disorders. Notably, the craving for cocaine replaces hunger; rats given unlimited cocaine will choose it over food until they starve to death. Thus, unlike marijuana use, cocaine use has major nutritional consequences.

Drug Abuse, in General The effects of other addictive drugs vary in degree but are similar in kind to those caused by cocaine. Drug abusers face the multiple nutrition problems listed in the margin. During withdrawal from drugs, an important part of treatment is to identify and correct these nutrition problems.

Alcohol Abuse Sooner or later all teenagers face the decision whether to drink alcohol. The law forbids the sale of alcohol to people under 21, but most adolescents who seek alcohol can obtain it. Four out of five high school students have had at least one alcoholic beverage; about half drink regularly; and one in three students drinks heavily (defined as five or more drinks on at least one occasion in the previous month).[31]

Nutrition in Practice 8 describes how alcohol affects nutrition status. To sum it up, alcohol provides energy but no nutrients, and it can displace nutritious foods from the diet. Alcohol alters nutrient absorption and metabolism, so imbalances develop.

WWW. **ncadd.org**
National Council on Alcoholism and Drug Dependence

WWW. **lungusa.org**
American Lung Association

Smoking The prevalence of cigarette smoking among U.S. adolescents is on the rise.[32] Cigarette smoking is a pervasive health problem causing thousands of people to suffer from cancer and diseases of the cardiovascular, digestive, and respiratory systems. These effects are beyond the scope of nutrition, but smoking cigarettes does influence hunger, body weight, and nutrient status.

Smoking a cigarette eases feelings of hunger. When smokers receive a hunger signal, they can quiet it with cigarettes instead of food. Such behavior ignores body signals and postpones energy and nutrient intake. Studies on rats confirm that nicotine reduces food intake and increases the rate of energy expenditure, causing weight loss.[33]

Indeed, smokers tend to weigh less than nonsmokers and to gain weight when they stop smoking. Weight gain is often a concern for people contemplating giving up cigarettes. They should know that the average person who quits smoking gains less than 10 pounds. Smokers wanting to quit need to prepare for this possibility and adjust their diet and activity habits so as to maintain weight during and after quitting. Smoking cessation programs need to include strategies for weight management.

Nutrient intakes of smokers and nonsmokers differ. Smokers tend to have lower intakes of dietary fiber, vitamin A, beta-carotene, folate, and vitamin C.[34] The association between smoking and low vitamin intake may be noteworthy, considering the altered metabolism of vitamin C in smokers and the protective effect of vitamin A and beta-carotene against lung cancer.

The vitamin C RDA for people who regularly smoke cigarettes is 100 mg/day.

Research shows that compared to nonsmokers, smokers require almost twice as much vitamin C to maintain steady body pools. Oxidants in cigarette smoke accelerate vitamin C metabolism and deplete smokers' body stores of this antioxidant; this depletion is even evident to some degree in nonsmokers who are exposed to passive smoke.[35]

Beta-carotene enhances the immune response and protects against some cancer activity.[36] Specifically, the risk of lung cancer is greatest for smokers who have the lowest intakes of carotene. Of course, such evidence should not be misinterpreted. It does not mean that as long as people eat their carrots, they can safely use tobacco. Nor does it mean that beta-carotene supplements would be beneficial. In fact, as mentioned in Chapter 7, some research shows beta-carotene supplements may have adverse effects in smokers.[37] Smokers are ten times more likely to get lung cancer than nonsmokers. Both smokers and nonsmokers can, however, reduce their cancer risks by eating fruits and vegetables rich in carotene (see Nutrition in Practice 26 for details on antioxidant nutrients and disease prevention).

Smokeless Tobacco Nationwide, one in ten high school students reports having used smokeless tobacco products.[38] Like cigarettes, smokeless tobacco use is linked to many health problems, from minor mouth sores to tumors in the nasal cavities, cheeks, gums, and throat. The risk of mouth and throat cancers is even greater than for smoking tobacco. Other drawbacks to tobacco chewing and snuff dipping include bad breath, stained teeth, and blunted senses of smell and taste. Tobacco chewing also damages the gums, tooth surfaces, and jawbones, making it likely that users will lose their teeth in later life.

To review, nutrient needs rise dramatically as children enter the rapid growth phase of the teen years. The busy lifestyles of teenagers add to the challenge of meeting their nutrient needs—especially for iron and calcium. In addition to making wise food choices, teenagers need to refrain from using substances that will impair their health—including illicit drugs, tobacco, and alcohol.

Nutrition Assessment

Assessment of nutrition status in healthy children and teenagers can confirm that development is normal or can catch potential problems early. As with pregnant women and infants, careful histories and measures of growth are most useful. In assessing, focus on the following:

Chapters 15 and 16 offer details about nutrition assessment.

- The history should reveal whether a child is eating appropriately from all food groups (recall Table 13.1). How much is the child snacking? Are snacks nutritious? Is the child receiving adequate iron from the foods typically eaten? Is milk intake adequate? If milk must be omitted from the diet, are appropriate substitutes in use? The history should discover both home and school food intakes, as well as those in any other place where the child spends significant amounts of time. Aberrant food patterns may suggest food allergies or intolerances and a need for intervention.
- The socioeconomic history may reveal a need for food assistance programs. Are any risk factors for lead poisoning present?
- By adolescence, healthful physical activity habits should be apparent. The history should also reveal any factors that might interfere with adequate food intakes such as inappropriate dieting, bizarre diets, alcohol or drug use, or smoking.
- Height and weight plotted over time will produce a smooth growth curve if development is normal. If significant obesity or underweight is apparent in children, intervention is in order. During the adolescent growth spurt, wide variations in height and weight-gain patterns are expected.
- Physical examination can reveal many clues to nutrition status as shown in Table 13.3. If any signs suggest possible malnutrition, further investigation is appropriate.

The nutrition lifestyle choices people make as children, teenagers, and young adults have long-term, as well as immediate, effects on their health. The challenge for parents and health care professionals concerned with young people's well-being is to motivate them to make sound choices by modeling good habits early and then by showing them how such choices relate to their own interests.

Self Check

1. A general rule appropriate from age two to the teen years is that a portion of meat, fruit, and vegetables for children is _____ tablespoon(s) per year.
   a. 1
   b. 2
   c. 3
   d. 4
2. Children who are hungry may be irritable or apathetic because:
   a. they need vitamin D.
   b. they have had too many sweets.
   c. their blood glucose is low.
   d. their blood glucose is high.
3. Two infant feeding practices that have improved the iron status of older children in the United States are:
   a. feeding solids at an early age and giving iron-fortified milk.
   b. giving iron supplements and increasing milk intake.
   c. feeding infants liver and starches.
   d. breastfeeding and iron-fortified formulas.
4. Three symptoms of lead toxicity are:
   a. low blood sugar, hair loss, and skin rash.
   b. diarrhea, irritability, and fatigue.
   c. increased heart rate, hyperactivity, and dry skin.
   d. bleeding gums, brittle fingernails, and swollen glands.
5. Allergic reactions to foods are most often caused by:
   a. eggs, peanuts, or milk.
   b. corn, rice, or meats.
   c. seafood, dark greens, or lactose.
   d. red meats, milk, or MSG.
6. Children who watch a lot of television are likely to become obese because:
   a. their parents are obese.
   b. they eat the foods most often advertised on television.
   c. they spend less time being physically active.
   d. b and c.
7. When introducing new foods to children:
   a. offer one new food at the end of the meal.
   b. offer many choices to encourage variety.
   c. offer one new food at the beginning of the meal.
   d. reward children as they try new foods.
8. During the growth spurt of adolescence:
   a. similarities in body composition between males and females become apparent.
   b. females gain more weight than males.
   c. differences in body composition between males and females become apparent.
   d. males gain more fat, proportionately, than females.
9. Two nutrients that are usually lacking in teens' diets are:
   a. vitamin A and riboflavin.
   b. iron and calcium.
   c. protein and thiamin.
   d. zinc and fat.
10. Smoking increases the need for:
    a. vitamin E.
    b. vitamin C.
    c. folate.
    d. iron.

Answers to these questions appear in Appendix H.

Notes

1. S. Shea and coauthors, Variability and self-regulation of energy intake in young children in their everyday environment, *Pediatrics* 90 (1992): 542–546.

2. L. L. Birch and coauthors, Effects of a non-energy fat substitute on children's energy and macronutrient intake, *American Journal of Clinical Nutrition* 58 (1993): 326–333; L. L. Birch and coauthors, The variability of young children's energy intake, *New England Journal of Medicine* 324 (1991): 232–235.

3. V. Matkovic and J. Z. Ilich, Calcium requirements for growth: Are current recommendations adequate? *Nutrition Reviews* 51 (1993): 171–180.

4. R. E. Klesges and coauthors, Parental influence on food selection in young children and its relationships to childhood obesity, *American Journal of Clinical Nutrition* 53 (1991): 859–864.

5. E. Pollitt, Iron deficiency and cognitive function, *Annual Review of Nutrition* 13 (1993): 521–537.

6. E. Pollitt, Iron deficiency and educational deficiency, *Nutrition Reviews* 55 (1997): 133–141.

7. M. C. Holst, Developmental and behavioral effects of iron deficiency anemia in infants, *Nutrition Today* 33 (1998): 27–36.

8. R. A. Goyer, Toxic and essential metal interactions, *Annual Review of Nutrition* 17 (1997): 37–50; P. Mushak and A. F. Crocetti, Lead and nutrition: Biologic interactions of lead with nutrients, *Nutrition Today* 31 (1996): 12–17.

9. R. Yip, The interaction of lead and iron, in *Dietary Iron: Birth to Two Years,* ed. L. J. Filer (New York: Raven Press, 1989), pp. 179–181.

10. S. Piomelli and J. A. Wolff, Childhood lead poisoning in the '90s, *Pediatrics* 93 (1994): 508–510; H. Needleman and coauthors, The long-term effects of exposure to low doses of lead in childhood: An 11-year follow-up report, *New England Journal of Medicine* 322 (1990): 83–88.

11. FDA seeks lower lead levels in food additives and GRAS ingredients, *Journal of the American Dietetic Association* 94 (1994): 495.

12. Update: Blood lead levels—United States, 1991–1994, *Morbidity and Mortality Weekly Report* 46 (1997): 141–146.

13. Childhood lead poisoning: A disease for the history texts, *American Journal of Public Health* 81 (1991): 685.

14. H. A. Sampson and D. D. Metcalfe, Food allergies, *Journal of the American Medical Association* 268 (1992): 2840–2844.

15. S. A. Bock and F. M. Atkins, Patterns of food hypersensitivity challenges, *Journal of Pediatrics* 117 (1990): 561–567.

16. J. W. White and M. Wolraich, Effect of sugar and mental performance, *American Journal of Clinical Nutrition* 62 (1995): 242S–249S; M. L. Wolraich and coauthors, Effects of diets high in sucrose or aspartame on the behavior and cognitive performance of children, *New England Journal of Medicine* 330 (1994): 301–307.

17. C. Evers, Empower children to develop healthful eating habits, *Journal of the American Dietetic Association* 97 (1997): S116; E. Satter, *How to Get Your Kid to Eat . . . But Not Too Much* (Palo Alto, Calif.: Bull Publishing Company, 1987), pp. 13–28.

18. M. Nahikian-Nelms, Influential factors of caregiver behavior at mealtime: A study of child-care programs, *Journal of the American Dietetic Association* 97 (1997): 505–509.

19. Position of The American Dietetic Association: Nutrition standards for child care programs, *Journal of the American Dietetic Association* 94 (1994): 323–328.

20. Prevalence of overweight among adolescents, United States, 1988–1991, *Morbidity and Mortality Weekly Report* 43 (1994): 819–821.

21. A. D. Martin and coauthors, Bone mineral and calcium accretion during puberty, *American Journal of Clinical Nutrition* 66 (1997): 611–615.

22. S. I. Barr, Associations of social and demographic variables with calcium intakes of high school students, *Journal of the American Dietetic Association* 94 (1994): 260–266, 269.

23. G. M. Chan, K. Hoffman, and M. McMurry, Effects of dairy products on bone and body composition in pubertal girls, *Journal of Pediatrics* 126 (1995): 551–556.

24. A. M. Fehily and coauthors, Factors affecting bone density in young adults, *American Journal of Clinical Nutrition* 56 (1992): 579–586.

25. J. G. Dausch and coauthors, Correlates of high-fat/low-nutrient-dense snack consumption among adolescents: Results from two national health surveys, *American Journal of Health Promotion* 10 (1995): 85–88.

26. S. M. Hoerr and V. A. Louden, Can nutrition information increase sales of healthful vended snacks? *Journal of School Health* 63 (1993): 386–390.

27. International Food Information Council, Caffeine and health: Clarifying the controversies, *IFIC Review,* May 1993.

28. B. H. Lin, J. Guthrie, and J. R. Blaylock, *The Diets of America's Children—Influences of Dining Out, Household Characteristics, and Nutrition Knowledge* (Washington, D.C.: U.S. Department of Agriculture, December 1996).

29. L. Kann and coauthors, Youth risk behavior surveillance—United States, 1993, *Journal of School Health* 65 (1995): 163–171.

30. Kann and coauthors, 1995.

31. Kann and coauthors, 1995.

32. Centers for Disease Control and Prevention, Tobacco use among high school students—United States, 1997, *Journal of the American Medical Association* 279 (1998): 1250.

33. S. R. Schwid, M. D. Hirvonen, and R. E. Keesey, Nicotine effects on body weight: A regulatory perspective, *American Journal of Clinical Nutrition* 55 (1992): 878–884.

34. T. A. B. Sanders and coauthors, Essential fatty acids, plasma cholesterol, and fat-soluble vitamins in subjects with age-related maculopathy and matched control subjects, *American Journal of Clinical Nutrition* 57 (1993): 428–433; A. F. Subar, L. C. Harlan, and M. E. Mattson, Food and nutrient intake differences between smokers and non-smokers in the US, *American Journal of Public Health* 80 (1990): 1323–1329.

35. D. L. Tribble, L. J. Giuliano, and S. P. Fortmann, Reduced plasma ascorbic acid concentrations in nonsmokers regularly exposed to environmental tobacco smoke, *American Journal of Clinical Nutrition* 58 (1993): 886–890.

36. G. van Poppel, S. Spanhaak, and T. Ockhuizen, Effect of ß-carotene on immunological indexes in healthy male smokers, *American Journal of Clinical Nutrition* 57 (1993): 402–407.

37. K. Smigel, Beta-carotene fails to prevent cancer in two major studies: CARET intervention stopped, *Journal of the National Cancer Institute* 88 (1996): 145; G. S. Omenn and coauthors, Effects of a combination of beta carotene and vitamin A on lung cancer and cardiovascular disease, *New England Journal of Medicine* 334 (1996): 1150–1155.

38. Kann and coauthors, 1995.

Nutrition in Practice

▪ Childhood Obesity and the Early Development of Chronic Diseases ▪

Disease of the heart and blood vessels, known as cardiovascular disease, or CVD, is the number one killer of adults in the United States and Canada, but CVD begins in childhood. Over the past three decades, researchers have been observing how changes in body weight, blood lipids, blood pressure, and individual behaviors correlate with the development of CVD over time—from infancy to childhood through adolescence and into young adulthood. Some major findings have emerged from this research:

- Changes inside the arteries—changes predictive of CVD—are evident in childhood.
- Obesity in children affects these changes.
- Behaviors that influence the development of obesity and of CVD are learned and begin early in life. These behaviors include overeating, eating high-fat foods, physical inactivity, and cigarette smoking.

This nutrition in practice focuses on efforts to prevent childhood obesity and CVD, but the benefits extend to cancer, diabetes, and other chronic diseases as well. The years of childhood are emphasized here, for the earlier in life health-promoting habits become established, the better they will stick.

WWW.
amhrt.org
American Heart Association

What about genetics, though? Don't some people inherit the tendency to develop CVD regardless of the lifestyle habits they adopt?

Genetics does not appear to play a *determining* role in CVD; that is, a person is not simply destined at birth to develop CVD.[1] Instead, genetics appears to play a *permissive* role—the potential is inherited, and then CVD will develop, if given a push by poor health choices such as excessive weight gain, poor diet, sedentary lifestyle, and cigarette smoking.

How does CVD develop, and when does its development begin?

Most CVD involves atherosclerosis—the accumulation of cholesterol and other blood lipids along the walls of the arteries (see the accompanying glossary for atherosclerosis and related terms). Frequently, atherosclerosis alters the flow of blood to the heart and can lead to hypertension and coronary heart disease (CHD) which, in turn, raises the likelihood of a heart attack. When atherosclerosis alters blood flow to the brain, a stroke can result. Infants are born with healthy, smooth, clear arteries, but within the first decade of life, fatty streaks may begin to appear. During adolescence, these fatty streaks may begin to turn to fibrous plaques (Chapter 26 shows the formation of plaques in atherosclerosis). By early adulthood, the fibrous plaques may begin to calcify and become raised lesions, especially in boys and young men. As the lesions grow more numerous and enlarge, the heart disease rate begins to rise, and the rise becomes dramatic at about age 45 in men and 55 in women. From this point on, arterial damage and blockage progress rapidly, and heart attacks and strokes threaten life. In short, the consequences of atherosclerosis, which become apparent only in adulthood, have their beginnings in the first decades of life.[2]

Atherosclerosis is not inevitable; people can grow old with relatively clear arteries. Early lesions may either progress or regress, depending on several factors, many of which reflect lifestyle behaviors. Smoking, for example, is strongly associated with the prevalence of raised lesions, even in young adults.

Tell me more about blood cholesterol and atherosclerosis. Parents don't need to worry about their children's blood cholesterol, do they?

Atherosclerotic lesions reflect blood cholesterol: as blood cholesterol increases, lesion coverage increases. Cholesterol values at birth are similar in all popula-

CHAPTER 13

Nutrition in Practice

Glossary

atherosclerosis (ath-er-oh-scler-OH-sis): a type of artery disease characterized by accumulations of lipid-containing material on the inner walls of the arteries.
athero = porridge or soft
scleros = hard
osis = condition

cardiovascular disease (CVD): a general term for all diseases of the heart and blood vessels. Atherosclerosis is the main cause of CVD. When the arteries that carry blood to the heart muscle become occluded, the heart suffers damage known as **coronary heart disease (CHD).**
cardio = heart
vascular = blood vessels

fatty streaks: an accumulation of cholesterol and other lipids along the walls of the arteries.

fibrous plaques: mounds of lipid material, mixed with smooth muscle cells and calcium, which develop in the artery walls in atherosclerosis.

tions; differences emerge in early childhood. In countries where the adults have high blood cholesterol and high rates of CVD, the children also tend to have high blood cholesterol. Conversely, in countries where the adults have low blood cholesterol and low rates of CVD, the children tend to have low blood cholesterol, suggesting that adult heart disease tracks early trends and that early preventive efforts might reduce the incidence of later CVD.

Such is the case among populations, but individual cholesterol status also becomes established in childhood, as early as one year.[3] The best predictors of a person's blood cholesterol are that person's earlier baseline values: childhood values correlate with values in young adulthood.[4] Quite simply, if you want to know a child's future cholesterol, measure it now. Standard values for cholesterol screening in children and adolescents are listed in Table NP13.1.[5]

Table NP13.1 Cholesterol Values for Children and Adolescents

Disease Risk	Total Cholesterol (mg/dL)	LDL Cholesterol (mg/dL)
Acceptable	<170	<110
Borderline	170–199	110–129
High	≥200	≥130

Note: Adult values appear in Chapter 26. A deciliter (dL) is one-tenth of a liter or 100 milliliters.

I know that hypertension is a risk factor for heart disease or stroke in adults. Is this a concern for children and teenagers as well?

Pediatricians routinely monitor blood pressure in children and adolescents.[6] High blood pressure may signal an underlying disease or the early onset of hypertension. Hypertension accelerates the development of atherosclerosis. Standard values for hypertension screening in children and adolescents are given in Table NP13.2 (p. 352).[7]

Like atherosclerosis and high blood cholesterol, hypertension may develop in the first decades of life.[8] Children can control their hypertension by participating in regular aerobic activity and by losing weight or maintaining their weight as they grow taller.[9] No evidence suggests a benefit of restricting sodium to lower blood pressure in children and adolescents.[10]

You've talked about blood cholesterol and blood pressure as risk factors for CVD in children and adolescents and you also mentioned obesity. Is obesity a problem among children today?

Many experts agree that preventing or treating obesity in childhood will reduce the rate of CVD in adulthood. Without intervention, overweight children become overweight adolescents who become overweight adults, and being overweight exacerbates every chronic disease that adults face.[11]

Children are heavier today than they were 20 or so years ago. Since the late 1970s, the prevalence of overweight has almost doubled for children—and more than doubled for adolescents.[12] This pattern is a

Table NP13.2 Hypertension Standards for Children and Adolescents: Systolic over Diastolic Pressure (mm Hg)

	6 to 9 yr	10 to 12 yr	13 to 15 yr	16 to 18 yr
High normal	111–121 over 70–77	117–125 over 75–81	124–135 over 77–85	127–141 over 80–91
Significant hypertension	122–129 over 70–85	126–133 over 82–89	136–143 over 86–91	142–149 over 92–97
Severe hypertension	>129 over >85	>133 over >89	>143 over >91	>149 over >97

secular trend—that is, one that cannot be explained by genetics. Diet and physical activity must be responsible.

I guess children are eating more food and more fat than ever before. Is this true?

No. Children's energy intakes have remained relatively stable over the past 15 years. There has even been a slight decline in fat intake, from 38 to 36 percent of kcalories from fat daily.[13] This slight decline in dietary fat is not enough, however, to have influenced body weight; nor is it enough to meet current dietary recommendations.

Children's dietary fat intakes vary, of course, and some children do eat high-fat diets. Children who prefer high-fat foods tend to consume a relatively large percentage of their energy intake from fat.[14] They also tend to be more overweight than their peers. Particularly noteworthy is the finding that the children's fat preferences and consumption correlate with their parents' obesity as well. Such findings confirm the significant roles parents play—teaching children about healthy food choices, providing children with low-fat selections, and serving as role models.

If diet alone is not to blame for the increasing prevalence of obesity among the young, then what is?

Most likely, children have grown more overweight because of their lack of physical activity.[15] An inactive child can become obese even while eating less food than an active child. Today's children are more sedentary and less physically fit than children were 20 years ago.

Watching television accounts for some 24 hours a week of sedentary behavior. Beyond these 24 hours, children spend more sedentary time sitting at computers and playing video games. Both obesity and blood cholesterol correlate with hours of television viewed.[16]

Just as blood cholesterol and obesity track over the years, so does a person's level of physical activity. Researchers studying almost 1000 teenagers found that over half of those who were initially described as inactive remained inactive six years later.[17] Similarly, almost half of those who were physically active remained so. Compared with inactive teens, those who were physically active weighed less, smoked less, ate a diet lower in saturated fats, and had a better blood lipid profile. The message is clear: physical activity offers numerous health benefits, and children who are active today are most likely to be active for years to come.

So, encouraging children and teens to be physically active is an important step toward preventing obesity. What else can concerned adults do to help prevent childhood obesity?

In light of all these findings, parents and teachers of children are encouraged to make major efforts to prevent childhood obesity. Suggestions include the following: encourage children to eat slowly, to pause and enjoy their table companions, and to stop eating when they are full. Teach them how to select low-fat snacks and to serve themselves appropriate portions. Never force children to clean their plates.

Above all, be sensitive in teaching children nutrition principles that can help to prevent obesity. Children can easily get the idea that their worth is tied to their body weight. Some parents fail to realize that society's ideal of slimness can be perilously close to starvation, and that a child encouraged to "diet" cannot obtain the energy and nutrients required for normal growth and development. Even healthy children without diagnosable eating disorders have been observed to limit their growth through "dieting." Weight gain in truly overweight children can be controlled safely without compromising growth, but should be overseen by a health care professional.

I have a niece who is sedentary and overweight, and I'm concerned that her cholesterol may be high. Will

her pediatrician check her cholesterol on a routine visit?

Many children in the United States are not only overweight but also have high blood cholesterol.[18] These children are quite likely to have parents who developed CVD early.[19] For this reason, selective screening is recommended for children and adolescents whose parents or grandparents have CVD; those whose parents have elevated blood cholesterol; and those whose family history is unavailable, especially if other risk factors are evident.[20] Since blood cholesterol in children is a good predictor of adult values, some experts recommend universal screening to identify all children with high blood cholesterol.[21] They note that many children who have high blood cholesterol would be missed under current screening criteria. [22]

Among those children who may have high blood cholesterol, but may not meet screening criteria, are those who are overweight.[23] The incidence of high blood cholesterol in obese children with no other criteria is similar to that of nonobese children with family histories of CVD. In addition to overweight, health care professionals should consider whether children consume a high-fat diet.[24]

Early—but not advanced—atherosclerotic lesions are reversible, making screening and education a high priority. Both those with family histories of CVD and those with multiple risk factors need intervention. Children with the highest risks of developing CVD are sedentary and obese, with high blood pressure and high blood cholesterol. In contrast, children with the lowest risks of heart disease are physically active and of normal weight, with low blood pressure and favorable lipid profiles. Routine pediatric care should identify these known risk factors and provide intervention when needed.

I know that dietary recommendations for adults are aimed at reducing the risks of obesity and CVD. Are the same recommendations appropriate for children?

Regardless of family history, all children over age two should eat a variety of foods and maintain desirable weight. Children should receive 20 to 30 percent of total energy from fat, less than 10 percent from saturated fat, and less than 300 milligrams of cholesterol per day.[25]

Recommendations limiting fat and cholesterol are not intended for infants or children under two years old. Infants and toddlers need a higher percentage of fat to support their rapid growth.

Healthy children over age two can begin the transition to eating according to recommendations by eating fewer high-fat foods, replacing some high-fat foods with low-fat choices, and selecting more fruits and vegetables.[26] All high-fat foods need not be eliminated, though. Healthy meals can still include moderate amounts of a child's favorite foods, even if they are high-fat selections such as french fries and ice cream.[27] Without such additions, diets might be too low in fat, not to mention unappetizing and boring.

Balanced meals need to provide lean meat, poultry, fish, and vegetable sources of protein; fruits and vegetables; whole grains; and low-fat milk products. Such meals can provide enough food energy and nutrients to support growth and maintain blood cholesterol within a healthy range.[28]

Pediatricians warn parents to avoid extremes; they caution that while intentions may be good, excessive food restriction may create nutrient deficiencies and impair growth. Furthermore, parental control over eating may instigate battles and foster attitudes about foods that can lead to inappropriate eating behaviors.

Is there anything else parents or caregivers can do to help children reduce their risks of CVD?

Even though the focus of this text is nutrition, another risk factor for CVD that starts in childhood and carries over into adulthood must also be addressed—cigarette smoking. Each day 5000 children light up for the first time—typically, in grade school. Among high school students, two out of three have tried smoking, and one in seven smokes regularly.[29] Approximately 90 percent of all adult smokers began smoking before the age of 18.[30]

Efforts to teach children about the dangers of smoking need to be aggressive. Children are not likely to consider the long-term health consequences of tobacco use. They are more likely to be struck by the immediate health consequences, such as shortness of breath when playing sports, or social consequences, such as having bad breath. Whatever the context, the message to all children and teens should be clear: don't start smoking. If you've already started, quit.

In conclusion, *adult* CVD is a major *pediatric* problem.[31] Without intervention, some 60 million children are destined to suffer its consequences within the next 30 years. Optimal prevention efforts focus on children, especially on those who are overweight.

Just as young children receive vaccinations against infectious diseases, they need screening for, and education about, CVD. Many health education programs have been implemented in schools around the country. These programs are most effective when they include education in the classroom, heart-healthy meals in the cafeteria, fitness activities on the playground, and parental involvement at home.

Notes

1. W. B. Kannel, R. B. D'Agostino, and A. Belanger, Concept of bridging the gap from youth to adulthood—The Framingham Study, an address presented at the Recognition and Prevention of Heart Disease: State of the Art Conference, New Orleans, Louisiana, April 27 and 28, 1994.

2. H. C. McGill, Childhood nutrition and adult cardiovascular disease, *Nutrition Reviews* 55 (1997): S2–S11.

3. M. J. T. Kallio and coauthors, Tracking of serum cholesterol and lipoprotein levels from the first year of life, *Pediatrics* 91 (1993): 949–954.

4. S. Guo and coauthors, Serial analysis of plasma lipids and lipoproteins from individuals 9–21 years of age, *American Journal of Clinical Nutrition* 58 (1993): 61–67.

5. American Academy of Pediatrics, Committee on Nutrition, Cholesterol in childhood, *Pediatrics* 101 (1998): 141–147.

6. National High Blood Pressure Education Program Working Group on Hypertension Control in Children and Adolescents, Update on the 1987 Task Force Report on High Blood Pressure in Children and Adolescents: A working group report from the National High Blood Pressure Education Program, *Pediatrics* 98 (1996): 649–658.

7. American Academy of Pediatrics, Committee on Sports Medicine and Fitness, Athletic participation by children and adolescents who have systemic hypertension, *Pediatrics* 99 (1997): 637–638.

8. A. R. Sinaiko, Hypertension in children, *New England Journal of Medicine* 335 (1996): 1968–1973.

9. S. Shea and coauthors, The rate of increase in blood pressure in children 5 years of age is related to changes in aerobic fitness and body mass index, *Pediatrics* 94 (1994): 465–470.

10. B. Falkner and S. Michel, Blood pressure response to sodium in children and adolescents, *American Journal of Clinical Nutrition* 65 (1997): 618S–621S.

11. S. S. Guo and coauthors, The predictive value of childhood body mass index values for overweight at age 35 y, *American Journal of Clinical Nutrition* 59 (1994): 810–819.

12. Update: Prevalence of overweight among children, adolescents, and adults—United States, 1988–1994, *Morbidity and Mortality Weekly Report* 46 (1997): 199–202.

13. T. A. Nicklas and coauthors, Secular trends in dietary intakes and cardiovascular risk factors of 10-year-old children: The Bogalusa Heart Study (1973–1988), *American Journal of Clinical Nutrition* 57 (1993): 930–937.

14. J. O. Fisher and L. L. Birch, Fat preferences and fat consumption of 3- to 5-year-old children are related to parental obesity, *Journal of the American Dietetic Association* 95 (1995): 759–764.

15. S. A. Schlicker, S. T. Borra, and C. Regan, The weight and fitness status of United States children, *Nutrition Reviews* 52 (1994): 11–17.

16. E. Obarzanek and coauthors, Energy intake and physical activity in relation to indexes of body fat: The National Heart, Lung, and Blood Institute Growth and Health Study, *American Journal of Clinical Nutrition* 60 (1994): 15–22.

17. O. T. Raitakari and coauthors, Effects of persistent physical activity on coronary risk factors in children and young adults: The Cardiovascular Risk in Young Finns Study, *American Journal of Epidemiology* 140 (1994): 195–205.

18. G. S. Berenson, S. R. Srinivasan, and L. S. Webber, Cardiovascular risk prevention in children: A challenge or a poor idea? *Nutrition, Metabolism and Cardiovascular Diseases* 4 (1994): 46–52.

19. W. Bao and coauthors, Longitudinal changes in cardiovascular risk from childhood to young adulthood in offspring of parents with coronary artery disease: The Bogalusa Heart Study, *Journal of the American Medical Association* 278 (1997): 1749–1754.

20. American Academy of Pediatrics, 1998.

21. L. Van Horn and P. Greenland, Prevention of coronary artery disease is a pediatric problem, *Journal of the American Medical Association* 278 (1997): 1779–1780; Berenson, Srinivasan, and Webber, 1994.

22. S. J. Wadowski and coauthors, Family history of coronary artery disease and cholesterol: Screening children in disadvantaged inner-city population, *Pediatrics* 93 (1994): 109–113; K. Resnicow and D. Cross, Are parents' self-reported total cholesterol levels useful in identifying children with hyperlipidemia? An examination of current guidelines, *Pediatrics* 92 (1993): 347–354.

23. M. S. Glassman and S. M. Schwarz, Cholesterol screening in children: Should obesity be a risk factor? *Journal of the American College of Nutrition* 12 (1993): 270–273.

24. American Academy of Pediatrics, 1998.

25. American Academy of Pediatrics, 1998.

26. L. B. Dixon and coauthors, The effect of changes in dietary fat on the food group and nutrient intake of 4- to 10-year old children, *Pediatrics* 100 (1997): 863–872.

27. M. Sigman-Grant, S. Zimmerman, and P. M. Kris-Etherton, Dietary approaches for reducing fat intake of preschool-age children, *Pediatrics* 91 (1993): 955–960.

28. Is there a relationship between dietary fat and stature or growth in children three to five years of age? *Pediatrics* 92 (1993): 579–586.

29. L. Kann and coauthors, Youth risk behavior surveillance—United States, 1993, *Journal of School Health* 65 (1995): 163–171.

30. Tobacco use and usual source of cigarettes among high school students—United States, 1995, *Journal of the American Medical Association* 276 (1996): 184–185.

31. L. Van Horn and P. Greenland, Prevention of coronary artery disease is a pediatric problem, *Journal of the American Medical Association* 278 (1997): 1779–1780.

14 Nutrition for Older Adults

WWW.

ncoa.org
National Council on Aging

cdc.gov/nccdphp/sgr/sgr.htm
Physical Activity and Health: A Report of the Surgeon General

The last two chapters were devoted to stages of the life cycle that require special nutrition attention: pregnancy, lactation, infancy, childhood, and adolescence. Much of the text before that focused on nutrition to support wellness during adulthood. This chapter describes the special nutrition needs of the later adult years.

The most urgent nutrition need of older people, however, is to have made good food choices in the past! All of life's nutrition choices incur health consequences for the better or for the worse. A single day's intakes of nutrients may exert only a minute effect on body organs and their functions, but over years and decades their repeated effects accumulate to have major impacts. This being the case, it is of great importance for everyone, of every age, to pay close attention today to nutrition.

The U.S. population is graying. The majority of citizens are now middle-aged, and the ratio of old people to young is becoming greater, as Figure 14.1 shows. Our society uses the arbitrary age of 65 to define the transition point between middle age and old age, but growing "old" happens day by day, with changes occurring gradually over time. Since 1950 the population of people over 65 has more than doubled. Remarkably, the fastest-growing age group is people over 85 years (see Figure 14.2).[1] The U.S. Bureau of the Census projects that by the year 2040 there will be more than a million Americans 100 years old or older.

life expectancy: the average number of years lived by people in a given society.

longevity: long duration of life.

life span: the maximum number of years of life attainable by a member of a species.

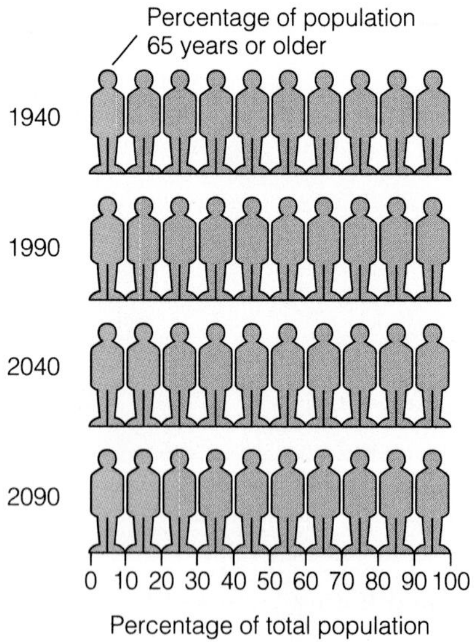

Life expectancy for women in the United States is 79 years; for men it is 73 years—both up from about 47 years in 1900.[2] Advances in medical science—antibiotics and other treatments—are largely responsible for almost doubling the life expectancy in this century. Improved nutrition and an abundant supply of food have also contributed to lengthening life expectancy.[3] Still, there appears to be an upper limit on human longevity that even nutrition cannot extend. The human life span is about 120 years and has not changed much over the years.

The study of the aging process is among the youngest of the scientific disciplines. It is only in this century that human beings have achieved a life expectancy worthy of a science devoted to studying it. The idea that nutrition can influence the way human bodies age is particularly appealing, since nutrition is a factor that people can control and change.

Figure 14.1
The Aging of the U.S. Population
In 1940, 6.8 percent of the population was 65 or older. In 1990, 12.7 percent of us had reached age 65; by 2040, 21.7 percent will have reached age 65; and a century from now, nearly one out of four Americans will be 65 or older. An estimated 25,000 Americans now living are 100 years old or older.

Nutrition and Longevity

What has been learned so far about the effects of nutrition and environment on longevity provides incentive for researchers to keep asking questions about how and why human beings age. Among their questions are:

- To what extent is aging inevitable, and can it be slowed through changes in lifestyle and environment?
- What roles does nutrition play in aging, and what roles can it play in retarding aging?

With respect to the first question, aging is an inevitable, natural process, programmed into the genes at conception. People can, however, slow the process within the natural limits set by heredity. They can adopt healthy lifestyle habits such as eating nutritious food and engaging in physical activity. With respect to the second question, clearly, good nutrition can retard and ease the aging process in many significant ways.

Slowing the Aging Process

One approach researchers use to search out the secret of long life has been to study older people. Some people are young for their ages, others old for their ages. What makes the difference?

Healthy Habits Six lifestyle habits seem to have a profound influence on people's health and therefore on their physiological age:[4]

physiological age: a person's age as estimated from her or his body's health and probable life expectancy.

- Sleeping regularly and adequately.
- Eating regular meals, including breakfast.
- Keeping weight under control.
- Engaging in regular physical activity.
- Not smoking.
- Not using alcohol, or using it in moderation.

Over the years, the effects of these lifestyle choices accumulate—that is, those who follow all of the practices are in better health, even if older in chronological age, than people who fail to do so. In fact, the physical health of people who report all positive health practices is comparable to that of people *30 years younger* who follow few or none. Other studies have confirmed that these health habits both extend longevity and support independence in later life.[5] The findings suggest that even though people cannot alter the years of their births, they can alter the probable lengths and quality of their lives. Physical activity seems to be most influential in preventing or slowing the many changes that many people seem to accept as an inevitable consequence of old age. In other words, physical activity and long life seem to go together.[6]

chronological age: a person's age in years from his or her date of birth.

Physical Activity The many and remarkable benefits of regular physical activity are not limited to the young: older adults who are active weigh less and have greater flexibility, more endurance, and better balance than those who are inactive.[7] They reap additional benefits as well; for example, evening activity helps to eliminate late night trips to the bathroom, moderate endurance activities improve the quality of sleep, and strength training significantly improves mobility and resistance to injury.[8] In fact, regular physical activity is the most powerful predictor of a person's mobility in the later years.[9]

Loss of muscle mass, strength, and quality is called *sarcopenia*.

Muscle mass and muscle strength tend to decline with aging, making older people vulnerable to falls and immobility. Falls are a major cause of fear, injury, disability, dependence, and even death among older adults. Regular physical activity tones, firms, and strengthens muscles, helping to improve confidence, reduce the risk of falling, and minimize the risk of injury should a fall occur. Strength training, even in frail, elderly people over 85 years of age, has been shown not only to improve muscle strength and mobility but to increase energy expenditure and energy intake, thereby enhancing nutrient intakes.[10] This finding highlights another reason to be physically active: a person spending energy on physical activity can afford to eat more food and with it, more nutrients. People who are committed to an ongoing fitness program have higher energy and nutrient intakes than more sedentary people.[11]

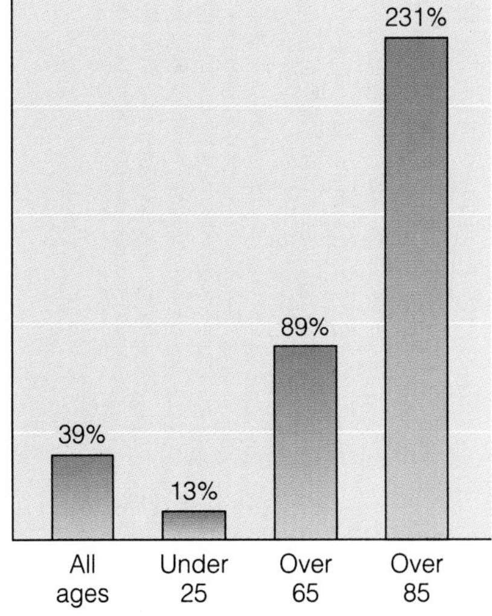

Figure 14.2

U.S. Population Growth, 1960 to 1990

The "oldest old"—those over 85 years—are the fastest-growing age group in the United States. Between 1960 and 1990, the U.S. population grew 39 percent, but the population of those over 85 more than doubled.

Activities of all kinds are recommended to maintain and promote health. Strength training improves muscle strength, which enhances a person's ability to perform many of life's daily tasks such as climbing stairs and carrying packages.[12] Aerobic activity can improve cardiorespiratory endurance and lower blood lipid concentrations.[13] Although aging affects both speed and endurance to some degree, older adults can still train and achieve exceptional performances.

Ideally, physical activity should be part of each day's schedule and should be intense enough to prevent muscle atrophy and to speed up the heartbeat and respiration rate. Healthy older adults who have not been active can ease into a suitable routine. They can start by walking short distances until they can walk at least a mile three times a week, and then they can gradually increase their pace to achieve a 20- to 25-minute mile.[14]

One expert suggests the following physical activity program to maintain good health and function in older adults:[15]

Strength training promotes strong muscles and bones and healthy appetites.

- *Every day.* Spend 60 minutes in some physical activity: gardening, walking, climbing stairs, or simply moving about. This can be for 5 minutes at a time, 12 times a day; 12 minutes at a time, 5 times a day; or any combination of activity to total 60 minutes.
- *Three days a week.* Spend 30 to 45 minutes in vigorous and continuous physical activity, such as swimming, dancing, rowing, or brisk walking.

With persistence, people can achieve great improvements at any age. Training not only benefits physical health but also increases the blood flow to the brain, thereby improving mental ability.

Restriction of kCalories Another approach to prevention of aging has been to manipulate animals' diets and look for effects on longevity. These studies have produced some interesting and suggestive findings.[16] For example, rats live longer when their food intake is restricted in the early weeks of their lives, or even when it is restricted after they are mature. Extensive research shows that it is the restriction of food energy rather than restriction of a specific nutrient that exerts the anti-aging action.[17]

Several mechanisms to explain how energy restriction prolongs life in rats have been proposed but not proved. Research suggests that food restriction may extend the life span by delaying age-related diseases and preventing damaging lipid oxidation.[18]

Experiments with food restriction and longevity in rats have *not* suggested any direct applications to human nutrition, though, and the animals given restricted feedings have suffered some distinct disadvantages. For example, the food restriction was so severe that half of the restricted animals died *very* early. The average length of life for the restricted rats was long because the few survivors lived a long time. Extreme starvation to extend life, like any extreme, is not worth the price.

One group of researchers studied the relationship between *moderate* energy restriction (80 percent of usual intake) and aging retardation in human beings.[19] Sixteen middle-aged, nonobese men were studied during ten weeks of energy restriction. Eight matched controls continued their usual diets. The energy-restricted men lost weight (mostly due to loss of fat), their blood pressures dropped significantly, and their HDL cholesterol concentrations rose significantly. Energy restriction had no adverse effects on mental and physical performance. For these men, moderate energy restriction favorably affected disease risk factors such as obesity, blood pressure, and cholesterol.

Nutrition and Disease Prevention

Nutrition alone, even if ideal, cannot ensure a long and robust life. Nevertheless, nutrition clearly affects aging and longevity in human beings by way of its role in disease prevention.

Among the better-known relationships between nutrition and disease are the following:

- Appropriate energy intake helps prevent *obesity, diabetes,* and related *cardiovascular diseases* such as atherosclerosis and hypertension (Chapters 9, 25, and 26) and may influence the development of some forms of *cancer* (Chapter 24).
- Adequate intakes of essential nutrients prevent *deficiency diseases* such as scurvy, goiter, anemia, and the like (Chapters 7 and 8).
- Variety in food intake, as well as ample intakes of certain vegetables, may be protective against certain types of *cancer* (Chapter 24).
- Moderation in sugar intake helps prevent *dental caries* (Nutrition in Practice 2).

- Appropriate fiber intakes help prevent malfunctions of the digestive tract such as *constipation, diverticulosis,* and possibly *colon cancer* (Chapters 2 and 19).
- Moderate sodium intake and adequate intakes of potassium, calcium, and other minerals help prevent *hypertension,* at least in people who are genetically predisposed to it (Chapters 8 and 26).
- An adequate calcium intake throughout life helps protect against *osteoporosis* (Chapter 8).

Figure 14.3 illustrates these relationships between diet and degenerative diseases. Note that the figure includes some risk factors such as heredity and age that cannot be modified, but notice how many can. Other, less well-established links between nutrition and disease are being discovered each day. For example, research efforts to uncover clues about the relationship between nutrition and immunity become more and more urgent as AIDS takes its toll on human life.

In summary, life expectancy for U.S. adults has almost doubled in the twentieth century. Advances in medical science, as well as improved nutrition, are largely responsible. Many of the health problems older people experience are currently attributed to normal, age-related processes, perhaps to a greater extent than is valid. Lifestyle factors that can slow the aging process include eating nutritious

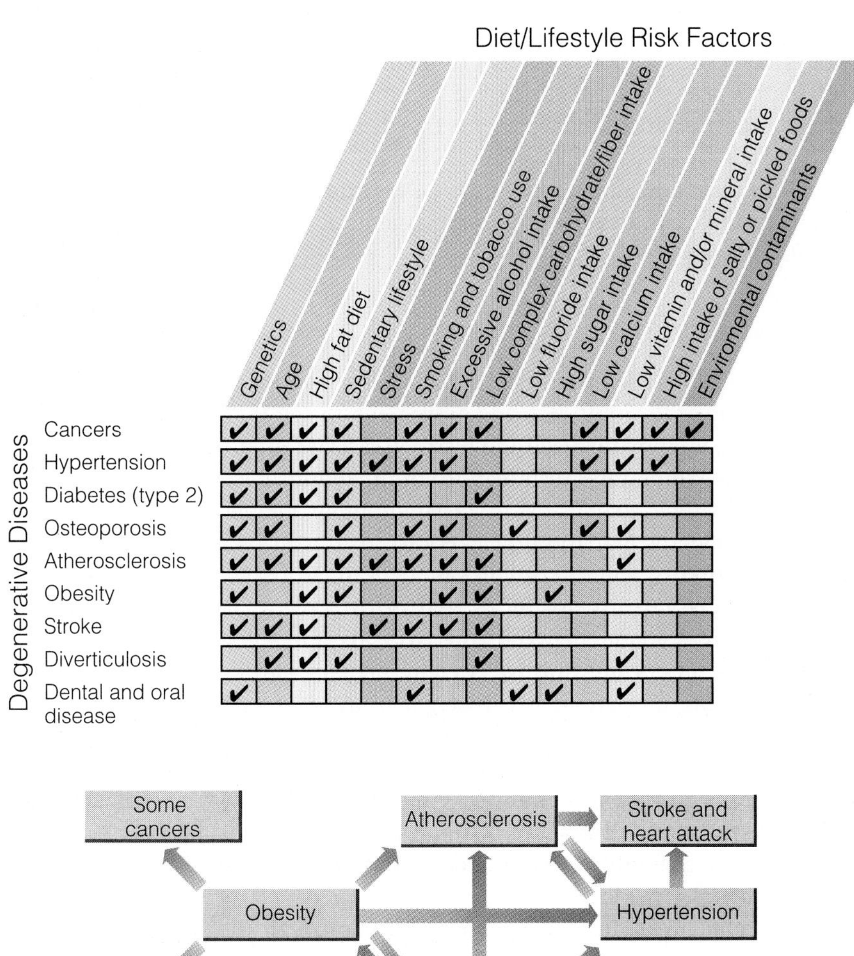

Figure 14.3
Risk Factors and Degenerative Diseases
The chart at the top shows that the same risk factor can affect many chronic diseases. Notice, for example, how many diseases have been linked to a high-fat diet. The chart also shows that a particular disease, such as atherosclerosis, may have several risk factors.

The flow chart at the bottom shows that many of these conditions are themselves risk factors for other chronic diseases. For example, a person with diabetes is likely to develop atherosclerosis and hypertension. These two conditions, in turn, worsen each other. Notice how all these diseases are linked to obesity.

food and being physically active. Research that focuses on how life factors affect aging and disease processes is vital to ensuring that more and more people can look forward to long, healthy lives.

Nutrition-Related Concerns during Late Adulthood

Nutrition through the prime years may play a greater role than has been realized in preventing many changes once thought to be inevitable consequences of growing older. The following discussions of cataracts, arthritis, and the aging brain show that nutrition may provide at least some protection against some of the conditions associated with aging.

Cataracts

cataracts: thickenings of the eye lenses that impair vision and can lead to blindness.

Cataracts are age-related thickenings in the lenses of the eye that impair vision. If not surgically removed, they ultimately lead to blindness. Cataracts occur even in well-nourished individuals due to ultraviolet light exposure, oxidative damage, injury, viral infections, toxic substances, and genetic disorders. Many cataracts, however, are vaguely called senile cataracts—meaning "caused by aging." In the United States, some 46 percent of people between the ages of 75 and 85 have cataracts, compared to only 5 percent of those between the ages of 52 and 64.[20]

WWW.

ascrs.org
American Society of Cataract and Refractive Surgery

rap.nas.edu/lab/NIH/EYE
National Eye Institute

Oxidative stress appears to play a significant role in the development of cataracts, and the antioxidant nutrients may help minimize the damage. Studies have reported an inverse relationship between cataracts and dietary intakes of vitamin C, vitamin E, and carotenoids.[21] One study found that people who had no cataracts took significantly more supplements of vitamins C and E than those who had cataracts.[22]

Arthritis

arthritis: inflammation of a joint, usually accompanied by pain, swelling, and structural changes.

osteoarthritis: a painful, chronic disease of the joints caused when the cushioning cartilage in a joint breaks down; joint structure is usually altered, with loss of function; also called *degenerative arthritis.*

The most common type of arthritis that disables older people is osteoarthritis, a painful swelling of the joints. During movement, the ends of bones are normally protected from wear by cartilage and by small sacs of fluid that act as a lubricant. With age, bones sometimes disintegrate, and the joints become malformed and painful to move. Osteoarthritis afflicts millions of people around the world, especially the elderly. Nutrition quackery to treat arthritis is abundant, but no known diet prevents, relieves, or cures it. Table 14.1 presents some of the many *non*effective dietary treatments for osteoarthritis.

A known connection between osteoarthritis and nutrition is overweight. Weight loss is important for overweight people with osteoarthritis, partly because the joints affected are often weight-bearing joints that are stressed and irritated by having to carry excess poundage. Interestingly, though, weight loss often relieves the worst pain of osteoarthritis in the hands as well, even though they are not weight-bearing joints. Jogging and other weight-bearing activities do not worsen osteoarthritis, even in marathon runners. In fact, both aerobic activity and weight training offer modest improvements in physical performance and pain relief.[23]

WWW.

arthritis.org
Arthritis Foundation

hypercon.com/evolve/oars.htm
Osteoarthritis Research Society

rheumatoid arthritis: a disease of the immune system involving painful inflammation of the joints and related structures.

Another type of arthritis, known as rheumatoid arthritis, has a possible link to diet through the immune system. In rheumatoid arthritis, the immune system mistakenly attacks the bone coverings as if they were made of foreign tissue. The integrity of the immune system depends on adequate nutrition, and a poor diet may worsen arthritis. It is also possible that in some individuals, certain foods may stimulate the immune system to attack. For example, milk and milk products seem to aggravate arthritis in some people.

Table 14.1 *None*ffective Dietary Strategies for Arthritis	
■ Alfalfa tea	■ Garlic
■ Aloe vera liquid	■ Honey
■ Amino acid supplements	■ Inositol
■ Blackstrap molasses	■ Kelp
■ Burdock root	■ Lecithin
■ Calcium	■ Para-amino benzoic acid (PABA)
■ Celery juice	■ Raw liver
■ Cod liver oil	■ Superoxide dismutase (SOD)
■ Copper supplements	■ Vitamin D
■ Dimethyl sulfoxide (DMSO)	■ Vitamin megadoses
■ Fasting	■ Watercress
■ Fresh fruit	■ Yeast

Another nutrient linked to arthritis is the omega-3 fatty acid found in fish oil, eicosapentaenoic acid (EPA). Research shows that the same diet recommended for heart health—one low in saturated fat from meats and milk products and high in oils from fish—helps prevent or reduce the inflammation in the joints that makes arthritis so painful. Researchers theorize that EPA probably interferes with the action of prostaglandins, compounds involved in inflammation.

WWW

rheumatology.org
American College of Rheumatology

rap.nas.edu/lab/NIH/ARTHRITIS
National Institute of Arthritis and Musculoskeletal and Skin Diseases

Another possible link between nutrition and rheumatoid arthritis involves lipid peroxidation. Lipid peroxidation of the membranes within joints causes inflammation and swelling.[24] Vitamin E helps prevent peroxidation, but it has not improved active cases of rheumatoid arthritis. This is not surprising, though, because the vitamin's role in lipid peroxidation is preventive, not restorative.

Medications used to relieve arthritis can impose nutrition risks. Many medications affect appetite and alter the body's use of nutrients.

The Aging Brain

The brain, like all of the body's organs, responds to both inherited and environmental factors that can enhance or diminish its amazing capacities. One of the challenges researchers face when studying aging of the brain in human beings is to distinguish among normal, age-related, physiological changes, changes caused by diseases, and changes that result from cumulative, extrinsic factors such as diet.

The brain normally changes in some characteristic ways as it ages. For one thing, its blood supply decreases. For another, the number of neurons, the brain cells that specialize in transmitting information, diminishes as people age. When the number of nerve cells in one part of the cerebral cortex diminishes, hearing and speech are affected. Losses of neurons in other parts of the cortex can impair memory and cognitive function. When the number of neurons in the hindbrain diminishes, balance and posture are affected. Losses of neurons in other parts of the brain affect still other functions.

neuron: a nerve cell; the structural and functional unit of the nervous system. Neurons initiate and conduct nerve transmissions.

cerebral cortex: the outer surface of the cerebrum.

Clinicians now recognize that much of the cognitive loss and forgetfulness generally attributed to aging are due in part to extrinsic, and therefore controllable, factors such as nutrient deficiencies. In some instances, the degree of cognitive loss is extensive and attributable to a specific disorder such as a brain tumor. In cases such as Alzheimer's disease, deterioration may be genetically determined and will not yield to external approaches.

Alzheimer's Disease Much attention has focused on the *abnormal* deterioration of the brain called senile dementia of the Alzheimer's type (SDAT), which affects 5 percent of U.S. adults by age 65 and 20 percent of those over 80.[25]

senile dementia: the loss of brain function beyond the normal loss of physical adeptness and memory that occurs with aging.

senile dementia of the Alzheimer's type (SDAT): a degenerative disease of the brain involving memory loss and major structural changes in neuron networks.

WWW. **alz.org**
Alzheimer's Association

WWW. **alzheimers.org**
Alzheimer's Disease Education and Referral Center

Diagnosis of SDAT depends on its characteristic symptoms: the victim gradually loses memory and reasoning, the ability to communicate, physical capabilities, and eventually life itself. Nerve cells in the brain die and communication between the cells breaks down.

The causes of SDAT continue to elude researchers, although genetic factors are apparently involved. A newly established connection to a human gene that makes part of a lipoprotein has sparked new hope for prevention of SDAT.[26*] Ultimately, researchers hope to use the genetic findings to develop early detection tests and a cure for this devastating disease.

Treatment involves providing care to clients and support to their families. One medication (trade-named Tacrine) seems to slow the advance of the disease in about 20 percent of those who use it, but it does not reverse the damage already done. Meanwhile, some medications seem to favorably influence the ability to remember and so hold promise for improving the lives of those with SDAT. Other medications may be used to control depression or behavior problems.

Most people have heard of an association between aluminum and the development of SDAT, although a causal connection seems unlikely. Brain concentrations of aluminum in SDAT people exceed normal brain concentrations by some 10 to 30 times, but blood and hair aluminum remains normal, indicating that the accumulation is caused by something in the brain itself, not by an overload of aluminum in the body. Thus the high brain aluminum must be at least partly a result, rather than a cause, of SDAT. Researchers are still investigating the relationship between dietary aluminum and SDAT in individuals, and the question is still open whether aluminum cookware, which slightly increases the aluminum content of foods, significantly affects the progress of SDAT.

Maintaining appropriate body weight may be the most important nutrition concern for the person with SDAT. Depression and forgetfulness can lead to poor food intake, and restlessness may increase energy needs. Perhaps the best that a caregiver can do nutritionally for an SDAT client is to supervise food planning and mealtimes. Providing well-liked and well-balanced meals and snacks in a cheerful atmosphere encourages food consumption. To minimize confusion, offer a few ready-to-eat foods, in bite-size pieces, with seasonings and sauces. To avoid mealtime disruptions, control distractions such as television, children, and the telephone.

SDAT is an identifiable disease, the course of which is probably not influenced by nutrition. But poor nutrition in general affects the brain in other ways.

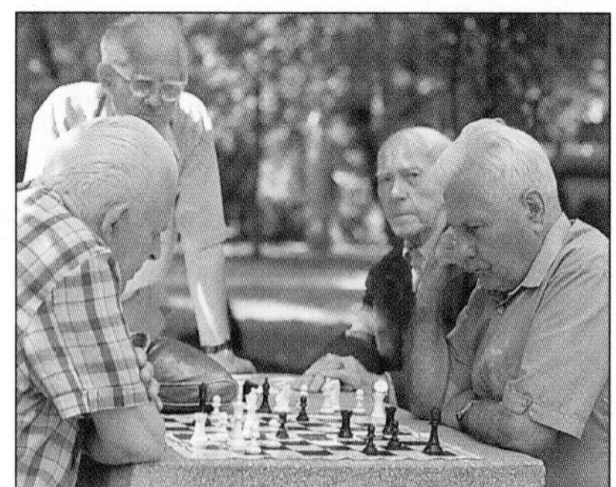
The brain is nourished by both foods and mental challenges.

Nutrient Deficiencies and Brain Function Moderate, long-term nutrient deficiencies may contribute to the loss of memory and cognition that some older adults experience.[27] Table 14.2 summarizes some of the better-known connections between impaired brain function and severe nutrient deficiencies. If long-term, moderate nutrient deficiencies influence the loss of cognitive function that accompanies aging, then the loss may be preventable or at least diminished or delayed through diet.

In summary, these brief discussions of cataracts, arthritis, and the aging brain show that nutrition may provide at least some protection against certain conditions associated with aging. In fact, it is beginning to look as though nutrition through the prime years may play a greater role than has been realized in preventing many changes once thought to be inevitable consequences of growing older.

*The gene associated with SDAT is apo E4, one of three apolipoprotein E varieties. A report on the genetic and other aspects of SDAT is available from Alzheimer's Disease Education and Referral Center, P.O. Box 8250, Silver Springs, MD 20907-8250.

Table 14.2 Summary of Nutrient-Brain Relationships

Brain Function	Nutrient Deficiency
Short-term memory loss	Vitamin B_{12}, vitamin C
Poor performance in problem-solving tests	Riboflavin, folate, vitamin B_{12}, vitamin C
Dementia	Thamin, niacin, zinc
Cognition	Folate, vitamin B_6, vitamin B_{12}, iron
Degeneration of brain tissue	Vitamin B_6

Energy and Nutrient Needs during Late Adulthood

Knowledge about the nutrient needs and nutrition status of older adults has grown considerably in the last decade or so. The RDA revision that is presently under way includes changes in age groupings (see inside front cover). Older adults are grouped into two age categories—51 to 70 years old, and 71 and older.[28]* After all, the needs of people 50 to 60 years old may be very different from those of people over 80. The need for more age-specific recommendations is becoming more and more urgent as the population ages.

2.nas.edu/fnb
Dietary Reference Intakes (from the Food and Nutrition Board)

Setting standards for older people is difficult, though, because individual differences become more pronounced as people grow older. One person may tend to omit vegetables from his diet, and by the time he is old, he will have an associated set of nutrition problems. Another may have omitted milk and milk products all her life—her nutrition problems will be different. Also, as people age, they may suffer different chronic diseases and take different medications—both having impacts on nutrient needs. Even before all this, people start out with different genetic predispositions and ways of handling nutrients, and the effects of these become magnified with the years. Researchers have difficulty even defining "healthy aging," a prerequisite to developing recommendations that are designed to meet the "needs of practically all healthy persons."[29] Still, some generalizations are valid. The next sections give special attention to a few nutrients of concern.

Energy and Energy Nutrients

Energy needs decline with advancing age. As a general rule, adult energy needs decline an estimated 5 percent per decade. For one thing, as people age, they usually reduce their physical activity, although they need not do so. For another, lean body mass diminishes, slowing the basal metabolic rate. One researcher contends, however, that if older adults (up to age 75) could avoid changes in body composition (by way of physical activity), they would show little, if any, decrease in basal metabolic rate.[30] Physical activity not only increases energy expenditure but, along with sound nutrition, enhances bone density and supports many body functions as well.[31]

The lower energy expenditures of many older adults require that they eat less food energy to maintain their weights. Accordingly, the energy RDA for adults decreases slightly, beginning at age 51. Energy intakes typically decline in parallel with needs. Still, many older adults are overweight, indicating that their food intakes do not decline enough to compensate for their reduced energy expenditure.

*The Canadian RNI divide older people into two age groups—50 to 74, and 75 and older.

Table 14.3 Daily Food Plan for Older Adults
2 to 3 servings (2 oz each) of protein-rich food (lean meat, poultry, fish, eggs, dried beans and peas, nuts)
2 to 3 servings (1 c each) of milk, cheese, or yogurt[a]
6 or more servings of whole-grain breads or cereals
2 to 4 servings (½ c each) of fruits
3 to 5 servings (½ c each) of vegetables

[a]Women should aim for 3 servings of milk, cheese, or yogurt.
Source: Adapted from A. Greeley, Nutrition and the elderly, *FDA Consumer,* October 1990, pp. 25–28.

On limited energy allowances, people must select mostly nutrient-dense foods. There is little leeway for sugars, fats, oils, or alcohol. Table 14.3 offers a dietary framework for older adults.

Protein The protein needs of older adults appear to be about the same as, or even greater than, those of younger people.[32] Since energy needs decrease, however, the protein has to be obtained from low-kcalorie sources of high-quality protein, such as lean meats, poultry, fish, and eggs; nonfat and low-fat milk products; and legumes and grains.

Carbohydrate and Fiber As always, abundant carbohydrate is needed to protect protein from being used as an energy source. Complex carbohydrate foods such as vegetables, whole grains, and fruits are also rich in fiber and essential vitamins and minerals.

The combination of ample water and high-fiber foods can alleviate constipation—a condition common among older adults, and especially among nursing home residents. Physical inactivity and medications probably contribute to the high incidence of constipation, but lack of water and fiber does, too. In fact, average fiber intakes among older adults are lower than current recommendations.

It has been estimated that as many as 50 percent of nursing home residents may be malnourished and underweight.[33] For these people, a diet that emphasizes fiber-rich foods such as whole grains, fruits, and vegetables may be too low in concentrated protein and energy. Protein- and energy-dense snacks such as hard-boiled eggs, tuna fish and crackers, peanut butter on graham crackers, and homemade soups are valuable additions to the diets of underweight or malnourished older adults.

Fat As is true for people of all ages, fat needs to be limited in the diets of most older adults. Cutting fat may help retard the development of cancer, atherosclerosis, and other degenerative diseases. For some older adults, though, limiting fat intake too severely may lead to nutrient deficiencies and weight loss—two problems that carry greater health risks in the elderly than overweight.[34]

Water

Dehydration is a risk for older adults, who may not notice or pay attention to their thirst, or who find it difficult and bothersome to get a drink or to get to a bathroom. Older adults who have lost bladder control may be afraid to drink too much water. Despite real fluid needs, older people do not seem to feel thirsty or notice mouth dryness.[35] Many nursing home employees say it is hard to persuade their elderly clients to drink enough water and fruit juices.

Water recommendation for adults:
1 to 1½ oz/kg actual body weight.

Total body water decreases as people age, so even mild stresses such as fever or hot weather can precipitate rapid dehydration in older adults. Dehydrated older adults seem to be more susceptible to urinary tract infections, pneumonia, pressure ulcers, and confusion and disorientation.[36] An intake of 6 to 8 glasses of water a day is recommended. Milk and juices may replace some of this water, but beverages containing alcohol or caffeine cannot because of their diuretic effect.

Vitamins and Minerals

As research reveals more about how specific vitamins and minerals influence disease prevention and age-related physiological changes affect nutrient metabolism, optimal intakes of vitamins and minerals for different groups of older adults are being defined. As mentioned earlier, new recommendations group older adults into two age categories rather than one.

Vitamin A Vitamin A stands alone in that it is absorbed and stored more efficiently by the aging GI tract and liver, although its processing within the body slows slightly. Several studies have reported that healthy older adults have normal concentrations of plasma vitamin A even when their dietary intakes fall below the RDA, suggesting that the current RDA may be too high.[37] The Committee on Dietary Allowances has hesitated to lower recommendations, however, recognizing both the need to prevent vitamin A deficiency and the possibility that the vitamin A precursor beta-carotene might prevent or delay the onset of some age-related diseases.

Vitamin A RDA during late adulthood:
1000 μg RE/day (men).
800 μg RE/day (women).

Vitamin D Older adults face a greater risk of vitamin D deficiency than younger people do. Only vitamin D–fortified milk provides significant vitamin D, and many older adults drink little or no milk. Consequently, many older adults have vitamin D intakes of less than half of recommendations. Further compromising the vitamin D status of many older people, especially those in nursing homes, is their limited exposure to sunlight. Finally, aging reduces the skin's capacity to make vitamin D and the kidneys' ability to convert it to its active form. Some research indicates that a vitamin D intake greater than the 1989 RDA may be necessary to prevent bone loss and to maintain vitamin D status in older people, especially in those who engage in minimal outdoor activity.[38] For this reason, the revised recommendations for vitamin D doubled the 1989 RDA from 5 to 10 micrograms per day.[39]

Vitamin D AI during late adulthood:
10 μg/day.

Vitamin B_6 Studies on vitamin B_6 reveal that its metabolism is altered with age, resulting in a higher requirement. Many older adults consume far less than the RDA for vitamin B_6.[40] Research suggests that immune response is impaired with vitamin B_6 deficiency. Such findings may have important implications for older adults because the aging process itself seems to be accompanied by a decline in immune function.[41]

Vitamin B_6 RDA during late adulthood:
1.7 μg/day (men).
1.5 μg/day (women).

Another approach to determining the vitamin B_6 requirements of older adults, as well as the requirements for vitamin B_{12} and folate, focuses on the amino acid homocysteine. An elevated homocysteine level is recognized as an independent risk factor for heart disease and stroke in the United States.[42] Homocysteine concentrations rise with vitamin B_6, vitamin B_{12}, or folate deficiencies (see Chapter 26).

atrophic gastritis: a condition characterized by chronic inflammation of the stomach accompanied by a diminished size and functioning of the mucosa and glands.

Vitamin B_{12} The prevalence of a condition known as atrophic gastritis among those 60 years of age and over is high. People with atrophic gastritis are particularly vulnerable to vitamin B_{12} deficiency for two reasons. First, digestion in the inflamed stomach is inefficient. Second, the abundant bacteria that accompany this condition use the vitamin. As with vitamin B_6, many older adults have vitamin B_{12} intakes below the RDA.[43]

Vitamin B_{12} RDA during late adulthood:
2.4 μg/day.

Folate As is true of vitamin B_6 and vitamin B_{12}, folate intakes of older adults fall short of recommendations.[44] Folate absorption, too, may be compromised by

Folate RDA during late adulthood: 400 μg/day.

atrophic gastritis, although increases in folate production in the small intestine may partially compensate.[45]

Many older adults take medications that influence folate absorption, use, and excretion. Antacids, diuretics, and anti-inflammatory drugs are among the medications that may alter folate metabolism.

Iron RDA during late adulthood: 10 mg/day.

Iron Among the minerals, iron deserves first mention. Iron-deficiency anemia is less common in older adults than in younger people, but it still occurs in some, especially in those with low food energy intakes. Aside from diet, other factors in many older people's lives make iron deficiency likely: chronic blood loss from disease conditions and medicines, and poor iron absorption due to reduced stomach acid secretion and antacid use. Anyone concerned with older people's nutrition should keep these possibilities in mind.

Zinc RDA during late adulthood:
15 mg/day (men).
12 mg/day (women).

Zinc Zinc intake is commonly low in older people, with many receiving less than half of the recommended amount.[46] In addition, older adults may absorb zinc less efficiently than younger people do. A number of different factors, including medications that older adults commonly use, can impair zinc absorption or enhance its excretion and thus lead to deficiency.[47] Older adults who do not make special efforts to eat zinc-rich foods such as meats, fish, and poultry will no doubt fail to meet the zinc RDA. Some of the symptoms of zinc deficiency resemble symptoms associated with aging—for example, decline in taste acuity and dermatitis. Whether these symptoms are attributable to zinc deficiency remains unclear.

Calcium AI during late adulthood: 1200 mg/day.

Calcium The appropriate calcium intake for older adults remains controversial. A National Institutes of Health panel has concluded that women over 50 who are not on estrogen replacement therapy and all adults over 65 should receive 1500 milligrams of calcium daily.[48] The revised recommendations for calcium are higher for both men and women than previous recommendations, but are still lower than the National Institutes of Health recommendation.

As researchers attempt to reach agreement about the calcium requirements of older adults, especially those of women, one thing is clear: the calcium intakes of many people, especially women, in the United States are well below recommended intakes.

Nutrient Supplements for Older Adults

People judge for themselves how to manage their nutrition, and some turn to supplements. Advertisers target older people with appeals to take supplements and eat "health" foods, claiming that these products prevent disease and promote longevity. About half of all women over 65 years of age take some type of nutrient supplement, while about one-fifth of older men do. Quite often those who take supplements are not deficient in the nutrients being supplemented.[49] Certain diseases or health problems may necessitate the taking of supplements, but quite often supplements have not been prescribed by health care professionals and are inappropriate.[50]

When recommended by a physician, vitamin D and calcium supplements for osteoporosis or iron for iron-deficiency anemia may be beneficial. In most cases, though, the money spent on supplements would be better spent on nutritious foods. Older adults with food energy intakes less than about 1500 kcalories should probably take the once-daily type of vitamin-mineral supplements.

People with small energy allowances would do well to become more active and earn the right to eat more food. Food is the best source of nutrients for everybody. Supplements are just that—supplements to foods, not substitutes for

Table 14.4 Strategies for Growing Old Gracefully

- Choose nutrient-dense foods.
- Maintain appropriate body weight.
- Reduce stress.
- For women, consult a physician about estrogen replacement to protect against osteoporosis.
- For people who smoke, quit.
- Expect to enjoy sex, and learn new ways of enhancing it.
- Use alcohol only moderately, if at all; use medications only as prescribed.
- Take care to prevent accidents.
- Expect good vision and hearing throughout life; obtain glasses and hearing aids if necessary.
- Be alert to confusion as a disease symptom, and seek diagnosis.
- Control depression through activities and friendships.
- Drink 8 glasses of water every day.
- Practice mental skills. Keep on solving math problems and crossword puzzles, playing cards or other games, reading, writing, imagining, and creating.
- Make financial plans early to ensure security.
- Accept change. Work at recovering from losses; make new friends.
- Cultivate spiritual health. Cherish personal values. Make life meaningful.
- Go outside for sunshine and fresh air as often as possible.
- Be physically active. Walk, run, dance, swim, bike, row, or climb for aerobic activity. Lift weights, do calisthenics, or pursue some other activity to tone, firm, and strengthen muscles. Change activities to suit changing abilities and tastes.
- Be socially active—play bridge, join an exercise group, take a class, teach a class, eat with friends, volunteer time to help others.
- Stay interested in life—pursue a hobby, spend time with grandchildren, take a trip, read, cultivate a garden, or go to the movies.
- Enjoy life.

them. For anyone who is motivated to obtain the best possible health, it is never too late to learn to eat well, become physically active, and adopt other lifestyle changes such as quitting smoking, moderating alcohol use, and the like. Table 14.4 offers strategies for growing old gracefully.

Table 14.5 (p. 370) summarizes the nutrient concerns of aging. Although some nutrients need special attention in the diet, supplements are not routinely recommended.

The Effects of Drugs on Nutrients

As people grow older, the use of medicines—from over-the-counter types such as aspirin and laxatives to prescription medications of all kinds—becomes commonplace. People over the age of 65 use more medications than any other age group; they take about 25 percent of all the over-the-counter and prescription medications sold. Most medications interact with one or more nutrients in several ways, usually resulting in greater-than-normal needs for these nutrients. Chapter 15 describes nutrient-medication interactions and Table 15.3 lists classes of medications that affect nutrition status.

The most common drug that can affect nutrition in older people is alcohol. A recent estimate sets the incidence of alcoholism in people over 60 in our society at 2 to 10 percent. The effects of alcohol on people of all ages are explained in Nutrition in Practice 8.

Table 14.5 Summary of Nutrient Concerns in Aging

Nutrient	Effect of Aging	Comments
Water	Lack of thirst and decreased total body water make dehydration likely	Mild dehydration is a common cause of confusion. Difficulty obtaining water or getting to the bathroom may compound the problem.
Energy	Need decreases.	Physical activity moderates the decline.
Fiber	Likelihood of constipation increases with low intakes and changes in the GI tract.	Inadequate water intakes and lack of physical activity, along with some medications, compound the problem.
Protein	Needs stay the same.	Low-fat, high-fiber legumes and grains meet both protein and other nutrient needs.
Vitamin A	Absorption increases.	RDA may be high.
Vitamin D	Increased likelihood of inadequate intake; skin synthesis declines.	Daily limited sunlight exposure may be of benefit.
Iron	In women, status improves after menopause; deficiencies are linked to chronic blood losses and low stomach acid output.	Adequate stomach acid is required for absorption; antacid or other medicine use may aggravate iron deficiency; vitamin C and meat increase absorption.
Zinc	Intakes may be low and absorption reduced; but needs may also decrease.	Medications interfere with absorption; deficiency may depress appetite and sense of taste.
Calcium	Intakes may be low; osteoporosis common.	Stomach discomfort commonly limits milk intake; calcium substitutes are needed.

Food Choices and Eating Habits of Older Adults

To provide any benefit, strategies and interventions to improve people's nutrition status must be based on knowledge of their food preferences and eating patterns. Menus and feeding programs for older adults must take into consideration not only the food likes and dislikes but also the living conditions, economic status, and medical conditions of this diverse group of people. If nutrition intervention is to be successful, it is essential to know what foods people will eat, in what settings they like to eat these foods, and whether they can buy and prepare meals.

Many factors affect food choices, eating habits, and the nutrition status of older adults. Information about specific subgroups of older people is lacking, making it difficult to interpret existing research. For instance, nutrition surveys do not always differentiate among older people living alone, those living with others, and those in institutions.

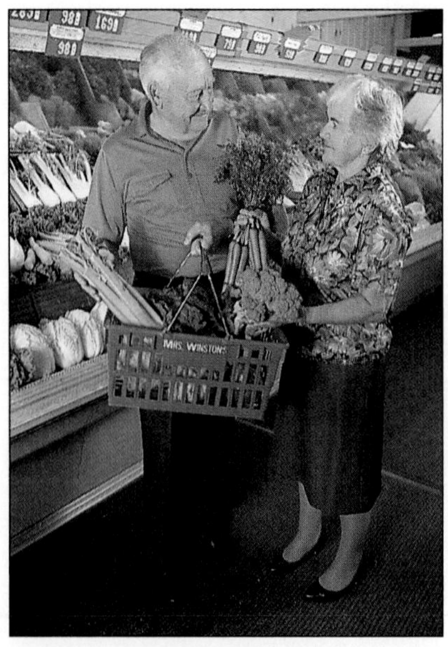

Taking time to nourish your body well is a gift you give yourself.

Older people are, for the most part, independent, socially sophisticated, mentally lucid, fully participating members of society who report themselves to be happy and healthy. In fact, chronic disabilities among the elderly have declined dramatically over the past decade.[51] Older people spend more money per person on foods to eat at home than other age groups and less money on foods away from home. Manufacturers would be wise to cater to the preferences of older adults by providing good-tasting, nutritious foods in easy-to-open, single-serving packages with labels that are easy to read. Such services enable older adults to maintain their independence; most of them want to take care of themselves and need to feel a sense of control and involvement in their own lives. As discussed earlier, another way older adults can take care of themselves is by remaining or becoming physically active. Physical activity helps preserve one's ability to perform daily tasks and so promotes independence.[52]

Individual Preferences Familiarity, taste, and health beliefs are most influential on older people's food choices. Eating foods that are familiar, especially those that recall family meals and pleasant times, can be comforting. The importance of diet and health beliefs in food selection is evidenced by surveys indicating that older adults are choosing low-fat poultry and fish, low-fat milk and milk products, and high-fiber breads and grains.[53] Few older adults however, consume the recommended amounts of milk products.[54]

Meal Setting Research shows that the settings in which older adults eat lead to significant differences in their food practices.[55] For example, men living alone are most likely to be poorly nourished.[56] Older adults who live alone do not make poorer food choices than those who live with companions; rather they consume too little food: loneliness is directly related to inadequacies, especially of energy intakes.[57]

Shared meals can brighten the day and enhance the appetite.

Depression Another factor affecting food intake and appetite in older people is depression (see Nutrition in Practice 15). Though not an inevitable component of aging, depression is more common with advancing age. Loss of appetite and motivation to cook or even to eat frequently accompanies depression. An overwhelming feeling of grief and sadness at the death of a spouse, friend, or family member may leave many people, particularly elderly people, with a feeling of powerlessness to overcome the depression. The support and companionship of family and friends, especially at mealtimes, can help overcome depression and enhance appetite. The case study on the next page presents a man who has some of these problems. Use the suggestions here, and in the nutrition in practice that follows this chapter, to help develop solutions.

In summary, food choices of older adults are affected by health beliefs, tastes, changed life circumstances, and meal settings. Intervention programs for older adults must consider these factors.

Nutrition Assessment of Older Adults

Nutrition assessment of older adults must focus on the past as well as the present. Each person's long prior history of eating and health habits has contributed to the present nutrition status.

- The history should reveal any current habits that may jeopardize adequate nutrient intakes: Does the person skip meals or routinely exclude any food groups? Can the person procure and prepare meals?
- The health and drug histories are particularly relevant to an older person's nutrition status. What are the health problems? What medications does the person take, and what are the effects on nutrition status? Does the person use alcohol? How much?
- Socioeconomic factors that affect nutrition status may also be present and may indicate the need for food assistance programs.
- The physical examination should be made with an eye to nutrition status. Hair, skin, and nails may offer clues to imbalances, excesses, or deficiencies. Inadequate fluid intake may be apparent from a high body temperature, a swollen tongue, lowered blood pressure, sunken eyeballs, or a reduced urine output. Are any chronic diseases, illnesses, or physical disabilities present?
- Oral health should be assessed, as it may affect food and nutrient intake. Can the person chew and swallow foods satisfactorily? Does the person have dentures or need them? Are there any lesions in the mouth, on the lips, or on the tongue?
- Anthropometric measures such as height, weight, and fatfold measures can reveal altered body composition that may indicate malnutrition. Loss of

Case Study

Elderly Man with a Poor Diet

Mr. Brezenoff is a 75-year-old man who lives alone. He has been losing weight slowly since his wife died a year ago. At 5 feet 8 inches tall, he currently weighs 124 pounds. His previous weight was 150 pounds. In talking with Mr. Brezenoff, you realize that he doesn't even like to talk about food, let alone eat it. "My wife always did the cooking before, and I ate well. Now I just don't feet like eating." You manage to find out that he skips breakfast, has soup and bread for lunch, and sometimes eats a cold-cut sandwich or a TV dinner for supper. He seldom sees friends or relatives. Mr. Brezenoff has also lost several teeth and doesn't eat any raw fruits or vegetables because he finds them hard to chew. He lives on a meager but adequate income.

- Is Mr. Brezenoff's weight within the range of suggested weights for adults his height (see Table 9.1)?
- Is his weight loss significant?
- What factors are contributing to his poor food intake?
- What nutrients are probably deficient in his diet?
- Look at Mr. Brezenoff as an individual and suggest ways he can improve his diet and his lifestyle.
- What other aspects of Mr. Brezenoff's physical and mental health should you consider in helping him to improve his food intake?

WWW **aafp.org/nsi/index.html**
Nutrition Screening Initiative

Risk factors for malnutrition in older adults:

- *Disease.*
- *Eating poorly.*
- *Tooth loss or oral pain.*
- *Economic hardship.*
- *Reduced social contact.*
- *Multiple medications.*
- *Involuntary weight loss or gain.*
- *Needs assistance with self-care.*
- *Elderly person older than 80 years.*

lean tissue is an early indicator of protein-energy malnutrition (PEM). Obesity suggests a heightened risk of chronic diseases.

- Biochemical measures may be used to follow up on other assessment steps as appropriate and as described in Chapter 16. These measures can confirm that iron deficiency or other conditions are present.

The Nutrition Screening Initiative is part of a national effort to identify and treat nutrition problems in older persons; it uses a screening checklist (see Table 14.6). To *determine* the risk of malnutrition in older clients, health care providers can keep in mind the characteristics listed in the margin.[58]

Many of the chapters to come describe diseases that affect the elderly and offer details on appropriate assessment measures.

Table 14.6 Nutrition Screening Initiative Checklist

Circle the number to the right if the statement applies to you.

Statement	Yes
I have an illness or condition that made me change the kind and/or amount of food I eat.	2
I eat fewer than 2 meals per day.	3
I eat few fruits or vegetables or milk products.	2
I have 3 or more drinks of beer, liquor, or wine almost every day.	2
I have tooth or mouth problems that make it hard for me to eat.	2
I don't always have enough money to buy the food I need.	4
I eat alone most of the time.	1
I take 3 or more different prescribed or over-the-counter medications a day.	1
Without wanting to, I have lost or gained 10 pounds in the last 6 months.	2
I am not always physically able to shop, cook, and/or feed myself.	2
Total	

SCORE:

0–2: Good. Recheck your score in 6 months.

3–5: Moderate nutritional risk. Visit your local office on aging, senior nutrition program, senior citizens center, or health department for tips on improving eating habits.

6 or more: High nutritional risk. See your doctor, dietitian, or other health care professional for help in improving your nutrition status.

Self Check

1. The fastest-growing age group in the United States is:
   a. people 21 years of age.
   b. people 30 to 45 years of age.
   c. people 50 to 70 years of age.
   d. people over 85 years of age.
2. Which of the following lifestyle habits can enhance the length and quality of people's lives?
   a. regular physical activity
   b. moderate smoking and alcohol intake
   c. adequate sleep
   d. a and c
3. Among the better-known relationships between nutrition and disease prevention are:
   a. appropriate energy intake helps prevent diabetes and cardiovascular disease.
   b. moderate sugar intake helps prevent hypertension.
   c. appropriate fiber intake helps prevent goiter.
   d. moderate sodium intake helps prevent obesity.
4. A disease of the immune system that involves painful inflammation of the joints is:
   a. osteoarthritis. b. sarcopenia.
   c. rheumatoid arthritis. d. senile dementia.
5. Examples of low-kcalorie, high-quality protein include:
   a. green and yellow vegetables and citrus fruits.
   b. cottage cheese, sour cream, and eggs.
   c. potatoes, rice, pasta, and whole-grain breads.
   d. lean meats, poultry, fish, legumes, nonfat milk, and eggs.
6. Dehydration is a risk for older adults because:
   a. they do not seem to feel thirsty.
   b. total body water decreases with age.
   c. they may find it difficult to get a drink.
   d. all of the above.
7. For malnourished and underweight people, protein- and energy-dense snacks include:
   a. whole grains and high-fiber legumes.
   b. yogurt and cottage cheese.
   c. hard-boiled eggs and peanut butter and crackers.
   d. fresh fruits and vegetables.
8. Inadequate milk intake and limited exposure to sunlight contribute to older adults' risk of:
   a. vitamin A deficiency. b. vitamin D deficiency.
   c. riboflavin deficiency. d. vitamin B_6 deficiency.
9. Two risk factors for malnutrition in older adults are:
   a. loneliness and multiple medication use.
   b. decreased mineral absorption and antioxidant intake.
   c. increased energy needs and lack of fiber.
   d. high carbohydrate intake and lack of physical activity.
10. Two strategies to improve nutrition status when growing old include:
    a. eat at least one big meal per day and drink at least 10 glasses of water daily.
    b. choose nutrient-dense foods and maintain appropriate weight.
    c. increase vitamin A intake and exercise 30 minutes daily.
    d. avoid high-fiber foods and take a daily vitamin-mineral supplement.

Answers to these questions appear in Appendix H.

Notes

1. Position of The American Dietetic Association: Nutrition, aging, and the continuum of care, *Journal of the American Dietetic Association* 96 (1996): 1048–1052; R. Chernoff, Demographics of aging, in *Geriatric Nutrition: The Health Professional's Handbook,* ed. R. Chernoff (Gaithersburg, Md.: Aspen Publishers, 1991), pp. 1–9.

2. Centers for Disease Control and Prevention, National Center for Health Statistics, *Monthly Vital Statistics Report,* October 1996, p. 4.

3. K. G. Kinsella, Changes in life expectancy 1900–1990, *American Journal of Clinical Nutrition* 55 (1992): 1196S–1202S; S. Kobayashi, A scientific basis for the longevity of Japanese in relation to diet and nutrition, *Nutrition Reviews* 50 (1992): 353–359.

4. L. Breslow and N. Breslow, Health practices and disability: Some evidence from Alameda County, *Preventive Medicine* 22 (1993): 86–95.

5. A. J. Vita and coauthors, Aging, health risks, and cumulative disability, *New England Journal of Medicine* 338 (1998): 1035–1041; A. Z. LaCroix and coauthors, Maintaining mobility in late life: Smoking, alcohol consumption, physical activity, and body mass index, *American Journal of Epidemiology* 137 (1993): 858–869.

6. U. M. Kujala and coauthors, Relationship of leisure-time, physical activity, and mortality: The Finnish twin cohort, *Journal of the American Medical Association* 279 (1998): 440–444; S. N. Blair and coauthors, Changes in physical fitness and all-cause mortality: A prospective study of healthy and unhealthy men, *Journal of the American Medical Association* 273 (1995): 1093–1098; R. S. Paffenbarger and coauthors, The association of changes in physical-activity level and other lifestyle characteristics with mortality among men, *New England Journal of Medicine* 328 (1993): 538–545.

7. American College of Sports Medicine, Position Stand, Exercise and physical activity for older adults, *Medicine and Science in Sports and Exercise* 30 (1998): 992–1008; L. E. Voorrips and coauthors, The physical condition of elderly women differing in habitual physical activity, *Medicine and Science in Sports and Exercise* 25 (1993): 1152–1157.

8. A. C. King and coauthors, Moderate-intensity exercise and self-rated quality of sleep in older adults, *Journal of the American Medical Association* 277 (1997): 32–37; W. Evans, Functional and metabolic consequences of sarcopenia, *Journal of Nutrition* 127 (1997): 998S–1003S.

9. LaCroix and coauthors, 1993.

10. King and coauthors, 1997; M. A. Fiatarone and coauthors, Exercise training and nutritional supplementation for physical frailty in very elderly people, *New England Journal of Medicine* 330 (1994): 1769–1775; W. W. Campbell and coauthors, Increased energy requirements and changes in body composition with resistance training in older adults, *American Journal of Clinical Nutrition* 60 (1994): 167–175.

11. D. E. Butterworth and coauthors, Exercise training and nutrient intake in elderly women, *Journal of the American Dietetic Association* 93 (1993): 653–657.

12. W. J. Evans and D. Cyr-Campbell, Nutrition, exercise, and healthy aging, *Journal of the American Dietetic Association* 97 (1997): 632–638.

13. J. S. Green and S. F. Crouse, The effects of endurance training on functional capacity in the elderly: A meta-analysis, *Medicine and Science in Sports and Exercise* 27 (1995): 920–926.

14. J. Posner, M. D., Professor of Medicine and Chief of the Divisions of Geriatric Medicine at the Medical College of Pennsylvania in Philadelphia, as cited in C. L. Pollock, Breaking the risk of falls, *The Physician and Sports Medicine* 20 (1992): 146–156.

15. P. Astrand, Physical activity and fitness, *American Journal of Clinical Nutrition* (supplement) 55 (1992): 1231–1236.

16. E. J. Masoro, Retardation of aging processes by food restriction: An experimental tool, *American Journal of Clinical Nutrition* (supplement) 55 (1992): 1250–1252.

17. Masoro, 1992.

18. R. Weindruch, Caloric restriction and aging, *Scientific American,* January 1997, pp. 46–52; Masoro, 1992.

19. E. J. M. Velthuis-te Wierik and coauthors, Energy restriction, a useful intervention to retard human ageing? Results of a feasibility study, *European Journal of Clinical Nutrition* 48 (1994): 138–148.

20. G. E. Bunce, J. Kinoshita, and J. Horwitz, Nutritional factors in cataract, *Annual Review of Nutrition* 10 (1990): 233–254.

21. G. E. Bunce, Antioxidant nutrition and cataract in women: A prospective study, *Nutrition Reviews* 51 (1993): 84–86; P. F. Jacques and L. T. Chylack, Epidemiologic evidence of a role for the antioxidant vitamins and carotenoids in cataract prevention, *American Journal of Clinical Nutrition* 53 (1991): 352S–355S.

22. P. F. Jacques and coauthors, Long-term vitamin C supplement use and prevalence of early age-related lens opacities, *American Journal of Clinical Nutrition* 66 (1997): 911–916; J. M. Robertson, A. P. Donner, and J. R. Trevithick, A possible role for vitamins C and E in cataract prevention, *American Journal of Clinical Nutrition* 53 (1991): 346S–351S.

23. W. H. Ettinger and coauthors, A randomized trial comparing aerobic exercise and resistance exercise with a health education program in older adults with knee osteoarthritis: The Fitness Arthritis and Seniors Trial (FAST), *Journal of the American Medical Association* 277 (1997): 25–31.

24. P. Merry and coauthors, Oxidative damage to lipids within the inflamed human joint provides evidence of radical-mediated hypoxic-reperfusion injury, *American Journal of Clinical Nutrition* 53 (1991): 362S–369S.

25. R. N. Butler, Senile dementia of the Alzheimer type (SDAT), in *The Merck Manual of Geriatrics* (Rahway, N.J.: Merck & Co., 1990), pp. 933–938.

26. E. M. Reiman and coauthors, Preclinical evidence of Alzheimer's disease in persons homozygous for the ϵ4 allele for apolipoprotein E, *New England Journal of Medicine* 334 (1996): 752–758; National Institute of Aging, *Progress Report on Alzheimer's Disease, 1994,* NIH publication number 94-3885 (Washington, D.C.: Government Printing Office, 1994).

27. A. La Rue and coauthors, Nutritional status and cognitive functioning in a normally aging sample: A 6-y reassessment, *American Journal of Clinical Nutrition* 65 (1997): 20–29.

28. J. Blumberg, Nutrient requirements of the elderly—Should there be specific RDAs? *Nutrition Reviews* 52 (1994): S15–S18; R. M. Russell and P. M. Suter, Vitamin requirements of elderly people: An update, *American Journal of Clinical Nutrition* 58 (1993): 4–14.

29. A. Bendich, Criteria for determining recommended dietary allowances for healthy older adults, *Nutrition Reviews* 53 (1995): S105–S110.

30. J. V. G. A. Durnin, Energy metabolism in the elderly, in *Nutrition of the Elderly,* eds. H. Munro and G. Schlierf (New York: Vevey/Raven Press, 1992), pp. 51–63.

31. L. DiPietro, The epidemiology of physical activity and physical function in older people, *Medicine and Science in Sports and Exercise* 28 (1996): 596–660; Durnin, 1992.

32. W. D. Campbell, Dietary protein requirements of older people: Is the RDA adequate? *Nutrition Today* 31 (1996): 192–197; D. Kritchevsky, Protein requirements of the elderly, in *Nutrition of the Elderly,* eds. H. Munro and G. Schlierf (New York: Vevey/Raven Press, 1992), pp. 109–118.

33. A. A. Abbase and D. Rudman, Undernutrition in the nursing home: Prevalence, consequences, causes and prevention, *Nutrition Reviews* 52 (1994): 113–122.

34. P. J. Nestel, Dietary fat for the elderly: What are the issues? in *Nutrition of the Elderly,* eds. H. Munro and G. Schlierf (New York: Vevey/Raven Press, 1992), pp. 119–127.

35. B. J. Rolls and P. A. Phillips, Aging and disturbances of thirst and fluid balance, *Nutrition Reviews* 48 (1990): 137–144.

36. J. C. Chidester and A. A. Spangler, Fluid intake in the institutionalized elderly, *Journal of the American Dietetic Association* 97 (1997): 23–28; S. A. Gilmore and coauthors, Clinical indicators associated with unintentional weight loss and pressure ulcers in elderly residents of nursing facilities, *Journal of the American Dietetic Association* 95 (1995): 984–992.

37. Russell and Suter, 1993.

38. Russell and Suter, 1993.

39. Committee on Dietary Reference Intakes, *Dietary Reference Intakes for calcium, phosphorus, magnesium, vitamin D, and fluoride* (Washington, D.C.: National Academy Press, 1997).

40. K. Tucker, Micronutrient status and aging, *Nutrition Reviews* (II) 53 (1995): 9–15.

41. A. L. de Weck, Immune response and aging, in *Nutrition of the Elderly,* eds. H. Munro and G. Schlierf (New York: Vevey/Raven Press, 1992), pp. 89–97.

42. J. Selhub and coauthors, Association between plasma homocysteine concentrations and extracranial carotid-artery stenosis, *New England Journal of Medicine* 332 (1995): 286–291.

43. Tucker, 1995.

44. Tucker, 1995.

45. R. M. Russell and coauthors, Folic acid malabsorption in atrophic gastritis: Possible compensation by bacterial folate synthesis, *Gastroenterology* 91 (1986): 1476–1478, as cited in B. A. Bowman, I. H. Rosenberg, and M. A. Johnson, Gastrointestinal function in the elderly, in *Nutrition of the Elderly,* eds. H. Munro and G. Schlierf (New York: Vevey/Raven Press, 1992), pp. 43–50.

46. R. J. Wood, P. M. Suter, and R. M. Russell, Mineral requirements of elderly people, *American Journal of Clinical Nutrition* 62 (1995): 493–505.

47. G. J. Fosmire, Trace mineral requirements, in *Geriatric Nutrition: The Health Professional's Handbook,* ed. R. Chernoff (Gaithersburg, Md.: Aspen Publishers, 1991), pp. 77–105.

48. D. V. Porter, Washington update: NIH consensus development conference statement optimal calcium intake, *Nutrition Today,* September/October 1994, pp. 37–40.

49. H. Payette and K. Gray-Donald, Do vitamin and mineral supplements improve the dietary intake of elderly Canadians? *Canadian Journal of Public Health* 82 (1993): 58–60; W. A. McIntosh and coauthors, The relationship between beliefs about nutrition and dietary practices of the elderly, *Journal of the American Dietetic Association* 90 (1990): 671–675.

50. F. Tripp, The use of dietary supplements in the elderly: Current issues and recommendations, *Journal of the American Dietetic Association* 97 (1997): S181–S183.

51. K. G. Manton, L. Corder, and E. Stallard, Chronic disability trends in elderly United States population: 1982–1994, *Proceedings of the National Academy of Sciences of the USA* 94 (1997): 2593–2598.

52. DiPietro, 1996.

53. Are older Americans making better food choices to meet diet and health recommendations? *Nutrition Reviews* 51 (1993): 20–22.

54. J. G. Fischer and coauthors, Dairy product intake of the oldest old, *Journal of the American Dietetic Association* 95 (1995): 918–921.

55. J. V. White and coauthors, Consensus of the Nutrition Screening Initiative: Risk factors and indicators of poor nutritional status in older Americans, *Journal of the American Dietetic Association* 91 (1991): 783–787.

56. I. Darnton-Hill, Psychosocial aspects of nutrition and aging, *Nutrition Reviews* 50 (1992): 476–479; M. A. Davis and coauthors, Living arrangements and dietary quality of older U.S. adults, *Journal of the American Dietetic Association* 90 (1990): 1667–1672.

57. D. Walker and R. E. Beauchene, The relationship of loneliness, social isolation, and physical health to dietary adequacy of independently living elderly, *Journal of the American Dietetic Association* 91 (1991): 300–304.

58. J. Dwyer and coauthors, Screening older Americans' nutritional health: Future possibilities, *Nutrition Today,* September/October 1991, pp. 21–24.

CHAPTER 14

Nutrition in Practice

▪ Food for Singles ▪

Singles of all ages face difficulties in purchasing, storing, and preparing food. Large packages of meat and vegetables are often intended for families of four or more, and even a head of lettuce can spoil before one person can use it all. Many singles live in small dwellings and have little storage space for foods. A limited income presents additional obstacles. The following ideas can help to solve some of these problems.

My grandmother is on a fixed income. Once the rent, utilities, and other bills are paid, she doesn't have much money for groceries. What advice can I give her?

First, make sure your grandmother knows what food assistance programs she can turn to. Such programs are available to older adults who need help obtaining nourishing meals because of financial or other difficulties. Table NP14.1 summarizes food assistance programs for the elderly.

People who have the means to shop and cook for themselves can cut their food bills just by being wise shoppers. The first decision a person with a tight grocery budget must make is where to shop. Large supermarkets are usually less expensive than convenience stores, but the cost of transportation to the market is a consideration.

Once a person decides where to shop, a grocery list that includes specials and coupons will help reduce impulse buying. Specials and coupons are a bargain only when the items featured are those that the shopper needs and uses. Foods that are almost always good buys include rice and nonfat dry milk, which can be stored on a shelf for months at room temperature; whole pieces of cheese, rather than sliced or shredded cheese; fresh produce in season; variety meats such as chicken livers; cereals that require cooking, instead of ready-to-serve cereals; and dried beans and peas.

All foods, whether featured as specials or not, are bargains only when they are available in quantities that can be used without waste or spoilage. For example, turkey may be on sale and is one of the most economical meats for the nutrients it offers, but only a person with ample storage space for leftovers would benefit from buying a whole turkey.

Table NP14.1 Food Assistance Programs for Older Adults

Program
Elderly Nutrition Program
▪ *Services:* Congregate and home-delivered meals to improve older people's nutrition status. Supportive services include transportation to congregate meal sites; shopping assistance; information and referral; and, to some extent, nutrition counseling and education. ▪ *Impact:* The Elderly Nutrition Program improves the nutrient content of high-risk older adults' diets and offers socialization and recreation. Many of the nutrition programs around the country go above and beyond federal requirements of congregate and home meals by offering lunch clubs, ethnic meals, and meals for older homeless people. An estimated 25 percent of our nation's elderly poor benefit from these meals.[a]
Food Stamps
▪ *Services:* Income supplement for low-income households in the form of coupons to purchase food. ▪ *Impact:* Food stamps serve more as an income supplement for some elderly participants than as a device to improve nutrition status. For other elderly food stamp participants, nutrient intakes are higher than those of nonparticipants with similar incomes.
Meals on Wheels
▪ *Services:* Direct meal delivery to the homebound elderly, integrated into the meal delivery services provided by the Elderly Nutrition Program. ▪ *Impact:* Meals on Wheels focuses on filling the need for weekend and holiday meals for homebound elderly people, a service that is limited in the Elderly Nutrition Program.

[a]Federal program nourishes poor elderly, *Journal of the American Medical Association* 278 (1997): 1301.

WWW

usda.gov/fcs/fs.htm
Food Stamp Program

That's a big problem for my grandmother. How can she buy small quantities when so many foods are packaged for families?

All singles face this problem. Packages of meat and fresh vegetables often come already wrapped in large servings. Even milk is often available only in gallons or half-gallons and can spoil before one person can use it all. People living alone can try these hints.

First, in the milk and milk product category, buy fresh milk in the size best suited for you. If your grocer doesn't carry pints or half-pints, try a nearby service station or convenience store. Pint-size and even cup-size boxes of heat-treated milk are also available and can be stored unopened on a shelf for up to three months without refrigeration. Dry powdered milk can be stored for months before it is reconstituted. You can use only the amount you need and store the rest.

Next, among the meats and meat alternates, buy only what you will use. Ask the grocer to break open a package of wrapped meat and rewrap the portion that you need. Buy eggs by the half-dozen—break the carton of a dozen eggs in half. Eggs do keep for long periods, though, if stored in the refrigerator and are such a good source of high-quality protein that you will probably use a dozen before they lose their freshness. Dried beans and peas offer high-quality protein, fiber, and many other nutrients for practically pennies and have a long shelf life.

If you have ample freezer space, you can buy large packages of meat, such as pork chops, ground beef, or chicken, when they are on sale. Then, immediately divide the package into individual servings. Wrap them in aluminum foil, not freezer paper: the foil can become the liner for the pan in which you bake or broil the meat, thus saving work over the sink. Don't label these individually; just put them all in a brown bag marked "hamburger" or "chicken thighs" or whatever, along with the date. The bag is easy to locate in the freezer, and you'll know when your supply is running low.

Among fruits and vegetables, purchase fresh ones individually. Buy only three pieces of each kind of fresh fruit: a ripe one, a semiripe one, and a green one. Eat the first right away, the second soon after, and let the last one ripen on your windowsill. If vegetables are packaged in large quantities, ask the grocer to break open the package so that you can buy what you need. Buy small cans of fruits and vegetables even though they are more expensive per unit. Remember, it is expensive to buy a regular-size can and let the unused portion spoil in the refrigerator. If you have space in your freezer, buy frozen vegetables in large bags rather than in small boxes. You can take out the exact amount you need and close the bag tightly with a rubber band. If you return the package quickly to the freezer each time, the vegetables will stay fresh for a long time.

Finally, breads and cereals usually must be purchased in larger quantities. When you buy bread, take out the amount you will use in a few days and store the rest in the freezer. Cereal grains (rice, barley, oatmeal) and pastas, like the dried beans mentioned earlier, generally have a long shelf life if you keep them sealed in jars.

Those are great suggestions. I know my grandmother can use them, and so can I. But sometimes you just have to buy more food than you can use. Do you have hints for what to do then?

One thing you can do is to make mixtures of leftovers that you have on hand. A thick stew prepared from leftover green beans, carrots, cauliflower, broccoli, and any meat with added onion, pepper, celery, and potatoes makes a complete and balanced meal—except for milk, but then you can add powdered milk to your stew.

You can also set aside a shelf in your kitchen for rows of glass jars containing staple items that you can't buy in single-serving quantities—rice, tapioca, lentils or other dry beans, flour, cornmeal, nonfat dry milk, pasta, cereal, or coconut, to name only a few possibilities. Freeze each filled jar for one night first to kill any insect eggs that might be present. The jars will then keep bugs out of the food indefinitely. They make an attractive display and will remind you of different choices you can make to vary your menus. Cut the directions-for-use labels from the packages and tape them on the jars.

Creative chefs think of various ways to use foods when only large amounts are available. For example, a head of cauliflower can be divided into thirds. Cook one-third and eat it as a hot vegetable. Put the other two-thirds into a vinegar and oil marinade for use as an appetizer or in a salad. You can keep half a package of frozen vegetables with other vegetables to be used in soup or stew.

Also, when you cook a lot of food, invite someone to share it with you. Next thing you know, that person will invite you back, and you'll get to enjoy a meal you wouldn't have thought to cook for yourself.

CHAPTER 14

Nutrition in Practice

My grandmother never drinks milk. What can I suggest?

She can try using nonfat dry milk. It is the greatest convenience food there is. Dry milk can be used in just about everything: hamburgers, gravies, soups, casseroles, sauces, and beverages, such as iced coffee. The taste is negligible, but 5 heaping tablespoons are the equivalent of a cup of fresh milk.

Your grandmother can also increase her calcium intake by making soup stock from pork and chicken bones soaked in vinegar. The bones release their calcium into the acid medium, and the vinegar boils off when the stock is boiled. One tablespoon of such stock may contain over 100 milligrams of calcium. Something can then be cooked in this stock every day: vegetables, rice, and stews. And, of course, this stock can be used as a soup base.

Sometimes my grandmother doesn't feel like cooking a meal for herself. I'm afraid that she may not be getting the nutrients she needs.

When your grandmother does feel like cooking, she can cook several meals at a time. For example, she can boil three potatoes with skins. She can eat one hot with margarine and chives. When the others have cooled, she can use one to make a potato-cheese casserole ready to be put into the oven for the next evening's meal. Then she can slice the third one into a covered bowl and pour the juice from pickles over it. The pickled potato will keep several days in the refrigerator and can be used later in a salad.

Depending on her freezer space, she can make double or triple portions of a dish that takes time to prepare: a casserole, vegetable pie, or meat loaf. The little aluminum trays from frozen foods can be saved and used to freeze the extra servings. Your grandmother needs to be sure to date the packages and use the oldest first. The work will seem worthwhile when several meals are prepared at once.

An occasional frozen TV dinner can also be useful—although expensive—if it makes the difference between a person's eating and not eating. Many such dinners that are now available are low in kcalories and nutritious. Adding a fresh salad, a whole-wheat roll, and a glass of milk can make a nice meal.

But encourage your grandmother to socialize, too. Sometimes, the loneliness that results from a single person's isolation can impair the person's appetite and motivation to cook appetizing meals.

One more suggestion for those who are alone at mealtime is this: make it a special occasion. One way to do this is to set the table with a tablecloth, a napkin, a full set of utensils, and fresh flowers. Set a pot of stew or homemade soup with vegetables and fresh herbs on low heat to cook, and make a salad. Get comfortable in a stuffed chair, and enjoy a book or some soothing music until the rich aroma of your simmering dinner beckons. After serving your plate, light a candle, dim the lights, savor the food, and relish some of the best company you will ever have—your own.

15 Nutrition Assessment: Health, Drug, Socioeconomic, and Diet Histories

The first 14 chapters of this book have described the basics of nutrition science and have shown how attention to nutrition supports a physically fit body and an alert mind throughout life. Turning now to clinical nutrition, this chapter outlines the approach health care professionals use to apply nutrition principles to help clients maintain optimal health. The chapter then shows how health care professionals use health, drug, socioeconomic, and diet histories to assess nutrition status. The next chapter describes how anthropometric data, biochemical tests, and physical examinations help complete the nutrition assessment process.

WWW.

eatright.org
The American Dietetic Association

vh.org/Providers/ClinGuide/Prevention Practice/ACounseling/55
The Virtual Hospital,University of Iowa, College of Medicine

clinnutr.org
The American Society for Parenteral and Enteral Nutrition

Nutrition in Health Care

In today's cost-conscious health care environment, alert health care professionals who learn to think "nutrition" can significantly affect their clients' health and quality of life. They recognize medical nutrition therapy as a medical service that can minimize health care costs by preventing, forestalling, and treating diseases and their complications. (Nutrition in Practice 16 addresses issues related to the delivery of nutrition services in a cost-conscious health care system.)

medical nutrition therapy: a term introduced by the American Dietetic Association in 1994 to emphasize the role of nutrition in medical care. In this book, the terms *medical nutrition therapy* and *diet therapy* are used interchangeably.

Malnutrition is a common problem, particularly among people who are hospitalized. Compared to well-nourished people, people with malnutrition are more likely to experience more complications, longer hospital stays, and higher mortality rates.[1] By identifying clients with existing malnutrition or those at risk for malnutrition, clinicians can provide appropriate nutrition therapy, avert negative health consequences, and potentially reduce health care costs.

Responsibility for Nutrition Care Although registered dietitians hold the primary responsibility for addressing the nutrition needs of clients, effective nutrition care requires the involvement of many health care professionals. Dietetic technicians directly assist dietitians. Physicians, nurses, nursing assistants, and home health care aides are instrumental in identifying clients with nutrition problems, alerting the appropriate health care professional to these problems, clarifying nutrition information, and sharing relevant observations about clients' health status or personal histories. In health care facilities that do not employ dietitians, nurses or physicians most often assume the responsibility for nutrition care. Social workers, pharmacists, physical therapists, and occupational therapists may also make valuable contributions to the nutrition care process (see Nutrition in Practice 21). To the extent that all members of the health care team understand the importance of nutrition to health and apply their nutrition knowledge, technical skills, and interpersonal skills, nutrition needs will be addressed in an efficient and realistic manner.

health care team: a group of professionals representing different disciplines who work together to promote their clients' health.

The Nutrition Care Process To deliver nutrition care efficiently and effectively, health care professionals use a systematic and logical approach that addresses both the person's nutrient and nutrition education needs. Appropriate medical nutrition therapy and skillful communication are the two complementary parts of nutrition care.

The nutrition care process consists of four steps:

- Assess nutrition status. (This chapter and the next describe nutrition assessment.)
- Develop a plan of action (a nutrition care plan) for meeting nutrition needs, including client education.

nutrition care process: an organized approach to nutrition intervention that consists of assessing, planning, implementing, and evaluating. The nutrition care process parallels the nursing care process except that it focuses on nutrition concerns.

nutrition status: the state of the body's nutritional health.

- Implement the nutrition care plan.
- Evaluate the effectiveness of the nutrition care plan through ongoing assessment, and make appropriate changes.

Ideally, health care professionals address the nutrition needs of all people seen in any type of health care facility. Even healthy persons who appear to have no nutrition-related problems can benefit from a review of their dietary habits and nutrition education aimed at promoting lifelong health. More often, however, nutrition intervention is provided primarily for clients with poor nutrition status and those with health problems that affect nutrient needs. In some cases, nutrition intervention is incorporated into a total medical care plan developed by the health care team. Such plans, called clinical pathways, critical pathways, or care maps, are charts or tables that outline a plan of care for a specific diagnosis, procedure, or treatment, with a goal of providing the best outcome at the lowest cost (see also Nutrition in Practice 16).

Active client participation ensures the success of the nutrition care process. In cases where active participation is not possible, such as for infants and young children, people who are very ill or unconscious, people with mental disabilities, or people who are uncooperative, health care professionals should enlist the involvement of family members or other support people.

nutrition care plan: a plan that translates nutrition assessment data into a strategy for meeting a client's nutrient and nutrition education needs.

clinical pathways critical pathways, or **care maps:** charts or tables that outline a plan of care for a specific diagnosis, procedure, or treatment, with a goal of providing the best outcome at the lowest cost. The plan, developed by the health care team after a careful study of each facility's unique client population, is regularly reassessed and improved.

nutrition assessment: the evaluation of many factors that influence or reflect nutritional health; the tools used for nutrition assessment include historical information, anthropometric measurements, physical findings, and laboratory tests.

nutrition screening: a tool for quickly identifying clients who need complete nutrition assessments.

Nutrition Assessment

Time constraints in health care facilities often make it difficult to perform nutrition assessments on each client. Instead, most hospitals and extended care facilities, such as nursing homes and some mental health facilities, use screening procedures to quickly identify clients who are at risk for poor nutrition status or whose diseases or symptoms require immediate attention. Ideally, nutrition screenings take place within 24 hours of admission. The exact criteria that determine nutrition risk vary from facility to facility. Clients determined to be at nutrition risk receive nutrition assessments and appropriate intervention as soon as possible. All clients are reevaluated at intervals, because their nutrition status may change. Chapter 17 provides additional information about nutrition screening.

A nutrition assessment evaluates the client's current nutrition status and determines how the client's age, health, lifestyle, and socioeconomic status affect nutrient needs and the ability to understand and follow nutrition advice. Nutrition assessments provide the foundation for nutrition care plans. By reassessing nutrition status at intervals, the assessor can determine how well the plan is working and make appropriate changes. Thus accurate and thorough assessments are pivotal to ensuring high-quality, cost-effective care. Assessors rely on many sources of data including:

- Historical information (described in this chapter).
- Physical examinations (Chapter 16).
- Anthropometric measurements (Chapter 16).
- Laboratory tests (Chapter 16).

By carefully interpreting each finding in relation to the others, the assessor obtains the basis for a meaningful evaluation. Computer programs often assist assessors in performing mathematical calculations and checking the results against standards. As you read through this chapter and the next, keep in mind that using every technique for assessing nutrition status is not practical, cost-effective, or necessary. Instead, health care facilities rely on the techniques that best identify clients at risk for poor nutrition status.

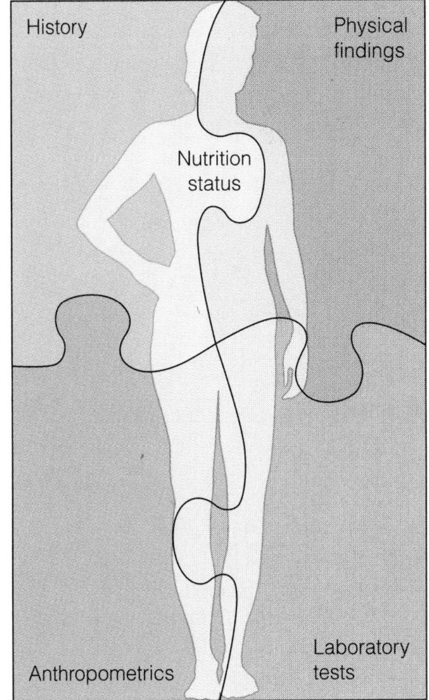

Taken as a whole, the information gathered during a nutrition assessment helps define a person's nutrition status.

Table 15.1 Historical Data Used in Nutrition Assessments	
Type of History	**What It Identifies**
Health history	Health factors that directly and indirectly affect nutrient needs and nutrition status
Drug history	Medications (prescription and over-the-counter), nutrient supplements, and illegal drugs that affect nutrition status
Socioeconomic history	Personal, financial, and environmental influences on food intake, food availability, nutrient needs, and diet therapy options
Diet history	Nutrient intake excesses or deficiencies and the reasons for imbalances

Historical Information

Assessors use the medical record, communicate with other health care professionals, and talk with clients and caregivers to gather the health, drug, socioeconomic, and diet history information summarized in Table 15.1. Although the following sections describe various histories separately, they are interrelated, as Figure 15.1 illustrates.

When talking with clients, an adept history taker uses the interview not only to gather facts, but also to establish rapport with the client and to assess motivation, education, and ability level. Interviewers who establish a caring and nonjudgmental relationship with the client are more likely to obtain accurate information about the client's medical history and personal facts. The accompanying box offers suggestions for conducting interviews.

The forms professionals use to record historical information vary and may be specific for the type of clients they serve. The form used by a dietitian in a maternal and infant health clinic will differ from the form used by a health care team that serves clients with swallowing disorders. Form 15.1 on pp. 384–385 provides an example.

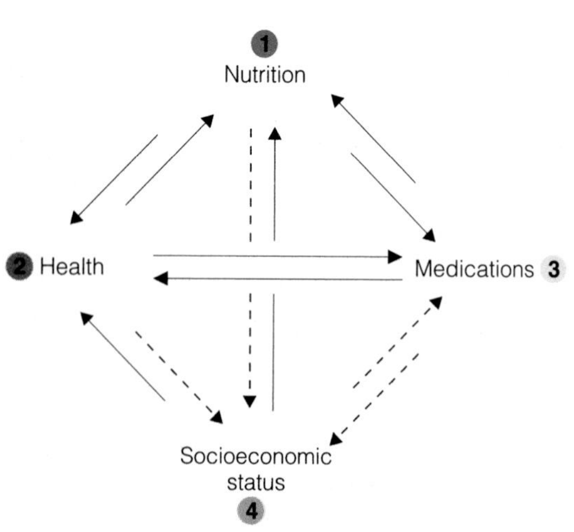

1 Nutrition status can affect a client's health status, which, in turn, can affect the need for medications. Nutrition status and food intake can also alter the way medications are digested, absorbed, metabolized, and excreted. Although many socioeconomic factors are independent of nutrition status, poor nutrition and health status can affect a person's quality of life and have an impact on lifestyle. The costs of some nutrition therapies as well as a person's quality of life can interfere with a person's ability to maintain an optimal economic status.

2 Health status can alter nutrient requirements, alter nutrition status, and directly affect the need for medications. Health care costs and the effects of health status on quality of life can alter lifestyle and seriously affect economic status.

3 Medications can alter nutrient requirements and affect nutrition status. Medications can have positive effects, negative effects, or both positive and negative effects on health. The costs of medications and their side effects can affect economic status, lifestyle, and quality of life.

4 Socioeconomic status can have an impact on nutrient needs, food availability, and food choices. Socioeconomic factors can affect health status by altering nutrition status. Socioeconomic factors can also increase the predisposition to some diseases, such as hypertension and diabetes, and can affect people's ability to pay for health care, as well as their attitudes and beliefs about seeking health care and/or following health care recommendations.

Figure 15.1
Interrelationships among the Various Histories
Attention to the interrelationships among the components of a client's history helps health care professionals provide high-quality, cost-effective, and realistic care. Here the solid lines indicate direct relationships, and the dashed lines indicate indirect relationships.

How to

Conduct Successful Interviews

To solicit accurate and reliable information from clients, the successful interviewer respects and shows genuine concern for the individual. Techniques to help clients feel comfortable and communicate openly and freely include:

- Each time you visit a client, introduce yourself, verify the client's name, and explain the purpose of your visit. In so doing, you confirm that you are talking with the right person, you set the tone for the discussion, and you let the person know who you are and what you expect.
- Arrange for privacy and reassure the client that all the information will be treated confidentially.
- Be sure that the client is comfortable and that the physical environment is conducive to an interview. If the client is in pain or is tired, it may be best to arrange another time for the interview. To facilitate communication, you might modify the physical environment by, for example, adjusting the lighting or turning off the television.
- Position yourself so that you are comfortable and can maintain eye contact. Sit down, if necessary. If you are uncomfortable or if you stand while the client sits or lies in bed, you may nonverbally communicate an unfriendly, overpowering, or hurried feeling.
- Allow adequate time for the interview. Hurrying through an interview conveys to the client that your discussion is not very important and that you really do not care enough to get all the facts.
- Allow others to be themselves. Accept and value people for who and what they are. If an interviewer reacts with advice, criticism, or judgment, the client may withhold information, not wanting to risk belittlement. To avoid such pitfalls, make nonjudgmental responses and ask open-ended questions that allow a wide choice of answers. For example,

Consider the differences between judgmental versus nonjudgmental responses:

Helper: Do you take any type of vitamin or mineral supplement?
Client: I take a vitamin E capsule and 2 grams of vitamin C every day.
Judgmental helper response: You know, of course, that there is no reason for taking these supplements.
Nonjudgmental helper response: For what reasons do you take these supplements?

Examples of open- versus closed-ended questions include:

Open-ended: When do you usually eat for the first time during the day?
Closed-ended: What do you usually eat for breakfast?
Open-ended: What foods do you usually eat at that time?
Closed-ended: Do you usually have cereal or eggs for breakfast?

- Be an active listener. Let the client do most of the talking and control the direction of the conversation. Don't interrupt the person's thoughts. Try not to follow a set format for asking important questions; rather, use forms as guides, and ask questions based on the client's responses.
- Avoid giving diet advice when seeking information. Reserve nutrition education sessions for a later time.
- Regularly provide the client with feedback to make sure you understand each other correctly.
- Prior to ending an interview, let the client know what to expect next. For example, "I'll come back tomorrow to talk with you about your eating plan."

While interviews serve as a tool for gathering information, they also provide an opportunity for establishing a relationship with the client. That relationship will influence all future communications with the client.

Health History

Physical and mental health both affect and reflect nutrition status; the history taker should be alert to conditions that place a client at risk for malnutrition (see Table 15.2 on p. 385). Figure 15.2 on p. 386 illustrates some of the relationships between illness and nutrition.

An assessor is wise to use the medical record to review the client's health, or medical, history before visiting the client. During the interview, the assessor can then keep in mind the factors that may affect the person's nutrition status. Conversations with the client can also uncover valuable health-related information that might otherwise be overlooked because no one thought to ask.

health history: an account of the client's current and past health status and risk factors for disease. Traditionally, the health history has been called the *medical history.* The term *health history* now seems more appropriate, however, because the contents describe the client's health status, and current trends in the medical profession emphasize health promotion and disease prevention.

Form 15.1 Historical Data That Affect the Nutrition Care Plan

Name ______________________ Date ______________________

Address ______________________ Phone ______________________

______________________ Reason for admission/visit ______________________

______________________ ______________________

Health History

1. Known health disorders (check any that apply and record type):

___ Alcoholism
___ Cancer
___ Diabetes
___ GI disorders
___ Heart disease
___ HIV infection
___ Hypertension
___ Kidney disease
___ Liver disease
___ Lung disease
___ Mental health disorder
___ Pancreatic disease
___ Physical disability
___ Recent major illness, infection, trauma, or surgery
___ Ulcers
___ Other

2. Symptoms (check any that apply and record reason if known):

___ Chronic fatigue
___ Constipation
___ Depression
___ Diarrhea
___ Difficulty chewing or swallowing
___ Fever
___ Indigestion
___ Lack of appetite
___ Nausea
___ Vomiting

3. Tobacco use ___ Yes ___ No Type ______________ Amount ______________

Drug History

1. Prescription and over-the-counter medications:

Medication	Reason for taking	Dose and frequency	Duration of intake

2. Alternative therapies (describe) ______________________

3. Side effects noted: ______________________

4. Vitamin/mineral supplements ___ Yes ___ No Type __________ Amount __________ Reason __________

Socioeconomic History

1. Age __________ Ethnic group __________ Religious affiliation __________

Last grade completed __________ Occupation __________

2. Other (check all that apply and describe where appropriate):

___ In school (grade) ______________
___ Employed (job) ______________
___ Others in household (number) ______________
___ Other ______________
___ Refrigerator
___ Stove
___ Access to grocery stores

Form 15.1 Historical Data That Affect the Nutrition Care Plan *(continued)*

Diet History

1. Height ____________ Weight ____________ Usual body weight ____________
2. Recent weight loss or gain ______ Over what length of time ______ Reason, if known ____________
3. Food likes and dislikes ____________
4. Food allergies or sensitivities ____________
5. Special diet ___ Restrictive diet ___ Monotonous diet ___ Client___ Household members___

   Type______ How long?____________
6. Alcohol use ____ Yes ____ No Amount ____________
7. Frequently eats out ____ Omits food groups (describe) ____________
8. Physical activity: Type ________ Intensity ________ Duration ________ Frequency ________
9. Other: ____________

Note: Use the appropriate form to record food intake data.

Appetite and Food Intake Loss of appetite commonly accompanies illness. A child with a fever frequently is unable to eat; so is an adult with cancer. Nausea, mouth dryness, problems with chewing or swallowing, and obstructions in the digestive tract can all lead to malnutrition.

The secondary effects of illness can also affect nutrient intake. Pain and anxiety associated with illness can make it difficult or impossible to eat; pain in the

Table 15.2 Health Factors That Can Affect or Reflect Nutrition Status

- Alcoholism
- Alzheimer's disease
- Anorexia (lack of appetite)
- Anorexia nervosa
- Bulimia
- Cachexia
- Cancer
- Celiac disease
- Chewing or swallowing difficulties (including poorly fitted dentures, dental caries, missing teeth, and mouth ulcers)
- Chronic obstructive pulmonary disease
- Circulatory problems
- Coma
- Constipation
- Crohn's disease
- Cystic fibrosis
- Decubitis ulcers
- Dementia
- Depression
- Diabetes mellitus
- Diarrhea
- Drug addiction
- Dysphagia
- Fever
- Heart disease
- HIV infection
- Hormonal imbalance
- Hyperlipidemia
- Hypertension
- Infection
- Kidney disease
- Liver disease
- Lung disease
- Malabsorption
- Mental illness
- Mental retardation or deterioration
- Multiple pregnancies
- Nausea
- Neurologic disorders
- Organ failure
- Overweight
- Pancreatic insufficiency
- Paralysis
- Physical disability
- Pneumonia
- Pregnancy
- Recent major illness
- Recent major surgery
- Recent weight loss or gain
- Short-bowel syndrome
- Surgery of the GI tract
- Tobacco use
- Trauma
- Ulcerative colitis
- Ulcers
- Underweight
- Vomiting

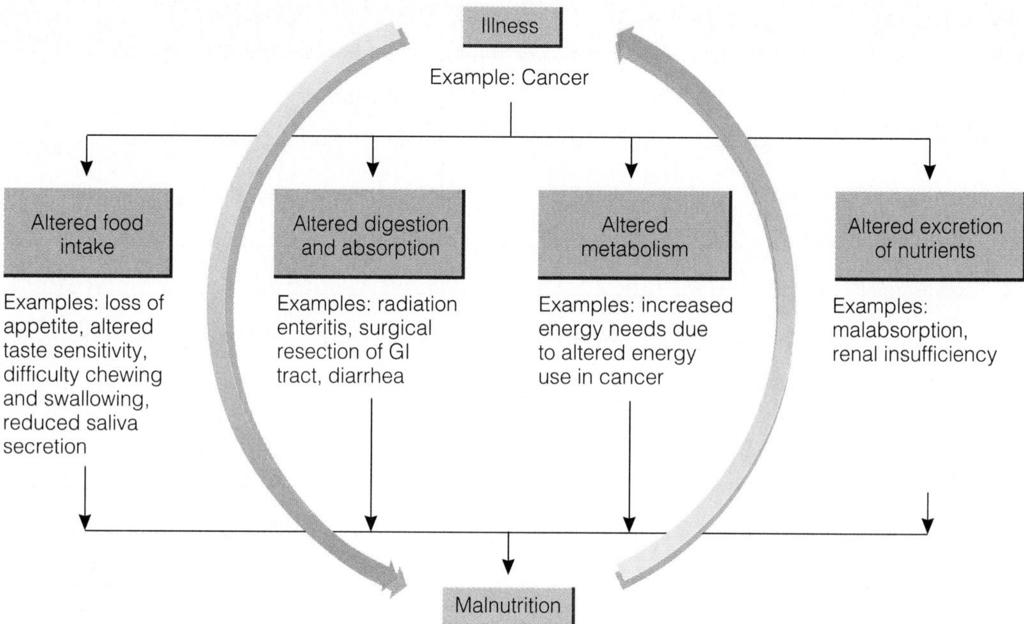

Figure 15.2
Relationships between Illness and Nutrition
(Nutrition for people with cancer is discussed in Chapter 24.)

mouth or throat is especially problematic. Anxiety stimulates stress hormone activity and suppresses digestive activity, making clients lose their appetites. Medical treatments and procedures may require that a person not eat at the very time when nutrient needs are especially high due to illness.

Digestion and Absorption Illnesses that interfere with digestion or absorption usually affect several nutrients and can cause nutrition status to deteriorate rapidly. Examples include cystic fibrosis, pancreatitis, and inflammatory bowel diseases. (Chapter 20 describes these disorders in more detail.)

Metabolism Illnesses can directly alter metabolism (Chapters 23–25) and change nutrient needs. They can also alter organ function and thereby alter metabolism indirectly; examples are renal failure (Chapter 27) and liver failure (Chapter 28).

Excretion Illnesses can also interfere with the excretion of nutrients. The result can be either excessive retention of nutrients, as in renal failure, or excessive loss of nutrients, as in malabsorption (Chapter 20) or nephrotic syndrome (Chapter 27).

Nutrition in Practice 8 describes nutrition concerns associated with alcohol abuse (one type of substance abuse), and Chapter 9 explores eating disorders and their nutrition consequences.

Mental Health The effects of mental illnesses on nutrition status may be less readily apparent than the effects of physiological illnesses, but they are no less important. Nutrition in Practice 15 describes some major interactions between emotional health and nutrition.

Drug History

drug history: a record of all the medications (over-the-counter and prescribed), illicit drugs, and nutrient supplements that a person takes routinely.

Medical drugs are often used in the treatment of illness, and nearly every medication affects nutrition status to some degree. Medications in wide use today are frequently potent and often have an extended duration of action, which increases the likelihood of nutrient-medication interactions.[2] In addition, as people are living longer, medication use is also increasing. Therefore, obtaining a drug his-

Table 15.3 Classes of Medications That Can Affect Nutrition Status

- Amphetamines and other stimulants
- Analgesics
- Antacids
- Antibiotics
- Anticonvulsants
- Antidepressants
- Antidiabetic agents
- Antidiarrheals
- Antifungal agents
- Anithyperlipemics
- Antihypertensives
- Antineoplastics
- Antiulcer agents
- Antiviral agents
- Catabolic steroids
- Diuretics
- Hormonal agents
- Immunosuppressive agents
- Laxatives
- Oral contraceptives
- Vitamin and other nutrient preparations

Note: Specific examples, medication by medication, appear in Appendix E, Table E.1.

Among those at high risk for nutrient-medication interactions are the elderly; alterations in organ function as well as changes in mental and emotional health increase the risks of malnutrition, poor health, inappropriate medication use, and altered drug metabolism. In addition, the elderly are more likely to be taking several medications over long periods of time.

tory is an increasingly important part of the assessment process. This discussion focuses on medical drugs, both prescription drugs and nonprescription over-the-counter (OTC) drugs. Illicit drugs, nutrient supplements, and herbal preparations are also important to consider. If a person is taking any drug routinely, the assessor records the name of the drug or supplement with the dose, frequency, and duration of intake; the reason for taking it; and signs of any adverse or positive effects (see Form 15.1).

Adverse nutrient-medication interactions are most likely to occur if medications are taken over long periods, if several medications are taken, if nutrition status is poor or deteriorating, or if nutrient needs are high. Understandably, then, elderly people, people with chronic diseases, people with malnutrition, and people with illnesses that markedly raise nutrient needs are most at risk. Studies of institutionalized elderly people suggest that multiple medication use may significantly affect nutrition status in this population.[3]

The trend toward making medications that once were available only by prescription available over-the-counter underscores the need to ask clients about all the nonprescription medications they may be taking.

Nutrients and medications may interact in many ways:

- Medications can alter food intake and the absorption, metabolism, and excretion of nutrients.
- Foods and nutrients can alter the absorption, metabolism, and excretion of medications.

Table 15.3 lists the general classes of medications notable for their interactions with nutrients. The following sections describe these interactions, and Table 15.4 on p. 388 summarizes this information and provides specific examples. Table E.1 in Appendix E provides details on specific interactions, medication by medication.

Medications and Food Intake Medications can alter food intake by suppressing the appetite, altering taste sensations or the sense of smell, drying the mouth, or leading to mouth lesions that make eating difficult. Amphetamines used to treat attention deficient hyperactivity disorder provide an example: they may effectively improve concentration and behavior, but they also suppress appetite, alter taste perceptions, dry the mouth, and cause nausea. Conversely, some medications stimulate the appetite and lead to undesirable weight gain. An example is thioridazine hydrochloride (Mellaril), an antipsychotic agent. Furthermore, sedatives and medications that cause excessive drowsiness may make a person too tired to eat.

Table 15.4 Mechanisms and Examples of Food-Medication Interactions

Medications Can Alter Food Intake by:

- Altering the appetite (amphetamines suppress the appetite).
- Interfering with taste or smell (methotrexate changes taste perceptions).
- Inducing nausea or vomiting (digitalis can do both).
- Changing the oral environment (phenobarbital can cause dry mouth).
- Irritating the GI tract (cyclophosphamide induces mucosal ulcers).
- Causing sores or inflammation of the mouth (methotrexate can cause painful mouth ulcers).

Medications Can Alter Nutrient Absorption by:

- Changing the acidity of the digestive tract (antacids can interfere with iron absorption).
- Altering digestive juices (cimetidine can improve fat absorption).
- Altering motility of the digestive tract (laxatives speed motility, causing the malabsorption of many nutrients).
- Inactivating enzyme systems (neomycin may reduce lipase activity).
- Damaging mucosal cells (chemotherapy can damage mucosal cells).
- Binding to nutrients (some antacids bind phosphorus).

Foods Can Alter Medication Absorption by:

- Changing the acidity of the digestive tract (candy can change the acidity, thereby causing slow-acting asthma medication to dissolve too quickly).
- Stimulating secretion of digestive juices (griseofulvin is absorbed better when taken with foods that stimulate the release of digestive enzymes).
- Altering rate of absorption (aspirin is absorbed more slowly when taken with food).
- Binding to drugs (calcium binds to tetracycline, limiting drug absorption).
- Competing for absorption sites in the intestines (dietary amino acids interefere with levodopa absorption this way).

Medications and Nutrients Can Interact and Alter Metabolism by:

- Acting as structural analogs (as anticoagulants and vitamin K do).
- Competing with each other for metabolic enzyme systems (as phenobarbital and folate do).
- Altering enzyme activity and contributing pharmacologically active substances (as monoamine oxidase inhibitors and tyramine do).

Medications Can Alter Nutrient Excretion by:

- Altering reabsorption in the kidneys (some diuretics increase the excretion of sodium and potassium).
- Displacing nutrients from their plasma protein carriers (aspirin displaces folate).

Foods Can Alter Medication Excretion by:

- Changing the acidity of the urine (vitamin C can alter urinary pH and limit the excretion of aspirin).

Absorption and Medications Foods, food components, and nutrients frequently affect medication absorption. Some medications are absorbed better with foods than without them. Foods reduce the absorption of one antihypertensive drug, captopril, and improve the absorption of another, propranolol. In some cases, foods delay a medication's absorption, but do not reduce absorption. An aspirin taken on an empty stomach works faster than when it is given with food, but because aspirin can irritate the GI tract, taking it with food can reduce nausea. Individual nutrients and nonnutrients in foods can also affect drug absorption. Among the more common substances in foods that can bind with drugs and reduce their absorption are minerals, phytates, and oxalates (see Table E.1 in Appendix E for examples).

phytate (FIGH-tate): a nonnutrient found in the husks of grains, legumes, and seeds.

oxalate (OX-a-late): a nonnutrient found in significant amounts in rhubarb, spinach, beets, nuts, chocolate, tea, wheat bran, and strawberries.

Laxatives provide an example of how medications can interfere with nutrient absorption. Some laxatives cause foods to move rapidly through the intestine, reducing the time available for nutrient absorption. Laxatives can reduce nutrient absorption for other reasons as well. For example, fat-soluble vitamins (notably, vitamin D) dissolve in and are excreted along with mineral oil, an indigestible oil that is sometimes used as a laxative. Calcium, too, is excreted. An added danger with all laxatives is that a person who uses them daily for a long time may find that the intestines can no longer function without them. The more often laxatives are used, the more likely that nutrient deficiencies will develop.

Nutrients and medications can interact and reduce the absorption of both. A classic example is the interaction between the antibiotic tetracycline and the minerals calcium and iron. When either of these minerals and tetracycline are taken at the same time, the mineral binds to the tetracycline and both are excreted. To circumvent this problem, clients are instructed to take tetracycline on an empty stomach at least one hour before or two hours after meals or after using milk, milk products, calcium-containing antacids, and iron supplements.

Metabolism and Medications Nutrients and nonnutrients can significantly affect a medication's metabolism. Some medications, for example, resemble vitamins in structure, and it is this property that makes them effective. Vitamin K and the anticlotting medication warfarin (Coumadin) provide an example. Because warfarin resembles vitamin K, warfarin interferes with the synthesis of clotting factors that require vitamin K. The warfarin dose clients take depends, in part, on how much vitamin K is in their diets. If a person's vitamin K intake changes, as may happen in summer when lettuce and greens are in season, then the physician has to alter the medication dose. Another example is methotrexate, which resembles folate in structure (see Figure 15.3). People who take methotrexate to treat certain cancers or rheumatoid arthritis can develop severe folate deficiencies.

Aspirin can also alter folate metabolism but in a different way. Aspirin competes with folate for its protein carrier, thus hindering the body's use of the vitamin. When aspirin is used over long periods of time, health care professionals should ensure that either the diet or supplements are supplying sufficient folate to meet the added demands.

Tyramine, a substance found in some foods and in monoamine oxidase (MAO) inhibitors, medications prescribed to treat certain forms of severe depression, provide an example of a potentially fatal food-medication interaction. MAO inhibitors block the action of the enzyme in the brain that normally inactivates tyramine. When people who take the drug consume large amounts of tyramine, tyramine remains active and stimulates the release of the neurotransmitter norepinephrine. Severe headaches and hypertension can result, and if blood pressure rises high enough, it can be fatal. For this reason, people taking MAO inhibitors are advised to restrict their intakes of foods rich in tyramine (see Table 15.5 on p. 390).

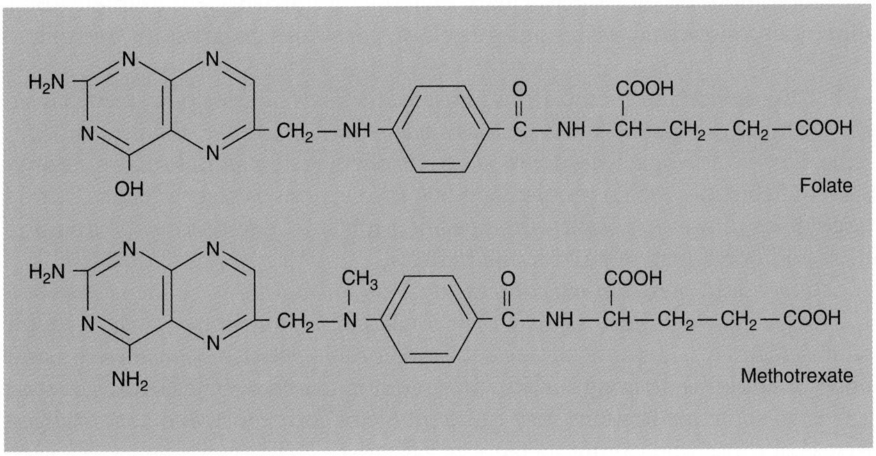

Figure 15.3
Folate and Methotrexate
By competing for the enzyme that activates folate, methotrexate can create a secondary deficiency of folate.

Table 15.5 Foods Restricted in a Tyramine-Controlled Diet	
Beverages:	Red wines including chianti, sherry[a]
Cheeses:	Aged cheeses, American, camembert, cheddar, gouda, gruyère, mozzarella, parmesan, provolone, romano, roquefort, stilton[b]
Meats:	Liver; dried, salted, smoked, or pickled fish; sausage; pepperoni; salami; dried meats
Vegetables:	Fava beans; Italian broad beans; sauerkraut; snow peas; fermented pickles and olives
Other:	Brewer's yeast;[c] all aged and fermented products; soy sauce in large amounts; cheese-filled breads, crackers, and desserts; salad dressings containing cheese

Note: The tyramine contents of foods vary from product to product depending on the methods used to prepare, process, and store the food. In some cases, as little as 1 ounce of cheese can cause a severe hypertensive reaction in people taking monoamine oxidase inhibitors. In general, the following foods contain small enough amounts of tyramine that they can be consumed in small quantities: ripe avocado, banana, yogurt, sour cream, acidophilus milk, buttermilk, raspberries, and peanuts.

[a]Most wine and domestic beer can be consumed in small quantities.

[b]Unfermented cheeses, such as ricotta, cottage cheese, and cream cheese, are allowed.

[c]Products made with baker's yeast are allowed.

In some cases, nutrients must be available to ensure a medication's maximum effectiveness. The medication aldendronate sodium (Fosamax), used to increase bone mass and prevent osteoporosis in postmenopausal women, for example, depends on an adequate supply of vitamin D and calcium, either from the diet or from supplements.

Among the notable foods and food components that can affect drug metabolism are grapefruit juice (but not other citrus juices), caffeine, and natural licorice. Grapefruit juice enhances the bioavailability of some drugs, including some calcium channel blockers used to treat hypertension, the antihistamine terfenadine, the hormone ethinyl estradiol, the sedative medazolam, the antiviral agent saquinavir, and the immunosuppressant cyclosporine.[4] Caffeine, which acts as a central nervous system stimulant, diuretic, and muscle relaxant can potentiate the actions of some medications. Natural licorice can complicate drug therapy using diuretics and antihypertensive agents because it promotes sodium retention and potassium excretion.

Excretion and Medications Urinary acidity affects the reabsorption of medications from the kidneys back into the blood. An acidic urine limits the excretion of acidic drugs like aspirin. Large doses of vitamin C taken along with aspirin increase the urine's acidity, and aspirin remains in the blood longer.

Medications can also alter urinary excretion of nutrients. For example, some diuretics accelerate the excretion of calcium, potassium, magnesium, and zinc.

Other Ingredients in Medications Besides the active ingredients, medications may contain other substances such as sugar, sorbitol, lactose, sodium, and caffeine. For most people who use medications on occasion and in small amounts, such ingredients pose no problems. When medications are taken regularly or in large doses, however, people with specific problems may need to be aware of these additional ingredients and their effects.

Many liquid preparations contain sugar or sorbitol to make them taste better. For people who must regulate their intakes of carbohydrates, such as people with diabetes, the amount of sugar in medications must be considered. Large doses of liquids containing sorbitol may result in diarrhea. The lactose added as a filler to some medications may cause problems for people who cannot digest lactose or those who cannot metabolize galactose.

Reminder: Lactose is composed of the two simple sugars, glucose and galactose.

Antibiotics and antacids often contain sodium. People who take Alka-Seltzer may not realize that a single two-tablet dose may exceed their safe sodium intakes for a whole day. Medications given by vein provide water and frequently provide sodium, potassium, and other electrolytes, or dextrose (the name for glucose in intravenous solutions). Assessors must consider these contributions when clients' diets must be modified in any of these nutrients. Administering drugs through a feeding tube requires additional precautions (see Chapter 21).

A Note to Assessors Hundreds of nutrient-medication interactions have been identified, and information continues to accumulate. It would be difficult, if not impossible, to remember every nutrient-medication interaction and its potential effects on nutrition status and medication effectiveness. Instead, assessors serve their clients best if they:

- Keep in mind that nutrient-medication interactions can and do occur.
- Record the complete drug and diet histories of clients and review these records with potential interactions in mind.
- Keep a nutrient-medication interaction reference handy and check it frequently.
- Be aware of groups of people who are likely to develop nutrient-medication interactions.
- Reassess nutrition status frequently for high-risk clients.
- Become familiar with the nutrient interactions of the medications commonly used to treat the disorders of their clients. For example, nurses working with people who have heart disease should become familiar with the nutrient interactions of medications used to treat that condition.

Information from the socioeconomic and diet histories, described in the next sections, can help assessors identify clients with nutrition problems that might raise the possibility of nutrient-medication interactions.

Socioeconomic History

Socioeconomic factors can affect food availability and food choices and, thus, nutrition status (see Table 15.6 and Form 15.1). One socioeconomic factor, age, affects nutrient requirements as well as food choices (see Chapters 12–14). Infants, for example, must meet their nutrient needs for growth primarily from breast milk or an infant formula. Infants and children depend on caregivers to provide nutritious and acceptable foods; so do adults who are unable to care for themselves. Assessors must therefore sometimes evaluate caregivers, as well as clients.

socioeconomic history: a record of a person's social and economic background, including such factors as age, education, income, and ethnic identity.

Occupation provides clues to the person's education and income. It can also reveal certain eating habits and physical activity levels. One job, for example, may entail deskwork and eating out; another may require vigorous physical activity and allow only a short lunch break.

The ethnic background, religious affiliation, and educational level of both the client and the other members of the household often influence food availability, food choices, health care decisions, and the ability to pay for medical services. These factors also suggest how the interviewer should word questions, interpret answers, and plan for nutrition education. The community environment may also influence the client's nutrition status. The interviewer should be familiar with the food habits of the major ethnic and religious groups in the locale, regional food preferences, local crops, and the nutrition resources and programs available in the community. Local health departments and social agencies can often provide such information.

Income level also influences the diet. In general, diet quality declines as income falls; an inadequate income puts an adequate diet out of reach. Agencies use poverty indexes to identify people at risk for poor nutrition status and to

Table 15.6 Socioeconomic Factors That Can Affect Nutrition Status

Access to grocery stores
Activities
Age
Education
Ethnic identity
Income
Kitchen facilities
Number of people in household
Occupation
Religious affiliation

Table 15.7 Dietary Factors That Can Affect Nutrition Status

Deficient or excessive food intake
Frequently eating out
Intravenous fluids (other than total parenteral nutrition) for 7 or more days
No intake for 5 or more days
Omission from diet of any food group (for example, vegetables)
Poor appetite
Restrictive or fad diets
Monotonous diet (lack of variety)

qualify people for government food assistance programs. Nutrition in Practice 17 addresses additional issues regarding poverty and hunger.

Low income affects not only the power to purchase foods but also the ability to shop for, store, and cook them. A skilled assessor will note whether a person has transportation to a grocery store that sells a sufficient variety of low-cost foods, and whether the person has access to a refrigerator and stove.

Diet History

diet history: a record of eating behaviors and the foods a person eats.

A diet history provides a record of eating habits and food intake and can help identify possible nutrient imbalances (see Table 15.7). Information about the person's eating habits provides the background for developing realistic and attainable nutrition goals.

Constructing an accurate diet history requires skill. Eating habits are an important part of lifestyle and often reflect a person's philosophy. By encouraging trust, the assessor enhances the likelihood of obtaining accurate information.

Form 15.1 on pp. 384–385 includes questions about eating habits and lifestyle that can provide assessors with clues to possible nutrient imbalances and factors that affect food intake. In addition to determining food habits, assessors evaluate food intake using various tools such as the 24-hour recall, the usual intake record, the food record, and direct observation of food intake. The assessor relies on clinical judgment to select the best tool or tools to obtain the needed information. Each tool depends on an accurate account of portion sizes.[5] Food models, photos, pictures, and measuring devices can help clients identify the types of foods and quantities consumed.[6] The assessor also needs to know how the foods are prepared and when they are eaten. In addition to asking about foods, assessors ask about beverage consumption, including beverages containing alcohol or caffeine.

24-hour recall: a record of foods eaten by a person during one 24-hour period.

24-Hour Recall The 24-hour recall provides data for one day only and is commonly used in nutrition surveys to obtain estimates of the typical food intakes for a population. For individuals, the assessor uses the 24-hour recall to get a general picture of eating habits and meal times. The assessor asks the client to recount everything eaten or drunk in the past 24 hours or for the previous day. Form 15.2 shows a typical 24-hour recall form.

An advantage of the 24-hour recall is that it is easy to obtain. It is also more likely to provide accurate data, at least about the past 24 hours, than a person's estimates of average intakes over long periods. It does not, however, provide enough information to allow accurate generalizations about an individual's usual food intake. Only when 24-hour recalls are collected on several nonconsecutive days, including both weekdays and weekend days, is this limitation overcome.

Usual Intake To obtain data about a person's usual intake, an inquiry might begin with "What is the first thing you usually eat or drink during the day?" Similar questions follow until a typical daily intake pattern emerges. The client may state, for example, that he usually eats cold cereal with skim milk and a glass of orange juice for breakfast, but that twice a week he eats 2 fried eggs, 2 strips of bacon, and juice instead. This method uses the same form as the 24-hour recall (Form 15.2) and can be useful, especially in verifying food intake when the past 24 hours have been atypical. It also helps the assessor verify usual eating habits. For example, one person may always eat an afternoon snack; another may never eat breakfast. A person whose intake varies widely from day to day, however, may find it difficult to answer such general questions.

Food Records A food record maintained over several days or more can be a valuable tool for gathering food intake data, monitoring a client's response to

Form 15.2 Food Intake for a 24-Hour Recall or Usual Intake Pattern

Name and address ______________________________ Date ______________

____

____

Did or do you take vitamin-mineral supplements? ______________________________

If yes, what kind? ______________________________ Dose ______________

Please record the type and amount of foods and beverages consumed today. [Or: Please record the types and amounts of foods and beverages you typically consume each day.]

Time of day	Food	Amount (c, tbs, or ounces)	Description (how cooked, how served)

and compliance with medical nutrition therapy, and ascertaining a client's tolerance for certain foods. The assessor instructs the person keeping the record to write down all foods and beverages consumed, the times foods are eaten, the amounts consumed, and methods of preparation. Often the person is asked to record other information as well, depending on the purpose of the food record. When the purpose of the record is to help a person change eating behaviors and lose weight, the record might also include information about the person's mood, the occasion (party, holiday, family meal), behaviors associated with eating food (watching TV, driving in the car while on the way to work, sitting at the table with the family), and physical activity. Figure 9.2 in Chapter 9 provided an example of this type of record. When the purpose of the food record is to establish blood glucose control (see Chapter 25), records include details of drug administration, physical activity, and the results of blood glucose monitoring. When the purpose of the record is to establish food tolerances (such as the amount of lactose a person can handle), food records also include symptoms associated with eating (for example, cramps, diarrhea, nausea, or hives).

food record: a log of all foods eaten over a period of time that may also include records of behaviors associated with eating, physical activity, medications, and symptoms associated with eating.

Food records, when carefully kept, provide an accurate record of food intake, food behaviors, and food tolerances. The assessor can use the record to identify problem food patterns or associated behaviors and find solutions. The record keeper assumes an active role and learns to take responsibility for personal food choices and eating habits.

A disadvantage of food records is that they take time to complete, and the client must be highly motivated to keep them. Another drawback is that the clients may either consciously or unconsciously change their eating behaviors while keeping the records. If clients understand their diets but are not following them, the clients may record what they believe they *should* be eating, rather than what they *are* eating.[7]

Observing Food Intake Direct observation of clients' food intakes is possible in health care facilities such as hospitals or nursing homes. Dietitians, dietary technicians, nurses, and nursing assistants frequently work together to keep records of the kinds and quantities of foods a client receives and leaves on the plate. From these records, the dietitian deduces what has been eaten and estimates nutrient intakes as described in the next section. Often direct observations are used to generate an estimate of a client's current energy and protein intake, and the procedure is simply called a kcalorie count.

Analysis of Food Intake Data After collecting food intake data, the assessor estimates nutrient intakes. Formal estimates of nutrient intakes can be obtained either by calculating them manually (by looking up each food in a table of food composition, recording its nutrients, and adding them up) or by using a computer diet analysis program. Often, though, assessors rely on informal estimates of energy and energy nutrients by using the exchange system (see Chapter 25) and of vitamins and minerals by using food guides. Informal evaluations are possible only if the assessor has enough prior experience with formal calculations to be able to "see" nutrient amounts in reported food intakes without doing calculations.

Once the nutrient intakes are obtained, the assessor compares them to standards, usually recommendations such as the RDA, DRI, or dietary guidelines, to determine how closely the diet meets the standards. Are the types and amounts of proteins, carbohydrates (including fiber), and fats (including cholesterol) appropriate? Are all food groups included in appropriate amounts? Is caffeine or alcohol consumption excessive? Are intakes of any vitamins or minerals (such as sodium and iron) excessive or deficient?

Limitations of Food Intake Analysis Food intake data can be informative, but the skillful assessor also keeps their limitations in mind. A computer diet analysis tends to imply greater accuracy than is possible to obtain from data as uncertain as the starting information. Nutrient contents of foods listed in tables of food composition or stored in computer databases are averages and, for some nutrients, are incomplete. In addition, the available data on nutrient contents of foods do not reflect the amounts of nutrients a person actually absorbs. Iron is a case in point: its availability from a given meal may vary from as high as 50 percent to below 2 percent. (Chapter 8 explains how to calculate iron absorption from a meal.)

Furthermore, reported portion sizes may not be correct. The person who reports eating "a serving" of greens may not distinguish between ¼ cup and 2 whole cups. Children tend to remember the serving sizes of foods they like as being larger than to serving sizes of foods they dislike.

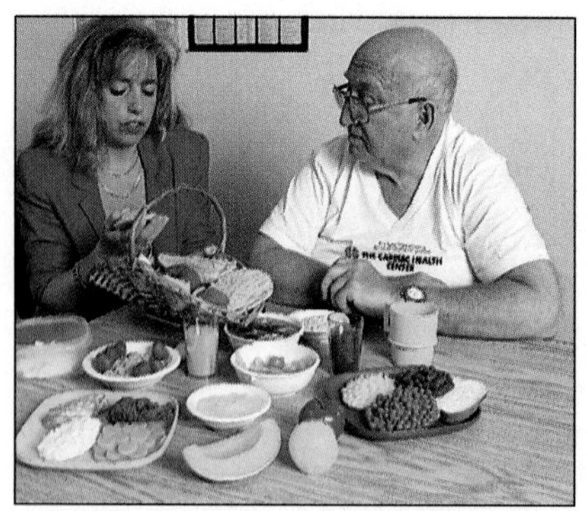

Help clients estimate food sizes by using food models, photos, pictures, or measuring utensils. Comparisons can also be helpful; for example, a small chicken leg is about 2 ounces, and a slice of luncheon meat is about 1 ounce.

Interpretation of Food Intake Data The assessor must remember that adequate nutrient *intakes* do not guarantee adequate nutrient *status* for an individual. Likewise, insufficient intakes do not always indicate deficiencies, but instead alert the assessor to possible problems. Each person digests, absorbs, metabolizes, and excretes nutrients in a unique way; individual needs vary. Intakes of nutrients identified by diet histories are only pieces of a puzzle that must be put together with other indicators of nutrition status in order to extract meaning.

Histories alert health care professionals to potential nutrition problems. Health and medication histories identify health factors and medications that can lead to malnutrition or alter nutrient requirements. Diet and socioeconomic histories pinpoint food intake, eating behaviors, and nutrition status. To substantiate findings, other assessment tools including anthropometric measurements, physical examinations, and biochemical tests (described in the next chapter) are useful.

Self Check

1. The term *medical nutrition therapy* is used interchangeably with:
   a. nutrition care. b. diet therapy.
   c. nutrition support. d. nutrition care process.
2. The person primarily responsible for assessment of nutrition status is the:
   a. dietitian. b. social worker.
   c. pharmacist. d. physician.
3. A plan that translates nutrition assessment data into a strategy for meeting a client's nutrient and nutrition education needs is called a:
   a. nutrition assessment plan. b. nutrition care process.
   c. nutrition diagnosis plan. d. nutrition care plan.
4. The primary purpose of nutrition screening is to:
   a. define nutrition status.
   b. identify clients who need complete nutrition assessments.
   c. determine nutrient needs.
   d. provide a basis for the nutrition care plan.
5. Which of the following is **not** a routine source of data for nutrition assessments?
   a. historical information b. biochemical analysis
   c. cost of care d. anthropometric data
6. Which of the following is **not** one of the purposes of an interview?
   a. to gather facts
   b. to establish rapport with the client
   c. to assess education, motivation, and ability level of the client
   d. to provide nutrition education
7. An effective communication technique to use while interviewing a client is to:
   a. give the client advice when she describes her eating habits.
   b. ask closed-ended questions so that the client will not have to think too hard.
   c. let the client do most of the talking.
   d. use and follow a guide for asking important questions.
8. A client who experiences loss of appetite, difficulty swallowing, and a reduced secretion of saliva is at risk of malnutrition due to:
   a. altered digestion and absorption.
   b. altered metabolism.
   c. altered food intake.
   d. altered excretion of nutrients.
9. Adverse nutrient-medication interactions are more likely to occur if:
   a. nutrient requirements are very high.
   b. one drug is taken exclusively.
   c. nutrition status is adequate.
   d. medication is used for a short time.
10. A client is taking the anticoagulant warfarin. The health care professional should periodically evaluate the client's intake of:
    a. folate. b. vitamin K.
    c. vitamin D. d. calcium.
11. People taking MAO inhibitors should restrict their intakes of foods rich in:
    a. sodium. b. phosphorus.
    c. calcium. d. tyramine.
12. Which foods and medications should not be taken together by a client using the antibiotic tetracycline:
    a. milk, calcium-containing antacids, iron supplements.
    b. milk, calcium-containing antacids, folate supplements.
    c. grapefruit juice, calcium-containing antacids, iron supplements.
    d. grapefruit juice, milk, and calcium-containing antacids.
13. Which foods and food components frequently affect the absorption of several medications?
    a. phytates and oxalates
    b. vitamin D
    c. vitamin K and vitamin B_{12}
    d. folate
14. Factors such as a person's age, education, and ethnic identity are part of the ____________ history:
    a. socioeconomic b. health
    c. diet d. drug
15. Techniques for determining food intake include all of the following **except:**
    a. kcalorie counts. b. food diaries.
    c. 24-hour recalls. d. food guides.
16. An advantage of food intake records is that:
    a. the client may change eating behaviors during the times records are kept.
    b. the client may also record behaviors associated with eating, symptoms associated with eating foods, and physical activities.
    c. the client must diligently maintain the record.
    d. the records are easy to keep.

17. Which of the following statements about food intake analysis is/are true?
    a. Reported portion sizes are generally accurate.
    b. Nutrient contents of foods in nutrient databases are averages.
    c. Data on nutrient contents of foods reflect the amount of the nutrient actually absorbed.
    d. People accurately report their intakes when they use food records.

Clinical Applications

1. Considering the many factors in a client's history that can affect or reflect nutrition status, explain why a systematic approach to nutrition care that involves many health care professionals improves the quality of care while minimizing health care costs. Describe some ways in which a physician, nurse, dietetic technician, nursing assistant, home health care aide, pharmacist, and social worker might assist the dietitian in the assessment process.
2. Describe the possible nutrition implications of these findings from a client's history: age 73, lives alone, recently lost spouse, uses a walker, no teeth, history of hypertension and diabetes, several medications prescribed.
3. Nurses and nursing assistants may shoulder much of the responsibility for collecting food intake data for kcalorie counts because they often deliver and retrieve food trays. Why is it important for a nurse or assistant to verify and record both what the client receives (both the foods and the amounts) and the foods that remain uneaten? When might clients be enlisted to aid in the collection of food intake data, and when might such a course be unwise?
4. Drawing upon the discussion of laxatives (and especially mineral oil) on p. 388, discuss why elderly people are at particular risk for both developing nutrient-medication interactions and suffering more serious consequences. Consider the following to guide your thinking: elderly people are more likely to use laxatives, elderly people are more likely to have chronic diseases and take many medications, elderly people are more likely to be malnourished, and elderly people are more likely to develop osteoporosis.

Notes

1. P. Charney, Nutrition assessment in the 1990s: Where are we now? *Nutrition in Clinical Practice* 10 (1995): 131–139.

2. J. A. Thomas, Drug-nutrient interactions, *Nutrition Reviews* 53 (1995): 271–282.

3. C. W. Lewis, E. A. Frogillo, and D. A. Roe, Drug-nutrient interactions in long-term care facilities, *Journal of the American Dietetic Association* 95 (1995): 309–315; R. N. Varma, Risk for drug-induced malnutrition is unchecked in elderly patients in nursing homes, *Journal of the American Dietetic Association* 94 (1994): 192–194.

4. E. B. Feldman, How grapefruit juice potentiates drug bioavailability, *Nutrition Reviews* 55 (1997): 398–400.

5. L. R. Young and M. Nestle, Portion sizes in dietary assessment: Issues and policy implications, *Nutrition Reviews* 53 (1995): 149–158.

6. V. S. Cypel, P. M. Guenther, and G. J. Petot, Validity of portion-size measurement aids: A review, *Journal of the American Dietetic Association* 97 (1997): 289–292.

7. D. J. Mela and J. I. Aaron, Honest but invalid: What subjects say about recording their food intakes, *Journal of the American Dietetic Association* 97 (1997): 791–793.

CHAPTER 15

Nutrition in Practice

▪ NUTRITION AND MENTAL HEALTH ▪

Mental and nutritional health go together. Mentally healthy people have the capacity to feed themselves well. Well-nourished people experience none of the nutrient deficiencies that might impair their mental health. By the same token, when either type of health is impaired, both may be affected. People with mental and emotional problems often have poor diets; and people who are malnourished are often in poor mental health. It is important to understand these connections, for they affect everyone from the man or woman on the street to the hospitalized person with a severe psychiatric illness. The health professional who recognizes the nutrition implications of mental and emotional disorders can sometimes offer effective help.

I'm interested in hearing more about the man or woman on the street. Are you saying that mental state affects everyone's nutritional health?

It often does. Consider what ordinary anxiety does to your own eating habits. You may be unable to eat at all, you may eat the wrong things, or you may vastly overeat, depending on your personality type. Temporary anxiety may have little effect on nutrition, but if anxiety becomes prolonged or chronic, the resulting changes in eating habits can lead to underweight or cause nutrient deficiencies, imbalances, or obesity. Emotional states that are more difficult to overcome, such as depression, are more likely to lead to nutrition problems.

Why are people who are depressed more likely to develop nutrition problems?

Depression can have profound effects on nutrition status. People who are depressed lose interest in caring for themselves and in usual activities such as eating, socializing with friends, hobbies, and entertainment. But when people fail to care for themselves and cut themselves off from pleasurable activities and friendships, depression deepens and becomes a self-aggravating condition. People with depression may feel worthless, hopeless, and drained of energy and be unable to sleep and concentrate. Thus depression is likely to interfere with eating and nutrition status.

The pain, loss of physical independence, and economic hardships imposed by serious illness can lead to feelings of hopelessness and depression. The side effects of some medications used in the treatment of illness also lead to depression. Depression can also result from loneliness associated with the loss of loved ones or loss of a job or sense of purpose. Although depression and its consequences affect people of all ages, depression among the elderly is pervasive and is a common cause of weight loss among the elderly in nursing homes.[1] Many authorities believe that among the elderly, loneliness is particularly relevant to depression and malnutrition.

WWW

nimh.nih.gov
National Institute of Mental Health

In what ways does loneliness lead to malnutrition in the elderly?

For human beings, eating is as much a social and psychological event as a biological one. Without companionship, appetite falters.

Some 6 million adults over age 65 live alone. Their most pressing need seems to be for companionship; food takes second place. Social interaction is important to emotional health, and elderly people of all classes in our society, both the financially secure and the poverty-stricken, tend to become isolated.

Jack Weinberg, professor of psychiatry at the University of Illinois, wrote perceptively of this problem:

> *In our efforts to provide the aged with a proper diet, we often fail to perceive it is not what the older person eats but with whom that will be the deciding factor in proper care for him. The oft-repeated complaint of the older patient that he has little incentive to prepare food for only himself is not merely a statement of fact but also a rebuke to the questioner for failing to perceive his isolation and aloneness and to realize that food . . . for one's self lacks the condiment of another's presence which can transform the simplest fare to the ceremonial act with all its shared meaning.*[2]

CHAPTER 15

Nutrition in Practice

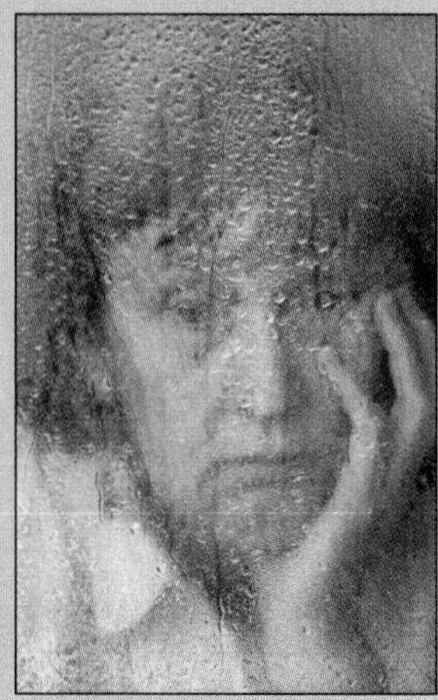

Depression and loneliness can profoundly affect nutrition status.

A sad spiral can set in when a lonely person begins to neglect to eat well. Malnutrition worsens the apathy felt due to loneliness. Then the person has even less energy with which to secure nourishment. Watch for this spiral in all people, and especially in elderly people who live alone and in those who have recently lost a spouse or other loved one and are grieving.

Can something be done to prevent such a spiral?

Health care professionals must be alert to signs of depression in a person of any age and be aware of its potential influence on nutrition status. Recognizing that depression is not a normal consequence of aging is particularly important, because health care professionals sometimes overlook depression in the elderly.[3] When it is recognized early, the person can receive appropriate treatment before health and nutrition status markedly deteriorate. Counselors and social workers can help elderly clients work through depression and find solutions to their loneliness. Dietitians can help clients understand how depression affects food intake and health and how eating a well-balanced diet can prevent further health and nutrition problems. Meal plans that emphasize foods that are easy to prepare and eat can help some clients sustain their nutrient needs when they lack the motivation to eat.

In what ways can nutrient deficiencies affect mental health?

People with B vitamin deficiencies often exhibit "mental" symptoms ranging from confusion, apathy, fatigue, memory deficits, and irritability to delirium and psychoses. Severe niacin deficiencies, for example, can lead to dementia.[4] Deficiencies of folate and vitamin B_{12} are also associated with depression and dementia.[5] Folate shows particularly strong correlations with depression. Depression is a common manifestation of folate deficiency, and conversely, people diagnosed with depressive disorders frequently have low serum or red blood cell folate levels.[6] In the latter case, it is not clear whether folate deficiency leads to depression or depression leads to low folate levels by altering food intake.

Is malnutrition responsible for the senility sometimes seen in the elderly?

Sometimes yes, sometimes no. Both folate and vitamin B_{12} deficiencies are associated with dementia and memory impairment.[7] Sometimes the confusion caused by a vitamin or mineral deficiency is incorrectly diagnosed as senility. An elderly person may even be wrongly confined to a nursing home. The story is told of a woman who exhibited the classic signs of senility—mental confusion, inability to make decisions, and forgetting to perform important tasks, such as turning off a stove burner. The woman's family decided to move her into a nursing home. While she was waiting for a place there, her family took her into their own home. After several weeks of eating good meals and enjoying social stimulation, the woman became her old self again and was able to return to her home. This story has been repeated with many variations and serves to remind us to think about loneliness and nutrition before concluding that a person is senile and needs institutional care. What harm could there be in first trying good, balanced meals served with tender, loving care?

Suppose a person is truly mentally ill. Then what nutrition considerations apply?

Nutrition in psychiatric care is a specialty all its own because there are so many connections. Mental illnesses characterized by depression, illogical thinking, dementia, paranoia, delusions, and inappropriate eating habits can alter food intake and thus interfere with nutrition status. Such disorders include schizophrenia, Alzheimer's disease, mood disorders, substance abuse, and eating disorders. The accompanying glossary defines terms related to mental illness.

As already described, people with mental disorders characterized by depression risk poor nutrition status. People with mental illnesses characterized by illogical thinking or dementia may have little interest in food or may be unable to make appropriate food choices. Those who are paranoid may believe that foods are

Glossary

Alzheimer's (ALTZ-high-mers) **disease:** a degenerative disease of the brain involving memory loss and major structural changes in the brain's nerve cells.

delusions: inappropriate beliefs not consistent with the individual's own knowledge and experience.

dementia (dee-MEN-she-ah): irreversible loss of mental function.

mood disorders: mental illnesses characterized by episodes of severe depression or excessive excitement (mania) or both.

paranoia (PARA-NOY-ah): mental illness characterized by delusions of persecution.

schizophrenia (SKITS-oh-FREN-ee-ah): mental illness characterized by an altered concept of reality and, in some cases, delusions and hallucinations.

being used to poison them. People suffering from delusions may attribute magical powers to certain foods and insist on eating only those foods. Medications used in the treatment of mental illnesses can also interact with nutrients in significant ways (see Appendix E).

What are the relationships between mental health and malnutrition in children?

Protein-energy malnutrition in pregnant women can cause mental retardation in their children. Children malnourished early in life show behavioral and social deficits as well as physical retardation. Also, children who are neglected early in life show a greater tendency to suffer from severe malnutrition than children who receive love and attention. Wherever you see abnormal nutrition status in a child, ask yourself if the child requires emotional as well as physical support. And wherever you see emotional illness, look to the child's nutrition, too.

What mental disorders affecting nutrition are especially common?

Alcoholism is one, and it is treated elsewhere in this book (see Nutrition in Practice 8). You almost always see abnormal energy balance and vitamin and mineral deficiencies in cases of severe alcoholism. Anorexia nervosa and bulimia nervosa are other examples (see Nutrition in Practice 9).

Nutrition affects the brain and the mind, and the brain and the mind affect the way people eat. They all work together, and the wise health care professional will keep these interrelationships in focus.

Notes

1. G. J. Kennedy, The geriatric syndrome of late-life depression, *Psychiatric Services* 46 (1995): 43–48.

2. J. Weinberg, Psychological implications of the nutritional needs of the elderly, *Journal of the American Dietetic Association* 60 (1972): 293–296.

3. C. Ryan and M. E. Shea, Recognizing depression in older adults: The role of the dietitian, *Journal of the American Dietetic Association* 96 (1996): 1042–1044.

4. J. E. Morley, Nutritional modulation of behavior and immunocompetence, *Nutrition Reviews* (supplement 2) 52 (1994): 6–8.

5. T. Bottiglieri, Folate, vitamin B_{12}, and neuropsychiatric disorders, *Nutrition Reviews* 54 (1996): 382–390.

6. J. E. Alpert and M. Fava, Nutrition and depression: The role of folate, *Nutrition Reviews* 55 (1997): 145–149.

7. Bottiglieri, 1996.

16 Nutrition Assessment: Physical Measurements and Observations

Chapter 15 introduced the nutrition care process and nutrition assessments and then showed how the assessor can use historical information to uncover the health, medication, socioeconomic, and diet factors that affect a client's nutrient needs, as well as lifestyle habits that will shape medical nutrition therapy. This chapter shows how physical measurements and observations of the body further aid in the assessment process.

Body Measurements

anthropometric (an-throw-poe-MEH-trick): relating to measurement of the physical characteristics of the body, such as height and weight.
anthropos = human
metric = measure

To evaluate nutrition status, assessors can use measures of body composition and development (anthropometric measurements) or measures of how well the body performs certain tasks (functional tests of nutrition status). Table 16.1 lists anthropometric measurements and functional tests useful in nutrition assessments and indicates what each measure reflects. This section describes anthropometric measurements, which serve three main purposes: first, to evaluate the progress of growth in pregnant women, infants, children, and adolescents; second, to detect undernutrition and overnutrition in all age groups; and third, to measure changes in body composition over time. Assessors use this information to help estimate an individual's energy and energy nutrient needs.

Assessors compare anthropometric measurements taken on an individual with population standards specific for gender and age to see how body composition compares to norms. Assessors may also take measurements periodically and compare them with previous measurements to detect changes in an individual's status and to determine whether estimated nutrient needs are appropriate.

Height and weight are well-recognized anthropometrics; others include fatfold measurements and various measures of lean tissue. Still other measures are useful in specific situations. A head circumference measurement may help to assess brain development in an infant, and an abdominal girth measurement supplies information about abdominal fluid retention or enlargement of abdominal organs.

Mastering the techniques for taking anthropometric measurements requires proper instruction and practice. Once the correct techniques are learned, however, taking the measurements is easy and generally requires minimal equipment.

Measures of Growth and Development

Height and weight are routinely measured in virtually all health care settings and are an important component of both nutrition screenings and nutrition assessments. Length measurements for infants and children up to age three and height measurements for children over three are particularly valuable in assessing growth, which depends on adequate nutrition. Poor growth in children indicates malnutrition. For adults, height measurements alone do not reflect current nutrition status but help with estimating desirable weight, interpreting other assessment data, and estimating energy needs. Once adult height has been reached, changes in body weight may reveal either overnutrition or undernutrition.

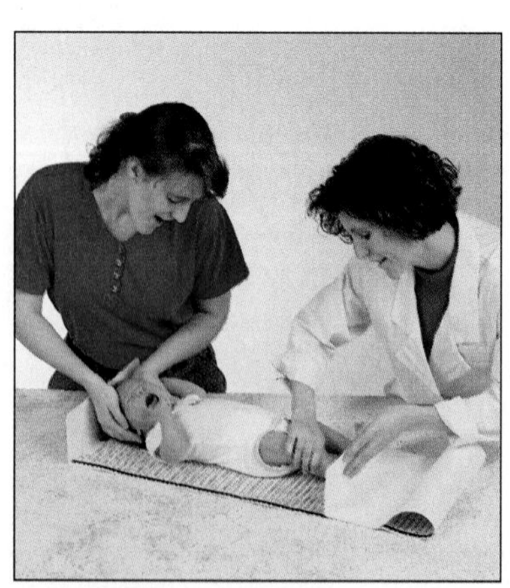

Lying still with legs straight for a length measurement can be a trying experience for an infant.

Length and Height For infants and children younger than three, health care professionals may use a measuring board that has a fixed headboard and movable footboard to measure length. The assessor places the barefoot infant on the measuring board with the infant's head against the headboard, straightens the infant's legs, and moves the footboard to the bottom of the infant's feet. This method provides the most accurate measure possible, but many health care pro-

Table 16.1 Body Measurements Used in Nutrition Assessments

Measurement	What It Reflects
Abdominal girth measurement	Abdominal fluid retention and abdominal organ size
Hand grip strength	Ability of muscles to perform work (protein status)
Height-weight	Overnutrition and undernutrition; growth in children
%IBW, %UBW,[a] recent weight change	Overnutrition and undernutrition
Head circumference	Brain growth and development in infants and children under age two
Fatfold	Subcutaneous and total body fat
Midarm muscle circumference	Skeletal muscle mass (protein status)
Skin test	Immune function (protein status)
Waist-to-hip ratio	Abdominal fat

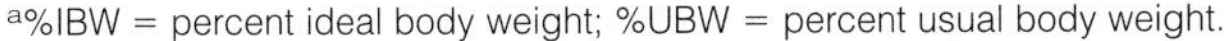

[a]%IBW = percent ideal body weight; %UBW = percent usual body weight.

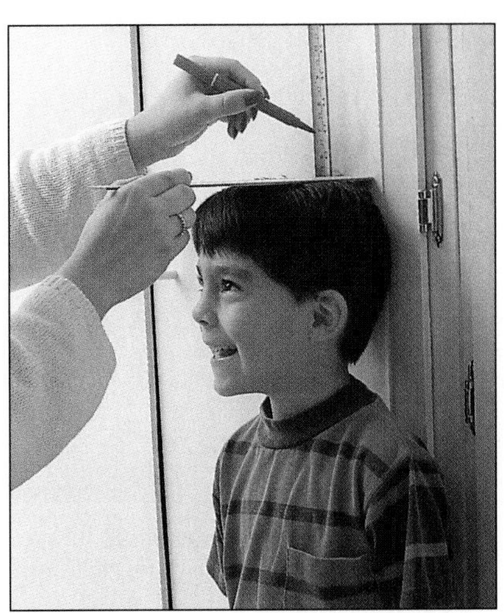

Standing erect allows for an accurate height measurement.

fessionals use a less exacting method. They simply hold the infant straight with its head against the headboard or other vertical support, mark the blanket with a chalk or pen at the infant's heel, and then measure the distance from the headboard to the mark. Even more informally and less accurately, they may lay the infant on a flat surface and extend a nonstretchable measuring tape along the side of the infant from the top of the head to the heel of the foot.

Once a child can stand erect, height measurements are made using the same procedure that is used to measure adults. The best way to measure standing height is with the person's back against a flat wall to which a nonstretchable measuring tape or stick has been fixed. The person stands erect, without shoes, with heels together. The person's line of sight should be horizontal, with the heels, buttocks, shoulders, and head touching the wall. The assessor places a ruler, book, or other stiff object on top of the head at a right angle to the wall, carefully checks the height measurement, and records it immediately in either inches or centimeters. Immediate recording prevents the assessor from forgetting the correct measurement.

The measuring rod of a scale is commonly used to measure height but is less accurate because it bends easily. The assessor follows the same general procedure, asking the person to face away from the scale and to take extra care to stand erect.

Many health care professionals merely ask clients how tall they are rather than measuring their height. Self-reported height is often inaccurate and should be used only as a last resort when measurement is impractical, as in the case of an uncooperative client or an emergency admission.

Weight Valid weight measurements require functional scales that have been carefully maintained, calibrated, and checked for accuracy at regular intervals. Beam balance and electronic scales are the most accurate types of scales. Special scales and hospital beds with built-in scales help weigh clients who are bedridden. Bathroom scales are inaccurate and inappropriate for use in professional settings.

To measure an infant's weight, assessors use special scales that allow the infant to lie down or sit. Weighing infants without clothes or diapers is standard procedure. Children who can stand are weighed in the same way as adults. The usefulness of repeated weight measurements can be improved by weighing the

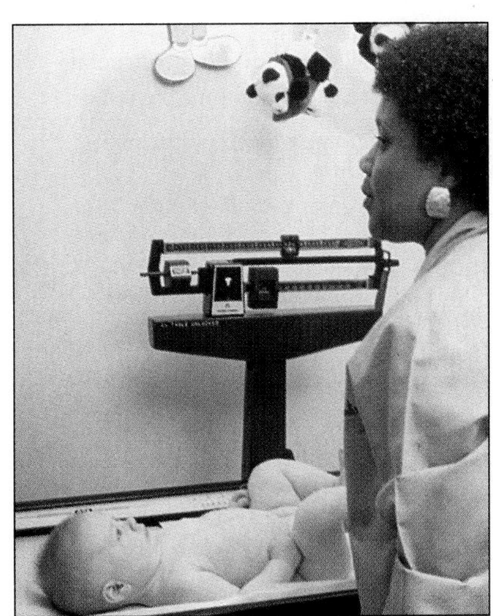

Infants are measured on scales that allow them to sit or lie down while they are being weighed.

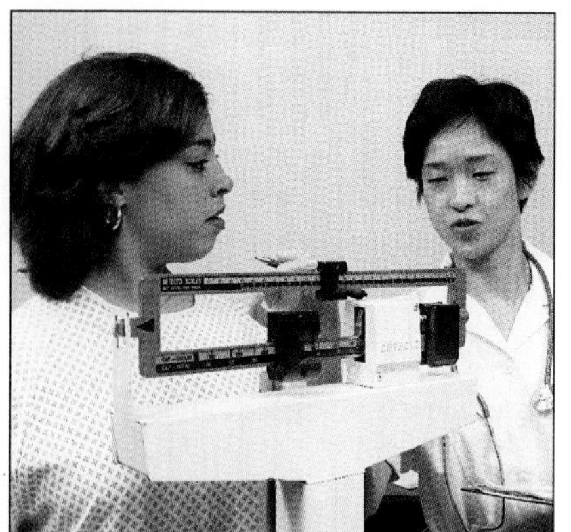

Beam balance scales provide accurate weight measurements for older children and adults.

client at the same time of day (preferably before breakfast), in the same amount of clothing (without shoes), after the person has voided, and on the same scale. As with all measurements, the assessor records observed weight immediately in either pounds or kilograms.

Head Circumference Assessors may also measure head circumference in infants and young children to confirm that growth is proceeding normally or to help detect protein-energy malnutrition (PEM) and evaluate the extent of its impact on brain size. To measure head circumference, the assessor places a nonstretchable tape so that it encircles the largest part of the head: just above the eyebrow ridges, just above the point where the ears attach, and around the occipital prominence at the back of the head. The assessor immediately notes the measure in either inches or centimeters.

Analysis of Measures in Infants and Children Health professionals evaluate the physical development of infants and children by monitoring the growth of a child over time and plotting the data on standard growth charts (see Appendix E for more information). Growth charts compare weight to age, height to age, and weight to height; ideally, height and weight are roughly in the same percentile. Although individual growth patterns may vary, a child's growth curve will generally stay at about the same percentile throughout childhood. Measurements below this percentile for height, weight, or head circumference in infants and young children indicate growth retardation, which is an important sign of malnutrition.

In children whose growth has been retarded, nutrition rehabilitation aims to increase height and weight to higher percentiles. In overweight children, the goal is for weight to remain stable as height increases, until weight becomes appropriate for height.

Head circumference is a useful indicator of brain growth in children under two years of age. Since the brain grows rapidly before birth and during early infancy, chronic malnutrition during these times can impair brain development. Head circumference percentiles should be similar to the child's weight and height percentiles. Nonnutritional factors, such as certain disorders and genetic variation, can also influence head circumference.

WWW

niddk.nih.gov/health/nutrit/win.htm
Weight-Control Information Network

niddk.nih.gov/health/nutrit/pubs/unders.htm
Understanding Adult Obesity

Analysis of Weight Measures for Adults Chapter 9 (pp. 206–207) discussed healthy body weight and described the controversies regarding weight standards for adults. Health care professionals typically compare weights with weight-for-height standards. Each facility or practitioner decides which standard to use based on a review of the literature and clinical judgment. Some of the commonly used standards for nutrition assessment include:

1. The body mass index (BMI) and tables based on the BMI observed to be consistent with health (see p. 207 and the inside back cover).
2. The Suggested Weights for Adults table (see p. 207).
3. A quick method of estimating desirable body weight described in the box on p. 405.

Reminder: The *body mass index (BMI)* is an index of a person's weight in relation to height, determined by dividing the weight in kilograms by the square of the height in meters:

$$\text{BMI} = \frac{\text{weight (kg)}}{\text{height (m)}^2}.$$

An alternate formula for determining BMI using pounds and inches:

$$\text{BMI} = \frac{\text{weight (lb)}}{\text{height}^2\text{ (in)}} \times 705.$$

The standard chosen can be used to compare a person's actual weight with desirable weight, a figure commonly known as the percent ideal body weight (%IBW). Assessors use the %IBW to help determine an appropriate energy intake for a healthy individual or for a person with stable chronic diseases.

Current weight can also be compared to usual body weight to generate the percent usual body weight (%UBW), which considers what is normal for a particular individual. The %UBW is particularly useful for evaluating the degree of nutrition risk associated with acute illness. For example, in cases where an overweight individual becomes acutely ill and is rapidly losing weight, the health care

Quickly Estimate Desirable Body Weight

Professionals sometimes bypass standard weight-for-height tables and use a quick estimate of desirable body weight based on height and gender. The assessor considers 106 pounds a reasonable weight for a man who is 5 feet tall and then adds 6 pounds for each inch over 5 feet (or subtracts 6 pounds for each inch under 5 feet). For example, the estimate for a man who is 5 feet 8 inches tall would be:

106 lb for the first 5 ft =	106 lb.
6 lb for each inch over 5 ft (8 × 6 lb) =	48 lb.
Total =	154 lb.

A reasonable estimate for this man is 154 pounds. Large-framed individuals may need to add 10 percent more, whereas small-framed individuals may need to subtract 10 percent. Thus the range for a man of this height is 139 to 169 pounds. This estimated range is close enough to the range of 125 to 164 pounds in the "suggested weights" table on p. 207 to serve for most purposes.

Similarly, the assessor considers 100 pounds a reasonable weight for a woman who is 5 feet tall and then adds 5 pounds for each inch over 5 feet (or subtracts 5 pounds for each inch under 5 feet). The calculation for a woman who is 5 feet 5 inches tall would be:

100 lb for the first 5 ft =	100 lb.
5 lb for each inch above 5 ft (5 × 5 lb) =	25 lb.
Total =	125 lb.

A reasonable estimate for this woman is 125 pounds. Because large-framed individuals may need to add 10 percent, and small-framed individuals may need to subtract 10 percent, the estimated range for women of this height is 112 to 138 pounds. Again this range is similar to the suggested range of 114 to 150 pounds (p. 207).

professional relying on the %IBW may inadvertently overlook significant weight loss. The box on p. 406 and Table 16.2 show how to calculate and evaluate %IBW and %UBW.

Assessors consider not only the *amount* of weight loss, but also the *rate* of weight loss. A person who loses less than 5 percent UBW over a six-month period is at minimal risk for poor nutrition status. A loss of 5 to 10 percent UBW over six months is significant, and a loss of greater than 10 percent UBW is highly significant.[1]

Reminder: Weight-for-height tables suggest a weight range, rather than pinpointing one weight—a helpful reminder that desirable weights are estimates at best.

Although the term *ideal body weight* is a misnomer, it is the term most likely to be used in health care settings, and so it is used here.

Weight Change during Pregnancy One of the anthropometric measures most predictive of an infant's birthweight is the mother's amount and pattern of weight gain during pregnancy. Chapter 12 describes normal weight gains related to pregnancy, and Appendix E provides an example of a chart used to monitor weight gain during pregnancy. Patterns of weight gain that deviate from these require further investigation.

Measures of Body Fat and Lean Tissue

Significant weight changes in both children and adults can reflect overnutrition or undernutrition with respect to energy and protein. To estimate the degree to

Table 16.2 Weight as an Indicator of Nutrition Status

%IBW	%UBW	Nutrition Status
>120	—	Obese
110–120	—	Overweight
90–109	—	Adequate
80–89	85–95	Mildly underweight
70–79	75–84	Moderately underweight
<70	<75	Severely underweight

Estimate %IBW and %UBW

To estimate %IBW, compare the individual's current weight with the desirable body weight from standard weight-for-height tables:

$$\%IBW = \frac{\text{actual body weight}}{\text{desirable (ideal) weight}} \times 100.$$

For example, suppose you wish to calculate %IBW for a man who is 5 feet 8 inches tall and weighs 115 pounds. For desirable (ideal) weight, use the midpoint of the weight range in Table 9.1 on p. 207. In this example, the desirable weight is 144 pounds.

$$\%IBW = \frac{115 \text{ lb}}{144 \text{ lb}} \times 100 = 80\%.$$

The man in this example is at 80 percent of his ideal body weight. A look at Table 16.2 indicates that at 80 percent IBW he is mildly underweight.

This man has lost 15 pounds in the last month. To calculate his %UBW, use this formula:

$$\%UBW = \frac{\text{actual weight}}{\text{usual weight}} \times 100.$$

Calculate the usual weight (130 pounds) by adding the weight loss (15 pounds) to the current body weight (115 pounds).

$$\%UBW = \frac{115 \text{ lb}}{130 \text{ lb}} \times 100 = 88\%.$$

The man is at 88 percent of his usual body weight. A look at Table 16.2 reveals that a person at 88 percent UBW is mildly underweight. Based on %UBW, the degree of underweight is less severe than the %IBW implied, because this man has consistently weighed less than standard weight. Nevertheless, his recent rate of weight change is very significant: he has lost almost 4 pounds per week.

which fat stores or lean tissues are affected by weight changes, several anthropometric measurements are useful. Note that these measurements are most likely to be used for clients determined to be at significant risk for poor nutrition status.

WWW **ama-assn.org/insight/gen_hlth/pernutri/checkeat.htm**

American Medical Association's Health Insight Personal Nutritionist

Fatfold Measures Approximately half the fat in the body is found in subcutaneous fat—the fat located directly beneath the skin—and its thickness reflects total body fat. In some parts of the body, this fat is loosely attached; a person can pull it up between the thumb and forefinger and obtain a measure of fatfold thickness. Although fatfold measures can be taken from a variety of body sites, most clinicians measure the triceps fatfold. The upper arm is easily accessible, and its measurement involves little inconvenience to the client. The proper techniques for measuring the triceps fatfold, as well as standards for comparison, are given in Appendix E.

Midarm Circumferences Just as subcutaneous fat provides an indirect estimate of total body fat, measurable muscles provide an indirect measure of protein status. To determine whether a person has depleted muscle mass, an indirect measure of arm muscle size is useful: the *midarm muscle circumference.* This measure is estimated by subtracting the area of fat on the arm from the total area (derived from its circumference). Appendix E shows how this is done and presents standards for comparison.

Waist-to-Hip Ratio Chapter 9 described how fat distribution correlates with health risks and mentioned that the waist-to-hip ratio is a valuable indicator of fat distribution. To calculate the waist-to-hip ratio, divide the number of inches (or centimeters) around the waistline by the number of inches (or centimeters) around the hips. For example, a person with a 28-inch waist and 38-inch hips would have a ratio of 28/38 or 0.74. Women with a ratio of 0.80 or higher and men with a ratio of 0.95 or higher are considered to be at greater risk for developing obesity-related health problems.

Analysis of Measures The reliability of anthropometric measurements is limited by the skills of the measurer and the accuracy of the equipment used for measurement. Sometimes taking measurements is difficult for physical reasons, such as when a person cannot be moved to be weighed or measured for height. Triceps fatfolds and midarm circumferences are sometimes difficult to measure in people with wounds or loose, hanging skin on the arms. In the elderly, the distribution of fat and the compressibility of the skin change, complicating the measurement of fatfolds and midarm circumferences.

Even when measurements are taken accurately, interpreting them can present problems. A person's state of hydration, for example, significantly influences anthropometric measurements because body composition reflects total body water as well as lean body mass and body fat. Diseases or medications that cause fluid retention can mask significant weight loss. Dehydration affects measurements of weight, fatfold, and midarm circumference. Besides the state of hydration, exercise alters anthropometric measurements. Exercise enhances muscle size, and lack of exercise may diminish muscle mass independently of nutrition factors.

Accurate interpretations of anthropometric meaurements are also confounded by problems with the standards used for comparisons. The controversies surrounding weight standards have already been described in Chapter 9. Another difficulty is that fatfold and muscle circumference standards were developed for specific populations, often of healthy persons in their middle years of life. The application of such standards to people who are ill or are not within the age groups studied requires cautious clinical judgment.

Another limitation is the inability of anthropometrics to describe small changes in body composition that occur over short periods of time. Thus anthropometrics are of limited value in quantifying the nutrition effects of acute illnesses or in assessing the immediate effects of nutrition therapy. Furthermore, additional research is needed to determine if internal fat stores change at the same rate as subcutaneous fat during illnesses that markedly raise the metabolic rate.

With all these limitations, is taking anthropometrics even useful? The answer is an emphatic *yes*! It is not essential to define weight, fatfolds, or arm circumference down to a precise measurement. Rather, assessors look for patterns—obviously high or low measurements and changes that occur in the individual. They use the measures to estimate energy and energy nutrient needs and then use them again to see whether their estimates require further refinements.

Other Measures of Body Composition Estimates of body composition can also be obtained using other techniques including hydrodensitometry (underwater weighing) and bioelectrical impedance analysis. Underwater weighing generates a good estimate of body fat, but requires bulky, expensive, nonportable equipment. Furthermore, submerging some people (especially those who are very young, very old, ill, or fearful) under water is generally not practical. For these reasons, underwater weighing is more likely to be used at research centers and private health clubs than in health care facilities.

In addition to anthropometric measurements, various other methods including isotope studies, X rays, ultrasonography, and computerized axial topography (CAT scans) have been used to determine body composition. The expense of these tests limits their clinical use, although they are quite useful in research.

Unlike underwater weighing, bioelectrical impedance analysis provides a good estimate of body composition using portable, less-expensive equipment that incurs minimal inconvenience to the client. To measure body fat using the bioelectrical impedance technique, an electrical current of very low intensity is briefly sent through the body by way of electrodes placed on the wrist and ankle. Since electrolyte-containing fluids, which readily conduct electrical current, are found primarily in lean body tissues, the leaner the person, the less resistance to the current. The measurement of electrical resistance is then used in a mathematical equation to estimate the percentage of body fat. When bioelectrical impedance analysis is performed under standardized conditions and the correct mathematical equations are applied, body composition estimates are reasonably accurate in healthy populations and in some people with stable chronic diseases.[2]

WWW **www.nlm.nih.gov/pubs/cbm/bioelimp.html**
Bioelectric Impedance Analysis

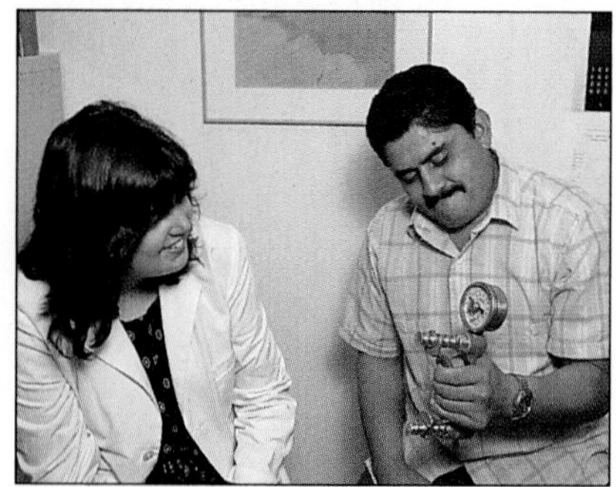

An instrument called a dynamometer measures hand grip strength.

For people with critical illnesses or those with unstable diseases, the value of bioelectrical impedance is unclear.

Functional Tests of Nutrition Status

Rather than looking at a fixed body measurement, functional tests of nutrition status measure the body's ability to perform specific tasks. Two of the most commonly used functional measures include hand grip strength and skin tests. Although these tests are not anthropometric measurements, they also require physical measurements and so are described here.

Hand Grip Strength The measurement of hand grip strength assesses nutrition status by measuring muscle function. The assessor asks the person to grip an instrument (called a dynamometer) as tightly as possible. Low grip strength (weak muscle function) suggests poor nutrition status. People with severe arthritis or muscular disorders may have low grip strengths unrelated to nutrition status. Grip strengths may also be deceptively low in people taking sedatives or those who are poorly motivated to grip the dynamometer as tightly as possible. These drawbacks can be overcome by using special equipment that measures muscle function without requiring a client to grip a dynamometer.[3]

Skin Tests Just as hand grip strength measures muscle function, skin tests measure immune function. Organisms (usually three or four kinds) that produce an immune reaction in most people are injected just under the skin.* Raised, hardened areas (indurations) appear after 24 to 48 hours in well-nourished people, but are minimal or absent in people with PEM. Many factors other than nutrition interfere with immune responses, however, and skin testing alone is an insensitive test of nutrition status.

induration: a raised, hardened area of skin.
durus = hard

Conditions other than PEM that can affect skin tests include metabolic stress, liver disease, kidney disease, and the use of many drugs including corticosteroids and general anesthesia.

Biochemical Analysis

Biochemical analyses or laboratory tests can provide information about protein-energy balance, vitamin and mineral status, fluid status, body composition, organ function, and metabolic status and can help determine if nutrition therapy is appropriate or if a person is complying with a special diet. Common tests are based on analysis of blood and urine samples, which contain nutrients, enzymes, and metabolites. Some tests useful in assessing nutrition status and response to diet therapy are described in this section. Other tests, such as serum glucose or cholesterol or tests that define fluid and electrolyte balance, acid-base balance, and organ function, help pinpoint disease-related problems with nutrition implications. Table 16.3 lists some biochemical tests with nutrition implications and indicates what these tests reflect. Tests important in specific disorders will be discussed in the appropriate chapters.

The analysis of several blood components from a single blood sample is referred to as **SMA (simultaneous multiple analysis).** SMA is followed by a number (for example, SMA-12) that indicates how many tests will be run.

The **serum** is the watery portion of the blood that remains after removal of the cells and clot-forming material; **plasma** is the fluid that remains when unclotted blood is centrifuged. Usually, serum and plasma concentrations are similar, but because plasma samples may clog mechanical blood analyzers, serum samples are preferred.

Limitations of Biochemical Tests

To interpret biochemical data for nutrition assessments, assessors must consider how a person's state of health affects test results independently of nutrition factors. The person's state of hydration, for example, greatly influences laboratory values, and indeed, laboratory tests are often used to detect dehydration and fluid retention. With dehydration, lab results may be deceptively high; with fluid retention, lab results may be deceptively low.

Recall that lab tests that confirm dehydration or fluid retention, including sodium, BUN (blood urea nitrogen), hemoglobin, and hematocrit, alert the assessor to interpret anthropometric measurements cautiously.

*Typical antigens include *Candida*, mumps, purified protein derivative (PPD), streptokinase-streptodornase (SK-SD), and *Monilla*.

Table 16.3 Routine Hospital Laboratory Tests

Test	Uses
Hematology	
Hemoglobin (Hg)	To detect anemia and determine state of hydration.
Hematocrit (Hct)	To detect anemia and determine state of hydration.
White blood cells (WBC)	To detect infection and determine total lymphocyte count.
Mean corpuscular volume (MCV)	To detect anemia and determine its causes.
Mean corpuscular hemoglobin (MCH)	To detect anemia and determine its causes.
Mean corpuscular hemoglobin concentration (MCHC)	To detect anemia and determine its causes.
Blood Chemistry	
Proteins	
Total protein[a]	To detect PEM and various nutrient imbalances.
Albumin	To detect PEM and determine state of hydration.
Transferrin	To detect PEM and monitor response to feeding.
Prealbumin	To detect PEM and monitor response to feeding.
Electrolytes	
Sodium	To check state of hydration.
Potassium	To monitor acid-base balance and renal function and detect imbalances.
Chloride	To monitor acid-base balance and detect GI losses of chloride (from vomiting or nasogastric suctioning).
Carbon dioxide	To monitor acid-base balance.
Other	
Glucose	To detect diabetes mellitus, pancreatic tumors, and hypoglycemia and monitor glucose intolerance.
Glycated hemoglobin A_{1c}	To monitor blood glucose control over the past 2–3 months.
Blood urea nitrogen	To monitor renal function and determine state of hydration.
Calcium	To detect hormonal imbalances, certain malignancies, and calcium imbalances.
Phosphorus	To detect imbalances and PEM and monitor renal function and response to feeding.
Magnesium	To monitor renal function and response to feeding and detect PEM.
Cholesterol	To assess risk of heart disease and possibility of obstructive jaundice.
Uric acid	To detect gout and determine state of hydration.
Serum creatinine	To monitor renal function and determine state of hydration.
Serum enzymes	
Creatinine phosphokinase (CPK)	To monitor heart function and muscle damage.
Lactic dehydrogenase (LDH)	To monitor heart and renal function.
Alanine transaminase (ALT, formerly SGPT)	To monitor heart and liver function.
Aspartate transaminase (AST, formerly SGOT)	To monitor heart and liver function.
Alkaline phosphatase	To monitor liver function.
Serum amylase	To monitor pancreatic function.
Serum lipase	To monitor pancreatic function.

Note: This table presents a partial listing of the major uses of certain commonly performed lab tests that have implications for nutrition.

[a]More than half of the total protein is albumin.

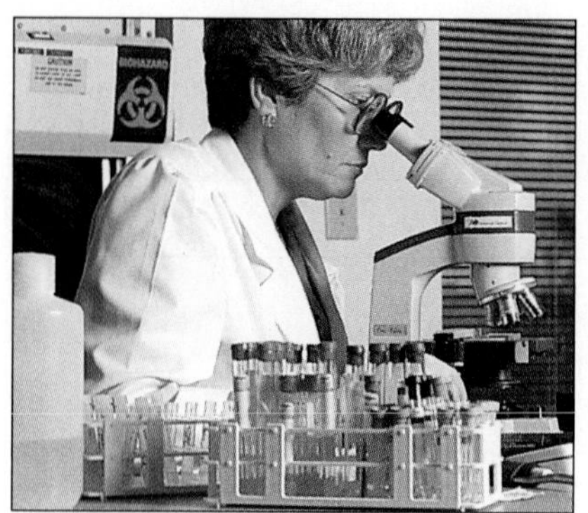
Blood and urine samples offer valuable clues for assessing nutrition status.

Assessors must also consider many other factors that influence test results. The low blood concentration of a nutrient, for example, may reflect a primary deficiency of that nutrient, but it may also be secondary to a deficiency of one or several other nutrients or to a condition unrelated to nutrition. Nutrient concentrations in the blood and urine sometimes reflect recent intakes rather than long-term intakes. Thus blood concentrations of a nutrient may be normal, even when tissue levels are deficient. Assessors who keep these limitations in mind and use lab tests along with other assessment data, however, can create a total picture that becomes meaningful with careful interpretation.

It is beyond the scope of this text to describe all lab tests used to assess nutrition status, define organ function, and develop and monitor care plans. Instead, the emphasis is on lab tests used to evaluate energy and protein status. Appendix E provides tables of lab tests useful in detecting various vitamin and mineral deficiencies, including those associated with nutrition-related anemias.

Biochemical Tests of Protein Status

Protein-energy malnutrition is pervasive in illness and can dramatically worsen outcome. Recall from Chapter 4 that body proteins maintain all metabolic activities including the management of fluid balances, digestion of nutrients, physical work, synthesis of hormones, maintenance of immune defenses, wound healing, and much more. Without adequate protein, the body is unable to support these vital functions.

Along with other assessment techniques, serum proteins can help evaluate a person's current energy-protein status and response to nutrition therapy. The larger the body's pool of a serum protein, and the slower the metabolism of that protein, the longer it takes for the protein to be affected by diet. While serum proteins reflect energy and protein intake (the availability of amino acids), they also reflect the distribution of proteins. During some severe stresses, such as extensive burn wounds, serum proteins may shift from the blood to extracellular fluid, for example. Thus serum proteins will be low, even though the body's pool may not be low. Furthermore, serum proteins are synthesized in the liver, so serum levels can reflect liver function. Table 16.4 provides standards for evaluating the serum proteins most widely used in nutrition assessments. Other biochemical tests of protein status and response to nutrition therapy include the total lymphocyte count, a measure of immune function; urinary nitrogen, used to evaluate nitrogen balance; and urinary creatinine, a measure of skeletal muscle mass.

Table 16.4 Relationship between Degree of Undernutrition and Serum Proteins

Indicator	Degree of Depletion			
	Normal	Mild	Moderate	Severe
Albumin (g/100 ml)	≥3.5	2.8–3.4	2.1–2.7	<2.1
Transferrin (mg/100 ml)	>200	150–200	100–149	<100
Transthyretin (mg/100 ml)	16–30	10–15	5–9	<5
Retinol-binding protein[a] (mg/100 ml)	2.6–7.6	—	—	—

Note: To convert albumin (g/100 ml) to standard international units (g/L), multiply by 100. To convert transferrin (mg/100 ml) to standard international units (g/L), multiply by 0.01.

[a]Levels less than normal suggest compromised protein status. The actual degree of depletion (mild, moderate, and severe) has not been defined.

Albumin Albumin is the most abundant serum protein and accounts for over 50 percent of the total serum proteins. Serum albumin is frequently used in nutrition screenings because it is an inexpensive test that is readily available. Serum albumin is slow to reflect changes in nutrition status because it is plentiful in the body and breaks down slowly. In people with chronic PEM, serum albumin levels remain normal despite a depletion of body proteins; the levels fall only after prolonged malnutrition. Likewise, albumin concentrations increase slowly with appropriate nutrition support, so albumin is not a sensitive indicator of response to nutrition therapy.

More often, albumin and other serum protein levels help assessors distinguish chronic malnutrition, acute malnutrition, and mixed malnutrition (see Table 16.5). The type of malnutrition helps guide nutrition therapy. Serum albumin levels appear to correlate directly with the development of complications and survival among people in the hospital.

Conditions other than PEM and severe stresses that can depress serum albumin include liver disease, kidney disease (nephrotic syndrome), and eclampsia.

Transferrin The serum protein transferrin transports iron, and its concentrations reflect both protein-energy and iron status. Like albumin, markedly reduced transferrin levels indicate severe PEM. Transferrin breaks down in the body more rapidly than albumin, but is still relatively slow to respond to changes in protein intake. Thus it may not be a sensitive indicator of response to nutrition therapy.[4] Furthermore, interpreting transferrin levels as a measure of energy-protein status is difficult when *iron* deficiency is present. Transferrin rises as iron deficiency grows worse and falls as iron status improves.

Conditions other than malnutrition that depress serum transferrin levels include liver disease, kidney disease (nephrotic syndrome), metabolic stress, and the use of some antibiotics. Pregnancy, iron deficiency, hepatitis, blood loss, and the use of oral contraceptives can raise transferrin levels.

Transthyretin and Retinol-Binding Protein Transthyretin and retinol-binding protein levels decrease rapidly during PEM and respond quickly to changes in protein intake. Lab tests of transthyretin and retinol-binding protein are more expensive than the relatively inexpensive serum albumin, which is routinely available for hospitalized clients. Therefore, tests of transthyretin and retinol-binding protein are often reserved for clients with disorders that markedly change metabolic rates and can rapidly and profoundly affect nutrition status.

Transthyretin is also known as **prealbumin** or **thyroxin-binding prealbumin.**

Conditions other than protein status that can lower transthyretin levels include metabolic stress, hemodialysis, and hypothyroidism; levels may be elevated in kidney disease and with the use of corticosteroids.

Conditions other than protein status that can lower retinol-binding protein levels include vitamin A deficiency, metabolic stress, hyperthyroidism, liver disease, and cystic fibrosis; levels may be elevated in kidney disease.

Other Serum Proteins Other serum proteins that may be useful in assessing the response to nutrition therapy include insulin-like growth factor (IGF-1) and fibronectin. These indicators of protein status are less practical and more expensive to measure than the other serum proteins and are not in common use.

Insulin-like growth factor (IGF-1) is also known as **somatomedin-C.**

Total Lymphocyte Count PEM compromises the immune system, reducing the number of white blood cells (lymphocytes), which are important in resisting and fighting infections. The total lymphocyte count, derived from the number of white blood cells and the percentage of lymphocytes (see the box on p. 412), is inexpensive and easy to obtain, but the many variables that affect this indicator limit its value in nutrition assessment.

Conditions other than protein status that affect the total lymphocyte count include metabolic stress (especially infection) and the use of chemotherapy, immunosuppressants, and corticosteroids.

Table 16.5 Distinguishing Characteristics of Different Types of PEM

	Weight and Fat Stores	Skeletal Protein	Blood Protein Status and Immune Function
Acute malnutrition (kwashiorkor)	Excessive or adequate	Rapidly becoming depleted	Depleted/compromised
Chronic malnutrition (marasmus)	Low/depleted	Depleted	Adequate/compromised
Mixed malnutrition	Low/depleted	Depleted	Depleted/compromised

How to

Calculate the Total Lymphocyte Count

To calculate the total lymphocyte count (TLC), look at the laboratory report and find the complete blood count (CBC). Two figures from this report, the number of white blood cells (WBC) in cubic millimeters (mm^3) and the percentage of lymphocytes, are used to determine the total lymphocyte count:

$$\text{TLC (mm}^3\text{)} = \text{WBC (mm}^3\text{)} \times \text{\% lymphocytes.}$$

A person with a WBC count of 7500 mm^3 with 15 percent lymphocytes would have a total lymphocyte count of:

$$\text{TLC (mm}^3\text{)} = 7500 \text{ mm}^3 \times 0.15 = 1125 \text{ mm}^3.$$

The standard lymphocyte count is 2500 mm^3. Values below 1500 mm^3 suggest mild depletion; below 1200 mm^3, moderate depletion; and below 800 mm^3, severe depletion. Thus the TLC of 1125 mm^3 in this example suggests moderate depletion.

Urinary Tests of Protein Status Two biochemical tests of protein status—urinary urea nitrogen (UUN) and urinary creatinine excretion—require the collection of urine over a 24-hour period. Both tests require normal kidney function for accurate results. Appendix E provides more information about these tests.

Laboratory tests help to define protein status, vitamin and mineral status, and alterations in metabolism or organ function with implications for nutrient needs. Physical findings, described next, help substantiate the findings.

Table 16.6 Physical Signs of Dehydration and Fluid Retention

Dehydration
- Sunken eyes
- Hollow cheekbones
- Dry mucous membranes
- Loss of skin turgor (elasticity)[a]
- Weak cry[b]
- Depression of the anterior fontanel[b]
- Deep, gasping respirations
- Weak, rapid pulse
- Thirst
- Reduced urinary output
- Weight loss

Fluid Retention
- Edema
- Ascites (abdominal fluid retention)
- Elevated blood pressure
- Increased urinary output
- Weight gain

[a]May not be a useful parameter in the elderly.
[b]Findings specific to infants.

Physical Examinations

One clinician astutely summarized the role of the physical examination in nutrition assessment this way: "To me, physical examination proves the saying, 'A picture is worth a thousand words.'"[5] Indeed, health care professionals can simply look at people and determine that they are overweight, underweight, lethargic, confused, or unable to feed themselves, to give a few examples.

With closer examination, a skilled assessor can use a physical examination to search for signs of nutrient deficiencies or toxicities. Signs of malnutrition appear most rapidly in parts of the body where cell replacement occurs at a high rate, such as the hair, skin, and digestive tract (including the mouth and tongue). The summary tables in Chapters 7 and 8 list physical signs of vitamin and mineral malnutrition.

Fluid Balance Among the most useful physical signs of nutrition status are those that reflect dehydration and fluid retention. Many illnesses upset fluid balances, and attention to physical signs and laboratory tests of fluid balance help with the interpretation of anthropometric and other biochemical tests and help guide therapy. Table 16.6 lists physical signs that may occur in dehydration and fluid retention. The causes of fluid imbalances vary, however, and the clinical manifestations vary as well.

Limitations of Physical Findings Like other assessment techniques, identifying and interpreting physical findings requires knowledge, skill, and clinical judgment. Many physical signs are nonspecific; they can reflect any of several nutrient deficiencies as well as conditions unrelated to nutrition (see Table 16.7).[6] For

Table 16.7 Physical Signs of Nutrient Deficiencies			
Body System	**Acceptable Findings**	**Malnutrition Findings**	**What the Findings May Reflect**
Hair	Shiny, firm in the scalp	Dull, brittle, dry, loose; falls out	PEM
Eyes	Bright, clear pink membranes; adjust easily to light	Pale membranes; spots; redness; adjust slowly to darkness	Vitamin A, the B vitamins, zinc and iron status
Teeth and gums	No pain or caries, gums firm, teeth bright	Missing, discolored, decayed teeth; gums bleed easily and are swollen and spongy	Mineral and vitamin C status
Face	Clear complexion without dryness or scaliness	Off-color, scaly, flaky, cracked skin	PEM, vitamin A, and iron status
Glands	No lumps	Swollen at front of neck	PEM and iodine status
Tongue	Red, bumpy, rough	Sore, smooth, purplish, swollen	B vitamin status
Skin	Smooth, firm, good color	Dry, rough, spotty; "sandpaper" feel or sores; lack of fat under skin	PEM, essential fatty acid deficiency, vitamin A, the B vitamins, and vitamin C status
Nails	Firm, pink	Spoon-shaped, brittle, ridged, pale	Iron status
Internal systems	Regular heart rhythm, heart rate, and blood pressure; no impairment of digestive function, reflexes, or mental status	Abnormal heart rate, heart rhythm, or blood pressure; enlarged liver, spleen; abnormal digestion; burning, tingling of hands, feet; loss of balance, coordination; mental confusion, irritability, fatigue	PEM and mineral status
Muscles and bones	Muscle tone; posture, long bone development appropriate for age	"Wasted" appearance of muscles; swollen bumps on skull or ends of bones; small bumps on ribs; bowed legs or knock-knees	PEM, mineral, and vitamin D status

example, cracked lips may be caused by sunburn, windburn, dehydration, or any of several B vitamin deficiencies. For this reason, physical findings are most valuable for identifying potential problems, which then may be confirmed by other assessment techniques. A diet history can provide further support for a suspected deficiency, for example. Weight measurements can confirm that a person is underweight and quantify the degree of underweight. Laboratory data can help verify a person's state of hydration or vitamin-mineral status.

The numerous markers of nutrition status described in this chapter and the last provide clinicians with a variety of tools for evaluating nutrition status, the risk of malnutrition, and the response to nutrition therapy. As Chapter 17 describes, cost considerations, staffing, and individual preferences determine which tests are routinely available in different facilities and which will be used under specific circumstances. One approach, the Subjective Global Assessment (SGA), relies on historical, anthropometric, and physical findings to assess nutrition status.[7] The case study that follows shows how the assessor uses assessment data. The next chapter provides the details of how nutrition assessment data are translated into nutrition care plans.

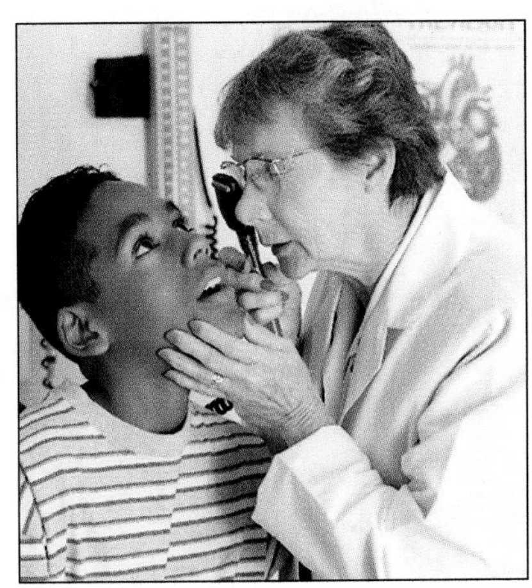

Physical signs of malnutrition appear in parts of the body where cells are replaced at a rapid rate.

Case Study

Nutrition Assessment of a Computer Scientist following a Car Accident

Ms. Green, a 38-year-old computer scientist, was admitted to the hospital for emergency surgery to repair a broken hip following a car accident. Other injuries included a wound over her left eye and bruises on her arm. After completing a nutrition screening, the nurse referred Ms. Green to a dietitian for further assessment. Before visiting the client, the dietitian reviewed the medical record and noted the following:

- Ms. Green has been in good health over the past ten years, although she has experienced a gradual weight gain (height, 5 feet, 7 inches; current weight, 150 pounds). She is in stable condition following surgery.
- Prior to hospitalization, Ms. Green was taking one multivitamin-mineral supplement daily and no medications. Morphine sulfate is now being administered for pain.
- Available information from the lab report shows a mildly depleted serum albumin (3 g/100 ml) and a moderately depleted total lymphocyte count (1000 mm^3).

From this information, the dietitian keeps these points in mind as he prepares to visit the client:

- Even though Ms. Green is overweight, her nutrient requirements are temporarily increased.
- Pain medication may make it difficult for Ms. Green to provide a detailed history.

Once in the client's room, the dietitian confirms Ms. Green's identity, introduces himself, and states the purpose of his visit. He takes time to establish rapport and get a sense of the client's ability to answer questions. Ms. Green appears to be alert. The dietitian learns that she lives alone and that her busy schedule seldom leaves time for her to prepare meals. Using a usual food intake, the dietitian finds that Ms. Green usually skips breakfast and often eats out. She enjoys foods from all food groups, although she seldom eats the recommended amounts of fruit and vegetables. For the past three weeks, she has been trying to lose weight, and her intake has been less than usual. She has lost 10 pounds during this time. Money for food and facilities for food preparation are not a problem. She appears to be overweight and pale.

What factors might have alerted the nurse to the need for nutrition assessment? What factors in Ms. Green's health, drug, socioeconomic, and diet histories or physical findings suggest possible nutrient imbalances?

Estimate Ms. Green's desirable body weight and calculate her percent ideal body weight and percent usual body weight. Consider Ms. Green's recent weight change. What is a safe rate of weight loss (see p. 405)? How does Ms. Green's weight loss compare with the safe rate? Does her recent weight change increase her risk for malnutrition? What do Ms. Green's lab values suggest with respect to her protein-energy balance? What factors besides protein-energy balance need to be considered when interpreting Ms. Green's lab test results?

Self Check

1. Health care professionals can use all of the following reasons for anthropometric measurements in nutrition assessment **except:**
   a. to evaluate growth in pregnant women and children.
   b. to detect undernutrition and overnutrition.
   c. to evaluate functional status of organs.
   d. to measure changes in body composition over time.

2. Both height and weight measurements:
   a. can be affected by fluid status.
   b. cannot be taken on people who are bedridden.
   c. are commonly used anthropometric measurements.
   d. require equipment that is not readily available in most health care facilities.

3. All of the following are true about growth measurements in infants and children **except:**
   a. height and weight stay at about the same percentile throughout childhood.
   b. for a malnourished child, the goal of nutrition therapy is for both height and weight to increase.
   c. for the overweight child, the goal of nutrition therapy is for weight to remain stable while height increases.
   d. for the malnourished child, the goal of nutrition therapy is for weight to increase while height remains stable.

4. The weight parameter that relates current weight to the weight that is normal for a particular individual is:
   a. percent usual body weight.
   b. percent ideal body weight.
   c. weight for height.
   d. weight for age.

5. The percent ideal body weight of a person who weighs 185 pounds and has a desirable body weight of 150 pounds is:
   a. 150 percent. b. 123 percent.
   c. 50 percent. d. 23 percent.

6. Problems that affect the reliability and use of anthropometric measurements include all of the following **except:**
   a. the accuracy of the equipment and the skill of the measurer.
   b. the use of standards that were derived for people who were ill.
   c. medications and illness that affect the state of hydration.
   d. physical problems, such as loose hanging skin, that can make some measurements difficult to make.

7. Hand grip strength is a(n):
   a. anthropometric measurement.
   b. biochemical test.
   c. functional test.
   d. body composition assessment technique.

8. A client has just begin to eat again after 10 days without significant amounts of foods. Which of the following laboratory tests would the dietitian expect to respond to changes in energy and protein intakes most quickly?
   a. albumin b. transferrin
   c. transthyretin d. total lymphocyte count

9. Urine tests used in nutrition assessments include:
   a. albumin and transferrin.
   b. urea nitrogen and creatinine.
   c. transthyretin and retinol-binding protein.
   d. fibronectin and somatomedin-C.

10. All of the following are true about physical findings used in nutrition assessments **except:**
    a. physical findings are highly specific.
    b. physical findings help confirm suspected nutrient deficiencies.
    c. physical findings can help detect dehydration and fluid retention.
    d. physical findings can reflect both nutrient imbalances and conditions unrelated to nutrition.

Clinical Applications

1. Calculate the percent ideal body weight and percent usual body weight for a man who is 5 feet 11 inches tall with a current weight of 160 pounds and a usual body weight of 180 pounds. State which standards you choose for determining desirable body weight. What additional information will be important for you to find out about this man's weight loss?

2. Recall that serum proteins are influenced by metabolic stress, including infection. With this in mind, what possible explanations can you suggest for these findings from a nutrition screening of a client who recently suffered a severe metabolic stress and now has an infection: 15-pound weight loss over the last four months (unintentional), 75 percent ideal body weight, depleted serum albumin, elevated total lymphocyte count. How might one sort through the possible explanations?

Notes

1. J. Hillhouse, Reliability of commonly used anthropometrics in adult hospitalized patients, *Support Line,* August 1996, pp. 9–13.

2. D. O. Jacobs, Bioelectrical impedance analysis: Implications for clinical practice, *Nutrition in Clinical Practice* 12 (1997): 204–210.

3. S. D. Brooks and P. J. Kearns, Muscle function analysis: An alternative nutrition assessment technique, *Support Line,* April 1997, pp. 12–15.

4. M. Russell, Serum proteins and nitrogen balance: Evaluating response to nutrition support, *Support Line,* February 1995, pp. 3–8.

5. K. Hammond, Nutrition focused physical assessment, *Support Line,* August 1996, pp. 1–4.

6. S. G. Morrisson, Clinical nutrition physical examination, *Support Line,* April 1997, pp. 16–18.

7. A. S. Detsky and coauthors, What is Subjective Global Assessment of nutrition status? *Journal of Parenteral and Enteral Nutrition* 11 (1987): 8–13.

CHAPTER 16

Nutrition in Practice

▪ Nutrition in a Cost-Conscious Health Care Environment ▪

Decades of medical research have resulted in an explosion of knowledge and technologies to diagnose and treat diseases. This astounding progress, however, has come at a tremendous financial cost. The United States spends more money on health care than any other nation, yet the high cost of health care has put medical care out of the reach of some citizens. Without attention to cost containment, the health status of the nation is threatened; sophisticated medical services do little good if people cannot use them. The medical community, which once embodied the idealistic approach of sparing no cost when it came to health care, has embraced the reality that cost is an element of quality.[1] To discuss all the ramifications of cost containment for health care delivery is beyond the scope of this discussion. Instead, this nutrition in practice describes major trends that affect medical services, including nutrition services, and some ways nutrition can help reduce health care costs.

How can health care costs be reduced without sacrificing quality of care?

Among other things, efforts to control health care costs aim to eliminate duplication of services, reduce the number of hours spent in client care, maximize the use of health care professionals, and limit access to unnecessary services. The task is to cut costs without compromising the quality of care. The changes in health care occurring today have affected the way that health care professionals function in their jobs and in the health care environment.

How have health care professionals' jobs changed as a result of cost-cutting measures?

One trend is the replacement of traditional health care delivery, which is discipline- or department-specific, with a multidisciplinary or team approach. By working together, health care teams can coordinate their services, eliminate unnecessary services, and solve problems in a timely and efficient manner. (Nutrition in Practice 21 provides additional information about health care teams.) Traditionally, for example, nurses performed nursing functions and left the responsibility for the nutrition care of clients to the dietitian. When a client was admitted to the hospital, for example, the nurse would perform a nursing assessment, and the dietitian, a nutrition assessment. Yet much of the information from these assessments overlaps. An alert nurse can use the nursing assessment to determine the client's nutrition risk and make appropriate referrals to the dietitian. Thus, in today's cost-conscious health care environment, the nurse often performs nutrition screenings and notifies the dietary department only when the screening reveals potential nutrition problems.

Another change is an increased use of skilled assistants in health care. A dietetic technician assisting a clinical dietitian, for example, may collect the data for a nutrition assessment, and the dietitian may analyze the data. Because the dietitian's time is saved for tasks that require a dietitian's unique skills, the dietitian can manage more clients, and money is saved.

The more health care professionals involved in nutrition care—or any medical service for that matter—the more important effective communications become. In the case of nutrition care, all persons involved must understand their roles and perform their functions effectively to ensure quality of care.

How do health care professionals measure quality of care?

In cost-conscious health care, quality of care is most often defined in terms of outcomes. Examples of desirable outcomes include the person's ability to function independently (the more care a person needs, the more expenses incurred), reduced number or length of hospital stays, prevention of diseases and complications, and extended survival time. Services are cost-justified when they have a reasonable chance of improving a client's outcome. Services that produce the best outcomes at the lowest costs are the most cost-effective. The more expensive the procedure or test, the more critical justifying its cost becomes. Thus it may not be cost-effective to perform a nutrition assessment on every client admitted to a hospital, whereas nutrition screening is cost-effective.

How can nutrition affect the quality of care?

A review of some of the desirable outcomes listed above reveals many ways that nutrition can affect quality of care and reduce health care costs. By providing clients with the nutrients they need to maintain both physical and mental health, nutrition can significantly improve the client's ability to function independently, prevent or forestall diseases, limit complications, and maintain quality of life. All of these factors can help prevent the need for medical treatments, hospitalization, or lengthy hospital stays. Both timely identification of nutrition problems and appropriate intervention are esential to improving outcomes and cutting costs. Malnutrition, a persistent and common problem in hospitalized clients, is associated with significantly longer hospital stays, higher costs, and the need for home health care.[2] Based on one analysis of data from several sources, researchers estimate that early attention to malnutrition in hospitalized clients can reduce length of stay and may result in cost savings of approximately $8300 per hospital bed per year.[3]

Medical nutrition therapy can also have a significant impact on the costs of common chronic diseases with a relationship to diet such as diabetes, hypertension, cardiovascular diseases, and cancer. The task for nutrition professionals is to document the cost savings of nutrition intervention and to ensure that nutrition is addressed in standards and treatment plans that guide the care of clients.

What do you mean?

Health care professionals and the organizations that represent them are working together to define outcome-oriented standards of practice and measures of care. The Joint Commission on Accreditation of Healthcare Organizations (JCAHO), an agency that oversees the accreditation of health care facilities, requires compliance with nutrition care standards to identify, address, and monitor each client's nutrition needs.[4] The standards promote coordination and communication among disciplines in an effort to improve quality of care in a cost-efficient manner.[5]

WWW.

jcaho.org
Joint Commission on Accreditation of Healthcare Organizations

eatright.org
American Dietetic Association

Cost-cutting measures have redefined the responsibilities of health care professionals as well as the settings in which they work.

Clinical pathways that focus on a desirable care plan for a specific diagnosis, procedure, or treatment are another way of providing quality care that is also cost-effective (see Chapter 15). Examples include clinical pathways for enteral nutrition (see Chapter 21), parenteral nutrition (Chapter 22), and the management of bone marrow transplantation. The health care team develops each pathway after carefully studying their unique client population. The plan defines a time frame for each intervention with a goal of providing the best outcome at the lowest cost. Once in place, each pathway is reassessed by studying unexpected outcomes or variances from the expected course of treatment. When necessary, the plan is adjusted. The active participation of nutrition professionals or nutrition advocates on the teams that develop and reassess the pathways helps ensure that clients' nutrition needs are promptly identified and addressed.

How has health care changed as a result of cost-cutting efforts?

One of the major changes has been a great reduction in the length of hospital stays. Early discharges lower hospital costs. As a result, hospital staffs have been reduced, and there is less time to identify nutrition problems. Clients may be discharged before they have regained health, and they may require additional medical and nutrition care at home or in other settings. For

Glossary

capitation: prepayment of a set fee per client in exchange for medical services.

health maintenance organization (HMO): a form of managed care that limits the subscriber's choice of health care professionals to those affiliated with the organization and controls access to services by directing care through a primary care physician.

indemnity insurance: traditional fee-for-service insurance.

managed care: a health care delivery system that aims to provide cost-effective health care by coordinating services and limiting access to services.

preferred provider organization: a form of managed care that encourages subscribers to select health care providers from a group that has contracted with the organization to provide services at lower costs.

this reason, it has become more important that clients have access to nutrition services in other settings.

The availability of nutrition services in other settings is affected by another major change in health care delivery—the transformation from the traditional fee-for-service system to managed care (see the accompanying glossary). A traditional health insurance plan allows clients to use the physicians and health care facilities of their choice and then reimburses the clients for a percentage of the services covered by the plan. Managed care organizations, which include health maintenance organizations (HMOs) and preferred provider organizations (PPOs), strive to provide high-quality, low-cost care by controlling the access to and cost of services. Each client selects a primary care physician, who then determines what services the client needs. Managed care is moving toward a capitated system; that is, a system where the health care provider agrees to supply all the services a client needs for a set monthly fee. Because the providers assume the financial responsibility if they fail to operate within a predetermined budget, they have a strong incentive to control costs. Managed care organizations, which were rare just 15 years ago, now account for over 40 percent of the insurance provided by employers, and that percentage appears to be rising.[6]

How do nutrition services fit into a managed care organization?

In most managed care settings, medical nutrition therapy is treated as specialty care that requires referral by the primary care physician. Medical nutrition therapy is a relatively inexpensive service, and when appropriately applied, it can help prevent, forestall, and treat diseases and their complications. Thus it is a potentially cost-effective service. In representing nutrition professionals, the American Dietetic Association takes the position that managed care organizations and integrated health delivery systems should provide medical nutrition therapy as an essential component of health care and that it should be provided by qualified nutrition professionals.[7]

Health care professionals are challenged to ensure that home care agencies and managed care organizations understand the cost-effectiveness of nutrition services in helping clients maintain health and quality of life. To this end, nutrition professionals must conduct research to document the ways in which timely and appropriate nutrition services improve clients' health outcomes in a cost-efficient manner. Nutrition advocates must take a proactive stand in ensuring that nutrition services are included in all settings.

The full implications of cost containment and its impact on the nation's health remain to be seen. In the words of one clinician, "Our American society is in the midst of the most far-reaching and profoundly disturbing uncontrolled study in the history of health care. We are experiencing major changes in the way we practice and pay for health care with very little evidence that these changes will achieve the desired outcomes."[8] As changes occur, health care professionals, as client advocates, are challenged to ensure that high-quality care is not sacrificed in an effort to control costs.

CHAPTER 16

Nutrition in Practice

Notes

1. A. Bothe, Consensus: We should not lower quality to cut costs, *Nutrition in Clinical Practice* (supplement) 10 (1995): 1–7.

2. C. S. Chima and coauthors, Relationship of nutritional status to length of stay, hospital costs, and discharge status of patients hospitalized in the medicine service, *Journal of the American Dietetic Association* 97 (1997): 975–978.

3. H. N. Tucker and S. G. Miguel, Cost containment through nutrition intervention, *Nutrition Reviews* 54 (1996): 111–121.

4. D. Dougherty and coauthors, Nutrition care given new importance in JCAHO standards, *Nutrition in Clinical Practice* 10 (1995): 26–31.

5. Bothe, 1995.

6. J. K. Iglehart, The American health care system: Introduction, *New England Journal of Medicine* 326 (1992): 962–967; D. A. August, Creation of a specialized nutrition support outcomes research consortium: If not now, when? *Journal of Parenteral and Enteral Nutrition* 20 (1996): 394–400.

7. Position of The American Dietetic Association: Nutrition services in managed care, *Journal of the American Dietetic Association* 96 (1996): 391–395.

8. J. R. Wesley, Managing the future of nutrition support, *Journal of Parenteral and Enteral Nutrition* 20 (1996): 383–384.

17 Nutrition Screening and Intervention

This chapter describes the procedures used to identify clients at risk for poor nutrition status and then describes the process used to correct the nutrition problems identified through nutrition assessments. Assuring that a person's nutrient needs are met is a key part of this process, so the chapter also describes medical nutrition therapy and looks at how clients' needs are communicated among health care team members.

Nutrition Screening

Now that you have reviewed the basics of nutrition assessments, you can see why conducting a complete assessment on every client may not be possible or cost-effective. Instead, screening procedures help identify clients at risk for poor nutrition status, and these clients receive in-depth assessments.

Often nutrition screens are incorporated into the nursing admission assessment and are completed by registered nurses. Nutrition screening techniques vary from facility to facility, but typically include the following:

- *Assessment of health history* Does the person's health history reveal risk factors for poor nutrition status (review Table 15.2 on p. 385)? Do the current medical problems include metabolic stress, malabsorption, cachexia (wasting), swallowing problems (dysphagia), or altered organ function? Do the person's current symptoms include nausea, vomiting, pain, confusion, or injuries that may interfere with eating?
- *Assessment of diet history* Has the person been eating less than half of his or her normal intake for five or more days? Has the person been following an extremely restricted diet?
- *Assessment of height and weight data* Does the person weigh less than 80 percent of ideal body weight? Has the person lost or gained weight? How much? How fast?
- *Assessment of lab data, when available* Do serum proteins or the total lymphocyte count suggest malnutrition?

Clients at risk for malnutrition are then referred to the dietary department for further evaluation. Nutrition screens are repeated at regular intervals throughout hospitalization, because nutrition status may change. In addition, health care professionals should take these steps to remain alert to other conditions that may place the client at nutrition risk:

- Check the client's tray to see if food is being eaten.
- If the client is to receive no food or is unable to eat, find out how long it has been since the client has eaten. Ask if the client is expected to be able to eat soon.
- If the client is unable to eat, determine whether adequate nutrients are being delivered by tube (Chapter 21) or by vein (Chapter 22).

Regardless of the health care setting in which a client is seen, the health care professional communicates any problems discovered, records the problem in the medical record, and follows up to make sure the problem is being addressed.

Nutrition Care Plans

After completing a nutrition assessment, the health care professional studies the nutrition assessment data. What are the potential nutrition problems? Is body weight appropriate? Are lab values normal? Are there physical signs of malnutrition? Are nutrient needs altered due to growth, illness, or medications? Will long-term dietary adjustments be necessary? Answers to questions like these enable the health care professional to generate a nutrition problem list—the basis of the nutrition care plan. The problems listed may be past, current, or future conditions that may impair nutrition status or alter nutrient needs. The nutrition problem list is to the nutrition care process what the *nursing diagnosis* is to the nursing care process.

The next step is to develop a plan of action that identifies the client's immediate and long-term nutrition needs and sets forth a strategy for meeting those needs. The plan specifies the goals of dietary recommendations, the content of counseling sessions, and a tentative time frame for accomplishing goals. To ensure that the nutrition care plan will be realistic, the health care professional takes into account food habits, lifestyle, and socioeconomic factors identified from the food intake data and diet histories generated during nutrition assessment. Form 17.1 provides a sample nutrition assessment and care plan summary.

When establishing goals for the plan, the health care professional states them in terms of measurable outcomes, such as target ranges for body weight or blood glucose levels, because the success of therapy can more readily be determined if goals are measurable. Consider, for example, a problem of diarrhea. If a medication is causing the diarrhea, an appropriate strategy might be to prevent dehydration by giving ample fluids and electrolytes while the physician tries substituting other drugs. The client should be informed of this plan. Measurable

The word **illness** as used in this book refers to any medical condition that alters nutrient needs. Not all such conditions are diseases. For example, major surgery can have a significant impact on nutrition status. It is not a disease, but it is a stress that alters nutrient needs.

The North American Nursing Diagnosis Association (NANDA) has developed the following nursing diagnoses that relate to food intake:

- *Altered nutrition: At high risk for more than body requirements.*
- *Altered nutrition: More than body requirements.*
- *Altered nutrition: Less than body requirements.*

Objectives of the nutrition care plan:

- *To meet the client's nutrition needs.*
- *To meet the client's needs for nutrition information.*

The nutrition care plan specifies:

- *Goals.*
- *Areas of content.*
- *Tentative time frame.*

Form 17.1 Sample Nutrition Care Plan Summary

After analyzing assessment data, the dietitian may use a form such as this one to list problems, set goals, and record the plan for meeting goals.

Client ______________________ Diet Order ______________________

Problem List

1. ______________________ 4. ______________________
2. ______________________ 5. ______________________
3. ______________________ 6. ______________________

Goals

1. ______________________ 4. ______________________
2. ______________________ 5. ______________________
3. ______________________ 6. ______________________

Nutrition Care Plan Summary

Plan of nutrition care ______________________ Compliance/understanding ______________________

Other therapy ______________________ Follow-up ______________________

Education ______________________ Date ______________________

Dietitian ______________________

WWW.nanda.org
North American Nursing Diagnosis Association

goals might include a target range for serum electrolytes, urinary output, and blood pressure, as well as production of stools of normal volume and consistency. If the diarrhea is caused by a milk allergy, then the problem-solving strategy is to eliminate milk and milk products. Nutrition education is a key part of the plan. The plan will include counseling sessions to provide the client with instructions and suggestions for a nutritionally adequate diet that eliminates milk and milk products. The health care professional might measure goals for nutrition education by ensuring that the client can verbally identify foods containing milk or milk products, plan a day's menus appropriate for the restrictions, and find sources of milk on food labels.

Nutrient Needs

Dietary recommendations provide an estimate of energy, protein, vitamin, and mineral intakes for healthy people. They do not apply to people with health problems. When an illness requires supplemented or restricted intakes, the health care team makes educated estimates of the client's nutrient needs. Always remember that these estimates are just that—estimates. An individual's response to the nutrition care plan reveals the person's needs better than estimates can.

Energy If an adult has maintained a desirable weight over a period of time, then that person's energy needs can be estimated from habitual food intake. The person is already consuming the amount of energy needed. For otherwise healthy adults who are not at a desirable weight, energy needs can be calculated as detailed in Chapter 6 on p. 130. As for infants, pregnant women, children, and others, Chapters 12–14 described their changing energy needs. For people with severe stresses, energy needs can be estimated using indirect calorimetry or formulas (see Chapter 23). Later chapters describe energy needs imposed by different states of health.

indirect calorimetry: an indirect estimate of resting energy needs made by measuring the ratio of carbon dioxide expired to the amount of oxygen inspired.

Protein, Vitamin, and Mineral For people who are ill, especially those with severe stresses, protein needs can be estimated by conducting nitrogen balance studies or monitoring the response of serum proteins such as transthyretin and retinol-binding protein. Exact vitamin and mineral needs during illness are largely unknown and can only be estimated. Protein, vitamin, and mineral needs vary among individuals and change in different stages of the life cycle and for different medical conditions, as later chapters describe.

Nutrition Education Needs

In addition to providing for the client's nutrient needs, the nutrition care plan also considers the client's needs for nutrition information and education. Data gathered during nutrition assessments help the health care professional to determine the best way to present nutrition information, the amount of information the client will be able to handle, and the level of the client's interest in nutrition and health. Such plans are estimates as well. The nutrition education plan must be flexible so that it can be adjusted to accommodate the client's goals, understanding of the information presented, and motivation to practice the suggestions offered.

For people on medications, nutrition education includes advice on potential nutrient-medication interactions. Health care professionals should provide clients with advice on how to take medications in relation to food or specific nutrients, signs of potential nutrient deficiencies or nutrition-related problems, and actions to take if problems arise.

Case Study

COMPUTER SCIENTIST WITH CAR ACCIDENT INJURIES

To practice using nutrition assessment data to develop a care plan, answer the questions that follow drawing upon the information presented about Ms. Green in the case study on p. 414. The dietitian determines that Ms. Green's current medical status indicates a need for sufficient kcalories and protein to minimize weight loss and the depletion of serum proteins. Once Ms. Green's medical condition has stabilized and she has recovered from her injuries, the dietitian will confer with Ms. Green's physician and recommend a safe weight-loss diet.

Contrast Ms. Green's current nutrient needs with her long-term needs. How will the nutrition care plan reflect these changing needs? Describe methods the health care professional can use to determine if the nutrition care plan is effective while Ms. Green is in the hospital.

Think ahead to Ms. Green's nutrient needs once she leaves the hospital. What factors in Ms. Green's history need to be considered in developing a realistic weight-loss plan? Other than adjusting kcalories, might the health care professional recommend any changes in Ms. Green's usual eating habits?

Once Ms. Green is discharged from the hospital, it is unlikely that she will see the person who provided the instructions again. Consider ways that follow-up care might be arranged.

Implementation of Nutrition Care Plans

Once a care plan is developed, the next step is to implement it by providing both the appropriate diet and education. In an in-patient health care facility (such as a hospital, nursing home, or in-patient mental health care facility), the diet part appears deceptively simple: appropriate foods are delivered to clients. Behind the scenes, however, the dietitian carefully plans the diet for each client, the dietitian or dietetic technician checks the client's menus, and the food service department assures that the appropriate foods are carefully prepared and delivered. Nutrition in Practice 19 describes how foodservice systems operate. Equally important, once delivered, the food must be eaten by the client. Whenever possible, health care professionals check with clients throughout their stay to make sure they understand their diets.

Nurses, thanks to their frequent daily contact with clients, can offer important support in the educational aspect of client care. Clients often think of questions long after the dietitian has left, and they most often ask the nurse. The nurse who is confident of the answers should provide them. If not sure of the answers, the nurse should express honest uncertainty and reassure the client that the information will be provided. The nurse should then inform the dietitian that the client needs a follow-up visit.

Ongoing Evaluation

While the planned strategies are being implemented, the health care professional must keep track of how they are working. If, for example, a client on a weight-reduction diet fails to lose weight, a change may be needed. Is the client eating too much? Is the client too inactive? Perhaps the client should keep a food and activity record to help identify problems with the weight-loss plan.

If a client's situation changes, so may nutrition status and nutrient needs. For example, when a pregnant woman delivers her baby, she will need instructions on how to feed her infant. She will also need information on how to revise her diet to support lactation (if she is breastfeeding) or to return to a healthy weight (if she is bottle feeding). The accompanying case study affords an opportunity to apply some of the planning principles introduced here.

A care plan may be ideal, but may still fall short of meeting goals if a client is unable or unwilling to comply with it. If the client is unable to comply, reassessing communication techniques may help. Perhaps the level of instruction needs to be simplified or cultural differences addressed. If the client is unwilling, despite the best efforts of health care professionals, little can be done except to try later when the client may be more receptive.

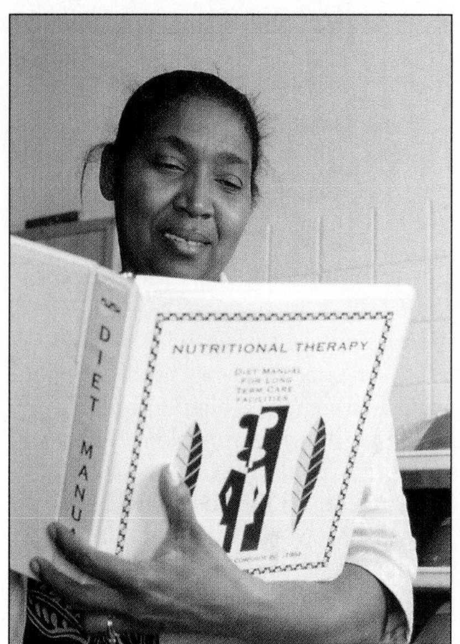

Health care professionals consult diet manuals to clarify what foods are included on or excluded from different diets.

Medical Nutrition Therapy

An essential component of every nutrition care plan is medical nutrition therapy. Medical nutrition therapy strives to provide the appropriate amounts of energy, protein, carbohydrate, fat, vitamins, major minerals, trace elements, and water in whatever form best meets the client's needs. For example, a person who cannot chew needs soft foods; a person who is in a coma may need to have a formula delivered by tube into the GI tract; a person with diabetes mellitus needs a special diet.

Medical nutrition therapy often complements other therapies. For example, a diet plan for an individual with type 1 diabetes mellitus is coordinated with a physical activity program and the insulin-delivery schedule.

diet order: a physician's written statement in the medical record of what diet a client should receive.

Diet Orders In facilities that serve food, the physician prescribes the client's diet and writes the diet order in the medical record. The physician often relies on the dietitian or health care team to suggest a diet prescription or make recommendations when changes in the diet order appear warranted or when clarity is lacking.

To avoid confusion, physicians should provide clear and precise diet orders. For example, a "low-sodium diet" order should specify the amount of sodium; otherwise, "low-sodium" could be interpreted to mean any amount from 500 to 4000 milligrams. For uncomplicated diets, such as low-sodium diets, the food service department often sends a preselected diet to the client until orders are clarified, and the level of restriction is recorded in the medical record. For more complicated diets, such as renal diets, meals will not be sent until the order is clarified.

Occasionally, diet orders may be inappropriate. For example, the physician may describe an obese individual as "well-nourished" and order a regular diet. If this occurs, the client will receive an inappropriate diet and no nutrition advice. As another example, a physician may order that a client receive no food or fluids after midnight for a lab test to be conducted in the morning. The doctor assumes that the order will be discontinued after the test, but it may not be. The client may miss several meals before someone notices the error. These examples illustrate situations where communication between health care professionals can make a difference in client care. Whenever you notice inappropriate diet orders, contact the dietitian or alert the physician.

An order to give a client nothing orally (including food, beverages, and medications) reads **NPO,** the abbreviation for *non per os,* which means "nothing by mouth." **PO** stands for *per os,* which means "by mouth" or "orally."

Diet Manuals The exact foods excluded from or included on a specific modified diet, and even the name given to the diet, may differ among health care facilities, generally in minor ways. These variations reflect different schools of thought regarding diet; institutional diet manuals are consulted as a standard of practice.

In large facilities, the staff of dietitians may compile a diet manual, subject to approval by the hospital administrator, several physicians, and representatives of the nursing service. Small facilities may adopt the diet manual of another hospital or an organization such as the state dietetic association. The diet manual describes the foods allowed and not allowed on each diet, outlines the rationale and indications for use of each diet, provides information on the nutritional adequacy of the diets, and offers sample menus. The dietary department uses the manual to design menus for each diet.

diet manual: a book that describes the foods allowed and restricted on a diet, outlines the rationale and indications for use of each diet, and provides sample menus.

WWW **cc.nih.gov/nutr/dietmanual/intro.html**
National Institutes of Health Clinical Center Nutrition Department Diet Manual

Standard and Modified Diets *Standard* or *regular* diets include all foods and provide all the nutrients in amounts appropriate for healthy people. Modified diets are used when standard diets fail to meet the specific needs of clients. Modifying the standard diet is much like tailoring a suit. A tailored suit is the same suit after alterations—only it fits better. In the case of a modified diet, the tailoring may involve changing the consistency; adjusting the amounts of energy, individual nutrients, or fluid; altering the number of meals; or including or eliminating certain foods. Nutrition in Practice 19 shows how a hospital menu can be modified for different diets. Table 17.1 on pp. 428–429 gives examples of modified diets used to treat diseases involving different organ systems. These diets are described further in later chapters.

It is helpful to think about modified diets in terms of the symptoms or conditions they relieve rather than in terms of disorders. Two people with the same disorder may need two different diets. Conversely, people with two different disorders may benefit from the same diet. Consider two people with cancer: one may need a diet that will help control nausea; the other may need a high-kcalorie diet. Now consider a pregnant woman with nausea. She may benefit from the same recommendations as those for the first person with cancer.

standard or **regular diet:** a diet that includes all foods and meets the nutrient needs of a healthy person.

modified or **therapeutic diet:** a regular diet that is adjusted to meet special nutrient needs. Such diets can be adjusted in consistency, level of energy and nutrients, amount of fluid, or number of meals, or by the inclusion or elimination of certain foods.

Medical Records

Maintaining strong professional communication networks benefits both health care professionals and their clients. Conversely, miscommunication between professionals can result in inappropriate therapy with serious consequences for clients' health. Professionals have many opportunities to discuss client concerns and can record these concerns in the medical record.

Medical records are legal documents that record a client's history; the assessment, diagnosis, and prognosis of medical problems; the measures being taken to treat those problems; and the results of tests and therapy. Writing in the medical records allows health care professionals to document the actions taken to comply with physicians' orders, the client's responses to those actions, and recommendations. This information helps determine if medical orders are being followed and directs future care.

medical record: a continuous written account of a client's health history, diagnosis, prognosis, and therapy.

diagnosis: the disease or condition a person has or is thought to have.
dia = through
gnosis = knowing

prognosis: the predicted course and outcome of a disease or condition.
pro = ahead of time, before

Types of Medical Records Medical records can be organized in many ways. In today's cost-conscious health care environment, time constraints compel health care professionals to minimize the time spent in documenting medical care.[1] Traditional problem-oriented medical records (POMR), which focus on a client's medical problems and the strategies being used to address those problems, are gradually being replaced by outcome- or goal-oriented medical records, which focus on observable medical goals. A problem-oriented medical record for a person with diabetes mellitus, for example, would list diabetes mellitus as a problem and then define the strategies used to control the diabetes, such as a 1600-kcalorie diet, oral hypoglycemic agents, and the like. A goal-oriented medical record for the same client would list the client's goal weight and target blood glucose measurements and then would show how these values have responded to therapy.

The Medical Record and Nutrition Care Learn how to effectively use the record in the facility where you work. Regardless of the approach used, be sure the client's medical record includes important nutrition-related information. Examples of important information include:

- Documentation of nutrition screening.
- Nutrition assessment data, including evaluation of the client's diet.
- Recommended medical nutrition therapy, including goals.

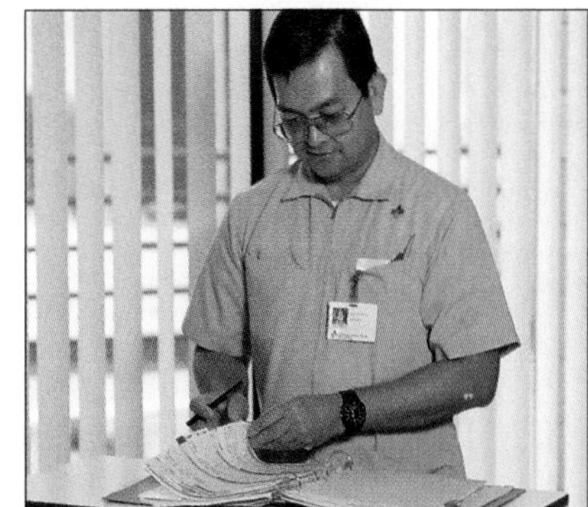
Take time to record important nutrition information in the client's medical record.

Table 17.1 Summary of Modified Diets by Organ System

Disorders	Possible Diet Modifications[a]
Conditions Affecting or Involving the GI Tract, Liver, and Exocrine Pancreas[b]	
Blind loop syndrome	Fat-restricted, fluid and electrolyte replacement, folate supplements, vitamin B_{12} injections
Broken jaw	Mechanical soft; liquid
Celiac disease	Gluten-restricted
Cirrhosis	High-kcalorie, high-protein, protein-restricted, sodium-restricted, fluid-restricted
Constipation	High-fiber, increased fluids
Cystic fibrosis	High-kcalorie, high-protein
Delayed gastric emptying	Liquid, low-fiber, tube feeding, total parenteral nutrition (TPN)
Dental caries	Mechanical soft
Diarrhea	Liquid, low-fiber, high-fiber lactose-free, regular, fluid and electrolyte replacement
Difficulty swallowing (dysphagia)	Mechanical soft, tube feeding, TPN
Diverticulitis	Low-fiber
Diverticulosis	High-fiber
Dry mouth	Mechanical soft
Dumping syndrome	Carbohydrate-restricted, no concentrated sugars, frequent small feedings, fluid and electrolyte replacement
Gastritis	Low-fiber, bland
Hepatic coma	Protein-restricted, sodium-restricted, fluid-restricted
Hepatitis	Regular, high-kcalorie, high-protein
Hiatal hernia	Frequent small feedings, fat-restricted, bland, kcalorie-restricted
Ill-fitting dentures	Mechanical soft
Indigestion (dyspepsia)	Low-fiber, bland, frequent small feedings
Inflammatory bowel disease	Low-fiber, fat-restricted, high-kcalorie, high-protein, fluid and electrolyte replacement, lactose-restricted, tube feeding, TPN
Irritable bowel syndrome	High-fiber, fat-restricted
Lactose intolerance	Lactose-restricted
Malabsorption	Fat-restricted, high-kcalorie, high-protein, fluid and electrolyte replacement
Missing teeth	Mechanical soft
Nausea	Low-fiber, bland, frequent small feedings, no liquids with meals
Oral surgery	Mechanical soft
Pancreatitis	Fat-restricted, regular, frequent small feedings, tube feeding, TPN
Peptic ulcer	Bland
Periodontal disease	Mechanical soft
Plastic surgery of head or neck	Mechanical soft, tube feeding, TPN
Reflux esophagitis	Frequent small feedings, fat-restricted, bland, kcalorie-restricted
Short-bowel syndrome	Fat-restricted, high-kcalorie, high-protein, fluid and electrolyte replacement
Ulcers of mouth or gums	Mechanical soft, avoid spicy foods and foods with seeds
Vomiting	Fluid and electrolyte replacement; NPO

[a]Diet modifications vary for each disease and sometimes depend on medical therapy.
[b]The pancreas produces both external (exocrine) and internal (endocrine) secretions. The external secretions (enzymes) play an important role in the digestion of food; the internal secretions (insulin and other hormones) play a primary role in the regulation of glucose metabolism.

- Acceptance and tolerance of current diet.
- Problems with food intake.
- Documentation of diet counseling, including the client's response to counseling.
- Any planned follow-up or referral to another person or agency.

Disorders	Possible Diet Modifications[a] *(continued)*
Conditions Affecting the Endocrine Pancreas[a]	
Diabetes mellitus	Carbohydrate-controlled, kcalorie-controlled, fat-restricted, high-fiber, sodium-restricted
Hypoglycemia	Carbohydrate-controlled, limited simple sugars, frequent small feedings
Conditions Affecting the Blood Vessels, Heart, and Lungs	
Atherosclerosis	Fat-restricted, low-cholesterol, kcalorie-restricted, sodium-restricted, high-fiber
Congestive heart failure	Sodium-restricted, kcalorie-restricted, low-fiber, bland, frequent small feedings, fluid-restricted, caffeine-restricted
Coronary heart disease	Fat-restricted, low-cholesterol, kcalorie-restricted, sodium-restricted, high-fiber
Hypertension	Sodium-restricted, kcalorie-restricted, high-potassium, fat-restricted
Myocardial infarction	Sodium-restricted, kcalorie-restricted, bland, frequent small feedings, moderate-temperature foods, fat-restricted, caffeine-restricted
Pulmonary disease	High-kcalorie, high-protein
Conditions Affecting the Kidneys	
Acute renal disease	Protein-restricted, high-kcalorie, fluid-controlled, sodium-controlled, potassium-controlled, fat-restricted, carbohydrate-controlled
Chronic renal disease	Protein-restricted, low-sodium, fluid-restricted, potassium-restricted, phosphorus-restricted, fat-restricted
Kidney stones	Increased fluid intake, calcium-controlled, oxalate-restricted, purine-restricted, methionine-restricted
Nephrotic syndrome	High-kcalorie, protein-restricted, sodium-restricted
Conditions Affecting Many Organ Systems	
Burns	High-kcalorie, high-protein, increased fluid intake
Cancer	High-kcalorie, high-protein (see also specific related conditions such as *Nausea*)
Food sensitivities	Elimination of offending substance
Galactosemia	Galactose-restricted
Human immunodeficiency virus (HIV) infection	High-kcalorie, high-protein, fat-restricted, fluid and electrolyte replacement, lactose-restricted, caffeine-restricted, mechanical soft, tube feeding, TPN (see also specific related conditions such as *Ulcers of mouth*)
Obesity, overweight	kCalorie-restricted, fat-restricted, high-fiber
Phenylketonuria (PKU)	Phenylalanine-restricted
Stroke	Mechanical soft, regular, tube feeding, fat-restricted, low-sodium, high-potassium
Surgery	Regular, high-kcalorie, high-protein, increased fluids
Underweight	High-kcalorie, high-protein

Effective nutrition care addresses the unique needs of the individual, framing nutrition needs in the context of the person's educational, socioeconomic, and medical needs. This chapter has described techniques for building effective nutrition care plans. The remaining chapters describe how therapeutic diets meet nutrient needs as they change as a result of illness.

Self Check

1. All of the following are true about nutrition screenings **except:**
   a. they should be conducted routinely, as soon as possible after admission to a hospital or long-term care facility.
   b. they cannot be completed until laboratory data are available.
   c. they identify clients at risk for poor nutrition status.
   d. they identify clients who need complete nutrition assessments.
2. Which of the following provides the foundation of the nutrition care plan?
   a. nutrition problem list or nursing diagnosis
   b. nutrition education needs
   c. analysis of assessment data
   d. nutrient needs
3. Diet manuals:
   a. explain which foods are included on and excluded from diets.
   b. explain how to provide nutrition education for special diets.
   c. are compiled by the nursing department.
   d. can only be found in the foodservice department.
4. All of the following statements are true about diet orders **except:**
   a. they are always appropriate.
   b. they are written by the physician.
   c. restricted diet orders should specify the level of restriction.
   d. the health care team may recommend an appropriate diet order to the physician.
5. The factors that determine how nutrition information will be presented include:
   a. the client's nutrient needs.
   b. the client's goals, ability level, and motivation.
   c. the client's lifestyle.
   d. the medical record.
6. Once a nutrition care plan is implemented:
   a. the health care professional must reevaluate nutrient needs.
   b. the health care professional performs a complete nutrition assessment.
   c. the health care professional's involvement with the client is over.
   d. the health care professional evaluates how well the plan is working.
7. All of the following are true about medical records **except:**
   a. they are legal documents.
   b. they provide information about a client's medical problems and the measures being taken to address those problems.
   c. they all conform to the same format.
   d. they demand the time of health care professionals.

Answers to these questions appear in Appendix H.

Clinical Applications

1. A client who recently suffered a heart attack is on a special diet. The client tells you that the dietitian has talked to him about his diet and that he is totally confused. He confides that diet is the last thing on his mind right now. What actions should you take?
2. An initial nutrition screening of an elderly client admitted to the hospital for surgery revealed that the client was not at risk for poor nutrition status. The client was not referred for nutrition assessment and has not seen a dietitian. Following surgery, the client developed several complications, and recovery has been slower than expected. The nurse working with the client notices that the client has eaten only minimal amounts of food for several days. What steps, if any, should be taken to uncover and address problems the client might be having with food?
3. Assume that the client in Clinical Application 2 loses a considerable amount of weight and the physician orders a kcalorie count. The physician's written order is recorded in the medical record, as is a note from the dietitian verifying the procedure. Considering the communication channels described in this chapter, what steps might the dietitian take to ensure that the client's intake will be recorded through all shifts?

Note

1. C. J. Klein, J. B. Bosworth, and C. E. Wiles, Physicians prefer goal-oriented note format more than three to one over other outcome-focused documentation, *Journal of the American Dietetic Association* 97 (1997): 1306–1310.

CHAPTER 17

Nutrition in Practice

▪ U.S. AND WORLD HUNGER ▪

Chapters 15 through 17 showed how to detect malnutrition and make plans for correcting nutrition problems. This nutrition in practice focuses on global and societal problems that can affect the availability of food and thus cause widespread malnutrition. It also describes strategies for tackling global malnutrition.

In the early 1990s, one person in every ten worldwide was experiencing hunger—not the healthy hunger we all feel, which leads us to sit down and eat a hearty meal, but chronic, painful hunger people feel when no food is available. Today, hundreds of millions of people are suffering from chronic hunger, both in the developing world and at home in the United States. Many, including children, are dying of starvation: tens of thousands die each day, one every two seconds.[1]

WWW

brown.edu/Departments/World_Hunger_Program
HungerWeb

bread.org
Bread for the World

oxfamamerica.org
Oxfam America

In the United States, an estimated 30 million Americans, including 12 million children, cannot afford to buy enough food to maintain good health.[2] Malnutrition and other health problems associated with chronic hunger—stunted growth, failure to thrive, low-birthweight babies, infant mortality, and anemia—are declining more slowly than in earlier decades; some problems are growing worse. The causes of hunger and malnutrition are many; this discussion focuses on three major causes—poverty, environmental degradation, and overpopulation.

I can understand how poverty affects food availability, but I find it hard to believe that 30 million people in the United States suffer from poverty and hunger. Who exactly is affected?

In the United States, poverty and hunger reach into all segments of society—not only the chronic poor (migrant workers, the unskilled and unemployed, the homeless, and some elderly) but also the so-called new poor. Some are displaced farm families. Some are former blue-collar and white-collar workers forced out of their trades and professions into minimum-wage jobs. These new poor, who outnumber the chronic poor, are not on welfare; they have jobs, but the pay is low. Families with incomes below a certain level are simply unable to buy sufficient amounts of nourishing foods, even if they are wise food shoppers.

Is anything being done to remedy the problem?

At present, many programs aimed at preventing or remediating malnutrition and hunger are in effect in the United States. To what extent federal programs will continue to feed those who are hungry is unknown, given the current political climate; many federal programs are being targeted in cost-saving measures.

Among the food assistance programs are programs for children such as the school lunch, breakfast, and child care food programs; programs to supply nourishing food to low-income pregnant women, mothers, and their young children; and food assistance programs for older adults such as congregate meals and Meals on Wheels. Another program aimed directly at the poor is the Food Stamp Program; the largest of the federal food assistance programs, both in dollars spent and in number of people participating, it is administered by the U.S. Department of Agriculture (USDA).

WWW

usda.gov/fcs/fs.htm
Food Stamp Program

Despite assistance programs, hunger continues to plague the United States. Of the estimated 2 million homeless people in the United States who are eligible for food assistance, only 15 percent of single adults and 50 percent of families receive food stamps.

To supplement federal programs, private efforts have sprung up in many communities, where concerned citizens work through local agencies and churches to help feed the hungry. Community-based soup kitchens and shelters generally provide good-quality meals, but

CHAPTER 17

Nutrition in Practice

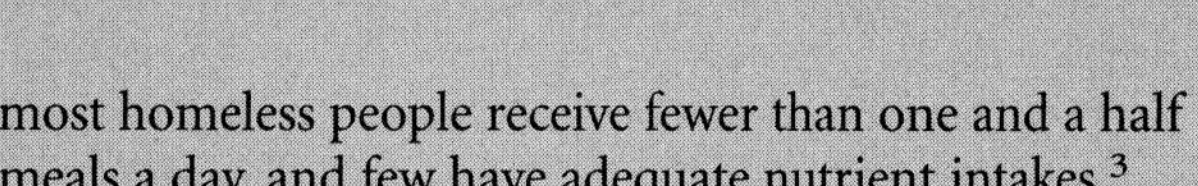

Feeding the hungry—in Sarajevo.

Feeding the hungry—in the United States.

most homeless people receive fewer than one and a half meals a day, and few have adequate nutrient intakes.[3]

How do hunger problems in developing countries compare to those in the United States?

Developing countries face more serious hunger problems than the United States, and the causes are more diverse. The primary cause of hunger is still poverty, but the poverty is more extreme. Most people find it almost impossible to comprehend the severity of poverty in the developing world. One-fifth of the world's 5 billion people have no land and no possessions *at all.* They survive on less than a dollar a day each, they lack water that is safe to drink, and they cannot read or write.[4] The average U.S. house cat eats twice as much protein every day as one of these people, and the cost of keeping the cat is greater than such a person's annual income.[5] Another important aspect of the hunger-poverty problem is its relationships to the environment.

How does the hunger problem relate to the environment?

Environmental degradation is beginning to threaten the world's ability to produce enough food to feed its people. Environmental problems include soil erosion, deforestation, water scarcity, air pollution, changes in climate, deteriorating rangelands, and diminishing fisheries.

Soil erosion is reducing agricultural productivity in every nation on earth. The resulting crop losses are estimated at 6 percent per year.[6] Another environmental problem, deforestation, leads to soil erosion, droughts, and floods. When areas of forests are cleared, water runs off the land instead of soaking into the ground; the result is soil erosion and flooding. The loss of trees, which normally transpire groundwater to the air and recycle rainwater, and the inability of the land to hold water contribute to droughts. Water supplies are threatened as they are increasingly used to irrigate crops.

Damage to crops from air pollution is now measurable in the car-centered societies of western Europe and the United States and in societies that burn coal to generate electricity—notably, eastern Europe and China. In the United States, the most damaging air pollutants are ground-level ozone, sulfur dioxide, and nitrous oxide, which come from the burning of fossil fuels. Crops are especially sensitive to ground-level ozone concentrations, which increasingly are being detected in rural as well as urban areas in amounts that reduce crop yields.

Not only is ground-level ozone pollution reducing agricultural outputs, but outer-atmosphere ozone depletion is doing so, too—especially to radiation-sensitive crops such as soybeans. For each 1 percent loss of outer-atmosphere ozone, the amount of damaging ultraviolet radiation reaching the earth increases by 2 percent. Based on studies of experimental plots, soybean yields fall 1 percent for each 1 percent rise in radiation. Soybeans are the world's leading protein crop, and at last report, no one was monitoring radiation-induced yield losses.

Crop yields may also be seriously affected by climate change caused by increased atmospheric concentrations of heat-trapping carbon dioxide, produced by fossil fuels. If hot summers become hotter, droughts during the growing season may become more common. In 1994, new heat records were set throughout the western United States, northern Europe, the Baltic, and Japan.[7] A rise of only a degree or so in average global temperature may reduce soil moisture, impair pollination of major staple food crops such as rice and corn, slow growth, weaken disease resistance, and disrupt many other factors affecting crop yields.

Environmental pollution and overuse also threaten the rangelands that feed the grazing animals that are eventually used to supply meats. Environmental pollution and overfishing are reducing the yield of fish from the ocean. Worse yet, environmental problems are disrupting food supplies at the same time that the world's population is increasing.

How is the increasing world population affecting food availability?

Population growth contributes to poverty and hunger, for the more mouths there are to feed, the worse the poverty and hunger become. Ironically, poverty and hunger lead to population growth. Lack of education regarding birth control and high childhood mortality rates contribute to this problem. People are willing to risk having fewer children only if they are sure that their children will live.

As the population grows, more forests are cut down and more agricultural lands are displaced by growing cities and industry. This shift forces people onto marginal land, where they cannot produce sufficient food for themselves. The sheer magnitude of the earth's annual population increase of 90 million people—or just over 10,000 people born every hour—is difficult to comprehend.[8] The rising population threatens the world's capacity to produce adequate food without degrading the environment further and threatening food supplies even more seriously than today.

What steps are being taken to tackle the world's hunger problem?

An important step was taken when the United Nations Convention of the Rights of the Child was ratified by more than a hundred nations. Significantly, for the first time in world history, the convention cited *nutrition* as an internationally recognized human right.[9]

To ensure this right, both rich and poor nations must contribute to solving the world's hunger, environmental, and poverty problems. Poor nations need to gain control of their rampaging population growth and to slow and reverse the destruction of their environmental resources: forests, waterways, and soil. To do this, they must, among other things, relieve their people's poverty. The rich nations need to stem their wasteful and polluting uses of resources and energy, which are contributing to global environmental degradation. They must also become willing to help relieve debtor nations of their poverty in ways that effectively reach the poor.

What role can the United States play in solving these problems?

The challenges for the United States are many. For one, we are being asked to reduce our consumption of fossil fuel and thereby reduce our disproportionate contribution to global environmental degradation. The willingness of U.S. consumers to take responsibility for their individual shares in solving global problems could make a substantial contribution to the quality of life for future generations.

Also, the developing countries need relief from the gigantic interest payments they have been making to U.S. and international banks. Debtor nations sell a bounty of cash crops, such as cotton, tobacco, coffee, sugar, and palm oil, to the developed world, but they cannot put the proceeds back into their economies. Instead, all the money goes to pay the interest on their loans. Furthermore, they must use the land for cash crops to pay back debts, rather than for food crops to feed their own people.

Debt relief must reach those who need it, and not just the wealthy. Relieving hunger requires land reform—returning sufficient land to the dispossessed so that they can live and grow food on it and become able to support themselves permanently.

The idea behind these measures is that relieving poverty will help relieve environmental degradation and hunger. To rephrase a well-known adage, if you give a man a fish, he will eat for a day. If you teach him to fish, he will eat for a lifetime. Unlike food giveaways and money doles, which are only stopgap measures, social programs that permanently better the lot of the poor can permanently solve the hunger problem.

What can individuals do to help?

All individuals can help by understanding the problems and helping others to understand. They can "talk it up," urging their friends and relatives in powerful positions to assist in the global effort to bring about a sustainable economy. They can support individuals, organizations, and laws that promote needed changes in environmental and economic policies.

Nutrition in Practice

Most importantly, individuals can try to make lifestyle choices that consider the environmental consequences. Several possible choices relating to typical U.S. foodways are presented in Table NP17.1. These suggested changes can easily be extended from food to other areas. All aspects of our lifestyles relate to global problems. Nutrition in Practice 11 recommended personal actions: reduce, reuse, recycle, and cut energy use for cars and homes. Admittedly, these approaches to solving today's global problems seem simplistic, but because we number 5 billion plus, individual actions can add up to exert an immense impact. As Margaret Mead said, "Never doubt that a small group of thoughtful, committed people can change the world. Indeed, it is the only thing that ever has."

Emphasis on personal lifestyle choices is important because it raises awareness and paves the way for larger actions. Individual choices are, however, only part of the solution to today's problems. Institutional changes are the other part—changes in the way agriculture, industry, and governments do business domestically and internationally. Consumers can be involved in promoting both kinds of changes: make personal lifestyle changes, and then vote for government changes and the officials who support them.

Finally, it makes sense for everyone in this world, rich or poor, in the United States or in any other country, to plan on bearing no more than one or two children. For those who want large families, there are plenty of children to adopt. And for those who love children and want to help them in other ways, there are numerous opportunities to play with, teach, and nurture the world's children, from the community center downtown to the remotest primitive village on the globe.

Table NP17.1 Environmentally Conscious Foodways

Food production taxes environmental resources and causes pollution. Consumers can make environmentally conscious choices at every step from food shopping to cooking and use of kitchen appliances to serving, cleanup, and waste disposal.

Food Shopping

Transportation:

- Whenever possible, walk or ride a bicycle; use car pools and mass transit.
- Shop only once a week, share trips, or take turns shopping for each other.
- When buying a car, choose an energy-efficient one.

Food choices:

- Eat low on the food chain; that is, eat plants, rather than animals that eat plants (this suggestion complements the Food Guide Pyramid recommendations for eating for good health).
- Avoid buying canned beef products (many of these foods come at the expense of cleared rainforest land).
- Eat small portions of meat; select range-fed beef, buffalo, poultry, and fish.
- Select local foods (they are transported shorter distances and less fuel is required to pack them, label them, and keep them cold if they are fresh).

Food packages:

- Whenever possible, select foods with no packages; next best are minimal, reusable, or recyclable ones.
- Buy juices and sodas in large glass or recyclable plastic bottles (not small individual cans or cartons); grains in bulk (not separate little packages); and eggs in pressed fiber cartons (not foam, unless it is recycled locally).
- Carry reusable shopping bags; alternatively, ask for plastic bags if they are recyclable.

Cooking Food

- Cook foods quickly in a pressure cooker or microwave oven.
- When using the oven, bake a lot of food at one ti me and keep the door closed tightly.
- Refuse throwaway utensils.
- Avoid spray products.

Kitchen Appliances

- Do without small electrical appliances such as can openers, mixers, knife sharpeners, and food processors.
- When buying a refrigerator, choose an energy-efficient one.
- Consider the possibility of using solar energy to meet home electrical needs.
- Set the water heater at 130°F (54°C), no hotter; put it on a timer; wrap it and the hot-water pipes in insulation; install water-saving faucets.

Food Serving, Dish Washing, and Waste Disposal

- Use "real" plates, cups, and glasses instead of disposable ones.
- Use cloth towels and napkins, reusable storage containers with lids, and dishcloths instead of paper towels, plastic wrap, plastic storage bags, and sponges.
- Run the dishwasher only when it is full.
- Recycle all glass, plastic, and aluminum.
- Compost all vegetable scraps, fruit peelings, and leftover plant foods.

CHAPTER 17

Nutrition in Practice

"Be part of the solution, not part of the problem," an old adage says. In other words, don't waste time or energy moaning and groaning about how rough things are; do something to improve them. This adage is applicable to today's global problems. They are our problems: human beings created them, and human beings must solve them.

Notes

1. L. N. Burby, *World Hunger* (San Diego, Calif.: Lucent Books, 1995), pp. 13– 16; P. L. Kutzner, *World Hunger: A Reference Handbook* (Santa Barbara, Calif.: ABC-CL10, 1991), pp. 158–159.

2. *Tallahassee Democrat,* October 14, 1994; P. Univ, The state of world hunger, *Nutrition Reviews* 52 (1994): 151–161.

3. J. C. Wolgemuth and coauthors, Wasting malnutrition and inadequate nutrient intakes identified in a multiethnic homeless population, *Journal of the American Dietetic Association* 92 (1992): 834–839; M. A. Drake, The nutritional status and dietary adequacy of single homeless women and their children in shelters, *Public Health Reports* 107 (1992): 312–319; B. E. Cohen, N. Chapman, and M. R. Burt, Food sources and intake of homeless persons, *Journal of Nutrition Education* (1 supplement) 24 (1992): 45–51.

4. World Bank, *World Development Report 1991* (New York: W. W. Norton, 1992), pp. 3–8.

5. L. Timberlake, *Only One Earth,* cited in Food for thought, *Seeds,* Sprouts edition, 1988.

6. L. R. Brown and J. E. Young, Feeding the world in the nineties, in L. R. Brown, *State of the World 1990* (New York: W. W. Norton, 1990).

7. L. R. Brown and coauthors, *State of the World 1995* (New York: W. W. Norton, 1995), p. 191.

8. Centers for Disease Control, Population based mortality assessment: Baidoa and Afgoi, Somalia, 1992, *Journal of the American Medical Association* (1993), as cited in L. R. Brown and H. Kane, *Full House* (New York: W. W. Norton, 1994), pp. 49–61.

9. S. Lewis, Food security, environment, poverty, and the world's children, *Journal of Nutrition Education* (1 supplement) 24 (1992): 3–5.

Consistency-Modified, Soft, and Bland Diets for Upper GI Tract Disorders

As Chapters 12 through 14 demonstrate, nutrient needs fluctuate in response to a multitude of physical and physiological changes that occur throughout life. Illnesses also alter nutrient needs. The remaining chapters of this book explore the many ways that modified diets assist in treating illnesses and their symptoms.

Just as the assimilation of nutrients begins with physical processes—chewing, swallowing, and the passage of food through the GI tract—your study of therapeutic nutrition begins by discussing foods and diets that affect these physical processes. This chapter describes the consistency-modified, soft, and bland diets that help treat conditions that primarily affect the upper GI tract. The next chapter describes diets used to treat lower GI tract disorders. Table 18.1 lists conditions for which consistency-modified, soft, and bland diets are indicated.

Consistency-Modified and Soft Diets

Consistency-modified diets provide foods that are modified in texture. Such foods may be liquid, pureed, chopped, tender-cooked, or simply whole foods like bananas and yogurt that are easy to eat. Soft diets also provide soft-textured foods, while omitting foods that are likely to cause indigestion.

Consistency-modified and soft diets aid in the treatment of conditions that affect the upper GI tract. This makes sense, because the consistency of a food and

Table 18.1 Indications for Consistency-Modified, Soft, and Bland Diets

Liquid Diet (Temporary)		
Following surgery Following complete bowel rest	Diarrhea Following a myocardial infarction (heart attack)	Refeeding in protein-energy malnutrition
Mechanical Soft Diet		
Severe dental caries Missing or no teeth Ill-fitting dentures Dry mouth	Periodontal disease Ulcers of the mouth or gums Surgery of the mouth, head, or neck	Broken jaw Dysphagia Stroke HIV infection
Soft Diet		
Temporary diet (see *Liquid Diet*) Indigestion Nausea	Esophageal varices Gastritis Diarrhea	Myocardial infarction Congestive heart failure
Bland Diet		
Ulcers of the mouth or gums Nausea Reflux esophagitis	Hiatal hernia Gastritis Peptic ulcers	Myocardial infarction Congestive heart failure

its effects on the stomach make a difference only in the upper GI tract. Once nutrients reach a functional intestine, they can be digested and absorbed, regardless of which foods originally provided them.

Liquid Diets

Two liquid diets—clear liquid and full liquid—are routine hospital diets. Liquids are often the first foods offered following surgery or when a person who has not been eating foods orally for any period of time begins to eat again.

Clear-Liquid Diets Clear-liquid diets consist of foods that are liquid and transparent at body temperature, such as gelatin, tea, and broth. Clear liquids provide minimal stimulation to the GI tract, and they are easily and almost completely absorbed. Table 18.2 lists the foods allowed on clear-liquid diets, and the sample menu on p. 440 shows an example of a day's meals.

clear liquids: foods that are liquid and transparent at body temperature.

Standard clear-liquid diets are temporary diets; they are deficient in kcalories and essential nutrients. Clear-liquid diets help determine if the digestive system is working well enough to handle more complex foods. Once the person tolerates clear liquids, the diet progresses to full liquids, then on to soft foods, and then regular foods. In addition to their use in progressive diets, clear-liquid diets may be used prior to surgery or diagnostic tests of the intestine, or when a person is experiencing gastrointestinal disturbances, such as vomiting or diarrhea.

progressive diet: a diet that progresses as a client's tolerances permit; such diets often progress from clear liquids to full liquids to soft foods to regular foods.

Full-Liquid Diets A full-liquid diet includes both clear and opaque liquid foods and semiliquid foods (see Table 18.2 and the sample menu on p. 440). Full-liquid diets may be used as the second diet as a person progresses from clear liquids to regular foods, or for people who are unable to chew or swallow regular foods for medical reasons or because they are too ill to eat. Like clear-liquid diets, unsupplemented, standard full-liquid diets are deficient in many nutrients and are temporary diets.

full liquids: foods that are liquid or semiliquid at body temperature.

Table 18.2 Foods Included on Liquid Diets

Clear-Liquid Diets	Full-Liquid Diets
Bouillon	All clear liquids
Broth, clear	Butter
Carbonated beverages	Commercially prepared liquid formulas (all)
Coffee, regular and decaffeinated	Cooked cereals, strained
Commercially prepared clear-liquid formulas	Cream
Fruit drinks	Custard
Fruit ices	Ice cream, plain
Fruit juices, strained	Instant breakfast drinks
Gelatin	Margarine
Hard candy	Milk, all types
Honey	Pudding
Lemonade	Sherbet
Popsicles	Soups, strained vegetable, meat, or cream
Salt	Sour cream
Salt substitutes	Vegetable juices, strained
Sugar	Vegetable purees, diluted in cream soups
Sugar substitutes	Yogurt
Tea, regular and decaffeinated	

Sample Clear-Liquid Diet Menu

Breakfast	Lunch	Supper	Between Meals
Strained orange juice	Bouillon	Bouillon	Soft drinks
Flavored gelatin	Apple juice	Cranberry juice	Gelatin
Ginger ale	Flavored gelatin	Fruit ice	Fruit juices
Coffee or tea	Coffee or tea	Flavored gelatin	
Sugar	Sugar	Coffee or tea	
		Sugar	

Cautious Use of Liquid Diets Liquids are offered in small amounts at first to make sure the person can tolerate them. Many clients willingly accept liquid diets because they are too ill to eat. When left on the diet for any length of time, however, people may understandably find these foods unappetizing and boring. Even more disturbing, liquid diets are deficient in energy and most nutrients, especially in relation to the high energy and nutrient needs often imposed by illness. To meet the energy and nutrient needs of people who must stay on liquid diets for more than a few days, health care professionals use formulas. (Chapter 21 describes formula diets in detail.) Formulas are of a known nutrient composition, they can be given in quantities that meet all nutrient needs, and they can be selected to meet a variety of medical needs and to reduce the likelihood of GI problems. Almost all formulas are lactose-free, which accommodates the loss of lactose tolerance often seen following GI tract disuse or in GI dysfunction.

Solid Foods and Diet Progression

Once a person is tolerating liquids, the next step is solid foods. The exact diet used as the first solid diet depends on the client's medical condition. Some clients begin receiving regular foods, but sometimes clients better tolerate the shift from liquid diets to solid diets if they first receive soft diets.

Soft Diets Soft diets provide soft but solid foods (such as tender meats and soft fresh fruit) that are lightly seasoned and moderate in fiber. Generally, soft foods

Sample Full-Liquid Diet Menu

Breakfast	Lunch	Supper	Between Meals
Orange juice	Apricot nectar	Apple juice	Milk shakes (made with plain ice cream), ice cream, eggnog, pudding, custard, or gelatin
Strained oatmeal	Yogurt, plain	Creamed soup	
Milk	Pudding	Custard	
Sugar	Milk	Milk	
Margarine	Coffee or tea	Coffee or tea	
	Sugar	Sugar	

Table 18.3 Foods Included on Soft Diets
Meat and Meat Alternates
Tender, moist meats, fish, or poultry; mild cheeses, creamy peanut butter, and eggs
Milk and Milk Products
Milk, milk products, yogurt without seeds or nuts
Fruits and Vegetables
Cooked, canned, or soft fresh fruits such as melons; fruit juice; soft-cooked vegetables except those likely to produce gas (see Table 19.4 in Chapter 19)
Grains
Refined white or light rye bread, rolls, or crackers; cooked or ready-to-eat cereals without nuts or seeds
Miscellaneous
Mild condiments; salt; sugar; mildly seasoned broths or soups; all nonalcoholic beverages; desserts without nuts, seeds, or coconut

are offered in frequent small meals. As Table 18.3 shows, soft diets include foods that are easy to chew, digest, and absorb. Soft diets limit high-fiber foods, nuts, foods with seeds, and coconut. By limiting highly seasoned foods that might irritate the GI tract and high-fiber foods that are more likely to produce gas, soft diets minimize the risk of indigestion, nausea, distension, cramping, and other GI upsets. In addition to their use in a progressive diet, soft diets also benefit people with temporary indigestion or nausea. Clients tolerating soft diets are then progressed to regular diets that provide the full spectrum of nutrients and include all foods.

Progressive Diets As previously described, progressive diets are used to slowly reintroduce foods after periods of GI tract disuse or dysfunction. Such diets can be adapted to include other dietary modifications such as sodium and lactose restrictions. The person who needs a progressive low-sodium diet, for example, might first receive low-sodium liquids; next, low-sodium, soft foods; and, finally, a regular low-sodium diet.

Advancing the Diet The diet order for a progressive diet may simply say "progress diet from clear liquids to full liquids to soft foods to regular foods as tolerated." It is up to the nurse or dietitian to keep track of the client's tolerance and readiness to advance the diet. Indigestion, nausea, vomiting, diarrhea, cramping, or other GI upsets indicate intolerance. At each progressive step, a client may be intolerant to particular foods (orange juice or milk, for example), rather than to the diet itself. In such a case, the offending food is withheld temporarily.

Sometimes the physician inadvertently forgets to advance the diet; the alert health care professional will then make the recommendation. Other times a person truly cannot tolerate enough food to maintain nutrition status. In those cases, liquid formulas, tube feedings, or intravenous nutrition may be necessary. The following section describes the use of progressive diets for people undergoing surgery.

Surgery

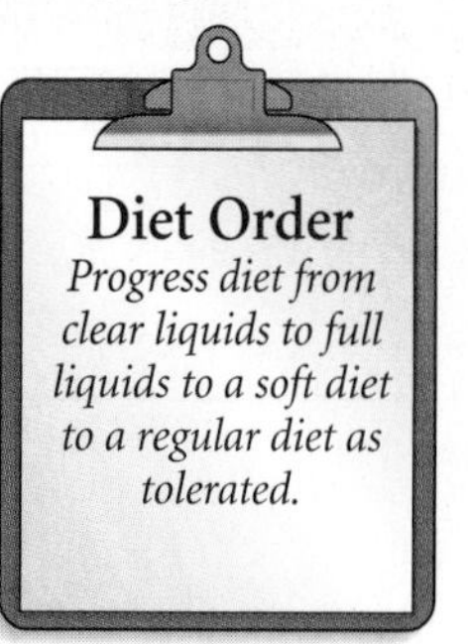

Diet Order
Progress diet from clear liquids to full liquids to a soft diet to a regular diet as tolerated.

Ideally, all clients enter surgery at the appropriate weight with full nutrient reserves. Such optimal status may be difficult to achieve, however. The illness that necessitates the surgery, the medications used in its treatment, or the psychological stress associated with it may interfere with a person's food intake or with the body's use of nutrients. Additionally, the person may have to fast or follow a nutritionally inadequate liquid diet before undergoing diagnostic and laboratory tests, further taxing nutrient status. The health care professional who appreciates the vital roles nutrition plays in resistance and recovery will watch for signs of nutrition risk and encourage clients to eat as well as possible, when possible.

This section describes the nutrient needs of well-nourished clients who undergo uncomplicated surgery. Malnourished clients or those who undergo extensive surgery are severely stressed and need the kcalorie-modified, high-protein diets described in Chapter 23.

aspiration: the flux of food or liquid into the lungs.

Immediate Presurgery Diet Physicians generally order all foods and fluids withheld for at least eight hours before surgery. This helps to prevent regurgitation and aspiration, which can occur during anesthesia or recovery. In the special case of GI surgery, people receive liquid or very-low-residue diets (see Chapter 19) for two or three days before surgery.

Immediate Postsurgery Support In the immediate postsurgical period, the most important nutrition-related task for the medical team is to maintain fluid and electrolyte balances. Clients lose blood, fluid, and electrolytes during surgery, and they may lose additional fluids thereafter from fever, draining wounds, vomiting, and diarrhea. Excessive losses without replacement can lead to dehydration and shock.

Immediately following surgery, gastric motility is often slowed, and the person is given nothing to eat or drink orally. Clients receive intravenous infusions (see Chapter 22) to maintain fluid and electrolyte balances. Physicians determine fluid and electrolyte needs based on the person's blood pressure, pulse rate, urinary output, level of consciousness, breathing patterns, body temperature, and laboratory test results.

Postsurgical Diets When GI tract activity resumes, the postoperative client is ready to begin a progressive diet. Liquids are offered first to see if the client is able to tolerate oral feedings. A well-nourished person often tolerates solid foods and is ready for a regular hospital diet within a short period of time. Clients undergoing certain gastrointestinal surgeries may be advanced to low-fiber or low-residue diets (see Chapter 19) rather than soft or regular diets. Clients undergoing surgery of the mouth or esophagus may advance from liquid diets to the mechanical soft diets described next.

Mechanical Soft Diets

mechanical soft diet: a diet that excludes all foods that are difficult to chew or swallow; also called a **dental soft diet.**

Another type of consistency-modified diet is the mechanical soft diet. Mechanical soft diets provide foods modified in texture to make them easy to chew and swallow. Unlike the soft diets described above, mechanical soft diets include all foods and seasonings. Individual tolerances determine if foods should be provided in liquid, pureed, chopped, or tender-cooked form. Foods that are naturally soft are provided as tolerated.

Table 18.4 Conditions That May Interfere with Chewing and Swallowing

Achalasia	Ill-fitting dentures
Alzheimer's disease	Missing teeth
Broken jaw	Multiple sclerosis
Cancer	Myasthenia gravis
Chemotherapy	Parkinson's disease
Congenital defects of upper GI tract	Periodontal disease
Dental caries	Radiation therapy of the head and neck
Dryness of mouth	Sensitivity of mouth to hot or cold
Dysphagia	Strokes
Guillain-Barré syndrome	Surgery of the mouth, head, or neck
Head injury	Ulcers of mouth, gums, or esophagus
HIV Infection	

Many conditions can temporarily or permanently interfere with chewing or swallowing (see Table 18.4). Although it is possible to meet all nutrient needs using a mechanical soft diet, people with long-term conditions that necessitate the use of liquid and pureed foods only may become bored with the diet and eat too little, lose too much weight, and suffer the consequences of deteriorating nutrition status.

Individualizing the Diet People's tolerances of food consistencies vary greatly. Following surgery to repair a broken jaw, for example, a person may be able to consume liquids only. With time, however, the person may progress to a pureed diet, then to a diet of chopped or ground foods, and then to a regular diet.

In another case, a person without teeth may need to follow a mechanical soft diet permanently. Such a person may be able to eat soft natural foods such as fish or chopped or ground meats. Moist, soft-textured foods, such as casseroles and foods prepared with sauces and gravies, often work well for clients who can handle some solid foods. Drinking liquids along with meals makes it easier to chew and swallow.

A person without teeth is **edentulous** (ee-DENT-you-lus).

Dietitians work carefully with clients and caregivers to determine which foods a client can handle. The goal is to provide a wide variety of foods that are as similar as possible to those of a regular diet. Therefore, soft natural foods and chopped foods are provided whenever possible; only when the person cannot chew or swallow adequate amounts of these foods are pureed foods provided. Such a strategy enhances appetite and minimizes the likelihood of nutrient deficiencies.

Diet Adjustments for Mouth Ulcers The mechanical soft diet for people with mouth ulcers provides moist, soft-textured foods and eliminates spicy, salty, or acidic foods (such as citric fruits and tomatoes) that may be painful to eat. In addition, nuts or seeds in foods (such as sesame or poppy seeds in breads) can become trapped in mouth ulcers and cause discomfort. Heat may intensify pain, too. Clients with mouth ulcers often prefer cold foods and beverages.

mouth ulcers: lesions or sores in the lining of the mouth. Certain drugs, radiation therapy, and some disorders, such as oral herpes virus infections, can cause mouth ulcers.

Diet Adjustment for Mouth Dryness People with mouth dryness due to a reduced flow of saliva often tolerate moist, soft foods. Clients can moisten foods with sauces and gravies. Salty foods and snacks dry the mouth and should be avoided. Encourage these clients to practice good oral hygiene; when salivary flow is reduced, the mouth is poorly defended against dental caries. Clients can increase salivary secretions by sucking on sugarless candy, sugarless chewing gum, or lemon or lime ices or by using drugs that stimulate the flow of saliva.

Pureed Foods For clients who have chewing and swallowing problems and cannot handle any solid foods, pureed diets provide foods that are near liquid in consistency, yet offer a wider variety of foods than full-liquid diets. Pureed foods can meet all nutrient needs, but for clients who can tolerate only pureed foods, the monotony of eating food of the same consistency for long periods of time can create a psychological block to eating, and malnutrition can follow. Foodservice departments in nursing homes, rehabilitation centers, and other long-term care facilities that must provide many clients with pureed foods over long periods of time face a difficult challenge. Fortunately, a variety of commercial products are available to thicken and shape pureed foods to give them an appetizing appearance. Clinicians should also be mindful that other factors may interfere with eating for clients in long-term care facilities. Such clients lack the comforts of home and may be overwhelmed by their medical conditions, isolation, and loss of control. Addressing the client's emotional needs is critical to stimulating the appetite. The box on p. 445 offers suggestions for improving the acceptance of pureed foods.

Difficulties Swallowing

dysphagia (dis-FAY-gee-ah): difficulty in swallowing.
dys = bad
phagein = to eat

Dysphagia that occurs when the cardiac sphincter fails to relax and allow swallowed food to enter the stomach is known as **achalasia** (ACK-ah-LA-zee-ah).

Diet Order
Liquids and semisolid foods as tolerated.

Problems in swallowing, known as dysphagia, can arise from many causes, including aging, nervous system diseases, injuries, developmental disabilities, and strokes. A person with dysphagia may be unable to initiate swallowing, to chew foods and mix them with saliva, or to push foods to the back of the throat and into the esophagus. Some forms of dysphagia specifically interfere with the muscular contractions of the esophagus.

Subtle and Dangerous Dysphagia sometimes goes undiagnosed, especially when caused by aging, because symptoms are not always obvious. All people "catch food in their throat" at one time or another, so a person who gradually does this more and more frequently may consider the condition normal and ignore it. Eventually, the person may eat less and develop nutrient deficiencies.

Dysphagia can be dangerous when it goes undetected. Unlike healthy people, people with dysphagia may fail to cough in response to the presence of food in the trachea. Food particles may then pass unnoticed into the lungs (aspiration), allowing bacteria to multiply. Such "silent" aspiration, which has been observed in people following strokes, carries with it the risk of serious pneumonia and death.

Signs of Dysphagia Health care professionals should be alert to subtle symptoms of dysphagia including an unexplained decline in food intake or repeated bouts of pneumonia. Other symptoms include pain on swallowing, weight loss, a fear of eating certain foods or any food at all, a feeling that food is sticking in the throat, a tendency to hold food in the mouth rather than swallowing it, coughing or choking during meals, frequent throat clearing, drooling, or a change in voice quality. Depending on the cause of dysphagia, the voice may be hoarse or nasal or have a "wet" sound. Diagnosis is based on extensive testing that may include cranial nerve assessment, X rays, fluoroscopy, and measurements of esophageal sphincter pressure and esophageal peristalsis.

WWW **dysphagia.com**
Dysphagia Resource Center

Diet for Dysphagia The mechanical soft diet for dysphagia leaves little room for error or experimentation. Speech pathologists, dietitians, physicians, and nurses

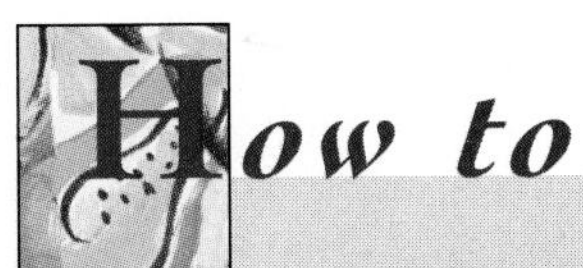

How to

Improve Acceptance of Pureed Foods

Take a moment to think about a meal of pureed foods. A typical dinner of baked chicken, mashed potatoes, and green beans is blenderized to white mush, more white mush, and a green blob. The foods may taste great, but on seeing the plate, the person may be reluctant to try the first bite. To stimulate the appetite, use creative techniques such as these for preparing and serving food:

- Encourage clients and their caregivers to prepare a variety of favorite foods and blenderize them to a tolerable consistency. The smell of favorite foods and the thought of consuming them stimulate the appetite.
- Consider color when planning meals. The meal of baked chicken, mashed potatoes, and green beans, described above, can be made more appealing by substituting mashed sweet potatoes for the white potatoes. Arranging foods attractively on a plate with appropriate garnishes also adds color and eye appeal.
- Serve foods at the right temperature and puree them so that they are smooth and thick—not watery and thin. Commercially available thickeners add shape, texture, and even nutrients to food.
- Experiment with seasonings and spices to enliven food flavors, excluding only those that the person cannot tolerate for personal or medical reasons. Seasoning food to accommodate personal tastes adds flavor, allows for variety, and improves the appetite.
- Supplement the diet with nutritious liquids such as milk, instant breakfast drinks, or liquid formulas (see Chapter 21).

Efforts to improve the acceptance of pureed foods can go a long way toward helping people to eat and maintain or improve their weights.[a] When efforts to improve a client's intake of pureed foods are unsuccessful, feeding the person by tube becomes an option (see Chapter 21).

[a] D. Cassens, E. Johnson, and S. Keelan, Enhancing taste, texture, appearance, and presentation of pureed food improved resident quality of life and weight status, *Nutrition Reviews* (supplement) 11 (1996): 51–54.

work together to assess the person's swallowing abilities and design an individualized diet. Often the person can handle only semisolid foods or thickened liquids, which flow slowly enough to allow time to coordinate swallowing movements. Smooth solids such as puddings and smooth yogurts are good choices; commercial thickeners, pudding, yogurt, or baby cereal can be used to thicken liquids. The suggestions in the box apply to diets for dysphagia.

With time, swallowing function may improve. The health care team continuously monitors the person and expands the diet to include additional food as tolerated. Ideally, the diet is progressed to a solid diet, although this is not always possible.

Thickeners enhance the eye appeal of pureed foods and may ease the task of swallowing for people with dysphagia.

Tube Feedings Feedings by tube (discussed in Chapter 21) are beneficial for malnourished individuals with dysphagia who are unable to take adequate nourishment orally and those whose swallowing function continues to deteriorate. Tube feedings delivered into the stomach, however, may be contraindicated due to the high risk of aspiration pneumonia in some people with dysphagia. Tube feedings delivered into the intestine often provide a safer alternative.

Consistency-modified diets test the person's ability to handle foods and thus ease the transition from GI tract disuse or dysfunction to recovery. Mechanical soft diets provide foods the person can chew and swallow without difficulty.

Bland Diets

bland diet: a diet that eliminates foods known to stimulate gastric acid secretion, irritate the gastric mucosa, or cause indigestion.

Bland diets eliminate foods that stimulate gastric acid secretions or those that irritate the gastric mucosa. Table 18.5 shows the recommendations that often serve as a starting point for a bland diet. Health care professionals may begin with these suggestions and then individualize the diet based on each person's tolerances. Practitioners frequently prescribe bland diets for reflux esophagitis, gastritis, and ulcers.

WWW.

gerd.com
GERD (Gastroesophageal Reflux Disease) Information Resource Center

gutfeelings.com
Gut Feelings (from South Denver Gastroenterology)

acg.gi.org
American College of Gastroenterology

Alcohol, Caffeine, Coffee, and Tea Bland diets include the recommendation to limit alcohol, caffeine, and all coffee and tea (even decaffeinated). Alcohol directly damages the gastric lining. Caffeine stimulates gastric acid secretions. Research has shown that decaffeinated coffee and tea stimulate gastric acid secretion to the same degree as caffeinated coffee and tea; therefore, all coffee and tea are restricted.

Spices Spices may cause discomfort for some people, but spicy foods do not damage the stomach lining. Spices frequently reported to cause discomfort include black pepper, red pepper, and chili powder.

Reflux Esophagitis

reflux esophagitis (eh-sof-ah-JYE-tis): the backflow or regurgitation of gastric contents from the stomach into the esophagus, causing inflammation of the esophagus; also called **gastroesophageal reflux** or **acid indigestion.**
re = back
fluxus = flow

heartburn: a burning pain felt behind the sternum caused by stomach acids irritating the esophagus.

esophageal stricture: narrowing of the inner diameter of the esophagus from scar tissue.

esophageal ulcers: lesions or sores in the lining of the esophagus.

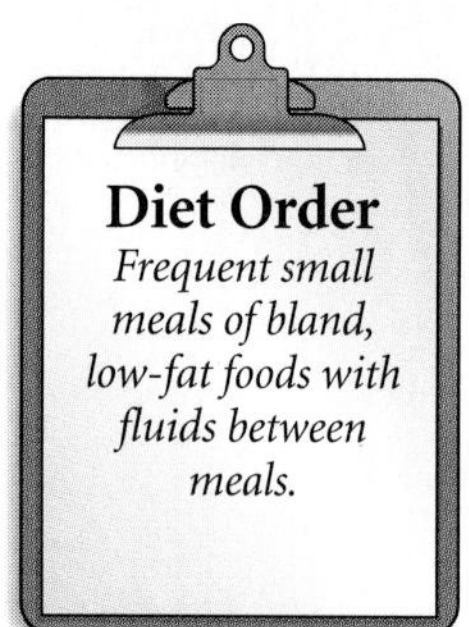

Diet Order
Frequent small meals of bland, low-fat foods with fluids between meals.

In reflux esophagitis, the cardiac sphincter fails to close tightly, allowing the highly acidic fluids from the stomach to splash backward into the esophagus. Stomach acids irritate the lining of the esophagus and cause the pain called heartburn. Although most people experience the discomfort of "acid indigestion" from time to time, when the problem becomes chronic, the esophagus develops a painful inflammation called reflux esophagitis. Severe inflammation and scarring may narrow the inner diameter of the esophagus. Dysphagia and its potential complications, esophageal ulcers, and bleeding are possible consequences.

Certain drugs, aging, and the use of feeding tubes that pass from the nose through the cardiac sphincter are associated with an increased risk of reflux esophagitis. Most often, however, reflux esophagitis develops as a consequence of a hiatal hernia.

Hiatal Hernia The esophagus joins the stomach at the cardiac sphincter. Normally, this sphincter sits right in the hiatus in the diaphragm and is reinforced by it. The esophagus lies completely above, and the stomach completely

Table 18.5 The Bland Diet

A bland diet includes all foods except:

- Those an individual identifies as irritating to the GI tract.
- Alcohol.
- Caffeine and caffeine-containing beverages (including cola beverages, cocoa, coffee, and tea).
- Decaffeinated coffee and tea.
- Pepper and spicy foods except as tolerated.

Note: The liberal bland diet shown here has replaced the earlier "traditional bland diet," which was invalidated by research.

Table 18.6 Substances That Relax the Cardiac Sphincter

Alcohol	Cigarette smoking	Onions
Anticholinergic agents	Diazepam	Peppermint and spearmint oils
Calcium channel blockers	Garlic	Theophylline
Chocolate	High-fat foods	
	Meperidine	

below, the diaphragm. Sometimes, however, the diaphragm weakens, and a portion of the stomach slides up through it, an abnormality called a hiatal hernia. Once a hiatal hernia forms, the cardiac sphincter, which holds foods in the stomach, no longer is reinforced by the surrounding diaphragm, and it becomes easy for acidic gastric juices to reflux into the esophagus. Because reflux occurs more readily in people with hiatal hernias, they frequently suffer from heartburn and esophagitis. Figure 18.1 shows the normal relationship of the upper GI tract to the diaphragm and the changes that occur with reflux and with a hernia.

hiatal hernia: a protrusion of a portion of the stomach through the esophageal hiatus of the diaphragm.

esophageal hiatus (high-AY-tus): the opening in the diaphragm through which the esophagus passes.
hiatus = to yawn

Dietary Prevention and Treatment Treatment for active reflux esophagitis aims to prevent reflux and reduce inflammation of the esophagus and its associ-

Figure 18.1
The Upper GI Tract and Acid Reflux

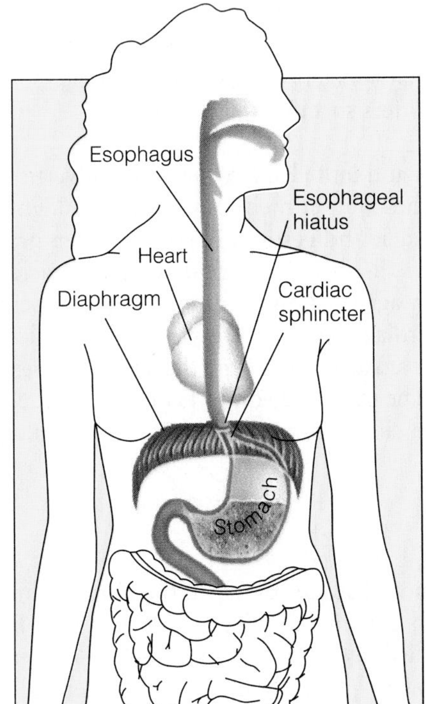

The stomach normally lies below the diaphragm, and the esophagus passes through the esophageal hiatus. The cardiac sphincter prevents reflux of stomach contents.

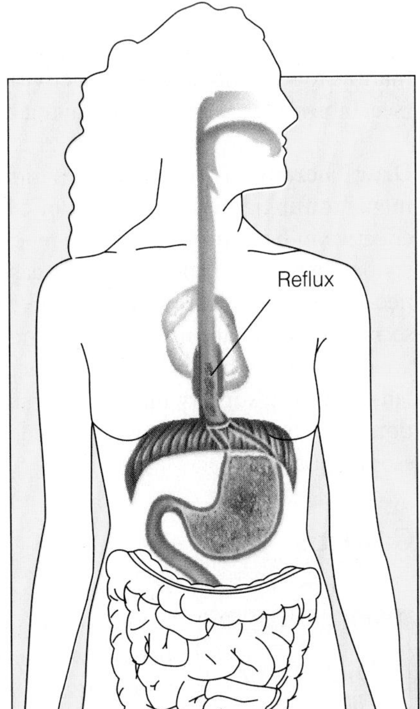

Whenever the pressure in the stomach exceeds the pressure in the esophagus, as can occur with overeating and overdrinking, the chance of reflux increases. The resulting "heartburn" is so-named because it is felt in the area of the heart.

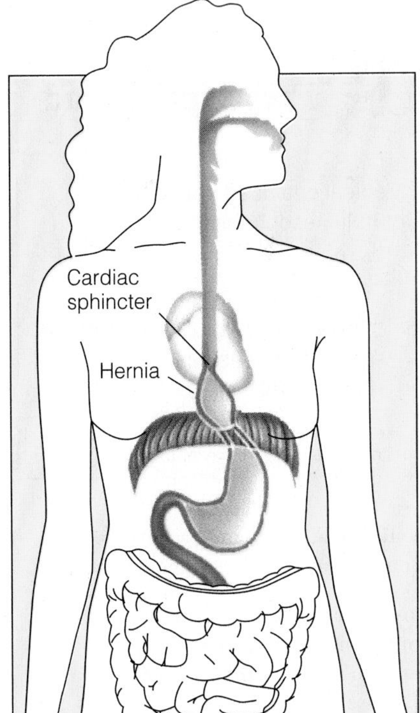

Acid reflux often occurs as a consequence of a hiatal hernia. The most common type, a sliding hiatal hernia, results when part of the stomach, with the cardiac sphincter, slip through the diaphragm.

How to

Prevent and Treat Reflux Esophagitis

To prevent and treat reflux esophagitis and its associated discomfort, recommend that clients:

- Eat small meals and drink liquids one hour before or one hour after meals to avoid distending the stomach. A distended stomach exerts pressure on the cardiac sphincter.
- Relax during mealtimes, eat foods slowly, and chew them thoroughly to avoid swallowing air and distending the stomach.
- Limit foods that weaken cardiac sphincter pressure or increase gastric acid secretion, including fat, alcohol, caffeine, decaffeinated coffee and tea, chocolate, spearmint, and peppermint.
- Avoid foods and beverages that irritate the esophagus, such as citrus fruits and juices, tomatoes and tomato-based products, pepper, spices, and very hot or very cold foods, according to individual tolerances. If a client cannot tolerate citrus fruits and juices, ensure that the diet provides adequate vitamin C from other foods or supplements.
- Lose weight, if necessary. Overweight tends to increase abdominal pressure.
- Refrain from lying down, bending over, and wearing tight-fitting clothing or belts, particularly after eating, to avoid increasing pressure in the stomach.
- Elevate the head of the bed by 4 to 6 inches. Keeping the chest higher than the stomach helps prevent reflux.
- Refrain from smoking cigarettes. Smoking relaxes the cardiac sphincter.

Individual tolerances of types and amounts of foods and spices vary markedly especially during periods of active esophagitis. Health care professionals can help clients pinpoint food intolerances by advising them to keep a record of the types and amounts of foods and beverages consumed, the time of consumption, GI symptoms, and time of occurrence. Assessment of the record by a dietitian or health care professional provides the basis for determining the types and amounts of food and food components that the client can handle without discomfort.

ated pain by reducing gastric acidity, reducing pressure in the stomach, and eliminating foods, substances, or activities that weaken cardiac sphincter pressure (see Table 18.6). The accompanying box offers specific suggestions.

Drug Therapy Turn on the television set and quite likely you will see advertisements touting a new generation of over-the-counter medications that are highly effective in both preventing and relieving acid indigestion. Physicians often prescribe these and other drugs to ease the discomfort of acid reflux. Antacids neutralize gastric acidity, and antisecretory agents suppress or inhibit gastric acid secretion. Medications that strengthen cardiac sphincter pressure (cholinergics) are prescribed when other measures prove unsuccessful. If medical management fails, however, surgery may be indicated. The case study on p. 449 provides questions that review the nutrition needs of a client with a hiatal hernia and reflux esophagitis.

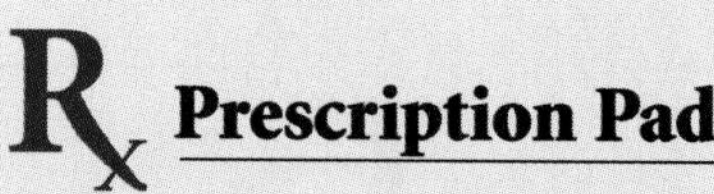

Prescription Pad

Medications used in the treatment of reflux esophagitis may include:

- Antacids
- Antisecretory agents
- Cholinergics (metoclopramide and bethanecol, see *anti-nauseants)*

See Appendix E for timing with meals and nutrition-related side effects.

Gastritis

Diet Order
Bland diet as tolerated.

Gastritis is a common disorder in which the mucosal lining of the stomach becomes inflamed and painful. The person with gastritis may complain of anorexia, nausea, vomiting, and epigastric pain. Although gastritis generally resolves with treatment, unresolved gastritis can lead to ulcers (described in the next section), hemorrhage, shock, obstruction, perforation, and gastric cancer.

Acute Gastritis Acute gastritis most often follows the repeated use of aspirin or other medications that irritate

antacids: acid-buffering agents used to counter excess acidity in the stomach.

antisecretory agents: drugs that suppress or inhibit gastric acid secretion; also called **anti-GERD (GastroEsophageal Reflux Disease).**

gastritis: inflammation of the stomach lining.

Case Study

ACCOUNTANT WITH REFLUX ESOPHAGITIS

Mrs. Scarlatti, a 49-year-old accountant, recently underwent a complete physical examination. She told her physician that she had been feeling fairly well until she began experiencing heartburn, which has become more frequent and more painful. The heartburn usually occurs after she has eaten a large meal, particularly when she lies down after eating. After inspecting the interior of her esophagus, stomach, and duodenum with a long tube equipped with an optical device called a gastroscope, the physician diagnosed a sliding hiatal hernia and reflux esophagitis.

Mrs. Scarlatti's past health history shows no indication of significant health problems. During her last physical, the physician did advise her to stop smoking cigarettes and to lose 20 pounds, which she has yet to do. During a diet history taken by the nurse, Mrs. Scarlatti mentioned that her life is pretty hectic because it is the middle of the tax season. She usually does not have time for breakfast, eats lunch hurriedly while continuing to work, and eats a large dinner around 8:00 P.M. She generally drinks one or two alcoholic beverages before going to sleep. Her current height and weight are 5 feet 6 inches and 170 pounds.

Explain to Mrs. Scarlatti what a sliding hiatal hernia is, and describe how the hernia leads to gastric reflux and heartburn.

From the brief history provided, list the factors and behaviors that increase Mrs. Scarlatti's chances of experiencing gastric reflux and the pain of heartburn. What recommendations can you make to help her change these behaviors? What drugs might the physician prescribe and why?

the gastric mucosa. Alcohol abuse, food irritants, food allergies, food poisoning, radiation therapy, metabolic stress, and bacterial infections can also cause gastritis.

For the person with gastritis who cannot eat because of nausea or vomiting, foods are generally withheld for a day or two. The diet then progresses from liquids to a bland diet and then a regular diet as tolerated. Antacids, antisecretory agents, and antibiotics (see margin note on p. 450) may also be prescribed.

Chronic Gastritis Chronic gastritis may be associated with aging, peptic ulcer disease (described in the next section), and conditions that cause the chronic reflux of basic fluids from the duodenum into the stomach (gastric reflux). Many people with chronic gastritis have pernicious anemia, and, indeed, pernicious anemia may lead to the diagnosis. In some cases the cause of chronic gastritis is unknown.

The symptoms of chronic gastritis are similar to those of acute gastritis, but they persist over time. As gastritis progresses, the gastric cells gradually shrink; and gastric secretions decline.

Early diagnosis and treatment help prevent further damage to the gastric cells. Interventions need to begin early enough to prevent complications, including malnutrition. An individualized bland diet may help to relieve GI symptoms in some cases. When intrinsic factor is lacking, vitamin B_{12} is given by injection to bypass the need for absorption.

epigastric: the region of the body just above the stomach.
epi = above
gastric = stomach

pernicious anemia: anemia caused by a lack of intrinsic factor and the consequent malabsorption of vitamin B_{12}.

In **atrophic gastritis,** all of the layers of the stomach's mucosal cells are inflamed.

The bacterial infection associated with both gastritus and ulcers is caused by *Helicobacter pylori.*

WWW.
helico.com
The Helicobacter Foundation

Ulcers

The term *ulcers* brings to mind the image of a frantic businessman rushing through the day with coffee in hand, gulping down high-fat, spicy foods, and working until midnight while suppressing the pain of open sores in the stomach or small intestine caused by excessive gastric acid. But like other aspects of ulcer prevention and treatment, this stereotype has fallen by the wayside. Neither a stressful lifestyle nor male gender typifies the person with ulcers; ulcers occur in stressed and unstressed men and women alike.

Diet Order
Bland diet as tolerated.

ulcer: an open sore or lesion. A **peptic ulcer** is an erosion of the top layer of cells from the mucosa of the stomach (**gastric ulcer**) or duodenum (**duodenal ulcer**). Ulcers may also develop in the mouth, esophagus, and other parts of the intestine, or on the skin.

Ulcers can develop both inside and outside the body, but the term *ulcer* used alone generally refers to a *peptic ulcer*—an erosion of the top layer of cells from the GI tract lining. This erosion leaves the underlying layers of cells exposed to gastric juices. When the gastric juices reach the capillaries, the ulcer bleeds, and when they reach the nerves, they cause pain.

The drugs associated with ulcers are nonsteroidal anti-inflammatory agents such as ibuprofen and naproxen.

Zollinger-Ellison syndrome: marked hypersecretion of gastric acid and consequent peptic ulcers caused by a tumor of the pancreas, which releases gastrin.

Causes of Ulcers Three major causes of ulcers have been identified: bacterial infection, the use of certain anti-inflammatory medications, and disorders that cause excessive gastric acid secretion. One such disorder, the Zollinger-Ellison syndrome, results from a tumor of the pancreas that produces gastrin, which, in turn, stimulates the production of gastric acid. Severe peptic ulcer disease follows.

A recent study of over 45,000 men suggests that dietary factors may aid in the prevention of ulcers.[2] High intakes of vitamin A, fruits and vegetables, and dietary fiber were associated with a lower incidence of duodenal ulcers during the six year study period. Further research is necessary to determine if other closely related nutrition factors may actually account for the protective effects observed.

Treatment for Ulcers Treatment aims at relieving pain, healing the ulcer, and minimizing the likelihood of recurrence. Drug therapy plays the primary role in treatment. Antibiotics are used to treat bacterial infections. Antiulcer agents include antisecretory agents to suppress or inhibit gastric acid secretion and other medications that protect the stomach and duodenal lining from acid erosion. As for diet, the client may be advised to follow a bland diet during periods of active disease activity, eliminating those foods that cause pain or discomfort.

Rx Prescription Pad

Medications used in the treatment of gastritis or ulcers may include:

- Antacids
- Antibiotics
- Antisecretory agents
- Antiulcer agents

See Appendix E for timing with meals and nutrition-related side effects.

Consistency-modified diets serve as progressive diets after surgery or when people have not been eating oral foods for a period of time. Mechanical soft diets provide foods for people who cannot chew or swallow. Bland diets help treat disorders characterized or affected by excessive gastric acid secretion. The nutrition assessment checklist reminds health care professionals of factors to assess and monitor for clients with disorders of the upper GI tract.

Nutrition Assessment Checklist

For People with Upper GI Tract Disorders

Medical Assess the client's health history for conditions and treatments that interfere with chewing and swallowing, delay gastric emptying, or affect gastric acid secretions. People with conditions that require consistency-modified diets for long periods of time risk malnutrition.

Medication Review the client's medications for possible nutrient-medication interactions, especially when medications are taken for long periods of time. People with reflux esophagitis, gastritis, and ulcers may use antacids and antisecretory agents over long periods of time. Aluminum-containing antacids may lead to phosphorus deficiencies. Magnesium-containing antacids may cause diarrhea. Calcium- and sodium-containing antacids contribute calcium and sodium to the diet, respectively. Some antisecretory agents may interfere with iron absorption; when iron supplements are necessary, they should be given two hours before or after taking these medications.

Food Intake Regularly assess tolerances for food following periods of GI tract disuse, and progress the diet as appropriate. For people with long-term problems with chewing or swallowing, regularly assess intake to uncover tolerable food consistencies, and quickly address problems with failing appetite and possible nutrient deficiencies. Nutritionally complete formulas, appropriate nutrient supplements, or tube feedings should be recommended for people who cannot eat adequate amounts of table foods or beverages. People on bland diets may need help pinpointing foods that cause pain or worsen symptoms. Food records can help.

Anthropometric Take accurate baseline height and weight measurements. Be alert to weight loss for people with long-term problems tolerating regular food consistencies or those who have severe abdominal pain that interferes with appetite. Timely dietary adjustments can help prevent malnutrition and help support growth in children and desirable weight in adults. People with reflux esophagitis benefit from a safe weight-loss plan if they are overweight.

Laboratory Monitor changes in serum albumin for people with long-term problems with food consistencies. Serum electrolytes, blood urea nitrogen, and hemoglobin and hematocrit help detect electrolyte and fluid imbalances, especially for people unable to drink adequate liquids or those with persistent diarrhea or vomiting. Nutrition-related anemias are common problems for people with chronic gastritis and possible problems for people whose diets are nutrient deficient or those who take antisecretory agents for long periods of time. Laboratory tests indicative of general anemia include hemoglobin and hematocrit. When assessors suspect that anemia is due to nutrition-related causes, mean corpuscular volume, serum ferritin, total iron-binding capacity, serum iron, serum folate, serum vitamin B_{12}, and tests of vitamin B_{12} absorption can help determine whether anemia is caused by iron, folate, or vitamin B_{12} deficiency (the most common types of nutrition-related anemias.)

Physical Check for physical signs of excessive weight loss, nutrient deficiencies, and dehydration. Anemia, smooth tongue, and fatigue may suggest vitamin B_{12} deficiency, a possible complication of gastritis. For people on long-term consistency-modified diets, look for signs of excessive fatigue and poor mental health.

Self Check

1. Standard liquid diets provide all of the following **except:**
   a. a test of a person's ability to tolerate oral foods.
   b. a diet that can be eaten with minimal effort.
   c. a diet that is nutritionally complete.
   d. a source of fluid and electrolytes.

2. A soft diet for a client with indigestion would include the following advice **except:**
   a. limit foods that produce gas.
   b. limit highly seasoned foods.
   c. use foods that are chopped or pureed.
   d. limit high-fiber foods.

3. For the well-nourished client, nutrition support immediately following uncomplicated surgery to remove an appendix may include:
   a. NPO with intravenous fluids and electrolytes.
   b. a diet that progresses from clear liquids to full liquids to soft foods to regular foods as tolerated.
   c. a diet that progresses from clear liquids to full liquids to pureed foods to mechanical soft foods and then a regular diet as tolerated.
   d. a and b.

4. Mechanical soft diets:
   a. provide only pureed foods to minimize the risk of dysphagia.
   b. aim to provide foods that resemble the consistency and variety of a regular diet.
   c. restrict highly seasoned foods.
   d. lack nutrients and are used only as temporary diets.
5. For a person with dysphagia:
   a. the diet requires a great deal of experimentation to uncover which foods the person can tolerate.
   b. the diet is based on a health care team's assessment of the person's swallowing abilities and close monitoring of the person's ability to handle different foods.
   c. nutrient needs should never be met through tube feedings due to the risk of aspiration.
   d. coughing during meals indicates that the person is able to clear the throat and is not at risk for aspiration.
6. Bland diets eliminate all of the following foods **except:**
   a. foods that cause indigestion.
   b. foods that significantly increase gastric acid production.
   c. foods that irritate the lining of the esophagus and stomach.
   d. foods that delay gastric emptying.
7. Reflux esophagitis is:
   a. an inflammation of the esophagus caused by the backflow of acidic gastric fluids from the stomach.
   b. a protuberance of a portion of the stomach above the cardiac sphincter.
   c. an erosion of the lining of the esophagus.
   d. an obstruction of the lower esophagus that results in dysphagia.
8. The health care professional is working with a client who has a hiatal hernia and reflux esophagitis. The health care professional recognizes that the client understands her diet when she says:
   a. "I need to eat three meals a day and drink liquids with my meals."
   b. "I need to eat foods slowly and relax during mealtimes and lay down after meals."
   c. "I need to drink more citrus juices and eat more tomato-based products."
   d. "I need to limit my intake of fat, alcohol, caffeine, and decaffeinated coffee and tea."
9. Nutrition concerns most commonly associated with chronic gastritis include:
   a. malnutrition and pernicious anemia.
   b. iron-deficiency anemia and protein malabsorption.
   c. delayed gastric emptying and pernicious anemia.
   d. dehyration and electrolyte imbalances.
10. Causes of ulcers may include all of the following **except:**
    a. bacterial infections.
    b. increased pressure in the stomach.
    c. anti-inflammatory medications.
    d. tumors of the pancreas.
11. A bland diet and antisecretory agents are recommended for a client with peptic ulcers. The client complies with the suggestions in Table 18.5, but continues to complain of indigestion and gastric pain. The most appropriate advice would be to:
    a. eliminate all highly seasoned food.
    b. take foods in frequent small meals.
    c. keep a record of food intake, gastric symptoms, and when symptoms occur.
    d. drink milk with each meal.

Clinical Applications

1. People on mechanical soft diets differ in the kinds of foods they can handle, in the lengths of time that diet modifications remain necessary, and in the help they need from health care professionals. Think about the difference between working with a person who has had no teeth for years and a person who recently had mouth surgery and is just beginning to eat again. Describe some nutrition-related concerns you might have for the person who has been following a mechanical soft diet for years. How would these concerns differ for a person who needs the mechanical soft diet only temporarily? Contrast the amounts of time a nurse or dietitian might need to spend with the two types of clients.
2. Many of the diets described in this chapter are highly individualized. A particular food may cause discomfort for one person and have no effect on another. Describe practical ways to keep track of food intolerances.

Note

1. W. H. Aldoori and coauthors, Prospective study of diet and the risk of duodenal ulcer in men, *American Journal of Epidemiology* 145 (1997): 42–50.

CHAPTER 18

Nutrition in Practice

▪ Helping People with Feeding Disabilities ▪

Chapter 18 described problems encountered when people have difficulties with chewing and swallowing. This nutrition in practice discusses a broader problem faced by thousands of people who must cope with disabilities that interfere with the process of eating. These obstacles can arise at any time in a person's life and from any number of conditions. An infant may be born with a physical impairment such as cleft palate; an adolescent may suffer nerve damage from injuries sustained in a car accident; a middle-aged adult may lose motor control following a stroke; an older adult may struggle with the pain of arthritis or the mental deterioration of dementia. Table NP18.1 lists some of the conditions that may lead to feeding problems.

Disabilities may also have nutrition-related consequences beyond their effects on feeding. As one example, a disability may make it difficult for a person to engage in enough physical activity to support a healthy appetite or, conversely, to burn energy. Then the person's nutrient intake may either fail to meet nutrient needs or result in excessive weight gain. As another example, a person who has lost a limb to amputation has altered energy needs. Energy needs are reduced in proportion to the weight and metabolism represented by the missing limb, but may be increased if extra energy is necessary to do ordinary things—such as walking on crutches. As still another example, people with involuntary motor activity may have exceedingly high energy needs.

The dietitian, nurse, and occupational therapist most often become involved with feeding disabilities. They can help people achieve as much independence in eating as possible and can teach caregivers to help.

WWW.

blvd.com
The Boulevard

disability.com
Solutions

nichcy.org
National Information Center for Children and Youth with Disabilities

How do disabilities impair eating?

An amazing number of individual coordinated motions are required to get food from the table to the stomach. Consider what an infant experiences when learning to feed himself. At first, the infant cannot sit upright and finds it impossible to hold a spoon. Every single little action—from picking up the utensils, to biting and chewing, to swallowing—requires coordinated movements. Any injury or disability that interferes with these movements can lead to feeding problems.

Other disabilities do not involve oral-motor skills. For example, a person who has problems with sight, or who cannot drive or walk or carry groceries, or who cannot plan meals and think through what to buy, has a disability that affects eating. Disabilities of any type can cause people to have trouble maintaining adequate nutrition status.[1] Their number one nutrition problem is inadequate food intake, which leads to malnutrition, underweight, and, in children, poor growth.[2] Many conditions that lead to feeding problems require medications, which may further affect nutrition status.

Table NP18.1 Conditions That May Lead to Feeding Problems

The following conditions may lead to feeding problems by interfering with a person's ability to suck, bite, chew, swallow, or coordinate hand-to-mouth movements:

- Accidents
- Amputations
- Arthritis
- Birth defects
- Cerebral palsy
- Cleft palate
- Down's syndrome
- Head injuries
- Huntington's chorea
- Hydrocephalia
- Language, visual, or hearing impairment
- Microcephalia
- Multiple sclerosis
- Muscle weakness
- Muscular dystrophy
- Neuromotor dysfunction
- Parkinson's disease
- Polio
- Spinal cord injuries
- Stroke

Nutrition in Practice

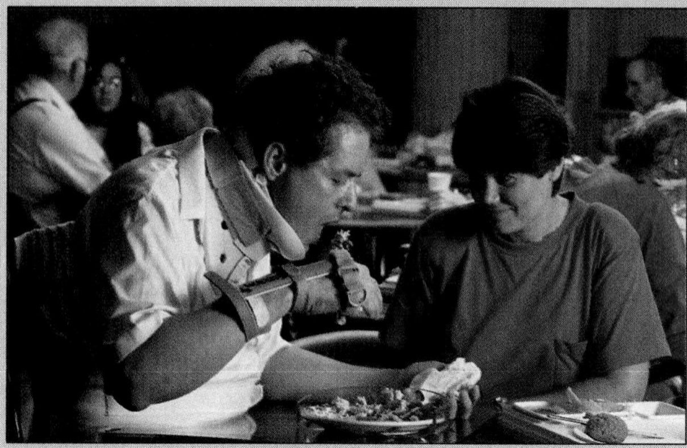

Adaptive feeding equipment can help clients with disabilities gain independence.

On top of nutrition-related problems, people who have difficulty eating often encounter emotional and social problems. For example, children fail to receive the social training that mealtimes provide, and older people miss the social stimulation that goes with eating in the company of others.

How can health care professionals promote independent eating for people with disabilities?

The evaluation and treatment of feeding problems ideally involve the joint efforts of several health care professionals, including a dietitian, a psychologist, an occupational therapist, a physical therapist, a speech-language pathologist, a dentist, and one or several nurses. Table NP18.2 on p. 455 provides a checklist of observations health care professionals use to assess feeding skills and nutrition needs. Because each case is different, health care professionals with a thorough understanding of the many variables that affect eating can best provide effective intervention.

Malnutrition and growth failure can be significant problems in children and adolescents with severe disabilities. Early identification of feeding disabilities and appropriate intervention are especially important for these groups.[3]

What roles do the various team members play in dealing with feeding problems?

The physician diagnoses the client's medical problems, prescribes treatments and medications, and coordinates the health care team. The dietitian assesses the client's nutrition status, plans a diet, monitors medications for possible nutrient-medication interactions, and provides nutrition counseling.[4] The nutrition assessment checklist on p. 456 focuses on topics of notable concern for people with feeding disabilities.

A speech-language pathologist most often evaluates a client's ability to chew and swallow and trains the client to use lips, tongue, and throat for eating and speaking. The occupational therapist may work with the client to evaluate the need for special feeding devices and then shows the client how to use appropriate devices. A dentist may evaluate the client's dental health and provide instructions for maintaining oral hygiene. The health care team faces many challenges in solving feeding problems, but the personal satisfaction of seeing someone gain control of eating can be very rewarding.

Do clients receive behavioral training, such as practice eating sessions?

Yes, such training sessions are essential. Knowing how is not enough to achieve good nutrition; doing is the key. Direct observation of the client during mealtimes allows the health care team to teach the appropriate feeding techniques, monitor the client's or caregiver's understanding of the techniques, and evaluate how well the strategies are working.

Some of the strategies health care professionals can use to help clients develop feeding skills are shown in Table NP18.3 on p. 457. For example, a child with a feeding problem may be overly sensitive to oral stimulation. In this case, the helper uses strategies to desensitize the client to oral stimulation. The helper may start by gently stroking the client's face with a hand, washcloth, or soft pliable toy. (This can be done playfully, making a game of it.) When the child can tolerate touch on less-sensitive areas of the face such as the forehead, cheeks, and lips, then the helper may begin to slowly and firmly rub the gums, palate, and tongue. With time, the client may be better able to tolerate the presence of food in the mouth.

What special feeding devices are available for helping people to eat?

Figure NP18.1 on p. 458 provides examples of special feeding devices and describes some of their uses. These

Table NP18.2 Feeding Evaluation

Oral Ability

- Sucking
- Oral prehension
- Swallowing
- Breathing and swallowing coordinated
- Drooling
- Lips
- Tongue size, thrust, mobility
- Biting
- Munching
- Chewing
- Drinking
- Response to input

Dental Health

- Structural malformation
- Impaired oral-motor ability
- Delayed weaning
- Use of cariogenic foods or medications
- Occlusion
- Teeth
- Caries
- Gingiva
- Oral hygiene
- Palate
- Pain on exam
- Hypersensitivity
- Teething stage

Body Position

- General tone and movement
- Reflex activity
- Head control
- Sitting balance
- Placement of feet
- Usual feeding position

Hand Use

- Palmar grasp
- Pincer grasp
- Opposition finger/thumb
- Hand-to-mouth control

Developmental Feeding

- Breast _________ bottle _________ weaned _________
- Baby food _________ junior food _________
- mashed table food _________ minced foods _________
- Cut table foods _________ regular table foods _________
- Closes hands in on bottle
- Hand to mouth/sucks on fingers
- Teething biscuit, holds and brings to mouth
- Finger feeds
- Opposes lips to rim of cup
- Attempts to grasp spoon
- Grasps spoon
- Dips spoon in dish
- Brings spoon to mouth
- Holds bottle and drinks independently
- Grasps cup
- Raises cup to mouth
- Lifts cup, drinks, and replaces
- Scoops well with spoon
- Feeds independently with spoon
- Drinks with straw
- Spears with fork
- Spreads with knife

Feeding Environment

- Time of feedings (note number and length of feedings)
- Atmosphere of feedings (tense, pleasant, unpleasant)
- Person responsible for feeding
- Parental and/or caregiver's attitude toward feeding
- Past successful and unsuccessful methods
- Identify positive and negative reinforcing behaviors
- Behavioral problems

Diet History, Including:

- Total fluid intake
- Types of foods consumed
- Method of feeding
- Texture modifications
- Food aversions, intolerances, or allergies
- Use of food as rewards

Medications

- Type
- Dosage
- Time given

Bowel Concerns

- Regular
- Constipation
- Diarrhea

Sources: Adapted from J. J. Cafferky, Nutrition assessment of children with developmental disabilities: Special considerations, *Support Line* (a newsletter of Dietitians in Nutrition Support) June 1993; R. B. Howard, Nutritional support of the developmentally disabled child, *Textbook of Pediatric Nutrition*, ed. R. M. Suskind (New York: Raven Press, 1981), pp. 577–582.

Nutrition Assessment Checklist

For People with Feeding Disabilities

Medical Determine if the client's primary medical problem imposes further dietary changes.

Medication Be aware that anticonvulsant medications used to treat some disorders may induce folate deficiency, impair vitamin D status, and raise blood cholesterol; medications prescribed for attention deficit hyperactivity disorder may suppress appetite and slow growth. Some medications may slow GI tract motility, contributing to constipation; others may cause gastric irritability, drowsiness, nausea, and altered taste sensations—factors that diminish appetite and food intake. Remember that clients on long-term drug therapy and those taking multiple medications are particularly vulnerable to nutrient imbalances.

Food Intake When taking a diet history, interview all parents and caregivers responsible for feeding the client. Take into account the differences between food offered and food eaten; spills can contribute to substantial food losses.

Anthropometric Obtain anthropometric measures as accurately as possible. In cases where standing height cannot be measured, use arm span length instead (measured from the tip of the middle finger on one hand to the tip of the middle finger on the other hand with arms stretched out as far as possible). Use standards specific to the medical disorder if available.

Laboratory Monitor serum albumin to ensure adequate protein status, serum ferritin to detect iron deficiency, and vitamin and mineral status as needed.

Physical Use Table NP18.2 to assess the client's feeding disabilities. Note any dental problems that may interfere with food intake.

devices can make a remarkable difference in a person's ability to eat independently. For a person who cannot grasp an ordinary fork, for example, a special fork may be the key to future health.

What happens if the person with a feeding disability is unable to eat enough food after training?

Sometimes, despite the best efforts of all involved, the client still can't eat enough food by mouth or may frequently experience complications such as choking or vomiting. Feeding through tubes placed directly in the stomach (gastrostomy) can improve nutrition status and ease problems with choking or vomiting.[5]

With so many things to do, how do clients and caregivers cope?

You have identified an area of great concern regarding feeding disabilities. Eating is only one of many routine tasks that clients with disabilities face. The time and patience required to handle these tasks are often a great source of frustration and distress for the client and the caregiver. Consider an example. Children with cerebral palsy take ten times longer to eat than children without disabilities. Some mothers have reported spending up to seven hours a day feeding these children.[6] Besides feeding, the caregiver must often cope with many other tasks that also require a great deal of time. Caregivers may feel they have little time to care for themselves and other family members. Psychologists can offer counseling to clients or caregivers to help them adjust; all members of the health care team can offer emotional support and practical suggestions to ease frustrations.

Successful therapy for feeding disabilities requires the involvement of many health care professionals and depends on the accurate identification of impaired feeding skills and appropriate interventions. Some peo-

Table NP18.3 Areas of Concern and Suggested Strategies for Developing Feeding Skills

Inability to Suck

- Use cold substances around lips to stimulate sucking.
- Use a cloth soaked with water for child to suck.
- Try different types of nipples.
- As child's ability begins to improve, change to nipple with a smaller hole.

Inability to Chew

- Place a small amount of food between back teeth and move jaw up and down. A mirror may help demonstrate and point out various body parts.
- Place foods such as peanut butter on lips and encourage client to wash lips with tongue. Gradually change from pureed foods to solid foods (sprinkle crackers in soup, etc.).

Inability to Swallow

- Close jaw and lips of client together (swallowing is easiest when the mouth is closed).
- Stroke throat upward under chin.
- Offer next bite of food only after client swallows.
- Demonstrate—let client feel *you* swallow.

Inability to Grasp

- Allow client to finger feed.
- Guide client in exploring mouth.
- Cut food into small pieces.
- Place your hand over client's hand and help client to grasp spoon.
- Use adaptive equipment (plastic spoon, etc.).
- Make sure bowl is stabilized (suction, tape).
- Use plates with high straight sides, or build higher edge using aluminum foil.

Poor Hand-Mouth Coordination

- Pour sand, etc.
- Exercise with ball.
- Exercise with push-pull objects.
- Study body parts with client, if appropriate.

Impaired Vision

- Place meats and vegetables consistently in same areas of plate.

Overweight

- Cut down snacks and high-kcalorie foods.
- Refrain from rewarding with food.
- Increase exercise and leisure-time activities.

Underweight

- Increase number of meals per day.
- Include high-kcalorie foods, especially liquid supplements.
- Encourage proper exercise.

Lack of Nutrition Education

- Work with families.
- Stress the importance of proper nutrition for *all* family members.
- Teach proper feeding environment (good eating habits, eating positions).
- Provide nutrition-instruction materials.

Source: Adapted with permission from S. Calvert and F. Davies, Nutrition of children with handicapping conditions, *Dietetic Currents* 4 (1977): 13–17.

ple with disabilities attain total independence with training—they are able to prepare, serve, and eat nutritionally adequate food daily without help. In other cases, these goals can be attained with the help of caregivers. The combined efforts of the health care team can support both clients and caregivers in enhancing quality of life and in achieving independence to the greatest degree possible.

Notes

1. Position of The American Dietetic Association: Nutrition in comprehensive program planning for persons with developmental disabilities, *Journal of the American Dietetic Association* 92 (1992): 613–615; Position of The American Dietetic Association: Nutrition services for children with special health needs, *Journal of the American Dietetic Association* 95 (1995): 809–812.

2. R. D. Stevenson, Feeding and nutrition in children with developmental disabilities, *Pediatric Annals* 24 (1995): 255–260;

Nutrition in Practice

Figure NP 18.1
Examples of Special Feeding Devices

Utensils

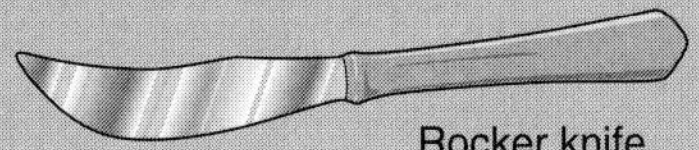

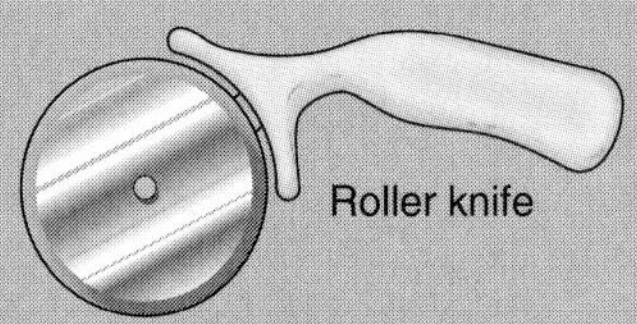

People with only one arm or hand may have difficulty cutting foods and may appreciate using a *rocker knife* or a *roller knife.*

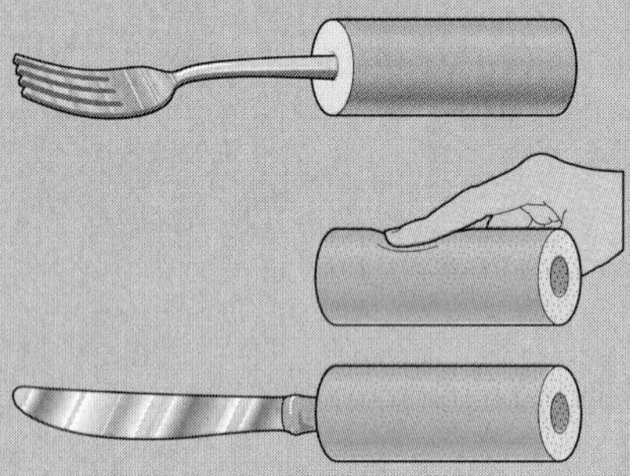

People with a limited range of motion can feed themselves better when they use *flatware with built-up handles.*

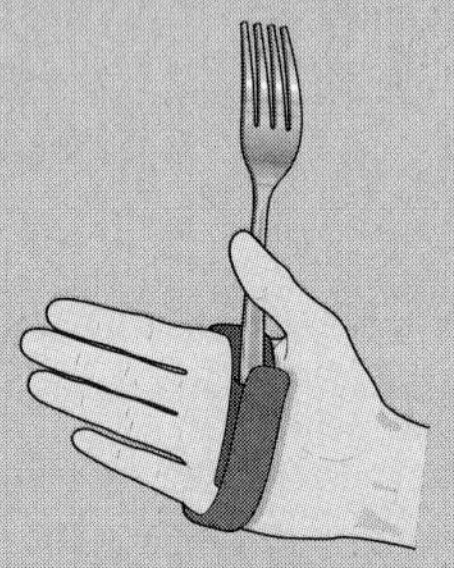

People with extreme muscle weakness may be able to eat with a *utensil holder.*

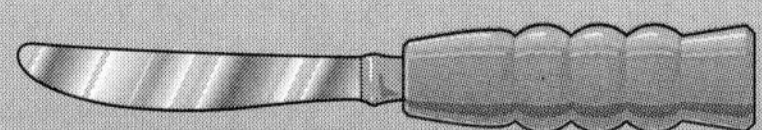

For people with tremors, spasticity, and uneven jerky movements, *weighted utensils* can aid the feeding process.

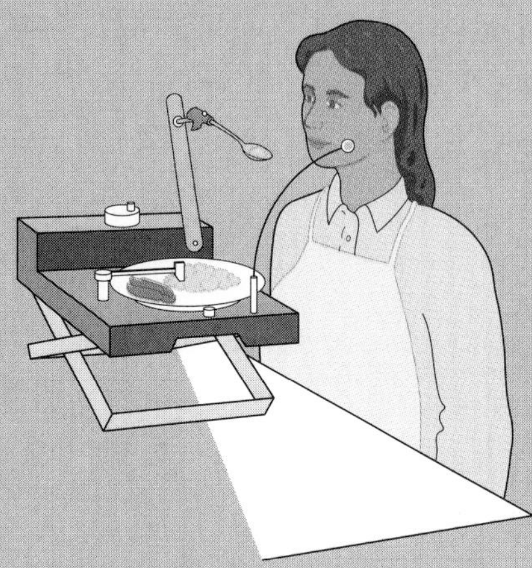

Battery-powered feeding machines enable people with severe limitations to eat with less assistance from others.

Plates

People who have limited dexterity and difficulty maneuvering food find *scoop dishes* or *food guards* useful.

People with uncontrolled or excessive movements might move dishes around while eating and may benefit from using *unbreakable dishes with suction cups.*

Cups

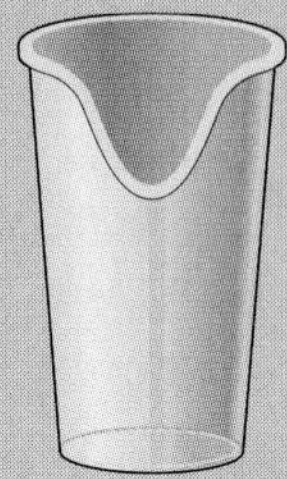

People with limited neck motion can use a *cutout plastic cup*.

Two-handed cups enable people with moderate muscle weakness to lift a cup with two hands.

People with uncontrolled or excessive movements might prefer to drink liquids from a *covered cup* or glass with a *slotted opening* or *spout*.

A soft, flexible long plastic straw may also ease the task of drinking.

CHAPTER 18

Nutrition in Practice

M. Thommessen and coauthors, Energy and nutrient intakes of disabled children: Do feeding problems make a difference? *Journal of the American Dietetic Association* 91 (1991): 1522–1525; R. K. Johnson and M. Maeda, Establishing outpatient nutrition services for children with cerebral palsy, *Journal of the American Dietetic Association* 89 (1989): 1504–1507; L. M. Thommessen and coauthors, Nutrition and growth retardation in 10 children with cogenital deaf-blindness, *Journal of the American Dietetic Association* 89 (1989): 69–73.

3. J. A. Amundson and coauthors, Early identification and treatment necessary to prevent malnutrition in children and adolescents with severe disabilities, *Journal of the American Dietetic Association* 94 (1994): 880–883.

4. H. H. Cloud, Expanding roles for dietitians working with persons with developmental disabilities, *Journal of the American Dietetic Association* 97 (1997): 129–130.

5. R. Tawfik and coauthors, Caregivers' perceptions following gastrostomy in severely disabled children with feeding problems, *Developmental Medicine and Child Neurology* 39 (1997): 746–751; J. S. Isaacs and coauthors, Weight gain and triceps skinfolds fat mass after gastrostomy placement in children with developmental disabilities, *Journal of the American Dietetic Association* 94 (1994): 849–854.

6. Thommessen and coauthors, 1991.

Fiber- and Residue-Modified Diets for Lower GI Tract Disorders

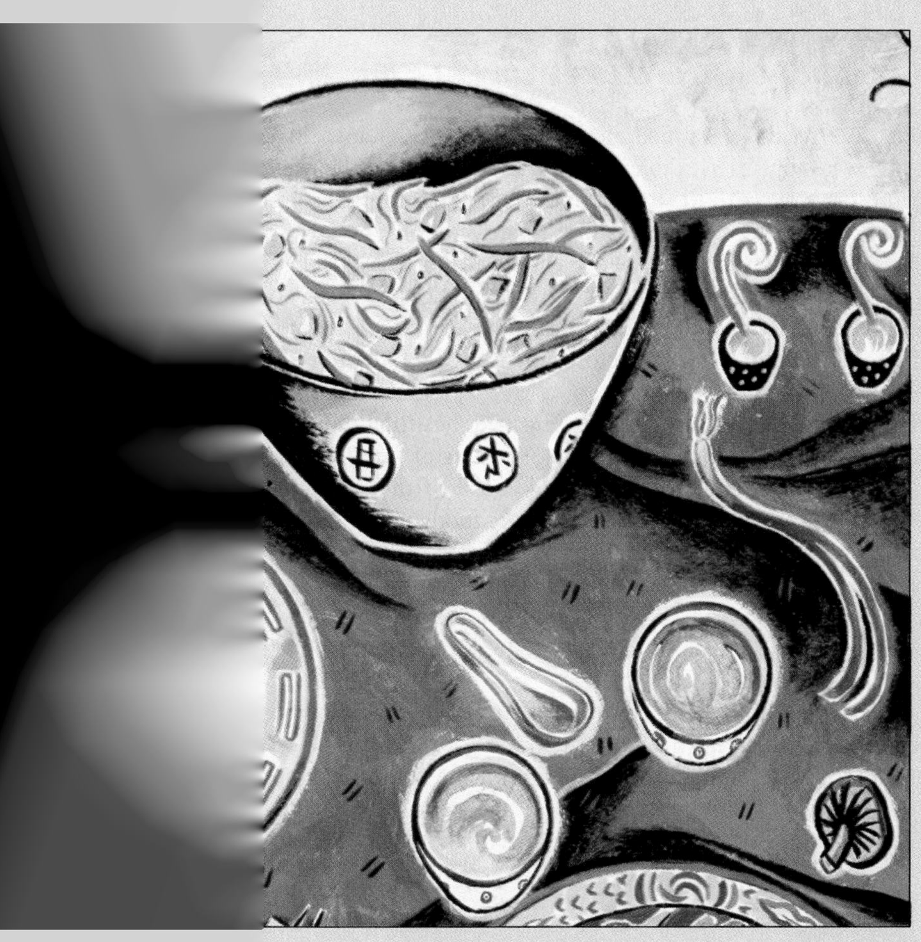

Table 19.1 Indications for Fiber- and Residue-Modified Diets
High-Fiber Diet
Constipation
Irritable bowel syndrome
Diverticulosis
Weight-loss efforts
Obesity
Heart disease
Low-Fiber Diet
Postsurgical diet
Diarrhea
Diverticulitis
Inflammatory bowel disease
Myocardial infarction (heart attack)
Congestive heart failure

Unlike the consistency-modified, soft, and bland diets used to treat conditions of the upper GI tract, fiber- and residue-modified diets serve primarily to treat conditions that affect the lower GI tract. This chapter describes fiber- and residue-modified diets and explains their uses for disorders characterized by altered intestinal motility and for conditions where minimal fecal bulk aids recovery. The next chapter describes disorders that affect the absorption of nutrients.

Fiber- and Residue-Modified Diets

Fiber- and residue-modified diets include high-fiber, low-fiber/low-residue, and very-low-residue diets. Table 19.1 lists indications for the use of these diets. The term *residue* refers to the total amount of material in the colon; it includes fiber and undigested food as well as intestinal secretions, bacteria, and shed mucosal cells. Foods containing fiber necessarily contribute residue, because dietary fibers cannot be digested by enzymes in the human digestive tract. Some foods, including milk and connective tissue from meats, are low in fiber; but they may contribute to fecal mass and, therefore, are limited on low-fiber/low-residue diets.

residue: the total amount of material in the colon; it includes dietary fiber and undigested food, intestinal secretions, bacterial cell bodies, and cells shed from the intestinal mucosa.

High-Fiber Diets As Chapter 2 described, health authorities recommend that people adopt high-fiber diets (20 to 35 grams of fiber) by increasing their consumption of whole-grain breads and cereals, fruits and vegetables, and dried beans and peas. Table 19.2 shows specific foods noted for their high-fiber contents. Because fibers draw fluids from the intestinal contents along with them as they pass through the intestine, health care professionals encourage people on high-fiber diets to drink plenty of fluids. The sample menu shows a day's meals for a high-fiber diet, and the box on p. 463 suggests ways to help clients increase their fiber intakes.

High-fiber diets help maintain GI tract function by adding volume and weight to the stool, normalizing the transit of undigested materials through the intestine, and minimizing pressure within the colon. Such diets help maintain intestinal health and are of particular benefit to people with constipation and diverticulosis, disorders that affect intestinal transit time and pressure in the

WWW wheatfoods.org
Wheat Foods Council

Table 19.2 High-Fiber Foods
Whole-Grain Breads and Cereals
Whole wheat, pumpernickel, rye, wheat germ, bran, bulgur, oatmeal, brown or wild rice
Fruits
Apples, berries, figs, papaya, prunes, pears
Vegetables
Artichokes, broccoli, brussels sprouts, raw carrots, chicory, kohlrabi, legumes, shítake mushrooms, sweet potatoes, rutabagas, turnips, yams
Other
Peanut butter, popcorn

How to

Increase Fiber Intake

During the first few weeks on a high-fiber diet, a person may feel bloated, pass gas frequently, or experience heartburn. These suggestions can help:

- *Go slow.* Add high-fiber foods gradually and in small portions at first. Increase portion sizes and add foods as tolerance improves.
- *Add fluids.* Fiber attracts water as it moves through the intestine.
- *Experiment.* Try small servings of various fiber-containing foods at first and adopt those that are most pleasing.
- *Mix high-fiber foods with other foods.* Sprinkle bran flakes, wheat germ, or raisins on salads or applesauce. Add bran or mashed legumes to meat loaf. Add legumes and other high-fiber vegetables to soups and salads.

colon, respectively. High-fiber diets may also help in weight-loss efforts (Chapter 9), in the regulation of blood lipids for people at risk for heart disease (Chapter 26), and in reducing the risk of certain cancers.

Low-Fiber and Low-Residue Diets Low-fiber and low-residue diets serve to reduce the total fecal volume. Low-fiber diets restrict high-fiber foods and tough meats. Low-residue diets limit these foods and also limit milk and milk products. Table 19.3 on p. 464 lists foods low in fiber. The sample menu shows a day's meals for a low-fiber/low-residue diet.

Very-low-residue diets restrict all high-fiber foods including all vegetables and fruits (including canned and tender-cooked) but allow vegetable and fruit juices (except prune). Very-low-residue diets are usually temporary; when required for long periods of time, they must be supplemented with nutrient-rich low-residue formulas or vitamin and mineral supplements.

Low-fiber/low-residue foods are least likely to obstruct an intestinal tract that is narrowed by inflammation or scarring or in which GI motility is slow. Thus low-fiber/low-residue diets are often used before and after intestinal surgery. People at risk for small bowel obstructions or those with diverticulitis, delayed gastric emptying, small bowel fistulas, or inflammatory bowel diseases may also benefit from low-fiber/low-residue diets. Physicians rely on clinical judgment to determine which level of restriction is appropriate for the person's medical condition and individual tolerances.

The **soft diet** described in Chapter 18, **low-fiber diets,** and **low-residue diets** are similar. Of the three, soft diets allow the most fiber, low-residue diets the least. Soft diets limit highly seasoned foods, and low-fiber diets limit serving sizes of allowed fruits and vegetables. Low-residue diets provide the same foods as low-fiber diets, but restrict milk and milk products. In facilities that serve meals, menus for these three diets are the same.

Sample High-Fiber Diet Menu

Breakfast	Lunch	Supper	Snack
1 c multigrain cereal	1 c black bean soup	3 oz baked fish	3 c popcorn
½ c strawberries	3 oz broiled chicken	½ c brown rice	1 c tomato juice
1 c nonfat milk	½ c steamed broccoli	½ c peas	
2 slices whole-wheat toast	½ c baked sweet potatoes	1 whole-wheat dinner roll	
2 tbs peanut butter	1 fresh pear	2 tsp margarine	
1 c coffee	1 whole-wheat dinner roll	1 piece carrot cake	
	1 tsp margarine	1 c nonfat milk	

Table 19.3 Low-Fiber Foods

Meat
Tender meat, poultry, seafood, and eggs
Breads and cereals
Refined breads, cereals, rice, and pasta
Fruits
Cooked or canned peeled apples, apricots, peaches, pineapple, plums; bananas, cherries, cranberry sauce, mandarin oranges, pomegranates, tangerines; fruit juice
Vegetables
Cooked bean sprouts, cabbage, onions, summer squash; celery, endive, lettuce, tomato paste, tomato puree, vegetable juice, water chestnuts, watercress
Other
Avocado, nuts

WWW

niddk.nih.gov/health/digest/pubs/gas/gas.htm
Gas in the Digestive Tract (from the National Digestive Diseases Clearinghouse)

Gas-Forming Foods Undigested and unabsorbed dietary fibers and foods may pass into the colon where bacteria may metabolize them and produce gas in the process. Thus many high-fiber foods, including legumes, some grains, fruits, and vegetables, are also gas-forming foods (see Table 19.4 on p. 465). Besides high-fiber foods, high-fat foods, fructose, and sugar alcohols (sorbitol, mannitol, xylitol, and maltitol) taken in large amounts may be incompletely absorbed and cause gas for some people. Unabsorbed lactose can cause gas for people with lactose intolerance.

Although expelling intestinal gas can be a humiliating experience, it is quite normal in healthy people. Excessive gas production, however, can cause abdominal pain, a bloated feeling, and distension. For people with uncomfortable or serious GI symptoms such as indigestion, nausea, or malabsorption, gas-forming foods can contribute to undesirable GI symptoms and are thus avoided.

Motility Disorders

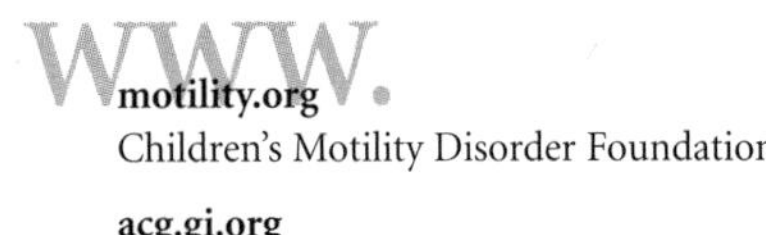

motility.org
Children's Motility Disorder Foundation

acg.gi.org
American College of Gastroenterology

Disorders that cause alterations in the transit of foods through the GI tract can result in symptoms such as delayed gastric emptying, constipation, diarrhea, and irritable bowel syndrome. Because dietary fibers affect GI transit time and stool volume and composition, fiber-modified diets are used in the treatment of motility disorders.

Sample Low-Fiber/Low-Residue Diet Menu

Breakfast	*Lunch*	*Supper*	*Snack*
Orange juice	Baked fish	Roast beef	Applesauce
Soft-cooked egg	White rice	Mashed potatoes	Vanilla wafers
Puffed rice cereal	Green beans	Cooked carrots	
White bread toast	Small banana	Canned peaches	
Coffee or tea	Roll	Roll	
Milk for cereal	Margarine	Margarine	
Creamer	Coffee or tea	Coffee or tea	
	Sugar	Sugar	
	Creamer	Creamer	

Delayed Gastric Emptying

Diet Order
Low-fiber, low-fat diet given in frequent small meals.

Although this chapter focuses on disorders of the lower GI tract, delayed gastric emptying is described here because its treatment often includes a fiber-modified diet. The rate at which the stomach empties can be delayed temporarily (following surgery, for example) or chronically (as a complication of diabetes mellitus). Table 19.5 on this page lists conditions that may delay gastric emptying. In some disorders, delayed gastric emptying occurs because the pyloric sphincter fails to open completely and the narrowed opening interferes with the passage of food into the intestine. Plant matter from food (fibers, leaves, and skins) may stagnate in the stomach and form masses called phytobezoars.

Symptoms of Delayed Gastric Emptying People experiencing delayed gastric emptying often feel full after eating small amounts of food and may also experience nausea, vomiting, abdominal pain, and bloating. The discomfort of eating can lead to anorexia and weight loss.

Treatments for Delayed Gastric Emptying Treatments aim to control the underlying medical disorder and may include medications to stimulate gastric motility. Diet therapy aims to eliminate symptoms and provide adequate nutrition. Often a low-fiber, low-fat diet given in frequent small meals is helpful. Fiber and fat slow the rate at which the stomach empties, and large volumes of food distend the stomach and increase the likelihood of GI upsets. In severe cases, the person tolerates only liquids in small amounts. In such cases, formula diets provide nutrients. If there is a complete obstruction, or if the person is unable to drink enough liquids, feeding by tube below the stomach or intravenous feedings may be necessary.

Table 19.4 Foods That May Produce Gas

Apples	High-fat meats
Asparagus	Honey
Beer	Kohlrabi
Bran	Legumes (dried beans and peas)
Broccoli	Mannitol
Brussels sprouts	Milk and milk products
Cabbage	Nuts
Carbonated beverages	Onions
Cauliflower	Prunes
Cream sauces	Radishes
Cucumbers	Raisins
Fried foods	Sorbitol
Fructose	Soybeans
Gravy	Wheat

phytobezoar (FIGH-toh-BEE-zor): a mass of plant matter (fibers, leaves, and skins) that forms a ball and may block the outlet from the stomach to the intestine. **Trichobezoars** contain hair and nails. People with some psychiatric disorders may swallow their hair and nails.

Table 19.5 Conditions That May Lead to Delayed Gastric Emptying

Temporary Delayed Gastric Emptying
Diabetic ketoacidosis
Electrolyte imbalances
Hyperglycemia
Infection
Surgery
Use of some drugs (examples: alcohol, nicotine, aluminum hydroxide antacids, opiates, levodopa, doxepin hydrochloride)
Chronic Delayed Gastric Emptying
Connective tissue disorders
Diabetic neuropathy
Gastric surgery
Neurological disorders
Peptic ulcer disease
Reflux esophagitis
Severe PEM

Constipation

Constipation describes a symptom, not a disease. People with constipation may pass stools that are difficult or painful to excrete or have infrequent bowel movements. Abdominal discomfort and headaches sometimes accompany constipation. Nutrition in Practice 5 discussed the constipation that may occur in any person from time to time and suggested strategies for establishing regular bowel habits.

Causes of Constipation Constipation may be associated with fluid and electrolyte and hormonal imbalances, with certain diseases of the GI tract, with chronic laxative abuse, with stress, and with a variety of medications, including narcotic analgesics and anticholinergics. Many elderly people self-report problems with constipation and take laxatives to treat it, although objective measurements of stool consistency and frequency fail to confirm constipation.[1]

Treatment of Constipation When constipation occurs as a result of an underlying medical condition, treatment of that disorder may alleviate the constipation. A high-fiber diet, drinking plenty of fluids, eating prunes, drinking prune juice, and engaging in regular physical activity can help, too. Laxatives, especially bulk-forming agents (hydrophilic colloids), may be prescribed in some cases.

hydrophilic colloids: laxatives composed of fibers that work like fibers from food; they attract water in the intestine to form a bulky stool, which then stimulates peristalsis. Metamucil and Fiberall are examples of hydrophilic colloids.

Diarrhea and Dehydration

Diet Order
Caffeine-restricted,
lactose-restricted,
low-fat, low-fiber
diet.

Diarrhea refers to an increased frequency or volume of stools. Diarrhea occurs either when fluids are drawn from the GI tract lining and added to the food residue or when the intestinal contents move so quickly through the GI tract that fluids are not absorbed. In both cases, the result is frequent, watery bowel movements.

Diarrhea is not a disease, but a symptom of many medical conditions and a complication of some medical treatments, including many medications. It can be acute, lasting less than two weeks, or chronic, lasting longer. Mild diarrhea that remits in 24 to 48 hours is seldom a cause for concern unless the person is already dehydrated. A person with severe, persistent diarrhea may rapidly become dehydrated, lose weight, and develop multiple nutrient deficiencies. A child or infant can lose proportionately more fluid and weight and can develop severe dehydration and malnutrition in a short time. Table 16.6 in Chapter 16 listed findings associated with dehydration.

Secretory diarrhea results from an accelerated movement of fluids and electrolytes from the intestinal capillaries into the lumen of the intestine.

Osmotic diarrhea results from an increase in the osmolarity of the intestinal contents due to unabsorbed nutrients or drugs.

Severe, chronic diarrhea that does not respond to treatment is **intractable diarrhea.**

Causes of Diarrhea Acute diarrhea that occurs abruptly in a healthy person frequently results from viral, bacterial, or protozoal infections or as a side effect of medications. It can also occur in the person who begins to eat foods or begins a tube feeding after a period of fasting, starvation, or intravenous nutrition. Infants frequently develop diarrhea when given formulas their immature GI tracts cannot tolerate or when they are ill. When used in large quantities, food ingredients such as sorbitol and olestra may cause diarrhea in some people.

Chronic diarrhea can occur as a result of disorders that alter GI tract motility, such as irritable bowel syndrome (discussed later in this chapter); disorders that cause malabsorption (Chapter 20); food intolerances (such as lactose intolerance); and some infections, including some parasitic infections and human immunodeficiency virus (HIV).

Treatment of Diarrhea The treatment of diarrhea requires treatment of the primary medical condition. If a food is responsible, that food must be omitted from the diet. If a medication is responsible, a different medication or form of medication (injectable versus oral, for example) may alleviate the problem. Infections are treated with appropriate anti-infective agents. Medications that slow GI motility are often recommended along with other therapies to treat diarrhea.

Until the diarrhea is resolved, treatment includes replacement of lost fluids and electrolytes to prevent dehydration. For mild cases of diarrhea, fluids and electrolytes can be replaced using fruit juices, sports drinks, caffeine-free carbonated beverages, tea, and broth with crackers. In mild-to-moderate cases of diarrhea, oral rehydration formulas—simple solutions of water, salts, and sugar—provide needed fluids and electrolytes. In severe cases, it may become necessary to stop placing demands on the GI tract by withholding all foods and beverages until the diarrhea remits, usually in a day or two. During bowel rest, intravenously administered fluids and electrolytes replace losses.

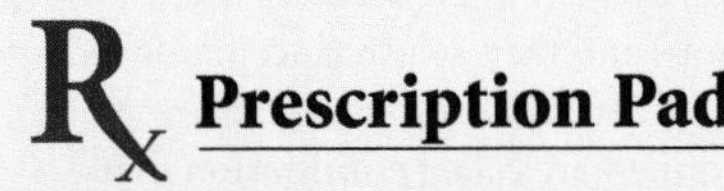

Medications used in the treatment of diarrhea may include:

- Antidiarrheals
- Anti-infective agents

See Appendix E for timing with meals and nutrition-related side effects.

Drugs used to treat diarrhea are called **antidiarrheal agents** (see Appendix E). Antidiarrheal agents are generally contraindicated for diarrhea caused by infectious agents, because they slow GI motility and prolong the time that the toxin remains in contact with GI cells.

Oral Diets Often people with diarrhea can tolerate solid foods, although they may benefit from temporarily avoiding highly seasoned foods, foods high in fat, foods that cause gas, lactose-containing foods, caffeinated beverages, and any food that aggravates the diarrhea. In other cases, including clients who begin eating after complete bowel rest, clients may be advised to drink only clear liquids (including oral rehydration formulas) to avoid irritating the GI tract. Once the diarrhea remits, the client may gradually advance to full liquids, to soft foods, and then to regular foods as tolerated. The diet temporarily excludes lactose and caffeine and any foods believed to have irritated the GI tract. Frequent small meals are easiest to tolerate at first. Permanent dietary changes may be necessary for diarrhea caused by food sensitivities or allergies. For some people with diarrhea, adding fiber, particularly soluble fiber, to the diet may reduce diarrhea by regulating the transit time of ingested materials through the intestine.

Reminder: In general, fruits, oats, barley, and legumes contain higher concentrations of soluble fibers.

Alternate Feedings Clients who are still unable to tolerate adequate amounts of foods after a few days on an oral diet may benefit from nutritionally complete, lactose-free liquid formulas (see Chapter 21), provided orally, if the client can drink them, or by tube, if oral intake remains inadequate. For people with severe diarrhea who are unable to tolerate any type of oral or tube feeding, intravenous nutrition (see Chapter 22) is indicated, and nothing is given by mouth (NPO).

Irritable Bowel Syndrome

Irritable bowel syndrome is a common motility disorder characterized by chronic diarrhea, constipation, or alternating episodes of both. The person may also experience indigestion, nausea, abdominal pain, bloating, and flatulence. Symptoms frequently occur shortly after a person eats and often resolve temporarily following a bowel movement.

Diet Order
High-fiber, fat-restricted diet. Avoid highly seasoned foods.

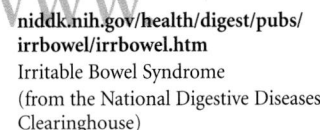

WWW **niddk.nih.gov/health/digest/pubs/irrbowel/irrbowel.htm**
Irritable Bowel Syndrome (from the National Digestive Diseases Clearinghouse)

Causes of Irritable Bowel Syndrome The cause of irritable bowel syndrome remains elusive, but stress and anxiety are believed to be contributing factors. Food intolerances may also be a factor in some cases. Many foods have been associated with irritable bowel syndrome; lactose, fructose, sorbitol, gluten, coffee, and tea are some of the most commonly cited. Irritable bowel syndrome occurs more frequently in women, and episodes of active symptoms often decrease with age.

Dietary Treatments Medical therapy for irritable bowel syndrome often includes stress management along with dietary modifications. If food intolerances

Diets that eliminate several foods or food groups and then reintroduce foods individually so that food intolerances can be identified are called **elimination diets.**

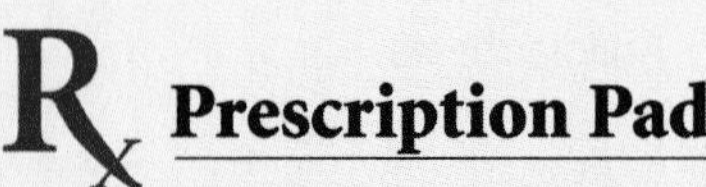

Medications used in the treatment of irritable bowel syndrome may include:

- Antianxiety agents
- Anticholinergics
- Antidepressants
- Antidiarrheals
- Antiemetics
- Antiflatulents
- Laxatives

See Appendix E for timing with meals and nutrition-related side effects.

An intestinal obstruction is also known as **ileus** (ILL-ee-us). Ileus caused by a physical obstruction is called **mechanical ileus.** When the intestinal muscles fail to function, the resulting obstruction is called **adynamic** or **paralytic ileus.**

intestinal lumen: the inner open space within the intestine.

A hole that develops in an organ is known as a **perforation.**

peritonitis (pear-ih-toe-NIGH-tus): infection and inflammation of the membrane lining the abdominal cavity caused by leakage of infectious organisms through a perforation in an abdominal organ.

septicemia (sep-tih-SEE-me-ah): the spread of an infection from a local area into the blood.

are suspected, the client avoids foods commonly thought to trigger irritable bowel syndrome. If symptoms subside, foods are added back to the diet individually to identify those that cause uncomfortable symptoms.

Many clients benefit from a high-fiber, fat-restricted diet. Some clients avoid uncomfortable symptoms by limiting highly seasoned food as well. High-fiber diets may also cause gas and abdominal pain, however, so it is important to add fiber gradually. If a client experiences gas and abdominal pain, it may be helpful to pinpoint which foods cause problems and then avoid those foods and ingredients.

People can use bran or hydrophilic colloids to help relieve constipation. Bran is likely to produce gas, though, so people who also have gas should use colloids only until symptoms resolve. Drug therapy is usually reserved for cases that do not respond to diet and stress management therapy.

Intestinal Obstructions

Diet Order
Low-fiber, low-residue diet in frequent small meals.

Partial or complete blockages of the intestinal lumen can occur in both the small and large intestine. Inflammation, scar tissue, tumors, or congenital malformations can physically block the movement of foods and fluids through the intestine. Sometimes the muscles of the intestine fail to function, often as a result of abdominal surgery. In this case, too, movement through the intestine is obstructed.

Consequences of Intestinal Obstructions The severity of complications depends on the degree and type of obstruction. Adynamic ileus may resolve without special treatment in a few days, for example, but complete obstructions can result in life-threatening complications in a matter of hours.

When the intestine becomes obstructed, its contents collect above the site of the blockage. Peristalsis increases as the body attempts to push the intestinal contents past the obstruction. Injury to the intestinal cells and intestinal distension follow. The distension cuts off blood flow to the area, which, in turn, can lead to malabsorption, acid-base imbalances, and dehydration. Damage to the intestinal lining can be so severe that a hole may form in the intestine and lead to peritonitis and septicemia.

People with mechanical obstructions of the small intestine may develop severe pain, abdominal distension, nausea, vomiting, and constipation. If the obstruction is complete, the person may vomit fecal material. People with mechanical obstructions of the large intestine often develop constipation and may suffer sudden, intermittent, and severe abdominal pain. People with nonmechanical obstructions also experience abdominal pain and vomiting, but the expelled material usually consists of gastric fluids and bile, rather than fecal material.

Treatments for Intestinal Obstructions In many cases, surgery is needed to remove the obstruction or correct the underlying medical condition. Most often, people with complete obstructions cannot be fed by mouth or by tube. If they are malnourished or at risk for malnutrition, they may receive intravenous nutrition.

People with partial obstructions and those with adynamic ileus can sometimes be treated without surgery. In these cases, a tube is inserted from the nose to the stomach and used to suction fluids and gas to help relieve pressure and reduce abdominal distension. Intravenous fluids help restore fluid, electrolyte, and acid-base balances. Antibiotics are given to prevent and treat infections.

Depending upon the degree of obstruction, some people with partial or intermittent obstructions can eat an oral diet of low-fiber, low-residue foods in frequent small meals. When a person's medical condition suggests that obstructions are likely to recur, a very-low-residue diet may be appropriate.

Disorders of the Large Intestine

Fiber-modified diets play a role in the management of diseases involving the colon. In diverticular disease, diet is important to both prevention and treatment. Treatment of some disorders of the large intestine necessitate surgery to remove a portion of the colon; here, too, diet plays a role in management.

Diverticular Disease

Sometimes pouches of the intestinal wall (called diverticula) bulge out through the muscles surrounding the large intestine, often at points where blood vessels enter the muscles (see Figure 19.1 on this page). Evidence suggests that the pouches result from intense pressure in the intestinal lumen combined with weakness of the supporting muscles in the intestinal wall. Strong intestinal contractions pinch off segments of the intestine; pressure then builds in the segments and forces parts of the membrane to balloon outward through the muscle layer. Diverticular disease often occurs with aging.

WWW **niddk.nih.gov/health/digest/pubs/divert/divert.htm**
Diverticulosis and Diverticulitis (from the National Digestive Diseases Clearinghouse)

Diverticulosis and Diverticulitis People with diverticulosis are frequently symptom-free and unaware of the disorder. In some people, however, fecal material gets trapped in the diverticula. A localized area of inflammation and infection develops, a condition called diverticulitis. People with diverticulitis may suffer from abdominal pain and distension, alternating episodes of diarrhea and constipation, indigestion, flatus, and fever. Occasionally, a diverticulum ruptures, causing a localized or sometimes life-threatening infection (peritonitis). If the diverticula become inflamed repeatedly, the intestinal wall can form scar tissue and thicken (fibrosis), narrowing the intestinal lumen and creating an obstruction. An inflamed bowel segment can also stick to other pelvic organs, forming a fistula.

Reminder: The term *diverticulosis* describes the condition of having diverticula. The term *diverticulitis* describes the condition when diverticula become inflamed.

fistula: an abnormal opening between two organs or from an organ to the skin.

Dietary Treatments For many years, health care professionals advised people with diverticulosis to adopt low-fiber diets in the belief that dietary fibers tended to become trapped in the diverticula and cause irritation. In recent years, however, this advice has changed dramatically. Now health care professionals believe a high-fiber diet may actually reduce the incidence of diverticulosis by reducing pressure in the colon and stimulating intestinal peristalsis. A study of over 45,000 people suggests that high-fiber, low-fat diets are associated with a lower incidence of symptomatic diverticular disease.[2] People with diverticular disease may be advised to avoid foods with seeds such as okra and strawberries, however, because the seeds may get trapped in the diverticula and cause irritation.

During periods of active diverticulitis, however, fiber restrictions are appropriate. The level of restriction is guided by the client's medical condition. A person with severe and active diverticulitis, for example, might receive a very-low-residue diet. As diverticulitis begins to resolve, the person might be switched to a low-fiber, low-residue diet. Finally, once diverticulitis has resolved, the person will be advised to gradually increase the intake of fiber-containing foods. The case study on the next page presents a client with diverticular disease.

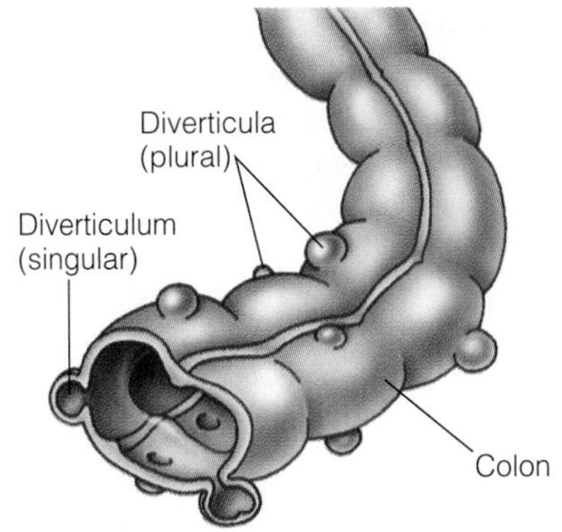

Figure 19.1
Diverticula
Outpocketings of intestinal linings that balloon through weakened muscles of the intestinal wall are known as **diverticula** (dye-ver-TIC-you-la); the singular is diverticulum.

Case Study

Retiree with Diverticular Disease

Mr. Stavros is a 69-year-old retired schoolteacher. He has recently been experiencing abdominal pain and fever. He told his doctor that he is frequently constipated, although he occasionally experiences diarrhea. He further complained of indigestion and a bloated feeling. Mr. Stavros was admitted to the hospital. After an examination that included X rays of the intestine, the physician made a diagnosis of diverticular disease, for which Mr. Stavros was treated. No food or drink was allowed by mouth, and a tube was inserted to suction gastric contents. An intravenous (IV) solution of fluid and electrolytes was started to prevent dehydration and maintain electrolyte balance. Antibiotics were given to treat the infection, and analgesics were given for pain and to lower Mr. Stavros's fever. After several days, suction was discontinued, and oral intake was initiated. Once Mr. Stavros was tolerating adequate liquids orally, the IV was discontinued. He is now symptom-free and ready for discharge from the hospital.

Describe diverticular disease to Mr. Stavros. Be sure to distinguish between diverticulosis and diverticulitis. How do diverticula form, and what consequences may follow?

Are Mr. Stavros's symptoms typical of people with diverticulitis? Do all people with diverticular disease have symptoms?

What diet would you recommend for Mr. Stavros to treat diverticular disease? What advice would you give him about adjusting to such a diet?

Resections of the Large Intestine

Diet Order
Progress diet from a low-fiber, bland diet to a diet as tolerated. Encourage fluids.

WWW 207.158.234.82
International Ostomy Association

For some clients, intestinal obstructions, lesions, or tumors necessitate the removal of all or a portion of the large intestine. In a colostomy, a segment of the colon or rectum (or both) is removed, and the remaining portion is then brought out through the abdominal wall via a stoma to allow for defecation (see Figure 19.2 on p. 471). In an ileostomy, both the entire colon and the rectum are removed, and the ileum becomes the terminal GI segment. In an alternative to an ileostomy, the ileal pouch/anal anastomosis, the surgeon removes the diseased colon and rectal tissue and connects the ileum to the anus. Thus defecation can occur through the anus, rather than through a stoma. A temporary ileostomy is made at the time of surgery to give the intestine time to heal. After about two to three months, the ileostomy is closed. In this section the word *ostomy* is used to refer to both colostomies and ileostomies.

stoma (STOH-ma): a surgically formed opening. After a colostomy or ileostomy, a stoma is formed by bringing the cut end of the intestine through the abdominal wall, rerouting the excretion of wastes.
stoma = window

The consistency of the stools following colostomies and ileostomies varies depending on both the length and the portion of the resected bowel. In general, ileostomies result in watery stools and colostomies in more formed stools.

Nutrition Support following Surgery Once solid foods are permitted following surgery, people who have undergone colostomies or ileostomies often receive low-fiber, bland diets to prevent obstructing the ostomy, help promote healing of the stoma, and prevent GI upsets. Encourage people to try other foods as soon as possible, however. They should add foods one at a time and in small amounts so that their effects can be assessed. If the added food presents problems, the person can try it again in a few weeks or months.

Preventing Obstructions Some foods are more likely than others to be incompletely digested and obstruct the stoma. These include stringy foods such as celery, spinach, and bean sprouts; foods with tough skins such as dried fruits, raw apples, and corn; foods with seeds; mushrooms; and nuts. Practitioners report that some of these foods can be used if the client cuts the food into small pieces and chews them thoroughly. An undigested mushroom, for example, may act as a plug and obstruct a stoma, but if it is cut into small pieces, it may be tolerated.

Encouraging Fluids People with ostomies need extra fluids because they are absorbing less fluid from the large intestine. They may tend to restrict their fluid intakes, however, for fear of aggravating diarrhea. Explain that drinking fluids helps prevent constipation and dehydration, and that excess fluid taken above and beyond the amount lost through the stoma will be absorbed and excreted by the kidneys; fluids will not aggravate diarrhea.

Controlling Diarrhea People with ostomies may benefit from foods that thicken the stool and help control diarrhea. These foods include applesauce, bananas, cheese, creamy peanut butter, and starchy foods such as breads, rice, and potatoes. Foods that may aggravate diarrhea include apple, grape, and prune juices; highly seasoned foods; and caffeine. The foods listed here are suggestions only; what works for the individual is determined by trial and error.

Reducing Gas and Odors People with ostomies are often concerned about gas and odors associated with foods. Gas-forming foods in general were listed in Table 19.4, but certain foods in particular seem to cause gas and odors for people with ostomies: asparagus, beans, beer, broccoli, brussels sprouts, cabbage, carbonated beverages, cauliflower, eggs, fish, garlic, and onions. Foods thought to reduce odors include buttermilk, cranberry juice, parsley, and yogurt.

Emotional Support Emotional support for people who need ostomies should begin before surgery. Meeting emotional needs helps alleviate people's fear and prepares them for the adjustments they face following surgery. People ready to undergo surgery are often frightened of the impact of physical changes to their

Figure 19.2
Colostomy and Ileostomy

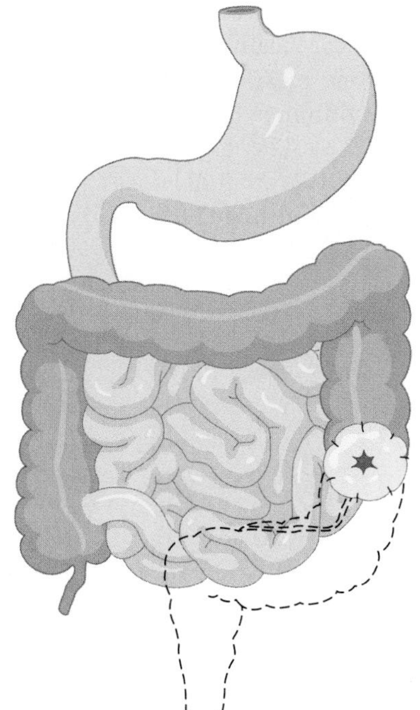

In a colostomy, the rectum and anus are removed, and the stoma is formed from the remaining colon.

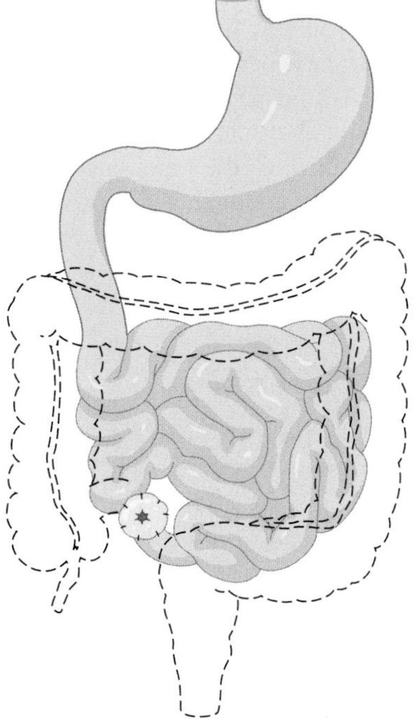

In an ileostomy, the entire colon, rectum, and anus are removed, and the stoma is formed from the ileum.

Nutrition Assessment Checklist

FOR PEOPLE WITH LOWER GI TRACT DISORDERS

Medical Review the client's medical record for test results that pinpoint the causes of problems such as diarrhea and for complications that alter diet therapy such as fever, intestinal obstructions, or fistulas.

Medication Review the client's drug therapy for possible nutrient-medication interactions. Many antibiotics have potential interactions. Laxatives may cause nausea, interfere with the absorption of fat-soluble vitamins, and result in laxative dependence. Some antidiarrheals may cause nausea or constipation.

Food Intake Assess food intake to determine individual food intolerances for people with diarrhea and irritable bowel syndrome. Assess fiber and fluid intake for people with constipation and diverticular disease. Assess foods and their effects on symptoms such as diarrhea, constipation, gas, and odors for people with colostomies and ileostomies.

Anthropometric Adjust diet to support growth in children and desirable weight in adults.

Laboratory Monitor lab values for signs of dehydration and electrolyte imbalances (electrolytes, blood urea nitrogen, hemoglobin, and hematocrit). Breath hydrogen tests help identify lactose intolerance.

Physical Check for physical signs of dehydration and nutrient imbalances.

body. They often feel they have lost control over a basic and private function. The training required to care for the ostomy and maintain bowel function may be difficult. Many people with osotomies also fear that loved ones, particularly spouses, may find them unattractive. The health care team should work closely with each person and the family to help everyone make the necessary adjustments and resume normal activities. Most communities have ostomy support groups that meet regularly to discuss mutual concerns. People who have already been through the experience can offer practical solutions to problems. The nutrition assessment checklist provides guidelines for monitoring the nutrition status of people with lower GI tract disorders.

Self Check

1. The nurse is instructing a client with diverticulosis about a high-fiber diet. The nurse emphasizes all of the following foods **except:**
   a. whole-grain breads and cereals.
   b. milk and milk products.
   c. legumes.
   d. fresh fruits and vegetables.
2. Intestinal residue is composed of:
   a. undigested foods and fibers.
   b. intestinal secretions.
   c. shed intestinal cells.
   d. all of the above.
3. A health care professional working with a client following a low-fiber diet learns that the clients eats a sandwich and piece of fruit and drinks a glass of milk for lunch. Considering lunch only, what other information will help the health care professional assess the client's compliance with the diet?

a. The type of bread and fruit the client uses.
b. The type of milk (nonfat, lowfat, or whole) and the condiments the client uses.
c. The time the client eats in relation to other meals.
d. The length of time the client has been on the diet.

4. Diet therapy for people with delayed gastric emptying most often includes:
   a. high-fiber foods given in frequent small meals.
   b. formulas provided by tube.
   c. low-fiber, low-fat foods given in frequent small meals.
   d. very-low-residue foods.

5. Treatments for constipation include:
   a. limited fluids.
   b. low-fiber foods.
   c. rest after meals.
   d. high-fiber foods.

6. Which of the following describe(s) diarrhea?
   a. It refers to an increased frequency or volume of stools.
   b. It is a medical disease.
   c. It always requires treatment with antidiarrheal agents.
   d. It always requires treatment with antimicrobial agents.

7. A common disorder characterized by a disturbance in the motility of the GI tract with symptoms that frequently occur shortly after a person eats is called:
   a. constipation.
   b. diarrhea.
   c. irritable bowel syndrome.
   d. diverticular disease.

8. Treatments for partial intestinal obstructions may include all of the following **except:**
   a. laxatives to help propel the intestinal contents.
   b. insertion of a tube from the nose to the stomach to help relieve pressure and distension.
   c. low-fiber/low-residue diets or very-low-residue diets.
   d. antibiotics.

9. Long-term management of diverticular disease includes a:
   a. high-fiber diet that omits foods with seeds.
   b. low-fiber diet that omits foods with seeds.
   c. high-fiber diet.
   d. low-residue diet.

10. Clients undergoing colostomies and ileostomies may temporarily receive a low-fiber, bland diet to:
   a. increase intestinal motility.
   b. help the stoma heal.
   c. prevent malabsorption.
   d. prevent mineral deficiencies.

Clinical Applications

1. Many symptoms and disorders of the GI tract described in this chapter and the last are associated with aging. Review both chapters and list these symptoms and disorders. Referring to Chapter 14, describe the effects of aging on the GI tract and describe how these changes might relate to the symptoms and disorders you find.

2. The last chapter discussed how an obstruction of the esophagus might occur as a consequence of chronic reflux esophagitis, and this chapter described how the stomach might become obstructed as a result of defective pyloric sphincter function. Contrast complete obstructions of the upper GI tract with complete intestinal obstructions. In which case would providing feedings by tube be possible and why?

Notes

1. D. Harari and coauthors, Bowel habit relation to age and gender: Findings from the National Health Interview Survey and clinical implications, *Archives of Internal Medicine* 156 (1996): 315–320; C. S. Probert and coauthors, Evidence for the ambiguity of the term constipation: The role of irritable bowel syndrome, *Gut* 35 (1994): 1455–1458.

2. W. H. Aldoori and coauthors, A prospective study of diet and the risk of symptomatic diverticular disease in men, *American Journal of Clinical Nutrition* 60 (1994): 757–764.

CHAPTER 19

Nutrition in Practice

■ Food and Foodservice in the Hospital ■

When medical conditions require treatment in a hospital or long-term care, foods are prepared for and provided to the clients. Many of these clients have conditions that interfere with the appetite either directly or through emotional distress. The hospital's foodservice department, under the direction of an administrative dietitian or foodservice manager, faces a challenge in planning, producing, and delivering meals designed to accommodate dozens of special diets and food preferences. This nutrition in practice addresses the special problems inherent in foodservice delivery and describes how foodservice systems work. Note that this discussion focuses on hospitals, but much of it applies to any health care facility that serves foods to large groups of people, including nursing homes, assisted living centers, rehabilitation centers, and residential mental health care facilities. Although people in hospitals may eat poorly, they can make up for nutrient deficits by eating well when they return home. The resident of a long-term care facility, however, does not have this option. For this reason, foodservice departments in long-term care facilities must make even greater efforts to ensure that their clients receive nutritious and appetizing foods.

WWW.

ashfsa.org
American Society for Healthcare Food Service Administrators of the American Hospital Association

ashe.org
American Society for Healthcare Engineering

eatright.org
American Dietetic Association

I don't know very many people who have been in the hospital, but those I do know always seem to have a comment about the food. Why?

Eating offers clients familiarity in an otherwise strange environment. Most people generally look forward to eating, and for many people in the hospital, a healthy appetite signals a return to health. Eating may become even more enjoyable than usual. It is also one of the few hospital experiences where clients have a choice. Consider that clients usually cannot choose when they will receive tests, how many times blood will be drawn, what nurse will care for them, or what time they will have surgery. But they usually can select their meals, and they can use those meals to exercise control or express their feelings: they can eat or refuse to eat!

When a client complains about hospital food, the complaint may have little to do with the food itself, but instead serves as a way to vent fear, frustration, anger, and physical pain. Clients need opportunities to express their feelings, and often you may find that a problem can be resolved simply by listening and providing emotional support.[1]

Don't clients sometimes have real complaints about food?

Yes, of course. Many disorders, medications, and treatments such as radiation therapy can dramatically alter taste perceptions and lead to complaints about food.[2] In addition, the hospital may not prepare foods in the same way the client does at home—a considerable problem when the client must eat three meals a day for many days in the hospital. Unfortunately, too, hot foods may not be hot and cold foods may not be cold by the time they arrive in the client's room. In addition, the client receives meals at specified times regardless of hunger and often must eat in bed without companionship, which can be more of a chore than a pleasurable experience. Meals may be unwelcome if the person is in pain.

What can be done to help correct these problems?

The majority of people in the hospital will eat adequate amounts of foods, even though they complain about it. If their intakes decrease somewhat, the deficit will be easy to correct once they are at home eating familiar foods. Chapter 21 provides suggestions for helping people to eat. For people in pain, administering pain

Figure NP19.1
Sample Lunch Menus

Lunch

REGULAR **SUNDAY**

Meats

Baked chicken❤ — Fried fish
Hamburger on bun with chips (with lettuce and tomato)

Starchy Vegetables

Cornbread dressing — Parsleyed potatoes❤

Vegetables

Baby carrots❤ — Stewed tomatoes

Soup/Salad/Juice	**Dressings**
Coleslaw	French
Clam chowder	Thousand Island
Gelatin	Italian
Tossed salad❤	Diet Thousand Island❤

Desserts

Apple pie — Butterscotch pudding
Fresh fruit❤

Breads

Dinner roll	Bran bread❤
White bread	Crackers
Wheat bread	

Beverages & Condiments

Coffee	Sugar
Decaf. coffee	Sugar substitute
Hot tea	Herb seasoning
Decaf. hot tea	Creamer
Iced tea	Lemon
Whole milk	Mustard
Buttermilk	Mayonnaise
2% milk	Catsup
Skim milk❤	Margarine
Chocolate milk	

PLEASE DO NOT LEAVE MENU ON THE TRAY

NAME ________________ **ROOM** ______

People on regular diets select the foods of their choice. The regular menu may also be used for high-kcalorie, high-protein diets. The menu items marked with a heart, guide people in selecting foods that are lower in fat, cholesterol, sodium, and caffeine or higher in fiber than other menu choices.

Lunch

SOFT/BLAND/LOW RESIDUE **SUNDAY**

Meats

Baked chicken — Baked fish(cod)
Hamburger on bun

Starchy Vegetables

Rice — Boiled potatoes

Vegetables

Baby carrots — Green beans

Soup/Salad/Juice	**Dressings**
Gelatin	Mayonnaise
Lemonade	Catsup
Tomato soup	

Desserts

Apple pie — Pears

Breads

Dinner roll	Crackers
White bread	

Beverages & Condiments

Decaf. coffee	Sugar
Decaf. hot tea	Sugar substitute
Decaf. iced tea	Creamer
Hot chocolate	Lemon
Whole milk	Margarine
2% milk	
Buttermilk	
Skim milk	

NO PEPPER
PLEASE DO NOT LEAVE MENU ON THE TRAY

NAME ________________ **ROOM** ______

Foods for soft/bland/low-residue diets are similar to those for regular diets. Foods from the regular menu that are not appropriate have been eliminated from the menu, and substitutes have been made. For clients on bland diets, decaffeinated coffee and tea would be crossed off the menu.

Lunch

KCALORIE RESTRICTED, DIABETIC
1200 **CALORIES** **SUNDAY**

LF = Low Fat LSLF = Low Sodium, Low Fat

Meat Exchange (Select 1)

LSLF Baked chicken (2oz) — LSLF Baked fish (2oz)
LSLF Hamburger on bun (with lettuce and tomato, 2oz meat) omit 2 starches

Starch Exchange (Select 1)

Clam chowder (1 c)	LF Dinner roll (1)
LSLF Rice (⅓c)	White bread (1 slice)
LSLF Boiled potatoes (½c)	Wheat bread (1 slice)
Angel food cake (1" slice)	Bran bread (1 slice)
	Crackers (6)

Vegetable Exchange (Select 2)

LSLF Baby carrots (½ c) — LSLF Green beans (½ c)

Fruit Exchange (Select 1)

Diet pears (½ c) — Fresh fruit

Milk Exchange (Select 1)

Whole milk (1 c) omit 2 fats	Buttermilk (1 c)
2% Milk (1 c) omit 1 fat	Skim milk (1 c)

Fat Exchange (Select 1)

Margarine (1 tsp)	Creamer (1 = ½ fat)
Diet mayonnaise (½ oz)	

Calorie-free Foods

Coffee	LSLF Coleslaw (½ c)
Decaf. coffee	Tossed salad (1 c)
Hot tea	Diet gelatin (½ c)
Decaf. hot tea	Diet French
Iced tea	Diet Thousand Island
Sugar substitute	Diet Italian
Lemon	Mustard
Herb seasoning	Diet catsup

PLEASE DO NOT LEAVE MENU ON THE TRAY

NAME ________________ **ROOM** ______

For kcalorie-restricted and diabetic diets, the number of exchanges allowed is written on the menu beforehand. (This example uses a 1200-kcalorie diet.) Note that the meat exchange is written in 2-ounce portions so that 1 serving = 2 exchanges. (Chapter 25 describes the exchange system.)

medications so that they will be effective during meals can be helpful. For people with altered taste perceptions, the dietitian can work closely with them to uncover the tastes and food preferences that they can tolerate and enjoy.

In some cases, the foodservice department must be contacted to solve food-related problems. In one study, researchers found that clients expressed the most satisfaction with meals that were served attractively, tasted good, and offered cold foods at the correct temperature.[3] Some problems can be handled directly by the person caring for the client. For these problems, understanding how the foodservice system works can help the caregiver handle problems more efficiently.

Who is responsible for problems directly related to the foodservice department?

The responsibility of budgeting, planning, preparing, and serving appropriate meals rests with either a chief administrative dietitian or a foodservice manager. In some facilities, foodservice companies from outside the hospital perform these duties.

Clinical dietitians work directly with clients to assess their nutrition status, plan appropriate diets, and provide nutrition education. In some facilities, dietetic technicians assist dietitians in both administrative and clinical responsibilities. Other dietary employees include clerks, porters, and other assistants. Keep in

Lunch

LOW-FAT/LOW CHOLESTEROL/ CARDIAC — **SUNDAY**

LF = Low Fat LSLF = Low Sodium, Low Fat

Meats
LSLF Baked chicken — LSLF Baked fish (cod)
LSLF Hamburger on bun (with lettuce and tomato)

Starchy Vegetables
LSLF Rice — LSLF Boiled potatoes

Vegetables
LSLF Baby carrots — LSLF Green beans

Soup/Salad/Juice
LSLF Coleslaw
Gelatin
Tomato soup
LS Chicken broth
Tossed salad

Dressings
Diet French
Diet Thousand Island
Diet Italian

Desserts
Pears — Angel food cake
Fresh fruit

Breads
LF Dinner roll — Bran bread
White bread — Crackers
Wheat bread — LS Crackers

Beverages & Condiments
Coffee — Creamer
Decaf. coffee — Sugar
Hot tea — Sugar substitute
Decaf. hot tea — Herb seasoning
Iced tea — Lemon
Buttermilk — Margarine
Skim milk — Mustard
Diet mayonnaise
Catsup

PLEASE DO NOT LEAVE MENU ON THE TRAY

NAME ____________ **ROOM** ______

People on low-fat, low-cholesterol diets who also need kcalorie restriction receive a kcalorie-restricted menu to control portion sizes and number of servings. Both menus provide low-fat, low-cholesterol foods. Foods not appropriate for a low-fat, low-cholesterol diet, such as whole milk, would be crossed off the menu beforehand.

Lunch

LOW SODIUM — **SUNDAY**

LF = Low Fat LSLF = Low Sodium, Low Fat

Meats
LSLF Baked chicken — LSLF Baked fish (cod)
LSLF Hamburger on bun (with lettuce and tomato)

Starchy Vegetables
LSLF Rice — LSLF Boiled potatoes

Vegetables
LSLF Baby carrots — LSLF Green beans

Soup/Salad/Juice
LSLF Coleslaw
LS Chicken broth
Apple juice
Tossed salad

Dressings
Diet French
Diet Thousand Island
Diet Italian

Desserts
Angel food cake — Pears
Fresh fruit

Breads
Dinner roll — Bran bread
White bread — LS Crackers
Wheat bread

Beverages & Condiments
Coffee — Sugar
Decaf. coffee — Sugar substitute
Hot tea — Creamer
Decaf. hot tea — Lemon
Iced tea — Herb seasoning
Whole milk — Margarine
2% Milk — Diet mustard
Skim milk — Diet mayonnaise
Diet catsup

NO SALT

PLEASE DO NOT LEAVE MENU ON TRAY

NAME ____________ **ROOM** ______

Low-sodium menus are similar to those provided for low-fat, low-cholesterol diets, but they eliminate high-sodium foods, such as tomato soup. The person on a low-sodium, low-fat, low-cholesterol diet selects foods from a low-fat menu with high-sodium foods crossed off beforehand. If the person is also on a low-kcalorie diet, foods would be selected from a low-kcalorie menu with high-sodium foods crossed off the menu beforehand.

Lunch

RENAL — **SUNDAY**

LF = Low Fat LSLF = Low Sodium, Low Fat

Meats (2 oz)
LSLF Baked chicken — LSLF Baked fish
LSLF Hamburger on bun (with lettuce)

Starchy Vegetables
LSLF Rice — LSLF Dialyzed potatoes

Vegetables
LSLF Baby carrots — LSLF Green beans

Soup/Salad/Juice
Lemonade
LSLF Coleslaw
Tossed salad (no tomato)

Dressings
Diet French
Diet Thousand Island
Diet Italian

Desserts
Pears — Apple pie

Breads
Dinner roll — Bran bread
White bread — LS Crackers
Wheat bread

Beverages & Condiments
Coffee — Sugar
Decaf. coffee — Sugar substitute
Hot tea — Creamer
Decaf. hot tea — Lemon
Iced tea — Margarine
Diet mustard
Mayonnaise

NO SALT

PLEASE DO NOT LEAVE MENU ON THE TRAY

NAME ____________ **ROOM** ______

Renal diets must be highly individualized, and the person checking the menu has to carefully consider the client's selections and make appropriate changes when necessary.

mind that many dietary employees do not have formal education in nutrition, and their ability to interpret diet orders and provide accurate information is limited.

How does the foodservice department know what foods to serve each client?

Most hospitals provide menus from which clients can select their meals. A client who must follow a modified diet receives menus that list only foods specified in the hospital's diet manual for that particular diet. By allowing a choice, this system helps to ensure that clients receive foods they enjoy and will eat. An added advantage for people on special diets is that they become familiar with their diets by marking appropriate menus.

Although procedures vary somewhat between facilities, generally dietary employees deliver menus to each client's room early in the day and pick them up again later in the day. Each menu shows the client's name and room number, as well as the name of the meal, the type of diet, and the day the menu will be served. Generally, the client makes selections for the next day or for the next few days to give the foodservice department time to collect the menus and estimate the amount and type of food to prepare. Menus are usually color coded by diet. Color coding helps ensure that foodservice

CHAPTER 19

Nutrition in Practice

Foodservice departments prepare foods to accommodate dozens of special diets and hundreds of food preferences.

employees put the right types of foods on food trays and helps the person delivering the tray to quickly determine if the right diet has been delivered. Figure NP19.1 shows lunch menus for several different diet menus and explains how each menu might be used.

If clients are making their own selections, how can they be receiving foods they don't like?

Several problems with menus can occur. Clients may not receive foods they enjoy if they fail to mark the menus or make the wrong food selections. Consider these possible scenarios:

- A client may have trouble seeing, reading, understanding, or physically marking menus.
- Clients may not understand that their selections will be for the *next* (or another) day.
- Clients may be out of their rooms (for tests, procedures, or physical activity) or asleep when the menus arrive; when the clients return or wake up, they may not see the menus or may have missed the menu pickup time.
- Clients may be too ill or too disinterested in food to make menu selections.

If menus are not marked or if a menu is lost, the client receives meals preselected by the foodservice department.

Occasional problems with menu selections can usually be corrected simply by explaining the menu system to clients or taking extra time to help them mark menus. If clients continue to complain about food selections, contact the dietetic technician or dietitian.

What happens to the menus once they are collected?

Once food selections have been made and menus collected, menus are often checked by a member of the foodservice staff (usually, a dietetic technician or dietitian) to make sure that selections are appropriate. Completed menus can provide valuable clues about a person's usual eating habits or understanding of a modified diet. In checking menus, for example, the technician may notice that one person on a regular diet is selecting very little or that another is selecting too much. In another case, the technician may notice that a person on a kcalorie-restricted diet is not selecting the appropriate number of servings from each food group. Such problems suggest the need for further intervention.

Do all facilities offer selective menus?

Not all. Those that do not offer a selective menu serve a standard house diet, adjusting the menu for individual food preferences. For example, clients can request simple changes, such as the substitution of one vegetable for another.

How do foodservice departments prepare foods to meet the needs of a variety of diets?

The logistics of preparing foods tailored to each modified diet can be overwhelming. For this reason, foodservice departments use systems designed to limit costs and minimize errors. Foods prepared for regular and soft/bland/low-fiber/low-residue diets are prepared with some fat and salt, because these dietary components are not restricted on such diets. Note that the other diet menus shown in Figure NP19.1 provide a number of low-fat (LF) or low-sodium (LS) or low-sodium, low-fat (LSLF) foods.

If the foodservice department were to prepare a food (baked chicken, for example) for each different diet, it would have to prepare regular baked chicken, low-fat baked chicken, low-sodium baked chicken, and low-sodium, low-fat baked chicken. Using the system illustrated in the menus of Figure NP19.1, only two types of baked chicken need to be prepared—one with some fat and salt, and the other without fat or salt.

Keep in mind that baked chicken is only one of many menu items in a day, and you can see why preparing individual foods for each diet is not feasible. Instead, clients can add allowed ingredients to food.

For example, the person on a low-salt diet could add margarine to a serving of vegetables; the person on a low-fat diet could add salt to a serving of rice.

How does the foodservice delivery system work?

Sometimes foods are prepared in a main kitchen, assembled on trays, and heated in areas close to the clients' rooms. In other cases, foods are delivered directly from the main kitchen. In either case, foodservice personnel deliver food carts directly to the nursing unit; then nursing or foodservice personnel take a tray to each client. Efficient delivery of foods to the nursing unit and then to the client helps ensure that clients receive foods at the appropriate temperature.

Once the client is finished eating, the tray is returned to the food cart. Foodservice personnel pick up the carts and return them to the foodservice department.

How can I use this information to help clients eat better?

You can help clients greatly—and save yourself needless aggravation and time—by learning about the foodservice system in the health care facility where you work. Better yet, ask to spend a few hours or days working with different foodservice employees to see firsthand how the department operates and what problems they encounter. If that is not possible, learn the facility's procedures for ordering diets, making diet changes, reporting problems with a client's tray, or making special requests. Remember that requests are not simply made by one individual to another. Often many people are involved in processing a single request, and the number of requests made during any one meal can be considerable. Translating requests (for example, preparing another tray) takes time, and delays are unavoidable. The best strategy is prevention—make sure that clients mark menus and that you call in requests as early as possible to allow the foodservice department the time to process the request.

One of the most important things to know about your facility's foodservice system is the time when meals are actually assembled, so that you can call in requests well before this time. Once tray assembly begins, dietary employees are extremely busy, and requests will be difficult to process.

With so many people and steps involved in foodservice, and so many clients with individual dietary needs and food preferences, it is easy to see many opportunities for problems to arise. Once you understand how the foodservice department operates, you can use the system to tackle problems efficiently and avoid needless frustration for your client and yourself.

Notes

1. M. Bélanger and L. Dubé, The emotional experience of hospitalization: Its moderation and its role in patient satisfaction with foodservice, *Journal of the American Dietetic Association* 96 (1996): 354–360.

2. M. A. Hess, Taste: The neglected nutritional factor, *Journal of the American Dietetic Association* (supplement 2) 97 (1997): 205–207.

3. P. A. O'Hara and coauthors, Taste, temperature, and presentation predict satisfaction with foodservices in a Canadian continuing-care hospital, *Journal of the American Dietetic Association* 97 (1997): 401–405.

20 Carbohydrate-, Fat-, and Protein-Modified Diets for Malabsorption Syndromes

Chapters 18 and 19 have shown how medical nutrition therapy aids in the treatment of disorders that affect the intake of food and its transit through the GI tract, as well as disorders in which diets help alleviate GI symptoms and complications. This chapter describes diets that help treat disorders that lead to malabsorption. Malabsorption can arise as a consequence of many disorders, treatments, and medications and can involve one or many nutrients. Table 20.1 lists indications for carbohydrate-, fat-, and protein-modified diets for malabsorption syndromes.

Table 20.1 Indications for the Use of Carbohydrate-, Fat-, and Protein-Modified Diets for Malabsorption Syndromes

Carbohydrate-Modified Diets
Dumping syndrome
Lactose intolerance
Fat-Modified Diets
Bacterial overgrowth of stomach and small intestine
Bile duct obstruction
Blind loop syndrome
Cystic fibrosis
HIV infection
Inflammatory bowel disease
Liver disease
Pancreatitis
Short-bowel syndrome
Protein-Modified Diets
Celiac disease

Carbohydrate-Modified Diets for Malabsorption Syndromes

Carbohydrate-modified diets used in the treatment of malabsorption syndromes include the postgastrectomy diet and the lactose-restricted diet. The postgastrectomy diet serves to control malabsorption that often occurs following surgery that removes a portion of the stomach. Lactose-restricted diets prevent symptoms that accompany lactose intolerance.

Postgastrectomy Diets

Postgastrectomy diets control the total amount of carbohydrate and carefully limit simple sugars (see Table 20.2 on p. 483). To provide energy and slow the passage of foods through the stomach, the diet emphasizes foods high in protein and provides additional energy from fat. A sample postgastrectomy diet menu is shown on p. 483.

gastrectomy (gas-TREK-tah-mee): surgery that removes all (total gastrectomy) or part (partial or subtotal gastrectomy) of the stomach.

pyloroplasty (pie-LOOR-o-PLAS-tee): surgery that enlarges the pyloric sphincter.

vagotomy (vay-GOT-o-mee): surgery that severs the nerves that stimulate gastric acid secretion.

gastric partitioning: surgery for severe obesity that limits the functional size of the stomach.

Gastric Surgery Several surgical procedures affect the functions of the stomach. During a gastrectomy, the surgeon removes either a portion or all of the stomach. Figure 20.1 illustrates three common gastrectomy procedures. Another type of gastric surgery, pyloroplasty, enlarges the pyloric sphincter (the sphincter at the junction of the stomach and small intestine) so that the basic intestinal fluids reflux into the stomach and neutralize gastric acidity. During a vagotomy, the surgeon severs the nerves that stimulate gastric acid production. A vagotomy may accompany either a gastrectomy or a pyloroplasty in some cases. In gastric partitioning, a treatment for severe obesity, the stomach remains intact, but all or a portion of the stomach is bypassed.

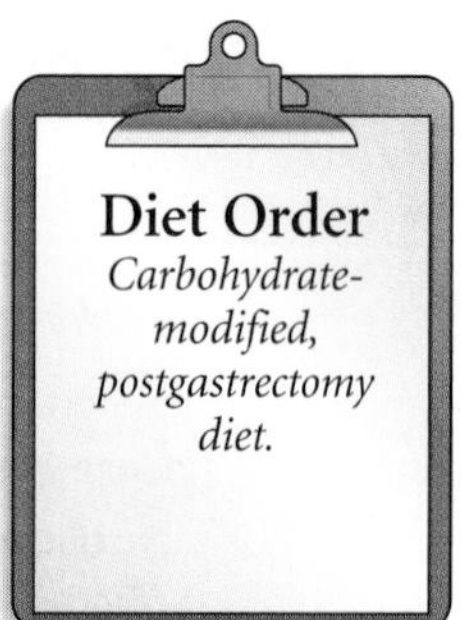
Diet Order
Carbohydrate-modified, postgastrectomy diet.

Postsurgical Care In the immediate postsurgical period, the client receives no foods or fluids by mouth. Health care professionals monitor fluid and electrolyte balances carefully, and the physician corrects imbalances promptly. After several days, liquids and then solids are gradually introduced in frequent small amounts. The postgastrectomy diet aims to prevent and control the dumping syndrome, described next.

Dumping Syndrome When the portion of the stomach containing the pyloric sphincter has been removed, bypassed, or disrupted, one problem that may occur is the dumping syndrome (see Figure 20.2 on p. 484). Without a functional pyloric sphincter, the stomach loses control over the rate at which food enters the intestine. Instead, food gets "dumped" rapidly into the jejunum. (The duodenum is short, and even if it has not been bypassed surgically, food passes quickly through it into the jejunum.)

Table 20.2 Postgastrectomy Diet[a]
Meat and Meat Alternatives
Any type allowed.
Milk and Milk Products
Withheld initially and then gradually introduced as tolerated.
Grains and Starchy Vegetables
Allowed (up to 5 servings per day): Plain breads, crackers, rolls, unsweetened cereal, rice, pasta, corn, lima beans, parsnips, peas, white potatoes, sweet potatoes, pumpkin, yams, winter squash. *Excluded:* Sweetened cereal; cereal containing dates, raisins, or brown sugar.
Nonstarchy Vegetables
Allowed (unlimited): Cabbage, celery, Chinese cabbage, cucumbers, lettuce, parsley, radishes, watercress. *Allowed* (up to two ½ c servings per day as individual tolerances permit): Asparagus, bean sprouts, beets, broccoli, brussels sprouts, carrots, cauliflower, eggplant, green pepper, greens, mushrooms, okra, onions, rhubarb, sauerkraut, string beans, summer squash, tomatoes, turnips, zucchini. *Excluded:* Vegetables prepared with sugar or creamed.
Fruits
Allowed (up to 3 servings per day): Unsweetened fruits and fruit juices. *Excluded:* Sweetened fruits and fruit juices, dates, raisins.
Fats
Any type allowed.
Beverages
Allowed: Coffee, tea, artificially sweetened drinks. *Excluded:* Alcohol; sweetened milk, beverages, and fruit drinks; cocoa.
Other
Excluded: Cakes, cookies, ice cream, sherbet, honey, jam, jelly, syrup, and sugar.

[a]Clients with dumping syndrome unable to tolerate a sufficient variety or volume of foods over long periods of time often require nutrient supplements.

Sample Postgastrectomy Diet Menu

Breakfast
1 scrambled egg
1 slice toast
1 tsp butter
Coffee (take 30–60 minutes after meal)

Midmorning Snack
¼ c cottage cheese
1 graham cracker

Lunch
2 oz hamburger patty
¼ c mashed potatoes
1 tsp margarine
½ small banana
Iced tea (take 30–60 minutes after meal)

Midafternoon snack
2 tbs peanut butter
3 saltine crackers

Supper
2 oz boiled ham
¼ c rice
½ c carrots
2 tsp butter
¼ c unsweetened peach slices
Tea (take 30–60 minutes after meal)

Evening Snack
¼ c tuna
1 tsp mayonnaise
1 slice bread

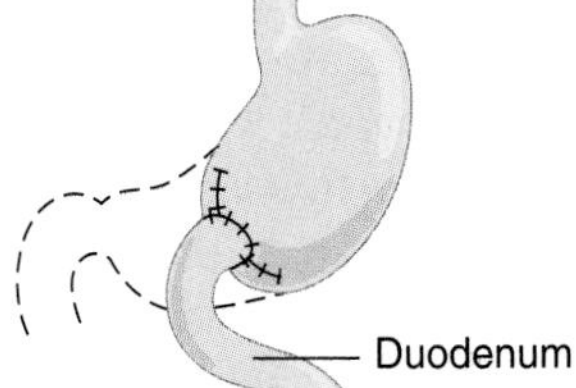

Gastroduodenostomy

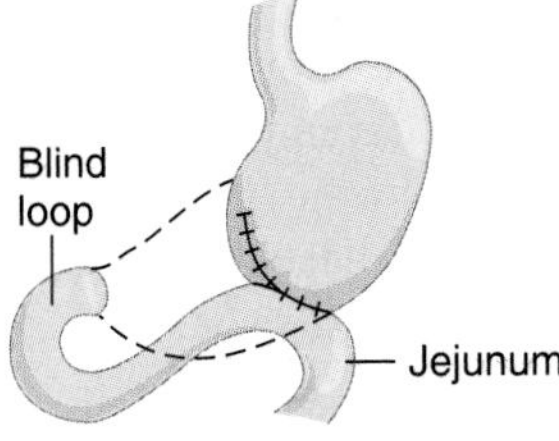

Gastrojejunostomy

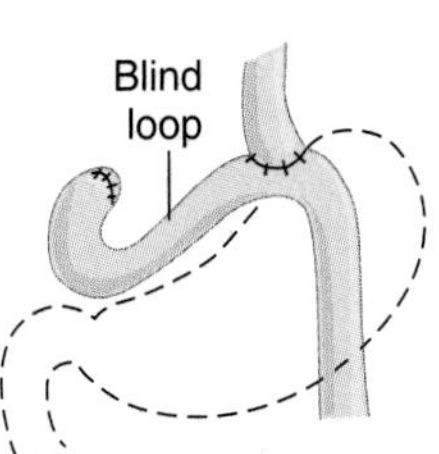

Total gastrectomy

Figure 20.1
Typical Gastric Surgery Resections
In a gastric resection, part or all of the stomach is surgically removed. The dashed lines show the removed section.

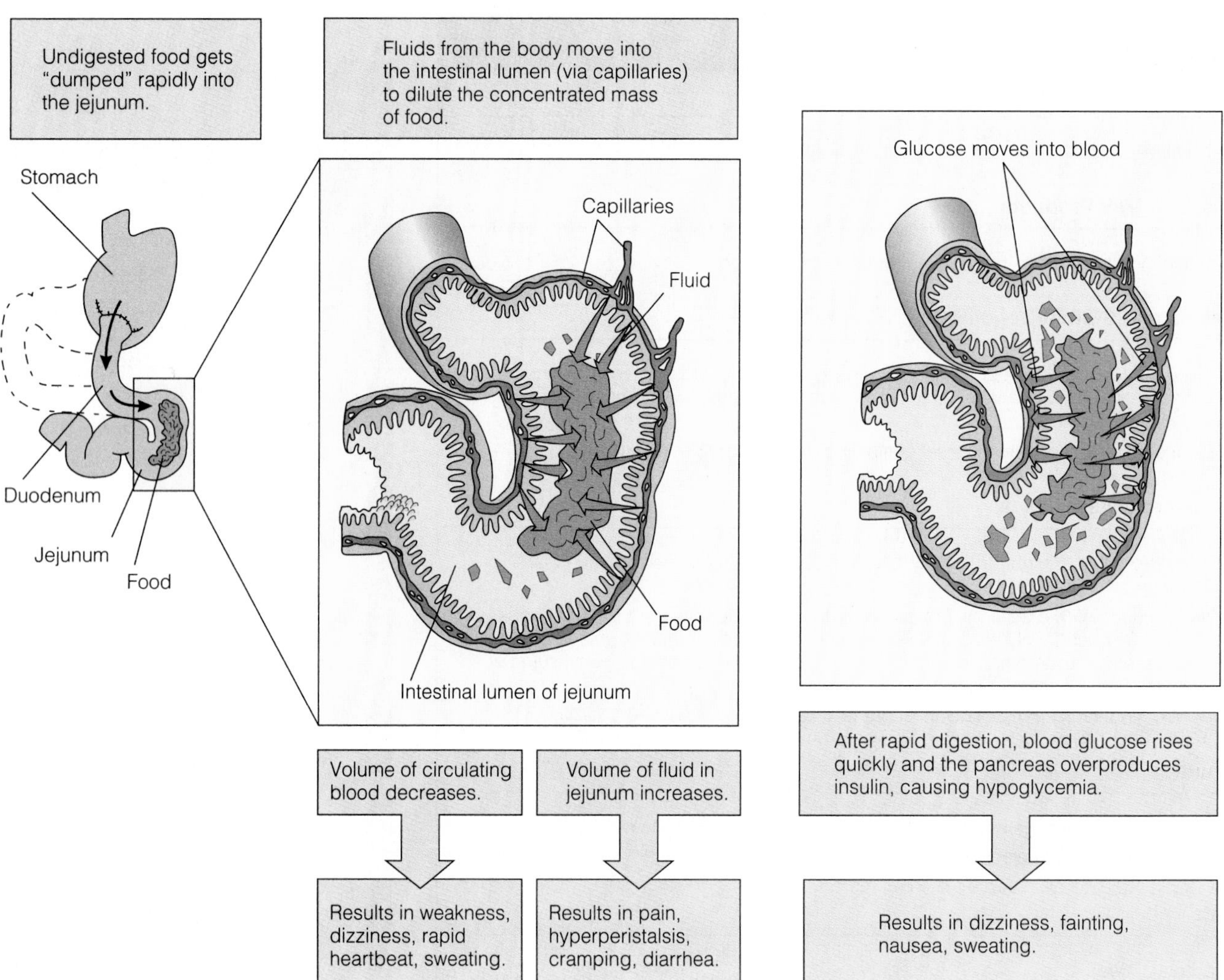
Undigested food gets "dumped" rapidly into the jejunum.
Fluids from the body move into the intestinal lumen (via capillaries) to dilute the concentrated mass of food.
Glucose moves into blood
Stomach
Capillaries
Fluid
Duodenum
Jejunum
Food
Food
Intestinal lumen of jejunum
Volume of circulating blood decreases.
Volume of fluid in jejunum increases.
After rapid digestion, blood glucose rises quickly and the pancreas overproduces insulin, causing hypoglycemia.
Results in weakness, dizziness, rapid heartbeat, sweating.
Results in pain, hyperperistalsis, cramping, diarrhea.
Results in dizziness, fainting, nausea, sweating.

Figure 20.2

Dumping Syndrome

When partially digested food rapidly enters the jejunum, it is quickly digested and creates a concentrated mass. Fluid from the intestinal capillaries enters the jejunum, diminishing blood volume and stimulating peristalsis. The result: low blood pressure and diarrhea.

As the mass of food is digested, the intestinal contents rapidly become concentrated (hypertonic). Fluids from the body move into the intestinal lumen to dilute the concentration. Consequently, the volume of circulating blood diminishes rapidly, causing weakness, dizziness, and a rapid heartbeat. The large volume of hypertonic fluid and unabsorbed material in the jejunum causes pain and hyperperistalsis, and diarrhea results.

Two to three hours later, many of the same symptoms may occur again and others may appear: dizziness, fainting, nausea, and sweating. This time the cause is different. The intestines efficiently absorbed so much glucose from the meal that blood glucose rose quickly. The pancreas responded by overproducing insulin, which made the blood glucose *fall* quickly. Now low blood glucose (hypoglycemia) is causing the symptoms.

Not all people experience the diarrhea of the dumping syndrome following gastric procedures. Even fewer develop hypoglycemia. Many people who initially experience the dumping syndrome gradually adapt to a fairly regular diet. Nevertheless, a postgastrectomy diet serves as a preventive measure in the immediate postsurgical period and benefits clients with prolonged or severe problems.

Diet Adjustment Diet advice to offer the client along with the postgastrectomy diet is provided in the next box. Dietitians carefully tailor the postgastrectomy diet to meet the person's tolerances. Initially, the dietitian visits the client after each meal to check the client's tolerance for food. With time, the symptoms of

How to

Adjust Meals to Prevent the Dumping Syndrome

The person on a postgastrectomy diet benefits from the following suggestions:

- Avoid concentrated sweets (sugar, cookies, cakes, pies, or soft drinks) because the body digests and absorbs these carbohydrates rapidly and breaks them down into many particles that draw fluids from the body into the intestine.
- Eat frequent small meals to fit the reduced storage capacity of the stomach.
- Drink liquids in small amounts about 45 minutes before or after meals, not with them. This precaution prevents overloading the stomach's reduced storage capacity and slows the transit of food from the stomach to the intestine.
- Lie down immediately after eating to help slow the transit of food to the intestine. Clients who experience esophageal reflux, however, should not lie down after eating.
- Be aware that lactose intolerance (described in a later section) may develop and add to the problem of diarrhea and abdominal pain. Enzyme-treated milk and milk products should also be avoided because the enzymes break down lactose to glucose and galactose—simple sugars that further concentrate the intestinal contents and promote dumping. Discontinue all types of milk and milk products until recovery is under way. Then introduce them gradually and in small amounts.

the dumping syndrome often resolve or improve. Gradually, most people begin to tolerate limited amounts of concentrated sweets, larger quantities of food, and some liquids with meals. Sometimes adding pectin and guar gum (types of dietary fibers) to the diet can help the dumping syndrome to resolve. If dietary and medical management of the dumping syndrome fail to adequately resolve the problem, additional surgery may be necessary.

Rx Prescription Pad

Medications used in the treatment of the dumping syndrome may include:

- Anticholinergics (atropine, see *antidiarrheals*)
- Antihistamines (cyprohepatine, see *miscellaneous*)
- Hormones (octreotide, see *miscellaneous*)

See Appendix E for timing with meals and nutrition-related side effects.

Weight Loss and Malabsorption After gastric surgery, many people experience significant weight loss and develop nutrient deficiencies. Limited food intake can occur as a consequence of early satiety, postsurgical pain, the desire to prevent the symptoms of the dumping syndrome, reflux esophagitis, and, sometimes, dysphagia.

In addition to the malabsorption that occurs whenever foods pass rapidly through the intestine, fat malabsorption (described later in this chapter) can occur whenever clients have had total gastrectomies or surgery that bypasses the duodenum. When food bypasses the duodenum, the trigger for the release of hormones that mediate the secretion of enzymes and bile to aid fat digestion and absorption is also bypassed, and fat malabsorption results.

Reduced gastric acid secretion can lead to bacterial overgrowth (described later) in the stomach or upper small intestine, which may also contribute to the malabsorption of fat, fat-soluble vitamins (especially vitamin D), folate, vitamin B_{12}, and calcium.

Anemia Iron-deficiency anemia is another common problem following gastric surgery, although it may take several years to develop. When iron's exposure to gastric acid is limited, less iron is converted to its absorbable form. The rapid transit of food through the duodenum (where 50 percent of iron absorption normally occurs), inadequate iron intake, and blood loss contribute to the problem. An iron supplement helps correct the deficiencies.

Inadequate intake and malabsorption can also lead to anemia caused by folate deficiency and, less often, by vitamin B_{12} deficiency. To correct deficiencies, clients receive supplements.

Although one might expect vitamin B_{12} deficiency to be common after gastric surgery because intrinsic factor production might be affected, surgeons often avert this problem by leaving intact a small part of the stomach that produces intrinsic factor.

Reminder: *Osteomalacia* is a bone disease characterized by softening of the bones.

Bone Disease People who experience fat malabsorption following gastric surgery also malabsorb vitamin D and calcium. After many years, a significant number of people who have undergone gastrectomies develop a bone disease similar to osteomalacia.[1] Vitamin D and calcium supplements are frequently provided, although the bone disease is often resistant to treatment.

As mentioned, temporary or permanent lactose intolerance may develop following gastric surgery. The next section describes the lactose-restricted diet.

Lactose-Restricted Diets

Lactose-restricted diets are highly individualized diets that most often limit, but do not exclude, milk and milk products. People test their tolerance for lactose by consuming small amounts of lactose-containing foods throughout the day and gradually increasing serving sizes up to the point that they develop symptoms of lactose intolerance (described later). Studies suggest that many lactose-intolerant individuals can tolerate up to 1 or 2 cups of milk a day without significant symptoms, provided that the milk is taken with food and intake is divided throughout the day.[2] Often people can eat cheeses, particularly aged cheeses.[3] Some tolerate chocolate milk better than plain milk. Most tolerate yogurt well, although they may need to experiment with different brands to find the one that works best.

WWW **niddk.nih.gov/** health/digest/pubs/lactose/lactose.htm Lactose Intolerance

People can also add a lactase enzyme preparation to milk before they drink it or take enzyme tablets whenever they eat lactose-containing foods. Milk and milk products already treated with lactase are also available. However, because people can usually tolerate a fair amount of milk, products to aid lactose digestion are probably unnecessary.

People with temporary lactose intolerance, described below, are advised to temporarily restrict all milk and milk products. Then lactose-containing foods are gradually reintroduced in small amounts.

Lactose Intolerance Lactose intolerance results from a deficiency of the digestive enzyme lactase, the enzyme that splits lactose to glucose and galactose in the intestine. In rare cases, a person is born with a lactase deficiency; more often, lactase activity gradually diminishes with age. Lactose intolerance is prevalent among people in certain ethnic groups including those of Mediterranean origin, African Americans, Asians, Jews, Mexicans, and Native Americans. Permanent or temporary lactase deficiencies can develop as a consequence of any disorder or condition that damages the delicate intestinal microvilli including malnutrition, radiation therapy (see Chapter 24), and many of the conditions described later in this chapter including inflammatory bowel diseases, GI tract surgery, and celiac disease.

Strong advertising campaigns to promote products for lactose intolerance have led many people to believe they have lactose intolerance, when, in fact, they do not.[4] Some believe they are intolerant to even the smallest amounts of lactose. These people may have underlying flatulence that they incorrectly attribute to lactose intolerance. Breath hydrogen tests (see Nutrition in Practice 20) confirm a diagnosis of lactose intolerance.

Symptoms of Lactose Intolerance When lactose absorption in the small intestine is blocked, the intestinal contents become hypertonic, causing cramps, distension, and diarrhea. Bacteria in the intestine metabolize the undigested sugars to irritating acids and gases, further contributing to cramping and diarrhea and causing flatulence. The severity of the symptoms varies depending on the amount of lactose eaten and the degree of lactose intolerance.

Preventing Calcium and Vitamin D Deficiencies People who restrict milk and milk products because they have lactose intolerance, or believe they have it, risk calcium and vitamin D deficiencies. To help prevent such deficiencies, encourage people to include milk and milk products in their diets to the extent that they can tolerate them. People who need to include more milk and milk products than they can tolerate, as well as those who are truly intolerant to even small amounts of lactose, can use enzyme-treated products, add enzymes to milk and milk beverages, or take enzyme tablets when they consume lactose-containing foods. People who fail to receive adequate amounts of calcium from milk and milk products should be encouraged to eat other food sources of calcium. Getting enough calcium from foods other than milk and milk products is difficult, however, because most other foods do not contain significant amounts of calcium, or the calcium they contain is poorly absorbed. Calcium supplements may be indicated, particularly for children, adolescents, and pregnant, lactating, or postmenopausal women. Vitamin D is not a problem if the person gets regular exposure to sunlight; otherwise, a supplement may be necessary.

People who avoid milk and milk products should be encouraged to use other food sources of calcium such as calcium-fortified juices, calcium-fortified cereals, broccoli, mustard greens, kale, and sardines.

Carbohydrate-modified diets for malabsorption syndromes help prevent symptoms associated with the dumping syndrome or lactose intolerance. The disorders described in the next section may include lactose intolerance as a symptom, but their nutrition consequences are related to fat malabsorption.

Fat-Modified Diets for Malabsorption Syndromes

As Chapter 18 described, dietary fats delay the rate at which the stomach empties, and for that reason, low-fat diets may be used to treat disorders that delay gastric emptying or for reflux esophagitis. The fat-modified diets described here help solve nutrition problems caused by fat malabsorption in the intestine.

Fat Malabsorption

Disorders of the stomach, intestine, pancreas, and liver can lead to fat malabsorption and the consequent loss of kcalories and many other nutrients. In addition to accelerating nutrient losses, malabsorption syndromes and their treatments may lead to malnutrition by reducing nutrient intakes and raising nutrient needs (see Table 20.3)—all of which can profoundly threaten nutrition status.

Unabsorbed fat is excreted from the body in the stools, causing the type of diarrhea called steatorrhea. Protein may also be malabsorbed, although usually

steatorrhea (stee-ah-toe-REE-ah): fatty diarrhea characterized by loose, foamy, foul smelling stools.

Table 20.3 Possible Causes of Malnutrition in Malabsorption Syndromes

Reduced Nutrient Intake	Excessive Nutrient Losses	Raised Nutrient Needs
Abdominal pain	Blood loss	High basal energy expenditure
Anorexia	Diarrhea	Infection
Bowel rest	Fistulas	Medications
Emotional stress	General malabsorption	Surgery
Food intolerance	Intestinal losses of serum proteins	
Indigestion	Medications	
Medications	Steatorrhea	
Nausea	Vomiting	
Obstructions		

to a lesser extent than fat. The loss of fat in the stools means that valuable food energy, essential fatty acids, fat-soluble vitamins, and some minerals are lost as well (see Figure 20.3).

Essential Fatty Acid Deficiency Among the fats lost in steatorrhea are the essential fatty acids. Studies show that many people with severe fat malabsorption develop essential fatty acid deficiencies.[5]

Vitamin and Mineral Malabsorption Fat-soluble vitamins are excreted in the stools along with unabsorbed fat. Minerals normally are absorbed in the colon, but when fat malabsorption occurs, unabsorbed fatty acids form soaps with calcium and magnesium and are not available for absorption. Vitamin D losses further aggravate calcium malabsorption.

soaps: chemical compounds formed between a basic mineral (such as calcium) and unabsorbed fatty acids. Soaps give steatorrhea its foamy appearance.

Oxalate Stones The binding of calcium to fatty acids can cause another problem. Oxalate, which is present in some foods, normally binds with some of the calcium in the gut and is excreted with it. But when fatty acids bind the calcium, the oxalate remains unbound. The intestine absorbs the unbound oxalate, but the body cannot metabolize it, and so excretes it in the urine. High urinary oxalate favors the formation of kidney stones (see Chapter 27).

Treatments for Malabsorption Syndromes

To treat steatorrhea successfully, the underlying disorder must be diagnosed and treated. Drug therapy, and sometimes surgery, may be necessary. Dietary fat is often modified in type or amount; otherwise enzyme replacements (described later) are given to aid absorption. The following diet-planning principles apply to all fat malabsorption syndromes:

- Foods are provided in frequent small meals because fat is best tolerated in small amounts at a time.

Figure 20.3
The Effects of Fat Malabsorption

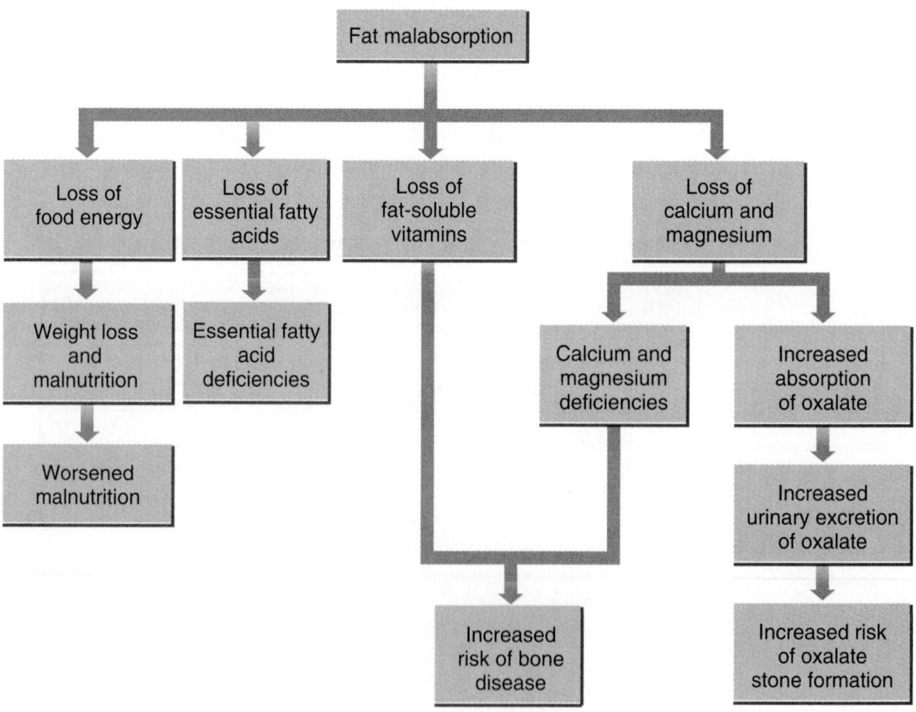

- Part of the fat may be given as medium-chain triglycerides because these fats are easier to digest and absorb than long-chain triglycerides.
- Fat-soluble vitamins are provided in a water-miscible form when fat malabsorption is severe. Water-miscible, fat-soluble vitamins mix readily with water and can be absorbed without fat.

Most naturally occurring fats are LCT; LCT contain side chains with at least 14 carbon atoms, and they require lipase and bile for digestion. MCT contain fatty acid side chains with 8 to 12 carbon atoms, and they require only minimal lipase and no bile for absorption.

Fat-Modified Diets for Malabsorption Individual tolerances determine the specific amount of fat prescribed, but usually it ranges from 20 to 40 percent of the total energy. Typically, the person with malabsorption begins with a diet containing about 35 to 40 grams of fat per day. If the person is tolerating this level of fat, additional fat may be provided gradually to meet energy needs. Table 20.4 provides instructions for a fat-restricted diet, and a sample fat-restricted diet menu is shown on p. 490.

Medium-Chain Triglycerides Note that the diet is not severely restricted in fat for any longer than is necessary; the person needs food energy. If the person cannot tolerate an adequate amount of fat, an alternative is to provide some fat from

Table 20.4 Fat-Restricted Diet (35 grams)

Use:

1. Nonfat milk, cheeses and yogurt made from nonfat milk, sherbet, and fruit ices.
2. Low-fat egg substitutes and up to three regular eggs per week.
3. Up to 6 oz of lean meats and poultry without skin daily.
4. Up to 3 servings of fat daily. One serving is any one of the following:
   1 tsp butter, margarine, shortening, oil, or mayonnaise
   1 strip crisp bacon
   1 tbs salad dressing
   ⅛ avocado
   2 tbs cream (half and half)
   10 small nuts
   8 large olives
   If fat is used to cook or season food, it must be taken from this allowance.
5. All vegetables prepared without fat.
6. All fruits prepared without fat.
7. Plain white or whole-grain bread; nonfat cereals, pasta, rice, noodles, and macaroni.
8. Clear soups.
9. Angel food cake and fruit whips made with gelatin, sugar, and egg-white meringues.
10. Jelly, jam, honey, gumdrops, jelly beans, and marshmallows.

Do not use:

1. Whole milk, chocolate milk, whole-milk cheeses, and ice cream.
2. Pastries, cakes, pies, sweet rolls, breads, or vegetables made with fat.
3. More than one egg a day, fried or fatty meats (sausage, luncheon meats, spareribs, frankfurters), duck, goose, or tuna packed in oil (unless well drained).
4. More than 3 servings of fat.
5. Desserts, candy, or anything made with chocolate, nuts, or foods not allowed.
6. Creamed soups made with whole milk.

Suggestions:

1. To make the diet still lower in fat, reduce the fat and meat (and egg) servings.
2. To raise the fat content, give additional fat or meat servings.
3. To improve acceptance of the diet, check the fat content of a well-liked food and allow that food if possible. Use the exchange system fat list for alternate suggestions for fat servings (see Appendix C).

Note: The box on p. 490 provides additional tips for lowering fat in the diet.

How to

Improve Acceptance of Fat-Restricted Diets

Fat-restricted diets can be difficult to follow. Fats give flavors, aromas, and textures to foods—characteristics that some people may miss.[a] Unlike some diets that can be introduced gradually, the fat-restricted diet for malabsorption must be implemented right away without giving the person time to adapt to the changes. These suggestions may help:

- Provide clients with tips for making foods palatable while lowering fat intake, such as those found in the box on p. 69.
- Remind clients that new fat-free and low-fat products appear on market shelves daily, and most people find these products very acceptable. Unless research proves otherwise, caution clients to avoid products containing fat substitutes (olestra), however. Even in healthy people, fat substitutes are associated with digestive problems and fat-soluble vitamin malabsorption.

People who use MCT need additional advice:

- Advise clients to add MCT to the diet gradually. Nausea, vomiting, diarrhea, abdominal pain, and distension can result from using too much MCT all at once.
- Recommend that clients improve the palatability of MCT oil by substituting it for regular oil in salad dressing and for cooking and baking and by adding it to beverages, desserts, and other dishes.
- Warn clients that MCT products are expensive, and explain that these products can be purchased at pharmacies and are sometimes covered by medical insurance.

[a]A. Drewnowski, Why do we like fat? *Journal of the American Dietetic Association* (supplement) 97 (1997): 58–62.

medium-chain triglycerides (MCT) rather than long-chain triglycerides (LCT). Products made from MCT oil and formulas containing MCT supply almost as many kcalories as regular fats, but people who cannot digest and absorb LCT can digest and absorb MCT. MCT oil does not contain essential fatty acids, however, so the diet must include some LCT. The accompanying box offers suggestions for helping people accept fat-restricted diets and includes tips for using MCT. Appendix G shows that many formulas provide some fat from MCT.

Sample Fat-Restricted Diet Menu

Breakfast	Lunch	Supper	Snack
Soft-cooked egg	3 oz broiled chicken	3 oz lean roast beef	Fruit ice
Dry cereal	Rice	Mashed potatoes	
Orange juice	Green beans	Peas	
Coffee, sugar	Tossed salad	Bread	
Whole-wheat toast	1 tbs low-fat French dressing	1 tsp margarine	
½ tsp margarine	Fresh apple	Peaches	
Nonfat milk	Iced tea, sugar	Nonfat milk	
	1 tsp margarine		

All foods are prepared without added fat.

Water-Miscible, Fat-Soluble Vitamins Most often, the person with malabsorption absorbs enough fat-soluble vitamins that a standard supplement can be given, if necessary. For people who fail to maintain adequate vitamin pools, however, fat-soluble vitamins can be supplemented in a water-miscible form that facilitates absorption.

Oxalate-Restricted Diet To reduce the risk of oxalate stones, clients with fat malabsorption may be advised to limit foods high in oxalate. Foods notable for their high oxalate content include spinach, rhubarb, beets, nuts, chocolate, tea, wheat bran, and strawberries.

Enzyme Replacement Therapy Enzyme replacements are used when the person suffers malabsorption related to chronic and severe damage to the pancreas or whenever steatorrhea is severe. Enzyme replacements, like naturally occurring enzymes, work best in a basic pH. Because a severely damaged pancreas also fails to secrete enough bicarbonate to provide an optimal pH for enzymes to function, people with pancreatic insufficiency often take antisecretory agents to limit gastric acid production. At best, enzyme replacements taken with meals lessen the malabsorption of protein and fat, but may not fully correct it.

enzyme replacements: extracts of pork or beef pancreatic enzymes that are taken as supplements to aid digestion.

Fat-modified diets are useful in the treatment of many disorders characterized by fat malabsorption. Because the causes of malabsorption vary, diet therapy varies somewhat as well.

Pancreatitis

Diet Order
Progress diet from clear liquids to a fat-restricted diet provided in frequent small meals to a regular diet as tolerated.

Pancreatic secretions contain many enzymes necessary for the digestion of protein, fat, and carbohydrate, together with bicarbonate-rich juices that provide the optimal pH necessary to activate these enzymes. Normally, the pancreas stores digestive enzymes in an inactive form to protect itself from digestion. In pancreatitis, however, digestive enzymes are activated within the pancreas and begin to damage the organ itself. The blood picks up some of these enzymes; thus elevated serum amylase and lipase serve as indicators of acute pancreatitis.

pancreatitis: inflammation of the pancreas.

Consequences of Pancreatitis Typical symptoms of pancreatitis include severe abdominal pain, nausea, and vomiting. Some people develop adynamic ileus. In some cases, pancreatitis can lead to tissue death and life-threatening complications including shock, renal failure (Chapter 27), respiratory failure, pancreatic hemorrhages, fistulas, and abscesses.[6] Chronic pancreatitis often leads to fat malabsorption. Pancreatitis most often develops as a consequence of gallstones or alcoholism; sometimes, though, the reasons are unclear because a variety of medical conditions and some drugs can also precipitate pancreatitis.[7]

Reminder: *Adynamic ileus* is a temporary paralysis of the intestine.

Reminder: A *fistula* is an abnormal opening between two organs or from an organ to the skin.

abscess: an accumulation of pus, caused by local infection, that builds up and may eventually burst.

Treatment of Acute Pancreatitis Initial therapy aims to suppress pancreatic secretions and to treat the underlying cause of pancreatitis. Food is withheld, because food stimulates pancreatic activity. For some people, a tube inserted into the stomach to suction gastric secretions helps treat ileus, relieves pain, or alleviates severe vomiting. Intravenous fluids and electrolytes help maintain fluid and electrolyte balances. Analgesics are provided to relieve pain.

Diet Therapy for Acute Pancreatitis In most cases, pancreatitis resolves in less than a week; the person can begin oral intake when abdominal pain subsides and

℞ Prescription Pad

Medications used in the treatment of pancreatitis may include:

- Analgesics
- Antisecretory agents
- Insulin
- Pancreatic enzyme replacements

See Appendix E for timing with meals and nutrition-related side effects.

serum amylase returns to normal or near normal. The diet progresses from liquids to a fat-restricted diet and, finally, to a regular diet as tolerated. If eating aggravates the pain, or if serum amylase rises, food is withheld; when these signs and symptoms subside, food can again be reintroduced.

If pancreatitis is severe or if complications arise, and if oral food intake fails to meet nutrient needs for more than a week, tube feeding or intravenous nutrition is indicated.[8] Studies suggest that feedings of easy-to-absorb (hydrolyzed) formulas delivered by tube directly into the jejunum do not significantly stimulate pancreatic secretions.[9] Tube feedings are preferred, but if abdominal pain, edema, vomiting, or other complications preclude the use of enteral nutrition, then intravenous nutrition is appropriate.

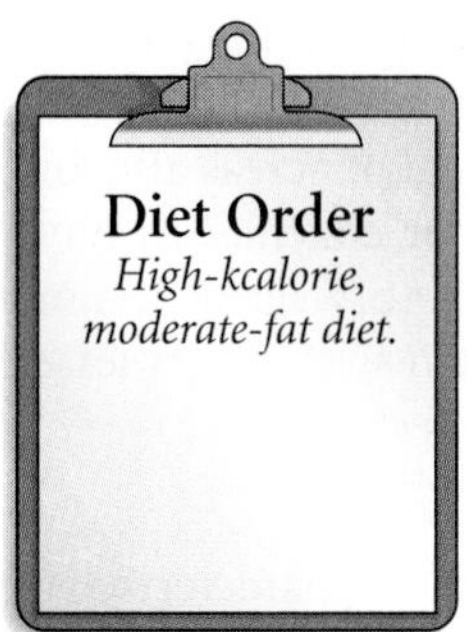
Diet Order
High-kcalorie, moderate-fat diet.

Chronic Pancreatitis If an acute episode of pancreatitis does not subside or if episodes recur at frequent intervals, the pancreatic cells can be permanently destroyed, leading to chronic pancreatitis. With extensive degeneration of the pancreas, digestion, especially of fat, becomes permanently impaired. Abdominal pain is often severe and unrelenting, vomiting is frequent, and severe weight loss is common.

Diet Therapy for Chronic Pancreatitis The goals of diet therapy for chronic pancreatitis are to maintain optimal nutrition status, reduce steatorrhea (if present), minimize pain, and avoid subsequent attacks of acute pancreatitis. Health care professionals recommend a moderate fat-restricted diet for chronic pancreatitis; restricting fat too severely makes it difficult for the person to gain or maintain weight. Enzyme replacements taken with meals help the person digest and absorb protein and fat. Absolutely no alcohol is permitted.

Complicating Conditions During active attacks of pancreatitis, diet therapy reverts to that described earlier for acute pancreatitis. Sometimes pancreatitis damages the cells that produce insulin and glucagon. In these cases, clients become glucose intolerant, as in diabetes (see Chapter 25), and must follow a diet for diabetes. Deficiencies of the hormone glucagon, which lead to hypoglycemia, complicate the task of regulating blood glucose.

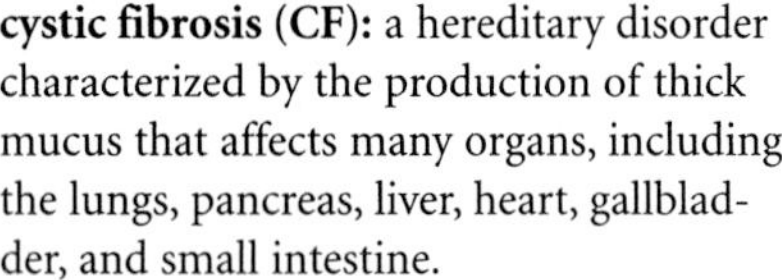
cystic fibrosis (CF): a hereditary disorder characterized by the production of thick mucus that affects many organs, including the lungs, pancreas, liver, heart, gallbladder, and small intestine.

WWW.
cff.org
Cystic Fibrosis Foundation

Cystic Fibrosis

Diet Order
High-kcalorie, high-protein diet.

People with cystic fibrosis, the most common fatal genetic disorder of the white population, produce secretions of thick, sticky mucus that may seriously impair the function of many organs, most notably the lungs and pancreas.[10] Just a few decades ago, few infants born with cystic fibrosis survived to adulthood. Today, thanks to advances in medical therapy and nutrition care, the outlook is much brighter, with some surviving into their forties and even fifties.

Cystic fibrosis has three major consequences: chronic lung disease, pancreatic insufficiency and malabsorption, and abnormally high electrolyte concentrations in the sweat. Chronic lung disease develops because the airways become congested with mucus, causing breathing to be labored. As the thick mucus stagnates in the bronchial tubes, bacteria can multiply there. Lung infections occur frequently and are the usual cause of death in people with cystic fibrosis.

Cystic fibrosis probably causes some degree of pancreatic insufficiency in all cases, with about 90 percent of cases serious enough to require enzyme replacement therapy. With time, damage to the pancreas worsens. The thick mucus obstructs the small pancreatic ducts and interferes with the secretion of digestive enzymes, pancreatic juices, and pancreatic hormones. Malabsorption of many nutrients, including fat, protein, vitamins, and minerals, often leads to malnutrition. Additionally, the secretion of insulin may be affected, resulting in glucose intolerance and diabetes.

Treatment of Cystic Fibrosis Therapy for cystic fibrosis aims to promote growth and development and prevent respiratory failure and other complications. Treatment includes respiratory, diet, and drug therapy.

Energy and Nutrient Needs Nutrient needs for people with cystic fibrosis range between 120 and 150 percent of the RDA for sex and age. Reduced food intake, nutrient losses through malabsorption, and heightened nutrient needs due to frequent infections, rapid turnover of protein and essential fatty acids, intense protein catabolism, and high basal energy expenditure all raise nutrient requirements. Extra energy is needed simply to breathe. Dietitians estimate individual energy requirements based on basal metabolic rate, activity level, pulmonary function, and degree of malabsorption.

Obtaining enough energy, however, is often complicated by a loss of appetite that may accompany repeated infections, emotional stress, and drug therapy. Coughing to clear the lungs may trigger vomiting or reflux of food from the stomach. Thus the person with cystic fibrosis may find it difficult to eat enough foods to meet nutrient needs.

Dietary Fat and Enzyme Replacements With such high energy needs, fat restrictions are inappropriate. Instead, enzyme replacements are used to control steatorrhea and relieve abdominal pain.[11]

Feeding Infants All infants with cystic fibrosis must be closely monitored to ensure that their diets meet their high energy and nutrient needs. Breastfeeding can sustain normal growth and development for infants with cystic fibrosis provided that enzyme replacements are given.[12] Additionally, the breastfed infant with cystic fibrosis needs ⅛ to ¼ teaspoon of table salt daily, given in water, to replace sweat-induced electrolyte losses.

Infants who are not breastfed can usually tolerate regular infant formulas, which can be provided in a more concentrated form to maximize nutrient intake. Infants who cannot tolerate regular formulas often receive easy-to-digest (hydrolyzed) formulas. Regardless of the type of feeding—human milk, standard infant formula, or hydrolyzed formula—enzyme replacements are always given as well.

Feeding Children and Adults A child with cystic fibrosis is weaned from breast milk or infant formula and thereafter requires a high-kcalorie, nutritionally balanced diet carefully tailored to food tolerances. People with cystic fibrosis need regular nutrition assessments to ensure that their diets are supporting their nutrient needs and well-being. Height and weight measurements are particularly relevant.

For children, every effort should be made to maintain weight at greater than 90 percent of that appropriate for height, gender, and age.[13] For adults, the goal is to maintain a healthy weight for height. If weight falls between 85 and 90 percent of desirable weight, diets should also include high-kcalorie snacks and formula supplements to meet energy needs.[14] Tube feedings are indicated for people with cystic fibrosis whose weights consistently remain below 85 percent of desirable weight. Weights below 75 percent indicate advanced malnutrition that necessitates either tube feedings or intravenous nutrition. Clients may benefit from home nutrition programs that include oral diets during waking hours and tube feedings or parenteral nutrition at night.

Rx Prescription Pad

Medications used in the treatment of cystic fibrosis may include:

- Antibiotics
- Antisecretory agents
- Bronchodilators (to ease breathing)
- Insulin
- Mucolytics (to thin mucous secretions)
- Pancreatic enzyme replacements

See Appendix E for timing with meals and nutrition-related side effects.

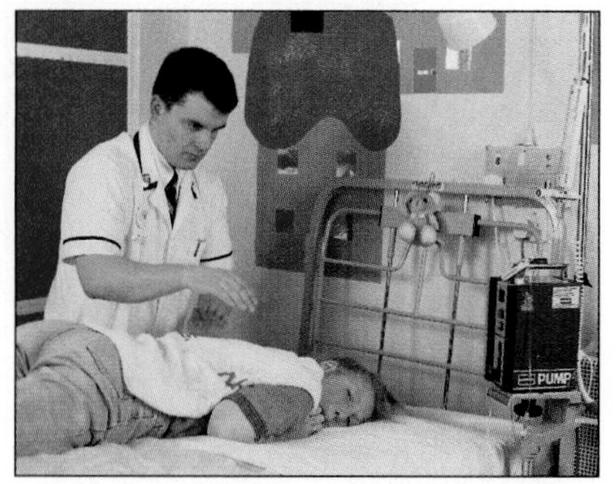

Chest physical therapy (postural drainage) is an important treatment in the management of cystic fibrosis.

Nutrient Supplementation Multivitamin and fat-soluble vitamin supplements help meet the nutrient demands imposed by a high-energy, high-protein intake and malabsorption. People with severe malabsorption may need to receive fat-soluble vitamins in a water-miscible form.

As for minerals, as mentioned earlier, abnormally high concentrations of electrolytes (sodium and chloride) in sweat are characteristic of cystic fibrosis. Fever, high environmental temperatures, vomiting, and malabsorption can further deplete electrolytes and lead to dehydration. The liberal use of table salt and fluids is encouraged.

Emotional Support People with cystic fibrosis and their caregivers are often acutely aware of the many aspects of care important to survival, of which nutrition is only one. They must often manage daily chest physical therapy treatments, enzyme replacements, and, often, antibiotics and other medications. Caregivers must also help people with cystic fibrosis to adjust to their disease, manage their social life, keep up with work or schoolwork, and assume as much responsibility for their health as possible. Health care professionals who are mindful of these challenges are best able to offer encouragement and support.

Disorders involving the pancreas can result in malabsorption by limiting the supply of digestive enzymes and other pancreatic secretions. Disorders affecting the intestine, described in the next sections, can cause malabsorption by altering either the absorptive area or the function of the intestinal cells.

Inflammatory Bowel Diseases

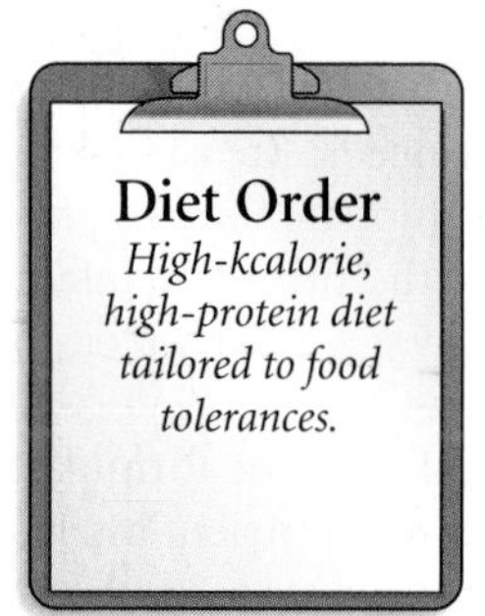

Diet Order
High-kcalorie, high-protein diet tailored to food tolerances.

The two most prevalent disorders characterized by inflammation of the intestine are Crohn's disease and ulcerative colitis. Inflammatory bowel diseases (IBD) share some clinical features but are distinct conditions. Their causes remain unknown, although heredity, environment, and immune factors are thought to be contributing factors.[15]

inflammatory bowel diseases (IBD): diseases characterized by inflammation of the intestine.

Crohn's disease: inflammation and ulceration along the length of the GI tract, often with granulomas.

ulcerative colitis (ko-LYE-tis): inflammation and ulceration of the colon.

granuloma (gran-you-LOH-mah): a tumor or growth that contains foreign organisms surrounded by immune system cells and covered with a fibrous coat.

Crohn's Disease In Crohn's disease, cracklike ulcers and sometimes granulomas accompany inflammation of the bowel. Crohn's disease most often affects the ileum and colon, but can affect the entire GI tract. The person often complains of intermittent fatigue, abdominal pain, diarrhea, and weight loss. There is no medical cure for Crohn's disease, and even if acute symptoms resolve, recurrences are likely.

Ulcerative Colitis Unlike Crohn's disease, which can occur anywhere along the GI tract, ulcerative colitis develops only in the large intestine. It causes severe bloody diarrhea, cramping, abdominal pain, anorexia, and weight loss. Diarrhea can be almost continuous, resulting in malabsorption and great losses of fluids and electrolytes. Anemia may develop due to the bleeding and malabsorption. For people with active ulcerative colitis who fail to respond to medical therapy, surgery to remove the colon and rectum is often recommended. Unlike intestinal surgery for Crohn's disease, which fails to cure the disorder, removal of the colon and rectum does cure ulcerative colitis.

WWW.
ccfa.org
Crohn's and Colitis Foundation of America

Consequences of Inflammatory Bowel Diseases As an inflammatory bowel disease progresses, fibrous tissue forms in the intestine, reducing its absorptive ability, narrowing the intestinal lumen, and sometimes causing an obstruction.

Localized infections can form. The intestine may also rupture, which can lead to peritonitis.

Reminder: *Peritonitis* is a serious, and sometimes fatal, infection of the lining surrounding the abdominal organs.

Fistulas may develop if an inflamed loop of intestine sticks to another loop of intestine, to another organ, or to the skin and gradually erodes. If a fistula forms between the stomach or the upper portion of the small intestine and the colon, ingested food is shunted directly into the colon, and malabsorption worsens. Bacteria from the colon can then invade the stomach or upper small intestine, contributing to further malabsorption (see "Bacterial Overgrowth" later in this chapter), increasing the risk of serious infections, and causing severe inflammation, nausea, and vomiting. If a fistula forms between the small intestine and the skin, significant malabsorption can occur if large volumes of fluid are lost through the fistula. Surgery may be necessary to remove a diseased or obstructed portion of the intestine or to repair a fistula.

Nutrition Status Nutrition status is severely threatened in people with inflammatory bowel diseases. The person often experiences emotional stress, anorexia, weight loss, fever, diarrhea, malabsorption, and cramping abdominal pain—all of which can lead to nutrient deficiencies. Oral intake may be withheld so that the bowel can rest, particularly when the disease is active or an obstruction develops or a fistula forms. In addition, bleeding from the lesions can lead to anemia, and inflammation can cause the secretion and loss of serum proteins, with resulting hypoalbuminemia. The accompanying drug therapy can further impair nutrition status. If the person requires surgery or develops an infection, nutrient needs become even greater. Because surgery removes a portion of the bowel, the procedure itself may contribute to malabsorption (see "Short-Bowel Syndrome" later in this chapter and ileostomies and colostomies in Chapter 19).

Because children with inflammatory bowel diseases need additional nutrients to support growth and maturation, nutrition problems are compounded. Growth failure is common and may be the manifestation of the disease that leads to its diagnosis.

Restoring and maintaining nutrition status can be a challenging task. Protein-energy malnutrition (PEM) and deficiencies of calcium, magnesium, zinc, iron, vitamin B_{12}, folate, vitamin C, and fat-soluble vitamins are commonly reported. Low serum albumin and multiple nutrient deficiencies threaten immune function and may reduce the effectiveness of drug therapy.

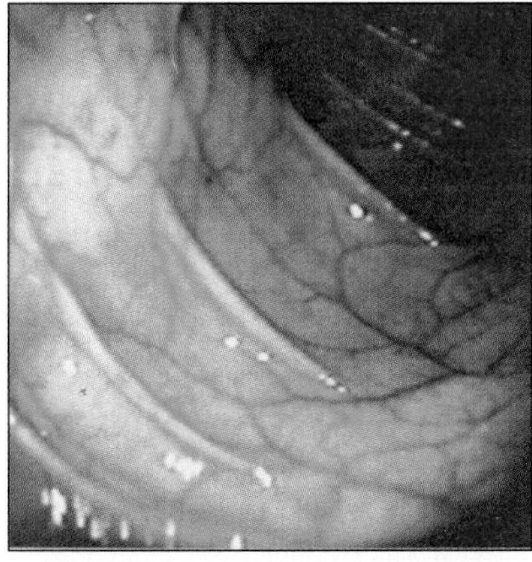

The normal colon has a smooth, shiny surface with a visible pattern of fine blood vessels.

Nutrition Support during Active Episodes For people with active Crohn's disease who have intestinal obstructions, all foods and fluids are withheld, nasogastric suction is initiated, and fluids and electrolytes are replaced intravenously. Diarrhea is a problem for people with active inflammatory bowel diseases, especially those with ulcerative colitis or Crohn's disease affecting the colon. A primary concern, then, is to ensure adequate intakes of fluids and electrolytes.

For people with active ulcerative colitis, no dietary interventions seem to lessen disease activity. People with severe abdominal pain and diarrhea need complete bowel rest with no enteral stimulation.

For people with Crohn's disease, enteral nutrition is the preferred feeding route, and easy-to-absorb (hydrolyzed) formulas may offer some advantages over standard formulas or table foods.[16] Hydrolyzed formulas may be unpalatable to some people, however, and if the formula cannot be taken orally, it can be fed by tube. In some cases, feeding tubes can be placed so as to bypass fistulas or partial obstructions, allowing enteral feeding without adding to the risk of complications.

Intravenous nutrition is used to deliver nutrients when oral or tube feedings significantly aggravate pain and diarrhea; when the bowel is obstructed; when complete bowel rest might help a fistula to close; or when oral or tube feedings cannot meet nutrient requirements.

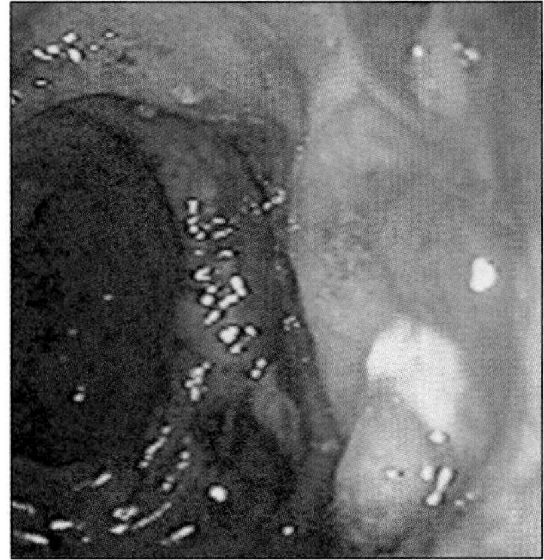

In ulcerative colitis, the colon appears inflamed and reddened, and ulcers are visible.

Rx Prescription Pad

Medications used in the treatment of inflammatory bowel diseases may include:

- Analgesics
- Antibiotics (metronidazole, sulfasalazine)
- Antidiarrheals
- Anti-inflammatory agents (prednisone)

See Appendix E for timing with meals and nutrition-related side effects.

Oral Diets As the acute stage of inflammatory bowel disease resolves, the person gradually progresses, as tolerance permits, to an oral diet, often a high-kcalorie, high-protein diet. Low-fiber diets are sometimes recommended for people with partial obstructions of the intestine. Fat-restricted diets may be necessary for people with fat malabsorption. Fish oils may help modulate the inflammatory responses that occur in inflammatory bowel disease. Some studies suggest that fish oil supplements may help prevent recurrences of active Crohn's disease.[17] Studies of the effects of fish oil for people with ulcerative colitis have shown modest benefits, but limitations of the studies make the findings difficult to interpret.[18] Further research is needed to clarify whether fish oil can be an effective treatment for inflammatory bowel diseases.

Clients may be intolerant to specific foods or food components, including lactose, and these should be identified and eliminated from the diet. Vitamin-mineral supplements are frequently prescribed. Encourage clients to eat a nutrient-rich, well-balanced diet, and reassess nutrition status frequently to ensure that nutrient needs are being met. The case study on p. 497 presents a person with Crohn's disease.

Bacterial Overgrowth

Diet Order
Progress diet from liquids to a fat-restricted diet to a regular diet as tolerated.

Although the colon normally houses a bacterial population, the small intestine is protected from bacterial overgrowth by gastric acid, which kills bacteria, and peristalsis, which flushes microorganisms through the small intestine before they can multiply. Conditions that disrupt these protective mechanisms can result in bacterial overgrowth in the stomach and small intestine. Gastric surgeries, for instance, can interfere with gastric acid secretions and alter peristalsis. In some types of gastric surgery, a portion of the small intestine is bypassed, causing stasis in that portion of the intestine and allowing bacteria to flourish (see Figure 20.1 on p. 483). The bypassed portion is called a blind loop, and the symptoms associated with bacterial overgrowth are called the blind loop syndrome.

blind loop syndrome: the problems of fat malabsorption and vitamin B_{12} and folate deficiencies that result from the overgrowth of bacteria in a bypassed segment of the intestine.

Other conditions that can lead to bacterial overgrowth by significantly reducing gastric acid secretions include chronic gastritis (see p. 449), medications (antisecretory agents), and HIV infections (see Chapter 24). Small bowel obstructions and nerve dysfunction associated with diabetes (Chapter 25) can lead to bacterial overgrowth by altering peristalsis.

Consequences of Bacterial Overgrowth Bacteria in the small intestine partly dismantle the bile salts, which are essential for fat digestion and absorption. Fat malabsorption and its related consequences occur as a result. Figure 20.4 repeats the figure from Chapter 5 (p. 108) that shows how bile prepares fat for digestion, this time illustrating how bacteria interfere with that process. The bacteria also compete with the body for vitamin B_{12} and folate, limiting the available supply and leading to vitamin B_{12} and folate deficiencies.

Treatment of Bacterial Overgrowth To control bacterial overgrowth, physicians prescribe antibiotics. Along with drug therapy, clients receive fat-restricted diets, parenterally administered vitamin B_{12}, and oral folate supplements. If medical treatment fails, people with surgically created blind loops may need additional surgery to remove the blind loop.

Case Study

College Student with Crohn's Disease

Lilinoe, a 19-year-old college student, was admitted to the hospital to manage an active episode of Crohn's disease. When Lilinoe was first diagnosed with Crohn's disease, she was 18 years old, weighed 120 pounds, and was 5 feet 7 inches tall. Her normal weight had been 130 pounds until the disease symptoms began to appear. Since then, she has been hospitalized several times for recurrent attacks of Crohn's disease. As anticipated from her weight history, Lilinoe's nutrition assessment shows that she is suffering from PEM. In talking with Lilinoe, you discover that she is very particular about the foods she eats and often simply does not eat. Her physician has ordered an easy-to-absorb formula diet to be fed to Lilinoe by tube.

Review Lilinoe's weight history. What is her desirable weight? What measures could have been taken to help Lilinoe avoid weight loss?

What possible benefits might an easy-to-absorb formula offer Lilinoe? Why might the physician prefer a tube feeding rather than oral feedings? How might Lilinoe be fed if she develops an obstruction or fistula?

Consider Lilinoe's long-term dietary management. What type of diet should she eventually follow? What goals should be set for weight gain? Why is it important to reassess her nutrition status regularly?

Given Lilinoe's age and stage of development, what emotional concerns might she be experiencing? What interventions might be planned to enhance her emotional health and well-being?

Short-Bowel Syndrome

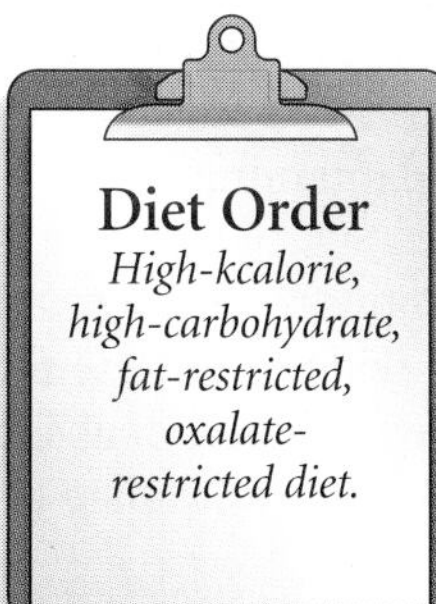

Diet Order
High-kcalorie, high-carbohydrate, fat-restricted, oxalate-restricted diet.

Short-bowel or short-gut syndrome is characterized by diarrhea, weight loss, muscle wasting, bone disease, hypocalcemia, hypomagnesemia, and anemia. It can occur whenever the absorptive surface of the intestine is significantly reduced. Short-bowel syndrome frequently results from surgery to remove a significant portion of the small intestine, which may be necessary in the treatment of inflammatory bowel diseases, cancer of the

short-bowel or **short-gut syndrome:** severe malabsorption that may occur when the absorptive surface of the small bowel is reduced, resulting in diarrhea, weight loss, bone disease, hypocalcemia, hypomagnesemia, and anemia.

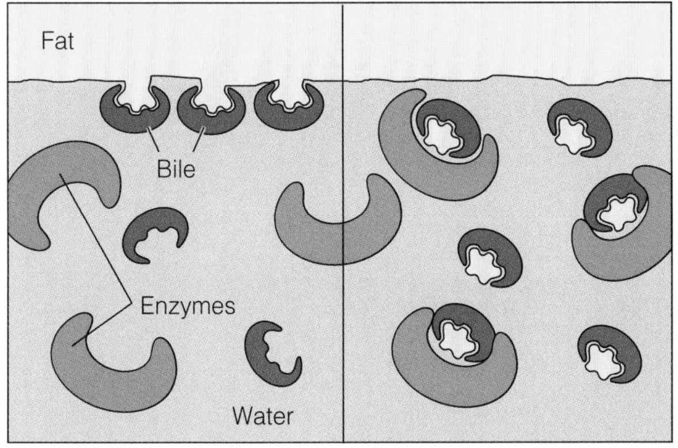

When fat enters the small intestine, bile arrives. Bile has an affinity for both fat and water, so it can bring the fat into solution in the water.

After emulsification, the fat is mixed in the water solution, so the enzymes have access to it.

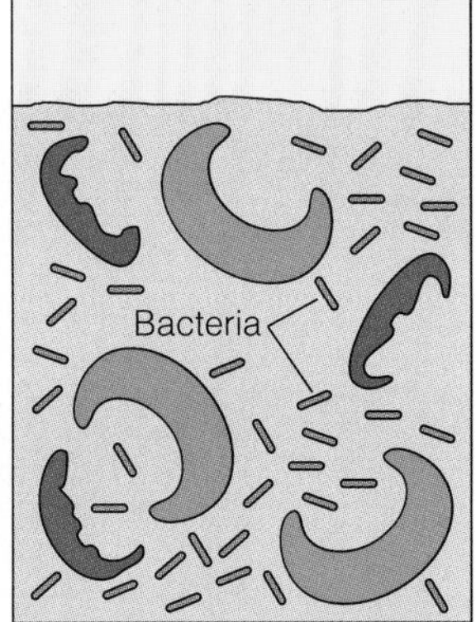

Bacterial overgrowth damages the bile, so that it is ineffective in fat digestion and absorption. The result is fat malabsorption and steatorrhea.

Figure 20.4
Bacterial Overgrowth and Steatorrhea

In general, fat malabsorption is not a major problem for people with ileostomies and colostomies (provided the small intestine is left intact) because most nutrients are absorbed in the upper small intestine. Fluid and electrolyte imbalances, however, can be a significant problem, as Chapter 19 described.

intestine, intestinal obstructions, fistulas, diverticulitis, or impaired blood supply to the intestine. The length, location, and health of the remaining intestine determine the degree to which nutrient absorption is affected. Figure 20.5 reviews nutrient absorption in the GI tract and describes how absorption is affected by surgical resection.

Extent and Location of the Resection Generally, up to 50 percent of the intestine can be resected without serious nutrition consequences. Remarkably, even resections of up to 80 percent may be well tolerated, provided that the terminal ileum, the ileocecal valve, and the colon remain intact.[19] When the ileum has been resected, however, the absorption of fat, protein, carbohydrate, fat-soluble vitamins, vitamin B_{12}, calcium, and magnesium can be impaired. The ileum is also where bile salts are normally reabsorbed. Without bile salt reabsorption, the body's pool of bile salts diminishes, and fat malabsorption worsens.

The Ileocecal Valve The ileocecal valve controls the rate at which the intestinal contents move from the small to the large intestine. Without the valve, transit time through the small intestine is rapid, and the time available for nutrient absorption is limited. Consequently, the colon receives large volumes of unabsorbed nutrients, fluids, electrolytes, and bile salts. Nutrient absorption is impaired and diarrhea results. If the colon is resected as well, severe fluid and electrolyte imbalances threaten health.

The Colon Recent studies suggest that an intact colon plays a significant role in reducing carbohydrate and, to a lesser extent, protein malabsorption following intestinal resections.[20] Bacteria in the colon salvage energy from some of the unabsorbed carbohydrate by metabolizing it to short-chain fatty acids, which can then be absorbed and utilized for energy. This salvage function appears to be enhanced in people with intestinal resections—a form of adaptation.

Adaptation After an intestinal resection, a remarkable adaptive response occurs in the portion that remains: it gets longer, thicker, and wider, and it either absorbs nutrients more efficiently or begins to absorb nutrients it did not absorb before. The presence of nutrients in the remaining gut appears to stimulate this

Figure 20.5
Nutrient Absorption and Consequences of Intestinal Surgeries
About 90 to 95 percent of nutrient absorption takes place in the first half of the small intestine. After a resection, nutrient absorption may be reduced.

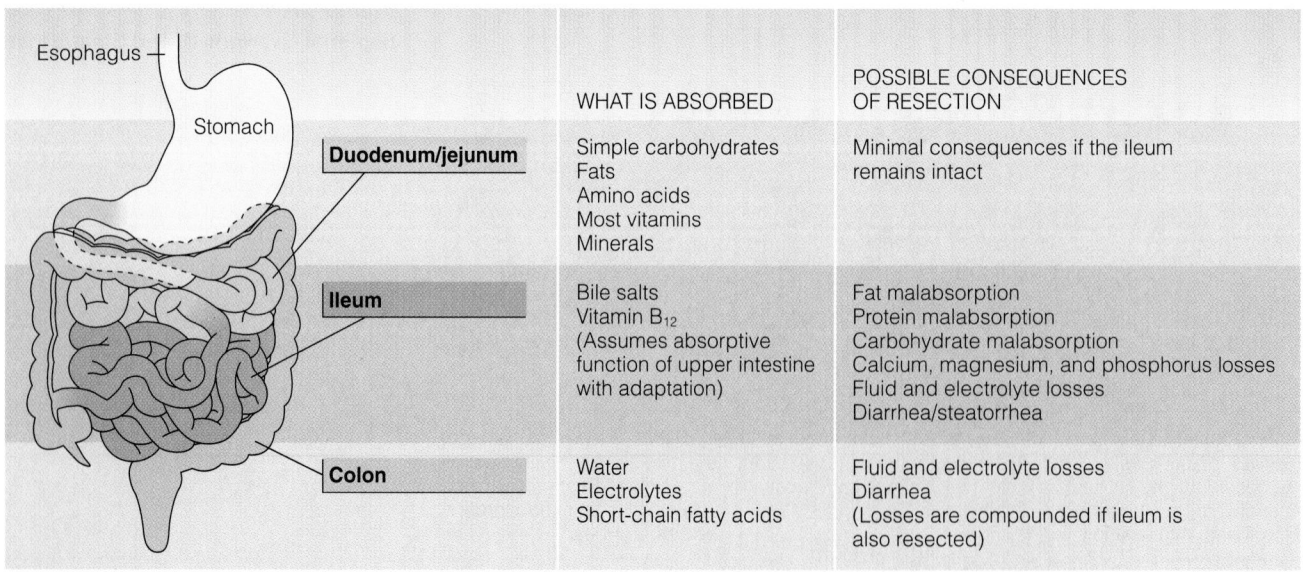

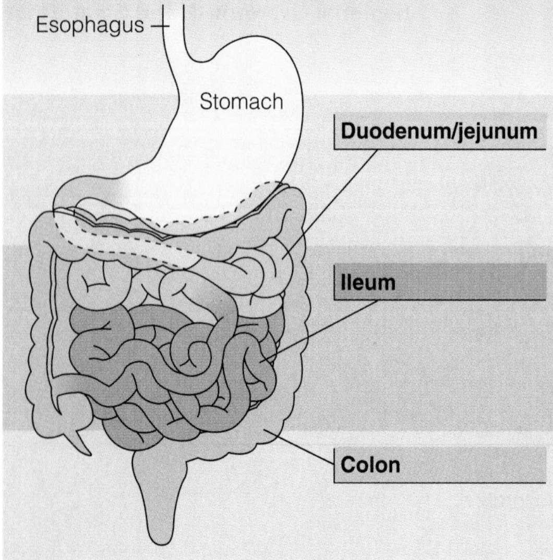

	WHAT IS ABSORBED	POSSIBLE CONSEQUENCES OF RESECTION
Duodenum/jejunum	Simple carbohydrates Fats Amino acids Most vitamins Minerals	Minimal consequences if the ileum remains intact
Ileum	Bile salts Vitamin B_{12} (Assumes absorptive function of upper intestine with adaptation)	Fat malabsorption Protein malabsorption Carbohydrate malabsorption Calcium, magnesium, and phosphorus losses Fluid and electrolyte losses Diarrhea/steatorrhea
Colon	Water Electrolytes Short-chain fatty acids	Fluid and electrolyte losses Diarrhea (Losses are compounded if ileum is also resected)

adaptation—a good reason to begin enteral nutrition as early as possible. Specific dietary constituents, such as the amino acid glutamine, short-chain fatty acids, fiber, and growth hormone, may also aid in this adaptation. With an extensive bowel resection, however, even adaptation will fail to compensate for the reduced surface area.

Nutrition Support Immediately after surgery, the primary nutrition concern is to maintain fluid and electrolyte balances. For resections of less than 50 percent of the intestine, oral nutrition begins a few days after surgery. For more extensive resections, parenteral nutrition is often provided initially to ensure that nutrient needs are met until adaptation is under way. Enteral nutrition (usually by tube feeding) is initiated as early as possible to stimulate adaptation.

Oral diets for people whose colons remain intact emphasize foods high in complex carbohydrates (60 percent of the kcalories from carbohydrate), restricted in fat (20 percent), and low in oxalate.[21] With time, fat intake can be liberalized if additional energy is needed and if the fat does not precipitate steatorrhea or diarrhea. People whose colons do not remain intact following surgery often have difficulty absorbing energy from either carbohydrate or fat. These people are more likely to rely permanently on intravenous nutrition to supply all or part of their nutrient needs.

Rx Prescription Pad

Medications used in the treatment of short-bowel syndrome may include:

- Antidiarrheals
- Antisecretory agents
- Calcium carbonate antacids (to bind oxalate)

See Appendix E for timing with meals and nutrition-related side effects.

Protein-Modified Diets for Celiac Disease

One type of protein-modified diet, the gluten-free diet, helps reverse malabsorption caused by celiac disease. In celiac disease, the intestinal mucosal cells become sensitive to certain fractions of proteins found in some grains, including wheat, rye, barley, and possibly oats. Gluten is a protein found in wheat, and gliadin is the fraction of gluten that causes sensitivity in celiac disease. Other grains have corresponding protein fractions that can cause sensitivity in celiac disease. These protein fractions act as toxic substances, damaging the intestinal villi and leading to malabsorption. The prevalence of celiac disease has been increasing, and the incidence may be as high as one in 250 people.[22] Clinical manifestations of the disease may range from simple iron and folate deficiencies to serious weight loss, fatigue, and severe anemia.[23] Bone mineral density may be reduced, and bone diseases are common.[24] People with unexplainable bone disease may actually have celiac disease that has gone undiagnosed because GI symptoms are either mild or absent.[25]

Diet Order
Gluten-free diet.

celiac (SEE-lee-ack) **disease:** a sensitivity to a part of the protein gluten that causes flattening of the intestinal villi and generalized malabsorption; also called **gluten-sensitive enteropathy** (EN-ter-OP-ah-thee) or **celiac sprue.**

gluten (GLUE-ten): a vegetable protein found in wheat; **gliadin** is the protein fraction of gluten that causes the toxic effects in celiac disease. Corresponding protein fractions in barley, rye, and possibly oats also have these effects.

Effects on Nutrition Status Although the severity of malabsorption varies, people with celiac disease may malabsorb many nutrients, notably, fat, protein, carbohydrate, vitamin K, folate, vitamin B_{12}, iron, and calcium. Consequently, people with celiac disease may experience steatorrhea, diarrhea, weight loss, and malnutrition. Lactose intolerance is common. Anemia may occur as a result of iron, folate, or vitamin B_{12} deficiency. A vitamin K deficiency may precipitate clotting abnormalities, and the person may bleed easily. Calcium deficiency can result in bone diseases, tetany, and bone pain.

Treatment for Celiac Disease A gluten-free diet serves as the primary treatment for celiac disease. Once the person follows a gluten-free diet for a few weeks, the intestinal changes reverse almost completely. With time, measurements of body composition show marked improvement.[26] With early diagnosis and treatment, children and adolescents with celiac disease who adhere to a gluten-free diet can achieve normal bone mass and avoid bone diseases.[27] In adults, bone demineralization is not always reversible.[28] Lactose intolerance may be permanent. Lifelong adherence to the diet is necessary to prevent the return of symptoms.

Gluten-Free Diets The treatment for celiac disease sounds deceptively simple: eliminate gluten. Actually, the diet eliminates not only wheat (because it contains gluten), but also traditionally eliminates rye, barley, and oats because protein fractions from these grains may also damage the intestinal cells. Wheat, rye, barley, and oats are common components of many foods (see Table 20.5), and many foods contain hidden sources of these grains, some of which are not listed on food labels.

Although gluten-free diets traditionally eliminate oats, this practice is controversial.[29] Studies suggest that moderate amounts of oats may be consumed without adverse effects.[30] However, oats may also be contaminated with wheat, and many practitioners believe that more evidence is needed before the inclusion of oats on gluten-free diets can be recommended.

Grains that can be included on gluten-free diets include corn and rice; products made from these grains can be consumed freely. Acceptable substitutes for wheat flour in recipes include tapioca and soybean, arrowroot, buckwheat, and potato flours. A low-gluten wheat starch flour is also available, but these products may contain small amounts of gliadin, and should be avoided.[31]

Table 20.5 Gluten-Free Diet

Meat and Meat Alternates
Any allowed except those that are breaded, prepared with bread crumbs, or creamed.
Milk and Milk Products
Any allowed if client is not intolerant to lactose except milk mixed with Ovaltine, commercial chocolate milk with a cereal additive, milk beverages flavored with malt, pudding thickened with wheat flour, or ice cream or sherbet containing gluten stabilizers.
Fruits and Vegetables
Any allowed except those that are breaded, prepared with bread crumbs, or creamed.
Starches and Grains
Allowed: Bread, cereal, or dessert products made from cornmeal, soybean flour, rice flour, or potato flour; tapioca; cornmeal, popcorn, and hominy; rice, cream of rice, puffed rice, and rice flakes; potato chips. *Not allowed:*[a] Bread, cereal, or dessert products made from wheat, rye, barley, and oats; commercially prepared mixes for biscuits, cornbread, muffins, pancakes, cakes, cookies, or waffles; bran; pasta, macaroni, and noodles; malt; pretzels; wheat germ; doughnuts; ice cream cones; matzo.
Other
Not allowed: Beer; ale; certain whiskeys (Canadian rye); alcohol-based extracts; cereal beverages (Postum); root beer; commercial salad dressings that contain gluten stabilizers; distilled white vinegar; soy sauce; soups containing any ingredients not allowed (such as barley or noodles); products made with hydrolyzed vegetable protein.

[a]Note that many special products made with allowed ingredients are available. Gluten-free pastas and macaroni, for example, are acceptable substitutes for regular pastas and macaroni.

Nutrition Assessment Checklist

For People with Malabsorption Syndromes

Medical Review the client's medical record for diagnoses associated with malabsorption (including lactose intolerance), and check the record regularly for improvements (such as tolerance to a liquid diet) or complications (such as obstructions or fistulas) in the medical condition that might call for dietary adjustments.

Medication Check the client's medications for possible nutrient-medication interactions. Many antibiotics have potential interactions. Anti-inflammatory agents may cause nausea, esophagitis, abdominal pain, fluid retention (may mask weight loss), and glucose intolerance. People taking antimucolytics for cystic fibrosis need adequate fluids to help liquify thick mucous secretions. Narcotic analgesics may cause nausea and drowsiness and interfere with eating.

Food Intake Assess energy intake in relation to weight changes and adjust the diet appropriately. Determine individual tolerances for people with the dumping syndrome, lactose intolerance, and inflammatory bowel diseases. For people with steatorrhea, assess the effects of fat-restricted diets and/or enzyme replacements on reducing the volume and frequency of steatorrhea. Various nutrient deficiencies frequently accompany malabsorption syndromes—assess the diet and supplemental intakes of these nutrients.

Anthropometric Regularly assess measures of growth and development in children and weight in adults with malabsorption syndromes to provide the information necessary for diet adjustments and to prevent deterioration of nutrition status. Remember to interpret anthropometric measurements cautiously for people with dehydration or edema.

Laboratory Monitor lab values for signs of PEM, dehydration, anemia, and nutrient deficiencies. Table 20.6 shows laboratory tests useful in assessing problems associated with malabsorption syndromes.

Physical Watch for physical signs of dehydration and of nutrient deficiencies commonly associated with malabsorption syndromes; these may include deficiencies of energy, protein, essential fatty acids, fat-soluble vitamins, folate, vitamin B_{12}, calcium, and iron. Assess the client's energy level and emotional state.

People who are lactose intolerant need to exclude milk and milk products. A family with a member who has celiac disease needs a lot of support from the health care team. Dietitians and support groups can offer tips about reading food labels, hidden sources of restricted grains, books, recipes, and products that can help clients manage the diet restrictions.

WWW.

celiac.com
Celiac Support Page

csaceliacs.org
Celiac Sprue Association

primenet.com/cdf
Celiac Disease Foundation

Table 20.6 Laboratory Tests Useful in Assessing Malabsorption Syndromes

- Direct stool examinations: Stool checked for weight (greater than normal weight suggests malabsorption) and oily materials (excess fat in stool suggests steatorrhea).
- Chemical analysis of fecal fat: Fecal fat of greater than 7 g/day when the diet includes 100 g of fat/day indicates fat malabsorption.
- Serum carotene: Low serum levels accompany steatorrhea.
- Serum calcium: Low levels seen in calcium or vitamin D malabsorption. (Recall that steatorrhea can lead to calcium and vitamin D malabsorption.)
- D-xylose test: Test of carbohydrate absorption.
- Chemical analysis of fecal nitrogen: Normal fecal nitrogen is less than 2 g/day.
- Schilling test: Identifies vitamin B_{12} malabsorption.

This chapter has described a variety of diets used to treat many disorders that result in malabsorption. The accompanying nutrition assessment checklist highlights factors of importance in assessing the nutrition status of people with malabsorption syndromes. In some cases, people may have difficulty eating enough foods to maintain nutrition status. In those cases, the formula diets described in the next chapter, fed orally or by tube, can help prevent nutrient deficiencies.

Self Check

1. A client receiving a post gastrectomy diet would like a snack. Which of the following snacks would be an appropriate choice?
   a. milkshake
   b. cheese and crackers
   c. cookies
   d. granola
2. Foods containing protein and fat are emphasized in the postgastrectomy diet because they:
   a. are digested more slowly than carbohydrate.
   b. give a greater feeling of fullness than carbohydrate.
   c. attract fluid into the intestine more rapidly than foods containing carbohydrate.
   d. contain pectin and guar gum.
3. The nurse interviewing a client who has done well for 3 years following a gastrectomy should be alert to signs of:
   a. dumping syndrome
   b. blind loop syndrome
   c. iron-deficiency anemia
   d. vitamin C deficiency
4. For people with permanent lactose intolerance, lactose-restricted diets:
   a. strictly limit lactose from all sources.
   b. require the use of lactase-containing digestive aids.
   c. often include some milk and milk products.
   d. may lead to vitamin D deficiencies if the person is overexposed to sunlight.
5. Nutrition problems associated with fat malabsorption syndromes include all of the following **except:**
   a. essential amino acid deficiencies.
   b. severe weight loss and PEM.
   c. bone diseases.
   d. oxalate kidney stones.
6. Which strategy is safe for increasing the energy intake for people on fat-restricted diets?
   a. Gradually add additional fat as tolerance permits.
   b. Provide additional fat as MCT.
   c. Provide additional fat along with enzyme replacements.
   d. All of the above.
7. For the person with acute pancreatitis, oral diets can begin when abdominal pain subsides and:
   a. steatorrhea resolves.
   b. when serum amylase and lipase levels return to normal or near-normal levels.
   c. gastric suction is no longer necessary.
   d. liquids are well tolerated.
8. Diet therapy for chronic pancreatitis may include all of the following **except:**
   a. a fat-restricted diet.
   b. enzyme replacements.
   c. small meals.
   d. a diet to control blood glucose.
9. Current dietary recommendations for people with cystic fibrosis include:
   a. limited use of table salt.
   b. a high-kcalorie, high-protein diet.
   c. fat restrictions.
   d. fluid restrictions.
10. For children and adults with cystic fibrosis, severe malnutrition is associated with a weight:
    a. at 85 to 90 percent of desirable body weight.
    b. at or below 65 percent of desirable body weight.
    c. at 90 to 95 percent of desirable body weight.
    d. at or below 75 percent of desirable body weight.
11. All of the following may affect the nutrient needs of people with inflammatory bowel diseases **except:**
    a. steatorrhea.
    b. medications.
    c. dumping syndrome.
    d. fistulas.
12. In comparing the nutrition-related problems associated with Crohn's disease and ulcerative colitis:
    a. the person with Crohn's disease is more likely to be lactose intolerant.
    b. the person with ulcerative colitis is more likely to have problems maintaining fluid and electrolyte balances.

c. the person with Crohn's disease is more likely to develop bone disease.
d. the person with ulcerative colitis is more likely to develop bone disease.

13. The primary nutrition problems associated with bacterial overgrowth in the stomach and small intestine include all of the following **except:**
    a. fat malabsorption.
    b. folate and vitamin B_{12} malabsorption.
    c. loss of bile salts.
    d. permanent lactose intolerance.

14. A complex of symptoms that may occur whenever the absorptive surface of the small intestine is reduced is referred to as:
    a. blind loop syndrome.
    b. dumping syndrome.
    c. short-bowel syndrome.
    d. Zollinger-Ellison syndrome.

15. Diets for people who have undergone extensive intestinal resections but whose colons remain intact should be:
    a. low in carbohydrate to prevent lactose intolerance.
    b. high in complex carbohydrates and restricted in fat.
    c. low in simple sugars and high in protein and fat.
    d. gluten-free.

16. In celiac disease, a fraction of the protein gluten in wheat and corresponding protein fractions in barley, oats, and rye:
    a. act as toxic substances and cause the intestinal villi to lose their absorptive capacity.
    b. bind fats and calcium so that they cannot be absorbed.
    c. allow oxalate to be absorbed and contribute to a higher incidence of oxalate kidney stones.
    d. are clearly marked on food labels when they are used in food products.

17. A gluten-free diet excludes:
    a. wheat, corn, and oats.
    b. barley, soybeans, oats, and corn.
    c. wheat, barley, rye, and oats.
    d. wheat, barley, rice, and oats.

Answers to these questions appear in Appendix H.

Clinical Applications

1. Using Table 20.4 as a guide, plan a day's menu for a diet containing 35 grams of fat. Take care to make the menu both palatable and nutritious. How can this menu be improved using the suggestions on p. 490?
2. Treatments for a disease often alleviate disease symptoms rather than the disease itself. With this in mind, describe the similarities in the treatment of chronic pancreatitis and cystic fibrosis. In what ways do the treatments differ and why?
3. As stated in this chapter, treatment of celiac disease is deceptively simple—eliminate gluten. Take a trip to the grocery store and randomly select ten of your favorite snack and convenience foods. Check the labels of these products and see if they are allowed on gluten-restricted diets. (As you do this part of the assignment, keep in mind that the labels may not list all offending ingredients.) Find acceptable substitutes for the products that are not allowed. No doubt, this will be a tough assignment.

Notes

1. J. Grant, G. Chapman, and M. K. Russell, Malabsorption associated with surgical procedures and its treatment, *Nutrition in Clinical Practice* 11 (1996): 43–52.

2. F. L. Suarez, D. A. Savaiano, and M. D. Levitt, A comparison of symptoms after the consumption of milk or lactose-hydrolyzed milk by people with self-reported severe lactose intolerance, *New England Journal of Medicine* 333 (1997): 1–4; F. L. Suarez and coauthors, Tolerance to the daily ingestion of two cups of milk by individuals claiming lactose intolerance, *American Journal of Clinical Nutrition* 65 (1997): 1502–1506.

3. S. R. Hertzler and coauthors, How much lactose is low lactose? *Journal of the American Dietetic Association* 96 (1996): 243–246.

4. Suarez and coauthors, 1997.

5. P. B. Jeppesen and coauthors, Essential fatty acid deficiency in patients with severe fat malabsorption, *American Journal of Clinical Nutrition* 65 (1997): 837–843.

6. W. Steinberg and S. Tenner, Acute pancreatitis, *New England Journal of Medicine* 330 (1994): 1198–1210.

7. J. Hurst and A. L. Gallagher, Pathophysiology and nutrition management in acute pancreatitis, *Support Line,* December 1994, pp. 6–11.

8. A.S.P.E.N. Board of Directors, Practice guidelines: Pancreatitis, *Journal of Parenteral and Enteral Nutrition* (supplement) 17 (1993): 16.

9. S. A. McClave and coauthors, Comparison of the safety of early enteral *vs.* parenteral nutrition in mild acute pancreatitis, *Journal of Parenteral and Enteral Nutrition* 21 (1997): 14–20; S. Marulendra and D. F. Kirby, Nutrition support in pancreatitis, *Nutrition in Clinical Practice* 10 (1995): 45–53.

10. C. E. Beck and coauthors, Improvement in the nutritional and pulmonary profiles of cystic fibrosis patients undergoing bilateral sequential lung and heart-lung transplantation, *Nutrition in Clinical Practice* 12 (1997): 216–221.

11. S. Creveling and coauthors, Cystic fibrosis, nutrition, and the health care team, *Journal of the American Dietetic Association* (supplement 2) 97 (1997): 186–191.

12. J. Dowsett, Nutrition in the management of cystic fibrosis, *Nutrition Reviews* 54 (1996): 31–33.

13. B. W. Ramsey and coauthors, Nutritional assessment and management in cystic fibrosis: A consensus report, *American Journal of Clinical Nutrition* 55 (1992): 108–116.

14. A.S.P.E.N. Board of Directors, Practice guidelines: Cystic fibrosis, *Journal of Parenteral and Enteral Nutrition* (supplement) 17 (1993): 44.

15. Y. Kim, Can fish oil maintain Crohn's disease in remission? *Nutrition Reviews* 54 (1996): 248–257.

16. M. H. Giaffer, G. North, and C. D. Holdsworth, Controlled trial of polymeric versus elemental diet in treatment of active Crohn's disease, *Lancet* 335 (1990): 816–819.

17. A. Belluzzi and coauthors, Effect of an enteric-coated fish-oil preparation on relapses in Crohn's disease, *New England Journal of Medicine* 334 (1996): 1557–1560.

18. Kim, 1996.

19. Presented by W. D. Heizer, Short bowel syndrome: Treatment strategies, *Fourth Annual Advances and Controversies in Clinical Nutrition,* sponsored by the Mayo Clinic, Jacksonville, Fla., April 18, 1994.

20. I. Nordgaard, B. S. Hansen, and P. B. Mortensen, Importance of colonic support for energy absorption as small-bowel failure proceeds, *American Journal of Clinical Nutrition* 64 (1996): 222–231.

21. T. A. Byrne and coauthors, A new treatment option of patients with short bowel syndrome: Bowel rehabilitation with growth hormone, glutamine, and a modified diet, *Support Line,* February 1996, pp. 1–7.

22. T. Not and coauthors, Celiac disease risk in the USA: High prevalence of antiendomysium antibodies in healthy blood donors, *Scandinavian Journal of Gastroenterology* 33 (1998): 494–498.

23. J. R. Saltzman and B. D. Clifford, Identification of the triggers of celiac sprue, *Nutrition Reviews* 52 (1994): 317–319.

24. S. Mora and coauthors, Reversal of low bone density with a gluten-free diet in children and adolescents with celiac disease, *American Journal of Clinical Nutrition* 67 (1998): 477–481; G. R. Corazza and coauthors, Propeptide of type I procollagen is predictive of posttreatment bone mass gain in adult celiac disease, *Gastroenterology* 113 (1997): 67–71.

25. J. L. Shaker and coauthors, Hypocalcemia and skeletal disease as presenting features of celiac, *Archives of Internal Medicine* 157 (1997): 1013–1016.

26. E. Smecuol and coauthors, Longitudinal study on the effect of treatment on body composition and anthropometry of celiac disease patients, *American Journal of Gastroenterology* 92 (1997): 639–643.

27. Mora and coauthors, 1998.

28. Corazza and coauthors, 1997.

29. T. Thompson, Do oats belong in a gluten-free diet? *Journal of the American Dietetic Association* 97 (1997): 1413–1416.

30. E. K. Janatuinen and coauthors, A comparison of diets with and without oats in adults with celiac disease, *New England Journal of Medicine* 333 (1995): 1033–1037.

31. L. J. Chartrand, Wheat starch intolerance in patients with celiac disease, *Journal of the American Dietetic Association* 97 (1997): 612–618.

CHAPTER 20

Nutrition in Practice

▪ Nutrition and Diagnostic Tests ▪

Diagnostic tests play a pivotal role in the identification and treatment of many disorders. Diagnostic tests include laboratory analyses of the blood, stool, urine, and breath and other procedures such as X rays, endoscopy, and other imaging techniques. Nutrition may influence the results of these tests and their interpretation. Some blood tests, for example, require that a person fast overnight; some require the consumption of a specific substance; others are not affected by diet.

Why must a person sometimes fast before a blood test?

Meals change the concentration of nutrients in the blood, and this may affect test results. After a meal, the concentration of nutrients in the blood rises—a reflection of recent nutrient absorption. These levels may affect the test results for some blood constituents directly; for example, vitamin and mineral tests often reflect recent intake.

Sometimes nutrient levels indirectly affect blood test results. Fasting blood samples are often recommended when multiple tests will be performed on one blood sample because after a meal, triglyceride levels rise and cause the blood sample to become cloudy. The cloudiness interferes with many chemical reactions that affect test results.[1]

In some cases, the expected effect of a food on a lab value is known, and it may even be desirable to study that effect. Such is the case with tests for fat malabsorption, which require that a certain amount of fat be eaten prior to and during testing.

What does the test used to diagnose fat malabsorption involve?

To diagnose fat malabsorption and determine its severity, clinicians aim to provide about 100 grams (900 kcalories) of fat per day for two to three days; during the same period, the person's stools are collected.[2]

That seems like a lot of fat. Is it safe to give a person that much fat?

Providing 900 kcalories from fat goes against traditional nutrition wisdom, but the diet is temporary, and therefore the risk is minimal. Some people, however, simply cannot eat that much fat, particularly when they are ill. In those cases, diets can be planned to include from 60 to 80 grams of fat, and different standards are used to determine malabsorption. To ensure that the data will be useful, the clinician should take care to monitor the actual fat intake, and not just present the client with a high-fat diet.

When must foods be restricted so that a test will be valid?

Several tests require dietary restrictions in order to obtain valid results. One example is a stool test used to detect GI bleeding, an aid to the detection of cancer of the colon. The person often takes this test at home as part of a routine physical examination. For 48 to 72 hours before the test, and during the times stools are collected, the person should follow a high-fiber diet and eat no red meats, poultry, fish, turnips, and horseradish. The high-fiber foods speed up intestinal transit time and maximize the likelihood of detecting significant blood loss if it is present. Omitting red meats and the other foods listed above helps prevent false positive results—that is, results that suggest significant GI blood loss, when in fact there is none.

Vitamin C and iron supplements are restricted during this GI test. Vitamin C supplements (more than 500 milligrams per day) can cause false normal test results, even when the person is experiencing significant GI bleeding. In contrast, iron supplements may produce some GI bleeding, even though there is no lesion.

Another example is a urine test used to detect certain tumors by looking for a compound called 5-hydroxyindoleacetic acid (5-HIAA). If present, 5-HIAA signifies abnormal production of the brain neurotransmitter serotonin. Some foods contain significant amounts of serotonin and can interfere with test results. For 24 hours preceding the test, clients are instructed not to eat avocados, bananas, kiwis, pineapples, plantains, plums, eggplant, tomatoes, butternuts, pecans, and walnuts. Clients must also abstain from alcohol because it suppresses 5-HIAA levels.

Nutrition in Practice

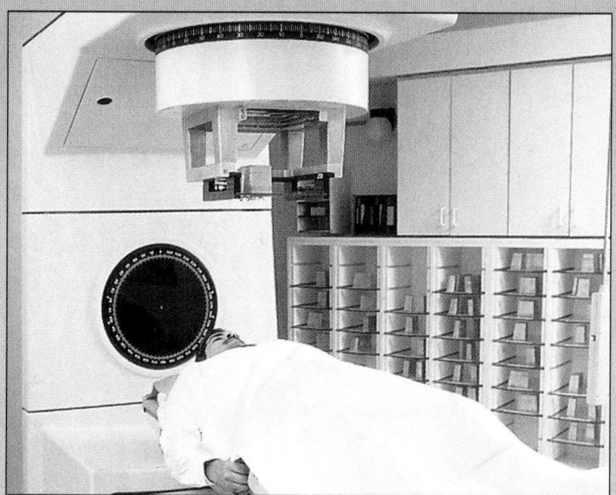

Clients must often adjust their diets in preparation for diagnostic tests.

Even breath tests can be affected by diet. An analysis of hydrogen in the breath is used to detect lactose intolerance. Breads and pastas made from wheat flour and legumes can contribute significant quantities of hydrogen to the breath, so the person is cautioned not to eat these foods the evening before the test. The person must fast from at least midnight until the next morning when breath hydrogen is to be collected to obtain a background level; then the person is given an oral dose of lactose. Breath hydrogen is measured thereafter at intervals.

So far all the tests you have mentioned have been lab tests. How is nutrition related to other diagnostic tests?

X rays, endoscopy, and other imaging techniques sometimes require special dietary restrictions. X rays of the intestinal tract, for example, may require that the person fast from midnight the day before the test. Likewise, endoscopy procedures frequently require fasting.

Who is responsible for making sure the diet rules are followed?

This is an important question. Like so many elements of client care, the tests require cooperation and communication between various health care professionals. First, the physician ordering the test must order the test diet. The diet order may read simply "diet for breath hydrogen test." In the hospital, the foodservice department is then responsible for sending the appropriate diet, which is specified in the diet manual. The nurse working with the client can check the diet manual to see what the diet entails and instruct the client, explaining how long the special diet (or fast) will last. The laboratory analyzing the test may also provide information about testing procedures. To make sure the test is valid, clients should be advised to eat only those foods provided by the health care facility.

For tests that require ingestion of a measured amount of a nutrient, such as the fat malabsorption tests, nurses and dietitians frequently share responsibility for monitoring and recording the client's intake. The dietitian is often responsible for calculating expected fecal fat excretion when intakes fall below desirable levels. Nurses are responsible for stool and urine collections; laboratory technicians draw blood for blood tests.

Extra care must be taken when clients are being tested in an outpatient setting. Clients who are at home during the test period have free access to a wide variety of foods, nutrient supplements, and medications. Like foods and nutrients, medications can interfere with some test results. The dietitian or nurse is wise to provide clearly written instructions and ask clients to keep food and medication records so that their test results can be accurately assessed.

Finally, it is important to note that diagnostic procedures change from time to time. When new tests are developed, precautions may change as well. Regular communication between the department performing the test and the physician, nursing service, and foodservice departments can help keep everyone up to date.

Notes

1. *Diagnostic Tests Handbook* (Springhouse, Pa.: Springhouse Corporation, 1993), p. xvi.

2. J. K. Nelson and coauthors, *Mayo Clinic Diet Manual* (St. Louis: Mosby, 1994), p. 413.

21 *Specialized Nutrition Support: Enteral Nutrition*

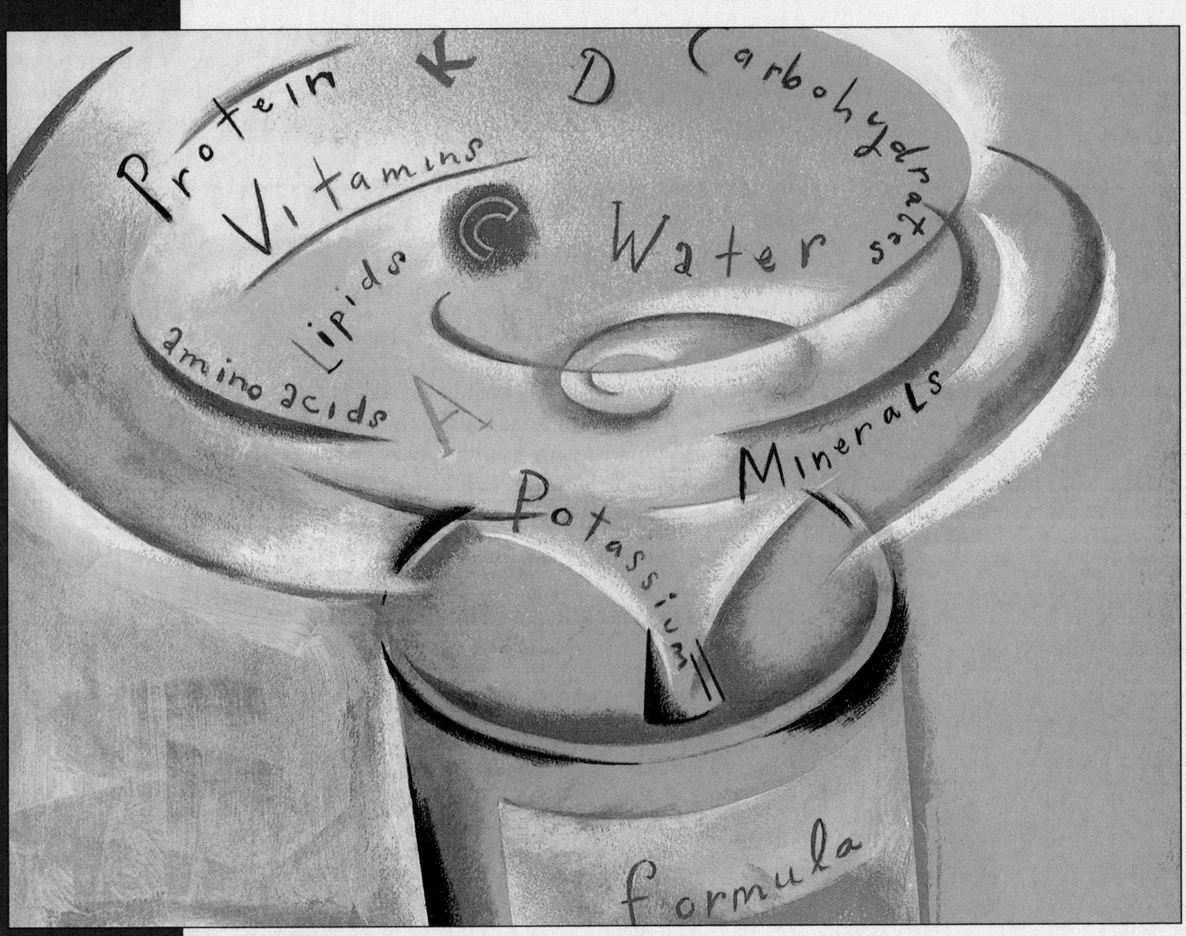

To meet nutrient needs, a person must be able to eat, digest, and absorb nutrients in the amounts necessary to satisfy metabolic demands. Most people, whether healthy or ill, can meet these needs with conventional foods. As Chapters 18, 19, and 20 have shown, illnesses may interfere with eating, digestion, and absorption to such a degree that conventional foods fail to deliver necessary nutrients. If poor appetite is the primary nutrition problem, health care professionals can use the strategies described in the accompanying box to encourage clients to eat. Alternatively, liquid formulas given orally can help clients meet nutrient needs, if the clients can drink them in sufficient amounts.

clinnutr.org
American Society for Parenteral and Enteral Nutrition

magi.com/~cpena
Canadian Parenteral-Enteral Nutrition Association

bapen.org.uk
British Association for Parenteral and Enteral Nutrition

For clients who cannot eat or drink, however, it may be necessary to deliver nutrients by tube or by vein. Feedings provided either orally or by tube are *enteral* feedings, the subject of this chapter. Enteral feedings are possible whenever a client can digest and absorb nutrients via the GI tract. Otherwise, feedings are given by vein as *parenteral* feedings, the subject of the next chapter. Figure 21.1 summarizes some of the factors involved in deciding the most appropriate way to feed a client.

Enteral Formulas

enteral formulas: liquid diets intended for oral use or for tube feedings.
enteron = intestine

The number of enteral formulas on the market is staggering (some are listed in Appendix G). Most formulas are available in ready-to-use form or in powdered form. They are designed to meet a variety of medical and nutrition needs and can be used alone or given along with other foods. The health care team identifies each client's nutrient needs through a careful medical and nutrition assessment and then selects a formula that meets these needs.

Whenever formula is the primary source of nutrients, complete formulas are necessary. Such is the case when a client is on a tube feeding or an oral liquid diet

Figure 21.1
Selecting a Feeding Method

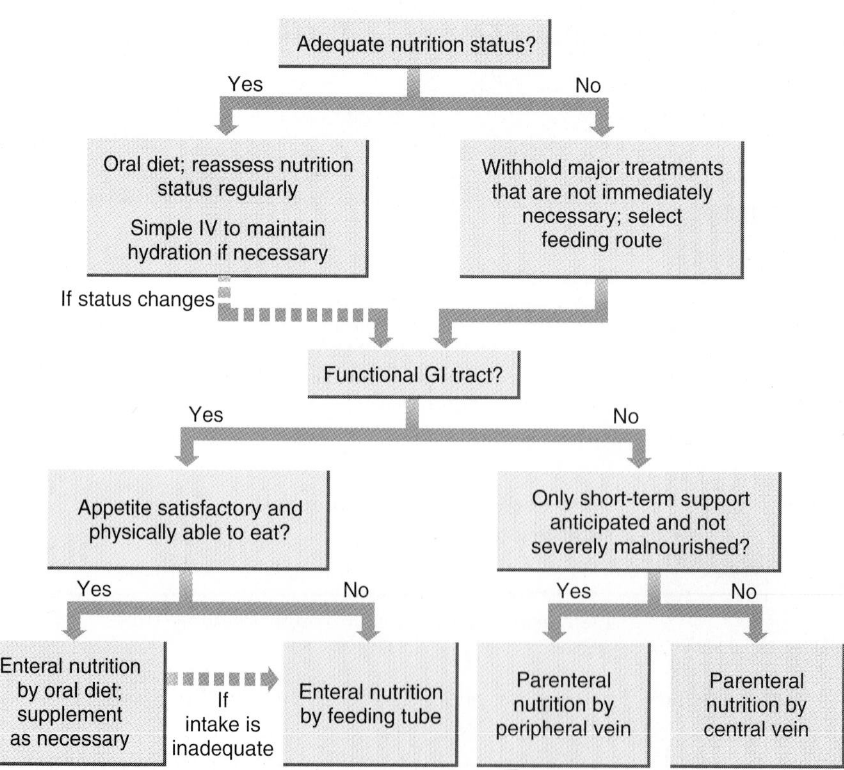

How to

Help the Hospitalized Client Meet Nutrient Needs with Oral Diets

1. Empathize. If the person is frightened, angry, or confused, show that you care and are there to help. Imagine feeling too sick to move or too tired to sit up. Show that you understand how difficult eating may be.
2. Motivate. Be sure the client understands how important nutrition is to recovery.
3. Help clients select foods they like and mark menus appropriately. Call the dietitian if the client needs extra help. When appropriate and permissible, let a friend or family member bring in favorite foods from outside the hospital. This may be especially helpful for clients with strong ethnic, religious, or personal food preferences.
4. Solve eating problems. Encourage clients who feel full after a short time to eat the most nutritious foods first and save liquids until after meals. For clients who are weak or tired, suggest foods that require little effort to eat. Eating a roast beef sandwich, for example, requires less effort than cutting and eating a steak; drinking soup is easier than eating it with a spoon. For clients who either fill up quickly when eating or are weak or tired, smaller meals combined with snacks, such as a sandwich at bedtime and milk shakes or instant breakfast drinks between meals, can improve intake considerably.
5. Suggest that clients add extra energy to the foods they eat by using extra sugar or fats. Dry milk powder added to milk-based drinks, soups, and casseroles boosts nutrient intake.
6. Help clients prepare for meals. Encourage clients to wash their hands and faces and to brush their teeth or rinse their mouths before eating. Help them get comfortable, either in bed or in a chair. Adjust the extension table to a comfortable distance and height, and make sure it is clean. A clean, odor-free room also helps. Take these steps before the tray arrives, so the meal can be served promptly and at the right temperature.

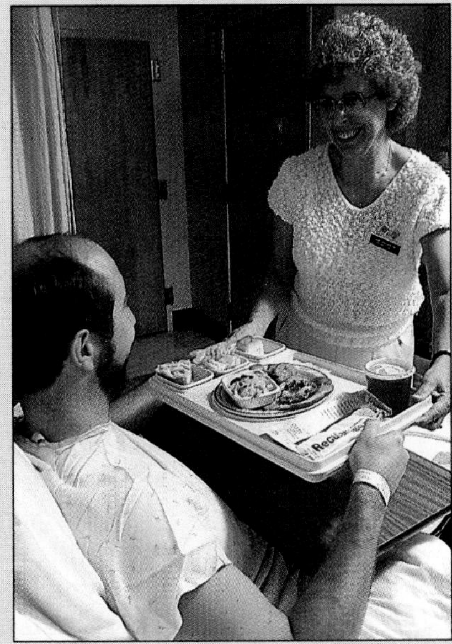

People enjoy eating when they feel comfortable and cared for.

7. Check for accuracy and appearance. When the food cart arrives, check the client's tray. Confirm that the client is receiving the right diet, that the foods on the tray are the ones the client marked on the menu, and that the foods look appealing. Order a new tray if foods are not appropriate.
8. Help with eating. Help clients who need assistance in opening containers or cutting foods and those who are unable to feed themselves.
9. Take a positive attitude toward the hospital's food. Never say something like "I couldn't eat this stuff either." Instead, say, "The foodservice department really tries to make foods appetizing. I'm sure we can find a solution."

for more than a few days. Complete formulas, when given in appropriate amounts, supply all the nutrients a client needs. Complete formulas can also be (and often are) used in smaller quantities to supplement table foods.

complete formulas: enteral formulas designed to supply all needed nutrients when provided in sufficient volume.

Types of Formulas

Formulas are classified in many ways, but for purposes of this book, it is reasonable to think of two major kinds categorized by the type of protein they supply. Standard, or intact, formulas contain complete proteins, whereas hydrolyzed formulas contain small fragments of proteins, which may include free amino acids, dipeptides, and tripeptides.

standard formula: a liquid diet that contains complete molecules of proteins; also called **intact** or **polymeric formula.**

protein isolate: a protein that has been separated from a food. Examples include casein from milk and albumin from egg.

hydrolyzed formula: a liquid diet that contains broken-down molecules of protein, such as amino acids and short peptide chains; also called **monomeric formula.**

modules: formulas or foods that provide primarily a single nutrient and are designed to be added to other formulas or foods to alter nutrient composition; they can also be combined together to create a highly individualized formula.

Caution: Although formulas designed to be delivered intravenously can also be delivered enterally, the reverse is not true. Enteral formulas cannot be delivered by vein without serious consequences.

Standard, or Intact, Formulas Standard, or intact, formulas are appropriate for people who are able to digest and absorb nutrients without difficulty. Most standard formulas contain one or a combination of protein isolates (purified proteins). A few formulas derive their protein primarily from pureed meat (a mixture of whole proteins).

Hydrolyzed Formulas To simplify the body's digestive work, a complete protein can be hydrolyzed—that is, partially broken down to yield small peptides. Alternatively, a formula can be made from free amino acids. In this text, we call both types *hydrolyzed* for simplicity. Hydrolyzed formulas are often also low in fat because fat is difficult to digest and absorb. People who cannot digest nutrients well may benefit from hydrolyzed formulas.

Modular Formulas Unlike complete formulas, a few formulas, called modules, provide essentially a single nutrient (protein, carbohydrate, or fat). In addition to commercial modular formulas, intravenous nutrients and even table foods (vegetable oil or corn syrup, for example) can serve as modules; modules can then be added to enteral formulas to alter nutrient composition (for example, to add kcalories or protein). Modules can also be combined with other modules and liquid vitamin and mineral preparations to construct individualized formulas for clients with unique nutrient needs. Designing, preparing, and delivering such a formula is a challenge that requires an in-depth knowledge of nutrition and the skills of a committed nutrition support team (see Nutrition in Practice 21).

Distinguishing Characteristics

Formulas vary not only in the form of protein they contain but in other characteristics as well. The physician or dietitian considers these characteristics in selecting a formula for individual clients.

For practical purposes, 1 ml (milliliter) is equivalent to 1 cc (cubic centimeter).

Nutrient Composition Standard formulas provide about 1.0 kcalorie per milliliter. High–nutrient density formulas, which provide 1.2 to 2.0 kcalories per milliliter, meet energy and nutrient needs in a smaller volume. Thus high–nutrient density formulas may be useful for clients with high nutrient needs or those requiring fluid restrictions. Formulas also vary in the percentage of energy from protein, fat, and carbohydrate.

Formulas derive their nutrients from different sources. One formula may derive its protein from milk, another from soy, and still others from free amino acids or combinations of protein sources. Some formulas provide all of their fat from long-chain triglycerides (LCT); others provide varying amounts of medium-chain triglycerides (MCT) in addition to LCT. Sources of carbohydrate also vary; one client may benefit from a formula containing higher amounts of complex carbohydrates, while another may prefer the taste of a formula sweetened with simple sugars.[1]

Percentages of vitamins and minerals also vary from one formula to the next, although most standard formulas supply all known essential nutrients in a smaller volume than most clients receive. (Appendix G lists the volume of individual formulas that meets the RDA.) Formulas designed for special purposes may contain altered levels of specific vitamins and minerals. Formulas designed to support wound healing, for example, may contain added amounts of vitamins A and C and zinc; formulas for renal failure may provide lower amounts of electrolytes than the standard formulas.

Residue and Fiber The positive health effects of dietary fibers suggest that fiber-enriched formulas would be the best choice for most people. Why, then, do many standard formulas have a low-to-moderate residue content? The answer is that formulas are most often used for relatively short periods of time and low-to-moderate residue formulas are least likely to cause gas and abdominal distension and thus are often well tolerated by many who need them: people with GI tract disorders (inflammatory bowel diseases or partial obstructions, for example), those who have undergone surgeries of the GI tract, or those beginning enteral nutrition after periods of GI tract disuse.

With that said, people with disorders that benefit from high-fiber diets may also benefit from fiber-enriched formulas, and many fiber-enriched formulas are available. Thus people who depend on tube feedings for long periods of time, those with constipation, and some people with short-bowel syndrome (see Chapter 20) are more likely than others to benefit from fiber-enriched formulas. One type of dietary fiber, pectin, may be effective in controlling some forms of diarrhea in people receiving enteral formulas.[2]

Reminder: Chapter 19 describes residue and fiber (see p. 462).

Since hydrolyzed formulas are almost completely absorbed, they leave little residue in the gut.

Recall from Chapter 20 that pectins are sometimes used to control diarrhea associated with the dumping syndrome.

Osmolality Osmolality is a measure of the concentration of molecular and ionic particles in a solution. A formula that approximates the osmolality of the blood serum (about 300 milliosmoles per kilogram) is referred to as an isotonic formula. A hypertonic formula has a higher osmolality than serum.

Standard formulas are isotonic or only moderately hypertonic. High–nutrient density formulas, because they contain less water than the same volume of a standard formula, and hydrolyzed formulas, because they contain more particles than standard formulas, tend to have a greater osmolality. A formula's osmolality does not generally affect the selection of a formula, because most people tolerate both isotonic and hypertonic formulas without difficulty. Osmolality may affect the way a formula is provided, however. When hypertonic formulas are delivered directly into the intestine, the hyperosmolar load can result in diarrhea, a situation analogous to the dumping syndrome (see Chapter 20). To prevent this problem, hypertonic formulas delivered into the intestine are initially delivered at a slow, even rate.

osmolality (OZ-mow-LAL-eh-tee): a measure of the concentration of particles in a solution, expressed as the number of milliosmoles (mOsm) per kilogram.

isotonic formula: a formula with an osmolality similar to that of blood serum (300 mOsm/kg).

iso = the same
ton = tension

hypertonic formula: a formula with an osmolality greater than that of blood serum.

hyper = greater, more

Cost Costs of individual products vary greatly in different parts of the country and in different facilities. As a general rule, however, hydrolyzed formulas and products formulated for specific disorders (renal failure, respiratory failure, diabetes mellitus, or HIV infection, for example) are more expensive than standard formulas.

Formula Choices Although formulas differ, many fit into general categories, and some can be used interchangeably. For example, several standard formulas are isotonic and provide similar amounts of energy, protein, carbohydrate, fat, and other nutrients. They may derive their protein from different sources, but the sources are all high-quality proteins. One formula may provide more vitamins in a smaller volume, but that may be of little consequence if both formulas meet nutrient needs in the volume the person is receiving. In some cases, however, such differences can be significant. A client on a fluid-restricted diet, for example, may need a formula of high nutrient density to meet nutrient needs in a smaller volume.

Like table foods, enteral formulas can meet a variety of dietary needs. With that in mind, it is important to identify which people benefit from such formulas.

Enteral Formulas: Who Needs What?

All people with functional GI tracts who cannot get the nutrients they need from table foods can potentially benefit from enteral formulas. They could also get nutrients from parenteral nutrition, but enteral nutrition is preferable whenever

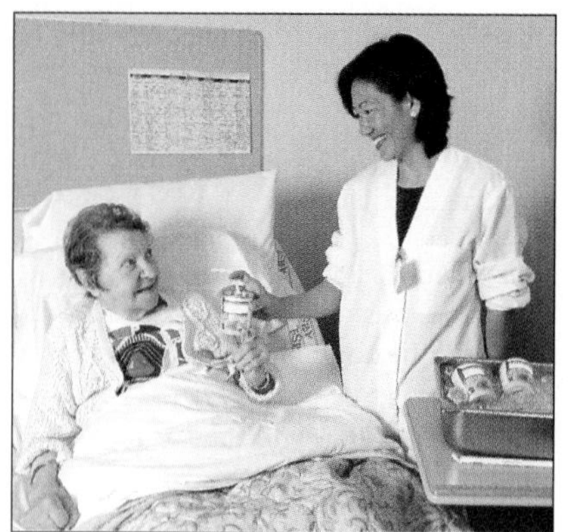

With help from caring professionals, a client can sometimes meet nutrient needs with formulas provided orally.

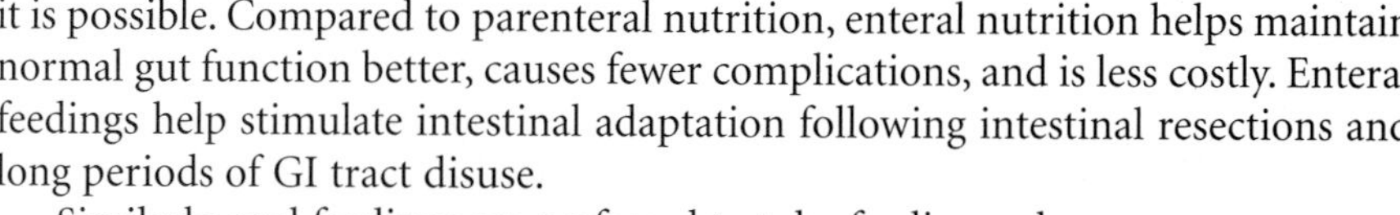

it is possible. Compared to parenteral nutrition, enteral nutrition helps maintain normal gut function better, causes fewer complications, and is less costly. Enteral feedings help stimulate intestinal adaptation following intestinal resections and long periods of GI tract disuse.

Similarly, oral feedings are preferred to tube feedings whenever a person can drink the formula and drink enough of it. In so doing, the client avoids the stress of the procedure to insert a feeding tube. Compared to feeding by tube, orally provided formulas are less costly and less likely to result in complications.

Enteral Formulas Provided Orally

For people who can tolerate only liquids for long periods and those who need hydrolyzed formulas, formulas provided orally can meet all nutrient needs in ways that foods cannot. If people can drink enough of the formula, they can avoid being fed by tube.

More often, formulas provided orally supplement a conventional diet. Some people can eat table foods, but not in sufficient quantities to meet their nutrient needs. Enteral formulas provide a reliable source of nutrients and work particularly well for adding energy and protein to the diet. Psychologically, liquids seem less filling than foods, and they are easier for debilitated, weak, or tired clients to handle.

When used as an oral diet, the formula's taste must be acceptable to the individual. As a general guide, the more hydrolyzed the formula and the lower its fat content, the less tasty it will be. Remember, however, that people's likes and dislikes vary greatly, and what is unpalatable to one person might be acceptable to another. Limited evidence suggests that hydrolyzed formulas may become more acceptable over time.[3] Allowing clients to sample different products and different flavors of formula and select the ones they like best is helpful in promoting acceptance.

Tube Feedings

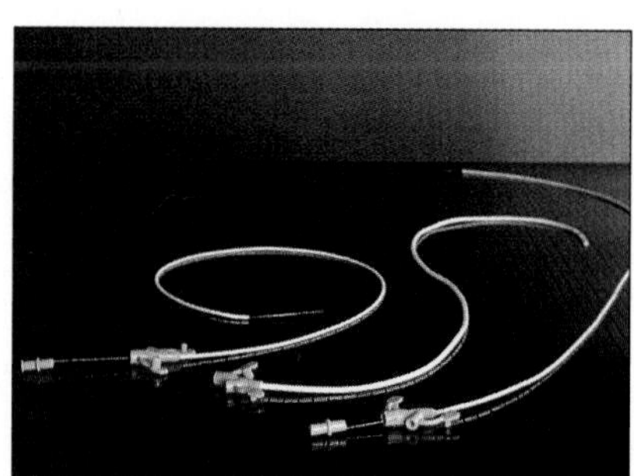

Feeding tubes provide access to the stomach and intestine for clients who cannot eat oral diets.

Tube feedings are simply complete formulas delivered by tube into the stomach or intestine. An individual who has a functional GI tract but is unable to ingest enough nutrients (or the appropriate type) by mouth to meet present needs is a candidate for a tube feeding.[4] Such a person may have physical problems that make chewing and swallowing difficult; have no appetite for an extended time; have a partial obstruction, fistula, or altered motility in the upper GI tract; be in a coma; have very high nutrient requirements; or be unable to ingest a hydrolyzed formula orally. Where the tube feeding is delivered depends on the client's medical needs.

Feeding Tube Placement Feeding tubes are inserted into different locations along the GI tract depending on the client's medical problems and the estimated length of time that the feeding will be required. Figure 21.2 shows various feeding tube placement sites, and the accompanying glossary describes these sites.

For infants, feeding tubes are often passed from the mouth to the stomach before each feeding and removed after each feeding to allow the infant to breathe easily and reduce the risk of regurgitation.

The final location of the feeding tube determines the type of feeding. If a feeding tube is passed through a gastrostomy into the duodenum or jejunum, the feeding is intestinal rather than gastric.

Tube Insertion When clients are not expected to be on tube feedings for more than about four weeks, feeding tubes are frequently inserted through the nose and passed into the stomach or intestine.[5] When a client will be on a tube feeding for a longer period, or when a feeding tube cannot be passed through the nose, esophagus, or stomach due to an obstruction or for other medical reasons, an opening can be made into the stomach or jejunum. Tube enterostomies can be made either surgically or nonsurgically using local anesthesia. Table 21.1 on p. 514 compares some of the features of various tube feeding sites. The box on p. 515 suggests ways to reduce anxiety for clients beginning a tube feeding.

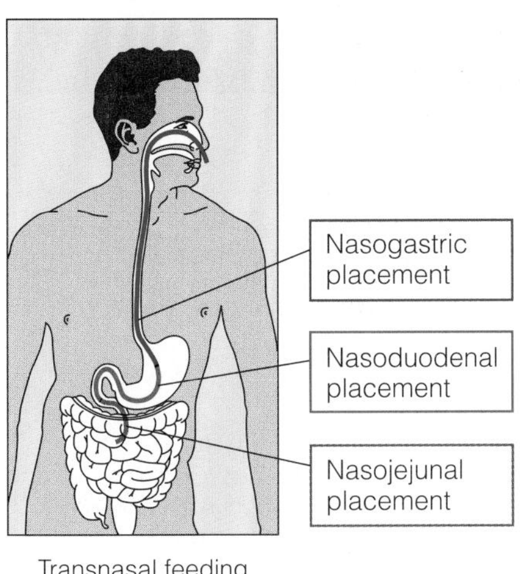

Transnasal feeding tube placements

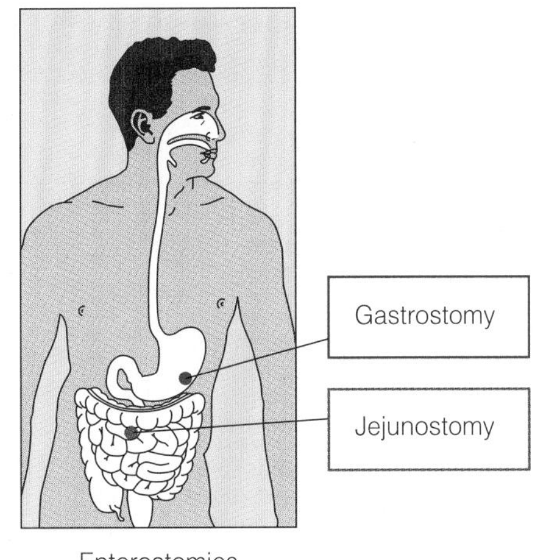

Enterostomies

Figure 21.2
Feeding Tube Placement Sites

Feeding Tubes Feeding tubes are soft and flexible and come in a variety of diameters and lengths. Many have special characteristics that make them desirable for specific purposes. For example, tubes with double lumens allow for gastric decompression and intestinal feedings at the same time.

Which feeding tube is appropriate depends on the client's age, medical condition, how the tube will be placed (transnasally or through an enterostomy), how far the final placement will be from the insertion site (from the nose to the stomach or intestine, for example), and its inner diameter. Once the appropriate length is selected, the smallest tube through which the formula will flow without clogging the tube is selected. Unclogging a tube is a difficult procedure that interrupts the feeding schedule and is not always successful.[6] Insertion of a new tube can cause undue stress and anxiety for the client and raises the cost of the feeding.

The outer diameter of feeding tubes is measured using the French scale. Each unit on the French scale is about one-third of a millimeter. Thus the diameter of a 10 French feeding tube is a little over 3 mm.

Glossary of Feeding Tube Placement Sites

These terms are listed in order from the nose to lower organs of the digestive system.

transnasal: through the nose. A **transnasal feeding tube** is one that is inserted through the nose.
naso = nose

nasogastric (NG): from the nose to the stomach.

nasoenteric: from the nose to the stomach or intestine. *Nasoenteric feedings* include nasogastric, nasoduodenal, and nasojejunal feedings. Most clinicians use nasoenteric to refer to nasoduodenal and nasojejunal feedings only.

nasoduodenal (ND): from the nose to the duodenum.

nasojejunal (NJ): from the nose to the jejunum.

orogastric: from the mouth to the stomach. This method is often used to feed infants because they breathe through their noses, and tubes inserted through the nose can hinder the infant's breathing. The tube is inserted before, and removed after, each feeding.

enterostomy (EN-ter-OSS-toe-mee): a gastric or jejunal opening made surgically or under local anesthesia through which a feeding tube can be passed.

gastrostomy (gas-TROSS-toe-mee): an opening in the stomach made surgically or under local anesthesia through which a feeding tube can be passed. The technique for creating a gastrostomy under local anesthesia is called **percutaneous endoscopic gastrostomy,** or **PEG** for short. When the feeding tube is guided from such an opening into the jejunum, the procedure is called **percutaneous endoscopic jejunostomy (PEJ),** a misnomer because the enterostomy site is in the stomach.

jejunostomy (JEE-ju-NOSS-toe-mee): an opening in the jejunum made surgically or under local anesthesia through which a feeding tube can be passed. The technique for creating a jejunostomy under local anesthesia is called a **direct endoscopic jejunostomy (DEJ).** Note: Some clinicians also refer to this procedure as a PEJ, which is more accurate use of the term than the more common use described above.

Table 21.1 Comparison of Feeding Tube Sites[a]

Insertion Method and Feeding Site	Advantages	Disadvantages
Transnasal	Does not require surgery or incisions for placement.	Easy to remove by disoriented clients; long-term use may irritate the nasal passages, throat, and esophagus.
Nasogastric	Easiest to insert and confirm placement; feedings can often be given intermittently and without an infusion pump.	Highest risk of aspiration in compromised clients.
Nasoduodenal and nasojejunal	Lower risk of aspiration in compromised clients; allow for enteral nutrition earlier than gastric feedings following severe stress; may allow for enteral feeding when partial obstructions, fistulas, or other medical conditions prevent gastric feeding.	More difficult to insert and confirm placement; feedings require an infusion pump for administration; may take longer to reach nutrition goals.
Tube enterostomies	Allow gastroesophageal sphincter to remain closed, lowering the risk of aspiration; more comfortable than transnasal insertion for long-term use.	May require general anesthesia for insertion; require incisions; greater risk of complications from the insertion procedure; greater risk of infection; may cause skin irritation around the insertion site.
Gastrostomy	Feedings can often be given intermittently and without a pump; easier to insert than a jejunostomy.	Moderate risk of aspiration in high-risk clients.
Jejunostomy	Lowest risk of aspiration; allows for enteral nutrition earlier following severe stress; may allow for enteral feeding when partial obstructions, fistulas, or medical conditions prevent gastric feeding.	Most difficult to insert; feedings require an infusion pump for administration; may take longer to reach nutrition goals.

[a]Relative to each tube feeding site. The actual advantages and disadvantages of different insertion procedures depend on the person's medical condition.

Formula Selection

The health care team uses a logical approach to sort through the many available formulas and find an appropriate choice. Figure 21.3 on p. 516 shows some considerations involved. In a nutshell, the formula that meets the client's medical and nutrient needs with the lowest risk of complications and at the lowest cost is the best choice. If no formula can be found that meets the client's needs, then modules can be used to create an appropriate formula.

Factors to consider in selecting a formula include:

1. *The client's ability to digest nutrients.* The person with a functional, but impaired, GI tract may benefit from hydrolyzed formulas. Otherwise, standard formulas are appropriate.
2. *Nutrient requirements.* Nutrient requirements are estimated based on a careful nutrition assessment. The client's age, medical condition, nutrition status, and metabolic rate are all important considerations in estimating nutrient requirements. If a nutrient must be restricted, a formula must be selected that delivers just the prescribed amount of that nutrient per day.
3. *Residue or fiber modifications.* The choice of formulas is narrowed when a person needs a low-residue or a high-fiber diet.
4. *Individual tolerances (food allergies and sensitivities).* Most formulas are lactose-free because temporary and permanent lactose intolerances are common problems for people who need enteral formulas.

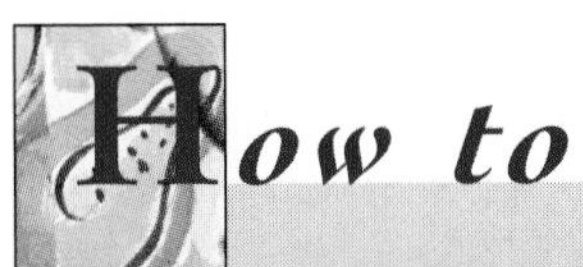

How to

Help Clients Cope with Tube Feedings

The thought of being "force-fed" is frightening to many people. One person may envision a large feeding tube and fear that the procedure will be extremely painful. Another may have heard about tube feedings only from the popular press and associate them with irreversible comas. All clients benefit when they understand the insertion procedure, the expected duration of the tube feeding, and the strategic role that nutrition plays in recovery from disease. These pointers can help health care professionals prepare clients for transnasal tube feedings:

- Allow clients to see and touch the feeding tube. Seeing firsthand that the tube is soft and narrow (only about half the diameter of a pencil) often alleviates anxiety. Show clients how the feeding apparatus is attached to the feeding tube, and explain how the feeding will work. Use dolls or stuffed toys to demonstrate tube insertion and feeding procedures to a young child.
- Explain that the client remains fully alert during the procedure and helps pass the tube by swallowing. A numbing solution sprayed on the back of the throat minimizes discomfort and prevents gagging during the procedure.
- Tell the client that once the tube has been inserted, most people become accustomed to its presence within a few hours. In most cases, the client can easily swallow foods and liquids with the tube in place. If permitted, favorite foods or beverages can still be enjoyed.
- Assure the client that the tube feeding will be temporary, if such assurance is appropriate.

Although a tube feeding may be frightening for some, for others, it is a relief. People who understand that they should eat, but cannot do so, may be relieved to receive sound nutrition without any effort. As they feel better and begin to eat again, the volume of the tube feeding can often be reduced and then discontinued when oral intake is adequate.

Some people feel a loss of control over their lives; others feel self-conscious about how the feeding tube looks or awkward about moving around with the equipment. A few simple measures can help:

- Involve older children, teens, and adults in the decision-making and care process whenever possible. Clients can help arrange daily feeding schedules, and some can also perform many of the feeding procedures themselves.
- Show clients how to manipulate the feeding equipment so that they can get out of bed and move around.
- Recommend that clients walk around and socialize with other clients, if permissible.
- Recommend that clients maintain contact with friends and keep busy with hobbies and activities they enjoy. This measure is especially important for children, teens, and those on long-term feedings.
- For infants and children, keep the developmental age of the child in mind and work with parents to ensure that appropriate feeding skills are mastered (see Nutrition in Practice 18). For infants, providing a pacifier during feedings helps maintain the associations between sucking, swallowing, eating, and fullness. When possible, some of the tube feeding formula may be provided by bottle or by spoon to further develop skills.

The more complex the procedure that a health care professional is responsible for, the easier it becomes to focus on the procedure and forget about the client's emotions. No matter how many technicalities you have to keep in mind, remember to stay focused on the person receiving your care.

5. *Availability.* Health care facilities cannot stock all formulas, so formula selection is limited by availability.

In the final analysis, the dietitian or physician can make only an educated guess about the best formula for an individual. The health care team monitors each person's nutrition status and responses to the formula to help ensure that individual needs are being met.

Once a feeding route has been selected and the feeding tube inserted, attention turns to delivering the formula. Thereafter, faithful and frequent monitoring helps ensure success. Using the safest methods of preparing and administering the formula helps prevent complications and ensures that the tube feeding will successfully support nutritional health.

Figure 21.3
Selecting a Formula

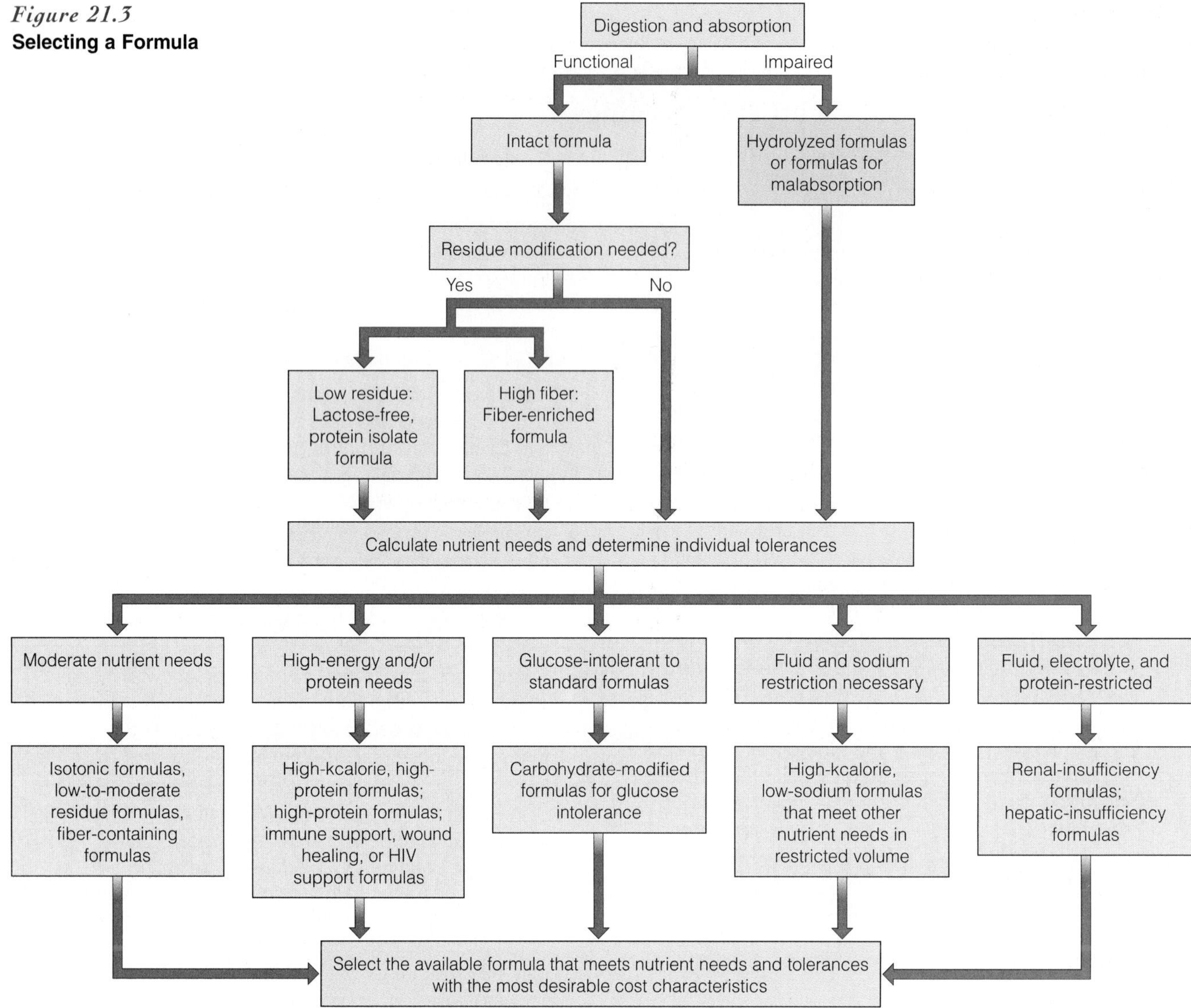

Tube Feedings: Preparation and Administration

Many people beginning a tube feeding are seriously ill or malnourished, and feedings must be handled carefully to avoid placing further stress on the person. The techniques described in the next sections help minimize the risk of complications, which can interfere with the attainment of medical and nutrition goals. Although the following sections specifically address the preparation and administration of tube feedings in health care institutions, the principles apply to people who receive tube feedings at home as well.

Formula Preparation

People who are ill or malnourished may have suppressed immune systems that make them vulnerable to infection from food-borne illness. To prevent contamination, all personnel involved in preparing or delivering formulas should handle them only in clean environments, using clean equipment and clean hands.

Formulas that must be mixed or diluted are most often prepared and packaged in the foodservice department or pharmacy. To prevent bacterial contamination, some institutions use sterile water for formula preparation. Once a formula has been mixed or diluted, the container is labeled with the client's name, room, date, and time of preparation and then sent to the nursing station. Ready-to-use cans of formula are often sent unopened to the nursing station; they, too, should be labeled with the client's name and room number.

At the Nursing Station Once a formula reaches the nursing station, the nursing staff assumes responsibility for its safe handling. The following steps reduce the likelihood of formula contamination:

- Before opening a can of formula, carefully clean the can opener and the lid. If you do not use the entire can at one feeding, label the can with the time it was opened.
- Cover opened cans. Store mixed or diluted formulas in clean, closed containers. Refrigerate the unused portion of formula promptly.
- Discard unlabeled or improperly labeled containers and all opened containers of formula not used within 24 hours.

At Bedside Prevent the risk of bacterial infections by following these procedures:

- Change the feeding bag or bottle and the tubing attached to it (except the feeding tube itself) every 24 hours.
- For intermittent feedings (described in the next section), rinse the feeding container and the attached tubing with water and allow them to air dry following each feeding. Never add fresh formula to formula still in the container.
- Flush the feeding tube with water before and after each use.

Tube feedings sometimes come prepackaged in closed containers that can be connected directly to the feeding tube without having to be transferred to another feeding container. Such systems save nursing time and significantly reduce the risk of bacterial contamination.

Formula Administration

Recommended formula administration schedules vary between institutions, and many protocols are based on clinical judgment rather than research. Almost all people can receive undiluted formula (either isotonic or hypertonic) at the start of a feeding.[7] People under severe stress, those who have not eaten for several weeks, or those receiving intestinal feedings may not be able to tolerate large volumes of hypertonic formulas initially. In such cases, formulas may need to be given slowly at first, at about 25 to 50 milliliters per hour. If the person tolerates the formula, the rate of feeding can be increased by about 25 milliliters per hour every 4 to 12 hours (see the margin note), depending on the location of the feeding tube (gastric or intestinal) as well as the person's medical condition. If the new rate is not tolerated, back up and proceed more slowly, giving the person more time to adapt. In a few cases, formulas may need to be diluted at first, the strength increased gradually, and then the rate advanced.

As an example of volume progression for a tube feeding, start the feeding at 50 ml/hr at full strength and then progress as follows:

- *After 6 hours: 75 ml/hr.*
- *After 12 hours: 100 ml/hr.*
- *After 18 hours: 125 ml/hr.*

Supplemental Water In addition to the formula itself, water can also be provided through the feeding tube. Using water to flush the feeding tube before and after a feeding or when the feeding apparatus is being changed helps prevent clogged feeding tubes and keeps the client hydrated. (Water can also be given orally if the person can drink it.) Supplemental water is often needed to meet the

Note: Fluid status must be carefully monitored in infants, the elderly, and people who are unconscious.

Formulas themselves contain considerable amounts of water. A standard formula (1.0 kcal/ml) contains about 850 ml of water per liter of formula. Higher-kcalorie formulas contain less water: formulas that contain 1.5 kcal/ml or 2.0 kcal/ml provide about 775 ml and 600 ml of water per liter of formula, respectively.

client's daily fluid requirements. As a guideline, adults require about 2000 milliliters (approximately 2 quarts) of water daily. Fever, excessive sweating, severe vomiting, diarrhea, blood loss, and burns raise water requirements. In kidney, liver, and heart diseases, water may need to be restricted.

Attention to indicators of body water balance can help determine how much additional water an individual needs. In alert adults, thirst is a good indicator of water needs; a person complaining of thirst generally needs water. In the elderly, however, thirst may be slow to develop in response to dehydration. Other clues to dehydration include unexplained weight loss, high serum electrolytes or hematocrit, and low blood pressure.

Delivery Techniques When people are receiving formula, they should not be lying flat; the risk of aspiration is too great. Whatever method is used to deliver a formula, elevate the client's upper body to at least a 45-degree angle during the feeding and for 30 minutes after the feeding whenever possible.[8]

A can of ready-to-use formula typically contains 240 ml of formula, and feedings are often divided so that one can of formula can be given at each feeding.

A day's volume of a formula can be given either intermittently or continuously over a period of 8 to 24 hours. Each method has specific uses, advantages, and disadvantages.

Delivery of no more than 250 ml of formula over 30 minutes is sometimes called an **intermittent feeding.**

Intermittent Feedings Intermittent feedings are best tolerated when they are delivered into the stomach and no more than 250 to 400 milliliters is given in 20 to 30 minutes using the gravity drip method or an infusion pump. The larger the volume of formula needed to meet nutrient needs, the more frequently feedings are delivered. (People who have very high nutrient needs benefit from either high–nutrient density formulas or continuous feedings.) Rapid delivery (in 10 minutes or less) of a large volume of formula (300 to 400 milliliters)—called bolus feeding—often leads to complaints of abdominal discomfort, nausea, fullness, and cramping. This makes sense. After all, people do not gobble down a meal in just a few minutes, especially when they are not feeling well.

Delivery of about 300 to 400 ml of formula over 10 minutes or less is called a **bolus feeding.**

gastric residual: the volume of formula that remains in the stomach from a previous feeding. It is measured by gently withdrawing the gastric contents through the feeding tube using a syringe. If the measured gastric residual is acceptable, the residual is returned to the client through the feeding tube.

Nurses measure gastric residuals to ensure that the stomach is emptying properly and to prevent nausea, vomiting, and possible aspiration of formula into the lungs. For intermittent feedings, the gastric residual is measured before each feeding. The gastric residual that is considered excessive varies among facilities, but ranges from about 75 to 150 milliliters. If the residual is excessive, the feeding is held for about an hour, and then the residual is rechecked. If excessive residuals persist, the physician may withhold the feeding, reduce the rate of administration, or begin drug therapy to stimulate gastric emptying. Studies suggest, however, that the formula's composition and the administration rate have a significant effect on gastric emptying, and that higher gastric residuals may be acceptable in some cases.[9]

Intermittent feedings work well for clients able to tolerate them. Often clients gradually adapt to larger volumes of formula given over shorter periods of time. Such feedings mimic the usual pattern of eating and allow the client freedom of movement between meals. They also require less time, making them less costly and easier for people to use at home.

Caution: Young children may be attracted to the bright lights, interesting sounds, and many controls of an infusion pump. Keep infusion pumps at a safe distance to prevent children from changing the flow rate or toppling the pump or intravenous pole and possibly injuring themselves or damaging the equipment.

Continuous Feedings Continuous feedings are delivered slowly and in constant amounts over a period of 8 to 24 hours. Such feedings benefit people who have received no food through the GI tract for a long time, those who are hypermetabolic, and those receiving intestinal feedings. Infusion pumps help ensure accurate and constant flow rates. For people receiving continuous feedings, gastric residuals are measured every 4 to 6 hours and should not exceed the volume of formula infused during the preceding 2 hours. The accompanying box explains several ways to plan tube feeding schedules.

How to

Plan a Tube Feeding Schedule

After selecting a formula that meets the client's medical and nutrient needs, the planner, usually a dietitian, determines the volume of formula per day that meets those needs. Consider a client who needs 2000 milliliters of formula per day. If the client is to receive the formula intermittently six times a day, he needs about 330 milliliters of formula at each feeding (2000 ml ÷ 6 feedings = 333 ml/feeding). Alternatively, if he is to receive the same volume of formula eight times a day, then he needs 250 milliliters (or about one can of ready-to-feed formula) at each feeding (2000 ml ÷ 8 feedings = 250 ml/feeding). He will probably tolerate this volume of formula best if it is given to him over 20 to 30 minutes at each feeding. If the client is to receive the formula continuously over 24 hours, he needs about 85 milliliters of formula each hour (2000 ml ÷ 24 hr = 83 ml/hr).

Delivering Medications through Feeding Tubes

Clients receiving tube feedings are often quite ill, and they are also likely to be receiving numerous medications. Often these medicines are delivered through feeding tubes, and in some cases, complications can occur.

Keep in mind that enteral formulas can interact with medications in the same ways that foods can. The health care team must consider the effects of drug therapy on nutrient requirements, the effects of the formula on drug absorption, the effects of medications on physical properties of formulas, and the prevention of complications.

Medication Forms A medication may come in any of several forms including tablets, liquid, injectable, and intravenous. People on tube feedings have functional GI tracts, and oral medications are less costly and easier to deliver than injectable or intravenous forms. Thus clinicians often prefer to use oral drugs for clients on tube feedings. The following guidelines may be helpful in preventing medication-medication interactions, medication-formula interactions, or clogged feeding tubes when medications are delivered through feeding tubes:

- Give medications by mouth instead of by tube whenever possible.
- Do not mix medications together or mix medications with the formula. Instead, stop the feeding temporarily and give each medication individually. Flush the feeding tube with warm water before and after administering each medication.
- Deliver liquid medications through the tube using a syringe, if possible. If liquid medications are thick or sticky, dilute them with water first.
- Consider using the injectable or intravenous form if the medication is not available in liquid form. Ideally, tablets should not be crushed and administered through feeding tubes. If using tablets is unavoidable, crush the tablets to a fine powder and mix with water before administering them. Do not crush tablets intended to release their contents slowly; in these cases, another medication form must be given.
- Avoid medications known to be incompatible with formulas, such as those listed in Table 21.2 on p. 520.

Additional Considerations The location of the feeding tube (whether gastric or intestinal) is also relevant when administering medications. Some medications

Table 21.2 Selected Medications That Are Incompatible with Some Formulas

Aluminum hydroxide	MCT oil
Chlorpromazine concentrate	Mellaril concentrate
Cibalith-S syrup	Mellaril oral solution
Cimetidine	Paregoric elixir
Dimetane elixir	Potassium chloride
Dimetapp elixir	Reglan syrup
Feosol elixir	Riopan
Fleet's phosphosoda	Robitussin expectorant
Gevrabon liquid	Sudafed syrup
Klorvess syrup	Thorazine concentrate
Mandelamine Forte suspension	Zinc sulfate capsules

Note: These substances may be compatible with some formulas and not others.

Sources: P. E. Burns, L. McCall, and R. Wirsching. Physical compatibility of enteral formulas with various common medications, *Journal of the American Dietetic Association* 88 (1988): 1094–1096; A. J. Cutle, E. Altman, and L. Lenkel, Compatibility of enteral products with commonly employed drug additives, *Journal of Parenteral and Enteral Nutrition* 7 (1983): 186–191; Z. M. Pronsky, *Food-Medication Interactions,* 9th ed. (Pottstown, Pa.: Food-Medication Interactions, 1995).

are designed to dissolve in the stomach's acidic environment. Such medications may not be readily absorbed if delivered directly into the duodenum or jejunum. Similarly, a medication that is optimally absorbed in the duodenum may be poorly absorbed in the jejunum. In such cases, oral, intravenous, or injectable forms of the medication should be provided.

In some cases, formulas alter medication absorption. One example is phenytoin, a medication used to control seizures. Absorption of phenytoin may be markedly reduced for a person who is on continuous tube feedings. Although opinions of the best way to handle this problem vary, some clinicians suggest that for most clients on either intermittent or continuous feedings, the feeding should be stopped for 2 hours before and 2 hours after giving phenytoin.[10] For clients requiring continuous feedings, the rate of delivery is increased during the times the feeding is given to ensure that nutrient needs are met.

Some clients have a specific type of feeding jejunostomy, called a *needle catheter jejunostomy,* through which phenytoin cannot be delivered. In such cases, phenytoin is given intravenously.

Addressing Tube Feeding Complications

Table 21.3 summarizes complications associated with tube feedings and shows that many problems can be prevented or corrected by selecting the formula and feeding route wisely, preparing the formula correctly, and delivering it appropriately. Attention to the person's primary medical condition and medications is important as well. Table 21.4 on p. 522 provides a monitoring schedule that helps ensure early detection of problems that may be encountered when clients are fed by tube.

Types of Complications Failure to estimate nutrient needs correctly or to ensure that the selected formula meets these needs limits the client's ability to achieve or maintain adequate nutrition status. Mechanical problems, such as a clogged feeding tube, a malfunctioning feeding pump, or a tube that becomes dislodged from its appropriate location, can interrupt the feeding schedule. Other complications related to the formula or its administration can result in nausea, vomiting, diarrhea, cramps, constipation, delayed gastric emptying, abdominal distension, and aspiration.

Metabolic complications such as dehydration, electrolyte imbalance, and elevated blood glucose can also occur. When complications arise, they can place further stress on the client and make it more difficult to achieve medical and nutrition goals.

Table 21.3 Causes and Prevention or Correction of Tube Feeding Complications		
Complications	**Possible Causes**	**Preventive/Corrective Measures**
Aspiration of formula	Compromised gastro-esophageal sphincter, delayed gastric emptying	Use nasoenteric, gastrostomy, or jejunostomy feedings in high-risk clients; elevate head of bed during and 45 minutes after feeding; check gastric residuals.
Clogged feeding tube	Formula too thick for tube	Select appropriate tube size; flush tubing with water before and after giving formula; use infusion pump to deliver thick formulas; remedies reported to help unclog feeding tubes include cola, cranberry juice, meat tenderizer, and pancreatic enzymes.
	Medications delivered through feeding tube	Use oral, liquid, or injectable medications whenever possible; dilute thick or sticky liquid medications with water before administering; crush tablets to a fine powder and mix with water; flush tubing with water before and after medications are given; give medications individually; do not mix medications with formula.
Constipation	Low-fiber formula	Provide additional fluids; use high-fiber formula.
	Lack of exercise	Encourage walking and other activities, if appropriate.
Dehydration and electrolyte imbalance	Excessive diarrhea	See items under *Diarrhea.*
	Inadequate fluid intake	Provide additional fluid.
	Carbohydrate intolerance	Use continuous drip administration of formula; monitor blood glucose; select a formula with a lower amount or different type of carbohydrate.
	Excessive protein intake	Monitor blood electrolyte levels; reduce protein intake.
Diarrhea, cramps, abdominal distension	Bacterial contamination	Use fresh formula every 24 hours; store opened or mixed formula in a refrigerator; rinse feeding bag and tubing before adding fresh formula; change feeding apparatus every 24 hours; prepare formula with clean hands using clean equipment in a clean environment.
	Lactose intolerance	Use lactose-free formula in lactose-intolerant and high-risk clients.
	Hypertonic formula	Use small volume of formula and increase volume gradually.
	Rapid formula administration	Slow administration rate or use continuous drip feedings.
	Malnutrition/low serum albumin	Use small volume of dilute formula and increase volume and concentration gradually.
Hyperglycemia	Diabetes, hypermetabolism, drug therapy	Check blood glucose; slow administration rate; provide adequate fluids; select a formula with a lower amount or different type of carbohydrate.
Nausea and vomiting	Obstruction	Discontinue tube feeding.
	Delayed gastric emptying	Check gastric residual; slow administration rate, use continuous drip feedings, or discontinue tube feeding.
	Intolerance to concentration or volume of formula	Use small volume of formula and increase volume and concentration gradually; use continuous drip feedings.
	Psychological reaction to tube feeding	Address client's concerns.
Skin irritation at enterostomy site	Leakage of GI secretions and friction caused by the tube	Keep site clean; inspect area for redness, tenderness, and drainage; use protective skin cream.

Note: Many of the complications presented here can be caused by the client's primary disorder or drug therapy rather than the tube feeding itself. In such a case, the corrective measure would include treatment of the disorder or a change in drug therapy. Additionally, other corrective measures require a physician's order and are not shown here.

What to Chart Chapter 17 emphasized the importance of the medical record as a legal document and communication tool. Before reimbursing for the cost of tube feedings, many types of insurance (including Medicare) and managed care organizations require that the physician appropriately document the client's need for a tube feeding as well as justification for using an infusion pump or a special formula, when appropriate. Medicare does not cover the cost of tube feeding unless the physician believes the condition necessitating the tube feeding

Table 21.4 Checklist for Monitoring Clients Recently Placed on Tube Feedings

Before starting a new feeding:	Complete a nutrition assessment.
	Check tube placement.
Before each intermittent feeding:	Check gastric residual.
Every half hour:	Check gravity drip rate, when applicable.
Every hour:	Check pump drip rate, when applicable.
Every 4 hours:	Check vital signs, including blood pressure, temperature, pulse, and respiration.
Every 6 hours:	Check blood glucose; monitoring blood glucose can be discontinued after 48 hours if test results are consistently negative in a nondiabetic client.
Every 4 to 6 hours of continuous feeding:	Check gastric residual.
Every 8 hours:	Check intake and output.
	Check specific gravity of urine.
	Check tube placement.
	Chart client's total intake of, acceptance of, and tolerance to tube feeding.
Every day:	Weigh client.
	Change feeding container and attached tubing.
	Clean feeding equipment.
Every 7 to 10 days:	Reassess nutrition status.
As needed:	Observe client for any undesirable responses to tube feeding; for example, delayed gastric emptying, nausea, vomiting, or diarrhea.
	Check nitrogen balance.
	Check laboratory data.
	Chart significant details.

will last for at least three months. Furthermore, Medicare does not cover the cost of tube feeding for clients with functioning GI tracts who are unable to eat due to a lack of appetite.

Health care professionals should routinely document the following information for clients on tube feedings:

- The nutrition goals for the client.
- Where (gastric or intestinal) and how (transnasal or enterostomy) the feeding tube is placed and the type and size of the feeding tube.
- The formula selected to meet nutrition goals and its nutrient composition.
- The recommended administration schedule (concentration and rate) and method of delivery (intermittent or continuous, gravity drip or infusion pump).
- Education of the client regarding the nutrition goals and the tube feeding procedure.
- The client's physical and emotional responses to the tube insertion and tube feeding procedure.
- The client's tolerances to the formula and administration schedule, complications (if any), and corrective actions recommended.
- Documentation of all substances delivered through the feeding tube including the formula, additional water, medications, and any substances used to unclog the feeding tube (see Table 21.3).

Case Study

Graphics Designer Requiring Enteral Nutrition

Mrs. Innis is a 24-year-old graphics designer who suffered multiple fractures when she fell from a cliff while hiking. She has been in the hospital for seven days and has no appetite. Mrs. Innis has lost 8 pounds over the course of her hospitalization. Due to the nature of her injuries, Mrs. Innis is in traction and is immobile, although the head of her bed can be elevated to 30 degrees. From the history, the dietitian determined that Mrs. Innis's nutrition status was adequate prior to hospitalization. The health care team agrees that a nasoduodenal tube feeding should be instituted before nutrition status deteriorates further. The standard formula selected for the feeding is lactose-free, and Mrs. Innis's nutrient requirements can be met with 2200 milliliters of the formula per day.

What steps can the health care team take to prepare Mrs. Innis for tube feeding? Why might nasoduodenal placement of the feeding tube be preferred to nasogastric placement for Mrs. Innis? Based on the limited information available, is the choice of formula appropriate?

The physician's orders specified that the feeding should be given continuously over 18 hours. Develop a tube feeding schedule for Mrs. Innis.

What parameters should be monitored to ensure that Mrs. Innis's fluid needs are being met? How can additional fluids be given? Describe precautions that should be taken if Mrs. Innis is to receive medications through the feeding tube.

After three days of feeding, Mrs. Innis develops diarrhea. Look at Table 21.3 on p. 521 to determine the possible causes. What measures can be taken to correct the various causes of diarrhea?

What tube feeding information should be charted in Mrs. Innis's medical record? When Mrs. Innis is ready to eat table foods again, what steps will the health care team take?

- Confirmation that measures for monitoring the client have been taken.
- Reasons why a tube feeding was interrupted or could not be delivered, if necessary.

The medical record serves as both a legal record and an invaluable source of information for investigating problems so that corrective measures can be taken promptly.

From Tube Feedings to Table Foods

Once the problem creating the need for a tube feeding resolves, the client can gradually shift to an oral diet as the volume of formula is tapered off. The client should be eating about two-thirds of the estimated nutrient needs by mouth before the tube feeding is discontinued.[11] In many cases, the person can begin to drink the same formula that is being delivered by tube. As clients begin to take more food or formula orally, they receive less of the formula by tube. Some people cannot make the transition to oral intake for medical reasons and go home on tube feedings. Chapter 22 discusses how home tube feedings work. The accompanying case study helps you to consider the many factors involved in tube feedings, and the nutrition assessment checklist reviews key points for assessing nutrition status in people receiving tube feedings.

Tube feeding is a practical solution to feeding a person who is unable to consume adequate nutrients by mouth. However, a person without a functional GI tract cannot benefit from a tube feeding. In such a case, intravenous nutrition (the subject of the next chapter) can be a lifesaving treatment option.

Nutrition Assessment Checklist

For People Receiving Tube Feedings

Medical Review the client's medical record for the information necessary to determine nutrient needs, and select the formula, feeding site (gastric versus intestinal), and placement technique (transnasal or enterostomy). Regularly review the medical record for information regarding changes in medical status and symptoms associated with the client's medical condition and therapy (especially drug) that suggest the need to alter the formula selection or delivery techniques or that require prompt intervention.

Medication Review the client's medications for possible nutrient-medication interactions and GI side effects that may affect the client's tolerance for the tube feeding. If the feeding tube is used to deliver drugs, follow the precautions on p. 519.

Nutrient/Food Intake Assess the client's nutrient needs and ensure that the formula, when given as prescribed, meets those needs. Check the medical or nurses' records to be sure that the formula is being delivered as prescribed, and take corrective actions as needed. For clients beginning to eat, determine the degree to which nutrient needs are being met by table foods or formula taken orally, and reduce the volume of the tube feedings accordingly.

Anthropometric Assess the client's weight daily to determine whether the selected formula is meeting the client's energy needs.

Laboratory Monitor serum and urine lab values for signs of fluid and electrolyte imbalances and glucose intolerance. Check serum protein levels to ensure that they are improving or being maintained. When available, assess nitrogen balance to determine whether the tube feeding is meeting the client's protein needs. Check other lab values as appropriate for the client's medical condition.

Physical Check gastric residual for signs of delayed gastric emptying to prevent GI complications and reduce the risk of aspiration. Check tube placement and gravity drip rate or infusion pump drip rate as needed (see Table 21.4). Assess blood pressure, temperature, pulse, and respiration every 4 hours. Look for physical signs of malnutrition or dehydration.

Self Check

1. A client has had a poor appetite for several days. Strategies the health care professional can use to improve a client's intake include all of the following **except:**
   a. help the client prepare for meals.
   b. provide foods that are easy to eat if the client is weak or tired.
   c. tell the client that if she cannot eat enough food she will need to be fed by tube.
   d. make sure that the client receives the right tray with the foods she selected.
2. Which of the following statements is correct?
   a. Standard formulas contain whole proteins or protein isolates.
   b. Standard formulas contain free amino acids or small peptide chains.
   c. Hydrolyzed formulas are made from pureed meats.
   d. Hydrolyzed formulas may contain protein isolates or whole proteins.
3. When a client cannot meet nutrient needs from table foods:
   a. enteral formulas provided by tube are preferred to formulas provided orally.
   b. tube feedings are preferred to parenteral nutrition.
   c. parenteral nutrition is preferred to enteral nutrition.
   d. parenteral nutrition is preferred to enteral formulas provided orally.
4. Which of the following statements is correct?
   a. Enterostomies often serve clients who need short-term tube feedings.
   b. Nasogastric and gastrostomy feedings carry a higher risk of regurgitation of formula than nasojejunal or jejunostomy feedings do.
   c. Feeding tubes inserted through enterostomies are more easily removed by uncooperative or disoriented clients than transnasal feeding tubes are.
   d. General anesthesia and surgery are always needed to create enterostomies.

5. The health care professional selecting an appropriate enteral formula for a client, considers all of the following **except:**
   a. the formula's osmolality.
   b. the client's ability to digest and absorb nutrients.
   c. the client's nutrient needs.
   d. the formula's cost.
6. To prevent bacterial contamination of tube feeding formulas, you would:
   a. discard all opened containers of formula not used in 48 hours.
   b. change the feeding bag and attached tubing every 12 hours.
   c. deliver the formula continuously.
   d. prepare the formula with clean hands in a clean environment.
7. A client needs 1800 milliliters of formula a day. If the client is to receive the formula intermittently every 4 hours, she will need _____ milliliters of formula at each feeding.
   a. 225
   b. 300
   c. 400
   d. 425
8. Factors to consider when medications are delivered through feeding tubes include:
   a. enteral formulas do not interact with medications in the same way that foods do.
   b. thick or sticky liquid medications and tablets that are not crushed to a fine powder can clog feeding tubes.
   c. medications can be added directly to formulas.
   d. medications as well as tube feedings can result in GI complaints.
9. Tube feedings can gradually be discontinued when:
   a. the client is able to eat foods or drink formula in sufficient amounts.
   b. as soon as the medical condition resolves.
   c. discharge planning begins.
   d. the client experiences hunger.

Answers to these questions appear in Appendix H.

Clinical Applications

1. Complex procedures, such as those necessary to deliver tube feedings, require attention to many technical details, making it easy to focus on the procedure and forget about the client. Imagine that you need a transnasal tube feeding. How might you react to the news that you need the feeding and to the insertion procedure? What things would you miss most about eating table foods? Think about ways health care professionals might help you deal with these feelings.
2. Take a look at the checklist for monitoring people on tube feedings (Table 21.4 on p. 522). You can see that the person on a tube feeding requires a great deal of care. Discuss the advantages of a nutrition support team in monitoring clients on tube feedings. What contributions might various team members make in working with tube-fed clients (see Nutrition in Practice 21)?
3. Review Chapters 18, 19, and 20 and note the symptoms and disorders that may require the use of tube feedings. For each, consider when and why a tube feeding might be appropriate and which conditions might require hydrolyzed formulas. Also note those conditions associated with esophageal reflux that might preclude the use of nasogastric feedings. What alternatives are possible in those cases?

Notes

1. H. M. Storm and P. Lin, Forms of carbohydrate in enteral formulas, *Support Line,* June 1996, pp. 7–9.

2. E. A. Emergy and coauthors, Banana flakes control diarrhea in enterally fed patients, *Nutrition in Clinical Practice* 12 (1997): 72–75.

3. K. Teahon and coauthors, Practical aspects of enteral nutrition in the management of Crohn's disease, *Journal of Parenteral and Enteral Nutrition* 19 (1995): 365–368.

4. V. M. Herrmann and coauthors, Who and when to enterally tube feed, *Nutrition in Clinical Practice* (supplement) 12 (1997): 59.

5. F. W. Clevenger and D. J. Rodriguez, Decision-making for enteral feeding administration: The why behind the where and how, *Nutrition in Clinical Practice* 10 (1995): 104–113; Q. Duh, Decision tree for route of enteral nutrition support: Placement techniques, A summary, in *Enteral Nutrition Support,* Report of the First Ross Conference on Enteral Devices, Ross Laboratories, 1996.

6. K. Sriram and coauthors, Prophylactic locking of enteral feeding tubes with pancreatic enzymes, *Journal of Parenteral and Enteral Nutrition* 21 (1997): 353–356.

7. G. P. Zaloga, Enteral nutrition in hospitalized patients: A summary, in *Enteral Nutrition Support for the 1990s: Innovations in Nutrition, Technology, and Techniques,* Report of the Twelfth Ross Roundtable on Medical Issues, Ross Laboratories, 1992.

8. E. H. Elpern, Pulmonary aspiration in hospitalized adults, *Nutrition in Clinical Practice* 12 (1997): 5–13.

9. H. C. Lin and G. W. Van Citters, Stopping enteral feeding for arbitrary gastric residual volume may not be physiologically sound: Results of a computer simulation model, *Journal of Parenteral and Enteral Nutrition* 21 (1997): 286–289.

10. J. Hatton and B. Magnuson, How to minimize interaction between phenytoin and enteral nutrition: Two approaches, *Nutrition in Clinical Practice* 11 (1996): 28–31.

11. T. C. Lykins, Nutrition support clinical pathways, *Nutrition in Clinical Practice* 11 (1996): 16–20.

CHAPTER 21

Nutrition in Practice

▪ The Health Care Team and Nutrition Care ▪

Teamwork in health care can promote high-quality care in a more efficient manner than individual professionals working solo. In today's cost-conscious health care environment, the value of health care teams in maintaining the highest-quality care while avoiding unnecessary duplication of services is increasingly recognized.

Each professional on a health care team possesses specialized knowledge. By integrating their strengths, team members can provide the client with the benefits of their combined expertise. By working together, each team member learns how the others contribute to the care process and discovers ways to provide services efficiently.

Who serves on a health care team?

The number and types of health care teams vary from one institution to the next. A health care team consists of a group of professionals specializing in a particular area of medicine. For example, the cardiac rehabilitation team works with clients recovering from heart attacks or cardiac surgery. Many health care teams include physicians, nurses, dietitians, pharmacists, and mental health specialists as their core members. Other members vary according to the specialty area. A respiratory therapist would be a primary member of a pulmonary rehabilitation team, for example. Physical therapists and occupational therapists frequently participate on rehabilitation or burn teams. Dietitians participate on health care teams that recognize medical nutrition therapy as essential to client care. Some examples include teams that specialize in disorders of the heart, blood vessels, lungs, endocrine system (such as diabetes mellitus), gastrointestinal tract, liver, and kidneys. Dietitians also serve on teams where multiple organ systems are involved such as severe trauma, burns, wound care, HIV infection, cancer, and inborn errors of metabolism.

Some institutions have nutrition support teams. Nutrition support teams develop standards and protocols for nutrition screening and nutrition assessment, identify people who need nutrients from enteral formulas or parenteral nutrition, and oversee the care of these people.

What do team members do?

Team members maintain records, review the progress of each client in their care, and meet frequently to discuss concerns affecting individual clients, problems with procedures, or issues appropriate to their specialty area. They often serve as consultants to the rest of the institution's staff, providing information, answering questions, and solving problems that arise. Team members also carry responsibility for keeping abreast of new developments in their respective fields. They analyze new products, review current research, and communicate their findings to other team members.

The team physician assumes primary responsibility for client care: diagnosing the client's medical problems, performing medical procedures, and coordinating and prescribing appropriate therapy. The physician frequently supervises the activities of other team members, monitors complications that arise, and makes the final decisions on what steps to take to correct problems.

Team physicians also direct the development and approval of guidelines and protocols pertinent to the team's area of specialization. They oversee in-hospital educational and training programs and act as consultants to other physicians who require their professional expertise.

What roles do nurses play?

The team nurse plays a central role in client care management and communications with the client, family, and nursing staff. The nurse often explains medical procedures and treatment plans to clients and their caregivers. In addition, the team nurse supervises other nurses who care for each client to assure that they are delivering appropriate and optimal care. The team nurse teaches the staff nurses, as a group and individually, appropriate procedures and their rationales. The

Nutrition in Practice

The team approach helps to ensure safe and cost-effective care.

nurse also assists in educating physicians and other health care professionals.

The team nurse often coordinates discharge from the hospital. Discharge responsibilities include discussing with clients any steps they will need to follow at home, making sure clients have written instructions, providing them with appropriate supplies and equipment, and arranging for follow-up care. (The nurse may contact the social worker to provide solutions to financial problems and coordinate home care programs.)

WWW.

ana.org
American Nurses Association

What about the dietitian and the pharmacist?

The dietitian, as the team's nutrition expert, assesses the client's nutrition status, determines the client's nutrition requirements, recommends appropriate medical nutrition therapy, and translates diet orders into foods or formulas. The dietitian calculates nutrient intake data and nitrogen balance and measures energy needs using indirect calorimetry. In addition, the dietitian visits clients regularly to monitor nutrient intakes and nutrition status and to help them understand and cope with their unique feeding situations. The dietitian also instructs clients about their diets and prepares them to follow special diets at home, when necessary.

Dietitians provide in-service training on nutrition-related topics in their specialty area for other dietitians and health care professionals. If particular nutrition problems arise, the dietitian actively investigates the possible causes, devises appropriate solutions, and sees that they are carried out. The team dietitian also acts as a liaison between the team and the foodservice department.

The pharmacist assists the physician in managing the client's drug therapy; alerts team members to possible interactions of medications with other medications, nutrients, or nutrient solutions; identifies complications that may be medication related; and recommends solutions to medication-related problems. The pharmacist may recommend an optimal medication administration schedule and educates clients about the proper use and possible side effects of their medications. The pharmacist also serves as a liaison between the pharmacy and the health care team.

WWW.

eatright.org/dpg24.html
American Dietetic Association Dietetic Practice Group—Dietitians in Nutrition Support

faseb.org/ascn
American Society for Clinical Nutrition

clinnutr.org
American Society for Parenteral and Enteral Nutrition

ashp.com
American Society of Health-System Pharmacists

Describe how a health care team solves nutrition problems.

Using the core members of a typical nutrition support team as an example, Figure NP21.1 shows the unique responsibilities of individual team members as well as responsibilities shared by the team. Team members communicate with each other both informally and formally. They may share office space and see each other throughout the day, but their primary avenue for managing team responsibilities is during rounds, when the entire team discusses each client in the team's care.

During rounds, the physician oversees the discussions and assesses recommendations. Sometimes the team discusses problems that affect many clients. For example, while discussing a client who has undergone a nitrogen balance study, the dietitian may mention that problems have occurred repeatedly with this procedure in the last month. The nurse then recalls that many of the new nurses are having difficulty with the procedure and suggests an in-service training session on the proper techniques for nitrogen balance studies. The nurse agrees to conduct the in-service training, and the dietitian joins in to explain how nitrogen balance studies help determine protein needs.

Nutrition in Practice

The physician
- Diagnoses medical problems
- Performs medical procedures
- Coordinates and prescribes therapy
- Directs and supervises team
- Approves guidelines and protocols
- Consults with other physicians

The nurse
- Assesses nursing needs
- Performs direct client care
- Explains medical procedures and treatment plans
- Instructs clients regarding medical care
- Acts as a liaison between team and nursing staff
- Coordinates discharge plans

All team members
- Review current research
- Analyze new products
- Develop guidelines
- Provide in-service training
- Monitor clients
- Correct problems
- Educate clients
- Evaluate the outcome of the care provided and cost savings
- Promote the appropriate use of nutrition support
- Improve communications among team members and between the team and other health care professionals

The dietitian
- Assesses nutrition status
- Determines clients' nutrient needs
- Recommends appropriate diet therapy
- Reevaluates clients regularly
- Instructs clients about their diets
- Acts as a liaison between the team and the dietary department

The pharmacist
- Recommends appropriate drug therapy
- Identifies medication-medication and nutrient-medication interactions
- Identifies medication-related complications
- Educates clients about their medications
- Acts as a liaison between the team and the pharmacy

Figure NP21.1
The Nutrition Support Team

During rounds, team members may bring up current research that may affect the care they give clients, or they may share information about new products they may wish to consider for their clients. The team then decides if further action is warranted.

Sometimes discussions focus on isolated problems. For example, the dietitian may express concern for a client who is unable to be weaned from tube feedings because she is experiencing pain that interferes with her ability to eat. The pharmacist recommends a pain medication and administration schedule that will optimize pain relief at mealtimes. The physician sees no problem with making this change and writes the medication order. The nurse makes sure that staff nurses working with the client understand the new medication schedule and its rationale. Working together, the team has efficiently identified and solved a problem that might otherwise go unresolved and result in poorer-quality care and increased costs.

Could such a simple step really improve the quality of care and reduce health care costs?

Yes, it can. Provided that an effective pain management therapy can be found, the client will not only feel better, but may also be able to begin eating foods orally. The sooner the person is able to take adequate foods by mouth, the sooner the tube feeding can be discontinued. Removal of the tube benefits the client's well-being. As Chapter 21 described, tube feedings necessitate equipment, formulas, and more time from physicians, nurses, and dietitians—all of which raise costs. The longer the feeding continues, then, the more time the client is burdened with the feeding equipment and the more expensive the care.

Even when the amount of money saved on one case is relatively small, small interventions can add up to substantial savings because institutions house many clients. Nutrition support teams can help save costs in many ways that can have a significant impact on health care costs.

Can you give me some examples?

A few of the ways nutrition support teams have been shown to improve the quality of care and save health care costs include:

- Identifying malnutrition and instituting appropriate treatments. One analysis suggests that early recognition and treatment of malnutrition can result in cost savings of about $8300 per hospital bed per year.[1]
- Preventing the overuse of parenteral nutrition, which is far more costly than enteral nutrition. In one hospital that implemented processes to reduce the inappropriate use of parenteral nutrition, quality nutrition care could be achieved with an estimated cost savings of over $225,000 for one year.[2]
- Selecting the most cost-effective enteral and parenteral formulas and supplies.[3]
- Developing standards and protocols that reduce the risk of complications. In one study, researchers found that developing and adhering to a strict protocol for inserting and caring for parenteral nutrition catheters and caring for the catheter insertion site resulted in potential cost savings of $14,500 to $35,000 per month.[4]

To appreciate the team approach, remember the adage "Two heads are better than one." In this case, several heads are better still. Clients benefit when many eyes are noting problems and many brains are searching for solutions. In short, teamwork works.

Notes

1. H. N. Tucker and S. G. Miguel, Cost containment through nutrition intervention, *Nutrition Reviews* 54 (1996): 111–121.

2. N. M. Pace and coauthors, Performance model anchors successful nutrition support protocol, *Nutrition in Clinical Practice* 12 (1997): 274–279.

3. As cited in J. R. Wesley, Nutrition support teams: Past, present, and future, *Journal of Parenteral and Enteral Nutrition* 10 (1995): 219–228.

4. W. C. Faubion and coauthors, TPN catheter sepsis: Impact of the team approach, *Journal of Parenteral and Enteral Nutrition* 10 (1986): 642–645.

22 Specialized Nutrition Support: Parenteral Nutrition

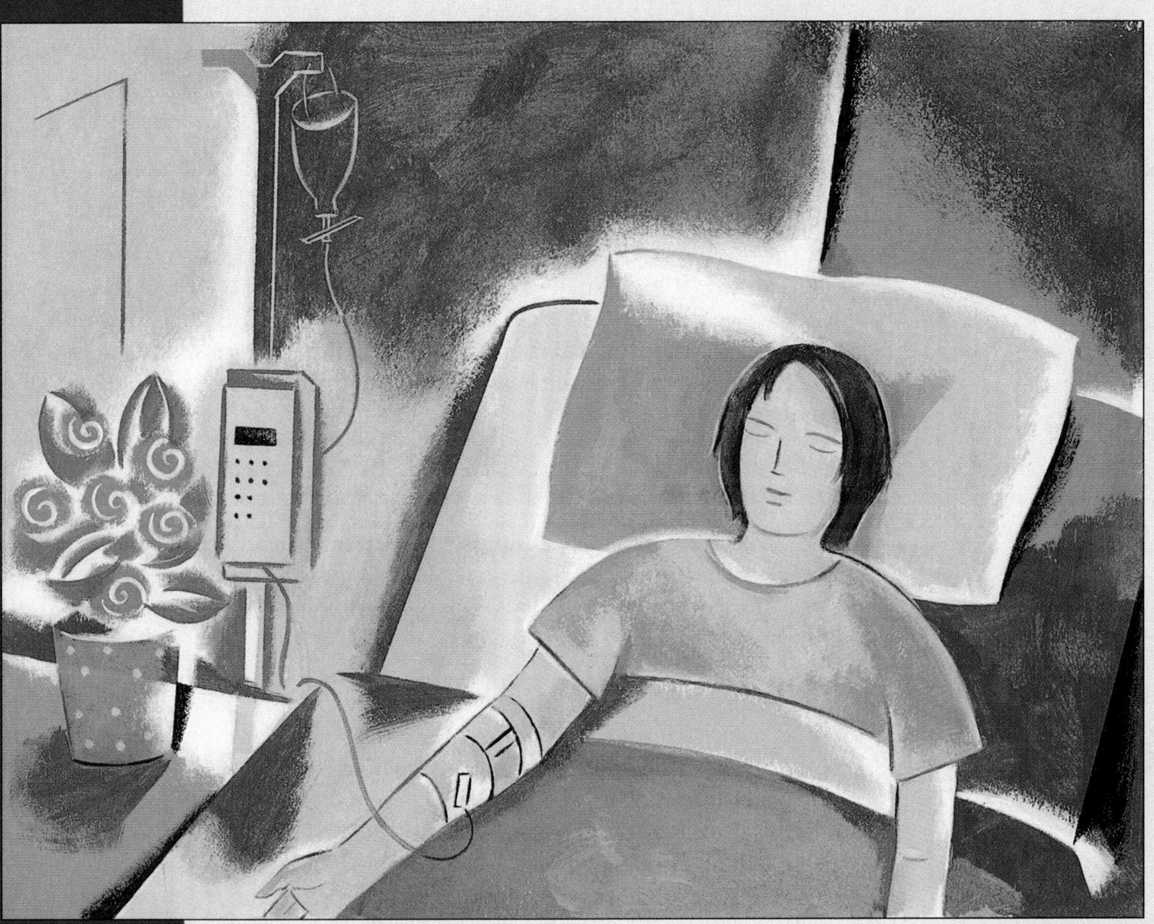

WWW.clinnutr.org
American Society for Parenteral and Enteral Nutrition

magi.com/~cpena
Canadian Parenteral-Enteral Nutrition Association

bapen.org.uk
British Association for Parenteral and Enteral Nutrition

parenteral nutrition: delivery of nutrient solutions directly into a vein, bypassing the intestine.
para = outside
enteron = intestine

intravenous (IV): through a vein.
intra = within
vena = vein

The science of medical nutrition as we know it today was shaped tremendously by the demonstration in 1968 that all nutrient needs could be met by vein.[1] Practitioners now had the means to feed people who otherwise might have died from malnutrition. With time, clinicians learned more and more about the solutions and delivery techniques to provide parenteral nutrition. They also discovered that while parenteral nutrition is a lifesaving procedure, it is also very costly and is associated with serious complications including liver dysfunction, progressive kidney problems, bone disorders, and many nutrient deficiencies. These findings prompted a renewed appreciation for the GI tract, and clinicians today subscribe to the adage "If the GI tract works, use it." Only when people cannot meet their nutrient needs using the enteral route should they receive total parenteral (or intravenous) nutrition support.

Intravenous Nutrition

As is true of all medical nutrition therapy, the decision to use intravenous (IV) nutrition, the method of delivery, and the type and amount of nutrients to provide are based on a thorough assessment of the client's medical condition and nutrient needs. Infusion of IV nutrients immediately changes blood levels of fluids, electrolytes, and other nutrients and, therefore, requires vigilant attention to the individual's responses.

Intravenous Solutions

A variety of nutrients can be administered by vein. These IV solutions may contain any or all of the essential nutrients: water, amino acids, carbohydrate, fat, vitamins, and minerals. Skilled pharmacists compound the individualized IV solutions prescribed by the physician.

Amino Acids Intravenous amino acid solutions usually contain both essential and nonessential amino acids to meet the body's need for protein. Special products that contain only essential amino acids or have large amounts of certain amino acids and small amounts of others are also available for specific medical conditions. Products designed for liver failure, for example, may contain more branched-chain amino acids and fewer aromatic amino acids (see Chapter 28).

Sometimes a nonessential amino acid may be omitted from standard parenteral solutions because it does not mix well or is unstable in the solution. For example, glutamine, which may be a conditionally essential amino acid for some clients, is not stable in IV solutions. Providing glutamine as a dipeptide solves the instability problem, and studies suggest that short-chain peptides can be digested to free amino acids by enzymes bound to cell membranes.[2]

Glutamine may be a conditionally essential amino acid following intestinal resections and during recovery from severe stress.

Carbohydrate Standard IV solutions provide carbohydrate as dextrose (glucose). Because the form of dextrose in IV solutions contains some water, dextrose provides only 3.4 kcalories per gram, whereas glucose provides 4.

dextrose monohydrate: a form of glucose that contains some water and is stable in IV solutions. Dextrose solutions provide 3.4 kcal/g, whereas glucose provides 4 kcal/g.

Lipid Intravenous lipid emulsions are the vehicle for fat in IV solutions. Intravenous fats are provided either daily or periodically (once or twice a week). If provided daily, IV fat serves as a concentrated source of energy; if offered less often, it serves primarily as a source of essential fatty acids. A 500-milliliter bottle of a 10 percent fat emulsion provides 550 kcalories (1.1 kcalories per milliliter). The same volume of a 20 percent fat emulsion provides 1000 kcalories (2.0 kcalories per milliliter). The accompanying box explains how to calculate the nutrient content of IV solutions.

How to

Calculate the Nutrient Content of IV Solutions

You can have confidence in IV solutions if you know what they contain. The basic thing to remember is that the percentage of a substance in solution tells you how many grams of that substance are present in 100 milliliters. For example, a 5 percent dextrose solution contains 5 grams of dextrose per 100 milliliters. A 3.5 percent amino acid solution contains 3.5 grams of amino acids per 100 milliliters. A 0.9 percent normal saline solution contains 0.9 gram of sodium chloride per 100 milliliters.

Suppose a person is receiving 3 liters (3000 milliliters) of an IV solution containing 1500 milliliters of 50 percent dextrose and 1500 milliliters of 7 percent amino acids. For dextrose, the person would get:

$$\frac{50 \text{ g dextrose}}{100 \text{ ml}} = \frac{X \text{ g dextrose.}}{1500 \text{ ml}}$$

$$\frac{50 \text{ g} \times 1500 \text{ ml}}{100 \text{ ml}} = 750 \text{ g dextrose.}$$

And for amino acids:

$$\frac{7 \text{ g amino acids}}{100 \text{ ml}} = \frac{X \text{ g amino acids.}}{1500 \text{ ml}}$$

$$\frac{7 \text{ g} \times 1500 \text{ ml}}{100 \text{ ml}} = 105 \text{ g amino acids.}$$

To calculate the total kcalories in 3000 milliliters of the solution, simply multiply by kcalories per gram and add the totals:

$$\begin{aligned} 750 \text{ g dextrose} \times 3.4 \text{ kcal/g} &= 2550 \text{ kcal} \\ 105 \text{ g amino acids} \times 4.0 \text{ kcal/g} &= \underline{420 \text{ kcal}} \\ \text{Total} &= 2970 \text{ kcal.} \end{aligned}$$

Intravenous fat emulsions are contraindicated for some newborns with markedly elevated bilirubin levels, people with some types of hyperlipidemia, people with severe liver disease, and those with severe egg allergies. Cautious use of IV fats is recommended for people with atherosclerosis, moderate liver disease, blood coagulation disorders, pancreatitis, and some types of lung problems. After long-term administration, brown pigments may accumulate in certain liver cells, but these pigments disappear after parenteral therapy is discontinued; their effects on liver function are unknown. Prolonged IV lipid use may also enlarge the liver and spleen and reduce the number of blood platelets and white blood cells.

bilirubin: a pigment in the bile whose concentration in the blood may rise as a result of some disorders.

Hyperlipidemia and atherosclerosis are discussed in Chapter 26. Liver disorders are the subject of Chapter 28.

Micronutrients Vitamins, electrolytes (minerals), and trace elements may all be used in IV solutions. Currently available IV multivitamin formulations for adults meet the recommendations of the Nutrition Advisory Group of the American Medical Association. These recommendations do not include vitamin K, because administering vitamin K to clients who are also receiving anticoagulants can lead to potentially serious interactions.[3] For clients not receiving anticoagulants, vitamin K is provided in one weekly dose. Pediatric multivitamin solutions contain vitamin K.

Failure to include vitamins in parenteral nutrition solutions can result in serious nutrient deficiencies and even death. During 1988, for example, a national shortage of parenteral multivitamin formulations led to the omission of vitamins in parenteral solutions and resulted in severe thiamin deficiencies and three deaths.[4]

Some electrolytes (particularly, calcium and phosphorus) can precipitate with other IV solution components, potentially resulting in life-threatening complications. As an indication of the seriousness of this problem, the Food and Drug Administration recently warned health care professionals that a precipitate of calcium phosphate might have been responsible for two deaths and at least two cases of respiratory distress in one institution. Pharmacists skilled in

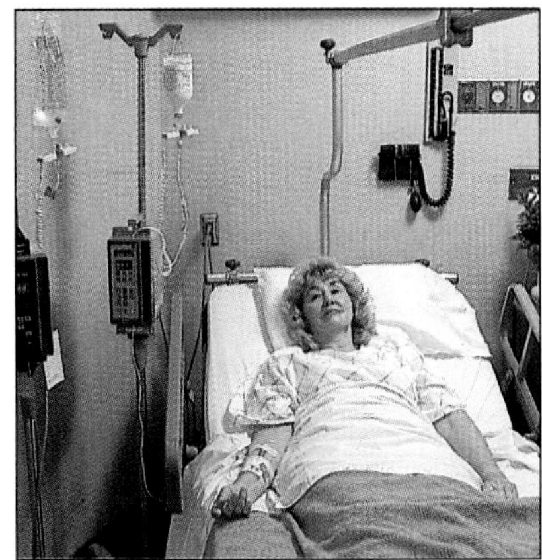

Intravenous lipid emulsions, made from egg phospholipids and plant-derived oils, provide energy and essential fatty acids and can be easily identified by their milky white color.

compounding IV solutions learn to mix solutions to minimize the risk of precipitation and protect the stability of IV solutions.

Other Additives Intravenous medications are sometimes added directly to the IV solution or infused into IV tubing through a separate port. Common examples include heparin, insulin, cimetidine, ranitidine, and famotidine. Providing medications along with parenteral nutrition solutions saves time and avoids the need for a separate infusion site. Interactions between medications and IV solutions, however, can and do occur. Only medications proved to be physically compatible with, and biochemically stable in, IV solutions are safe to administer. When medications are added directly to the IV solution, the health care team must remember that if the total volume of solution is not infused, then the client may not receive the full dose of medication. Conversely, if the addition of a medication to an IV solution is not noted on the medication record, the physician may inadvertently reorder the medication, and the client may suffer potentially severe consequences.

Types of Intravenous Feedings

Intravenous solutions can be delivered in different ways. The method used depends on the person's immediate medical and nutrient needs, nutrition status, and the anticipated length of time on IV nutrition support.

Simple IV solutions typically contain 5% dextrose and normal saline. Other electrolytes or salts may be added as needed. Often 3 liters of the solution are provided daily and deliver about 150 g of dextrose or about 500 kcal per day.

IV catheter: a thin tube inserted into a vein. Additional IV tubing connects the IV solution to the catheter.

peripheral veins: the small-diameter veins that carry blood from the extremities (arms and legs).

Simple IV Administration Simple IV solutions are used routinely in hospitals to provide water, dextrose, and electrolytes, and they help to maintain or restore fluid and electrolyte and acid-base balances. Simple IV solutions are delivered through catheters inserted into the small-diameter peripheral veins, such as those in the forearm and back of the hand. Most people are expected to be able to eat within a few days following surgery or illness, and simple IV solutions often meet their needs satisfactorily.

Meeting All Nutrient Needs by Vein Simple IV solutions fall far short of meeting the nutrient needs of people who cannot use their GI tracts for a long time, especially those who are malnourished or have high nutrient requirements. The volume of simple IV solutions required to meet energy needs exceeds the volume of fluid the body can safely handle. If higher concentrations of dextrose are provided through peripheral veins, the veins become irritated and eventually collapse. Two options remain: peripheral parenteral nutrition and central parenteral nutrition.

peripheral parenteral nutrition (PPN): the use of the peripheral veins to provide a solution that meets nutrient needs.

Peripheral Parenteral Nutrition (PPN) For some people, nutrient needs can be met using the peripheral veins to deliver IV solutions that provide lipids in addition to dextrose, amino acids, vitamins, minerals, and trace elements—*peripheral parenteral nutrition (PPN).* Intravenous lipid emulsions make it possible to deliver needed nutrients because they provide a concentrated source of kcalories in a form that is isotonic to blood and less irritating to the blood vessels than highly concentrated dextrose solutions. A typical PPN solution delivers about 2500 kcalories per day and provides about 150 grams of amino acids; IV lipid emulsions contribute more than half of the total kcalories.

Peripheral parenteral nutrition best suits people with normal renal function who need only short-term nutrition support (about 7 to 14 days), people who need additional nutrients temporarily to supplement an oral diet or tube feeding, or those in whom inserting an IV catheter into a central vein might be difficult.[5] People with very high energy requirements, people with weak peripheral veins that collapse easily, and those with fluid restrictions are not candidates for PPN.

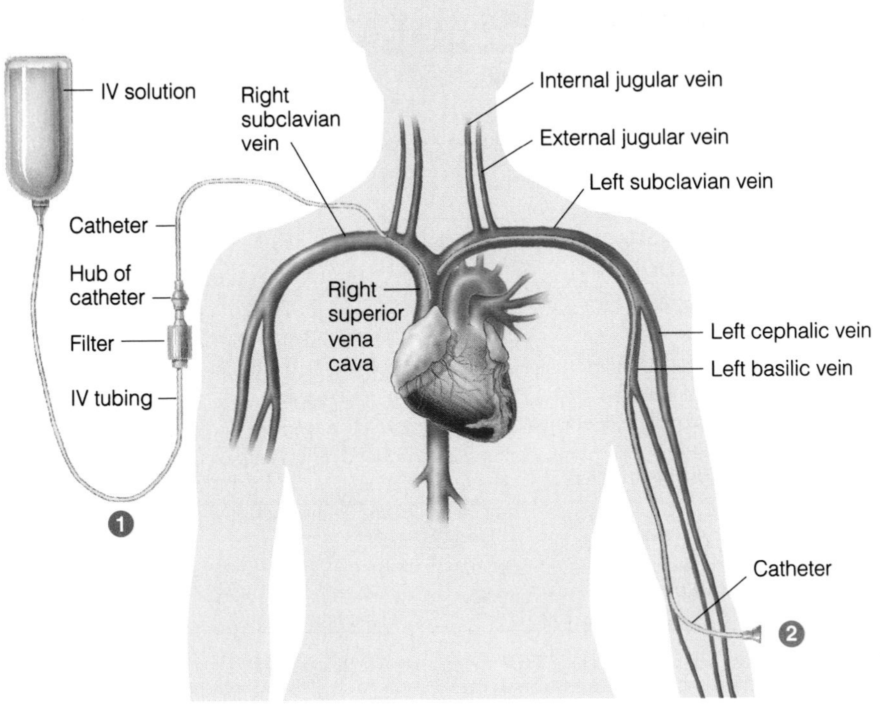

Figure 22.1
The Central Veins Used for TPN
1 Traditionally, TPN catheters enter the circulation at the right subclavian vein and are threaded into the superior vena cava with the tip of the catheter lying close to the heart. Sometimes catheters are threaded into the superior vena cava from the left subclavian vein, the internal jugular veins, or the external jugular veins.
2 Peripherally inserted central catheters usually enter the circulation at the basilic or cephalic vein and are guided up toward the heart so that the catheter tip rests in a central vein, often the superior vena cava.

Total Parenteral Nutrition (TPN) by Central Vein Another method used to meet all nutrient needs by vein is *central total parenteral nutrition,* or *TPN* for short. In TPN, the tip of the IV catheter is either placed directly in a large-diameter central vein (see Figure 22.1) or threaded into a central vein through a peripheral vein. Almost a gallon of blood rushes through one such central vein, the right superior vena cava, each minute, so highly concentrated solutions quickly become diluted. By the time these solutions reach the peripheral veins, they are no longer concentrated enough to irritate the blood vessels.

central total parenteral nutrition (TPN): a method for meeting all nutrient needs by infusing formula either directly into a large-diameter central vein or threading a catheter through a peripheral vein and into a central vein.

central veins: the large-diameter veins located close to the heart (see Figure 22.1).

TPN is indicated whenever long-term parenteral nutrition will be required, when nutrient requirements are high, or when people are severely malnourished (see Table 22.1 on p. 536). People who need TPN for weeks or months, but risk serious complications if a catheter is inserted directly into a central vein, may be candidates for peripherally inserted central catheters.[6]

peripherally inserted central catheter (PICC): a catheter inserted into a peripheral vein and advanced into a central vein.

Regardless of how the catheter is placed, TPN should be initiated before nutrition status is severely compromised. It is much easier to maintain nutrition status than to try to replenish lost nutrient stores.

Composition of TPN Solutions The actual concentrations of amino acids, dextrose, and lipids that compose the TPN solution are determined by each person's unique medical and nutrient needs. TPN solutions meet energy needs primarily from dextrose. Providing too much dextrose, however, can result in hyperglycemia, a common metabolic complication associated with TPN. Clinicians recommend that the solution provide not more than 4 to 5 milligrams of dextrose per minute per kilogram of body weight.[7]

One liter of a typical central TPN solution contains 25% dextrose and 3.5% amino acids. Often 3 liters of the solution are given daily and provide about 3000 kcal and 105 g protein.

To prevent hyperglycemia, provide no more than 4 to 5 mg dextrose/min/kg body weight.

If additional energy is needed, IV lipids can provide as much as 50 to 60 percent of the total daily energy requirement for an adult who is not severely stressed.[8] During stress, clinicians frequently restrict fat to 30 percent of the total daily energy allowance (see Chapter 23).[9]

Providing some energy from fat helps to minimize hyperglycemia in people sensitive to high glucose loads and may also help prevent respiratory acidosis in

Table 22.1 Possible Indications for TPN by Central Vein

Acquired immune deficiency syndrome (AIDS)
Extensive small bowel resections
Radiation enteritis (inflammation of intestine caused by radiation)
Intractable diarrhea
Intractable vomiting
Severe GI tract obstructions
Bone marrow transplants
Severe acute pancreatitis
Severe malnutrition if surgical or intensive medical intervention is necessary
Hypermetabolic disorders, or major surgery, when it is anticipated that the GI tract will be unusable for more than 2 weeks
High-output enterocutaneous fistulas
Severe nausea and vomiting associated with pregnancy (hyperemesis gravidarum) when they last for more than 14 days
Low birthweight with necrotizing enterocolitis (severe GI inflammatory disease) or bronchopulmonary dysplasia (chronic lung disease)
When it is anticipated that adequate enteral nutrition cannot be established within 14 days of hospitalization

Note: In all cases, parenteral nutrition is indicated only when enteral nutrition is contraindicated. If short-term parenteral nutrition support is anticipated (less than 14 days), PPN might be preferred.

Source: Adapted from A.S.P.E.N. Board of Directors, Guidelines for the use of parenteral and enteral nutrition in adult and pediatric patients, *Journal of Parenteral and Enteral Nutrition* (supplement) 17 (1993): 1–49.

respiratory acidosis: a condition of too much acid in the blood caused by failure of the lungs to expel carbon dioxide properly. Excess carbon dioxide is normally released from the lungs during exhalation; failing lungs, however, are unable to perform this function rapidly enough.

people with respiratory failure. People with respiratory failure are unable to expel carbon dioxide efficiently, and fat oxidation produces less carbon dioxide than glucose oxidation does. If lipids are not used as an energy source, essential fatty acid requirements may be met by giving IV lipids periodically (two to three times per week).

Researchers are actively working to identify the best types and amounts of amino acids, carbohydrates, and lipids for TPN solutions, as well as the factors that affect the bioavailability of vitamins, minerals, trace minerals, and medications. Many of these studies are particularly relevant for clients with severe stresses and will be described in Chapter 23.

Intravenous Nutrition Techniques

Intravenous nutrition is like tube feedings in that careful attention to formula selection, preparation, and delivery helps support nutrition status while minimizing the risks of complications. To prevent bacterial contamination and ensure the stability of IV solutions, they should be shielded from light and refrigerated until used. As Table 22.2 shows, many of the risks associated with IV nutrition are more serious than those associated with enteral nutrition.

Insertion and Care of the TPN Catheter

Insertion of a catheter for PPN is the same as for simple IV solutions, but for central TPN, it often requires a surgical procedure. Skilled nurses can place peripherally inserted catheters for PPN or TPN, but a catheter for direct central vein access is inserted surgically by a qualified physician, either at bedside or in an operating room. The client may be awake for the procedure, but is given a local anesthetic. Unnecessary apprehension can be avoided by explaining the procedure to the client.

Table 22.2 Complications Associated with TPN

Catheter- or Care-Related Complications
Fluid in the chest (hydrothorax)
Air or gas in the chest (pneumothorax)
Blood in the chest (hemothorax)
Catheter tip broken off, obstructing blood flow (catheter embolism)
Air leaking into catheter, obstructing blood flow (air embolism)
Hole or tear in heart made by catheter tip (myocardial perforation)
Catheter inadvertently placed in subclavian artery (arterial puncture)
Improperly positioned catheter tip
Sepsis
Blood clot (thrombosis)
Infusion pump malfunctions
Metabolic or Nutrition-Related Complications
Elevated blood glucose (hyperglycemia)
Low blood glucose (hypoglycemia)
Dehydration
Fluid overload
Coma from excessive glucose load (hyperosmolar, hyperglycemic, nonketotic coma)
Electrolyte imbalances
Essential fatty acid deficiency
Vitamin and mineral deficiencies
Trace element deficiencies
High blood ammonia levels (hyperammonemia)
Acid-base imbalances
Elevated liver enzymes
Fatty liver
Bone demineralization

Maintaining the integrity of peripheral veins is often a problem with PPN. Veins may become inflamed and sometimes infected. Often the infusion catheter must be removed and reinserted at a new site; consequently, long-term feedings are difficult and rarely indicated. Peripherally inserted central catheters are less irritating to the veins and can often remain in place longer than PPN catheters.

Infections can develop at the catheter site in both PPN and central TPN. Compared with peripherally inserted catheters (for either PPN or TPN), though, central TPN presents a greater risk of introducing disease-causing microorganisms into the bloodstream because the catheter is inserted so near the heart. Health care workers must inspect the catheter site regularly and change the dressing frequently to keep the site clean.

Reminder: The presence of disease-causing bacteria in the blood is **sepsis**—a major complication that may occur as a consequence of TPN.

Administration of the TPN Solution

Just as a tube feeding is started slowly to give the GI tract time to adapt to the formula, a central TPN feeding is started slowly to allow time for the body to adapt to the high glucose concentration and osmolality of the TPN solution. Typically, 1 liter of TPN solution is infused at a constant rate (about 40 milliliters per hour) during the first 24 hours. An infusion pump ensures an accurate and steady delivery rate. Electrolytes and blood glucose are monitored periodically. If tests indicate electrolyte imbalances or unacceptably high blood glucose, the causes are investigated and treated. After the first 24 hours, the infusion rate is often

Recall from Chapter 21 that infusion pumps may seem like toys to young children and must be kept at a safe distance from the bed so that the child will not change the flow rate or topple the pump and IV pole.

increased by 1 liter a day until the desired volume of solution is being given every 24 hours.

Rapid changes in the infusion rate can cause severe hyperglycemia or hypoglycemia, which can lead to coma, convulsions, and even death, so all changes must be made gradually and cautiously. Problems are more likely to occur in people with organ dysfunction or in infants with immature organ systems. When the administration of solution gets behind or ahead of schedule, the drip rate should be adjusted to the correct hourly infusion rate, but no attempt should be made to speed up or slow down the rate to meet the originally ordered volume. When a person is being taken off TPN, the infusion rate of the solution must be tapered off gradually to prevent hypoglycemia. Table 22.3 provides guidelines for monitoring clients on TPN.

Peripheral TPN Infusion Unlike central TPN solutions, peripheral TPN does not have to be increased gradually when feedings are initiated or tapered off gradually when feedings are discontinued. Peripheral TPN solutions do not have the high dextrose concentration or osmolality of central TPN solutions and do not present the associated problems.

IV Lipid Infusion Traditionally, IV fat emulsions are infused separately from the base TPN solution containing amino acids, dextrose, and micronutrients (see the photo on p. 533). Occasionally, people experience adverse reactions to IV lipid emulsions, particularly when the IV lipids are given in large amounts or administered too rapidly. Immediate reactions may include fever, warmth, chills, backache, chest pain, allergic reactions, palpitations, rapid breathing, wheezing, cyanosis, nausea, and an unpleasant taste in the mouth. To guard against adverse reactions, the client receives only small amounts of lipid emulsion over the first 15 to 30 minutes. After that time, the rate can be increased.

When IV lipid emulsions are used as an energy source, they are often added directly to the base solution and infused along with it. The use of total nutrient admixtures for clients in the hospital as well as at home has grown dramatically.[10] Total nutrient admixtures must be compounded carefully, refrigerated prior to use, and mixed gently before they are infused.

cyclic parenteral nutrition: the continuous administration of TPN solutions for 8 to 12 hours with time periods when no nutrients are infused.

Cyclic Infusion A person on cyclic parenteral nutrition receives a TPN solution at a constant rate 8 to 12 hours a day. The infusion can be given during the night to allow freedom for routine daytime activities and, therefore, is often used for long-term TPN. When a person receives a TPN solution continuously, insulin

Table 22.3 Guidelines for Monitoring People on TPN

Before starting TPN:	Complete nutrition assessment.
	Confirm placement of catheter tip by X ray.
	Check blood glucose, electrolytes, chemical profile, and complete blood count.
Every 4 to 6 hours:	Check blood glucose.
	Monitor vital signs.
	Check pump infusion rate.
Daily:	Monitor weight changes.
	Record intake and output.
	Check urine specific gravity.
Weekly:	Reassess nutrition status.
	Monitor serum proteins, ammonia, and triglycerides.
	Check the complete blood count.
As needed:	Monitor serum electrolytes, calcium, magnesium, phosphorus, and blood urea nitrogen.

levels stay high. As a result, the person cannot mobilize fat stores for energy or for essential fatty acids; eventually, fat may be deposited in the liver. Cyclic TPN reverses these problems.[11] Additionally, fewer kcalories seem to be effective in maintaining nitrogen balance, probably because the person uses body fat for energy. Some people, however, cannot tolerate the delivery of a day's volume of solution over a shorter period of time.

From Parenteral to Enteral Feedings

Once the problem causing the need for IV nutrition resolves, the client can gradually shift to an enteral diet while the volume of the IV feeding is tapered off. The transition requires careful planning. During long periods of disuse, the intestinal villi shrink and lose some of their function. Reintroducing nutrients to the GI tract at the appropriate rate and volume will stimulate the progressive restoration of the villi's normal structure and function and prevent malabsorption and other GI discomforts.

Transitional Feedings The transition from IV feeding to an enteral diet can be accomplished in different ways and often involves a combination of feeding methods. One way is to start an oral diet while the person is still on IV nutrition. The diet is often progressive, beginning with liquids provided in small amounts. If the person cannot eat enough food to meet at least 50 percent of daily nutrient needs within a few days, and intake does not seem to be improving, tube feedings should be considered.[12]

Whether a person is given a tube feeding or provided an oral diet, the volume of the IV solution is reduced as the volume of enteral feeding is increased. The person who cannot tolerate enteral feedings can still rely on TPN to meet nutrient needs. Parenteral nutrition can be discontinued when at least some 70 to 75 percent of estimated energy needs are being met by oral intake, tube feeding, or a combination of the two.[13] Chapter 21 described the transition from tube feedings to table foods.

Psychological Effects Returning to oral intake after having been fed either intravenously or by tube can have a variety of psychological effects. Some people may be extremely eager to eat again, and food can be an important morale booster. Others may be apprehensive about eating, particularly if they have had extensive GI problems. Appetite may be slow to return for some. In such circumstances, all members of the health care team can support the successful reintroduction of food. Recognize the person's concerns, and provide reassurances that help will be available throughout the process. The many decisions surrounding the provision of TPN require careful consideration. The case study on p. 540 presents an example for review.

Specialized Nutrition Support at Home

WWW **wizvax.net/oleyfdn/index.html**
The Oley Foundation

dataphone.se/~hpn
Swedish Association for Children with Home Parenteral Nutrition

Occasionally, a client must continue to receive nutrition support (tube feedings or parenteral nutrition) after the primary medical condition has stabilized. In such a case, home nutrition support might be an option.

Since the first report of a person sent home successfully on TPN in 1969,[14] the use of home nutrition support has expanded rapidly. According to recent

Case Study

Mail Carrier Requiring Parenteral Nutrition

Mr. Rossi, a 37-year-old mail carrier, has been admitted to the hospital for Crohn's disease (see Chapter 20). He has been steadily losing weight and appears emaciated. A thorough examination indicates that Mr. Rossi needs surgery as soon as possible to remove a portion of his small intestine. In the meantime, he cannot be placed on enteral feedings. The nutrition assessment reveals protein-energy malnutrition.

Mr. Rossi is placed on central TPN before surgery. He progresses well, gains weight, and undergoes surgery one week after admission.

What factors in Mr. Rossi's history indicate the need for central TPN? How would you explain the need for TPN to Mr. Rossi?

Describe the components of a typical TPN solution. Calculate the energy and protein that 1 liter of this solution provides.

Consider some of the physiological and psychological problems Mr. Rossi might face when enteral nutrition is reintroduced. How will the health care team know when it will be safe to take Mr. Rossi off central TPN? Describe the different ways the transition from TPN to enteral feedings can be accomplished.

estimates, about 40,000 people receive parenteral nutrition at home, and more than 150,000 receive enteral nutrition at home.[15] As the number of people benefiting from home nutrition programs continues to grow, health care professionals who work with these programs are gaining valuable experience and improving the quality of home nutrition support. Medical supply companies provide the equipment, formulas, and service necessary to support home nutrition care.

The Basics of Home Programs

As with tube feedings and TPN in the hospital, the main objective of home nutrition support is to maintain or achieve adequate nutrition status. Nutrition support at home, however, has an added dimension: it permits a person to continue nutrition care in familiar surroundings. If you have ever been in the hospital, or taken a long trip for that matter, you probably remember the comfort you experienced when you returned to your own bed, knew where things were, and could get things when you needed them.

Candidates for Home Nutrition Support The nutrition support team most frequently determines whether a client is a candidate for a home nutrition program. In addition to medical considerations, the candidate for home nutrition support and those who care for that person must have rational, stable personalities so they can successfully handle the problems that arise. They must be capable of learning the necessary techniques and of dealing with complications. They must be willing to comply with recommendations. They must also have adequate financial resources and access to the equipment, supplies, and professional support that are integral components of a successful home program.

Roles of Health Care Professionals Health care teams who work with home nutrition programs must understand the regulations that govern payment for home nutrition programs by both government and private insurance. They must use this knowledge to document the need for services to ensure maximum reimbursement for the client.

Once a home nutrition program is initiated, the physician directs the home care program and monitors the client's care. Nurses train clients in the appropriate techniques, dietitians monitor nutrition status, pharmacists coordinate the delivery of formulas, medications, and supplies with the home care company, and social workers provide insurance assistance. All team members are expected to answer questions and provide emotional support. Home visits by a nurse, and sometimes the dietitian, help ensure that the client is able to comply with the demands of the home care program. A qualified nurse, dietitian, or physician must be available to answer questions and handle problems as they arise.

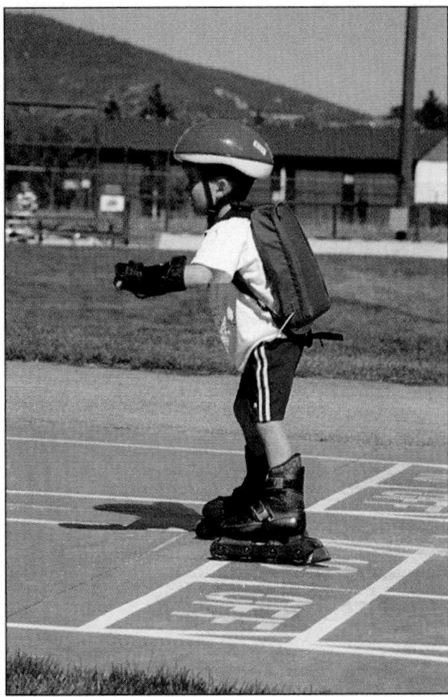

Portable pumps and convenient carrying cases allow clients who require nutrition support at home to move about freely.

Cost Considerations When home nutrition programs first began, dramatic cost savings were reported as clients and their caregivers assumed the responsibilities formerly performed by hospital staff. Costs for home care are rising, however, because of the increasingly complex care available for home clients and the strict regulation of the home care industry.[16] As a result, more skilled professionals and additional time are required, and both increase costs. The cost of home parenteral nutrition programs (solutions and supplies) ranges from \$51,000 to \$110,000 per year.[17] Although costs for enteral nutrition programs (\$5,100 to \$11,000 per year) are about 90 percent less, even these costs can be quite substantial.[18] Insurance often pays for a portion of the costs, but people on home programs are also more likely than others to have high medical costs for managing their diseases as well. The economic impact of home nutrition programs is often the primary concern for people on these programs.

Adjustments for Clients While home nutrition programs can extend and improve the quality of life, clients may struggle with the many lifestyle adjustments the programs entail. In addition to the economic impact of the therapy, clients may find the demands of the home schedule to be frustrating, inconvenient, and monotonous. Although a home nutrition program has a big impact on the person's lifestyle, clients report that they feel their lives have improved with home nutrition therapy.[19] Many people who require nutrition support at home can also eat some foods by mouth, and some resume activities, such as going to work, driving, and playing sports.

How Home Enteral and Parenteral Nutrition Programs Work

People on home enteral nutrition programs commonly have cancer or neurological disorders that interfere with swallowing. People on home TPN often have acquired immune deficiency syndrome (AIDS), cancer, Crohn's disease, or other intestinal disorders. Sometimes home TPN is relatively short term, lasting for weeks or months as opposed to lifelong.

Home Enteral Nutrition Both transnasal tubes and tube enterostomies provide access to the GI tract for home care tube feedings (see Chapter 21). When possible, intermittent feeding schedules are arranged so that clients are free to move around between meals. Some people are able to meet some of their nutrient needs by eating during the day and using tube feedings at night. Clients on continuous feedings who must use pumps can obtain small pumps that are easily concealed in clothing and allow freedom of movement.

Home TPN Different types of home TPN programs are currently in use. Ideally, clients are given as much responsibility for their own care as they can handle. For example, a client who is capable of changing the catheter dressings is trained to do so. Typically, caregivers also learn the techniques so that they can assist as needed.

Special catheters designed for long-term use are often inserted for home TPN. Many times these catheters are tunneled under the skin so that the catheter exit site lies in the abdominal area where the client can access and care for it easily. The day's volume of TPN solution is frequently delivered within 8 to 12 hours using an infusion pump. The client infuses the solution while sleeping or at some other convenient time and thus is free to move about unencumbered for much of the day.

For people who cannot tolerate an infusion rate of 8 to 12 hours, nutrients are infused over 24 hours. Those who are ambulatory can benefit from a lightweight carrying case that holds a small pump and IV bags. This system allows the client to move around freely with little inconvenience.

Unquestionably, specialized nutrition support provides a lifesaving feeding alternative for people who cannot eat traditional diets either in the hospital or at home. The nutrition assessment checklist reviews areas of concern for people on parenteral nutrition support.

Nutrition Assessment Checklist

For People on Parenteral Nutrition Support

Medical Review the client's medical record for information necessary to select the appropriate feeding site (peripheral versus central) and insertion method (PPN, peripherally inserted central catheter, or central TPN catheter) and to determine the composition of the IV solution. Check the medical record for the development of undesirable symptoms related to IV fluids—these require prompt treatment.

Medication Assure that medications delivered along with an IV solution are physically compatible with the solution and that their activity will not be adversely affected. Note any medications being infused along with IV solutions. Remember that all of the medication dosage will not be delivered if the medication has been added to the solution and the infusion is stopped for any reason.

Nutrient/Food Intake Ensure that IV solutions are being delivered as prescribed. For clients beginning enteral nutrition, determine the degree to which nutrient needs are being met by enteral formulas (orally or by tube) or by table foods, and reduce the volume of nutrient IV solutions accordingly.

Anthropometric Assess the client's weight daily to make sure that nutrient goals are being met.

Laboratory Monitor serum and urine lab values for signs of glucose intolerance and fluid and electrolyte imbalances. Check serum protein levels to ensure that they are rising or being maintained. When available, assess nitrogen balance to determine whether the IV solution is meeting the client's protein needs. Check the results of laboratory tests of vitamin and mineral status whenever possible.

Physical Inspect catheter insertion sites for signs of infection or inflammation. Check the infusion pump drip rate as needed. Assess blood pressure, temperature, pulse, and respiration about every 4 hours. Look for physical signs of malnutrition and dehydration.

Self Check

1. All of the following can be delivered intravenously **except:**
   a. hydrolyzed enteral formulas.
   b. carbohydrate.
   c. amino acids.
   d. dextrose.
2. The health care professional recognizes that a simple IV solution is most appropriate for:
   a. people who can't eat for long periods of time.
   b. people who are malnourished.
   c. people who are well nourished and are expected to eat in a few days.
   d. people who have high nutrient needs.
3. All of the following may benefit from peripheral parenteral nutrition (PPN) **except:**
   a. those with normal renal function who need short-term intravenous support.
   b. those who need long-term IV nutrition support.
   c. those in whom inserting an IV catheter into a central vein would be difficult.
   d. those on oral or tube feedings who need additional nutrients temporarily.
4. A typical PPN solution delivers about _____ kcalories per day.
   a. 1000 b. 1500
   c. 2000 d. 2500
5. Compared to PPN solutions, TPN solutions provide:
   a. more vitamins and minerals.
   b. more fat.
   c. more dextrose.
   d. a greater osmolality.
6. The nurse working with a client receiving TPN recognizes that the intravenous lipid emulsions provided two or three times a week serve primarily as a source of:
   a. concentrated energy. b. essential fatty acids.
   c. fat-soluble vitamins. d. vitamin C.
7. Central TPN is indicated for people who need IV nutrients in all of the following situations **except:**
   a. when short-term parenteral nutrition is required.
   b. when nutrient requirements are high.
   c. when people are severely malnourished.
   d. when fluid restriction prevents the use of PPN.
8. The solutions infused through peripherally inserted central catheters are most similar to those infused for:
   a. simple IVs. b. PPN.
   c. TPN. d. enteral feedings.
9. Which type of IV insertion procedure is associated with the highest risk of sepsis?
   a. peripheral catheter for PPN
   b. central TPN catheter
   c. peripherally inserted central catheter
   d. peripheral catheter for simple IVs
10. Which of the following is true?
    a. Central TPN solutions must be started slowly and stopped gradually to prevent hyperglycemia and hypoglycemia.
    b. PPN solutions must be started slowly and stopped gradually to prevent hyperglycemia and hypoglycemia.
    c. Intravenous fat emulsions must be started slowly at first and discontinued gradually to prevent adverse reactions.
    d. Simple IV solutions must be started slowly and stopped gradually to prevent hyperglycemia and hypoglycemia.
11. A gradual transition from an IV feeding to an oral diet is primarily designed to:
    a. improve the appetite.
    b. prevent hypoglycemia.
    c. ensure that nutrient needs will continue to be met.
    d. prevent apprehension about eating.
12. Primary concerns of the health care team evaluating a client's ability to manage a home enteral or parenteral nutrition program include all of the following **except:**
    a. The client's financial resources.
    b. The client and caregivers' abilities to learn the necessary techniques and follow medical instructions.
    c. The client's psychological status.
    d. The exact composition of the enteral formula or intravenous solution.

Answers to these questions appear in Appendix H.

Clinical Applications

1. One liter of a TPN solution contains 500 milliliters of 50 percent dextrose and 500 milliliters of 8.5 percent amino acids. Determine the daily kcalorie and protein intakes of a person who receives 2 liters of such a TPN solution. Calculate the average daily energy intake if the person also receives 500 milliliters of a 20 percent fat emulsion three times a week.
2. Consider what it must be like to be on a home TPN program with no foods allowed by mouth. What would be the advantages of being at home instead of in the hospital? What would be the disadvantages? Think of how you would manage feedings. How would you feel about the time, costs, and commitment required to maintain this therapy? How would you feel about not being able to eat after a long time? How would you handle holidays and special occasions that often center around food?

Notes

1. D. W. Wilmore and S. J. Dudrick, Growth and development of an infant receiving all nutrients exclusively by way of the vein, *Journal of the American Medical Association* 203 (1968): 860–864.

2. M. S. Dahn, Intravenous peptides, *Nutrition in Clinical Practice* 8 (1993): 93–94; J. A. Vazquez, H. Daniel, and S. A. Adibi, Dipeptides in parenteral nutrition: From basic science to clinical applications, *Nutrition in Clinical Practice* 8 (1993): 95–105; P. Fürst and P. Stehle, The potential use of parenteral dipeptides in clinical nutrition, *Nutrition in Clinical Practice* 8 (1993): 106–114.

3. National Advisory Group on Standards and Practice Guidelines for Parenteral Nutrition, *Journal of Parenteral and Enteral Nutrition* 22 (1998): 49–66.

4. As cited in National Advisory Group on Standards and Practice Guidelines for Parenteral Nutrition, 1998.

5. M. A. Stokes and G. L. Hill, Peripheral parenteral nutrition: A preliminary report on its efficacy and safety, *Journal of Parenteral and Enteral Nutrition* 17 (1993): 145–147.

6. J. Z. Rogers, K. McKee, and E. McDermott, Peripherally inserted central venous catheters, *Support Line,* October 1995, pp. 6–9; S. C. Loughran and M. Borzatta, Peripherally inserted central catheters: A report of 2506 catheter days, *Journal of Parenteral and Enteral Nutrition* 19 (1995): 133–136.

7. D. K. Rosmarin, G. M. Warlaw, and J. Mirtallo, Hyperglycemia associated with high, continuous infusion rates of total parenteral nutrition dextrose, *Nutrition in Clinical Practice* 11 (1996): 151–156; A.S.P.E.N. Board of Directors, Guidelines for the use of total parenteral nutrition in adult and pediatric patients, *Journal of Parenteral and Enteral Nutrition* (supplement) 17 (1993): 21.

8. A. M. Karch, *Lippincott's Nursing Drug Guide* (Philadelphia: Lippincott-Raven Publishers, 1997), p. 468.

9. R. G. Barton, Nutrition support in critical illness, *Nutrition in Clinical Practice* 9 (1994): 127–139.

10. D. F. Driscoll, Total nutrient admixtures: Theory and practice, *Nutrition in Clinical Practice* 10 (1995): 114–119.

11. L. M. Gramlich and B. Bistrian, Cyclic parenteral nutrition: Considerations of carbohydrates and lipid metabolism, *Nutrition in Clinical Practice* 9 (1994): 127–139.

12. M. F. Winkler and coauthors, Transitional feeding: The relationship between nutritional intake and plasma protein components, *Journal of the American Dietetic Association* 89 (1989): 969–970.

13. T. C. Clark, Nutrition support clinical pathways, *Nutrition in Clinical Practice* 11 (1996): 16–20; The 1995 A.S.P.E.N. standards for nutrition support: Hospitalized patients, *Nutrition in Clinical Practice* 10 (1995): 206–207.

14. M. E. Shils and coauthors, Long-term parenteral nutrition through an external arteriovenous shunt, *New England Journal of Medicine* 283 (1970): 341–344.

15. As cited in M. Evans, Home nutrition support materials, *Nutrition in Clinical Practice* 10 (1995): 37–39.

16. K. S. Crocker, Current status of home infusion therapy, *Nutrition in Clinical Practice* 7 (1992): 256–263.

17. D. G. Kelly, Home parenteral nutrition, presented at the *Eighth Annual Advances and Controversies in Clinical Nutrition,* sponsored by the Mayo Clinic Foundation, Dallas, Texas, April 5–7, 1998.

18. J. A. Weckwerth, Selected topics in home enteral nutrition, presented at the *Eighth Annual Advances and Controversies in Clinical Nutrition,* sponsored by the Mayo Clinic Foundation, Dallas, Texas, April 5–7, 1998.

19. M. Malone, Effect of home nutrition support on patient's lifestyle (abstract), *Journal of Parenteral and Enteral Nutrition* (supplement) 19 (1995): 23.

CHAPTER 22

Nutrition in Practice

▪ Ethical Issues in Nutrition Care ▪

Enteral and parenteral feedings can meet nutrient needs and support recovery in many cases. In other cases, such support can improve the quality of a declining life. Sometimes, however, nutrition support can prolong life by merely delaying death, and the remaining life may be of low quality. Thus the very availability of special nutrition support forces health care professionals and society to face ethical issues. (Terms relating to ethical issues appear in the glossary on p. 546.)

Is it ever morally and legally appropriate to withhold or withdraw nutrition support?

In attempting to answer a question such as this one, ethics experts are guided by the following principles:

- *Autonomy*—the client's right to make decisions concerning his or her own well-being.
- *Beneficence*—the treatment or its withdrawal will do more good than harm, and the caregivers will promote well-being and act without selfish intent.
- *Justice*—the actions are based on fairness, honesty, and loyalty to agreements.

Although these principles can help sift through ethical dilemmas, answers rarely come easily. It may be difficult to determine whether clients truly comprehend the complexity and potential finality of their decisions, and it may be equally as difficult to determine whether a treatment will do more harm than good. In reality, the answers often lie tangled in personal values, charged emotions, and legal conflict.

WWW.

aslme.org
American Society of Law, Medicine, and Ethics

asbh.org
American Society for Bioethics & Humanities

Isn't there a moral obligation to give nutrition support whenever it will prolong life?

Not necessarily. Health care professionals readily recommend whatever form of nutrition is necessary to support clients who have any reasonable chance of recovering from a disease. Clearly, health care professionals cannot rightfully withhold nutrition support because of poor judgment or negligence. The decision of whether to feed a client becomes less clear, however, when clients are terminally ill or in persistent vegetative states, or when they simply refuse specialized nutrition support because they feel the quality of their lives are poor. Do we (as a society) allow them such choices? Are health care professionals morally and legally obligated to comply with, or to deny, such requests? Furthermore, who determines when clients are competent to speak for themselves? If a client is incompetent to make such a decision, who, if anyone, should be allowed to make such life-and-death decisions for the client? These unanswered questions are but a few that have evolved along with advances in medical technology and nutrition support; a discipline known as "medical ethics" has developed out of the need to discuss and solve problems such as these.

Who decides what will be done in cases where there are ethical dilemmas?

Most often the client (when competent) and the client's family work out their own answers based on extensive consultation with the physician and the health care team. Health care professionals must provide decision makers with the information they need to fully understand the disease, its treatments, the potential benefits of treatments, and the risks of treatments in order for them to make a truly informed choice. When the client or client's family makes a decision that the health care team or the facility housing the client believes may make the team or facility liable for malpractice, the conflict of interests may give rise to court cases. Then the court is charged with defining the problems and solving them.

How have courts resolved nutrition support issues?

One of the most widely publicized cases involving nutrition support issues was that of Nancy Cruzan. Nancy Cruzan was a young woman who suffered

Glossary

advance directive: the means by which competent adults record their preferences for future medical interventions. The living will and durable power of attorney are types of advance directives.

artificial feeding: parenteral and enteral nutrition; feeding by a route other than the normal ingestion of food.

comatose: in a state of deep unconsciousness from which the person cannot be aroused.

competent: having sufficient mental ability to understand a treatment, weigh its risks and benefits, and comprehend the consequences of refusing or accepting the treatment.

durable power of attorney: a legal document in which one competent adult authorizes another competent adult to make decisions for her or him in the event of incapacitation. The phrase "durable power" means that the agent's authority survives the client's incompetence; "attorney" refers to an attorney-in-fact (not an attorney-at-law).

ethical: in accordance with moral principles or professional standards. Socrates described *ethics* as "how we ought to live."

legal: established by law.

living will: a document signed by a competent adult that specifically states whether the person wishes aggressive treatment in the event of terminal illness or irreversible coma from which the person is not expected to recover.

persistent vegetative state: exhibiting motor reflexes but without the ability to regain cognitive behavior, communicate, or interact purposefully with the environment.

terminal illness: a progressive, irreversible disease that will lead to death in the near future.

permanent and irreversible brain damage after a car crash in 1983. For eight years, she was in a persistent vegetative state—awake but unaware. Her physicians and parents held no hope for her recovery, yet given food and water, she might have lived for another 30 years. Her parents requested permission to discontinue tube feeding, but their request was rejected by the Missouri Supreme Court in 1987. The court held that Cruzan never definitively stated her "right to die" wishes and that her parents had no legal right to make such a request for her. The court stated that preserving life, no matter what its quality, takes precedence over all other considerations.

Cruzan's parents appealed the decision, and in 1989 the U.S. Supreme Court agreed to hear their arguments. The Cruzans tried to convince the Supreme Court that their once independent and vivacious daughter would not want to live in a vegetative state. No one questioned that Cruzan's parents knew their daughter's wishes better than anyone and had the highest and most loving motives. The question for the Court was whether families (or anyone) can make life-and-death decisions on behalf of incompetent persons. The Supreme Court recognized that competent adults have the right to stop life-sustaining treatment, but said that these treatments could not be withdrawn without "clear and convincing" evidence that the incompetent person would refuse treatment.[1] The Court, therefore, upheld the Missouri Supreme Court's decision. It took still another round of court battles before additional evidence convinced the Missouri Supreme Court of Nancy Cruzan's wishes. Finally, her feeding tube was removed, and she died from dehydration two weeks later.

The cost of maintaining Cruzan must have been astronomical.

It was indeed, both financially and emotionally. In terms of money, health care costs to support her ran about $130,000 per year (paid by the state). The emotional costs are more difficult to calculate, of course. Cruzan's parents were first faced with the initial shock of their daughter's accident. Then, for several years, they maintained hope that with continued care she would survive and regain consciousness. Finally, they endured court battles over their child's fate—a fate that meant grief whatever the courts decided.

Are such cases unusual?

Unfortunately, no. Some 30,000 other families of people who live in a persistent vegetative state face the same dilemma. The High Court's decision had widespread implications for these people and the health

Nutrition in Practice

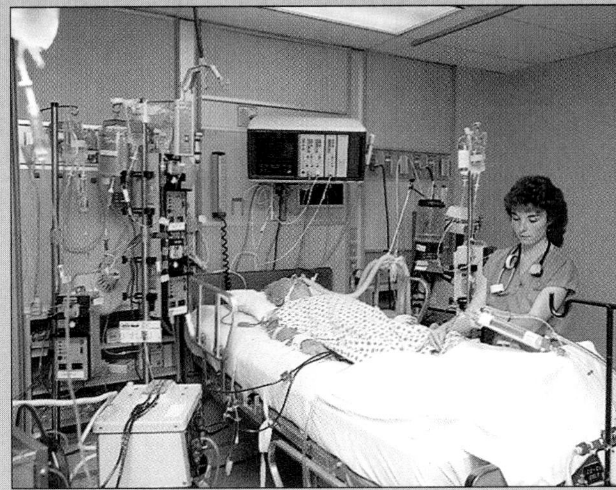
When is it morally and legally appropriate to use special nutrition support?

professionals who take care of them. Ethical concerns are more likely to arise as the population of elderly clients grows and new medical advances capable of sustaining life are developed. End-of-life decisions touch all of us, because they influence the extent to which our society views life-sustaining treatment as optional not only for our clients, but for ourselves and our families.[2] They decide how we may be allowed to die.[3]

What is the prevailing attitude about a person's right to make decisions about life-sustaining treatments?

The emerging ethical, medical, and legal consensus seems to support the view that individual rights outweigh those of the state. Competent individuals have a legal right to refuse medical treatment—including nourishment and hydration—even when medical experts consider that treatment necessary to sustain life. In other words, even when treatment is lifesaving and its refusal may bring an earlier death, clients' rights remain paramount.

WWW.
soros.org/death.html
The Project on Death in America

How can people protect their rights to retain or refuse treatment in the event that the are unable to make a decision when life-sustaining treatment is necessary?

Health care professionals should routinely encourage their competent clients to use advance directives to express ahead of time their preferences regarding medical treatments, including artificial feedings, should terminal illness, coma, or incompetency develop.[4] Each client's preferences should be noted in the medical records. Clients should expect that physicians and facilities will comply with their preferences and not merely tolerate them grudgingly.

Health care professionals should reassure clients that their decisions to refuse life-sustaining treatments, including nutrition support, will not mean that other care will be withheld. The client who makes such a choice still deserves meticulous physical care, pain management, and compassionate emotional support.[5] Any health care professional who is unwilling to abide by the client's stated preferences should arrange for continuing care by another equally qualified professional and then withdraw from that client's care.[6]

One form of an advance directive is a living will (see Form NP22.1 on p. 548). Another is the durable power of attorney (see Form NP22.2 on p. 549). All states have statutes governing the use of advance directives; health care professionals should be aware of the regulations of the states where they work.[7] Some states' laws specifically address the conditions under which nutrition support can be withheld or withdrawn.

What is a living will?

A living will allows a competent adult to express clear directions regarding medical treatment in the event that the person is unable to make the necessary decisions at that time. The living will may specify that no extraordinary treatments should be administered, or alternatively, it may declare that every effort should be made to maintain life. People who prefer that nourishment and hydration be continued or discontinued need to write this specification into their living wills to ensure that their wishes are known. Physicians seem more likely to comply with a *specific* living will than with a standard one.[8]

WWW.
choices.org
Choice in Dying

What is a durable power of attorney?

In many states, the durable power of attorney offers clients a way to ensure that their wishes will be carried out. Some states have enacted statutes authorizing the use of durable powers of attorney specifically for health

Form NP22.1 An Example of a Living Will

DECLARATION TO MY FAMILY, MY PHYSICIAN, MY LAWYER, AND MY SPIRITUAL ADVISER

If the time arrives when I can no longer take part in decisions for my own future, this statement and Declaration shall stand as the expression of my wishes.

I recognize that death is as much a reality as birth, growth, maturity, and old age. It is but one phase in the cycle of life and is the only certainty. I do not fear death as much as I fear there is no reasonable expectation of my recovery from physical or mental disability. I wish to be allowed to die and not to be kept alive by artificial means or heroic measures, but wish only that drugs be mercifully administered to me for terminal suffering, even if they hasten the moment of my death.

I recognize that my wishes place a heavy burden of responsibility upon you, and I therefore make the following declaration with the intention of sharing this responsibility and this decision with you and of mitigating any feelings of guilt that you may have:

THIS DECLARATION is made this ______ day of ______________, 19______.

I, ________________________________, willfully and voluntarily make known my desire that my dying not be artificially prolonged under the circumstances set forth below, and I do hereby declare:

If at any time I should have a terminal condition and if my attending physician has determined that there can be no recovery from such condition and that my death is imminent, I direct that life-prolonging procedures be withheld or withdrawn when the application of such procedures would serve only to prolong artificially the process of dying, and that I be permitted to die naturally with only the administration of medication or the performance of any medical procedures deemed necessary to provide me with comfort care or to alleviate pain. I desire that nutrition and hydration (food and water) be withheld or withdrawn when the application of such procedures would serve only to prolong artificially the process of dying.

In the absence of my ability to give directions regarding the use of such life-prolonging procedures, it is my intention that this declaration be honored by my family and physician as the final expression of my legal right to refuse medical or surgical treatment and to accept the consequences for such refusal.

I understand the full import of this declaration, and I am emotionally and mentally competent to make this declaration.

____
(signature)

The declarant is known to me, and I believe him/her to be of sound mind.

____
Witness

____
Witness

The foregoing instrument was acknowledged before me this ______ day of __________, 19___, by ____________.

____
Notary Public

Form NP22.2 Durable Power of Attorney

I, ______________________ now residing at ______________________ hereby constitute and appoint ______________________ as my true and lawful attorney-in-fact for me and in my name, place and stead, giving and granting unto my said attorney full power and authority to do and perform every act as fully as I might do if personally present, with full power of substitution and revocation. I hereby ratify and confirm all that my attorney shall lawfully do or cause to be done pursuant to this power.

This Power includes, but is not limited to, the right to encumber, assign or convey realty, including homestead realty. In addition, my attorney-in-fact is authorized to arrange for and consent to medical, therapeutical and surgical procedures for me as principal, including the administration of drugs.

I have executed this Power while in command of my faculties and with knowledge of the consequences, both legal and practical.

This Durable Power of Attorney shall not be affected by my disability as principal except as provided by statute.

IN WITNESS WHEREOF, I have hereunto set my hand and seal this ______ day of ____________, 19 ______.

WITNESSES:

______________________ ______________________ (SEAL)

____

STATE OF ______________________
COUNTY OF ______________________

I HEREBY CERTIFY that on this day, before me, a Notary Public duly authorized in the State and County named above to take acknowledgments, personally appeared ______________________ to me known to be the person described in and who executed the foregoing DURABLE POWER OF ATTORNEY, and said individual acknowledged before me that execution of this instrument was for the uses and purposes herein expressed.

WITNESS my hand and official seal in the State and County named above this ______ day of ____________, 19____.

(SEAL)

____
NOTARY PUBLIC

care. A durable power of attorney allows a competent adult to designate another competent adult (usually a relative or close friend) as an agent to make health care decisions in the event of incapacitation. In essence, it says, "I give this person the right to make health care decisions on my behalf should I become unable to make them."

How well do advance directives hold up? Do physicians and hospitals honor them?

Most health care professionals advocate the use of advance directives and agree that clients' wishes should be honored, but actual medical care may not always reflect these beliefs. In practice, physicians are more likely to follow family directives concerning tube feedings and other life support than instructions in living wills.[9]

When people give another person a durable power of attorney, they should appoint a person who understands their health care preferences and will make decisions that reflect those preferences. Clearly, this is not always the case. When faced with hypothetical situations, clients and those who would have to make health care decisions for them agree on treatment only 70 percent of the time.[10] Such discrepancy is natural because each views the other's experience as most important when making decisions. Clients want to avoid burdening their families, and families want to provide any treatment that offers a possible cure or relief from pain.

If people wish to ensure that they receive medical care consistent with their individual wishes, they need to discuss their living wills and hypothetical scenarios with family members and the person who holds the durable power of attorney before medical conditions arise that will necessitate others making decisions for them. The person who plans ahead for future care in the case of a terminal illness or irreversible state of unconsciousness relieves others of the guilt and some of the anxiety of having to make decisions. Imagine the anguish a family member may go through in directing the health care team to stop nutrition support, knowing that doing so will hasten death. That decision is a little easier if the family member knows it is what the individual would want, or if a legal document takes the decision out of the family's hands altogether.

The questions raised in this nutrition in practice have no easy answers, yet decisions must be made. Each case must be carefully considered individually. Most hospitals and extended care facilities have established ethics committees to deal with issues such as those presented here. Health care team members should ensure that their disciplines are represented on such committees and become familiar with their profession's ethics policies and guidelines.[11]

Notes

1. *Cruzan v. Director, Missouri Department of Health,* 497 U.S. 261, 110 S.Ct. 2841, 111 L.Ed.2d 224 (1990).

2. M. Angell, Prisoners of technology: The case of Nancy Cruzan, *New England Journal of Medicine* 322 (1990): 1226–1228; B. Lo, F. Rouse, and L. Dornbrand, Family decision making on trial—Who decides for incompetent patients? *New England Journal of Medicine* 322 (1990): 1228–1232.

3. The Court and Nancy Cruzan, *Hastings Center Report,* January/February 1990, pp. 38–50.

4. *Advance Directives: The Role of Health Care Professionals* (Columbus, Ohio: Ross Products Division, Abbott Laboratories, 1996).

5. V. M. Herrmann, Ethics in nutrition support, presented at the *Eighth Annual Advances and Controversies in Clinical Nutrition,* sponsored by the Mayo Clinic Foundation in association with the American Society for Parenteral and Enteral Nutrition, April 5–7, 1998.

6. C. R. Gallagher-Allred, Managing ethical issues in nutrition support of terminally ill patients, *Nutrition in Clinical Practice* 6 (1991): 113–116.

7. B. Dorner and coauthors, The "to feed or not to feed" dilemma, *Journal of the American Dietetic Association* (supplement 2) 97 (1997): 172–176.

8. J. W. Ely and coauthors, The physician's decision to use tube feedings: The role of the family, living will, and the *Cruzan* decision, *Journal of the American Geriatric Society* 40 (1992): 471–475.

9. Ely and coauthors, 1992.

10. J. Hare, C. Pratt, and C. Nelson, Agreement between patients and their self-selected surrogates on difficult medical decisions, *Archives of Internal Medicine* 152 (1992): 1049–1054.

11. American Dietetic Association, Position of The American Dietetic Association: Legal and ethical issues in feeding permanently unconscious patients, *Journal of the American Dietetic Association* 95 (1995): 231–234; American Academy of Pediatrics Committee on Bioethics, Guidelines on forgoing life-sustaining medical treatment, *Pediatrics* 93 (1994): 532–536; A.S.P.E.N. Board of Directors, Ethical and legal issues in specialized nutrition support, *Journal of Parenteral and Enteral Nutrition* (supplement) 17 (1993): 50–52; American Dietetic Association, Position of The American Dietetic Association: Issues in feeding the terminally ill adult, *Journal of the American Dietetic Association* 92 (1992): 996–1005.

23 *Energy-Modified, High-Protein Diets for Severe Stresses*

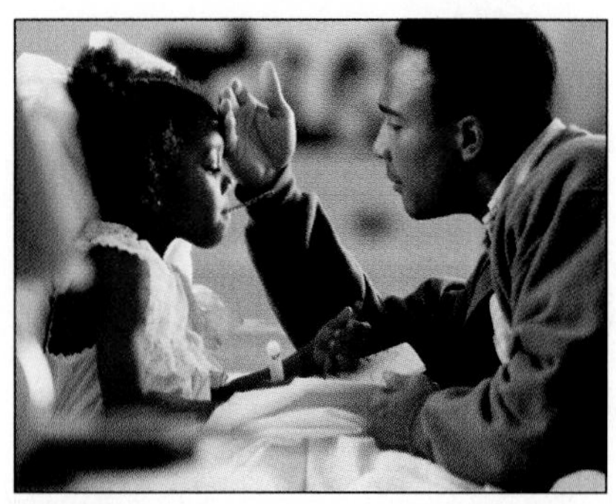

Although all illnesses threaten the body's internal balances, the body adapts to minor stresses quickly and efficiently.

Reminder: Any threat to a person's well-being is a *stress.* Some stresses fall within the body's normal and healthy functioning and are known as **physiological stresses.** Outside of these limits, additional stresses imposed by disease or tissue injury are **pathological stresses.** The term **severe stresses** is used here to refer to pathological stresses that rapidly and markedly raise the body's metabolic rate and significantly upset its normal internal balance.

stress response: an elaborate series of cardiovascular and metabolic events orchestrated by the body in response to *stressors* such as infection, surgery, burns, fractures, and deep, penetrating wounds (gunshot wounds, fistulas, or surgical incisions). These stressors may lead to tissue injury or necrosis (death), inflammation, and shock.

Diets modified in energy include both low-kcalorie and high-kcalorie diets. Chapter 9 discussed the use of these diets for overweight and underweight. This chapter and the next describe the use of energy-modified, high-protein diets in acute and chronic conditions, respectively.

Severe Stress

Diet Order
Energy-modified, high-protein diet.

All illnesses threaten the body and impair nutrition status to some extent, but the burden of acute stresses that involve extensive infection or severe tissue damage is especially great. When confronted with such a severe stress, the body employs complicated mechanisms to reestablish its balance. These mechanisms, collectively called the stress response, constitute an adaptation that is sustained for as long as is necessary or until the body can no longer sustain it and exhaustion leads to death.

The Body's Responses to Stress

In response to severe stresses, the body alters blood flow to the site of injury to prevent or halt the spread of infection and speeds up its metabolic rate (hypermetabolism) to mobilize nutrients into glucose and amino acid pools. From these pools the body can synthesize the special factors it needs to limit and repair damage and to regain homeostasis. Immune system factors and hormones mediate the response, which ultimately affects many body systems.

The exact factors the body makes depend on the type of stress. Mending a broken bone requires different factors than does healing a skin wound. Researchers are working to elucidate the body's complex and interrelated responses to inflammation and infection. Much of that research focuses on *cytokines*—proteins that enhance immune defenses by orchestrating changes in the cardiovascular and nervous systems and stimulating the production of the many cells necessary to attack and destroy foreign organisms. The accompanying glossary defines terms related to severe stress.

Immune System Responses The body's natural system of defense against pathogens—the immune system—enables the body to fight off infectious agents. The immune system defends the body so alertly and silently that most healthy people are unaware that thousands of microbes mount attacks against them every day. Occasionally, though, an infection succeeds in making a person ill, and the immune system must then mount a more vigorous counterattack. Serious infections greatly tax the body, and if the counterattack fails, death follows.

Tissues damaged by trauma, heat, chemicals, or infectious agents render the body vulnerable to invading organisms. The body's response to tissue damage—the *inflammatory response*—serves to repair the tissue damage and inactivate and remove foreign particles. Inflammatory response factors alter blood flow to the injured area, sealing it off from the rest of the body while tissue repair is under way. Meanwhile immune factors disarm any invaders that may have gained entry into the site. Together these responses protect the body from the spread of infection.

If the immune factors fail to disarm the invaders but succeed in preventing an infection from spreading, an abscess or granuloma may form. If local defenses are overwhelmed and the invaders gain entry into the bloodstream, sepsis develops. As microbes enter the bloodstream, more immune system factors are produced. The organisms, surrounded by immune system cells, may attach to and damage blood vessel walls and stimulate further inflammatory responses. The

Glossary

abscess: an accumulation of pus that contains live microorganisms and immune system cells; treatment involves draining the abscess.

counterregulatory hormones: hormones such as glucagon, cortisol, and catecholamines that oppose insulin's actions and promote catabolism.

cytokines (SIGH-toe-kynes): immune system factors that help regulate inflammatory responses. Some cause anorexia, fever, and discomfort.

granuloma: a grainlike growth that contains live microorganisms and immune system cells enclosed within a fibrous coat.

inflammatory response: the changes orchestrated by the immune system when tissues are injured by such forces as blows, wounds, foreign bodies (chemicals, microorganisms), heat, cold, electricity, or radiation.

sepsis: the presence of infectious microorganisms in the bloodstream.

shock: a sudden drop in blood volume that disrupts the supply of oxygen to the tissues and organs and the return of blood to the heart. Shock is a critical event that must be corrected immediately.

systemic inflammatory response syndrome (SIRS): the complex of symptoms (see the text) that occur as a result of immune and inflammatory factors in response to tissue damage. In severe cases SIRS may progress to multiple organ failure.

resulting alterations in circulation can disrupt blood flow to organs and may eventually lead to organ failure and shock.

Hormonal Responses The hormonal changes characteristic of the immediate stress response shift the balance between insulin, which promotes the storage of carbohydrate and lipid and the synthesis of protein, and the counterregulatory hormones, which promote the breakdown of glycogen, the mobilization of fatty acids from lipids, and the synthesis of glucose from protein (see Table 23.1). Although insulin levels rise, the rise in the level of counterregulatory hormones is greater still. Important changes occur in the body's use of protein. While proteins are broken down from skeletal muscle, connective tissue, and the gut, protein synthesis increases in the liver and the tissues that produce immune cells. Thus

Table 23.1 Hormonal Changes That Occur during Severe Stress

Hormone	Alteration	Metabolic Effect
Catecholamines	Increase	Glucagon release increases.
		Insulin-to-glucagon ratio decreases.[a]
		Glycogen breakdown increases.
		Glucose production from amino acids increases.
		Mobilization of free fatty acids increases.
Cortisol	Increases	Mobilization of free fatty acids increases.
		Glucose production from amino acids increases.
Glucagon	Increases	Insulin-to-glucagon ratio decreases.[a]
		Glucose production from amino acids increases.
		Glycogen breakdown increases.
		Storage of glucose, amino acids, and fatty acids decreases.
Antidiuretic hormone	Increases	Retention of water increases.
Aldosterone	Increases	Retention of sodium increases.
		Excretion of potassium increases.

Note: These changes are part of the immediate stress response. As adaptation occurs and recovery is in progress, hormone levels gradually return to normal.

[a]The net effect of a decrease in the ratio of insulin to glucagon is that catabolism predominates.

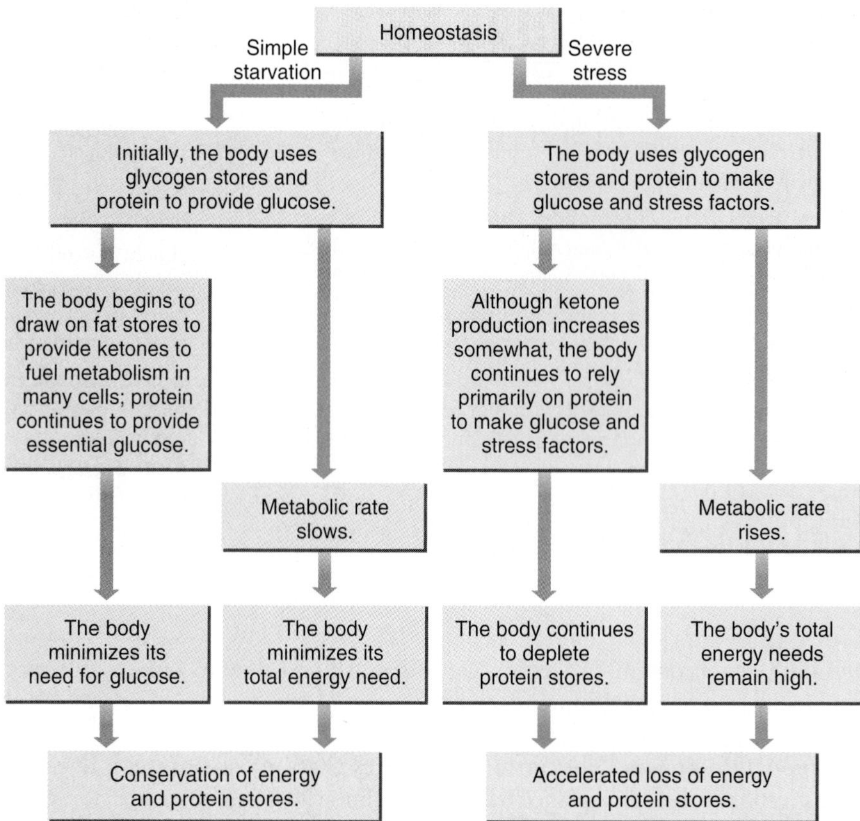

Figure 23.1
Metabolic Responses to Fasting and Stress

Reminder: Not all amino acids can be used to make glucose. The most important amino acids in glucose production are alanine and glutamine; they are synthesized in the body from branched-chain amino acids.

The **acute,** or **flow, phase** of the body's response to stress is the catabolic period immediately following the onset of stress. During the **adaptive phase,** the body begins to recover from the stress. If adaptation succeeds, **recovery** follows. If adaptation fails, **exhaustion** follows.

Cytokines are believed to contribute to the metabolic and GI tract changes as well as to the anorexia, fever, and malaise that accompany the immune response.

protein from skeletal muscle, connective tissue, and the gut is sacrificed to supply the glucose and amino acids necessary to support the synthesis of stress factors and immune cells.[1]

As a result of the hormonal changes, the metabolic rate rises, and the body mobilizes energy stores and elevates blood glucose at the expense of protein tissue. Fat metabolism increases as well, but the high insulin levels suppress the mobilization of fat from body stores. Plasma levels of essential fatty acids fall dramatically, and clinical signs of deficiency may develop in as few as 10 days.[2] Figure 23.1 illustrates the differences between the metabolic responses to simple fasting and to metabolic stress to show how severe stresses rapidly deplete nutrient stores.

Other hormones promote the retention of water and sodium and the excretion of potassium. Hypermetabolism generally peaks at about 3 to 4 days and subsides in about 7 to 10 days. With recovery, hormone levels gradually return to normal.

Clinical Signs and Symptoms The inflammatory and immune responses to stressors result in redness, swelling, heat, and pain at the injury site. Body temperature, heart rate, and respiratory rate increase and anorexia develops. The person often experiences malaise—a feeling of discomfort and uneasiness. Laboratory tests reveal elevated blood glucose (hyperglycemia), elevated blood urea nitrogen (from protein catabolism), negative nitrogen balance, an increased retention of fluid and sodium, and an increased excretion of potassium. Blood concentrations of albumin, iron, and zinc fall; white blood cell counts are often markedly elevated. All of these changes are believed to assist the body in fighting infection.

GI Tract Responses Severe stresses reduce blood flow to the GI tract and slow GI tract motility. Altered blood flow to the stomach may hinder the stomach's ability to protect itself from gastric acid, and ulcers may form in the stomach or small intestine. Motility returns more quickly to the intestine than it does to the stomach. Additionally, the cells of the intestinal tract gradually shrink and lose some of their absorptive and immune functions as gut proteins are sacrificed to support the body's defenses.

Some severe stresses directly impair GI tract functions. Wounds or trauma to the GI tract, including surgical incisions or resections, can interfere with GI tract function. Nutrition in Practice 23 explores the immune functions of the GI tract and describes how impaired GI immunity might contribute to sepsis and multiple organ failure that sometimes accompany severe stress.

Although the effects of severe stresses may have wide-reaching effects on many body systems, it is important to remember that these changes are necessary to defend the body successfully. The following sections describe the many ways that stress affects nutrition status. Clearly, the better nourished people are at the onset of a period of stress, the better able they will be to carry the metabolic burdens stress imposes.

Figure 23.2
Illness, Malnutrition, and Immunity
Regardless of where a person enters the spiral, the effects of stress, malnutrition, and impaired immunity can interact to worsen malnutrition and compromise recovery.

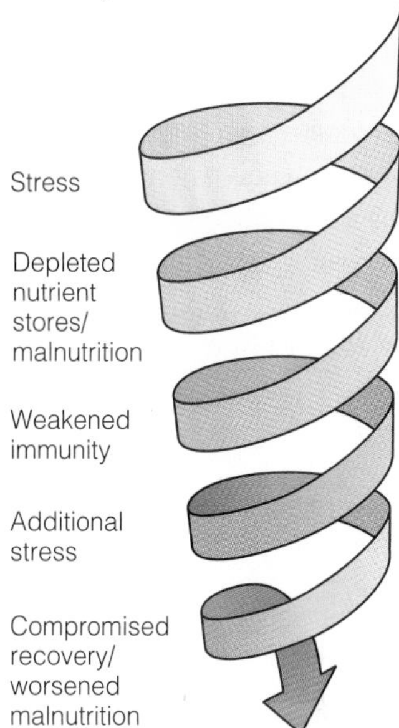

Effects on Nutrition Status

Hypermetabolic illnesses rapidly deplete energy reserves and break down protein tissues. Consequently, they may lead to acute malnutrition in a previously healthy person or worsen preexisting protein-energy malnutrition (PEM). PEM and stress then interact in a deadly cycle: stress worsens malnutrition, and malnutrition hinders the stress response (see Figure 23.2). The combination complicates treatment and hinders recovery.

To understand the relationship of PEM to severe stress, think of energy stores and protein status as money in the bank. Severe stress can be compared to a major expense that arises unexpectedly. The person who has saved enough money can pay off the expense without too much difficulty. If more and more expenses arise, however, the money may run out. The person with little or no savings may be unable to pay even the first expense.

Energy and Protein Stores Unlike glucose and fat, which are held in storage until they are needed, all of the body's proteins are already in use as skeletal muscle, cell structures, enzymes, hormones, immune system factors, and other blood proteins and body components. For a well-nourished person, the period of negative nitrogen balance (loss of protein) is acceptable, because sufficient protein will remain to support defense systems and maintain vital functions. If the person is malnourished or the stress is extreme or of long duration, however, the loss of vital proteins can compromise the function of the immune system, heart, lungs, kidneys, and GI tract. Once organ systems begin to fail, recovery is severely compromised.

pressure sores: the breakdown of skin and underlying tissues due to constant pressure and lack of oxygen to the affected area; also called **decubitus** (dee-CUE-bih-tis) **ulcers** or **bedsores.**

Appetite and Eating Although nutrient needs are high, the effects of stress on the GI tract and appetite limit the oral intake of foods. Pain, malaise, anxiety, and medications further limit food intake. The location of an injury may also interfere with eating. A person with burned hands, for example, may be unable to hold utensils. A person with surgical incisions near the waist my find sitting up to eat uncomfortable.

Immobility and Pressure Sores Severe stresses often necessitate bed rest and immobility. Immobility further compromises nutritional health. Without the muscle tension and weight load incurred by normal activity, the muscles and bones begin to lose nitrogen and calcium, respectively. In prolonged immobilization, calcium losses may cause urinary calcium to rise so high that calcium stones form in the bladder and kidneys.

Immobility, severe stress, and poor food intake are all associated with the development of pressure sores.[3] Pressure sores can form wherever there is constant pressure on the skin. The elderly and people who are unable to respond to pain or change body positions are at great risk for developing pressure sores.

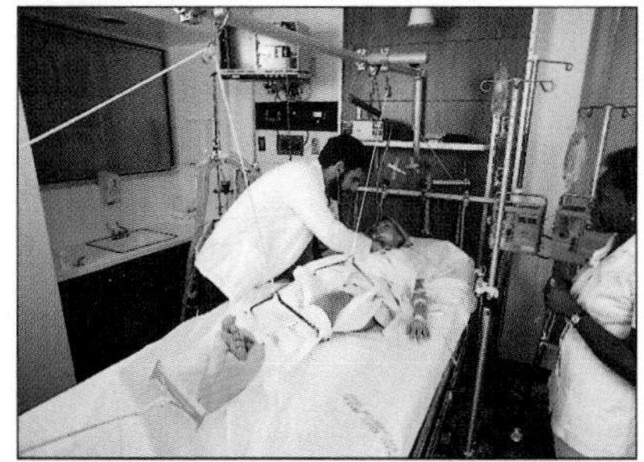

Prolonged immobilization can further impair nutrition status.

R Prescription Pad

Medications used in the treatment of severe stresses may include:

- Analgesics
- Antidiarrheals
- Anti-infective agents
- Anti-inflammatory agents
- Antisecretory agents
- Immunosuppressants

See Appendix E for timing with meals and nutrition-related side effects.

These sores can be extremely painful and are an open invitation for infections, which can further tax the body and nutrient stores.

Nutrient-Medication Interactions Drug therapy, critical in the treatment of many stresses and illnesses, may further tax nutrition status. When people who are severely stressed or malnourished are given multiple medications, as frequently occurs in hypermetabolic illnesses, the likelihood of nutrient-medication interactions increases, and serious deficiencies may result. Furthermore, the impairment of intestinal function seen in both stress and malnutrition can hinder the absorption of both medications and nutrients.

Severe stresses and malnutrition can also interfere with the metabolism and excretion of medications. Many medications are transported in the blood bound to blood proteins such as albumin, and low blood albumin is a symptom of both stress and PEM. Without sufficient carriers, medications may be slow to reach their sites of action. Once medications do arrive at their target cells, the lack of carriers may delay the medications' transport to the liver and kidneys, where they are often detoxified and excreted. Thus medications may take a long time to work and then may remain active for too long, making side effects more likely.

Nutrition Support following Stress

Following stress, the immediate concerns of the health care team are to restore blood flow and oxygen transport and to prevent or treat infection. Possible measures include providing IV solutions to correct fluid and electrolyte imbalances, giving transfusions, removing dead tissues, draining abscesses, and administering antibiotics. Nutrition support following stress helps to minimize nutrient loses, preserve organ function, and maintain immune defenses—all of which aid the client's recovery. Once hypermetabolism subsides, nutrition support aims to restore nutrient deficits and promote positive energy and nitrogen balance.

The removal of dead tissues resulting from burns and other wounds, called **debridement** (dee-BREED-ment), speeds healing and helps prevent infection.

Nourishment must be introduced cautiously to the stressed person. If nutrients are introduced too rapidly, they constitute an additional stress to the body, and severe complications, including malabsorption, cardiac insufficiency, respiratory distress, congestive heart failure, convulsions, coma, and even death, can result. Collectively, these complications are called the refeeding syndrome.

refeeding syndrome: a set of physiologic and metabolic complications associated with reintroducing adequate nutrition too rapidly. These complications can include malabsorption, cardiac insufficiency, respiratory distress, congestive heart failure, convulsions, coma, and possibly death.

Nutrient Needs

Health care professionals caring for severely stressed individuals face a formidable challenge. Providing enough nourishment aids recovery, but providing too much compromises recovery.

Fluids and Electrolytes To restore circulation and prevent dehydration and electrolyte imbalances, the medical team must act quickly to stabilize the body's blood volume and fluid and electrolyte balances. This step is critical, for without adequate blood volume, oxygen, nutrients, and medications cannot be delivered to the cells, organ systems cannot function properly, and waste products cannot be eliminated from the body.

The physician determines the person's fluid needs based on clinical measures such as blood pressure, heart rate, respiratory rate, urinary output, level of consciousness, and body temperature. Serum electrolytes are closely monitored, and adjustments are made as necessary.

With catabolism, blood levels of electrolytes normally concentrated in the intracellular fluids (potassium, phosphorus, magnesium, and calcium) rise.

Once catabolism subsides, these electrolytes move back into the cells along with glucose and amino acids to begin rebuilding tissues. Without careful attention to replacement, blood levels of these electrolytes can plummet, resulting in life-threatening complications.

Energy While supplying too little energy compromises recovery, supplying too much energy contributes to an elevated metabolic rate, which increases oxygen use and carbon dioxide production. Then the heart and lungs, already working hard as a consequence of stress, must work even harder to keep the body's gases in balance. If these vital organs are weakened by preexisting malnutrition, the stress is even greater. Overfeeding can also worsen hyperglycemia, which can adversely affect fluid and electrolyte balances and may also increase the risk of infections in stressed people.[4]

Indirect calorimetry is widely believed to provide the best estimate of energy needs during stress when measurements are carefully taken and cautiously interpreted.[5] Many facilities, however, lack the equipment to perform this measurement, so clinicians often rely on various formulas to estimate energy needs. Among the most widely used formulas to predict energy needs is the Harris-Benedict equation (see Table 23.2).

Reminder: *Indirect calorimetry* is the estimation of energy output from measures of the amount of oxygen used and carbon dioxide eliminated.

Although providing adequate energy following stress is important, resting energy needs are not markedly elevated in the majority of cases. Energy needs can generally be met with diets that provide about 100 percent to 120 percent of those estimated using the Harris-Benedict equation.[6] Exceptions include energy needs for people with burns (which may range from 120 percent to over 200 percent, depending on the severity of the burn)[7] and people with severe head injuries (which may range from 130 percent to 150 percent).[8] Some research suggests that providing less than the estimated energy needs might improve the utilization of lipids for energy, reduce the rate of catabolism, and limit nitrogen losses.[9]

Note that a diet that provides energy at 100 to 120% of the basal energy expenditure is not a high-kcalorie diet. The primary aim of the energy-modified diet for severe stress is to provide adequate kcalories from appropriate sources.

The effects of obesity on the heart, lungs, and immune function, as well as the insulin resistance typical of obesity, are additional concerns for the obese person who suffers a severe stress. While it is no less important that timely and appropriate nutrition support be given to the person who is obese, researchers report that obese people may benefit by providing energy at about half of their estimated needs, provided that protein needs are met.[10] The energy equation shown in Table 23.3 can be used to estimate energy needs for people who are obese.

WWW.

ameriburn.org
American Burn Association Gateway

burnrehab.com
Alberta Burn Rehabilitation Society

aarbf.org
Alisa Ann Ruch Burn Foundation

Table 23.2 The Harris-Benedict Equation for Estimating Basal Energy Expenditure (BEE)[a]

Women:

$$\text{BEE} = 655 + (9.6 \times \text{wt}^{b} \text{ in kg}) + (1.7 \times \text{ht in cm}) - (4.7 \times \text{age in yr})$$

Men:

$$\text{BEE} = 66 + (13.7 \times \text{wt}^{b} \text{ in kg}) + (5 \times \text{ht in cm}) - (6.8 \times \text{age in yr})$$

Add to BEE for stress:
- 0–20% for most stresses
- 30–50% for head injuries
- 20–100% for severe burns

Note: wt = weight; ht = height; yr = years; kg = kilograms; cm = centimeters.

[a]Basal metabolic rate (BMR, described on p. 128) and BEE express the same thing: basal energy need. The equation for BMR is traditionally used in physiology and fitness laboratories; that for BEE, in hospitals. The two equations yield slightly different results, each suitable for the purposes intended. All results are approximations; all require judgment in their application.

[b]Use actual body weight, not ideal body weight.

Table 23.3 The Ireton-Jones Energy Equation[a]	
Estimated energy expenditure (EEE) = 629 + 11(A) + 25(W) − 609(O)	
EEE =	kcal/day
A =	age
W =	body weight
O =	obesity is defined as >30% above ideal body weight from the 1959 Metropolitan Life Insurance weight tables (present = 1, absent = 0)

[a]This equation is appropriate for people who can breathe without the assistance of a ventilator.

Source: Adapted from C. Ireton-Jones, Energy expenditure assessment: Predictive equations, *Support Line,* February 1997, pp. 14–16.

No consensus has been reached on the best formulas for estimating energy needs, and it is important to remember that formulas yield estimates only. Clinical judgment and continual monitoring of nutrition status are always necessary to confirm that energy needs are being met without overfeeding.

Protein To meet protein needs, stressed people with normal kidney and liver function generally need between 1.0 and 1.5 grams of protein per kilogram of body weight per day. People with severe burns may require up to 3 grams of protein per kilogram of body weight per day.[11] Only after the hypermetabolic stage of severe stress subsides can negative nitrogen balance be fully corrected. The box on p. 559 shows an example of how to estimate energy and protein needs.

Amino Acids In recent years, increasing emphasis has been placed on supplying higher amounts of specific amino acids during stress, rather than simply supplying protein. One strategy is to supplement branched-chain amino acids (leucine, isoleucine, and valine) to minimize negative nitrogen balance, but research has been unable to confirm or refute a benefit to this therapy.

An amino acid receiving wide attention in relation to stress is glutamine. Glutamine provides fuel for the intestinal cells and helps maintain their structure and function. Glutamine also plays important roles in maintaining immune function and promoting wound healing.[12] Glutamine is a nonessential amino acid in healthy people, but the body's demands for glutamine during stress may exceed the body's ability to synthesize glutamine in adequate amounts. Under these conditions, glutamine may become a conditionally essential amino acid.[13] Although research has been promising, and glutamine supplementation is widely applied in clinical practice, a consistent benefit to supplementing glutamine in severely stressed clients has not been confirmed.[14] Some research supports the view that given adequate protein, the stressed body meets its need for glutamine without additional supplements.[15]

Other nitrogen-containing substances, including the amino acid arginine and nucleotides, may also be important. Studies suggest that supplemental arginine may help minimize negative nitrogen balance, improve wound healing, and stimulate the immune system. Nucleotides may help improve the function of certain immune cells. Studies are sparse, however, and their results are inconclusive.

nucleotides: nitrogen-containing components of RNA and DNA. Nucleotides can be synthesized in the bodies of healthy people, but the demands for nucleotides during severe stress may make them conditionally essential.

Carbohydrate and Lipids Nonprotein energy sources spare protein, and the amount of energy provided as carbohydrate or lipid has ramifications for stressed people. Carbohydrate provides a readily usable source of energy, but the body can metabolize only a fixed amount (about 500 grams per day) of glucose during stress. Excess glucose serves no benefit, but contributes to hyperglycemia and its consequences. Lipids provide energy and essential fatty acids, but given in excess, they can also tax metabolic functions and hamper immune responses.

How to

Estimate Energy and Protein Needs following Severe Stress

Bernadette is a 39-year-old female, who is 5 feet 3 inches tall and weighs 130 pounds. She recently underwent extensive surgery. Her energy needs can be estimated using the Harris-Benedict equation as follows:

Weight in kilograms = 130 lb ÷ 2.2 kg/lb = 59 kg.
Height in centimeters = 63 in × 2.54 cm/in = 160 cm.
BEE = 655 + (9.6 × wt in kg) + (1.7 × ht in cm) − (4.7 × age in yr).
655 + (9.6 × 59 kg) + (1.7 × 160 cm) − (4.7 × 39)
= 655 + 566 + 272 − 183 = 1310 kcal.

Next add 20% × BEE for surgery:

1310 kcal × 20% = 262 kcal.
1310 + 262 = 1572 kcal.

Bernadette needs about 1572 kcalories to meet her BEE and additional energy needs due to surgery; clinicians monitor weight changes to determine if actual needs are higher or lower. Her energy needs will change as stress resolves.

Protein needs for Bernadette can be estimated at 1.0 to 1.5 grams of protein per kilogram of body weight per day. Use her weight of 59 kilograms to make the calculation:

59 kg × 1.0 g/kg = 59 g protein.
59 kg × 1.5 g/kg = 89 g protein.

Bernadette needs an estimated 59 to 89 grams of protein daily. Some clinicians subtract the kcalories provided by protein from the total energy needs and then provide the remaining calories as carbohydrate and fat.[a] Others meet the energy needs with carbohydrate and fat only and do not consider the kcalories contributed by protein. In either case, clinicians can monitor serum proteins (see Chapter 16) or use nitrogen balance studies to determine whether the estimate is meeting actual protein needs.

[a]J. M. Miles and J. A. Klein, Should protein be included in calorie calculations for a TPN prescription? *Nutrition in Clinical Practice* 11 (1996): 204–206.

Clinicians often supply nonprotein kcalories through a mixture of about 70 percent carbohydrate and 30 percent lipids.[16]

Fatty Acids Intravenous lipid emulsions and enteral formulas are rich sources of omega-6 fatty acids. When given in excess of essential fatty acid requirements, omega-6 fatty acids may impair immune function and, therefore, may not be ideal for severely stressed clients. Alternate lipid sources such as fish oils (a rich source of omega-3 fatty acids) are currently under investigation.[17] Structured lipids—triglycerides chemically modified to contain both long- and medium-chain fatty acids—may also be advantageous during stress.[18]

Micronutrients Vitamin and mineral needs in stress are highly variable, and specific requirements are unknown. The needs for many B vitamins increase when energy and protein intakes increase. Some micronutrients act as cofactors in the many metabolic reactions that are occurring, so their levels may dwindle quickly. Other micronutrients play specific roles in healing wounds and mending broken bones. Levels of antioxidant nutrients fall, and although research is ongoing, supplementing these nutrients is believed to be beneficial.[19]

Growth Hormone and Insulin-like Growth Factor Researchers have begun to study nondietary factors that might improve nitrogen balance and lessen the impact of severe stress on the host. Two such factors are growth hormone and insulin-like growth factor-1 (IGF-1), which stimulate the growth of intestinal cells. Although research is limited, growth hormone has been shown to improve nitrogen balance in both animals and people.[20] Similarly, IGF-1 improves nitrogen balance in animals.[21]

Case Study

Journalist with a Third-Degree Burn

Mr. Sampson, a 48-year-old journalist, has been admitted to the emergency room. He suffered a severe burn covering over 40 percent of his body when he was trapped in a burning building. His wife told the nurse that Mr. Sampson's height is 6 feet and that he weighs about 175 pounds. The physician ordered lab work, including serum proteins; the results are not back yet.

Identify Mr. Sampson's immediate postinjury needs. How can these needs be met?

Do you have enough information to determine Mr. Sampson's preinjury, preburn nutrition status? If not, what information would be useful? Is information about preburn nutrition status important in this case? Why or why not?

Considering Mr. Sampson's condition, what problems might the health care team encounter in getting information from him about his preburn nutrition status?

Calculate Mr. Sampson's energy and protein needs to support burn healing (use 2 × the BEE for energy and 2 to 3 grams of protein per kilogram of body weight). What other nutrients might be of concern?

Describe the possible benefits of early enteral nutrition to Mr. Sampson. How might these benefits be particularly important following a severe burn injury?

Delivery of Nutrients following Stress

Selecting the appropriate amounts and types of nutrients to help people recover from stress is only part of medical nutrition therapy. Just as important is supplying nutrients in a form that best serves the body's ability to recover.

Oral Diets Well-nourished clients who are expected to eat within 7 to 10 days following stress receive simple IV solutions to maintain fluid and electrolyte balance and provide minimal kcalories. Once GI tract motility returns, they begin an oral diet that often progresses from clear liquids to full liquids and on to soft and then regular foods as tolerated (see Chapter 18).

WWW. **clinnutri.org**
American Society for Enteral and Parenteral Nutrition

Tube Feedings and Parenteral Nutrition People with severe malnutrition or those who are not expected to be able to eat within 7 to 10 days benefit from tube feedings or parenteral nutrition. Oral or gastric feedings have to wait until gastric motility is restored to prevent abdominal distension, nausea, vomiting, and the possible aspiration of foods or formula into the lungs. Peristalsis returns more quickly to the small intestine than to the stomach, however, and feeding formula directly into the small intestine through a tube is not only possible, but provides advantages over parenteral nutrition.[22] Early feeding (often initiated within about 36 hours following stress) stimulates intestinal blood flow, function, and adaptation and may minimize hypermetabolism and help prevent bacterial translocation (see Nutrition in Practice 23). Most significantly, however, early enteral feeding improves recovery following stress by reducing septic complications.[23] Early enteral feedings are not possible in cases where blood flow to the intestine is severely disrupted, however. Additionally, some clients may need both enteral feedings and parenteral nutrition until they are able to meet all nutrient needs orally. Once oral feeding is possible, tube feedings or parenteral nutrition is gradually discontinued (see Chapters 21 and 22).

Stress Formulas Clinicians eager to improve a client's outcome often rely on formulas designed to meet nutrient needs during stress. Many such formulas are of high nutrient density. Some contain extra vitamins A and C, zinc, and other

Nutrition Assessment Checklist

For People Experiencing Severe Stress

Medical Use the medical record to determine the degree and type of stress and to help estimate preexisting nutrition status. Find the information necessary to estimate energy needs. For severely stressed clients, indirect calorimetry may provide a more accurate assessment of energy needs.

Medication Assess the client's medication history for nutrient-medication interactions that might alter nutrient needs. For people able to eat food by mouth, consider adjusting pain medications so that they are most effective during meals. Narcotic analgesics can cause nausea and may make the person too tired to eat. Many anti-infective agents can affect nutrient intake and nutrition status.

Nutrient/Food Intake Calculate nutrient intake from parenteral and enteral formulas and oral diets to determine if intake is meeting calculated needs. If not, investigate the cause and take corrective actions, when possible. For clients on oral diets, a careful history of food preferences will be invaluable in encouraging adequate oral intake.

Anthropometric Interpret anthropometrics cautiously in the immediate poststress period. Weights may reflect the infusion of fluids or edema and can be deceptively high. The location of injuries may make anthropometric measurements impossible.

Laboratory Anticipate low serum albumin and transferrin in stressed clients, especially those with burns or deep or extensive wounds. (Remember that plasma proteins leak through the capillaries to the injury site.) Use nitrogen balance studies for a more accurate assessment of protein needs for severely stressed clients when necessary. Check blood glucose at regular intervals and treat hyperglycemia according to the hospital's protocol or physician's orders. Monitor electrolytes to replace losses immediately after stress and to prevent metabolic complications once hypermetabolism subsides.

Physical Check for physical signs of PEM, fatty acid deficiencies, dehydration, and nutrient deficiencies. Assess energy level and emotional state. Regular assessment of blood pressure, pulse, and fluid intake and urinary output records can help prevent dehydration or overhydration.

nutrients designed to promote wound healing. Formulas designed to improve immune function often contain added glutamine, arginine, nucleotides, and omega-3 fatty acids. Although such formulas appear to be beneficial for specific situations, further research is necessary to determine if their impact on recovery justifies their use.[24] The accompanying case study of a client with burns tests your knowledge of nutrition and severe stress and the nutrition assessment checklist summarizes information necessary to monitor the nutrition care of stressed clients.

Stresses marked by extensive tissue damage and hypermetabolism can place tremendous demands on the body. Then the body uses its arsenal of defenses to survive and regain health. Recovery depends, in part, on the body's receiving the energy and nutrients required to mount a defense, repair damaged tissues, and replenish nutrient reserves. The next chapter describes the impact of chronic conditions that tax nutrient stores.

Self Check

1. The severe stresses described in this chapter are characterized by:
   a. chronic malnutrition.
   b. tissue damage and hypermetabolism.
   c. reduced protein synthesis in the liver.
   d. hormonal changes that protect skeletal muscle.

2. The health care team recognizes that the complication most likely to lead to multiple organ failure during severe stress is:
   a. essential fatty acid deficiency.
   b. low serum albumin and transferrin.
   c. catabolism of protein.
   d. sepsis.

3. Cytokines are proteins that:
   a. orchestrate the inflammatory response to stress.
   b. oppose insulin's action and result in the catabolism of protein in skeletal muscle, connective tissue, and the gut.
   c. destroy microorganisms.
   d. repair damaged tissues.

4. All of the following metabolic changes accompany severe stress **except:**
   a. protein synthesis in the liver increases.
   b. proteins are broken down in skeletal muscle, connective tissue, and the gut.
   c. protein synthesis in the liver decreases.
   d. the body fails to conserve protein as it does in simple fasting.

5. All of the following statements with respect to nutrition and severe stress are true **except:**
   a. a previously well-nourished person can develop acute malnutrition if the stress is extreme or prolonged.
   b. a person with chronic malnutrition who suffers a severe stress requires immediate attention to nutrient needs.
   c. a person with chronic malnutrition who suffers a severe stress or a person who develops acute malnutrition as a result of stress generally does not need tube feedings or TPN unless he or she will be unable to eat for 7 to 10 days.
   d. a person with either acute or chronic malnutrition may not have the energy reserves or protein needed to respond successfully to stress.

6. The use of oral diets in the immediate poststress period is prohibited because severe stress:
   a. results in increased blood flow to the GI tract.
   b. slows gastric motility.
   c. decreases the appetite.
   d. alters plasma amino acid levels.

7. The health care professional may use all of the following parameters to assess a person's fluid needs during stress **except:**
   a. blood pressure. b. urinary output.
   c. iron status. d. body temperature.

8. Which of the following statements is true about the energy-modified diet for severe stresses?
   a. The diet is high in kcalories.
   b. The amounts of carbohydrates, lipids, and proteins that supply energy make little difference.
   c. Although the diet must supply adequate energy, energy needs are generally not high, except for some people with extensive burns or head injuries.
   d. The diet has little effect on blood glucose levels or respiratory rate.

9. Depending on the type of stress, the amount of protein a stressed person who weighs 150 lbs needs can range from ________ to ________ grams of protein per day.
   a. 68–136 b. 68–200
   c. 136–200 d. 136–270

10. The major reason early enteral nutrition following a severe stress must be introduced by tube is that:
    a. slowed gastric motility prohibits the use of the stomach.
    b. appetite is depressed.
    c. oral diets cannot meet nutrient needs.
    d. hydrolyzed diets are necessary.

Answers to these questions appear in Appendix H.

Clinical Applications

1. Returning to Bernadette from the box on p. 559 and assuming that she can tolerate an intact enteral formula, find at least three formulas in Appendix G that the health care team might select if she needed a tube feeding. Determine the volume of each formula that would be needed to meet Bernadette's energy and protein needs. Would this volume also meet the recommendations for vitamins and minerals?
2. Bennie is a well-nourished seven-year-old who develops the flu and has a fever of 101°F for two days. Describe how this stress could temporarily affect his nutrition status. How would your concerns differ if Bennie were a seven-year-old hospitalized for injuries suffered in a car accident, who develops the flu and a fever and is unable to eat for several days?
3. Susan Griff is a 28-year-old woman admitted to the hospital following a car accident in which she broke several bones, ruptured a portion of her small intestine, and suffered a severe burn. She has been in the hospital for several weeks and is now eating table foods. Aside from the nutrient demands imposed by the stresses she withstood, describe how the following factors can impair her nutrition status:
   - Susan's injuries are painful.
   - Susan's medications cause extreme drowsiness.
   - Susan is depressed.
   - Susan is often out of her room for X rays and other diagnostic tests when her menus and food trays arrive.
   - Susan's food intake is often restricted for diagnostic tests she will be receiving.

   How might these problems be resolved to improve Susan's ability to eat?

Notes

1. L. L. Moldawer, Cytokines and the cachexia response to acute inflammation, *Support Line,* April 1996, pp. 1–6.

2. R. G. Barton, Nutrition support in critical illness, *Nutrition in Clinical Practice* 9 (1994): 127–139.

3. B. J. Braden, Using the Braden Scale for predicting pressure sore risk, *Support Line,* August 1996, pp. 14–17.

4. M. M. McMahon, Nutrition support of hospitalized patients, presented at the *Eighth Annual Advances and Controversies in Clinical Nutrition,* Mayo Clinic Foundation, April 5–7, 1998; M. O. Kwoun and coauthors, Immunologic effects of acute hyperglycemia in nondiabetic rats, *Journal of Parenteral and Enteral Nutrition* 21 (1997): 91–95; J. J. Pomposelli and coauthors, Early postoperative glucose control predicts nosocomial infection rate in diabetic patients, *Journal of Parenteral and Enteral Nutrition* 22 (1998): 77–81.

5. C. Porter and N. H. Cohen, Indirect calorimetry in critically ill patients: Role of the clinical dietitian in interpreting results, *Journal of the American Dietetic Association* 96 (1996): 49–54; L. E. Matarase, Indirect calorimetry: Technical aspects, *Support Line,* February 1997, pp. 6–12; B. J. Osborne and coauthors, Clinical comparison of three methods to determine resting energy expenditure, *Nutrition in Clinical Practice* 9 (1994): 241–246.

6. As cited by J. M. Miles in J. M. Miles and J. A. Klein, Should protein be included in calorie calculations for a TPN prescription? *Nutrition in Clinical Practice* 11 (1996): 204–206; McMahon, 1998.

7. D. J. Rodriguez, Nutrition in major burn patients: State of the art, *Support Line,* August 1995, pp. 1–8.

8. E. Weekes and M. Elia, Observations on the patterns of 24-hour energy expenditure changes in body composition and gastric emptying in head-injured patients receiving nasogastric tube feeding, *Journal of Parenteral and Enteral Nutrition* 20 (1996): 31–37.

9. D. C. Frankenfield, S. Smith, and R. N. Cooney, Accelerated nitrogen loss after traumatic injury is not attenuated by achievement of energy balance, *Journal of Parenteral and Enteral Nutrition* 21 (1997): 324–329.

10. J. C. Burge and coauthors, Efficacy of hypocaloric total parenteral nutrition in hospitalized obese patients: A prospective, double-blind randomized trial, *Journal of Parenteral and Enteral Nutrition* 18 (1994): 203–207; P. S. Choban, J. C. Burge, and L. Flancbaum, Nutrition support of obese hospitalized patients, *Nutrition in Clinical Practice* 12 (1997): 149–154.

11. Rodriguez, 1995.

12. R. G. Barton, Immune-enhancing enteral formulas: Are they beneficial in critically ill patients? *Nutrition in Clinical Practice* 12 (1997): 51–62.

13. G. K. Savy, Enteral glutamine supplementation, *Nutrition in Clinical Practice* 12 (1997): 259–262.

14. S. Klein and coauthors, Nutrition support in clinical practice: Review of published data and recommendations for future research directions, *Journal of Parenteral and Enteral Nutrition* 21 (1997): 133–156.

15. C. L. Long and coauthors, Glutamine supplementation of enteral nutrition: Impact on whole body protein kinetics and glucose metabolism in critically ill patients, *Journal of Parenteral and Enteral Nutrition* 19 (1995): 470–475.

16. McMahon, 1998.

17. S. J. Bell and coauthors, The new dietary fats in health and disease, *Journal of the American Dietetic Association* 97 (1997): 280–286; J. D. Palombo and coauthors, Cyclic *vs* continuous enteral feeding with ω-3 and γ-linolenic fatty acids: Effects on modulation of phospholipid fatty acids in rat lung and liver immune cells, *Journal of*

Parenteral and Enteral Nutrition 21 (1996): 123–132; M. Roulet and coauthors, Effects of intravenously infused fish oil on platelet fatty acid phospolipid composition and on platelet function in postoperative trauma, *Journal of Parenteral and Enteral Nutrition* 21 (1997): 296–301.

18. Bell and coauthors, 1997; R. Sandström and coauthors, Structured triglycerides were well-tolerated and induced increased whole body fat oxidation compared with long-chain triglycerides in postoperative patients, *Journal of Parenteral and Enteral Nutrition* 19 (1995): 381–386.

19. V. Sardesai, Role of antioxidants in health maintenance, *Nutrition in Clinical Practice* 10 (1995): 19–25; C. J. Schorah and coauthors, Total vitamin C, ascorbic acid, and dehydroascorbic acid concentrations in plasma of critically ill patients, *American Journal of Clinical Nutrition* 63 (1996): 760–765.

20. K. Takagi and coauthors, Recombinant human growth hormone and protein metabolism of burned rats and esophagectomized patients, *Nutrition* 11 (1995): 22–26; M. Jeevanandam and coauthors, Adjuvant recombinant human growth hormone normalizes plasma amino acids in parenterally fed trauma patients, *Journal of Parenteral and Enteral Nutrition* 19 (1995): 137–144.

21. T. Inaba and coauthors, Effects of growth hormone and insulin-like growth factor-1 (IGF-1) treatments on the nitrogen metabolism and hepatic IGF-1-messenger RNA expression in postoperative parenterally fed rats, *Journal of Parenteral and Enteral Nutrition* 20 (1996): 325–331.

22. W. W. Souba, Nutritional support, *New England Journal of Medicine* 336 (1997): 41–48; B. Beier, E. A. Bergman, and M. J. Morrissey, Factors related to the use of early postoperative enteral feedings in thoracic and abdominal surgery patients in the United States, *Journal of the American Dietetic Association* 97 (1997): 293–295; S. Trice, G. Melnik, and C. P. Page, Complications and costs of early postoperative parenteral versus enteral nutrition in trauma patients, *Nutrition in Clinical Practice* 12 (1997): 114–119.

23. K. A. Kudsk, Immunologic support: Enteral vs parenteral feeding, in *Enteral Nutrition Support,* Report of the First Ross Conference on Enteral Devices, Ross Laboratories, 1996, pp. 70–74.

24. Barton, 1997.

CHAPTER 23

Nutrition in Practice

▪ GI Tract Immunity and Stress ▪

The significant role of the GI tract in preventing foreign invaders from entering the body has only recently received the attention it deserves. This discussion describes GI tract immunity and dietary factors that may help protect this immunity.

Controversies abound in this area, partly because much of the research has been conducted on animals, and an animal's GI tract responds differently to stress and lack of enteral nutrition than a person's GI tract does. From what is known, however, it seems clear that preserving GI tract immune function has far-reaching effects on health and nutrition status.

What role does the GI tract play in immunity?

A protective coating of mucus lines the entire GI tract. The mucus contains antimicrobial chemicals and enzymes to destroy foreign bodies and forms a slippery coat that prevents them from attaching to the lining of the GI tract. To reach the intestine, invaders must also avoid destruction by the highly acidic contents of the stomach. Invaders that enter the intestine directly (through breaks in tissue) or avoid destruction by mucus or gastric acidity encounter formidable obstacles in the intestinal tract.

How is the intestinal tract protected?

Recall from Chapter 5 that the absorptive surface of the intestine is lined with fingerlike projections called villi. Healthy villi are crowded close together, forming a physical barrier that prevents anything from passing between them. Consequently, substances can pass from the intestine to the inside of the body only by crossing the cells' membranes, and the cells are remarkably efficient at keeping foreign materials out.

Interspersed among the villi are mucus-secreting cells and lymph tissues that house immune cells to fend off invaders (see Figure NP23.1 on p. 566). To appreciate the vast importance of the intestinal lymph tissue in fighting foreign invaders, consider that of all the body's immunologic-secreting cells, 70 to 80 percent are located within the intestine.[1]

The large intestine also supports a bacterial population that inhibits the growth of harmful bacteria by competing with them for nutrients and space.[2] The normal bacterial flora also produce short-chain fatty acids that prevent harmful microbes from sticking to the intestinal surface.

How can these functions be disrupted by stress?

Conditions that compromise the GI tract's barrier function, alter the normal bacterial flora, or compromise the function of its immune cells may allow infectious agents to cross the intestinal barrier and enter the body—a process called translocation.[3] As Table NP23.1 shows, many of the conditions that increase the likelihood of translocation exist during severe stress or malnutrition. In small amounts, the translocation of infectious agents may stimulate the immune system, but extensive translocation may cause serious infection and even death. Research suggests that translocation may be an important factor in the development of sepsis and multiple organ failure that may accompanying severe stress.[4]

Table NP23.1 Conditions That Increase the Likelihood of Translocation

Altered structure and function of GI tract barrier:
Prolonged fasting or lack of enteral nutrients
Injury to the GI tract
Inflammatory responses
Malnutrition
Changes in bacterial flora:
Lack of enteral nutrients
Decreased GI tract motility
Use of broad-spectrum antibiotics
Compromised function of immune factors:
Malnutrition
Hypermetabolism

Source: Adapted from M. T. DeMeo, The role of enteral nutrition in maintaining the structural and functional integrity of the gastrointestinal tract, in *Enteral Nutrition Support,* Report of the First Ross Conference on Enteral Devices, Ross Laboratories, 1996, pp. 4–8.

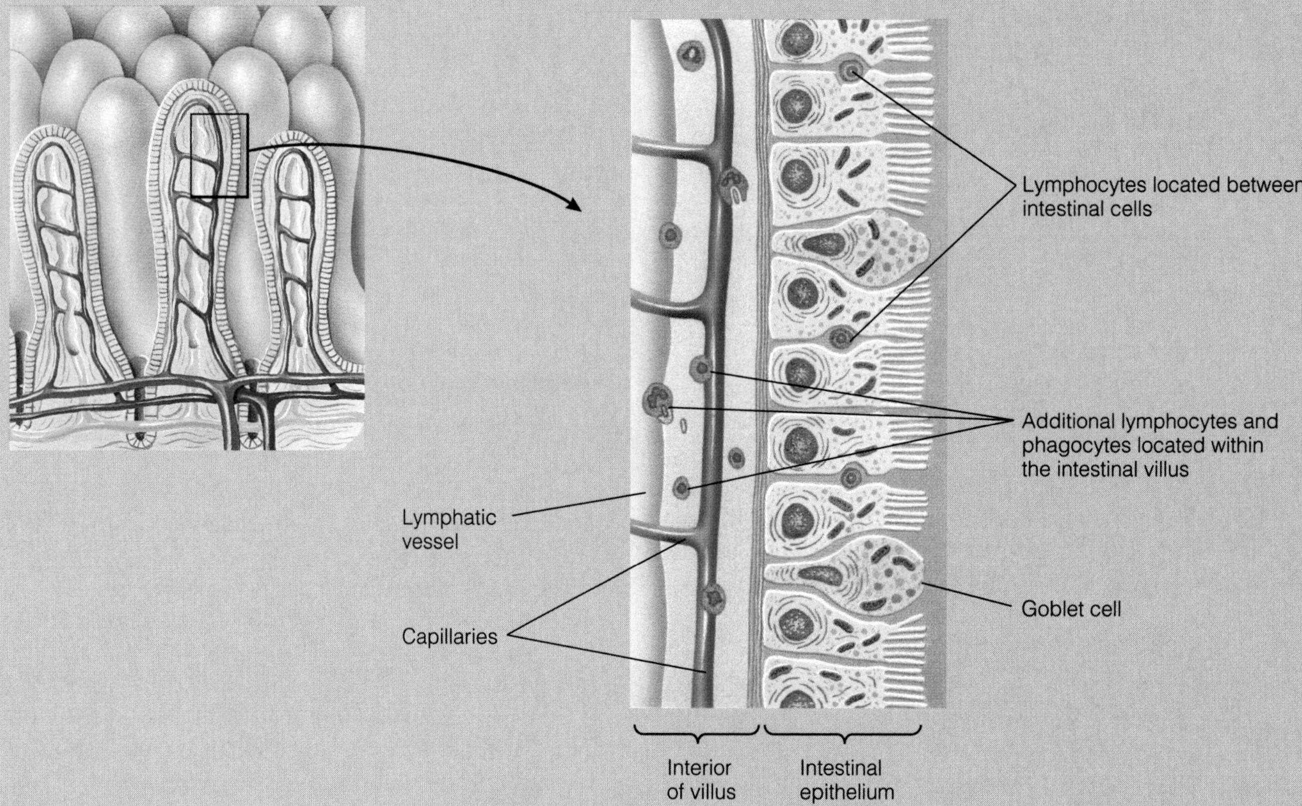

Figure NP23.1
Immune Cells That Protect the Intestinal Villi

How is translocation associated with multiple organ failure?

Multiple organ failure has characteristics that are remarkably similar in most cases, even though the original stresses are different. This suggests that common factors may be involved. For one thing, multiple organ failure does not develop until days or weeks following the initial stress. People who develop multiple organ failure develop sepsis initially, and the first organs to fail are the lungs, followed by the liver and kidneys. Some of the infectious agents associated with sepsis and multiple organ failure arise from the intestinal tract.[5] Clinical evidence to support a role for intestinal translocation as a primary factor, or even one of the factors, leading to sepsis and multiple organ failure is lacking, however, and the theory remains unproved.[6] In practice, however, measures to support intestinal integrity and maintain immune function, including early enteral nutrition (see Chapter 23), are widely accepted and utilized by clinicians.

In what ways can early enteral nutrition protect GI tract integrity and immune function?

Animal studies indicate that when nutrients are not present in the GI tract, intestinal cells atrophy and gradually become dysfunctional, even when nutrients are being delivered by vein. Most clinicians believe that early enteral nutrition following stress is superior to parenteral nutrition in maintaining the structure and functions of the intestinal tract, although this has not been proved in humans.[7] A primary objective of early enteral feeding is to preserve intestinal immunity, but it is not clear whether preserving GI tract integrity can prevent extensive translocation. In addition to early enteral nutrition, researchers are looking for specific dietary constituents that might protect the health of intestinal cells and GI immunity. Among these dietary constituents, glutamine is the most widely studied.

Nutrition in Practice

What is glutamine?

Glutamine is an amino acid abundant both in food and in the blood. Healthy people synthesize glutamine in skeletal muscle from branched-chain amino acids. Glutamine provides fuel for the intestinal cells as well as for some immune cells. After the intestinal cells have metabolized glutamine, the liver uses the end products, alanine and ammonia, to make glucose and urea, respectively. Thus glutamine is important in gluconeogensis (the synthesis of glucose). Glutamine also serves in the replication of all body cells, which use it to synthesize nucleotides (the basic units of RNA and DNA) as well as other amino acids.

During stress, the body's demands for glutamine may exceed the body's ability to synthesize it, and glutamine may become a conditionally essential amino acid. Intravenous feeding solutions traditionally do not contain glutamine, and most enteral formulas contain only small amounts. Studies have shown that adding glutamine to standard TPN solutions can help preserve intestinal structure.[8] Other research suggests that glutamine-supplemented TPN can prevent atrophy of lymph tissue in the intestine.[9] People undergoing bone marrow transplants who received glutamine-supplemented TPN had significantly fewer infections, better nitrogen balances, and shorter hospital stays.[10] While studies of parenteral glutamine supplementation have shown promise, data to support a benefit to glutamine-supplemented enteral nutrition are lacking.[11]

Recall from Chapter 23 that some stress formulas contain higher amounts of branched-chain amino acids than standard formulas, and that branched-chain amino acids can be used by the body to synthesize glutamine. Other stress formulas contain nucleotides, which can be synthesized from glutamine. Studies to date have failed to confirm a benefit to these formulas either in sparing glutamine or through other effects on immunity.

Evidence to support the theory that a breach in the intestinal barrier can be a significant source of infection in severe stress is mounting, and findings on the use of specific dietary interventions to maintain GI tract integrity are encouraging. Clearly, studies of the possible role of early enteral nutrition and glutamine in severe stress are promising. Although many practitioners are currently using these strategies to improve their clients' outcomes, further studies are needed to confirm the benefits. More work must be done to determine if a breakdown of the intestinal barrier is truly a major route of infection during stress. Carefully controlled studies in human beings are lacking, and little work has been done to determine under which conditions, and in what amounts, glutamine might be safe and effective. Hopes are high, however. The prospects of supporting the GI immune system and preventing further decline in stressed people are inviting, and the potential benefits may prove to be lifesaving.

Notes

1. P. Brandtzaeg and coauthors, Immunobiology and immunopathology of human gut mucosa: Humoral immunity and intraepithelial lymphocytes, *Gastroenterology* 97 (1989): 1562–1584.

2. M. B. Roberfroid and coauthors, Colonic microflora: Nutrition and health, *Nutrition Reviews* 53 (1995): 127–130.

3. M. T. DeMeo, The role of enteral nutrition in maintaining the structural and functional integrity of the gastrointestinal tract, in *Enteral Nutrition Support,* Report of the First Ross Conference on Enteral Devices, Ross Laboratories, 1996, pp. 4–8.

4. DeMeo, 1996; E. V. Shronts, Enteral versus parenteral nutrition: A clinical review, *Support Line,* June 1996, pp. 10–13; K. A. Kudsk, Clinical applications of enteral nutrition, *Nutrition in Clinical Practice* 9 (1994): 165–171.

5. Shronts, 1996.

6. T. O. Lipman, Bacterial translocation and enteral nutrition in humans: An outsider looks in, *Journal of Parenteral and Enteral Nutrition* 19 (1995): 156–165.

7. T. O. Lipman, Grains or veins: Is enteral nutrition really better than parenteral nutrition? A look at the evidence, *Journal of Parenteral and Enteral Nutrition* 22 (1998): 167–182.

8. R. R. J. Van Der Hulst and coauthors, Glutamine and the preservation of gut integrity, *Lancet* 222 (1993): 243–255.

9. J. Li and coauthors, Glycyl-L-glutamine-enriched total parenteral nutrition maintains small intestine gut-associated lymphoid tissue and upper respiratory tract immunity, *Journal of Parenteral and Enteral Nutrition* 22 (1998): 31–36.

10. T. R. Ziegler and coauthors, Clinical and metabolic efficacy of glutamine supplemented parenteral nutrition after bone marrow transplantation: A randomized, double-blind controlled study, *Annals of Internal Medicine* 116 (1992): 821–828.

11. R. G. Barton, Immune-enhancing enteral formulas: Are they beneficial in critically ill patients? *Nutrition in Clinical Practice* 12 (1997): 51–62; J. Lipp, Glutamine and immune function: Theory and clinical applications, *Support Line,* October 1997, pp. 14–17.

24 High-kCalorie, High-Protein Diets for Wasting Syndromes

Chapter 23 described the use of energy-modified, high-protein diets in acute stresses, the kind that place immediate demands on nutrient stores. This chapter focuses on high-kcalorie, high-protein diets for chronic stresses noted for their association with wasting, including chronic obstructive pulmonary diseases (COPD), congestive heart failure (CHF), cancer, and human immunodeficiency virus (HIV) infection. Other disorders that often lead to wasting, including malabsorption syndromes, liver disease, and renal failure, are examined in appropriate chapters.

High-kCalorie, High-Protein Diets in Wasting versus Stress

High-kcalorie, high-protein diets for chronic disorders and those for acute stresses share certain similarities. In both cases, loss of appetite, altered metabolism, and accelerated nutrient losses compromise nutrient stores, thereby reducing the supply of nutrients at a time when they are greatly needed. In both cases, diet therapy aims to limit the loss of lean body mass and to preserve nutrient stores and organ function. Acute stresses, however, rapidly alter nutrient needs in short periods of time and then resolve or lead to exhaustion, whereas chronic conditions develop gradually and persist for long periods. Chronic conditions require long-term treatment (including medications) and continue to tax nutrient stores over the entire period. The body weakened by chronic stresses is also more prone to acute stresses, most notably, infections. Whereas an acute stress often strikes the body with little warning, the wasting associated with the disorders described in this chapter frequently does not develop until the disease progresses to its end stages. Alert health care professionals institute nutrition therapy early to improve the quality of life and prevent early death associated with malnutrition.

Reminder: *Cytokines* are proteins that mediate the body's immune and inflammatory responses.

As Chapter 23 explained, cytokines mediate the body's response to acute stress. Although less research has been conducted on their part in chronic conditions, studies suggest that cytokines may play a role in the development of protein-energy malnutrition (PEM) that accompanies chronic obstructive pulmonary disease, congestive heart failure, cancer, and HIV infection.[1] Like severe stresses, these disorders trigger inflammatory and immune responses.

Energy and Protein Energy and protein needs for a wasting disorder vary depending on the stage of the disorder, its treatment, related complications, and the person's current nutrition status. When the person's disease has progressed to the point of unintentional weight loss, the prescribed diet often provides energy at about 150 percent of the basal energy expenditure (see Table 23.2 on p. 557) and 1.5 to 2.0 grams of protein per kilogram of body weight.

Vitamins and Minerals Vitamin and mineral needs for wasting syndromes are highly variable, and little information is available concerning specific needs. Often vitamin and mineral supplements are prescribed to ensure an adequate intake.

Wasting in Chronic Heart Failure and Chronic Obstructive Pulmonary Disease

Chronic disorders of the heart and lungs interfere with the delivery of oxygen to the body's tissues, which depend on oxygen to process the energy nutrients they need to function. Labored breathing, fatigue, and repeated infections typically accompany both disorders, and further compromises nutrition status.

Chronic Heart Failure (CHF)

While heart attacks represent acute heart failure, heart failure can also occur gradually. Coronary heart disease, hypertension (see Chapter 26), and obesity are common causes. A person may develop chronic heart failure following a heart attack or a severe stress. The conditions that lead to chronic, or *congestive,* heart failure (CHF) cause the heart muscle to work unusually hard. As a result, the heart muscle enlarges and gradually weakens as it strains to supply adequate blood to the tissues.

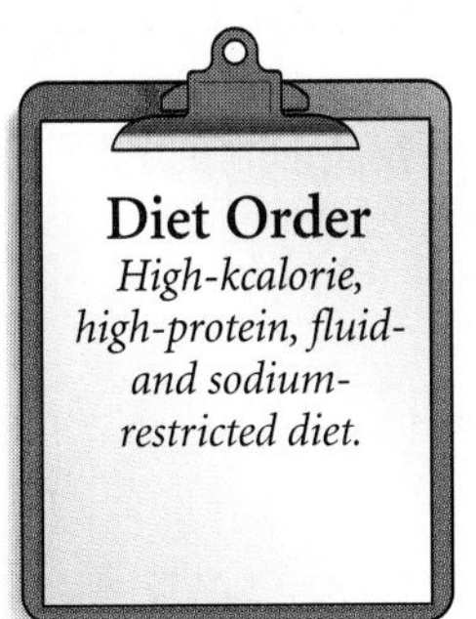
Diet Order
High-kcalorie, high-protein, fluid- and sodium-restricted diet.

chronic or **congestive heart failure (CHF):** a syndrome in which the heart can no longer pump blood through the circulatory system.

Enlargement of the heart is **cardiomegaly** (CAR-dee-oh-MEG-ah-lee).
cardio = heart
mega = large

Consequences of CHF As heart failure progresses, reduced blood flow impairs the function of all organs. Reduced blood flow to the kidneys triggers the retention of fluid, further stressing the heart and compounding the stagnation of fluid in the body. Peripheral, pulmonary, and hepatic edema may develop as the person becomes increasingly "congested" with excess fluids. Pulmonary edema increases the likelihood of respiratory infections, which can further stress the heart and lungs.

CHF and Nutrition Status As CHF progresses, energy needs increase, because organ systems, particularly the heart and lungs, must work extra hard to maintain their functions. At the same time, the disrupted blood flow limits the supply of nutrients and oxygen to the organs and tissues. Repeated respiratory infections can further tax nutrition status. People with CHF are often unable to eat enough to meet energy and protein demands; oral intake may be limited due to anorexia, altered taste sensitivity, intolerance to food odors, physical exhaustion, the diet used for treatment (described later), and medications. Weight loss in people with CHF may go unnoticed until it has progressed considerably, because edema masks their underweight condition. Thus severe PEM is a frequent consequence, particularly in the later stages of the disease. Malnutrition can further contribute to the weakness of heart muscle and lungs and the development of respiratory infections.

Chronic PEM that develops as a consequence of heart disease is called **cardiac cachexia** (ka-KEKS-ee-ah).

Treatment of CHF Drug therapy for CHF includes diuretics to reduce the fluid volume and cardiac glycosides to increase the strength of heart muscle contractions. People taking thiazide or loop diuretics and cardiac glycosides are at high risk for potassium deficiency and may be prescribed a potassium supplement as well. Stool softeners may be prescribed, particularly for elderly clients who frequently experience constipation, because straining to empty the bowels can stress the heart. Initially, bed rest helps reduce the heart's workload. Once recovery is under way, the person must rest frequently and avoid overexertion.

Rx Prescription Pad

Medications used in the treatment of CHF may include:

- Antihypertensives (vasodilators)
- Cardiac glycosides
- Diuretics
- Laxatives
- Potassium supplements

See Appendix E for timing with meals and nutrition-related side effects.

Diet Therapy The person newly diagnosed with CHF who is overweight benefits from a safe weight-loss program. Weight loss relieves strain on the heart and slows the progression of the disease.

Diet therapy for the wasting associated with CHF aims to preserve or restore nutrition status and reduce the work of the heart. Providing adequate energy is vital, but providing too much energy increases the body's metabolic rate, stressing the heart. Likewise, giving too much fluid and sodium expands the body's fluid volume, which also taxes the heart.

With respect to sodium and fluid, the treatment of CHF is similar to the treatment of acute heart failure (heart attack). If CHF develops as a consequence of coronary heart disease, the diet also restricts fat. Chapter 26 provides the details of sodium- and fat-restricted diets. Dietary fiber is carefully adjusted: the goal is to provide some fiber to prevent constipation, but to avoid amounts and types of fibers that produce gas and abdominal distension.

Chronic Obstructive Pulmonary Disease (COPD)

chronic obstructive pulmonary disease (COPD): one of several disorders, including emphysema and bronchitis, that interfere with respiration.

emphysema (EM-fe-SEE-ma): a type of COPD in which the lungs lose their elasticity and the victim has difficulty breathing; often occurs along with bronchitis.

bronchitis (bron-KYE-tis): inflammation of the lungs' air passages.

Chronic obstructive pulmonary disease (COPD) is a term that describes several conditions characterized by persistent obstruction of air flow through the lungs. Unlike acute respiratory failure, which may occur suddenly, lung failure resulting from COPD is gradual. The two major types of COPD are emphysema and chronic bronchitis. COPD ranks as the fourth leading cause of death in the United States.[2] Smoking is a primary risk factor for COPD, and most people with COPD are smokers. Other risk factors include exposure to environmental pollution (including exposure of nonsmokers to cigarette smoke) and, possibly, repeated respiratory tract infections.

Experts believe that COPD may be largely preventable by encouraging people not to smoke and controlling environmental pollution. Although COPD has no cure, people with COPD who smoke can help preserve lung function if they quit smoking before the disorder has progressed too far.

Consequences of COPD Regardless of the type of COPD, the lungs gradually lose their functional surface area and strength, making it difficult for them to deliver oxygen to the blood and to remove carbon dioxide from it. As lung function becomes increasingly compromised, pulmonary infections, respiratory failure, and heart failure can follow.

lungusa.org
American Lung Association

lung.ca
Canadian Lung Association

COPD and Nutrition Status People with COPD frequently experience weight loss, PEM, and infections and account for many cases of malnutrition in hospitals. The extent of malnutrition is associated with the severity of the pulmonary disease. Body weight is a predictor of survival in people with COPD, and low body weight is associated with reduced muscle mass, respiratory function, and immune competence.[3] Weight loss, which may be rapid and dramatic, may occur for many reasons, including the following:

- Anorexia and poor food intake. These may be caused by depression and anxiety; chronic mouth breathing, which can alter tastes for foods; difficulty breathing while preparing and eating food; and gastric discomfort from swallowing air while eating.[4]
- High energy expenditures associated with labored breathing.
- Medications, including anti-inflammatory agents, diuretics, and antibiotics, which alter nutrient requirements and compromise nutrition status (see Table E.1 in Appendix E).
- Use of oxygen masks, because people cannot eat while using them.
- Repeated infections, which raise nutrient needs and deplete nutrient stores. Poor nutrition status, in turn, opens the way for further infection, a vicious cycle.

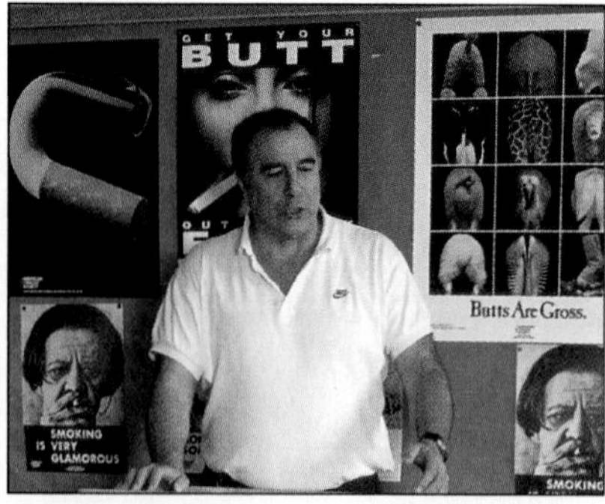

Health care professionals urge clients not to smoke to reduce the incidence of COPD or preserve lung function in those who already have the disorder.

Treatment of COPD The main treatment for COPD is to stop smoking. People with COPD are encouraged to receive vaccinations to prevent influenza (flu) and pneumonia, because infections significantly impair lung function. Likewise, clients are often encouraged to take antibiotics at the first sign of bacterial infection. Medications for COPD frequently include bronchodilators to limit respiratory muscle spasms and corticosteroids to reduce inflammation of lung tissue. People with severe COPD may need long-term oxygen therapy to help relieve symptoms.

Diet Therapy Like people with CHF, people newly diagnosed with COPD or those with mild cases who are also overweight benefit from safe weight-loss programs. Excessive body weight increases the work of the lungs in maintaining respiration. As the disease progresses, unintentional weight loss is more likely to occur. At this stage, maintaining or repleting nutrient stores helps to maintain lung function and prevent lung infections. Malnourished clients replete nutrient stores best when refed gradually, with a goal of providing a high-kcalorie, high-protein diet without stressing the lungs. Overfeeding people with COPD and PEM, however, can be as harmful as underfeeding them. Recall from Chapter 23 that overfeeding and excessive carbohydrate intakes produce high carbon dioxide levels, which the already stressed lungs must work hard to expel.

Meeting energy and protein needs from easy-to-eat foods provided in frequent small meals works best. When appetite is poor, clients may benefit from enteral formulas provided either orally or by tube. Pulmonary formulas that provide more kcalories from fat and fewer from carbohydrate than standard formulas are available, but there is little evidence to suggest that pulmonary formulas are superior to standard enteral formulas for managing COPD.[5]

CHF and COPD often lead to PEM, and PEM itself can cause further decline and impair immunity. The wasting associated with cancer and HIV infection shares these consequences, as the next sections describe.

℞ Prescription Pad

Medications used in the treatment of COPD may include:

- Antibiotics
- Anticholinergics
- Anti-inflammatory agents (corticosteroids)
- Bronchodilators
- Diuretics

See Appendix E for timing with meals and nutrition-related side effects.

Wasting in Cancer and HIV Infection

Although cancer and HIV infection are distinct disorders, from a nutrition standpoint, they share many similarities. Both disorders are associated with severe wasting characterized by anorexia, inadequate nutrient intake, altered metabolism, and excessive nutrient losses and are also associated with treatments that compound the nutrition problems. In both cases, unintentional weight loss and loss of lean body tissues result in complications that impair the quality of life and shorten life expectancy. In both cases, death may result from wasting, rather than from the disease process itself.[6] Both affect many organ systems, and in both, nutrition care is individualized based on the symptoms and organ systems involved. Likewise, in neither case can nutrition support reverse the course of the disease, but in both cases, attention to nutrition can improve the quality of life and possibly lengthen survival. Figure 24.1 illustrates the many factors that can lead to anorexia and wasting in people with cancer and HIV infection.

Cancer

Diet Order
High-kcalorie, high-protein diet.

Cancer is not a single disorder; instead, there are many different cancers. They have different characteristics, occur in different locations in the body, take different courses, and require different treatments. Whereas an isolated, nonspreading type of skin cancer may be removed in a physician's office with no observable effect on nutrition status, other cancers can seriously impair nutrition status.

How Cancer Develops Cancers develop from mutations in the genes that normally regulate cell division. The

cancers: diseases that arise from the unchecked growth of malignant tumors.

tumor: a new growth of tissue forming an abnormal mass with no function; also called a **neoplasm** (NEE-oh-plazm). Tumors that multiply out of control, threaten health, and require treatment are **malignant** (ma-LIG-nant). Tumors that stop growing without intervention or can be removed surgically and pose no threat to health are **benign** (bee-NINE).
malignus = of bad kind
benign = mild

A cancer that spreads from one part of the body to another is said to **metastasize** (me-TAS-tah-size).

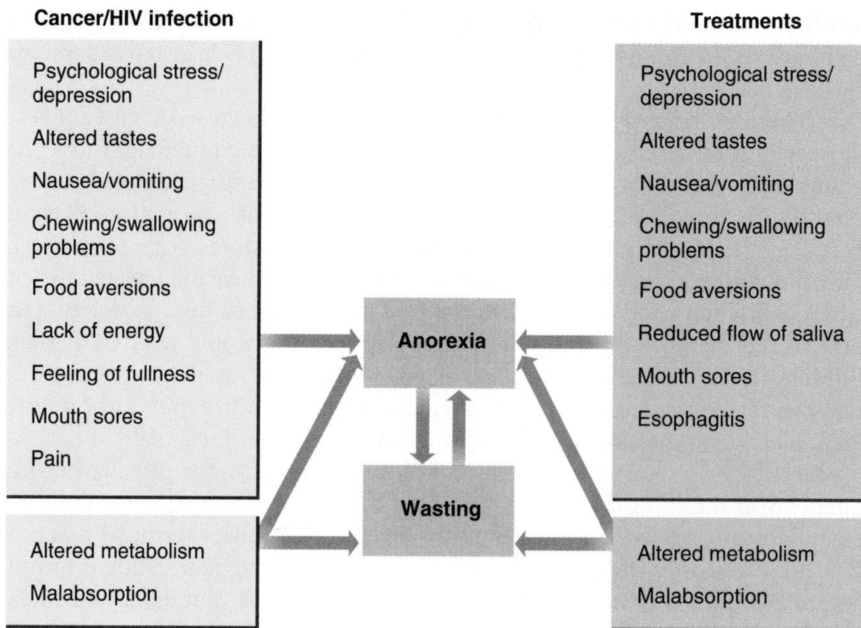

Figure 24.1

Causes of Anorexia and Wasting in Cancer and HIV Infection

Anorexia and wasting contribute to each other. Cancer and HIV infection and their available treatments make both problems worse.

radiation therapy: the use of radiation to arrest or destroy cancer cells.

chemotherapy: the use of drugs to arrest or destroy cancer cells. Drugs used for chemotherapy are called **chemotherapeutic** or **antineoplastic agents.**
chemo = chemical

bone marrow transplant: the replacement of diseased bone marrow in a recipient with healthy bone marrow from a donor; used as a treatment for breast cancer, leukemia, lymphomas, and certain blood disorders.

tissue rejection: destruction of healthy donor cells by the recipient's immune system, which recognizes the donor cells as foreign; also called **graft-versus-host disease (GVHD).**

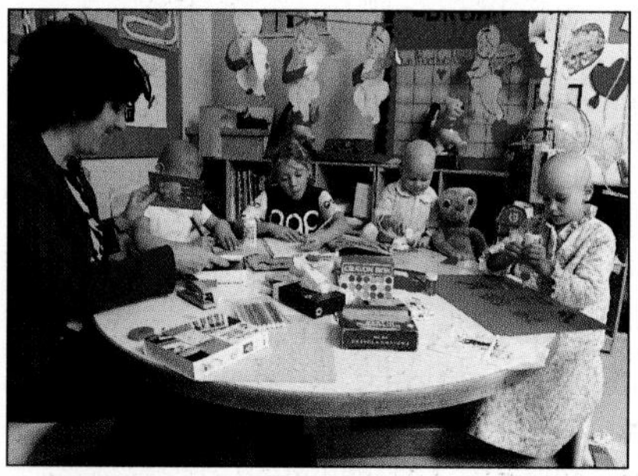

People with cancer take comfort from the support of others and from the knowledge that medical science is waging an unrelenting battle to find better treatments and possible cures.

affected cells seemingly have no brakes to halt cell division. As the abnormal mass of cells, called a malignant *tumor,* or *neoplasm,* grows, blood vessels form to supply the tumor with the nutrients it needs to support its growth. Eventually, the tumor invades healthy tissue and may spread.

Although the thought of cancer often strikes fear in people, the prognosis for most people with cancer today is far brighter than in the past. New techniques for detecting cancers early and innovative therapies to treat cancers offer hope and encouragement. In one of the latest avenues of study, drugs that prevent tumors from developing their own blood supplies were highly effective in halting tumor development in mice with no obvious side effects.[7] Although researchers are cautiously optimistic, they warn that the drugs may not have similar effects in human beings.

Treatments for Cancer The primary treatments for cancer aim to annihilate cancer cells, relieve pain, and prevent tumor growth. They include radiation therapy, chemotherapy, surgery, or any combination of the three. Table 24.1 summarizes the nutrition-related side effects of radiation therapy and chemotherapy, and Table 24.2 shows how various cancer surgeries can affect nutrition status. The use of bone marrow transplants to treat certain cancers has grown markedly.[8] To prepare a person for a bone marrow transplant, high doses of chemotherapy and sometimes whole-body radiation are used to eradicate cancer cells. Thus the nutrition-related side effects of these treatments apply to bone marrow transplants as well. In addition, immunosuppressants, given to help prevent tissue rejection following a bone marrow transplant, have multiple effects on nutrition status (see Appendix E).

Depending on the organ systems affected by cancer and by the side effects of treatments, many medications may be used in treatment. In addition to antineoplastic agents (chemotherapy), medications commonly used to treat symptoms of cancer include antinausea agents, antidiarrheals, analgesics, and sedatives. A later section describes medications that may be prescribed specifically to alleviate the wasting associated with cancer. People who feel they are making little progress in their fight against cancer or who think conventional medicine offers little hope of recovery may turn to alternative therapies, the subject of Nutrition in Practice 24.

Table 24.1 Possible Causes of Wasting Associated with Radiation and Chemotherapy			
	Reduced Nutrient Intake	**Accelerated Nutrient Losses**	**Altered Metabolism**
Radiation	Anorexia Damage to teeth and jaws Esophagitis Mouth ulcers Nausea Reduced salivary secretions Taste alterations Thick salivary secretions Vomiting	Chronic blood loss from intestine and bladder Diarrhea Fistula formation Intestinal obstructions Malabsorption Radiation enteritis Vomiting	Secondary effects of malnutrition or infection
Chemotherapy	Abdominal pain Anorexia Mouth ulcers Nausea Taste alterations Vomiting	Diarrhea Intestinal ulcers Malabsorption Vomiting	Fluid and electrolyte imbalances Hyperglycemia Interference with vitamins or other metabolites Negative nitrogen and calcium balance Secondary effects of malnutrition or infection

Wasting Associated with Cancer The likelihood of wasting varies with the type of cancer and its severity. Wasting occurs in as many as 80 percent of people with cancer before they die.[9] Malnutrition may be the cause of death in as many as 22 percent of deaths associated with cancer.[10]

Table 24.2 Possible Effects of Surgery for Cancer on Nutrition Status	
Head and Neck Resection	
Difficulty in chewing/swallowing	Inability to chew/swallow
Esophageal Resection	
Diarrhea Fistula formation Reduced gastric acid secretion	Reduced gastric motility Steatorrhea (fat malabsorption) Stenosis (constriction)
Gastric Resection	
Dumping syndrome General malabsorption Hypoglycemia	Lack of gastric acid Vitamin B_{12} malabsorption
Intestinal Resection	
Blind loop syndrome Diarrhea Fluid and electrolyte imbalance	General malabsorption Hyperoxaluria Steatorrhea
Pancreatic Resection	
Diabetes mellitus	General malabsorption

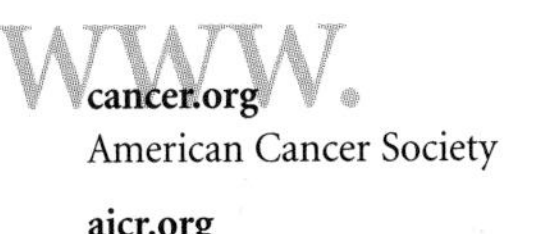

cancer.org
American Cancer Society

aicr.org
American Institute for Cancer Research

Wasting in cancer appears to be tumor derived; that is, the tumor itself causes the changes that lead to wasting, and when the tumor can be successfully removed, wasting can be reversed. Weight loss is often evident at the time cancer is diagnosed. Anorexia, weight loss, and depletion of lean body mass and serum proteins typify the wasting associated with cancer, often called the *cancer cachexia syndrome.*

R_x Prescription Pad

Medications used in the treatment of cancer may include:

- Analgesics
- Antidiarrheals
- Anti-inflammatory agents
- Antinauseants
- Antineoplastics
- Appetite stimulants
- Sedatives

See Appendix E for timing with meals and nutrition-related side effects.

cancer cachexia (ka-KEKS-ee-ah) **syndrome:** a syndrome that frequently accompanies many types of cancer; characterized by anorexia, inadequate intake of food, malnutrition, accelerated metabolism and wasting, and general ill health.

Anorexia and Reduced Food Intake Anorexia is a major contributor to wasting associated with cancer. Cytokines appear to play a role in the anorexia associated with cancer. Some other factors that can contribute to anorexia or otherwise reduce food intake in the person with cancer include:

- *Chronic nausea and early satiety* People with cancer frequently experience nausea and a premature feeling of fullness after eating small amounts of food.
- *Fatigue* People with cancer often tire easily and lack energy to prepare and eat meals. Once wasting is evident, these tasks become even more difficult for the person to handle.
- *Pain* People in pain may have little interest in eating, particularly if eating makes the pain worse.
- *Psychological stress* The very diagnosis of cancer can cause so much distress that eating becomes unimportant. Stress can be compounded by the person's anxiety about the medical, personal, and financial concerns created by the diagnosis. Once wasting is evident, people may become depressed by their inability to perform routine tasks and by their physical appearance.
- *Obstructions* A tumor may partially or completely obstruct any portion of the GI tract and interfere with chewing and swallowing; cause delayed gastric emptying, early satiety, nausea, or vomiting; or make oral diets impossible.
- *Cancer therapy* Therapy for cancer including medications, chemotherapy, radiation therapy, surgery, and bone marrow transplants can dramatically affect food intake by causing nausea, vomiting, altered taste perceptions, diminished taste sensitivity (mouth blindness), inflammation of the mouth (stomatitis) and esophagus (esophagitis), mouth ulcers, reduced flow of saliva, food aversions (strong dislike for certain foods), and depression.

Metabolic Alterations and Nutrient Losses Metabolic pathways in people with cancer are also altered so that they use nutrients in inefficient ways that demand more energy and waste vital protein tissues. This explains why some people with cancer fail to regain lean body mass even when they are receiving adequate energy and nutrients. Some people with cancer are hypermetabolic and have high nutrient needs; others may become hypermetabolic as a consequence of surgery or infection. Chemotherapy can interfere with normal metabolic pathways and create nutrient deficiencies as well.

Like anorexia, nutrient losses can develop as a consequence of cancer itself or the treatment for it. Causes of nutrient losses include inadequate digestion, malabsorption, vomiting, and radiation enteritis. Cancers of the pancreas, liver, or small intestine frequently lead to malabsorption. Radiation therapy to the small intestine can lead to radiation enteritis, which can lead to malabsorption, chronic blood loss, fluid and electrolyte imbalances, and, sometimes, intestinal obstructions and fistulas. Intestinal function may return after radiation therapy ends, but for some the changes are permanent. Severe diarrhea and malabsorption, with fluid losses often exceeding 10 liters a day, can occur in people who undergo bone marrow transplants and then reject the transplanted tissue.

radiation enteritis: inflammation and scarring of the intestinal cells caused by exposure to radiation.

HIV Infection

For many years, the devastating effects of infection by the human immunodeficiency virus (HIV), the infection that eventually causes acquired immune deficiency syndrome (AIDS), seemed unstoppable. Although the disease still has no cure, remarkable progress has been made in extending the life expectancy of people with HIV infections. Researchers remain optimistic that a vaccine to prevent HIV infection or treatments to arrest the disease may someday be possible.

human immunodeficiency virus (HIV): the virus that causes AIDS. The infection progresses until it seriously hampers the function of the immune system and leaves its victims defenseless against numerous infections and cancer.

acquired immune deficiency syndrome (AIDS): the end stage of HIV infection, in which severe complications are manifested.

Once a person has been infected with HIV, it may take several weeks before laboratory tests can confirm a diagnosis. Because most people remain symptom-free or have only mild, vague symptoms for years after the initial infection, people may have the virus for many years before it is detected. During this time, the person may spread the infection to others. Thus early detection to prevent the spread of HIV infection and to ensure early treatment for the infected person are critical concerns of health care professionals around the world.

WWW.

thebody.com
The Body: A Multimedia AIDS and HIV Information Resource

aegis.com
AIDS Education Global Information System

Course of HIV Infection HIV infection attacks the immune system and leaves its victims defenseless against opportunistic infections—infections from which most people are protected. HIV infections progress in stages. The virus gradually destroys cells with a specific protein called CD4+ on their surfaces. Among the cells most affected are the CD4+ lymphocytes, essential components of the immune system. At first, the number of CD4+ lymphocytes declines gradually, and the HIV-infected individual remains symptom-free. As the infection progresses, significant depletion of CD4+ lymphocytes greatly impairs immune function. Early symptoms may include fatigue, skin rashes, fevers, diarrhea, joint pain, night sweats, weight loss, oral lesions and infections, and other opportunistic infections that are not life-threatening. In the final stages, the person suffers frequent and eventually fatal complications, such as severe weight loss; tuberculosis; recurrent bacterial pneumonia; serious infections of the central nervous system, GI tract, and skin; hepatitis; cancers; and severe diarrhea. On average, it takes about ten years for an HIV infection to progress to AIDS (the final stage of infection). Clinicians monitor the progress of HIV infection by measuring concentrations of CD4+ lymphocytes and circulating virus (viral load) as well as by monitoring clinical symptoms.

opportunistic infections: infections from microorganisms that normally do not cause disease in the general population but can cause great harm in people once their immune systems are compromised.

CD4+ lymphocyte: a type of circulating white blood cell that has the CD4+ protein on its surface and is a necessary component of the immune system.

The cluster of symptoms that sometimes occur before AIDS develops is called **AIDS-related complex (ARC).**

Treatments for HIV Infection Treatments for HIV infection focus on slowing the course of the infection, controlling symptoms, and alleviating pain. Drug therapy often includes a combination of AZT (zidovudine) and other medications called protease inhibitors that prevent the virus from replicating. The use of medications to help alleviate wasting associated with HIV infection is described later. Like people with cancer, people with HIV infections may turn to alternative therapies to deal with their disorders (see Nutrition in Practice 24).

The antiviral drugs used in the treatment of HIV infection include zidovudine (AZT), didanosine (ddl), zalcitabine (ddC), stavudine (d4T), lamivudine (3TC), nevirapine, indinavir, ritonavir, and saquinavir.

Wasting and HIV Infection Weight loss in people with HIV infection often begins early in the progression of the disease, and severe wasting often occurs in the later stages. Among the clinical conditions that define AIDS, AIDS-related wasting is the second most frequently reported condition.[11] Slow, progressive weight loss is often associated with reduced food intake and GI complications, whereas rapid weight loss is most often associated with infections.[12] The strongest predictors of weight loss and depletion of lean body mass and fat stores in people with HIV infection include anorexia, diarrhea, and other infections.[13]

The countless lives touched by AIDS serve as a potent reminder of the need to continue to search for a cure.

Causes of Wasting in HIV Infection Much as in cancer, the wasting associated with HIV infection is multifactorial: anorexia and inadequate food intake, altered metabolism, excessive nutrient losses, and nutrient-medication interactions. Likewise, the exact factors that affect nutrition status depend on the complications that arise in each case.

Anorexia and Reduced Nutrient Intake People with HIV infections frequently suffer from anorexia and inadequate nutrient intakes, particularly as the disease progresses. Recall that slow, progressive weight loss in people with HIV infection is often associated with reduced food intake and gastrointestinal complications. HIV-related causes of anorexia and reduced food intake include:

thrush: a fungal infection of the mouth and esophagus caused by *Candida albicans;* the technical term for this infection is **candidiasis.** Thrush is characterized by a thick white coating of the tongue that alters taste sensations and causes pain on chewing and swallowing.

herpes virus: a virus that can lead to mouth lesions and may also affect the lower GI tract, causing diarrhea.

Kaposi's (cap-OH-seez) **sarcoma:** a type of cancer rare in the general population but common in people with HIV infection.

- *Psychological stress and pain* As in cancer, depression and anxiety over the HIV diagnosis, the prognosis, and the medical, personal, and financial problems that lie ahead, as well as the pain associated with complications of the disorder, can destroy the appetite.
- *Oral infections* Infections and fever cause anorexia. In addition, oral infections associated with HIV infection cause further problems. Thrush, a common oral infection associated with HIV infection, can alter taste sensitivity, reduce the flow of saliva, and cause pain on swallowing. Oral infections caused by herpes virus can cause painful mouth ulcers that interfere with chewing and swallowing.
- *Respiratory infections* Pneumonia and tuberculosis, frequent complications associated with HIV infection, cause fever and pain that contribute to anorexia. The person who must use an oxygen mask to improve breathing may find it difficult to eat.
- *GI tract complications and altered organ function* In addition to oral infections, people with HIV infection may experience belching, reflux esophagitis, and heartburn that may cause nausea and interfere with eating. Intestinal complications and altered organ function contribute to anorexia, early satiety, and food aversions.[14]
- *Lethargy and dementia* In the later stages of HIV infection, lethargy and dementia frequently occur and may interfere with food intake. The individual may be chronically exhausted and may not care or even remember to eat.
- *Cancer* As previously described, cancer often leads to anorexia. Kaposi's sarcoma, a cancer associated with HIV infection, can cause lesions and obstructions in the esophagus that make eating painful.
- *Medical treatments* Medications used to treat HIV infection, associated infections, and cancer often cause anorexia, taste alterations, nausea, and vomiting and reduce food intake. Food aversions may also arise.

The diarrhea and malabsorption associated with HIV infection is called **HIV enteropathy** (EN-ter-OP-a-thee).

Metabolic Alterations and Nutrient Losses The metabolic alterations associated with repeated infections and cancer contribute to wasting in people with HIV infections. In addition to anorexia and altered metabolism, significant nutrient losses can occur as a consequence of diarrhea and malabsorption, which often occur late in the course of HIV infection. From 50 to 90 percent of people with AIDS experience significant diarrhea and malabsorption. Both the structure and the function of the intestinal cells are altered as a consequence of HIV infection and other GI tract infections (see Table 24.3). The prolonged use of antibiotics to treat infections and antisecretory agents and antacids to relieve nausea can lead to bacterial overgrowth in the upper small intestine, further contributing to malabsorption. Food and enteral formulas may serve as a source of infectious agents, and people with advanced HIV infection are highly susceptible to food-borne illness. In most cases, diarrhea is recurrent, and typical losses average less than one liter of diarrheal fluids daily. Diarrhea caused by parasites, however,

Table 24.3 Causes of GI Infections in AIDS	
Bacterial	Protozoan
Clostridium dificile	*Cryptosporidium* species
Mycobacterium avium-intracellulare	*Giardia lamblia*
Mycobacterium tuberculosis	*Isosporia belli*
Salmonella species	*Microsporidium* species
Fungal	*Pneumocystitis carinii*
Candida albicans	*Toxoplasma gondii*
Cryptococcus neoformans	Viral
Parasitic	AIDS enteropathy
Nonpathogenic amoeba	*Cytomegalovirus* species
Entamoeba histolytica	Epstein-Barr
	Herpes simplex

may be severe and unresponsive to medications—the person may lose 10 or more liters of diarrheal fluids daily.

Treatments common among people with HIV infections, especially anti-infective agents, chemotherapy, and radiation therapy, can accelerate nutrient losses due to vomiting, diarrhea, and malabsorption. Megadoses of vitamin C and other home remedies that some people with HIV infections use may also cause diarrhea. Once malnutrition is under way, it, too, contributes to malabsorption.

Nutrition Support for People with Cancer or HIV Infections

In an era of improved treatments for people with cancer and HIV infections, measures that improve the quality of life assume great importance. Attention to nutrition can bolster the immune system and offer an improved quality of life and a reduced risk of complications. Although nutrition cannot change the ultimate outcome of cancer or HIV infection, compared to the malnourished person with these diseases, the person in good nutrition status:

- Feels better.
- Functions better.
- Is more active.
- Is stronger.
- Eats more.
- Resists infections better.
- Enjoys a better quality of life.

At a minimum, meeting nutrient needs eliminates the additional stresses imposed by malnutrition.

Nutrition Concerns Applicable to Both Cancer and HIV Infection

Although cancer and HIV infection are distinct disorders, they involve similar nutrition care because both are characterized by anorexia and are associated with multiple problems that reduce the intake of food. Considering the many adverse

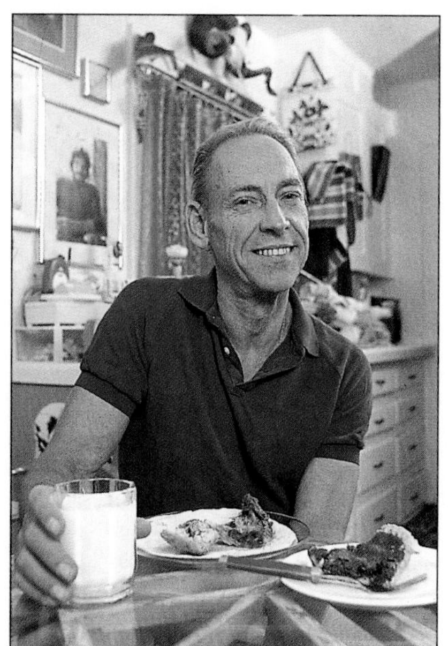

Nutrition provides an edge in maintaining quality of life and encouraging independence.

effects that threaten the nutrition status of people with cancer or HIV infections, health care professionals face an enormous challenge in helping these individuals meet their nutrient needs.

Early Nutrition Intervention For people with cancers associated with wasting or those with a positive test for HIV infection, early nutrition intervention takes a high priority. The initial nutrition assessment evaluates the individual's current nutrition status and establishes baseline parameters from which to monitor changes. Early intervention may detect and correct deficiencies before they become severe and helps to prepare the person for the stresses ahead. Clinicians can begin to encourage gradual improvements in eating habits before the person becomes debilitated and the task becomes monumental. For people with HIV infections, nutrition therapy may be most effective in the early stages when reduced food intake is more likely to be the cause of malnutrition than in the later stages when repeated infections and hypermetabolism quickly deplete available nutrients.[15]

Oral Diets The appropriate diet for a person with cancer or HIV infection, as for anyone, is an oral diet for as long as possible. A thorough nutrition assessment uncovers specific problems that each person is experiencing, and the nutrition care plan addresses these problems. The box on pp. 581–582 outlines some strategies for improving oral intake. The suggestions are numerous and detailed, reflecting both the complexity of the problems and the importance of offering specific suggestions to deal with specific problems.

Formula Supplements Clients with cancer or HIV infection who are unable to eat adequate amounts of table foods can often benefit from nutrient-dense enteral formulas. Provided early in the course of HIV infection, nutrition counseling, a high-kcalorie, high-protein oral diet, and enteral formula supplements have been successful in halting weight loss and restoring weight in people without secondary infections.[16] Limited research suggests that immune-enhancing formulas containing omega-3 fatty acids, arginine, and nucleotides may improve nutrition status early in the course of HIV infection.[17] Other researchers report that supplementing the diet with hydrolyzed protein and fish oil (rich in omega-3 fatty acids) may help prevent weight loss and reduce the frequency of hospitalizations in the early stages of HIV infection.[18]

Medications to Combat Wasting Several medications may be useful in the treatment of anorexia and wasting.[19] The most promising medication, megestrol acetate, stimulates the appetite and promotes weight gain (primarily as body fat). Preliminary studies suggest that growth hormone and insulin-like growth factor can promote weight gain, particularly a gain in lean body mass. Dronabinol (a medication containing the principal psychoactive ingredient in marijuana) works as both an appetite stimulant and an antiemetic and may be useful in some cases.

Ethica Issues Every malnourished person with cancer or HIV infection who cannot eat an adequate diet orally is a potential candidate for a tube feeding or TPN. The next sections describe uses of tube feedings or TPN as they are applied to people with cancer and HIV infections who have a chance of recovery (for cancer) or a reasonable life expectancy. When incurable cancer or HIV infection has reached its final stages, however, some important ethical issues should be considered by the person, caregivers, and the health care team before a tube feeding or parenteral nutrition support is undertaken (see Nutrition in Practice 22). What is the prognosis if the wasting can be corrected or reversed? Will the person survive longer? Will quality of life improve? Will treatments have a better chance of success? Does the client understand that nutrition support will not

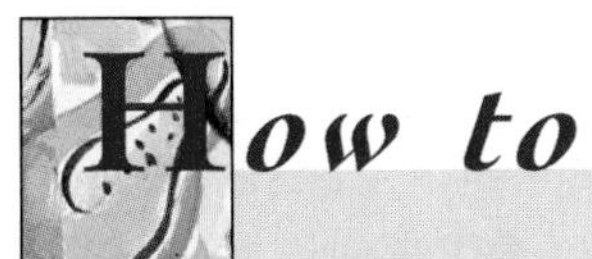

How to

Help Clients Handle Food-Related Problems

For each problem, find a solution using these suggestions.

1. To improve nutrient intake:
   - Explain why eating is important.
   - Encourage clients to eat the most when they feel the best.
   - Encourage people to eat extra food between chemotherapy or radiation treatments.
   - Recommend that clients use nutrient-dense foods (see item 7) for meals and snacks.
   - Suggest that clients eat nutrient-dense foods first.
   - Recommend indulging in favorite foods throughout the day.
   - Encourage clients to eat with family and friends.
   - Recommend smaller, more frequent meals.
   - Advise clients to avoid drinking large amounts of liquids with meals.
   - Work out a medication schedule that allows the client to take pain or antinausea medications at times when they will be effective during meals.
   - Advise clients to eat in a pleasant and relaxed environment.
   - Encourage clients or caregivers to serve foods attractively.
   - Reassess clients regularly to identify and solve problems as they arise.
2. To save energy:
   - Recommend that others prepare foods.
   - Suggest foods that are easy to prepare and eat.
   - Recommend that clients eat nutrient-dense foods first.
   - Encourage the use of time-saving appliances for food preparation.
3. To combat bitter or metallic taste perceptions:
   - Advise clients to brush their teeth or use a mouthwash before eating.
   - Recommend adding sauces and seasonings to meats.
   - Suggest that meats be served cold or at room temperature.
   - Encourage clients to try using eggs, fish, poultry, and dairy products instead of meats.
   - Encourage clients to try new foods and experiment with herbs and spices.
   - Recommend plastic, rather than metal, feeding utensils.
4. To control nausea and vomiting:
   - Give antinausea drugs at times when they will be effective during meals.
   - Recommend small, frequent meals.
   - Advise clients to avoid spicy and high-fat foods.
   - Suggest that clients avoid food odors that cause nausea. It may help to have others prepare meals, if possible.
   - Encourage clients to save most liquids for after meals. Clear liquids or popsicles after meals help prevent dehydration.
   - Suggest that clients keep the head of the bed elevated after eating, rest after meals, and loosen tight clothing. Some clients may benefit from getting fresh air after eating.
5. To prevent food aversions:
   - Suggest that clients save favorite foods for times when they are feeling relatively good.
   - Advise clients to avoid their favorite foods during the times of day when they usually experience nausea or vomiting.
   - Suggest that clients maintain a food-free "window" of an hour or so before and after treatment times, if the treatments cause nausea or vomiting.
6. To alleviate problems with chewing and swallowing:
   - Work with clients to find the consistency of food that is easiest to handle. Thin liquids, true solids, and sticky foods are often difficult to swallow.
   - Recommend that clients add sauces and gravies to dry foods.
   - Provide fluids with meals to ease chewing and swallowing.
   - Advise clients with mouth sores to try eating cold or frozen foods They are often soothing.
   - Recommend that clients with mouth sores avoid foods that are spicy, acidic, or coarse; foods that contain seeds that can be trapped in an ulcer; or sticky foods such as peanut butter that may be difficult to swallow.
   - Recommend that clients with mouth sores try local anesthetic solutions before eating to reduce pain.
   - Recommend that clients experiment with tilting the head forward and backward to see if swallowing is easier with the head positioned differently.
   - Suggest trying a straw for drinking liquids.
   - Encourage clients who suffer from a reduced flow of saliva to rinse the mouth frequently. Artificial saliva from the pharmacy can also help. Sour candy or gum can stimulate the flow of saliva.
   - Encourage good oral and dental hygiene to prevent cavities and oral infections.
7. To add kcalories and protein:
   - Add milk powder to liquid milk, meat loaves, casseroles, soups, puddings, and cereals.
   - Add ground meats, chicken, fish, or grated cheeses to sauces, soups, casseroles, or vegetables.

Help Clients Handle Food-Related Problems *(continued)*

- Eat peanut butter on fruit, celery, or crackers, if chewing or swallowing is not a problem.
- Use plenty of butter, margarine, mayonnaise, cream cheese, oil, and salad dressings on breads, sandwiches, potatoes, vegetables, salads, pasta, and rice.
- Use yogurt, sour cream, or a sour cream dip with vegetables.
- Add whipping cream to desserts and hot chocolate, or use it to lighten coffee.
- Have nutrient-dense snacks available at all times.
- Add nuts and dried fruits such as raisins to desserts, cereals, or salads.
- Use cream instead of milk with cereal.
- Try commercially available liquid supplements or instant breakfast mixes for milk shakes, meals, or between-meal snacks.
- Use whole milk and regular yogurt instead of low-fat or nonfat milks and yogurts.

change the ultimate outcome of the disease? Does the client understand the costs of nutrition support? Does the client want aggressive nutrition support? Special nutrition support should be undertaken only if the client clearly understands the benefits and risks and if the client chooses it.

Nutrition Concerns Specific to Cancer

The need to address specific problems that limit food intake is similar for both people with cancer and those with HIV infections. For people with cancer, additional dietary adjustments may apply based on the organ systems involved and the type and severity of the cancer (see Table 24.4 on p. 583). Some other nutrition problems and considerations that apply specifically to people with cancer are described next.

Tube Feedings and TPN In general, a tube feeding or TPN is not routinely recommended for an adequately nourished or mildly malnourished person with cancer who is unable to eat.[20] Special nutrition support is indicated when anorexia persists or when a person is severely malnourished and is about to undergo aggressive cancer therapy. Each case is decided individually, and the use of special nutrition support is more likely when the person's chances of recovery or of significant response to treatment are good, or when the type of cancer is associated with a high risk of death from malnutrition.[21] People requiring head and neck resections, for example, may need long-term tube feedings and may need to continue tube feedings at home. People with severe radiation enteritis may require home TPN.

Nutrition Support and Bone Marrow Transplants The person undergoing a bone marrow transplant routinely receives TPN before and after the transplant, because the GI tract is severely compromised by the preparatory procedure. Researchers have found that adding glutamine to the TPN solution results in fewer infections and shorter hospital stays for people undergoing bone marrow transplants.[22]

When GI tract function returns, the person begins to receive foods orally, while TPN is gradually tapered off (see Chapter 22). After a bone marrow transplant, nutrition complications can be severe and debilitating, especially for people who reject the transplant and develop GI complications. Early oral feedings often start with lactose-free, low-residue, low-fat liquids to maximize absorption and minimize nausea, vomiting, and fat malabsorption. Gradually, solid foods

Table 24.4 Dietary Considerations for Various Cancers	
Cancer Sites	**Dietary Considerations**
Brain	Physical feeding disabilities (see Nutrition in Practice 18); chewing and swallowing problems (see Chapter 18).
Head/neck	Chewing and swallowing problems.
Mouth/esophagus	Chewing and swallowing problems; if obstructed, tube feeding below the obstruction may be necessary.
Stomach	Nausea, vomiting; if obstructed, tube feeding below the obstruction or TPN may be necessary; if resection is performed, a postgastrectomy diet (see Chapter 20) may be needed; nutrient deficiencies due to bacterial overgrowth (Chapter 20) may occur.
Intestine	If obstructed, tube feeding or TPN may be necessary; resections or inflammation may cause multiple nutrition problems (see Chapter 20); fat- and lactose-restricted diet may be useful.
Liver	Protein-, sodium-, and fluid-restricted diet may be necessary (see Chapter 28).
Pancreas	Fat-restricted diet and enzyme replacements may be necessary (see Chapter 20); diabetic diet may be necessary if insulin production is affected (see Chapter 25).
Kidneys	Protein-, electrolyte-, and fluid-controlled diet may be necessary (see Chapter 27).

Note: The considerations listed here are specific to the type of cancer; they do not include other nutrition-related concerns, such as anorexia, nausea, and vomiting.

are introduced. For about three months after the transplant, the diet excludes most fresh fruits and vegetables, undercooked meats, poultry and eggs, and ground meats to minimize the risk of food-borne infections. Fiber, lactose, and fat are gradually added to the diet as individual tolerances allow. Because the transplant recipient also receives immunosuppressants, which often incur negative nitrogen and calcium balances, the final goal is to provide a high-kcalorie, high-protein, high-calcium diet. Clients are advised to follow safe food handling and preparation precautions (see pp. 273–278). In addition, physicians often prescribe calcium and vitamin D supplements. Individuals with persistent diarrhea are encouraged to eat high-potassium foods (see p. 178).

Nutrition Concerns Specific to HIV Infection

Severe diarrhea and malabsorption commonly occur as a consequence of HIV infection. This section describes the treatment of severe diarrhea and the use of tube feedings and TPN in people with HIV infections.

Treatment of HIV-Related Diarrhea and Malabsorption To prevent diarrhea from food-borne microorganisms, clients are provided with instructions for the safe handling and preparation of foods (see pp. 273–278). Treatment of HIV-related diarrhea depends on its cause and the extent to which the intestine is affected. Although diarrhea is sometimes unresponsive to therapy, often a pathogen can be identified. Appropriate medications along with the provision of adequate fluids and electrolytes are at the core of treatment. Drinking plenty of fluids is essential. The liberal use of table salt, salty broths, and high-potassium foods and juices can help replace electrolytes. Other dietary modifications may include lactose and fat restrictions.

For people with severe diarrhea, commercially prepared oral rehydration formulas (see Chapter 19) may be useful. TPN is often used when GI tract function is seriously impaired. Limited research suggests that for people with severe HIV-related malabsorption, orally administered hydrolyzed formulas containing

Case Study

Travel Agent with HIV Infection

Mr. Sands, a travel agent, sought medical help at age 34 when he began feeling run-down and developed a painful white coating over his mouth and tongue. The presence of thrush and anemia alerted Mr. Sands's physician to the possibility of an HIV infection. When Mr. Sands tested positive for an HIV infection, he and his family and friends were devastated by the news. Fortunately, those closest to him have been supportive during this difficult time, and he has a strong desire to live out his life as independently as possible. Four months after the diagnosis of HIV infection, Mr. Sands developed a serious, continuous diarrhea that required hospitalization to classify and control. Since the diagnosis of HIV infection was made, Mr. Sands has lost 10 pounds. At 6 feet tall, he currently weighs 158 pounds.

Describe how HIV infection can lead to reduced food intake, nutrient losses, and hypermetabolism.

From the limited information given here, what factors could have contributed to Mr. Sands's weight loss? Is his weight loss significant? What is Mr. Sands's percent ideal body weight (%IBW)? What steps could prevent further weight loss?

Discuss nutrition strategies for dealing with thrush and diarrhea.

What additional nutrition consequences might be anticipated if Mr. Sands develops cancer?

medium-chain triglycerides and supplemented with glutamine are as effective as TPN in reversing weight loss and wasting.[23] Although neither type of nutrition support alters intestinal function, the use of oral formulas is significantly less expensive and far less difficult for the person to manage than TPN.

Tube Feedings and TPN People with HIV infections who are unable to consume oral diets that are sufficient to prevent nutrition complications and unintentional weight loss need aggressive nutrition support. Tube feedings are preferred whenever the GI tract is functional. Tube feedings can be given at night to supplement oral diets during the day. Preventing bacterial contamination of the formula is particularly important because of the susceptibility of HIV-infected individuals to GI infections.

TPN is generally reserved for people with HIV infections who are unable to tolerate enteral nutrition, but need to maintain their nutrition status while undergoing a therapy that is expected to improve their condition. People with GI tract obstructions, severe vomiting, or GI infections affecting the entire small intestine may benefit from TPN. As with enteral formulas, the concern for infection in people with HIV infection is magnified. The accompanying case study discusses nutrition concerns of a person with HIV infection.

CHF, COPD, cancer, and HIV infection are all associated with wasting and severe malnutrition. Health care providers who work with people with these disorders serve their clients best by identifying nutrition problems early (see the accompanying nutrition assessment checklist) and offering solutions before nutrition status seriously deteriorates.

Nutrition Assessment Checklist

FOR PEOPLE WITH CHF, COPD, CANCER, AND HIV INFECTION

Medical Check the client's medical history for the type of COPD or cancer or stage of HIV infection, associated complications, medical therapy, and symptoms.

Medication Review the client's medications for possible nutrient-medication interactions and nutrient-related complications. Anti-inflammatory agents, antineoplastic agents, antiviral agents, and antibiotics can significantly and adversely affect nutrition status. People with CHF who are taking loop diuretics and cardiac glycosides risk potassium deficiencies. Medications used to treat wasting can promote weight gain for people with cancer or HIV infection. Ask clients if they are using alternative therapies, including megadoses of vitamins and herbal preparations.

Food Intake Determine if anorexia or nutrition-related complications are interfering with the client's ability to eat. Aggressively work with clients to improve nutrient intake and preserve nutrition status before severe wasting occurs. For clients with pain or nausea, check the timing of the administration of analgesics and antinausea agents to be sure they are given at times when they will improve appetite.

Anthropometric Take baseline height and weight measurements, and record weight at regular intervals to detect early signs of wasting. Recall that weight measurements may be a misleading indicator of nutrition status in people with CHF or COPD who are retaining fluids.

Laboratory Assess laboratory data for changes in nutrition status, fluid and electrolyte balance, organ function, and response to therapy. Low serum protein levels are common in later stages of CHF, COPD, cancer, and HIV infection.

Physical Check for physical signs of nutrient deficiencies, energy level, fluid status (especially for those with fever, diarrhea, or vomiting, or for people with CHF or COPD), and mouth ulcers. Muscle weakness, numbness and tingling sensations, and irregular heartbeats can be signs of potassium deficiencies in people with CHF.

Self Check

1. With respect to nutrient needs, chronic stresses **differ** from acute stresses because chronic stresses:
   a. are less serious than acute stresses.
   b. continue to tax nutrient stores over long periods of time.
   c. alter nutrient intake.
   d. alter the metabolism and excretion of nutrients.
2. Energy and protein needs for people with wasting syndromes are often about _____ of the basal energy expenditure (BEE) and _____ grams of protein per kilogram of body weight per day.
   a. 100 percent; 1.5 to 2.0 b. 120 percent; 1.0 to 1.2
   c. 150 percent; 1.5 to 2.0 d. 200 percent; 1.0 to 1.2
3. All of the following factors contribute to the development of wasting in CHF **except:**
   a. anorexia. b. severe nausea and vomiting.
   c. medications. d. respiratory infections.
4. Diet therapy for the person with CHF may include:
   a. a high-kcalorie, high-protein, low-sodium diet.
   b. fiber restrictions.
   c. fat restrictions.
   d. potassium restrictions.
5. Wasting in people with COPD is associated with all of the following **except:**
   a. anorexia and reduced food intake.
   b. medications used in treatment.
   c. use of oxygen masks.
   d. diarrhea and malabsorption.
6. Principles of feeding the person with COPD include:
   a. providing a diet high enough in energy to promote weight gain or weight maintenance without overtaxing the lungs.
   b. providing a low-fat diet.
   c. providing a potassium-restricted diet.
   d. providing a low-carbohydrate diet.
7. All of the following statements describe the wasting associated with cancer **except:**
   a. altered metabolism plays a role in its development.
   b. anorexia is a major factor in its development.

c. cytokines may play a role in its development.
d. it accompanies all forms of cancer.

8. Practical advice for a person with cancer or HIV infection who has trouble preparing and eating foods due to fatigue might include:
   a. prepare fresh vegetables every day.
   b. make homemade ice cream for snacks.
   c. keep premixed breakfast drinks or enteral formulas in the refrigerator for snacks.
   d. prepare dinner for friends.
9. Mouth sores in people with HIV infections are most frequently due to:
   a. malabsorption.
   b. food-borne illnesses.
   c. oral infections.
   d. dehydration.
10. Diets for people who begin oral diets immediately after bone marrow transplants:
   a. limit calcium.
   b. avoid fresh fruits and vegetables.
   c. test tolerance for lactose.
   d. avoid high-protein foods.
11. A tube feeding or TPN is most likely to benefit people with cancer or HIV infection if:
   a. they cannot tolerate further medical treatments.
   b. malnutrition may have an undesirable effect on their ability to receive additional treatments.
   c. they do not wish to prolong their lives.
   d. they feel they have no other alternatives.
12. All of the following statements are true with respect to the malabsorption that frequently occurs in people with AIDS **except:**
   a. HIV and secondary infections play a role in its development.
   b. use of antisecretory agents can contribute to bacterial overgrowth.
   c. use of antibiotics can contribute to bacterial overgrowth.
   d. all forms can be corrected with appropriate nutrition therapy.
13. People with wasting disorders who are most prone to infections arising from foods, enteral formulas, and TPN include:
   a. people with COPD.
   b. people who have undergone radiation therapy.
   c. people with CHF.
   d. people with bone marrow transplants and those with HIV infections.

Answers to these questions appear in Appendix H.

Clinical Applications

1. Many disorders can lead to wasting. For some of these disorders, such as fat malabsorption (Chapter 20), diet is a cornerstone of treatment. For others, such as CHF, COPD, cancer, and HIV infection, nutrition plays a supportive role. What determines whether nutrition plays a major or supportive role in the treatment of a disorder? Review the effects of PEM on pp. 85–88 and pp. 410–411. Carefully consider how severe malnutrition can further debilitate people with CHF, COPD, cancer, or HIV infection.
2. The suggestions for handling food-related problems in the box on pp. 581–582 appear simple enough, but many of the suggestions may be difficult to implement in some cases. What suggestions to control nausea and vomiting contradict suggestions to add kcalories and protein? What other contradictions can you find? How might a health care provider deal with such contradictions?
3. Consider problems associated with nutrition in a 36-year-old woman with a malignant brain tumor affecting her ability to move the right side of her body (including the tongue) and to speak coherently. She has an expected length of survival of six months and is taking a pain medication that makes her nauseated and sleepy. What would be a realistic goal of nutrition support? If she is right-handed, how can her impairment interfere with eating? What suggestions might you have for overcoming this problem? How might nutrition be affected by her problems with communication? Describe ways that the medications she is taking can affect her nutrition status. Would tube feedings or TPN be appropriate for this woman? Why or why not?

Notes

1. B. R. Bistrian, Cytokines in clinical nutrition, presented at the *Eighth Annual Advances and Controversies in Clinical Nutrition,* Mayo Clinic Foundation, Dallas, Texas, April 5–7, 1998; T. Cederholm and coauthors, Enhanced generation of interleukin 1B and 6 may contribute to the cachexia of chronic disease, *American Journal of Clinical Nutrition* 65 (1997): 876–882; S. D. Katz and coauthors, Pathophysiological correlates of increased tumor necrosis factor in patients with congestive heart failure, *Circulation* 90 (1994): 12–16; J. Falconer and coauthors, Cytokines, the acute phase response, and resting energy expenditure in cachectic patients with pancreatic cancer, *Annals of Surgery* 219 (1994): 325–331; S. Bell and coauthors, Dietary fish oil and cytokine and eicosanoid production during human immunodeficiency virus infection, *Journal of Parenteral and Enteral Nutrition* 20 (1996): 43–49.

2. T. L. Petty and G. G. Weinmann, Building a national strategy for the prevention and management of and research in chronic obstructive pulmonary disease, *Journal of the American Medical Association* 277 (1997): 246–253.

3. K. Gray-Donald and coauthors, Nutritional status and mortality in chronic obstructive pulmonary disease, *American Journal of Respiratory and Critical Care Medicine* 153 (1996): 961–966.

4. K. M. Chapman and L. Winter, COPD: Using nutrition to prevent respiratory function decline, *Geriatrics* 51 (1996): 37–42.

5. A. M. Malone, Is a pulmonary enteral formula warranted for patients with pulmonary dysfunction? *Nutrition in Clinical Practice* 12 (1997): 168–171.

6. C. Grunfeld, Therapy for treatment of the wasting syndrome in cancer and AIDS: What can we do and what should we do? *Nutrition in Clinical Practice* 12 (1997): 99–100.

7. M. S. O'Reilly and coauthors, Angiostatin induces and sustains dormancy of primary tumors in mice, *Nature Medicine* 2 (1996): 689–692.

8. T. Duell and coauthors, Health and functional status of long-term survivors of bone marrow transplantation, *Annals of Internal Medicine* 126 (1997): 182–184.

9. As cited in A. M. Herrington, J. D. Herrington, and C. A. Church, Pharmacologic options for the treatment of cancer, *Nutrition in Clinical Practice* 12 (1997): 101–113.

10. C. L. Loprinzi and coauthors, Alleviation of cancer anorexia and cachexia: Studies of the Mayo Clinic and the North Central Treatment Group, *Seminars in Oncology* 17 (1990): 8–12.

11. As cited in J. S. Young, HIV and medical nutrition therapy, *Journal of the American Dietetic Association* (supplement 2) 97 (1997): 161–166.

12. D. C. Macallan and coauthors, Prospective analysis of weight change in stage IV human immunodeficiency virus infection, *American Journal of Clinical Nutrition* 58 (1993): 417–424.

13. A. Schwenk and coauthors, Clinical risk factors in HIV-1-infected patients, *AIDS* 7 (1993): 1213–1219.

14. C. Fields-Gardner, A review of the mechanisms of wasting in HIV disease, *Nutrition in Clinical Practice* 10 (1995): 167–176.

15. J. A. Stack and coauthors, High-energy, high-protein, oral, liquid, nutrition supplementation in patients with HIV infections: Effect on weight status in relation to incidence of secondary infection, *Journal of the American Dietetic Association* 96 (1996): 337–341.

16. Stack and coauthors, 1996.

17. U. Süttmann and coauthors, Incidence and prognostic value of malnutrition and wasting in human immunodeficiency virus–infected outpatients, *Journal of Acquired Immune Deficiency Syndromes and Human Retrovirology* 8 (1995): 239–246.

18. R. T. Chelowski and coauthors, Long-term effects of early nutrition support with new enterotropic peptide-based formula vs. standard enteral formula in HIV-infected patients: Randomized prospective study, *Nutrition* 9 (1993): 507–512.

19. Herrington, Herrington, and Church, 1997.

20. A.S.P.E.N. Board of Directors, Practice guidelines: Cancer, *Journal of Parenteral and Enteral Nutrition* (supplement) 17 (1993): 12–13.

21. M. Marian, Cancer cachexia: Prevalence, mechanisms, and interventions, *Support Line,* April 1998, pp. 3–12.

22. T. R. Ziegler and coauthors, Clinical and metabolic efficacy of glutamine-supplemented parenteral nutrition after bone marrow transplantation, *Annals of Internal Medicine* 116 (1992): 821–828; P. R. Schloerb and M. Amare, Total parenteral nutrition with glutamine in bone marrow transplantation and other clinical applications (randomized, double-blind study), *Journal of Parenteral and Enteral Nutrition* 17 (1993): 407–413.

23. D. P. Kotler, L. Fogelman, and A. R. Tierney, Comparison of total parenteral nutrition and an oral, semielemental diet on body composition, physical function, and nutrition-related costs of patients with malabsorption due to acquired immunodeficiency syndrome, *Journal of Parenteral and Enteral Nutrition* 22 (1998): 120–126.

CHAPTER 24

Nutrition in Practice

▪ Alternative Therapies ▪

Alternative therapies have become increasingly popular in recent years. Sales of herbal products in the United States amounted to $3.24 billion in 1997.[1] Some states license alternative therapy practitioners in areas such as chiropractic, acupuncture, homeopathy, and naturopathy.[2] While many mainstream health care professionals remain skeptical about alternative therapies, many consumers have become skeptical of, and feel overwhelmed by, the costly and high-tech diagnostic tests and treatments that conventional medicine offers. They want to try a simpler approach to health care, and they want to take more responsibility for maintaining their own health and finding cures for their own diseases, especially when traditional medical therapies prove ineffective or offer little hope for marked improvement. This nutrition in practice explores alternative therapies in search of their possible benefits and with an awareness of their potential harms.

WWW **altmed.od.nih.gov**
National Institutes of Health Office of Alternative Medicine

Can you give some examples of alternative therapies?

Table NP24.1 lists selected fields of alternative medicine, and the accompanying glossary defines terms. Notice that most alternative medicines fall outside the field of nutrition, but that nutrition itself can be an alternative therapy.

Furthermore, many alternative therapies prescribe specific dietary regimens. The many dietary recommendations presented throughout this text are based on scientific evidence and do not fall into the alternative category; strategies that are still experimental, however, do. For example, alternative therapists may recommend megadoses of antioxidant supplements or macrobiotic diets to help prevent chronic diseases, whereas most registered dietitians would advise people to eat at least five servings of vegetables and fruits daily instead.

In what ways are alternative therapies being used?

People may use alternative therapies to foster good health or to prevent and treat diseases and symptoms ranging from anxiety and headaches to cancer and HIV infection. Most often, people use alternative therapies in addition to, rather than in place of, conventional

Table NP24.1 Fields of Alternative Medicine and Selected Examples

Mind-body interventions
Biofeedback
Faith healing
Hypnotherapy
Imagery
Meditation
Bioelectromagnetic applications in medicine
Electroacupuncture
Microwave resonance therapy
Alternative systems of medical practice
Acupuncture
Ayurveda
Homeopathic medicine
Naturopathic medicine
Manual healing methods
Biofield therapeutics
Chiropractic
Massage therapy
Pharmacological and biological treatments
Cartilage therapy
Chelation therapy
Ozone therapy
Herbal medicine
Diet and nutrition in the prevention and treatment of chronic disease
Macrobiotic diets
Orthomolecular medicine

Source: Alternative Medicine: Expanding Medical Horizons, A report to the National Institutes of Health and Alternative Medical Systems and Practices in the United States (Washington, D.C.: Government Printing Office, 1992).

CHAPTER 24

Nutrition in Practice

Glossary

acupuncture (AK-you-PUNK-cher): a technique that involves piercing the skin with long thin needles at specific anatomical points to relieve pain or illness. Acupuncture sometimes uses heat, pressure, friction, suction, or electromagnetic energy to stimulate the points.

alternative therapies: approaches to medical diagnosis and treatment that are not fully accepted by the established medical community; as such, they are not widely taught at U.S. medical schools or practiced in U.S. hospitals; also called adjunctive, unconventional, or unorthodox therapies.

aroma therapy: a technique that uses oil extracts from plants and flowers (usually applied by massage or baths) to enhance physical, psychological, and spiritual health.

ayurveda (EYE-your-VAY-dah): a traditional Hindu system of improving health by using herbs, diet, meditation, massage, and yoga to stimulate the body to make its own natural drugs.

bioelectromagnetic medical applications: the use of electrical energy, magnetic energy, or both to stimulate bone repair, wound healing, and tissue regeneration.

biofeedback: the use of special devices to convey information about heart rate, blood pressure, skin temperature, muscle relaxation, and the like to enable a person to learn how to consciously control these medically important functions.

biofield therapeutics: a manual healing method that directs a healing force from an outside source (commonly God or another supernatural being) through the practitioner and into the client's body; commonly known as "laying on of hands."

cartilage therapy: the use of cleaned and powdered connective tissue, such as collagen, to improve health.

chelation therapy: the use of ethylene diamine tetraacetic acid (EDTA) to bind with metallic ions, thus healing the body by removing toxic metals.

chiropractic (KYE-roe-PRAK-tik): a manual healing method of manipulating vertebrae to relieve musculoskeletal pain suspected of causing problems with internal organs.

complementary therapy: therapy that incorporates both traditional medical and alternative treatments.

DHEA (dehydroepiandrosterone): a hormone secreted by the adrenal glands. DHEA is available without prescription and is sold as an anti-aging remedy to improve energy, strength, and immunity. Proof of safety or effectiveness is lacking.

faith healing: healing by invoking divine intervention without the use of medical, surgical, or other traditional therapy.

garlic oil: extract of garlic. Proof of effectiveness is lacking.

hemlock: a poisonous herb having finely divided leaves and small white flowers.

herbal medicine: the use of plants to treat disease or improve health; also known as botanical medicine or phytotherapy.

homeopathic (home-ee-OP-ah-thick) **medicine:** a practice based on the theory that "like cures like," that is, that substances that cause symptoms in healthy people can cure those symptoms when given in very dilute amounts.
homeo = like
pathos = suffering

hypnotherapy: a technique that uses hypnosis and the power of suggestion to improve health behaviors, relieve pain, and heal.

imagery: a technique that guides clients to achieve a desired physical, emotional, or spiritual state by visualizing themselves in that state.

iridology: the study of changes in the iris of the eye and their relationships to disease.

macrobiotic diet: a diet consisting of brown rice, miso soup, sea vegetables, and other traditional Japanese foods.

massage therapy: a healing method in which the therapist manually kneads muscles to reduce tension, increase blood circulation, improve joint mobility, and promote healing of injuries.

meditation: a self-directed technique of relaxing the body and calming the mind.

melatonin: a hormone secreted by the pineal gland believed to help regulate the body's daily rhythms and promote sleep. Proof of safety or effectiveness is lacking.

naturopathic medicine: a system that integrates traditional medicine with botanical medicine, clinical nutrition, homeopathy, acupuncture, East Asian medicine, hydrotherapy, and manipulative therapy.

orthomolecular medicine: the use of large doses of vitamins to treat chronic disease.

ozone therapy: the use of ozone gas to enhance the body's immune system.

Nutrition in Practice

Digoxin, a medication commonly prescribed for abnormal heart rhythms, derives from the foxglove plant.

therapies.[3] Thus the term *complementary therapy*, often used interchangeably with alternative therapy, really describes therapy that integrates both types of therapy. Most people seem to seek alternative therapies for nonserious medical conditions or health promotion. They simply want to feel better and access is easy. Sometimes their symptoms are chronic and subjective, such as pain and fatigue, and difficult to treat. In these cases, the chances of finding relief are often as good with an alternative therapy as they are with a placebo, standard medical intervention, or even nonintervention.

Why are many mainstream health care professionals skeptical about alternative therapies?

Unlike conventional medical therapies, alternative therapies are not widely taught in medical schools, are not practiced in hospitals, and are usually not reimbursed by medical insurance companies.[4] Most importantly, while some alternative therapies have been proved to be safe and effective, most information on alternative therapies comes from folklore, tradition, and testimonial accounts. In short, many practitioners are unfamiliar with alternative therapies, and scientific evidence proving their safety and effectiveness is lacking. Some say alternative therapies simply do not work; others argue that the established medical community has not given these therapies a fair trial.

Health care professionals are also skeptical about some alternative therapies because they can be dangerous. If people delay or discontinue conventional treatments that have been proved effective for remedies that tout "miracle cures," the consequences can be severe and irreversible. Furthermore, some alternative therapies, particularly those that include megadoses of vitamins and herbal remedies, can interact with medications, and some can have potentially serious consequences. Garlic and ginger, for example, can magnify the anticlotting effects of the medication warfarin; high doses of vitamin K can reduce warfarin's effectiveness.

One thing is clear—people are using alternative therapies whether or not mainstream medical practitioners approve. One of the most disturbing facts about alternative therapies is that 70 percent of people using them have never mentioned that fact to their physicians.[5] Furthermore, most physicians don't ask.[6] Health care professionals serve their clients best by inquiring about alternative therapies the client may be using.

Are alternative therapies safe?

Many alternative therapies can be safe if they are used by the right people for the right reason and in the right amounts. Herbal remedies, for example, may be appropriate for minor ailments—a cup of chamomile tea to ease gastric discomfort or the gel of an aloe vera plant to soothe a sunburn, for example—but not for major health problems such as cancer.

Ideally, an alternative therapy provides benefits with little or no risk. The use of ginseng as an adjunct in the management of type 2 diabetes (Chapter 25), for example, correlates with improved measures of blood glucose control without adverse side effects.[7] Though still in the experimental stage, such findings, if replicated, hold promise that this alternative therapy may one day become accepted medical practice.

Some alternative therapies are innocuous, providing little or no benefit for little or no risk. Sipping a cup of warm tea with a pleasant aroma, for example, won't cure heart disease, but it may improve the person's mood and help relieve tension. Given no physical hazard and little financial risk, such therapies are acceptable. In contrast, other products and procedures are downright dangerous, posing great risks while providing no benefits. One example is the folk practice of geophagia (eating earth or clay), which can cause GI impaction and impair iron absorption. Clearly, such therapies are too harmful to be used. Figure NP24.1 shows how the benefit-risk relationship can help clients and health care professionals determine whether or not an alternative therapy is safe.

Among the most popular alternative therapies are the use of herbal remedies to treat diseases and symptoms. Herbal remedies, however, carry some additional risks to health.

WWW

herbalgram.org
American Botanical Council

Nutrition in Practice

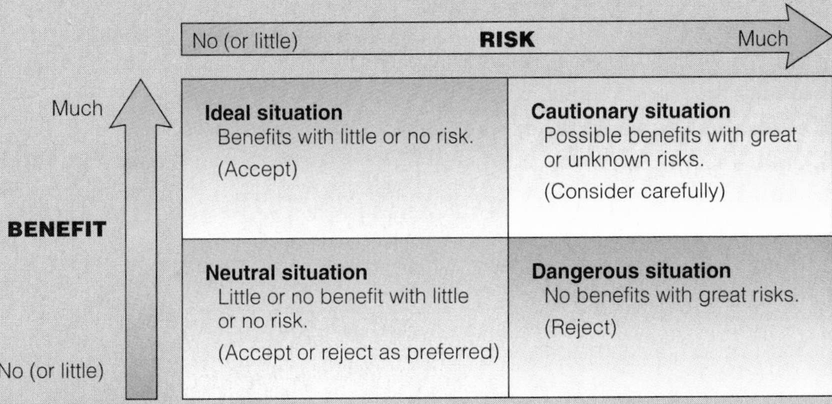

Figure NP24.1
Risk-Benefit Relationships

In what ways do herbal remedies carry additional risks?

Before medications became widely available, a myriad of herbs and plants were used to cure aches and ills with varying degrees of success. Upon scientific scrutiny, dozens of these folk remedies reveal their secrets. For example, myrrh, a plant resin used as a painkiller in ancient times, does indeed have an analgesic effect.[8] The herb valerian, which has long been used as a tranquilizer, contains oils that have a sedative effect. Senna leaves, brewed as a laxative tea, produce compounds that act as a potent cathartic drug. The compounds that plants make are so beneficial that today they contribute to more than half of our modern medicines.

Thus herbs are drugs, even though they are often less potent than manufactured drugs. In the United States, however, herbs are not classified as drugs, but rather as dietary supplements. Under the Dietary Supplement Health and Education Act, herbs can be marketed without prior approval from the Food and Drug Administration (FDA), and unlike foods and food additives, which require the manufacturer to provide proof of safety, the FDA has the burden of proving that an herbal product is not safe.[9]

Some factors that can affect an individual's risk of adverse effects from herbal supplements include the person's age, genetic makeup, nutrition, state of health, and concurrent use of medications, as well as how much of the supplement the person takes and how long it is used. Some issues that relate to the safety of herbal preparations include:

- *True identification of herbs* Most mint teas are safe, for instance, but some varieties contain the highly toxic pennyroyal oil. Mistakenly used to treat colic, mint tea laden with pennyroyal has been blamed for the liver and neurological injuries of at least two infants, one of whom died.[10] In another case, neurological and cardiovascular complications developed in three children, and hepatitis was reported in three adults after ingesting the Chinese herb *jin bu huan.*[11] Chemical analysis of the herbal preparation revealed that its active ingredient was actually from another herb.
- *Mislabeling of active ingredients* Chemical analysis of the Chinese herbal preparation *tung shueh,* which had been linked to acute renal failure, revealed the presence of two synthetically manufactured drugs, mefenamic acid (a nonsteroidal anti-inflammatory agent) and diazepam (an antianxiety agent) that were not listed as active ingredients on the label.[12]
- *Purity of herbal preparations* Potentially toxic quantities of arsenic and mercury have been detected in traditional Chinese herbal balls used to treat fever, rheumatism, and cataracts.[13]
- *Adverse reactions and toxicity levels of herbs* As is true of all drugs, herbs may produce undesirable reactions. Herbal preparations containing ephedrine, commonly known as *ma huang* and used to promote weight loss, act as strong central nervous system stimulants, causing rapid heart rate, nervousness, headaches, insomnia, and even death.
- *Safe dosages of herbs* Herbs that are effective contain active ingredients that need to be administered in proper doses. Each of these active ingredients has a different potency, time of onset, duration of activity, and consequent effects, making the plant itself too unpredictable to be useful. Consider that labels of over-the-counter bronchodilators containing ephedrine at doses of from 12.5 to 25.0 milligrams instruct users to take one dose two or three times per day, yet chemical analysis of one herbal preparation containing ephedrine revealed that each tablet

contained 45 milligrams of ephedrine and 20 milligrams of caffeine (another stimulant) and the label instructed the user to take five tablets per dose.[14]

- *Interactions of herbs with medicines and other herbs* Like medications, herbs may interfere with, or potentiate, the effects of other herbs and medications. Chewing foxglove leaves when taking the medication digoxin, for example, could be catastrophic.

Not only are herbal preparations not strictly regulated, but their labels may carry unsubstantiated health claims as long as the following disclaimer also appears: "Has not been evaluated by the Food and Drug Administration." Consumers who decide to use herbs do so at their own risk, but it is important for health care professionals to help clients sift through the information that is available so that consumers can learn to make informed choices.

What can health care professionals do to help?

First of all, it is important that health care professionals maintain open communications with their clients about alternative therapies they may be using or are considering using. To do so, health care professionals need to remain nonjudgmental, culturally sensitive, respectful of the clients' beliefs, and open-minded. Working together with their clients, health care professionals can develop medical plans that include both conventional and alternative therapies.

For a client who is considering an alternative therapy, discuss what is known about the therapy's safety and effectiveness. Recommend that the client try only one alternative therapy at a time, so the effectiveness of the measure can be evaluated. One clinician recommends that the client keep a symptom record, logging the severity and frequency of symptoms as well as factors that make the symptom worse or better.[15] The client and health care professional can then review the record and learn how the treatment is working.

The public's growing interest in unorthodox remedies, coupled with the soaring costs of traditional medical approaches, opened the way for alternative medicine to prove itself. In 1992, Congress passed legislation requiring the National Institutes of Health (NIH) to create an Office of Alternative Medicine. Its task is "to more adequately explore unconventional medical practices."[16] What the results will be remain to be seen. Alternative therapies come in a variety of shapes and sizes. Both their benefits and their risks may be either small, none, or great. Accept the beneficial, or even neutral, practices with an open mind and reject only those practices known to cause harm. Making healthful choices requires knowing what all the choices are.

Notes

1. As cited in C. L. Bartels and S. J. Miller, Herbal and related remedies, *Nutrition in Clinical Practice* 12 (1998): 5–19.

2. D. Shattuck, Complementary medicine: Finding a balance, *Journal of the American Dietetic Association* 97 (1997): 1367–1369.

3. D. M. Eisenberg and coauthors, Unconventional medicine in the United States: Prevalence, costs, and patterns of use, *New England Journal of Medicine* 328 (1993): 246–252.

4. Office of Alternative Medicine Clearinghouse, http://altmed.of.nih.gov., visited on June 9, 1998.

5. Eisenberg and coauthors, 1993.

6. D. M. Eisenberg, Advising patients who seek alternative medical therapies, *Annals of Internal Medicine* 127 (1997): 61–69.

7. E. A. Sotaniemi, E. Haapakoski, and A. Rautio, Ginseng therapy in noninsulin-dependent diabetic patients: Effects on psychophysical performance, glucose homeostasis, serum lipids, serum aminoterminalpropeptide concentration, and body weight, *Diabetes Care* 10 (1995): 1373–1375.

8. P. Lipkin, An ancient salve dampens pain, *Science News* 149 (1996): 20.

9. Bartels and Miller, 1998.

10. J. A. Bakerink and coauthors, Multiple organ failure after ingestion of pennyroyal oil from herbal tea in two infants, *Pediatrics* 98 (1996): 944–947.

11. R. S. Horowitz and coauthors, The clinical spectrum of jin bu huan toxicity, *Archives of Internal Medicine* 156 (1996): 899–903.

12. A. B. Abt and coauthors, Chinese herbal medicine induced acute renal failure, *Archives of Internal Medicine* 155 (1995): 211–212.

13. E. O. Espinoza, M. J. Mann, and B. Bleasdell, Arsenic and mercury in traditional Chinese herbal balls, *New England Journal of Medicine* 333 (1995): 803–804.

14. As cited in Bartels and Miller, 1998.

15. Eisenberg, 1997.

16. Alternative Medicine: Expanding Medical Horizons, A report to the National Institutes of Health and Alternative Medical Systems and Practices in the United States (Washington, D.C.: Government Printing Office, 1992).

25 Carbohydrate-Modified Diets for Diabetes and Hypoglycemia

As earlier chapters have shown, diet planners use carbohydrate-modified diets to prevent and treat a variety of disorders and symptoms. This chapter describes the use of carbohydrate-modified diets to help regulate blood glucose levels for people with diabetes mellitus and hypoglycemia.

Diabetes Mellitus

diabetes (DYE-uh-BEET-eez) **mellitus** (MELL-ih-tus or mell-EYE-tus): a group of metabolic disorders of glucose regulation and utilization.
diabetes = passing through (the body)
mellitus = honey-sweet (sugar)

Reminder: *Insulin* is the hormone that, among other things, enables many cells to take up glucose from the blood and store energy fuels. Insulin is produced by the beta cells of the islets of Langerhans—a specific type of endocrine cell in the pancreas.

WWW.
diabetes.org
American Diabetes Association

joslin.org
Joslin Diabetes Center

aadenet.org
American Association of Diabetes Educators

Diabetes mellitus describes a group of metabolic disorders characterized by elevated blood glucose and altered energy metabolism and caused by defective insulin secretion, defective insulin action, or a combination of the two. The classic symptoms of untreated diabetes are increased urination, increased thirst, and unexplained weight loss. Insulin, which signals the body to store energy fuels following meals, is produced by certain endocrine cells in the pancreas—specifically, the beta cells of the islets of Langerhans.

Types of Diabetes

There are two major types of diabetes, and their distinguishing features are summarized in Table 25.1. A later section describes the special case of gestational diabetes. Some other types of diabetes can occur as a consequence of genetic disorders, diseases of the exocrine pancreas (chronic pancreatitis and cystic fibrosis, for example), hormonal imbalances, drugs or chemicals, certain infections, and immune system disorders.

Diagnosis of Diabetes The primary criteria used to confirm a diagnosis of diabetes include:

- A random blood glucose sample (one taken without regard to food intake) that exceeds 200 milligrams per 100 milliliters in a person with symptoms of diabetes.
- A blood glucose of 126 milligrams per 100 milliliters or greater in a person who has been fasting for at least 8 hours.

Some clinicians prefer to test a person's fasting blood glucose and then retest it several times after a specific amount of glucose is given orally—a procedure called a glucose tolerance test.

type 1 diabetes mellitus: the less common type of diabetes in which the person produces no insulin at all.

Immune system disorders in which the body destroys its own cells or tissues are called **autoimmune disorders.**

type 2 diabetes mellitus: the more common type of diabetes that develops gradually and is associated with insulin resistance.

insulin resistance: the condition in which the cells fail to respond to insulin as they do in healthy people.

Type 1 Diabetes In type 1 diabetes, the less common of the two major types of diabetes (about 5 to 10 percent of diagnosed cases), the pancreas cannot synthesize insulin. Blood glucose rises abnormally high, but cannot enter the cells that need it for energy. Without insulin, the body's energy metabolism is dramatically altered with such severe consequences that people with type 1 diabetes cannot survive unless they inject insulin regularly. In many cases of type 1 diabetes, the individual appears to inherit a defect in which immune cells mistakenly attack and destroy insulin-producing pancreatic cells.

Type 2 Diabetes The predominant type of diabetes mellitus (90 to 95 percent of cases), and the type most likely to go undiagnosed, is type 2 diabetes. The pancreas produces insulin, and the cells respond to it, but with less sensitivity. As blood glucose rises, the pancreas makes more insulin, and blood insulin rises to abnormally high levels (hyperinsulinemia). During this period of impaired glucose tolerance, the body is able to maintain blood glucose within a fairly normal range but at a cost. Gradually, the chronic demand for insulin exhausts the beta

Table 25.1 Features of Type 1 and Type 2 Diabetes	Type 1	Type 2
Other names	IDDM[a] Juvenile-onset diabetes Ketosis-prone diabetes Brittle diabetes	NIDDM[a] Adult-onset diabetes Ketosis-resistant diabetes Lipoplethoric diabetes Stable diabetes
Age of onset	<20 (mean age, 12)	>40
Associated conditions	Viral infection, heredity	Obesity, heredity
Insulin required?	Yes	Sometimes
Cell response to insulin	Normal	Resistant
Symptoms	Relatively severe	Relatively moderate
Prevalence in diabetic population	5 to 10%	90 to 95%

[a]The names IDDM (insulin-dependent diabetes mellitus) and NIDDM (noninsulin-dependent diabetes mellitus) frequently appear in references and generally describe type 1 and type 2 diabetes, respectively.

cells of the pancreas, and finally insulin production falters as the disease progresses. Although the exact causes are unknown, impaired glucose tolerance and type 2 diabetes appear to be associated with obesity, abdominal fat, and physical inactivity.[1] Obesity aggravates insulin resistance: as body fat increases, body tissues become less and less able to respond to insulin.

About 8 million people in the United States have been diagnosed with diabetes, and estimates suggest that another 8 million people may have type 2 diabetes that goes undiagnosed.[2] This is especially dangerous because the damaging effects of diabetes on the circulatory system (described later) may begin to develop years before the symptoms of diabetes appear.[3] A nationwide multicenter study (the Diabetes Prevention Program) is currently under way to determine whether early intervention for people with impaired glucose tolerance can prevent type 2 diabetes.[4] People with impaired glucose tolerance may have fasting glucose levels higher than normal but not high enough to diagnose diabetes; others may have normal blood glucose levels most of the time, but when given a large amount of glucose, their blood glucose rises too high. Those most likely to develop impaired glucose tolerance include people who are obese (greater than 120 percent of desirable body weight or a body mass index of 27 or higher), have immediate family members with diabetes, are members of high-risk ethnic populations (Native Americans, Hispanic Americans, and African Americans), are over age 45, and women who have given birth to babies weighing over 9 pounds or have been diagnosed with diabetes while pregnant. (Gestational diabetes is described in a later section.)

impaired glucose tolerance: inability of the body to adequately regulate its blood glucose concentration in response to either the intake of dietary carbohydrate or the release of glucose from cells during fasting or metabolic stress; sometimes called **borderline diabetes.**

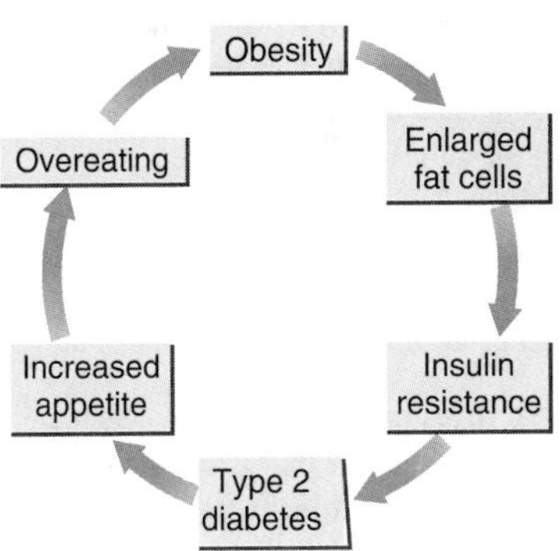

Acute Complications of Diabetes

Figure 25.1 on p. 596 presents an overview of the metabolic changes and acute complications that occur in uncontrolled diabetes. The metabolic consequences of type 1 diabetes are more immediate and severe than those of type 2 diabetes because in type 1 diabetes no glucose enters the cells. Disruption of energy metabolism and exposure of the tissues to high glucose concentrations result in both acute and chronic complications. The accompanying glossary defines diabetes-related symptoms and complications.

Hyperglycemia, Dehydration, and Glycosuria With insufficient or ineffective insulin, blood glucose rises and hyperglycemia results. High blood glucose creates

Figure 25.1

Metabolic Consequences and Acute Clinical Manifestations of Untreated Type 1 and Type 2 Diabetes

As you can see, when glucose cannot enter the cells, a cascade of metabolic changes follows. In type 2 diabetes, some glucose enters the cells. Because the cells are not "starved" for glucose, the body does not shift into the metabolism of fasting (losing weight and producing ketones).

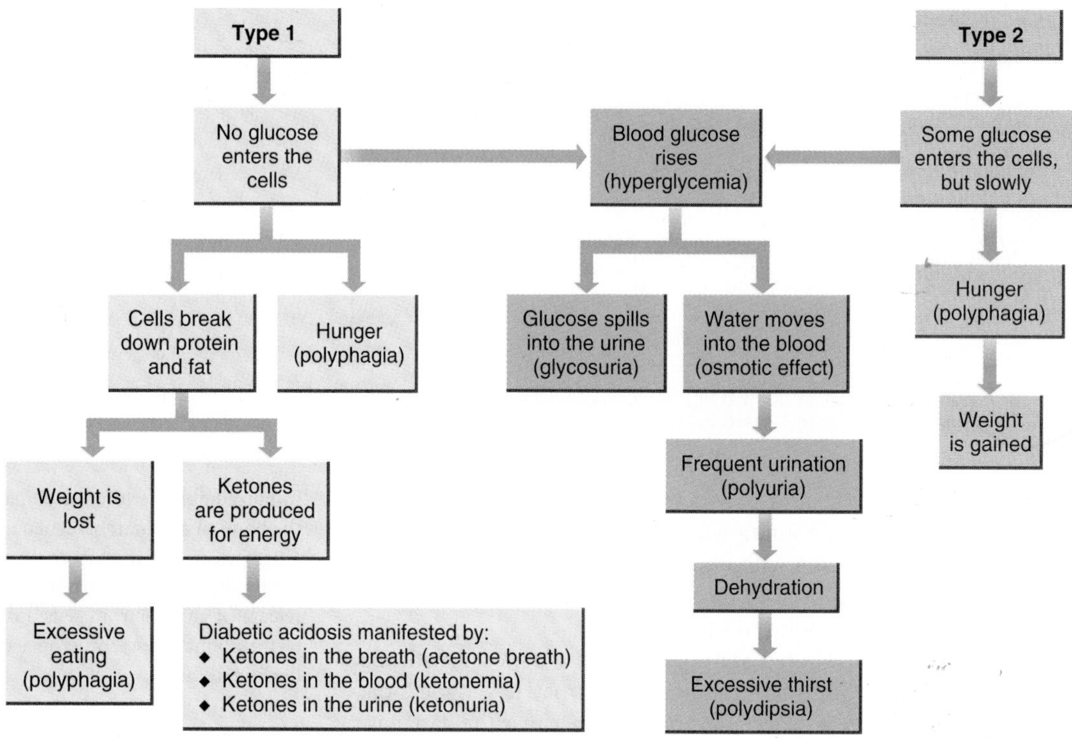

Symptoms of hyperglycemia:

- *Intense thirst and, sometimes, hunger.*
- *Increased urination.*
- *Blurred vision.*
- *Fatigue.*
- *Acetone breath.*
- *Glycosuria.*
- *Labored breathing.*

renal threshold: the point at which a blood constituent that is normally reabsorbed by the kidneys reaches a level so high the kidneys cannot reabsorb it. The renal threshold for glucose generally is reached when blood glucose rises above 180 mg/100 ml.

Reminder: Ketone bodies are produced by the incomplete breakdown of fat when glucose is not available to the cells.

Reminder: The hormonal changes that accompany severe stress result in insulin resistance.

an osmotic effect, drawing water from the tissues into the blood. The high blood glucose concentration eventually exceeds the kidneys' ability to reabsorb glucose (the renal threshold), and the excess glucose is excreted in the urine along with fluids and electrolytes. This series of events explains why people with uncontrolled diabetes produce excessive urine and are thirsty.

Ketosis and Coma Type 1 diabetes continuously deprives cells of the energy fuels they need, and the body responds by mobilizing protein and fat for energy. The liver responds to the mobilization of fatty acids by making ketone bodies, which accumulate in the blood and urine and result in acidosis. A fruity odor on the breath of a person with poorly controlled diabetes reflects the presence of the ketone acetone. If the combination of hyperglycemia, dehydration, and acidosis becomes severe enough, a potentially fatal coma may follow. The person with type 2 diabetes is usually not prone to ketosis and its consequences, because some glucose enters the cells. However, ketosis and coma can occur in the person with type 2 diabetes who suffers a severe stress or an infection.

Nonketotic Coma People with type 2 diabetes can develop another kind of coma caused by extremely high blood glucose and dehydration. This problem is more common in the elderly because they may not recognize thirst and drink enough water to compensate for high blood glucose levels. This kind of coma is termed hyperosmolar hyperglycemic nonketotic coma.

Glossary of Diabetes-Related Symptoms and Complications

acetone breath: a distinctive fruity odor that can be detected on the breath of a person who is experiencing ketosis.

diabetic coma: unconsciousness precipitated by hyperglycemia, dehydration, ketosis, and acidosis in people with diabetes who require insulin.

gangrene: death of tissue due to a deficient blood supply and/or infection.

gastroparesis: delayed gastric emptying.

glycosuria (GLY-ko-SUE-ree-ah) or **glucosuria** (GLUE-ko-SUE-ree-ah): glucose in the urine, which generally occurs when blood glucose exceeds 180 mg/100 ml.

hyperglycemia: elevated blood glucose. Normal fasting blood glucose is less than 110 mg/100 ml. Fasting blood glucose between 110 and 125 mg/100 ml suggests impaired glucose tolerance; values of 126 mg/100 ml or higher suggest diabetes.

hyperosmolar hyperglycemic nonketotic coma: coma that occurs in uncontrolled type 2 diabetes precipitated by the presence of hypertonic blood and dehydration.

hypoglycemia: low blood glucose.

ketonemia: ketones in the blood.

ketonuria: ketones in the urine.

macroangiopathies: disorders of the large blood vessels.

microangiopathies: disorders of the capillaries.

nephropathy: a disorder of the kidneys.

neuropathy: a disorder of the nerves.

polydipsia (POLL-ee-DIP-see-ah): excessive thirst.

polyphagia (POLL-ee-FAY-gee-ah): excessive eating.

polyuria (POLL-ee-YOU-ree-ah): excessive urine production.

retinopathy: a disorder of the retina.

Thinness in Type 1 Diabetes With the loss of glucose and ketone bodies (both energy sources) in the urine, combined with protein breakdown, serious weight loss may follow. The person with uncontrolled or poorly controlled type 1 diabetes is likely to be thin despite eating apparently adequate amounts of foods.

Weight Gain in Type 2 Diabetes In type 2 diabetes, people are also hungry and overeat, but because insulin is present, they may eventually store fat from the excess energy they consume and gain weight. The person with type 2 diabetes is likely to become or remain overweight.

Hypoglycemia Hypoglycemia is a consequence, not of untreated diabetes, but rather of inappropriate management. It can result from too much insulin or glucose-lowering medications, strenuous physical activity, skipped or delayed meals, inadequate food intake, vomiting, or severe diarrhea. The mental confusion and shakiness that are symptoms of hypoglycemia (see the margin) may make it difficult for the person to recognize the problem and take corrective actions.

Left untreated, hypoglycemia can lead to loss of consciousness, brain damage, and even death. Some research suggests that repeated episodes of hypoglycemia might permanently impair cognitive function.[5] People who have had diabetes for a long time risk severe hypoglycemia because the warning signs become less noticeable over time.

Notice that many of the symptoms of hypoglycemia are those of alcohol intoxication. If the true problem goes unrecognized, the person may die. To prevent such a tragic mistake, advise every person with diabetes to wear identification in the form of a bracelet or necklace.

Symptoms of hypoglycemia:

- *Hunger.*
- *Headache.*
- *Sweating.*
- *Shakiness.*
- *Nervousness.*
- *Confusion.*
- *Disorientation.*
- *Slurred speech.*

Chronic Complications of Diabetes

Chronic hyperglycemia damages the structures of the blood vessels and nerves (see glossary). Circulation becomes poor and nerve function falters. Infections are more likely to occur due to poor circulation coupled with glucose-rich blood and urine.

Infections may go undetected due to impaired nerve function; gangrene may follow. People with diabetes must pay special attention to hygiene and keep alert for early signs of infection.

Elevated blood glucose may have damaging effects on blood vessels and nerves even before a diagnosis of diabetes is made.[6] The combination of insulin resistance, secretion of more and more insulin to maintain blood glucose levels, obesity, hypertension, elevated LDL and triglycerides, and lowered HDL (see Chapter 26) is frequently observed in people with both type 2 diabetes and cardiovascular disease and is called *syndrome X.*

syndrome X: the combination of insulin resistance, hyperinsulinemia, obesity, hypertension, elevated LDL and triglycerides, and lowered HDL that is frequently associated with type 2 diabetes and cardiovascular disease.

A heart attack that goes unnoticed is called a **silent heart attack.**

Cardiovascular Diseases Atherosclerosis (see Chapter 26) tends to develop early, progress rapidly, and be more severe in people with diabetes. More than 80 percent of people with diabetes die as a consequence of cardiovascular diseases, especially heart attacks. If nerve function is also impaired, the person may suffer a heart attack and not even realize it.

Small Blood Vessel Disorders Disorders of the smallest blood vessels—the capillaries—may also develop and lead to loss of kidney function and retinal degeneration and loss of vision. About 85 percent of people with diabetes have nephropathy, retinopathy, or both. Consequently, diabetes is a leading cause of both kidney failure and blindness.

Neuropathy Nerve tissues may also deteriorate, resulting in neuropathy. At first, the person may experience a painful prickling sensation, often in the arms and legs, which progresses until the person loses sensations in the hands and feet. Injuries to these areas may go unnoticed, and infections can progress rapidly. If tissues die as a consequence, amputation of the affected limb (usually the toes, feet, or legs) may be necessary.

Neuropathy can also delay gastric emptying. When the stomach empties slowly after a meal (gastroparesis), the person may experience a premature feeling of fullness, nausea, vomiting, weight loss, and poor blood glucose control due to irregular nutrient absorption.

Treatment of Diabetes Mellitus: Coordinating Diet, Physical Activity, and Medications

A diagnosis of diabetes can be devastating. The parents of a young child with type 1 diabetes may feel overwhelmed, angry, anxious, and even guilty. A teenager may feel that it is the end of the world. An adult with type 2 diabetes may fear possible complications and resent having to readjust lifestyles and adopt a new diet. To control blood glucose successfully, the person must master the complex task of coordinating diet, physical activity, and medications. On the bright side, however, is that with such mastery the person can live a full and active life and significantly reduce the risk of chronic complications. Nutrition in Practice 25 describes how health care professionals can assist in this process.

Treatment Goals The goals of medical and nutrition therapy for diabetes are to maintain blood glucose within a fairly normal range, achieve optimal blood lipid levels (see Chapter 26), control blood pressure (Chapter 26), support health and well-being, and treat complications. The most important of these goals is blood glucose control. A major multicenter clinical trial (the Diabetes Control and

Complications Trial, or DCCT) clearly showed that tightly controlling blood glucose reduces the risks of onset and progression of nephropathy, retinopathy, and neuropathy by about 50 percent.[7] Although this study specifically addressed the management of diabetes for people who use insulin, most clinicians believe that the results apply to people with type 2 diabetes as well.

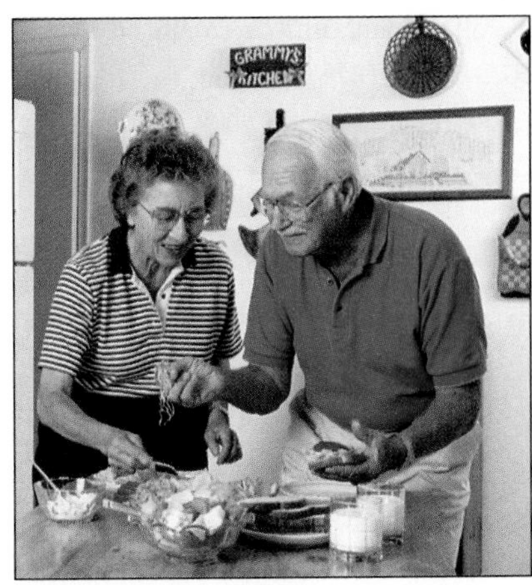

The diet for diabetes emphasizes a consistent intake of carbohydrate from well-balanced meals spaced evenly throughout the day.

Treatment Plans Because many factors influence blood glucose levels and clients must often make many lifestyle changes in controlling diabetes and its complications, accurate assessment and monitoring of a person's diet, physical activity, and health status are vital in helping people with diabetes work out acceptable goals for managing their diseases. To promote success, the health care team plans and adjusts therapy for the client's medical needs, motivational level, educational ability, and lifestyle. Successful diabetes education takes time and must be flexible to accommodate changing needs.

Diet for Diabetes

The diet for diabetes parallels a healthy diet for all people in both amounts and types of nutrients. Attention to all energy nutrients is important: controlling carbohydrate prevents hyper- and hypoglycemia; controlling protein can help prevent loss of kidney function; controlling fat helps prevent cardiovascular complications.

Energy The diet for diabetes first focuses on providing adequate food energy to achieve or maintain a healthy and realistic body weight and to support growth in children and pregnant women. To determine whether energy intake is appropriate, the planner uses the energy RDA as a guide, takes height and weight measures periodically, and adjusts the diet as necessary.

Weight-loss diets are often prescribed for people with type 2 diabetes. Even moderate weight loss (10 to 20 pounds) can help reverse insulin resistance, improve the blood lipid profile, and reduce blood pressure. Weight-loss diets that provide at least 10 kcalories per pound of body weight allow a safe and gradual weight loss. In some cases, very-low-kcalorie diets (less than 800 kcalories per day) may help establish blood glucose control, but the value of such diets is controversial.[8]

Protein Protein provides about 10 to 20 percent of the total kcalories in the diet for diabetes. Providing adequate, but not excessive protein, may help delay the onset or progression of kidney disease (see Chapter 27). At the first sign of kidney disease, people with diabetes may need to restrict protein to 0.8 gram per kilogram of body weight (the same as the RDA).[9]

An early sign of impending kidney disease is the excretion of small amounts of albumin in the urine or **microalbuminuria** (see Chapter 27).

After considering the energy provided by protein, the remaining kcalories are distributed between carbohydrate and fat. The next sections describe some of the considerations that apply.

Carbohydrate To control blood glucose, the person with diabetes needs to have glucose available throughout the day, but not so much at one time that blood glucose levels rise too high. Carbohydrates have the greatest effect of all energy nutrients on blood glucose. Once eaten, carbohydrates raise blood glucose in about an hour. Some protein from a meal may also be converted to glucose, but very slowly and in lesser amounts than carbohydrate. Fat is not converted to glucose in significant amounts. Thus the diet for diabetes emphasizes a consistent carbohydrate intake from day to day and at each meal and snack. Typically, diet plans provide from 45 to 60 percent of the total kcalories from carbohydrate. Eating about the same amount of carbohydrate at about the same time each day helps the person avoid hyperglycemia and hypoglycemia and eases the task of coordinating medications and food intake.

Recall that authorities recommend 20 to 35 g of dietary fiber a day.

Complex versus Simple Sugars Encourage clients to select foods rich in complex carbohydrates: whole-grain breads and cereals, legumes, fruits, and vegetables. In addition to carbohydrates, these foods provide fiber, vitamins, and minerals and offer many health benefits (see Chapter 2).

Traditionally, concentrated sweets were strictly excluded from the diet for diabetes, but now they are restricted only to the same extent as they are for all people. Health care professionals recognize that *total* carbohydrate is of greater concern in diabetes than the *type* of carbohydrate.[10] The person with diabetes can consume concentrated sweets as a limited part of a healthy diet, as long as they are counted as part of the carbohydrate allowance. Artificial sweeteners that contain minimal kcalories, and products made from them, can be used in place of sugar, especially for the person with diabetes who must lose weight.

Fat People with diabetes who have acceptable blood lipid concentrations benefit from a fat intake consistent with the *Dietary Guidelines for Americans* (30 percent or less of kcalories from fat, less than 10 percent of kcalories from saturated fat, and less than 300 milligrams of cholesterol). People with diabetes who have elevated LDL may need to restrict saturated fat intake to 7 percent or less of total kcalories and cholesterol to less than 200 milligrams daily (see Chapter 26). To lower fat and cholesterol intakes, clients can use low-fat and nonfat milk and milk products and lean meats, among other strategies (see the box on p. 69).

When clients need to lower their total fat, the percentage of kcalories from carbohydrate increases. Some people with diabetes, especially those with type 2 diabetes, may have difficulty controlling their blood glucose and blood lipid levels when they increase their intake of carbohydrates. High-carbohydrate diets raise triglyceride and lower HDL levels. Others who are accustomed to eating a higher-fat diet may have difficulty complying with a diet plan that strictly limits fats. Mounting evidence suggests that clients can control their blood glucose and improve their blood lipid levels with higher fat intakes, provided that monounsaturated fats replace saturated fats in the diet.[11] When compared with diets that provided 55 percent of the kcalories from carbohydrate and 30 percent from fat, diets that provided 40 percent of the kcalories from carbohydrate and 45 percent from fat (25 percent monounsaturated fat) improved blood glucose, insulin, and triglycerides in people with type 2 diabetes.[12]

Sodium People with diabetes frequently develop hypertension and are likely to be salt-sensitive. Practitioners generally advise all clients with diabetes to limit sodium to less than 3000 milligrams per day.[13] People with diabetes and hypertension may need to restrict sodium to 2400 milligrams or less per day. Chapter 26 provides more information about sodium-restricted diets.

One drink is defined as 1½ oz distilled liquor or 5 oz wine or 12 oz beer. Note: light beer contains the same amount of alcohol as regular beer, with half the carbohydrate. When counting kcalories, 1 drink = 2 fat exchanges.

oral antidiabetic agents: medications taken by mouth to lower blood glucose levels in people with type 2 diabetes.

Alcohol The person whose blood glucose is well controlled can usually include some alcoholic beverages with the consent of the physician. Because alcohol can cause hypoglycemia in any person, however, people who also take insulin are advised to drink only moderate amounts (no more than two drinks per day), with meals, and in addition to the usual meal plan. Remember, too, that the person with hypoglycemia may appear to be intoxicated and alcohol use can add confusion to a potentially dangerous situation.

Alcohol intake is discouraged for people with a history of alcohol abuse; those with pancreatitis, abnormal blood lipids, or neuropathy; and pregnant women. The combination of alcohol and some oral antidiabetic agents may cause flushing of the skin and a rapid heartbeat. Alcohol use is also discouraged for people who are overweight; if it is used, alcohol should be substituted for fat exchanges. Drinks that contain simple sugars (mixers, sweet wines, and liqueurs) are best avoided. If they are used, the person must count their carbohydrate contents as part of the daily carbohydrate allowance.

Timing and Composition of Meals and Snacks For people with diabetes, consistent timing and composition of meals and snacks from day to day improve glucose control. An evening snack is especially important because it helps sustain the person's blood glucose through the night. A person with a regular physical activity program who takes a prescribed dose of insulin at a set time and then eats about the same amount of carbohydrate at about the same time each day knows that glucose and insulin will be available to the body when they are needed. Diet-related behaviors that may improve blood glucose control include:

- Adherence to the meal plan.
- Appropriate treatment of hypoglycemia.
- Prompt treatment of hyperglycemia.
- Consistent and appropriate bedtime snacking.[14]

Dietitians teach the diet in stages, starting first with simple concepts and progressing to more difficult ones as the client's abilities and needs dictate.

Meal-Planning Strategies No single approach to diet therapy meets everyone's needs, and diet planners use several approaches to help clients achieve blood glucose control. Some diet strategies teach clients to use food guides or simple menus to plan diets. Traditionally, however, diet planners use the exchange system to help people with diabetes plan their diets.

Exchange Lists The exchange system sorts foods into three main groups by their proportions of carbohydrate, fat, and protein. The foods in these three groups—the carbohydrate group, the fat group, and the meat and meat substitutes group (protein)—are then organized into several exchange lists.

Appendix B includes the Canadian system, and Appendix C includes the U.S. exchange system.

The carbohydrate group includes these exchange lists:

- Starch (cereals, grains, pasta, breads, crackers, snacks, starchy vegetables, and dried beans, peas, and lentils).
- Fruit.
- Milk and some milk products (nonfat and very low-fat, low-fat, and whole).
- Other carbohydrates (desserts and snacks with added sugars and fats).
- Vegetables.

The fat group includes typical fats such as butter, margarine, oil, and salad dressing as well as high-fat foods such as nuts, olives, bacon, avocado, coconut, and cream cheese. The meat and meat substitutes group includes high-protein foods. Figure 25.2 on pp. 602–603 shows examples of foods and portion sizes for each group of foods; and Table 25.2 on p. 604 shows the energy and energy nutrients found in each exchange list.

By strictly defining portion sizes, all of the foods in a given exchange list provide approximately the same amounts of energy nutrients (carbohydrate, fat, and protein) and the same number of kcalories. Any food on a list can then be exchanged, or traded, for any other food on that same list without affecting a plan's balance or total kcalories. The box on pp. 605–606 shows how to plan a diet for diabetes using exchange lists.

The exchange system helps clients learn to recognize sources and types of fat. By including items like bacon and avocados on the fat list, the exchange system alerts users to foods that are unexpectedly high in fat. The fat list also shows which fats are monounsaturated, polyunsaturated, and saturated to help clients easily plan lipid-lowering diets. The exchange system encourages users to think of nonfat milk as milk and of whole milk as milk with added fat; and to think of very lean meats as meats and of lean, medium-, and high-fat meats as meats with

Figure 25.2

The Exchange System: Examples of Foods, Portion Sizes, and Energy-Nutrient Contributions

The Carbohydrate Group

Starch

1 starch exchange is like:

1 slice bread.

¾ c ready-to-eat cereal.

½ c cooked pasta, rice noodles, or bulgar.

⅓ c cooked rice.

½ c cooked legumes.[a]

½ c corn, peas, or yams.

1 small (3 oz) potato.

½ bagel, English muffin, or bun.

1 tortilla, waffle, roll, taco, or matzoh.

(1 starch = 15 g carbohydrate, 3 g protein, 0–1 g fat, and 80 kcal.)

Vegetables

1 vegetable exchange is like:

½ c cooked carrots, greens, green beans, brussels sprouts, beets, broccoli, cauliflower, or spinach.

1 c raw carrots, radishes, or salad greens.

1 lg tomato.

(1 vegetable = 5 g carbohydrate, 2 g protein, and 25 kcal.)

Fruits

1 fruit exchange is like:

1 small banana, nectarine, apple, or orange.

½ large grapefruit, pear, or papaya.

½ c orange, apple, or grapefruit juice.

17 small grapes.

⅓ cantaloupe (or 1 c cubes).

2 tbs raisins.

1½ dried figs.

3 dates.

1½ carambola (star fruit).

(1 fruit = 15 g carbohydrate and 60 kcal.)

[a]Count as 1 very lean meat exchange *plus* 1 starch exchange.

The Meat and Meat Substitutes Group (Protein)

Meat and substitutes (very lean)

1 very lean meat exchange is like:

1 oz chicken (white meat, no skin).

1 oz cod, flounder, or trout.

1 oz tuna (canned in water).

1 oz clams, crab, lobster, scallops, shrimp, or imitation seafood.

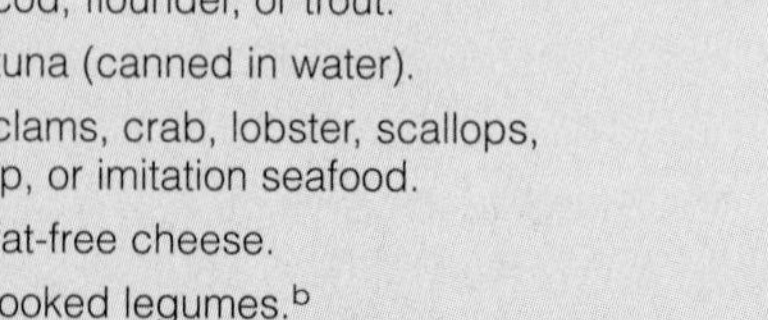

1 oz fat-free cheese.

½ c cooked legumes.[b]

¼ c nonfat or low-fat cottage cheese.

2 egg whites (or ¼ c egg substitute).

(1 very lean meat = 7 g protein, 0–1 g fat, and 35 kcal).

Meats and substitutes (lean)

1 lean meat exchange is like:

1 oz beef or pork tenderloin.

1 oz chicken (dark meat, no skin).

1 oz herring or salmon.

1 oz tuna (canned in oil, drained).

1 oz low-fat cheese or luncheon meats.

(1 lean meat = 7 g protein, 3 g fat, and 55 kcal.)

Meats and substitutes (medium-fat)

1 medium-fat meat exchange is like:

1 oz ground beef.

1 oz pork chop.

1 egg (high in cholesterol, limit 3 per week).

¼ c ricotta.

4 oz tofu.

(1 medium-fat meat = 7 g protein, 5 g fat, and 75 kcal.)

[b]Count as 1 very lean meat exchange *plus* 1 starch exchange.

Other carbohydrates

1 other carbohydrates exchange is like:

2 small cookies.

1 small brownie or cake.

5 vanilla wafers.

1 granola bar.

½ c ice cream.

(1 other carbohydrate = 15 g carbohydrate and may be exchanged for 1 starch, 1 fruit, or 1 milk. Because many items on this list contain added sugar and fat, their fat and kcalorie values vary, and their portion sizes are small.)

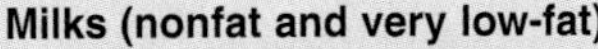

Milks (nonfat and very low-fat)

1 nonfat milk exchange is like:

1 c nonfat or 1% milk.

¾ c nonfat yogurt, plain.

1 c nonfat or low-fat buttermilk.

½ c evaporated nonfat milk.

⅓ c dry nonfat milk.

(1 nonfat milk = 12 g carbohydrate, 8 g protein, 0–3 g fat, and 90 kcal.)

Milks (low-fat)[c]

1 reduced-fat milk exchange is like:

1 c 2% milk.

¾ c low-fat yogurt, plain.

(1 reduced-fat milk = 12 g carbohydrate, 8 g protein, 5 g fat, and 120 kcal.)

Milks (whole)

1 whole-milk exchange is like:

1 c whole milk.

½ c evaporated whole milk.

(1 whole milk = 12 g carbohydrate, 8 g protein, 8 g fat, and 150 kcal.)

[c]The higher the fat content of milk and milk products, the greater the amount of saturated fat and cholesterol.

Meats and substitutes (high-fat)[d]

1 high-fat meat exchange is like:

1 oz pork sausage.

1 oz luncheon meat (such as bologna).

1 oz regular cheese (such as cheddar or swiss).

1 slice bacon

1 small hot dog (turkey or chicken).[e]

2 tbs peanut butter.[f]

(1 high-fat meat = 7 g protein, 8 g fat, and 100 kcal.)

[d]These foods are high in saturated fat, cholesterol, and kcalories.
[e]A beef or pork hot dog counts as 1 high-fat meat exchange *plus* 1 fat exchange.
[f]Count as 1 high-fat meat exchange *plus* 1 fat exchange.

The Fat Group

Fats

1 fat exchange is like:

1 tsp butter.[f]

1 tsp margarine or mayonnaise (1 tbs reduced fat).

1 tsp any oil.

1 tbs salad dressing (2 tbs reduced fat).

8 large black olives.

10 large peanuts.

⅛ medium avocado.

1 slice bacon.[f]

2 tbs shredded coconut.[f]

1 tbs cream cheese (2 tbs reduced fat).[f]

(1 fat = 5 g fat and 45 kcal.)

[f]Butter, bacon, coconut, and cream cheese contain saturated fats.

Table 25.2 The Exchange Lists: Energy Nutrients per Serving

Group/Lists	Carbohydrate(g)	Protein (g)	Fat (g)	Energy (kcal)
Carbohydrate Group				
Starch	15	3	1 or less	80
Fruit	15	—	—	60
Milk				
Nonfat and very low-fat	12	8	0–3	90
Low-fat	12	8	5	120
Whole	12	8	8	150
Other carbohydrates	15	varies	varies	varies
Vegetable	5	2	—	25
Meat and Meat Substitutes Group				
Meat				
Very lean	—	7	0–1	35
Lean	—	7	3	55
Medium-fat	—	7	5	75
High-fat	—	7	8	100
Fat Group				
Fat	—	—	5	45

added fat. To that end, foods on the milk and meat lists are separated into categories based on their fat contents. Meat exchanges that are particularly high in cholesterol are noted. The starch list also specifies grain products that contain added fat.

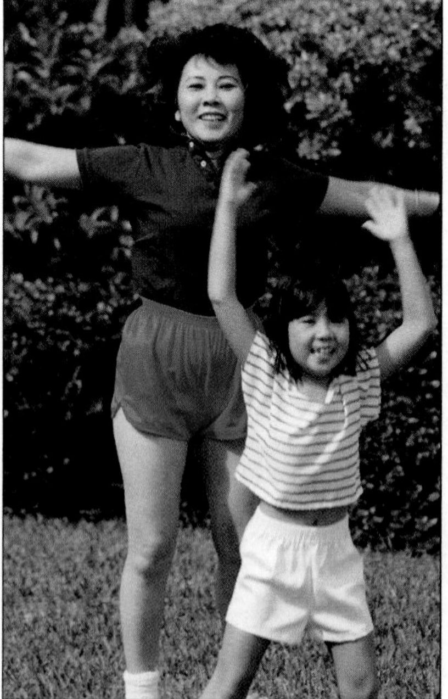

Physical activity plays an important role in the management of diabetes.

Carbohydrate Counting Another diet strategy that is gaining wide use is called *carbohydrate counting.* It teaches clients to focus mainly on the carbohydrate contents of foods. Clients learn to eat consistent amounts of carbohydrates at meals and snacks. They must have the motivation to monitor their blood glucose and keep records of their glucose levels, the time and amount of carbohydrate they eat at each meal, and the time and amount of insulin or oral antidiabetic agents they use. They must also be able to perform the mathematical operations necessary to calculate their carbohydrate intakes. The box on p. 607 provides more information about carbohydrate counting. Regardless of the diet strategy, all clients receive instructions on planning well-balanced and healthy meals, eating consistent amounts of foods at regular times, and maintaining a desirable weight.

Physical Activity

The client with diabetes should be carefully evaluated to determine a safe type and amount of physical activity. For people with type 2 diabetes, a regular program of physical activity improves blood glucose control, contributes to weight loss, improves blood lipid levels, and lowers blood pressure. Although physical activity has not been shown to aid blood glucose control in people with type 1 diabetes, its value lies in its benefits to the cardiovascular system. People who use insulin, however, need to take special precautions when exercising.

How to

Plan a Diet for Diabetes Using Exchange Lists

The diet planner (usually a dietitian) carefully considers the client's lifestyle and medical goals in planning a diet for diabetes. Planning a diet using exchange lists takes time at first, but with practice, the planning process becomes routine. This box describes a simplified diet plan.

The first step is to assess each individual to determine what weight is reasonable and how many kcalories will be necessary to achieve or maintain that body weight. Chapter 16 described ways of estimating desirable body weights, and the box on p. 130 in Chapter 6 showed how to estimate energy needs based on body weight and physical activity levels. Chapter 9 described kcalorie needs for safe weight loss (see p. 215) and weight gain (see pp. 220–221). The growth charts in Appendix E can be used to estimate desirable weights for children, and Chapter 13 (see p. 330) described their energy needs. Remember, though, that desirable weights and calculated energy needs are estimates only.

For this example, we will use a man who is 6 feet tall and is comfortable with the weight of 178 pounds that he has maintained throughout his adult life. From an assessment of food intake, the dietitian estimates that the man has maintained his weight on about 2900 kcalories per day with 25 percent of kcalories from protein, 45 percent from carbohydrate, and 30 percent from fat.

1. The first step is to determine the grams of protein, carbohydrate, and fat recommended for a diet for diabetes.
   - 10 to 20% of the kcalories from protein.
   - 45 to 60% of the kcalories from carbohydrate.
   - 30% or less of the kcalories from fat.

   For 2900 kcalories, this division of nutrients translates into grams as follows:
   - Protein:

     10% × 2900 kcal = 290 kcal.
     290 kcal ÷ 4 kcal/g = 73 g.

     20% × 2900 kcal = 580 kcal.
     580 kcal ÷ 4 kcal/g = 145 g.

   Thus the man needs between 73 and 145 grams of protein.
   - Carbohydrate:

     45% × 2900 kcal = 1305 kcal.
     1305 kcal ÷ 4 kcal/g = 326 g.

     60% × 2900 kcal = 1740 kcal.
     1740 kcal ÷ 4 kcal/g =435 g.

   Thus the man needs between 326 and 435 grams of carbohydrate.
   - Fat:

     30% × 2900 kcal = 870 kcal.
     870 kcal ÷ 9 kcal/g = 97 g.

   Thus the man needs about 97 grams of fat or less.
2. The dietitian recognizes that the client will need to make dietary changes to conform to a healthy eating plan. To minimize the changes the client must make, the dietitian decides to plan the diet to include 20 percent protein or 580 kcalories. Thus 80 percent or 2320 kcalories remain for carbohydrate and fat. After reviewing information about the man's blood lipids, which are within acceptable limits, the dietitian plans the diet to keep fat at the current level of 30 percent (870 kcalories). This means that 50 percent of the kcalories (1450 kcalories) remain for carbohydrate.

   2900 total kcal − 580 protein kcal − 870 fat kcal =
   1450 carbohydrate kcal.
3. To translate the kcalories from fat and carbohydrate to grams:

   870 fat kcal ÷ 9 kcal/g =
   96.6 g (round down to 96 to limit fat).
   1450 carbohydrate kcal ÷ 4 kcal/g =
   362.5 g (round up to 363).

   Thus the diet will provide 2900 kcalories with a distribution of 20 percent protein (145 grams or 580 kcalories), 50 percent carbohydrate (363 grams or 1450 kcalories), and 30 percent fat (96 grams or 870 kcalories).
4. Now it is time to translate the diet prescription into a meal plan. Table 25.2 on p. 604 shows the grams of carbohydrate, protein, and fat and the energy value in each serving on an exchange list. Using this table and the client's food intake record as a guide, the dietitian first plans servings of foods that contain carbohydrate, then protein, and finally fat, trying to match foods as closely as possible to the client's usual food intake. This process takes practice and requires some adjusting based on trial and error. Most often, the final result does not fit the meal plan exactly, but comes close. Table 25.3 on p. 606 shows how the dietitian might plan a day's exchanges for the man in this example. Note that the plan falls within the guidelines of the Daily Food Guide on pp. 14–15. The plan uses nonfat milk and lean meat exchanges for calculations. Lower-fat foods are encouraged. If the client occasionally

(continued)

Plan a Diet for Diabetes Using Exchange Lists *(continued)*

Table 25.3 A Day's Exchange for a Sample 2900-kCalorie Diet

Exchange Group/List	Number of Exchanges	Carbohydrate (g)	Protein (g)	Fat[a] (g)
Carbohydrate group[b]				
Starch	13	195	39	0
Fruit	7	105	—	—
Milk	3	36	24	0
Vegetable	6	30	12	—
Meat and meat substitutes group				
Lean	10	—	70	30
Fat group	13	—	—	65
Total grams		366	145	95
Total kcalories		1464	580	855
% kcalories		50.5	20	29.5

[a]To ease calculation, exchanges from the carbohydrate groups are assumed to have 0 grams fat. If the client uses a fat-containing exchange, the fat can be deducted from the daily fat allowance.

[b]Foods from the "other carbohydrates" list can be substituted for a starch, fruit, or milk list exchange. Any fat in the selected food is then deducted from the daily fat allowance.

chooses to use another type of milk or meat, the number of fat servings must be adjusted accordingly. For example, if the client eats 4 ounces of a high-fat meat (32 grams of fat) instead of lean meat (12 grams of fat), he must then use four fewer fat exchanges during the day (20 grams of fat). The plan shown in Table 25.3 does not include the "other carbohydrates" list; starches and other foods (described later) can be substituted for foods on this list.

5. Distribute foods into meals that fit the client's usual eating patterns. Table 25.4 shows how the day's exchanges might be divided for the man in this example. With this information in hand, the dietitian and client can begin to fill in the plan with real foods to create a sample menu such as the one shown on p. 607. The client is reminded to eat about the same amount of carbohydrate at about the same time each day.
6. Teach clients how to tailor the diet to meet their own preferences. For example, foods from the starch, fruit, milk, and other carbohydrate lists contain similar amounts of energy and carbohydrate and can be substituted for one another from time to time. Regular substitution is discouraged, however, because each list makes unique contributions to other nutrient needs. A client who regularly substitutes fruit for milk, for example, may not be getting enough calcium. The client who regularly substitutes milk for a starch or fruit may not be getting enough fiber.[a]
7. Several servings of free foods can be included as long as they are spread throughout the day. Free foods contain up to 20 kcalories and 5 grams of carbohydrate per serving.

[a]M. Wheeler, M. J. Franz, and P. Barrier, Helpful hints: Using the 1995 exchange lists for meal planning, *Diabetes Spectrum* 8 (1995): 325–326.

Table 25.4 Translating a Day's Exchanges into Meals

Exchange Group/List	Number of Exchanges[a]	Breakfast	Lunch	Midafternoon Snack	Supper	Bedtime Snack
Carbohydrate group						
Starch	13	3	3	2	3	2
Fruit	7	2	1	1	1	2
Milk	3	1	1		1	
Vegetable	6		3		3	
Meat and meat substitutes group						
Lean	10		3	1	4	2
Fat group	13	3	3	2	3	2

[a]From Table 25.3.

How to

Help Clients Count Carbohydrates

1. Begin by calculating how much carbohydrate the person usually eats at each meal and snack. For an example, assume that the client is a woman who maintains her weight and controls her blood glucose levels on an intake of 1700 kcalories. Use exchanges to determine her usual carbohydrate intake. In this example, the client and dietitian determined that her usual eating pattern includes:
   - Breakfast: 2 starch, 1 fruit, 1 milk.
   - Lunch: 2 starch, 1 fruit, 2 vegetables.
   - Midafternoon snack: 1 fruit, 1 milk.
   - Supper: 2 starch, 1 fruit, 2 vegetables.
   - Bedtime snack: 1 starch, 1 fruit.
2. Next, use Table 25.2 to determine the grams of carbohydrate in each meal. Breakfast, for example, contains:

   2 starch = 2 × 15 g carbohydrate = 30 g
   1 fruit = 1 × 15 g carbohydrate = 15 g
   1 milk = 1 × 12 g carbohydrate = 12 g
   Total = 57 g carbohydrate.

   Using the same method to calculate the carbohydrate from other meals and snacks, the carbohydrate contents are as follows:
   - Lunch: 55 g.
   - Midafternoon snack: 27 g.
   - Supper: 55 g.
   - Bedtime snack: 30 g.
3. Teach the client to include these same amounts of carbohydrate consistently at each meal and snack. Clients can use exchanges, food labels, and food composition tables to determine the carbohydrate contents of the foods they eat.
4. Encourage clients to carefully weigh and measure all foods initially. Once clients are familiar with portion sizes, they can weigh and measure foods less often, but should still check portion sizes occasionally.
5. Remind clients that they still need to be aware of the total amount of food they eat as well as the type and amount of fat they are eating.

Clients who use insulin learn how much insulin they need to cover their carbohydrate intakes. Then, if they change their carbohydrate intakes on occasion, they can adjust their insulin dose accordingly.

Sample 2900-kCalorie Diet Menu

Breakfast	*Lunch*	*Midafternoon Snack*	*Supper*	*Bedtime Snack*
½ c bran cereal	1 cup pasta served with: ½ c spaghetti sauce 2 oz meatballs (medium fat) 2 tbs grated parmesan cheese	1 fat-free granola bar	4 oz grilled salmon	¼ c lowfat cottage cheese
1 c nonfat milk	1½ c broccoli	2 tbs peanut butter spread over slices of 1 small apple	⅔ c rice	1 c fruit cocktail
1 bagel (2 oz)	2 tsp margarine		1 whole-wheat roll	1 small frosted cupcake
3 tsp margarine	17 small grapes		1 c cooked carrots	1 c nonfat milk
1 banana	1 c nonfat milk		1 tsp margarine	
Coffee			1 c mixed green salad	
			Salad dressing made with 2 tsp olive oil and wine vinegar	
			1 c diced cantaloupe	
			1 c nonfat milk	

Physical Activity and Blood Glucose Levels People who need insulin should check their blood glucose before and after engaging in physical activity to help them maintain their blood glucose levels. Physical activity should not be undertaken if blood glucose is too low (less than 100 milligrams per 100 milliliters) or too high (greater than 300 milligrams per 100 milliliters). If blood glucose is too low, hypoglycemia can quickly develop. If blood glucose is too high, exercising can cause blood glucose levels to rise even higher.

Physical Activity and Food Intake All persons, especially those with diabetes, need to make sure they are adequately hydrated before and during physical activity by drinking liquids throughout the day. The person who needs insulin may need to eat before, during, and after vigorous physical activity. Especially important is carbohydrate, which is readily available from fruits, fruit juices, yogurt, crackers, and other starches. The amount of carbohydrate the client needs depends on the type of activity, its duration, the client's individual responses, and the results of blood glucose tests.

Drug Therapy for Diabetes

People with type 2 diabetes can sometimes control their blood glucose without medications by using a combination of diet and physical activity. When these measures fail, one or a combination of oral antidiabetic agents may be prescribed. Oral antidiabetic agents do not replace diet and physical activity in diabetes management; advise clients to continue these therapies as instructed. Oral antidiabetic agents have a maximum dose; if blood glucose cannot be controlled at the maximum dose, the physician may prescribe insulin or insulin analogs, alone or in combination with oral antidiabetic agents. Estimates suggest that about 29 percent of people with type 2 diabetes require insulin.[15] All people with type 1 diabetes need insulin and do not benefit from oral antidiabetic agents.

Oral antidiabetic agents include sulfonylureas, metformin, acarbose, troglitazone, and replaglinide. Sulfonylureas and replaglinide are also called **hypoglycemic agents** because they stimulate insulin release and can cause hypoglycemia. Sulfonylureas are the type of antidiabetic agent that can react with alcohol and cause flushing of the skin and a rapid heartbeat. Acarbose is the type of antidiabetic agent that reduces the rate of digestion of complex carbohydrate.

Oral Antidiabetic Agents The number of oral agents available to treat people with type 2 diabetes has grown markedly in recent years.[16] Because some oral agents work by stimulating the release of insulin from the beta cells, people who take them may develop hypoglycemia. Other oral agents work by reducing insulin resistance and depressing the liver's production of glucose without raising insulin levels. Still other oral agents reduce the rate of complex carbohydrate and sucrose digestion in the intestine, consequently slowing the rate of glucose absorption. People using oral agents that slow the rate of carbohydrate absorption must use glucose to treat episodes of hypoglycemia.

Of note among the newest classes of oral antidiabetic agents is troglitazone, which works primarily by lessening peripheral insulin resistance.[17] Troglitazone also lowers triglyceride levels and raises HDL and has been shown to improve insulin resistance and reduce blood pressure in people with impaired glucose tolerance. For these reasons, troglitazone may prove valuable in preventing or delaying the development of type 2 diabetes or the complications associated with insulin resistance, and it is being included as an intervention in the Diabetes Prevention Program.[18]

Although the availability of insulin has been lifesaving, insulin therapy cannot achieve the same degree of blood glucose control as a body that produces its own insulin.

Insulin and Insulin Analogs For people who need insulin, commercial insulin is available in different forms that act with different timings so that it can be delivered in a manner that mimics the body's normal insulin actions as closely as possible. As Figure 25.3 shows, insulins can be either rapid acting (regular), intermediate acting (NPH and lente), or long acting (ultralente). Insulin analog (lispro) is a rapid-acting insulin whose amino acid composition has been modified so that it works faster and has a shorter duration of action.[19] As a result, lispro reduces after-meal hyperglycemia to a greater extent than regular insulin and is also associated with a lower risk of hypoglycemia between meals and during the night.

Insulin Delivery People who need insulin must inject it or use pumps to deliver the insulin they need. Insulin is a protein, so if it is taken orally, digestive enzymes reduce it to amino acids. Because the pancreas of a person who needs insulin cannot synthesize insulin or cannot synthesize enough of it from amino acids, it would not be available to the body. The person who chooses injections often receives a mixture of two or more types of insulin three or more times daily; single injections are seldom effective.

External pumps, about the size of a beeper, hold enough insulin to meet needs for two or three days. From the pump, insulin enters the body through tubing and a needle inserted into the abdominal area. Personal preferences, motivational level, and financial considerations guide clients in deciding which delivery system works best for them.

Some people experience a temporary remission from diabetes after their initial treatment with insulin—a time referred to as the "honeymoon phase." Perhaps with insulin treatment and relief from hyperglycemia, the insulin-producing cells of the pancreas become able to function again—but only temporarily.

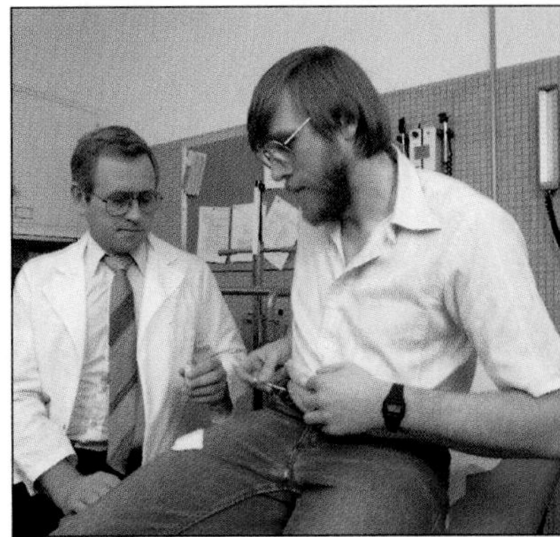

Injections are one option for delivering insulin for people with diabetes.

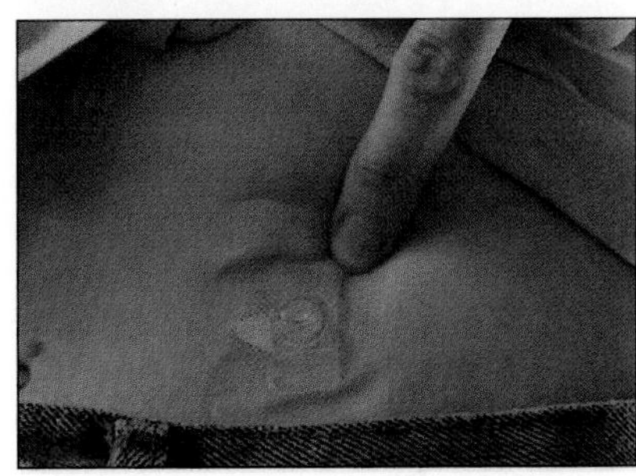

For people who use insulin pumps, insulin passes from the pump through tubing that enters the body through the abdomen as shown.

Insulin and Food Intake Normally, the body secretes a constant, baseline amount of insulin at all times and more after blood glucose rises following a meal. The person with type 1 diabetes often receives NPH (intermediate-acting) insulin to meet baseline needs and regular (rapid-acting) insulin or insulin analogs to process energy nutrients after meals. People with type 2 diabetes may be treated with insulin alone, or they may use it in combination with oral antidiabetic agents. Often a single injection of NPH (intermediate-acting) insulin is given at bedtime.[20] Insulin analogs may also be used in some cases.

Insulin and Physical Activity Generally, insulin should be taken more than an hour before physical activity. Vigorous physical activity and warm temperatures speed blood flow, increase the rate of insulin absorption, and set the stage for a hypoglycemic reaction, which may occur after several hours. Reducing the insulin dose before and after the activity by up to 30, or even 50, percent can help to prevent this sequence of events.

multiple daily injections: delivery of different types of insulin by injection three or more times daily.

Pancreas Transplants For people with type 1 diabetes who encounter serious problems managing their diseases with insulin, pancreas transplants have been

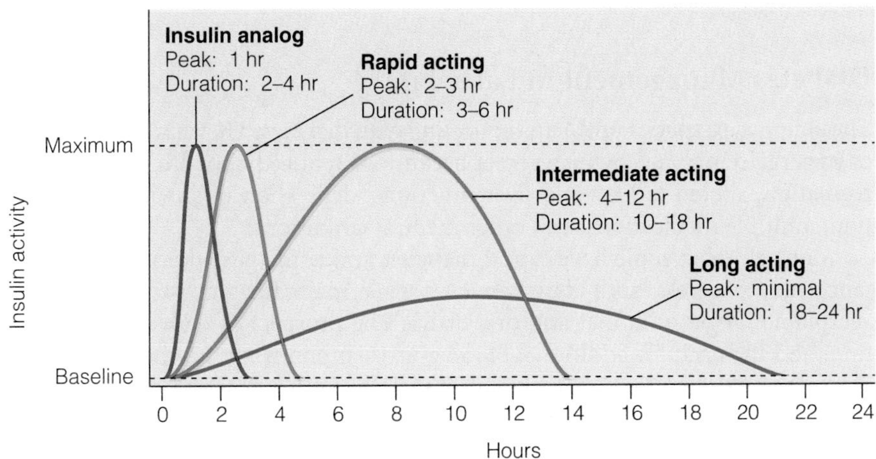

Figure 25.3
Actions of Insulin Types

successful in providing functional, insulin-producing beta cells. Most often, a pancreas transplant is combined with a kidney transplant, because the person often has significant kidney problems as well. A combination pancreas-kidney transplant can eliminate the need for insulin and for dialysis (see Chapter 27) and greatly enhance the person's quality of life. One year after surgery, 75 percent of people who receive a pancreas-kidney transplant do not need insulin.[21] For people undergoing a pancreas transplant only, 50 percent do not need insulin after one year.

Children with Diabetes

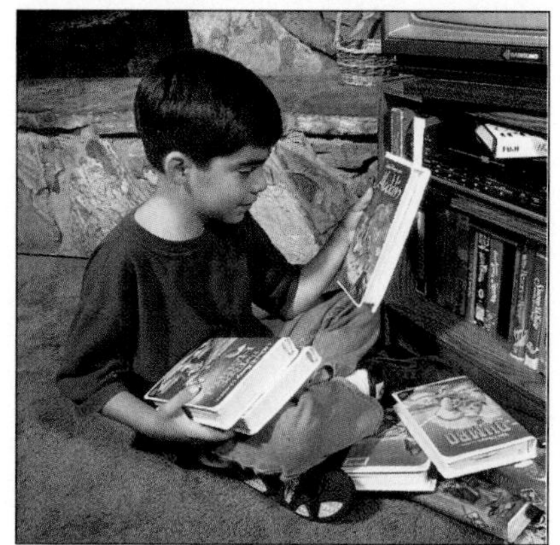

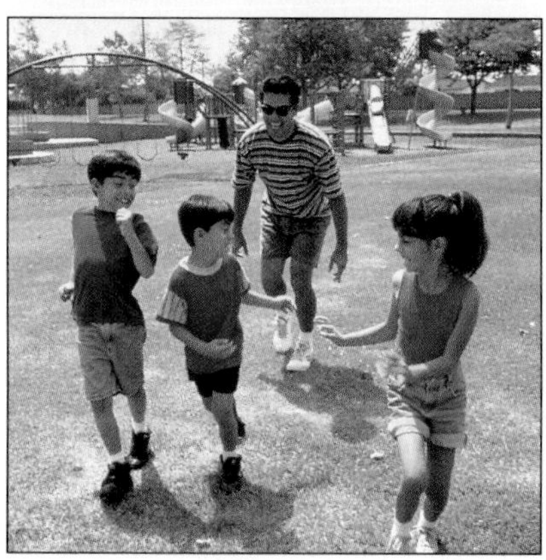

Because a child's activities vary from day to day, food and insulin needs may also change.

The overall approach to diabetes remains the same throughout life, but special problems may become apparent at different ages. Consider some of the problems of children with diabetes. Like those of all children, the energy and nutrient needs of children with diabetes keep changing throughout the growing years. Children's appetites and activities vary widely from day to day. A child may eat like a horse one day and like a mouse the next. A teen may spend hours walking around the mall one day and spend the next day watching TV. Growth and activity influence the needs for food and insulin, and management must adjust to meet those needs.

Meal Plans To support growth and development, children with diabetes need flexible, balanced meals and snacks that offer wide varieties of foods from each of the food groups. Dietitians often teach children and caregivers carbohydrate counting to provide flexibility from day to day. Concentrated sweets are allowed within the context of a healthy diet. Caregivers need not force children to finish meals, but should encourage them not to skip meals either, because hypoglycemia can result, particularly if the child uses insulin. Meals are best taken at about the same times each day, and children with diabetes can eat the same foods as the rest of the family.

Family Lifestyles Successful diet management incorporates prescribed meals into existing family lifestyles and eating patterns. For children who take insulin, a typical eating plan may include three meals with two to three snacks a day. Snacks before bedtime help prevent hypoglycemia while the child is sleeping, especially if the child engages in strenuous activity late in the day. Caregivers should vary meals and snacks to prevent boredom, provide enough to share with friends, and avoid identifying foods as "good" or "bad." Such connotations create unrealistic expectations or fears and invite the development of manipulative eating behaviors. The case study on p. 611 presents a child with type 1 diabetes.

Diabetes Management in Later Life

The elderly face special problems in dealing with diabetes. They have greater risks of hyperglycemia and hypoglycemia because of reduced appetite, altered thirst regulation, altered kidney and liver functions, depression or mental deterioration, multiple medications, and other medical conditions.

Many elderly people have type 2 diabetes, and with aging their insulin resistance may progress until they can no longer maintain glucose levels within acceptable ranges with diet and oral drugs. The prospect of daily insulin injections and the need for additional blood glucose monitoring may be overwhelming to the elderly client. Those who have also suffered a loss of vision as a consequence of aging or diabetes may have great difficulty drawing correct insulin doses, giving self-injections, or using glucose strips or meters to monitor blood glucose. The inability to perform these necessary tasks may make it impossible for the person to live independently.

Case Study

Child with Type 1 Diabetes

One year ago, Yusuf, a 12-year-old boy, was diagnosed with type 1 diabetes. The initial diagnosis was made after Yusuf's parents became concerned when he began to lose weight, urinate excessively, and complain of thirst. Aware of a family history of diabetes, the parents quickly sought medical help. Since that time, Yusuf's diabetes has been well controlled. Recently, however, Yusuf was admitted to the emergency room, complaining of nausea, vomiting, and intense thirst. He had a fever, and his blood glucose records from the previous day showed that his blood glucose was high throughout the day. The physician observed that Yusuf was confused and breathing with difficulty and also noted the smell of acetone on his breath. Urine tests were positive for glycosuria and ketonuria, and Yusuf's blood glucose was 400 milligrams per 100 milliliters. The diagnosis was ketosis and acidosis.

Describe the metabolic events that led to the symptoms associated with diabetes (before diagnosis), as well as those associated with ketosis and acidosis. Were Yusuf's physical symptoms and laboratory tests consistent with this diagnosis? How can you distinguish between diabetic ketoacidosis and hypoglycemia?

When Yusuf recovers, what advice can you offer him to prevent future incidents of ketosis and acidosis? Assume that Yusuf had instructions for a diet for diabetes. What dietary modifications would you advise him to make?

Think about and discuss the influence of Yusuf's age on his outlook and ability to cope with diabetes. What problems does his age pose? Consider some ways you might help him deal with these problems. Regarding his future, describe the possible role of diet in preventing the chronic complications of diabetes.

The elderly may also lack the financial resources or social support necessary to help them cope with their diabetes. Caring health care professionals address these problems and help elderly clients find solutions.

Mastering Glucose Control

The health care team determines acceptable fasting and after-meal blood glucose goals for each client. Motivated clients who wish to tightly control their blood glucose levels keep accurate records of food intake, physical activity, blood glucose measurements, medication administration, and illnesses. People with type 1 diabetes learn how their blood glucose changes in response to foods and activity schedules and learn how to alter their insulin doses when their normal routines are disrupted.

Monitoring Diabetes Management

Clients with diabetes and the health care team depend on several tests to see how well they are managing their disorders. At medical appointments, the health care team reviews the client's test results and records to monitor the client's progress and look for patterns that suggest the need for adjustments in the treatment plan. Early intervention can avoid unnecessary complications.

Blood Glucose Monitoring To monitor blood glucose, the client pricks a finger to get a blood sample and transfers the blood to a disposable strip. Most often clients rely on computerized strips to measure the glucose concentration from the blood sample. Less accurate and less costly are paper strips that change to different colors, depending on the blood glucose concentration.

At first, the person performs blood tests at least seven times a day: before each meal, two hours after each meal, and at bedtime. As people learn how their blood glucose responds to food intake and physical activity, they can test less often—about three or four times daily.

The **carbohydrate-to-insulin ratio** identifies the number of units of insulin a person needs per gram of carbohydrate. The lower the ratio, the more insulin a person needs to cover carbohydrate intake.

Although blood glucose goals for individuals may vary, in general glucose before meals should be between 80 and 120 mg per 100 ml (normal, <110 mg/100 ml), and glucose at bedtime should be between 100 and 140 mg (normal, <120 mg/100 ml). In the Diabetes Control and Complications Trial, study groups receiving intensive therapy achieved an average fasting blood glucose of 155 mg/100 ml.

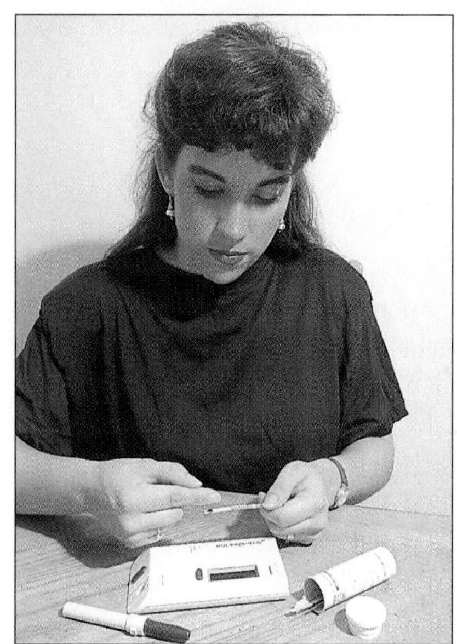

Blood glucose monitoring helps people with diabetes maintain blood glucose in a safe range.

Glycated hemoglobin is also called **glycosylated hemoglobin.** The type of glycated hemoglobin most commonly measured to monitor blood glucose control is **hemoglobin A_{1c}**. In the Diabetes Control and Complications Trial, groups receiving intensive treatment achieved an average hemoglobin A_{1c} of 7.2% (normal, <6%).

dawn phenomenon: early morning hyperglycemia that develops in response to increased levels of counterregulatory hormones that act to raise blood glucose during an overnight fast. Without adequate insulin, the glucose cannot enter the cells and remains in the blood.

rebound hyperglycemia: hyperglycemia resulting from excessive secretion of counterregulatory hormones in response to excessive insulin and consequent low blood glucose levels; also called the **Somogyi** (so-MOHG-yee) **effect.**

Glycated Hemoglobin In addition to checking blood glucose records, physicians monitor blood glucose control by evaluating the percentage of hemoglobin A_{1c}. As blood glucose rises, glucose attaches to amino acids on hemoglobin molecules and remains there until the red blood cells that carry the hemoglobin die (about 120 days). Hemoglobin A_{1c} reflects diabetes control over the past two to three months, rather than just prior to the test.

Urinary Ketones Health care professionals often recommend that clients with consistently high blood glucose also monitor ketones in the urine, especially during illness. As described earlier, high blood glucose may predispose the person to ketosis and coma. Such a course is especially likely during illness.

Other Measures The health care team also monitors the client's weight, blood lipid levels, blood pressure, and reflexes and checks for early signs of complications. Urine tests can help detect the early signs of kidney disease, eye exams help detect the early signs of retinopathy, and foot exams help detect early signs of infection.

Managing Hyperglycemia and Hypoglycemia

Table 25.5 summarizes strategies for adjusting treatment plans to correct problems with hyperglycemia and hypoglycemia. Except for adjusting insulin doses, the strategies shown in the table apply to people with either type 1 or type 2 diabetes. For people with type 2 diabetes who have consistent problems with hyper- or hypoglycemia, the physician may add, change, or adjust the medications the client uses.

Hyperglycemia A pattern of hyperglycemia before lunch or dinner in a person with a consistent carbohydrate intake and activity schedule clues the health care professional to look back to the previous meal and make corrections. (Recall that it takes time for food to be digested and absorbed before blood glucose rises.) Hyperglycemia can occur whenever counterregulatory hormone levels become elevated. Thus hyperglycemia can be present when the person arises, as a consequence of an overnight fast; when blood glucose levels are high and the person engages in strenuous physical activity; when a person uses too much insulin; or when the person develops an illness or infection.

Severe Hyperglycemia Even a minor illness such as a cold or flu can cause blood glucose to rise dramatically. During illness, clients may be advised to increase their doses of insulin, reduce carbohydrate intakes somewhat, or a combination of both. A major concern is the prevention of dehydration and vomiting.

Severe hyperglycemia and ketosis in the person with diabetes constitute a medical emergency that can lead to coma and death. The physician treats severe hyperglycemia and ketoacidiosis in the hospital by carefully administering insulin and correcting fluid and electrolyte and acid-base imbalances using IV fluids.

Enteral and Parenteral Formulas The indications for enteral and parenteral nutrition for people with diabetes are the same as for other people (see Chapters 21 and 22). However, when people with diabetes require enteral or parenteral formulas, adjustments may need to be made for the large amounts of carbohydrates the formulas contain. Often health care professionals provide or adjust insulin doses to cover the higher carbohydrate loads.[22] If insulin fails to control hyperglycemia, people on parenteral nutrition may need to receive less energy from dextrose and more from IV lipid emulsions. People who cannot tolerate the carbohydrate in standard enteral formulas may benefit from specially designed formulas that contain less total carbohydrate (see Appendix G).

Table 25.5 Strategies for Managing Hyperglycemia and Hypoglycemia	
Problem	**Possible Solutions**[a]
Hyperglycemia	
Before breakfast	▪ Adjust dose of intermediate-acting insulin at bedtime.[b]
Before lunch	▪ Adjust morning dose of rapid-acting insulin.[b]
	▪ Reduce amount of carbohydrate at breakfast.
	▪ Reduce or omit midmorning snack.
	▪ Change time of breakfast or midmorning snack.
	▪ Add physical activity after breakfast.
Before dinner	▪ Adjust afternoon dose of rapid-acting insulin.[b]
	▪ Reduce carbohydrate at lunch.
	▪ Reduce or omit midafternoon snack.
	▪ Change time of lunch or midafternoon snack.
	▪ Add physical activity between lunch and dinner.
At bedtime	▪ Adjust insulin dose before dinner.[b]
	▪ Reduce amount of carbohydrate at dinner.
	▪ Reduce or omit evening snack.
	▪ Add physical activity after dinner.
Hypoglycemia	
Before breakfast	▪ Adjust dose of intermediate- or long-acting insulin at bedtime.[b]
	▪ Add carbohydrate at evening snack.
	▪ Avoid strenuous activity late in the day.
Before lunch	▪ Adjust morning dose of rapid-acting insulin.[b]
	▪ Add carbohydrate at breakfast.
	▪ Add a morning snack.
	▪ Change time of breakfast, lunch, or morning snack.
	▪ Adjust physical activity schedule.
Before dinner	▪ Adjust afternoon dose of rapid-acting insulin.[b]
	▪ Add carbohydrate at lunch.
	▪ Add an afternoon snack.
	▪ Change time of lunch, dinner, or afternoon snack.
	▪ Adjust physical activity schedule.
At bedtime	▪ Adjust insulin dose before dinner.[b]
	▪ Add carbohydrate at dinner.
	▪ Add an evening snack.
	▪ Change time of dinner or evening snack.

[a]Skilled health care professionals gather additional data to find the best solution for problems with blood glucose control. Is the problem an isolated occurrence or a pattern? Has food intake changed? If yes, why? Has the activity level changed? Has illness been a problem? Whenever altering the diet or physical activity plan is difficult for the client, insulin is adjusted to correct problems, if possible.
[b]Insulin doses can be adjusted in amount or timing or both.

Hypoglycemia Tightly controlling blood glucose reduces the risk of chronic complications, but increases the risk of hypoglycemia. Hypoglycemia can result from an overdose of insulin or some oral agents, strenuous physical activity, skipped meals, or inadequate food intake. People with diabetes and those who spend time with them need to learn to recognize the symptoms of hypoglycemia (see the margin on p. 597).

Hypoglycemia in a person who uses insulin is also called an **insulin reaction** or **insulin shock.**

Judicious treatment of hypoglycemia prevents overtreatment and subsequent hyperglycemia. As soon as the symptoms are observed, the person needs 10 to 15 grams of carbohydrate. Except for people with type 2 diabetes who take oral agents that interfere with the digestion of sucrose and complex carbohydrates, any carbohydrate source that is readily available and easy to eat is a good choice.[23] It is best to avoid foods that also contain fat because fat slows the absorption of carbohydrate. Blood glucose is then checked within 15 to 20 minutes to see if it has risen to an acceptable level. If not, an additional 10 to 15 grams of carbohydrate are given, and blood glucose is rechecked. The procedure continues until blood glucose returns to an acceptable range. Advise clients to carry some convenient source of carbohydrates with them at all times, so they can act immediately when hypoglycemic symptoms occur.

Easy-to-eat sources of carbohydrate (10 to 15 g per serving):
- *2 to 3 tsp honey.*
- *4 to 5 candies (such as Lifesavers).*
- *5 to 6 large jelly beans.*
- *4 to 6 oz sweetened soft drink.*
- *4 oz orange or other fruit juice.*
- *1 tbs icing from a can or tube.*
- *Glucose gel or tablets (check label for amount).*

nocturnal hypoglycemia: hypoglycemia that occurs while a person is sleeping.

If the person frequently experiences hypoglycemia, treatment can lead to excessive weight gain. Repeated hypoglycemic episodes must be carefully investigated and corrected. People prone to nocturnal hypoglycemia may be advised to wake up during the night and test their blood glucose. Strategies that may help eliminate the problem are to snack consistently at bedtime, undertake strenuous activity earlier in the day, or reduce the insulin dose following evening physical activity.

Missed Meals Associated with Illness When people with diabetes are ill, they need to have some carbohydrate to forestall hypoglycemia. If appetite is poor, people can use juice, flavored gelatin, soft drinks, or frozen juice bars to meet their carbohydrate needs. In the hospital, if a person with diabetes misses a meal, different procedures may be employed. One procedure provides at least half the prescribed carbohydrate and kcalories within three hours of the missed meal. If this cannot be done, the physician may change the insulin schedule, give IV dextrose, and/or change the diet prescription to include more simple carbohydrates.

Severe Hypoglycemia In severe cases of hypoglycemia, the person may be disoriented, unable to recognize a hypoglycemic reaction, and unable to swallow safely. In such cases, the person needs to receive IV glucose or the hormone glucagon or both to counteract the insulin reaction. Without treatment, the person may lapse into shock and die.

Diabetes Management in Pregnancy

Pregnancy in all women elevates blood insulin and alters insulin resistance. Blood insulin begins to rise soon after conception, and the cells respond by storing energy nutrients to provide for the developing fetus. Later in pregnancy, insulin remains high, but the cells become insulin resistant. Levels of hormones that act antagonistically to insulin rise. This hormonal shift signals the body to stop storing energy fuels and to begin allowing the fetus to rapidly take up energy nutrients. Because pregnancy stresses the glucose regulatory system in these ways, women with diabetes should expect control to become more difficult during pregnancy.

The hormones that oppose the action of insulin during late pregnancy are placental lactogen, cortisol, prolactin, and progesterone.

Health Risks Associated with Diabetes during Pregnancy Women with diabetes who are contemplating pregnancy should know that uncontrolled diabetes presents risks for both mother and infant.[24] Women with gestational diabetes have a higher risk than women without diabetes of requiring a cesarean delivery or experiencing pregnancy-induced hypertension. Infants may suffer increased mortality and morbidity, congenital abnormalities, and other complications

Reminder: High blood pressure that develops during pregnancy is known as *pregnancy-induced hypertension* and may signal the onset of other complications (see Chapter 12).

such as severe hypoglycemia or respiratory distress, both of which can be fatal. Women need to receive preconception care, which aims to achieve the best possible blood glucose control before pregnancy, and continued prenatal care to maintain blood glucose control during pregnancy.

Gestational Diabetes Women who have never had diabetes or never knew that they had it may be diagnosed with diabetes during pregnancy (gestational diabetes). Gestational diabetes is the most common medical complication of pregnancy.[25] The American Diabetes Association recommends that women be screened for diabetes between 24 and 28 weeks of gestation, with the exception of women who meet all of these criteria: less than 25 years of age, normal body weight, no first-degree relatives with diabetes, and not of a high-risk ethnic origin including Hispanic, Native American, and African American.[26]

Blood Glucose Monitoring Obstetricians recommend blood glucose monitoring for all pregnant women with any type of diabetes. Establishing blood glucose control is important to the health of both mother and infant.

Diet Therapy For the pregnant woman with diabetes, a diet tailored to meet the increased nutrient demands of pregnancy and carefully coordinated with insulin therapy (when necessary) is central to therapy. The diet plan aims to provide adequate but not excessive kcalories to support appropriate weight gain (see the weight-gain recommendations in the margin on p. 305). Carbohydrate is often provided in lower amounts (40 to 45 percent of the total kcalories) than in the usual diet for diabetes to keep blood glucose levels from rising too high after meals.[27] Limiting carbohydrate to about 15 to 30 grams at breakfast helps maintain morning blood glucose levels in an acceptable range until counterregulatory hormone levels diminish. Frequent small meals help assure an ongoing supply of glucose without inducing hyperglycemia. A bedtime snack is recommended to prevent nocturnal hypoglycemia in the mother and to provide fuel to the developing fetus.

Oral antidiabetic agents are not used to treat diabetes during pregnancy.

Preventive Measures after Gestational Diabetes For most women with gestational diabetes, glucose tolerance returns to normal after pregnancy. Nevertheless, women with gestational diabetes are at high risk for developing type 2 diabetes later in life, especially if they are overweight. In women with previous gestational diabetes, being 20 percent overweight doubles the risk of developing type 2 diabetes. For this reason, health care professionals encourage clients with gestational diabetes to avoid excessive weight gain during pregnancy and to achieve or maintain a healthy weight thereafter.

Poorly controlled diabetes is associated with many acute and chronic complications. To minimize the risk of complications, people with diabetes learn to carefully control their blood glucose levels. The carbohydrate-controlled diet is only one part of the total management of diabetes. Meeting all the goals requires support from the diabetes care team—physicians, nurses, dietitians, counselors, and physical therapists. All members of the team coordinate instructions with other team members so that clients receive consistent advice that will not overwhelm or confuse them. People with diabetes and those who work with them should realize that the learning process takes time. After their initial introduction to the world of diabetes, clients benefit from frequent follow-up visits to address problems, expand their knowledge, and promote independence.

Hypoglycemia

Diet Order
Carbohydrate-modified diet with frequent small meals.

Strictly speaking, the term *hypoglycemia* simply means "low blood glucose" and refers not to a disease, but to a symptom of an alteration in carbohydrate metabolism. Ordinarily, blood glucose initially rises and then falls after eating; in healthy people, the decline is gradual, glucose remains in the normal range, and the transition occurs without notice. In people with hypoglycemia, however, blood glucose falls too low as the body shifts from the fed to the fasting state. Hypoglycemia may or may not be accompanied by other symptoms, and the symptoms may or may not be uncomfortable. There are two major types of hypoglycemia, reactive and fasting.

Reactive Hypoglycemia

reactive hypoglycemia: hypoglycemia experienced simultaneously with epinephrine-release symptoms one to three hours after a meal; also called **postprandial hypoglycemia.**

Reactive hypoglycemia occurs within an hour or two after eating and is related to the release of the hormone epinephrine in response to rapidly falling blood glucose. Hypoglycemia may occur, for example, after gastric surgery (see Chapter 20), when TPN solutions are discontinued too quickly (see Chapter 22), or early in type 2 diabetes when insulin levels are elevated. In most cases, the reason for the rapid decline in blood glucose is unknown. This section describes reactive hypoglycemia of unknown origin.

Symptoms of Reactive Hypoglycemia The symptoms of reactive hypoglycemia are similar to the symptoms of an anxiety attack: weakness, rapid heartbeat, sweating, anxiety, hunger, and trembling. These symptoms are not surprising since they are caused by the "emergency hormone," epinephrine.

Diagnosis of Reactive Hypoglycemia True reactive hypoglycemia can be identified by directly testing blood glucose at intervals after a meal. Low blood glucose and the simultaneous presence of symptoms confirm a diagnosis of reactive hypoglycemia. True reactive hypoglycemia is rare, although it is often misdiagnosed and has been the subject of much misguided advice. Many dishonest or ill-informed practitioners "diagnose" hypoglycemia on the basis of their clients' verbal reports alone and prescribe all sorts of "remedies" for it with transparently thin rationale. People also "diagnose" themselves so commonly that physicians have identified a special category for their condition: *non*hypoglycemia.

Carbohydrate-Modified Diet for Reactive Hypoglycemia For people who experience true reactive hypoglycemia, a judicious carbohydrate-modified diet may bring relief. Avoiding both low-carbohydrate dieting and sudden large carbohydrate doses may be all that is required. The remedy, then, is similar to the diet for diabetes: eat a consistent amount of carbohydrate from balanced meals at regular times. If several average-sized meals fail to relieve symptoms, smaller meals eaten more frequently may help.

Fasting Hypoglycemia

fasting hypoglycemia: hypoglycemia that develops gradually and primarily affects the brain and central nervous system.

A person who has symptoms while well advanced into the fasting state (for example, overnight) is experiencing a different kind of hypoglycemia. So is the person whose symptoms occur because insulin has driven too much glucose into the cells. Fasting hypoglycemia arises from medically diverse disorders, such as diabetes or tumors of the pancreas or liver, that interfere with normal blood glucose regulation. Table 25.6 lists the major distinguishing characteristics of reactive and fasting hypoglycemia.

Symptoms of Fasting Hypoglycemia The symptoms of fasting hypoglycemia differ from those of reactive hypoglycemia because they are not related to epinephrine release. Instead, blood glucose falls slowly, and the major effect is on the brain and central nervous system. The symptoms include headaches, blurred vision, mental dullness, fatigue, confusion, amnesia, and even seizures and unconsciousness.

Carbohydrate-Modified Diets for Fasting Hypoglycemia Earlier sections of this chapter described how an evenly spaced, consistent carbohydrate intake can help prevent fasting hypoglycemia in people with diabetes and also discussed how carbohydrates are used to treat hypoglycemia. Surgery is the primary treatment for tumors that precipitate fasting hypoglycemia, although carbohydrate-modified diets (as described for reactive hypoglycemia) may be used temporarily.

Carbohydrate-modified diets help control the symptoms and prevent the complications of hyperglycemia and hypoglycemia. In both cases, diets are individualized to meet each person's needs; no single diet is appropriate for everyone. The nutrition assessment checklist highlights areas of concern for people with diabetes and hypoglycemia. The accompanying nutrition in practice explains how diets can be planned around clients' lifestyles, health needs, and willingness to follow diet advice.

Table 25.6 Characteristics of Reactive and Fasting Hypoglycemia

	Reactive Type	Fasting Type
Onset of symptoms	Sudden; occurs 1 to 3 hours after meals	Gradual
Type of symptoms	Anxiety, weakness, sweating, rapid heartbeat, hunger, trembling	Headache, mental dullness, fatigue, confusion, amnesia, seizures, unconsciousness
Duration of symptoms	Transient	Persistent
Possible causes	Early type 2 diabetes, gastric surgery, TPN	Hormonal imbalance, diabetes, medications, tumors
Clinical course	Less serious; treat with diet	Can be serious; treat underlying problems

Nutrition Assessment Checklist

For People with Diabetes and Hypoglycemia

Medical For people with diabetes, use the medical record to determine the person's type of diabetes, its duration, acute and chronic complications, and other medical conditions that may affect nutrient needs. For people with hypoglycemia, check the medical record for the cause, if known, and for a confirmation of the diagnosis.

Medication For clients with preexisting diabetes who use insulin, note the types of insulin and schedule of administration. Note medications, if any, for clients with type 2 diabetes. Remind clients who use oral antidiabetic agents that slow carbohydrate digestion to keep glucose tablets or gel on hand to treat hypoglycemia. Check for other drug therapy including antilipemics (to lower blood lipids) and antihypertensives (to reduce blood pressure) and note possible nutrient-medication interactions.

Food Intake Obtain an accurate diet and physical activity record from the person with diabetes from which to devise an acceptable plan and to coordinate medications. During reassessment, use food intake and physical activity records along with records of blood glucose monitoring to evaluate the effectiveness of the treatment plan and to help make acceptable adjustments, when necessary. An assessment of food intake for the person with hypoglycemia helps pinpoint foods or amounts of foods that cause undesirable symptoms.

Anthropometric Take accurate height and weight measurements and determine desirable weight. Initial doses of insulin and calculated energy needs rely on body weight. Food energy intakes and insulin doses may need to be adjusted regularly to account for weight changes, particularly for growing children.

Laboratory Monitor blood glucose, hemoglobin A_{1c}, and blood lipids regularly for people with diabetes. Check results of urine tests for microalbuminuria when available.

Physical Check results of eye and foot exams and monitor blood pressure. Look for physical signs of dehydration, especially in elderly people with diabetes.

Self Check

1. Diabetes mellitus describes a group of disorders characterized primarily by:
   a. altered fat metabolism and hyperglycemia.
   b. altered fluid and electrolyte imbalances.
   c. altered energy metabolism and hyperglycemia.
   d. insulin resistance.
2. The classic symptoms of undiagnosed diabetes include:
   a. increased urination, increased thirst, and unexplained weight loss.
   b. headache, shakiness, and irritability.
   c. blurred vision, labored breathing, and fatigue.
   d. nausea, vomiting, and diarrhea.
3. Which of the following is characteristic of type 1 diabetes?
   a. Insulin secretion is ineffective in preventing hyperglycemia.
   b. It is the predominant form of diabetes.
   c. It frequently goes undiagnosed.
   d. It is associated with ketosis and acidosis.
4. Which of the following is characteristic of type 2 diabetes?
   a. Chronic complications may have begun to develop before it is diagnosed.
   b. Immune factors play a role.
   c. No insulin is synthesized by the beta cells.
   d. Ketosis and acidosis are common complications.
5. The chronic complications associated with diabetes result from:
   a. alterations in kidney function.
   b. damage to blood vessels and nerves.
   c. infections that deplete nutrient reserves.
   d. weight gain and hypertension.
6. The health care professional recognizes that the primary goal of medical and nutrition therapy in diabetes is to:
   a. support health and well-being.
   b. maintain near-normal blood glucose levels.
   c. control blood lipids and blood pressure.
   d. encourage weight loss.

7. The nurse working with a client with diabetes emphasizes that the diet provide:
   a. more protein than regular diets.
   b. a consistent carbohydrate intake from day to day and at each meal and snack.
   c. a restricted fat intake.
   d. a restricted intake of simple sugars and concentrated sweets.
8. Which of the following is true regarding the use of alcohol in a diet for diabetes?
   a. A serving of alcohol should always be exchanged for two fat exchanges.
   b. Alcohol can cause hyperglycemia in all people, including those with diabetes.
   c. Some types of insulin can cause flushing of the skin and a rapid heartbeat in people with type 2 diabetes.
   d. People with well-controlled blood glucose levels can use alcohol in moderation.
9. The meal-planning strategy that is most effective for the person with diabetes is:
   a. the exchange list system.
   b. carbohydrate counting.
   c. food guides and sample menus.
   d. the one that best helps the client control blood glucose levels.
10. All of the following statements are true regarding physical activity and diabetes **except:**
    a. people with type 2 diabetes can improve control of blood glucose through a regular program of physical activity.
    b. people with type 1 diabetes can improve blood glucose control through a regular program of physical activity.
    c. people with type 1 diabetes should not exercise if their blood glucose is too high.
    d. people with type 1 diabetes should not exercise if their blood glucose is too low.
11. Which of the following best describes insulin therapy in a person receiving both NPH insulin and an insulin analog?
    a. NPH insulin covers the carbohydrate from meals while the insulin analog covers basal insulin needs.
    b. Since NPH insulin is of long duration, there is no need for the insulin analog.
    c. NPH insulin covers basal insulin needs while the insulin analog covers the carbohydrate from meals.
    d. Since glucose is not available from food between meals, insulin analog alone would cover the person's insulin needs.
12. Which of the following describes the treatment plans for a person with type 1 diabetes who practices intensive therapy?
    a. The person uses one type of insulin one or two times daily.
    b. The person must monitor blood glucose several times a day.
    c. After learning to control blood glucose levels, the person can quit monitoring blood glucose levels.
    d. The health care team monitors blood glucose levels to evaluate long-term control.
13. Hyperglycemia in a person who is eating a consistent amount of carbohydrate from day to day can be precipitated by:
    a. infections or illnesses.
    b. conditions that lower levels of counterregulatory hormones.
    c. undertreatment of hypoglycemia.
    d. alcohol ingestion.
14. Repeated episodes of hypoglycemia in the person with diabetes can result in all of the following complications **except:**
    a. excessive weight gain.
    b. retinopathy.
    c. possible impairment of cognitive function.
    d. shock and death.
15. Women with pregnancies complicated by diabetes:
    a. need more carbohydrate than women with diabetes who are not pregnant.
    b. generally benefit from larger meals and a snack at bedtime.
    c. often need less carbohydrate at breakfast.
    d. benefit from diet therapy that provides kcalories to support a weight gain of about 30 pounds.
16. Which of the following is true regarding hypoglycemia?
    a. Fasting hypoglycemia is more likely to result from serious medical conditions.
    b. Fasting hypoglycemia is more likely to benefit from a diet that provides a consistent amount of carbohydrate at regular times, without additional treatment.
    c. Symptoms of fasting hypoglycemia are associated with the release of epinephrine.
    d. Symptoms of reactive hypoglycemia include confusion and headaches.

Answers to these questions appear in Appendix H.

Clinical Applications

1. Using the box on pp. 605–606, plan a diet using the exchange lists for a sedentary woman with type 1 diabetes who is 5 feet 9 inches tall and weighs 160 pounds. Assume that the distribution of kcalories will be 55 percent from carbohydrate, 20 percent from protein, and 25 percent from fat. Develop a sample menu.
2. An important part of learning is being able to apply knowledge and guidelines to real-life situations. Using Table 25.5 on p. 613 as a guide, think about the possible remedies for either hyperglycemia or hypoglycemia. Describe at least one situation when it might be preferable to alter the insulin dose and when it might be preferable to alter the carbohydrate intake.
3. Take a trip to a pharmacy and price these items: blood glucose meter, glucose test strips, lancets, insulin, and syringes. Determine the approximate cost of insulin for a person who uses 14 units of regular insulin and 26 units of NPH insulin in three injections daily. Then estimate the cost of testing blood glucose four times daily. How does using a meter affect the cost of glucose test strips? Consider how an external pump might affect the total cost of managing diabetes. How does the need for a well-balanced diet influence the cost of diabetes care?
4. You have been asked to instruct a client on a diet for hypoglycemia. You read the medical chart and realize that the client was never given the proper diagnostic tests. After talking with the client, you discover that he is convinced that he has hypoglycemia and expects the proper diet will cure his mood swings, fatigue, and depression. Should you give the diet instructions? Are there any risks in following a diet for hypoglycemia?

Notes

1. M. I. Harris, NIDDM: Epidemiology and scope of the problem, *Diabetes Spectrum* 9 (1996): 26–29.

2. Harris, 1996.

3. The Expert Committee on the Diagnosis and Classification of Diabetes Mellitus, Report of the Expert Committee on the diagnosis and classification of diabetes mellitus, *Diabetes Care* (supplement 1) 21 (1998): 5–19.

4. W. Y. Fujimoto, A national multicenter study to learn whether type II diabetes can be prevented: The Diabetes Prevention Program, *Clinical Diabetes* 15 (1997): 13–15.

5. I. J. Deary, Hypoglycemia-induced cognitive decrements in adults with Type I: A case to answer? *Diabetes Spectrum* 10 (1997): 13–15.

6. The Expert Committee on the Diagnosis and Classification of Diabetes Mellitus, 1998.

7. American Diabetes Association, Implications of the Diabetes Control and Complications Trial, *Diabetes Care* 16 (1993): 1517–1520.

8. R. R. Wing, Use of very-low-kcalorie diets in the treatment of persons with non-insulin-dependent diabetes mellitus, *Journal of the American Dietetic Association* 95 (1995): 569–672.

9. American Diabetes Association, Nutrition recommendations and principles for people with diabetes mellitus, *Diabetes Care* 17 (1994): 519–522.

10. American Diabetes Association, 1994.

11. J. P. Barnett and A. Garg, Medical nutrition therapy for patients with diabetes mellitus: Role of dietary fats, *On the Cutting Edge: Diabetes Care and Education,* Fall 1997, pp. 5–6; A. M. Coulston, Monounsaturated fats for people with diabetes, *On the Cutting Edge: Diabetes Care and Education,* Fall 1997, pp. 14–16.

12. A. Garg and coauthors, Effects of varying carbohydrate content of diet in patients with non-insulin-dependent diabetes mellitus, *Journal of the American Medical Association* 271 (1994): 1421–1428.

13. M. Karlsen, D. Khakpour, and L. L. Thomson, Efficacy of medical nutrition therapy: Are your patients getting what they need? *Clinical Diabetes* 14 (1996): 54–60.

14. American Diabetes Association, 1994.

15. As cited in J. R. White, The pharmacologic reduction of blood glucose in patients with type 2 diabetes mellitus, *Clinical Diabetes* 16 (1998): 58–67.

16. J. R. White, Combination oral agent/insulin therapy in patients with type II diabetes mellitus, *Clinical Diabetes* 15 (1997): 102–112.

17. S. V. Edelman, Troglitazone: A new and unique oral antidiabetic agent for the treatment of Type II diabetes and the insulin resistance syndrome, *Clinical Diabetes* 15 (1997): 60–65.

18. T. Antonucci and coauthors, Impaired glucose tolerance normalized by treatment with the thiozolidinedone troglitazone, *Diabetes Care* 20 (1997): 188–193.

19. American Diabetes Association, Lispro: A new fast-acting insulin option, *Diabetes Spectrum* 9 (1996): 253.

20. White, 1997.

21. J. D. Pirsch and coauthors, Pancreas transplant for diabetes mellitus, *American Journal of Kidney Diseases* 27 (1996): 132–138.

22. P. J. Charney, Nutrition support in patients with diabetes mellitus, *Support Line,* April 1993, pp. 1–4.

23. M. Franz and coauthors, Who, what, and where—questions from "Maximizing the role of nutrition in diabetes management" continuing education program, *Diabetes Spectrum* 8 (1995): 369–374.

24. American Diabetes Association, Position statement: Preconception care of women with diabetes, *Diabetes Care* (supplement 1) 21 (1998): 56–59.

25. D. B. Carr and S. Gabbe, Gestational diabetes: Detection, management, and implications, *Clinical Diabetes* 16 (1998): 4–11.

26. The Expert Committee on the Diagnosis and Classification of Diabetes Mellitus, 1998.

27. C. Fagen, J. D. King, and M. Erick, Nutrition management in women with gestational diabetes mellitus, *Diabetes Care* 16 (1993): 1146–1157.

CHAPTER 25

Nutrition in Practice

▪ Living with Diabetes ▪

A healthy person goes about daily activities with little thought to how the body will react to everyday routines or disruptions to those routines. If you usually eat breakfast at 8:00 A.M., you may sleep in and choose not to eat breakfast on weekends without a second thought. If your friend asks you to play tennis and the match interferes with dinner, you simply eat later. If you get hungry during the day, you eat a snack. If you're not hungry at your usual dinner hour, you wait and eat later. For people with diabetes, even such simple variations in a daily schedule require thought and adjustment. They need to learn facts, master techniques, and develop new attitudes and behaviors that will provide for a healthy life. Ideally, clients learn to manage their disease, rather than allow their disease to control their lives.

WWW.

diabetes.org
American Diabetes Association

joslin.org
Joslin Diabetes Center

aadenet.org
American Association of Diabetes Educators

How can health care professionals help clients take control of their disorders?

Health care professionals who simply "prescribe" remedies and then expect their clients to comply with those remedies fail to consider the impact that lifestyle changes have on a person's quality of life. Clients can easily be overwhelmed, and their motivation and compliance may be poor. This nutrition in practice describes a different educational approach—client empowerment—that is directed by the client. To use this approach, health care professionals provide clients with the information and skills they need to make decisions about their treatment plans and manage their diseases.

The combined expertise of many health care professionals including physicians, nurses, dietitians, counselors, and physical therapists or exercise physiologists enhances diabetes management. Along with the client, these professionals form the health care team. Throughout this discussion, keep in mind that the client is the central member of the team.

What are some of the things clients must understand in order to manage diabetes?

In order to take control of their care, clients with diabetes need to learn a lot about it so they can manage their disease and prevent complications:

- *Medication* Clients need to learn the appropriate type, dose, and schedule. Clients on insulin need to learn how to draw insulin, give themselves an injection, and rotate injection sites. Clients who use external pumps need to know how to operate and maintain them and have them refilled. Clients on oral agents need to learn when to take their medications and about possible interactions and precautions.
- *Blood testing* Clients need to learn how to administer the tests, record and interpret results, and bring glucose levels within a desirable range.
- *Diet* Clients need to know how to schedule their meals, distribute carbohydrate throughout the day, and control portion sizes.
- *Changes to accommodate physical activity, missed meals, or illness* Clients need to know how to meet these demands.
- *Complications associated with diabetes* Clients need to learn how to prevent complications and how to recognize and treat them when they occur.
- *Record keeping* Clients need to learn how to keep accurate food, activity, and insulin administration records so that they can understand how their bodies respond to diabetes and how they can gain control over their disease.

For clients with diabetes and their families, all this new information and simply the time required to manage the disease can be overwhelming. They may find the diagnosis of diabetes and all it entails difficult to grasp and accept. Clients need a great deal of support to help

Nutrition in Practice

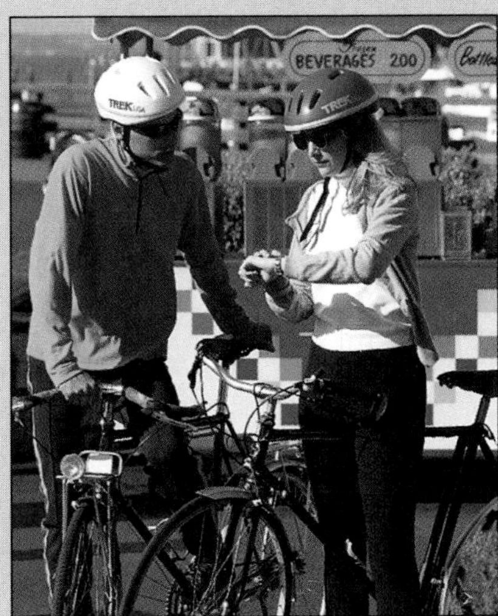

For a person with diabetes, even simple changes in routine require planning and adjustments.

them cope with their fears, stay motivated when they feel overwhelmed, gain confidence in their abilities to manage the disorder, and live a high-quality life.

How can health care professionals leave important health decisions up to clients without endangering their health?

Health care professionals cannot knowingly encourage clients to practice behaviors that are medically harmful. Nor can they "force" clients to follow advice. Some aspects of disease management are critical for survival. Other aspects may be beneficial but less critical. Still other aspects may be ideal but less pressing when considered in the total context of a treatment plan.

The health care team must use clinical judgment in assigning priorities to various treatments. A person with type 1 diabetes, for example, must take insulin or face death. The person's goals, motivation, finances, and ability level help determine if intensive therapy or traditional therapy is more appropriate. A 24-year-old client with type 1 diabetes may be eager to learn intensive therapy, while an 80-year-old client with type 2 diabetes may refuse insulin therapy, even if it means poor blood glucose control. Health care professionals who practice client empowerment explain the ramifications of treatment choices, but respect the individual's right to make health care decisions without judgment. In addition, the health care team cannot merely "abandon" the client. The team members work to promote health to the highest degree possible.

What can health care professionals do to help the client design a management plan that meets the client's needs?

Health care professionals facilitate the learning process by working out a highly individualized plan that carefully considers the client's motivation, goals, and ability to grasp new concepts and make lifestyle changes. The plan must be flexible to accommodate changing needs; a person who is highly motivated at one point, for example, may become totally discouraged at another time and need to restructure goals temporarily. At each step, the health care team balances medical needs with the individual's goals and motivation.

All of this seems very time-consuming. Do health care professionals have the time to individualize plans?

Health care professionals recognize that all newly diagnosed clients and their families need intensive diabetes care training and counseling and that the learning process takes time. Even highly motivated clients who listen attentively need several counseling sessions to learn the basics of diabetes management. Inevitably, efforts at in-depth, short-term counseling will meet with failure. Diabetes education is a continuous process that is a routine part of diabetes management.

How does the educational process begin?

Health care professionals can help clients make lifestyle changes by using a stepwise process that includes assessment, goal setting, intervention, and evaluation.[1] As Chapter 25 describes, a complete assessment serves as the first step in formulating a treatment plan.[2] The more detailed the assessment, the more closely the plan can be tailored to the individual's needs and the better the chances for success. Using assessment data, the health care team sets long-term medical goals; to enable the team and the client to measure the success of therapy, the goals are stated in terms of measurable outcomes such as target ranges for blood glucose, blood lipids, and body weight.

Health care professionals also work with the client to negotiate short-term goals geared toward making lifestyle adjustments. For short-term goals to be successful, they must incorporate clients' personal health goals and include only those actions clients are willing and ready to take to reach their goals.

Can you give me an example?

Let's look at an example of a 55-year-old woman newly diagnosed with type 2 diabetes. After a complete nutrition assessment, the dietitian may find that in addition to the elevated blood glucose, blood lipids are also elevated; the client is 30 pounds overweight, does not have a regular physical activity plan, skips breakfast, has a large dinner, eats many fried and high-fat foods and snacks, and seldom eats vegetables. The dietitian recognizes that several dietary changes are warranted, but talks with the client to determine how she perceives her health situation and what steps she thinks she can make to improve her diet. The client tells the dietitian that she wants to lose weight, but she feels very stressed and doesn't know how to get started. With the guidance of the dietitian, the woman sets these goals: she will eat a consistent amount of carbohydrate three times a day, with careful attention to portion sizes, and begin a physical activity program. By involving the client in goal setting, the dietitian improves the chances for success and encourages the client to take responsibility for her health.

What does the dietitian do next?

Once goals have been set, the next step is intervention. What specific activities can help the client meet her goals? Returning to our example, the dietitian and client might discuss a diet plan that includes appropriate portion sizes for breakfast, lunch, and dinner; review several menu options; practice weighing and measuring foods; and develop a physical activity plan in which the client agrees to walk after dinner for 20 minutes three times a week. The dietitian would like records of food, activity, and blood glucose, but the client feels she cannot handle that task right now. Instead, she will monitor blood glucose as instructed by the nurse. The dietitian must wait until the client returns for her next appointment to decide what to do next.

How will the dietitian know what to do next?

Once the client tries the plan, the next step is evaluation.[3] Which strategies were successful and which were not? Suppose the client in our example returns for her next appointment. She has been successful at eating breakfast, inconsistent about reducing her portion sizes at lunch and dinner, and has managed to walk for 20 minutes only once or twice a week. From the client's records, the dietitian sees that her blood glucose levels have improved somewhat and that she has lost half a pound. The dietitian reinforces the value of the positive changes the client has made and praises her efforts. At this point, the dietitian, keeping in mind the client's medical goals, must reassess the client's motivation and decide what steps to take next. The client may want to continue the plan and agree to renew her commitment to control portion sizes and to walk more frequently. Alternatively, she may be so pleased with how she is feeling that she is ready to do more. She may agree not only to continue her original plan, but also to keep food and activity records and limit servings of fried foods to two a week, for example.

To help people adjust to the physiological demands imposed by diabetes, health care professionals guide clients through management plans with measurable goals set by the clients. Clients' responses to the interventions and level of motivation dictate future actions. Health care professionals who are aware of the psychological burdens associated with diabetes are better equipped to support their clients' emotional health—an important factor in diabetes management.

How do emotional issues enter into diabetes management?

Consider that in the example just discussed, the plan focused only on the nutrition component of diabetes education. Clients have many other diabetes-related tasks to master. In addition, they have responsibilities related to work, family, and community. It should come as little surprise that even when clients know what to do and why they should do it, they may be unable to carry out the plan at times.[4] Many people with diabetes report feeling overwhelmed and frustrated by the multitude of self-care demands.[5] In the words of one diabetes educator: "No one but another person who has diabetes can fully appreciate the demands of diabetes. It is 24 hours a day, 365 days a year (except on Leap Year when you get an extra day of diabetes). It involves all manner of imposition and deprivation. And even when you do everything right, there are no guarantees."[6]

People with diabetes often feel that health care professionals, family, and friends "blame" them for having diabetes-related problems and complications.[7] They may feel guilt and remorse because their efforts at diabetes control were not good enough to prevent complications. Health care professionals are wise to remain nonjudgmental when working with clients with diabetes and to recognize that diabetes control must be balanced with quality of life. Clients also need to know that they may experience complications even when they are doing everything possible. Insisting on perfection can only lead to failure.

Teenagers often have intense difficulty accepting the initial diagnosis of diabetes. At a time when they are striving to develop their identity with a group and to be as similar to their peers as possible, they are faced with an unwelcome diagnosis and new rules they are expected to follow; their response may be denial and refusal to cooperate. Yet their refusal to cooperate might result in serious consequences. The person who appreciates a teen's special views on life is best prepared to help with the adjustment. Adolescents especially need to know that they can manage the disease themselves—that it won't turn them back into dependent children.

Parents and other family members also face the challenge of living with diabetes. The intensity of the situation can either reinforce or disrupt family unity. Parents may resent the demands of caring for a child with a chronic illness and then may experience guilt for having those feelings. They may feel anxious and be reluctant to allow their child to follow the diabetes care plan without their constant assistance. Especially in the case of an older child, they may press their care and control on a child who needs to develop autonomy and self-care. Parents may also become emotionally upset when they see their child feeling anxious, depressed, or withdrawn. Parents need time to work through these feelings. They might want to attend meetings for parents of children with diabetes. Such meetings offer opportunities to share feelings, ideas, and frustrations with others in similar situations. Sometimes just knowing that you're not alone helps.

Are there any other resources people with diabetes can turn to for help?

After an initial introduction to the world of diabetes, many hospitals and medical centers offer clients educational programs designed to expand their knowledge and promote independence. Some programs encourage children to bring friends, which makes the experience more comfortable and fun. Some programs are designed specifically for parents, grandparents, and other caregivers.

Children can combine education and summer vacation at camps designed especially for children with diabetes. These camps offer the chance to learn more about diabetes while "living" the lifestyle with companions under supervision. Children trade snack ideas, try new recipes, and help prepare meals. Older children assist younger ones, and all benefit.

The results of the Diabetes Control and Complications Trial clearly show that tightly managing diabetes can dramatically reduce the complications associated with it. Helping clients with diabetes make the necessary adjustments is a continuous process that balances medical and individual needs.

The principles described in this Nutrition in Practice apply to any chronic disease that requires diet and other lifestyle changes to ensure health. Remember that even relatively minor lifestyle changes can be important to the individual. A person with a hiatal hernia, for example, may be unwilling to give up coffee, even though the consequences include the pain of heartburn and worsening esophagitis. The biggest change the person may be willing to make may be to reduce coffee intake from 6 cups to 2 cups a day or to agree to drink coffee only along with foods. Clients with extremely serious disorders who lack motivation to change their lifestyles may benefit from professional counseling along with the encouragement of health care professionals, family, and friends, but in the end, only the clients can determine the course that they can accept.

Notes

1. American Diabetes Association and American Dietetic Association, *Facilitating Lifestyle Change: A Resource Manual* (Alexandria, Va.: American Diabetes Association, 1996).

2. J. G. Pastors, Nutrition assessment for diabetes medical nutrition therapy, *Diabetes Spectrum* 9 (1996): 99–103.

3. M. Peyrot, Evaluation of patient education programs: How to do it and how to use it, *Diabetes Spectrum* 8 (1996): 86–93.

4. M. M. Funnell and R. M. Anderson, Judge not: Lessons learned from simulated diabetes regimens, *Diabetes Spectrum* 8 (1995): 328–329.

5. W. H. Polonsky, Listening to our patients' concerns: Understanding and addressing diabetes-specific emotional distress, *Diabetes Spectrum* 9 (1996): 8–11.

6. R. R. Rubin, Life's work they have not chosen, *Diabetes Spectrum* 8 (1995): 308.

7. K. F. McFarland, The power of words, *Diabetes Spectrum* 8 (1995): 308.

26 Fat- and Mineral-Modified Diets for Cardiovascular Diseases

Cardiovascular disease (CVD) is a general term for all diseases of the heart and blood vessels. **Atherosclerosis,** the most common cause of CVD, is characterized by plaques along the inner walls of the arteries, which occlude the affected artery and restrict blood flow.

plaques (PLACKS): mounds of lipid material that form within certain immune system cells (macrophages) that are embedded in artery walls. If atherosclerosis progresses to an advanced stage, the plaques may become hardened with fibrous connective tissue and calcium deposits.
placken = patch or plate

coronary heart disease (CHD): damage to the heart muscle that results from obstruction of blood flow to the heart muscle.

angina: a painful feeling of tightness or pressure, felt in the area in and around the heart, often radiating to the back, neck, and arms, caused by a lack of oxygen to an area of heart muscle.

platelets: tiny, disc-shaped bodies in the blood that are important in clot formation.

thrombosis: the formation of a **thrombus,** a blood clot that may obstruct a blood vessel, causing gradual tissue death. A **coronary thrombosis** blocks blood flow through an artery that feeds the heart. A **cerebral thrombosis** blocks blood flow through an artery that feeds the brain.
thrombo = clot

embolism: the obstruction of a blood vessel by an **embolus** (EM-boh-luss), or traveling clot, causing sudden tissue death.
embol = to insert, plug

Table 26.1 Indications for the Use of Fat-Restricted, Fat-Modified Diets

Atherosclerosis
Chronic renal disease
Congestive heart failure
Diabetes mellitus
Hyperlipidemia
Hypertension
Nephrotic syndrome
Post cerebrovascular accident
Post myocardial infarction

The fat-modified diets described in this chapter limit both the amount and type of fat (see Table 26.1). Such diets are prescribed to prevent and treat cardiovascular disease (CVD)—the leading cause of death around the world today. In the United States, death rates from CVD in men aged 35 to 50 are three times greater than in women of the same age, but in later years (65 to 74), the incidence is similar. The consequences of CVD are usually heart disease and strokes, the first and third leading causes of death for adults, respectively. Most CVD involves atherosclerosis and hypertension. Each of these conditions has its own set of risk factors, and each condition makes the other one worse.

This chapter examines the factors that lead to the development of atherosclerosis and hypertension and the diets used in their treatment. Then it describes the major consequences of these conditions—heart attacks and strokes.

Atherosclerosis

Atherosclerosis usually begins with the accumulation of soft fatty streaks, called plaques, along the inner arterial walls, especially at branch points (see Figure 26.1). These plaques may gradually enlarge and become hardened with fibrous connective tissue and calcium. Most people have well-developed plaques by age 30, and preventive efforts focus on delaying or reversing the progression of existing plaques. Mounting evidence suggests that highly unstable and potentially toxic molecules, called *free radicals,* that form in the body during the course of normal metabolism play a role in the progression of plaques.[1] Nutrition in Practice 26 provides more information about free-radical formation and how antioxidant nutrients might protect the body from atherosclerosis.

Consequences of Atherosclerosis

Atherosclerosis directly and indirectly obstructs blood flow through the arteries, damages tissues, and raises blood pressure. When blood flow in the arteries feeding the heart is obstructed, coronary artery disease results. When coronary artery disease damages the heart muscle (coronary heart disease, or CHD), the person may experience pain and pressure in and around the area of the heart (angina). If blood flow to the heart is cut off, that area of heart muscle dies, and a heart attack results. When blood flow to the brain is obstructed, a transient ischemic attack or stroke may result.

Blood Clots Form Small, cell-like bodies in the blood, known as platelets, cause clots to form whenever they encounter injuries in blood vessels. Clots normally form and dissolve in the blood all the time, but in atherosclerosis, clots form faster than they dissolve, because the platelets respond to plaques as if they were injuries. A blood clot may stick to a plaque and gradually grow large enough to close off a blood vessel (thrombosis). A clot may also break free from the artery wall and travel through the circulatory system until it lodges in a small artery and suddenly shuts off the blood flow to that area (embolism). The gradual or sudden loss of blood flow to the portion of tissue supplied by the clotted artery robs the tissue of oxygen and nutrients, and the tissue may eventually die.

Blood Pressure Rises The heart must create enough pressure to push blood through the circulatory system. When arteries are narrowed by plaques, clots, or both, blood flow is restricted, so the heart must generate more pressure to deliver

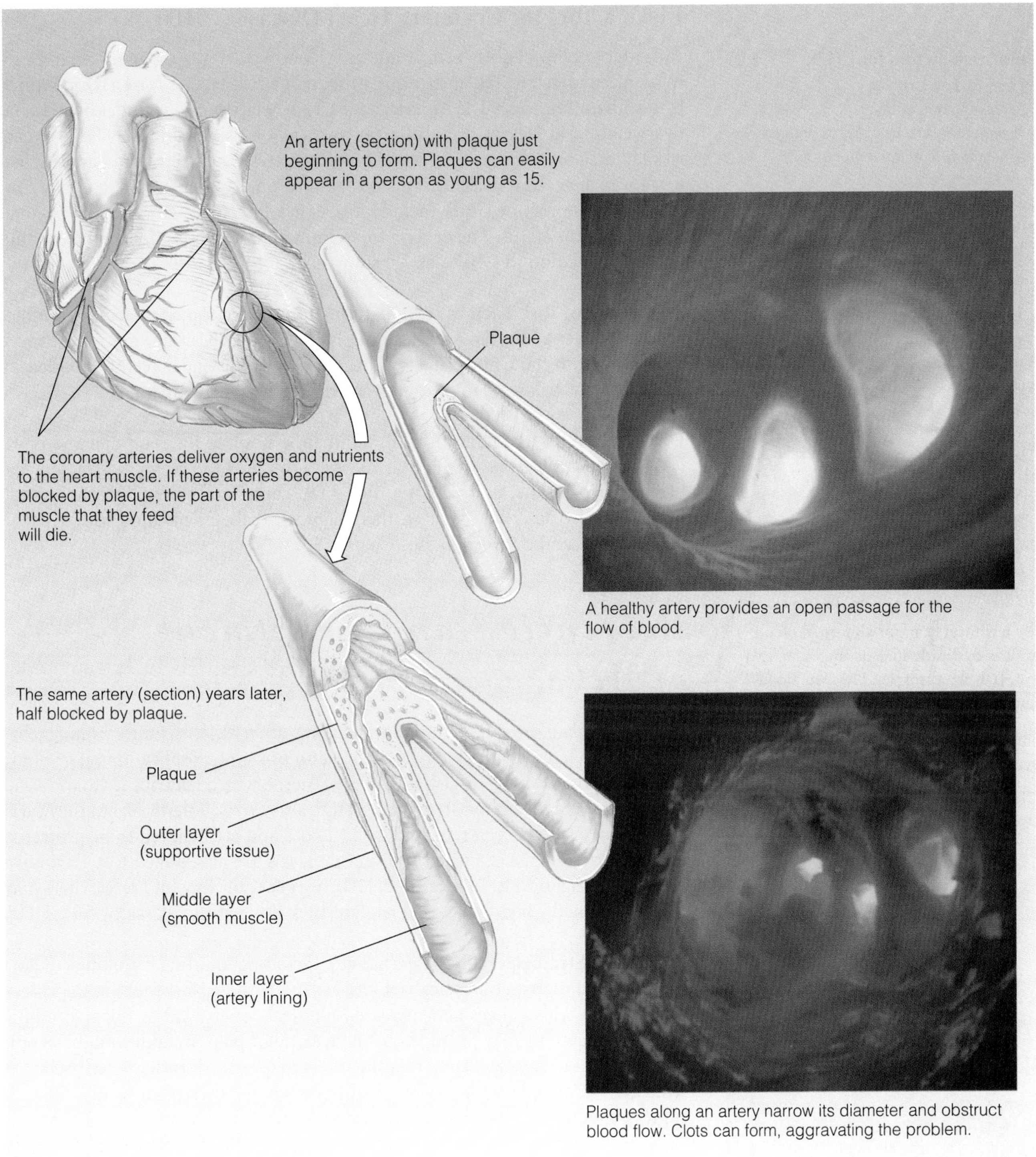

An artery (section) with plaque just beginning to form. Plaques can easily appear in a person as young as 15.

Plaque

The coronary arteries deliver oxygen and nutrients to the heart muscle. If these arteries become blocked by plaque, the part of the muscle that they feed will die.

A healthy artery provides an open passage for the flow of blood.

The same artery (section) years later, half blocked by plaque.

Plaque

Outer layer (supportive tissue)

Middle layer (smooth muscle)

Inner layer (artery lining)

Plaques along an artery narrow its diameter and obstruct blood flow. Clots can form, aggravating the problem.

Figure 26.1

The Formation of Plaques in Atherosclerosis

When plaques have covered 60 percent of the coronary artery walls, the critical phase of heart disease begins.

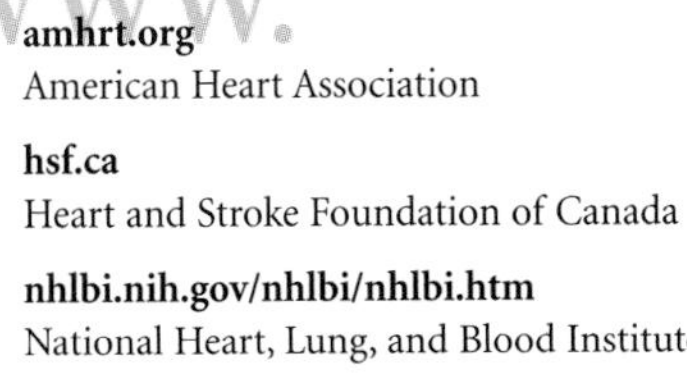

amhrt.org
American Heart Association

hsf.ca
Heart and Stroke Foundation of Canada

nhlbi.nih.gov/nhlbi/nhlbi.htm
National Heart, Lung, and Blood Institute

blood to the tissues. This higher blood pressure further damages the artery walls, and plaques and clots are especially likely to form at damage points. Thus the development of atherosclerosis is a self-accelerating process.

Risk Factors for Coronary Heart Disease (CHD)

Major risk factors for CHD:

- *High LDL cholesterol.*
- *Male, 45 years or older.*
- *Female, 55 years or older, or with premature menopause and not on estrogen replacement therapy.*
- *Low HDL cholesterol. (Subtract 1 risk factor if HDL cholesterol ≥60 mg/dL.)*
- *Hypertension.*
- *Smoking.*
- *Diabetes mellitus.*
- *Family history of heart attacks or sudden death prior to age 55 in a male parent or sibling or prior to age 65 in a female parent or sibling.*

Reminder: Cholesterol is carried in several lipoproteins, chief among them LDL and HDL (see Chapter 5 for details). Remember them this way:

- ***HDL** = **H**igh-density lipoprotein = **H**ealthy.*
- ***LDL** = **L**ow-density lipoprotein = **L**ess healthy.*

The technical term for abnormal blood lipids is **dyslipidemia;** elevated LDL and low HDL are examples. Elevated blood lipids may also be called **hyperlipidemia.**

Nutrition in Practice 13 discusses dietary recommendations for preventing atherosclerosis in children.

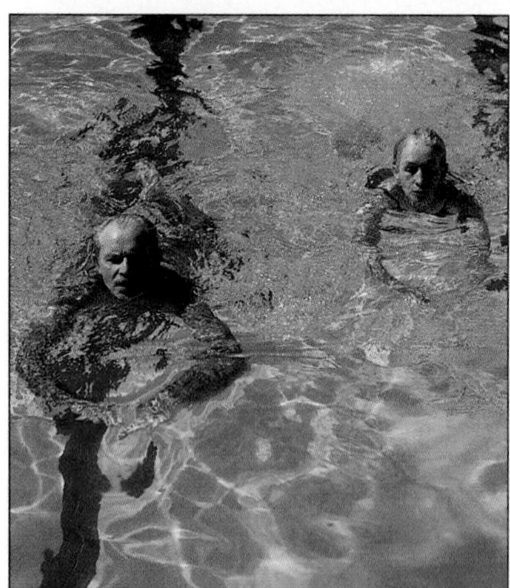

Regular aerobic exercise can help to defend against heart disease by strengthening the heart muscle, promoting weight loss, reducing blood pressure, and improving blood lipid and blood glucose levels.

Although atherosclerosis can invade any blood vessel, the coronary arteries are most often affected. The margin lists the major risk factors for CHD identified by health authorities.[2] The American Heart Association also includes obesity and overweight and lack of physical activity as major risk factors for CHD.[3] Many risk factors are interrelated—lack of physical activity and overweight are risk factors for diabetes as well as heart disease, for example. Other factors that may contribute to the risk of CHD include a person's life stress, behavior habits, and socioeconomic status. The criteria for defining blood lipids, blood pressure, and obesity in relation to CHD risk are shown in Table 26.2.

Prevalence of Risk Factors Population studies have found that most middle-aged and older adults have at least one risk factor for CHD and many have more than one. Both the United States and Canada recommend screening to identify individuals at risk so as to offer preventive advice and treatment.

Diet-Related Risk Factors It befits a nutrition book to focus on diet and physical activity strategies to reduce CHD risk. Several of the major risk factors can be modified by diet and activity: high LDL cholesterol, low HDL cholesterol, hypertension (described later in this chapter), diabetes mellitus (discussed in Chapter 25), and overweight (see Chapter 9).

Prevention and Treatment of Atherosclerosis

Treatment plans for preventing and treating atherosclerosis aim to normalize blood lipids, alter modifiable risk factors, and prevent complications. The blood cholesterol linked to CHD risk is LDL cholesterol, which contributes to the accumulation of plaque on the artery walls. HDL also carry cholesterol, but they carry cholesterol that has been removed from tissues and are believed to be protective against heart disease. Elevated triglyceride levels are also linked to CHD; however, when triglycerides are elevated, HDL are generally low, and researchers have not determined whether elevated triglycerides alone are a risk factor for CHD.[4]

Treatment Strategies Lifestyle changes including quitting smoking, dietary changes, and increased physical activity are the first step in reducing the risk of CHD. If lifestyle changes and increased physical activity fail to normalize lipid levels, medications may be added to the treatment plan. In addition, for people with diabetes or hypertension, treatment of the specific disorder is an important strategy. For people with advanced atherosclerosis, therapy may also include surgery to restore blood flow to the affected organ.

Physical Activity Physical activity deserves attention in any program to reduce CHD risk or treat heart disease. Physical activity can help speed weight loss and loss of body fat, strengthen the cardiovascular system, raise HDL, reduce hypertension, and improve glucose tolerance in people with type 2 diabetes.[5] Physical activity can also improve mental health. Frequent and sustained aerobic, endurance-type activities, such as brisk walking, undertaken faithfully for 30 minutes or more as a daily or every-other-day routine, may be most effective.

Table 26.2 Standards for CHD Risk Factors

LDL Cholesterol[a]	Total Cholesterol[b]
<130 mg/dL = desirable	<200 mg/dL = desirable
130–159 mg/dL = borderline high	200–239 mg/dL = borderline high
≥160 mg/dL = high	≥240 mg/dL = high

HDL Cholesterol	Triglycerides (Fasting)[d]
HDL ≤35 mg/dL indicates risk[c]	<200 mg/dL = desirable
LDL-to-HDL ratio:	200–400 mg/dL = borderline high
Men: >5.0 indicates risk	400–1000 mg/dL = high
Women: >4.5 indicates risk	>1000 mg/dL = very high

Blood Pressure	Systolic pressure		Diastolic pressure
Optimal	<120	and	<80
Normal	<130	and	<85
High-normal	130–139	or	85–89
Hypertension			
Mild (stage 1)	140–159	or	90–99
Moderate (stage 2)	160–179	or	100–109
Severe (stage 3)	≥180	or	≥110

Weight (Body Mass Index)	
Overweight:	BMI 25–29.9
Obese:	BMI ≥30

[a]For people with existing CHD, desirable LDL cholesterol values are lower (≤100 mg/dL).
[b]To convert cholesterol (mg/dL) to standard international units (mmol/L), multiply by 0.02586. For cholesterol values for children and adolescents, see Table NP13 on p. 351.
[c]This HDL value may be too low for women; no alternative value has yet been proposed. NIH Consensus Development Panel on Triglyceride, High-Density Lipoprotein, and Coronary Heart Disease, Triglyceride, high-density lipoprotein, and coronary heart disease, *Journal of the American Medical Association* 269 (1993): 505–510.
[d]High triglycerides alone normally do not indicate *direct* risk, but may reflect lipoprotein abnormalities associated with CHD. The risk of CHD increases as triglyceride levels increase in people with other risk factors. High triglycerides also occur in conditions such as kidney disease and diabetes, which suggest a high CHD risk.

Sources: Blood lipid standards adapted from The Expert Panel, Summary of the second report of the National Cholesterol Education Program (NCEP) Expert Panel on Detection, Evaluation, and Treatment of High Blood Cholesterol in Adults (Adult Treatment Panel II), *Journal of the American Medical Association* 269 (1993): 3015–3023; hypertension standards adapted from the Sixth Report of the Joint National Committee on Prevention, Detection, Evaluation, and Treatment of High Blood Pressure, National High Blood Pressure Education Program, National Heart, Lung, and Blood Institute, National Institutes of Health, November 1997, p. 11.

Diet Therapy

As Table 26.3 on p. 632 shows, the goals of diet therapy for preventing and treating CHD focus on reducing LDL cholesterol. To that end, people are advised to control their body weights and their intakes of total fat, saturated fat, and dietary cholesterol. A panel of experts recommends a two-step plan, shown in Table 26.4.[6] If blood lipids do not improve on the Step I diet, therapy moves on to Step II. People with existing CHD or those who have already been following a fat-modified diet prior to seeking treatment start on the Step II diet.

Control Weight Dietary plans to reduce LDL cholesterol and triglycerides include attention to weight reduction or weight maintenance (see Chapter 9).

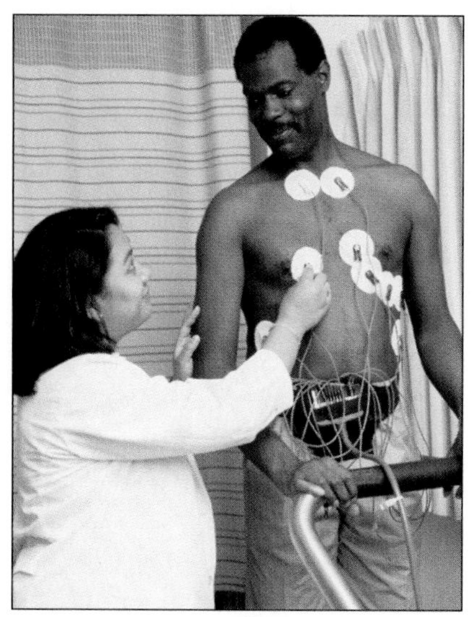
A stress test is one tool clinicians use to evaluate cardiovascular fitness.

Table 26.3 Dietary Treatment Guidelines Based on LDL Cholesterol

Risk Factor Status	Cholesterol Values When Diet Therapy Should Begin	Goals for Cholesterol Values with Diet Therapy
■ Without existing CHD and with one risk factor	≥160 mg/dL[a]	<160 mg/dL
■ Without existing CHD and with two or more risk factors	≥130 mg/dL[b]	<130 mg/dL
■ With existing CHD	>100 mg/dL	≤100 mg/dL

[a]Note that ≥160 mg/dL indicates high risk.
[b]Note that ≥130 mg/dL indicates borderline-high risk.
Source: Adapted from The Expert Panel, Summary of the second report of the National Cholesterol Education Program (NCEP) Expert Panel on Detection, Evaluation, and Treatment of High Blood Cholesterol in Adults (Adult Treatment Panel II), *Journal of the American Medical Association* 269 (1993): 3015–3023.

Obesity, especially abdominal obesity, is associated with high blood lipids, hypertension, and diabetes. With weight loss, heart disease risk factors improve: blood pressure, LDL cholesterol, and triglycerides decline. Weight loss also improves insulin resistance in people with type 2 diabetes.

Notice that the fat-modified diet described here controls both the amount of fat and the amount of saturated, monounsaturated, and polyunsaturated fat. By comparison, the fat-modified diet described in Chapter 20's discussion of fat malabsorption controls the amount of fat and sometimes includes fat from medium-chain triglycerides (MCT).

Reduce Fat, Especially Saturated Fat The Step I diet recommends a total fat intake of less than 30 percent of daily kcalories, with saturated fat no more than one-third of that and dietary cholesterol less than 300 milligrams a day. Such a diet restricts obvious sources of fat, saturated fat, and cholesterol and can be accomplished by following the practical suggestions for reducing fat outlined in Chapter 3 on p. 69.

The Step II diet further reduces saturated fat to 7 percent of daily kcalories and cholesterol to less than 200 milligrams per day. For persons requiring a Step II diet, the help of a registered dietitian can ensure that saturated fat and cholesterol are reduced without sacrificing nutritional quality of the diet.

Other Fats In addition to the dietary recommendations of the expert panel, other fat-related strategies for reducing CHD risk are under investigation. Some

Table 26.4 Diet Strategies to Reduce LDL Cholesterol

	Step I	Step II
Energy	Adequate to achieve or maintain desirable weight	Adequate to achieve or maintain desirable weight
Total fat[a]	≤30%	≤30%
Saturated fat[a]	<10%	<7%
Polyunsaturated fat[a]	<10%	<10%
Monounsaturated fat[a]	5–15%	5–15%
Cholesterol	<300 mg/day	<200 mg/day

[a]All fats except cholesterol are expressed as percentages of total food energy.
Source: Adapted from E. J. Schaefer, New recommendations for the diagnosis and treatment of plasma lipid abnormalities, *Nutrition Reviews* 51 (1993): 246-252.

researchers have shown, for example, that for people with type 2 diabetes, a diet that provides up to 45 percent of total kcalories from fat can effectively lower LDL and triglycerides, provided that the additional fat comes from monounsaturated sources.[7] As an added benefit, the diet does not appear to lower HDL, as low-fat diets sometimes do.

Fish oils, rich in omega-3 fatty acids, have received much attention with respect to CHD risk. A large and carefully conducted study failed to show a correlation between fish oil consumption and CHD risk.[8] The authors noted, however, that very few men in the study group ate no fish at all, and eating fish one or two times per week may confer the same benefits as eating fish five to six times per week. Although eating one or two fish meals per week may be beneficial and is certainly safe, there is not enough evidence to recommend the use of fish oil supplements, which are associated with side effects and may be toxic (see p. 60).[9]

Margarine is often substituted for butter in fat-modified diets because butter is a rich source of both saturated fat and cholesterol. Margarines, while lower in saturated fat and cholesterol, contain *trans*-fatty acids (see Chapter 3), which tend to raise total cholesterol and LDL and lower HDL. When people limit their total fat intakes, it is unlikely that they will eat excessive *trans*-fatty acids. However, it may be prudent to recommend that people use margarine in tub or liquid form, because the more liquid the margarine, the less *trans*-fatty acids it contains.

Include Complex Carbohydrates Encouraging people on fat-modified diets to include fiber-rich foods may have additional cholesterol-lowering and other benefits. When used together with a diet low in both fat and saturated fat, the fiber found in oats, barley, and pectin-rich fruits and vegetables may help reduce blood lipids further.[10] Furthermore, people with elevated triglycerides should avoid simple sugars, which often cause triglyceride levels to rise.

In addition, high-fiber diets tend to be higher in folate and other B vitamins, which may help lower blood homocysteine levels. Studies have shown a positive association between elevated blood homocysteine and the risk of CHD.[11] How homocysteine may increase the risk of CHD is unknown, but it may promote atherosclerosis by damaging the artery walls or promoting thrombosis. Although folate and the other B vitamins can reduce homocysteine levels, the question remains to be answered whether such a reduction also reduces CHD risk.[12] For most people, selecting at least five servings of folate-rich fruits and vegetables should be sufficient to normalize homocysteine concentrations.

See p. 155 in Chapter 7 for information on foods rich in folate.

Antioxidant Nutrients A discussion of dietary factors to reduce CHD risk must include the antioxidant nutrients—a hot topic in both the popular press and scientific journals. Evidence continues to mount that antioxidant nutrients, particularly vitamin E, may help protect against CHD by preventing the toxic effects of free radicals. Nutrition in Practice 26 provides more information about free radicals and antioxidant nutrients in heart disease.

Alcohol Research suggests that a *moderate* consumption of alcohol (no more than two drinks a day) may reduce the risk of CHD by raising HDL cholesterol and preventing clot formation.[13] Beneficial effects of alcohol are more apparent in people over age 50, in those with other risk factors, and in those with high LDL.[14] These findings pose a dilemma for health care professionals who are well aware of the potentially damaging effects of alcohol on many body systems. Furthermore, alcohol can raise triglycerides, and people with elevated triglycerides are advised to restrict alcohol.

Vegetarian and Meat-Restricted Diets When compared to people who eat meat regularly, vegetarians generally maintain a healthier body weight; eat less saturated fat and cholesterol; eat more dietary fiber; have higher intakes of folate, other B vitamins, and vitamins C, A, and E (antioxidants); and enjoy a lower incidence of cardiovascular disease. Whether the vegetarian diet is responsible for protection from cardiovascular disease is difficult to determine, because people who adopt vegetarian diets are also more likely than others to avoid smoking, to use alcohol in moderation (if at all), and to be physically active.

People in Mediterranean countries also experience a low incidence of cardiovascular disease. Mediterranean diets are low in saturated fat and rich in monounsaturated fat (olive oil), dietary fiber, folate, other B vitamins, and antioxidants. Again, however, the lifestyles of people living in Mediterranean countries differ from lifestyles of people in other regions, and it is difficult to sort through the many factors that might have the greatest effect on cardiovascular health.

℞ Prescription Pad

Medications used in the treatment of cardiovascular disease may include:

- Anticoagulants (including aspirin)
- Antihypertensives
- Antilipemics
- Diuretics (isorbide nitrate)
- Nitroglycerin

See Appendix E for timing with meals and nutrition-related side effects.

Drug Therapy

When used together with diet therapy and a physical activity program, drug therapy can effectively lower blood lipids. Lipid-lowering medications carry health risks, however, and are costly. Therefore, physicians as a rule do not prescribe medications to treat hyperlipidemia until after a six-month trial of intensive diet therapy and physical activity has proved unsuccessful in lowering blood lipid concentrations. For people with very high LDL cholesterol (greater than 220 milligrams per deciliter), a shorter diet trial may be considered. Besides lipid-lowering medications, the treatment of atherosclerosis may include aspirin and anticoagulants to prevent clot formation, antihypertensives and diuretics to reduce blood pressure, and nitroglycerin to alleviate angina. All of these medications are associated with significant risks and nutrition-related side effects, a problem compounded by the fact that drug therapy often includes multiple medications and continues for many years or even life.

Hypertension

Diet Order
kCalorie- and fat-modified, high-potassium, no-added-salt diet

Chronic elevated blood pressure, or hypertension, is a major CHD risk factor; it is believed to affect some 43 million people in the United States, more than a quarter of the adult population.[15] Hypertension damages the kidneys and contributes to over a million heart attacks and half a million strokes each year. The higher the blood pressure above normal, the greater the risk of heart disease. People cannot feel the physical effects of high blood pressure, but it can impair life's quality and end life prematurely.

Optimal resting blood pressure for adults averages about 120 over 80 millimeters of mercury (mm Hg). Readings of 130 over 85 mm Hg are considered to be high-normal. At readings of 140 over 90 mm Hg or higher, the risks of heart attacks and strokes increase in direct proportion to increasing blood pressure.

Blood Pressure Regulation The body's ability to maintain blood pressure is vital to life. The heart's pumping action must create enough force to push the blood through the major arteries into the smaller arteries and finally into tiny capillaries, whose thin porous walls permit fluid exchange between the blood and tissues (see Figure 26.2). The nervous system helps maintain blood pressure by adjusting the size of the blood vessels and by influencing the heart's pumping action. The kidneys help regulate blood pressure by setting in motion mechanisms that change the blood volume. The narrower the blood vessels or the greater the volume of blood in the circulatory system, the harder the heart must pump (and the more pressure the heart must create) to feed the tissues.

peripheral resistance: resistance to the flow of blood caused by the reduced diameter of the vessels at the periphery of the body—the smallest arteries and capillaries.

Obesity, Insulin Resistance, and Hypertension Obesity can contribute to the development of hypertension or make it worse. Obesity can also increase insulin

Figure 26.2
How Normal Blood Pressure Supports Fluid Exchange
At the same time the heart pushes blood into an artery, the small-diameter arteries and capillaries at its other end resist the blood's flow (peripheral resistance). Both actions contribute to the pressure inside the artery. Another determining factor is the volume of fluid in the circulatory system, which depends in turn on the number of dissolved particles in that fluid.

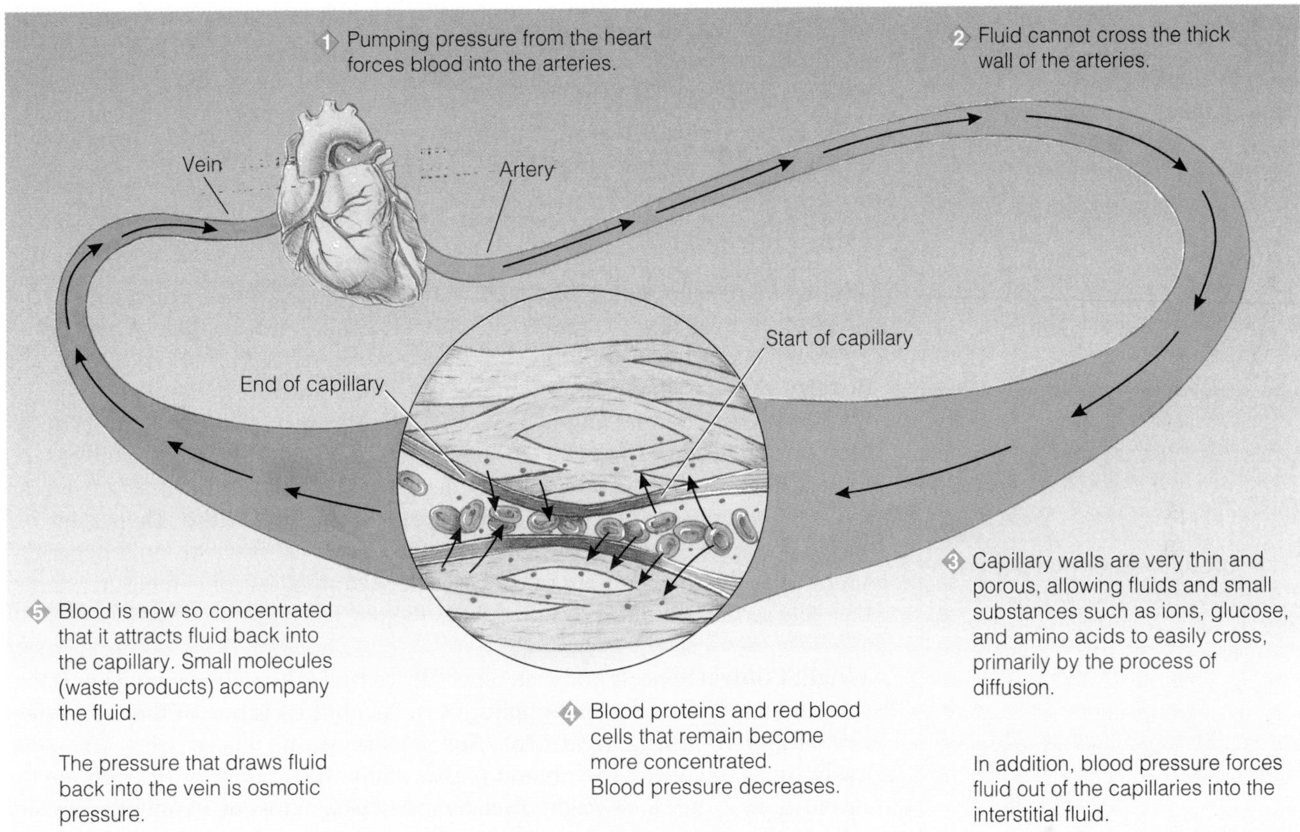

1 Pumping pressure from the heart forces blood into the arteries.
2 Fluid cannot cross the thick wall of the arteries.
Vein
Artery
Start of capillary
End of capillary
3 Capillary walls are very thin and porous, allowing fluids and small substances such as ions, glucose, and amino acids to easily cross, primarily by the process of diffusion.
In addition, blood pressure forces fluid out of the capillaries into the interstitial fluid.
4 Blood proteins and red blood cells that remain become more concentrated.
Blood pressure decreases.
5 Blood is now so concentrated that it attracts fluid back into the capillary. Small molecules (waste products) accompany the fluid.
The pressure that draws fluid back into the vein is osmotic pressure.

resistance, and both are associated with both type 2 diabetes mellitus and hypertension. The interrelationships among type 2 diabetes, obesity, hypertension, and atherosclerosis may explain why 80 percent of people with diabetes die from cardiovascular diseases.

Consequences of Hypertension Strain on the heart's pump, the left ventricle, can enlarge and weaken it until finally it fails (Chapter 24 described chronic heart failure). Constant high pressures in an artery may cause it to balloon out and burst (aneurysm). Aneurysms that go undetected can lead to massive bleeding and death, particularly when a large vessel such as the aorta is affected. In the small arteries of the brain, an aneurysm may lead to stroke, and in the eye, it may lead to blindness. Likewise the kidneys may be damaged (kidney disease) when the heart is unable to adequately pump blood through them. Hypertension injures the artery linings and accelerates plaque formation, thus setting the stage for atherosclerosis or worsening existing atherosclerosis. Then the plaques and reduced blood flow induce a further rise in blood pressure, and hypertension and atherosclerosis become mutually aggravating conditions.

Prevention Significant risk factors for the development of hypertension are listed in the margin. The most effective single step people can take against hypertension is to find out whether they have it: know your blood pressure. A major

Major risk factors for hypertension:

- Age. *Most people develop hypertension in their 50s and 60s.*
- Heredity. *A family history of cardiovascular disease raises the risk of hypertension.*
- Race. *Hypertension develops earlier, is twice as common, and is more severe among African Americans than among whites.*
- Obesity. *Obese people are more likely to develop hypertension.*
- Diabetes mellitus. *People with type 2 diabetes are more likely to develop hypertension.*

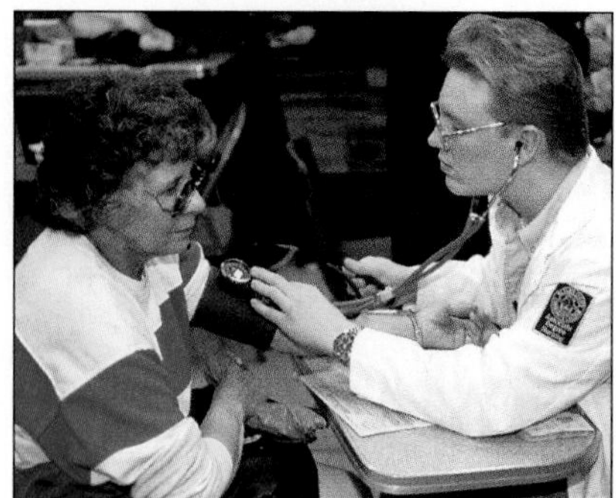

Screening people for hypertension is a first step toward early detection of hypertension and prevention of complications.

WWW dash.bwh.harvard.edu
DASH Diet

Table 26.5 The Dietary Approaches to Stop Hypertension (DASH) Diet

Food Group	Servings per Day[a]
Grains	7–8
Vegetables	4–5
Fruits and juices	4–5
Milk,nonfat or low-fat	2–3
Meats, poultry, fish	2 or less
Nuts, seeds, legumes	½–1
Fats and oils	2–3

[a]Servings for a person consuming a 2100-kcalorie diet.

Source: Adapted from L. J. Appel and coauthors, A clinical trial of the effects of dietary patterns on blood pressure, *New England Journal of Medicine* 336 (1997): 1117–1124.

People with chronic renal disease or diabetes, people with one or both parents who have hypertension, African Americans, and people over age 50 are most likely to be salt-sensitive—that is, their blood pressure responds to different amounts of dietary salt.

Caution: Many salt substitutes contain potassium and should be avoided by people with kidney disease and those on potassium-sparing diuretics.

national effort to identify and treat hypertension is currently under way. Even mild hypertension can be serious; early treatment promotes health and a higher-quality, longer life.

Treatments for Hypertension

The treatment of hypertension focuses on diet, physical activity, and, sometimes, medications. For people with high-normal or mild hypertension, a trial of diet and exercise is often the treatment choice. Medications are often necessary for people with moderate-to-severe hypertension or those who have preexisting cardiovascular disease. Treatment of underlying disorders such as atherosclerosis and diabetes is also important.

Diet Therapy

Traditional diets to reduce blood pressure have typically included recommendations to control weight, reduce salt intakes, and limit alcohol. While health authorities continue to recommend these modifications, they also recommend a diet that contains adequate amounts of potassium, calcium, and magnesium and that limits saturated fat and cholesterol.[16] Results of the Dietary Approaches to Stop Hypertension (DASH) trial show that a diet rich in fruits, vegetables, and low-fat dairy products and with a reduced total fat and saturated fat intake can lower blood pressure to a significant degree.[17] For people without hypertension, such a diet may be able to prevent hypertension. For those with mild hypertension, such a diet may prevent or delay the need for medications. Table 26.5 shows the daily servings of foods on the DASH diet.

Weight Control Excess body fat, especially central fat, can precipitate hypertension and thus raise risks of complications. Weight loss is one of the most effective and long-lasting treatments for hypertension. Those who are using medications to control their blood pressure can often reduce or discontinue the medications if they lose weight. Even a modest weight loss of 10 pounds may significantly lower blood pressure.

Minerals The minerals that have been associated with hypertension include sodium, potassium, calcium, and magnesium. Although researchers have suspected a link between these minerals and hypertension for some time, data from studies have been inconclusive and difficult to interpret. Perhaps the effect of each mineral is too small to detect, but the combined effect of all minerals together results in a significant effect.[18] The DASH trial did not try to sort out how individual nutrients affect blood pressure, but rather how dietary patterns affect it.

Sodium in combination with chloride (table salt) may aggravate hypertension in some people who are salt-sensitive.[19] Mounting evidence suggests a link between salt sensitivity and the adequacy of potassium, calcium, and magnesium in the diet. For people who regularly consume adequate intakes of these minerals, high-salt diets may not be associated with elevated blood pressure. Current recommendations suggest that salt be restricted to about 6 grams a day (2.4 grams of sodium) and that people obtain adequate potassium (3.5 grams), calcium, and magnesium from food sources, rather than supplements. The recommendations do not include specific recommended intakes of calcium and magnesium. For people following the DASH diet at energy levels of about 2000 kcalories, the diet provided about 4.4 grams of potassium, 1250 grams of calcium, 480 milligrams of magnesium, and about three grams of sodium per day.

Potassium may be of concern for people taking diuretics to treat hypertension. A later section describes how these diuretics can lead to potassium imbalances.

Sample Fat-Modified, No-Added-Salt, High Potassium Diet Menu

Breakfast	Lunch	Snack	Supper
¾ c cantaloupe	Broiled chicken breast, 3 oz	Popcorn	Baked flounder, 3 oz
½ c all-bran cereal	Baked beans	Grapefruit juice	Brown rice
Nonfat milk	Carrots		Broccoli
2 slices whole-wheat toast	Tossed salad		Hard roll
Margarine	Hard roll		Margarine
½ c orange juice	Low-fat dressing		1 medium orange
Coffee or tea	Nonfat milk		Coffee or tea
Sugar (optional)	1 banana		

Foods for this menu are prepared with little or no salt and a minimal amount of fat. The fats used for cooking or for flavor would be monounsaturated and polyunsaturated.

Helping Clients Adjust Their Diets The person who normally uses salt liberally may find unsalted foods unpalatable. Diet compliance will probably be poor if the person is handed an instruction sheet with no additional encouragement or guidance. The suggestions in the box on p. 638 can help clients make the necessary changes.

Reminder: 6 g table salt (slightly more than a teaspoon) = 2400 mg sodium.

Salt substitutes and low-sodium products may be helpful for some people, but palatable diets can be planned without them. Caution clients that some of these products contain some salt and cannot be used freely. Furthermore, tell clients not to heat salt substitutes containing potassium because they can turn bitter.

With respect to potassium, calcium, and magnesium, remind clients to follow a well-balanced diet that includes all food groups; emphasizes grains, fruits, vegetables, and legumes; and provides at least two servings of low-fat milk and milk products. Chapter 8 and Appendix A provide more information about the mineral contents of various foods. The accompanying menu illustrates a sample fat-modified, no-added-salt, high-potassium diet.

Dietary Fats Modifying fat intake does not appear to affect blood pressure directly, but because altered blood lipids are risk factors for CHD, reducing total fat and saturated fat is often recommended for the person with hypertension. The DASH diet was low in both total fat and saturated fat, but the trial design did not directly address the effects of fats on blood pressure.

Alcohol Alcohol, especially if consumed in large amounts, is associated with hypertension. An expert committee recommends that if people with hypertension do drink, they should do so in moderation. Moderation means no more than two drinks a day for men and one drink per day for women. People who need to lose weight must consider the kcalories contributed by alcohol as well.

Other Therapies

Some other therapies that may be used in the treatment of hypertension include physical activity and medications. For people who smoke cigarettes, quitting can reduce the risk of cardiovascular disease and is recommended.

How to

Help Clients Cut Their Salt Intake

Most people can significantly lower their sodium intakes by avoiding highly salted foods and removing the saltshaker from the table. Foods eaten without salt may seem less tasty at first, but with repetition, people can learn to enjoy the natural flavors of many unsalted foods. Specific strategies to cut salt intake include:

- Cook without, or with only small amounts of, added salt.
- Prepare foods with sodium-free spices such as basil, bay leaves, curry, garlic, ginger, lemon, mint, oregano, pepper, rosemary, and thyme.
- Add little or no salt at the table.
- Read labels with an eye open for salt. (See Table 1.6 on pp. 23–24 for terms used to describe the sodium contents of foods on labels.)
- Find acceptable low-salt or salt-free products and use them whenever possible.

Use these high-salt foods sparingly if at all:

- Foods prepared in brine, such as pickles, olives, and sauerkraut.
- Cured or smoked meats, such as beef jerky, bologna, corned or chipped beef, frankfurters, ham, lunch meats, salt pork, sausage, and smoked tongue.
- Salty or smoked fish, such as anchovies, caviar, salted and dried cod, herring, sardines, and smoked salmon.
- Snack items such as potato chips, pretzels, salted popcorn, salted nuts, and crackers.
- Bouillon cubes; seasoned salts; soy, steak, Worcestershire, and barbecue sauces.
- Cheeses, especially processed types.
- Canned and instant soups.
- Prepared horseradish, catsup, and mustard.

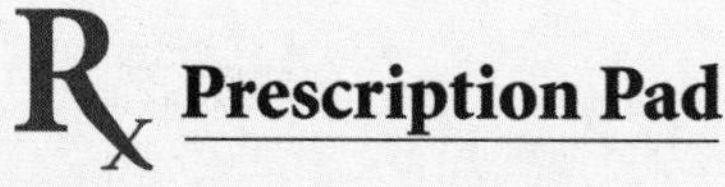

Medications used in the treatment of hypertension may include:

- Antihypertensives
- Diuretics
- Potassium supplements

See Appendix E for timing with meals and nutrition-related side effects.

Physical Activity Not only does physical activity help with weight control, but moderate aerobic activity also helps to lower blood pressure directly. Moderate activities, such as brisk walking for 30 to 45 minutes, can be helpful when undertaken daily or most days of the week.

Medications Drug therapy for hypertension may include one or a combination of antihypertensive agents and diuretics. Many clients may also be receiving medications to treat CHD, diabetes mellitus, and, sometimes, congestive heart failure. The risk of side effects, including medication-medication and nutrient-medication interactions, is heightened because several medications may be necessary, and they may be necessary for long periods of time.

Diuretics can cause potassium imbalances and lead to muscle weakness, unexplained numbness and tingling sensations, irregular heartbeats, and cardiac arrest. Thiazide and loop diuretics increase the urinary excretion of potassium, and clients on these diuretics need to include rich sources of potassium daily. Potassium-sparing diuretics, on the other hand, can lead to potassium retention. Clients taking potassium-sparing diuretics should be cautioned to avoid excessive potassium intakes, potassium supplements, and salt substitutes that contain potassium.

Heart Attacks and Strokes

When atherosclerosis and hypertension run their courses, the consequences can be fatal. Most often, these conditions lead to heart attacks, strokes, or congestive heart failure. Chapter 24 describes congestive heart failure because this chronic disorder frequently leads to severe wasting.

heart attack: sudden tissue death caused by blockages of vessels that feed the heart muscle; also called **myocardial infarction (MI)** or **cardiac arrest.**
myo = muscle
cardial = heart
infarct = tissue death

How to Manage Diets after a Myocardial Infarction (MI)

- Offer nothing by mouth until shock resolves.
- After several hours, give a 1000- to 1200-kcalorie diet that progresses from low-sodium liquids to low-sodium soft foods of moderate temperature in frequent small feedings.
- After five to ten days, adjust the diet to meet individual needs, generally in three meals a day.
- Although somewhat controversial, many practitioners restrict caffeine completely during the first few days after an MI and generally recommend a moderate restriction (no more than 3 cups of a caffeine-containing beverage per day) thereafter.

Heart Attacks

The heart receives nutrients and oxygen, not from inside its chambers, but from arteries on its surface (see Figure 26.1 on p. 629). As mentioned earlier, a heart attack, or myocardial infarction (MI), occurs when the supply of blood to the heart muscle is cut off. Treatment aims to relieve pain, stabilize the heart rhythm, and reduce the heart's workload.

Immediate Care Like an accident victim, a heart attack victim is in shock at first, and diet therapy cannot begin until shock resolves. After several hours of observation, the person can usually begin to eat again, as outlined in the accompanying box. The diet aims to reduce the work of the heart and therefore restricts total energy, the amount of food or drink at each feeding, sodium, and caffeine. Because nausea is a common problem following an MI, liquids are offered first. Low-sodium foods prevent fluid retention, and soft foods can help prevent nausea and abdominal distension, which can push the diaphragm up toward the heart and stress the heart muscle. Temperature extremes can stimulate nerves that slow the heart rate, so foods are offered at moderate temperatures. Caffeine stimulates the metabolic rate and is usually restricted.

Long-Term Diet Therapy After the person is out of immediate danger (in about five to ten days), the diet is tailored to meet individual needs and to deal with conditions such as hyperlipidemia, hypertension, obesity, and diabetes. Whatever else may be necessary, a fat-modified diet (see Table 26.4) that moderately restricts sodium is appropriate for reducing the risk of further CHD and hypertension. Such a diet can be planned to provide three meals a day, but people who still have chest pain after a heart attack may continue to benefit from eating frequent small meals. Advise clients to eat slowly and to avoid strenuous physical activity before and after meals.

Encourage Lifestyle Changes Often an MI victim eagerly applies helpful advice, so use the opportunity to offer sound and useful counsel. Continue to offer support and encouragement for as long as possible. Clients may readily return to their old habits when their symptoms disappear if they haven't fully incorporated new healthful behaviors into their lives. To make permanent lifestyle changes, clients can benefit from the techniques described in Nutrition in Practice 25.

Case Study

History Professor with Cardiovascular Disease

Mr. Jablonski, a 48-year-old history professor, has a blood lipid profile that includes elevated LDL cholesterol. He is 5 feet 7 inches tall and weighs 200 pounds. Mr. Jablonski has a family history of CHD. His diet history shows excessive intakes of food energy, cholesterol, total fat, saturated fat, and salt. He smokes a pack of cigarettes a day, and his lifestyle leaves him little time for physical activity. Mr. Jablonski also has hypertension, for which diuretics have been prescribed. He frequently forgets to take his pills, though, and his blood pressure is often quite high.

Name the risk factors for CHD and hypertension in Mr. Jablonski's history. Which of them can he control? Which can be helped by diet? What complications might you expect if his condition goes untreated?

What type of diet, if any, would you recommend to treat Mr. Jablonski's high LDL cholesterol? Explain the rationale for each diet change. How will his current diet change? Prepare a day's menus for Mr. Jablonski. What type of diet, if any, would you recommend for Mr. Jablonski's hypertension? What suggestions might you offer to help him make the necessary diet changes? Are these diet changes consistent with those you would recommend for high LDL cholesterol?

What laboratory and clinical tests would you expect to see monitored regularly? Why?

Name at least three ways in which Mr. Jablonski could benefit from losing weight. How does physical activity fit into a weight-loss plan? Describe other ways Mr. Jablonski might benefit from a physical activity program.

Discuss nutrition considerations if Mr. Jablonski should suffer a heart attack or stroke. Describe the relationships of these disorders to elevated blood lipids and hypertension.

Strokes

transient ischemic attack (TIA): a temporary reduction in blood flow to the brain that causes temporary symptoms that depend on the part of the brain that is affected. Some common symptoms include light-headedness, visual disturbances, paralysis, staggering, numbness, or dysphagia.

stroke: an event in which the blood flow to a part of the brain is cut off; also called a **cerebral vascular accident (CVA).**
cerebro = brain
vascular = blood vessels

Reminder: *Dysphagia* refers to an inability to coordinate swallowing appropriately.

Temporary interference with blood flow to the brain may result in a transient ischemic attack (TIA), a condition that causes changes in mental status that may last for a few minutes or a few hours. People who experience a TIA may or may not develop a total blockage of blood flow to a portion of the brain that results in a stroke, or cerebral vascular accident (CVA). Most strokes occur as a consequence of atherosclerosis, hypertension, or a combination of the two.

Complications Affecting Food Intake For some stroke victims, recovery is unremarkable; others may require months of rehabilitative therapy. Victims may suffer from temporary or permanent problems that interfere with the ability to communicate. The inability to communicate effectively makes it difficult for them to tell health care professionals about foods they can or would like to eat or about problems they may be having with swallowing foods.

Dysphagia Dysphagia (see p. 444 in Chapter 18) affects many stroke victims. The stroke victim may fail to respond to the presence of food in the trachea with a coughing response, and food may enter the lungs. "Silent" aspiration has been documented in a number of people who develop dysphagia following a stroke. Tube feedings may be indicated initially until the person is able to work with a speech pathologist and dietitian to determine which foods can safely be chewed and swallowed. (Chapter 21 described cautions for feeding clients with a high risk for aspiration.)

Physical Problems Some stroke victims have problems with the physical process of eating. Such problems can include an inability to grasp utensils or coordinate movements that bring foods or liquids from the table to the mouth. Nutrition in Practice 18 described ways to handle such problems.

Long-Term Diet Therapy Long-term diet therapy for stroke victims depends on the underlying medical condition. A diet that restricts kcalories, fat, and sodium is often appropriate. Food energy may need to be limited due to inactivity. A client who must relearn how to walk or use other muscle groups can best do so if not overweight. Underweight can also hinder physical rehabilitation, another reason to ensure that the person does not become malnourished.

The accompanying case study presents a client with CHD. Carefully consider the questions posed to review the information presented in this chapter.

Cardiovascular diseases gradually lead to loss of organ function and can eventually progress to acute events that are fatal. Health care professionals seek to identify people at risk for cardiovascular disorders and encourage clients to modify behaviors known to increase that risk. The nutrition assessment checklist helps to identify nutrition-related factors that play a role in the prevention and treatment of cardiovascular disorders.

Nutrition Assessment Checklist

FOR PEOPLE WITH DISORDERS OF THE HEART AND BLOOD VESSELS

Medical Use the health history to note risk factors for or the presence of cardiovascular diseases, complications associated with those disorders, and other medical conditions that might affect nutrition status.

Medication Assess the client's medications for nutrient-medication interactions. Note that the likelihood of adverse effects on nutrition status is heightened when people use multiple medications over long periods of time. Keep in mind the special potassium needs of those taking diuretics and/or cardiac glycosides (used in the treatment of congestive heart failure).

Food Intake For clients with cardiovascular diseases, assess nutrient intake for total energy; saturated, monounsaturated, and polyunsaturated fats; and cholesterol. In addition, check intakes for salt, potassium, calcium, magnesium, and alcohol.

Anthropometric Assess weight, body fat, and central fat for people with cardiovascular diseases. Weight measurements must be interpreted cautiously in people who develop congestive heart failure (see Chapter 24). For people with atherosclerosis and hypertension, achieving or maintaining a healthy body weight reduces the risk of complications and prevents the additional stress of obesity on the cardiovascular system.

Laboratory Monitor serum lipids in people at risk for CHD and in those with hyperlipidemia. Monitor serum potassium in people taking thiazide or loop diuretics and cardiac glycosides.

Physical Monitor blood pressure. Watch for physical signs of potassium imbalances (muscle weakness, numbness and tingling sensations, and irregular heartbeats). Clients with heart disorders may experience shortness of breath and chest pain, especially when they are physically active.

Self Check

1. Which of the following is the leading cause of death around the world today?
   a. cardiovascular disease
   b. atherosclerosis
   c. coronary heart disease
   d. coronary artery disease
2. Major risk factors for CHD include:
   a. obesity, diabetes, low LDL, and high HDL.
   b. elevated LDL, low HDL, and stress.
   c. obesity, diabetes, elevated LDL, and low HDL.
   d. hypertension and sodium intake.
3. A health care professional working with a client with elevated LDL recommends which of the following diet modifications?
   a. Low total fat, low saturated fat, and high potassium.
   b. Low total fat, low monounsaturated fat, and high carbohydrate.
   c. Low total fat, low saturated fat, and low salt.
   d. Low total fat, low saturated fat, and low cholesterol.

4. The Step II diet differs from the Step I diet in its recommendations for:
   a. weight loss.
   b. saturated fat and cholesterol.
   c. total fat, monounsaturated fat, and cholesterol.
   d. complex carbohydrates.
5. Diet-related components that may be protective against CHD include all of the following **except:**
   a. fiber.
   b. B vitamins, particularly folate.
   c. vitamin K.
   d. antioxidant nutrients, particularly vitamin E.
6. Which of the following statements is true?
   a. The wider the diameter of the arteries, the greater the blood pressure.
   b. Obesity, central obesity, and insulin resistance are associated with the development of hypertension.
   c. Hypertension aggravates atherosclerosis, but atherosclerosis does not aggravate hypertension.
   d. Low blood pressure reduces life expectancy.
7. The single nutrition-related strategy that appears to be most effective in lowering blood pressure is:
   a. salt restriction.
   b. fat restriction.
   c. weight control.
   d. potassium supplementation.
8. For people receiving drug therapy for either hyperlipidemia or hypertension:
   a. diet restrictions and physical activity programs are no longer effective.
   b. the risk of nutrient-medication interactions is high.
   c. the risk of potassium imbalances is lower for people taking loop or thiazide diuretics than for those taking potassium-sparing diuretics.
   d. physical activity progams must be restricted.
9. Which of the following statements about alcohol and cardiovascular diseases might a health care professional relay to a client who wishes to start using alcohol to lower their risks of CHD?
   a. Risks associated with alcohol use may outweigh its benefits in protecting against CHD.
   b. alcohol lowers blood pressure.
   c. alcohol intake should not exceed more than two drinks a day, for both men and women.
   d. a reduced risk of CHD is observed at all levels of alcohol intake.
10. Diet strategies to help a person recover from a myocardial infarction include:
    a. high-kcalorie, high-protein, low-sodium liquids at first to speed repair of the heart and prevent fluid retention.
    b. foods of moderate temperatures to avoid slowing the heart rate.
    c. high-fiber foods to prevent constipation.
    d. three larger meals to allow the heart more time to rest between meals.
11. Following a stroke:
    a. diets must be consistency modified to prevent dysphagia.
    b. weight-loss diets are inappropriate.
    c. physical problems may interfere with eating.
    d. fat- and sodium-modified diets are ineffective.

Answers to these questions appear in Appendix H.

Clinical Applications

1. Consider the list of major risk factors for CHD and include obesity and lack of physical activity as well. Describe possible interrelationships among the factors. For example, a female over age 55 is also at higher risk of diabetes; a person with diabetes is more likely to have hypertension.
2. Consider the DASH diet shown in Table 26.5 that was found to significantly lower blood pressure. In what ways might the DASH diet also be protective against CHD? To guide your thinking, consider these components of the DASH diet: total fat, saturated fat, fiber, folate, other B vitamins, and antioxidants.

Notes

1. B. Halliwell, Antioxidants and human disease: A general introduction, *Nutrition Reviews* (supplement 2) 55 (1997): 44–52.

2. The Expert Panel, Summary of the second report of the National Cholesterol Education Program (NCEP) Expert Panel on Detection, Evaluation, and Treatment of High Blood Cholesterol in Adults (Adult Treatment Panel II), *Journal of the American Medical Association* 269 (1993): 3015–3023.

3. American Heart Association, Risk factors and coronary heart disease: AHA Scientific position, http://www.amhrt.org/Heart_and_Stroke_A_Z_Guide/riskfact.ht, site visited on June 17, 1998.

4. S. M. Grundy and coauthors, Primary prevention of coronary heart disease: Guidance from Framingham, *Circulation* 97 (1998): 1876–1887.

5. G. F. Fletcher and coauthors, Statement on exercise: Benefits and recommendations for physical activity programs for all Americans, *Circulation* 94 (1996): 857–862.

6. The Expert Panel, 1993.

7. A. Garg and coauthors, Effects varying carbohydrate content of diet in patients with non-insulin dependent diabetes, *Journal of the American Medical Association* 271 (1994): 1421–1428.

8. A. Ascherio and coauthors, Marine *n*-3 fatty acids, fish intake, and the risk of coronary disease among men, *New England Journal of Medicine* 332 (1995): 977–982.

9. N. J. Stone, Fish consumption, fish oil, lipids, and coronary heart disease, *Circulation* 94 (1996): 2337–2340.

10. L. Van Horn, Fiber, lipids, and coronary heart disease, *Circulation* 95 (1997): 2701–2704.

11. M. J. Stampfer and coauthors, A prospective study of plasma homocyst(e)ine and risk of myocardial infarction in US physicians, *Journal of the American Medical Association* 268 (1992): 877–880; E. Arnesen and coauthors, Serum total homocysteine and coronary heart disease, *International Journal of Epidemiology* 24 (1995): 704–709.

12. M. J. Stampfer and E. B. Rimm, Folate and cardiovascular disease, *Journal of the American Medical Association* 275 (1996): 1929–1930.

13. J. M. Gaziano and coauthors, Moderate alcohol intake, increased levels of high-density lipoprotein and its subfractions, and decreased risk of myocardial infarction, *New England Journal of Medicine* 329 (1993): 1829–1834; P. R. Ridker and coauthors, Association of moderate alcohol consumption and plasma concentrations of endogenous tissue-type plasminogen activator, *Journal of the American Medical Association* 272 (1994): 929–933.

14. C. S. Fuchs and coauthors, Alcohol consumption and mortality among women, *New England Journal of Medicine* 332 (1995): 1245–1250; H. O. Hein, P. Suadicani, and F. Gyntelberg, Alcohol consumption, serum low density lipoprotein cholesterol concentration and risk of ischaemic heart disease: Six year follow up in the Copenhagen male study, *British Journal of Medicine* 312 (1996): 731–736.

15. V. L. Burt and coauthors, Prevalence of hypertension in the US adult population: Results from the Third National Health and Nutrition Examination Survey, 1988–1991, *Hypertension* 25 (1995): 305–331.

16. The sixth report of the National Committee on Prevention, Detection, Evaluation, and Treatment of High Blood Pressure, *Archives of Internal Medicine* 21 (1997): 2413–2446.

17. L. J. Appel and coauthors, A clinical trial of the effects of dietary patterns on blood pressure, *New England Journal of Medicine* 336 (1997): 1117–1124.

18. M. B. Zemel, Dietary pattern and hypertension: The DASH study, *Nutrition Reviews* 55 (1997): 303–308.

19. T. A. Kotchen and J. M. Kotchen, Dietary sodium and blood pressure: Interactions with other nutrients, *American Journal of Clinical Nutrition* (supplement) 65 (1997): 708–711.

CHAPTER 26

Nutrition in Practice

▪ Free Radicals, Antioxidants, and CHD ▪

Health-conscious consumers around the country have eagerly been swept up in the mad dash for antioxidants, which they have been led to believe can prevent cardiovascular disease, cancer, and aging. Each bit of fresh news about antioxidants convinces more people to use antioxidant supplements, yet few people truly understand what antioxidants are or how they work. The purpose of this nutrition in practice is to explain how antioxidants work and how one antioxidant, vitamin E, might function in preventing CHD. Such a discussion necessarily includes an explanation of free-radical formation—the process through which antioxidant nutrients exert their protective effects.

WWW

amhrt.org
American Heart Association

hsf.ca
Heart and Stroke Foundation of Canada

nhlbi.nih.gov/nhlbi/nhlbi.htm
National Heart, Lung, and Blood Institute

What are free radicals?

Many types of free radicals exist, but the ones that are most important in the human body derive from oxygen. Oxygen is basic to life, and yet, ironically, it can also be toxic.[1] Most of the oxygen we breathe in is used by the body's cells to produce energy. A small amount of oxygen, however, is not used to produce energy, but instead reacts with electrons generated during energy production to form free radicals—molecules with unpaired electrons. A molecule with a free electron is chemically unstable and highly reactive; free radicals quickly react with stable molecules to capture another electron and return to a stable state. In so doing, the stable molecule loses an electron and becomes a free radical itself, and a chain reaction is set in motion that damages other molecules.

In addition to free radicals generated during metabolism, free radicals can also form in the body from exposure to sunlight, ozone, radiation, tobacco smoke, and environmental pollution. Finally, sometimes the immune system intentionally generates free radicals as a means of killing bacteria and fungi and inactivating viruses. With free radicals being generated by these various sources, virtually all of the body's cells are under constant attack from free radicals.

How do we survive such a hostile environment?

The body would not survive without defense mechanisms to keep free-radical formation in check. One line of defense is enzymes, which intercept free radicals and convert them to less toxic substances. Many of these enzymes depend on nutrient cofactors, including copper, zinc, manganese, and selenium. In addition, the body relies on a variety of antioxidants that are not enzymes, including nutrients like vitamins A, C, and E and a variety of nonnutrients.

How can free radicals harm the body if the body's defenses are countering their effects?

Although the defense systems keep free-radical formation in check, free-radical damage is not completely prevented.[2] The term *oxidative stress* describes a serious imbalance between free radicals and antioxidant defenses, such that antioxidant defenses are overwhelmed. Sometimes cells under oxidative stress adapt, and mild oxidative stress may actually make cells more resistant to oxidative damage. But if the oxidative stress is severe or prolonged, damage is self-aggravating. Free radicals can damage lipid and protein molecules in cell membranes and even DNA.[3] As a result, cells may be injured or lose their function, and mutations can arise in DNA. Tissue damage activates the immune system (see Chapter 23), which generates more free radicals. In addition, when cells break down, they release metals (mainly iron and copper) that can catalyze chemical reactions that make a particularly reactive free radical, and the damage becomes progressive. Aging is associated with both an increased production of free radicals and the accumulation of unrepaired damage from free radicals. Thus free-radical formation has been impli-

Nutrition in Practice

Vegetable oils and margarines are the most commonly eaten sources of vitamin E—the antioxidant most closely related to a reduced risk of CHD.

cated as a factor in aging and in the development of chronic diseases including CHD and cancer, inflammatory diseases (such as Crohn's disease and arthritis), and degenerative diseases (such as Alzheimer's and Parkinson's diseases).

How exactly do free radicals relate to CHD?

As Chapter 26 describes, high LDL cholesterol is a major risk factor for CHD. Data suggest that the oxidation of LDL by free radicals promotes the adhesion of certain immune system cells (macrophages) to the artery walls and exposes receptors that allow the macrophages to fill with oxidized LDL.[4] The immune system cells undergo changes and eventually become the fatty streaks characteristic of plaques. Oxidized LDL also damage the blood vessel walls, allowing the entry of other plaque-forming factors and promoting clot formation.

Interestingly, homocysteine appears to stimulate the oxidation of LDL. Recall from Chapter 26 that elevated homocysteine may be a risk factor for CHD.

Is there any way to prevent oxidative stress?

This is perhaps the million dollar question. With free radicals implicated in so many diseases, researchers are busy trying to find an answer to this question. Not surprisingly, many investigations focus on antioxidant nutrients. With respect to cardiovascular disease, vitamin E has shown the strongest correlation. Vitamin E is packaged along with cholesterol in LDL. Epidemiological studies suggest a negative correlation between vitamin E status and death rates from heart disease.[5] Researchers selected groups of men in 16 European regions where rates of death from heart disease varied sixfold. The researchers measured plasma vitamin E, cholesterol, and blood pressure in men from each region. When the groups were compared, high death rates from heart disease correlated more strongly with low vitamin E concentrations than with either cholesterol or blood pressure. The authors cautioned that while the evidence is suggestive of a protective effect of vitamin E in heart disease, it is by no means conclusive. Other factors or a combination of factors not controlled for in the studies may be responsible for the findings.

Other studies suggest that vitamin E may slow the progression of plaques that have already formed in the arteries or may lower the risk of developing a heart attack in people with existing CHD.[6] To date there has been no carefully designed study to compare the effects of vitamin E and a placebo in altering complications or death rates from CHD.[7]

Might there be any harm in taking vitamin E to prevent CHD until more is known?

It is certainly safe to eat a diet rich in natural sources of vitamin E (vegetable oils, polyunsaturated margarines, some nuts, and wheat germ). However, the amount of vitamin E that appears to be protective against CHD is difficult to consume from a typical diet, particularly from a diet that obtains less than 30 percent of its energy from fat. Therefore, vitamin E supplements may be necessary to exert an effect on CHD risk. When provided in large amounts, vitamin E may have pharmacological, rather than nutritional, effects. The impact of taking a single antioxidant for many years is unknown. Such a practice could possibly interfere with normal physiological processes, perhaps immune defenses.[8] Until further studies are completed to verify whether vitamin E can protect against heart disease and to determine what dose is effective and safe, most authorities do not recommend vitamin E supplements.

Another safe strategy to try is to include a wide variety of fruits and vegetables—rich sources of folate, other B vitamins, phytochemicals, and fiber. Folate and other B vitamins can lower homocysteine levels, and homocysteine stimulates the oxidation of LDL. Although studies have not confirmed that lowering homocysteine levels by eating adequate amounts of folate and other B vitamins reduces CHD risk, including fruits and vegetables presents no health risks.

Much remains to be discovered about cardiovascular diseases and the nutrition-related and other factors that can prevent or delay their development. Health care professionals have the responsibility for staying abreast of new developments so that they can help clients balance fact and fiction and thus protect their health.

CHAPTER 26

Nutrition in Practice

Notes

1. B. Halliwell, Antioxidants and human disease: A general introduction, *Nutrition Reviews* (supplement 2) 55 (1997): 44–52.

2. Halliwell, 1997.

3. V. M. Sardesai, Role of antioxidants in health maintenance, *Nutrition in Clinical Practice* 10 (1995): 19–25.

4. B. Frei, Reactive oxygen species and antioxidant vitamins: Mechanisms of action, *American Journal of Medicine* (supplement 3A) 97 (1994): 5–13.

5. K. F. Gey and coauthors, Inverse correlation between plasma vitamin E and mortality from ischemic heart disease in cross-cultural epidemiology, *American Journal of Clinical Nutrition* 53 (1991): 326–334; K. F. Gey and coauthors, Increased risk of cardiovascular disease at suboptimal plasma concentrations of essential antioxidants: An epidemiological update with special attention to carotene and vitamin C, *American Journal of Clinical Nutrition* 57 (1993): 333–336.

6. H. N. Hodis and coauthors, Serial coronary andiographic evidence that antioxidant vitamin intake reduces progression of coronary artery atherosclerosis, *Journal of the American Medical Association* 273 (1995): 1849–1854; N. G. Stephens and coauthors, Randomised controlled trial of vitamin E in patients with coronary disease: Cambridge Heart Antioxidant Study (CHAOS), *Lancet* 347 (1996): 781–786.

7. P. O. Kwiterovich, The effect of dietary fat, antioxidants, and pro-oxidants on blood lipids, lipoproteins, and atherosclerosis, *Journal of the American Dietetic Association* (supplement) 97 (1997): 31–41.

8. Sardesai, 1995.

27 *Protein-, Mineral-, and Fluid-Modified Diets for Renal Diseases*

WWW.

kidney.org
National Kidney Foundation

kidney.ca
Kidney Foundation of Canada

niddk.nih.gov
National Institute of Diabetes and Digestive and Kidney Diseases

renalnet.org/renalnet/renalnet.cfm
Renalnet Kidney Information Clearinghouse

aakp.org
American Association of Kidney Patients

Diets modified in protein, minerals, and fluids serve as a primary treatment of diseases that alter renal function. This chapter begins by examining the principles that underlie diets for renal diseases. It then shows how these principles are applied in the specific cases of the nephrotic syndrome, acute renal failure, and chronic renal failure. Finally, the chapter describes the dietary management of kidney stones. (Although kidney stones do not alter renal function, they arise in the kidneys and can lead to complications that affect the kidneys.)

Figure 27.1 shows the location of the kidneys and the structure of a nephron—one of the kidneys' functional units. Notice the glomerulus, a key part of the nephron, which serves as a gate through which blood passes into the nephron to be cleansed and returned to the body. Healthy kidneys, by continuously filtering the blood, maintain the body's fluid and electrolyte and acid-base balances and eliminate metabolic waste products. They also help regulate blood pressure and produce the hormone erythropoietin, which stimulates red blood cell production. Additionally, the kidneys convert vitamin D to its most active form, and they play an important role in maintaining healthy bone tissue. Failure of the kidneys, then, can affect any or all of these functions. The glossary defines terms related to kidney function.

Diets for Altered Renal Function

Diets for altered renal function aim to restore or maintain nutrition status and to prevent the progression of renal disease, the buildup of metabolic waste prod-

Glossary

active vitamin D: the 1,25-dihydroxy form of vitamin D that promotes calcium balance and bone mineralization.

dialysis (dye-AL-ih-sis): removal of waste from the blood through a semipermeable membrane using the principles of simple diffusion and osmosis. The two main types of dialysis are hemodialysis and peritoneal dialysis (see Nutrition in Practice 27).

ehrythropoietin (eh-RITH-row-POY-eh-tin): a hormone secreted by the kidneys in response to oxygen depletion or anemia that stimulates the bone marrow to produce red blood cells.

erthro = red
poi = to make

glomerular filtration rate (GFR): the rate at which the kidneys form filtrate. Normally, the GFR is about 120 ml/min. Falling GFR signifies deteriorating kidney function. When the GFR falls drastically, to about 10 ml/min, dialysis is usually initiated.

glomerulus (glow-MARE-you-lus): a cup-shaped membrane enclosing a tuft of capillaries within a nephron. (The plural is *glomeruli.*)

nephrons (NEF-rons): the working units of the kidneys; each nephron consists of a glomerulus and a tubule.

renal: pertaining to the kidneys.

renal failure: failure of the kidneys to maintain normal function to a degree that requires dialysis or kidney transplantation for survival.

renal filtrate: the fluid that passes from the blood through the capillary walls of the glomeruli, eventually forming urine.

renal insufficiency: reduced renal function but not to the degree that requires dialysis or kidney transplantation.

renin (REN-in): an enzyme secreted by the kidneys in response to a reduced blood flow that triggers the release of the hormone aldosterone from the adrenal glands. Aldosterone, in turn, signals the kidneys to retain sodium and fluid.

tubule: a tubelike structure that surrounds the glomerulus and descends through the nephron. A pressure gradient between the glomerular capillaries and the tubule returns needed materials to the blood and moves wastes into the tubule to be excreted in the urine.

uremia (you-REE-me-ah): abnormal accumulation of nitrogen-containing substances, especially urea, in the blood; also called **azotemia** (AZE-oh-TEE-me-ah). Normal BUN (blood urea nitrogen) levels are 10 to 20 mg/dL. A BUN of 50 to 150 mg/dL indicates serious impairment of renal function. BUN may rise to as high as 150 to 250 mg/dL in end-stage renal disease.

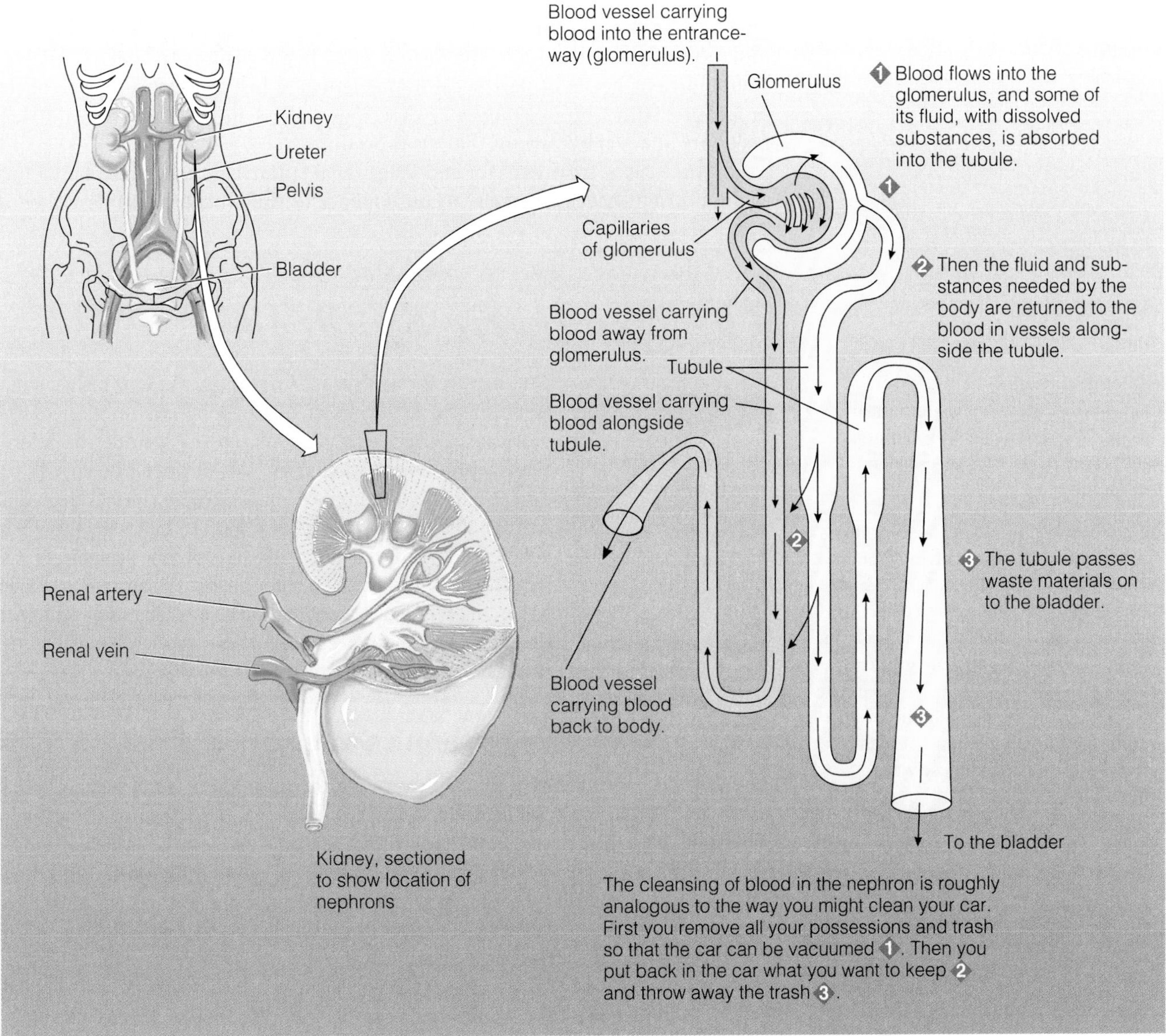
A nephron (a working unit of the kidney)
Blood vessel carrying blood into the entrance-way (glomerulus).
Glomerulus
1 Blood flows into the glomerulus, and some of its fluid, with dissolved substances, is absorbed into the tubule.
Kidney
Ureter
Pelvis
Bladder
Capillaries of glomerulus
2 Then the fluid and substances needed by the body are returned to the blood in vessels alongside the tubule.
Blood vessel carrying blood away from glomerulus.
Tubule
Blood vessel carrying blood alongside tubule.
3 The tubule passes waste materials on to the bladder.
Renal artery
Renal vein
Blood vessel carrying blood back to body.
To the bladder
Kidney, sectioned to show location of nephrons
The cleansing of blood in the nephron is roughly analogous to the way you might clean your car. First you remove all your possessions and trash so that the car can be vacuumed 1. Then you put back in the car what you want to keep 2 and throw away the trash 3.

Figure 27.1
A Nephron, One of the Kidneys' Many Functioning Units

ucts or nutrient excesses, and complications associated with renal disease. Some of the complications with nutrition implications include growth failure and wasting, hypertension, edema, elevated blood lipids, congestive heart failure, bone disease, and anemia. The diet prescription specifies amounts of energy, protein, lipids, electrolytes, and fluid; the exact amounts depend on the type of renal disease, its severity, and its treatment.

Diet and the Type and Severity of Renal Disease The kidneys may fail suddenly or deteriorate gradually. As renal function deteriorates, the composition of the blood and urine changes. The rate at which the kidneys form filtrate, the glomerular filtration rate (GFR), declines, upsetting fluid and electrolyte and acid-base balances. The body's principal nitrogen-containing metabolic waste products—blood urea nitrogen (BUN), creatinine, and uric acid—accumulate in the blood. The degree of renal function, as well as complications such as malnutrition or severe stress, determines what diet will best preserve nutrition status while preventing the buildup of waste products.

WWW. kdf.org.sg
Kidney Dialysis Foundation

Diet and Treatment for Renal Disease If diet alone cannot control the accumulation of waste products, the person must begin dialysis to survive. Dialysis alters nutrient needs in part because nutrients are lost as a consequence of the procedure. Nutrition in Practice 27 describes dialysis and explains how nutrient needs are affected by various dialysis procedures.

In some cases, treatment for end-stage renal failure is a renal transplant. Once again, nutrient needs change as a consequence of the procedure, as described in a later section.

The Nephrotic Syndrome

nephrotic syndrome: the complex of symptoms that occur when glomerular function falters; it includes proteinuria, low serum proteins, edema, and elevated blood lipids.

The loss of protein in the urine is **proteinuria,** and the loss of the protein albumin in the urine is **albuminuria.** Sensitive laboratory tests allow clinicians to detect **microalbuminuria,** the loss of albumin in the urine in quantities that are greater than normal, but not enough to precipitate symptoms. Tests for microalbuminuria are routinely performed in people at high risk for renal disease, such as those with diabetes.

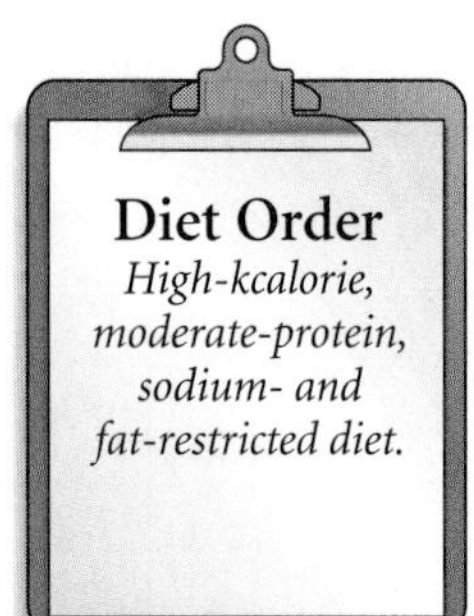

Diet Order
High-kcalorie, moderate-protein, sodium- and fat-restricted diet.

The nephrotic syndrome is not a disease, but rather a distinct cluster of symptoms caused by dysfunction of the glomerular capillaries. The symptoms include the loss of plasma proteins into the urine, low serum albumin, edema, and elevated blood lipids. Damage to the kidneys from diabetic nephropathy, infections, blood clots in the renal veins, and some drugs and toxins can precipitate the nephrotic syndrome. The nephrotic syndrome is sometimes an early sign of renal failure, especially in people with diabetes (see Chapter 25). In other cases, treatment of the underlying condition can correct the disorder.

Consequences of the Nephrotic Syndrome

The major consequences of the nephrotic syndrome include protein-energy malnutrition (PEM), anemia, blood coagulation disorders, occlusion of blood vessels from clots in the lungs and legs, infection, and accelerated atherosclerosis. Many of the consequences of this disorder are the same as those of malnutrition, which is not surprising because both alter protein status. If the nephrotic syndrome progresses to renal failure, the person develops uremia and other manifestations as a later section describes.

Loss of Blood Proteins As plasma proteins escape through the urine, blood proteins plummet. Figure 27.2 shows the consequences of falling levels of various proteins.

Edema While plasma proteins are lost as a consequence of the nephrotic syndrome, sodium is reabsorbed by the nephrons in greater amounts than normal.[1] Sodium retention results in edema.

Altered Blood Lipids Elevated cholesterol, triglycerides, LDL, and VLDL and low HDL are characteristic of the nephrotic syndrome. People with the nephrotic syndrome have a high risk of cardiovascular disease and stroke.

℞ Prescription Pad

Medications used in the treatment of the nephrotic syndrome may include:

- Antibiotics
- Anticoagulants
- Anti-inflammatory agents (corticosteroids)
- Diuretics
- Immunosuppressants

See Appendix E for timing with meals and nutrition-related side effects.

Treatment of the Nephrotic Syndrome

Medical treatment of the nephrotic syndrome first requires treatment of the underlying disorder and then drug and diet therapy to resolve symptoms. Diet is central to preventing protein malnutrition and alleviating edema.

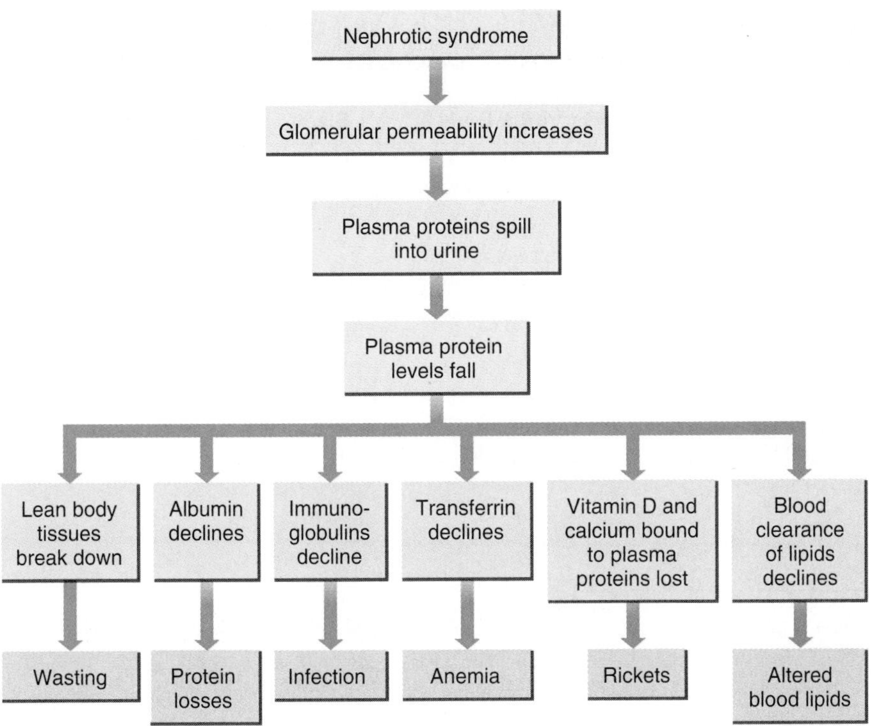

Figure 27.2
Consequences of Urinary Protein Losses in the Nephrotic Syndrome

Energy A diet adequate in energy (about 35 kcalories per kilogram of body weight per day) sustains weight and spares protein. Weight loss or infections signal the need for additional kcalories. People who are obese may benefit from lowering their energy intakes to help control blood lipid levels.

Protein In the past, high-protein diets (about 120 grams per day) were often prescribed for people with the nephrotic syndrome, stemming from the belief that high dietary protein would compensate for protein losses. High-protein diets accelerate protein synthesis, but they also incur greater urinary protein losses.[2] Furthermore, extra dietary protein may accelerate deterioration of renal function.[3] For these reasons, protein is provided at about 0.8 to 1.0 gram per kilogram of body weight per day, about the RDA.

Fat A low-fat diet such as the plan shown in Chapter 26 (less than 30 percent of kcalories from fat) can help control the elevated blood lipids associated with the nephrotic syndrome. Often, however, people with the nephrotic syndrome are unable to control blood lipids adequately using diet alone, and physicians may prescribe medications as well.

Sodium Because a person with the nephrotic syndrome avidly retains sodium, sodium is generally restricted. Although the level of sodium restriction varies depending on the individual's response to diuretics, the diet generally provides about 2 grams of sodium. Table 27.1 on p. 652 shows the foods allowed on a 2-gram sodium-restricted diet. Clients placed on thiazide or loop diuretics should be encouraged to select foods rich in potassium (see Chapter 26).

Acute Renal Failure

In acute renal failure, the nephrons suddenly lose function and are unable to maintain homeostasis. The degree of renal dysfunction varies from mild to severe. With prompt treatment, acute renal failure may be reversible in some cases, but in others, the damage is permanent.

Table 27.1 Two-Gram (2600 mg) Sodium-Restricted Diet

Foods Restricted	Number of Servings Daily	Serving Size	Sodium per Serving (mg)
Regular breads and cooked cereals	4	1 slice bread or ½ c cereal	125
Fresh, frozen, or canned vegetables without salt: artichokes; beets; carrots; celery; beet, collard, dandelion, mustard, and turnip greens; kale; swiss chard; white turnips; low-sodium vegetable juice	Avoid excessive use	½ cup	50
Canned or frozen vegetables with salt; frozen corn, lima beans, mixed vegetables, and peas	2	½ cup	250
Regular nonfat, whole, and evaporated milk and milk products	2	8 oz	120
Fresh and fresh frozen meats, poultry, and freshwater fish; low-sodium canned meats and fish, peanut butter, cheese; unsalted soybeans, textured vegetable protein, and cottage cheese	8	1 oz	25
Eggs	1	1	70
Regular butter and margarine	6	1 tsp	50

Foods Allowed

1. Low-sodium breads, bread products, and cereals; bread products made without salt and with low-sodium baking powder; puffed rice and wheat and shredded wheat cereals; rice; pasta.
2. Fresh, unsalted frozen, and low-sodium canned vegetables (except those listed above); low-sodium tomato juice.
3. All fruits and fruit juices.
4. Unsalted butter, margarine, nuts, and gravy; low-sodium salad dressings and mayonnaise; shortening.
5. Low-sodium catsup, mustard, and tabasco sauce.
6. Soups, casseroles, and recipes made with allowed foods and food ingredients.

Foods Not Allowed (Unless Calculated into the Diet)

1. Table, celery, garlic, and onion salts; reduced-sodium salts; regular catsup, mustard, and tabasco sauce; monosodium glutamate; Worcestershire, barbeque, and soy sauces; baking powder and soda.
2. Instant and quick-cooking hot cereals; commercial bread products made from self-rising flour or cornmeal, salt, baking powder, or baking soda; salted snack foods such as potato chips, corn chips, tortilla chips, popcorn, and pretzels.
3. Sauerkraut, pickles, and salted vegetable juices.
4. Maraschino cherries; crystallized or glazed fruits, and dried fruits with sodium sulfite added.
5. Buttermilk, chocolate milk, instant milk mixes, regular cheeses, and prepared pudding mixes; commercial ice cream, sherbet, and frozen desserts.
6. Cured, canned, salted, or smoked meats, poultry, and fish such as bacon, luncheon meats, corned beef, kosher meats, and canned tuna and salmon; imitation fish products; salted textured vegetable protein; regular peanut butter; salted nuts.
7. Salt pork and bacon fat; commercial salad dressings and mayonnaise; olives; regular gravy.
8. Regular canned soups and bouillon.

Reduced blood flow to the kidneys is a **prerenal** cause of acute renal failure.

A urinary tract obstruction is a **postrenal** cause of acute renal failure.

Damage to the kidneys' cells is an **intrarenal** cause of acute renal failure.

Acute renal failure frequently develops when blood flow to the kidneys suddenly declines, often as a result of a severe stress such as heart failure, shock, or severe blood loss after surgery or trauma. In these cases, acute renal failure is imposed on catabolic illness, and mortality rates are high. When acute renal failure develops for other reasons, such as urinary tract obstructions, recovery rates are much better. Infections, toxins, and some drugs can directly damage the kidneys' cells and lead to acute renal failure.

Consequences of Acute Renal Failure

Acute renal failure is characterized by a sudden and precipitous drop in the GFR and urine output. As the nephrons fail, the nitrogen-containing waste products accumulate in the blood (uremia). Consider that the person who is also catabolic is in negative nitrogen balance and has large amounts of nitrogen to excrete. It is easy to see how uremia and its clinical manifestations—the uremic syndrome—can quickly develop. The margin lists the symptoms of the uremic syndrome, which affects virtually every organ system. Hemorrhages may be visible on the skin, and GI bleeding sometimes occurs.

Electrolyte Levels Rise As renal function falters, blood electrolyte levels rise and can result in serious consequences. In a person who is catabolic, electrolyte levels can rise rapidly as the body's cells break down and release potassium, phosphorus, and magnesium. Elevated potassium (hyperkalemia) is of particular concern because potassium imbalances can alter the heart rate and even lead to heart failure. The problems of renal failure are multiplied as blood flow to the kidneys is compromised.

Blood Volume Changes In the early stages of acute renal failure, blood pressure may rise dangerously high as fluids accumulate. Later in the course of acute renal failure, the kidneys cannot conserve water, and the person begins to excrete large amounts of fluid and electrolytes. If recovery occurs, kidney function gradually normalizes.

uremic (you-REE-mic) **syndrome:** the many symptoms that accompany the buildup of toxic waste products in the blood including:

- *Fatigue.*
- *Weakness.*
- *Diminished mental alertness.*
- *Agitation.*
- *Muscle twitches.*
- *Muscle cramps.*
- *Anorexia.*
- *Nausea.*
- *Vomiting.*
- *Stomatitis.*
- *Unpleasant taste in the mouth.*
- *Diarrhea.*

The early phase of acute renal failure, when urine volume is reduced, is the **oliguric phase;** the phase characterized by large fluid and electrolyte losses in the urine is the **diuretic phase;** the gradual return of renal function marks the **recovery phase.**

Treatment of Acute Renal Failure

The primary goal in acute renal failure is to treat the underlying cause in order to prevent permanent or further damage to the kidneys. For example, if severe blood loss is the problem, a blood transfusion may be needed to restore blood volume. Diet therapy, drug therapy, and dialysis may be undertaken to restore fluid and electrolyte balances and minimize blood concentrations of toxic waste products.

Energy If the person in acute renal failure receives too little energy, body proteins break down, raising blood urea and potassium even higher and taxing the kidneys further. Thus wasting and malnutrition complicate recovery. The person's energy needs depend on the rate of catabolism, so indirect calorimetry provides the best estimate of energy needs.[4] When indirect calorimetry is not available, energy needs may be estimated at about 30 to 45 kcalories per kilogram of body weight per day.[5]

Protein Although protein contributes nitrogen and can tax the kidneys, protein is necessary to prevent complications such as impaired wound healing, infections, muscle wasting, and negative nitrogen balance, and some of these complications may prove fatal. Accordingly, clinicians often provide a higher protein intake, even if it means the person will need dialysis. Actual protein needs of people with acute renal failure depend on the degree of renal function, the metabolic rate, and nutrition status. People who are not on dialysis receive about 0.6 to 1.0 gram of protein per kilogram of body weight per day. If dialysis is begun, a more liberal protein intake of 1.1 to 2.5 grams of protein per kilogram of body weight per day is provided because nitrogenous waste products are removed and some amino acids are lost during the dialysis procedure (see Nutrition in Practice 27).[6]

Fluids Fluid balances are carefully restored in clients who are either overhydrated or dehydrated. Thereafter, health care professionals determine fluid needs by measuring urine output and then adding about 500 milliliters for water lost through the skin, lungs, and perspiration. The person who is vomiting, has diarrhea, has a high fever, or otherwise loses fluids has greater fluid needs. In the oliguric stage, the person needs small amounts of fluids. In the diuretic stage, urine volume may increase significantly, and large amounts of fluids may have to be provided.

Electrolytes Sodium may be restricted (to 500 to 1000 milligrams) in the oliguric phase, but this may change as the person enters the diuretic phase. Generally, potassium, phosphorus, and magnesium are restricted; however, lab values must be monitored carefully, especially in those who are catabolic.

American Society for Enteral and Parenteral Nutrition

Enteral and Parenteral Nutrition Because clients with acute renal failure are often severely stressed, they frequently receive their nutrients from tube feedings or TPN. Special enteral and parenteral formulas for renal failure meet energy and protein needs in small volumes. Compared with standard enteral formulas, renal formulas have less protein, fewer electrolytes, and more kcalories per milliliter (see Appendix G). TPN formulas are compounded with mixtures of both nonessential and essential amino acids at lower concentrations and dextrose at higher concentrations than in standard TPN solutions. Electrolytes are added in appropriate amounts.

The carbohydrate-dense enteral and parenteral formulas used for acute renal failure may exacerbate the hyperglycemia and insulin resistance that frequently accompany both severe stress and renal failure. Fat can be used to add kcalories, reduce the need for carbohydrate, and limit the glucose load. Insulin may be provided to lower blood glucose.

Rx Prescription Pad

Medications used in the treatment of acute renal failure may include:

- Anti-infective agents
- Diuretics
- Exchange resins (sodium polystyrene)
- Insulin

See Appendix E for timing with meals and nutrition-related side effects.

Drug Therapy In the oliguric phase of acute renal failure, diuretics may be used to mobilize fluids. Medications called exchange resins must sometimes be used to treat hyperkalemia. These medications, provided by mouth or through an enema, cause sodium to be exchanged for potassium in the colon, and the potassium is then excreted in the stool.

As mentioned, insulin may be provided to help lower blood glucose. Insulin also lowers blood potassium and phosphorus in two ways. First, as insulin moves glucose into the cells, potassium and phosphorus follow. Second, as an anabolic hormone, insulin minimizes tissue breakdown, and consequently, the cells retain potassium and phosphorus. The accompanying case study helps direct your thoughts toward the needs of a client with acute renal failure.

Chronic Renal Failure

Chronic renal failure [is like] a downhill course for your patient; acute renal failure is like seeing him go over a cliff no one quite knew was there.

—J. L. Stark

Unlike acute renal failure, in which the GFR drops suddenly and sharply, in chronic renal failure the GFR gradually deteriorates, and renal function is irreversibly altered. Chronic renal failure develops from conditions that progressively and permanently damage the kidneys, notably certain infections, renal artery obstructions, diabetic nephropathy (Chapter 25), hypertension (Chapter 26), atherosclerosis (Chapter 26), and congestive heart failure (Chapter 24). Recall that some people who develop the nephrotic syndrome eventually develop chronic renal failure. In a few cases, chronic renal failure results from a disorder that rapidly causes irreversible kidney damage, as may occur following acute renal failure.

Consequences of Chronic Renal Failure

The capacity of the kidneys to function despite loss of some nephrons is referred to as **renal reserve.**

end-stage renal disease: the severe stage of renal failure in which dialysis or a renal transplant is necessary to sustain life.

In the early stages of chronic renal failure, the body compensates for the loss of some nephron function by enlarging the remaining functional nephrons. The hypertrophied nephrons work so efficiently that the GFR may fall to 75 percent of its normal rate before the symptoms of renal failure appear.

The body eventually exhausts the overworked nephrons, and renal function deteriorates further. (This effort is similar to that of the pancreatic beta cells in type 2 diabetes, which at first produce more and more insulin in response to high blood glucose and later become exhausted and unable to produce adequate insulin.) In end-stage renal disease, the GFR drops below 20 percent of normal.

Case Study

Store Manager with Acute Renal Failure

Mrs. Calley is a 35-year-old woman admitted to the hospital's intensive care unit. She was first seen in the emergency room after she sustained multiple and severe injuries in an auto accident. She had lost so much blood she almost died before reaching the hospital. Her injuries include a fractured leg, broken ribs, a collapsed lung, and internal bleeding. Following emergency surgery to stop the internal bleeding and repair injuries, she developed acute renal failure. Mrs. Calley is 5 feet 3 inches tall and weighs 125 pounds.

Mrs. Calley has a urine volume of less than 50 ml/day and a BUN of 75 mg/dL. A test of GFR could not be performed due to the low volume of urine she was excreting.

Describe the most probable reason why Mrs. Calley developed acute renal failure. What other problems can cause acute renal failure? Describe the phases of acute renal failure.

What are Mrs. Calley's dietary needs in the early phase? What waste products and electrolytes are of greatest concern? Why? What factors do you have to keep in mind in determining Mrs. Calley's energy and protein needs during acute renal failure? How will these needs change if dialysis is begun?

How will Mrs. Calley's nutrient needs change as she progresses to the second stage of acute renal failure? Why? As you read through the discussion of chronic renal failure, consider how Mrs. Calley's needs would change if her renal failure became chronic.

Uremic Syndrome As renal function deteriorates, nitrogen-containing waste products accumulate in the blood, and the uremic syndrome (see p. 653) develops. The skin becomes dry and scaly, and the person may itch uncomfortably. Skin hemorrhages may be visible. In the later stages, urea (which can be excreted through the sweat) may crystallize on the skin, a symptom known as uremic frost. Nausea, vomiting, diarrhea, gastritis, and GI bleeding frequently accompany the uremic syndrome.

Reminder: The nitrogen-containing waste products that accumulate in renal failure include blood urea nitrogen (BUN), creatinine, and uric acid.

uremic frost: the appearance of urea crystals on the skin.

Blood Chemistry Alterations In addition to the retention of nitrogen-containing waste products, the body retains excess fluids and electrolytes and acids produced during metabolism that are normally excreted in the urine. As a consequence, the person develops edema and acidosis, which stress the cardiovascular and pulmonary systems. Elevated blood potassium can trigger irregular heartbeats that further stress the heart and can cause the heart to fail. Elevated phosphorus upsets the body's balance of phosphorus and calcium, often leading to bone disease.

Hormonal balances are also upset. Altered hormone levels together with the retention of waste products lead to hypertension, hyperglycemia, elevated blood lipids (especially triglycerides), low serum albumin, anemia, and altered bone metabolism.

Reminder: *Acidosis* is the condition of having too much acid in the blood.

renal osteodystrophy (OS-tee-oh-DIS-tro-fee): a bone disorder resulting from calcium and phosphorus imbalances in renal disease.

Other Complications Accelerated atherosclerosis and cardiovascular diseases frequently accompany renal failure, and they can aggravate hypertension and lead to congestive heart failure, heart attacks, or pulmonary edema. Anemia may result from depressed erythropoietin synthesis by the damaged kidneys, restrictive diets, nausea and vomiting, GI blood losses, and blood losses through dialysis and from frequent blood testing.

Growth Failure and Wasting Both children and adults with chronic renal disease frequently develop wasting and PEM. Nutrition status becomes more difficult to maintain as renal disease progresses. Table 27.2 on p. 656 summarizes the causes of wasting associated with renal failure. Children with renal disease need nutrition intervention before the end of puberty if they are to make up growth deficits. Adults with renal disease can maintain or restore their nutrition status, avoid complications, and improve quality of life by attending to diet as well.

Table 27.2 Possible Causes of Wasting in Renal Failure		
Reduced Nutrient Intake	**Excessive Nutrient Losses**	**Raised Nutrient Needs**
Anorexia	Dialysis	Hormonal alterations
Fatigue	Diarrhea	Infection
Medications	GI bleeding	Medications
Nausea	Medications	
Pain	Numerous blood tests	
Restrictive diet	Poor absorption	
Taste alterations	Vomiting	

Treatment of Chronic Renal Failure

Treatment of chronic renal failure includes diet, medications, dialysis, and, sometimes, kidney transplants. Table 27.3 summarizes nutrient needs for chronic renal failure and shows how they are affected by hemodialysis and peritoneal dialysis. The complexity of the renal diet, as well as its critical role in the treatment of renal disease, underscores the need for a specialist, a renal dietitian, to educate clients and provide diet plans. Other health care professionals do not need to know the specifics of diet therapy, but they must understand the general concepts in order to communicate effectively with clients.

Rx Prescription Pad

Medications used in the treatment of chronic renal failure may include:

- Antiemetics
- Antihypertensives
- Antilipemics
- Antisecretory agents
- Cardiac glycosides
- Diuretics
- Insulin
- Phosphate binders

See Appendix E for timing with meals and nutrition-related side effects.

To evaluate dialysis and guide diet therapy, renal teams often use a mathematical model that takes into account the kidneys' ability to clear urea and the person's protein catabolic rate. The technique is called **urea kinetic modeling.**

Energy All people with renal failure need adequate energy (see Table 27.3) to maintain a desirable weight and prevent protein catabolism. For children, an intake of 100 or even more kcalories per kilogram of body weight is desirable, but 80 kcalories per kilogram of body weight is considered a reasonable intake. At the minimum, energy intake should meet the RDA.[7]

Protein Providing the right amount of protein to the person in renal failure is like walking a tightrope. Too little protein, and the person develops malnutrition. Too much protein, and blood urea (the toxic waste product of protein metabolism) rises.

For people with renal insufficiency, restricting protein may help protect the remaining nephrons, although human studies have not clearly demonstrated a beneficial effect.[8] Clinicians generally recommend a protein-restricted diet that derives about two-thirds of its protein from sources such as eggs, milk, meat, poultry, and fish. The remaining protein is derived from plant sources. A diet that includes protein from both animal and plant sources may be best because it combines the high quality of animal proteins with the low–saturated fat, low-cholesterol nature of plant proteins.

As renal failure progresses, some clinicians prescribe very-low-protein diets supplemented with essential amino acids or essential amino acid precursors (keto acids). Once the client begins dialysis, protein restrictions can be relaxed somewhat because dialysis removes nitrogenous waste products and incurs protein losses (see Nutrition in Practice 27).

Lipid The ideal renal diet restricts total fat, saturated fat, and cholesterol to help control elevated blood lipid levels. A later section ("Diet Planning") explains the difficulty in meeting this dietary goal.

Carbohydrate A diet rich in complex carbohydrates helps minimize elevated blood glucose and triglycerides. People on peritoneal dialysis (see Nutrition in Practice 27) may need to further restrict their intake of total carbohydrate, and

Table 27.3 Nutrient Needs in Chronic Renal Failure

Nutrients[a]	Renal Insufficiency (Predialysis)	Hemodialysis	Peritoneal Dialysis
Energy (kcal/kg)	35–40	30–35	25–35
Protein (g/kg)	0.6–0.8	1.2–1.4	1.2–1.5
Fluid (ml)	Typically not restricted	500–750 plus daily urine output, or 1000 if anuric	≥2000
Sodium (g)	2–4	2–3	2–4
Potassium (g/kg)	Typically not restricted	3–4	Typically not restricted
Phosphorus (mg/g protein)	10–12[b]	12–15[b]	12–15[b]
Supplements			
Calcium (mg)	1000–1500	1000–1500	1000–1500
Folate (mg)	1	1	1
Vitamin B_6 (mg)	5	10	10
Vitamin D	As appropriate	As appropriate	As appropriate

Note: The actual amounts of these nutrients in the diet must be highly individualized based on each person's responses. For example, calcium supplementation may be as high as 3000 milligrams per day.
[a]Besides the specific nutrients listed, all others should meet recommended amounts.
[b]The extent of phosphorus restriction depends on serum phosphorus. The goal is to maintain serum phosphorus between 4.5 and 6.0 milligrams per deciliter. Often, phosphate binders are useful for this purpose.

Sources: Adapted from Meeting the challenge of the renal diet: A preview of the "National Renal Diet" educational series, *Journal of the American Dietetic Association* 93 (1993): 637–639; J. A. Beto, Which diet for which renal failure: Making sense of the options, *Journal of the American Dietetic Association* 95 (1995): 898–903.

especially simple carbohydrates, because they absorb a considerable amount of glucose as a consequence of the dialysis procedure.

Sodium and Fluids As renal failure progresses, the person excretes less urine and cannot handle normal amounts of sodium and fluids. At this point, limiting sodium and fluids helps to prevent hypertension, edema, and heart failure. Individual needs for sodium and fluids are determined by carefully monitoring each person's weight, blood pressure, urine output, and blood electrolyte levels. A rapid rise in body weight and blood pressure suggests that the person is retaining sodium and fluid; conversely, a rapid decline in body weight and blood pressure (a desirable outcome of dialysis) represents fluid loss.

Fluids are not restricted in renal insufficiency until urine output decreases. For the person who is neither dehydrated nor overhydrated, daily fluid needs amount to the daily urine output plus 500 to 750 milliliters to provide for insensible water losses.

Once a person is on dialysis, sodium and fluid intakes are controlled to allow a weight gain of about 2 pounds (of fluid) between dialysis treatments, although larger weight gains are common.[9] The margin lists foods considered part of the fluid allowance.

Foods and other substances that are considered part of the fluid allowance on fluid restricted diets include:

- *Cream*
- *Frozen yogurt*
- *Fruit ice*
- *Gelatin*
- *Ice*
- *Ice cream*
- *Ice milk*
- *Liquid medications*
- *Popsicles*
- *Sherbet*
- *Soup*

Potassium Most people with renal insufficiency and those on peritoneal dialysis can handle typical intakes of potassium.[10] People on hemodialysis, however, may develop hyperkalemia between dialysis treatments. When hyperkalemia is a problem, potassium may be moderately restricted to about 2 to 3 grams per day. Remember, however, that individual needs vary. People taking thiazide and loop diuretics may need to adjust their potassium intakes accordingly.

Caution: People with renal failure who must restrict potassium should avoid using low-sodium products and salt substitutes that contain potassium.

Phosphorus Dietary phosphorus restrictions help control rising blood phosphorus and may help slow the progression of renal failure. Fortunately, when a

Foods such as milk and milk products, eggs, cheese, peanut butter, bran cereal, sardines, and legumes are high in phosphorus and so must be restricted in a renal diet. See Appendix A for other foods high in phosphorus.

People with renal failure should avoid aluminum- and magnesium-containing antacids and laxatives or enemas containing magnesium; aluminum toxicity or serious hypermagnesemia can result.

The active form of supplemental vitamin D, or **calcitriol,** can be given orally or intravenously.

client follows a protein-restricted diet, phosphorus is restricted as well. In addition, the person must limit foods particularly high in phosphorus. The margin photo shows some examples of foods high in phosphorous.

Physicians may also prescribe medications that bind phosphorus in the GI tract, thus making it unavailable for absorption. Calcium salts—calcium carbonate and calcium acetate—are the preferred phosphate binders for people with end-stage renal disease. Aluminum and magnesium salts also bind phosphate, but they carry a risk of aluminum toxicity and are therefore avoided.

Calcium Supplements Impaired calcium absorption due to the lack of active vitamin D and limited calcium intake may contribute to bone disease in people with renal failure. Most people with renal disease need calcium supplements; some, however, develop hypercalcemia. The renal team monitors serum calcium closely to prevent both low and high blood calcium. Calcium-containing phosphate binders provide some calcium, but the absorption of calcium from these products varies widely.

Water-Soluble Vitamins People with renal failure frequently develop vitamin B_6 and folate deficiencies because of the restrictive diet, loss of vitamins during dialysis, drug therapy, and altered metabolism. For these reasons, clients receive generous amounts of vitamin B_6 and folate, along with the recommended amounts of the remaining water-soluble vitamins. Intakes of vitamin C from both the diet and supplements should be limited to less than 100 milligrams per day to prevent the formation of oxalate stones (see Chapter 20).[11]

Fat-Soluble Vitamins Supplemental vitamin D in its active form can help maintain blood calcium and prevent bone disease. The doses and methods of administering vitamin D supplements must be carefully adjusted for each client to maintain serum calcium levels within the normal range. Supplementation of fat-soluble vitamins other than vitamin D is usually not necessary.

Trace Minerals The administration of human erythropoietin along with the iron needed to synthesize hemoglobin is effective in treating iron-deficiency anemia, once a common and persistent problem in people with chronic renal disease. Clients should be cautioned to avoid iron supplements that also contain vitamin C.

People on dialysis frequently complain of anorexia and altered taste perceptions (dysgeusia), symptoms typical of zinc deficiency. Clients who have these symptoms may need zinc supplements if their serum zinc levels are inadequate. Recall that children have particularly high needs for zinc, and zinc deficiencies can contribute to growth retardation.

Enteral and Parenteral Nutrition Enteral and parenteral nutrition can provide nutrients to people with chronic renal failure who are unable to eat adequate amounts of foods. Formulas designed for use in renal failure are described on p. 654.

Diet Planning The ideal renal diet presents a challenge: provide adequate energy, but restrict protein, fat, and sometimes simple carbohydrate. Complex carbohydrates, which might appear to be the ideal energy source, are often also rich sources of fiber; because fiber absorbs fluids, which are often restricted, complex carbohydrates must be used cautiously. Diet planners must accept that under such circumstances, no diet is truly ideal. They must recognize that the need to adjust protein and electrolytes outweighs the need to restrict fat and simple carbohydrate.

To help meet energy needs, clients include as many complex carbohydrates as their diet plans allow. They supplement their meals with formulas high in kcalo-

How to

Help Clients Comply with a Renal Diet

The following suggestions can assist clients in complying with the renal diet:

1. To keep track of fluid intake:
   - Fill a container with an amount of water equal to your total fluid allowance. Each time you use a liquid food or beverage, discard an equivalent amount of water from the container. The amount remaining in the container will show you how much fluid you have left for the day.
   - Be sure to save enough fluid to take medications.
2. To help control thirst:
   - Chew gum or suck hard candy.
   - Freeze fluids so they take longer to consume.
   - Add lemon juice to water to make it more refreshing.
   - Gargle with refrigerated mouthwash.
3. To prevent the diet from becoming monotonous:
   - Experiment with new combinations of allowed foods.
   - Use favorite foods whenever possible.
   - Substitute nondairy products for regular dairy products. Nondairy products are lower in protein, phosphorus, and potassium than regular dairy foods, and they can substitute for milk and add energy to the diet.
   - Add zest to foods by seasoning with garlic, onion, chili, curry powder, oregano, pepper, or lemon juice.
   - Consult a dietitian when you want to eat restricted foods. Many restricted foods can be used occasionally and in small amounts if the diet is carefully adjusted.

ries but restricted in protein and electrolytes and use foods such as sugars (hard candy and jelly) and fats (margarine, butter, and oil) freely. The person with elevated blood lipids is advised to restrict fat and modify the type of fat to whatever extent is possible. The person with diabetes or hyperglycemia is advised to eat a consistent carbohydrate intake at regular intervals and to adjust insulin to cover carbohydrate intake.

To help individuals on renal diets find foods they will accept and enjoy, food lists similar to the exchange system are available. Whereas the exchange system for diabetes groups foods by their energy, carbohydrate, protein, and fat contents, renal food lists group foods by their energy, protein, sodium, potassium, and phosphorus contents. The box above provides suggestions to ease the task of complying with a renal diet. A sample renal diet menu is shown on p. 660.

Diet Compliance The challenges dietitians face in designing renal diets pale in comparison with those encountered by clients and caregivers who must follow a complicated medical plan of which diet is only a part. To help clients understand their medical plans and support their efforts, the members of the health care team must effectively communicate with their clients and, just as important, listen to them. Within a tangled web of grossly altered metabolic processes, dialysis lines, and toxic waste products is a person, often a frightened or discouraged one. All members of the health care team need to understand the plan if they are to offer the most effective support.

A nutrition education approach that emphasizes self-management appears to be a useful strategy for helping clients comply with a renal diet.[12] This approach (see Nutrition in Practice 25) guides clients in identifying goals, selecting strategies to meet goals, and evaluating progress. Both psychosocial and behavioral factors play key roles in helping clients adhere to protein-restricted diets.[13]

Psychosocial factors include:

- *Knowledge.*
- *Attitude.*
- *Support.*
- *Satisfaction.*
- *Self-perception of success.*

Behavioral factors include:

- *Self-monitoring of protein intake.*
- *Provision of feedback by a dietitian.*

Kidney Transplants and Diet

A preferable alternative to dialysis in end-stage renal disease is a kidney transplant. Kidney transplants can successfully restore kidney function and promote

Sample Renal Diet Menu

Breakfast	Lunch	Supper	Snacks
1 egg, fried with margarine	Sandwich with 2 oz turkey	3 oz roast beef	Hard candy, gumdrops, marshmallows
1 slice toast	2 slices bread	½ c rice	Carbonated beverages
Margarine	Mayonnaise	Margarine	
Jelly	Lettuce leaf	½ c mushrooms sautéed in olive oil and seasonings	
½ c grape juice	1 c green beans	½ c applesauce	
Coffee	Margarine	Iced tea	
Nondairy creamer	½ c strawberries	Sugar	
Sugar	Whipping cream		
	Sugar		
	½ c milk		

This diet menu provides 60 grams of protein and controls phosphorus, potassium, and sodium intake. To increase the kcalories in this diet, prepare food with oil or unsalted margarine (regular margarine, if allowed) and use additional fat, sugar, or syrup whenever possible. For example, canned fruit packed in heavy syrup, rather than juice, adds kcalories.

normal growth. For this reason, transplants are particularly desirable in children. Given a choice, many would prefer transplants, but suitable kidney donors cannot always be found.

Immunosuppressive Drug Therapy After receiving a new kidney, the person must take very large doses of immunosuppressive drugs to prevent rejection (see Table E.1 in Appendix E for nutrient-medication interactions). Muscular weakness, GI bleeding, protein catabolism, carbohydrate intolerance, sodium retention, fluid retention, hypertension, weight gain, and a characteristic puffy-faced appearance commonly accompany immunosuppressant therapy. Infections and increased susceptibility to malignant tumors are also common. Diuretics are frequently prescribed to promote the excretion of fluid and sodium and to prevent hypertension. The kidney may be rejected or fail to function, in which case the renal diet and dialysis must be reinstituted either temporarily, until another kidney can be found, or permanently.

Table 27.4 Dietary Guidelines following Kidney Transplant

Energy: Adequate to achieve or maintain desirable body weight.

Protein: 1 gram per kilogram of body weight. (Adjust based on renal function tests.)

Fat: ≤30 percent of total kcalories; ≤300 milligrams cholesterol.

Sodium: 3 to 4 grams per day.

Potassium: Adjust according to diuretic therapy.

Source: Adapted from J. A. Beto, Which diet for which renal failure: Making sense of the options, *Journal of the American Dietetic Association* 95 (1995): 898–903.

Diet Interventions Once recovery is under way, the degree of renal function guides diet therapy. Typical post-transplant diet modifications appear in Table 27.4. Protein is provided in amounts adequate to prevent protein catabolism that immunosuppressants may incur, but not so much as to tax renal function. Because blood lipids are frequently elevated, clients are advised to follow a fat-modified diet. Sodium restrictions help prevent fluid retention and hypertension. Depending on the type of diuretic prescribed and the person's responses, potassium intake may have to be adjusted.

As mentioned, a person with a kidney transplant may reject the new kidney either temporarily or permanently. During these times, the person must return to the prescribed diet for renal failure. Clients may find this regression difficult to accept and should be prepared for the possibility before it occurs.

WWW.

ohf.org
Oxalosis and Hyperoxaluria Foundation

afud.org
American Foundation for Urologic Disease

Kidney Stones

Unlike most renal diseases, which are rare, kidney stones are common. About one in every ten adults in the United States suffers from kidney stones, a painful, although rarely fatal, condition. Most of these people are men over 25 years old. Stones often occur only once; recurrences may be preventable.

Kidney stones form when stone constituents become concentrated in the urine and form crystals that grow. The stone constituents vary, but about three-fourths of them contain calcium as calcium oxalate. Less commonly, stones are composed of calcium phosphate, uric acid, the amino acid cystine, or magnesium ammonium phosphate (known as sturvite). Table 27.5 lists some disorders associated with stone formation.

Table 27.5 Conditions Associated with Kidney Stones

Cystinuria
Fat malabsorption
Glucocorticoid excess
Gout
Hyperparathyroidism
Hyperthyroidism
Immobilization
Malignancies (some types)
Osteoporosis
Paget's disease
Recurrent urinary tract infections
Renal tubular acidosis
Vitamin D toxicity

Consequences of Kidney Stones

In most cases, kidney stones pose no problems, especially when they are few and small. Small stones (less than one-fifth of an inch in diameter) may readily pass through the ureters and out of the body via the urine with minimal treatment (see Figure 27.3). Large stones, on the other hand, cannot pass easily through the ureters. When a large stone enters a ureter, it produces a sharp, stabbing pain, called renal colic. Typically, the pain starts suddenly in the back and intensifies as the stone follows the ureter's course down the abdomen toward the groin. The intense pain is often accompanied by nausea and vomiting. When the stone reaches the bladder, the pain subsides abruptly. Large stones that fail to pass through the ureter may cause a urinary tract obstruction or infection and serious bleeding. Symptoms may include frequent urination, urgency of urination, dysuria, and hematuria.

renal colic: the severe pain that accompanies the movement of a kidney stone through the ureter to the bladder.

hematuria (HE-mah-TOO-ree-ah): blood in the urine.

dysuria (dis-YOU-ree-ah): painful or difficult urination.

Treatment of Kidney Stones

Clients with a small kidney stone may more readily pass the stone if they drink plenty of fluids (more than 3 liters a day). They may also need antimicrobials if an infection is present and may sometimes need diuretics to help maintain urine output so that more stones do not have a chance to form. Large stones that block the flow of urine or cause an infection require removal either surgically or, more commonly by using shock waves to break the stone into pieces small enough to pass through the urinary tract.

Treatment of the underlying medical condition is necessary to help prevent recurrences of kidney stones. Dietary measures vary according to the composition of the stone, but all advice to prevent stones includes this recommendation: increase fluid intake to dilute the urine. People who have had kidney stones need to drink enough fluid (mostly water) to maintain a urine volume of at least 2 liters per day. Generally, this requires a total intake of about 3 to 4 liters of fluid throughout the day. People who are physically active or live in warm climates may need additional fluids. People with fevers, diarrhea, or vomiting also need additional fluids until these conditions resolve.

Dietary therapies for specific types of kidney stones are described next. Phosphate stones are not responsive to diet therapy and are treated with medications or surgery.

People with calcium stones treated with thiazide diuretics benefit from diets that moderately restrict salt. Such a diet helps reduce the urinary excretion of calcium and potassium.

hypercalciuria (HIGH-per-kal-see-YOU-ree-ah): excessive urinary excretion of calcium.

Calcium Stones About half of the people with calcium stones excrete normal amounts of calcium in the urine, and the other half excrete excess calcium (hypercalciuria). People with hypercalciuria are either more efficient at absorbing calcium from the intestine or more wasteful in their excretion of calcium than most people. Diet therapy for people with hypercalciuria limits calcium to the DRI appropriate for age and sex. It is important that calcium intakes not fall too low, however. Those with hypercalciuria who follow a low-calcium diet generally excrete more calcium than they ingest, indicating that they are losing calcium from their bones.

Calcium restriction also increases urinary oxalate excretion, which poses problems for people with calcium oxalate stones. Hyperoxaluria increases the

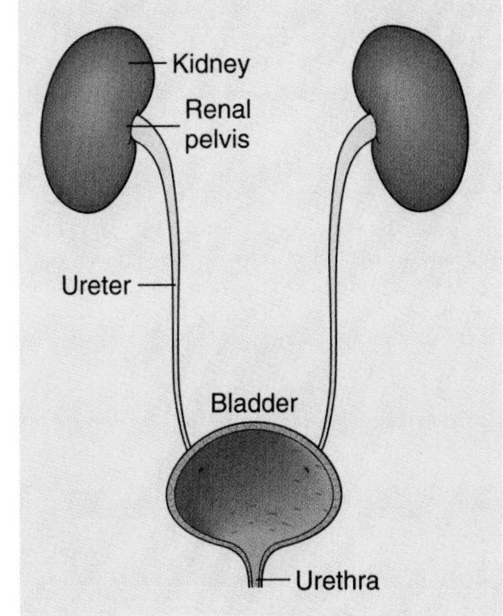

Figure 27.3
The Urinary Tract

hyperoxaluria (HIGH-per-OX-all-YOU-ree-ah): excessive urinary excretion of oxalate.

likelihood of calcium oxalate stone formation even more than hypercalciuria does. Calcium restriction is not appropriate in the treatment of calcium oxalate stones caused by fat malabsorption (see Chapter 20).

People with calcium oxalate stones, including people with fat malabsorption, are advised to limit their intakes of foods high in oxalate—spinach, rhubarb, beets, nuts, chocolate, tea, wheat bran, and strawberries. Some clinicians report that most people need to restrict only nuts (and peanut butter). Megadoses of vitamin C over long times raise urinary oxalate concentrations, so people at risk for oxalate stones are advised to avoid vitamin C supplements.

In addition to calcium and oxalate, attention to salt intake may also be important in preventing calcium oxalate stones. Excess salt increases urinary calcium excretion in all people, but causes a proportionately greater amount of calcium to be excreted in people with hypercalciuria.[14] Further research is necessary to determine whether moderate salt restrictions can prevent calcium stones from forming.

gout: a metabolic disorder that results in excess uric acid in the blood and sometimes in the urine; characterized by acute arthritis and inflammation of the joints.

Uric Acid Stones Uric acid stones are frequently associated with gout, a metabolic disorder characterized by elevated levels of uric acid in the blood and urine. Uric acid stones form when the urine becomes persistently acid, contains excessive uric acid, or both. Purine-restricted diets are commonly prescribed to prevent uric acid stones. A purine-restricted diet limits red meats, particularly organ meats, anchovies, sardines, and meat extracts. The benefits of such a diet are unproven, but avoiding excessive protein may be useful; health care professionals recommend a diet limited to no more than 100 grams of protein per day. Alcohol intake is also limited. Medications (allopurinol) are often prescribed to inhibit uric acid production, thus reducing both uric acid concentrations and urinary acidity.

cystinuria (SIS-tih-NEW-ree-ah): the presence of cystine in the urine; a symptom of an inherited metabolic disorder in which large amounts of the amino acids cystine, lysine, arginine, and ornithine are excreted in the urine.

Cystine Stones Cystine stones form when an inherited disorder of amino acid metabolism (cystinuria) causes the abnormal excretion of cystine in the urine. For cystine stones, health care professionals prescribe a diet restricted in the amino acid methionine, because the body makes cystine from methionine. Medications to reduce urinary acidity may also be beneficial.

For people with the nephrotic syndrome or renal failure, malnutrition can develop as a consequence of the disorder and worsen its progression. Use the nutrition assessment checklist to review assessment findings that can help detect malnutrition before it progresses too far.

Nutrition Assessment Checklist

FOR PEOPLE WITH RENAL DISORDERS

Medical Use the medical record to evaluate the cause of renal insufficiency, renal failure, or the type of kidney stone and the treatment plan. Determine whether other conditions such as severe stress, diabetes mellitus, hyperlipidemia, hypertension, or congestive heart disease are present and may alter nutrient needs.

Medication Review the client's medications for possible nutrient-medication interactions, particularly for clients with chronic renal diseases that will require long-term use of various medications including antilipemics, antihypertensives, diuretics, immunosuppressants, phosphate binders, and potassium exchange resins. Assess the client's over-the-counter drug use for sources of electrolytes.

Food Intake For people with renal failure, assess food intake to evaluate usual intake of energy, protein, fluids, sodium, potassium, phosphorus, calcium, vitamins, and minerals. For people with kidney stones, assess intake of fluids and of calcium, oxalate, sodium, protein, and methionine as appropriate to the type of stone. Use food records to assess the client's compliance with diet therapy and to make additional suggestions.

Anthropometric Monitor changes in height and weight in children and weight in adults. Interpret anthropometric measurements cautiously in people with oliguria, anuria, or edema because anthropometrics may be deceptively normal or high due to fluid retention. For people on dialysis, the weight measured immediately after a treatment most accurately reflects the person's true weight and is sometimes called the "dry weight." Rapid weight gain between dialysis treatments often reflects fluid retention. Weight loss should be expected following dialysis treatment.

Laboratory Check serum protein levels, which are often low in people with nephrotic syndrome and renal failure and may be even lower if malnutrition complicates renal disease. Review laboratory measurements of GFR, electrolytes, BUN, and creatinine, which are commonly used to determine dietary and medical treatments. Note whether the person has hyperglycemia or if serum lipids are elevated.

Physical Check for physical signs of fluid retention or dehydration, iron deficiency (pale skin and conjunctiva), uremia (fatigue, mental confusion, dry itchy skin, skin hemorrhages, nausea, vomiting, altered taste perceptions), osteodystrophy (bone pain, fractures, and bowed legs in children), hyperkalemia (arrhythmias and muscle weakness), and zinc deficiencies (altered taste perceptions, growth retardation in children). Record blood pressure and vital signs.

Self Check

1. All of the following are functions of the kidneys **except:**
   a. maintenance of fluid and electrolyte balance.
   b. activation of vitamin K.
   c. elimination of metabolic waste products.
   d. maintenance of acid-base balance.

2. A person with the nephrotic syndrome is prone to infections due to losses of:
   a. albumin. b. transferrin.
   c. immunoglobulins. d. lean body mass.

3. Diet recommendations for the nephrotic syndrome include:
   a. high protein intake. b. sodium restriction.
   c. phosphorus restriction. d. low protein intake.

4. The energy needs for a woman who weighs 57 kilograms and is in acute renal failure may range from:
   a. 2000 to 3000 kcalories. b. 1500 to 2700 kcalories.
   c. 3000 to 4000 kcalories. d. 1700 to 2600 kcalories.

5. The electrolytes that may rise rapidly in people with acute renal failure who are catabolic include:
   a. potassium, phosphorus, and magnesium.
   b. potassium, phosphorus, and calcium.
   c. sodium, potassium, and phosphorus.
   d. sodium, phosphorus, and calcium.

6. The health care team estimates fluid requirements for clients with acute renal failure by adding _____ milliliters to the amount of urine output.
   a. 100 b. 300
   c. 500 d. 750

7. The nurse recognizes dry, flaky, scaly, and itchy skin of a person with chronic renal failure as symptoms of:
   a. the uremic syndrome. b. the nephrotic syndrome.
   c. uremic frost. d. essential fatty acid deficiency.

8. Complications commonly associated with chronic renal failure may include:
   a. growth failure, altered taste sensations, and renal colic.
   b. GI bleeding, bone disease, and cardiovascular disease.
   c. anemia, edema, and potassium deficiencies.
   d. nausea, vomiting, and reflux esophagitis.
9. The renal diet intentionally restricts all of the following nutrients **except:**
   a. fluid.
   b. protein.
   c. phosphorus.
   d. calcium.
10. Anemia may occur in chronic renal failure due to all of the following **except:**
    a. GI bleeding.
    b. restrictive diets.
    c. increased life span of red blood cells.
    d. decreased production of erythropoietin.
11. Compared to the person with renal failure who is not on dialysis, the diet of a person on dialysis is:
    a. lower in protein.
    b. higher in protein.
    c. higher in potassium and phosphorus.
    d. lower in potassium and phosphorus.
12. Which nutrient should be evaluated in a client with chronic renal failure who complains of altered taste perceptions?
    a. iron
    b. phosphorus
    c. potassium
    d. zinc
13. Dietary guidelines following a kidney transplant include all of the following **except:**
    a. sodium restriction.
    b. low fat.
    c. potassium in accordance with individual responses.
    d. high protein.
14. Treatment for all kidney stones includes:
    a. calcium intake that meets but does not exceed the DRI.
    b. fluid intake to maintain a urine volume of at least 2 liters a day.
    c. a protein-restricted diet.
    d. a methionine-restricted diet.
15. People with calcium oxalate stones may benefit from diets that restrict:
    a. oxalate and sodium.
    b. calcium and potassium.
    c. protein and methionine.
    d. calcium and phosphorus.

Answers to these questions appear in Appendix H.

Clinical Applications

1. Consider that a person with chronic renal failure may need multiple medications to control disease progression and treat symptoms and complications. For people with diabetes and hyperlipidemia who develop renal failure, medications might include insulin, antihypertensives, diuretics, antilipemics, antiemetics (for nausea), antisecretory agents, and phosphate binders. Review the nutrition-related side effects of these medications in Appendix E. Describe the ways that medications can make it harder for people to maintain nutrition status.
2. Using the box on p. 659 as a guide, give suggestions for helping people adjust to different aspects of their renal diets. Can you think of any other suggestions?

Notes

1. S. R. Orth and E. Ritz, The nephrotic syndrome, *New England Journal of Medicine* 338 (1998): 1202–1211.

2. R. Rodrigo and M. Pino, Proteinuria and albumin homeostasis in the nephrotic syndrome: Effect of dietary protein intake, *Nutrition Reviews* 54 (1996): 337–347.

3. G. A. Kaysen, Nutritional management of nephrotic syndrome, *Journal of Renal Nutrition* 2 (1992): 50–58.

4. D. J. Rodriguez and W. M. Sandoval, Nutrition support in acute renal failure patients: Current perspectives, *Support Line,* December 1997, pp. 3–7.

5. J. D. Kopple, The nutrition management of the patient with acute renal failure, *Journal of Parenteral and Enteral Nutrition* 20 (1996): 3–12.

6. Kopple, 1996.

7. A.S.P.E.N. Board of Directors, Practice guidelines: Kidney failure—Pediatric, *Journal of Parenteral and Enteral Nutrition* (supplement) 17 (1993): 43.

8. The Modification of Diet in Renal Disease Study Group, The effects of dietary protein restriction and blood pressure control on the progression of chronic renal disease, *New England Journal of Medicine* 330 (1994): 877–884.

9. J. A. Beto, Which diet for which renal failure: Making sense of the options, *Journal of the American Dietetic Association* 95 (1995): 898–903.

10. Beto, 1995.

11. Beto, 1995.

12. B. P. Gillis and coauthors, Nutrition intervention program of the Modification of Diet in Renal Disease Study: A self-management approach, *Journal of the American Dietetic Association* 95 (1995): 1288–1294.

13. N. C. Milas and coauthors, Factors associated with adherence to the dietary protein intervention in the Modification of Diet in Renal Disease Study, *Journal of the American Dietetic Association* 95 (1995): 1295–1300; T. Coyne and coauthors, Dietary satisfaction correlated with adherence in the Modification of Diet in Renal Disease Study, *Journal of the American Dietetic Association* 95 (1995): 1301–1306.

14. W. J. Burtis and coauthors, Dietary hpercalciuria in patients with calcium oxalate stones, *American Journal of Clinical Nutrition* 60 (1994): 424–429.

CHAPTER 27

Nutrition in Practice

▪ DIALYSIS AND NUTRITION ▪

No perfect substitute for one's own kidneys has been found, although dialysis is a life-sustaining treatment for people with end-stage renal disease either as a permanent treatment or as a temporary measure until a suitable kidney donor can be found. Dialysis also benefits the person with acute renal failure who needs immediate help in restoring normal blood balances. All health care professionals need to understand why renal diseases alter nutrient needs and how dialysis will affect those needs. Those who routinely work with clients with renal diseases, however, need to understand more about dialysis procedures. This nutrition in practice describes the different types of dialysis procedures and explains why different procedures affect nutrient needs differently.

WWW.
kdf.org.sg
Kidney Dialysis Foundation

What is dialysis?

Dialysis removes excess fluids and wastes from the blood by employing the principles of simple diffusion and osmosis across a semipermeable membrane (see the accompanying glossary). For hemodialysis and peritoneal dialysis, a solution similar in composition to normal blood plasma, called the *dialysate,* is placed on one side of the semipermeable membrane; the person's blood flows by on the other side. A semipermeable membrane is like a filter; different types of membranes have pores of different sizes—the size of the pores determines which molecules will pass through the membrane and which will not. Small molecules like urea and electrolytes cross the membrane freely; large molecules like proteins are less likely to cross the membrane.

How do the dialysate and semipermeable membrane work together to remove wastes?

Wastes are removed from the blood by altering the composition of the dialysate. When the dialysate contains a limited amount of electrolytes, for example, electrolytes (which are small molecules) cross the membrane and diffuse out of the blood because they are more concentrated in the blood than in the dialysate. The dialysate can also be used to add needed components back into the blood. For a person with acidosis, for example, bases such as bicarbonate can be added to the dialysate. The base moves by diffusion into the person's blood to ease acidosis.

Excess fluids are removed from the blood by creating a pressure gradient between the blood and the dialysate. Sodium moves into the dialysate along with

Glossary

continuous renal replacement therapy (CRRT): a method for removing wastes from the blood that uses a semipermeable membrane to slowly remove all blood constituents that will cross the membrane; selected blood components are then returned to the client intravenously.

dialysate (dye-AL-ih-sate): a solution used during dialysis to draw wastes and fluids from the blood.

dialysis (dye-AL-ih-sis): removal of wastes and fluids from the blood using the principles of simple diffusion and osmosis through a semipermeable membrane.

dialyzer (dye-ah-LYES-er): the machine used for hemodialysis; also called an **artificial kidney.**

diffusion: movement of solutes from an area of high concentration to one of low concentration.

hemodialysis: removal of wastes and fluids from the blood by passing it through a dialyzer.

osmosis: movement of water from an area of low solute concentration to one of high solute concentration.

peritoneal dialysis: removal of wastes and fluids from the body using the peritoneal membrane as a semipermeable membrane.

water. Dialysis can only filter excess substances from the blood. Thus electrolytes like potassium and phosphorus that reside mainly in the intracellular fluid are more difficult to control with dialysis.[1]

What is the difference between hemodialysis and peritoneal dialysis?

In hemodialysis, the blood enters a machine called a *dialyzer* or *artificial kidney.* The dialyzer houses a series of tubes made from synthetic semipermeable membranes. Blood is pumped out of the body and into the dialyzer where it flows between the tubing that carries the dialysate. Wastes are extracted and blood is returned to the body.

In peritoneal dialysis, the peritoneal membrane (the membrane that covers the abdominal membranes) serves as the semipermeable membrane. The dialysate is infused directly through a tube into the peritoneal space—the space within the person's abdomen that overlays the intestine. Fluid is kept in the abdomen for about 30 minutes and then is drained from the abdomen through a tube by gravity. The procedure is repeated with fresh fluids. One complete exchange takes about 45 minutes, and the treatment requires about six exchanges.

For people in acute renal failure, another procedure, called continuous renal replacement therapy (CRRT) allows removal of fluids and wastes.[2] This method uses no dialysate. Instead blood is circulated through a filter where pressure gradients force all blood components small enough to pass through the pores of a semipermeable membrane out of the blood. Then just those components needed by the client are provided intravenously. If a small amount of blood containing electrolytes and wastes is removed and replaced with a very dilute dextrose solution, for example, eventually the concentration of electrolytes and wastes in the blood is lowered. This method is usually reserved for people who cannot tolerate hemodialysis or peritoneal dialysis for medical reasons.

How do the different forms of dialysis alter nutrient needs?

Compared to people with renal failure who are not on dialysis, people on dialysis receive diets higher in pro-

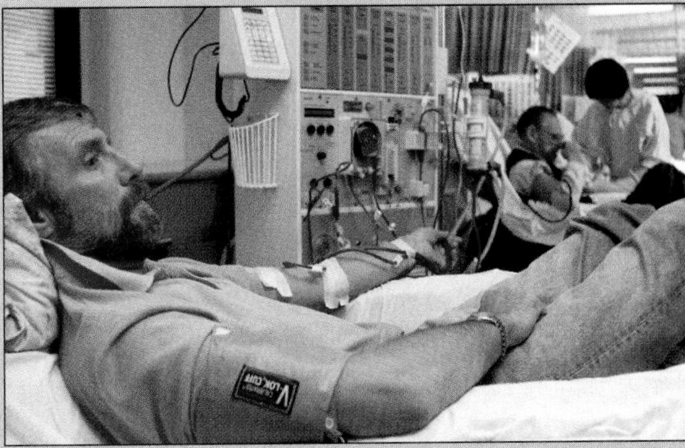

During hemodialysis, shown here, blood passes through a dialyzer where wastes are extracted. The cleansed blood is then returned to the body.

tein. In part this is because dialysis can remove the nitrogen-containing waste products of protein metabolism. In addition, some amino acids and smaller proteins may be lost in the dialysate. Protein losses from peritoneal dialysis are slightly higher than for hemodialysis, because more blood proteins pass into the dialysate through the peritoneal membrane than through the synthetic membrane used for hemodialysis.

For people on peritoneal dialysis, dextrose (glucose) is added to the dialysate. Dextrose helps draw fluid into the dialysate through osmosis. Gradually, however, dextrose is slowly absorbed across the peritoneal membrane; this makes the dialysate less efficient at drawing fluid out of the blood and also means that the person may absorb significant amounts of glucose.

Another difference between hemodialysis and peritoneal dialysis with respect to nutrition concerns potassium. In hemodialysis, the dialysate contains potassium because as blood is pumped through the dialyzer, potassium levels can drop rapidly, and hypokalemia can interfere with heart function. In peritoneal dialysis, the exchange of potassium between the blood and the dialysate is much slower, and blood potassium does not change rapidly. Thus potassium can be left out of the dialysate altogether. Since peritoneal dialysis can remove more potassium, the person on peritoneal dialysis generally does not have to restrict dietary sources of potassium. The person on hemodialysis, however, does need to moderately restrict potassium between dialysis treatments.

Dialysis takes over two important functions of the kidneys: waste removal and fluid removal. In so doing, dialysis reduces the symptoms of uremia, hypertension, and edema, as well as the risk of congestive heart failure. The hormonal functions of the kidneys, however, are not restored.

Notes

1. D. I. Charney, Medical treatment in renal disease: Basic concepts in dialysis, *Support Line,* February 1998, pp. 3–7.

2. D. J. Rodriguez and W. M. Sandoval, Nutrition support in acute renal failure patients: Current perspectives, *Support Line,* December 1997, pp. 3–7.

28 *Protein-Modified Diets for Liver Diseases*

As Chapter 27 showed, protein-modified diets for renal diseases are among the most difficult diets to plan and follow. This chapter describes the protein-modified diets used to treat liver diseases, which aim to provide enough protein to maintain nutrition status, repair damaged liver cells, and preserve liver function. Nutrition in Practice 28 shows how a protein-modified diet serves in the treatment of PKU—an inborn error of metabolism.

The liver is the metabolic crossroads of the body, and its health is crucial to every body function. The liver receives nutrients and then metabolizes, packages, stores, or ships them out for use by other organs. It manufactures bile, which the body uses to prepare fat for digestion and absorption. It synthesizes albumin and other proteins. The liver also detoxifies drugs, packages excess nitrogen in urea so that it can be excreted by the kidneys, and participates in iron recycling and the manufacture of red blood cells. No wonder then that liver disorders can profoundly affect both nutrition and general health status.

WWW

sadieo.ucsf.edu/alf/alffinal/homepagealf.html
American Liver Foundation

liver.ca
Canadian Liver Foundation

hepar-sfgh.ucsf.edu
American Association for the Study of Liver Disease

livertx.org
Children's Liver Alliance

Fatty Liver and Hepatitis

Two of the more common disorders of the liver that sometimes require a protein-modified diet are fatty liver and hepatitis. Dietary factors may play a role in the development of both disorders, although both may also arise from causes unrelated to diet.

Fatty Liver

fatty liver: an accumulation of triglycerides in the liver resulting from many disorders, including malnutrition and alcoholic liver disease; also called **hepatic steatosis.**

Fatty liver is not a disease, but rather a clinical finding common to many conditions. Fatty liver may develop from damage to liver cells or from altered nutrient availability to the liver. Most commonly, fatty liver develops from the liver's exposure to toxic substances such as alcohol (described in Nutrition in Practice 8), from an inadequate intake of protein (as in protein-energy malnutrition, or PEM), or as the result of an infection or malignant disease. Fatty liver can also develop as a complication of drug therapy (such as therapy with corticosteroids or tetracycline), obesity, long-term TPN, or small bowel bypass surgery.

Fatty liver develops when the liver either synthesizes too much fat, oxidizes too little, takes up too much from the blood, releases too little back to the blood, or any combination of these errors. Whatever the reasons, triglycerides accumulate in the liver and cause it to enlarge. In severe cases, liver weight may increase from a normal weight of about 3 pounds to as much as 11 pounds, with triglycerides increasing from 5 percent to as much as 40 percent of liver weight. Laboratory findings associated with fatty liver may include elevated serum transaminases (ALT and AST), alkaline phosphatase, and bilirubin.

The two transaminase enzymes that may be elevated in liver disease are **alanine transaminase (ALT)** and **aspartate transaminase (AST).**

Consequences of Fatty Liver Fatty liver alone usually causes no harm. Fatty liver associated with TPN, for example, may resolve as the feeding continues, when the feeding is changed to a cyclic infusion, or when TPN is discontinued. In other cases, however, the liver's accumulation of fat suggests the presence of an underlying disorder that can progress to permanent liver damage, metabolic and blood coagulation disturbances, renal failure, and death, if left untreated.

Treatment of Fatty Liver The appropriate therapy for fatty liver focuses on eliminating the cause and reversing its effects. Fatty liver caused by malnutrition requires a gradual introduction of a high-kcalorie, high-protein diet adequate in all other nutrients. Fatty liver due to alcohol abuse requires abstinence from alcohol and an adequate diet to replenish nutrient stores. Fatty liver related to obesity requires weight reduction. Fatty liver caused by drug therapy requires alternative medications or other therapies.

Hepatitis

In hepatitis, the destruction of liver cells causes inflammation and enlargement of the liver. Most often hepatitis results from a viral infection. Of the five types of viruses known to cause hepatitis (A, B, C, D, and E), type A is highly contagious and can be spread through contaminated foods and water. In other cases, hepatitis occurs as a consequence of chronic and excessive alcohol ingestion, certain medications, or toxins.

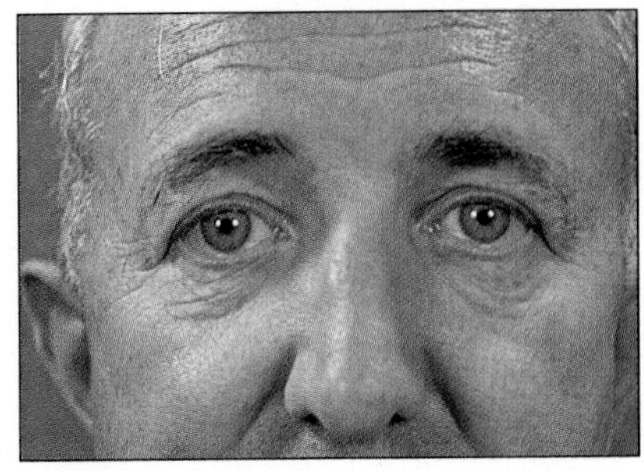

Jaundice is easy to see in the whites of the eyes.

Symptoms of Hepatitis During the early stages of hepatitis, the person may develop fatigue, joint and muscle pain, anorexia, nausea, vomiting, diarrhea or constipation, and a mild, flu-like respiratory illness with fever. In some cases, the symptoms are mild and may go unnoticed. In other cases, as hepatitis progresses, yellow bile pigments accumulate in the inflamed liver and spill into the blood, causing jaundice and producing a dark urine. Serum transaminase levels (AST and ALT) rise, and the liver becomes enlarged and tender.

jaundice (JON-dis): a characteristic yellowing of the skin, whites of the eyes, mucous membranes, and body fluids resulting from the accumulation of bile pigments (bilirubin) in the blood.

Consequences of Hepatitis The origin and type of hepatitis, the extent of liver damage, and the person's response to treatment all determine how seriously the disease will affect health. In many cases, liver cells gradually regenerate and liver function recovers. Hepatitis A, for example, is often mild, although relapses may occur after apparent recovery. Sometimes chronic hepatitis develops. Liver cancer may develop after infection with hepatitis B or C. Far less frequently, severe hepatitis progresses rapidly and leads to acute liver failure and death.

Treatment of Hepatitis Liver cells need nutrients to help them recover from hepatitis. Recovery also aims to reduce further insult to the liver cells; thus the person must abstain from alcohol. The person with hepatitis who is in good nutrition status receives a regular, well-balanced diet. The malnourished person with hepatitis receives a high-kcalorie, high-protein diet to replenish nutrient stores. For the person with mild anorexia, frequent small meals, enteral formula supplements, or both may be helpful. For the person with persistent vomiting, parenteral nutrition is an alternative.

WWW **scn.org/health/hepatitis/index.htm**
Hepatitis Education Project

hepfi.org
Hepatitis Foundation International

www2.hepb.org/hepb
Hepatitis B Foundation

Cirrhosis

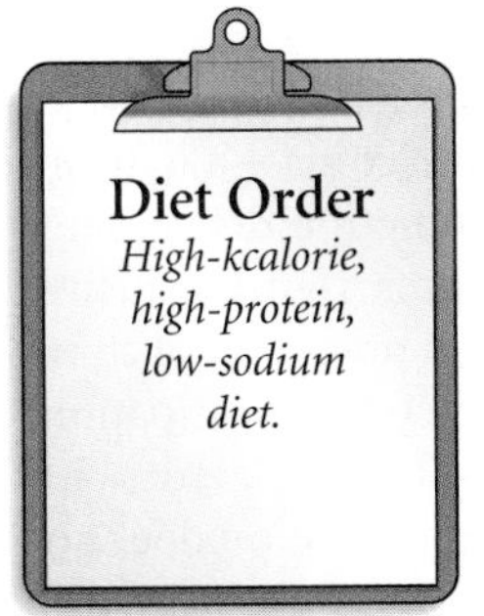

Diet Order
High-kcalorie, high-protein, low-sodium diet.

In cirrhosis, a serious and chronic form of liver disease, liver cells are damaged to the point that they cannot regenerate. Instead scar tissue (fibrosis) forms within the liver, altering the structure of the liver and the blood flow through it.

Chronic alcohol abuse is the most common cause of cirrhosis in the United States, although not all people with cirrhosis are alcohol abusers, and not all alcohol abusers develop cirrhosis. Other causes of cirrhosis include infections (including hepatitis), prolonged obstruction or diseases of the bile ducts, heart disease, some inherited disorders, severe reactions to medications, and exposure to toxic chemicals.

cirrhosis (sih-ROW-sis): an advanced form of liver disease in which scar tissue replaces liver cells that have permanently lost their function. The type of cirrhosis associated with alcohol abuse is called **Laennec's** (LAY-eh-necks) **cirrhosis. Postnecrotic cirrhosis** develops as a complication of chronic hepatitis. **Biliary cirrhosis** can develop from a prolonged inflammation or obstruction of the bile ducts. **Cardiac cirrhosis** is associated with congestive heart failure. In still other cases, cirrhosis may be **idiopathic,** having no identifiable cause.

Consequences of Cirrhosis

Unlike healthy liver tissue, which is soft and flexible, scar tissue is unyielding, a difference that leads to major consequences. Many people with cirrhosis have no symptoms in the early stages. As the disorder progresses, symptoms may include nausea, vomiting, weight loss, liver enlargement, jaundice, mental disturbances, and itching from the accumulation of bile pigments under the skin.

WWW **nhtech.demon.co.uk/pbc/index.html**
Primary Biliary Cirrhosis Foundation

Reminder: The *portal vein* is the blood vessel that carries nutrients from the GI tract to the liver. The *hepatic vein* returns blood from the liver to the heart. The *hepatic artery* delivers oxygen-rich blood from the heart back to the liver.

portal hypertension: elevated blood pressure in the portal vein caused by obstructed blood flow through the liver.

collaterals: small blood vessels that develop to divert blood flow away from an obstructed organ; also called **shunts.**

varices (VAIR-ih-seez): blood vessels that have become distended and twisted.

Portal Hypertension The portal vein and the hepatic artery carry 1½ quarts of blood every minute to the miles of intermeshed capillaries and arterioles within the liver. This huge volume of blood cannot pulse easily through the scarred tissue of a cirrhotic liver. Consequently, blood flow to and through the liver decreases, blood backs up, and pressure in the portal vein rises sharply, causing portal hypertension. Figure 28.1 shows the consequences of portal hypertension that the following paragraphs describe.

Esophageal Varices With normal blood flow through the portal vein obstructed, pressure forces some of the blood from the portal vein to take a detour through the smaller vessels around the liver. These collaterals, or shunts, can develop throughout the GI tract, but often occur in the area of the esophagus. With portal hypertension, the collaterals become enlarged and twisted, forming varices. Esophageal varices bulge into the lumen of the esophagus, and they can rupture and lead to massive bleeding that can be fatal. In addition, blood from ruptured varices can travel to the intestine, where it can serve as a source of ammonia (described later).

ascites (ah-SIGH-teez): a type of edema characterized by the accumulation of fluid in the abdominal cavity.

Ascites The rising pressure in the portal vein forces plasma out of the liver's blood vessels into the abdominal cavity, causing the abdomen to swell. This accumulation of fluid in the abdominal cavity is called ascites. In addition, the diseased liver is unable to synthesize adequate amounts of albumin, which normally creates a pressure that draws fluids from tissues back into the blood. Ascites tends to be a self-aggravating condition. Because less blood reaches the kidneys, the body responds by making more aldosterone and antidiuretic hormone, hormones that expand the body's blood volume by triggering retention of sodium and water. As a result of sodium and water retention, ascites worsens, and edema may spread to all body compartments (peripheral edema). To make matters worse, the diseased liver cannot dispose of aldosterone as it usually does, so aldosterone levels remain high. Ascites frequently causes early satiety and nausea and also increases the basal metabolic rate.

An elevated blood ammonia level is called **hyperammonemia.** Normal blood ammonia levels are less than 50 μg/100 ml.

Elevated Blood Ammonia Levels Normally, the healthy liver removes ammonia from circulation and converts it to urea, but a severely diseased liver fails at this task, and blood ammonia rises. Even if the liver does handle the ammonia it receives, some ammonia-laden blood bypasses the liver by way of the collaterals. Elevated blood ammonia disrupts central nervous system function.

hepatic encephalopathy: mental changes associated with liver disease that may include irritability, short-term memory loss, and an inability to concentrate.

hepatic coma: a state of unconsciousness that results from severe liver disease.

Hepatic Encephalopathy and Hepatic Coma Liver diseases can cause changes in mental function (encephalopathy) and can lead to hepatic coma—a dangerous complication that develops in some people when liver function deteriorates significantly. The causes of encephalopathy and hepatic coma remain elusive, although elevated blood ammonia plays a role. Events that can precipitate a coma include GI bleeding, electrolyte imbalances, renal failure, infection, constipation, and the use of sedatives. Elevated blood ammonia is associated with encephalopathy and hepatic coma, but the degree of ammonia elevation does not correlate with the severity of symptoms. A possible explanation for this poor correlation is that blood ammonia does not always parallel brain ammonia concentration. When brain ammonia is high, the body produces greater quantities of two substances (glutamine and ketoglutarate), and the degree of their elevation tends to correlate with the severity of symptoms.

Phenylalanine and tyrosine are the aromatic amino acids; leucine, isoleucine, and valine are the branched-chain amino acids.

Blood amino acid patterns also change in the later stages of liver disease; aromatic amino acid levels rise, and branched-chain amino acid levels fall. These alterations may add ammonia to the blood and may contribute to encephalopathy and hepatic coma in some people.

The person with hepatic encephalopathy may suffer from irritability, short-term memory loss, and an inability to concentrate. The person with impending

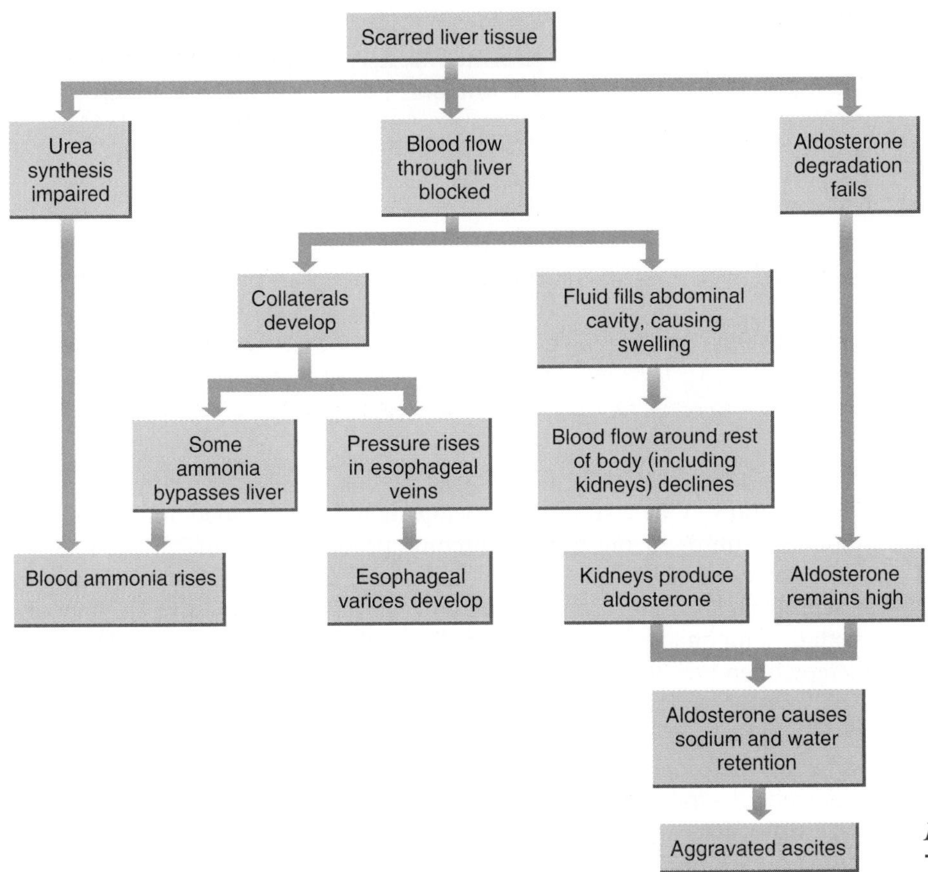

Figure 28.1
The Consequences of Cirrhosis

hepatic coma exhibits mental disturbances such as changes in judgment, personality, or mood. The person may be unable to draw even a simple shape, such as a star. A sweet, musty, or pungent odor may develop on the breath. Flapping tremor may also develop in the precoma state. Just before passing into a coma, the person becomes very difficult to arouse.

The odor that may develop in people with impending hepatic coma is called **fetor hepaticus.**

flapping tremor: uncontrolled movement of the muscle group that causes the outstretched arm and hand to flap like a wing; occurs in hepatic coma and other diseases that cause encephalopathy; also called **asterixis** (AS-ter-ICK-sis).

Insulin Resistance Although the reasons remain unclear, insulin resistance, hyperinsulinemia, and hyperglycemia develop in many people with chronic liver disease. In some cases, the pancreas is unable to produce enough insulin to overcome the insulin resistance, and diabetes mellitus develops.

Wasting As liver deterioration progresses, malnutrition and wasting become evident. Table 28.1 summarizes possible causes of wasting in people with liver diseases, which can include anorexia and reduced food intake, altered metabolism, and malabsorption. Fat malabsorption can occur if the liver disease interferes with bile salt production or the person develops pancreatic insufficiency.

Other Consequences When liver function deteriorates significantly, the alterations in fluid and electrolyte balance as well as the buildup of waste products in the blood can precipitate kidney failure. Bone disorders that are generally unresponsive to vitamin D and calcium supplementation can also become a problem.

hepatorenal syndrome: the symptoms of renal failure that occur as a consequence of liver failure.

Treatment of Cirrhosis

Treatment of cirrhosis aims to preserve remaining organ function, control the clinical manifestations of the disease, and prevent hepatic coma. When liver failure has progressed to a severe stage, liver transplantation becomes a treatment option.

Table 28.1 Possible Causes of Wasting in Cirrhosis

Reduced Nutrient Intake	Excessive Nutrient Losses	Raised Nutrient Needs
Abdominal pain	Blood loss through GI bleeding	Ascites (raises metabolic rate)
Anorexia	Diarrhea	Infections
Early satiety	Malabsorption	Malnutrition
Esophagel varices	Medications	
Nausea	Steatorrhea	
Restrictive diet	Vomiting	
Vomiting		

℞ Prescription Pad

Medications used in the treatment of cirrhosis may include:

- Antibiotics (neomycin)
- Antidiabetic agents (glipzide and glyburide)
- Antihypertensives
- Appetite stimulants
- Cholestyramine (to control excessive itching; see *antilipemics*)
- Diuretics
- Laxatives (lactulose)

See Appendix E for timing with meals and nutrition-related side effects.

Avoiding severe stress is critical for a person with compromised liver function. The impaired liver is already taxed in maintaining homeostasis. The added metabolic demands of stress can easily overwhelm the liver and lead to death.

Drug Therapy One goal of drug therapy is to control blood ammonia levels. About two-thirds of the body's ammonia production occurs in the intestine where bacteria make ammonia from undigested dietary proteins as well as from protein from shed mucosal cells and GI tract bleeding. Digestive enzymes also produce ammonia in the process of dismantling dietary proteins. Thus drug therapy often includes antibiotics to limit the growth of intestinal bacteria and laxatives to speed intestinal transit time.

In addition to these medications, antihypertensives may be used to control portal hypertension, and diuretics may be used to reduce fluid retention and prevent ascites.[1] Oral antidiabetic agents may be used to control hyperglycemia. Megestrol acetate and dronabinol (see Chapter 24, p. 580) may help improve the appetite and stimulate weight gain.[2]

Energy In providing diet therapy for cirrhosis, health care professionals must carefully consider each client's needs for energy, protein, sodium, and fluid (see Table 28.2). Energy needs vary depending on the complications each person exhibits. Energy needs are increased for people with ascites, malabsorption, or infections. For people with ascites, energy needs are based on the person's desirable or estimated dry weight to avoid overestimating energy needs.[3]

Protein For people with cirrhosis who do not show signs of impending coma, protein restrictions are unneccessary, and the diet should supply at least 1.0 to 1.5 grams of protein per kilogram of body weight per day.[4] For people who develop signs of impending coma, the cause is investigated and the underlying condition is treated. In a small number of people (7 to 9 percent), dietary protein can precipitate hepatic encephalopathy.[5] For people who are protein sensitive, the diet plan is low in protein at first, and protein is gradually increased. The final amount of protein that is optimal depends on the person's nutrition status and individual tolerance to protein. For a person who is malnourished, up to 1.5 grams of protein per kilogram of body weight is the final goal. The final level of protein the person tolerates may be less than this amount, however. Sometimes a higher protein level can be attained by adjusting medications to control symptoms.

In some cases, the person who is protein sensitive can tolerate only very low protein intakes (0.5 to 0.75 gram of protein per kilogram of body weight per day). Although research is limited and controversial, many clinicians recommend special enteral (or parenteral, when needed) formulas that are low in aromatic amino acids and high in branched-chain amino acids (see Appendix G).[6] Some studies suggest that vegetable proteins may be better tolerated than meat proteins, perhaps because vegetables contain fewer ammonia-forming constituents and aromatic amino acids and more branched-chain amino acids than meats do. In addition, the fiber from plant foods speeds intestinal transit time, reducing the amount of time for ammonia absorption from the intestine.

Table 28.2 Diet Guidelines for Cirrhosis

Energy

- Without ascites, malnutrition, or infection: 120% of basal energy expenditure (BEE); see Table 23.2 on p. 557.
- With ascites, malnutrition, or infection: 150–175% of BEE.

Protein

- Not sensitive to protein: 1.0 to 1.5 g protein/kg body weight/day.
- Protein sensitive: Start with 0.5 to 0.7 g protein/kg body weight/day; increase gradually to level of tolerance with a goal of at least 1.0 g protein/kg body weight/day and up to 1.5 g protein/kg body weight/day if the person is malnourished.
- Very sensitive to protein (unable to tolerate at least 0.8 g protein/kg body weight/day): Consider vegetable-based diets or enteral supplements enriched with branched-chain amino acids and restricted in aromatic amino acids.

Carbohydrate

- Not restricted.
- For people with insulin resistance or diabetes, provide up to 50 to 60% of kcalories from carbohydrates (mainly complex carbohydrates) with a consistent carbohydrate intake from day to day and at each meal and snack.

Fat

- Not restricted unless fat malabsorption is present.
- With fat malabsorption: Restrict fat only as necessary to control steatorrhea (see Chapter 20); use medium-chain triglycerides (MCT) to increase kcalories as necessary.

Sodium

- Restrict only as necessary to control ascites, but not less than 2 g sodium/day in most cases.

Fluid

- Restrict to 1.0 to 1.5 liters per day for people with ascites who also have low blood sodium levels.

Vitamins and minerals

- Ensure adequate intake from diet or supplements based on individual needs.

Source: Adapted from J. Hasse and coauthors, Nutrition therapy for end-stage liver disease: A practical approach, *Support Line*, August 1997, pp. 8–15.

If a person falls into a coma, all protein is withheld temporarily until the cause is determined and corrective measures are taken. For protein-sensitive people, protein is then gradually reintroduced in small amounts and increased gradually as tolerated.

Nonprotein Energy Sources In a diet for liver disease, carbohydrate generally provides 50 to 60 percent of energy needs. People with hyperglycemia should follow the guidelines for diets in diabetes; that is, eat mostly complex carbohydrates and eat them at consistent times throughout the day. A high-carbohydrate (50 grams) snack at bedtime may help prevent excessive breakdown of fat and protein during an overnight fast.[7]

Because fat helps make foods appetizing and delivers energy efficiently, fat plays an important role in the diet of a person with cirrhosis. Fat is restricted only if the person develops fat malabsorption. In these cases, medium-chain triglycerides (MCT) can provide additional kcalories.

The foods shown here provide about 60 grams of protein, the entire day's protein allowance for some people with liver disease.

Sodium and Fluid For people with ascites, the diet restricts sodium (see Table 28.2). Fluids are restricted only in people who have low blood sodium levels—a

Low blood sodium is **hyponatremia.** In liver disease, the total amount of sodium in the body is excessive, even though blood levels are sometimes low. Thus adding salt to the diet is inappropriate.

The nutrition assessment checklist describes how to measure abdominal girth.

sign that the person is overloaded with fluid. In these cases, fluids may need to be restricted to about 1.0 to 1.5 liters per day. To assess changes in fluid balance, health care professionals monitor weight changes and measure abdominal girth. Rapid weight gain and an increasing abdominal girth indicate fluid retention; sudden weight loss and decreasing abdominal girth indicate successful fluid excretion.

The lower the sodium intake, the more quickly ascites resolves. Very-low-sodium diets are unpalatable to many people, however, so most clinicians recommend a diet that allows from 2 to 4 grams of sodium (see Chapter 27, p. 652).

Alcohol To protect the liver from further injury, clients with cirrhosis must abstain completely from alcohol use. A cirrhotic liver exposed to the toxic effects of alcohol will be further impaired.

Vitamins The liver's central role in the metabolism and storage of vitamins and minerals, combined with coexisting conditions (such as malabsorption, alcoholism, and malnutrition), explains why nutrient deficiencies commonly occur in people with liver diseases. Virtually all people with advanced liver disease require supplements of some vitamins, as well as minerals and trace elements. Physicians determine which nutrients to supplement by monitoring serum levels and checking for clinical signs of deficiencies.

The B vitamins serve as cofactors for the liver's many metabolic reactions and repair work; deficiencies of thiamin, vitamin B_6, riboflavin, and folate are common. Fat-soluble vitamins may be malabsorbed if steatorrhea develops. If the diseased liver fails to synthesize adequate amounts of retinol-binding protein, body tissues may not receive the vitamin A they need. Vitamin D deficiencies may develop from inability of the liver to activate vitamin D for the body's use. Although bone diseases are generally not responsive to vitamin D supplements, it is important to ensure adequate vitamin D intakes to prevent further problems. Vitamin K deficiencies can prolong the time it takes for blood to clot, a dangerous complication that increases the risk of massive bleeding from the GI tract.

One laboratory test that evaluates the time it takes for blood to clot is called the **prothrombin time.** Both vitamin K deficiency and liver disease can prolong the prothrombin time.

Minerals Calcium deficiencies can develop from three causes: steatorrhea, low serum albumin (albumin, which carries calcium in the blood, is manufactured in the liver), and impaired vitamin D metabolism. Thus it is important that the diet or supplements provide adequate calcium to prevent deficiencies. Limited research suggests that zinc deficiencies may be implicated in the development of hepatic encephalopathy.[8]

Enteral and Parenteral Nutrition People with cirrhosis face many difficulties in eating enough food to maintain nutrition status. The accompanying box offers tips on how to help people with cirrhosis cope with the demands of their diets. If the person with cirrhosis cannot take enough food or formula by mouth, tube feedings or TPN is indicated. As mentioned earlier, enteral and parenteral formulas designed for liver disease provide fewer aromatic and more branched-chain amino acids than standard formulas (see Appendix G). Both enteral and parenteral nutrition have been used successfully in people with cirrhosis.

People with bleeding esophageal varices will be unable to consume food by mouth and are often given simple intravenous solutions to maintain fluid and electrolyte balance. Parenteral nutrition should be considered if the person is malnourished or the health care team anticipates that oral intake will not resume for an extended time. The case study on p. 678 asks you to use clinical knowledge and judgment in answering questions about cirrhosis.

How to

Help the Person with Cirrhosis Eat Enough Food

People with cirrhosis often have difficulty eating enough food to prevent malnutrition and its consequences. Ascites and gastrointestinal symptoms such as nausea and vomiting can interfere with food intake. The person with encephalopathy may be confused about what to eat or have little interest in eating. People with sodium restrictions may have difficulty adjusting to a low-salt diet. To facilitate diet compliance:

- Individualize the diet plan based on each person's symptoms and responses. The diet should restrict protein, fat, sodium, or fluid only when such restrictions are warranted.
- Recommend the suggestions for improving nutrient intake, controlling nausea and vomiting, and adding kcalories and protein found in the box on pp. 581–582.
- Advise clients to eat frequent small meals and to eat a high-carbohydrate bedtime snack. Eating frequent small meals can help with nausea and can also improve glucose tolerance. For people who are protein sensitive, eating protein frequently throughout the day improves tolerance.
- Recommend the suggestions in the box on p. 659 for people who must restrict fluids and proteins.
- Point out that there is probably no food that cannot be incorporated into the diet on special occasions and in limited amounts. Show clients how to change their meal plans so that favorite foods can be incorporated from time to time.

The low-sodium diet used in the treatment of cirrhosis may be more restrictive than the low-salt diet recommended for hypertension. Table 27.1 on p. 652 shows the foods allowed and not allowed on a 2-gram sodium-restricted diet. Some suggestions that can help clients adhere to their sodium restrictions include:

- Suggest that clients replace the salt they use for cooking and seasoning with herbs or spices like basil, bay leaves, curry, garlic, ginger, lemon, mint, oregano, rosemary, and thyme.
- Suggest that clients experiment with low-sodium products to find the ones they like. (People on potassium-sparing diuretics should be cautioned to avoid salt substitutes that replace sodium with potassium.)
- Advise clients to check food labels to learn the sodium content of the foods they eat. They may be able to find similar products with lower sodium contents.

Continue to offer support and encouragement to the client with cirrhosis. Severe weight loss is less likely to occur if nutrition intervention is provided before problems progress too far.

Liver Transplantation

When liver failure progresses to a severe and irreversible state, liver transplantation may be an option. In some liver transplant cases, the graft fails to function, and retransplantation forestalls an otherwise inevitable death.

Nutrition before Transplantation In severe liver failure, malnutrition has often progressed for some time. Clinicians report malnutrition in over 70 percent of liver transplant recipients and note that malnutrition increases the risk of complications following a liver transplant.[9] A liver transplant candidate must often wait for a liver donor before surgery is possible. Wise health care professionals use this time to identify and correct nutrient imbalances whenever possible. Often TPN is used to provide nutrients prior to a transplant. The person equipped with adequate nutrient stores faces the transplant better prepared to fight infections, heal wounds, and mount a stress response.

Nutrition following Transplantation Following liver transplantation, liver function determines nutrient needs. All people are hypermetabolic after surgery, and energy needs must be met. Immunosuppressant drugs given to prevent tissue rejection can contribute to nutrient imbalances by causing nausea, vomiting, diarrhea, and mouth sores. The nutrition support team often uses indirect calorimetry to estimate energy needs and carefully monitors clinical and laboratory data so that specific nutrient recommendations can be made.

Case Study

Carpenter with Cirrhosis

Mr. Sloan, a 48-year-old carpenter, has been hospitalized many times. He recognizes his problem with alcohol abuse and has entered alcohol rehabilitation programs several times over the last few years. Nevertheless, he is still drinking. Mr. Sloan was recently admitted to the hospital, and a diagnosis of alcoholic cirrhosis has been confirmed. At 5 feet 7 inches tall, Mr. Sloan, who once weighed 150 pounds, now weighs 120 pounds. He is jaundiced and looks thin, although his abdomen is distended with ascites. He has advanced liver disease and is showing signs of impending hepatic coma. Laboratory findings include elevated AST, ALT, alkaline phosphatase, and blood ammonia. Compare these findings with Table 28.3 to determine if they are consistent with liver disease.

Can you explain to Mr. Sloan what cirrhosis is and what its consequences are? From the limited information available, what can you determine about Mr. Sloan's nutrition status? What medical problem makes it difficult to interpret Mr. Sloan's actual weight? How can his weight measurements help determine if his condition is improving? Calculate Mr. Sloan's energy needs. What factors influence protein needs for a person with impending hepatic coma? What signs suggest that a person is in a precoma state?

Why is Mr. Sloan's abdomen distended? Explain the development of ascites in liver disease and how diet is adjusted.

Would you expect Mr. Sloan's blood ammonia levels to be high? Why or why not?

Describe portal hypertension, jaundice, and esophageal varices. How would Mr. Sloan's diet be changed if he were found to have bleeding esophageal varices?

Although TPN has been the traditional source of nutrients in the posttransplant period, researchers report equal success with intestinal tube feedings.[10] Early enteral nutrition may reduce the incidence of infection, a particularly important consideration for people with suppressed immune systems.[11]

Halting or delaying the progression of liver diseases depends in large part on attention to nutrition and nutrition assessment parameters. The accompanying

Table 28.3 Standards for Tests Used to Diagnose and Monitor Liver Disease

Test	Normal Values	Values in Liver Disease
Albumin	3.5–5.0 g/100 ml	Decreased
Alkaline phosphatase	Varies[a]	Normal or elevated
ALT (formerly SGPT)[b]	Varies[a]	Elevated
Ammonia	<50 μg/100 ml	Elevated
AST (formerly SGOT)[b]	Varies[a]	Elevated
Bilirubin (direct)	0.1–0.3 mg/100 ml	Elevated
Prothrombin time		Prolonged

Note: To convert albumin (g/100 ml) to standard international (SI) units, multiply by 10; to convert ammonia (μg/100 ml) to SI units, multiply by 0.5872; to convert bilirubin (mg/100 ml) to SI units (μmol/L), multiply by 17.10.

[a]Reference ranges vary depending on the test used. Consult laboratory report for normal ranges.
[b]ALT = alanine transaminase; SGPT = serum glutamic pyruvic transaminase; AST = aspartate transaminase; SGOT = serum glutamic oxaloacetic transaminase.

nutrition assessment checklist reviews important points to keep in mind when assessing the nutrition status of people with liver diseases.

This chapter brings to a close your introduction to normal and clinical nutrition. Congratulations! You have received an abundance of information since you first began your study. The normal nutrition chapters of this text provided you with current recommendations to promote optimal health. You learned how the body transforms foods into nutrients and how those nutrients support the body's well-being. The clinical chapters addressed you as a future health care professional, concerned with the well-being of others during times of illness.

We hope this text has served you well and that you will remember, when selecting food for yourself or when making recommendations for others, to honor the body.

Nutrition Assessment Checklist

For People with Liver Disorders

Medical Review the medical record to determine the type and cause of liver disease, as well as any history of alcohol abuse, hepatitis, or biliary tract obstruction. Recognize that the effects of advanced liver disease and malnutrition are often difficult to distinguish.

Medication Assess medications for possible nutrient-medication interactions. Lactulose can cause nausea, abdominal distension, and diarrhea. Cholestyramine can cause nausea and may reduce the absorption of fat and fat-soluble vitamins, folate, calcium, iron, zinc, and magnesium. In addition, cholestyramine can lead to constipation, which can precipitate hepatic coma. Spironolactone, a diuretic frequently used to treat ascites, is a potassium-sparing diuretic. People on spironolactone should avoid high-potassium foods and salt substitutes that contain potassium.

Food Intake Obtain an accurate diet history to determine nutrition status, calculate nutrient requirements, and identify inadequate nutrient intake. Assess current intake to help pinpoint tolerance for protein in people with hepatic encephalopathy who are sensitive to protein. People who abuse alcohol normally derive much of their daily energy intake from alcohol. Without that energy source, as occurs during hospitalization, special care must be taken to ensure that the diet supplies sufficient energy.

Anthropometric Interpret anthropometric data cautiously in people with edema and ascites. Use desirable or dry weights for calculating energy and protein needs. In advanced liver disease, weight is measured daily to assess changes in fluid status. To measure abdominal girth, place a tape measure around the back and over the person's abdomen directly over the umbilicus. Abdominal girth measurements are also useful for assessing fluid status.

Laboratory Anticipate low serum protein levels in people with advanced liver diseases. Many laboratory tests, including serum albumin, reflect liver function as well as nutrition status. Supplying adequate protein may fail to raise serum proteins if the liver is unable to synthesize them. AST, ALT, ammonia, and bilirubin levels increase with deteriorating liver function.

Physical Note physical signs of altered liver function including ascites, edema, and jaundice. Gynecomastia (abnormally large mammary tissue) and testicular atrophy may be present in men with liver failure. Other physical signs of liver dysfunction include angiomas (masses of dilated capillaries and arterioles) and distended abdominal blood vessels. Flapping tremor suggests impending hepatic coma.

Self Check

1. Diet strategies that may reverse fatty liver include all of the following **except:**
   a. low-kcalorie diets for weight loss.
   b. high-kcalorie diets for malnutrition.
   c. continuous delivery of TPN solutions.
   d. abstinence from alcohol.
2. Which of the following statements about hepatitis is true?
   a. Hepatitis A is often mild and can be spread through contaminated foods and water.
   b. Regardless of the type of hepatitis, symptoms are severe.
   c. People with hepatitis always need high-kcalorie, high-protein diets.
   d. Hepatitis is never the cause of cirrhosis.
3. The consequences of cirrhosis are primarily due to:
   a. chronic malnutrition.
   b. chronic alcohol abuse.
   c. destruction of liver cells, which alters their structure and function.
   d. elevated blood ammonia levels.
4. The health care professional recognizes that all of the following factors may contribute to ascites in cirrhosis **except:**
   a. portal hypertension.
   b. low serum albumin levels.
   c. increased secretion of aldosterone and antidiuretic hormone.
   d. rising blood ammonia levels.
5. Esophageal varices are a dangerous complication of liver disease primarily because:
   a. they can rupture and cause massive bleeding.
   b. they can interfere with food intake.
   c. they divert blood flow from the GI tract.
   d. they cause portal hypertension.
6. Contributors to rising blood ammonia levels in cirrhosis include all of the following **except:**
   a. inability of the liver to make urea.
   b. detour of ammonia-laden blood from the intestine through collaterals.
   c. increased levels of branched-chain amino acids.
   d. blood loss and its subsequent digestion in the intestine.
7. For the person with cirrhosis, the nurse recognizes irritability, short-term memory loss, and an inability to concentrate as possible signs of:
   a. hepatic coma.
   b. hepatic encephalopathy.
   c. hyperammonemia.
   d. fetor hepaticus.
8. Diets for people with cirrhosis are:
   a. restricted in protein, sodium, and fluid.
   b. high in kcalories and protein.
   c. restricted in fat.
   d. highly individualized based on symptoms and responses to treatment.
9. All of the following statements regarding diet and cirrhosis are true **except:**
   a. people with steatorrhea may need to restrict fat.
   b. people with ascites need to restrict sodium.
   c. people who are protein sensitive need to restrict protein to 0.5 to 0.75 gram of protein per kilogram of body weight per day.
   d. people with ascites and low blood sodium levels need to restrict sodium and fluid.
10. With respect to vitamins and minerals, people with cirrhosis:
    a. seldom require nutrient supplements.
    b. may develop clotting abnormalities associated with vitamin A deficiency.
    c. frequently develop vitamin C deficiencies.
    d. may develop calcium deficiencies.
11. All of the following are true regarding liver transplants **except:**
    a. immunosuppressant drugs given to prevent tissue rejection can alter nutrient needs.
    b. enteral nutrition is inappropriate in the posttransplant period.
    c. attention to nutrition before a transplant improves chances of recovery.
    d. before a transplant, nutrition status is difficult to assess.

Answers to these questions appear in Appendix H.

Clinical Applications

1. Think about the problems a person might encounter when following an 1800-kcalorie diet that is restricted to 40 grams of protein. On such a diet, the total protein allowance could be used up by just one scrambled egg, 3 ounces of meat, a cup of milk, and two slices of bread. Using Appendix A and Figure 25.2 on pp. 602–603 and Table 25.2 on p. 604 for reference, calculate the kcalories these foods provide. Then consider what other foods to add to the diet to meet energy needs without contributing additional protein.

Now compare the result with the Daily Food Guide on pp. 14–15. Which food groups have you offered in the recommended quantities? Which food groups are in short supply on such a diet? Which nutrients might be low? How might fats and sugars be useful in such a diet?

2. Chapter 23 described severe stresses and discussed how the combination of hypermetabolism and infection can lead to multiple organ failure. The sequence of multiple organ failure often includes liver failure and kidney failure. On a sheet of paper make three columns, one for stress, one for liver failure, and one for kidney failure. Under each column, list the energy, protein, fat, fluid, and other nutrient considerations for each disorder individually. Which nutrient modifications are common to all disorders? Do some of the modifications necessary for one disorder conflict with those for another? If yes, describe how the final decision might be made for the most appropriate diet.

Notes

1. J. S. Crippin, Medical management of end-stage liver disease: A bridge to transplantation, *Support Line,* August 1997, pp. 3–7.

2. C. Gurk-Turner, Management of the metabolic complications of liver disease: An overview of commonly used pharmacologic agents, *Support Line,* August 1997, pp. 17–19.

3. J. Hasse and coauthors, Nutrition therapy for end-stage liver disease: A practical approach, *Support Line,* August 1997, pp. 8–15.

4. C. Corish, Nutrition and liver disease, *Nutrition Reviews* 55 (1997): 17–20.

5. As cited in Hasse and coauthors, 1997.

6. A. Fabri and coauthors, Overview of randomized clinical trials of oral branched-chain amino acid treatment in chronic hepatic encephalopathy, *Journal of Parenteral and Enteral Nutrition* 20 (1996): 159–164.

7. W. Chang and coauthors, Effects of extra-carbohydrate supplementation in the late evening on energy expenditure and substrate oxidation in patients with liver cirrhosis, *Journal of Parenteral and Enteral Nutrition* 21 (1997): 96–99.

8. Hasse and coauthors, 1997.

9. J. Hasse, Nutrition and transplantation, *Nutrition in Clinical Practice* 8 (1993): 3–4; J. Pikul and coauthors, Degree of preoperative malnutrition is predictive of postoperative morbidity and mortality in liver transplant recipients, *Transplantation* 57 (1994): 469–472.

10. C. Wicks and coauthors, Comparison of enteral feeding and total parenteral nutrition after liver transplantation, *Lancet* 344 (1994): 837–840; J. M. Hasse, Early enteral nutrition support in patients undergoing liver transplantation, *Journal of Parenteral and Enteral Nutrition* 19 (1995): 437–443.

11. Hasse, 1995.

CHAPTER 28

Nutrition in Practice

Inborn Errors of Metabolism

The discussion in Chapter 28 of the metabolic consequences of liver disorders sets the stage for a closer look at metabolic disorders caused by genetic errors in protein synthesis. Each inborn error affects metabolism in a unique way and has specific implications for diet.

What is an inborn error of metabolism?

An inborn error of metabolism is a genetic error that alters the production of a protein. In many cases, the protein is an enzyme. When the body fails to make an enzyme, makes an enzyme in insufficient amounts, or makes an enzyme with an abnormal structure, body functions that depend on that enzyme cannot proceed. For example, if an enzyme is missing or malfunctioning in the metabolic pathway that converts compound A to compound B, then compound A accumulates and compound B becomes deficient. Both the excess of compound A and the lack of compound B can lead to a variety of problems and, in many cases, can cause death. Furthermore, this imbalance creates excesses and deficiencies in other metabolic pathways that present another array of problems. The accompanying glossary defines related terms.

Do all inborn errors cause severe illness or death?

No, many do not. In some instances, the accumulated compound is not toxic, and the deficient compound is not essential, so individuals experience no problem. They may not even know about the error. In other cases, however, inborn errors have severe consequences, including possible mental retardation. Without prompt diagnosis and treatment, they can be lethal.

What are the treatments for inborn errors?

The primary treatment for many inborn errors is nutrition intervention. With an understanding of the biochemical pathway involved, a clinician can often manipulate the diet to compensate for excesses and inadequacies. Management involves restricting dietary precursors that occur prior to the error in the metabolic pathway, replacing needed products that fail to be produced, or both. The goal of therapy is to:

- Prevent the accumulation of toxic metabolites.
- Replace essential nutrients that are deficient as a result of the defective metabolic pathway.

Glossary

carrier: an individual who possesses one dominant and one recessive gene for a genetic trait, such as an inborn error of metabolism. When a trait is recessive, a carrier may show no signs of the trait but can pass it on.

dominant gene: a gene that has an observable effect on an organism. If an altered gene is dominant, it has an observable effect even when it is paired with a normal gene; see also *recessive gene.*

galactosemia (ga-LAK-toe-SEE-me-ah): an inborn error of metabolism in which enzymes that normally metabolize galactose to compounds the body can handle are missing and an alternative metabolite accumulates in the tissues, causing damage.

genes: the basic units of hereditary information, made of DNA, that are passed from parent to offspring in the chromosomes. A pair of genes codes for each genetic trait.

inborn error of metabolism: an inherited flaw evident as a metabolic disorder or disease present from birth.

mutation: an alteration in a gene such that an altered protein is produced.
muta = change

PKU, phenylketonuria (FEN-el-KEY-toe-NEW-ree-ah): an inborn error of metabolism in which phenylalanine, an essential amino acid, cannot be converted to tyrosine. Alternative metabolites of phenylalanine (phenylketones) accumulate in the tissues, causing damage, and overflow into the urine.

recessive gene: a gene that has no observable effect on an organism. If an altered gene is recessive, it has no observable effect as long as it is paired with a normal gene that can produce a normal product. In this case, the normal gene is said to be *dominant.*

CHAPTER 28
Nutrition in Practice

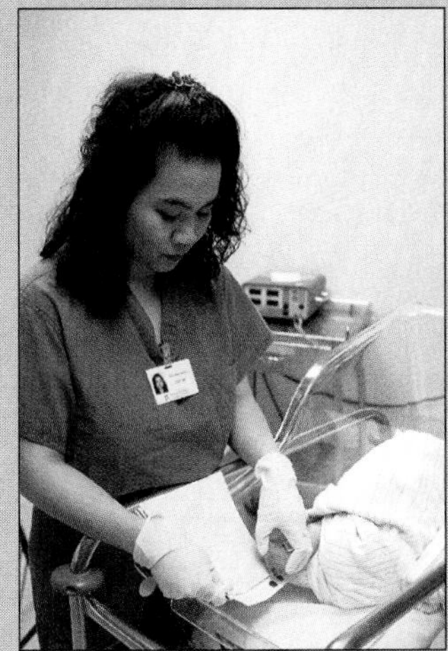

A simple blood test screens newborns for PKU—the most common inborn error of metabolism.

- Provide a diet that supports normal growth, development, and maintenance.

Meeting these three objectives is a major challenge that was previously unattainable. New knowledge about the body's many biochemical pathways, coupled with current technology for synthesizing formulas with specific nutrient compositions, has greatly enhanced the treatment of inborn errors.

WWW **wolfenet.com/~kronmal** National PKU News

Can you give me a specific example?

A classic example that illustrates the principles of treatment is the most common inborn error of metabolism: phenylketonuria (PKU). PKU is one of many inborn errors that affect amino acid metabolism. Other disorders affect not only amino acid metabolism but also carbohydrate, lipid, vitamin, and mineral metabolism.

PKU affects approximately 1 out of every 10,000 newborns in the United States each year. The ability to detect and treat PKU has saved and significantly improved the lives of many people; its example offers hope to those suffering from other inborn errors.

What causes PKU?

Classic PKU results from a deficiency of an enzyme that converts the essential amino acid phenylalanine to tyrosine (see Figure NP28.1 on p. 684). Without the enzyme, abnormally high concentrations of phenylalanine and other related compounds accumulate and damage the developing nervous system. Simultaneously, the body cannot make tyrosine or other compounds (such as the neurotransmitter epinephrine) that normally derive from tyrosine. Under these conditions, tyrosine becomes an essential amino acid; that is, the body cannot make it, and so the diet must supply it.

How is PKU diagnosed?

PKU is a hidden disease that cannot be seen at birth, yet diagnosis and treatment beginning in the first few days of life can prevent its devastating effects. For these reasons, and because PKU is the most common inborn error of metabolism, all newborns in the United States receive a screening test for PKU.[1] The test must be conducted after the infant has consumed several meals containing protein (usually after 24 hours and before seven days). Before screening became routine, an infant with PKU would suffer the dire consequences of uncorrected high phenylalanine concentrations. At first, the only signs are a skin rash and light skin pigmentation. Between three and six months, signs of developmental delay begin to appear. By one year, irreversible brain damage is clearly evident.

What is the dietary treatment for PKU?

Essentially, the diet restricts phenylalanine and supplements tyrosine intake to maintain blood concentrations within a safe range. The diet's effectiveness is remarkable; in almost every case, it can prevent the devastating array of symptoms described. As most dietitians can attest, though, the diet is more easily described than designed.

Because phenylalanine is an essential amino acid, the diet cannot exclude it completely. If phenylalanine intake is too low, children suffer bone, skin, and blood disorders; growth and mental retardation; and death. Therefore, the diet must strike a balance, providing enough phenylalanine to support normal growth and health but not enough to cause harm. The problem is not that children with PKU require less phenylalanine than other children, but that they cannot handle excesses without detrimental effects. To ensure that blood phenylalanine and tyrosine concentrations remain within a safe range, children with PKU receive blood tests periodically and alterations in their diets when necessary. With a controlled phenylalanine intake, children with PKU can lead normal, happy lives.

Special low-phenylalanine formulas are the primary source of energy and protein for children with PKU. Their diets exclude high-protein foods such as meat, fish, poultry, cheese, eggs, milk, nuts, and dried beans and peas. Also excluded are commercial breads and pastries made from regular flour, which has a high phenylalanine content. Basically, the diet allows foods

Nutrition in Practice

Figure NP28.1
The Biochemical Pathway in PKU

Normal:

Normally, the amino acid phenylalanine follows two pathways, one in the liver, the other in the kidneys. In the liver, the enzyme phenylalanine hydroxylase adds a hydroxyl group (OH) to produce the amino acid tyrosine. Tyrosine, in turn, produces melanin, the pigmented compound found in skin and brain cells; the neurotransmitters epinephrine and norepinephrine; and the hormone thyroxin. In the kidneys, enzymes convert phenylalanine to by-products that are excreted.

In the liver:

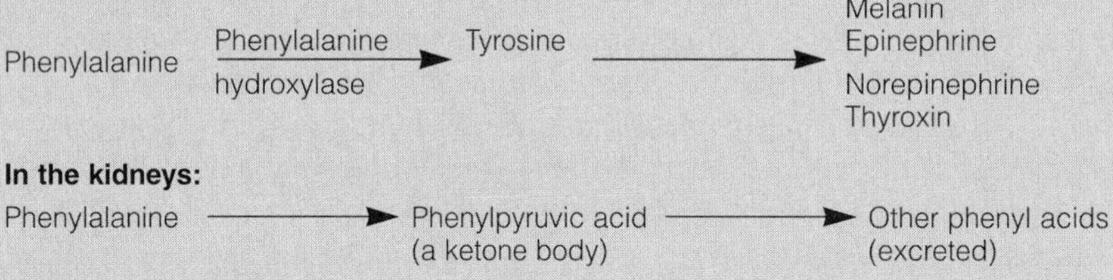

In PKU:

Individuals with PKU lack the liver enzyme phenylalanine hydroxylase, impairing conversion of phenylalanine to tyrosine. Phenylalanine accumulates in the liver and blood, reaching the kidneys in abnormally high concentrations. In the kidneys, an aminotransferase enzyme converts phenylalanine to the ketone body phenylpyruvic acid, which spills into the urine—thus the name phenylketonuria.

In the liver:

Phenylalanine (accumulates) —Phenylalanine hydroxylase (deficient)→ Tyrosine (deficient)

In the kidneys:

Phenylalanine (accumulates) → Phenylpyruvic acid (accumulates) → Other phenyl acids (accumulate)

that contain some phenylalanine, such as fruits, vegetables, and cereals, and those that contain none, such as fats, sugars, jellies, and some candies. Clearly, it is impossible to create such a diet using only whole, natural foods, but children who depend primarily on a formula for their nourishment risk multiple trace mineral deficiencies.[2] Health care professionals monitor trace mineral status and supplement as needed. The accompanying menu provides a sample phenylalanine-restricted diet for a child with PKU.

Do people with PKU have to stay on the diet for life?

The answer to that question is unclear. During the early years of central nervous system development, the diet is clearly critical to preventing irreversible mental retardation. Less certain is the length of time the nervous system is vulnerable to the PKU defect. Until the late 1970s, researchers assumed that the child with PKU could abandon the special diet after the first few years of life when the central nervous system had completed its development. Unfortunately, however, elevated phenylalanine concentrations in the older child do cause problems such as short attention span, poor short-term memory, and poor eye-to-hand coordination, although the damage is less severe than at an earlier age. A child with PKU who has discontinued the controlled diet may experience problems in school performance, mood, and behavior. For these reasons, clinicians generally encourage children to continue the low-phenylalanine diet indefinitely.[3] Convincing adolescents to return to the phenylalanine-restricted diet after several years of unrestricted diets requires intense education and reinforcement. Even then efforts are quite often unsuccessful. Reinstitution of a controlled diet, however, does improve blood phenylalanine concentrations, behavior, and IQ scores.

If a woman with PKU becomes pregnant, how does the disorder affect her fetus?

Her high blood phenylalanine presents a hostile environment to fetal development. The fetus's blood concentrations rise even higher than hers, and she may experience a spontaneous abortion; or her infant may suffer mental retardation, congenital heart disease, and low birthweight. For these reasons, women with elevated phenylalanine concentrations need counseling prior to pregnancy on the problems their condition may create for their children.

Dietary control of maternal PKU may protect the fetus, at least in part, if implemented early enough. Dietary control does not ensure a successful outcome of pregnancy, but the infants of women who follow a low-phenylalanine diet from at least one to two months prior to conception and continue it throughout pregnancy are more likely to have higher birthweights, larger head circumferences, fewer malformations, and higher scores on intelligence tests than the infants of women who begin diet therapy dur-

Sample Phenylalanine-Restricted Menu for a Child with PKU

Breakfast	Snack	Lunch	Snack	Supper	Snack
2 tbs raisins	4 oz orange juice	½ small banana	4 oz Lofenalac	2 tbs instant potatoes (without milk)	2 tbs raisins
5 tbs cream of rice		2 tbs tomato soup (without milk)	5 round butter crackers	3 tbs green beans	4 oz Lofenalac
2 tsp sugar		3 tbs rice		4 tbs vegetable and beef broth	
8 oz Lofenalac		1½ tsp margarine		1½ tsp margarine	
		8 oz Lofenalac		¾ c sliced peaches	
				8 oz Lofenalac	

Note: Lofenalac is a special low-phenylalanine formula that is commercially available.

ing their pregnancies or not at all.[4] Special formulas that meet the energy, protein, vitamin, and mineral needs of pregnant women with PKU are now available.

Are all inborn errors managed the same way as PKU?

In some ways, all are similar, but in other ways each is different. As an example, galactosemia is an inborn error of carbohydrate metabolism in which any one of three enzymes that convert galactose to glucose is missing or defective. When infants with galactosemia are given infant formula or breast milk (which contains a galactose unit in each molecule of lactose), they vomit and have diarrhea. The abnormal metabolism causes growth failure, liver enlargement, and other neurological abnormalities that lead to coma and death. Early introduction of a galactose-restricted diet prevents or minimizes most of these symptoms.

Dietary adjustment in galactosemia is simpler than in PKU for two reasons. First, unlike phenylalanine, galactose is not an essential nutrient. The galactosemia diet needs only to exclude galactose, not to provide a perfectly calculated dose. Second, galactose occurs primarily in lactose (the sugar in milk), so treatment depends chiefly on the careful elimination of all milk and milk products. This is not to say that the diet is easy to follow; many commercially prepared products contain milk. Still, milk is less widespread in the diet than the amino acid phenylalanine, which appears in all proteins.

As scientific understanding of human genetics and biochemistry increases, more and more inborn errors affecting enzyme function are being recognized. Understanding the roles of enzymes in metabolism sometimes makes it possible to compensate for these defects, which otherwise would destroy the quality of life. In such cases, diet can make a dramatic difference in people's lives.

WWW

miele-herndon.com/galactosemia/galactosemia.html
Galactosemia Resources and Information

Notes

1. Committee on Genetics, Newborn screening fact sheet, *Pediatrics* 98 (1996): 473–501.

2. C. Reilly and coauthors, Trace element nutrition status and dietary intake of children with phenylketonuria, *American Journal of Clinical Nutrition* 52 (1990): 159–165.

3. R. O. Fisch and coauthors, Phenylketonuria: Current dietary treatment practices in the United States and Canada, *Journal of the American College of Nutrition* 16 (1997): 147–151.

4. The Maternal Phenylketonuria Collaborative Study: A status report, *Nutrition Reviews* 52 (1994): 390–393.

Appendixes

A *Table of Food Composition*

B *Canada: Recommendations, Choice System, and Labels*

C *United States: Recommendations and Exchanges*
World Health Organization: Recommendations

D *Nutrition Resources*

E *Nutrition Assessment: Supplemental Information*

F *Aids to Calculation*

G *Enteral Formulas*

H *Answers to Self Check Questions*

Contents

A TABLE OF FOOD COMPOSITION

This edition of the table of food composition contains more complete values for several nutrients than any comparable table.[1] These include dietary fiber; saturated, monounsaturated, and polyunsaturated fat; vitamin B_6; vitamin E; folate; magnesium; and zinc. The table includes a wide variety of foods from all food groups and is updated yearly to reflect current food patterns. For example, this edition includes many new nonfat items; several new ethnic items such as adzuki beans, tahitian taro, and gai choy chinese mustard; and a new selection of vegetarian foods.

• ***Sources of Data*** • To achieve a complete and reliable listing of nutrients for all the foods, over 1200 sources of information are researched. Government sources are the primary base for all data for most foods. In addition to USDA data (from Release 12 and surveys), provisional USDA information—both published and unpublished—is included.

Even with all the government sources available, however, some nutrient values are still missing; and as the USDA updates various data, it sometimes reports conflicting values for the same items. To fill in the missing values and resolve discrepancies, other reliable sources of information are used. These sources include journal articles, food composition tables from Canada and England, information from other nutrient data banks and publications, unpublished scientific data, and manufacturers' data.

The data for brand foods are listed as provided by the food manufacturers and the food chain restaurants. This information changes often because recipes and formulations are modified to meet consumer preferences, and the data are usually limited to those nutrients required for food labels. To provide more complete information, values for several nutrients are often estimated based on known values for major ingredients.

• ***Accuracy*** • The energy and nutrients in recipes and combination foods vary widely, depending on the ingredients. The amounts of various fatty acids and cholesterol are influenced by the type of fat used (the specific type of oil, vegetable shortening, butter, margarine, etc.).

Estimates of nutrient amounts for foods and nutrients include all possible adjustments in the interest of accuracy. When multiple values are reported for a nutrient, the numbers are averaged and weighted with consideration of the original number of analyses in the separate sources. Whenever water percentages are available, estimates of nutrient amounts are adjusted for water content. When no water is given, water percentage is assumed to be that shown in the table. Whenever a reported weight appeared inconsistent, many kitchen tests were made, and the average weight of the typical product was given as tested.

When estimates of nutrient amounts in cooked foods are derived from reported amounts in raw foods, published retention factors are applied. Data for combination foods are modified to include newer data for major ingredients.

Considerable effort has been made to report the most accurate data available. The table is revised annually, and the authors welcome any suggestions or comments for future editions.

• ***Average Values*** • It is important to know that many different nutrient values can be reported for foods, even by reliable sources. Many factors influence the amounts of nutrients in foods, including the mineral content of the soil, the method of processing, genetics, the diet of the animal or the fertilizer of the plant, the season of the year, methods of analysis, the difference in moisture content of the samples analyzed, the length and method of storage, and methods of cooking the food. The mineral content of water also varies according to the source.

Although each nutrient is presented as a single number, each number is actually an average of a range of data. More detailed reports from the USDA, for example, indicate the number of samples and the standard deviation of the data. One can also find different reported values for foods as older data is replaced with newer data from more recent analyses using newer analytical techniques. Therefore, nutrient data should be viewed and used only as a guide, a close approximation of nutrient content.

• ***Dietary Fiber*** • There can be many different reported values for dietary fiber in foods because information depends on the type of analytical technique used. The fiber data in this table are primarily from the USDA/ARS Human Nutrition Information Service in Hyattsville, Maryland; Composition of Foods by Southgate and Paul (England); and many journal articles.

• ***Vitamin A*** • Vitamin A is reported in retinol equivalents (RE). The amount of vitamin A can vary by the season of the year and the maturity of the plant. Reported values in both dairy products and plants are higher in summer and early fall than in winter. The values reported here represent year-round averages. The organ meats of all animal products (liver especially) contain large amounts of vitamin A, which vary widely, depending on the background of the animal. The vitamin is also present in very small amounts in regular meat and is often reported as a trace.

• ***Vitamin E*** • Vitamin E is actually a combination of various forms of this nutrient, and the measure of alpha tocopherol equivalents (α-TE) summarizes the activity of the various types of tocopherols and tocotrienols into one measure.

• ***Fats*** • Total fats, as well as the breakdown of total fats to saturated, monounsaturated, and polyunsaturated fats, are listed in the table. The fatty acids seldom add up to the total due to rounding and to other fatty acid components that are not included in these basic categories, such as *trans*-fatty acids and glycerol. *Trans*-fatty acids can comprise a large share of the total fat in margarine and shortening (hydrogenated oils) and in any foods that include them as ingredients.

• ***Enrichment-Fortification*** • The mandatory enrichment values for foods are presented as appropriate, including the new values for folate enrichment folacin in grain products.

• ***Niacin*** • Niacin values are for preformed niacin and do not include additional niacin that may form in the body from the conversion of tryptophan.

• ***Using the Table*** • The items in this table have been organized into several categories, which are listed at the head of each right-hand page. As the key shows, each group has been color-coded to make it easier to find individual items.

In an effort to conserve space, the following abbreviations have been used in the food descriptions and nutrient breakdowns:

- diam = diameter
- ea = each
- enr = enriched
- f/ = from
- frzn = frozen
- g = grams
- liq = liquid
- pce = piece
- pkg = package
- w/ = with
- w/o = without
- t = trace
- 0 = zero (no nutrient value)
- blank space = information not available

• ***Caffeine Sources*** • Caffeine occurs in several plants, including the familiar coffee bean, the tea leaf, and the cocoa bean from which chocolate is made. Most human societies use caffeine regularly, most often in beverages, for its stimulant effect and flavor. Caffeine contents of beverages vary depending on the plants they are made from, the climates and soils where the plants are grown, the grind or cut size, the method and duration of brewing, and the amounts served. The accompanying table shows that in general, a cup of coffee contains the most caffeine; a cup of tea, less than half as much; and cocoa or chocolate, less still. As for cola beverages, they are made from kola nuts which contain caffeine, but most of their caffeine is added, using the purified compound obtained from decaffeinated coffee beans.

The FDA lists caffeine as a multipurpose GRAS substance that may be added to foods and beverages. Drug industries in developed countries use caffeine in many kinds of drugs: stimulants, pain relievers, cold remedies, diuretics, and weight-loss aids.

1. This food composition table has been prepared for West-Wadsworth Publishing Company and is copyrighted by ESHA Research in Salem, Oregon—the developer and publisher of the Food Processor®, Genesis® R&D, and the Computer Chef® nutrition software systems. The major sources for the data are from the USDA, supplemented by more than 1200 additional sources of information. Because the list of references is so extensive, it is not provided here, but is available from the publisher.

Caffeine Content of Beverages, Foods, and Over-the-Counter Drugs

Beverages and Foods	Average (mg)	Range (mg)
Coffee (5-oz cup)		
Brewed, drip method	130	110–150
Brewed, percolator	94	64–124
Instant	74	40–108
Decaffeinated, brewed or instant	3	1–5
Tea (5-oz cup)		
Brewed, major U.S. brand	40	20–90
Brewed, imported brands	60	25–110
Instant	30	25–50
Iced (12-oz can)	70	67–76
Soft drinks (12-oz can)		
Dr. Pepper	40	
Colas and cherry cola		
Regular		30–46
Diet		2–58
Caffeine-free		0–trace
Jolt	72	
Mountain Dew, Mello Yello	52	
Fresca, Hires Root Beer, 7-Up, Sprite, Squirt, Sunkist Orange	0	
Cocoa beverage (5-oz cup)	4	2–20
Chocolate milk beverage (8 oz)	5	2–7
Milk chocolate candy (1 oz)	6	1–15
Dark chocolate, semisweet (1 oz)	20	5–35
Baker's chocolate (1 oz)	26	
Chocolate flavored syrup (1 oz)	4	

Drugs[a]	Average (mg)
Cold remedies (standard dose)	
Dristan	0
Coryban-D, Triaminicin	30
Diuretics (standard dose)	
Aqua-ban, Permathene H_2Off	200
Pre-Mens Forte	100
Pain relievers (standard dose)	
Excedrin	130
Midol, Anacin	65
Aspirin, plain (any brand)	0
Stimulants	
Caffedrin, NoDoz, Vivarin	200
Weight-control aids (daily dose)	
Prolamine	280
Dexatrim, Dietac	200

NOTE: A pharmacologically active dose of caffeine is defined as 200 milligrams.
[a]Because products change, contact the manufacturer for an update on products you use regularly.

Table A-1

Food Composition (Computer code number is for West Diet Analysis program) (For purposes of calculations, use "0" for t, <1, <.1, <.01, etc.)

Computer Code Number	Food Description	Measure	Wt (g)	H_2O (%)	Ener (kcal)	Prot (g)	Carb (g)	Dietary Fiber (g)	Fat (g)	Fat Breakdown (g) Sat	Mono	Poly
	BEVERAGES											
	Alcoholic:											
	Beer:											
1	Regular (12 fl oz)	1½ c	356	92	146	1	13	1	0	0	0	0
2	Light (12 fl oz)	1½ c	354	95	99	1	5	0	0	0	0	0
1506	Nonalcoholic (12 fl oz)	1½ c	360	98	32	1	5	0	0	0	0	0
	Gin, rum, vodka, whiskey:											
3	80 proof	1½ fl oz	42	67	97	0	0	0	0	0	0	0
4	86 proof	1½ fl oz	42	64	105	0	<1	0	0	0	0	0
5	90 proof	1½ fl oz	42	62	110	0	0	0	0	0	0	0
	Liqueur:											
1359	Coffee liqueur, 53 proof	1½ fl oz	52	31	175	<1	24	0	<1	.1	t	.1
1360	Coffee & cream liqueur, 34 proof	1½ fl oz	47	46	154	1	10	0	7	4.5	2.1	.3
1361	Crème de menthe, 72 proof	1½ fl oz	50	28	186	0	21	0	<1	t	t	.1
	Wine, 4 fl oz:											
6	Dessert, sweet	½ c	118	72	181	<1	14	0	0	0	0	0
7	Red	½ c	118	88	85	<1	2	0	0	0	0	0
8	Rosé	½ c	118	89	84	<1	2	0	0	0	0	0
9	White medium	½ c	118	90	80	<1	1	0	0	0	0	0
1592	Nonalcoholic	1 c	232	98	14	1	3	0	0	0	0	0
1593	Nonalcoholic light	1 c	232	98	14	1	3	0	0	0	0	0
1409	Wine cooler, bottle (12 fl oz)	1½ c	340	90	169	<1	20	<1	<1	t	t	t
1595	Wine cooler, cup	1 c	227	90	113	<1	13	<1	<1	t	t	t
	Carbonated:											
10	Club soda (12 fl oz)	1½ c	355	100	0	0	0	0	0	0	0	0
11	Cola beverage (12 fl oz)	1½ c	372	89	153	0	39	0	0	0	0	0
12	Diet cola w/aspartame (12 fl oz)	1½ c	355	100	4	<1	<1	0	0	0	0	0
13	Diet soda pop w/saccharin (12 fl oz)	1½ c	355	100	0	0	<1	0	0	0	0	0
14	Ginger ale (12 fl oz)	1½ c	366	91	124	0	32	0	0	0	0	0
15	Grape soda (12 fl oz)	1½ c	372	89	160	0	42	0	0	0	0	0
16	Lemon-lime (12 fl oz)	1½ c	368	89	147	0	38	0	0	0	0	0
17	Orange (12 fl oz)	1½ c	372	88	179	0	46	0	0	0	0	0
18	Pepper-type soda (12 fl oz)	1½ c	368	89	151	0	38	0	<1	.1	0	0
19	Root beer (12 fl oz)	1½ c	370	89	152	0	39	0	0	0	0	0
20	Coffee, brewed	1 c	237	99	5	<1	1	0	<1	t	0	t
21	Coffee, prepared from instant	1 c	238	99	5	<1	1	0	<1	t	0	t
	Fruit drinks, noncarbonated:											
22	Fruit punch drink, canned	1 c	248	88	117	0	29	<1	<1	t	t	t
1358	Gatorade	1 c	241	93	60	0	15	0	0	0	0	0
23	Grape drink, canned	1 c	250	87	125	<1	32	<1	0	0	0	0
1304	Koolade sweetened with sugar	1 c	262	90	97	0	25	0	<1	t	t	t
1356	Koolade sweetened with nutrasweet	1 c	240	95	43	0	11	0	0	0	0	0
26	Lemonade,frzn concentrate (6-oz can)	¾c	219	52	396	1	103	1	<1	.1	t	.1
27	Lemonade, from concentrate	1 c	248	89	99	<1	26	<1	<1	t	t	t
28	Limeade, frzn concentrate (6-oz can)	¾c	218	50	408	<1	108	1	<1	t	t	.1
29	Limeade, from concentrate	1 c	247	89	101	0	27	<1	<1	t	t	t
24	Pineapple grapefruit, canned	1 c	250	88	118	<1	29	<1	<1	t	t	.1
25	Pineapple orange, canned	1 c	250	87	125	3	29	<1	0	0	0	0
	Fruit and vegetable juices: see Fruit and Vegetable sections											
	Ultra Slim Fast, ready to drink, can:											
30411	Chocolate Royale	1 ea	350	84	220	10	38	5	3	1	1	.5
30415	French Vanilla	1 ea	350	84	220	10	38	5	3	1	1.5	.5
30413	Strawberries n' cream	1 ea	350	83	220	10	42	5	3	1	1.5	.5
1357	Water, bottled: Perrier (6½ fl oz)	1 ea	192	100	0	0	0	0	0	0	0	0
1594	Water, bottled: Tonic water	1½ c	366	91	124	0	32	0	0	0	0	0
	Tea:											
30	Brewed, regular	1 c	237	100	2	0	1	0	<1	t	0	t
1662	Brewed, herbal	1 c	237	100	2	0	<1	0	<1	t	t	t
32	From instant, sweetened	1 c	259	91	88	<1	22	0	<1	t	t	t
31	From instant, unsweetened	1 c	237	100	2	<1	<1	0	0	0	0	0

Chol (mg)	Calc (mg)	Iron (mg)	Magn (mg)	Pota (mg)	Sodi (mg)	Zinc (mg)	VT-A (RE)	Thia (mg)	VT-E (a-TE)	Ribo (mg)	Niac (mg)	V-B6 (mg)	Fola (µg)	VT-C (mg)
0	18	.11	21	89	18	.07	0	.02	0	.09	1.61	.18	21	0
0	18	.14	18	64	11	.11	0	.03	0	.11	1.39	.12	14	0
0	25	.04	32	90	18	.04	0	.02	0	.09	1.63	.18	22	0
0	0	.02	0	1	<1	.02	0	<.01	0	<.01	<.01	0	0	0
0	0	.02	0	1	<1	.02	0	<.01	0	<.01	<.01	0	0	0
0	0	.02	0	1	<1	.02	0	<.01	0	<.01	<.01	0	0	0
0	1	.03	2	16	4	.02	0	<.01	0	.01	.07	0	0	0
7	8	.06	1	15	43	.07	20	0	.12	.03	.04	.01	0	0
0	0	.03	0	0	2	.02	0	0	0	0	<.01	0	0	0
0	9	.28	11	109	11	.08	0	.02	0	.02	.25	0	<1	0
0	9	.51	15	132	6	.11	0	.01	0	.03	.1	.04	2	0
0	9	.45	12	117	6	.07	0	<.01	0	.02	.09	.03	1	0
0	11	.38	12	94	6	.08	0	<.01	0	.01	.08	.02	<1	0
0	21	.93	23	204	16	.19	0	0	0	.02	.23	.05	2	0
0	21	.93	23	204	16	.19	0	0	0	.02	.23	.05	2	0
0	19	.92	18	152	29	.2	<1	.02	.02	.02	.15	.04	4	6
0	13	.61	12	102	19	.13	<1	.01	.02	.02	.1	.03	3	4
0	18	.04	4	7	75	.35	0	0	0	0	0	0	0	0
0	11	.11	4	4	15	.04	0	0	0	0	0	0	0	0
0	14	.11	4	0	21	.28	0	.02	0	.08	0	0	0	0
0	14	.14	4	7	57	.18	0	0	0	0	0	0	0	0
0	11	.66	4	4	26	.18	0	0	0	0	0	0	0	0
0	11	.3	4	4	56	.26	0	0	0	0	0	0	0	0
0	7	.26	4	4	40	.18	0	0	0	0	.05	0	0	0
0	19	.22	4	7	45	.37	0	0	0	0	0	0	0	0
0	11	.15	0	4	37	.15	0	0	0	0	0	0	0	0
0	18	.18	4	4	48	.26	0	0	0	0	0	0	0	0
0	5	.12	12	128	5	.05	0	0	0	0	.53	0	<1	0
0	7	.12	10	86	7	.07	0	0	0	<.01	.67	0	0	0
0	20	.52	5	62	55	.3	3	0	0	.06	.05	0	3	73
0	0	.12	2	26	96	.05	0	.01	0	0	0	0	0	0
0	7	.25	10	87	2	.07	<1	.02	0	.02	.25	.05	2	40
0	42	.13	3	3	37	.08	0	0	0	<.01	<.01	0	<1	31
0	17	.65	5	50	50	.26	2	.02	0	.05	.05	0	5	77
0	15	1.58	11	147	9	.17	22	.06	0	.21	.16	.05	22	39
0	7	.4	5	37	7	.1	5	.01	0	.05	.04	.01	5	10
0	11	.22	9	129	0	.09	0	.02	0	.02	.22	0	9	26
0	7	.07	2	32	5	.05	0	<.01	0	<.01	.05	0	2	7
0	17	.77	15	153	35	.15	9	.07	0	.04	.67	.1	26	115
0	12	.67	15	115	7	.15	13	.07	0	.05	.52	.12	27	56
5	400	2.7	140	530	220	2.24	525	.52	7	.59	7	.7	120	21
5	400	2.8	140	450	460	2.1	525	.52	7	.59	7	.7	120	21
5	400	2.7	140	450	460	2.24	525	.52	7	.59	7	.7	120	21
0	27	0	0	0	2	0	0	0	0	0	0	0	0	0
0	4	.04	0	0	15	.37	0	0	0	0	0	0	0	0
0	0	.05	7	88	7	.05	0	0	0	.03	0	0	12	0
0	5	.19	2	21	2	.09	0	.02	0	.01	0	0	1	0
0	5	.05	5	49	8	.08	0	0	0	.05	.09	<.01	10	0
0	5	.05	5	47	7	.07	0	0	0	<.01	.09	<.01	1	0

Table A-1

Food Composition (Computer code number is for West Diet Analysis program) (For purposes of calculations, use "0" for t, <1, <.1, <.01, etc.)

Computer Code Number	Food Description	Measure	Wt (g)	H_2O (%)	Ener (kcal)	Prot (g)	Carb (g)	Dietary Fiber (g)	Fat (g)	Fat Breakdown (g) Sat	Mono	Poly
	DAIRY											
	Butter: see Fats and Oils, #158,159,160											
	Cheese, natural:											
33	Blue	1 oz	28	42	99	6	1	0	8	5.2	2.2	.2
34	Brick	1 oz	28	41	104	6	1	0	8	5.3	2.4	.2
35	Brie	1 oz	28	48	93	6	<1	0	8	4.9	2.2	.2
36	Camembert	1 oz	28	52	84	6	<1	0	7	4.3	2	.2
37	Cheddar:	1 oz	28	37	113	7	<1	0	9	5.9	2.6	.3
38	1″ cube	1 ea	17	37	68	4	<1	0	6	3.6	1.6	.2
39	Shredded	1 c	113	37	455	28	1	0	37	24	10.6	1.1
1406	Low fat, low sodium	1 oz	28	65	48	7	1	0	2	1.2	.6	.1
	Cottage:											
984	Low sodium, low fat	1 c	225	83	162	28	6	0	2	1.4	.7	.1
40	Creamed, large curd	1 c	225	79	232	28	6	0	10	6.4	2.9	.3
41	Creamed, small curd	1 c	210	79	216	26	6	0	9	6	2.7	.3
42	With fruit	1 c	226	72	280	22	30	0	8	4.9	2.2	.2
43	Low fat 2%	1 c	226	79	203	31	8	0	4	2.8	1.2	.1
44	Low fat 1%	1 c	226	82	164	28	6	0	2	1.5	.7	.1
46	Cream	1 tbs	15	54	52	1	<1	0	5	3.3	1.5	.2
983	low fat	1 tbs	15	64	35	2	1	0	3	1.7	.9	.1
47	Edam	1 oz	28	42	100	7	<1	0	8	4.9	2.3	.2
48	Feta	1 oz	28	55	74	4	1	0	6	4.2	1.3	.2
49	Gouda	1 oz	28	41	100	7	1	0	8	4.9	2.2	.2
50	Gruyère	1 oz	28	33	116	8	<1	0	9	5.3	2.8	.5
51	Gorgonzola	1 oz	28	43	97	6	1	0	8	5		
1676	Limburger	1 oz	28	48	92	6	<1	0	8	4.7	2.4	.1
53	Monterey Jack	1 oz	28	41	104	7	<1	0	8	5.3	2.4	.3
54	Mozzarella, whole milk	1 oz	28	54	79	5	1	0	6	3.7	1.8	.2
55	Mozzarella, part-skim milk, low moisture	1 oz	28	49	78	8	1	0	5	3	1.4	.1
56	Muenster	1 oz	28	42	103	7	<1	0	8	5.3	2.4	.2
2422	Neufchatel	1 oz	28	62	73	3	1	0	7	4.1	1.9	.2
1399	Nonfat cheese (Kraft Singles)	1 oz	28	61	44	6	4	0	0	0	0	0
59	Parmesan, grated:	1 oz	28	18	128	12	1	0	8	5.5	2.4	.2
57	Cup, not pressed down	1 c	100	18	456	42	4	0	30	19.7	8.7	.7
58	Tablespoon	1 tbs	6	18	27	2	<1	0	2	1.2	.5	t
60	Provolone	1 oz	28	41	98	7	1	0	7	4.9	2.1	.2
61	Ricotta, whole milk	1 c	246	72	428	28	7	0	32	20.4	8.9	.9
62	Ricotta, part-skim milk	1 c	246	74	339	28	13	0	19	12.1	5.7	.6
63	Romano	1 oz	28	31	108	9	1	0	8	4.8	2.2	.2
64	Swiss	1 oz	28	37	105	8	1	0	8	5	2	.3
976	low fat	1 oz	28	60	50	8	1	0	1	.9	.4	t
	Pasteurized processed cheese products:											
65	American	1 oz	28	39	105	6	<1	0	9	5.5	2.5	.3
66	Swiss	1 oz	28	42	93	7	1	0	7	4.5	2	.2
67	American cheese food, jar	½ c	57	43	187	11	4	0	14	9	4.1	.4
68	American cheese spread	1 tbs	15	48	43	2	1	0	3	2	.9	.1
982	Velveeta cheese spread, low fat, low sodium, slice	1 pce	34	62	61	9	1	0	2	1.5	.7	.1
	Cream, sweet:											
69	Half & half (cream & milk)	1 c	242	81	315	7	10	0	28	17.3	8	1
70	Tablespoon	1 tbs	15	81	19	<1	1	0	2	1.1	.5	.1
71	Light, coffee or table:	1 c	240	74	468	6	9	0	46	28.8	13.4	1.7
72	Tablespoon	1 tbs	15	74	29	<1	1	0	3	1.8	.8	.1
73	Light whipping cream, liquid:	1 c	239	63	698	5	7	0	74	46.1	21.7	2.1
74	Tablespoon	1 tbs	15	63	44	<1	<1	0	5	2.9	1.4	.1
75	Heavy whipping cream, liquid:	1 c	238	58	821	5	7	0	88	54.7	25.5	3.3
76	Tablespoon	1 tbs	15	58	52	<1	<1	0	6	3.4	1.6	.2
77	Whipped cream, pressurized:	1 c	60	61	154	2	7	0	13	8.3	3.8	.5
78	Tablespoon	1 tbs	4	61	10	<1	<1	0	1	.6	.3	t
79	Cream, sour, cultured:	1 c	230	71	492	7	10	0	48	29.9	13.9	1.8
80	Tablespoon	1 tbs	14	71	30	<1	1	0	3	1.8	.8	.1

Chol (mg)	Calc (mg)	Iron (mg)	Magn (mg)	Pota (mg)	Sodi (mg)	Zinc (mg)	VT-A (RE)	Thia (mg)	VT-E (a-TE)	Ribo (mg)	Niac (mg)	V-B6 (mg)	Fola (μg)	VT-C (mg)
21	148	.09	6	72	391	.74	64	.01	.18	.11	.29	.05	10	0
26	189	.12	7	38	157	.73	85	<.01	.14	.1	.03	.02	6	0
28	51	.14	6	43	176	.67	51	.02	.18	.15	.11	.07	18	0
20	109	.09	6	52	236	.67	71	.01	.18	.14	.18	.06	17	0
29	202	.19	8	28	174	.87	85	.01	.1	.1	.02	.02	5	0
18	123	.12	5	17	106	.53	51	<.01	.06	.06	.01	.01	3	0
119	815	.77	31	111	702	3.51	342	.03	.41	.42	.09	.08	21	0
6	197	.2	8	31	6	.86	17	.01	.05	.01	.02	.02	5	0
9	137	.31	11	194	29	.85	25	.04	.25	.36	.29	.16	27	0
33	135	.31	12	190	911	.83	108	.05	.27	.37	.28	.15	27	0
31	126	.29	11	177	851	.78	101	.04	.26	.34	.26	.14	26	0
25	108	.25	9	151	915	.65	81	.04	.21	.29	.23	.12	22	0
19	155	.36	14	217	918	.95	45	.05	.13	.42	.32	.17	30	0
10	138	.32	12	193	918	.86	25	.05	.25	.37	.29	.15	28	0
16	12	.18	1	18	44	.08	66	<.01	.14	.03	.01	.01	2	0
8	17	.25	1	25	44	.11	33	<.01	.07	.04	.02	.01	3	0
25	205	.12	8	53	270	1.05	71	.01	.21	.11	.02	.02	5	0
25	138	.18	5	17	312	.81	36	.04	.01	.24	.28	.12	9	0
32	196	.07	8	34	229	1.09	49	.01	.1	.09	.02	.02	6	0
31	283	.05	10	23	94	1.09	84	.02	.1	.08	.03	.02	3	0
30	170	.18			280		43							0
25	139	.04	6	36	224	.59	88	.02	.18	.14	.04	.02	16	0
25	209	.2	8	23	150	.84	71	<.01	.09	.11	.03	.02	5	0
22	145	.05	5	19	104	.62	67	<.01	.1	.07	.02	.02	2	0
15	205	.07	7	26	148	.88	53	.01	.13	.1	.03	.02	3	0
27	201	.11	8	37	176	.79	88	<.01	.13	.09	.03	.02	3	0
21	21	.08	2	32	112	.15	74	<.01	.26	.05	.03	.01	3	0
4	221	0		81	427		126		0	.1				0
22	385	.27	14	30	521	.89	48	.01	.22	.11	.09	.03	2	0
79	1375	.95	51	107	1861	3.19	173	.04	.8	.39	.31	.1	8	0
5	82	.06	3	6	112	.19	10	<.01	.05	.02	.02	.01	<1	0
19	212	.15	8	39	245	.9	74	<.01	.1	.09	.04	.02	3	0
124	509	.93	28	258	207	2.85	330	.03	.86	.48	.26	.11	30	0
76	669	1.08	36	308	308	3.3	278	.05	.53	.45	.19	.05	32	0
29	298	.22	11	24	336	.72	39	.01	.2	.1	.02	.02	2	0
26	269	.05	10	31	73	1.09	71	.01	.14	.1	.03	.02	2	0
10	269	.05	10	31	73	1.09	18	.01	.05	.1	.02	.02	2	0
26	172	.11	6	45	400	.84	81	.01	.13	.1	.02	.02	2	0
24	216	.17	8	60	384	1.01	64	<.01	.19	.08	.01	.01	2	0
36	327	.48	17	159	678	1.7	125	.02	.4	.25	.08	.08	4	0
8	84	.05	4	36	202	.39	28	.01	.11	.06	.02	.02	1	0
12	233	.15	8	61	2	1.13	22	.01	.17	.13	.03	.03	3	0
89	254	.17	25	315	98	1.23	259	.08	.27	.36	.19	.09	6	2
6	16	.01	2	19	6	.08	16	<.01	.02	.02	.01	.01	<1	<1
159	231	.1	21	293	95	.65	437	.08	.36	.35	.14	.08	6	2
10	14	.01	1	18	6	.04	27	<.01	.02	.02	.01	<.01	<1	<1
265	166	.07	17	231	82	.6	705	.06	1.43	.3	.1	.07	9	1
17	10	<.01	1	14	5	.04	44	<.01	.09	.02	.01	<.01	1	<1
326	154	.07	17	179	89	.55	1001	.05	1.5	.26	.09	.06	9	1
21	10	<.01	1	11	6	.03	63	<.01	.09	.02	.01	<.01	1	<1
46	61	.03	6	88	78	.22	124	.02	.36	.04	.04	.02	2	0
3	4	<.01		6	5	.01	8	<.01	.02	<.01	<.01	<.01	<1	0
102	267	.14	26	331	123	.62	449	.08	1.3	.34	.15	.04	25	2
6	16	.01	2	20	7	.04	27	<.01	.08	.02	.01	<.01	2	<1

Table A-1

Food Composition (Computer code number is for West Diet Analysis program) (For purposes of calculations, use "0" for t, <1, <.1, <.01, etc.)

Computer Code Number	Food Description	Measure	Wt (g)	H_2O (%)	Ener (kcal)	Prot (g)	Carb (g)	Dietary Fiber (g)	Fat (g)	Fat Breakdown (g) Sat	Mono	Poly
	DAIRY—Continued											
	Cream products—imitation and part dairy:											
81	Coffee whitener, frozen or liquid	1 tbs	15	77	20	<1	2	0	1	1.4	t	0
82	Coffee whitener, powdered	1 tsp	2	2	11	<1	1	0	1	.6	t	0
83	Dessert topping, frozen, nondairy:	1 c	75	50	239	1	17	0	19	16.4	1.2	.4
84	Tablespoon	1 tbs	5	50	16	<1	1	0	1	1.1	.1	t
85	Dessert topping, mix with whole milk:	1 c	80	67	151	3	13	0	10	8.6	.7	.2
86	Tablespoon	1 tbs	5	67	9	<1	1	0	1	.5	t	t
88	Dessert topping, pressurized	1 c	70	60	185	1	11	0	16	13.3	1.3	.2
87	Tablespoon	1 tbs	4	60	11	<1	1	0	1	.8	.1	t
91	Sour cream, imitation:	1 c	230	71	478	6	15	0	45	40.9	1.3	.1
92	Tablespoon	1 tbs	14	71	29	<1	1	0	3	2.5	.1	t
89	Sour dressing, part dairy:	1 c	235	75	418	8	11	0	39	31.3	4.6	1.1
90	Tablespoon	1 tbs	15	75	27	1	1	0	2	2	.3	.1
	Milk, fluid:											
93	Whole milk	1 c	244	88	150	8	11	0	8	5.1	2.7	.3
94	2% reduced-fat milk	1 c	244	89	121	8	12	0	5	2.9	1.3	.2
95	2% milk solids added	1 c	245	89	125	9	12	0	5	2.9	1.4	.2
96	1% lowfat milk	1 c	244	90	102	8	12	0	3	1.6	.7	.1
97	1% milk solids added	1 c	245	90	104	9	12	0	2	1.5	.7	.1
98	Nonfat milk, vitamin A added	1 c	245	91	85	8	12	0	<1	.3	.1	t
99	Nonfat milk solids added	1 c	245	90	90	9	12	0	1	.4	.2	t
100	Buttermilk, skim	1 c	245	90	99	8	12	0	2	1.3	.6	.1
	Milk, canned:											
101	Sweetened condensed	1 c	306	27	982	24	166	0	27	16.8	7.4	1
103	Evaporated, nonfat	1 c	256	79	199	19	29	0	1	.3	.2	t
	Milk, dried:											
104	Buttermilk, sweet	1 c	120	3	464	41	59	0	7	4.3	2	.3
105	Instant, nonfat, vit A added (makes 1 qt)	1 ea	91	4	326	32	47	0	1	.4	.2	t
106	Instant nonfat, vit A added	1 c	68	4	243	24	35	0	<1	.3	.1	t
107	Goat milk	1 c	244	87	168	9	11	0	10	6.5	2.7	.4
108	Kefir	1 c	233	88	149	8	11	0	8			
	Milk beverages and powdered mixes:											
	Chocolate:											
109	Whole	1 c	250	82	209	8	26	2	8	5.3	2.5	.3
110	2% fat	1 c	250	84	179	8	26	1	5	3.1	1.5	.2
111	1% fat	1 c	250	84	158	9	26	1	2	1.5	.7	.1
	Chocolate-flavored beverages:											
112	Powder containing nonfat dry milk:	1 oz	28	1	101	3	22	<1	1	.7	.4	t
113	Prepared with water	1 c	275	86	138	4	30	3	2	.9	.5	t
114	Powder without nonfat dry milk:	1 oz	28	1	98	1	25	2	1	.5	.3	t
115	Prepared with whole milk	1 c	266	81	226	9	31	1	9	5.5	2.6	.3
116	Eggnog, commercial	1 c	254	74	343	10	34	0	19	11.3	5.7	.9
974	Eggnog, 2% reduced-fat	1 c	254	85	189	12	17	0	8	3.7	2.7	.7
1027	Instant Breakfast, envelope,powder only:	1 ea	37	7	131	7	24	<1	1	.3	.1	t
1028	Prepared with whole milk	1 c	281	77	280	15	36	<1	9	5.4	2.5	.3
1029	Prepared with 2% milk	1 c	281	78	252	15	36	<1	5	3.3	1.5	.2
1283	Prepared with 1% milk	1 c	281	79	233	15	36	<1	3	1.9	.9	.1
1284	Prepared with nonfat milk	1 c	282	80	216	16	36	<1	1	.7	.3	t
117	Malted milk, chocolate, powder:	3 tsp	21	1	79	1	18	<1	1	.5	.2	.1
118	Prepared with whole milk	1 c	265	81	228	9	30	<1	9	5.5	2.6	.4
1661	Ovaltine with whole milk	1 c	265	81	225	9	29	<1	9	5.5	2.5	.4
119	Malted mix powder, natural:	3 tsp	21	2	87	2	16	<1	2	.9	.4	.3
120	Prepared with whole milk	1 c	265	81	236	10	27	<1	10	6	2.8	.6
121	Milk shakes, chocolate	1 c	166	71	211	6	34	1	6	3.8	1.8	.2
122	Milk shakes, vanilla	1 c	166	75	184	6	30	1	5	3.1	1.4	.2
	Milk desserts:											
134	Custard, baked	1 c	282	79	296	14	30	0	13	6.6	4.3	1
1548	Low-fat frozen dessert bars	1 ea	81	72	88	2	19	0	1	.2	.1	.4
	Ice cream, vanilla (about 10% fat):											
123	Hardened: ½ gallon	1 ea	1064	61	2138	37	251	0	117	72.4	33.8	4.4
124	Cup	1 c	132	61	265	5	31	0	14	9	4.2	.5
126	Soft serve	1 c	172	60	370	7	38	0	22	12.9	6	.8

PAGE KEY: A–2 = Beverages A–4 = Dairy A–8 = Eggs A–8 = Fat/Oil A–12 = Fruit A–18 = Bakery A–26 = Grain A–30 = Fish A–32 = Meats A–36 = Poultry A–38 = Sausage A–40 = Mixed/Fast A–44 = Nuts/Seeds A–48 = Sweets A–50 = Vegetables/Legumes A–62 = Misc A–64 = Soups/Sauces A–66 = Fast A–82 = Frozen Entree A–86 = Baby foods

Chol (mg)	Calc (mg)	Iron (mg)	Magn (mg)	Pota (mg)	Sodi (mg)	Zinc (mg)	VT-A (RE)	Thia (mg)	VT-E (a-TE)	Ribo (mg)	Niac (mg)	V-B6 (mg)	Fola (µg)	VT-C (mg)
0	1	<.01		29	12	<.01	1	0	.24	0	0	0	0	0
0		.02		16	4	.01	<1	0	<.01	<.01	0	0	0	0
0	5	.09	1	14	19	.02	64	0	.14	0	0	0	0	0
0		.01		1	1	<.01	4	0	.01	0	0	0	0	0
8	72	.03	8	121	53	.22	39	.02	.11	.09	.05	.02	3	1
<1	5	<.01		8	3	.01	2	<.01	.01	.01	<.01	<.01	<1	<1
0	4	.01	1	13	43	.01	33	0	.12	0	0	0	0	0
0		<.01		1	2	0	2	0	.01	0	0	0	0	0
0	6	.9	15	370	235	2.71	0	0	.34	0	0	0	0	0
0		.05	1	22	14	.16	0	0	.02	0	0	0	0	0
13	266	.07	23	381	113	.87	5	.09	.29	.38	.17	.04	28	2
1	17	<.01	1	24	7	.06	<1	.01	.02	.02	.01	<.01	2	<1
33	290	.12	33	371	120	.93	76	.09	.24	.39	.2	.1	12	2
18	298	.12	33	376	122	.95	139	.09	.17	.4	.21	.1	12	2
18	314	.12	35	397	128	.98	140	.1	.17	.42	.22	.11	13	2
10	300	.12	34	381	123	.95	144	.09	.1	.41	.21	.1	12	2
10	314	.12	35	397	128	.98	145	.1	.1	.42	.22	.11	13	2
4	301	.1	28	407	126	.98	149	.09	.1	.34	.22	.1	13	2
5	316	.12	35	419	130	1	149	.1	.1	.43	.22	.11	13	2
9	284	.12	27	370	257	1.03	20	.08	.15	.38	.14	.08	12	2
104	869	.58	79	1135	389	2.88	248	.27	.65	1.27	.64	.16	34	8
9	742	.74	69	850	294	2.3	300	.11	.01	.79	.44	.14	22	3
83	1420	.36	132	1910	620	4.82	65	.47	.48	1.9	1.05	.41	57	7
17	1119	.28	106	1551	500	4.01	646	.38	.02	1.58	.81	.31	45	5
12	836	.21	80	1159	373	3	483	.28	.01	1.18	.61	.23	34	4
28	327	.12	34	498	122	.73	137	.12	.22	.34	.68	.11	1	3
		.3	33	373	107									
30	280	.6	32	418	149	1.03	72	.09	.23	.4	.31	.1	12	2
17	285	.6	33	423	151	1.03	143	.09	.13	.41	.31	.1	12	2
7	288	.6	33	425	152	1.03	148	.09	.06	.41	.32	.1	12	2
1	91	.33	23	199	141	.41	1	.03	.04	.16	.16	.03	0	<1
3	129	.47	33	270	198	.6	1	.04	.06	.21	.22	.04	0	1
0	10	.88	27	165	59	.43	1	.01	.11	.04	.14	<.01	2	<1
32	301	.8	53	497	165	1.28	77	.1	.21	.43	.32	.1	12	2
149	330	.51	47	419	138	1.17	203	.09	.58	.48	.27	.13	2	4
194	269	.71	32	367	155	1.26	197	.11	1.01	.55	.21	.15	30	2
4	105	4.74	84	350	142	3.16	554	.31	5.31	.07	5.27	.42	105	28
38	396	4.86	117	719	262	4.09	630	.41	5.51	.47	5.46	.52	118	31
23	401	4.86	118	726	264	4.12	693	.41	5.41	.48	5.46	.53	118	31
14	406	4.86	118	731	266	4.12	698	.41	5.36	.48	5.47	.53	118	31
9	407	4.83	112	755	268	4.14	703	.4	5.3	.42	5.47	.52	118	31
1	13	.48	15	130	53	.17	4	.04	.08	.04	.42	.03	4	<1
34	305	.61	48	498	172	1.09	79	.13	.26	.44	.62	.13	16	3
34	384	3.76	53	620	244	1.17	901	.73	.32	1.26	10.9	1.02	32	34
4	63	.15	19	159	104	.21	18	.11	.08	.19	1.1	.09	10	1
37	355	.26	53	530	223	1.14	95	.2	.32	.59	1.31	.19	22	3
22	188	.51	28	332	161	.68	38	.1	.11	.41	.27	.08	6	1
18	203	.15	20	289	136	.6	53	.07	.1	.3	.31	.09	5	1
245	316	.85	39	431	217	1.49	169	.09	.68	.64	.24	.14	28	1
1	82	.07	10	111	47	.26	38	.03	.07	.11	.06	.03	3	1
468	1361	.96	149	2117	851	7.34	1244	.44	0	2.55	1.23	.51	53	6
58	169	.12	18	263	106	.91	154	.05	0	.32	.15	.06	7	1
157	225	.36	21	304	105	.89	265	.08	.64	.31	.16	.08	15	1

Table A-1

Food Composition (Computer code number is for West Diet Analysis program) (For purposes of calculations, use "0" for t, <1, <.1, <.01, etc.)

Computer Code Number	Food Description	Measure	Wt (g)	H_2O (%)	Ener (kcal)	Prot (g)	Carb (g)	Dietary Fiber (g)	Fat (g)	Fat Breakdown (g) Sat	Mono	Poly
	DAIRY—Continued											
	Ice cream, rich vanilla (16% fat):											
127	Hardened: ½ gallon	1 ea	1188	60	2554	49	264	0	154	88.9	41.5	5.5
128	Cup	1 c	148	57	357	5	33	0	24	14.8	6.9	.9
1724	Ben & Jerry's	½ c	108		230	4	21	0	17	10		
	Ice milk, vanilla (about 4% fat):											
129	Hardened: ½ gallon	1 ea	1048	68	1456	40	238	0	45	27.7	12.9	1.7
130	Cup	1 c	132	68	183	5	30	0	6	3.5	1.6	.2
131	Soft serve (about 3% fat)	1 c	176	70	222	9	38	0	5	2.9	1.3	.2
	Pudding, canned (5 oz can = .55 cup):											
135	Chocolate	1 ea	142	69	189	4	32	1	6	1	2.4	2
136	Tapioca	1 ea	142	74	169	3	27	<1	5	.9	2.3	1.9
137	Vanilla	1 ea	142	71	185	3	31	<1	5	.8	2.2	1.9
	Puddings, dry mix with whole milk:											
138	Chocolate, instant	1 c	294	74	326	9	55	3	9	5.4	2.7	.5
139	Chocolate, regular, cooked	1 c	284	74	315	9	51	3	10	5.9	2.8	.4
140	Rice, cooked	1 c	288	72	351	9	60	<1	8	5.1	2.4	.3
141	Tapioca, cooked	1 c	282	74	321	8	55	0	8	5.1	2.3	.3
142	Vanilla, instant	1 c	284	73	324	8	56	0	8	4.9	2.4	.4
143	Vanilla, regular, cooked	1 c	280	75	311	8	52	0	8	5.1	2.4	.4
	Sherbet (2% fat):											
132	½ gallon	1 ea	1542	66	2127	17	469	8	31	17.9	8.3	1.2
133	Cup	1 c	198	66	273	2	60	1	4	2.3	1.1	.2
144	Soy milk	1 c	245	93	81	7	4	3	5	.7	1	2.6
2301	Soy milk, fortified, fat free	1 c	240	88	110	6	22	1	0	0	0	0
	Yogurt, frozen, low-fat											
1584	Cup	1 c	144	65	229	6	35	0	8	4.9	2.3	.3
1512	Scoop	1 ea	79	74	78	4	15	0	<1	.1	t	t
	Yogurt, lowfat:											
1172	Fruit added with low-calorie sweetener	1 c	241	86	122	12	19	1	<1	.2	.1	t
145	Fruit added	1 c	245	74	250	11	47	<1	3	1.7	.7	.1
146	Plain	1 c	245	85	155	13	17	0	4	2.4	1	.1
147	Vanilla or coffee flavor	1 c	245	79	209	13	34	0	3	2	.8	.1
148	Yogurt, made with nonfat milk	1 c	245	85	137	15	19	0	<1	.3	.1	t
149	Yogurt, made with whole milk	1 c	245	88	150	9	11	0	8	5.1	2.2	.2
	EGGS											
	Raw, large:											
150	Whole, without shell	1 ea	50	75	74	6	1	0	5	1.5	1.9	.7
151	White	1 ea	33	88	16	3	<1	0	0	0	0	0
152	Yolk	1 ea	17	49	61	3	<1	0	5	1.6	2	.7
	Cooked:											
153	Fried in margarine	1 ea	46	69	91	6	1	0	7	1.9	2.8	1.3
154	Hard-cooked, shell removed	1 ea	50	75	77	6	1	0	5	1.6	2	.7
155	Hard-cooked, chopped	1 c	136	75	211	17	2	0	14	4.4	5.5	1.9
156	Poached, no added salt	1 ea	50	75	74	6	1	0	5	1.5	1.9	.7
157	Scrambled with milk & margarine	1 ea	61	73	101	7	1	0	7	2.2	2.9	1.3
1681	Egg substitute, liquid:	½ c	126	83	106	16	1	0	4	.8	1.1	2
1254	Egg Beaters, Fleischmann's	½ c	122		60	12	2	0	0	0	0	0
1262	Egg substitute, liquid, prepared	½ c	105	80	100	14	1	0	4	.8	1.1	1.9
	FATS and OILS											
158	Butter: Stick	½ c	114	16	817	1	<1	0	92	57.7	27.4	3.4
159	Tablespoon:	1 tbs	14	16	100	<1	<1	0	11	7.1	3.4	.4
8025	Unsalted	1 tbs	14	18	100	<1	<1	0	11	7.1	3.4	.4
160	Pat (about 1 tsp)	1 ea	5	16	36	<1	<1	0	4	2.5	1.2	.2
1682	Whipped	1 tsp	3	16	21	<1	<1	0	2	1.5	.7	.1
	Fats, cooking:											
1363	Bacon fat	1 tbs	14		125	0	0	0	14	6.3	5.9	1.1
1362	Beef fat/tallow	1 c	205	0	1849	0	0	0	205	103	87.3	8.2
1364	Chicken fat	1 c	205		1845	0	0	0	205	61.1	91.6	42.8
161	Vegetable shortening:	1 c	205	0	1812	0	0	0	205	52.1	91.2	53.5
162	Tablespoon	1 tbs	13	0	115	0	0	0	13	3.3	5.8	3.4

PAGE KEY: A–2 = Beverages A–4 = Dairy A–8 = Eggs A–8 = Fat/Oil A–12 = Fruit A–18 = Bakery A–26 = Grain A–30 = Fish A–32 = Meats A–36 = Poultry A–38 = Sausage A–40 = Mixed/Fast A–44 = Nuts/Seeds A–48 = Sweets A–50 = Vegetables/Legumes A–62 = Misc A–64 = Soups/Sauces A–66 = Fast A–82 = Frozen Entree A–86 = Baby foods

Chol (mg)	Calc (mg)	Iron (mg)	Magn (mg)	Pota (mg)	Sodi (mg)	Zinc (mg)	VT-A (RE)	Thia (mg)	VT-E (a-TE)	Ribo (mg)	Niac (mg)	V-B6 (mg)	Fola (μg)	VT-C (mg)
1081	1556	2.49	143	2102	725	6.18	1829	.58	4.4	2.16	1.13	.57	107	9
90	173	.07	16	235	83	.59	272	.06	0	.24	.12	.06	7	1
95	150	.36			55		214		0					0
147	1456	1.05	157	2211	891	4.61	493	.61	0	2.78	.94	.68	63	8
18	183	.13	20	279	112	.58	62	.08	0	.35	.12	.09	8	1
21	276	.11	25	389	123	.93	51	.09	0	.35	.21	.08	11	2
4	128	.72	30	256	183	.6	16	.04	.18	.22	.49	.04	4	3
1	119	.33	11	148	168	.38	0	.03	.13	.14	.44	.14	4	1
10	125	.18	11	160	192	.35	9	.03	.18	.2	.36	.02	0	0
32	300	.85	53	488	835	1.23	62	.1	.18	.41	.28	.11	12	3
34	315	1.02	43	463	293	1.28	74	.09	.17	.49	.29	.1	11	2
32	297	1.09	37	372	314	1.09	58	.22	.17	.4	1.28	.1	11	2
34	293	.17	34	372	341	.96	76	.08	.23	.4	.21	.11	11	2
31	287	.2	34	364	812	.94	71	.09	.18	.39	.21	.1	11	2
34	300	.14	36	381	448	.98	76	.08	.17	.4	.21	.09	11	2
77	833	2.16	123	1480	709	7.4	216	.39	.88	1.05	1.48	.52	77	66
10	107	.28	16	190	91	.95	28	.05	.11	.13	.19	.07	10	9
0	10	1.42	47	345	29	.56	7	.39	.02	.17	.36	.1	4	0
0	400	1.44		20	60		0	.07		.1	3			0
3	206	.43	20	304	125	.6	82	.05	.07	.32	.41	.11	9	1
1	137	.07	13	175	53	.67	1	.03	<.01	.16	.08	.04	8	1
3	369	.61	41	550	139	1.83	6	.1	.17	.45	.5	.11	32	26
10	372	.17	36	478	143	1.81	27	.09	.07	.44	.23	.1	23	2
15	448	.2	43	573	172	2.18	39	.11	.1	.52	.28	.12	27	2
12	419	.17	40	537	161	2.03	32	.1	.08	.49	.26	.11	26	2
4	488	.22	47	625	187	2.38	5	.12	.01	.57	.3	.13	30	2
31	296	.12	28	380	114	1.45	73	.07	.22	.35	.18	.08	18	1
213	24	.72	5	60	63	.55	95	.03	.52	.25	.04	.07	23	0
0	2	.01	4	47	54	<.01	0	<.01	0	.15	.03	<.01	1	0
218	23	.6	2	16	7	.53	99	.03	.54	.11	<.01	.07	25	0
211	25	.72	5	61	162	.55	114	.03	.75	.24	.03	.07	17	0
212	25	.59	5	63	62	.52	84	.03	.52	.26	.03	.06	22	0
577	68	1.62	14	171	169	1.43	228	.09	1.43	.7	.09	.16	60	0
212	24	.72	5	60	140	.55	95	.02	.52	.21	.03	.06	17	0
215	43	.73	7	84	171	.61	119	.03	.8	.27	.05	.07	18	<1
1	67	2.65	11	416	223	1.64	272	.14	.61	.38	.14	<.01	19	0
0	80	2.16		170	200		80		.59					
1	63	2.51	10	394	211	1.55	258	.11	.58	.34	.12	<.01	13	0
250	27	.18	2	30	942	.06	860	.01	1.8	.04	.05	<.01	3	0
31	3	.02		4	116	.01	106	<.01	.22	<.01	.01	0	<1	0
31	3	.02		4	2	.01	106	<.01	.22	<.01	.01	0	<1	0
11	1	.01		1	41	<.01	38	0	.08	<.01	<.01	0	<1	0
7	1	<.01		1	25	<.01	23	0	.05	<.01	<.01	0	<1	0
14		0			76	<.01	0	0	.31	0	0	0	0	0
223	0	0	0		<1	0	0	0	3.08	0	0	0	0	0
174	0	0	0	0	0	0	0	0	5.54	0	0	0	0	0
0	0	0	0	0	0	0	0	0	17	0	0	0	0	0
0	0	0	0	0	0	0	0	0	1.08	0	0	0	0	0

Table A-1

Food Composition (Computer code number is for West Diet Analysis program) (For purposes of calculations, use "0" for t, <1, <.1, <.01, etc.)

Computer Code Number	Food Description	Measure	Wt (g)	H_2O (%)	Ener (kcal)	Prot (g)	Carb (g)	Dietary Fiber (g)	Fat (g)	Fat Breakdown (g) Sat	Mono	Poly
	FATS and OILS—Continued											
163	Lard:	1 c	205	0	1849	0	0	0	205	81.1	87	28.3
164	Tablespoon	1 tbs	13	0	117	0	0	0	13	5.1	5.5	1.8
	Margarine:											
165	Imitation (about 40% fat), soft:	1 c	232	58	800	1	1	0	90	17.9	36.4	32
166	Tablespoon	1 tbs	14	58	48	<1	<1	0	5	1.1	2.2	1.9
167	Regular, hard (about 80% fat):	½ c	114	16	820	1	1	0	92	18	40.8	29
168	Tablespoon	1 tbs	14	16	101	<1	<1	0	11	2.2	5	3.6
169	Pat	1 ea	5	16	36	<1	<1	0	4	.8	1.8	1.3
170	Regular, soft (about 80% fat):	1 c	227	16	1625	2	1	0	183	31.3	64.7	78.5
171	Tablespoon	1 tbs	14	16	100	<1	<1	0	11	1.9	4	4.8
2056	Saffola, unsalted	1 tbs	14	20	100	0	0	0	11	2	3	4.5
2057	Saffola, reduced fat	1 tbs	14	37	60	0	0	0	8	1.3	2.7	4.4
172	Spread (about 60% fat), hard:	1 c	227	37	1225	1	0	0	138	32	59	41.1
173	Tablespoon	1 tbs	14	37	76	<1	0	0	9	2	3.6	2.5
174	Pat	1 ea	5	37	27	<1	0	0	3	.7	1.2	1
175	Spread (about 60% fat), soft:	1 c	227	37	1225	1	0	0	138	29.3	71.5	31.3
176	Tablespoon	1 tbs	14	37	76	<1	0	0	9	1.8	4.4	1.9
2160	Touch of Butter (47% fat)	1 tbs	14		60	0	0	0	7	1.5	3.1	1.5
	Oils:											
1585	Canola:	1 c	218	0	1927	0	0	0	218	15.5	128	64.5
1586	Tablespoon	1 tbs	14	0	124	0	0	0	14	1	8.2	4.1
177	Corn:	1 c	218	0	1927	0	0	0	218	29.4	54.1	131
178	Tablespoon	1 tbs	14	0	124	0	0	0	14	1.9	3.5	8.4
179	Olive:	1 c	216	0	1909	0	0	0	216	29.4	159	21.3
180	Tablespoon	1 tbs	14	0	124	0	0	0	14	1.9	10.3	1.4
1683	Olive, extra virgin	1 tbs	14		126	0	0	0	14	2	10.8	1.3
181	Peanut:	1 c	216	0	1909	0	0	0	216	40	99.8	71.3
182	Tablespoon	1 tbs	14	0	124	0	0	0	14	2.6	6.5	4.6
183	Safflower:	1 c	218	0	1927	0	0	0	218	19.8	26.4	162
184	Tablespoon	1 tbs	14	0	124	0	0	0	14	1.3	1.7	10.4
185	Soybean:	1 c	218	0	1927	0	0	0	218	32	50.8	126
186	Tablespoon	1 tbs	14	0	124	0	0	0	14	2.1	3.3	8.1
187	Soybean/cottonseed:	1 c	218	0	1927	0	0	0	218	39.5	64.3	105
188	Tablespoon	1 tbs	14	0	124	0	0	0	14	2.5	4.1	6.7
189	Sunflower:	1 c	218	0	1927	0	0	0	218	25.3	42.5	143
190	Tablespoon	1 tbs	14	0	124	0	0	0	14	1.6	2.7	9.2
	Salad dressings/sandwich spreads:											
191	Blue cheese, regular	1 tbs	15	32	76	1	1	0	8	1.5	1.8	4.2
1040	Low calorie	1 tbs	15	79	15	1	<1	<1	1	.2	.5	.4
1684	Caesar's	1 tbs	12	36	56	1	<1	<1	5	1	3.7	.5
192	French, regular	1 tbs	16	38	69	<1	3	0	7	1.5	1.3	3.5
193	Low calorie	1 tbs	16	71	21	<1	3	0	1	.1	.2	.5
194	Italian, regular	1 tbs	15	40	70	<1	2	0	7	1	1.7	4.2
195	Low calorie	1 tbs	15	84	16	<1	1	<1	1	.2	.3	.9
	Kraft, Deliciously Right:											
2150	1000 Island	1 tbs	16		35	0	4	0	2	.5		
2153	Bacon & tomato	1 tbs	16		31	1	2	0	3	.5		
2154	Cucumber ranch	1 tbs	16		31	0	1	0	3	.5		
2151	French	1 tbs	16		25	0	3	0	1	.2		
2152	Ranch	1 tbs	16		52	0	3	0	5	.8		
199	Mayo type, regular	1 tbs	15	40	58	<1	4	0	5	.7	1.3	2.7
1030	Low calorie	1 tbs	14	54	36	<1	3	0	3	.4	.7	1.4
	Mayonnaise:											
197	Imitation, low calorie	1 tbs	15	63	35	<1	2	0	3	.5	.7	1.6
196	Regular (soybean)	1 tbs	14	17	100	<1	<1	0	11	1.7	3.2	5.8
1488	Regular, low calorie, low sodium	1 tbs	14	63	32	<1	2	0	3	.5	.6	1.4
1493	Regular, low calorie	1 tbs	15	63	35	<1	2	0	3	.5	.7	1.6
198	Ranch, regular	1 tbs	15	39	80	0	<1	0	8	1.2		
2251	Low calorie	1 tbs	14	69	30	<1	1	0	2	.5		
1685	Russian	1 tbs	15	34	74	<1	2	0	8	1.1	1.8	4.4
1502	Salad dressing, low calorie, oil free	1 tbs	15	88	4	<1	1	<1	<1	0	0	0

Chol (mg)	Calc (mg)	Iron (mg)	Magn (mg)	Pota (mg)	Sodi (mg)	Zinc (mg)	VT-A (RE)	Thia (mg)	VT-E (a-TE)	Ribo (mg)	Niac (mg)	V-B6 (mg)	Fola (µg)	VT-C (mg)
195		0			<1	.23	0	0	2.46	0	0	0	0	0
12		0			<1	.01	0	0	.16	0	0	0	0	0
0	41	0	4	59	2227	0	1853	.01	5.41	.05	.03	.01	2	<1
0	2	0		4	134	0	112	<.01	.33	<.01	<.01	<.01	<1	<1
0	34	.07	3	48	1075	0	911	.01	14.6	.04	.03	.01	1	<1
0	4	.01		6	132	0	112	<.01	1.79	<.01	<.01	<.01	<1	<1
0	1	<.01		2	47	0	40	<.01	.64	<.01	<.01	0	<1	<1
0	60	0	5	86	2447	0	1813	.02	27.2	.07	.04	.02	2	<1
0	4	0		5	151	0	112	<.01	1.68	<.01	<.01	<.01	<1	<1
	0	0			0		51							0
	0	0			115		51							0
0	47	0	4	68	2256	0	1813	.02	11.4	.06	.04	.01	2	<1
0	3	0		4	139	0	112	<.01	.7	<.01	<.01	<.01	<1	<1
0	1	0		1	50	0	40	0	.25	<.01	<.01	0	<1	<1
0	47	0	4	68	2256	0	1813	.02	20.5	.06	.04	.01	2	<1
0	3	0		4	139	0	112	<.01	1.26	<.01	<.01	<.01	<1	<1
0	0	0		0	110		100		1.27					0
0	0	0	0	0	0	0	0	0	45.8	0	0	0	0	0
0	0	0	0	0	0	0	0	0	2.94	0	0	0	0	0
0	0	0	0	0	0	0	0	0	46	0	0	0	0	0
0	0	0	0	0	0	0	0	0	2.95	0	0	0	0	0
0		.82		0	<1	.13	0	0	26.8	0	0	0	0	0
0		.05		0	<1	.01	0	0	1.74	0	0	0	0	0
									1.74					
0		.06			<1	.02	0	0	27.9	0	0	0	0	0
0		<.01			<1	<.01	0	0	1.81	0	0	0	0	0
0	0	0	0	0	0	0	0	0	94	0	0	0	0	0
0	0	0	0	0	0	0	0	0	6.03	0	0	0	0	0
0		.04		0	0	0	0	0	39.7	0	0	0	0	0
0		<.01		0	0	0	0	0	2.55	0	0	0	0	0
0	0	0	0	0	0	0	0	0	61.5	0	0	0	0	0
0	0	0	0	0	0	0	0	0	3.95	0	0	0	0	0
0	0	0	0	0	0	0	0	0	110	0	0	0	0	0
0	0	0	0	0	0	0	0	0	7.08	0	0	0	0	0
3	12	.03	0	6	164	.04	10	<.01	1.4	.01	.01	.01	1	<1
<1	13	.07	1	1	180	.04	<1	<.01	.13	.01	.01	<.01	<1	<1
12	23	.2	3	21	207	.13	7	<.01	.72	.02	.5	.01	2	1
0	2	.06	0	13	219	.01	21	<.01	1.35	<.01	0	<.01	1	0
0	2	.06	0	13	126	.03	21	0	.19	0	0	0	0	0
0	1	.03		2	118	.02	4	<.01	1.56	<.01	0	<.01	1	0
1		.03	0	2	118	.02	0	0	.22	0	0	0	0	0
2	0	0		27	160		0		.19					0
1	0	0		21	155		0		.75					0
0	0	0		10	232		0		.73					0
0	0	0		7	130		50		.42					0
0	0	0		5	165		0		1.31					0
4	2	.03		1	107	.03	13	<.01	.6	<.01	<.01	<.01	1	0
4	2	.03		1	99	.02	9	<.01	.6	<.01	0	<.01	1	0
4		0		1	75	.02	0	0	.96	0	0	0	0	0
8	3	.07		5	79	.02	12	0	1.65	0	<.01	.08	1	0
3	0	0	0	1	15	.01	1	0	.53	<.01	0	0	<1	0
4		0		1	75	.02	0	0	.96	0	0	0	0	0
5	0	0			105		0							0
5	5	0			120		0		.7					0
3	3	.09		24	130	.06	31	.01	1.53	.01	.09	<.01	2	1
0	1	.04	2	7	256	<.01	<1	0	0	0	<.01	<.01	<1	<1

Table A-1

Food Composition (Computer code number is for West Diet Analysis program) (For purposes of calculations, use "0" for t, <1, <.1, <.01, etc.)

Computer Code Number	Food Description	Measure	Wt (g)	H_2O (%)	Ener (kcal)	Prot (g)	Carb (g)	Dietary Fiber (g)	Fat (g)	Fat Breakdown (g) Sat	Mono	Poly
	FATS and OILS—Continued											
	Salad dressing, no cholesterol											
1605	Miracle Whip	1 tbs	15	57	48	0	2	0	4	1.1	1.1	2.1
203	Salad dressing, from recipe, cooked	1 tbs	16	69	25	1	2	0	2	.5	.6	.3
200	Tartar sauce, regular	1 tbs	14	34	74	<1	1	<1	8	1.5	2.6	4.1
1503	Low calorie	1 tbs	14	63	31	<1	2	<1	2	.4	.6	1.3
201	Thousand island, regular	1 tbs	16	46	60	<1	2	0	6	1	1.3	3.2
202	Low calorie	1 tbs	15	69	24	<1	2	<1	2	.2	.4	.9
204	Vinegar and oil	1 tbs	16	47	72	0	<1	0	8	1.5	2.4	3.9
	Wishbone:											
2180	Creamy Italian, lite	1 tbs	15		26	<1	2		2	.4	.9	.7
2166	Italian, lite	1 tbs	16	90	6	0	1		<1	0	.2	.1
8427	Ranch, lite	1 tbs	15	56	50	0	2	0	4	.7		
	FRUITS and FRUIT JUICES											
	Apples:											
	Fresh, raw, with peel:											
205	2¾" diam (about 3 per lb w/cores)	1 ea	138	84	81	<1	21	4	<1	.1	t	.1
206	3¼" diam (about 2 per lb w/cores)	1 ea	212	84	125	<1	32	6	1	.1	t	.2
207	Raw, peeled slices	1 c	110	84	63	<1	16	2	<1	.1	t	.1
208	Dried, sulfured	10 ea	64	32	156	1	42	6	<1	t	t	.1
209	Apple juice, bottled or canned	1 c	248	88	117	<1	29	<1	<1	t	t	.1
210	Applesauce, sweetened	1 c	255	80	194	<1	51	3	<1	.1	t	.1
211	Applesauce, unsweetened	1 c	244	88	105	<1	28	3	<1	t	t	t
	Apricots:											
212	Raw, w/o pits (about 12 per lb w/pits)	3 ea	105	86	50	1	12	3	<1	t	.2	.1
	Canned (fruit and liquid):											
213	Heavy syrup	1 c	240	78	199	1	52	4	<1	t	.1	t
214	Halves	3 ea	120	78	100	1	26	2	<1	t	t	t
215	Juice pack	1 c	244	87	117	2	30	4	<1	t	t	t
216	Halves	3 ea	108	87	52	1	13	2	<1	t	t	t
217	Dried, halves	10 ea	35	31	83	1	22	3	<1	t	.1	t
218	Dried, cooked, unsweetened, w/liquid	1 c	250	76	213	3	55	8	<1	t	.2	.1
219	Apricot nectar, canned	1 c	251	85	141	1	36	2	<1	t	.1	t
	Avocados, raw, edible part only:											
220	California	1 ea	173	73	306	4	12	8	30	4.5	19.5	3.5
221	Florida	1 ea	304	80	340	5	27	16	27	5.3	14.8	4.5
222	Mashed, fresh, average	1 c	230	74	370	5	17	11	35	5.6	22.1	4.5
	Bananas, raw, without peel:											
223	Whole, 8¾" long (175g w/peel)	1 ea	118	74	109	1	28	3	1	.2	t	.1
224	Slices	1 c	150	74	138	2	35	4	1	.3	.1	.1
1285	Bananas, dehydrated slices	½ c	50	3	173	2	44	4	1	.3	.1	.2
225	Blackberries, raw	1 c	144	86	75	1	18	8	1	t	.1	.3
	Blueberries:											
226	Fresh	1 c	145	85	81	1	20	4	1	t	.1	.2
227	Frozen, sweetened	10 oz	284	77	230	1	62	6	<1	t	.1	.2
228	Frozen, thawed	1 c	230	77	186	1	51	5	<1	t	t	.1
	Cherries:											
229	Sour, red pitted, canned water pack	1 c	244	90	88	2	22	3	<1	.1	.1	.1
230	Sweet, red pitted, raw	10 ea	68	81	49	1	11	2	1	.1	.2	.2
231	Cranberry juice cocktail,vitamin C added	1 c	253	85	144	0	36	<1	<1	t	t	.1
1411	Cranberry juice, low calorie	1 c	237	95	45	0	11	<1	0	0	0	0
232	Cranberry-apple juice,vitamin C added	1 c	245	83	164	<1	42	<1	0	0	0	0
233	Cranberry sauce, canned, strained	1 c	277	61	418	1	108	3	<1	t	.1	.2
234	Dates, whole, without pits	10 ea	83	22	228	2	61	6	<1	.2	.1	t
235	Dates, chopped	1 c	178	22	490	4	131	13	1	.3	.3	.1
236	Figs, dried	10 ea	190	28	485	6	124	18	2	.4	.5	1.1
	Fruit cocktail, canned, fruit and liq:											
237	Heavy syrup pack	1 c	248	80	181	1	47	2	<1	t	t	.1
238	Juice pack	1 c	237	87	109	1	28	2	<1	t	t	t
	Grapefruit:											
	Raw 3¾" diam (half w/rind = 241g)											
239	Pink/red, half fruit, edible part	1 ea	123	91	37	1	9	2	<1	t	t	t
240	White, half fruit, edible part	1 ea	118	90	39	1	10	1	<1	t	t	t
241	Canned sections with light syrup	1 c	254	84	152	1	39	1	<1	t	t	.1

PAGE KEY: A–2 = Beverages A–4 = Dairy A–8 = Eggs A–8 = Fat/Oil A–12 = Fruit A–18 = Bakery A–26 = Grain A–30 = Fish A–32 = Meats A–36 = Poultry A–38 = Sausage A–40 = Mixed/Fast A–44 = Nuts/Seeds A–48 = Sweets A–50 = Vegetables/Legumes A–62 = Misc A–64 = Soups/Sauces A–66 = Fast A–82 = Frozen Entree A–86 = Baby foods

Chol (mg)	Calc (mg)	Iron (mg)	Magn (mg)	Pota (mg)	Sodi (mg)	Zinc (mg)	VT-A (RE)	Thia (mg)	VT-E (a-TE)	Ribo (mg)	Niac (mg)	V-B6 (mg)	Fola (μg)	VT-C (mg)
0	0	<.01	0	0	102	0	2	0	.64	0	0	0	0	0
9	13	.08	1	19	117	.06	20	.01	.3	.02	.04	<.01	1	<1
7	3	.13		11	99	.02	9	<.01	2.24	<.01	0	<.01	1	<1
3	2	.09		5	83	.02	2	0	.83	<.01	.01	<.01	<1	<1
4	2	.1		18	112	.02	15	<.01	.18	<.01	<.01	<.01	1	0
2	2	.09		17	150	.02	14	<.01	.18	<.01	0	<.01	1	0
0	0	0	0	1	<1	0	0	0	1.41	0	0	0	0	0
<1	0	0			148		0	0	.56	0	0			0
0	1	0			255		2	0	.24	0	0			<1
2	0	0			120		0							0
0	10	.25	7	159	0	.05	7	.02	.44	.02	.11	.07	4	8
0	15	.38	11	244	0	.08	11	.04	.68	.03	.16	.1	6	12
0	4	.08	3	124	0	.04	4	.02	.09	.01	.1	.05	<1	4
0	9	.9	10	288	56	.13	4	0	.35	.1	.59	.08	0	2
0	17	.92	7	295	7	.07	<1	.05	.02	.04	.25	.07	<1	2
0	10	.89	8	156	8	.1	3	.03	.03	.07	.48	.07	2	4
0	7	.29	7	183	5	.07	7	.03	.02	.06	.46	.06	1	3
0	15	.57	8	311	1	.27	274	.03	.93	.04	.63	.06	9	10
0	22	.72	17	336	10	.26	295	.05	2.14	.05	.9	.13	4	7
0	11	.36	8	168	5	.13	148	.02	1.07	.03	.45	.06	2	4
0	29	.73	24	403	10	.27	412	.04	2.17	.05	.84	.13	4	12
0	13	.32	11	178	4	.12	183	.02	.96	.02	.37	.06	2	5
0	16	1.65	16	482	3	.26	253	<.01	.52	.05	1.05	.05	4	1
0	40	4.18	42	1222	7	.65	590	.01	1.25	.07	2.36	.28	0	4
0	18	.95	13	286	8	.23	331	.02	.2	.03	.65	.05	3	2
0	19	2.04	71	1096	21	.73	106	.19	2.32	.21	3.32	.48	113	14
0	33	1.61	103	1483	15	1.28	185	.33	2.37	.37	5.84	.85	162	24
0	25	2.35	90	1377	23	.97	140	.25	3.08	.28	4.42	.64	142	18
0	7	.37	34	467	1	.19	9	.05	.32	.12	.64	.68	22	11
0	9	.46	43	594	1	.24	12	.07	.4	.15	.81	.87	29	14
0	11	.57	54	746	1	.3	15	.09	0	.12	1.4	.22	7	3
0	46	.82	29	282	0	.39	23	.04	1.02	.06	.58	.08	49	30
0	9	.25	7	129	9	.16	14	.07	1.45	.07	.52	.05	9	19
0	17	1.11	6	170	3	.17	11	.06	2.02	.15	.72	.17	19	3
0	14	.9	5	138	2	.14	9	.05	1.63	.12	.58	.14	15	2
0	27	3.34	15	239	17	.17	183	.04	.32	.1	.43	.11	19	5
0	10	.26	7	152	0	.04	14	.03	.09	.04	.27	.02	3	5
0	8	.38	5	45	5	.18	1	.02	0	.02	.09	.05	<1	90
0	21	.09	5	52	7	.05	1	.02	0	.02	.08	.04	<1	76
0	17	.15	5	66	5	.1	1	.01	0	.05	.15	.05	<1	78
0	11	.61	8	72	80	.14	6	.04	.28	.06	.28	.04	3	6
0	27	.95	29	541	2	.24	4	.07	.08	.08	1.83	.16	10	0
0	57	2.05	62	1160	5	.52	9	.16	.18	.18	3.92	.34	22	0
0	274	4.24	112	1352	21	.97	25	.13	9.5	.17	1.32	.43	14	2
0	15	.72	12	218	15	.2	50	.04	.72	.05	.93	.12	6	5
0	19	.5	17	225	9	.21	73	.03	.47	.04	.95	.12	6	6
0	13	.15	10	159	0	.09	32	.04	.31	.02	.23	.05	15	47
0	14	.07	11	175	0	.08	1	.04	.29	.02	.32	.05	12	39
0	36	1.02	25	328	5	.2	0	.1	.63	.05	.62	.05	22	54

Table A-1

Food Composition (Computer code number is for West Diet Analysis program) (For purposes of calculations, use "0" for t, <1, <.1, <.01, etc.)

Computer Code Number	Food Description	Measure	Wt (g)	H_2O (%)	Ener (kcal)	Prot (g)	Carb (g)	Dietary Fiber (g)	Fat (g)	Fat Breakdown (g) Sat	Mono	Poly
	FRUITS and FRUIT JUICES											
	Grapefruit juice:											
242	Fresh, white, raw	1 c	247	90	96	1	23	<1	<1	t	t	.1
243	Canned, unsweetened	1 c	247	90	94	1	22	<1	<1	t	t	.1
244	Sweetened	1 c	250	87	115	1	28	<1	<1	t	t	.1
	Frozen concentrate, unsweetened:											
245	Undiluted, 6-fl-oz can	¾ c	207	62	302	4	72	1	1	.1	.1	.2
246	Diluted with 3 cans water	1 c	247	89	101	1	24	<1	<1	t	t	.1
	Grapes, raw European (adherent skin):											
247	Thompson seedless	10 ea	50	81	35	<1	9	<1	<1	.1	t	.1
248	Tokay/Emperor, seeded types	10 ea	50	81	35	<1	9	<1	<1	.1	t	.1
	Grape juice:											
249	Bottled or canned	1 c	253	84	154	1	38	<1	<1	.1	t	.1
	Frozen concentrate, sweetened:											
250	Undiluted, 6-fl-oz can, vit C added	¾ c	216	54	387	1	96	1	1	.2	t	.2
251	Diluted with 3 cans water, vit C added	1 c	250	87	128	<1	32	<1	<1	.1	t	.1
1410	Low calorie	1 c	253	84	154	1	38	<1	<1	.1	t	.1
252	Kiwi fruit, raw, peeled (88 g with peel)	1 ea	76	83	46	1	11	3	<1	t	t	.2
253	Lemons, raw, without peel and seeds (about 4 per lb whole)	1 ea	58	89	17	1	5	2	<1	t	t	.1
	Lemon juice:											
254	Fresh:	1 c	244	91	61	1	21	1	0	0	0	0
255	Tablespoon	1 tbs	15	91	4	<1	1	<1	0	0	0	0
256	Canned or bottled, unsweetened:	1 c	244	92	51	1	16	1	1	.1	t	.2
257	Tablespoon	1 tbs	15	92	3	<1	1	<1	<1	t	t	t
258	Frozen, single strength, unsweetened:	1 c	244	92	54	1	16	1	1	.1	t	.2
2298	Tablespoon	1 tbs	15	92	3	<1	1	<1	<1	t	t	t
	Lime juice:											
260	Fresh:	1 c	246	90	66	1	22	1	<1	t	t	.1
261	Tablespoon	1 tbs	15	90	4	<1	1	<1	<1	t	t	t
262	Canned or bottled, unsweetened	1 c	246	92	52	1	16	1	1	.1	.1	.2
263	Mangos, raw, edible part (300 g w/skin & seeds)	1 ea	207	82	135	1	35	4	1	.1	.2	.1
	Melons, raw, without rind and contents:											
264	Cantaloupe, 5″ diam (2⅓ lb whole with refuse), orange flesh	½ ea	276	90	97	2	23	2	1	.2	t	.3
265	Honeydew, 6½″ diam (5¼ lb whole with refuse), slice = ⅒ melon	1 pce	160	90	56	1	15	1	<1	t	t	.1
266	Nectarines, raw, w/o pits, 2¼″ diam	1 ea	136	86	67	1	16	2	1	.1	.2	.3
	Oranges, raw:											
267	Whole w/o peel and seeds, 2⅝″ diam (180 g with peel and seeds)	1 ea	131	87	62	1	15	3	<1	t	t	t
268	Sections, without membranes	1 c	180	87	85	2	21	4	<1	t	t	t
	Orange juice:											
269	Fresh, all varieties	1 c	248	88	112	2	26	<1	<1	.1	.1	.1
270	Canned, unsweetened	1 c	249	89	105	1	24	<1	<1	t	.1	.1
271	Chilled	1 c	249	88	110	2	25	<1	1	.1	.1	.2
	Frozen concentrate:											
272	Undiluted (6-oz can)	¾ c	213	58	339	5	81	2	<1	.1	.1	.1
273	Diluted w/3 parts water by volume	1 c	249	88	112	2	27	<1	<1	t	t	t
1345	Orange juice, from dry crystals	1 c	248	88	114	0	29	0	0	0	0	0
274	Orange and grapefruit juice, canned	1 c	247	89	106	1	25	<1	<1	t	t	t
	Papayas, raw:											
275	½″ slices	1 c	140	89	55	1	14	3	<1	.1	.1	t
276	Whole, 3½″ diam by 5⅛″ w/o seeds and skin (1 lb w/refuse)	1 ea	304	89	119	2	30	5	<1	.1	.1	.1
1031	Papaya nectar, canned	1 c	250	85	143	<1	36	1	<1	.1	.1	.1
	Peaches:											
277	Raw, whole, 2½″ diam, peeled, pitted (about 4 per lb whole)	1 ea	98	88	42	1	11	2	<1	t	t	t
278	Raw, sliced	1 c	170	88	73	1	19	3	<1	t	.1	.1
	Canned, fruit and liquid:											
279	Heavy syrup pack:	1 c	262	79	194	1	52	3	<1	t	.1	.1
280	Half	1 ea	98	79	72	<1	19	1	<1	t	t	t

Chol (mg)	Calc (mg)	Iron (mg)	Magn (mg)	Pota (mg)	Sodi (mg)	Zinc (mg)	VT-A (RE)	Thia (mg)	VT-E (a-TE)	Ribo (mg)	Niac (mg)	V-B6 (mg)	Fola (µg)	VT-C (mg)
0	22	.49	30	400	2	.12	2	.1	.12	.05	.49	.11	25	94
0	17	.49	25	378	2	.22	2	.1	.12	.05	.57	.05	26	72
0	20	.9	25	405	5	.15	0	.1	.12	.06	.8	.05	26	67
0	56	1.01	79	1001	6	.37	6	.3	.37	.16	1.6	.32	26	248
0	20	.35	27	336	2	.12	2	.1	.12	.05	.54	.11	9	83
0	5	.13	3	92	1	.02	3	.05	.35	.03	.15	.05	2	5
0	5	.13	3	92	1	.02	3	.05	.35	.03	.15	.05	2	5
0	23	.61	25	334	8	.13	3	.07	0	.09	.66	.16	7	<1
0	28	.78	32	160	15	.28	6	.11	.38	.2	.93	.32	9	179
0	10	.25	10	52	5	.1	2	.04	.12	.06	.31	.1	3	60
0	23	.61	25	334	8	.13	3	.07	0	.09	.66	.16	7	<1
0	20	.31	23	252	4	.13	14	.01	.85	.04	.38	.07	29	74
0	15	.35	5	80	1	.03	2	.02	.14	.01	.06	.05	6	31
0	17	.07	15	303	2	.12	5	.07	.22	.02	.24	.12	31	112
0	1	<.01	1	19	<1	.01	<1	<.01	.01	<.01	.01	.01	2	7
0	27	.32	19	249	51	.15	5	.1	.22	.02	.48	.1	25	60
0	2	.02	1	15	3	.01	<1	.01	.01	<.01	.03	.01	2	4
0	19	.29	19	217	2	.12	2	.14	.22	.03	.33	.15	23	77
0	1	.02	1	13	<1	.01	<1	.01	.01	<.01	.02	.01	1	5
0	22	.07	15	268	2	.15	2	.05	.22	.02	.25	.11	20	72
0	1	<.01	1	16	<1	.01	<1	<.01	.01	<.01	.01	.01	1	4
0	29	.57	17	185	39	.15	5	.08	.17	.01	.4	.07	19	16
0	21	.27	19	323	4	.08	805	.12	2.32	.12	1.21	.28	29	57
0	30	.58	30	853	25	.44	889	.1	.41	.06	1.58	.32	47	116
0	10	.11	11	434	16	.11	6	.12	.24	.03	.96	.09	10	40
0	7	.2	11	288	0	.12	101	.02	1.21	.06	1.35	.03	5	7
0	52	.13	13	237	0	.09	27	.11	.31	.05	.37	.08	40	70
0	72	.18	18	326	0	.13	38	.16	.43	.07	.51	.11	54	96
0	27	.5	27	496	2	.12	50	.22	.22	.07	.99	.1	75	124
0	20	1.1	27	436	5	.17	45	.15	.22	.07	.78	.22	45	86
0	25	.42	27	473	2	.1	20	.28	.47	.05	.7	.13	45	82
0	68	.75	72	1435	6	.38	60	.6	.68	.14	1.53	.33	330	294
0	22	.25	25	473	2	.12	20	.2	.47	.04	.5	.11	109	97
0	62	.2	2	50	12	.1	551	<.01	0	.04	0	0	143	121
0	20	1.14	25	390	7	.17	30	.14	.17	.07	.83	.06	35	72
0	34	.14	14	360	4	.1	39	.04	1.57	.04	.47	.03	53	86
0	73	.3	30	781	9	.21	85	.08	3.4	.1	1.03	.06	116	188
0	25	.85	7	77	12	.37	27	.01	.05	.01	.37	.02	5	7
0	5	.11	7	193	0	.14	53	.02	.69	.04	.97	.02	3	6
0	8	.19	12	335	0	.24	92	.03	1.19	.07	1.68	.03	6	11
0	8	.71	13	241	16	.24	86	.03	2.33	.06	1.61	.05	8	7
0	3	.26	5	90	6	.09	32	.01	.87	.02	.6	.02	3	3

Table A-1

Food Composition (Computer code number is for West Diet Analysis program) (For purposes of calculations, use "0" for t, <1, <.1, <.01, etc.)

Computer Code Number	Food Description	Measure	Wt (g)	H_2O (%)	Ener (kcal)	Prot (g)	Carb (g)	Dietary Fiber (g)	Fat (g)	Fat Breakdown (g) Sat	Mono	Poly
	FRUITS and FRUIT JUICES—Continued											
281	Juice pack:	1 c	248	87	109	2	29	3	<1	t	t	t
282	Half	1 ea	98	87	43	1	11	1	<1	t	t	t
283	Dried, uncooked	10 ea	130	32	311	5	80	11	1	.1	.4	.5
284	Dried, cooked, fruit and liquid	1 c	258	78	199	3	51	7	1	.1	.2	.3
	Frozen, slice, sweetened:											
285	10-oz package,vitamin C added	1 ea	284	75	267	2	68	5	<1	t	.1	.2
286	Cup, thawed measure, vitamin C added	1 c	250	75	235	2	60	4	<1	t	.1	.2
1032	Peach nectar, canned	1 c	249	86	134	1	35	1	<1	t	t	t
	Pears:											
	Fresh, with skin, cored:											
287	Bartlett, 2½" diam (about 2½ per lb)	1 ea	166	84	98	1	25	4	1	t	.1	.2
288	Bosc, 2⅛" diam (about 3 per lb)	1 ea	139	84	82	1	21	3	1	t	.1	.1
289	D'Anjou, 3" diam (about 2 per lb)	1 ea	209	84	123	1	32	5	1	t	.2	.2
	Canned, fruit and liquid:											
290	Heavy syrup pack:	1 c	266	80	197	1	51	4	<1	t	.1	.1
291	Half	1 ea	76	80	56	<1	15	1	<1	t	t	t
292	Juice pack:	1 c	248	86	124	1	32	4	<1	t	t	t
293	Half	1 ea	76	86	38	<1	10	1	<1	t	t	t
294	Dried halves	10 ea	175	27	459	3	122	13	1	.1	.2	.3
1033	Pear nectar, canned	1 c	250	84	150	<1	39	1	<1	t	t	t
	Pineapple:											
295	Fresh chunks, diced	1 c	155	86	76	1	19	2	1	t	.1	.2
	Canned, fruit and liquid:											
	Heavy syrup pack:											
296	Crushed, chunks, tidbits	½ c	127	79	99	<1	26	1	<1	t	t	.1
297	Slices	1 ea	49	79	38	<1	10	<1	<1	t	t	t
298	Juice pack, crushed, chunks, tidbits	1 c	250	83	150	1	39	2	<1	t	t	.1
299	Juice pack, slices	1 ea	47	83	28	<1	7	<1	<1	t	t	t
300	Pineapple juice, canned, unsweetened	1 c	250	85	140	1	34	<1	<1	t	t	.1
	Plantains, yellow flesh, without peel:											
301	Raw slices (whole=179 g w/o peel)	1 c	148	65	181	2	47	3	1	.2	t	.1
302	Cooked, boiled, sliced	1 c	154	67	179	1	48	4	<1	.1	t	.1
	Plums:											
303	Fresh, medium, 2⅛" diam	1 ea	66	85	36	1	9	1	<1	t	.3	.1
304	Fresh, small, 1½" diam	1 ea	28	85	15	<1	4	<1	<1	t	.1	t
	Canned, purple, with liquid:											
305	Heavy syrup pack:	1 c	258	76	230	1	60	3	<1	t	.2	.1
306	Plums	3 ea	138	76	123	<1	32	1	<1	t	.1	t
307	Juice pack:	1 c	252	84	146	1	38	3	<1	t	t	t
308	Plums	3 ea	138	84	80	1	21	1	<1	t	t	t
1698	Pomegranate, fresh	1 ea	154	81	105	1	26	1	<1	.1	.1	.1
	Prunes, dried, pitted:											
309	Uncooked (10 = 97 g w/pits, 84 g w/o pits)	10 ea	84	32	201	2	53	6	<1	t	.3	.1
310	Cooked, unsweetened, fruit & liq (250 g w/pits)	1 c	248	70	265	3	70	16	1	t	.4	.1
311	Prune juice, bottled or canned	1 c	256	81	182	2	45	3	<1	t	.1	t
	Raisins, seedless:											
312	Cup, not pressed down	1 c	145	15	435	5	115	6	1	.2	t	.2
313	One packet, ½ oz	½ oz	14	15	42	<1	11	1	<1	t	t	t
	Raspberries:											
314	Fresh	1 c	123	87	60	1	14	8	1	t	.1	.4
315	Frozen, sweetened	10 oz	284	73	293	2	74	12	<1	t	t	.3
316	Cup, thawed measure	1 c	250	73	258	2	65	11	<1	t	t	.2
317	Rhubarb, cooked, added sugar	1 c	240	68	278	1	75	5	<1	t	t	.1
	Strawberries:											
318	Fresh, whole, capped	1 c	144	92	43	1	10	3	1	t	.1	.3
	Frozen, sliced, sweetened:											
319	10-oz container	10 oz	284	73	273	2	74	5	<1	t	.1	.2
320	Cup, thawed measure	1 c	255	73	245	1	66	5	<1	t	t	.2
	Tangerines, without peel and seeds:											
321	Fresh (2⅜" whole) 116 g w/refuse	1 ea	84	88	37	1	9	2	<1	t	t	t
322	Canned, light syrup, fruit and liquid	1 c	252	83	154	1	41	2	<1	t	t	t

PAGE KEY: A–2 = Beverages A–4 = Dairy A–8 = Eggs A–8 = Fat/Oil A–12 = Fruit A–18 = Bakery A–26 = Grain A–30 = Fish A–32 = Meats A–36 = Poultry A–38 = Sausage A–40 = Mixed/Fast A–44 = Nuts/Seeds A–48 = Sweets A–50 = Vegetables/Legumes A–62 = Misc A–64 = Soups/Sauces A–66 = Fast A–82 = Frozen Entree A–86 = Baby foods

Chol (mg)	Calc (mg)	Iron (mg)	Magn (mg)	Pota (mg)	Sodi (mg)	Zinc (mg)	VT-A (RE)	Thia (mg)	VT-E (a-TE)	Ribo (mg)	Niac (mg)	V-B6 (mg)	Fola (μg)	VT-C (mg)
0	15	.67	17	317	10	.27	94	.02	3.72	.04	1.44	.05	8	9
0	6	.26	7	125	4	.11	37	.01	1.47	.02	.57	.02	3	4
0	36	5.28	55	1294	9	.74	281	<.01	0	.28	5.69	.09	<1	6
0	23	3.38	33	826	5	.46	52	.01	0	.05	3.92	.1	<1	10
0	9	1.05	14	369	17	.14	79	.04	2.53	.1	1.85	.05	9	268
0	7	.92	12	325	15	.12	70	.03	2.23	.09	1.63	.04	8	236
0	12	.47	10	100	17	.2	65	.01	.2	.03	.72	.02	3	13
0	18	.41	10	208	0	.2	3	.03	.83	.07	.17	.03	12	7
0	15	.35	8	174	0	.17	3	.03	.69	.06	.14	.02	10	6
0	23	.52	12	261	0	.25	4	.04	1.05	.08	.21	.04	15	8
0	13	.58	11	173	13	.21	0	.03	1.33	.06	.64	.04	3	3
0	4	.17	3	49	4	.06	0	.01	.38	.02	.18	.01	1	1
0	22	.72	17	238	10	.22	2	.03	1.24	.03	.5	.03	3	4
0	7	.22	5	73	3	.07	1	.01	.38	.01	.15	.01	1	1
0	59	3.68	58	933	10	.68	1	.01	0	.25	2.4	.13	0	12
0	12	.65	7	32	10	.17	<1	<.01	.25	.03	.32	.03	3	3
0	11	.57	22	175	2	.12	3	.14	.15	.06	.65	.13	16	24
0	18	.48	20	132	1	.15	1	.11	.13	.03	.36	.09	6	9
0	7	.19	8	51	<1	.06	<1	.04	.05	.01	.14	.04	2	4
0	35	.7	35	305	2	.25	10	.24	.25	.05	.71	.18	12	24
0	7	.13	7	57	<1	.05	2	.04	.05	.01	.13	.03	2	4
0	42	.65	32	335	2	.27	1	.14	.05	.05	.64	.24	58	27
0	4	.89	55	739	6	.21	167	.08	.4	.08	1.02	.44	33	27
0	3	.89	49	716	8	.2	140	.07	.22	.08	1.16	.37	40	17
0	3	.07	5	114	0	.07	21	.03	.4	.06	.33	.05	1	6
0	1	.03	2	48	0	.03	9	.01	.17	.03	.14	.02	1	3
0	23	2.17	13	235	49	.18	67	.04	1.81	.1	.75	.07	6	1
0	12	1.16	7	126	26	.1	36	.02	.97	.05	.4	.04	3	1
0	25	.86	20	388	3	.28	255	.06	1.76	.15	1.19	.07	7	7
0	14	.47	11	213	1	.15	139	.03	.97	.08	.65	.04	4	4
0	5	.46	5	399	5	.18	0	.05	.85	.05	.46	.16	9	9
0	43	2.08	38	626	3	.44	167	.07	1.22	.14	1.65	.22	3	3
0	57	2.75	50	828	5	.59	77	.06	<.01	.25	1.79	.54	<1	7
0	31	3.02	36	707	10	.54	1	.04	.03	.18	2.01	.56	1	10
0	71	3.02	48	1088	17	.39	1	.23	1.02	.13	1.19	.36	5	5
0	7	.29	5	105	2	.04	<1	.02	.1	.01	.11	.03	<1	<1
0	27	.7	22	187	0	.57	16	.04	.55	.11	1.11	.07	32	31
0	43	1.85	37	324	3	.51	17	.05	1.28	.13	.65	.1	74	47
0	37	1.63	32	285	2	.45	15	.05	1.13	.11	.57	.08	65	41
0	348	.5	29	230	2	.19	17	.04	.48	.05	.48	.05	13	8
0	20	.55	14	239	1	.19	4	.03	.2	.09	.33	.08	25	82
0	31	1.68	20	278	9	.17	6	.04	.4	.14	1.14	.08	42	118
0	28	1.5	18	250	8	.15	5	.04	.36	.13	1.02	.08	38	106
0	12	.08	10	132	1	.2	77	.09	.2	.02	.13	.06	17	26
0	18	.93	20	197	15	.6	212	.13	.86	.11	1.12	.11	12	50

Table A-1

Food Composition (Computer code number is for West Diet Analysis program) (For purposes of calculations, use "0" for t, <1, <.1, <.01, etc.)

Computer Code Number	Food Description	Measure	Wt (g)	H_2O (%)	Ener (kcal)	Prot (g)	Carb (g)	Dietary Fiber (g)	Fat (g)	Fat Breakdown (g) Sat	Mono	Poly
	FRUITS and FRUIT JUICES—Continued											
323	Tangerine juice, canned, sweetened	1 c	249	87	125	1	30	<1	<1	t	t	.1
	Watermelon, raw, without rind and seeds:											
324	Piece, 1⁄16 wedge	1 pce	286	91	91	2	20	1	1	.1	.3	.4
325	Diced	1 c	152	91	49	1	11	1	1	.1	.2	.2
	BAKED GOODS: BREADS, CAKES, COOKIES, CRACKERS, PIES											
326	Bagels, plain, enriched, 3½″ diam.	1 ea	71	33	195	7	38	2	1	.2	.1	.5
1663	Bagel, oat bran	1 ea	71	33	181	8	38	3	1	.1	.2	.3
	Biscuits:											
327	From home recipe	1 ea	60	29	212	4	27	1	10	2.6	4.2	2.5
328	From mix	1 ea	57	29	191	4	28	1	7	1.6	2.4	2.5
329	From refrigerated dough	1 ea	74	27	276	4	34	1	13	8.7	3.4	.5
330	Bread crumbs, dry, grated (see # 364, 365 for soft crumbs)	1 c	108	6	427	13	78	3	6	1.4	2.3	1.7
2087	Bread sticks, brown & serve	1 ea	57	34	150	7	28	1	1	.5	.5	.5
	Breads:											
331	Boston brown, canned, 3¼″ slice	1 pce	45	47	88	2	19	2	1	.1	.1	.3
332	Cracked wheat (¼ cracked-wheat & ¾ enr wheat flour): 1-lb loaf	1 ea	454	36	1180	39	225	25	18	4.2	8.6	3.1
333	Slice (18 per loaf)	1 pce	25	36	65	2	12	1	1	.2	.5	.2
334	Slice, toasted	1 pce	23	30	65	2	12	1	1	.2	.5	.2
335	French/Vienna, enriched: 1-lb loaf	1 ea	454	34	1243	40	236	14	14	2.9	5.5	3.1
337	Slice, 4¾ x 4 x ½″	1 pce	25	34	68	2	13	1	1	.2	.3	.2
336	French, slice, 5 x 2½″	1 pce	25	34	68	2	13	1	1	.2	.3	.2
	French toast: see Mixed Dishes, and Fast Foods, #691											
2083	Honey wheatberry	1 pce	38	38	100	3	18	2	1	0	.5	
338	Italian, enriched: 1-lb loaf	1 ea	454	36	1230	40	227	12	16	3.9	3.7	6.3
339	Slice, 4½ x 3¼ x ¾″	1 pce	30	36	81	3	15	1	1	.3	.2	.4
340	Mixed grain, enriched: 1-lb loaf	1 ea	454	38	1135	45	211	29	17	3.7	6.9	4.2
341	Slice (18 per loaf)	1 pce	26	38	65	3	12	2	1	.2	.4	.2
342	Slice, toasted	1 pce	24	32	65	3	12	2	1	.2	.4	.2
343	Oatmeal, enriched: 1-lb loaf	1 ea	454	37	1221	38	220	18	20	3.2	7.2	7.7
344	Slice (18 per loaf)	1 pce	27	37	73	2	13	1	1	.2	.4	.5
345	Slice, toasted	1 pce	25	31	73	2	13	1	1	.2	.4	.5
346	Pita pocket bread, enr, 6½″ round	1 ea	60	32	165	5	33	1	1	.1	.1	.3
	Pumpernickel (⅔ rye & ⅓ enr wheat flr):											
347	1-lb loaf	1 ea	454	38	1135	40	216	29	14	2	4.2	5.6
348	Slice, 5 x 4 x ⅜″	1 pce	26	38	65	2	12	2	1	.1	.2	.3
349	Slice, toasted	1 pce	29	32	80	3	15	2	1	.1	.3	.4
350	Raisin, enriched: 1-lb loaf	1 ea	454	34	1243	36	237	19	20	4.9	10.5	3.1
351	Slice (18 per loaf)	1 pce	26	34	71	2	14	1	1	.3	.6	.2
352	Slice, toasted	1 pce	24	28	71	2	14	1	1	.3	.6	.2
353	Rye, light (⅓ rye & ⅔ enr wheat flr): 1-lb loaf	1 ea	454	37	1175	39	219	26	15	2.9	6	3.6
354	Slice, 4¾ x 3¾ x 7⁄16″	1 pce	32	37	83	3	15	2	1	.2	.4	.3
355	Slice, toasted	1 pce	24	31	68	2	13	2	1	.2	.3	.2
356	Wheat (enr wheat & whole-wheat flour): 1-lb loaf	1 ea	454	37	1160	43	213	25	19	3.9	7.3	4.5
357	Slice (18 per loaf)	1 pce	25	37	65	2	12	1	1	.2	.4	.2
358	Slice, toasted	1 pce	23	32	65	2	12	1	1	.2	.4	.2
359	White, enriched: 1-lb loaf	1 ea	454	35	1293	36	225	9	26	5.4	5.9	12.6
360	Slice	1 pce	42	35	120	3	21	1	2	.5	.5	1.2
361	Slice, toasted	1 pce	38	29	119	3	21	1	2	.5	.5	1.2
366	Whole-wheat: 1-lb loaf	1 ea	454	38	1116	44	209	31	19	4.2	7.6	4.5
367	Slice (16 per loaf)	1 pce	28	38	69	3	13	2	1	.3	.5	.3
368	Slice, toasted	1 pce	25	30	69	3	13	2	1	.3	.5	.3
	Bread stuffing, prepared from mix:											
369	Dry type	1 c	200	65	356	6	43	6	17	3.5	7.6	5.2
370	Moist type, with egg and margarine	1 c	232	65	390	9	51	5	17	3.4	7.4	4.9

Chol (mg)	Calc (mg)	Iron (mg)	Magn (mg)	Pota (mg)	Sodi (mg)	Zinc (mg)	VT-A (RE)	Thia (mg)	VT-E (a-TE)	Ribo (mg)	Niac (mg)	V-B6 (mg)	Fola (μg)	VT-C (mg)
0	45	.5	20	443	2	.07	105	.15	.22	.05	.25	.08	11	55
0	23	.49	31	332	6	.2	106	.23	.43	.06	.57	.41	6	27
0	12	.26	17	176	3	.11	56	.12	.23	.03	.3	.22	3	15
0	52	2.53	21	72	379	.62	0	.38	.02	.22	3.24	.04	62	0
0	9	2.19	40	145	360	1.48	<1	.23	.17	.24	2.1	.14	57	<1
2	141	1.74	11	73	348	.32	14	.21	1.45	.19	1.77	.02	37	<1
2	105	1.17	14	107	544	.35	15	.2	.23	.2	1.72	.04	3	<1
5	89	1.64	9	87	584	.29	24	.27	.44	.18	1.63	.03	6	0
0	245	6.61	50	239	931	1.32	<1	.83	.95	.47	7.4	.11	118	0
0	60	2.7			290		0	.22		.1	1.6			0
<1	31	.94	28	143	284	.22	5	.01	.13	.05	.5	.04	5	0
														0
0	195	12.8	236	804	2442	5.63	0	1.63	2.56	1.09	16.7	1.38	277	
0	11	.7	13	44	135	.31	0	.09	.14	.06	.92	.08	15	0
0	11	.7	13	44	135	.31	0	.07	.14	.05	.83	.07	7	0
0	341	11.5	123	513	2764	3.95	0	2.36	1.07	1.49	21.6	.19	431	0
0	19	.63	7	28	152	.22	0	.13	.06	.08	1.19	.01	24	0
0	19	.63	7	28	152	.22	0	.13	.06	.08	1.19	.01	24	0
0	20	.72			200		0	.12	.24	.07	.8			0
0	354	13.3	123	499	2651	3.9	0	2.15	1.26	1.33	19.9	.22	431	0
0	23	.88	8	33	175	.26	0	.14	.08	.09	1.31	.01	28	0
0	413	15.8	241	926	2210	5.77	0	1.85	2.79	1.55	19.8	1.51	363	1
0	24	.9	14	53	127	.33	0	.11	.16	.09	1.14	.09	21	<1
0	24	.9	14	53	127	.33	0	.08	.16	.08	1.02	.08	16	<1
0	300	12.3	168	645	2719	4.63	9	1.81	1.56	1.09	14.3	.31	281	2
0	18	.73	10	38	162	.27	1	.11	.09	.06	.85	.02	17	<1
0	18	.73	10	38	163	.28	<1	.09	.09	.06	.77	.02	13	<1
0	52	1.57	16	72	322	.5	0	.36	.02	.2	2.78	.02	57	0
0	309	13	245	944	3046	6.72	0	1.48	2.3	1.38	14	.57	363	0
0	18	.75	14	54	174	.38	0	.08	.13	.08	.8	.03	21	0
0	21	.91	17	66	214	.47	0	.08	.17	.09	.89	.04	20	0
0	300	13.2	118	1030	1770	3.27	1	1.54	3.44	1.81	15.8	.31	395	2
0	17	.75	7	59	101	.19	<1	.09	.2	.1	.9	.02	23	<1
0	17	.76	7	59	102	.19	<1	.07	.2	.09	.81	.02	18	<1
0	331	12.8	182	754	2996	5.18	2	1.97	2.51	1.52	17.3	.34	390	1
0	23	.91	13	53	211	.36	<1	.14	.18	.11	1.22	.02	27	<1
0	19	.74	10	44	174	.3	<1	.09	.15	.08	.9	.02	17	<1
0	572	15.8	209	627	2447	4.77	0	2.09	3	1.45	20.5	.49	204	0
0	26	.83	11	50	133	.26	0	.1	.14	.07	1.03	.02	19	0
0	26	.83	11	50	132	.26	0	.08	.14	.06	.93	.02	15	0
14	259	13.5	86	663	1629	2.91	100	1.84	4.95	1.74	16.3	.23	413	1
1	24	1.25	8	61	151	.27	9	.17	.46	.16	1.51	.02	38	<1
1	24	1.24	8	61	150	.27	8	.13	.46	.14	1.35	.02	12	<1
0	327	15	390	1144	2392	8.81	0	1.59	4.72	.93	17.4	.81	227	0
0	20	.92	24	71	148	.54	0	.1	.29	.06	1.08	.05	14	0
0	20	.93	24	71	148	.54	0	.08	.23	.05	.97	.04	9	0
0	64	2.18	24	148	1086	.56	162	.27	2.8	.21	2.96	.08	202	0
0	148	3.8	35	304	1069	.74	160	.39	2.78	.33	3.69	.12	39	4

Table A-1

Food Composition (Computer code number is for West Diet Analysis program) (For purposes of calculations, use "0" for t, <1, <.1, <.01, etc.)

Computer Code Number	Food Description	Measure	Wt (g)	H_2O (%)	Ener (kcal)	Prot (g)	Carb (g)	Dietary Fiber (g)	Fat (g)	Fat Breakdown (g) Sat	Mono	Poly
	BAKED GOODS: BREADS, CAKES, COOKIES, CRACKERS, PIES—Continued											
	Cakes, prepared from mixes using enriched flour and veg shortening, w/frostings made from margarine:											
	Angel food:											
371	Whole cake, 9 ¾" diam tube	1 ea	340	33	877	20	197	5	3	.4	.2	1.2
372	Piece, 1⁄12 of cake	1 pce	28	33	72	2	16	<1	<1	t	t	.1
373	Boston cream pie, ⅛ of cake	1 pce	123	45	310	3	53	2	10	3.1	5.4	1.2
	Coffee cake:											
374	Whole cake, 7¾ x 5⅝ x 1¼"	1 ea	336	30	1068	18	177	4	32	6.3	13	10.7
375	Piece, ⅙ of cake	1 pce	56	30	178	3	30	1	5	1	2.2	1.8
	Devil's food, chocolate frosting:											
376	Whole cake, 2 layer, 8 or 9" diam	1 ea	1021	23	3747	42	557	29	167	47.9	91.9	19.5
377	Piece, 1⁄16 of cake	1 pce	64	23	235	3	35	2	10	3	5.8	1.2
378	Cupcake, 2½" diam	1 ea	42	23	154	2	23	1	7	2	3.8	.8
	Gingerbread:											
379	Whole cake, 8" square	1 ea	603	33	1863	24	306	7	61	15.8	34	8.1
380	Piece, ⅑ of cake	1 pce	67	33	207	3	34	1	7	1.8	3.8	.9
	Yellow, chocolate frosting, 2 layer:											
381	Whole cake, 8 or 9" in diam	1 ea	1024	22	3880	39	567	18	178	49	99	21.4
382	Piece, 1⁄16 of cake	1 pce	64	22	243	2	35	1	11	3.1	6.2	1.3
	Cakes from home recipes w/enr flour:											
	Carrot cake, made with veg oil, cream cheese frosting:											
383	Whole, 9 x 13" cake	1 ea	1776	21	7743	82	838	21	469	86.8	116	242
384	Piece, 1⁄16 of cake, 2¼ x 3¼" slice	1 pce	111	21	484	5	52	1	29	5.4	7.2	15.1
	Fruitcake, dark:											
385	Whole cake, 7½"diam tube, 2¼"high	1 ea	1376	25	4458	40	848	51	125	15.4	57.4	44.6
386	Piece, 1⁄32 of cake, ⅔" arc	1 pce	43	25	139	1	26	2	4	.5	1.8	1.4
	Sheet, plain, made w/veg shortening, no frosting:											
387	Whole cake, 9" square	1 ea	774	24	2817	35	433	3	108	29.9	51.5	25.5
388	Piece, ⅑ of cake	1 pce	86	24	313	4	48	<1	12	3.3	5.7	2.8
	Sheet, plain, made w/margarine, uncooked white frosting:											
389	Whole cake, 9" square	1 ea	576	22	2148	20	339	2	83	13.8	35.5	29.5
390	Piece, ⅑ of cake	1 pce	64	22	239	2	38	<1	9	1.5	3.9	3.3
	Cakes, commerical:											
	Cheesecake:											
401	Whole cake, 9" diam	1 ea	960	46	3081	53	245	4	216	111	74.4	13.2
402	Piece, 1⁄12 of cake	1 pce	80	46	257	4	20	<1	18	9.2	6.2	1.1
	Pound cake:											
393	Loaf, 8½ x 3½ x 3"	1 ea	340	25	1319	19	166	2	68	38.1	19	3.7
394	Slice, 1⁄17 of loaf, 2" slice	1 pce	28	25	109	2	14	<1	6	3.1	1.6	.3
	Snack cakes:											
395	Chocolate w/creme filling,Ding Dong	1 ea	50	20	188	2	30	<1	7	1.6	2.7	2.1
396	Sponge cake w/creme filling,Twinkie	1 ea	43	20	157	1	27	<1	5	1.2	1.9	1.5
1677	Sponge cake, 1⁄12 of 12" cake	1 pce	38	30	110	2	23	<1	1	.3	.4	.2
	White, white frosting, 2 layer:											
397	Whole cake, 8 or 9" diam	1 ea	1136	20	4260	37	716	11	153	68.2	60.1	15.4
398	Piece, 1⁄16 of cake	1 pce	71	20	266	2	45	1	10	4.3	3.8	1
	Yellow, chocolate frosting, 2 layer:											
399	Whole cake, 8 or 9" in diam	1 ea	1024	22	3880	39	567	18	178	49	99	21.4
400	Piece, 1⁄16 of cake	1 pce	64	22	243	2	35	1	11	3.1	6.2	1.3
1332	Bagel chips	5 pce	70	3	298	6	52	6	7	1.2	2	3.4
2225	Bagel chips, onion garlic, toasted	1 oz	28		193	5	31	3	8	1.7	5.2	0
1035	Cheese puffs/Cheetos	1 c	20	1	111	2	11	<1	7	1.3	4.1	1
	Cookies made with enriched flour:											
	Brownies with nuts:											
403	Commercial w/frosting, 1½ x 1¾ x ⅞"	1 ea	61	14	247	3	39	1	10	2.6	5.1	1.6
1902	Fat free fudge, Entenmann's	1 pce	40	24	110	2	27	1	0	0	0	0

Chol (mg)	Calc (mg)	Iron (mg)	Magn (mg)	Pota (mg)	Sodi (mg)	Zinc (mg)	VT-A (RE)	Thia (mg)	VT-E (a-TE)	Ribo (mg)	Niac (mg)	V-B6 (mg)	Fola (µg)	VT-C (mg)
0	476	1.77	41	316	2546	.24	0	.35	.34	1.67	3	.1	119	0
0	39	.15	3	26	210	.02	0	.03	.03	.14	.25	.01	10	0
45	28	.47	7	48	177	.2	28	.5	1.3	.33	.23	.03	18	<1
165	457	4.8	60	376	1414	1.51	134	.56	5.58	.59	5.11	.17	228	1
27	76	.8	10	63	236	.25	22	.09	.93	.1	.85	.03	38	<1
470	439	22.5	347	2042	3410	7.04	286	.28	17.3	1.36	5.89	.32	174	1
29	27	1.41	22	128	214	.44	18	.02	1.08	.08	.37	.02	11	<1
19	18	.92	14	84	140	.29	12	.01	.71	.06	.24	.01	7	<1
211	416	20	96	1453	2761	2.47	96	1.14	8.26	1.12	9.41	.23	60	1
23	46	2.22	11	161	307	.27	11	.13	.92	.12	1.05	.02	7	<1
563	379	21.3	307	1822	3450	6.35	276	1.23	27.6	1.61	12.8	.3	225	1
35	24	1.33	19	114	216	.4	17	.08	1.73	.1	.8	.02	14	<1
959	444	22.2	320	1989	4368	8.7	6819	2.42	74.9	2.77	17.9	1.35	213	19
60	28	1.39	20	124	273	.54	426	.15	4.68	.17	1.12	.08	13	1
69	454	28.5	220	2105	3715	3.72	261	.69	42.9	1.36	10.9	.63	261	5
2	14	.89	7	66	116	.12	8	.02	1.34	.04	.34	.02	8	<1
503	495	11.7	108	611	2322	2.74	372	1.24	11	1.39	10.1	.26	54	2
56	55	1.3	12	68	258	.3	41	.14	1.22	.15	1.12	.03	6	<1
323	357	6.16	35	305	1981	1.44	109	.58	10.9	.4	2.88	.2	156	1
36	40	.68	4	34	220	.16	12	.06	1.22	.04	.32	.02	17	<1
528	490	6.05	106	864	1987	4.9	1545	.27	10.1	1.85	1.87	.5	173	6
44	41	.5	9	72	166	.41	129	.02	.84	.15	.16	.04	14	<1
751	119	4.69	37	405	1353	1.56	530	.47	2.24	.78	4.45	.12	139	<1
62	10	.39	3	33	111	.13	44	.04	.18	.06	.37	.01	11	<1
8	36	1.68	20	61	213	.28	2	.11	1.01	.15	1.22	.01	14	<1
7	19	.55	3	39	157	.13	2	.07	.83	.06	.52	.01	12	<1
39	27	1.03	4	38	93	.19	17	.09	.17	.1	.73	.02	15	0
91	545	9.09	60	659	2658	1.76	368	1.14	20.4	1.48	10.2	.16	64	1
6	34	.57	4	41	166	.11	23	.07	1.28	.09	.64	.01	4	<1
563	379	21.3	307	1822	3450	6.35	276	1.23	27.6	1.61	12.8	.3	225	1
35	24	1.33	19	114	216	.4	17	.08	1.73	.1	.8	.02	14	<1
0	9	1.38	41	167	419	.9	0	.13	.46	.12	1.57	.19	58	0
0	0	2.52			490		0	.39	<.01	.24	3.5			0
1	12	.47	4	33	210	.08	7	.05	1.02	.07	.65	.03	24	<1
10	18	1.37	19	91	190	.44	12	.16	1.3	.13	1.05	.02	13	<1
0	0	1.08		90	140		0		.01					0

Table A-1

Food Composition (Computer code number is for West Diet Analysis program) (For purposes of calculations, use "0" for t, <1, <.1, <.01, etc.)

Computer Code Number	Food Description	Measure	Wt (g)	H_2O (%)	Ener (kcal)	Prot (g)	Carb (g)	Dietary Fiber (g)	Fat (g)	Fat Breakdown (g) Sat	Mono	Poly
	BAKED GOODS: BREADS, CAKES, COOKIES, CRACKERS, PIES—Continued											
	Chocolate chip cookies:											
405	Commercial, 2¼" diam	4 ea	60	12	275	2	35	2	15	4.5	7.8	1.6
406	Home recipe, 2¼" diam	4 ea	64	6	312	4	37	2	18	5.2	6.7	5.4
407	From refrigerated dough, 2¼" diam	4 ea	64	13	284	3	39	1	13	4.5	6.5	1.3
408	Fig bars	4 ea	64	16	223	2	45	3	5	.9	2.6	.8
2052	Fruit bar, no fat	1 ea	28		90	2	21	0	0	0	0	0
2162	Fudge, fat free, Snackwell	1 ea	16	14	53	1	12	<1	<1	.1	.1	t
409	Oatmeal raisin, 2⅝" diam	4 ea	60	6	261	4	41	2	10	1.9	4.1	3
410	Peanut butter, home recipe, 2⅝"diam	4 ea	80	6	380	7	47	2	19	3.5	8.7	5.8
411	Sandwich-type, all	4 ea	40	2	189	2	28	1	8	1.7	4.7	1.1
412	Shortbread, commercial, small	4 ea	32	4	161	2	21	1	8	2	4.3	1
413	Shortbread, home recipe, large	2 ea	22	3	120	1	12	<1	7	4.5	2.1	.3
414	Sugar from refrigerated dough, 2" diam	4 ea	48	5	232	2	31	<1	11	2.8	6.2	1.4
1874	Vanilla sandwich, Snackwell's	2 ea	26	4	109	1	21	1	2	.5	.8	.2
415	Vanilla wafers	10 ea	40	5	176	2	29	1	6	1.4	2.4	1.5
416	Corn chips	1 c	26	1	140	2	15	1	9	1.2	2.5	4.3
	Crackers (enriched):											
417	Cheese	10 ea	10	3	50	1	6	<1	3	.9	.9	.5
418	Cheese with peanut butter	4 ea	28	4	135	4	16	1	6	1.4	3.4	1.2
	Fat free, enriched:											
2161	Cracked pepper, Snackwell	1 ea	15	2	60	2	12	<1	<1	.1	t	.1
2159	Wheat, Snackwell	7 ea	15	1	60	2	12	1	<1	.1	.1	.1
2075	Whole wheat, herb seasoned	5 ea	14	5	50	2	11	2	0	0	0	0
2077	Whole wheat, onion	5 ea	14	4	50	2	11	2	0	0	0	0
419	Graham, enriched	2 ea	14	4	59	1	11	<1	1	.4	.7	.2
420	Melba toast, plain, enriched	1 pce	5	5	19	1	4	<1	<1	t	t	.1
1514	Rice cakes, unsalted, enriched	2 ea	18	6	70	1	15	1	<1	.1	.2	.2
421	Rye wafer, whole grain	2 ea	22	5	73	2	18	5	<1	t	t	.1
422	Saltine-enriched	4 ea	12	4	52	1	9	<1	1	.3	.8	.2
1971	Saltine, unsalted tops, enriched	2 ea	6		25	1	4	0	<1	0	0	0
423	Snack-type, round like Ritz, enriched	3 ea	9	3	45	1	5	<1	2	.4	1	.7
424	Wheat, thin, enriched	4 ea	8	3	38	1	5	<1	2	.7	.8	.2
425	Whole-wheat wafers	2 ea	8	3	35	1	5	1	1	.2	.8	.2
426	Croissants, 4½ x 4 x 1¾"	1 ea	57	23	231	5	26	1	12	6.7	3.2	.7
1699	Croutons, seasoned	½ c	20	4	93	2	13	1	4	1	1.9	.5
	Danish pastry:											
427	Packaged ring, plain, 12 oz	1 ea	340	21	1349	19	181	1	65	13.5	40.9	6.4
428	Round piece, plain, 4¼" diam, 1" high	1 ea	88	21	349	5	47	<1	17	3.5	10.6	1.6
429	Ounce, plain	1 oz	28	21	111	2	15	<1	5	1.1	3.4	.5
430	Round piece with fruit	1 ea	94	29	335	5	45		16	3.3	10.1	1.6
	Desserts, 3 x 3" piece:											
1348	Apple crisp	1 pce	78	61	127	1	25	1	3	.6	1.2	.9
1353	Apple cobbler	1 pce	104	57	199	2	35	2	6	1.2	2.8	2
1349	Cherry crisp	1 pce	138	77	146	2	24	1	5	.9	2.5	1.8
1352	Cherry cobbler	1 pce	129	66	198	2	34	1	6	1.2	2.8	1.9
1350	Peach crisp	1 pce	139	75	155	2	27	2	5	.9	2.5	1.7
1351	Peach cobbler	1 pce	130	64	204	2	36	2	6	1.2	2.8	1.9
	Doughnuts:											
431	Cake type, plain, 3¼" diam	1 ea	47	21	198	2	23	1	11	1.8	4.5	3.8
432	Yeast-leavened, glazed, 3¾" diam	1 ea	60	25	242	4	27	1	14	3.5	7.8	1.7
	English muffins:											
433	Plain, enriched	1 ea	57	42	134	4	26	2	1	.2	.2	.5
434	Toasted	1 ea	52	37	133	4	26	2	1	.1	.2	.5
1504	Whole wheat	1 ea	66	46	134	6	27	4	1	.2	.3	.6
1414	Granola bar, soft	1 ea	28	6	124	2	19	1	5	2	1.1	1.5
1415	Granola bar, hard	1 ea	25	4	118	3	16	1	5	.6	1.1	3
1985	Granola bar, fat free, all flavors	1 ea	42	10	140	2	35	3	0	0	0	0
	Muffins, 2½" diam, 1½" high:											
	From home recipe:											
435	Blueberry	1 ea	57	39	165	4	23	1	6	1.4	1.6	3.1
436	Bran, wheat	1 ea	57	35	164	4	24	4	7	1.5	1.8	3.6
437	Cornmeal	1 ea	57	32	183	4	25	2	7	1.6	1.8	3.5

Chol (mg)	Calc (mg)	Iron (mg)	Magn (mg)	Pota (mg)	Sodi (mg)	Zinc (mg)	VT-A (RE)	Thia (mg)	VT-E (a-TE)	Ribo (mg)	Niac (mg)	V-B6 (mg)	Fola (μg)	VT-C (mg)
0	9	1.45	21	56	196	.28	1	.07	1.74	.12	.97	.1	23	0
20	25	1.57	35	143	231	.59	105	.12	1.86	.11	.87	.05	21	<1
15	16	1.44	15	115	134	.32	11	.12	1.31	.12	1.27	.03	36	0
0	41	1.86	17	132	224	.25	3	.1	.45	.14	1.2	.05	17	<1
0	0	.36			95		0		.01					0
0	3	.29	5	26	71	.08	<1	.02	<.01	.02	.26	<.01		0
20	60	1.59	25	143	323	.52	98	.15	1.5	.1	.76	.04	18	<1
25	31	1.78	31	185	414	.66	125	.18	3.04	.17	2.81	.07	44	<1
0	10	1.55	18	70	242	.32	<1	.03	1.21	.07	.83	.01	17	0
6	11	.88	5	32	146	.17	4	.11	.98	.1	1.07	.01	19	0
20	4	.58	3	15	102	.09	67	.08	.18	.06	.64	<.01	2	0
15	43	.88	4	78	225	.13	5	.09	1.54	.06	1.16	.01	25	0
<1	17	.61	5	28	95	.16	<1	.05		.07	.69	.01		0
23	19	.95	6	39	125	.14	7	.11	.54	.13	1.24	.03	20	0
0	33	.34	20	37	164	.33	2	.01	.35	.04	.31	.06	5	0
1	15	.48	4	14	99	.11	3	.06	.1	.04	.47	.05	8	0
1	22	.82	16	69	278	.3	10	.11	1.24	.1	1.83	.42	25	0
<1	26	.73	4	19	148	.14	<1	.05		.06	.78	.01		<1
<1	28	.58	7	43	169	.21	<1	.04		.07	.73	.02		0
0	0				80		100							2
0	0	0			80		100							2
0	3	.52	4	19	85	.11	0	.03	.27	.04	.58	.01	8	0
0	5	.18	3	10	41	.1	0	.02	.01	.01	.21	<.01	6	0
0	2	.27	24	52	5	.54	1	.01	.02	.03	1.41	.03	4	0
0	9	1.31	27	109	175	.62	<1	.09	.44	.06	.35	.06	3	<1
0	14	.65	3	15	156	.09	0	.07	.2	.05	.63	<.01	15	0
0		.36		5	50				.1					
0	11	.32	2	12	76	.06	0	.04	.4	.03	.36	<.01	7	0
2	3	.28	5	16	70	.13	<1	.04	.02	.03	.34	.01	1	0
0	4	.25	8	24	53	.17	0	.02	.31	.01	.36	.01	3	0
43	21	1.16	9	67	424	.43	78	.22	.24	.14	1.25	.03	35	<1
1	19	.56	8	36	248	.19	1	.1	.32	.08	.93	.02	18	0
105	143	6.94	54	371	1261	1.87	20	.99	3.06	.75	8.5	.2	211	10
27	37	1.8	14	96	326	.48	5	.25	.79	.19	2.2	.05	55	3
9	12	.57	4	30	104	.15	2	.08	.25	.06	.7	.02	17	1
19	22	1.4	14	110	333	.48	24	.29	.85	.21	1.8	.06	31	2
0	22	.58	5	76	142	.12	24	.07		.06	.6	.03	4	2
1	21	.79	6	106	288	.16	76	.1	1.11	.09	.74	.04	3	<1
0	26	2.14	11	154	74	.15	150	.06	.93	.08	.6	.06	11	3
1	28	1.81	9	133	294	.2	135	.1	1.01	.11	.85	.05	9	2
0	20	.89	12	189	70	.19	108	.06	1.13	.05	1.06	.03	6	5
1	24	.91	10	159	291	.23	105	.09	1.16	.09	1.19	.03	6	3
17	21	.92	9	60	257	.26	8	.1	1.63	.11	.87	.03	22	<1
4	26	1.22	13	65	205	.46	6	.22	1.75	.13	1.71	.03	26	0
0	99	1.43	12	75	264	.4	0	.25	.07	.16	2.21	.02	46	<1
0	98	1.41	11	74	262	.39	0	.2	.07	.14	1.98	.02	38	<1
0	175	1.62	47	139	420	1.06	0	.2	.46	.09	2.25	.11	28	0
<1	29	.72	21	91	78	.42	0	.08	.34	.05	.14	.03	7	0
0	15	.74	24	84	73	.51	4	.07	.33	.03	.39	.02	6	<1
0	0	3.6			5		100							0
22	107	1.29	9	69	251	.31	16	.15	1.03	.16	1.26	.02	7	1
20	106	2.39	44	181	335	1.57	136	.19	1.31	.25	2.29	.18	30	4
26	147	1.49	13	82	333	.35	23	.17	1.08	.18	1.36	.05	10	<1

Table A-1

Food Composition (Computer code number is for West Diet Analysis program) (For purposes of calculations, use "0" for t, <1, <.1, <.01, etc.)

Computer Code Number	Food Description	Measure	Wt (g)	H_2O (%)	Ener (kcal)	Prot (g)	Carb (g)	Dietary Fiber (g)	Fat (g)	Fat Breakdown (g) Sat	Mono	Poly
	BAKED GOODS: BREADS, CAKES, COOKIES, CRACKERS, PIES—Continued											
	From commercial mix:											
438	Blueberry	1 ea	50	36	150	3	24	1	4	.7	1.8	1.5
439	Bran, wheat	1 ea	50	35	138	3	23	2	5	1.2	2.3	.7
440	Cornmeal	1 ea	50	30	161	4	25	1	5	1.4	2.6	.6
1864	Nabisco Newtons, fat free, all flavors	1 ea	23		69	1	16		0	0	0	0
	Pancakes, 4″ diam:											
441	Buckwheat, from mix w/ egg and milk	1 ea	30	54	62	2	8	1	2	.6	.6	.8
442	Plain, from home recipe	1 ea	38	53	86	2	11	1	4	.8	.9	1.7
443	Plain, from mix; egg, milk, oil added	1 ea	38	53	74	2	14	<1	1	.2	.3	.3
1468	Pan dulce, sweet roll w/topping	1 ea	79	21	291	5	48	1	9	2	3.9	2.7
	Piecrust,with enriched flour, vegetable shortening, baked:											
444	Home recipe, 9″ shell	1 ea	180	10	949	12	85	3	62	15.5	27.4	16.4
	From mix:											
445	Piecrust for 2-crust pie	1 ea	320	10	1686	21	152	5	111	27.6	48.6	29.2
446	1 pie shell	1 ea	160	11	802	11	81	3	49	12.3	27.7	6.2
	Pies, 9″ diam; pie crust made with vegetable shortening, enriched flour:											
447	Apple: Whole pie	1 ea	1000	52	2370	19	340	16	110	21.1	59.4	20.9
448	Piece, ⅙ of pie	1 pce	167	52	396	3	57	3	18	3.5	9.9	3.5
449	Banana cream: Whole pie	1 ea	1152	48	3098	51	379	8	157	43.3	65.9	38
450	Piece, ⅙ of pie	1 pce	192	48	516	8	63	1	26	7.2	11	6.3
451	Blueberry: Whole pie	1 ea	1176	51	2881	32	394	16	140	34.3	60.2	36.2
452	Piece, ⅙ of pie	1 pce	196	51	480	5	66	3	23	5.7	10	6
453	Cherry: Whole pie	1 ea	1140	46	3078	32	439	17	139	34.1	60.5	37.1
454	Piece, ⅙ of pie	1 pce	240	46	648	7	92	4	29	7.2	12.7	7.8
455	Chocolate cream: Whole pie	1 ea	1194	63	2150	49	281	12	97	35.5	38.2	18.6
456	Piece, ⅙ of pie	1 pce	199	63	358	8	47	2	16	5.9	6.4	3.1
457	Custard: Whole pie	1 ea	630	61	1323	35	131	10	73	17.5	36.3	12.1
458	Piece, ⅙ of pie	1 pce	105	61	221	6	22	2	12	2.9	6	2
459	Lemon meringue: Whole pie	1 ea	678	42	1817	10	320	8	59	10.6	24.6	19.6
460	Piece, ⅙ of pie	1 pce	113	42	303	2	53	1	10	1.8	4.1	3.3
461	Peach: Whole pie	1 ea	1111	45	2994	26	443	16	130	31.1	55.7	37.4
462	Piece, ⅙ of pie	1 pce	139	45	375	3	55	2	16	3.9	7	4.7
463	Pecan: Whole pie	1 ea	678	19	2712	27	388	24	125	25.5	73.2	20.1
464	Piece, ⅙ of pie	1 pce	113	19	452	5	65	4	21	4.2	12.2	3.4
465	Pumpkin: Whole pie	1 ea	654	58	1373	25	179	18	62	13.2	32.8	10.5
466	Piece, ⅙ of pie	1 pce	109	58	229	4	30	3	10	2.2	5.5	1.7
467	Pies, fried, commercial: Apple	1 ea	85	40	266	2	33	1	14	6.5	5.8	1.2
468	Pies, fried, commercial: Cherry	1 ea	128	38	404	4	54	3	21	3.1	9.5	6.9
	Pretzels, made with enriched flour:											
469	Thin sticks, 2¼″ long	1 oz	28	3	107	3	22	1	1	.2	.4	.3
470	Dutch twists	10 pce	60	3	229	5	47	2	2	.4	.8	.7
471	Thin twists, 3¼ x 2¼ x ¼″	10 pce	60	3	229	5	47	2	2	.4	.8	.7
	Rolls & buns, enriched, commercial:											
472	Cloverleaf rolls, 2½″ diam, 2″ high	1 ea	28	32	84	2	14	1	2	.5	1	.3
473	Hot dog buns	1 ea	43	34	123	4	22	1	2	.5	1.1	.4
474	Hamburger buns	1 ea	43	34	123	4	22	1	2	.5	1.1	.4
475	Hard roll, white, 3¾″ diam, 2″ high	1 ea	57	31	167	6	30	1	2	.3	.6	1
476	Submarine rolls/hoagies, 11¼ x 3 x 2½″	1 ea	135	31	392	12	75	4	4	.9	1.3	1.4
	Rolls & buns, enriched, home recipe:											
477	Dinner rolls 2½″ diam, 2″ high	1 ea	35	29	112	3	19	1	3	.7	1.1	.7
	Sports/fitness bar:											
2043	Forza energy bar	1 ea	70	18	231	10	45	4	1			
2042	Power bar	1 ea	65		230	10	45	3	2			
2041	Tiger sports bar	1 ea	65	17	229	11	40	4	2			
478	Toaster pastries, fortified (Poptarts)	1 ea	52	12	204	2	37	1	5	.8	2.1	2
2132	Toaster strudel pastry—cream cheese	1 ea	54	32	188	3	24	<1	9	2.7		
2134	Toaster strudel pastry—french toast	1 ea	54	32	188	3	24	<1	9	2.9		

PAGE KEY: A–2 = Beverages A–4 = Dairy A–8 = Eggs A–8 = Fat/Oil A–12 = Fruit A–18 = Bakery A–26 = Grain A–30 = Fish A–32 = Meats A–36 = Poultry A–38 = Sausage A–40 = Mixed/Fast A–44 = Nuts/Seeds A–48 = Sweets A–50 = Vegetables/Legumes A–62 = Misc A–64 = Soups/Sauces A–66 = Fast A–82 = Frozen Entree A–86 = Baby foods

Chol (mg)	Calc (mg)	Iron (mg)	Magn (mg)	Pota (mg)	Sodi (mg)	Zinc (mg)	VT-A (RE)	Thia (mg)	VT-E (a-TE)	Ribo (mg)	Niac (mg)	V-B6 (mg)	Fola (µg)	VT-C (mg)
23	12	.56	5	39	219	.19	11	.07	.7	.16	1.12	.04	5	<1
34	16	1.27	28	73	234	.57	15	.1	.75	.12	1.44	.09	8	0
31	37	.97	10	65	398	.32	22	.12	.75	.14	1.05	.05	5	<1
					77									
20	77	.56	17	70	160	.35	20	.05	.62	.08	.4	.04	5	<1
22	83	.68	6	50	167	.21	20	.08	.36	.11	.6	.02	14	<1
5	48	.59	8	66	239	.15	3	.08	.32	.08	.65	.03	3	<1
26	13	1.82	10	57	140	.35	87	.23	1.35	.21	1.98	.04	22	<1
0	18	5.2	25	121	976	.79	0	.7	9.94	.5	5.96	.04	121	0
0	32	9.25	45	214	1734	1.41	0	1.25	17.7	.89	10.6	.08	214	0
0	96	3.44	24	99	1166	.62	0	.48	8.83	.3	3.79	.09	19	0
0	110	4.5	70	650	2660	1.6	300	.28	16.5	.27	2.63	.38	220	32
0	18	.75	12	109	444	.27	50	.05	2.76	.04	.44	.06	37	5
588	864	12	184	1900	2764	5.53	806	1.6	16.9	2.38	12.1	1.53	311	18
98	144	2	31	317	461	.92	134	.27	2.82	.4	2.02	.25	52	3
0	82	14.5	94	588	2175	2.35	47	1.8	24.7	1.55	14	.4	270	8
0	14	2.41	16	98	363	.39	8	.3	4.12	.26	2.33	.07	45	1
0	114	21.1	103	878	2177	2.28	547	1.69	21.7	1.43	14.6	.39	308	11
0	24	4.44	22	185	458	.48	115	.35	4.56	.3	3.07	.08	65	2
109	1028	8.84	170	1705	2085	4.93	235	1.03	11.4	2.43	7.34	.37	59	6
18	171	1.47	28	284	348	.82	39	.17	1.9	.41	1.22	.06	10	1
208	504	3.65	69	668	1512	3.28	315	.25	7.5	1.31	1.84	.3	126	2
35	84	.61	12	111	252	.55	52	.04	1.25	.22	.31	.05	21	<1
305	380	4.14	102	603	990	3.32	353	.42	9.7	1.42	4.4	.2	88	22
51	63	.69	17	101	165	.55	59	.07	1.62	.24	.73	.03	15	4
0	59	12.1	79	1047	2025	1.88	386	1.41	26.2	1.19	15.3	.2	58	589
0	7	1.52	10	131	253	.23	48	.18	3.28	.15	1.91	.02	7	74
217	115	7.05	122	502	2874	3.86	319	.62	17.2	.83	1.69	.14	183	7
36	19	1.18	20	84	479	.64	53	.1	2.86	.14	.28	.02	30	1
131	392	5.17	98	1007	1844	2.94	3139	.36	10.5	1	1.22	.37	131	10
22	65	.86	16	168	307	.49	523	.06	1.75	.17	.2	.06	22	2
13	13	.88	8	51	325	.17	33	.1	.37	.08	.98	.03	4	1
0	28	1.56	13	83	479	.29	22	.18	.55	.14	1.83	.04	23	2
0	10	1.21	10	41	480	.24	0	.13	.06	.17	1.47	.03	48	0
0	22	2.59	21	88	1029	.51	0	.28	.13	.37	3.15	.07	103	0
0	22	2.59	21	88	1029	.51	0	.28	.13	.37	3.15	.07	103	0
<1	33	.88	6	37	146	.22	0	.14	.22	.09	1.13	.01	27	<1
0	60	1.36	9	61	241	.27	0	.21	.2	.13	1.69	.02	41	0
0	60	1.36	9	61	241	.27	0	.21	.2	.13	1.69	.02	41	0
0	54	1.87	15	62	310	.54	0	.27	.1	.19	2.42	.03	54	0
0	122	3.78	27	122	783	.85	0	.54	.1	.33	4.47	.05	40	0
13	21	1.04	7	53	145	.24	28	.14	.35	.14	1.21	.02	15	<1
0	300	6.3	160	220	65	5.25		1.5	20	1.7	20	2	400	60
0	300	5.4	140	150	110	5.25		1.5		1.7	20	2	400	60
	349	4.49	140	279	100		50	1.5	19.9	1.69	19.9	1.99	399	60
0	13	1.81	9	58	218	.34	149	.15	.97	.19	2.05	.2	34	<1
12	12	.97			217		17		1					0
12	12	.97			217		17		1					0

Table A-1

Food Composition (Computer code number is for West Diet Analysis program) (For purposes of calculations, use "0" for t, <1, <.1, <.01, etc.)

Computer Code Number	Food Description	Measure	Wt (g)	H_2O (%)	Ener (kcal)	Prot (g)	Carb (g)	Dietary Fiber (g)	Fat (g)	Fat Breakdown (g) Sat	Mono	Poly
	BAKED GOODS: BREADS, CAKES, COOKIES, CRACKERS, PIES—Continued											
	Tortilla chips:											
1271	Plain	10 pce	18	2	90	1	11	1	5	.9	2.8	.7
1036	Nacho flavor	1 c	26	2	129	2	16	1	7	1.3	3.9	.9
1037	Taco flavor	1 pce	18	2	86	1	11	1	4	.8	2.6	.6
	Tortillas:											
479	Corn, enriched, 6″ diam	1 ea	26	44	58	2	12	1	1	.1	.2	.3
480	Flour, 8″ diam	1 ea	49	27	159	4	27	2	3	.6	1.4	1.4
1301	Flour, 10″ diam	1 ea	72	27	234	6	40	2	5	.8	2.1	2
481	Taco shells	1 ea	14	4	63	1	9	1	3	.4	1.5	.6
	Waffles, 7″ diam:											
482	From home recipe	1 ea	75	42	218	6	25	1	11	2.2	2.6	5.1
483	From mix, egg/milk added	1 ea	75	42	218	5	26	1	10	1.7	2.7	5.2
1510	Whole grain, prepared from frozen	1 ea	39	43	107	4	13	1	5	1.6	1.9	.9
	GRAIN PRODUCTS: CEREAL, FLOUR, GRAIN, PASTA and NOODLES, POPCORN											
484	Barley, pearled, dry, uncooked	1 c	200	10	704	20	155	31	2	.5	.3	1.1
485	Barley, pearled, cooked	1 c	157	69	193	4	44	6	1	.1	.1	.3
2009	Breakfast bars, fat free, all flavors	1 ea	38	25	110	2	26	3	0	0	0	0
	Breakfast bar, Snackwell:											
2165	Apple-cinnamon	1 ea	37	16	119	1	29	1	<1	.1	t	.1
2164	Blueberry	1 ea	37	16	121	1	29	1	<1	t	t	.1
2163	Strawberry	1 ea	37	16	120	1	29	1	<1	t	t	.1
	Breakfast cereals, hot, cooked w/o salt added:											
	Corn grits (hominy) enriched:											
486	Regular/quick prep w/o salt, yellow:	1 c	242	85	145	3	31	<1	<1	.1	.1	.2
487	Instant, prepared from packet, white	1 ea	137	82	89	2	21	1	<1	t	t	.1
	Cream of wheat:											
488	Regular, quick, instant	1 c	239	87	129	4	27	1	<1	.1	.1	.3
489	Mix and eat, plain, packet	1 ea	142	82	102	3	21	<1	<1	t	t	.2
1664	Farina cereal, cooked w/o salt	1 c	233	88	117	3	25	3	<1	t	t	.1
490	Malt-O-Meal, cooked w/o salt	1 c	240	88	122	4	26	1	<1	.1	.1	t
494	Maypo	1 c	216	83	153	5	29	5	2	.4	.7	.8
	Oatmeal or rolled oats:											
491	Regular, quick, instant,nonfortified cooked w/o salt	1 c	234	85	145	6	25	4	2	.4	.7	.9
	Instant, fortified:											
492	Plain, from packet	½ c	118	85	70	4	12	2	1	.2	.4	.4
493	Flavored, from packet	½ c	109	76	106	3	21	2	1	.2	.5	.5
	Breakfast cereals, ready to eat:											
495	All-Bran	1 c	62	3	160	8	46	20	2	.4	.4	1.3
1306	Alpha Bits	1 c	28	1	110	2	24	1	1	.1	.2	.2
1307	Apple Jacks	1 c	33	3	120	2	30	1	<1	.1	.1	.2
1308	Bran Buds	1 c	90	3	240	8	72	36	2	.4	.4	1.4
1305	Bran Chex	1 c	49	2	156	5	39	8	1	.2	.3	.7
1309	Honey BucWheat Crisp	1 c	38	5	147	4	31	3	1	.2	.3	.6
1310	C.W. Post, plain	1 c	97	2	421	9	73	7	13	1.7	6	4.7
1311	C.W. Post, with raisins	1 c	103	4	446	9	74	14	15	11	1.7	1.4
496	Cap'n Crunch	1 c	37	2	147	2	32	1	2	.5	.4	.3
1312	Cap'n Crunchberries	1 c	35	2	140	2	30	1	2	.5	.3	.3
1313	Cap'n Crunch, peanut butter	1 c	35	2	146	3	28	1	3	.7	1.1	.7
497	Cheerios	1 c	23	3	84	2	17	2	1	.3	.5	.2
1314	Cocoa Krispies	1 c	41	2	159	3	36	1	1	.7	.2	.2
1316	Cocoa Pebbles	1 c	32	2	131	1	27	1	2	1.1	.4	.1
1315	Corn Bran	1 c	36	3	120	2	30	6	1	.3	.3	.4
1317	Corn Chex	1 c	28	2	110	2	25	<1	<1	t	t	t
498	Corn Flakes, Kellogg's	1 c	28	3	100	2	24	1	<1	.1	t	.1
499	Corn Flakes, Post Toasties	1 c	24	3	93	2	21	1	<1	t	t	t
1340	Corn Pops	1 c	31	3	120	1	28	<1	<1	.1	.1	t
1318	Cracklin' Oat Bran	1 c	65	4	252	5	48	8	8	3.4	3.8	.9
1038	Crispy Wheat 'N Raisins	1 c	43	7	150	3	35	3	1	.1	.1	.2

Chol (mg)	Calc (mg)	Iron (mg)	Magn (mg)	Pota (mg)	Sodi (mg)	Zinc (mg)	VT-A (RE)	Thia (mg)	VT-E (a-TE)	Ribo (mg)	Niac (mg)	V-B6 (mg)	Fola (μg)	VT-C (mg)
0	28	.27	16	35	95	.27	4	.01	.24	.03	.23	.05	2	0
1	38	.37	21	56	184	.31	11	.03	.35	.05	.37	.07	4	<1
1	28	.36	16	39	142	.23	16	.04	.24	.04	.36	.05	4	<1
0	45	.36	17	40	42	.24	6	.03	.04	.02	.39	.06	30	0
0	61	1.62	13	64	234	.35	0	.26	.62	.14	1.75	.02	60	0
0	90	2.38	19	94	344	.51	0	.38	.91	.21	2.57	.04	89	0
0	35	.36	15	34	25	.19	6	.04	.59	.02	.24	.04	4	0
52	191	1.73	14	119	383	.51	49	.2	1.73	.26	1.55	.04	34	<1
38	93	1.22	15	134	458	.35	19	.15	1.5	.19	1.23	.07	9	<1
39	84	.69	15	91	150	.45	25	.08	.53	.13	.75	.04	7	<1
0	58	5	158	560	18	4.26	4	.38	.26	.23	9.2	.52	46	0
0	17	2.09	34	146	5	1.29	2	.13	.08	.1	3.23	.18	25	0
0	20	.72			25		20							1
<1	17	5	6	68	103	3.88	260	.39		.44	5.2	.52		<1
<1	14	4.83	5	43	107	3.85	260	.39		.44	5.2	.52		<1
<1	14	4.82	6	47	102	3.83	260	.39		.44	5.2	.52		2
0	0	1.55	10	53	0	.17	14	.24	.12	.14	1.96	.06	75	0
0	8	8.19	11	38	289	.21	0	.15	.03	.08	1.38	.05	47	0
0	50	10.3	12	45	139	.33	0	.24	.03	0	1.43	.03	108	0
0	20	8.09	7	38	241	.24	125	.43	.02	.28	4.97	.57	101	0
0	5	1.17	5	30	0	.16	0	.19	.03	.12	1.28	.02	54	0
0	5	9.6	5	31	2	.17	0	.48	.03	.24	5.76	.02	5	0
0	112	7.56	45	190	233	1.34	633	.65	1.51	.65	8.42	.86	9	26
0	19	1.59	56	131	2	1.15	5	.26	.23	.05	.3	.05	9	0
0	109	4.2	28	66	190	.58	302	.35	.14	.19	3.65	.49	100	0
0	112	4.45	34	91	169	.66	306	.35	.14	.25	3.92	.51	100	<1
0	200	9	280	620	560	7.5	450	.75	1.14	.85	10	1	186	30
0	8	2.66	16	54	178	1.48	371	.36	.02	.42	4.93	.5	99	0
0	0	4.5	8	35	150	3.75	225	.38	.05	.43	5	.5	116	15
0	60	13.5	240	809	599	11.3	676	1.17	1.42	1.26	15	1.53	270	45
0	29	14	69	216	345	6.47	11	.64	.56	.26	8.62	.88	173	26
0	54	10.9	43	142	361	.68	913	.9	8.99	1.03	12.1	1.88	11	36
<1	47	15.4	67	198	167	1.64	1284	1.26	.68	1.46	17.1	1.75	342	0
<1	50	16.4	74	261	161	1.64	1363	1.34	.72	1.55	18.1	1.85	364	0
0	7	6.18	13	47	286	5.14	5	.51	.18	.58	6.85	.68	137	0
0	9	6.06	13	49	256	5.39	6	.5	.25	.57	6.72	.67	135	<1
0	3	5.85	24	80	264	4.87	5	.49	.19	.55	6.48	.65	130	0
0	42	6.21	25	68	218	2.88	288	.29	.16	.33	3.84	.38	77	11
0	0	2.38	11	79	278	1.97	298	.49	.19	.57	6.6	.66	123	20
0	5	2.02	13	53	180	1.7	424	.42	.04	.48	5.63	.58	113	0
0	27	10.1	19	75	338	5	5	.1	.19	.56	6.66	.67	134	0
0	3	8.01	4	23	306	.1	14	.36	.07	.07	4.93	.5	99	15
0	0	8.68	3	25	300	.17	225	.36	.03	.43	5	.5	99	15
0	1	.63	4	28	252	.07	318	.31	.06	.36	4.22	.43	85	0
0	0	1.86	2	25	120	1.55	225	.4	.03	.43	5.18	.5	109	15
0	26	2.41	79	305	226	1.95	299	.5	.43	.56	6.63	.66	181	20
0	54	3.52	33	180	223	.85	293	.29	.45	.33	3.91	.39	78	0

Table A-1

Food Composition (Computer code number is for West Diet Analysis program) (For purposes of calculations, use "0" for t, <1, <.1, <.01, etc.)

Computer Code Number	Food Description	Measure	Wt (g)	H_2O (%)	Ener (kcal)	Prot (g)	Carb (g)	Dietary Fiber (g)	Fat (g)	Fat Breakdown (g) Sat	Mono	Poly
	GRAIN PRODUCTS: CEREAL, FLOUR, GRAIN, PASTA and NOODLES, POPCORN—Continued											
1319	Fortified Oat Flakes	1 c	48	3	180	8	36	1	1	.2	.3	.4
500	40% Bran Flakes, Kellogg's	1 c	39	4	121	4	32	7	1	.2	.2	.5
501	40% Bran Flakes, Post	1 c	47	3	152	5	37	9	1	.1	.1	.4
502	Froot Loops	1 c	32	2	120	2	28	1	1	.4	.2	.3
518	Frosted Flakes	1 c	41	3	159	2	37	1	<1	.1	t	.1
1320	Frosted Mini-Wheats	1 c	51	5	170	5	41	5	1	.2	.1	.6
1321	Frosted Rice Krispies	1 c	35	2	135	2	32	<1	<1	.1	.1	.1
1324	Fruit & Fibre w/dates	1 c	57	9	193	5	43	8	3	.4	1.3	1
1322	Fruity Pebbles	1 c	32	3	130	1	28	<1	2	1.4	.1	.1
503	Golden Grahams	1 c	39	3	150	2	33	1	1	.2	.4	.2
504	Granola, homemade	½ c	61	5	285	9	32	6	15	2.9	4.8	6.5
505	Granola, low fat	½ c	47	3	181	5	38	3	3	0		
1670	Granola, low fat, commercial	½ c	45	5	165	4	35	2	2	.6	.7	.9
505	Grape Nuts	½ c	55	3	196	7	45	5	<1	t	t	.1
1326	Grape Nuts Flakes	1 c	39	3	144	4	32	4	1	.6	.1	.2
1665	Heartland Natural with raisins	1 c	110	5	468	11	76	6	16	4	4.2	6.2
1327	Honey & Nut Corn Flakes	1 c	37	2	148	3	31	1	2	.3	.7	.6
506	Honey Nut Cheerios	1 c	33	2	126	3	27	2	1	.3	.5	.2
1328	HoneyBran	1 c	35	2	119	3	29	4	1	.3	.1	.3
1329	HoneyComb	1 c	22	1	86	1	20	1	<1	.2	.1	.1
1330	King Vitaman	1 c	21	2	81	2	18	1	1	.2	.3	.2
1039	Kix	1 c	19	2	72	1	16	1	<1	.1	.1	t
1331	Life	1 c	44	4	167	4	35	3	2	.3	.6	.8
507	Lucky Charms	1 c	32	2	124	2	27	1	1	.2	.4	.2
1323	Mueslix Five Grain	1 c	82	5	279	7	63	7	3	.5	1	1.2
508	Nature Valley Granola	1 c	113	4	510	12	74	7	20	2.6	13.3	3.8
1666	Nutri Grain Almond Raisin	1 c	40	6	147	3	31	3	2	.1	1	1.2
1336	100% Bran	1 c	66	3	178	8	48	19	3	.6	.6	1.9
509	100% Natural cereal, plain	1 c	104	2	462	11	71	8	17	7.5	7.5	2.3
1337	100% Natural with apples & cinnamon	1 c	104	2	477	11	70	7	20	15.5	1.8	1.3
1338	100% Natural with raisins & dates	1 c	110	3	496	12	72	7	20	13.6	3.7	1.7
510	Product 19	1 c	33	4	110	2	28	1	<1	t	.2	.2
1339	Quisp	1 c	30	3	121	2	25	1	2	.5	.4	.2
511	Raisin Bran, Kellogg's	1 c	61	9	200	6	47	8	1	.1	.1	.4
512	Raisin Bran, Post	1 c	56	9	172	5	42	8	1	.2	.1	.5
1667	Raisin Squares	1 c	71	9	241	6	55	7	2	.2	.2	.6
1041	Rice Chex	1 c	33	3	130	2	29	1	<1	t	t	t
513	Rice Krispies, Kellogg's	1 c	28	2	111	2	25	<1	<1	t	t	t
514	Rice, puffed	1 c	14	4	54	1	12	<1	<1	t	t	t
515	Shredded Wheat	1 c	43	5	154	5	35	4	1	.1	.1	.4
516	Special K	1 c	31	3	110	6	22	1	<1	t	t	.2
517	Super Golden Crisp	1 c	33	1	123	2	30	<1	<1	.1	.1	.1
519	Honey Smacks	1 c	36	3	133	3	32	1	1	.4	.1	.3
1341	Tasteeos	1 c	24	2	94	3	19	3	1	.2	.2	.2
1342	Team	1 c	42	4	164	3	36	1	1	.1	.2	.3
520	Total, wheat, with added calcium	1 c	40	3	140	4	32	4	1	.2	.2	.1
521	Trix	1 c	28	2	114	1	24	1	2	.4	.9	.3
1344	Wheat Chex	1 c	46	2	169	5	38	4	1	.2	.1	.5
1043	Wheat cereal, puffed, fortified	1 c	12	4	44	2	9	1	<1	t	t	.1
522	Wheaties	1 c	29	3	106	3	23	2	1	.2	.2	.1
523	Buckwheat flour, dark	1 c	120	11	402	15	85	12	4	.8	1.1	1.1
525	Buckwheat, whole grain, dry	1 c	170	10	583	23	122	17	6	1.3	1.8	1.8
526	Bulgar, dry, uncooked	1 c	140	9	479	17	106	26	2	.3	.2	.8
527	Bulgar, cooked	1 c	182	78	151	6	34	8	<1	.1	.1	.2
	Cornmeal:											
528	Whole-ground, unbolted, dry	1 c	122	10	442	10	94	9	4	.6	1.2	2
530	Degermed, enriched, dry	1 c	138	12	505	12	107	10	2	.3	.6	1
38041	Degermed, enriched, baked	1 c	138	12	505	12	107	10	2	.3	.6	1
	Macaroni, cooked:											
532	Enriched	1 c	140	66	197	7	40	2	1	.1	.1	.4
533	Whole wheat	1 c	140	67	174	7	37	4	1	.1	.1	.3

Chol (mg)	Calc (mg)	Iron (mg)	Magn (mg)	Pota (mg)	Sodi (mg)	Zinc (mg)	VT-A (RE)	Thia (mg)	VT-E (a-TE)	Ribo (mg)	Niac (mg)	V-B6 (mg)	Fola (μg)	VT-C (mg)
0	68	13.7	58	228	220	2.54	636	.62	.34	.72	8.45	.86	169	0
0	0	10.9	81	229	309	5.03	505	.51	7.22	.58	6.71	.66	138	20
0	21	13.4	102	251	431	2.49	622	.61	.54	.7	8.27	.85	166	0
0	0	4.51	8	35	150	3.75	225	.37	.12	.43	5	.5	96	15
0	0	6.15	4	26	264	.2	298	.5	.05	.56	6.61	.66	123	20
0	0	15	60	170	0	1.5	0	.37	.46	.42	4.64	.5	102	0
0	0	2.42	8	27	256	.42	303	.49	.03	.56	6.72	.66	140	20
0	30	10.1	81	335	270	3.02	725	.75	1.32	.85	10.1	1	201	0
0	4	2.02	9	24	178	1.7	424	.42	.03	.48	5.63	.58	113	0
0	19	5.85	12	69	357	4.88	293	.49	.29	.55	6.51	.65	130	19
0	49	2.56	109	328	15	2.48	2	.45	7.87	.17	1.25	.19	52	1
0		2.71	36	143	90	5.64	226	.56	7.57	.64	7.52	.75	151	
0	15	1.35	30	127	101	2.84	169	.27	4.03	.31	3.74	.36	90	0
0	5	15.7	37	184	382	1.21	728	.71	.14	.82	9.68	.99	194	0
0	16	11.2	43	136	220	.78	516	.51	.1	.58	6.86	.7	138	0
0	66	4.02	141	415	226	2.83	7	.32	.77	.14	1.54	.2	44	1
0	0	3.03	3	40	249	.2	152	.26	.09	.3	3.37	.33	74	10
0	22	4.95	32	94	285	4.13	248	.41	.34	.47	5.51	.55	110	16
0	16	5.57	46	151	202	.9	463	.45	.81	.52	6.16	.63	23	19
0	4	2.09	7	25	124	1.17	291	.29	.09	.33	3.87	.4	78	0
0	3	5.92	18	58	176	2.65	212	.26	1.42	.3	3.53	.35	71	8
0	28	5.13	6	26	167	2.38	238	.24	.05	.27	3.17	.32	63	9
0	134	12.3	43	109	240	5.5	2	.55	.22	.62	7.35	.73	147	0
0	35	4.8	21	58	217	4	240	.4	.14	.45	5.34	.53	107	16
0	38	8.94	82	369	107	7.46	747	.75	8.94	.84	9.84	.99	197	1
0	85	3.53	107	375	183	2.27	0	.35	7.97	.12	1.25	.16	17	0
0	122	1.14	13	147	139	3.06	0	.32	4.38	.35	4.08	.41	80	0
0	46	8.12	312	652	457	5.74	0	1.58	1.53	1.78	20.9	2.11	47	63
1	100	3.11	109	457	28	2.5	1	.36	1.19	.17	1.84	.19	26	<1
0	157	2.89	72	514	52	2	6	.33	.73	.57	1.87	.11	17	1
0	160	3.12	124	538	47	2.11	7	.31	.77	.65	2.09	.16	45	0
0	0	19.8	18	55	308	16.5	248	1.65	24.4	1.88	22	2.21	429	66
0	6	5.1	15	40	216	4.26	4	.42	.15	.48	5.67	.56	113	0
0	40	4.5	80	350	390	3.75	225	.37	.56	.43	5	.5	122	0
0	26	8.9	95	345	365	2.97	741	.73	1.3	.84	9.86	1.01	198	0
0	0	21.7	54	335	4	1.99	0	.5	.38	.57	6.67	.64	142	0
0	5	9.44	8	38	276	.45	2	.43	.04	.01	5.81	.59	116	17
0	5	.7	12	27	206	.46	371	.52	.03	.59	6.92	.69	138	15
0	1	.41	4	16	1	.15	0	.06	.01	.01	.87	0	1	0
0	16	1.81	57	155	4	1.42	0	.11	.23	.12	2.26	.11	21	0
0	0	8.4	16	55	250	3.75	225	.53	.08	.59	7.01	.71	93	15
0	7	2.08	20	48	51	1.75	437	.43	.12	.49	5.81	.59	116	0
0	0	2.4	21	53	67	.4	300	.5	.18	.58	6.66	.68	133	20
0	11	6.86	26	71	183	.69	318	.31	.17	.36	4.22	.43	85	13
0	6	12	12	71	260	.58	556	.55	.1	.63	7.39	.76	7	22
0	344	24	43	129	265	20	500	2	31.3	2.27	26.8	2.67	533	80
0	30	4.2	3	16	184	3.5	210	.35	.56	.4	4.68	.47	93	14
0	18	13.2	58	173	308	1.23	0	.6	.17	.17	8.1	.83	162	24
0	3	.56	16	44	1	.37	<1	.05	.08	.03	1.43	.02	4	0
0	53	7.83	31	101	215	.68	218	.36	.36	.41	4.84	.48	97	14
0	49	4.87	301	692	13	3.74	0	.5	1.24	.23	7.38	.7	65	0
0	31	3.74	393	782	2	4.08	0	.17	1.75	.72	11.9	.36	51	0
0	49	3.44	230	574	24	2.7	0	.32	.22	.16	7.15	.48	38	0
0	18	1.75	58	124	9	1.04	0	.1	.05	.05	1.82	.15	33	0
0	7	4.21	155	350	43	2.22	57	.47	.82	.24	4.43	.37	31	0
0	7	5.7	55	224	4	.99	57	.99	.45	.56	6.94	.35	258	0
0	7	5.7	55	224	4	.99	57	.79	.5	.5	6.25	.32	181	0
0	10	1.96	25	43	1	.74	0	.29	.04	.14	2.34	.05	98	0
0	21	1.48	42	62	4	1.13	0	.15	.14	.06	.99	.11	7	0

Table A-1

Food Composition (Computer code number is for West Diet Analysis program) (For purposes of calculations, use "0" for t, <1, <.1, <.01, etc.)

Computer Code Number	Food Description	Measure	Wt (g)	H_2O (%)	Ener (kcal)	Prot (g)	Carb (g)	Dietary Fiber (g)	Fat (g)	Fat Breakdown (g) Sat	Mono	Poly
	GRAIN PRODUCTS: CEREAL, FLOUR, GRAIN, PASTA and NOODLES, POPCORN—Continued											
534	Vegetable, enriched	1 c	134	68	172	6	36	2	<1	t	t	.1
535	Millet, cooked	1 c	240	71	286	8	57	3	2	.4	.4	1.2
	Noodles (see also Pasta and Spaghetti):											
1507	Cellophane noodles, cooked	1 c	190	79	160	<1	39	<1	<1	t	t	t
1995	Cellophane noodles, dry	1 c	140	13	491	<1	121	1	<1	t	t	t
537	Chow mein, dry	1 c	45	1	237	4	26	2	14	2	3.5	7.8
536	Egg noodles, cooked, enriched	1 c	160	69	213	8	40	2	2	.5	.7	.7
538	Spinach noodles, dry	3½ oz	100	8	372	13	75	11	2	.2	.2	.6
1343	Oat bran, dry	¼ c	24	7	59	4	16	4	2	.3	.6	.7
	Pasta, cooked:											
1418	Fresh	2 oz	57	69	75	3	14	1	1	.1	.1	.2
1417	Linguini/Rotini	1 c	140	66	197	7	40	4	1	.1	.1	.4
	Popcorn:											
539	Air popped, plain	1 c	8	4	31	1	6	1	<1	t	.1	.2
1042	Microwaved, low fat, low sodium	1 c	6	3	25	1	4	1	1	.1	.2	.3
540	Popped in vegetable oil/salted	1 c	11	3	55	1	6	1	3	.5	.9	1.5
541	Sugar-syrup coated	1 c	35	3	151	1	28	2	4	1.3	1	1.6
	Rice:											
542	Brown rice, cooked	1 c	195	73	216	5	45	4	2	.4	.6	.6
2215	Mexican rice, cooked	1 c	226		820	16	180	6	30	4	1	1
2216	Spanish rice, cooked	1 c	246	85	130	3	28	2	1			
	White, enriched, all types:											
543	Regular/long grain, dry	1 c	185	12	675	13	148	2	1	.3	.4	.3
544	Regular/long grain, cooked	1 c	158	68	205	4	45	1	<1	.1	.1	.1
545	Instant, prepared without salt	1 c	165	76	162	3	35	1	<1	.1	.1	.1
	Parboiled/converted rice:											
546	Raw, dry	1 c	185	10	686	13	151	3	1	.3	.3	.3
547	Cooked	1 c	175	72	200	4	43	1	<1	.1	.1	.1
1486	Sticky rice (glutinous), cooked	1 c	174	77	169	4	37	2	<1	.1	.1	.1
548	Wild rice, cooked	1 c	164	74	166	7	35	3	1	.1	.1	.4
1700	Rice and pasta (Rice-a-Roni), cooked	1 c	202	72	246	5	43	1	6	1.1	2.3	1.9
549	Rye flour, medium	1 c	102	10	361	10	79	15	2	.2	.2	.8
1044	Soy flour, low-fat	1 c	88	3	325	45	30	9	6	.9	1.3	3.3
	Spaghetti pasta:											
550	Without salt, enriched	1 c	140	66	197	7	40	4	1	.1	.1	.4
551	With salt, enriched	1 c	140	66	197	7	40	2	1	.1	.1	.4
552	Whole-wheat spaghetti, cooked	1 c	140	67	174	7	37	6	1	.1	.1	.3
1302	Tapioca-pearl, dry	1 c	152	11	544	<1	135	1	<1	t	t	t
553	Wheat bran, crude	1 c	58	10	125	9	37	25	2	.4	.4	1.3
554	Wheat germ, raw	1 c	115	11	414	27	60	15	11	1.9	1.6	6.9
555	Wheat germ, toasted	1 c	113	6	432	33	56	15	12	2.1	1.7	7.5
1669	Wheat germ, with brown sugar & honey	1 c	113	3	420	30	66	11	9	1.5	1.2	5.5
556	Rolled wheat, cooked	1 c	240	84	149	5	33	4	1	.1	.1	.5
557	Whole-grain wheat, cooked	1 c	150	86	84	4	20	3	<1	.1	.1	.2
	Wheat flour (unbleached):											
	All-purpose white flour, enriched:											
558	Sifted	1 c	115	12	419	12	88	3	1	.2	.1	.5
559	Unsifted	1 c	125	12	455	13	95	3	1	.2	.1	.5
560	Cake or pastry, enriched, sifted	1 c	96	12	348	8	75	2	1	.1	.1	.4
561	Self-rising, enriched, unsifted	1 c	125	11	443	12	93	3	1	.2	.1	.5
562	Whole wheat, from hard wheats	1 c	120	10	407	16	87	15	2	.4	.3	.9
	MEATS: FISH and SHELLFISH											
1045	Bass, baked or broiled	4 oz	113	69	165	27	0	0	5	1.4	1.5	2.3
1046	Bluefish, baked or broiled	4 oz	113	63	180	29	0	0	6	1.4	2	2.7
1686	Catfish, breaded/flour fried	4 oz	113	49	325	21	14	1	20	5	9	4.7
	Clams:											
563	Raw meat only	1 ea	145	82	107	19	4	0	1	.3	.4	.7
564	Canned, drained	1 c	160	64	237	41	8	0	3	.7	.9	1.5
1290	Steamed, meat only	10 ea	95	64	141	24	5	0	2	.4	.5	.9

PAGE KEY: A–2 = Beverages A–4 = Dairy A–8 = Eggs A–8 = Fat/Oil A–12 = Fruit A–18 = Bakery A–26 = Grain A–30 = Fish A–32 = Meats A–36 = Poultry A–38 = Sausage A–40 = Mixed/Fast A–44 = Nuts/Seeds A–48 = Sweets A–50 = Vegetables/Legumes A–62 = Misc A–64 = Soups/Sauces A–66 = Fast A–82 = Frozen Entree A–86 = Baby foods

Chol (mg)	Calc (mg)	Iron (mg)	Magn (mg)	Pota (mg)	Sodi (mg)	Zinc (mg)	VT-A (RE)	Thia (mg)	VT-E (a-TE)	Ribo (mg)	Niac (mg)	V-B6 (mg)	Fola (µg)	VT-C (mg)
0	15	.66	25	41	8	.59	7	.15	.05	.08	1.43	.03	87	0
0	7	1.51	106	149	5	2.18	0	.25	.43	.2	3.19	.26	46	0
0	14	1	3	5	9	.23	0	.07	.06	0	.09	.02	1	0
0	35	3.04	4	14	14	.57	0	.21	.18	0	.28	.07	3	0
0	9	2.13	23	54	198	.63	4	.26	.07	.19	2.68	.05	40	0
53	19	2.54	30	45	11	.99	10	.3	.08	.13	2.38	.06	102	0
0	58	2.13	174	376	36	2.76	46	.37	.04	.2	4.55	.32	48	0
0	14	1.3	56	136	1	.75	0	.28	.41	.05	.22	.04	12	0
19	3	.65	10	14	3	.32	3	.12	.09	.09	.56	.02	36	0
0	10	1.96	25	43	1	.74	0	.29	.08	.14	2.34	.05	98	0
0	1	.21	10	24	<1	.27	2	.02	.01	.02	.15	.02	2	0
0	1	.14	9	14	29	.23	1	.02	.06	.01	.12	.01	1	0
0	1	.31	12	25	97	.29	2	.01	.03	.01	.17	.02	2	<1
2	15	.61	12	38	72	.2	3	.02	.42	.02	.77	.01	1	0
0	19	.82	84	84	10	1.23	0	.19	.53	.05	2.98	.28	8	0
0	300	9			2700		120							96
0		.72			1340									
0	52	7.97	46	213	9	2.02	0	1.07	.24	.09	7.75	.3	427	0
0	16	1.9	19	55	2	.77	0	.26	.08	.02	2.34	.15	92	0
0	13	1.04	8	7	5	.4	0	.12	.08	.08	1.45	.02	68	0
0	111	6.59	57	222	9	1.78	0	1.1	.24	.13	6.72	.65	427	0
0	33	1.98	21	65	5	.54	0	.44	.09	.03	2.45	.03	87	0
0	3	.24	9	17	9	.71	0	.03	.07	.02	.5	.04	2	0
0	5	.98	52	166	5	2.2	0	.08	.38	.14	2.12	.22	43	0
2	16	1.9	24	85	1147	.57	0	.25	.27	.16	3.6	.2	89	<1
0	24	2.16	76	347	3	2.03	0	.29	1.36	.12	1.76	.27	19	0
0	165	5.27	202	2261	16	1.04	4	.33	.17	.25	1.9	.46	361	0
0	10	1.96	25	43	1	.74	0	.29	.08	.14	2.34	.05	98	0
0	10	1.96	25	43	140	.74	0	.29	.38	.14	2.34	.05	98	0
0	21	1.48	42	62	4	1.13	0	.15	.07	.06	.99	.11	7	0
0	30	2.4	2	17	2	.18	0	.01	0	0	0	.01	6	0
0	42	6.15	354	686	1	4.22	0	.3	1.35	.33	7.89	.75	46	0
0	45	7.2	275	1025	14	14.1	0	2.16	20.7	.57	7.83	1.5	323	0
0	51	10.3	362	1070	5	18.9	0	1.89	20.5	.93	6.32	1.11	398	7
0	56	9.1	307	1089	12	15.7	11	1.51	24.9	.78	5.34	.56	376	0
0	17	1.49	53	170	0	1.15	0	.17	.48	.12	2.14	.17	26	0
0	9	.88	35	99	1	.73	0	.12	.3	.03	1.5	.08	12	0
0	17	5.34	25	123	2	.8	0	.9	.07	.57	6.79	.05	177	0
0	19	5.8	27	134	2	.87	0	.98	.07	.62	7.38	.05	193	0
0	13	7.03	15	101	2	.59	0	.86	.06	.41	6.52	.03	148	0
0	423	5.84	24	155	1587	.77	0	.84	.07	.52	7.29	.06	193	0
0	41	4.66	166	486	6	3.52	0	.54	1.48	.26	7.64	.41	53	0
98	116	2.16	43	515	102	.94	40	.1	.84	.1	1.72	.16	19	2
86	10	.7	47	539	87	1.18	156	.08	.71	.11	8.19	.52	2	0
92	41	1.44	34	376	598	1.05	33	.4	2.48	.18	3.37	.21	19	1
49	67	20.3	13	455	81	1.99	131	.12	1.45	.31	2.57	.09	23	19
107	147	44.8	29	1004	179	4.37	274	.24	3.04	.68	5.36	.18	46	35
64	87	26.6	17	597	106	2.59	162	.14	1.86	.4	3.18	.1	27	21

Table A-1

Food Composition (Computer code number is for West Diet Analysis program) (For purposes of calculations, use "0" for t, <1, <.1, <.01, etc.)

Computer Code Number	Food Description	Measure	Wt (g)	H_2O (%)	Ener (kcal)	Prot (g)	Carb (g)	Dietary Fiber (g)	Fat (g)	Fat Breakdown (g) Sat	Mono	Poly
	MEATS: FISH and SHELLFISH—Continued											
	Cod:											
565	Baked	4 oz	113	76	119	26	0	0	1	.2	.1	.5
566	Batter fried	4 oz	113	67	196	20	8	<1	9	2.2	3.6	2.6
567	Poached, no added fat	4 oz	113	77	116	25	0	0	1	.2	.1	.3
	Crab, meat only:											
1048	Blue crab, cooked	1 c	118	77	120	24	0	0	2	.3	.3	.8
1049	Dungeness crab, cooked	1 c	118	73	130	26	1	0	1	.2	.3	.5
568	Blue crab, canned	1 c	135	76	134	28	0	0	2	.4	.3	.6
1587	Crab, imitation, from surimi	4 oz	113	74	115	14	11	0	1	.3	.2	.8
569	Fish sticks, breaded pollock	2 ea	56	46	152	9	13	<1	7	1.8	2.8	1.8
572	Flounder/sole, baked	4 oz	113	73	132	27	0	0	2	.5	.4	.9
1599	Grouper, baked or broiled	4 oz	113	73	133	28	0	0	1	.4	.4	.6
573	Haddock, breaded, fried	4 oz	113	55	264	22	14	1	13	3.2	5.4	3.3
1050	Haddock, smoked	4 oz	113	71	131	28	0	0	1	.3	.3	.5
	Halibut:											
17291	Baked	4 oz	113	72	158	30	0	0	3	.7	1	1.4
1051	Smoked	4 oz	113	64	203	34			4	.6	1.2	1.5
1054	Raw	4 oz	113	78	124	23	0	0	3	.7	.8	1.1
575	Herring, pickled	4 oz	113	55	296	16	11	0	20	4.4	11	4.8
1052	Lobster meat, cooked w/moist heat	1 c	145	76	142	30	2	0	1	.2	.2	.5
1687	Ocean perch, baked/broiled	4 oz	113	73	137	27	0	0	2	.4	1	.7
576	Ocean perch, breaded/fried	4 oz	113	59	249	22	9	<1	13	3.2	5.7	3.4
1056	Octopus, raw	4 oz	113	80	93	17	2	0	1	.3	.2	.3
	Oysters:											
577	Raw, Eastern	1 c	248	85	169	17	10	0	6	2	.9	2.8
578	Raw, Pacific	1 c	248	82	201	23	12	0	6	1.3	.9	2.2
	Cooked:											
579	Eastern, breaded, fried, medium	5 ea	73	65	144	6	8	<1	9	2.7	1.7	4.6
580	Western, simmered	5 ea	125	64	204	24	12	0	6	1.9	.9	2.7
581	Pollock, baked, broiled, or poached	4 oz	113	74	128	27	0	0	1	.3	.2	.6
	Salmon:											
582	Canned pink, solids and liquid	4 oz	113	69	157	22	0	0	7	1.7	2.1	2.3
583	Broiled or baked	4 oz	113	62	244	31	0	0	12	2.2	6	2.7
584	Smoked	4 oz	113	72	132	21	0	0	5	1.2	2.3	1.1
585	Atlantic sardines, canned, drained, 2 = 24 g	4 oz	113	60	235	28	0	0	13	1.9	4.4	6.4
586	Scallops, breaded, cooked from frozen	6 ea	93	58	200	17	9	<1	10	2	2.5	5.3
1588	Scallops, imitation, from surimi	4 oz	113	74	112	14	12	0	1	.1	.1	.3
1688	Scallops, steamed/boiled	½ c	60	76	64	10	1	0	2	.3	.7	.6
	Shrimp:											
587	Cooked, boiled, 2 large = 11g	16 ea	88	77	87	18	0	0	1	.2	.2	.5
588	Canned, drained	½ c	64	73	77	15	1	0	1	.3	.3	.7
589	Fried, 2 large = 15 g, breaded	12 ea	90	53	218	19	10	<1	11	2	3	5.9
1057	Raw, large, about 7g each	14 ea	98	76	104	20	1	0	2	.3	.3	1
1589	Shrimp, imitation, from surimi	4 oz	113	75	114	14	10	0	2	.3	.2	1
1053	Snapper, baked or broiled	4 oz	113	70	145	30	0	0	2	.4	.4	.7
1060	Squid, fried in flour	4 oz	113	64	198	20	9	0	8	2.1	3.1	2.4
1590	Surimi	4 oz	113	76	112	17	8	0	1	.2	.2	.6
1058	Swordfish, raw	4 oz	113	76	137	22	0	0	5	1.3	1.7	1
1059	Swordfish, baked or broiled	4 oz	113	69	175	29	0	0	6	1.7	2.2	1.3
590	Trout, baked or broiled	4 oz	113	70	170	26	0	0	7	1.8	2	2.2
	Tuna, light, canned, drained solids:											
591	Oil pack	1 c	145	60	287	42	0	0	12	2.2	4.3	4.2
592	Water pack	1 c	154	74	179	39	0	0	1	.4	.2	.5
1061	Bluefin tuna, fresh	4 oz	113	68	163	26	0	0	6	1.4	1.8	1.6
	MEATS: BEEF, LAMB, PORK and others											
	BEEF, cooked, trimmed to ½" outer fat:											
	Braised, simmered, pot roasted:											
	Relatively fat, choice chuck blade:											
593	Lean and fat, piece 2½ x 2½ x ¾"	4 oz	113	47	393	30	0	0	29	13	14.8	1.2
594	Lean only	4 oz	113	55	297	35	0	0	16	7.3	8.2	.7

Chol (mg)	Calc (mg)	Iron (mg)	Magn (mg)	Pota (mg)	Sodi (mg)	Zinc (mg)	VT-A (RE)	Thia (mg)	VT-E (a-TE)	Ribo (mg)	Niac (mg)	V-B6 (mg)	Fola (µg)	VT-C (mg)
62	16	.55	47	276	88	.65	16	.1	.39	.09	2.84	.32	9	1
64	43	.92	36	443	124	.62	17	.13	.92	.13	2.58	.23	10	1
61	23	.54	41	496	69	.63	14	.09	.32	.08	2.48	.28	8	1
118	123	1.07	39	382	329	4.98	2	.12	1.18	.06	3.89	.21	60	4
90	70	.51	68	481	446	6.45	37	.07	1.33	.24	4.27	.2	50	4
120	136	1.13	53	505	450	5.43	2	.11	1.35	.11	1.85	.2	57	4
23	15	.44	49	102	950	.37	23	.04	.12	.03	.2	.03	2	0
63	11	.41	14	146	326	.37	17	.07	.77	.1	1.19	.03	10	0
77	20	.38	65	389	119	.71	12	.09	2.6	.13	2.46	.27	10	0
53	24	1.29	42	537	60	.58	56	.09	.71	.01	.43	.4	11	0
96	63	1.92	46	345	523	.59	33	.08	1.56	.14	4.49	.28	19	<1
87	55	1.58	61	469	862	.56	25	.05	.56	.05	5.73	.45	17	0
46	68	1.21	121	651	78	.6	61	.08	1.23	.1	8.05	.45	16	0
59	87	1.56	154	833	2260	.78	86	.11	1.11	.14	10.8	.64	22	0
36	53	.95	94	509	61	.47	53	.07	.96	.08	6.61	.39	14	0
15	87	1.38	9	78	983	.6	292	.04	1.81	.16	3.73	.19	3	0
104	88	.57	51	510	551	4.23	38	.01	2.1	.1	1.55	.11	16	0
61	155	1.33	44	396	108	.69	16	.15	1.84	.15	2.76	.3	12	1
71	136	1.57	38	323	431	.67	23	.14	2.41	.18	2.68	.24	15	1
54	60	5.99	34	396	260	1.9	51	.03	1.36	.04	2.37	.41	18	6
131	112	16.5	117	387	523	225	74	.25	1.98	.24	3.42	.15	25	9
124	20	12.7	55	417	263	41.2	201	.17	2.11	.58	4.98	.12	25	20
59	45	5.07	42	178	304	63.6	66	.11	1.66	.15	1.2	.05	23	3
125	20	11.5	55	378	265	41.5	183	.16	2.21	.55	4.53	.11	19	16
108	7	.32	82	437	131	.68	26	.08	.32	.09	1.86	.08	4	0
62	241	.95	38	368	626	1.04	19	.03	1.53	.21	7.39	.34	17	0
98	8	.62	35	424	75	.58	71	.24	1.42	.19	7.54	.25	6	0
26	12	.96	20	198	886	.35	29	.03	1.53	.11	5.33	.31	2	0
160	432	3.3	44	449	571	1.48	76	.09	.34	.26	5.93	.19	13	0
57	39	.76	55	310	432	.99	20	.04	1.77	.1	1.4	.13	34	2
25	9	.35	49	116	898	.37	23	.01	.12	.02	.35	.03	2	0
19	15	.15	32	168	246	.55	27	.01	.81	.04	.6	.08	7	1
172	34	2.72	30	160	197	1.37	58	.03	.66	.03	2.28	.11	3	2
111	38	1.75	26	134	108	.81	11	.02	.59	.02	1.77	.07	1	1
159	60	1.13	36	203	310	1.24	50	.12	1.35	.12	2.76	.09	7	1
149	51	2.36	36	181	145	1.09	53	.03	.8	.03	2.5	.1	3	2
41	21	.68	49	101	797	.37	23	.03	.12	.04	.19	.03	2	0
53	45	.27	42	590	64	.5	40	.06	.71	<.01	.39	.52	7	2
294	44	1.14	43	315	346	1.97	12	.06	2.09	.52	2.94	.07	16	5
34	10	.29	49	127	162	.37	23	.02	.28	.02	.25	.03	2	0
44	5	.91	30	325	102	1.3	41	.04	.56	.11	10.9	.37	2	1
56	7	1.18	38	417	130	1.66	46	.05	.71	.13	13.3	.43	3	1
78	97	.43	35	506	63	.58	17	.17	.57	.11	6.52	.39	21	2
26	19	2.02	45	300	513	1.31	33	.05	1.74	.17	18	.16	8	0
46	17	2.36	42	365	521	1.19	26	.05	.82	.11	20.5	.54	6	0
43	9	1.15	56	285	44	.68	740	.27	1.13	.28	9.77	.51	2	0
112	11	3.45	21	275	67	7.57	0	.08	.26	.27	3.54	.32	10	0
120	15	4.16	26	297	80	11.6	0	.09	.16	.32	3.02	.33	7	0

Table A-1

Food Composition (Computer code number is for West Diet Analysis program) (For purposes of calculations, use "0" for t, <1, <.1, <.01, etc.)

Computer Code Number	Food Description	Measure	Wt (g)	H_2O (%)	Ener (kcal)	Prot (g)	Carb (g)	Dietary Fiber (g)	Fat (g)	Fat Breakdown (g) Sat	Mono	Poly
	MEATS: BEEF, LAMB, PORK and others—Continued											
	Relatively lean, like choice round:											
595	Lean and fat, pce 4⅛ x 2½ x ¾"	4 oz	113	52	311	32	0	0	19	8.5	9.7	.8
596	Lean only	4 oz	113	57	249	36	0	0	11	4.8	5.4	.5
	Ground beef, broiled, patty 3 x ⅝":											
597	Extra lean, about 16% fat	4 oz	113	54	299	32	0	0	18	8	9	.8
598	Lean, 21% fat	4 oz	113	53	316	32	0	0	20	8.9	10.1	.8
	Roasts, oven cooked, no added liquid:											
	Relatively fat, prime rib:											
601	Lean and fat, piece 4⅛ x 2¼ x ½"	4 oz	113	46	425	25	0	0	35	15.8	17.9	1.5
602	Lean only	4 oz	113	58	271	31	0	0	16	7	7.9	.7
	Relatively lean, choice round:											
603	Lean and fat, piece 2½ x 2½ x ¾"	4 oz	113	59	272	30	0	0	16	7.1	8.1	.7
604	Lean only	4 oz	113	65	198	33	0	0	6	2.9	3.3	.3
1701	Steak, rib, broiled, lean	4 oz	113	58	250	32	0	0	13	5.7	6.4	.5
	Steak, broiled, relatively lean,											
606	choice sirloin, lean only	4 oz	113	62	228	34	0	0	9	4	4.6	.4
	Steak, broiled, relatively fat,											
	choice T-bone:											
1063	Lean and fat	4 oz	113	52	349	26	0	0	26	11.8	13.3	1.1
1064	Lean only	4 oz	113	61	232	30	0	0	11	5.1	5.8	.5
	Variety meats:											
1086	Brains, panfried	4 oz	113	71	221	14	0	0	18	6.7	7	3.9
599	Heart, simmered	4 oz	113	64	198	32	<1	0	6	2.9	1.5	1.5
600	Liver, fried	4 oz	113	56	245	30	9	0	9	3.1	1.8	1.9
1062	Tongue, cooked	4 oz	113	56	320	25	<1	0	23	10.1	10.7	.9
607	Beef, canned, corned	4 oz	113	58	283	31	0	0	17	7.5	8.5	.7
608	Beef, dried, cured	1 oz	28	56	46	8	<1	0	1	.5	.5	.1
	LAMB, domestic, cooked:											
	Chop, arm, braised (5.6 oz raw w/bone):											
609	Lean and fat	1 ea	70	44	242	21	0	0	17	7.8	7.3	1.4
610	Lean only	1 ea	55	49	153	19	0	0	8	3.6	3.4	.6
	Chop, loin, broiled (4.2 oz raw w/bone):											
611	Lean and fat	1 ea	64	52	202	16	0	0	15	6.8	6.4	1.2
612	Lean only	1 ea	46	61	99	14	0	0	4	2.1	1.9	.4
1067	Cutlet, avg of lean cuts, cooked	4 oz	113	54	330	28	0	0	23	10.9	10.2	1.9
	Leg, roasted, 3 oz = 4⅛ x 2¼ x ½":											
613	Lean and fat	4 oz	113	57	292	29	0	0	19	8.7	8.1	1.6
614	Lean only	4 oz	113	64	216	32	0	0	9	4.1	3.8	.7
615	Rib, roasted, lean and fat	4 oz	113	48	406	24	0	0	34	15.7	14.7	2.8
616	Rib, roasted, lean only	4 oz	113	60	262	30	0	0	15	7	6.6	1.2
1065	Shoulder, roasted, lean and fat	4 oz	113	56	312	25	0	0	23	10.5	9.8	1.9
1066	Shoulder, roasted, lean only	4 oz	113	63	231	28	0	0	12	5.7	5.3	1.1
	Variety meats:											
1069	Brains, panfried	4 oz	113	76	164	14	0	0	11	4.4	3.7	1.9
1068	Heart, braised	4 oz	113	64	209	28	2	0	9	3.9	2.7	1.1
1070	Sweetbreads, cooked	4 oz	113	60	264	26	0	0	17	8.1	6.5	1.4
1071	Tongue, cooked	4 oz	113	58	311	24	0	0	23	8.8	11.6	1.4
	PORK, cured, cooked (see also Sausages											
	and Lunch Meats)											
617	Bacon, medium slices	3 pce	19	13	109	6	<1	0	9	3.3	4.5	1.1
1087	Breakfast strips, cooked	2 pce	23	27	106	7	<1	0	8	2.9	3.8	1.3
618	Canadian-style bacon	2 pce	47	62	87	11	1	0	4	1.3	1.9	.4
	Ham, roasted:											
619	Lean and fat, 2 pces 4⅛ x 2¼ x ¼"	4 oz	113	64	201	25	0	0	10	3.4	5	1.7
620	Lean only	4 oz	113	68	164	24	2	0	6	2.1	3	.6
621	Ham, canned, roasted, 8% fat	4 oz	113	69	154	24	1	0	6	1.8	2.8	.5
	PORK, fresh, cooked:											
	Chops, loin (cut 3 per lb with bone):											
1291	Braised, lean and fat	1 ea	89	58	213	24	0	0	12	4.5	5.4	1
1292	Lean only	1 ea	80	61	163	23	0	0	7	2.7	3.3	.6
622	Broiled, lean and fat	1 ea	82	58	197	23	0	0	11	3.9	4.8	.8
623	Broiled, lean only	1 ea	74	61	149	22	0	0	6	2.2	2.7	.4

Chol (mg)	Calc (mg)	Iron (mg)	Magn (mg)	Pota (mg)	Sodi (mg)	Zinc (mg)	VT-A (RE)	Thia (mg)	VT-E (a-TE)	Ribo (mg)	Niac (mg)	V-B6 (mg)	Fola (µg)	VT-C (mg)
108	7	3.53	25	319	56	5.55	0	.08	.21	.27	4.21	.37	11	0
108	6	3.91	28	348	58	6.19	0	.08	.2	.29	4.61	.41	12	0
112	10	3.13	28	417	93	7.27	0	.08	.2	.36	6.61	.36	12	0
114	14	2.77	27	394	101	7.01	0	.07	.23	.27	6.75	.34	12	0
96	12	2.61	21	334	71	5.92	0	.08	.27	.19	3.8	.26	8	0
91	11	2.95	28	425	84	7.84	0	.09	.14	.24	4.64	.34	9	0
81	7	2.07	27	406	67	4.87	0	.09	.23	.18	3.92	.4	7	0
78	6	2.2	30	446	70	5.36	0	.1	.12	.19	4.24	.43	8	0
90	15	2.9	30	445	78	7.9	0	.11	.16	.25	5.42	.45	9	0
101	12	3.8	36	455	75	7.37	0	.15	.16	.33	4.84	.51	11	0
76	9	3.06	26	363	72	5.03	0	.1	.24	.24	4.46	.37	8	0
67	7	3.58	32	427	80	6	0	.12	.16	.28	5.23	.44	9	0
2254	10	2.51	17	400	179	1.53	0	.15	2.37	.29	4.27	.44	7	4
218	7	8.49	28	263	71	3.54	0	.16	.81	1.74	4.6	.24	2	2
545	12	7.1	26	411	120	6.16	12123	.24	.72	4.68	16.3	1.62	249	26
121	8	3.83	19	203	68	5.42	0	.03	.4	.4	2.43	.18	6	1
97	14	2.35	16	154	1136	4.03	0	.02	.17	.17	2.75	.15	10	0
12	2	1.26	9	124	972	1.47	0	.02	.04	.06	1.53	.1	3	0
84	17	1.67	18	214	50	4.26	0	.05	.1	.17	4.66	.08	13	0
67	14	1.49	16	186	42	4.02	0	.04	.1	.15	3.48	.07	12	0
64	13	1.16	15	209	49	2.23	0	.06	.08	.16	4.54	.08	11	0
44	9	.92	13	173	39	1.9	0	.05	.07	.13	3.15	.07	11	0
110	12	2.26	25	340	77	4.67	0	.12	.15	.32	7.48	.16	19	0
105	12	2.24	27	354	75	4.97	0	.11	.17	.3	7.45	.17	23	0
101	9	2.4	29	382	77	5.58	0	.12	.2	.33	7.16	.19	26	0
110	25	1.81	23	306	82	3.94	0	.1	.11	.24	7.63	.12	17	0
99	24	2	26	356	91	5.05	0	.1	.17	.26	6.96	.17	25	0
104	23	2.23	26	284	75	5.91	0	.1	.16	.27	6.95	.15	24	0
98	21	2.41	28	299	77	6.83	0	.1	.2	.29	6.51	.17	28	0
2308	14	1.9	16	232	151	1.54	0	.12	1.73	.27	2.79	.12	6	14
281	16	6.24	27	212	71	4.16	0	.19	.79	1.34	4.93	.34	2	8
452	14	2.4	21	329	59	3.03	0	.02	.78	.24	2.89	.06	15	23
214	11	2.97	18	179	76	3.38	0	.09	.36	.47	4.17	.19	3	8
16	2	.31	5	92	303	.62	0	.13	.1	.05	1.39	.05	1	0
24	3	.45	6	107	483	.85	0	.17	.08	.08	1.75	.08	1	0
27	5	.38	10	183	727	.8	0	.39	.15	.09	3.25	.21	2	0
67	9	1.51	25	462	1695	2.79	0	.82	.45	.37	6.95	.35	3	0
60	9	1.67	16	324	1359	3.25	0	.85	.29	.23	4.54	.45	3	0
34	7	1.04	24	393	1282	2.52	0	1.18	.29	.28	5.53	.51	6	0
71	19	.95	17	333	43	2.12	2	.56	.3	.23	3.93	.33	3	1
63	14	.9	16	310	40	1.98	2	.53	.3	.21	3.67	.31	3	<1
67	27	.66	20	294	48	1.85	2	.88	.27	.24	4.3	.35	5	<1
61	23	.63	20	278	44	1.76	2	.85	.31	.23	4.1	.35	4	<1

Table A-1

Food Composition (Computer code number is for West Diet Analysis program) (For purposes of calculations, use "0" for t, <1, <.1, <.01, etc.)

Computer Code Number	Food Description	Measure	Wt (g)	H_2O (%)	Ener (kcal)	Prot (g)	Carb (g)	Dietary Fiber (g)	Fat (g)	Fat Breakdown (g) Sat	Mono	Poly
	MEATS: BEEF, LAMB, PORK and others—Continued											
624	Panfried, lean and fat	1 ea	78	53	216	23	0	0	13	4.7	5.5	1.5
625	Panfried, lean only	1 ea	63	59	152	16	0	0	10	3.2	3.9	1.2
626	Leg, roasted, lean and fat	4 oz	113	55	308	30	0	0	20	7.3	8.9	1.9
627	Leg, roasted, lean only	4 oz	113	61	233	35	0	0	9	3.2	4.3	.9
628	Rib, roasted, lean and fat	4 oz	113	56	288	31	0	0	17	6.7	7.9	1.4
629	Rib, roasted, lean only	4 oz	113	59	252	32	0	0	13	4.9	5.9	1
630	Shoulder, braised, lean and fat	4 oz	113	48	372	32	0	0	26	9.6	11.8	2.6
631	Shoulder, braised, lean only	4 oz	113	54	280	36	0	0	14	4.7	6.5	1.3
1088	Spareribs, cooked, yield from 1 lb raw with bone	4 oz	113	40	449	33	0	0	34	12.5	15.3	3.1
1095	Rabbit, roasted (1 cup meat = 140 g)	4 oz	113	61	223	33	0	0	9	4	2	3
	VEAL, cooked:											
632	Cutlet, braised or broiled, 4⅛ x 2¼ x ½"	4 oz	113	52	321	34	0	0	19	7.6	7.6	1.3
633	Rib roasted, lean, 2 pieces 4⅛ x 2¼ x ¼"	4 oz	113	60	258	27	0	0	16	6.1	6.1	1.1
634	Liver, panfried	4 oz	113	67	186	24	3	0	8	3.3	1.9	2.2
1096	Venison (deer meat), roasted	4 oz	113	65	179	34	0	0	4	1.4	1	.7
	MEATS: POULTRY and POULTRY PRODUCTS											
	CHICKEN, cooked:											
	Fried, batter dipped:											
635	Breast	1 ea	280	52	728	69	25	1	37	9.9	15.3	8.6
636	Drumstick	1 ea	72	53	193	16	6	<1	11	3	4.7	2.7
637	Thigh	1 ea	86	51	238	19	8	<1	14	3.8	5.9	3.3
638	Wing	1 ea	49	46	159	10	5	<1	11	2.9	4.5	2.5
	Fried, flour coated:											
639	Breast	1 ea	196	57	435	62	3	<1	17	4.9	7.1	3.8
1212	Breast, without skin	1 ea	86	60	161	29	<1	<1	4	1.1	1.5	.9
640	Drumstick	1 ea	49	57	120	13	1	<1	7	1.8	2.7	1.6
641	Thigh	1 ea	62	54	162	17	2	<1	9	2.5	3.7	2.1
1099	Thigh, without skin	1 ea	52	59	113	15	1	<1	5	1.4	2	1.3
642	Wing	1 ea	32	49	103	8	1	<1	7	1.9	2.9	1.6
	Roasted:											
643	All types of meat	1 c	140	64	266	40	0	0	10	2.9	3.8	2.4
644	Dark meat	1 c	140	63	287	38	0	0	14	3.7	5.2	3.2
645	Light meat	1 c	140	65	242	43	0	0	6	1.8	2.2	1.4
646	Breast, without skin	1 ea	172	65	284	53	0	0	6	1.8	2.2	1.3
647	Drumstick, without skin	1 ea	44	67	76	12	0	0	2	.7	.8	.6
1703	Leg, without skin	1 ea	95	65	181	26	0	0	8	2.2	2.9	1.9
648	Thigh	1 ea	62	59	153	16	0	0	10	2.7	3.9	2.1
1100	Thigh, without skin	1 ea	52	63	109	13	0	0	6	1.6	2.2	1.3
649	Stewed, all types	1 c	140	67	248	38	0	0	9	2.6	3.5	2.2
656	Canned, boneless chicken	4 oz	113	69	186	25	0	0	9	2.5	3.6	2
1102	Gizzards, simmered	1 c	145	67	222	39	2	0	5	1.5	1.3	1.5
1101	Hearts, simmered	1 c	145	65	268	38	<1	0	11	3.3	2.9	3.3
2300	Liver, simmered: Ounce	3 oz	85	68	133	21	1	0	5	1.6	1.1	.8
1098	Liver, simmered: Piece = 20 g	6 ea	120	68	188	29	1	0	7	2.2	1.6	1.1
	DUCK, roasted:											
1293	Meat with skin, about 2.7 cups	½ ea	382	52	1287	73	0	0	108	36.9	49.3	13.9
651	Meat only, about 1.5 cups	½ ea	221	64	444	52	0	0	25	9.2	8.2	3.2
	GOOSE, domesticated, roasted:											
1294	Meat only, about 4.2 cups	½ ea	591	57	1406	173	0	0	75	23.6	40.2	10.9
1295	Meat with skin, about 5.5 cups	½ ea	774	52	2360	195	0	0	170	53.2	80.5	24.8
	TURKEY:											
	Roasted, meat only:											
652	Dark meat	4 oz	113	63	211	33	0	0	8	2.7	1.8	2.5
653	Light meat	4 oz	113	66	177	34	0	0	4	1.2	.6	1
654	All types, chopped or diced	1 c	140	65	238	42	0	0	7	2.3	1.5	2
1103	Ground, cooked	4 oz	113	59	266	31	0	0	15	4.1	5.5	3.6
1106	Gizzard, cooked	2 ea	134	65	218	39	1	0	5	1.5	1	1.5
1107	Heart, cooked	4 ea	64	64	113	17	1	0	4	1.1	.8	1.1
1108	Liver, cooked	1 ea	75	66	127	18	3	0	4	1.4	1.1	.8

PAGE KEY: A–2 = Beverages A–4 = Dairy A–8 = Eggs A–8 = Fat/Oil A–12 = Fruit A–18 = Bakery A–26 = Grain A–30 = Fish A–32 = Meats A–36 = Poultry A–38 = Sausage A–40 = Mixed/Fast A–44 = Nuts/Seeds A–48 = Sweets A–50 = Vegetables/Legumes A–62 = Misc A–64 = Soups/Sauces A–66 = Fast A–82 = Frozen Entree A–86 = Baby foods

Chol (mg)	Calc (mg)	Iron (mg)	Magn (mg)	Pota (mg)	Sodi (mg)	Zinc (mg)	VT-A (RE)	Thia (mg)	VT-E (a-TE)	Ribo (mg)	Niac (mg)	V-B6 (mg)	Fola (μg)	VT-C (mg)
72	21	.71	23	332	62	1.8	2	.89	.32	.24	4.37	.37	5	1
52	14	.67	16	230	49	2.44	1	.46	.3	.23	2.8	.26	3	<1
106	16	1.14	25	398	68	3.34	3	.72	.34	.35	5.16	.45	11	<1
108	8	1.29	33	442	73	3.4	3	.91	.46	.4	5.56	.38	3	<1
82	32	1.06	24	476	52	2.33	2	.82	.41	.34	6.92	.37	3	<1
80	29	1.11	25	494	53	2.41	2	.86	.55	.36	7.25	.38	3	<1
123	20	1.82	21	417	99	4.72	3	.61	.5	.35	5.89	.4	5	<1
129	9	2.2	25	458	115	5.62	3	.68	.58	.41	6.71	.46	6	<1
137	53	2.09	27	362	105	5.2	3	.46	.52	.43	6.19	.4	5	0
93	21	2.57	24	433	53	2.57	0	.1	.96	.24	9.53	.53	12	0
133	32	1.23	27	316	90	4.1	0	.04	.45	.34	10.2	.29	16	0
124	12	1.1	25	333	104	4.62	0	.06	.4	.3	7.89	.28	15	0
634	8	2.96	21	232	60	10.8	9095	.15	.42	2.19	9.58	.55	858	35
127	8	5.05	27	379	61	3.11	0	.2	.28	.68	7.58	.42	5	0
238	56	3.5	67	563	770	2.66	56	.32	2.97	.41	29.4	1.2	42	0
62	12	.97	14	134	194	1.68	19	.08	.88	.15	3.67	.19	13	0
80	15	1.25	18	165	248	1.75	25	.1	1.05	.19	4.92	.22	16	0
39	10	.63	8	68	157	.68	17	.05	.52	.07	2.58	.15	9	0
174	31	2.33	59	508	149	2.16	29	.16	1.12	.26	26.9	1.14	12	0
78	14	.98	27	237	68	.93	6	.07	.36	.11	12.7	.55	3	0
44	6	.66	11	112	44	1.42	12	.04	.41	.11	2.96	.17	5	0
60	9	.92	15	147	55	1.56	18	.06	.52	.15	4.31	.2	7	0
53	7	.76	13	135	49	1.45	11	.05	.3	.13	3.7	.2	5	0
26	5	.4	6	57	25	.56	12	.02	.18	.04	2.14	.13	2	0
125	21	1.69	35	340	120	2.94	22	.1	.58	.25	12.8	.66	8	0
130	21	1.86	32	336	130	3.92	31	.1	.81	.32	9.17	.5	11	0
119	21	1.48	38	346	108	1.72	13	.09	.37	.16	17.4	.84	6	0
146	26	1.79	50	440	127	1.72	10	.12	.66	.2	23.6	1.03	7	0
41	5	.57	11	108	42	1.4	8	.03	.25	.1	2.68	.17	4	0
89	11	1.24	23	230	86	2.72	18	.07	.55	.22	6	.35	8	0
58	7	.83	14	138	52	1.46	30	.04	.35	.13	3.95	.19	4	0
49	6	.68	12	124	46	1.34	10	.04	.3	.12	3.4	.18	4	0
116	20	1.64	29	252	98	2.79	21	.07	.42	.23	8.57	.36	8	0
70	16	1.79	14	156	568	1.59	38	.02	.24	.15	7.15	.4	5	2
281	14	6.02	29	260	97	6.35	81	.04	2.29	.35	5.77	.17	77	2
351	28	13.1	29	191	70	10.6	13	.1	2.32	1.07	4.06	.46	116	3
536	12	7.2	18	119	43	3.69	4176	.13	1.45	1.49	3.78	.49	655	13
757	17	10.2	25	168	61	5.21	5895	.18	2.04	2.1	5.34	.7	924	19
321	42	10.3	61	779	225	7.11	241	.66	2.5	1.03	18.5	.69	23	0
197	26	5.97	44	557	144	5.75	51	.57	1.55	1.04	11.3	.55	22	0
567	83	17	148	2293	449	18.7	71	.54	9.16	2.3	24.1	2.78	71	0
704	101	21.9	170	2546	542	20.3	163	.6	13.5	2.5	32.3	2.86	15	0
96	36	2.63	27	328	89	5.04	0	.07	.94	.28	4.12	.41	10	0
78	21	1.53	32	345	72	2.31	0	.07	.12	.15	7.73	.61	7	0
106	35	2.49	36	417	98	4.34	0	.09	.59	.25	7.62	.64	10	0
115	28	2.18	27	305	121	3.23	0	.06	.45	.19	5.45	.44	8	0
311	20	7.29	25	283	72	5.57	74	.04	.27	.44	4.11	.16	70	2
145	8	4.41	14	117	35	3.37	5	.04	.13	.56	2.08	.2	51	1
470	8	5.85	11	146	48	2.32	2805	.04	2.41	1.07	4.46	.39	500	1

Table A-1

Food Composition (Computer code number is for West Diet Analysis program) (For purposes of calculations, use "0" for t, <1, <.1, <.01, etc.)

Computer Code Number	Food Description	Measure	Wt (g)	H_2O (%)	Ener (kcal)	Prot (g)	Carb (g)	Dietary Fiber (g)	Fat (g)	Fat Breakdown (g) Sat	Mono	Poly
	MEATS: POULTRY and POULTRY PRODUCTS—Continued											
	POULTRY FOOD PRODUCTS (see also items in Sausages & Lunchmeats section):											
1567	Chicken patty, breaded, cooked	1 ea	75	49	213	12	11	<1	13	4.1	6.4	1.6
659	Turkey and gravy, frozen package	3 oz	85	85	57	5	4	<1	2	.8	.8	.4
	Turkey breast, Louis Rich:											
1104	Barbecued	2 oz	56	72	58	12	2	0	<1	.2	.2	.1
1943	Hickory smoked	1 pce	80		80	16	2	0	1	0		
1947	Honey roasted	1 pce	80		80	16	3	0	1	.5		
1945	Oven roasted	1 pce	80		70	16		0	1	0		
661	Turkey patty, breaded, fried	2 oz	57	50	161	8	9	<1	10	2.7	4.3	2.7
662	Turkey, frozen, roasted, seasoned	4 oz	113	68	175	24	3	0	7	2.1	1.4	1.9
1704	Turkey roll, light meat	1 pce	28	72	41	5	<1	0	2	.6	.7	.5
	MEATS: SAUSAGES and LUNCHMEATS (see also Poultry Food Products)											
1072	Beerwurst/beer salami, beef	1 oz	28	53	92	3	<1	0	8	3.6	3.9	.3
1074	Beerwurst/beer salami, pork	1 oz	28	61	67	4	1	0	5	1.8	2.5	.7
1075	Berliner sausage	1 oz	28	61	64	4	1	0	5	1.7	2.2	.4
	Bologna:											
1297	Beef	1 pce	23	55	72	3	<1	0	7	2.8	3.2	.3
2115	Beef, light, Oscar Mayer	1 pce	28	65	56	3	2	0	4	1.6	2	.1
663	Beef & pork	1 pce	28	54	88	3	1	0	8	3	3.7	.7
2155	Healthy Favorites	1 pce	23		22	3	1	0	<1	0		
1298	Pork	1 pce	23	61	57	4	<1	0	5	1.6	2.2	.5
2114	Regular, light, Oscar Mayer	1 pce	28	65	56	3	2	0	4	1.6	2	.4
664	Turkey	1 pce	28	65	56	4	<1	0	4	1.4	1.3	1.2
1970	Turkey, Louis Rich	1 pce	28	67	57	3	<1	0	5	1.5	1.8	1.3
665	Braunschweiger sausage	2 pce	57	48	205	8	2	0	18	6.2	8.5	2.1
1073	Bratwurst, link	1 ea	70	51	226	10	2	0	19	6.9	9.3	2
666	Brown & serve sausage links, cooked	2 ea	26	45	102	4	1	0	10	3.4	4.4	1
1089	Cheesefurter/cheese smokie	2 ea	86	52	281	12	1	0	25	9	11.8	2.6
2157	Chicken breast, Healthy Favorites	4 pce	52		40	9	1	0	0	0	0	0
1556	Chorizo, pork & beef	1 ea	60	32	273	15	1	0	23	8.6	11	2.1
1090	Corned beef loaf, jellied	1 pce	28	69	43	6	0	0	2	.7	.7	.1
	Frankfurters:											
1077	Beef, large link, 8/package	1 ea	57	55	180	7	1	0	16	6.9	7.9	.8
1078	Beef and pork, large link, 8/package	1 ea	57	54	182	6	1	0	17	6.2	8	1.6
667	Beef and pork, small link, 10/pkg	1 ea	45	54	144	5	1	0	13	4.9	6.3	1.2
668	Turkey frankfurter, 10/package	1 ea	45	63	102	6	1	0	8	2.7	2.5	2.2
1968	Turkey/chicken frank 8/pkg	1 ea	43		80	6	1	0	6	2		
	Ham:											
669	Ham lunchmeat, canned, 3 x 2 x ½"	1 pce	21	52	70	3	<1	0	6	2.3	3	.7
670	Chopped ham, packaged	2 pce	42	64	96	7	0	0	7	2.4	3.4	.9
2156	Honey ham, Healthy Favorites	4 pce	52	73	55	9	2	0	1	.4	.8	.1
2113	Oscar Mayer lower sodium ham	1 pce	21	73	23	3	1	0	1	.3	.4	.1
673	Turkey ham lunchmeat	2 pce	57	71	73	11	<1	0	3	1	.7	.9
1091	Kielbasa sausage	1 pce	26	54	81	3	1	0	7	2.6	3.4	.8
1092	Knockwurst sausage, link	1 ea	68	55	209	8	1	0	19	6.9	8.7	2
1093	Mortadella lunchmeat	2 pce	30	52	93	5	1	0	8	2.8	3.4	.9
1097	Olive loaf lunchmeat	2 pce	57	58	134	7	5	<1	9	3.3	4.5	1.1
1952	Turkey breast, fat free	1 pce	28	77	22	4	1	0	<1	.1	.1	t
1080	Turkey pastrami	2 pce	57	71	80	10	1	0	4	1	1.2	.9
1969	Turkey salami	1 pce	28	72	41	4	<1	0	3	.9	1	.8
1081	Pepperoni sausage	2 pce	11	27	55	2	<1	0	5	1.8	2.3	.5
1094	Pickle & pimento loaf	2 pce	57	57	149	7	3	<1	12	4.5	5.5	1.5
1082	Polish sausage	1 oz	28	53	91	4	<1	0	8	2.9	3.8	.9
674	Pork sausage, cooked, link, small	2 ea	26	45	96	5	<1	0	8	2.8	4.1	.8
1079	Pork sausage, cooked, patty	4 oz	113	45	417	22	1	0	35	12.1	17.7	3.3
675	Salami, pork and beef	2 pce	57	60	143	8	1	0	11	4.6	5.2	1.1
677	Salami, pork and beef, dry	3 pce	30	35	125	7	1	0	10	3.7	5.1	1
676	Salami, turkey	2 pce	57	66	112	9	<1	0	8	2.3	2.6	2
	Sandwich spreads:											
1300	Ham salad spread	2 tbs	30	63	65	3	3	0	5	1.5	2.2	.8
678	Pork and beef	2 tbs	30	60	70	2	4	<1	5	1.8	2.3	.8
1296	Chicken/turkey	2 tbs	26	66	52	3	2	0	4	.9	.8	1.6

Chol (mg)	Calc (mg)	Iron (mg)	Magn (mg)	Pota (mg)	Sodi (mg)	Zinc (mg)	VT-A (RE)	Thia (mg)	VT-E (a-TE)	Ribo (mg)	Niac (mg)	V-B6 (mg)	Fola (µg)	VT-C (mg)
45	12	.94	15	185	399	.78	22	.07	1.46	.1	5.04	.23	8	<1
15	12	.79	7	52	471	.59	11	.02	.3	.11	1.53	.08	3	0
25	14	.62	16	175	599	.59	0	.02		.06	5.35	.22	2	0
35	0	.72			1060		0							0
35	0	.72			940		0							0
35	0				910		0							0
35	8	1.25	9	157	456	.82	6	.06	1.36	.11	1.31	.11	16	0
60	6	1.84	25	337	768	2.87	0	.05	.43	.18	7.09	.3	6	0
12	11	.36	4	70	137	.44	0	.02	.04	.06	1.96	.09	1	0
17	3	.42	3	49	288	.68	0	.02	.05	.03	.95	.05	1	0
16	2	.21	4	71	347	.48	0	.15	.06	.05	.91	.1	1	0
13	3	.32	4	79	363	.69	0	.11	.06	.06	.87	.06	1	0
13	3	.38	3	36	226	.5	0	.01	.04	.02	.55	.03	1	0
13	4	.34	4	44	314	.53	0							0
15	3	.42	3	50	285	.54	0	.05	.06	.04	.72	.05	1	0
7		.18			255									
14	3	.18	3	65	272	.47	0	.12	.06	.04	.9	.06	1	0
15	14	.39	5	46	312	.45	0							0
28	23	.43	4	56	246	.49	0	.01	.15	.05	.99	.06	2	0
22	34	.45	5	51	242	.57	0	.01		.05	1.08	.05		0
89	5	5.34	6	113	652	1.6	2405	.14	.2	.87	4.77	.19	25	0
44	34	.72	11	197	778	1.47	0	.17	.19	.16	2.31	.09	3	0
16	2	.62	4	70	248	.3	0	.21	.06	.09	.96	.06	1	0
58	50	.93	11	177	931	1.94	33	.21	.27	.14	2.49	.11	3	0
25		.72			620									
53	5	.95	11	239	741	2.05	0	.38	.13	.18	3.08	.32	1	0
13	3	.57	3	28	267	1.15	0	0	.05	.03	.49	.03	2	0
35	11	.81	2	95	585	1.24	0	.03	.11	.06	1.38	.07	2	0
28	6	.66	6	95	638	1.05	0	.11	.14	.07	1.5	.07	2	0
22	5	.52	4	75	504	.83	0	.09	.11	.05	1.18	.06	2	0
48	48	.83	6	81	642	1.4	0	.02	.28	.08	1.86	.1	4	0
40	60	1.08			480		0							0
13	1	.15	2	45	271	.31	0	.08	.05	.04	.66	.04	1	<1
21	3	.35	7	134	576	.81	0	.26	.11	.09	1.63	.15	<1	0
24	6	.7	18	144	635	1.02	0							0
9	1	.3	5	197	174	.42	0							0
32	6	1.57	9	185	568	1.68	0	.03	.36	.14	2.01	.14	3	0
17	11	.38	4	70	280	.52	0	.06	.06	.06	.75	.05	1	0
39	7	.62	7	135	687	1.13	0	.23	.39	.09	1.86	.12	1	0
17	5	.42	3	49	374	.63	0	.04	.07	.05	.8	.04	1	0
22	62	.31	11	169	846	.79	34	.17	.14	.15	1.05	.13	1	0
9	3	.34	8	59	387	.24	0							0
31	5	.95	8	148	596	1.23	0	.03	.12	.14	2.01	.15	3	0
21	11	.35	6	61	281	.65	0							0
9	1	.15	2	38	224	.27	0	.03	.02	.03	.55	.03	<1	0
21	54	.58	10	194	792	.8	4	.17	.14	.14	1.17	.11	3	0
20	3	.4	4	66	245	.54	0	.14	.06	.04	.96	.05	1	<1
22	8	.33	4	94	336	.65	0	.19	.07	.07	1.18	.09	1	<1
94	36	1.42	19	408	1462	2.84	0	.84	.29	.29	5.11	.37	2	2
37	7	1.52	9	113	607	1.22	0	.14	.12	.21	2.02	.12	1	0
24	2	.45	5	113	558	.97	0	.18	.08	.09	1.46	.15	1	0
47	11	.92	9	139	572	1.03	0	.04	.32	.1	2.01	.14	2	0
11	2	.18	3	45	274	.33	0	.13	.52	.04	.63	.04	<1	0
11	4	.24	2	33	304	.31	3	.05	.52	.04	.52	.04	1	0
8	3	.16	3	48	98	.27	11	.01	.57	.02	.43	.03	1	<1

Table A-1

Food Composition (Computer code number is for West Diet Analysis program) (For purposes of calculations, use "0" for t, <1, <.1, <.01, etc.)

Computer Code Number	Food Description	Measure	Wt (g)	H_2O (%)	Ener (kcal)	Prot (g)	Carb (g)	Dietary Fiber (g)	Fat (g)	Fat Breakdown (g) Sat	Mono	Poly
	MEATS: SAUSAGES and LUNCHMEATS—Continued											
1084	Smoked link sausage, beef and pork	1 ea	68	52	228	9	1	0	21	7.2	9.7	2.2
1083	Smoked link sausage, pork	1 ea	68	39	265	15	1	0	22	7.7	9.9	2.6
1085	Summer sausage	2 pce	46	51	154	7	<1	0	14	5.5	6	.6
1076	Turkey breakfast sausage	1 pce	28	60	64	6	0	0	5	1.6	1.8	1.2
679	Vienna sausage, canned	2 ea	32	60	89	3	1	0	8	3	4	.5
	MIXED DISHES and FAST FOODS											
	MIXED DISHES:											
1445	Almond Chicken	1 c	242	77	275	20	18	4	14	2	5.3	5.8
1981	Baked beans, fat free, honey	½ c	120	73	110	7	24	7	0	0	0	0
1454	Bean cake	1 ea	32	23	130	2	16	1	7	1	2.9	2.6
680	Beef stew w/ vegetables, homemade	1 c	245	82	218	16	15	2	10	4.9	4.5	.5
1109	Beef stew w/ vegetables, canned	1 c	245	82	194	14	17	2	8	2.4	3.1	.3
1116	Beef, macaroni, tomato sauce casserole	1 c	226	76	255	16	26	2	10	3.8	4.1	.5
2295	Beef fajita	1 ea	223	63	409	17	46	4	17	5.1	7.6	3.9
1265	Beef flauta	1 ea	113	49	360	16	13	2	27	4.9	11.6	9.1
681	Beef pot pie, homemade	1 pce	210	55	517	21	39	3	30	8.4	14.7	7.3
1898	Broccoli, batter fried	1 c	85	74	123	3	9	2	9	1.3	2.2	4.9
1462	Buffalo wings/spicy chicken wings	2 pce	32	53	98	8	<1	<1	7	1.8	2.8	1.6
1675	Carrot raisin salad	½ c	88	58	204	1	21	2	14	2	3.9	7.3
2248	Cheeseburger deluxe	1 ea	219	52	563	28	38		33	15	12.6	2
682	Chicken à la king, homemade	1 c	245	68	468	27	12	1	34	12.7	14.3	6.2
683	Chicken & noodles, homemade	1 c	240	71	367	22	26	2	18	5.9	7.1	3.5
684	Chicken chow mein, canned	1 c	250	89	95	6	18	2	1	0	.1	.8
685	Chicken chow mein, homemade	1 c	250	78	255	31	10	1	10	2.4	4.3	3.1
1266	Chicken fajita	1 ea	223	61	405	22	50	4	13	2.4	6	3.5
1264	Chicken flauta	1 ea	113	52	343	14	13	2	27	4.3	11.1	9.6
686	Chicken pot pie, homemade (⅓)	1 pce	232	57	545	23	42	3	31	10.9	14.5	5.8
1672	Chili con carne	½ c	127	77	128	12	11	2	4	1.7	1.7	.3
1112	Chicken salad with celery	½ c	78	53	268	11	1	<1	25	3.1	4.5	15.8
1382	Chicken teriyaki, breast	1 ea	128	67	176	26	7	<1	4	.9	1	.9
687	Chili with beans, canned	1 c	256	75	287	15	30	11	14	6	6	.9
1479	Chinese pastry	1 oz	28	46	67	1	13	<1	1	.2	.4	.8
688	Chop suey with beef & pork	1 c	220	63	425	22	31	3	24	5	8.6	9.3
690	Coleslaw	1 c	132	74	195	2	17	2	15	2.1	3.2	8.5
689	Corn pudding	1 c	250	76	273	11	32	4	13	6.4	4.3	1.8
1110	Corned beef hash, canned	1 c	220	67	398	19	23	1	25	11.9	10.9	.9
1255	Deviled egg (½ egg + filling)	1 ea	31	69	62	4	<1	0	5	1.2	1.7	1.5
	Egg foo yung patty:											
1467	Meatless	1 ea	86	78	113	6	3	1	8	1.9	3.3	2.1
1458	With beef	1 ea	86	74	129	9	3	<1	9	2.2	3.2	2.4
1465	With chicken	1 ea	86	74	130	9	4	<1	9	2.1	3.1	2.5
1602	Egg roll, meatless	1 ea	64	70	102	3	10	1	6	1.2	2.5	1.6
1550	Egg roll, with meat	1 ea	64	66	114	5	9	1	6	1.5	2.7	1.5
1113	Egg salad	1 c	183	57	586	17	3	0	56	10.6	17.4	24.2
691	French toast w/wheat bread, homemade	1 pce	65	54	151	5	16	<1	7	2	3	1.7
1355	Green pepper, stuffed	1 ea	172	75	229	11	20	2	11	5	4.9	.5
1487	Hot & sour soup (Chinese)	1 c	244	88	133	12	5	<1	6	2	2.5	1
2242	Hamburger deluxe	1 ea	110	49	279	13	27		13	4.1	5.3	2.6
1997	Hummous/hummus	¼ c	62	65	106	3	12	3	5	.8	2.2	2
	Lasagna:											
1346	With meat, homemade	1 pce	245	67	382	22	39	3	15	7.7	5	.9
1111	Without meat, homemade	1 pce	218	69	298	15	39	3	9	5.4	2.4	.6
1117	Frozen entree	1 ea	340	75	390	24	42	4	14	6.7	5.5	.8
1606	Lo mein, meatless	1 c	200	82	134	6	27	3	1	.1	.1	.3
1607	Lo mein, with meat	1 c	200	70	285	17	31	2	10	1.9	2.9	4.5
692	Macaroni & cheese, canned	1 c	240	80	228	9	26	1	10	4.2	3.1	1.4
693	Macaroni & cheese, homemade	1 c	200	58	430	17	40	1	22	8.9	8.8	3.6
1115	Macaroni salad, no cheese	1 c	177	60	461	5	28	2	37	4	6	25.5
1120	Meat loaf, beef	1 pce	87	63	182	16	4	<1	11	4.4	4.7	.5
1119	Meat loaf, beef and pork (⅓)	1 pce	87	60	205	15	4	<1	14	5.2	6.3	.9
1303	Moussaka (lamb & eggplant)	1 c	250	82	237	16	13	4	13	4.6	5.4	1.9

PAGE KEY: A–2 = Beverages A–4 = Dairy A–8 = Eggs A–8 = Fat/Oil A–12 = Fruit A–18 = Bakery A–26 = Grain A–30 = Fish A–32 = Meats A–36 = Poultry A–38 = Sausage A–40 = Mixed/Fast A–44 = Nuts/Seeds A–48 = Sweets A–50 = Vegetables/Legumes A–62 = Misc A–64 = Soups/Sauces A–66 = Fast A–82 = Frozen Entree A–86 = Baby foods

Chol (mg)	Calc (mg)	Iron (mg)	Magn (mg)	Pota (mg)	Sodi (mg)	Zinc (mg)	VT-A (RE)	Thia (mg)	VT-E (a-TE)	Ribo (mg)	Niac (mg)	V-B6 (mg)	Fola (μg)	VT-C (mg)
48	7	.99	8	129	643	1.43	0	.18	.15	.12	2.2	.12	1	0
46	20	.79	13	228	1020	1.92	0	.48	.17	.17	3.08	.24	3	1
34	6	1.17	6	125	571	1.18	0	.07	.1	.15	1.98	.12	1	0
23	5	.51	6	75	188	.96	0	.03	.14	.08	1.4	.08	1	0
17	3	.28	2	32	305	.51	0	.03	.07	.03	.51	.04	1	0
35	81	2	59	551	615	1.54	75	.08	2.64	.19	8.59	.42	31	10
0	40	2.7			135		450							12
0	3	.67	6	57	55	.16	0	.07	1.14	.05	.55	.02	9	0
64	29	2.94	40	613	292	5.29	568	.15	.49	.17	4.66	.28	37	17
34	29	2.21	39	426	1006	4.24	262	.07	.34	.12	2.45	.2	31	7
39	26	2.7	40	522	862	3.14	97	.22	.57	.22	4.31	.29	20	14
26	76	3.69	38	427	850	2.38	52	.46	2.08	.3	4.73	.32	25	29
45	50	2.15	29	292	187	4.18	15	.07	4	.15	2.13	.25	10	14
44	29	3.78	6	334	596	3.17	519	.29	3.78	.29	4.83	.24	29	6
16	67	.94	20	242	62	.38	102	.08	2.1	.13	.75	.11	43	53
26	5	.4	6	59	61	.56	17	.01	.23	.04	2.06	.13	1	<1
10	26	.75	14	317	118	.19	1452	.08	5.03	.05	.64	.22	9	5
88	206	4.66	44	445	1108	4.6	129	.39	1.18	.46	7.38	.28	81	8
186	127	2.45	20	404	760	1.8	272	.1	.98	.42	5.39	.23	11	12
96	26	2.16	26	149	600	1.53	10	.05		.17	4.32	.19	10	0
7	45	1.25	14	418	725	1.3	28	.05	.05	.1	1	.09	12	12
77	57	2.5	28	473	718	2.12	50	.07	.75	.22	4.25	.41	19	10
41	83	3.7	51	532	439	1.77	55	.48	2.04	.37	6.64	.35	41	22
37	52	.97	27	243	189	1.18	21	.05	4.06	.1	3.21	.22	8	14
72	70	3.02	25	343	594	2	735	.32	3.25	.32	4.87	.46	29	5
67	34	2.6	23	347	505	1.79	84	.06	.81	.57	1.24	.16	23	1
48	16	.62	11	138	201	.79	31	.03	6.27	.07	3.28	.34	8	1
80	27	1.75	36	309	1866	1.94	16	.08	.35	.2	8.69	.46	13	3
43	120	8.78	115	934	1336	5.12	87	.12	1.88	.27	.92	.34	58	4
0	8	.51	6	28	3	.18	<1	.05	.25	<.01	.41	.02	1	0
46	39	4.16	54	515	818	3.52	134	.36	1.82	.37	5.63	.44	44	20
7	45	.96	12	236	356	.26	66	.05	5.28	.04	.11	.14	51	11
250	100	1.4	37	403	138	1.25	90	1.03	.52	.32	2.47	.29	63	7
73	29	4.4	36	440	1188	3.3	0	.02	.48	.2	4.62	.43	20	0
121	15	.35	3	36	94	.3	49	.02	.86	.14	.02	.05	13	0
184	31	1.04	12	118	310	.7	86	.04	1.57	.25	.44	.09	30	5
180	26	1.11	11	145	184	1.16	92	.05	1.79	.24	.74	.15	22	3
182	27	.86	12	144	187	.81	95	.05	1.87	.25	.96	.13	22	3
30	12	.76	9	98	306	.25	15	.08	.81	.1	.81	.05	12	3
37	13	.78	10	124	304	.46	14	.16	.78	.13	1.31	.1	9	2
574	74	1.8	13	180	665	1.42	260	.08	8.87	.66	.09	.46	62	0
76	64	1.09	11	86	311	.44	81	.13	.31	.21	1.06	.05	15	<1
34	16	1.77	20	233	201	2.28	44	.15	.75	.1	2.74	.3	17	55
23	29	1.83	27	351	1562	1.17	2	.19	.12	.22	4.58	.15	12	1
26	63	2.63	22	227	504	2.06	9	.23	.82	.2	3.69	.12	52	2
0	31	.97	18	108	151	.68	1	.06	.62	.03	.25	.25	37	5
56	258	3.22	50	461	745	3.25	158	.23	1.15	.33	3.97	.21	19	15
31	252	2.5	44	375	714	1.77	156	.22	1.07	.27	2.49	.17	17	15
55	263	3.44	64	752	823	3.7	248	.29	3.45	.39	5.07	.32	28	41
0	47	2.06	33	389	623	.92	130	.23	.35	.24	2.83	.19	49	12
30	25	2.11	39	246	276	1.63	6	.37	1.51	.24	4.25	.28	41	8
24	199	.96	31	139	730	1.2	73	.12	.14	.24	.96	.02	8	<1
42	362	1.8	37	240	1086	1.2	234	.2	.12	.4	1.8	.05	10	1
27	31	1.56	20	170	352	.53	44	.18	10.3	.1	1.43	.33	20	4
84	29	1.61	14	187	145	3.23	23	.05	.31	.22	2.61	.15	11	1
84	33	1.42	14	213	381	2.68	23	.19	.32	.22	2.68	.17	10	1
97	68	1.79	40	557	432	2.56	105	.15	.81	.31	4.14	.23	45	6

A

Table A–1

Food Composition (Computer code number is for West Diet Analysis program) (For purposes of calculations, use "0" for t, <1, <.1, <.01, etc.)

Computer Code Number	Food Description	Measure	Wt (g)	H_2O (%)	Ener (kcal)	Prot (g)	Carb (g)	Dietary Fiber (g)	Fat (g)	Fat Breakdown (g) Sat	Mono	Poly
	MIXED DISHES and FAST FOODS—Continued											
1899	Mushrooms, batter fried	5 ea	70	66	148	2	8	1	12	2.1	3	6.4
715	Potato salad with mayonnaise and eggs	½ c	125	76	179	3	14	2	10	1.8	3.1	4.7
1674	Pizza, combination, 1/12 of 12″ round	1 pce	79	48	184	13	21		5	1.5	2.5	.9
1673	Pizza, pepperoni, 1/12 of 12″ round	1 pce	71	46	181	10	20		7	2.2	3.1	1.2
694	Quiche Lorraine ⅛ of 8″ quiche	1 pce	176	54	508	20	20	1	39	17.6	13.8	4.9
1449	Ramen noodles, cooked	1 c	227	82	156	6	29	3	2	.4	.5	.5
1671	Ravioli, meat	½ c	125	68	194	10	18	1	9	2.9	3.6	1
1597	Fried rice (meatless)	1 c	166	68	264	5	34	1	12	1.7	3	6.3
2142	Roast beef hash	½ c	117	66	230	9	11	1	16	7	5.8	3.2
	Spaghetti (enriched) in tomato sauce											
	With cheese:											
695	Canned	1 c	250	80	190	5	38	2	1	0	.4	.5
696	Home recipe	1 c	250	77	260	9	37	2	9	2	5.4	1.2
	With meatballs:											
697	Canned	1 c	250	78	258	12	28	6	10	2.1	3.9	3.9
698	Home recipe	1 c	248	70	332	19	39	8	12	3.3	6.3	2.2
716	Spinach soufflé	1 c	136	74	219	11	3	3	19	9.5	5.7	2.2
1553	Sweet & sour pork	1 c	226	77	231	15	25	1	8	2.2	2.9	2.4
1263	Sweet & sour chicken breast	1 ea	131	79	117	8	15	1	3	.6	.8	1.5
1515	Three bean salad	1 c	150	82	139	4	13	3	8	1.2	1.9	4.9
717	Tuna salad	1 c	205	63	383	33	19	0	19	3.2	5.9	8.4
1121	Tuna noodle casserole, homemade	1 c	202	75	237	17	25	1	7	1.9	1.5	3.2
1270	Waldorf salad	1 c	137	58	408	4	13	2	40	4.1	7.3	27
	FAST FOODS and SANDWICHES (see end of this appendix for additional Fast Foods)											
699	Burrito, beef & bean	1 ea	116	52	255	11	33	3	9	4.2	3.5	.6
700	Burrito, bean	1 ea	109	52	225	7	36	4	7	3.5	2.4	.6
2106	Burrito, chicken con queso	1 ea	306	76	280	12	53	5	6	1.5		
701	Cheeseburger with bun, regular	1 ea	154	55	359	18	28		20	9.2	7.2	1.5
702	Cheeseburger with bun, 4-oz patty	1 ea	166	51	417	21	35		21	8.7	7.8	2.7
703	Chicken patty sandwich	1 ea	182	47	515	24	39	1	29	8.5	10.4	8.4
704	Corndog	1 ea	175	47	460	17	56		19	5.2	9.1	3.5
1922	Corndog, chicken	1 ea	113	52	271	13	26		13			
705	Enchilada	1 ea	163	63	319	10	28		19	10.6	6.3	.8
706	English muffin with egg, cheese, bacon	1 ea	146	49	383	20	31	1	20	9	6.8	2.1
	Fish sandwich:											
707	Regular, with cheese	1 ea	183	45	523	21	48	<1	28	8.1	8.9	9.4
708	Large, no cheese	1 ea	158	47	431	17	41	<1	23	5.2	7.7	8.2
709	Hamburger with bun, regular	1 ea	107	45	275	14	33	1	10	3.5	3.7	1.8
710	Hamburger with bun, 4-oz patty	1 ea	215	50	576	32	39		32	12	14.1	2.8
711	Hotdog/frankfurter with bun	1 ea	98	54	242	10	18		14	5.1	6.8	1.7
	Lunchables:											
2129	Bologna & American cheese	1 ea	128		450	18	19	0	34	15		
2130	Ham & cheese	1 ea	128		320	22	19	0	17	8		
2117	Honey ham & Amer. w/choc pudding	1 ea	176		390	18	34	<1	20	9		
2118	Honey turkey & cheddar w/Jello	1 ea	163		320	17	27	<1	16	9		
2131	Pepperoni & American cheese	1 ea	128		480	20	19	0	36	17		
2125	Salami & American cheese	1 ea	128		430	18	18	0	32	15		
2127	Turkey & cheddar cheese	1 ea	128		360	20	20	1	22	11		
712	Pizza, cheese, ⅛ of 15″ round	1 pce	63	48	140	8	20	1	3	1.5	1	.5
	SANDWICHES:											
	Avocado, chesse, tomato & lettuce:											
1276	On white bread, firm	1 ea	210	58	478	15	41	5	29	8.8	11.3	7.2
1278	On part whole wheat	1 ea	201	59	444	14	34	7	29	8.6	11.4	7.3
1277	On whole wheat	1 ea	214	58	468	16	40	8	30	8.7	11.6	7.5
	Bacon, lettuce & tomato sandwich:											
1137	On white bread, soft	1 ea	124	53	308	10	28	2	18	4.5	6.1	6.1
1139	On part whole wheat	1 ea	124	54	303	10	26	3	17	4.3	6.2	6.1
1138	On whole wheat	1 ea	137	53	328	12	32	4	18	4.4	6.4	6.3

Chol (mg)	Calc (mg)	Iron (mg)	Magn (mg)	Pota (mg)	Sodi (mg)	Zinc (mg)	VT-A (RE)	Thia (mg)	VT-E (a-TE)	Ribo (mg)	Niac (mg)	V-B6 (mg)	Fola (μg)	VT-C (mg)
14	54	.76	8	180	121	.42	10	.07	.92	.22	1.65	.05	8	1
85	24	.81	19	318	661	.39	41	.1	2.33	.07	1.11	.18	8	12
20	101	1.53	18	179	382	1.11	101	.21		.17	1.96	.09	32	2
14	65	.94	9	153	267	.52	55	.13		.23	3.05	.06	37	2
205	201	1.9	27	271	549	1.66	243	.23	1.91	.44	4.71	.19	17	3
38	20	1.89	24	51	1349	.76	204	.22	.09	.1	1.75	.06	9	<1
84	32	2.03	20	259	619	1.67	94	.15	1.52	.22	2.95	.14	14	11
42	30	1.84	24	134	286	.89	21	.21	2.46	.11	2.25	.15	22	4
40	10	.9	22	362	695	2.99	0	.09		.12	2.33	.3	12	0
7	40	2.75	21	303	955	1.12	120	.35	2.13	.27	4.5	.13	6	10
7	80	2.25	26	408	955	1.3	140	.25	2.75	.17	2.25	.2	8	12
22	52	3.25	20	245	1220	2.39	100	.15	1.5	.17	2.25	.12	5	5
74	124	3.72	40	665	1009	2.45	159	.25	1.64	.3	3.97	.2	10	22
184	230	1.35	38	201	763	1.29	675	.09	1.22	.3	.48	.12	80	3
38	28	1.36	34	390	1219	1.46	28	.55	.62	.21	3.6	.41	10	20
23	16	.79	21	187	732	.66	20	.06	.39	.08	3.06	.18	6	12
0	35	1.42	25	224	514	.54	23	.07	1.96	.09	.4	.04	53	4
27	35	2.05	39	365	824	1.15	55	.06	1.95	.14	13.7	.17	16	5
41	34	2.3	30	182	772	1.2	13	.18	1.18	.15	7.78	.2	10	1
21	43	.88	39	270	236	.63	39	.1	8.67	.05	.36	.36	27	6
24	53	2.46	42	329	670	1.93	32	.27	.7	.42	2.71	.19	58	1
2	57	2.27	44	328	495	.76	16	.32	.87	.3	2.04	.15	44	1
10	40	.72			600		40							15
52	182	2.65	26	229	976	2.62	71	.32	1.34	.23	6.38	.15	65	2
60	171	3.42	30	335	1050	3.49	65	.35		.28	8.05	.18	61	2
60	60	4.68	35	353	957	1.87	31	.33	.55	.24	6.81	.2	100	9
79	102	6.18	17	263	973	1.31	37	.28	.7	.7	4.17	.09	103	0
64					668									
44	324	1.32	50	240	784	2.51	186	.08	1.47	.42	1.91	.39	65	1
234	207	3.29	34	213	784	1.81	158	.48	.6	.53	3.93	.16	47	1
68	185	3.5	37	353	939	1.17	97	.46	1.83	.42	4.23	.11	91	3
55	84	2.61	33	340	615	.99	30	.33	.87	.22	3.4	.11	85	3
43	51	2.46	22	215	564	2.05	13	.26	.43	.32	4.7	.13	52	3
103	92	5.55	45	527	742	5.81	4	.34	1.61	.41	6.73	.37	84	1
44	23	2.31	13	143	670	1.98	0	.23	.27	.27	3.65	.05	48	<1
85	300	2.7			1620		60							0
60	300	1.8			1770		80							
55	250	2.7			1540		40							
50	20	6			1360		80							
95	250	2.7			1840		60							
80	250	2.7			1740		60							
70	300	1.8			1650		60							
9	117	.58	16	110	336	.81	74	.18		.16	2.48	.04	35	1
34	294	3.06	54	581	550	1.71	140	.37	4.55	.39	3.77	.32	80	11
31	291	3.1	67	617	525	1.91	140	.35	4.55	.39	3.94	.35	80	11
31	281	3.53	102	679	593	2.68	140	.36	4.17	.37	4.29	.42	92	11
21	52	1.99	20	233	590	.96	31	.35	2.34	.19	3.2	.14	34	12
20	63	2.27	33	283	604	1.19	31	.36	2.7	.21	3.62	.17	37	12
20	55	2.68	64	342	670	1.9	31	.37	2.36	.2	3.97	.24	48	12

Table A-1

Food Composition (Computer code number is for West Diet Analysis program) (For purposes of calculations, use "0" for t, <1, <.1, <.01, etc.)

Computer Code Number	Food Description	Measure	Wt (g)	H_2O (%)	Ener (kcal)	Prot (g)	Carb (g)	Dietary Fiber (g)	Fat (g)	Fat Breakdown (g) Sat	Mono	Poly
	MIXED DISHES and FAST FOODS—Continued											
	Cheese, grilled:											
1140	On white bread, soft	1 ea	119	37	400	18	30	1	24	13.2	7.5	2
1142	On part whole wheat	1 ea	119	37	396	18	28	3	24	13	7.6	2.1
1141	On whole wheat	1 ea	132	38	420	20	33	4	24	13.1	7.8	2.2
1596	Chicken fillet	1 ea	182	47	515	24	39	1	29	8.5	10.4	8.4
	Chicken salad:											
1143	On white bread, soft	1 ea	110	40	369	11	31	1	23	3.7	6.1	12.1
1145	On part whole wheat	1 ea	110	41	364	11	29	4	23	3.5	6.1	12.1
1144	On whole wheat	1 ea	123	41	387	13	34	5	23	3.6	6.3	12.2
1146	Corned beef & swiss on rye	1 ea	156	49	420	28	22	<1	26	9.4	7.4	6.3
	Egg salad:											
1147	On white bread, soft	1 ea	117	43	380	10	31	1	25	4.4	6.8	12
1149	On part whole wheat	1 ea	116	43	374	10	29	3	25	4.1	6.8	12
1148	On whole wheat	1 ea	130	43	400	12	35	5	25	4.3	7.1	12.2
	Ham:											
1279	On rye bread	1 ea	150	60	283	22	21	<1	13	2.4	3.8	6
1151	On white bread, soft	1 ea	157	55	334	22	30	1	14	3	4.4	5.9
1153	On part whole wheat	1 ea	156	55	328	22	28	3	14	2.7	4.4	6
1152	On whole wheat	1 ea	169	54	352	24	34	4	15	2.9	4.7	6.1
	Ham & cheese:											
1280	On white bread, soft	1 ea	157	49	403	23	30	1	22	8.4	5.9	6.1
1282	On part whole wheat	1 ea	156	50	397	23	28	3	21	8.1	5.9	6.2
1281	On whole wheat	1 ea	170	49	424	24	34	4	22	8.3	6.2	6.4
1150	Ham & swiss on rye	1 ea	150	54	339	22	22	<1	19	6.5	5.1	6
	Ham salad:											
1154	On white bread, soft	1 ea	131	47	362	11	37	1	20	4.8	6.7	7.4
1156	On part whole wheat	1 ea	131	48	357	11	35	3	20	4.5	6.8	7.5
1155	On whole wheat	1 ea	144	47	380	12	40	4	20	4.7	7	7.6
1157	Patty melt: Ground beef & cheese on rye	1 ea	182	46	561	37	22	3	37	13.2	11.7	8.4
	Peanut butter & jelly:											
1158	On white bread, soft	1 ea	101	26	351	12	47	3	15	3.1	6.7	3.9
1160	On part whole wheat	1 ea	101	27	346	12	45	5	15	2.9	6.7	4
1159	On whole wheat	1 ea	114	28	370	13	50	6	15	3	7	4.1
1161	Reuben, grilled: Corned beef, swiss cheese, sauerkraut on rye	1 ea	239	64	462	28	25	2	29	9.9	9.5	7.1
	Roast beef:											
713	On a bun	1 ea	139	49	346	21	33		14	3.6	6.8	1.7
1162	On white bread, soft	1 ea	157	46	404	29	34	1	17	3.4	4.2	8.2
1164	On part whole wheat	1 ea	156	47	398	29	32	3	17	3.2	4.3	8.3
1163	On whole wheat	1 ea	169	46	422	31	38	4	17	3.3	4.5	8.4
	Tuna salad:											
1165	On white bread, soft	1 ea	122	46	327	14	35	2	15	2.5	3.8	7.9
1167	On part whole wheat	1 ea	122	47	322	14	33	4	15	2.2	3.8	8
1166	On whole wheat	1 ea	135	46	346	16	39	5	15	2.3	4.1	8.1
	Turkey:											
1168	On white bread, soft	1 ea	156	54	346	24	29	1	15	2.4	3.2	8.3
1170	On part whole wheat	1 ea	155	54	338	24	27	3	14	2.1	3.2	8.3
1169	On whole wheat	1 ea	169	53	365	26	33	4	15	2.3	3.5	8.5
	Turkey ham:											
1272	On rye bread	1 ea	150	60	280	21	20	<1	14	2.5	2.8	6.9
1273	On white bread, soft	1 ea	156	55	331	21	29	1	14	3	3.4	6.8
1275	On part whole wheat	1 ea	156	56	326	21	28	3	14	3	4.2	5.6
1274	On whole wheat	1 ea	169	55	350	23	33	4	15	2.9	3.7	7
714	Taco	1 ea	171	58	369	21	27		20	11.4	6.6	1
	Tostada:											
1114	With refried beans	1 ea	144	66	223	10	26	7	10	5.4	3	.7
1118	With beans & beef	1 ea	225	70	333	16	30	4	17	11.5	3.5	.6
1354	With beans & chicken	1 ea	156	68	248	19	18	3	11	5.3	3.9	1.6
	NUTS, SEEDS, and PRODUCTS											
	Almonds:											
1365	Dry roasted, salted	1 c	138	3	810	22	33	19	71	6.7	46.2	14.9

Chol (mg)	Calc (mg)	Iron (mg)	Magn (mg)	Pota (mg)	Sodi (mg)	Zinc (mg)	VT-A (RE)	Thia (mg)	VT-E (a-TE)	Ribo (mg)	Niac (mg)	V-B6 (mg)	Fola (μg)	VT-C (mg)
55	399	1.81	25	154	1143	2.05	212	.24	1.13	.34	1.91	.06	24	<1
54	412	2.12	39	209	1160	2.31	212	.25	1.53	.36	2.39	.1	28	<1
54	402	2.54	73	271	1226	3.08	212	.26	1.14	.35	2.74	.17	40	<1
60	60	4.68	35	353	957	1.87	31	.33	.55	.24	6.81	.2	100	9
32	60	2.04	18	139	460	.8	24	.25	6.16	.18	3.68	.25	26	1
31	73	2.35	33	195	475	1.06	24	.26	6.57	.2	4.16	.29	30	1
30	63	2.79	68	259	543	1.85	24	.27	6.14	.19	4.5	.36	42	1
82	268	3.12	28	225	1392	3.65	81	.19	2.59	.33	2.72	.17	19	1
157	71	2.18	16	113	526	.74	76	.26	4.52	.31	1.98	.2	37	0
155	83	2.47	31	169	539	1	75	.27	4.91	.33	2.45	.23	41	0
155	73	2.92	66	234	611	1.8	76	.28	4.53	.32	2.82	.3	53	0
47	48	2.3	26	364	1566	2.11	8	.99	2.36	.31	5.45	.47	15	23
47	60	2.39	29	368	1619	2.04	8	1.02	2.34	.33	5.99	.47	24	22
45	71	2.68	43	421	1630	2.29	8	1.03	2.73	.35	6.45	.5	27	22
45	62	3.1	77	483	1696	3.05	8	1.04	2.34	.33	6.78	.57	39	22
61	232	2.28	30	315	1620	2.34	90	.76	2.53	.36	4.64	.36	25	15
59	244	2.58	45	368	1630	2.59	90	.77	2.93	.39	5.09	.39	28	15
59	236	3.02	79	432	1707	3.37	90	.78	2.56	.37	5.47	.46	40	15
57	258	2.25	29	344	1602	2.59	79	.72	2.52	.36	4.06	.35	16	15
30	56	2.07	19	159	921	1.06	8	.51	3.29	.22	3.25	.17	22	4
29	69	2.38	33	216	936	1.32	8	.52	3.69	.24	3.74	.21	26	4
29	59	2.81	69	279	1001	2.11	8	.53	3.29	.23	4.09	.28	38	4
113	222	4.19	36	391	701	7.11	123	.25	3.5	.46	6.14	.35	25	<1
2	60	2.25	56	245	293	1.07	<1	.27	.12	.17	5.33	.13	40	<1
0	72	2.55	70	299	308	1.32	<1	.28	.51	.19	5.8	.17	44	<1
0	63	2.97	104	361	375	2.09	<1	.29	.14	.17	6.14	.24	56	<1
80	288	4.24	38	361	1949	3.73	130	.21	4.46	.34	2.79	.27	38	13
51	54	4.23	31	316	792	3.39	21	.37	.19	.31	5.87	.26	57	2
45	60	3.98	28	432	1595	3.78	12	.29	3.3	.3	6.39	.39	30	12
43	72	4.27	42	485	1607	4.02	12	.31	3.69	.32	6.84	.42	34	12
43	62	4.7	77	547	1672	4.79	12	.31	3.31	.31	7.18	.49	45	12
14	60	2.24	22	161	567	.67	22	.25	2.72	.18	5.53	.11	25	1
13	73	2.55	37	217	582	.93	22	.26	3.12	.2	6	.16	29	1
12	63	2.98	72	280	649	1.72	22	.27	2.71	.19	6.34	.23	41	1
45	56	2	29	302	1585	1.33	12	.26	3.46	.23	8.97	.4	24	0
43	68	2.28	43	354	1589	1.57	11	.27	3.82	.25	9.38	.44	28	0
43	59	2.73	77	418	1665	2.35	12	.28	3.47	.23	9.77	.51	39	0
55	51	4.06	25	342	1185	3	8	.22	2.8	.33	4.3	.29	17	<1
55	62	4.09	28	346	1248	2.9	8	.27	2.75	.35	4.87	.28	25	0
53	74	4.37	42	400	1262	3.15	8	.28	3.57	.37	5.34	.32	29	0
53	65	4.81	76	462	1329	3.91	8	.29	2.76	.35	5.68	.39	41	0
56	221	2.41	70	474	802	3.93	147	.15	1.88	.44	3.21	.24	68	2
30	210	1.89	59	403	543	1.9	85	.1	1.15	.33	1.32	.16	43	1
74	189	2.45	67	491	871	3.17	173	.09	1.8	.49	2.86	.25	85	4
53	168	1.79	47	365	433	2.28	86	.11	1.87	.2	4.52	.32	53	3
0	389	5.24	420	1062	1076	6.76	0	.18	7.66	.83	3.89	.1	88	1

Table A-1

Food Composition (Computer code number is for West Diet Analysis program) (For purposes of calculations, use "0" for t, <1, <.1, <.01, etc.)

Computer Code Number	Food Description	Measure	Wt (g)	H_2O (%)	Ener (kcal)	Prot (g)	Carb (g)	Dietary Fiber (g)	Fat (g)	Fat Breakdown (g) Sat	Mono	Poly
	NUTS, SEEDS, and PRODUCTS—Continued											
	Almonds:											
718	Slivered, packed, unsalted	1 c	108	4	636	22	22	12	56	5.3	36.6	11.9
719	Whole, dried, unsalted	1 c	142	4	836	28	29	15	74	7	48.1	15.6
720	Ounce	1 oz	28	4	165	6	6	3	15	1.4	9.5	3.1
721	Almond butter:	1 tbs	16	1	101	2	3	1	9	.9	6.1	2
4572	Salted	1 tbs	16	1	101	2	3	1	9	.9	6.1	2
722	Brazil nuts, dry (about 7)	1 c	140	3	918	20	18	8	93	22.7	32.2	33.7
	Cashew nuts, dry roasted:											
723	Salted:	1 c	137	2	786	21	45	4	64	12.8	37.4	10.7
724	Ounce	1 oz	28	2	161	4	9	1	13	2.6	7.6	2.2
4621	Unsalted:	1 c	137	2	786	21	45	4	64	12.8	37.4	10.7
4621	Ounce	1 oz	28	2	161	4	9	1	13	2.6	7.6	2.2
725	Oil roasted:	1 c	130	4	749	23	37	5	63	12.6	36.9	10.6
726	Ounce	1 oz	28	4	161	5	8	1	13	2.7	7.9	2.3
4622	Unsalted:	1 c	130	4	749	21	37	5	63	12.6	36.9	10.6
4622	Ounce	1 oz	28	4	161	5	8	1	13	2.7	7.9	2.3
727	Cashew butter, unsalted	1 tbs	16	3	94	3	4	<1	8	1.6	4.7	1.3
4662	Cashew butter, salted	1 tbs	16	3	94	3	4	<1	8	1.6	4.7	1.3
728	Chestnuts, European, roasted (1 cup = approx 17 kernels)	1 c	143	40	350	5	76	7	3	.6	1.1	1.2
	Coconut, raw:											
729	Piece 2 x 2 x ½″	1 pce	45	47	159	2	7	4	15	13.5	.6	.2
730	Shredded/grated, unpacked	½ c	40	47	142	1	6	4	13	12	.6	.1
	Coconut, dried, shredded/grated:											
731	Unsweetened	1 c	78	3	515	6	19	13	50	45.1	2.1	.6
732	Sweetened	1 c	93	13	466	3	44	4	33	29.6	1.4	.4
733	Filberts/hazelnuts, chopped:	1 c	135	5	853	18	21	8	84	6.2	66.3	8.1
734	Ounce	1 oz	28	5	177	4	4	2	17	1.3	13.7	1.7
735	Macadamias, oil roasted, salted:	1 c	134	2	962	10	17	12	103	15.4	80.9	1.8
736	Ounce	1 oz	28	2	201	2	4	3	21	3.2	16.9	.4
1368	Macadamias, oil roasted, unsalted	1 c	134	2	962	10	17	12	103	15.4	80.9	1.8
	Mixed nuts:											
737	Dry roasted, salted	1 c	137	2	814	24	35	12	71	9.4	43	14.8
738	Oil roasted, salted	1 c	142	2	876	24	30	13	80	12.4	45	18.9
1369	Oil roasted, unsalted	1 c	142	2	876	27	30	14	80	12.4	45	18.9
	Peanuts:											
739	Oil roasted, salted	1 c	144	2	837	38	27	13	71	9.8	35.3	22.5
740	Ounce	1 oz	28	2	163	7	5	3	14	1.9	6.9	4.4
1370	Oil roasted, unsalted	1 c	144	2	837	38	27	10	71	9.8	35.3	22.5
741	Dried, salted	1 c	146	2	854	35	31	12	73	10.1	36.1	22.9
742	Ounce	1 oz	28	2	164	7	6	2	14	1.9	6.9	4.4
743	Peanut butter:	½ c	128	1	759	33	25	8	65	14.3	31.1	17.7
1371	Tablespoon	2 tbs	32	1	190	8	6	2	16	3.6	7.8	4.4
744	Pecan halves, dried, unsalted:	1 c	108	5	720	9	20	8	73	5.8	45.6	18
745	Ounce	1 oz	28	5	187	2	5	2	19	1.5	11.8	4.7
1372	Pecan halves, dry roasted, salted	¼ c	28	1	185	2	6	3	18	1.4	11.3	4.5
746	Pine nuts/piñons, dried	1 oz	28	6	176	3	5	3	17	2.6	6.4	7.2
747	Pistachios, dried, shelled	1 oz	28	4	162	6	7	3	14	1.8	9.2	2
1373	Pistachios,dry roasted,salted,shelled	1 c	128	2	776	19	35	14	68	8.8	45.7	10.2
748	Pumpkin kernels, dried, unsalted	1 oz	28	7	151	7	5	1	13	2.4	4	5.8
1374	Pumpkin kernels, roasted, salted	1 c	227	7	1184	75	30	9	96	18.1	29.7	43.6
749	Sesame seeds, hulled, dried	¼ c	38	5	223	10	4	3	21	2.9	7.9	9.1
	Sunflower seed kernels:											
750	Dry	¼ c	36	5	205	8	7	4	18	1.9	3.4	11.8
751	Oil roasted	¼ c	34	3	209	7	5	2	20	2	3.7	12.9
752	Tahini (sesame butter)	1 tbs	15	3	91	3	3	1	8	1.2	3.2	3.7
1334	Trail mix w/chocolate chips	1 c	146	7	707	21	66	8	47	9.3	19.8	16.5
753	Black walnuts, chopped:	1 c	125	4	759	31	15	6	71	4.8	15.9	46.9
754	Ounce	1 oz	28	4	170	7	3	1	16	1.1	3.6	10.5
755	English walnuts, chopped:	1 c	120	4	770	17	22	6	74	7.2	17	46.9
756	Ounce	1 oz	28	4	180	4	5	1	17	1.7	4	10.9

Chol (mg)	Calc (mg)	Iron (mg)	Magn (mg)	Pota (mg)	Sodi (mg)	Zinc (mg)	VT-A (RE)	Thia (mg)	VT-E (a-TE)	Ribo (mg)	Niac (mg)	V-B6 (mg)	Fola (μg)	VT-C (mg)
0	287	3.95	320	791	12	3.15	0	.23	25.9	.84	3.63	.12	63	1
0	378	5.2	420	1039	16	4.15	0	.3	34.1	1.11	4.77	.16	83	1
0	74	1.02	83	205	3	.82	0	.06	6.72	.22	.94	.03	16	<1
0	43	.59	48	121	2	.49	0	.02	3.25	.1	.46	.01	10	<1
0	43	.59	48	121	72	.49	0	.02	3.25	.1	.46	.01	10	<1
0	246	4.76	315	840	3	6.43	0	1.4	10.6	.17	2.27	.35	6	1
0	62	8.22	356	774	877	7.67	0	.27	.78	.27	1.92	.35	95	0
0	13	1.68	73	158	179	1.57	0	.06	.16	.06	.39	.07	19	0
0	62	8.22	356	774	22	7.67	0	.27	.78	.27	1.92	.35	95	0
0	13	1.68	73	158	4	1.57	0	.06	.16	.06	.39	.07	19	0
0	53	5.33	332	689	814	6.18	0	.55	2.03	.23	2.34	.32	88	0
0	11	1.15	71	148	175	1.33	0	.12	.44	.05	.5	.07	19	0
0	53	5.33	332	689	22	6.18	0	.55	2.03	.23	2.34	.32	88	0
0	11	1.15	71	148	5	1.33	0	.12	.44	.05	.5	.07	19	0
0	7	.8	41	87	2	.83	0	.05	.25	.03	.26	.04	11	0
0	7	.8	41	87	98	.83	0	.05	.25	.03	.26	.04	11	0
0	41	1.3	47	847	3	.81	3	.35	1.72	.25	1.92	.71	100	37
0	6	1.09	14	160	9	.49	0	.03	.33	.01	.24	.02	12	1
0	6	.97	13	142	8	.44	0	.03	.29	.01	.22	.02	11	1
0	20	2.59	70	424	29	1.57	0	.05	1.05	.08	.47	.23	7	1
0	14	1.79	46	313	244	1.69	0	.03	1.26	.02	.44	.25	8	1
0	254	4.41	385	601	4	3.24	9	.67	32.3	.15	1.54	.83	97	1
0	53	.92	80	125	1	.67	2	.14	6.69	.03	.32	.17	20	<1
0	60	2.41	157	441	348	1.47	1	.28	.55	.15	2.71	.26	21	0
0	13	.5	33	92	73	.31	<1	.06	.11	.03	.57	.05	4	0
0	60	2.41	157	441	9	1.47	1	.28	.55	.15	2.71	.26	21	0
0	96	5.07	308	818	917	5.21	1	.27	8.22	.27	6.44	.41	69	1
0	153	4.56	334	825	926	7.21	3	.71	8.52	.31	7.19	.34	118	1
0	153	4.56	334	825	16	7.21	3	.71	8.52	.31	7.19	.34	118	1
0	127	2.64	266	982	624	9.55	0	.36	10.7	.16	20.6	.37	181	0
0	25	.51	52	191	121	1.86	0	.07	2.07	.03	4	.07	35	0
0	127	2.64	266	982	9	9.55	0	.36	10.7	.16	20.6	.37	181	0
0	79	3.3	257	961	1186	4.83	0	.64	10.8	.14	19.7	.37	212	0
0	15	.63	49	184	228	.93	0	.12	2.07	.03	3.78	.07	41	0
0	49	2.36	204	856	598	3.74	0	.11	12.8	.13	17.2	.58	95	0
0	12	.59	51	214	149	.93	0	.03	3.2	.03	4.29	.14	24	0
0	39	2.3	138	423	1	5.91	14	.92	3.35	.14	.96	.2	42	2
0	10	.6	36	110	<1	1.53	4	.24	.87	.04	.25	.05	11	1
0	10	.61	37	104	218	1.59	4	.09	.84	.03	.26	.05	11	1
0	2	.86	65	176	20	1.2	1	.35	.98	.06	1.22	.03	16	1
0	38	1.9	44	306	2	.37	6	.23	1.46	.05	.3	.07	16	2
0	90	4.06	166	1241	998	1.74	31	.54	8.26	.31	1.8	.33	76	9
0	12	4.2	150	226	5	2.09	11	.06	.28	.09	.49	.06	16	1
0	98	33.8	1212	1829	1305	16.9	86	.48	2.27	.72	3.95	.2	130	4
0	50	2.96	132	155	15	3.91	3	.27	.86	.03	1.78	.05	36	0
0	42	2.44	127	248	1	1.82	2	.82	18.1	.09	1.62	.28	82	<1
0	19	2.28	43	164	1	1.77	2	.11	17.1	.09	1.4	.27	80	<1
0	21	.95	53	69	<1	1.58	1	.24	.34	.02	.85	.02	15	0
6	159	4.95	235	946	177	4.58	7	.6	15.6	.33	6.44	.38	95	2
0	72	3.84	253	655	1	4.28	37	.27	3.28	.14	.86	.69	82	4
0	16	.86	57	147	<1	.96	8	.06	.73	.03	.19	.15	18	1
0	113	2.93	203	602	12	3.28	14	.46	3.14	.18	1.25	.67	79	4
0	26	.68	47	141	3	.76	3	.11	.73	.04	.29	.16	18	1

Table A-1

Food Composition (Computer code number is for West Diet Analysis program) (For purposes of calculations, use "0" for t, <1, <.1, <.01, etc.)

Computer Code Number	Food Description	Measure	Wt (g)	H_2O (%)	Ener (kcal)	Prot (g)	Carb (g)	Dietary Fiber (g)	Fat (g)	Fat Breakdown (g) Sat	Mono	Poly
	SWEETENERS and SWEETS (see also Dairy [milk desserts] and Baked Goods)											
757	Apple butter	2 tbs	36	52	66	<1	17	<1	<1	t	t	t
1124	Butterscotch topping	2 tbs	41	32	103	1	27	<1	<1	t	t	0
1125	Caramel topping	2 tbs	41	32	103	1	27	<1	<1	t	t	0
	Cake frosting, creamy vanilla:											
1127	Canned	2 tbs	39	13	163	<1	27	<1	7	1.9	3.4	.9
1123	From mix	2 tbs	39	12	165	<1	28	<1	6	1.3	2.6	2.2
	Cake frosting, lite:											
2061	Milk chocolate	1 tbs	16	18	58	<1	11	<1	1	.4		
2062	Vanilla	1 tbs	16	15	60	0	12	<1	1	.4		
	Candy:											
1128	Almond Joy candy bar	1 oz	28	10	131	1	16	1	7	4.8	1.8	.4
2069	Butterscotch morsels	¼ c	43	1	246	0	31	0	12	12.5	0	0
758	Caramel, plain or chocolate	1 pce	10	8	38	<1	8	<1	1	.7	.1	t
1961	Chewing gum, sugarless	1 pce	3		6	0	2		0	0	0	0
	Chocolate (see also #784, 785, 971):											
	Milk chocolate:											
759	Plain	1 oz	28	1	144	2	17	1	9	5.2	2.8	.3
760	With almonds	1 oz	28	1	147	3	15	2	10	4.8	3.8	.6
761	With peanuts	1 oz	28	1	155	5	11	2	11	3.4	5.1	2.5
762	With rice cereal	1 oz	28	2	139	2	18	1	7	4.4	2.4	.2
763	Semisweet chocolate chips	1 c	168	1	805	7	106	10	50	29.9	16.8	1.6
764	Sweet dark chocolate (candy bar)	1 ea	41	1	226	2	25	2	13	8.3	4.6	.4
765	Fondant candy, uncoated (mints, candy corn, other)	1 pce	16	7	57	0	15	0	0	0	0	0
1697	Fruit Roll-Up (small)	1 ea	14	11	49	<1	12	<1	<1	.1	.2	.1
766	Fudge, chocolate	1 pce	17	10	65	<1	13	<1	1	.9	.4	.1
767	Gumdrops	1 c	182	1	703	0	180	0	0	0	0	0
768	Hard candy, all flavors	1 pce	6	1	22	0	6	0	<1	0	0	0
769	Jellybeans	10 pce	11	6	40	0	10	0	<1	t	t	t
1134	M&M's plain chocolate candy	10 pce	7	2	34	<1	5	<1	1	.9	.5	t
1135	M&M's peanut chocolate candy	10 pce	20	2	103	2	12	1	5	2.1	2.2	.8
1130	Mars almond bar	1 ea	50	4	234	4	31	1	11	2.7	5.5	2.8
1129	Milky Way candy bar	1 ea	60	6	254	3	43	1	10	4.7	3.6	.4
1708	Milk chocolate-coated peanuts	1 c	149	2	773	19	74	7	50	21.8	19.4	6.4
1709	Peanut brittle, recipe	1 c	147	2	666	11	102	3	28	7.4	12.5	6.9
1132	Reese's peanut butter cup	2 ea	50	3	271	5	27	2	16	5.5	6.5	2.8
1133	Skor English toffee candy bar	1 ea	39	3	217	2	22	1	13	8.5	4.3	.5
1131	Snickers candy bar (2.2oz)	1 ea	62	5	297	5	37	2	15	5.6	6.5	3
1482	Fruit juice bar (2.5 fl oz)	1 ea	77	78	63	1	16	0	<1	t	0	t
771	Gelatin dessert/Jello, prepared	½ c	135	85	80	2	19	0	0	0	0	0
1702	SugarFree	½ c	117	98	8	1	1	0	0	0	0	0
772	Honey:	1 c	339	17	1030	1	279	1	0	0	0	0
773	Tablespoon	1 tbs	21	17	64	<1	17	<1	0	0	0	0
774	Jams or preserves:	1 tbs	20	29	54	<1	14	<1	<1	0	t	t
775	Packet	1 ea	14	34	34	<1	9	<1	<1	t	t	0
776	Jellies:	1 tbs	19	28	51	<1	13	<1	<1	t	t	t
777	Packet	1 ea	14	28	38	<1	10	<1	<1	t	t	t
1136	Marmalade	1 tbs	20	33	49	<1	13	<1	0	0	0	0
770	Marshmallows	1 ea	7	16	22	<1	6	<1	<1	t	t	t
1126	Marshmallow creme topping	2 tbs	38	18	118	1	30	<1	<1	t	t	t
778	Popsicle/ice pops	1 ea	128	80	92	0	24	0	0	0	0	0
	Sugars:											
779	Brown sugar	1 c	220	2	827	0	214	0	0	0	0	0
780	White sugar, granulated:	1 c	200		774	0	200	0	0	0	0	0
781	Tablespoon	1 tbs	12		46	0	12	0	0	0	0	0
782	Packet	1 ea	6		23	0	6	0	0	0	0	0
783	White sugar, powdered, sifted	1 c	100		389	0	99	0	<1	t	t	t
	Sweeteners:											
1711	Equal, packet	1 ea	1	12	4	<1	1	0	<1	0	t	t
1712	Sweet 'N Low, packet	1 ea	1		0	0	1	0	0	0	0	0

PAGE KEY: A–2 = Beverages A–4 = Dairy A–8 = Eggs A–8 = Fat/Oil A–12 = Fruit A–18 = Bakery A–26 = Grain A–30 = Fish A–32 = Meats A–36 = Poultry A–38 = Sausage A–40 = Mixed/Fast A–44 = Nuts/Seeds A–48 = Sweets A–50 = Vegetables/Legumes A–62 = Misc A–64 = Soups/Sauces A–66 = Fast A–82 = Frozen Entree A–86 = Baby foods

Chol (mg)	Calc (mg)	Iron (mg)	Magn (mg)	Pota (mg)	Sodi (mg)	Zinc (mg)	VT-A (RE)	Thia (mg)	VT-E (a-TE)	Ribo (mg)	Niac (mg)	V-B6 (mg)	Fola (μg)	VT-C (mg)
0	2	.05	1	33	0	.02	0	<.01	.01	<.01	.03	.01	<1	1
<1	22	.08	3	34	143	.08	11	<.01	0	.04	.02	.01	1	<1
<1	22	.08	3	34	143	.08	11	<.01	0	.04	.02	.01	1	<1
0	1	.04		14	35	0	88	0	.79	<.01	<.01	0	0	0
0	4	.09	1	9	87	.04	42	.01	.79	.01	.13	<.01	0	0
0	1	.24			40		<1							0
0		.02			29		0							0
1	17	.39	18	69	41	.22	1	.01	.63	.04	.13	.02		<1
0	0	0		80	46		0	.03		.04	.03			0
1	14	.01	2	21	24	.04	1	<.01	.05	.02	.02	<.01	<1	<1
				0	0									
6	53	.39	17	108	23	.39	15	.02	.35	.08	.09	.01	2	<1
5	63	.46	25	124	21	.37	4	.02	.53	.12	.21	.01	3	<1
3	32	.52	34	150	11	.68	6	.08	1.3	.05	2.12	.04	23	0
5	48	.21	14	96	41	.31	3	.02	.35	.08	.13	.02	3	<1
0	54	5.26	193	613	18	2.72	3	.09	2	.15	.72	.06	5	0
<1	11	.98	47	139	3	.61	1	.01	.41	.1	.27	.02	1	0
0		.01		3	6	.01	<1	0	0	<.01	0	0	0	0
0	4	.14	3	41	9	.03	2	.01	.04	<.01	.01	.04	1	1
2	7	.08	4	17	10	.07	8	<.01	.02	.01	.02	<.01	<1	<1
0	5	.73	2	9	80	0	0	0	0	<.01	<.01	0	0	0
0		.02			2	<.01	0	0	0	0	0	0	0	0
0		.12		4	3	.01	0	0	0	0	0	0	0	0
1	7	.08	3	19	4	.07	4	<.01	.06	.01	.02	<.01	<1	<1
2	20	.23	12	69	10	.27	5	.03	.43	.04	.41	.02	7	<1
4	84	.55	36	163	85	.55	22	.02	.3	.16	.47	.03	9	<1
8	78	.46	20	145	144	.43	34	.02	.39	.13	.21	.03	6	1
13	155	1.95	134	748	61	2.8	0	.17	3.8	.26	6.33	.31	12	0
19	44	2.03	73	306	664	1.43	69	.28	2.41	.08	5.15	.15	103	0
2	39	.6	42	176	159	.7	9	.02	.66	.1	1.99	.04	27	<1
20	51	.02	13	93	108	.3	27	.01	.53	.13	.03	.01		<1
8	58	.47	42	209	165	.88	24	.13	.95	.1	2.26	.07	25	<1
0	4	.15	3	41	3	.04	2	.01	0	.01	.12	.02	5	7
0	3	.04	1	1	57	.04	0	0	0	<.01	<.01	<.01	0	0
0	2	.01	1	0	56	.03	0	0	0	<.01	<.01	<.01	0	0
0	20	1.42	7	176	14	.75	0	0	0	.13	.41	.08	7	2
0	1	.09		11	1	.05	0	0	0	.01	.02	<.01	<1	<1
0	4	.2	1	18	2	.01	<1	<.01	.02	.01	.04	<.01	2	<1
0	3	.07	1	11	6	.01	<1	0	0	<.01	<.01	<.01	5	1
0	2	.04	1	12	7	.01	<1	0	0	<.01	.01	<.01	<1	<1
0	1	.03	1	9	5	.01	<1	0	0	<.01	<.01	<.01	<1	<1
0	8	.03		7	11	.01	1	<.01	0	<.01	.01	<.01	7	1
0		.02			3	<.01	<1	0	0	0	<.01	0	<1	0
0	1	.08	1	2	17	.01	<1	0	0	0	.03	<.01	<1	0
0	0	0	1	5	15	.03	0	0	0	0	0	0	0	0
0	187	4.2	64	761	86	.4	0	.02	0	.01	.18	.06	2	0
0	2	.12	0	4	2	.06	0	0	0	.04	0	0	0	0
0		.01	0		<1	<.01	0	0	0	<.01	0	0	0	0
0		<.01	0		<1	<.01	0	0	0	<.01	0	0	0	0
0	1	.06	0	2	1	.03	0	0	0	0	0	0	0	0
0		<.01			<1	0	0	0	0	0	0	0	0	0
0	0	0			0		0							0

Table A-1

Food Composition (Computer code number is for West Diet Analysis program) (For purposes of calculations, use "0" for t, <1, <.1, <.01, etc.)

Computer Code Number	Food Description	Measure	Wt (g)	H_2O (%)	Ener (kcal)	Prot (g)	Carb (g)	Dietary Fiber (g)	Fat (g)	Fat Breakdown (g) Sat	Mono	Poly
	SWEETENERS and SWEETS—Continued											
	Syrups, chocolate:											
785	Hot fudge type	2 tbs	43	22	149	2	27	1	4	2.4	1.6	1.4
784	Thin type	2 tbs	38	29	93	1	25	1	<1	.3	.2	t
786	Molasses, blackstrap	2 tbs	41	29	96	0	25	0	0	0	0	0
1710	Light cane syrup	2 tbs	41	24	103	0	27	0	0	0	0	0
787	Pancake table syrup (corn and maple)	2 tbs	40	24	115	0	30	0	0	0	0	0
	VEGETABLES and LEGUMES											
788	Alfalfa sprouts	1 c	33	91	10	1	1	1	<1	t	.1	t
1815	Amaranth leaves, raw, chopped	1 c	28	92	7	1	1	<1	<1	t	t	t
1816	Amaranth leaves, raw, each	1 ea	14	92	4	<1	1	<1	<1	t	t	t
1817	Amaranth leaves, cooked	1 c	132	91	28	3	5	2	<1	.1	.1	.1
1987	Arugula, raw, chopped	½ c	10	92	2	<1	<1	<1	<1	t	t	t
789	Artichokes, cooked globe (300 g with refuse)	1 ea	120	84	60	4	13	6	<1	t	t	.1
1177	Artichoke hearts, cooked from frozen	1 c	168	86	76	5	15	8	1	.2	t	.4
1176	Artichoke hearts, marinated	1 c	130	81	128	3	10	6	10	1.5	2.3	5.9
2021	Artichoke hearts, in water	½ c	100	91	37	2	6	0	0	0	0	0
	Asparagus, green, cooked:											
	From fresh:											
790	Cuts and tips	½ c	90	92	22	2	4	1	<1	.1	t	.1
791	Spears, ½" diam at base	4 ea	60	92	14	2	3	1	<1	t	t	.1
	From frozen:											
792	Cuts and tips	½ c	90	91	25	3	4	1	<1	.1	t	.2
793	Spears, ½" diam at base	4 ea	60	91	17	2	3	1	<1	.1	t	.1
794	Canned, spears, ½" diam at base	4 ea	72	94	14	2	2	1	<1	.1	t	.2
795	Bamboo shoots, canned, drained slices	1 c	131	94	25	2	4	2	1	.1	t	.2
1795	Bamboo shoots, raw slices	1 c	151	91	41	4	8	3	<1	.1	t	.2
1798	Bamboo shoots, cooked slices	1 c	120	96	14	2	2	1	<1	.1	t	.1
	Beans (see also alphabetical listing this section):											
1990	Adzuki beans, cooked	½ c	115	66	147	9	28	1	<1	t	t	t
796	Black beans, cooked	½ c	86	66	114	8	20	7	<1	.1	t	.2
	Canned beans (white/navy):											
803	With pork and tomato sauce	½ c	127	73	124	7	25	6	1	.5	.6	.2
804	With sweet sauce	½ c	130	71	144	7	27	7	2	.7	.8	.2
805	With frankfurters	½ c	130	69	185	9	20	9	9	3.1	3.7	1.1
	Lima beans:											
797	Thick seeded (Fordhooks), cooked from frozen	½ c	85	73	85	5	16	5	<1	.1	t	.1
798	Thin seeded (Baby), cooked from frozen	½ c	90	72	94	6	18	5	<1	.1	t	.1
799	Cooked from dry, drained	½ c	94	70	108	7	20	7	<1	.1	t	.2
1998	Red Mexican, cooked f/dry	½ c	112	70	126	8	23	9	<1	.1	.1	.2
	Snap bean/green string beans cuts and french style:											
800	Cooked from fresh	½ c	63	89	22	1	5	2	<1	t	t	.1
801	Cooked from frozen	½ c	68	91	19	1	4	2	<1	t	t	.1
802	Canned, drained	½ c	68	93	14	1	3	1	<1	t	t	t
1713	Snap bean, yellow, cooked f/fresh	½ c	63	89	22	1	5	2	<1	t	t	.1
	Bean sprouts (mung):											
806	Raw	½ c	52	90	16	2	3	1	<1	t	t	t
807	Cooked, stir fried	½ c	62	84	31	3	7	1	<1	t	t	t
808	Cooked, boiled, drained	½ c	62	93	13	1	3	<1	<1	t	t	t
1788	Canned, drained	½ c	63	96	8	1	1	<1	<1	t	t	t
	Beets, cooked from fresh:											
809	Sliced or diced	½ c	85	87	37	1	8	2	<1	t	t	.1
810	Whole beets, 2" diam	2 ea	100	87	44	2	10	2	<1	t	t	.1
	Beets, canned:											
811	Sliced or diced	½ c	79	91	24	1	6	1	<1	t	t	t
812	Pickled slices	½ c	114	82	74	1	19	2	<1	t	t	t
813	Beet greens, cooked, drained	½ c	72	89	19	2	4	2	<1	t	t	.1

Chol (mg)	Calc (mg)	Iron (mg)	Magn (mg)	Pota (mg)	Sodi (mg)	Zinc (mg)	VT-A (RE)	Thia (mg)	VT-E (a-TE)	Ribo (mg)	Niac (mg)	V-B6 (mg)	Fola (μg)	VT-C (mg)
5	43	.52	21	92	56	.34	9	.01	0	.09	.09	.01	2	<1
0	5	5.17	25	183	58	.28	494	<.01	.01	.31	12.8	.01	2	<1
0	353	7.18	88	1021	23	.41	0	.01	0	.02	.44	.29	<1	0
0	68	1.76	100	376	6	.12	0	.03	0	.02	.08	.27	0	0
0		.04	1	1	33	.02	0	<.01	0	<.01	.01	0	0	0
0	11	.32	9	26	2	.3	5	.02	—	.04	.16	.01	12	3
0	60	.65	15	171	6	.25	82	.01	.22	.04	.18	.05	24	12
0	30	.32	8	85	3	.13	41	<.01	.11	.02	.09	.03	12	6
0	276	2.98	73	846	28	1.16	366	.03	.66	.18	.74	.23	75	54
0	16	.15	5	37	3	.05	24	<.01	.04	.01	.03	.01	10	1
0	54	1.55	72	425	114	.59	22	.08	.23	.08	1.2	.13	61	12
0	35	.94	52	444	89	.6	27	.1	.32	.26	1.54	.15	200	8
0	30	1.24	37	335	688	.41	21	.05	1.43	.13	1.06	.11	114	40
0	0	1.35		0	250		12							4
0	18	.66	9	144	10	.38	49	.11	.34	.11	.97	.11	131	10
0	12	.44	6	96	7	.25	32	.07	.23	.08	.65	.07	88	6
0	21	.58	12	196	4	.5	74	.06	1.13	.09	.94	.02	122	22
0	14	.38	8	131	2	.34	49	.04	.75	.06	.62	.01	81	15
0	11	1.32	7	124	207	.29	38	.04	.31	.07	.69	.08	69	13
0	10	.42	5	105	9	.85	1	.03	.5	.03	.18	.18	4	1
0	20	.75	5	805	6	1.66	3	.23	1.51	.11	.91	.36	11	6
0	14	.29	4	640	5	.56	0	.02	.8	.06	.36	.12	3	1
0	32	2.3	60	612	9	2.04	1	.13	.11	.07	.82	.11	139	0
0	23	1.81	60	305	1	.96	1	.21	.07	.05	.43	.06	128	0
9	71	4.17	44	381	559	7.44	15	.07	.69	.06	.63	.09	29	4
9	79	2.16	44	346	437	1.95	14	.06	.7	.08	.46	.11	49	4
8	62	2.25	36	306	559	2.43	19	.07	.61	.07	1.17	.06	39	3
0	19	1.16	29	347	45	.37	16	.06	.25	.05	.91	.1	18	11
0	25	1.76	50	370	26	.49	15	.06	.58	.05	.69	.1	14	5
0	16	2.25	40	478	2	.89	0	.15	.17	.05	.4	.15	78	0
0	42	1.86	48	369	240	.87	<1	.13	.08	.07	.37	.11	94	2
0	29	.81	16	188	2	.23	42	.05	.09	.06	.39	.03	21	6
0	33	.6	16	86	6	.33	27	.02	.09	.06	.26	.04	16	3
0	18	.61	9	74	178	.2	24	.01	.09	.04	.14	.02	22	3
0	29	.81	16	188	2	.23	5	.05	.18	.06	.39	.03	21	6
0	7	.47	11	77	3	.21	1	.04	.02	.06	.39	.05	32	7
0	8	1.18	20	136	6	.56	2	.09	.01	.11	.74	.08	43	10
0	7	.4	9	63	6	.29	1	.03	.01	.06	.51	.03	18	7
0	9	.27	6	17	88	.18	1	.02	.01	.04	.14	.02	6	<1
0	14	.67	20	259	65	.3	3	.02	.25	.03	.28	.06	68	3
0	16	.79	23	305	77	.35	4	.03	.3	.04	.33	.07	80	4
0	12	1.44	13	117	153	.17	1	.01	.24	.03	.12	.04	24	3
0	12	.47	17	169	301	.3	1	.01	.15	.05	.29	.06	30	3
0	82	1.37	49	654	174	.36	367	.08	.22	.21	.36	.09	10	18

Table A-1

Food Composition (Computer code number is for West Diet Analysis program) (For purposes of calculations, use "0" for t, <1, <.1, <.01, etc.)

Computer Code Number	Food Description	Measure	Wt (g)	H_2O (%)	Ener (kcal)	Prot (g)	Carb (g)	Dietary Fiber (g)	Fat (g)	Fat Breakdown (g) Sat	Mono	Poly
	VEGETABLES and LEGUMES—Continued											
	Broccoli, raw:											
817	Chopped	½ c	44	91	12	1	2	1	<1	t	t	.1
818	Spears	1 ea	31	91	9	1	2	1	<1	t	t	.1
	Broccoli, cooked from fresh:											
819	Spears	1 ea	180	91	50	5	9	5	1	.1	t	.3
820	Chopped	½ c	78	91	22	2	4	2	<1	t	t	.1
	Broccoli, cooked from frozen:											
821	Spear, small piece	½ c	92	91	26	3	5	3	<1	t	t	.1
822	Chopped	½ c	92	91	26	3	5	3	<1	t	t	.1
1603	Broccoflower, steamed	½ c	78	90	25	2	5	2	<1	t	t	.1
823	Brussels sprouts, cooked from fresh	½ c	78	87	30	2	7	2	<1	.1	t	.2
824	Brussels sprouts, cooked from frozen	½ c	78	87	33	3	6	3	<1	.1	t	.2
	Cabbage, common varieties:											
825	Raw, shredded or chopped	1 c	70	92	17	1	4	2	<1	t	t	.1
826	Cooked, drained	1 c	150	94	33	2	7	3	1	.1	t	.3
	Cabbage, Chinese:											
1178	Bok choy, raw, shredded	1 c	70	95	9	1	2	1	<1	t	t	.1
827	Bok choy, cooked, drained	1 c	170	96	20	3	3	3	<1	t	t	.1
1937	Kim chee style	1 c	150	92	31	2	6	2	<1	t	t	.2
828	Pe Tsai, raw, chopped	1 c	76	94	12	1	2	2	<1	t	t	.1
1796	Pe Tsai, cooked	1 c	119	95	17	2	3	3	<1	t	t	.1
	Cabbage, red, coarsely chopped:											
829	Raw	1 c	89	92	24	1	5	2	<1	t	t	.1
830	Cooked, drained	1 c	150	94	31	2	7	3	<1	t	t	.1
831	Cabbage, savoy, coarsely chopped, raw	1 c	70	91	19	1	4	2	<1	t	t	t
1785	Cabbage, savoy, cooked	1 c	145	92	35	3	8	4	<1	t	t	.1
1896	Capers	1 ea	5	86		<1	<1	<1	<1			
	Carrots, raw:											
832	Whole, 7½ x 1⅛"	1 ea	72	88	31	1	7	2	<1	t	t	.1
833	Grated	½ c	55	88	24	1	6	2	<1	t	t	t
	Carrots, cooked, sliced, drained:											
834	From raw	½ c	78	87	35	1	8	3	<1	t	t	.1
835	From frozen	½ c	73	90	26	1	6	3	<1	t	t	t
836	Carrots, canned, sliced, drained	½ c	73	93	17	<1	4	1	<1	t	t	.1
837	Carrot juice, canned	1 c	236	89	94	2	22	2	<1	.1	t	.2
	Cauliflower, flowerets:											
838	Raw	½ c	50	92	12	1	3	1	<1	t	t	t
839	Cooked from fresh, drained	½ c	62	93	14	1	3	2	<1	t	t	.1
840	Cooked, from frozen, drained	½ c	90	94	17	1	3	2	<1	t	t	.1
	Celery, pascal type, raw:											
841	Large outer stalk, 8 x 1½"(root end)	1 ea	40	95	6	<1	1	1	<1	t	t	t
842	Diced	1 c	120	95	19	1	4	2	<1	t	t	.1
1789	Celeriac/celery root, cooked	1 c	155	92	39	1	9	2	<1	.1	.1	.2
1179	Chard, swiss, raw, chopped	1 c	36	93	7	1	1	1	<1	t	t	t
1180	Chard, swiss, cooked	1 c	175	93	35	3	7	4	<1	t	t	t
1855	Chayote fruit, raw	1 ea	203	94	39	2	9	3	<1	.1	t	.1
1856	Chayote fruit, cooked	1 c	160	93	38	1	8	4	1	.2	.1	.3
	Chickpeas (see Garbanzo Beans #854)											
	Collards, cooked, drained:											
843	From raw	½ c	95	92	25	2	5	3	<1	t	t	.1
844	From frozen	½ c	85	88	31	3	6	3	<1	.1	t	.2
	Corn, yellow, cooked, drained:											
845	From raw, on cob, 5" long	1 ea	77	73	72	2	17	2	1	.1	.2	.3
846	From frozen, on cob, 3½" long	1 ea	63	73	59	2	14	2	<1	.1	.1	.2
847	Kernels, cooked from frozen	½ c	82	77	66	2	16	2	<1	.1	.1	.2
	Corn, canned:											
848	Cream style	½ c	128	79	92	2	23	2	1	.1	.2	.3
849	Whole kernel, vacuum pack	½ c	105	77	83	3	20	2	1	.1	.2	.2
	Cowpeas (see Black-eyed peas #814-816)											
850	Cucumber slices with peel	7 pce	28	96	4	<1	1	<1	<1	t	t	t
1948	Cucumber, kim chee style	1 c	150	91	31	2	7	2	<1	t	0	.1

Chol (mg)	Calc (mg)	Iron (mg)	Magn (mg)	Pota (mg)	Sodi (mg)	Zinc (mg)	VT-A (RE)	Thia (mg)	VT-E (a-TE)	Ribo (mg)	Niac (mg)	V-B6 (mg)	Fola (μg)	VT-C (mg)
0	21	.39	11	143	12	.18	68	.03	.73	.05	.28	.07	31	41
0	15	.27	8	101	8	.12	48	.02	.51	.04	.2	.05	22	29
0	83	1.51	43	526	47	.68	250	.1	3.04	.2	1.03	.26	90	134
0	36	.65	19	228	20	.3	108	.04	1.32	.09	.45	.11	39	58
0	47	.56	18	166	22	.28	174	.05	.95	.07	.42	.12	28	37
0	47	.56	18	166	22	.28	174	.05	1.52	.07	.42	.12	52	37
0	25	.55	16	251	18	.39	5	.06	.23	.07	.59	.14	38	49
0	28	.94	16	247	16	.26	56	.08	.66	.06	.47	.14	47	48
0	19	.58	19	254	18	.28	46	.08	.45	.09	.42	.22	79	36
0	33	.41	10	172	13	.13	9	.03	.07	.03	.21	.07	30	22
0	46	.25	12	146	12	.13	19	.09	.16	.08	.42	.17	30	30
0	73	.56	13	176	45	.13	210	.03	.08	.05	.35	.14	46	31
0	158	1.77	19	631	58	.29	437	.05	.2	.11	.73	.28	69	44
0	145	1.28	27	375	995	.35	426	.07	.24	.1	.75	.34	88	80
0	58	.24	10	181	7	.17	91	.03	.09	.04	.3	.18	60	20
0	38	.36	12	268	11	.21	115	.05	.14	.05	.59	.21	63	19
0	45	.44	13	183	10	.19	4	.04	.09	.03	.27	.19	18	51
0	55	.52	16	210	12	.22	4	.05	.18	.03	.3	.21	19	52
0	24	.28	20	161	20	.19	70	.05	.07	.02	.21	.13	56	22
0	43	.55	35	267	35	.33	129	.07	.15	.03	.03	.22	67	25
0	2	.05			105		1							0
0	19	.36	11	233	25	.14	2025	.07	.33	.04	.67	.11	10	7
0	15	.27	8	178	19	.11	1547	.05	.25	.03	.51	.08	8	5
0	24	.48	10	177	51	.23	1914	.03	.33	.04	.39	.19	11	2
0	20	.34	7	115	43	.17	1292	.02	.31	.03	.32	.09	8	2
0	18	.47	6	131	177	.19	1005	.01	.31	.02	.4	.08	7	2
0	57	1.09	33	689	68	.42	2584	.22	.02	.13	.91	.51	9	20
0	11	.22	7	152	15	.14	1	.03	.02	.03	.26	.11	28	23
0	10	.2	6	88	9	.11	1	.03	.02	.03	.25	.11	27	27
0	15	.37	8	125	16	.12	2	.03	.04	.05	.28	.08	37	28
0	16	.16	4	115	35	.05	5	.02	.14	.02	.13	.03	11	3
0	48	.48	13	344	104	.16	16	.05	.43	.05	.39	.1	34	8
0	40	.67	19	268	95	.31	0	.04	.31	.06	.66	.16	5	6
0	18	.65	29	136	77	.13	119	.01	.68	.03	.14	.04	5	11
0	102	3.96	151	961	313	.58	550	.06	3.31	.15	.63	.15	15	31
0	34	.69	24	254	4	1.5	12	.05	.24	.06	.95	.15	189	16
0	21	.35	19	277	2	.5	8	.04	.19	.06	.67	.19	29	13
0	113	.44	16	247	9	.4	297	.04	.84	.1	.55	.12	88	17
0	179	.95	25	213	42	.23	508	.04	.42	.1	.54	.1	65	22
0	2	.47	22	193	3	.48	16	.13	.07	.05	1.17	.17	23	4
0	2	.38	18	158	3	.4	13	.11	.06	.04	.96	.14	19	3
0	3	.29	16	121	4	.33	18	.07	.07	.06	1.07	.11	25	3
0	4	.49	22	172	365	.68	13	.03	.11	.07	1.23	.08	57	6
0	5	.44	24	195	286	.48	25	.04	.09	.08	1.23	.06	52	9
0	4	.07	3	40	1	.06	6	.01	.02	.01	.06	.01	4	1
0	13	7.23	12	176	1531	.76	49	.04	.24	.04	.69	.16	34	5

Table A-1

Food Composition (Computer code number is for West Diet Analysis program) (For purposes of calculations, use "0" for t, <1, <.1, <.01, etc.)

Computer Code Number	Food Description	Measure	Wt (g)	H_2O (%)	Ener (kcal)	Prot (g)	Carb (g)	Dietary Fiber (g)	Fat (g)	Fat Breakdown (g) Sat	Mono	Poly
	VEGETABLES and LEGUMES—Continued											
	Dandelion Greens:											
851	Raw	1 c	55	86	25	1	5	2	<1	.1	t	.2
852	Chopped, cooked, drained	1 c	105	90	35	2	7	3	1	.2	t	.3
853	Eggplant, cooked	1 c	99	92	28	1	7	2	<1	t	t	.1
1714	Endive, fresh, chopped	1 c	50	94	8	1	2	2	<1	t	t	t
856	Escarole/curly endive, chopped	1 c	50	94	8	1	2	2	<1	t	t	t
854	Garbanzo beans (chickpeas), cooked	1 c	164	60	269	14	45	12	4	.4	1	1.9
1939	Grape leaf, raw:	1 ea	3	73	3	<1	1	<1	<1	t	t	t
7914	Cup	1 c	14	73	13	1	2	2	<1	t	t	.1
855	Great northern beans, cooked	1 c	177	69	209	15	37	12	1	.2	t	.3
857	Jerusalem artichoke, raw slices	1 c	150	78	114	3	26	2	<1	0	t	t
1794	Jicama	1 c	120	90	46	1	11	6	<1	t	t	t
	Kale, cooked, drained:											
858	From raw	1 c	130	91	36	2	7	3	1	.1	t	.3
859	From frozen	1 c	130	90	39	4	7	3	1	.1	t	.3
860	Kidney beans, canned	1 c	256	77	218	13	40	16	1	.1	.1	.5
1181	Kohlrabi, raw slices	1 c	135	91	36	2	8	5	<1	t	t	.1
861	Kohlrabi, cooked	1 c	165	90	48	3	11	2	<1	t	t	.1
1183	Leeks, raw, chopped	1 c	89	83	54	1	13	2	<1	t	t	.1
1182	Leeks, cooked, chopped	1 c	104	91	32	1	8	1	<1	t	t	.1
862	Lentils, cooked from dry	1 c	198	70	230	18	40	16	1	.1	.1	.3
1288	Lentils, sprouted, stir-fried	1 c	124	69	125	11	26	5	1	.1	.1	.2
1289	Lentils, sprouted, raw	1 c	77	67	82	7	17	3	<1	t	.1	.2
	Lettuce:											
	Butterhead/Boston types:											
863	Head, 5″ diameter	¼ ea	41	96	5	1	1	<1	<1	t	t	t
864	Leaves, inner or outer	4 ea	30	96	4	<1	1	<1	<1	t	t	t
	Iceberg/crisphead:											
867	Chopped or shredded	1 c	55	96	7	1	1	1	<1	t	t	.1
865	Head, 6″ diameter	1 ea	539	96	65	5	11	8	1	.1	t	.5
866	Wedge, ¼ head	1 ea	135	96	16	1	3	2	<1	t	t	.1
868	Looseleaf, chopped	½ c	28	94	5	<1	1	1	<1	t	t	t
869	Romaine, chopped	½ c	28	95	4	<1	1	<1	<1	t	t	t
870	Romaine, inner leaf	3 pce	30	95	4	<1	1	1	<1	t	t	t
1930	Luffa, cooked (Chinese okra)	1 c	178	89	57	3	13	6	<1	.1	t	.1
	Mushrooms:											
871	Raw, sliced	½ c	35	92	9	1	2	<1	<1	t	t	.1
872	Cooked from fresh, pieces	½ c	78	91	21	2	4	2	<1	t	t	.1
1962	Stir fried, shitake slices	½ c	73	83	40	1	10	2	<1	t	t	t
873	Canned, drained	½ c	78	91	19	1	4	2	<1	t	t	.1
1951	Mushroom caps, pickled	8 ea	47	92	11	1	2	1	<1	t	t	.1
	Mustard greens:											
874	Cooked from raw	½ c	70	94	10	2	1	1	<1	t	.1	t
875	Cooked from frozen	½ c	75	94	14	2	2	2	<1	t	.1	t
876	Navy beans, cooked from dry	1 c	182	63	258	16	48	12	1	.3	.1	.4
	Okra, cooked:											
877	From fresh pods	8 ea	85	90	27	2	6	2	<1	t	t	t
878	From frozen slices	1 c	184	91	51	4	11	5	1	.1	.1	.1
1236	Batter fried from fresh	1 c	92	69	175	3	11	2	13	2.1	3.4	7.1
1930	Chinese, (Luffa), cooked	1 c	178	89	57	3	13	6	<1	.1	t	.1
	Onions:											
879	Raw, chopped	½ c	80	90	30	1	7	1	<1	t	t	t
880	Raw, sliced	½ c	58	90	22	1	5	1	<1	t	t	t
881	Cooked, drained, chopped	½ c	105	88	46	1	11	1	<1	t	t	.1
882	Dehydrated flakes	¼ c	14	4	49	1	12	1	<1	t	t	t
1934	Onions, pearl, cooked	½ c	93	87	41	1	9	1	<1	t	t	.1
883	Spring/green onions, bulb and top, chopped	½ c	50	90	16	1	4	1	<1	t	t	t
884	Onion rings, breaded, heated f/frozen	2 ea	20	28	81	1	8	<1	5	1.7	2.2	1
1917	Palm hearts, cooked slices	1 c	146	69	150	4	39	2	<1	.1	.1	t
885	Parsley, raw, chopped	½ c	30	88	11	1	2	1	<1	t	.1	t
888	Parsnips, sliced, cooked	½ c	78	78	63	1	15	3	<1	t	.1	t

Chol (mg)	Calc (mg)	Iron (mg)	Magn (mg)	Pota (mg)	Sodi (mg)	Zinc (mg)	VT-A (RE)	Thia (mg)	VT-E (a-TE)	Ribo (mg)	Niac (mg)	V-B6 (mg)	Fola (µg)	VT-C (mg)
0	103	1.71	20	218	42	.23	770	.1	1.38	.14	.44	.14	15	19
0	147	1.89	25	244	46	.29	1228	.14	2.63	.18	.54	.17	13	19
0	6	.35	13	246	3	.15	6	.07	.03	.02	.59	.08	14	1
0	26	.41	7	157	11	.39	103	.04	.22	.04	.2	.01	71	3
0	26	.41	7	157	11	.39	103	.04	.22	.04	.2	.01	71	3
0	80	4.74	79	477	11	2.51	5	.19	.57	.1	.86	.23	282	2
0	11	.08	3	8	<1	.02	81	<.01	.06	.01	.07	.01	2	<1
0	51	.37	13	38	1	.09	378	.01	.28	.05	.33	.06	12	2
0	120	3.77	88	692	4	1.56	<1	.28	.53	.1	1.21	.21	181	2
0	21	5.1	25	644	6	.18	3	.3	.28	.09	1.95	.12	20	6
0	14	.72	14	180	5	.19	2	.02	5.48	.03	.24	.05	14	24
0	94	1.17	23	296	30	.31	962	.07	1.11	.09	.65	.18	17	53
0	179	1.22	23	417	19	.23	826	.06	.23	.15	.87	.11	19	33
0	61	3.23	72	658	873	1.41	0	.27	.13	.22	1.17	.06	130	3
0	32	.54	26	473	27	.04	5	.07	.65	.03	.54	.2	22	84
0	41	.66	31	561	35	.51	7	.07	2.76	.03	.64	.25	20	89
0	52	1.87	25	160	18	.11	9	.05	.82	.03	.36	.21	57	11
0	31	1.14	15	90	10	.06	5	.03	.63	.02	.21	.12	25	4
0	38	6.59	71	731	4	2.51	2	.33	.22	.14	2.1	.35	358	3
0	17	3.84	43	352	12	1.98	5	.27	.11	.11	1.49	.2	83	16
0	19	2.47	28	248	8	1.16	4	.18	.07	.1	.87	.15	77	13
0	13	.12	5	105	2	.07	40	.02	.18	.02	.12	.02	30	3
0	10	.09	4	77	1	.05	29	.02	.13	.02	.09	.01	22	2
0	10	.27	5	87	5	.12	18	.02	.15	.02	.1	.02	31	2
0	102	2.7	48	852	48	1.19	178	.25	1.51	.16	1.01	.22	302	21
0	26	.67	12	213	12	.3	45	.06	.38	.04	.25	.05	76	5
0	19	.39	3	74	3	.08	53	.01	.12	.02	.11	.01	14	5
0	10	.31	2	81	2	.07	73	.03	.12	.03	.14	.01	38	7
0	11	.33	2	87	2	.07	78	.03	.13	.03	.15	.01	41	7
0	112	.8	101	570	420	.97	103	.23	1.22	.1	1.54	.33	81	29
0	2	.43	3	130	1	.26	0	.04	.04	.16	1.44	.03	7	1
0	5	1.36	9	278	2	.68	0	.06	.09	.23	3.48	.07	14	3
0	2	.32	10	85	3	.97	0	.03	.09	.12	1.1	.12	15	<1
0	9	.62	12	101	332	.56	0	.07	.09	.02	1.24	.05	10	0
0	2	.5	5	139	95	.28	0	.03	.05	.16	1.42	.03	6	1
0	52	.49	10	141	11	.08	212	.03	1.41	.04	.3	.07	51	18
0	76	.84	10	104	19	.15	335	.03	1.31	.04	.19	.08	52	10
0	127	4.51	107	670	2	1.93	<1	.37	.73	.11	.97	.3	255	2
0	54	.38	48	274	4	.47	49	.11	.59	.05	.74	.16	39	14
0	177	1.23	94	431	6	1.14	94	.18	1.27	.23	1.44	.09	269	22
15	104	.77	37	214	137	.5	43	.13	3.08	.1	.75	.13	37	10
0	112	.8	101	570	420	.97	103	.23	1.22	.1	1.54	.33	81	29
0	16	.18	8	126	2	.15	0	.03	.1	.02	.12	.09	15	5
0	12	.13	6	91	2	.11	0	.02	.07	.01	.09	.07	11	4
0	23	.25	12	174	3	.22	0	.04	.14	.02	.17	.13	16	5
0	36	.22	13	227	3	.26	0	.07	.19	.01	.14	.22	23	10
0	21	.22	10	154	218	.19	0	.04	.12	.02	.15	.12	14	5
0	36	.74	10	138	8	.19	19	.03	.06	.04	.26	.03	32	9
0	6	.34	4	26	75	.08	5	.06	.14	.03	.72	.01	13	<1
0	26	2.47	15	2636	20	5.45	10	.07	.73	.25	1.25	1.06	30	10
0	41	1.86	15	166	17	.32	156	.03	.54	.03	.39	.03	46	40
0	29	.45	23	286	8	.2	0	.06	.78	.04	.56	.07	45	10

Table A-1

Food Composition (Computer code number is for West Diet Analysis program) (For purposes of calculations, use "0" for t, <1, <.1, <.01, etc.)

Computer Code Number	Food Description	Measure	Wt (g)	H_2O (%)	Ener (kcal)	Prot (g)	Carb (g)	Dietary Fiber (g)	Fat (g)	Fat Breakdown (g) Sat	Mono	Poly
	VEGETABLES and LEGUMES—Continued											
	Peas:											
	Black-eyed, cooked:											
814	From dry, drained	½ c	86	70	100	7	18	6	<1	.1	t	.2
815	From fresh, drained	½ c	82	75	79	3	17	4	<1	.1	t	.1
816	From frozen, drained	½ c	85	66	112	7	20	5	1	.1	.1	.2
889	Edible pod peas, cooked	½ c	80	89	34	3	6	2	<1	t	t	.1
890	Green, canned, drained:	½ c	85	82	59	4	11	3	<1	.1	t	.1
5267	Unsalted	½ c	124	86	66	4	12	4	<1	.1	t	.2
891	Green, cooked from frozen	½ c	80	79	62	4	11	4	<1	t	t	.1
1786	Snow peas, raw	½ c	49	89	21	1	4	1	<1	t	t	t
1787	Snow peas, raw	10 ea	34	89	14	1	3	1	<1	t	t	t
892	Split, green, cooked from dry	½ c	98	69	116	8	21	8	<1	.1	.1	.2
1187	Peas & carrots, cooked from frozen	½ c	80	86	38	2	8	2	<1	.1	t	.2
1186	Peas & carrots, canned w/liquid	½ c	128	88	49	3	11	3	<1	.1	t	.2
	Peppers, hot:											
893	Hot green chili, canned	½ c	68	92	14	1	3	1	<1	t	t	t
894	Hot green chili, raw	1 ea	45	88	18	1	4	1	<1	t	t	t
1715	Hot red chili, raw, diced	1 tbs	9	88	4	<1	1	<1	<1	t	t	t
1988	Jalapeno, raw	1 ea	45	90	11	<1	2		<1			
895	Jalapeno, chopped, canned	½ c	68	89	18	1	3	2	1	.1	t	.3
1918	Jalapeno wheels, in brine (Ortega)	2 tbs	29		10	0	2		0	0	0	0
	Peppers, sweet, green:											
896	Whole pod (90 g with refuse), raw	1 ea	74	92	20	1	5	1	<1	t	t	.1
897	Cooked, chopped (1 pod cooked = 73g)	½ c	68	92	19	1	5	1	<1	t	t	.1
	Peppers, sweet, red:											
1286	Raw, chopped	½ c	75	92	20	1	5	1	<1	t	t	.1
1807	Raw, each	1 ea	74	92	20	1	5	1	<1	t	t	.1
1287	Cooked, chopped	½ c	68	92	19	1	5	1	<1	t	t	.1
	Peppers, sweet, yellow:											
1872	Raw, large	1 ea	186	92	50	2	12	2	<1	.1	t	.2
1873	Strips	10 pce	52	92	14	1	3	<1	<1	t	t	.1
898	Pinto beans, cooked from dry	½ c	85	64	116	7	22	7	<1	.1	.1	.2
1191	Poi, two finger	½ c	120	72	134	<1	33	<1	<1	t	t	.1
	Potatoes:											
	Baked in oven, 4¾"x2⅓" diam											
899	With skin	1 ea	202	71	220	5	51	5	<1	.1	t	.1
900	Flesh only	1 ea	156	75	145	3	34	2	<1	t	t	.1
901	Skin only	1 ea	58	47	115	2	27	5	<1	t	t	t
	Baked in microwave, 4¾"x 2⅓"dm:											
902	With skin	1 ea	202	72	212	5	49	5	<1	.1	t	.1
903	Flesh only	1 ea	156	74	156	3	36	2	<1	t	t	.1
904	Skin only	1 ea	58	63	77	3	17	3	<1	t	t	t
	Boiled, about 2½" diam:											
905	Peeled after boiling	1 ea	136	77	118	3	27	2	<1	t	t	.1
906	Peeled before boiling	1 ea	135	77	116	2	27	2	<1	t	t	.1
	French fried, strips 2–3½" long:											
907	Oven heated	10 ea	50	35	167	2	20	2	9	3	5.7	.7
908	Fried in vegetable oil	10 ea	50	38	158	2	20	2	8	1.9	4.7	.7
1188	Fried in veg and animal oil	10 ea	50	38	158	2	20	2	8	1.9	4.7	.7
909	Hashed browns from frozen	1 c	156	56	340	5	44	3	18	7	8	2.1
	Mashed:											
910	Home recipe with whole milk	½ c	105	78	81	2	18	2	1	.4	.2	.1
911	Home recipe with milk and marg	½ c	105	76	111	2	17	2	4	1.1	1.9	1.3
912	Prepared from flakes; water, milk, margarine, salt added	½ c	110	76	124	2	16	3	6	1.6	2.5	1.7
	Potato products, prepared:											
	Au gratin:											
913	From dry mix	½ c	123	79	114	3	16	1	5	3.5	1.5	.2
914	From home recipe, using butter	½ c	122	74	161	7	14	2	9	4.8	3.2	1.3
	Scalloped:											
915	From dry mix	½ c	122	79	113	3	16	1	5	3.2	1.5	.2
916	From home recipe, using butter	½ c	123	81	106	4	13	2	5	1.7	1.7	.9

PAGE KEY: A–2 = Beverages A–4 = Dairy A–8 = Eggs A–8 = Fat/Oil A–12 = Fruit A–18 = Bakery A–26 = Grain A–30 = Fish A–32 = Meats A–36 = Poultry A–38 = Sausage A–40 = Mixed/Fast A–44 = Nuts/Seeds A–48 = Sweets A–50 = Vegetables/Legumes A–62 = Misc A–64 = Soups/Sauces A–66 = Fast A–82 = Frozen Entree A–86 = Baby foods

Chol (mg)	Calc (mg)	Iron (mg)	Magn (mg)	Pota (mg)	Sodi (mg)	Zinc (mg)	VT-A (RE)	Thia (mg)	VT-E (a-TE)	Ribo (mg)	Niac (mg)	V-B6 (mg)	Fola (μg)	VT-C (mg)
0	21	2.16	46	239	3	1.11	2	.17	.24	.05	.43	.09	179	<1
0	105	.92	43	343	3	.84	65	.08	.18	.12	1.15	.05	104	2
0	20	1.8	42	319	4	1.21	7	.22	.33	.05	.62	.08	120	2
0	34	1.58	21	192	3	.3	10	.1	.31	.06	.43	.11	23	38
0	17	.81	14	147	214	.6	65	.1	.32	.07	.62	.05	38	8
0	22	1.26	21	124	11	.87	47	.14	.47	.09	1.04	.08	35	12
0	19	1.26	23	134	70	.75	54	.23	.14	.08	1.18	.09	47	8
0	21	1.02	12	98	2	.13	7	.07	.19	.04	.29	.08	20	29
0	15	.71	8	68	1	.09	5	.05	.13	.03	.2	.05	14	20
0	14	1.26	35	355	2	.98	1	.19	.38	.05	.87	.05	64	<1
0	18	.75	13	126	54	.36	621	.18	.26	.05	.92	.07	21	6
0	29	.96	18	128	333	.74	739	.09	.24	.07	.74	.11	23	8
0	5	.34	10	127	798	.12	41	.01	.47	.03	.54	.1	7	46
0	8	.54	11	153	3	.13	35	.04	.31	.04	.43	.12	10	109
0	2	.11	2	31	1	.03	97	.01	.06	.01	.09	.02	2	22
				2	2		30		.37					53
0	16	1.28	10	131	1136	.23	116	.03	.47	.03	.27	.13	10	7
0				55	390		10		.2					21
0	7	.34	7	131	1	.09	47	.05	.51	.02	.38	.18	16	66
0	6	.31	7	113	1	.08	40	.04	.47	.02	.32	.16	11	51
0	7	.34	7	133	1	.09	428	.05	.52	.02	.38	.19	16	143
0	7	.34	7	131	1	.09	422	.05	.51	.02	.38	.18	16	141
0	6	.31	7	113	1	.08	256	.04	.47	.02	.32	.16	11	116
0	20	.86	22	394	4	.32	45	.05	1.28	.05	1.66	.31	48	342
0	6	.24	6	110	1	.09	12	.01	.36	.01	.46	.09	13	96
0	41	2.22	47	398	2	.92	<1	.16	.8	.08	.34	.13	146	2
0	19	1.06	29	220	14	.26	2	.16	.22	.05	1.32	.33	26	5
0	20	2.75	54	844	16	.65	0	.22	.1	.07	3.33	.7	22	26
0	8	.55	39	610	8	.45	0	.16	.06	.03	2.18	.47	14	20
0	20	4.08	25	332	12	.28	0	.07	.02	.06	1.78	.36	12	8
0	22	2.5	54	903	16	.73	0	.24	.1	.06	3.45	.69	24	30
0	8	.64	39	641	11	.51	0	.2	.06	.04	2.54	.5	19	24
0	27	3.45	21	377	9	.3	0	.04	.02	.04	1.29	.28	10	9
0	7	.42	30	515	5	.41	0	.14	.07	.03	1.96	.41	14	18
0	11	.42	27	443	7	.36	0	.13	.07	.03	1.77	.36	12	10
0	6	.83	11	270	307	.2	0	.04	.25	.02	1.34	.11	11	3
0	9	.38	17	366	108	.19	0	.09	.25	.01	1.63	.12	14	5
6	9	.38	17	366	108	.19	0	.09	.25	.01	1.63	.12	14	5
0	23	2.36	26	680	53	.5	0	.17	.3	.03	3.78	.2	10	10
2	27	.28	19	314	318	.3	6	.09	.05	.04	1.18	.24	9	7
2	27	.27	19	303	310	.28	21	.09	.31	.04	1.13	.23	8	6
4	54	.24	20	256	365	.2	23	.12	.77	.05	.74	.01	8	11
18	102	.39	18	269	540	.29	38	.02	1.48	.1	1.15	.05	8	4
18	145	.78	24	483	528	.84	46	.08	.64	.14	1.21	.21	13	12
13	44	.46	17	248	416	.3	26	.02	.18	.07	1.26	.05	12	4
7	70	.7	23	465	412	.49	23	.08	.4	.11	1.29	.22	13	13

Table A-1

Food Composition (Computer code number is for West Diet Analysis program) (For purposes of calculations, use "0" for t, <1, <.1, <.01, etc.)

Computer Code Number	Food Description	Measure	Wt (g)	H_2O (%)	Ener (kcal)	Prot (g)	Carb (g)	Dietary Fiber (g)	Fat (g)	Fat Breakdown (g) Sat	Mono	Poly
	VEGETABLES and LEGUMES—Continued											
	Potato Salad (see Mixed Dishes #715)											
1192	Potato puffs, cooked from frozen	½ c	64	53	142	2	19	2	7	3.3	2.8	.5
918	Pumpkin, cooked from fresh, mashed	½ c	123	94	25	1	6	1	<1	t	t	t
919	Pumpkin, canned	½ c	123	90	42	1	10	4	<1	.2	t	t
1891	Radicchio, raw, shredded	½ c	20	93	5	<1	1	<1	<1	t	t	t
1894	Radicchio, raw, leaf	10 ea	80	93	18	1	4	1	<1	t	t	.1
920	Red radishes	10 ea	45	95	8	<1	2	1	<1	t	t	t
1793	Daikon radishes (Chinese) raw	½ c	44	95	8	<1	2	1	<1	t	t	t
921	Refried beans, canned	½ c	126	76	118	7	19	7	2	.6	.7	.2
1375	Rutabaga, cooked cubes	½ c	85	89	33	1	7	2	<1	t	t	.1
922	Sauerkraut, canned with liquid	½ c	118	92	22	1	5	3	<1	t	t	.1
923	Seaweed, kelp, raw	½ c	40	82	17	1	4	1	<1	.1	t	t
924	Seaweed, spirulina, dried	½ c	8	5	23	5	2	<1	1	.2	.1	.2
1866	Shallots, raw, chopped	1 tbs	10	80	7	<1	2	<1	<1	t	t	t
1557	Snow peas, stir-fried	½ c	83	89	35	2	6	2	<1	t	t	.1
925	Soybeans, cooked from dry	½ c	86	63	149	15	9	5	8	1.1	1.7	4.4
1996	Soybeans, dry roasted	½ c	86	1	387	34	28	7	19	2.7	4.1	10.6
	Soybean products:											
926	Miso	½ c	138	41	284	17	39	7	8	1.2	1.9	4.7
	Soy milk (see #144 and #2301 under Dairy)											
	Tofu (soybean curd):											
7540	Extra firm, silken	½ c	126	88	69	9	3	<1	2	.4	.4	1.3
7542	Firm, silken	½ c	126	87	77	9	3	<1	3	.5	.7	1.9
927	Regular	½ c	124	87	76	8	2	<1	5	.7	1	2.6
7541	Soft, silken	½ c	124	89	68	6	4	<1	3	.4	.6	1.9
	Spinach:											
928	Raw, chopped	½ c	28	92	6	1	1	1	<1	t	t	t
929	Cooked, from fresh, drained	½ c	90	91	21	3	3	2	<1	t	t	.1
930	Cooked from frozen (leaf)	½ c	95	90	27	3	5	3	<1	t	t	.1
931	Canned, drained solids:	½ c	107	92	25	3	4	3	1	.1	t	.2
5149	Unsalted	½ c	107	92	25	3	4	3	1	.1	t	.2
	Spinach soufflé (see Mixed Dishes)											
	Squash, summer varieties,cooked w/skin:											
932	Varieties averaged	½ c	90	94	18	1	4	1	<1	.1	t	.1
933	Crookneck	½ c	90	94	18	1	4	1	<1	.1	t	.1
934	Zucchini	½ c	90	95	14	1	4	1	<1	t	t	t
	Squash, winter varieties, cooked:											
	Average of all varieties, baked:											
935	Mashed	1 c	245	89	96	2	21	7	2	.3	.1	.6
936	Cubes	1 c	205	89	80	2	18	6	1	.3	.1	.5
937	Acorn, baked, mashed	½ c	123	83	69	1	18	5	<1	t	t	.1
1218	Acorn, boiled, mashed	½ c	122	90	41	1	11	3	<1	t	t	t
	Butternut squash:											
938	Baked cubes	½ c	103	88	41	1	11	3	<1	t	t	t
1219	Baked, mashed	½ c	103	88	41	1	11	3	<1	t	t	t
1193	Cooked from frozen	½ c	120	88	47	1	12	3	<1	t	t	t
1194	Hubbard, baked, mashed	½ c	120	85	60	3	13	3	1	.2	.1	.3
1195	Hubbard, boiled, mashed	½ c	118	91	35	2	8	3	<1	.1	t	.2
1196	Spaghetti, baked or boiled	½ c	77	92	22	<1	5	1	<1	t	t	.1
1189	Succotash, cooked from frozen	½ c	85	74	79	4	17	3	1	.1	.1	.4
	Sweet potatoes:											
939	Baked in skin, peeled, 5 x 2″ diam	1 ea	114	73	117	2	28	3	<1	t	t	.1
940	Boiled without skin, 5 x 2″ diam	1 ea	151	73	159	2	37	3	<1	.1	t	.2
941	Candied, 2½ x 2″	1 pce	105	67	144	1	29	3	3	1.4	.7	.2
	Canned:											
942	Solid pack	½ c	128	74	129	3	30	2	<1	.1	t	.1
943	Vacuum pack, mashed	½ c	127	76	116	2	27	2	<1	.1	t	.1
944	Vacuum pack, 3¾ x 1″	2 pce	80	76	73	1	17	1	<1	t	t	.1
1940	Taro shoots, cooked slices	1 c	140	95	20	1	4	1	<1	t	t	t
1941	Taro, tahitian, cooked slices	1 c	137	86	60	6	9	1	1	.2	.1	.4
	Tomatillos:											
1877	Raw, each	1 ea	34	92	11	<1	2	1	<1	t	.1	.1
1875	Raw, chopped	1 c	132	92	42	1	8	3	1	.2	.2	.6

Chol (mg)	Calc (mg)	Iron (mg)	Magn (mg)	Pota (mg)	Sodi (mg)	Zinc (mg)	VT-A (RE)	Thia (mg)	VT-E (a-TE)	Ribo (mg)	Niac (mg)	V-B6 (mg)	Fola (μg)	VT-C (mg)
0	19	1	12	243	477	.19	1	.12	.03	.05	1.38	.15	11	4
0	18	.7	11	283	1	.28	1330	.04	1.3	.1	.51	.05	10	6
0	32	1.71	28	253	6	.21	2713	.03	1.3	.07	.45	.07	15	5
0	4	.11	3	60	4	.12	1	<.01	.45	.01	.05	.01	12	2
0	15	.45	10	242	18	.5	2	.01	1.81	.02	.2	.05	48	6
0	9	.13	4	104	11	.13	<1	<.01	0	.02	.13	.03	12	10
0	12	.18	7	100	9	.07	0	.01	0	.01	.09	.02	12	10
10	44	2.09	42	336	377	1.47	0	.03	.39	.02	.4	.18	14	8
0	41	.45	20	277	17	.3	48	.07	.13	.03	.61	.09	13	16
0	35	1.73	15	201	780	.22	2	.02	.12	.03	.17	.15	28	17
0	67	1.14	48	36	93	.49	5	.02	.35	.06	.19	<.01	72	1
0	10	2.28	16	109	84	.16	5	.19	.4	.29	1.02	.03	8	1
0	4	.12	2	33	1	.04	125	.01	.01	<.01	.02	.03	3	1
0	36	1.73	20	166	3	.22	11	.11	.32	.06	.47	.13	28	42
0	88	4.42	74	443	1	.99	1	.13	1.68	.24	.34	.2	46	1
0	120	3.4	196	1173	2	4.1	2	.37	3.96	.65	.91	.19	176	4
0	91	3.78	58	226	5032	4.58	12	.13	.01	.34	1.19	.3	45	0
0	39	1.5	34	195	80	.76	0	.1	.18	.04	.3	.01		0
0	41	1.3	34	244	45	.77	0	.13	.24	.05	.31	.01		0
0	138	1.38	33	149	10	.79	1	.06	.01	.05	.66	.06	55	<1
0	38	1.02	36	223	6	.64	0	.12	.25	.05	.37	.01		0
0	28	.76	22	156	22	.15	188	.02	.53	.05	.2	.05	54	8
0	122	3.21	78	419	63	.68	737	.09	.86	.21	.44	.22	131	9
0	139	1.44	66	283	82	.66	739	.06	.91	.16	.4	.14	103	12
0	136	2.46	81	370	29	.49	939	.02	1.39	.15	.41	.11	105	15
0	136	2.46	81	370	29	.49	939	.02	1.39	.15	.41	.11	105	15
0	24	.32	22	173	1	.35	26	.04	.11	.04	.46	.06	18	5
0	24	.32	22	173	1	.35	26	.04	.11	.04	.46	.08	18	5
0	12	.31	20	228	3	.16	22	.04	.11	.04	.38	.07	15	4
0	34	.81	20	1070	2	.64	872	.21	.29	.06	1.72	.18	69	23
0	29	.68	16	896	2	.53	730	.17	.25	.05	1.44	.15	57	20
0	54	1.14	53	538	5	.21	53	.2	.15	.02	1.08	.24	23	13
0	32	.68	32	321	4	.13	32	.12	.15	.01	.65	.14	14	8
0	42	.62	30	293	4	.13	721	.07	.17	.02	1	.13	20	16
0	42	.62	30	293	4	.13	721	.07	.17	.02	1	.13	20	16
0	23	.7	11	160	2	.14	401	.06	.16	.05	.56	.08	20	4
0	20	.56	26	430	10	.18	725	.09	.14	.06	.67	.21	19	11
0	12	.33	15	253	6	.12	473	.05	.14	.03	.39	.12	11	8
0	16	.26	8	90	14	.15	8	.03	.09	.02	.62	.08	6	3
0	13	.76	20	225	38	.38	20	.06	.31	.06	1.11	.08	28	5
0	32	.51	23	397	11	.33	2487	.08	.32	.14	.69	.27	26	28
0	32	.85	15	278	20	.41	2574	.08	.42	.21	.97	.37	17	26
8	27	1.19	12	198	73	.16	440	.02	3.99	.04	.41	.04	12	7
0	38	1.7	31	269	96	.27	1936	.03	.35	.11	1.22	.3	14	7
0	28	1.13	28	396	67	.23	1013	.05	.32	.07	.94	.24	21	33
0	18	.71	18	250	42	.14	638	.03	.2	.05	.59	.15	13	21
0	20	.57	11	482	3	.76	7	.05	1.4	.07	1.13	.16	4	26
0	204	2.14	70	854	74	.14	241	.06	3.7	.27	.66	.16	10	52
0	2	.21	7	91	<1	.07	4	.01	.13	.01	.63	.02	2	4
0	9	.82	26	354	1	.29	14	.06	.5	.05	2.44	.07	9	15

Table A–1

Food Composition (Computer code number is for West Diet Analysis program) (For purposes of calculations, use "0" for t, <1, <.1, <.01, etc.)

Computer Code Number	Food Description	Measure	Wt (g)	H_2O (%)	Ener (kcal)	Prot (g)	Carb (g)	Dietary Fiber (g)	Fat (g)	Fat Breakdown (g) Sat	Mono	Poly
	VEGETABLES and LEGUMES—Continued											
	Tomatoes:											
945	Raw, whole, 2 ⅗" diam	1 ea	123	94	26	1	6	1	<1	.1	.1	.2
946	Raw, chopped	1 c	180	94	38	2	8	2	1	.1	.1	.2
947	Cooked from raw	1 c	240	92	65	3	14	2	1	.1	.2	.4
948	Canned, solids and liquid:	1 c	240	94	46	2	10	2	<1	t	t	.1
5741	Unsalted	1 c	240	94	46	2	10	2	<1	t	t	.1
1879	Tomatoes, sundried:	1 c	54	15	139	8	30	7	2	.2	.3	.6
1881	Pieces	10 pce	20	15	52	3	11	2	1	.1	.1	.2
1885	Oil pack, drained	10 pce	30	54	64	2	7	2	4	.6	2.6	.6
2020	Tomato, raw	1 ea	123	94	26	1	6	1	<1	.1	.1	.2
949	Tomato juice, canned:	1 c	243	94	41	2	10	1	<1	t	t	.1
5397	Unsalted	1 c	243	94	41	2	10	2	<1	t	t	.1
	Tomato products, canned:											
950	Paste, no added salt	1 c	262	74	215	10	51	11	1	.2	.2	.6
951	Puree, no added salt	1 c	250	87	100	4	24	5	<1	.1	.1	.2
952	Sauce, with salt	1 c	245	89	73	3	18	3	<1	.1	.1	.2
953	Turnips, cubes, cooked from fresh	1 c	156	94	33	1	8	3	<1	t	t	.1
	Turnip greens, cooked:											
954	From fresh, leaves and stems	1 c	144	93	29	2	6	5	<1	.1	t	.1
955	From frozen, chopped	1 c	164	90	49	6	8	6	1	.2	t	.3
956	Vegetable juice cocktail, canned	1 c	242	93	46	2	11	2	<1	t	t	.1
	Vegetables, mixed:											
957	Canned, drained	½ c	81	87	38	2	7	2	<1	t	t	.1
958	Frozen, cooked, drained	½ c	91	83	54	3	12	4	<1	t	t	.1
1818	Water chestnuts, Chinese, raw	½ c	62	73	60	1	15	2	<1	t	t	t
	Water chestnuts, canned:											
959	Slices	½ c	70	86	35	1	9	2	<1	t	t	t
960	Whole	4 ea	28	86	14	<1	3	1	<1	t	t	t
1190	Watercress, fresh, chopped	½ c	17	95	2	<1	<1	<1	<1	t	t	t
	VEGETARIAN FOODS:											
7509	Bacon strips, meatless	3 ea	15	49	46	2	1	<1	4	.7	1.1	2.3
1511	Baked beans, canned	½ c	127	73	118	6	26	6	1	.1	t	.2
7526	Bakon crumbles	¼ c	7	16	28	2	1	<1	2			
7548	Chicken, breaded, fried, meatless	1 pce	57	70	97	6	3	3	7	1	2.9	2.5
7547	Chicken slices, meatless	2 ea	60	59	132	10	4	3	8	1.3	2	4.4
7557	Chili w/meat substitute	½ c	107	64	141	19	15	4	2	.3	.6	.9
7549	Fish stick, meatless	2 ea	57	45	165	13	5	3	10	1.6	2.5	5.4
7550	Frankfurter, meatless	1 ea	51	58	102	10	4	2	5	.8	1.2	2.7
7504	GardenBurger, patty	1 ea	45	53	87	5	13	3	1	.4	.3	.7
7505	GardenSausage, patty	1 ea	35	15	117	4	22	5	1	.7	.3	t
7551	Luncheon slice, meatless	1 sl	67	46	188	17	6	3	11	1.7	2.6	5.6
7560	Meatloaf, meatless	1 ea	71	58	142	15	6	3	6	1	1.5	3.3
1171	Nuteena	1 ea	55	58	162	6	6	2	13	5.1	5.8	1.7
7556	Pot pie, meatless	1 ea	227	59	524	15	41	5	34	9.5	12.6	9.8
7554	Soyburger, patty	1 ea	71	58	142	15	6	3	6	1	1.5	3.3
7562	Soyburger w/cheese, patty	1 ea	135	50	316	21	29	4	13	4.2	3.9	3.7
7564	Tempeh	1 c	166	55	330	31	28	9	13	1.9	2.9	7.2
7670	Vegan burger, patty	1 ea	78	71	75	11	6	4	<1	.1	.3	.2
	Vegetarian foods, Green Giant:											
7677	Breakfast links	3 ea	68	65	114	12	5	4	5	.7	1.2	3.1
7676	Breakfast patties	2 ea	57	65	95	10	5	3	4	.6	1	2.6
	Burger, harvest, patty:											
7673	Italian	1 ea	90	65	139	17	8	5	4	1.4	.3	.4
7674	Original	1 ea	90	65	137	18	8	5	4	1.3	.1	.4
7675	Southwestern	1 ea	90	65	135	16	9	5	4	1.4	.2	.4
	Vegetarian foods, Loma Linda											
7727	Chik nuggets, frozen	5 pce	85	47	245	12	13	5	16	2.5	4	8.8
7753	Chik-fried, frozen	1 pce	57	51	178	11	1	1	15	1.9	3.7	8.7
7744	Franks, big, canned	1 ea	51	59	110	10	2	2	7	1.1	1.7	3.8
7747	Linketts, canned	1 ea	35	60	72	7	1		4	.7	1.2	2.5
1173	Redi-burger, patty	1 ea	85	59	172	16	5		10	1.5	2.4	5.8
7755	Swiss stake w/gravy, canned	1 pce	92	71	120	9	8	4	6	.8	1.5	3.3

Chol (mg)	Calc (mg)	Iron (mg)	Magn (mg)	Pota (mg)	Sodi (mg)	Zinc (mg)	VT-A (RE)	Thia (mg)	VT-E (a-TE)	Ribo (mg)	Niac (mg)	V-B6 (mg)	Fola (μg)	VT-C (mg)
0	6	.55	13	273	11	.11	76	.07	.47	.06	.77	.1	18	23
0	9	.81	20	400	16	.16	112	.11	.68	.09	1.13	.14	27	34
0	14	1.34	34	670	26	.26	178	.17	.91	.14	1.8	.23	31	55
0	72	1.32	29	530	355	.38	144	.11	.77	.07	1.76	.22	19	34
0	72	1.32	29	545	24	.38	144	.11	.91	.07	1.76	.22	19	34
0	59	4.91	105	1850	1131	1.07	47	.28	<.01	.26	4.89	.18	37	21
0	22	1.82	39	685	419	.4	17	.11	<.01	.1	1.81	.07	14	8
0	14	.8	24	470	80	.23	39	.06	.16	.11	1.09	.1	7	31
0	6	.55	13	273	11	.11	76	.07	.47	.06	.77	.1	18	23
0	22	1.41	27	535	877	.34	136	.11	2.21	.07	1.64	.27	48	44
0	22	1.41	27	535	24	.34	136	.11	2.21	.07	1.64	.27	48	44
0	92	5.08	134	2454	231	2.1	639	.41	11.3	.5	8.44	1	59	111
0	42	3.1	60	1065	85	.55	320	.18	6.3	.13	4.3	.38	27	26
0	34	1.89	47	909	1482	.61	240	.16	3.43	.14	2.82	.38	23	32
0	34	.34	12	211	78	.31	0	.04	.05	.04	.47	.1	14	18
0	197	1.15	32	292	42	.2	792	.06	2.48	.1	.59	.26	170	39
0	249	3.18	43	367	25	.67	1308	.09	4.79	.12	.77	.11	65	36
0	27	1.02	27	467	653	.48	283	.1	.77	.07	1.76	.34	51	67
0	22	.85	13	236	121	.33	944	.04	.49	.04	.47	.06	19	4
0	23	.75	20	154	32	.45	389	.06	.33	.11	.77	.07	17	3
0	7	.04	14	362	9	.31	0	.09	.74	.12	.62	.2	10	2
0	3	.61	3	83	6	.27	0	.01	.35	.02	.25	.11	4	1
0	1	.24	1	33	2	.11	0	<.01	.14	.01	.1	.04	2	<1
0	20	.03	4	56	7	.02	80	.01	.17	.02	.03	.02	2	7
0	3	.36	3	25	220	.06	1	.66	1.04	.07	1.13	.07	6	0
0	63	.37	41	376	504	1.78	22	.19	.67	.08	.54	.17	30	4
	8	.44	11	120	172	.25	0	.06		.02	.12	.02	7	0
0	13	.97	7	171	228	.37	0	.4	1.11	.27	2.68	.28	32	0
0	21	.78	10	198	474	.42	0	.66	1.61	.24	3.18	.42	46	0
0	53	4.24	36	362	527	1.26	78	.12	1.25	.07	1.21	.15	82	16
0	54	1.14	13	342	279	.8	0	.63	2.25	.51	6.84	.85	58	0
0	17	.92	9	76	219	.61	0	.56	.98	.61	8.16	.5	40	0
0	36	1.35		129	112		18	.05		.09		.06		<1
0	181	.33		307	78		3	.08		.2		.13		<1
0	27	1.54	15	188	576	1.07	0	.64	2.01	.37	7.37	.74	67	0
0	21	1.49	13	128	391	1.28	0	.64	1.23	.43	7.1	.85	55	0
0	9	.27	33	166	119	.46	0	.1		.35	1.04	.45	49	0
20	66	2.9	31	331	538	1.05	729	.65	4	.4	4.47	.31	40	10
0	21	1.49	13	128	391	1.28	0	.64	1.23	.43	7.1	.85	55	0
13	146	2.71	26	211	931	1.97	45	.77	1.43	.55	8.14	.86	69	1
0	154	3.75	116	609	10	3	115	.22	.03	.18	7.69	.5	86	0
0	79	2.66	15	398	351	.69	0	.23	.01	.51	3.78	.18	225	0
0	65	1.84			340	4.56	0	.18		.09	.27	.18		0
0	54	2			285	3.82	0	.15		.07	2.28	.15		0
0	74	2.61			374	6.93	3	.28		.14	4.05	.28		0
0	76	2.7			378	7.2	0	.29		.14	4.32	.29		0
0	71	2.52			371	6.66	3	.27		.13	4.05	.27		0
2	40	1.4		153	709	.43	0	.67		.3	2.89	.45		0
4	2	.63		76	503	.2	0	.98		.46	2.1	.35		0
2	8	.77		51	243	.89	0	.26		.46	1.98	.14		0
1	4	.39		29	160	.46	0	.13		.22	.64	.29		0
1	12	1.06	16	121	455	1.11	0	.14		.3	1.9	.51	21	0
2	24	.31		225	433	.41	0	1.25		.65	5.41	1		0

Table A–1

Food Composition (Computer code number is for West Diet Analysis program) (For purposes of calculations, use "0" for t, <1, <.1, <.01, etc.)

Computer Code Number	Food Description	Measure	Wt (g)	H_2O (%)	Ener (kcal)	Prot (g)	Carb (g)	Dietary Fiber (g)	Fat (g)	Fat Breakdown (g) Sat	Mono	Poly
	VEGETARIAN FOODS:—Continued											
1174	Vege-Burger, patty	1 ea	55	71	66	10	2	2	2	.4	.6	.5
	Vegetarian foods, Morningstar Farms:											
7672	Better-n-burgers, svg	1 ea	78	71	75	11	6	4	<1	.1	.3	.2
7766	Better-n-eggs	¼ c	57	88	23	5	<1	0	<1	.1	.1	.1
57436	Breakfast links	2 pce	45	60	63	8	2	2	2	.5	.7	1.3
7752	Breakfast strips	2 pce	16	43	56	2	2	<1	4	.7	1.1	2.6
7725	Burger crumbles, svg	1 ea	55	60	116	11	3	3	6	1.6	2.3	2.5
7726	Burger, spicy black bean	1 ea	78	60	113	11	15	5	1	.2	.3	.4
7665	Chik pattie	1 ea	71	51	177	7	15	2	10	1.3	2.6	5.9
7724	Frank, deli	1 ea	45	52	109	10	3	3	7	1	2.1	3.5
7722	Garden vege pattie	1 ea	67	60	104	11	9	4	4	.5	1.1	2.2
7746	Grillers	1 ea	64	55	139	14	5	3	7	1.7	2.2	3
7664	Prime pattie	1 ea	64	64	94	16	4	3	2	.2	.4	.6
	Vegetarian foods, Worthington:											
7634	Beef style, meatless, frzn	3 pce	55	58	113	9	4	3	7	1.2	2.7	2.6
7732	Burger, meatless, patty	¼ c	55	71	60	9	2	1	2	.3	.5	1.1
1846	Chik slices, canned	2 pce	60	78	62	6	1	1	4	.6	.9	2.3
1833	Chili, canned	½ c	106	73	136	9	10	4	7	1.1	1.7	4.1
1835	Choplets, slices, canned	2 pce	92	72	93	17	3	2	2	.9	.3	.3
7608	Corned beef style, meatless, frzn	4 pce	57	55	138	10	5	2	9	1.9	4.1	3.1
1831	Country stew, canned	1 c	240	81	208	13	20	5	9	1.6	2.3	4.8
7632	Egg roll, meatless, frzn	1 ea	85	53	181	6	20	2	8	1.7	4.5	2.3
1838	Numete, slices, canned	1 pce	55	58	132	6	5	3	10	2.4	4.4	2.7
1839	Prime stakes, slices, canned	1 pce	92	71	136	9	4	4	9	1.4	2.9	4.9
1840	Protose, slices, canned	1 pce	55	53	131	13	5	3	7	1	3	2.4
7606	Roast, dinner, meatless, frzn	1 ea	85	63	180	12	5	3	12	2.2	5	5.2
1842	Saucette links, canned	1 pce	38	62	86	6	1		6	1.1	1.6	3.8
1844	Savory slices, canned	1 pce	28	66	48	3	2	1	3	1.2	1.3	.6
7735	Stakelets, frzn	1 pce	71	58	145	12	6	2	8	1.4	2.7	3.9
1847	Turkee slices, canned	1 pce	33	64	68	5	1	1	5	.8	1.9	2.1
	MISCELLANEOUS											
	Baking powders for home use:											
	Sodium aluminum sulfate:											
962	With monocalcium phosphate monohydrate	1 tsp	5	2	6	<1	2	0	0	0	0	0
963	With monocalcium phosphate monohydrate, calcium sulfate	1 tsp	5	5	3	0	1	<1	0	0	0	0
964	Straight phosphate	1 tsp	5	4	3	<1	1	<1	0	0	0	0
965	Low sodium	1 tsp	5	6	5	<1	2	<1	<1	t	0	t
1204	Baking soda	1 tsp	5		0	0	0	0	0	0	0	0
966	Basil, dried	1 tbs	5	6	13	1	3	2	<1	t	t	.1
2068	Cajun seasoning	1 tsp	3	5	6	<1	1	<1	<1			
961	Carob flour	1 c	103	4	185	5	92	41	1	.1	.2	.2
967	Catsup:	¼ c	61	67	63	1	17	1	<1	t	t	.1
968	Tablespoon	1 tbs	15	67	16	<1	4	<1	<1	t	t	t
1200	Cayenne/red pepper	1 tbs	5	8	16	1	3	1	1	.2	.1	.4
969	Celery seed	1 tsp	2	6	8	<1	1	<1	<1	t	.3	.1
1203	Chili powder:	1 tbs	8	8	25	1	4	3	1	.2	.3	.6
970	Teaspoon	1 tsp	3	8	9	<1	2	1	<1	.1	.1	.2
	Chocolate:											
971	Baking, unsweetened, square	1 oz	28	1	146	3	8	4	15	9.1	5.2	.5
	For other chocolate items, see Sweeteners & Sweets											
972	Cilantro/coriander, fresh	1 tbs	1	93		<1	<1	<1	<1	0	t	0
2287	Cinnamon	1 tsp	2	10	5	<1	2	1	<1	t	t	t
1197	Cornstarch	1 tbs	8	8	30	<1	7	<1	<1	t	t	t
2239	Curry powder	1 tsp	2	10	6	<1	1	1	<1	t	.1	.1
1202	Dill weed, dried	1 tbs	3	7	8	1	2	<1	<1	t	.1	t
975	Garlic cloves	1 ea	3	59	4	<1	1	<1	<1	t	0	t
2238	Garlic powder	1 tsp	3	6	10	<1	2	<1	<1	t	t	t
977	Gelatin, dry, unsweetened: Envelope	1 ea	7	13	23	6	0	0	<1	t	t	t
978	Ginger root, slices, raw	2 pce	5	82	3	<1	1	<1	<1	t	t	t

Chol (mg)	Calc (mg)	Iron (mg)	Magn (mg)	Pota (mg)	Sodi (mg)	Zinc (mg)	VT-A (RE)	Thia (mg)	VT-E (a-TE)	Ribo (mg)	Niac (mg)	V-B6 (mg)	Fola (μg)	VT-C (mg)
0	8	.5	12	30	114	.58	0	.2		.25	.78	.31	15	0
0	79	2.66	15	398	351	.69	0	.23	.01	.51	3.78	.18	225	0
2	7	.63		68	90	.51	64	.01		.26	0	.11		0
1	15	2.14	16	59	338	.36	0	6.95		.22	5.19	.33	12	0
<1	7	.27		15	220	.05	0	.75		.04	.6	.07		0
0	40	3.2	1	89	238	.82	0	4.96	.34	.18	1.49	.27		0
1	56	1.84	44	269	499	.93	14	8.03	.36	.14	0	.21		0
1	11	1.02		163	536	.31	0	2.15		.16	1.51	.14		0
2	16	.26	4	50	524	.38	0	.14	1.26	.02	0	.01		0
1	34	.72	29	180	382	.59	20	6.47	.98	.1	0	0	29	0
2	43	1.16		127	256	.49	0	11.7		.24	2.99	.37		0
1	46	2.14		142	247	.74	0	.51		.25	.92	.41		2
0	4	2.63		44	624	.22	0	.89		.34	6.46	.56		0
0	4	1.73		25	269	.38	0	.13		.1	1.96	.24		0
1	9	.73		111	257	.26	0	.06		.05	.37	.08		0
0	20	1.49		195	523	.57	0	.02		.03	1.04	.31		0
0	6	.37		40	500	.65	0	.05		.05	0	.05		0
1	6	1.17		58	524	.26	0	10.6		.07	1.36	.3		0
2	51	5.09		270	826	1.03	216	1.85		.29	4.22	.86		0
1	15	.57		96	384	.31	0	1.22		.19	0	.03		0
0	10	1.12		155	272	.56	0	.08		.06	.54	.2		0
2	12	.38		82	445	.38	0	.12		.13	1.98	.38		0
<1	1	1.84		50	283	.7	0	.18		.13	1.34	.24		0
2	36	2.87		38	566	.64	0	2.13		.25	6.02	.6		0
1	9	1.15		25	205	.26	0	.59		.08	.09	.13		0
<1		.47		14	179	.08	0	.08		.06	.48	.1		0
2	49	.99		95	484	.5	0	1.51		.12	3.1	.26		0
1	3	.47		16	203	.11	0	1.13		.05	.39	.09		0
0	97	0		7	547	0	0	0	0	0	0	0	0	0
0	294	.55	1	1	530	<.01	0	0	0	0	0	0	0	0
0	368	.56	2		395	<.01	0	0	0	0	0	0	0	0
0	217	.41	1	505	4	.04	0	0	<.01	0	0	0	0	0
0	0	0	0	0	1368	0	0	0	0	0	0	0	0	0
0	106	2.1	21	172	2	.29	47	.01	.08	.02	.35	.06	14	3
				29	474									
0	358	3.03	56	852	36	.95	1	.05	.65	.47	1.96	.38	30	<1
0	12	.43	13	293	723	.14	62	.05	.9	.04	.84	.11	9	9
0	3	.1	3	72	178	.03	15	.01	.22	.01	.21	.03	2	2
0	7	.39	8	101	1	.12	208	.02	.24	.05	.43	.1	5	4
0	35	.9	9	28	3	.14	<1	.01	.02	.01	.06	.01	<1	<1
0	22	1.14	14	153	81	.22	279	.03	.08	.06	.63	.15	8	5
0	8	.43	5	57	30	.08	105	.01	.03	.02	.24	.06	3	2
0	21	1.77	87	233	4	1.12	3	.02	.34	.05	.31	.03	2	0
0	1	.02		5	<1	<.01	3	<.01	.02	<.01	.01	<.01	<1	<1
0	25	.76	1	10	1	.04	1	<.01	0	<.01	.03	<.01	1	1
0		.04			1	<.01	0	0	0	0	0	0	0	0
0	10	.59	5	31	1	.08	2	<.01	.01	.01	.07	.01	3	<1
0	53	1.46	13	99	6	.1	18	.01		.01	.08	.04		1
0	5	.05	1	12	1	.03	0	.01	0	<.01	.02	.04	<1	1
0	2	.08	2	33	1	.08	0	.01	0	<.01	.02	.08	<1	1
0	4	.08	2	1	14	.01	0	<.01	0	.02	.01	0	2	0
0	1	.02	2	21	1	.02	0	<.01	.01	<.01	.03	.01	1	<1

Table A-1

Food Composition (Computer code number is for West Diet Analysis program) (For purposes of calculations, use "0" for t, <1, <.1, <.01, etc.)

Computer Code Number	Food Description	Measure	Wt (g)	H_2O (%)	Ener (kcal)	Prot (g)	Carb (g)	Dietary Fiber (g)	Fat (g)	Fat Breakdown (g) Sat	Mono	Poly
	MISCELLANEOUS—Continued											
1198	Horseradish, prepared	1 tbs	15	85	7	<1	2	<1	<1	t	t	.1
1997	Hummous/hummus	1 c	246	65	421	12	50	12	21	3.1	8.8	7.8
1909	Mustard, country dijon	1 tsp	5		5	<1	<1	0	0	0	0	0
2019	Mustard, gai choy Chinese	1 tbs	16	94	3	<1	1		<1			
979	Mustard, prepared (1 packet = 1 tsp)	1 tsp	5	80	4	<1	<1	<1	<1	t	.2	t
	Miso (see #926 under Vegetables and Legumes, Soybean products)											
980	Olives, green	5 ea	20	78	23	<1	<1	<1	3	.3	1.9	.2
981	Olives, ripe, pitted	5 ea	22	80	25	<1	1	1	2	.3	1.7	.2
26008	Onion powder	1 tsp	2	5	7	<1	2	<1	<1	t	t	t
2237	Oregano, ground	1 tsp	2	7	6	<1	1	1	<1	.1	t	.1
2236	Paprika	1 tsp	2	10	6	<1	1	<1	<1	t	t	.2
887	Parsley, freeze dried	¼ c	1	2	3	<1	<1	<1	<1	t	t	t
	Parsley, fresh (see #885 and #886)											
985	Pepper, black	1 tsp	2	10	5	<1	1	1	<1	t	t	t
	Pickles:											
986	Dill, medium, 3¾ x 1¼" diam	1 ea	65	92	12	<1	3	1	<1	t	t	t
987	Fresh pack, slices, 1½" diam x ¼"	2 pce	15	79	11	<1	3	<1	<1	0	0	t
988	Sweet, medium	1 ea	35	65	41	<1	11	<1	<1	t	t	t
989	Pickle relish, sweet	1 tbs	15	63	21	<1	5	<1	<1	t	t	t
	Popcorn (see Grain Products #539-541)											
917	Potato chips:	10 pce	20	2	107	1	11	1	7	2.2	2	2.4
44076	Unsalted	1 oz	28	2	150	2	15	1	10	3.1	2.8	3.4
1201	Sage, ground	1 tsp	1	8	3	<1	1	<1	<1	.1	t	t
1347	Salsa, from recipe	1 tbs	15	93	3	<1	1	<1	<1	t	t	t
2218	Salsa, pico de gallo, medium	1 tbs	15	92	2	0	1	<1	0	0	0	0
990	Salt	1 tsp	6		0	0	0	0	0	0	0	0
	Salt Substitutes:											
1205	Morton, salt substitute	1 tsp	6			0	<1		0	0	0	0
1207	Morton, light salt	1 tsp	6			0	<1		0	0	0	0
2067	Seasoned salt, no MSG	1 tsp	5	5	4	<1	1	<1	<1			
991	Vinegar, cider	½ c	120	94	17	0	7	0	0	0	0	0
2172	Balsamic	1 tbs	15	64	21	0	4	0	0	0	0	0
2176	Malt	1 tbs	15	90	5	0	<1	0	0	0	0	0
2182	Tarragon	1 tbs	15	95	3	0	<1	0	0	0	0	0
2181	White wine	1 tbs	15	89	5	0	<1	0	0	0	0	0
	Yeast:											
992	Baker's, dry, active, package	1 ea	7	8	21	3	3	1	<1	t	.2	t
993	Brewer's, dry	1 tbs	8	5	23	3	3	3	<1	t	t	0
	SOUPS, SAUCES, and GRAVIES											
	SOUPS, canned, condensed:											
	Unprepared, condensed:											
1210	Cream of celery	1 c	251	85	181	3	18	2	11	2.8	2.6	5
1215	Cream of chicken	1 c	251	82	233	7	18	<1	15	4.2	6.5	3
1216	Cream of mushroom	1 c	251	81	259	4	19	1	19	5.1	3.6	8.9
1220	Onion	1 c	246	86	113	8	16	2	3	.5	1.5	1.3
	Prepared w/equal volume of whole milk:											
994	Clam chowder, New England	1 c	248	85	164	9	17	1	7	2.9	2.3	1.1
1209	Cream of celery	1 c	248	86	164	6	14	1	10	3.9	2.5	2.6
995	Cream of chicken	1 c	248	85	191	8	15	<1	11	4.6	4.5	1.6
996	Cream of mushroom	1 c	248	85	203	6	15	<1	14	5.1	3	4.6
1214	Cream of potato	1 c	248	87	149	6	17	<1	6	3.8	1.7	.6
1213	Oyster stew	1 c	245	89	135	6	10	0	8	5	2.1	.3
997	Tomato	1 c	248	85	161	6	22	3	6	2.9	1.6	1.1
	Prepared with equal volume of water:											
998	Bean with bacon	1 c	253	84	172	8	23	9	6	1.5	2.2	1.8
999	Beef broth/bouillon/consommé	1 c	240	98	17	3	<1	0	1	.3	.2	t
1000	Beef noodle	1 c	244	92	83	5	9	1	3	1.1	1.2	.5
1001	Chicken noodle	1 c	241	92	75	4	9	1	2	.7	1.1	.6
1002	Chicken rice	1 c	241	94	60	4	7	1	2	.5	.9	.4
1208	Chili beef	1 c	250	85	170	7	21	9	7	3.3	2.8	.3
1003	Clam chowder, Manhattan	1 c	244	92	78	2	12	1	2	.4	.4	1.3

Chol (mg)	Calc (mg)	Iron (mg)	Magn (mg)	Pota (mg)	Sodi (mg)	Zinc (mg)	VT-A (RE)	Thia (mg)	VT-E (a-TE)	Ribo (mg)	Niac (mg)	V-B6 (mg)	Fola (µg)	VT-C (mg)
0	8	.06	4	37	47	.12	<1	<.01	<.01	<.01	.06	.01	9	4
0	123	3.86	71	428	600	2.71	5	.23	2.46	.13	1.01	.98	146	19
0				10	120									
0	4	.1	2	6	63	.03	0	0	.09	0	0	<.01	0	0
0	12	.32	4	11	480	.01	6	0	.6	0	0	<.01	<1	0
0	19	.73	1	2	192	.05	9	<.01	.66	0	.01	<.01	0	<1
0	7	.05	2	19	1	.05	0	.01	<.01	<.01	.01	.03	3	<1
0	31	.88	5	33	<1	.09	14	.01	.03	.01	.12	.02	5	1
0	4	.47	4	47	1	.08	121	.01	.01	.03	.31	.04	2	1
0	2	.54	4	63	4	.06	63	.01	.06	.02	.1	.01	15	1
0	9	.58	4	25	1	.03	<1	<.01	.02	<.01	.02	.01	<1	<1
0	6	.34	7	75	833	.09	21	.01	.1	.02	.04	.01	1	1
0	5	.27	1	30	101	0	2	0	.02	<.01	0	<.01	0	1
0	1	.21	1	11	329	.03	5	<.01	.06	.01	.06	<.01	<1	<1
0	3	.12	1	30	107	.01	1	0	.02	<.01	0	0	0	1
0	5	.33	13	255	119	.22	0	.03	.98	.04	.77	.13	9	6
0	7	.46	19	357	2	.3	0	.05	1.37	.05	1.07	.18	13	9
0	16	.28	4	11	<1	.05	6	.01	.02	<.01	.06	.01	3	<1
0	1	.06	1	24	58	.02	22	.01	.04	<.01	.06	.01	2	5
0					130									
0	1	.02			2325	.01	0	0	0	0	0	0	0	0
	33			3018	<1									
	2		4	1560	1170									
				15	1542									
0	7	.72	26	120	1	0	0	0	0	0	0	0	0	0
	2	.07		10	3		<1	.07		.07	.07			<1
	2	.07		13	4		<1	.07		.07	.07			1
		.07		2	1		<1	.07		.07	.07			<1
	1	.07		12	1		<1	.07		.07	.07			<1
0	4	1.16	7	140	3	.45	<1	.16	.01	.38	2.79	.11	164	<1
0	17	1.38	18	151	10	.63	0	1.25		.34	3.03	.4	313	0
28	80	1.26	13	246	1900	.3	60	.06	.38	.1	.66	.02	5	<1
20	68	1.2	5	176	1972	1.26	113	.06	.33	.12	1.64	.03	3	<1
3	65	1.05	10	168	1736	1.19	0	.06	2.61	.17	1.62	.02	8	2
0	54	1.35	5	138	2115	1.23	0	.07	.57	.05	1.21	.1	30	2
22	186	1.49	22	300	992	.8	40	.07	.15	.24	1.03	.13	10	3
32	186	.69	22	310	1009	.2	67	.07	.97	.25	.44	.06	8	1
27	181	.67	17	273	1046	.67	94	.07	.24	.26	.92	.07	8	1
20	179	.59	20	270	918	.64	37	.08	1.34	.28	.91	.06	10	2
22	166	.55	17	322	1061	.67	67	.08	.1	.24	.64	.09	9	1
32	167	1.05	20	235	1041	10.3	44	.07	.49	.23	.34	.06	10	4
17	159	1.81	22	449	744	.29	109	.13	2.6	.25	1.52	.16	21	68
3	81	2.05	45	402	951	1.03	89	.09	.08	.03	.57	.04	32	2
0	14	.41	5	130	782	0	0	<.01	0	.05	1.87	.02	5	0
5	15	1.1	5	100	952	1.54	63	.07	<.01	.06	1.07	.04	19	<1
7	17	.77	5	55	1106	.39	72	.05	.07	.06	1.39	.03	22	<1
7	17	.75	0	101	815	.26	65	.02	.05	.02	1.13	.02	1	<1
12	42	2.13	30	525	1035	1.4	150	.06	.17	.07	1.07	.16	17	4
2	27	1.63	12	188	578	.98	98	.03	.73	.04	.82	.1	10	4

Table A-1

Food Composition (Computer code number is for West Diet Analysis program) (For purposes of calculations, use "0" for t, <1, <.1, <.01, etc.)

Computer Code Number	Food Description	Measure	Wt (g)	H_2O (%)	Ener (kcal)	Prot (g)	Carb (g)	Dietary Fiber (g)	Fat (g)	Fat Breakdown (g) Sat	Mono	Poly
	SOUPS, SAUCES, and GRAVIES—Continued											
1004	Cream of chicken	1 c	244	91	117	3	9	<1	7	2.1	3.3	1.5
1005	Cream of mushroom	1 c	244	90	129	2	9	<1	9	2.4	1.7	4.2
1006	Minestrone	1 c	241	91	82	4	11	1	3	.6	.7	1.1
1211	Onion	1 c	241	93	58	4	8	1	2	.3	.7	.7
1007	Split pea & ham	1 c	253	82	190	10	28	2	4	1.8	1.8	.6
1008	Tomato	1 c	244	90	85	2	17	<1	2	.4	.4	1
1009	Vegetable beef	1 c	244	92	78	6	10	<1	2	.9	.8	.1
1010	Vegetarian vegetable	1 c	241	92	72	2	12	<1	2	.3	.8	.7
	Ready to serve:											
1707	Chunky chicken soup	1 c	251	84	178	13	17	2	7	2	3	1.4
	SOUPS, dehydrated:											
	Prepared with water:											
1299	Beef broth/bouillon	1 c	244	97	19	1	2	0	1	.3	.3	t
1376	Chicken broth	1 c	244	97	22	1	1	0	1	.3	.4	.4
1013	Chicken noodle	1 c	252	94	53	3	7	1	1	.3	.5	.4
1122	Cream of chicken	1 c	261	91	107	2	13	<1	5	3.4	1.2	.4
1014	Onion	1 c	246	96	27	1	5	1	1	.1	.3	.1
1217	Split pea	1 c	255	87	125	7	21	3	1	.4	.7	.3
1015	Tomato vegetable	1 c	253	93	56	2	10	<1	1	.4	.3	.1
	Unprepared, dry products:											
1011	Beef bouillon, packet	1 ea	6	3	14	1	1	0	1	.3	.2	t
1012	Onion soup, packet	1 ea	39	4	115	5	21	4	2	.5	1.4	.3
	SAUCES											
	From dry mixes, prepared with milk:											
1016	Cheese sauce	1 c	279	77	307	17	23	1	17	9.3	5.3	1.6
1017	Hollandaise	1 c	259	84	240	5	14	<1	20	11.6	5.9	.9
1018	White sauce	1 c	264	81	240	10	21	<1	13	6.4	4.7	1.7
	From home recipe:											
1206	Lowfat cheese sauce	¼ c	61	73	85	6	4	<1	5	2.1	1.9	.9
1019	White sauce, medium	¼ c	72	77	102	2	6	<1	8	2.3	3.2	2
	Ready to serve:											
2202	Alfredo sauce, reduced fat	¼ c	69		170	5	16	0	10	6		
1020	Barbeque sauce	1 tbs	16	81	12	<1	2	<1	<1	t	.1	.1
1706	Chili sauce, tomato base	1 tbs	17	68	18	<1	4	<1	<1	t	t	t
2126	Creole sauce	¼ c	62	89	25	1	4	1	1	.1	.2	.3
2124	Hoisin sauce	1 tbs	17	47	35	<1	7	0	1	0		
2199	Pesto sauce	2 tbs	16		83	2	1	0	8	1.8	5.4	.7
1021	Soy sauce	1 tbs	16	71	8	1	1	<1	<1	t	t	t
2123	Szechuan sauce	1 tbs	16	71	21	<1	3	<1	1	.1	.3	.4
1380	Teriyaki sauce	1 tbs	18	68	15	1	3	<1	0	0	0	0
	Spaghetti sauce, canned:											
1377	Plain	1 c	249	75	271	5	40	8	12	1.7	6.1	3.3
1378	With meat	1 c	250	74	300	8	37	8	14	2.7	7	3.2
1379	With mushrooms	½ c	123	84	108	2	13	1	3	.4	1.5	.8
	GRAVIES											
	Canned:											
1022	Beef	1 c	233	87	123	9	11	1	5	2.7	2.2	.2
1023	Chicken	1 c	238	85	188	5	13	1	14	3.4	6.1	3.6
1024	Mushroom	1 c	238	89	119	3	13	1	6	1	2.8	2.4
1025	From dry mix, brown	1 c	258	92	75	2	13	<1	2	.8	.7	.1
1026	From dry mix, chicken	1 c	260	91	83	3	14	<1	2	.5	.9	.4
	FAST FOOD RESTAURANTS											
	ARBY'S											
1402	Bac'n cheddar deluxe	1 ea	231	59	512	21	39	<1	31	8.7	12.7	10.1
	Roast beef sandwiches:											
1403	Regular	1 ea	155	47	383	22	35	1	18	7	8	3.5
1404	Junior	1 ea	89	48	233	11	23	<1	11	4.1	5.2	2.5
1405	Super	1 ea	254	58	552	24	54	1	28	7.6	12.2	8.4
1407	Beef 'n cheddar	1 ea	194	50	508	25	43		26	7.7	12	6.8
1408	Chicken breast sandwich	1 ea	204	52	445	22	52	1	22	3	9.7	10.1
1412	Ham'n cheese sandwich	1 ea	169	54	355	25	34	<1	14	5.1	5.8	3.8
1726	Italian sub sandwich	1 ea	297	58	671	34	47		39	12.8	15.7	8.5

Chol (mg)	Calc (mg)	Iron (mg)	Magn (mg)	Pota (mg)	Sodi (mg)	Zinc (mg)	VT-A (RE)	Thia (mg)	VT-E (a-TE)	Ribo (mg)	Niac (mg)	V-B6 (mg)	Fola (μg)	VT-C (mg)
10	34	.61	2	88	986	.63	56	.03	.2	.06	.82	.02	2	<1
2	46	.51	5	100	881	.59	0	.05	1.24	.09	.72	.01	5	1
2	34	.92	7	313	911	.73	234	.05	.07	.04	.94	.1	36	1
0	26	.67	2	67	1053	.61	0	.03	.29	.02	.6	.05	15	1
8	23	2.28	48	400	1006	1.32	45	.15	.15	.08	1.47	.07	3	2
0	12	1.76	7	264	695	.24	68	.09	2.49	.05	1.42	.11	15	66
5	17	1.12	5	173	791	1.54	190	.04	.32	.05	1.03	.08	10	2
0	22	1.08	7	210	822	.46	301	.05	.79	.05	.92	.05	11	1
30	25	1.73	8	176	889	1	131	.08	.18	.17	4.42	.05	5	1
0	10	.02	7	37	1361	.07	<1	<.01	.02	.02	.36	0	0	0
0	15	.07	5	24	1483	.01	12	.01	.02	.03	.19	0	2	0
3	33	.5	8	30	1282	.2	5	.07	.02	.06	.88	.01	18	<1
3	76	.26	5	214	1184	1.57	123	.1	.15	.2	2.61	.05	5	1
0	12	.15	5	64	849	.06	<1	.03	.1	.06	.48	0	1	<1
3	20	.94	43	224	1147	.56	5	.21	.13	.14	1.26	.05	39	0
0	8	.63	20	104	1146	.17	20	.06	.81	.05	.79	.05	10	6
1	4	.06	3	27	1018	0	<1	<.01	.01	.01	.27	.01	2	0
2	55	.58	25	260	3493	.23	1	.11	.42	.24	1.99	.04	6	1
53	569	.28	47	552	1565	.97	117	.15	.33	.56	.32	.14	13	2
52	124	.9	8	124	1564	.7	220	.04	.26	.18	.06	.5	22	<1
34	425	.26	264	444	797	.55	92	.08	1.58	.45	.53	.07	16	3
11	166	.25	10	100	389	.73	58	.03	.55	.14	.16	.03	4	<1
8	75	.21	9	100	82	.26	89	.05	.98	.12	.28	.03	4	1
30	150	0	8	80	600		80	0		.1	0			0
0	3	.14	3	28	130	.03	14	<.01	.18	<.01	.14	.01	1	1
0	3	.14	2	63	227	.05	24	.01	.05	.01	.27	.02	1	3
0	35	.31	9	187	339	.1	24	.03	.61	.02	.53	.07	9	0
0	0	0			250		0							0
4	64	.09	6	15	129	.29	39	.01		.03	0	.02	<1	0
0	3	.32	5	29	914	.06	0	.01	0	.02	.54	.03	2	0
0	2	.12	2	13	218	.02	10	<.01	.07	<.01	.1	.01	1	<1
0	4	.31	11	40	690	.02	0	<.01	0	.01	.23	.02	4	0
0	70	1.62	60	956	1235	.52	306	.14	4.98	.15	3.76	.88	54	28
15	68	1.94	60	952	1179	1.37	577	.13	5.91	.17	4.51	.87	52	26
0	15	1	15	332	494	.34	241	.08	1.35	.08	.93	.16	12	9
7	14	1.63	5	189	1304	2.33	0	.07	.15	.08	1.54	.02	5	0
5	48	1.12	5	259	1373	1.9	264	.04	.37	.1	1.05	.02	5	0
0	17	1.57	5	252	1356	1.67	0	.08	.19	.15	1.6	.05	29	0
3	67	.23	10	57	1075	.31	0	.04	.05	.08	.81	0	0	0
3	39	.26	10	62	1133	.32	0	.05	.05	.15	.78	.03	3	3
38	110	4.32		491	1094	3	40	.34		.46	9.6			11
43	60	4.86	16	422	936	3.75	0	.28		.48	11	.2	14	1
22	40	2.7	8	201	519	1.5		.18		.25	6.6	.1	7	
43	90	6.48	25	533	1174	3.75	30	.39		.58	12.4	.3	21	9
52	150	6.12		321	1166	3		.42		.63	9.8			1
45	60	2.88	30	330	1019	.15		.22		.54	9	.38	18	5
55	170	2.7	31	382	1400	.9	40	.82		.37	7.8	.31	26	24
69	410	4.32		565	2062		100	.91		.49	8.2			11

Table A-1

Food Composition (Computer code number is for West Diet Analysis program) (For purposes of calculations, use "0" for t, <1, <.1, <.01, etc.)

Computer Code Number	Food Description	Measure	Wt (g)	H_2O (%)	Ener (kcal)	Prot (g)	Carb (g)	Dietary Fiber (g)	Fat (g)	Fat Breakdown (g) Sat	Mono	Poly
	FAST FOOD RESTAURANTS											
	ARBY'S—Continued											
1413	Turkey sandwich, deluxe	1 ea	195	69	260	20	33	<1	6	1.6	2.3	2.4
1680	Turkey sub sandwich	1 ea	277	62	486	33	46		19	5.3	6	7
	Milkshakes:											
1419	Chocolate	1 ea	340	74	451	10	76	<1	12	2.8	7	1.7
1420	Jamocha	1 ea	326	75	368	9	59	0	10	2.5	6.4	1.6
1421	Vanilla	1 ea	312	77	330	10	46	0	11	3.9	5.3	2.3
1728	Salad, roast chicken	1 ea	400	88	204	24	12		7	3.3	.9	.9
1729	Sports drink, Upper Ten	1 ea	358	88	169	0	42		0	0	0	0
	Source: Arby's											
	BURGER KING											
1423	Croissant sandwich, egg, sausage & cheese	1 ea	176	46	600	22	25	1	46	16		
	Whopper sandwiches:											
1425	Whopper	1 ea	270	58	640	27	45	3	39	11		
1426	Whopper with cheese	1 ea	294	57	730	33	46	3	46	16		
	Sandwiches:											
1629	BK broiler chicken sandwich	1 ea	248	59	550	30	41	2	29	6		
1432	Cheeseburger	1 ea	138	48	380	23	28	1	19	9		
1434	Chicken sandwich	1 ea	229	45	710	26	54	2	43	9		
1427	Double beef	1 ea	351	57	870	46	45	3	56	19		
1428	Double beef & cheese	1 ea	375	56	960	52	46	3	63	24		
1433	Double cheeseburger with bacon	1 ea	218	48	640	44	28	1	39	18		
1431	Hamburger	1 ea	126	48	330	20	28	1	15	6		
1437	Ocean catch fish fillet	1 ea	255	51	700	26	56	3	41	6		
1435	Chicken tenders	1 ea	88	50	230	16	14	2	12	3		
1439	French fries (salted)	1 svg	116	40	370	5	43	3	20	5		
1630	French toast sticks	1 svg	141	33	500	4	60	1	27	7		
1440	Onion rings	1 svg	124	51	310	4	41	6	14	2	8	4
1441	Milk shakes, chocolate	1 ea	284	75	320	9	54	3	7	4		
1442	Milk shakes, vanilla	1 ea	284	75	300	9	53	1	6	4		
1443	Fried apple pie	1 ea	113	47	300	3	39	2	15	3		
	Source: Burger King Corporation											
	CHICK-FIL-A											
	Sandwiches:											
69153	Chargrilled chicken	1 ea	150	54	280	27	36	1	3	1		
69152	Chicken	1 ea	167	61	290	24	29	1	9	2		
69155	Chicken salad	1 ea	167	55	320	25	42	1	5	2		
69154	Chicken salad club	1 ea	232	62	390	33	38	2	12	5		
	Salads:											
52139	Carrot and raisin	1 ea	76	53	150	5	28	2	2	0		
52136	Chicken plate	1 ea	468	85	290	21	40	6	5	0		
52134	Chicken garden, charbroiled	1 ea	397	89	170	26	10	5	3	1		
52135	Chick-n-strips	1 ea	451	86	290	32	21	5	9	2		
52138	Cole slaw	1 ea	79	70	130	6	11	1	6	1		
52137	Tossed salad	1 ea	130	85	70	5	13	1	0	0	0	0
15263	Chicken nuggets, svg	1 ea	110	51	290	28	12	60	12	3		
15262	Chicken-n-strips, svg	1 ea	119	59	230	29	10	0	8	2		
50885	Hearty breast of chicken soup, svg	1 ea	215	86	110	16	10	1	1	0		
7973	Waffle potato fries, svg	1 ea	85	28	290	1	49	0	10	4		
46489	Cheesecake, svg	1 ea	88	52	270	13	7	0	21	9		
49134	Fudge nut brownie, svg	1 ea	88	8	416	12	49	0	19	3.6		
20601	Icedream, svg	1 ea	127	74	140	11	16	0	4	1		
48214	Lemon pie, svg	1 ea	99	56	280	1	19	0	22	6		
	Source: Chick-Fil-A											

PAGE KEY: A–2 = Beverages A–4 = Dairy A–8 = Eggs A–8 = Fat/Oil A–12 = Fruit A–18 = Bakery A–26 = Grain A–30 = Fish A–32 = Meats A–36 = Poultry A–38 = Sausage A–40 = Mixed/Fast A–44 = Nuts/Seeds A–48 = Sweets A–50 = Vegetables/Legumes A–62 = Misc A–64 = Soups/Sauces A–66 = Fast A–82 = Frozen Entree A–86 = Baby foods

Chol (mg)	Calc (mg)	Iron (mg)	Magn (mg)	Pota (mg)	Sodi (mg)	Zinc (mg)	VT-A (RE)	Thia (mg)	VT-E (a-TE)	Ribo (mg)	Niac (mg)	V-B6 (mg)	Fola (µg)	VT-C (mg)
33	130	3.42	30	353	1262	1.5	40	.08		.41	15.4	.52	20	12
51	400	4.68		500	2033		20	13.2		.54	18.8			
36	250	.72	48	410	341	1.5	60	.12		.68	.8	.14	14	5
35	250	2.7	36	525	262	1.5	60	.12		.68	.8	.14	14	2
32	300	2.7	36	686	281	1.5	60	.12		.68	4	.14	37	2
43	170	1.98		877	508		485	.33		.54	5.6			51
0				0	40									
260	150	3.6			1140		80							0
90	80	4.5			870		100	.33		.41	7	.35		9
115	250	4.5			1350		150	.34		.48	7	.33		9
80	60	5.4			480		60							6
65	100	2.7			770		60							0
60	100	3.6			1400		0							0
170	80	7.2			940		100	.34		.56	10			9
195	250	7.2			1420		150	.35		.63	10			9
145	200	4.5			1240		80	.31		.42	6			0
55	40	1.8			530		20	.28		.31	4.89			0
90	60	2.7			980		20							1
35	0	.72			530		0							0
0	0	1.08			240		0							4
0	60	2.7			490		0							0
0	100	1.44			810		0							0
20	200	1.8			230		60	.13		.55	.13			0
20	300	0			230		60	.11		.57	.13			4
0	0	1.44			230		0							6
40					640									
50					870									
10					810									
70					980									
6					650									
35					570									
25					650									
20					430									
15					430									
0					0									
14					770									
20					380									
45					760									
5					960									
10					510									
36					773									
40					240									
5					550									

Table A-1

Food Composition (Computer code number is for West Diet Analysis program) (For purposes of calculations, use "0" for t, <1, <.1, <.01, etc.)

Computer Code Number	Food Description	Measure	Wt (g)	H_2O (%)	Ener (kcal)	Prot (g)	Carb (g)	Dietary Fiber (g)	Fat (g)	Fat Breakdown (g) Sat	Mono	Poly
	FAST FOOD RESTAURANTS—Continued											
	DAIRY QUEEN											
	Ice cream cones:											
1446	Small vanilla	1 ea	142	63	230	6	38	0	7	4.5		
1447	Regular vanilla	1 ea	213	64	350	8	57	0	10	7		
1448	Large vanilla	1 ea	253	65	410	10	65	0	12	8		
1450	Chocolate dipped	1 ea	234	59	510	9	63	1	25	13		
1453	Chocolate sundae	1 ea	241	62	410	8	73	0	10	6		
1455	Banana split	1 ea	369	67	510	8	96	3	12	8		
1456	Peanut buster parfait	1 ea	305	51	730	16	99	2	31	17		
1457	Hot fudge brownie delight	1 ea	305	52	710	11	102	1	29	14	12	2
1459	Buster bar	1 ea	149	45	450	10	41	2	28	12		
1645	Breeze, strawberry, regular	1 ea	383	70	460	13	99	1	1	1	0	0
1460	Dilly bar	1 ea	85	55	210	3	21	0	13	7	3	3
1461	DQ ice cream sandwich	1 ea	61	46	150	3	24	1	5	2		
1463	Milk shakes, regular	1 ea	397	71	520	12	88	<1	14	8	2	2
1464	Milk shakes, large	1 ea	461	71	600	13	101	<1	16	10	2	2
1466	Milk shakes, malted	1 ea	418	68	610	13	106	<1	14	8	2	2
1470	Misty slush, small	1 ea	454	88	220	0	56	0	0	0	0	0
2250	Starkiss	1 ea	85	75	80	0	21	0	0	0	0	0
	Yogurt:											
1641	Yogurt cone, regular	1 ea	213	66	280	9	59	0	1	.5		
1643	Yogurt sundae, strawberry	1 ea	255	69	300	9	66	1	<1	.5	0	0
	Sandwiches:											
1481	Cheeseburger, double	1 ea	219	55	540	35	30	2	31	16		
1480	Cheeseburger, single	1 ea	152	55	340	20	29	2	17	8		
1474	Chicken	1 ea	191	56	430	24	37	2	20	4		
1647	Chicken fillet, grilled	1 ea	184	64	310	24	30	3	10	2.5		
1475	Fish fillet sandwich	1 ea	170	57	370	16	39	2	16	3.5		
1476	Fish fillet with cheese	1 ea	184	56	420	19	40	2	21	6	7	8
1477	Hamburger, single	1 ea	128	56	269	16	27	2	11	4.6	5.6	.9
1478	Hamburger, double	1 ea	212	62	440	30	29	2	22	10		
	Hotdog:											
1483	Regular	1 ea	99	57	240	9	19	1	14	5		
1484	With cheese	1 ea	113	55	290	12	20	1	18	8	8	2
1485	With chili	1 ea	128	61	280	12	21	2	16	6		
1489	French fries, small	1 ea	71	39	210	3	29	3	10	2	5	3
1490	French fries, large	1 ea	128	40	390	5	52	6	18	4	8	6
1491	Onion rings	1 ea	85	46	240	4	29	2	12	2.5		
	Source: International Dairy Queen											
	HARDEES											
1734	Frisco burger hamburger	1 ea	242		760	36	43		50	18		
1736	Frisco grilled chicken salad	1 ea	278		120	18	2		4	1		
1737	Peach shake	1 ea	345		390	10	77		4	3		
	Source: Hardees											
	JACK IN THE BOX											
	Breakfast items:											
1492	Breakfast Jack sandwich	1 ea	121	49	300	18	30	0	12	5	5	2.5
1494	Sausage crescent	1 ea	156	39	580	22	28	0	43	16		
1495	Supreme crescent	1 ea	153	40	530	23	34	0	33	10	18.9	7.8
1496	Pancake platter	1 ea	231	45	610	15	87	0	22	9	7.6	3.5
1497	Scrambled egg platter	1 ea	213	52	560	18	50	0	32	9	16.6	4.4
	Sandwiches:											
1654	Bacon cheeseburger	1 ea	242	49	710	35	41	0	45	15	15.7	8.7
1499	Cheeseburger	1 ea	110	41	330	16	32	0	15	6	5.9	2.3
1739	Chicken caesar pita sandwich	1 ea	237	59	520	27	44	4	26	6		
1655	Chicken sandwich	1 ea	160	52	400	20	38	0	18	4		
1656	Chicken sandwich, sourdough ranch	1 ea	225		490	29	45	1	21	6		
1505	Chicken supreme	1 ea	245	55	620	25	48	0	36	11	14.8	11.4
1583	Double cheeseburger	1 ea	152	44	450	24	35	0	24	12	11.6	3.1

Chol (mg)	Calc (mg)	Iron (mg)	Magn (mg)	Pota (mg)	Sodi (mg)	Zinc (mg)	VT-A (RE)	Thia (mg)	VT-E (a-TE)	Ribo (mg)	Niac (mg)	V-B6 (mg)	Fola (μg)	VT-C (mg)
20	200	1.08		250	115		122	.05		.28				
30	300	1.8		390	170		150	.09		.38	.16	.13		2
40	350	1.8		451	200		200	.11		.4	.2			2
30	300	1.8		435	200		150	.09		.38	.16	.13		2
30	250	1.44		394	210		150	.08		.35	.4	.19		0
30	250	1.8		860	180		200	.15		.25	.4	.2		15
35	300	1.8		660	400		150	.15		.51	3	.22		1
35	300	5.4		510	340		80	.15		.68	.3	.18		1
15	150	1.08		400	280		80	.09		.17	3	.08		0
10	450	2.7		530	270		0	.13		.73				9
10	100	.36		170	75		60	.03		.14		.06		0
5	60	.72		105	115		40	.03		.25	.4	.05		0
45	400	1.44		570	230		80	.12		.59	.8	.19		<1
50	450	1.44		660	260		200	.15		.68	.8			<1
45	400	1.44		570	230		80	.12		.59	.8	.19		<1
0	0	0			20		0							0
0	0	0			10		0							0
5	300	1.8		285	170		0	.09		.38				2
5	300	1.8		352	180		0	.09		.49				6
115	250	4.5		426	1130		150	.29		.49	6.78			4
55	150	3.6		263	850		60	.29		.33	3.89			4
55	40	1.8		350	760		0	.37		.34	11			0
50	200	2.7		330	1040		0	.3		1.02	12			0
45	40	1.8		280	630		0	.3		.22	3			0
60	100	1.8		290	850		80	.3		.25	5			0
42	56	2.5		234	584		37	.27		.23	3.6			3
90	60	4.5		444	680		60	.32		.45	7.49			6
25	60	1.8		170	730		20	.22		.14	2			4
40	150	1.8		180	950		60	.22		.17	2			4
35	60	1.8		262	870		80	.23		.14	3			4
0	0	.72		430	115		0	.09		.03	2			5
0	0	1.44		780	200		0	.15		.07	3			9
0	0	1.08		90	135		0	.09		.05	.4			0
70					1280									
60					520									
25					290									
185	200	2.7		220	890		80	.47		.41	3			9
185	150	2.7		260	1010		100	.6		.51	4.6			0
210	150	3.6		270	930		150	.65		.54	4.2			12
100	100	1.8		310	890		80	.03		.85	7			6
380	150	4.5		450	1060		150			.66	5			9
110	250	5.4		540	1240		80	.24		.48	8.8	.39		9
35	200	2.7		200	510		60	.23		.23	3			1
55	250	2.7		490	1050		80							2
45	150	1.8		180	1290		40							0
65	150	1.8		340	1060									0
75	200	2.7		190	1520		100	.39		.32	11			2
75	250	3.6		320	900		100	.15		.34	6			0

Table A-1

Food Composition (Computer code number is for West Diet Analysis program) (For purposes of calculations, use "0" for t, <1, <.1, <.01, etc.)

Computer Code Number	Food Description	Measure	Wt (g)	H_2O (%)	Ener (kcal)	Prot (g)	Carb (g)	Dietary Fiber (g)	Fat (g)	Fat Breakdown (g) Sat	Mono	Poly
	FAST FOOD RESTAURANTS—Continued											
	JACK IN THE BOX—Continued											
1651	Grilled sourdough burger	1 ea	223	48	670	32	39	0	43	16	17.8	7.9
1498	Hamburger	1 ea	97	42	280	13	31	0	11	4	4.9	2
1500	Jumbo Jack burger	1 ea	229	55	560	26	41	0	32	10	13	8
1501	Jumbo Jack burger with cheese	1 ea	242	55	610	29	41	0	36	12	15	9
1740	Monterey roast beef sandwich	1 ea	238	57	540	30	40	3	30	9		
1508	Tacos, regular	1 ea	78	57	190	7	15	2	11	4		
1509	Tacos, super	1 ea	126	59	280	12	22	3	17	6		
	Teriyaki bowl:											
1679	Beef	1 ea	440	62	640	28	124	7	3	1		
1668	Chicken	1 ea	440	62	580	28	115	6	1			
1516	French fries	1 ea	109	38	350	4	45	4	17	4		
1517	Hash browns	1 ea	57	53	160	1	14	1	11	2.5	6.8	.3
1518	Onion rings	1 ea	103	34	380	5	38	0	23	6	15.2	.9
	Milkshakes:											
1519	Chocolate	1 ea	322	72	390	9	74	0	6	3.5	2.1	
1520	Strawberry	1 ea	298	74	330	9	60	0	7	4	2	
1521	Vanilla	1 ea	304	73	350	9	62	0	7	4	1.8	
1522	Apple turnover	1 ea	110	34	350	3	48	0	19	4	10.6	1.5
	Source: Jack in the Box Restaurant, Inc											
	KENTUCKY FRIED CHICKEN											
	Rotisserie gold:											
1472	Dark qtr, no skin	1 ea	117	66	217	27	0	0	12	3.5		
1473	Dark qtr, w/skin	1 ea	146	62	333	30	1		24	6.6		
1513	White qtr with wing, w/skin	1 ea	176	65	335	40	1		19	5.4		
1525	White qtr with wing, no skin	1 ea	117	63	199	37	0	0	6	1.7		
	Original Recipe:											
1253	Center breast	1 ea	103	52	260	25	9	<1	14	3.8	7.8	2
1251	Side breast	1 ea	83	47	245	18	9	<1	15	4.2	8.8	2.2
1250	Drumstick	1 ea	57	54	152	14	3	<1	8	2.2	4.1	1.3
1252	Thigh	1 ea	95	49	287	18	8	<1	21	5.3	9.4	3.1
1249	Wing	1 ea	53	45	172	12	5	<1	12	3	6	1.8
	Hot & spicy:											
1451	Center breast	1 ea	125	48	360	28	13		22	5		
1452	Side breast	1 ea	120	43	400	22	16		28	6		
1430	Thigh	1 ea	119	47	370	24	10		27	6		
1471	Wing	1 ea	61	38	220	14	5		16	4		
	Extra crispy recipe:											
1261	Center breast	1 ea	118	48	330	26	14	<1	20	4.8	10.8	2.1
1259	Side breast	1 ea	116	40	400	21	19	<1	27	5.5	12.9	2.3
1258	Drumstick	1 ea	65	49	190	14	6	<1	12	3.4	7.7	1.7
1260	Thigh	1 ea	109	43	380	23	7	<1	30	7.7	16	4.2
1257	Wing	1 ea	59	34	240	13	8	<1	17	4.4	10.7	2.5
1390	Baked beans	½ c	167	70	200	8	36	6	3	1.5	.7	.4
1526	Breadstick	1 ea	33	30	110	3	17	0	3	0		
1388	Buttermilk biscuit	1 ea	65	28	234	5	28	<1	13	3.4	6.2	2.3
1391	Chicken little sandwich	1 ea	47	35	169	6	14	<1	10	2	4.7	3.4
1269	Coleslaw	1 svg	90	75	114	1	13	<1	6	1	1.7	3.4
1527	Cornbread	1 ea	56	26	228	3	25	1	13	2		
1268	Corn-on-the-cob	1 ea	151	70	222	4	27	8	12	2	1	1.5
1429	Chicken, hot wings	1 svg	135	38	471	27	18		33			
1386	Kentucky fries	1 svg	77	42	228	3	26	3	12	3.2		
1381	Kentucky nuggets	6 ea	95	48	284	16	15	<1	18	4		
1534	Macaroni & cheese	1 svg	114	71	162	7	15	0	8	3		
1387	Mashed potatoes & gravy	1 svg	120	80	103	1	16	<1	5	.4	.5	.2
1530	Pasta salad	1 svg	108	78	135	2	14	1	8	1		
1389	Potato salad	½ c	188	74	271	5	27	3	16	3	4.2	7.3
1383	Potato wedges	1 svg	92	59	192	3	25	3	9	3		
1535	Red beans & rice	1 svg	112	76	114	4	18	3	3	1		
1529	Vegetable medley salad	1 ea	114	77	126	1	21	3	4	1		
	Source: Kentucky Fried Chicken Corp											

Chol (mg)	Calc (mg)	Iron (mg)	Magn (mg)	Pota (mg)	Sodi (mg)	Zinc (mg)	VT-A (RE)	Thia (mg)	VT-E (a-TE)	Ribo (mg)	Niac (mg)	V-B6 (mg)	Fola (µg)	VT-C (mg)
110	200	4.5		510	1140		150	.65		.48	8	.33		6
25	100	2.7		190	430		20	.15		.26	2			1
65	100	4.5		450	700		40	.36		.29	1.8			6
80	200	5.4		460	780		60	.36		.44	1.6			6
75	300	3.6		500	1270		80							5
20	100	1.08	35	240	410	1.2	0	.07		.17	1	.13		0
30	150	1.8	45	370	720	1.8	0	.12		.08	1.4	.18		2
25	150	4.5		430	930		1000							6
30	100	1.8		380	1220		1100							9
0	0	1.08		690	190		0	.18		.03	3.8			24
0	0	.36		190	310		0	.05			1			6
0	20	1.8		130	450		0	.29		.17	2.6			2
25	300	.72		680	210		0	.15		.6	.4			0
30	300	0		550	180		0	.15		.43	.4			0
30	300	0		570	180		0	.15		.34	.4			0
0	0	1.8		80	460		0	.2		.12	1.8			9
128	10	.18			772		15							1
163	10	.18			980		15							1
157	10	.18			1104		15							1
97	10	.18			667		15							1
92	30	.72			609		15	.09		.17	11.5			
78	68	1.2			604		15	.06		.13	6.9			
75	21	1.1			269		15	.05		.12	3.2			
112	40	1.08			591		31	.08		.3	5.5			
59	30	.54			383		15	.03		.08	3.7			
80	20	.72			750		15							6
80	40	1.08			850		15							6
100	20	1.08			670		15							6
65	20	.72			440		30							
75	33	.8			740		15	.11		.13	13.1			
75	20	.72			710		15	.09		.1	8.5			
65	20	.36			310		30	.06		.12	3.7			
90	49	1.2			520		30	.1		.21	6.5			
65	20	.36			320		30			.04	.06	3.3		
5	61	2.17	44	348	812	1.96	38	.09		.06	.76	.11	49	3
0	30	.18			15		0							0
3	43	1.92			565		28	.26		.2	2.77			
18	23	1.7			331		5	.16		.12	2.2			
5	30	.36			177		32	.03		.03	.2			27
42	60	.72			194		10							
0	0	.36			76		20	.14		.11	1.8			2
150	40	3.24			1230		15							6
4	11	.98			535		0							0
66	2	.1			865		15	.02		.02	1	.05		<1
16	120	.72			531		190							0
<1	20	.4			388		15			.04	1.2			
1	20	1.08			663		110							7
16	16	3.25	23	385	636	.44	120	.1		.03	.9	.29	11	
3					428		0							
4	10	.72			315									
0	20	.36			240		375							5

Table A–1

Food Composition (Computer code number is for West Diet Analysis program) (For purposes of calculations, use "0" for t, <1, <.1, <.01, etc.)

Computer Code Number	Food Description	Measure	Wt (g)	H_2O (%)	Ener (kcal)	Prot (g)	Carb (g)	Dietary Fiber (g)	Fat (g)	Fat Breakdown (g) Sat	Mono	Poly
	FAST FOOD RESTAURANTS—Continued											
	LONG JOHN SILVER'S											
1528	Chicken plank dinner, 3 piece	1 ea	399	56	890	32	101		44	9.5	24.8	9.4
1531	Clam chowder	1 ea	198	86	140	11	10	1	6	1.8	2.5	1.7
1532	Clam dinner	1 ea	361	46	990	24	114		52	10.9	31.4	9.9
	Fish, batter fried:											
1523	Fish & fryes (fries), 3 piece	1 ea	384	54	980	31	92		50	11.3	28.4	9.7
1524	Fish & fryes, 2 piece	1 ea	261	54	610	27	52		37	7.9	23.5	5.3
2240	Fish and lemon crumb dinner, 3 piece	1 ea	493	71	610	39	86		13	2.2	3.9	5.3
2241	Fish and lemon crumb dinner, 2 piece	1 ea	334	77	330	24	46		5	.9	1.6	1.2
1533	Fish & chicken dinner	1 ea	431	55	950	36	102		49	10.6	28.8	9.5
1537	Shrimp dinner, batter fried	1 ea	331	54	840	18	88		47	9.7	27.2	9.1
	Salads:											
1541	Cole slaw	1 ea	98	70	140	1	20	1	6	1	1.5	3.5
1539	Ocean chef salad	1 ea	234	89	110	12	13	2	1	.4	.4	.2
1540	Seafood salad	1 ea	278	79	380	15	12	2	31	5.1	8.2	17.5
1542	Fryes (fries) serving	1 ea	85	43	250	3	28	1	15	2.5	7.4	5.1
1543	Hush puppies	1 ea	24	40	70	2	10	<1	2	.4	1.3	.2
	Source: Long John Silver's, Lexington KY											
	McDONALD'S											
	Sandwiches:											
1221	Big mac	1 ea	216	53	510	25	46	3	26	9.3	7.5	4.1
1226	Cheeseburger	1 ea	122	46	318	15	36	2	13	5.6	3.8	1.1
1224	Filet-o-fish	1 ea	145	49	364	14	41	1	16	3.7	3.8	5.6
1225	Hamburger	1 ea	108	49	266	12	35	2	9	3.2	2.8	.9
1444	McChicken	1 ea	189	52	491	17	42	2	29	5.4	8.5	10.2
1591	McLean deluxe	1 ea	214	64	345	23	37	2	12	4.4	3.6	1.2
1438	McLean deluxe with cheese	1 ea	228	63	398	26	38	2	16	6.8	4.6	1.3
1222	Quarter-pounder	1 ea	171	52	415	23	36	2	20	7.8	6.7	1.3
1223	Quarter-pounder with cheese	1 ea	199	50	520	28	37	2	29	12.6	8.7	1.6
1227	French fries, small serving	1 ea	68	40	207	3	26	2	10	1.7	3.1	2.5
1228	Chicken McNuggets	4 pce	73	51	198	12	10	0	12	2.5	3.7	2.4
	Sauces (packet):											
1229	Hot mustard	1 ea	30	60	63	1	7	1	4	.5	1.1	2
1230	Barbecue	1 ea	32	58	53	<1	12	<1	<1	.1	.1	.2
1231	Sweet & sour	1 ea	32	57	55	<1	12	<1	<1	.1	.1	.3
	Low-fat (frozen yogurt) milk shakes:											
1232	Chocolate	1 ea	295		348	13	62	1	6	3.5	.1	.7
1233	Strawberry	1 ea	294		343	12	63	<1	5	3.4	.1	.6
1234	Vanilla	1 ea	293		308	12	54	<1	5	3.3	.1	.6
	Low-fat (frozen yogurt) sundaes:											
1237	Hot caramel	1 ea	182	56	307	7	62	1	3	2	.3	1
1235	Hot fudge	1 ea	179	60	293	8	53	2	5	4.7	.1	.4
1267	Strawberry	1 ea	178	65	239	6	51	1	1	.7	.1	.2
1238	Vanilla	1 ea	90	68	118	4	24	<1	1	.5	.2	t
1241	Cookies, McDonaldland	1 ea	56	3	258	4	41	1	9	1.7	6.4	.8
1242	Cookies, chocolaty chip	1 ea	56	3	282	3	36	1	14	3.9	4.4	1
1240	Muffin, apple bran, fat-free	1 ea	75	39	182	4	40	1	1	.2	.1	.3
1239	Pie, apple	1 ea	84	35	289	3	37	1	14	3.7	4.5	.3
	Breakfast items:											
1243	English muffin with spread	1 ea	63	33	189	5	30	2	6	2.4	1.5	1.3
1244	Egg McMuffin	1 ea	137	57	289	17	27	1	13	.7	4.5	1.6
1245	Hotcakes with marg & syrup	1 ea	222	44	557	8	100	2	14	2.4	4.6	5.8
1246	Scrambled eggs	1 ea	102	73	170	13	1	0	12	3.6	5.3	1.7
1247	Pork sausage	1 ea	43	45	173	6	<1	0	16	5.5	6.4	2.1
1248	Hashbrown potatoes	1 ea	53	55	130	1	13	1	8	1.3	2.3	1.9
1392	Sausage McMuffin	1 ea	112	42	361	13	26	1	23	8.3	8.2	2.8
1393	Sausage McMuffin with egg	1 ea	163	52	443	19	27	1	29	10	10.7	3.6
1394	Biscuit with biscuit spread	1 ea	76	32	260	4	32	1	13	3.8	3.7	.8

Chol (mg)	Calc (mg)	Iron (mg)	Magn (mg)	Pota (mg)	Sodi (mg)	Zinc (mg)	VT-A (RE)	Thia (mg)	VT-E (a-TE)	Ribo (mg)	Niac (mg)	V-B6 (mg)	Fola (μg)	VT-C (mg)
55	200	4.5		1170	2000	3	40	.52		.51	16			9
20	200	1.8		380	590	.6	150	.09		.25	2			
75	200	4.5		910	1830	3	40	.75		.42	12			12
70	200	4.5		1120	1530	3	40	.45		.42	8			15
60	40	1.8		900	1480	1.2		.37		.34	8			9
125	200	5.4		990	1420	2.25	700	.75		.59	24			6
75	80	1.8		440	640	.9	1000	.3		.25	14			18
75	200	4.5		1280	2090	3	40	.6		.59	14			9
100	200	3.6		840	1630	3	40	.45		.42	9			9
15	60	.72		190	260	.6	40	.06		.07	2			
40	100	3.6		95	730	.3	500	.12		.14	3			21
55	150	4.5		130	980	.9	200	.15		.25	3			21
0	200	.72		370	500	.3	0	.09			1.6			6
	40	.72		65	25	.3		.06		.03	.8			
76	202	4.32	46	456	932	4.81	66	.49	1.01	.44	6.08	.25	49	3
42	134	2.73	27	281	766	2.62	64	.33	.46	.31	3.81	.15	24	2
37	123	1.85	32	266	708	.7	21	.32	1.52	.23	2.58	.07	30	0
28	126	2.73	24	260	531	2.25	22	.33	.23	.26	3.81	.14	21	2
52	128	2.5	32	319	797	1.06	29	.91	6.16	.24	7.74	.38	37	1
59	131	4.29	40	537	811	4.9	74	.42	.63	.34	7.16	.29	44	8
73	139	4.29	43	559	1046	5.26	115	.42	.85	.39	7.16	.3	47	8
70	127	4.33	33	405	692	4.66	33	.39	.36	.32	6.78	.24	27	3
97	143	4.5			1160		115	.39	.81	.43	6.78	.26	33	3
0	9	.53	26	469	135	.32	0	.05	.83	0	1.94	.24	26	8
42	9	.65	17	210	353	.69	0	.08	.96	.11	5.15	.21		0
3	7	.78		29	85		4	.01		.01	.15			0
0	4	0		51	277		0	.01		.01	.17			4
0	2	.16		7	158		74	0		.01	.08			0
24	372	1.04		543	241		46	.12		.51	.4	.1		3
24	366	.29		542	170		46	.12		.51	.4	.11		3
24	360	.29		533	193		45	.12		.51	.31			3
7	246	.15		344	197		18	.09		.34	.27			2
5	258	.59		441	190		7	.09		.34	.29			2
5	221	.25		325	115		6	.06		.34	.25			2
3	132	.23		175	84		4	<.01		.01	.23			1
0	10	1.73	11	62	267	.38	0	.24	.99	.16	2.01	.03		0
3	28	1.78	24	142	229	.4	0	.14	.92	.16	1.48			0
0	34	1.29	13	77	215	.33	0	.14	0	.14	1.32	.03	5	1
0	17	1.23	7	69	221	.23		.19	1.5	.12	1.55	.04	9	27
13	103	1.59	13	69	386	.42	33	.25	.13	.31	2.61	.04	57	1
234	151	2.44	24	199	730	1.56	100	.49	.85	.45	3.33	.15	33	2
11	108	1.98	27	285	746	.53	119	.24	1.2	.26	1.86	.09	<1	<1
424	50	1.19	10	126	143	1.06	168	.07	.92	.51	.06	.12	44	0
33	7	.5	7	102	292	.78	0	.18	.26	.06	1.7	.09		0
0	7	.3	11	213	332	.15	0	.08	.58	.02	.9	.08	8	3
46	132	2.07	22	191	751	1.51	48	.56	.66	.27	3.76	.14	16	0
257	156	2.8	26	251	821	2.07	117	.59	1.11	.49	3.79	.19	33	0
0	68	1.85	9	105	836	.3	2	.29	.81	.23	2.23	.03	5	0

Table A-1

Food Composition (Computer code number is for West Diet Analysis program) (For purposes of calculations, use "0" for t, <1, <.1, <.01, etc.)

Computer Code Number	Food Description	Measure	Wt (g)	H_2O (%)	Ener (kcal)	Prot (g)	Carb (g)	Dietary Fiber (g)	Fat (g)	Fat Breakdown (g) Sat	Mono	Poly
	FAST FOOD RESTAURANTS—Continued											
	McDONALD'S—Continued											
1395	Biscuit with sausage	1 ea	119	37	433	10	32	1	29	8.6	10.1	2.8
1396	Biscuit with sausage & egg	1 ea	170	48	518	16	33	1	35	10.5	12.7	3.7
1397	Biscuit with bacon, egg, cheese	1 ea	152	46	450	17	33	1	27	8.7	8.9	2.3
	Salads:											
1398	Chef salad	1 ea	313	86	206	19	9	3	11	4.2	3	1.2
1400	Garden salad	1 ea	234	92	84	6	7	3	4	1.1	1.4	.7
1401	Chunky chicken salad	1 ea	296	87	164	23	8	3	5	1.3	1.6	1
	Source: McDonald's Corporation											
	PIZZA HUT											
	Pan pizza:											
1657	Cheese	2 pce	216	51	522	24	56	4	22	10	6.8	3.4
1658	Pepperoni	2 pce	208	49	531	22	56	4	24	8	9.9	3.7
1659	Supreme	2 pce	273	56	622	30	56	6	30	12	12	4.2
1660	Super supreme	2 pce	286	56	645	30	56	6	34	12		
	Thin 'n crispy pizza:											
1649	Cheese	2 pce	174	52	411	22	42	4	16	8	4.4	2.3
1623	Pepperoni	2 pce	168	48	431	22	42	2	20	8		
1622	Supreme	2 pce	232	57	514	28	42	4	26	10		
1620	Super supreme	2 pce	247	57	541	28	44	4	28	12		
	Hand tossed pizza:											
1619	Cheese	2 pce	216	53	470	26	58	4	14	7.9		
1618	Pepperoni	2 pce	208	51	477	24	58	4	16	8		
1648	Supreme	2 pce	273	56	568	32	60	6	24	10		
1617	Super supreme	2 pce	286	57	591	32	60	6	26	10		
	Personal pan pizza:											
1610	Pepperoni	1 ea	255	50	637	27	69	5	28	10	11.8	4.5
1609	Supreme	1 ea	327	57	721	33	70	6	34	12	14.7	5.6
	Source: Pizza Hut											
	SUBWAY											
	Deli style sandwich:											
69104	Bologna	1 ea	171	64	292	10	38	2	12	4		
69102	Ham	1 ea	171	69	234	11	37	2	4	1		
69103	Roast beef	1 ea	180	69	245	13	38	2	4	1		
69105	Seafood and crab:	1 ea	178	66	298	12	37	2	11	2		
69106	With light mayo	1 ea	178	68	256	12	37	2	7	2		
69108	Tuna:	1 ea	178		354	11	37	2	18	3		
69107	With light mayo	1 ea	178	67	279	11	38	2	9	2		
69101	Turkey	1 ea	180	69	235	12	38	2	4	1		
	Sandwiches, 6 inch:											
	B.L.T.:											
69135	On white bread	1 ea	191	67	311	14	38	3	10	3		
69136	On wheat bread	1 ea	198	65	327	14	44	3	10	3		
	Chicken taco sub:											
69131	On white bread	1 ea	286	70	421	24	43	3	16	5		
69132	On wheat bread	1 ea	293	69	436	25	49	4	16	5		
	Club :											
69117	On white bread	1 ea	246	73	297	21	40	3	5	1		
69118	On wheat bread	1 ea	253	71	312	21	46	3	5	1		
	Cold cut trio:											
69113	On white bread	1 ea	246	71	362	19	39	3	13	4		
69114	On wheat bread	1 ea	253	68	378	20	46	3	13	4		
	Ham:											
69115	On white bread	1 ea	232	73	287	18	39	3	5	1		
69115	On wheat bread	1 ea	239	71	302	19	45	3	5	1		

Chol (mg)	Calc (mg)	Iron (mg)	Magn (mg)	Pota (mg)	Sodi (mg)	Zinc (mg)	VT-A (RE)	Thia (mg)	VT-E (a-TE)	Ribo (mg)	Niac (mg)	V-B6 (mg)	Fola (μg)	VT-C (mg)
33	75	2.35	15	207	1128	1.08	2	.48	1.07	.29	3.93	.12	5	0
245	100	2.95	20	271	1199	1.61	59	.51	1.53	.55	3.96	.18	27	0
238	103	2.6	20	245	1315	1.64	99	.39	1.49	.57	3.32	.13	30	0
179	157	1.81	40	605	727	2.16	1179	.33	1.45	.37	4.32	.36	100	22
139	52	1.34	24	407	61	.73	1114	.12	.95	.24	.65	.16	96	22
76	54	1.62	44	673	318	1.52	1973	.51	1.28	.21	8.46	.52	83	30
50	288	3	63	337	1002	4.32	211	.6		.64	5.48	.18		7
48	206	3.21	55	399	1140	4.14	190	.62		.48	5.31	.16	0	8
60	234	4.6	81	620	1529	6	195	.86		.84	6.4	.33		11
68	236	4.39	80	592	1649	5.99	201	.83		.73	7.13			12
50	291	2.06	56	307	1070	4.23	217	.46		.46	5.65	.18		6
50	208	2.2	51	330	1255	4.02	199	.48		.49	5.97			7
62	238	3.6	79	631	1591	5.4	197	.7		.57	6.27			12
70	238	3.41	73	563	1762	5.47	208	.72		.53	6.57			9
50	284	3	71	388	1242	4.6	198	.48		.48	5.3			10
48	202	3.21	84	610	1380	6.01	187	.72		.56	7.59			13
60	232	4.6	87	589	1769	5.48	192	.82		.66	8.45			14
68	232	4.39	89	607	1889	5.65	198	.84		.68	8.71			14
55	250	4	60	406	1338	3.8	233	.56		.66	8.16	.2		10
66	276	5.19	74	603	1757	4.69	240	.73		.82	9.91	.4		14
20	39	3			744		113							14
14	24	3			773		113							14
13	23	3			638		113							14
17	24	3			544		113							14
16	24	3			556		118							14
18	26	3			557		116							14
16	26	3			583		126							14
12	26	3			944		113							14
16	27	3			945		120							15
16	33	3			957		120							15
52	118	4			1264		209							18
52	124	4			1275		209							18
26	29	4			1341		120							15
26	35	4			1352		120							15
64	49	4			1401		130							16
64	55	4			1412		130							16
28	28	3			1308		120							15
28	35	3			1319		120							15

Table A-1

Food Composition (Computer code number is for West Diet Analysis program) (For purposes of calculations, use "0" for t, <1, <.1, <.01, etc.)

Computer Code Number	Food Description	Measure	Wt (g)	H_2O (%)	Ener (kcal)	Prot (g)	Carb (g)	Dietary Fiber (g)	Fat (g)	Fat Breakdown (g) Sat	Mono	Poly
	FAST FOOD RESTAURANTS—Continued											
	SUBWAY—Continued											
	Italian B.M.T.											
69139	On white bread	1 ea	246	66	445	21	39	3	21	8		
69140	On wheat bread	1 ea	253	64	460	21	45	3	22	7		
	Meatball:											
69129	On white bread	1 ea	260	70	404	18	44	3	16	6		
69130	On wheat bread	1 ea	267	67	419	19	51	3	16	6		
	Melt with turkey, ham, bacon, cheese:											
69127	On white bread	1 ea	251	70	366	22	40	3	12	5		
69128	On wheat bread	1 ea	258	68	382	23	46	3	12	5		
	Pizza sub:											
69133	On white bread	1 ea	250	66	448	19	41	3	22	9		
69134	On wheat bread	1 ea	257	65	464	19	48	3	22	9		
	Roast beef:											
69121	On white bread	1 ea	232	72	288	19	39	3	5	1		
69122	On wheat bread	1 ea	239	70	303	20	45	3	5	1		
	Roasted chicken breast:											
69125	On white bread	1 ea	246	70	332	26	41	3	6	1		
69126	On wheat bread	1 ea	253	68	348	27	47	3	6	1		
	Seafood and crab:											
69145	On white bread:	1 ea	246	69	415	19	38	3	19	3		
69147	With light mayo	1 ea	246	72	332	19	39	3	10	2		
69146	On wheat bread:	1 ea	253	67	430	20	44	3	19	3		
69148	With light mayo	1 ea	253	70	347	20	45	3	10	2		
	Spicy italian:											
69123	On white bread	1 ea	232	64	467	20	38	3	24	9		
69124	On wheat bread	1 ea	239	62	482	21	44	3	25	9		
	Steak and cheese:											
69119	On white bread	1 ea	257	68	383	29	41	3	10	6		
69120	On wheat bread	1 ea	264	67	398	30	47	3	10	6		
	Tuna:											
69141	On white bread:	1 ea	246	62	527	18	38	3	32	5		
69143	With light mayo	1 ea	246	70	376	18	39	3	15	2		
69142	On wheat bread:	1 ea	253	62	542	19	44	3	32	5		
69144	With light mayo	1 ea	253	68	391	19	46	3	15	2		
	Turkey:											
69111	On white bread	1 ea	232	73	273	17	40	3	4	1		
69112	On wheat bread	1 ea	239	71	289	18	46	3	4	1		
	Turkey breast and ham:											
69137	On white bread	1 ea	232	73	280	18	39	3	5	1		
69138	On wheat bread	1 ea	239	71	295	18	46	3	5	1		
	Veggie delite:											
69109	On white bread	1 ea	175	71	222	9	38	3	3	0		
69110	On wheat bread	1 ea	182	69	237	9	44	3	3	0		
	Salads:											
52128	B.L.T.	1 ea	276	91	140	7	10	2	8	3		
52124	B.M.T., classic Italian	1 ea	331	86	274	14	11	1	20	7		
52127	Chicken taco	1 ea	370	87	250	18	15	2	14	5		
52115	Club	1 ea	331	91	126	14	12	1	3	1		
52120	Cold cut trio	1 ea	330	89	191	13	11	1	11	3		
52123	Ham	1 ea	316	91	116	12	11	1	3	1		
52129	Meatball	1 ea	345	88	233	12	16	2	14	5		
52131	Melt	1 ea	336	88	195	16	12	1	10	4		
52121	Pizza	1 ea	335	86	277	12	13	2	20	8		
52126	Roast beef	1 ea	316	92	117	12	11	1	3	1		
52119	Roasted chicken breast	1 ea	331	89	162	20	13	1	4	1		
52117	Seafood and crab:	1 5	331	88	244	13	10	2	17	3		
52116	With light mayo	1 5	331	90	161	13	11	2	8	1		
52130	Steak and cheese	1 ea	342	87	212	22	13	1	8	5		
52122	Tuna:	1 ea	331	84	356	12	10	1	30	5		
52118	With light mayo	1 ea	331	89	205	12	11	1	13	2		
52114	Turkey breast	1 ea	316	92	102	11	12	1	2	1		
52125	With ham	1 ea	316	92	109	11	11	1	3	1		

Chol (mg)	Calc (mg)	Iron (mg)	Magn (mg)	Pota (mg)	Sodi (mg)	Zinc (mg)	VT-A (RE)	Thia (mg)	VT-E (a-TE)	Ribo (mg)	Niac (mg)	V-B6 (mg)	Fola (μg)	VT-C (mg)
56	44	4			1652		151							15
56	50	4			1664		151							15
33	32	4			1035		142							16
33	39	4			1046		142							16
42	93	4			1735		155							15
42	100	3			1746		156							15
50	103	4			1609		238							16
50	110	3			1621		238							16
20	25	4			928		120							15
20	32	3			939		120							15
48	35	3			967		123							15
48	42	3			978		123							15
34	28	3			849		121							15
32	28	3			873		131							15
34	34	3			860		121							15
32	34	3			884		131							15
57	40	4			1592		169							15
57	47	4			1604		169							15
70	88	5			1106		175							18
70	95	5			1117		176							18
36	32	3			875		125							15
32	32	3			928		146							15
36	38	3			886		126							15
32	38	3			940		146							15
19	30	4			1391		120							15
19	37	3			1403		120							15
24	29	3			1350		120							15
24	36	3			1361		120							15
0	25	3			582		120							15
0	32	3			593		120							15
16	24	1			672		273							32
56	41	2			1379		303							32
52	115	3			990		361							35
26	26	2			1067		273							32
64	46	2			1127		282							33
28	25	2			1034		273							32
33	30	2			761		295							33
42	90	2			1461		308							32
50	100	2			1336		390							33
20	23	2			654		273							32
48	32	2			693		276							32
34	25	2			575		273							32
32	25	2			599		284							32
70	86	3			832		328							35
36	29	2			601		278							32
32	29	2			654		298							32
19	28	2			1117		273							32
24	27	2			1076		273							32

Table A-1

Food Composition (Computer code number is for West Diet Analysis program) (For purposes of calculations, use "0" for t, <1, <.1, <.01, etc.)

Computer Code Number	Food Description	Measure	Wt (g)	H_2O (%)	Ener (kcal)	Prot (g)	Carb (g)	Dietary Fiber (g)	Fat (g)	Fat Breakdown (g) Sat	Mono	Poly
	FAST FOOD RESTAURANTS—Continued											
	SUBWAY—Continued											
52113	Veggie delite	1 ea	260	94	51	2	10	1	1	0		
	Cookies:											
47662	Brazil nut and chocolate chip	1 ea	48	12	229	3	27	1	12	3.5		
47655	Chocolate chip:	1 ea	48	14	209	2	29	1	10	3.5		
47658	With M&M's	1 ea	48	14	209	2	29	1	10	3		
47659	Chocolate chunk	1 ea	48	14	209	2	29	1	10	3.5		
47656	Oatmeal raisin	1 ea	48	15	199	3	29	1	8	2		
47657	Peanut butter	1 ea	48	13	219	3	26	1	12	2.5		
47660	Sugar	1 ea	48	11	229	2	28	0	12	3		
47661	White chip macademia	1 ea	48	12	229	2	28	1	12	2.5		
	Source: Subway International											
	TACO BELL											
	Breakfast burrito:											
1601	Bacon breakfast burrito	1 ea	99	48	291	11	23		17	4		
1627	Country breakfast burrito	1 ea	113	55	220	8	26	2	14	5		
1626	Fiesta breakfast burrito	1 ea	92	44	280	9	25	2	16	6		
1625	Grande breakfast burrito	1 ea	177	56	420	13	43	3	22	7		
1604	Sausage breakfast burrito	1 ea	106	49	303	11	23		19	6		
	Burritos:											
1544	Bean with red sauce	1 ea	198	58	380	13	55	13	12	4		
1545	Beef with red sauce	1 ea	198	57	432	22	42	4	19	8	6.7	.7
1546	Beef & bean with red sauce	1 ea	198	57	412	17	50	5	16	6	6.1	2.1
1569	Big beef supreme	1 ea	298	64	520	24	54	11	23	10		
1552	Chicken burrito	1 ea	171	58	345	17	41		13	5		
1547	Supreme with red sauce	1 ea	248	64	428	16	50	10	18	7.8		
1571	7 layer burrito	1 ea	234	61	438	13	55	11	19	5.8		
1538	Chilito	1 ea	156	49	391	17	41		18	9		
1549	Chilito, steak	1 ea	257	62	496	26	47		23	10		
	Tacos:											
1551	Taco	1 ea	78	58	180	9	12	3	10	4		
1554	Soft taco	1 ea	99	63	242	10	13	3	11	4.4		
1536	Soft taco supreme	1 ea	128	64	234	11	21	3	13	6.3		
1568	Soft taco, chicken	1 ea	128	63	212	15	22	2	7	2.6		
1572	Soft taco, steak	1 ea	100	63	180	12	16	2	8	1.9		
1555	Tostada with red sauce	1 ea	156	67	264	9	27	11	13	4.4		
1558	Mexican pizza	1 ea	223	53	578	21	43	8	35	10.1		
1559	Taco salad with salsa	1 ea	585	71	923	33	70	17	56	16.3		
1560	Nachos, regular	1 ea	106	40	343	5	36	3	19	4.3		
1561	Nachos, bellgrande	1 ea	287	51	708	19	77	16	36	10.1		
1562	Pintos & cheese with red sauce	1 ea	128	68	203	10	19	11	10	4.3		
1563	Taco sauce, packet	1 ea	9	94	2	<1	<1	<1	<1	0	0	0
1564	Salsa	1 ea	10	28	27	1	6		<1	0	0	0
1565	Cinnamon twists	1 ea	35	6	175	1	24	0	7	0		
1628	Caramel roll	1 ea	85	19	353	6	46		16	4		
	Source: Taco Bell Corporation											
	WENDY'S											
	Hamburgers:											
1566	Single on white bun, no toppings	1 ea	133	44	360	24	31	2	16	6		
1570	Cheeseburger, bacon	1 ea	166	55	380	20	34	2	19	7	10.3	1.4
1730	Chicken sandwich, grilled	1 ea	189	62	310	27	35	2	8	1.5		
	Baked potatoes:											
1573	Plain	1 ea	284	71	310	7	71	7	0	0	0	0
1574	With bacon & cheese	1 ea	380	69	530	17	78	7	18	4	10.7	3.3
1575	With broccoli & cheese	1 ea	411	74	470	9	80	9	14	2.5	8	2.5
1576	With cheese	1 ea	383	68	570	14	78	7	23	8	9.2	4.8

Chol (mg)	Calc (mg)	Iron (mg)	Magn (mg)	Pota (mg)	Sodi (mg)	Zinc (mg)	VT-A (RE)	Thia (mg)	VT-E (a-TE)	Ribo (mg)	Niac (mg)	V-B6 (mg)	Fola (μg)	VT-C (mg)
0	23	1			308		136							32
10	32	1.99			115		0							0
10	16	1.99			139		0							0
15	16	1			139		0							0
10	16	1			139		0							0
15	32	1			159		0							0
0	16	1			179		0							0
20	0	.72			179		0							0
10	16	1			139		0							0
181	80	1.8			652		310							
195	80	1.08			690		250							0
25	80	.72			580		150							0
205	100	1.8			1050		500							0
183	80	1.8			661		320							
10	150	2.7		495	1100		450	.04		2.02	1.98	.31		0
57	160	3.96		380	1303		530	.4		2.14	3.44	.32		1
32	170	3.78	50	442	1221	2.67	450	.49		.41	3.09	.59	38	1
55	150	2.7			1520		600							5
57	140	2.52			854		440							1
34	146	8.75	48	410	1196		486	.39		2.04	2.81	.34		5
21	165	2.98			1058		248							5
47	300	3.06			980		950							
78	200	2.7			1313		970							2
25	80	1.08		159	330		100	.05		.14	1.2	.12		0
27	88	1.19		211	363		110	.42		.24	2.95	1.08		0
31	90	1.62			532		135							3
37	85	1.52			571		63							1
19	62	1.13			797		31							0
13	132	1.59		401	573		441	.05		.17	.63	.26		1
46	253	3.65	80	408	1054	5.37	405	.32		.33	2.96	1.12	60	5
65	326	6.84		1048	1931	1.67	1736	.51		.76	4.8	.56	10	26
5	107	.77		160	610	1.68	64	.17		.16	.68	.19	10	0
32	184	3.31		674	1205		138	.1		.34	2.17			3
16	160	1.92	110	384	693	2.17	267	.05		.15	.43	.21	68	0
0	0	.07		9	75		30	0			.02			<1
0	50	.6		376	709		168	.02		.14	0			10
0	0	.45		27	238		50	.1		.04	.71	.04		0
15	60	1.44			312		330							4
65	110	4.14		296	580		0	.43		.38	6.71			0
60	170	3.42	38	375	850	5.9	80	.3		.31	6.43	.26	28	6
65	100	2.7			790		40							6
0	30	3.78	75	1187	25	.74	0	.31	.14	.12	4.3	.8	31	36
20	180	4.32	87	1498	1390	2.75	100	.24		.19	5.04	.94	36	36
5	210	4.5	93	1745	470	.97	350	.34		.29	4.5	.97	74	72
30	380	4.14	85	1510	640	.67	200	.25		.28	3.6	.88	36	36

Table A-1

Food Composition (Computer code number is for West Diet Analysis program) (For purposes of calculations, use "0" for t, <1, <.1, <.01, etc.)

Computer Code Number	Food Description	Measure	Wt (g)	H_2O (%)	Ener (kcal)	Prot (g)	Carb (g)	Dietary Fiber (g)	Fat (g)	Fat Breakdown (g) Sat	Mono	Poly
	FAST FOOD RESTAURANTS—Continued											
	WENDY'S—Continued											
1577	With chili & cheese	1 ea	439	69	630	20	83	9	24	9		
1578	With sour cream & chives	1 ea	314	71	380	8	74	8	6	4		
1579	Chili	1 ea	227	81	210	15	21	5	7	2.5		
1582	Chocolate chip cookies	1 ea	57	6	270	3	36	1	13	6		
1580	French fries	1 ea	130	41	390	5	50	5	19	3	11.9	2.4
1581	Frosty dairy dessert	1 ea	227	68	330	8	56	0	8	5		
	Source: Wendy's International											
	CONVENIENCE FOODS and MEALS											
	BUDGET GOURMET											
1695	Chicken cacciatore	1 ea	312	80	300	20	27		13			
1692	Linguini & shrimp	1 ea	284	77	330	15	33		15			
1691	Scallops & shrimp	1 ea	326	79	320	16	43		9			
2245	Seafood newburg	1 ea	284	74	350	17	43		12			
1693	Sirloin tips with country gravy	1 ea	284	80	310	16	21		18			
1694	Sweet & sour chicken with rice	1 ea	284	72	350	18	53		7			
1689	Teriyaki chicken	1 ea	340	77	360	20	44		12			
1690	Veal parmigiana	1 ea	340	75	440	26	39		20			
1696	Yankee pot roast	1 ea	312	77	380	27	22		21			
	Source: The All American Gourmet Co.											
	HAAGEN DAZS											
1755	Ice cream bar, vanilla almond	1 ea	107		371	6	26		27	14.1	10	3
	Sorbet:											
1758	Lemon	½ c	113		140	0	35		0	0	0	0
1760	Orange	½ c	113		140	0	36		0	0	0	0
1759	Raspberry	½ c	113		110	0	27		0	0	0	0
	Yogurt, frozen:											
1753	Chocolate	½ c	98		171	8	26		4	2	2	0
1754	Strawberry	½ c	98		171	6	27		4	2	2	0
	Yogurt extra, frozen:											
1752	Brownie nut	½ c	101		220	8	29		9	4	4	1
1751	Raspberry rendezvous	½ c	101		132	4	26		2	1	1	0
	Source: Pillsbury											
	HEALTHY CHOICE											
	Entrees:											
2112	Fish, lemon pepper	1 ea	303	78	290	14	47	7	5	1		
1624	Lasagna	1 ea	383	76	390	26	60	9	5	2		
2111	Meatloaf, traditional	1 ea	340	79	320	16	46	7	8	4		
2104	Zucchini lasagna	1 ea	396	80	329	20	58	11	1	1		
2110	Dinner, pasta shells marinara	1 ea	340	74	360	25	59	5	3	1.5		
	Low-fat ice cream:											
973	Brownie	½ c	71	60	120	3	22	2	2	1	.3	.7
259	Butter pecan	½ c	71	60	120	3	22	1	2	1	.3	.7
650	Chocolate chip	½ c	71	62	120	3	21	<1	2	1	1	0
1608	Cookie & cream	½ c	71	62	120	3	21	<1	2	1.5	.5	0
650	Chocolate chip	½ c	71	62	120	3	21	<1	2	1	1	0
45	Rocky road	½ c	71	53	140	3	28	2	1	1	.5	0
1621	Vanilla	½ c	71	66	100	3	18	1	2	.5	1.5	0
391	Vanilla fudge	½ c	71	62	120	3	21	1	2	1.5		
	Source: ConAgra Frozen Foods, Omaha, NE											

Chol (mg)	Calc (mg)	Iron (mg)	Magn (mg)	Pota (mg)	Sodi (mg)	Zinc (mg)	VT-A (RE)	Thia (mg)	VT-E (a-TE)	Ribo (mg)	Niac (mg)	V-B6 (mg)	Fola (μg)	VT-C (mg)
40	330	5.04	122	1745	770	4.15	200	.33		.29	4.5	.99	55	36
15	80	4.32	71	1438	40	.91	300	.23		.14	3.04	.8	32	48
30	80	2.9		501	800		80	.11		.15	2.66			4
30	10	1.8	13	89	120	.41	0	.05		.06	.36	.03	5	0
0	20	1.08	55	845	120	.62	0	.18		.04	3.6	.33	40	6
35	310	1.08	46	544	200	.97	150	.11		.47	.32	.13	17	0
60	150	1.8			810		40	.23		.51	5			21
75	10	3.6			1250		1000	.3		.17	3			2
70	150	.72			690		150			.26	3			12
70	100	.72			660		40	.23		.26	2			
40	60	.36			570		150	.15		.17	4	.28		2
40	60	.72			640		80	.12		.34	3			2
55	80	1.4			610		300	.15		.34	6			12
165	30	4.5			1160		1000	.45		.6	6			6
70	150	1.8			690		600	.15		.43	7			6
90	161	.38		221	85		161			.18				
0				30	20									7
0				80	20									20
0				60	15									7
40	147	.71		241	45		20			.17				
50	147			141	45		20	.03		.17				5
55	152	.73		250	60		20			.14				
20	81			97	25		0			.1				5
25	20	1.08			360		100							30
15	150	3.6		500	550		100	.3		.26	2			6
35	40	1.8			460		150							54
10	199	2.69			309		249							0
25	400	1.8			390		100							4
2	80	0		268	55		40							0
2	100	0		211	60		40							0
2	100	0		240	50		40							0
2	100			254	90		60	.03		.15				2
2	100	0		240	50		40							0
2	100	0		168	60		40	.03		.15				0
5	100			254	50		60	.05		.22				2
2	100	0		296	50		40							0

Table A-1

Food Composition (Computer code number is for West Diet Analysis program) (For purposes of calculations, use "0" for t, <1, <.1, <.01, etc.)

Computer Code Number	Food Description	Measure	Wt (g)	H_2O (%)	Ener (kcal)	Prot (g)	Carb (g)	Dietary Fiber (g)	Fat (g)	Fat Breakdown (g) Sat	Mono	Poly
	CONVENIENCE FOODS and MEALS—Continued											
	HEALTH VALLEY											
	Soups, fat-free:											
2001	Beef broth, no salt added	1 c	240	98	18	5	0	0	0	0	0	0
2073	Beef broth, w/salt	1 c	240	98	30	5	2	0	0	0	0	0
2016	Black bean & vegetable	1 c	240	85	110	11	24	12	0	0	0	0
2017	Chicken broth	1 c	240	97	30	6	0	0	0	0	0	0
2018	14 garden vegetable	1 c	240	90	80	6	17	4	0	0	0	0
2015	Lentil & carrot	1 c	240	85	90	10	25	14	0	0	0	0
2014	Split pea & carrot	1 c	240	89	110	8	17	4	0	0	0	0
2013	Tomato vegetable	1 c	240	90	80	6	17	5	0	0	0	0
	Source: Health Valley											
	LA CHOY											
2100	Egg rolls, mini, chicken	1 svg	106	53	220	8	35	3	6	1.5		
2099	Egg rolls, mini, shrimp	1 svg	106	56	210	7	35	3	4	1		
	Source: Beatrice/Hunt Wesson											
	LEAN CUISINE											
	Dinners:											
1639	Baked cheese ravioli	1 ea	241	77	250	12	32	4	8	3	2	1
1632	Chicken chow mein	1 ea	255	81	210	13	28	2	5	1	2	1
1633	Lasagna	1 ea	291	79	270	19	34	5	6	2.5	1.5	.5
1634	Macaroni & cheese	1 ea	255	78	270	13	39	2	7	3.5	1.5	.5
1631	Spaghetti w/meatballs	1 ea	269	74	290	17	40	4	7	2	3	1.5
	Pizza:											
1636	French bread sausage pizza	1 ea	170	53	420	19	41	4	20	5	13.9	1.1
	Source: Stouffer's Foods Corp, Solon, OH											
	TASTE ADVENTURE SOUPS											
1905	Black bean	1 c	242		139	6	28	6	1			
1904	Curry lentil	1 c	241		138	6	30	5	1			
1906	Lentil chili	1 c	242		181	11	33	6	1			
1903	Split pea	1 c	244		140	5	27	5	1			
	Source: Taste Adventure Soups											
	WEIGHT WATCHERS											
	Cheese, fat-free slices:											
1978	Cheddar, sharp	2 pce	21	65	30	5	2	0	0	0	0	0
1980	Swiss	2 pce	21	65	30	5	2	0	0	0	0	0
1977	White	2 pce	21	65	30	5	2	0	0	0	0	0
1979	Yellow	2 pce	21	65	30	5	2	0	0	0	0	0
	Dinners:											
2029	Chicken chow mein	1 ea	255	81	200	12	34	3	2	.5		
1646	Oven fried fish	1 ea	218	78	230	15	25	2	8	2.5	5	2
1972	Margarine, reduced fat	1 tbs	14	49	59	0	0	0	7	1.5		
	Pizza:											
1653	Cheese	1 ea	163	48	390	23	49	6	12	4	3	1
1650	Deluxe combination pizza	1 ea	186	56	380	23	47	6	11	3.5	5	2
1652	Pepperoni pizza	1 ea	158	48	390	23	46	4	12	4	5	2
	Desserts:											
1644	Chocolate brownie	1 ea	182	75	190	6	35	4	4	1	2	1
2024	Chocolate eclair	1 ea	60	45	151	3	24	2	5	1.5		
2247	Chocolate mousse	1 ea	78	44	190	6	33	3	4	1.5		
1642	Strawberry cheesecake	1 ea	111	62	180	7	28	2	5	2	1	2

Chol (mg)	Calc (mg)	Iron (mg)	Magn (mg)	Pota (mg)	Sodi (mg)	Zinc (mg)	VT-A (RE)	Thia (mg)	VT-E (a-TE)	Ribo (mg)	Niac (mg)	V-B6 (mg)	Fola (µg)	VT-C (mg)
0				196	74						.98			
0	0	0		196	160		0				.98			5
0	40	3.6		676	280		2000	.34		.11	1.35	.22	135	9
0	20	1.8		147	170		0			.03	2.45			1
0	40	1.8		406	250		2000	.26		.08	2.25	.18	27	15
0	60	5.4		439	220		2000	.1		.16	5.63	.45	27	2
0	40	5.4		439	230		2000	.1		.16	5.63	.45		9
0	40	5.4		609	240		2000	.1		.08	2.25	.13	<1	9
5	20	1.44			460		20							0
5	20	1.44			510		20							0
55	200	1.08	42	400	500	1.5	150	.06		.25	1.2	.2	48	6
35	20	.36	30	300	510	1.1	20	.15		.17	5			6
25	150	1.8	44	620	560	2.9	100	.15		.25	3	.32		12
20	250	.72		170	550		20	.12		.25	1.2			0
30	100	2.7	47	480	520	2.5	80	.15		.25	3	.2		4
35	250	2.7	39	340	900	2.2	80	.45		.51	5	.07		6
				650	565									
				467	584									
				650	448									
				484	591									
0	99	0		64	306		56							0
0	99	0		74	276		56							0
0	99	0		64	306		56							0
0	99	0		64	306		56							0
25	40	.72		360	430		300							36
25	20	1.44		370	450		40	.09		.14	1.6			0
0	0	0		5	128		49							0
35	700	1.8		290	590		80	.3		.51	3	.06		6
40	500	3.6		370	550		150	.3		.51	3	.2		5
45	450	1.8		320	650		80	.23		.51	3			5
5	80	1.08		230	160		0	.06		.03	.2	.03		0
0	40	0		65	151		0							0
5	60	1.8		320	150		0							0
15	80	.36		115	230		40	.06		.07	1.6			2

Table A–1

Food Composition (Computer code number is for West Diet Analysis program) (For purposes of calculations, use "0" for t, <1, <.1, <.01, etc.)

Computer Code Number	Food Description	Measure	Wt (g)	H_2O (%)	Ener (kcal)	Prot (g)	Carb (g)	Dietary Fiber (g)	Fat (g)	Fat Breakdown (g) Sat	Mono	Poly
	CONVENIENCE FOODS and MEALS—Continued											
	WEIGHT WATCHERS—CONTINUED											
2027	Triple chocolate cheesecake	1 ea	89	52	199	7	32	1	5	2.5		
	Source: Weight Watchers											
	SWEET SUCCESS:											
	Drinks, prepared:											
1776	Chocolate chip	1 c	265	81	180	15	30	6	3	1.6		
1777	Chocolate fudge	1 c	265	81	180	15	30	6	2			
1774	Chocolate mocha	1 c	265	81	180	15	30	6	1	1		
1778	Milk chocolate	1 c	265	81	180	15	30	6	2	1		
1775	Vanilla	1 c	265	81	180	15	33	6	1	.6		
	Drinks, ready to drink:											
2147	Chocolate mint	1 c	297	82	187	11	36	6	3	0		
2148	Strawberry	1 c	265	82	167	10	32	5	3	0		
	Shakes:											
1771	Chocolate almond	1 c	250	82	158	9	30	5	2	0	2.1	.2
1773	Chocolate fudge	1 c	250	82	158	9	30	5	2	0	2.1	.2
1768	Chocolate mocha	1 c	250	82	158	9	30	5	2	0	.6	1.8
1769	Chocolate raspberry truffle	1 c	250	82	158	9	30	5	2	0	2.2	.2
1770	Vanilla creme	1 c	250	82	158	9	30	5	2	0	2.1	.3
	Snack bars:											
1767	Chocolate brownie	1 ea	33	9	120	2	23	3	4	2	.5	.6
1766	Chocolate chip	1 ea	33	9	120	2	23	3	4	2	.4	.5
1921	Oatmeal raisin	1 ea	33	9	120	2	23	3	4	2		
1765	Peanut butter	1 ea	33	9	120	2	23	3	4	2	.6	.6
	Source: Foodway National Inc, Boise, ID											
	BABY FOODS											
1720	Apple juice	½ c	125	88	59	0	15	<1	<1	t	t	t
1721	Applesauce, strained	1 tbs	16	89	7	<1	2	<1	<1	t	t	t
1716	Carrots, strained	1 tbs	14	92	4	<1	1	<1	<1	t	t	t
1718	Cereal, mixed, milk added	1 tbs	15	75	17	1	2	<1	1	.3		
1719	Cereal, rice, milk added	1 tbs	15	75	17	<1	3	<1	1	.3		
1723	Chicken and noodles, strained	1 tbs	16	88	8	<1	1	<1	<1	.1	.1	t
1722	Peas, strained	1 tbs	15	87	6	1	1	<1	<1	t	t	t
1717	Teething biscuits	1 ea	11	6	43	1	8	<1	<1	.2	.2	.1

PAGE KEY: A–2 = Beverages A–4 = Dairy A–8 = Eggs A–8 = Fat/Oil A–12 = Fruit A–18 = Bakery A–26 = Grain A–30 = Fish A–32 = Meats A–36 = Poultry A–38 = Sausage A–40 = Mixed/Fast A–44 = Nuts/Seeds A–48 = Sweets A–50 = Vegetables/Legumes A–62 = Misc A–64 = Soups/Sauces A–66 = Fast A–82 = Frozen Entree A–86 = Baby foods

A

Chol (mg)	Calc (mg)	Iron (mg)	Magn (mg)	Pota (mg)	Sodi (mg)	Zinc (mg)	VT-A (RE)	Thia (mg)	VT-E (a-TE)	Ribo (mg)	Niac (mg)	V-B6 (mg)	Fola (μg)	VT-C (mg)
10	80	1.08		169	199		0							0
6	500	6.3	140	600	288	5.25	350	.52	7.05	.59	7	.7	140	21
6	500	6.3	140	750	336	5.25	350	.52	7.05	.59	7	.7	140	21
6	500	6.3	140	800	336	5.25	350	.52	7.05	.59	7	.7	140	21
6	500	6.3	140	750	336	5.25	350	.52	7.05	.59	7	.7	140	21
6	500	6.3	140	830	312	5.25	250	.52	7.05	.59	7	.7	140	21
6	470	5.94	131	526	226	5.05	329	.5	6.56	.56	6.53	.65	131	20
5	419	5.3	117	310	175	4.51	294	.45	5.86	.5	5.83	.58	117	17
5	396	5	110	443	190	4.25	277	.42	5.53	.47	5.5	.55	110	16
5	396	5	110	443	175	4.25	277	.42	5.53	.47	5.5	.55	110	16
5	396	5	110	403	175	4.25	277	.42	5.53	.47	5.5	.55	110	16
5	383	5	110	443	175	4.25	277	.42	5.53	.47	5.5	.55	110	16
5	396	5	110	293	175	4.25	277	.42	5.53	.47	5.5	.55	110	16
3	150	2.71	60	140	45	.59	150	.22	3.01	.25	3	.3	60	9
3	150	2.71	60	110	40	.59	150	.22	3.01	.25	3	.3	60	9
3	150	2.71	60		30	.59	150	.22	3.01	.25	3	.3	60	9
3	150	2.71	60	125	35	.59	150	.22	3.01	.25	3	.3	60	9
0	5	.71	4	114	4	.04	2	.01	.75	.02	.1	.04	<1	72
0	1	.03		11	<1	<.01	<1	<.01	.1	<.01	.01	<.01	<1	6
0	3	.05	1	27	5	.02	160	<.01	.07	.01	.06	.01	2	1
2	33	1.56	4	30	7	.11	4	.06		.09	.87	.01	2	<1
2	36	1.83	7	28	7	.1	4	.07		.07	.78	.02	1	<1
3	4	.07	1	6	3	.05	18	<.01	.04	.01	.07	<.01	2	<1
0	3	.14	2	17	1	.05	8	.01	.08	.01	.15	.01	4	1
0	29	.39	4	35	40	.1	1	.03	.05	.06	.48	.01	5	1

Contents

B

CANADA: RECOMMENDATIONS, CHOICE SYSTEM, AND LABELS

Chapter 1 introduced recommended nutrient intakes, food guides, and food labels and Chapter 25 introduced the U.S. exchange system (see Appendix C). This appendix presents details for Canadians. Appendix D includes addresses of

Table B-1

Recommended Nutrient Intakes for Canadians, 1990

				Vitamins					
Age	**Sex**	**Weight (kg)**	**Protein (g/day)[a]**	**Vitamin A (RE/day)[b]**	**Vitamin E (mg/day)[c]**	**Vitamin C (mg/day)[d]**	**Iron (mg/day)**	**Iodine (μg/day)**	**Zinc (mg/day)**
Infants (months)									
0–4	Both	6	12[e]	400	3	20	0.3[f]	30	2[g]
5–12	Both	9	12	400	3	20	7	40	3
Children (years)									
1	Both	11	13	400	3	20	6	55	4
2–3	Both	14	16	400	4	20	6	65	4
4–6	Both	18	19	500	5	25	8	85	5
7–9	M	25	26	700	7	25	8	110	7
	F	25	26	700	6	25	8	95	7
10–12	M	34	34	800	8	25	8	125	9
	F	36	36	800	7	25	8	110	9
13–15	M	50	49	900	9	30	10	160	12
	F	48	46	800	7	30	13	160	9
16–18	M	62	58	1000	10	40	10	160	12
	F	53	47	800	7	30	12	160	9
Adults (years)									
19–24	M	71	61	1000	10	40	9	160	12
	F	58	50	800	7	30	13	160	9
25–49	M	74	64	1000	9	40	9	160	12
	F	59	51	800	6	30	13[h]	160	9
50–74	M	73	63	1000	7	40	9	160	12
	F	63	54	800	6	30	8	160	9
75+	M	69	59	1000	6	40	9	160	12
	F	64	55	800	5	30	8	160	9
Pregnancy (additional amount needed)									
1st trimester			5	0	2	0	0	25	6
2nd trimester			20	0	2	10	5	25	6
3rd trimester			24	0	2	10	10	25	6
Lactation (additional amount needed)			20	400	3	25	0	50	6

Note: Recommended intakes of energy and of certain nutrients are not listed in this table because of the nature of the variables upon which they are based. The figures for energy are estimates of average requirements for expected patterns of activity (see Table B-2). For nutrients not shown, the following amounts are recommended based on at least 2000 kcalories per day and body weights as given: thiamin, 0.4 milligram per 1000 kcalories (0.48 milligram/5000 kilojoules); riboflavin, 0.5 milligram per 1000 kcalories (0.6 milligram/5000 kilojoules); niacin, 7.2 niacin equivalents per 1000 kcalories (8.6 niacin equivalents/5000 kilojoules); vitamin B_6, 15 micrograms, as pyridoxine, per gram of protein. Recommended intakes during periods of growth are taken as appropriate for individuals representative of the midpoint in each age group. All recommended intakes are designed to cover individual variations in essentially all of a healthy population subsisting upon a variety of common foods available in Canada.

Source: Health and Welfare Canada, *Nutrition Recommendations: The Report of the Scientific Review Committee* (Ottawa: Canadian Government Publishing Centre, 1990), Table 20, p. 204.

[a]The primary units are expressed per kilogram of body weight. The figures shown here are examples.

[b]One retinol equivalent (RE) corresponds to the biological activity of 1 microgram of retinol, 6 micrograms of beta-carotene, or 12 micrograms of other carotenes.

[c]Expressed as δ-α-tocopherol equivalents, relative to which β- and γ-tocopherol and α-tocotrienol have activities of 0.5, 0.1, and 0.3, respectively.

[d]Cigarette smokers should increase intake by 50 percent.

[e]The assumption is made that the protein is from breast milk or has the same biological value as breast milk and that, between 3 and 9 months, adjustment for the quality of the protein is made.

[f]Based on the assumption that breast milk is the source of iron.

[g]Based on the assumption that breast milk is the source of zinc.

[h]After menopause, the recommended intake is 8 milligrams per day.

Canadian governmental agencies and professional organizations that may provide additional information.

RNI

As Chapter 1 mentioned, a major revision of the nutrient recommendations is underway in both the United States and Canada. The Dietary Reference Intakes (DRI) reports are replacing the 1989 RDA in the United States and the 1991 RNI in Canada. Recommendations from the DRI reports are presented on the inside front covers. For nutrients that do not yet have new values, the RNI will continue to serve health professionals in Canada (Tables B-1 and B-2).

Table B-2
Average Energy Requirements for Canadians

Age	Sex	Average Height (cm)	Average Weight (kg)	Requirements[a]					
				(kcal/kg)[b]	(MJ/kg)[b]	(kcal/day)	(MJ/day)	(kcal/cm)	(MJ/cm)
Infants (months)									
0–2	Both	55	4.5	120–100	0.50–0.42	500	2.0	9	0.04
3–5	Both	63	7.0	100–95	0.42–0.40	700	2.8	11	0.05
6–8	Both	69	8.5	95–97	0.40–0.41	800	3.4	11.5	0.05
9–11	Both	73	9.5	97–99	0.41	950	3.8	12.5	0.05
Children and Adults (years)									
1	Both	82	11	101	0.42	1100	4.8	13.5	0.06
2–3	Both	95	14	94	0.39	1300	5.6	13.5	0.06
4–6	Both	107	18	100	0.42	1800	7.6	17	0.07
7–9	M	126	25	88	0.37	2200	9.2	17.5	0.07
	F	125	25	76	0.32	1900	8.0	15	0.06
10–12	M	141	34	73	0.30	2500	10.4	17.5	0.07
	F	143	36	61	0.25	2200	9.2	15.5	0.06
13–15	M	159	50	57	0.24	2800	12.0	17.5	0.07
	F	157	48	46	0.19	2200	9.2	14	0.06
16–18	M	172	62	51	0.21	3200	13.2	18.5	0.08
	F	160	53	40	0.17	2100	8.8	13	0.05
19–24	M	175	71	42	0.18	3000	12.6		
	F	160	58	36	0.15	2100	8.8		
25–49	M	172	74	36	0.15	2700	11.3		
	F	160	59	32	0.13	1900	8.0		
50–74	M	170	73	31	0.13	2300	9.7		
	F	158	63	29	0.12	1800	7.6		
75+	M	168	69	29	0.12	2000	8.4		
	F	155	64	23	0.10	1500	6.3		

[a]Requirements can be expected to vary within a range of ±30 percent.
[b]First and last figures are averages at the beginning and end of the three-month period.
Source: Health and Welfare Canada, *Nutrition Recommendations: The Report of the Scientific Review Committee* (Ottawa: Canadian Government Publishing Centre, 1990), Tables 5 and 6, pp. 25, 27.

Choice System for Meal Planning

The *Good Health Eating Guide* is the Canadian choice system of meal planning.[1] It contains several features similar to those of the U.S. exchange system including the following:

[1] The tables for the Canadian choice system are adapted from the *Good Health Eating Guide Resource,* copyright 1994, with permission of the Canadian Diabetes Association.

- Foods are divided into lists according to carbohydrate, protein, and fat content.
- Foods are interchangeable within a group.
- Most foods are eaten in measured amounts.
- An energy value is given for each food group.

Tables B-3 through B-10 present the Canadian choice system.

Table B-3
Canadian Choice System: Starch Foods

1 starch choice = 15 g carbohydrate (starch), 2 g protein, 290 kJ (68 kcal)

Food	Measure	Mass (Weight)
Breads		
Bagels	½	30 g
Bread crumbs	50 mL (¼ c)	30 g
Bread cubes	250 mL (1 c)	30 g
Bread sticks	2	20 g
Brewis, cooked	50 mL (¼ c)	45 g
Chapati	1	20 g
Cookies, plain	2	20 g
English muffins, crumpets	½	30 g
Flour	40 mL (2½ tbs)	20 g
Hamburger buns	½	30 g
Hot dog buns	½	30 g
Kaiser rolls	½	30 g
Matzo, 15 cm	1	20 g
Melba toast, rectangular	4	15 g
Melba toast, rounds	7	15 g
Pita, 20-cm (8″) diameter	¼	30 g
Pita, 15-cm (6″) diameter	½	30 g
Plain rolls	1 small	30 g
Pretzels	7	20 g
Raisin bread	1 slice	30 g
Rice cakes	2	30 g
Roti	1	20 g
Rusks	2	20 g
Rye, coarse or pumpernickel	½ slice	30 g
Soda crackers	6	20 g
Tortillas, corn (taco shell)	1	30 g
Tortilla, flour	1	30 g
White (French and Italian)	1 slice	25 g
Whole-wheat, cracked-wheat, rye, white enriched	1 slice	30 g
Cereals		
Bran flakes, 100% bran	125 mL (½ c)	30 g
Cooked cereals, cooked	125 mL (½ c)	125 g
Dry	30 mL (2 tbs)	20 g
Cornmeal, cooked	125 mL (½ c)	125 g
Dry	30 mL (2 tbs)	20 g
Ready-to-eat unsweetened cereals	125 mL (½ c)	20 g
Shredded wheat biscuits, rectangular or round	1	20 g
Shredded wheat, bite size	125 mL (½ c)	20 g
Wheat germ	75 mL (⅓ c)	30 g
Cornflakes	175 mL (⅔ c)	20 g
Rice Krispies	175 mL (⅔ c)	20 g

B

Table B-3 (continued)
Canadian Choice System: Starch Foods

1 starch choice = 15 g carbohydrate (starch), 2 g protein, 290 kJ (68 kcal)

Food	Measure	Mass (Weight)
Cheerios	200 mL (¾ c)	20 g
Muffets	1	20 g
Puffed rice	300 mL (1¼ c)	15 g
Puffed wheat	425 mL (1⅔ c)	20 g
Grains		
Barley, cooked	125 mL (½ c)	120 g
Dry	30 mL (2 tbs)	20 g
Bulgur, kasha, cooked, moist	125 mL (½ c)	70 g
Cooked, crumbly	75 mL (⅓ c)	40 g
Dry	30 mL (2 tbs)	20 g
Rice, cooked, brown & white (short & long grain)	125 mL (½ c)	70 g
Rice, cooked, wild	75 mL (⅓ c)	70 g
Tapioca, pearl and granulated, quick cooking, dry	30 mL (2 tbs)	15 g
Couscous, cooked moist	125 mL (½ c)	70 g
Dry	30 mL (tbs)	20 g
Quinoa, cooked moist	125 mL (½ c)	70 g
Dry	30 mL (2 tbs)	20 g
Pastas		
Macaroni, cooked	125 mL (½ c)	70 g
Noodles, cooked	125 mL (½ c)	80 g
Spaghetti, cooked	125 mL (½ c)	70 g
Starchy Vegetables		
Beans and peas, dried, cooked	125 mL (½ c)	80 g
Breadfruit	1 slice	75 g
Corn, canned, whole kernel	125 mL (½ c)	85 g
Corn on the cob	½ medium cob	140 g
Cornstarch	30 mL (2 tbs)	15 g
Plantains	⅓ small	50 g
Popcorn, air-popped, unbuttered	750 mL (3 c)	20 g
Potatoes, whole (with or without skin)	½ medium	95 g
Yams, sweet potatoes, (with or without skin)	½	75 g

Food	Choices per serving	Measure	Mass (Weight)
NOTE: Food items found in this category provide more than 1 starch choice:			
Bran flakes	1 starch + ½ sugar	150 mL (⅔ c)	24 g
Croissant, small	1 starch + 1½ fats	1 small	35 g
Large	1 starch + 1½ fats	½ large	30 g
Corn, canned creamed	1 starch + ½ fruits and vegetables	12 mL (½ c)	113 g
Potato chips	1 starch + 2 fats	15 chips	30 g
Tortilla chips (nachos)	1 starch + 1½ fats	13 chips	20 g
Corn chips	1 starch + 2 fats	30 chips	30 g
Cheese twists	1 starch + 1½ fats	30 chips	30 g
Cheese puffs	1 starch + 2 fats	27 chips	30 g
Tea biscuit	1 starch + 2 fats	1	30 g
Pancakes, homemade using 50 mL (¼ c) batter (6" diameter)	1½ starches + 1 fat	1 medium	50 g
Potatoes, french fried (homemade or frozen)	1 starch + 1 fat	10 regular size	35 g
Soup, canned* (prepared with equal volume of water)	1 starch	250 mL (1 c)	260 g
Waffles, packaged	1 starch + 1 fat	1	35 g

*Soup can vary according to brand and type. Check the label for Food Choice Values and Symbols or the core nutrient listing.

B

Table B-4
Canadian Choice System: Fruits and Vegetables

1 fruits and vegetables choice = 10 g carbohydrate, 1 g protein, 190 kJ (44 kcal)

Food	Measure	Mass (Weight)
Fruits (fresh, frozen, without sugar, canned in water)		
Apples, raw (with or without skin)	½ medium	75 g
Sauce unsweetened	125 mL (½ c)	120 g
Sweetened	see *Combined Food Choices*	
Apple butter	20 mL (4 tsp)	20 g
Apricots, raw	2 medium	115 g
Canned, in water	4 halves, plus 30 mL (2 tbs) liquid	110 g
Bake-apples (cloudberries), raw	125 mL (½ c)	120 g
Bananas, with peel	½ small	75 g
Peeled	½ small	50 g
Berries (blackberries, blueberries, boysenberries, huckleberries, loganberries, raspberries)		
Raw	125 mL (½ c)	70 g
Canned, in water	125 mL (½ c), plus 30 mL (2 tbs) liquid	100 g
Cantaloupe, wedge with rind	¼	240 g
Cubed or diced	250 mL (1 c)	160 g
Cherries, raw, with pits	10	75 g
Raw, without pits	10	70 g
Canned, in water, with pits	75 mL (⅓ c), plus 30 mL (2 tbs) liquid	90 g
Canned, in water, without pits	75 mL (⅓ c), plus 30 mL (2 tbs) liquid	85 g
Crabapples, raw	1 small	55 g
Cranberries, raw	250 mL (1 c)	100 g
Figs, raw	1 medium	50 g
Canned, in water	3 medium, plus 30 mL (2 tbs) liquid	100 g
Foxberries, raw	250 mL (1 c)	100 g
Fruit cocktail, canned, in water	125 mL (½ c), plus 30 mL (2 tbs) liquid	120 g
Fruit, mixed, cut-up	125 mL (½ c)	120 g
Gooseberries, raw	250 mL (1 c)	150 g
Canned, in water	250 mL (1 c), plus 30 mL (2 tbs) liquid	230 g
Grapefruit, raw, with rind	½ small	185 g
Raw, sectioned	125 mL (½ c)	100 g
Canned, in water	125 mL (½ c), plus 30 mL (2 tbs) liquid	120 g
Grapes, raw, slip skin	125 mL (½ c)	75 g
Raw, seedless	125 mL (½ c)	75 g
Canned, in water	75 mL (⅓ c), plus 30 mL (2 tbs) liquid	115 g
Guavas, raw	½	50 g
Honeydew melon, raw, with rind	½	225 g
Cubed or diced	250 mL (1 c)	170 g
Kiwis, raw, with skin	2	155 g
Kumquats, raw	3	60 g
Loquats, raw	8	130 g
Lychee fruit, raw	8	120 g
Mandarin oranges, raw, with rind	1	135 g
Raw, sectioned	125 mL (½ c)	100 g
Canned, in water	125 mL (½ c), plus 30 mL (2 tbs) liquid	100 g
Mangoes, raw, without skin and seed	⅓	65 g
Diced	75 mL (⅓ c)	65 g
Nectarines	½ medium	75 g
Oranges, raw, with rind	1 small	130 g
Raw, sectioned	125 mL (½ c)	95 g

B

Table B-4 (continued)

Canadian Choice System: Fruits and Vegetables

1 fruits and vegetables choice = 10 g carbohydrate, 1 g protein, 190 kJ (44 kcal)

Food	Measure	Mass (Weight)
Papayas, raw, with skin and seeds	¼ medium	150 g
Raw, without skin and seeds	¼ medium	100 g
Cubed or diced	125 mL (½ c)	100 g
Peaches, raw, with seed and skin	1 large	100 g
Raw, sliced or diced	125 mL (½ c)	100 g
Canned in water, halves or slices	125 mL (½ c), plus 30 mL (2 tbs) liquid	120 g
Pears, raw, with skin and core	½	90 g
Raw, without skin and core	½	85 g
Canned, in water, halves	1 half plus 30 mL (2 tbs) liquid	90 g
Persimmons, raw, native	1	30 g
Raw, Japanese	¼	50 g
Pineapple, raw	1 slice	75 g
Raw, diced	125 mL (½ c)	75 g
Canned, in juice, diced	75 mL (⅓ c), plus 15 mL (1 tbs) liquid	55 g
Canned, in juice, sliced	1 slice, plus 15 mL (1 tbs) liquid	55 g
Canned, in water, diced	125 mL (½ c), plus 30 mL (2 tbs) liquid	100 g
Canned, in water, sliced	2 slices, plus 15 mL (1 tbs) liquid	100 g
Plums, raw	2 small	60 g
Damson	6	65 g
Japanese	1	70 g
Canned, in apple juice	2, plus 30 mL (2 tbs) liquid	70 g
Canned, in water	3, plus 30 mL (2 tbs) liquid	100 g
Pomegranates, raw	½	140 g
Strawberries, raw	250 mL (1 c)	150 g
Frozen/canned, in water	250 mL (1 c), plus 30 mL (2 tbs) liquid	240 g
Rhubarb	250 mL (1 c)	150 g
Tangelos, raw	1	205 g
Tangerines, raw	1 medium	115 g
Raw, sectioned	125 mL (½ c)	100 g
Watermelon, raw, with rind	1 wedge	310 g
Cubed or diced	250 mL (1 c)	160 g
Dried Fruit		
Apples	5 pieces	15 g
Apricots	4 halves	15 g
Banana flakes	30 mL (2 tbs)	15 g
Currants	30 mL (2 tbs)	15 g
Dates, without pits	2	15 g
Peaches	½	15 g
Pears	½	15 g
Prunes, raw, with pits	2	15 g
Raw, without pits	2	10 g
Stewed, no liquid	2	20 g
Stewed, with liquid	2, plus 15 mL (1 tbs) liquid	35 g
Raisins	30 mL (2 tbs)	15 g
Juices (no sugar added or unsweetened)		
Apricot, grape, guava, mango, prune	50 mL (¼ c)	55 g
Apple, carrot, papaya, pear, pineapple, pomegranate	75 mL (⅓ c)	80 g
Cranberry (see Sugars Section)		
Clamato (see Sugars Section)		

(continued on the next page)

Table B-4 (continued)
Canadian Choice System: Fruits and Vegetables

1 fruits and vegetables choice = 10 g carbohydrate, 1 g protein, 190 kJ (44 kcal)

Food	Measure	Mass (Weight)
Grapefruit, loganberry, orange, raspberry, tangelo, tangerine	125 mL (½ c)	130 g
Tomato, tomato-based mixed vegetables	250 mL (1 c)	255 g
Vegetables (fresh, frozen, or canned)		
Artichokes, French, globe	2 small	50 g
Beets, diced or sliced	125 mL (½ c)	85 g
Carrots, diced, cooked or uncooked	125 mL (½ c)	75 g
Chestnuts, fresh	5	20 g
Parsnips, mashed	125 mL (½ c)	80 g
Peas, fresh or frozen	125 mL (½ c)	80 g
Canned	75 mL (⅓ c)	55 g
Pumpkin, mashed	125 mL (½ c)	45 g
Rutabagas, mashed	125 mL (½ c)	85 g
Sauerkraut	250 mL (1 c)	235 g
Snow peas	250 mL (1 c)	135 g
Squash, yellow or winter, mashed	125 mL (½ c)	115 g
Succotash	75 mL (⅓ c)	55 g
Tomatoes, canned	250 mL (1 c)	240 g
Tomato paste	50 mL (¼ c)	55 g
Tomato sauce*	75 mL (⅓ c)	100 g
Turnips, mashed	125 mL (½ c)	115 g
Vegetables, mixed	125 mL (½ c)	90 g
Water chestnuts	8 medium	50 g

*Tomato sauce varies according to brand name. Check the label or discuss with your dietitian.

Table B-5
Canadian Choice System: Milk

Type of Milk	Carbohydrate (g)	Protein (g)	Fat (g)	Energy
Nonfat (0%)	6	4	0	170 kJ (40 kcal)
1%	6	4	1	206 kJ (49 kcal)
2%	6	4	2	244 kJ (58 kcal)
Whole (4%)	6	4	4	319 kJ (76 kcal)

Food	Measure	Mass (Weight)
Buttermilk (higher in salt)	125 mL (½ c)	125 g
Evaporated milk	50 mL (¼ c)	50 g
Milk	125 mL (½ c)	125 g
Powdered milk, regular	30 mL (2 tbs)	15 g
Instant	50 mL (¼ c)	15 g
Plain yogurt	125 mL (½ c)	125 g

Food	Choices per serving	Measure	Mass (Weight)
NOTE: Food items found in this category provide more than 1 milk choice:			
Milkshake	1 milk + 3 sugars + ½ protein	250 mL (1 c)	300 g
Chocolate milk, 2%	2 milks 2% + 1 sugar	250 mL (1 c)	300 g
Frozen yogurt	1 milk + 1 sugar	125 mL (½ c)	125 g

Table B-6
Canadian Choice System: Sugars

1 sugar choice = 10 g carbohydrate (sugar), 167 kJ (40 kcal)

Food	Measure	Mass (Weight)
Beverages		
Condensed milk	15 mL (1 tbs)	
Flavoured fruit crystals*	75 mL (⅓ c)	
Iced tea mixes*	75 mL (⅓ c)	
Regular soft drinks	125 mL (½ c)	
Sweet drink mixes*	75 mL (⅓ c)	
Tonic water	125 mL (½ c)	
*These beverages have been made with water.		
Miscellaneous		
Bubble gum (large square)	1 piece	5 g
Cranberry cocktail	75 mL (⅓ c)	80 g
Cranberry cocktail, light	350 mL (1⅓ c)	260 g
Cranberry sauce	30 mL (2 tbs)	
Hard candy mints	2	5 g
Honey, molasses, corn & cane syrup	10 mL (2 tsp)	15 g
Jelly bean	4	10 g
Licorice	1 short stick	10 g
Marshmallows	2 large	15 g
Popsicle	1 stick (½ popsicle)	
Powdered gelatin mix		
(Jello®) (reconstituted)	50 mL (¼ c)	
Regular jam, jelly, marmalade	15 mL (1 tbs)	
Sugar, white, brown, icing, maple	10 mL (2 tsp)	10 g
Sweet pickles	2 small	100 g
Sweet relish	30 mL (2 tbs)	

Food	Choices per Serving	Measures	Mass (Weight)
The following food items provide more than 1 sugar choice:			
Brownie	1 sugar + 1 fat	1	20 g
Clamato juice	1½ sugars	175 mL (⅔ c)	
Fruit salad, light syrup	1 sugar + 1 fruits & vegetables	125 mL (½ c)	130 g
Aero® bar	2½ sugars + 2½ fats	1 bar	43 g
Smarties®	4½ sugars + 2 fats	1 box	60 g
Sherbet	3 sugars + ½ fat	125 mL (½ c)	95 g

Table B-7
Canadian Choice System: Protein Foods

1 protein choice = 7 g protein, 3 g fat, 230 kJ (55 kcal)

Food	Measure	Mass (Weight)
Cheese		
Low-fat cheese, about 7% milk fat	1 slice	30 g
Cottage cheese, 2% milkfat or less	50 mL (¼ c)	55 g
Ricotta, about 7% milkfat	50 mL (¼ c)	60 g
Fish		
Anchovies (see *Extras,* Table B-9)		
Canned, drained (e.g., mackerel, salmon, tuna packed in water)	(⅓ of 6.5 oz can)	30 g
Cod tongues, cheeks	75 mL (⅓ c)	50 g
Fillet or steak (e.g., Boston blue, cod, flounder, haddock, halibut, mackerel, orange roughy, perch, pickerel, pike, salmon, shad, snapper, sole, swordfish, trout, tuna, whitefish)	1 piece	30 g
Herring	⅓ fish	30 g
Sardines, smelts	2 medium or 3 small	30 g
Squid, octopus	50 mL (¼ c)	40 g
Shellfish		
Clams, mussels, oysters, scallops, snails	3 medium	30 g
Crab, lobster, flaked	50 mL (¼ c)	30 g
Shrimp, fresh	5 large	30 g
Frozen	10 medium	30 g
Canned	18 small	30 g
Dry pack	50 mL (¼ c)	30 g
Meat and Poultry (e.g., beef, chicken, goat, ham, lamb, pork, turkey, veal, wild game)		
Back, peameal bacon	3 slices, thin	30 g
Chop	½ chop, with bone	40 g
Minced or ground, lean or extra-lean	30 mL (2 tbs)	30 g
Sliced, lean	1 slice	30 g
Steak, lean	1 piece	30 g
Organ Meats		
Hearts, liver	1 slice	30 g
Kidneys, sweetbreads, chopped	50 mL (¼ c)	30 g
Tongue	1 slice	30 g
Tripe	5 pieces	60 g
Soyabean		
Bean curd or tofu	½ block	70 g
Eggs		
In shell, raw or cooked	1 medium	50 g
Without shell, cooked or poached in water	1 medium	45 g
Scrambled	50 mL (¼ c)	55 g

Food	Choices per Serving	Measures	Mass (Weight)
Note: The following choices provide more than 1 protein exchange:			
Cheese			
Cheeses	1 protein + 1 fat	1 piece	25 g
Cheese, coarsely grated (e.g., cheddar)	1 protein + 1 fat	50 mL (¼ c)	25 g
Cheese, dry, finely grated (e.g., parmesan)	1 protein + 1 fat	45 mL	15 g
Cheese, ricotta, high fat	1 protein + 1 fat	50 mL (¼ c)	55 g
Fish			
Eel	1 protein + 1 fat	1 slice	50 g

Table B-7 (continued)
Canadian Choice System: Protein Foods

1 protein choice = 7 g protein, 3 g fat, 230 kJ (55 kcal)

Food	Choices per Serving	Measures	Mass (Weight)
Meat			
Bologna	1 protein + 1 fat	1 slice	20 g
Canned lunch meats	1 protein + 1 fat	1 slice	20 g
Corned beef, canned	1 protein + 1 fat	1 slice	25 g
Corned beef, fresh	1 protein + 1 fat	1 slice	25 g
Ground beef, medium-fat	1 protein + 1 fat	30 mL (2 tbs)	25 g
Meat spreads, canned	1 protein + 1 fat	45 mL	35 g
Mutton chop	1 protein + 1 fat	½ chop, with bone	35 g
Paté (see *Fats and Oils* group, Table B-8)			
Sausages, garlic, Polish or knockwurst	1 protein + 1 fat	1 slice	50 g
Sausages, pork, links	1 protein + 1 fat	1 link	25 g
Spareribs or shortribs, with bone	1 protein + 1 fat	1 large	65 g
Stewing beef	1 protein + 1 fat	1 cube	25 g
Summer sausage or salami	1 protein + 1 fat	1 slice	40 g
Weiners, hot dog	1 protein + 1 fat	½ medium	25 g
Miscellaneous			
Blood pudding	1 protein + 1 fat	1 slice	25 g
Peanut butter	1 protein + 1 fat	15 mL (1 tbs)	15 g

B

Table B-8
Canadian Choice System: Fats and Oils

1 fat choice = 5 g fat, 190 kJ (45 kcal)

Food	Measure	Mass (Weight)	Food	Measure	Mass (Weight)
Avocado*	⅛	30 g	Nuts (continued):		
Bacon, side, crisp*	1 slice	5 g	Sesame seeds	15 mL (1 tbs)	10 g
Butter*	5 mL (1 tsp)	5 g	Sunflower seeds		
Cheese spread	15 mL (1 tbs)	15 g	Shelled	15 mL (1 tbs)	10 g
Coconut, fresh*	45 mL (3 tbs)	15 g	In shell	45 mL (3 tbs)	15 g
Coconut, dried*	15 mL (1 tbs)	10 g	Walnuts	4 halves	10 g
Cream, Half and half			Oil, cooking and salad	5 mL (1 tsp)	5 g
(cereal), 10%*	30 mL (2 tbs)	30 g	Olives, green	10	45 g
Light (coffee), 20%*	15 mL (1 tbs)	15 g	Ripe black	7	57 g
Whipping, 32 to 37%*	15 mL (1 tbs)	15 g	Pâté, liverwurst,	15 mL (1 tbs)	15 g
Cream cheese*	15 mL (1 tbs)	15 g	meat spreads		
Gravy*	30 mL (2 tbs)	30 g	Salad dressing: blue,	10 mL (2 tsp)	10 g
Lard*	5 mL (1 tsp)	5 g	French, Italian,		
Margarine	5 mL (1 tsp)	5 g	mayonnaise,		
Nuts, shelled:			Thousand Island	5 mL (1 tsp)	5 g
Almonds	8	5 g	Salad dressing,	30 mL (2 tbs)	30 g
Brazil nuts	2	10 g	low-calorie		
Cashews	5	10 g	Salt pork, raw	5 mL (1 tsp)	5 g
Filberts, hazelnuts	5	10 g	or cooked*		
Macadamia	3	5 g	Sesame oil	5 mL (1 tsp)	5 g
Peanuts	10	10g	Sour cream		
Pecans	5 halves	5 g	12% milkfat	30 mL (2 tbs)	30 g
Pignolias, pine nuts	25 mL (5 tsp)	10 g	7% milkfat	60 mL (4 tbs)	60 g
Pistachios, shelled	20	10 g	Shortening*	5 mL (1 tsp)	
Pistachios, in shell	20	20 g			
Pumpkin and squash seeds	20 mL (4 tsp)	10 g			

*These items contain higher amounts of saturated fat.

B

Table B-9
Canadian Choice System: Extras

Extras have no more than 2.5 g carbohydrate, 60 kJ (14 kcal)

Vegetables 125 mL (½ c)

Artichokes
Asparagus
Bamboo shoots
Bean sprouts, mung or soya
Beans, string, green, or yellow
Bitter melon (balsam pear)
Bok choy
Broccoli
Brussels sprouts
Cabbage
Cauliflower
Celery
Chard
Cucumbers
Eggplant
Endive
Fiddleheads
Greens: beet, collard, dandelion, mustard, turnip, etc.
Kale
Kohlrabi
Leeks
Lettuce
Mushrooms
Okra
Onions, green or mature
Parsley
Peppers, green, yellow or red
Radishes
Rapini
Rhubarb
Sauerkraut
Shallots
Spinach
Sprouts: alfalfa, radish, etc.
Tomato wedges
Watercress
Zucchini

Free Foods (may be used without measuring)

Artificial sweetener, such as cyclamate or aspartame	Lime juice or lime wedges Marjoram, cinnamon, etc.
Baking powder, baking soda	Mineral water
Bouillon from cube, powder, or liquid	Mustard Parsley
Bouillon or clear broth	Pimentos
Chowchow, unsweetened	Salt, pepper, thyme
Coffee, clear	Soda water, club soda
Consommé	Soya sauce
Dulse	Sugar-free Crystal Drink
Flavorings and extracts	Sugar-free Jelly Powder
Garlic	Sugar-free soft drinks
Gelatin, unsweetened	Tea, clear
Ginger root	Vinegar
Herbal teas, unsweetened	Water
Horseradish, uncreamed	Worcestershire sauce
Lemon juice or lemon wedges	

Condiments

Food	Measure
Anchovies	2 fillets
Barbecue sauce	15 mL (1 tbs)
Bran, natural	30 mL (2 tbs)
Brewer's yeast	5 mL (1 tsp)
Carob powder	5 mL (1 tsp)
Catsup	5 mL (1 tsp)
Chili sauce	5 mL (1 tsp)
Cocoa powder	5 mL (1 tsp)
Cranberry sauce, unsweetened	15 mL (1 tbs)
Dietetic fruit spreads	5 mL (1 tsp)
Maraschino cherries	1
Nondairy coffee whitener	5 mL (1 tsp)
Nuts, chopped pieces	5 mL (1 tsp)
Pickles	
unsweetened dill	2
sour mixed	11
Sugar substitutes, granular	5 mL (1 tsp)
Whipped toppings	15 mL (1 tbs)

B

Table B-10
Canadian Choice System: Combined Food Choices

Food	Choices per serving	Measure	Mass (Weight)
Angel food cake	½ starch + 2½ sugars	1/12 cake	50 g
Apple crisp	½ starch + 1½ fruits & vegetables + 1 sugar + 1–2 fats	125 mL (½ c)	
Applesauce, sweetened	1 fruits & vegetables + 1 sugar	125 mL (½ c)	
Beans and pork in tomato sauce	1 starch + ½ fruits & vegetables + ½ sugar + 1 protein	125 mL (½ c)	135 g
Beef burrito	2 starches + 3 proteins + 3 fats		110 g
Brownie	1 sugar + 1 fat	1	20 g
Cabbage rolls*	1 starch + 2 proteins	3	310 g
Caesar salad	2–4 fats	20 mL dressing (4 tsp)	
Cheesecake	½ starch + 2 sugars + ½ protein + 5 fats	1 piece	80 g
Chicken fingers	1 starch + 2 proteins + 2 fats	6 small	100 g
Chicken and snow pea Oriental	2 starches + ½ fruits & vegetables + 3 proteins + 1 fat	500 mL (2 c)	
Chili	1½ starches + ½ fruits & vegetables + 3½ protein	300 mL (1¼ c)	325 g
Chips			
Potato chips	1 starch + 2 fats	15 chips	30 g
Corn chips	1 starch + 2 fats	30 chips	30 g
Tortilla chips	1 starch + 1½ fats	13 chips	
Cheese twist	1 starch + 1½ fats	30 chips	30 g
Chocolate bar			
Aero®	2½ sugars + 2½ fats	bar	43 g
Smarties®	4½ sugars + 2 fats	package	60 g
Chocolate cake (without icing)	1 starch + 2 sugars + 3 fats	1/10 of a 8″ pan	
Chocolate devil's food cake (without icing)	2 starches + 2 sugars + 3 fats	1/12 of a 9″ pan	
Chocolate milk	2 milks 2% + 1 sugar	250 mL (1 c)	300 g
Clubhouse (triple-decker) sandwich	3 starches + 3 proteins + 4 fats		
Cookies			
chocolate chip	½ starch + ½ sugar + 1½ fats	2	22 g
oatmeal	1 starch + 1 sugar + 1 fat	2	40 g
Donut (chocolate glazed)	1 starch + 1½ sugars + 2 fats	1	65 g
Egg roll	1 starch + ½ protein + 1 fat		75 g
Four bean salad	1 starch + ½ protein + 1 fat	125 mL (½ c)	
French toast	1 starch + ½ protein + 2 fats	1 slice	65 g
Fruit in heavy syrup	1 fruits & vegetables + 1½ sugars	125 mL (½ c)	
Granola bar	½ starch + 1 sugar + 1–2 fats		30 g
Granola cereal	1 starch + 1 sugar + 2 fats	125 mL (½ c)	45 g
Hamburger	2 starches + 3 proteins + 2 fats	junior size	
Ice cream and cone, plain flavour			
ice cream	½ milk + 2–3 sugars + 1–2 fats		100 g
cone	½ sugar		4 g
Lasagna			
regular cheese	1 starch + 1 fruits & vegetables + 3 proteins + 2 fats	3″ x 4″ piece	
low-fat cheese	1 starch + 1 fruits & vegetables + 3 proteins	3″ x 4″ piece	
Legumes			
Dried beans (kidney, navy, pinto, fava, chick peas)	2 starches + 1 protein	250 mL (1 c)	180 g
Dried peas	2 starches + 1 protein	250 mL (1 c)	210 g
Lentils	2 starches + 1 protein	250 mL (1 c)	210 g

* If eaten with sauce, add ½ fruits & vegetables exchange.

Table B-10 (continued)
Canadian Choice System: Combined Food Choices

Food	Choices per serving	Measure	Mass (Weight)
Macaroni and cheese	2 starches + 2 proteins + 2 fats	250 mL (1 c)	210 g
Minestrone soup	1½ starches + ½ fruits & vegetables + ½ fat	250 mL (1 c)	
Muffin	1 starch + ½ sugar + 1 fat	1 small	45 g
Nuts (dry or roasted without any oil added).			
Almonds, dried sliced	½ protein + 2 fats	50 mL (¼ c)	22 g
Brazil nuts, dried unblanched	½ protein + 2½ fats	5 large	23 g
Cashew nuts, dry roasted	½ starch + ½ protein + 2 fats	50 mL (¼ c)	28 g
Filbert hazelnut, dry	½ protein + 3½ fats	50 mL (¼ c)	30 g
Macadamia nuts, dried	½ protein + 4 fats	50 mL (¼ c)	28 g
Peanuts, raw	1 protein + 2 fats	50 mL (¼ c)	30 g
Pecans, dry roasted	½ fruits & vegetables + 3 fats	50 mL (¼ c)	22 g
Pine nuts, pignolia dried	1 protein + 3 fats	50 mL (¼ c)	34 g
Pistachio nuts, dried	½ fruits & vegetables + ½ protein + 2½ fats	50 mL (¼ c)	27 g
Pumpkin seeds, roasted	2 proteins + 2½ fats	50 mL (¼ c)	47 g
Sesame seeds, whole dried	½ fruits & vegetables + ½ protein + 2½ fats	50 mL (¼ c)	30 g
Sunflower kernel, dried	½ protein + 1½ fats	50 mL (¼ c)	17 g
Walnuts, dried chopped	½ protein + 3 fats	50 mL (¼ c)	26 g
Perogies	2 starches + 1 protein + 1 fat	3	
Pie, fruit	1 starch + 1 fruits & vegetables + 2 sugars + 3 fats	1 piece	120 g
Pizza, cheese	1 starch + 1 protein + 1 fat	1 slice (⅛ of a 12")	50 g
Pork stir fry	½ to 1 fruits & vegetables + 3 proteins	200 mL (¾ c)	
Potato salad	1 starch + 1 fat	125 mL (½ c)	130 g
Potatoes, scalloped	2 starches + 1 milk + 1–2 fats	200 mL (¾ c)	210 g
Pudding, bread or rice	1 starch + 1 sugar + 1 fat	125 mL (½ c)	
Pudding, vanilla	1 milk + 2 sugars	125 mL (½ c)	
Raisin bran cereal	1 starch + ½ fruits & vegetables + ½ sugar	175 mL (⅔ c)	40 g
Rice krispie squares	½ starch + 1½ sugars + ½ fat	1 square	30 g
Shepherd's pie	2 starches + 1 fruits & vegetables + 3 proteins	325 mL (1⅓ c)	
Sherbet, orange	3 sugars + ½ fat	125 mL (½ c)	
Spaghetti and meat sauce	2 starches + 1 fruits & vegetables + 2 proteins + 3 fats	250 mL (1 c)	
Stew	2 starches + 2 fruits & vegetables + 3 proteins + ½ fat	200 mL (¾ c)	
Sundae	4 sugars + 3 fats	125 mL (½ c)	
Tuna casserole	1 starch + 2 proteins + ½ fat	125 mL (½ c)	
Yogurt, fruit bottom	1 fruits & vegetables + 1 milk + 1 sugar	125 mL (½ c)	125 g
Yogurt, frozen	1 milk + 1 sugar	125 mL (½ c)	125 g

Food Labels

Consumers can gather a lot of information from a nutrition label. Figure B-1 demonstrates the reading of a food label and Table B-11 defines terms.

Figure B-1

Example of a Food Label

OUR COMMITMENT TO QUALITY

Kellogg's is committed to providing foods of outstanding quality and freshness. If this product in any way falls below the high standards you've come to expect from Kellogg's, please send your comments and both top flaps to:

Consumer Affairs
KELLOGG CANADA INC.
Etobicoke, Ontario M9W 5P2

IF IT DOESN'T SAY *Kellogg's* ON THE BOX, IT'S NOT *Kellogg's* IN THE BOX.
SI LE NOM *Kellogg's* N'EST PAS SUR LA BOÎTE, CE N'EST PAS *Kellogg's* DANS LA BOÎTE.

- HIGH IN FIBRE
- LOW IN FAT
- PRESERVATIVE FREE
- SOURCE ÉLEVÉE DE FIBRES
- FAIBLE EN MATIÈRES GRASSES
- SANS AGENT DE CONSERVATION

NUTRITION INFORMATION
APPORT NUTRITIONNEL

	Per 40 g serving cereal (175 mL, ¾ cup) / Par ration de 40 g de céréale (175 mL, ¾ tasse)	Per 40 g serving cereal with 125 mL Partly Skimmed Milk (2%) / Par ration de 40 g de céréale avec 125 mL de lait partiellement écrémé (2,0 %)	
ENERGY	130Cal	195Cal	ÉNERGIE
	540kJ	810kJ	
PROTEIN	3.0g	7.3g	PROTÉINES
FAT	0.4g	2.9g	MATIÈRES GRASSES
CARBOHYDRATE	32g	38g	GLUCIDES
SUGARS*	11g	18g	*SUCRES
STARCH	16g	16g	AMIDON
DIETARY FIBRE	4.6g	4.6g	FIBRES ALIMENTAIRES
SODIUM	235mg	300mg	SODIUM
POTASSIUM	240mg	440mg	POTASSIUM
% of Recommended Daily Intake / % de l'apport quotidien conseillé			
VITAMIN A	0%	7%	VITAMINE A
VITAMIN D	0%	23%	VITAMINE D
VITAMIN B1	62%	66%	VITAMINE B1
VITAMIN B2	3%	16%	VITAMINE B2
NIACIN	13%	18%	NIACINE
VITAMIN B6	13%	16%	VITAMINE B6
FOLACIN	11%	14%	FOLACINE
VITAMIN B12	0%	25%	VITAMINE B12
PANTOTHENATE	9%	15%	PANTOTHÉNATE
CALCIUM	1%	15%	CALCIUM
PHOSPHORUS	12%	23%	PHOSPHORE
MAGNESIUM	20%	27%	MAGNÉSIUM
IRON	38%	39%	FER
ZINC	16%	22%	ZINC

*Approximately half of the sugars occur naturally in the raisins.
Environ la moitié des sucres se retrouvent à l'état naturel dans les fruits.

Canadian Diabetes Association Food Choice Values: 40 g (175 mL, ¾ cup) cereal. Système des choix d'aliments de l'Association canadienne du diabète : 40 g (175 mL, ¾ tasse) céréale = 1 [■] + ½ [●] + ½ [✱] choices/choix

INGREDIENTS / INGRÉDIENTS

WHOLE WHEAT, RAISINS (COATED WITH SUGAR, HYDROGENATED VEGETABLE OIL), WHEAT BRAN, SUGAR/GLUCOSE-FRUCTOSE, SALT, MALT (CORN FLOUR, MALTED BARLEY), VITAMINS (THIAMIN HYDROCHLORIDE, PYRIDOXINE HYDROCHLORIDE, FOLIC ACID, d-CALCIUM PANTOTHENATE), MINERALS (IRON, ZINC OXIDE).

BLÉ ENTIER, RAISINS SECS (ENROBÉS DE SUCRE, D'HUILE VÉGÉTALE HYDROGÉNÉE), SON DE BLÉ, SUCRE/GLUCOSE-FRUCTOSE, SEL, MALT (FARINE DE MAÏS, ORGE MALTÉ), VITAMINES (CHLORHYDRATE DE THIAMINE, CHLORHYDRATE DE PYRIDOXINE, ACIDE FOLIQUE, PANTOTHÉNATE DE d-CALCIUM), MINÉRAUX (FER, OXYDE DE ZINC).

Made by / Produit par
KELLOGG CANADA INC.
ETOBICOKE, ONTARIO
CANADA M9W 5P2
*Registered trademark of / *Marque déposée de
KELLOGG CANADA INC. © 1994
00094

WHAT YOU WILL FIND ON A LABEL:

Nutrition Claims

- in Canada, it is optional for a company to decide to use claims,
- when claims appear on a label, they must follow government laws

Nutrition Information

- gives detailed nutrition facts about the product, including serving size and core list
- does not have to appear by law on food products in Canada
- refers to the food as packaged, so if you add milk, eggs or other food, the nutritional content of the food you eat can be very different

Serving Size

- the amount of food for which the information is given
- check the serving size: the serving size on the label may not be the same as the serving size you would actually eat (for example, the serving size of cereal may be ¾ cup, much smaller than your regular serving

Core List

- the energy (in Calories and kilojoules), grams of protein, fat and carbohydrate for each serving
- some products break down fat into monounsaturates, polyunsaturates, saturates, and cholesterol (to find out what these mean, look at the Fats & Oils section)
- carbohydrates may include the amount of sugars, starch and fibre, or may list these items separately

Sodium and Potassium (in milligrams)

Vitamins and Minerals (as percent of your recommended daily intake)

Canadian Diabetes Association Food Choice Values and Symbols

- the Values and Symbols are tools to help you fit the food into your meal plan, they are not an endorsement by CDA
- it is up to the food company to decide if they want their foods analyzed and assigned symbols
- when they are on a label, they have been assigned by a dietitian working for CDA, so you can be sure the information is correct

Ingredients

- must be found on all food labels by law
- ingredients are listed in decreasing order by weight, so what you see first is what you get the most of

Table B-11
Terms on Food Labels

Energy

kcalorie reduced: 50% or fewer kcalories than the regular version.

light: term may be used to describe anything (for example, light in colour, texture, flavour, taste, or kcalories); read the label to find out what is "light" about the product.

low kcalorie: kcalorie-reduced and no more than 15 kcalories per serving.

Fat and Cholesterol

low cholesterol: no more than 3 mg of cholesterol per 100 g of the food and low in saturated fat; *does not* always mean low in total fat.

low fat: no more than 3 g of fat per serving; *does not* always mean low in kcalories.

lower fat: at least 25% less fat than the comparison food; be aware that 80% fat-free still means the food is 20% fat.

Carbohydrates: Fibre and Sugar

carbohydrate reduced: not more than 50% of the carbohydrate found in the regular version; *does not* always mean the product is lower in kcalories because other ingredients such as fat may have increased.

source of dietary fibre: a product that provides 2–4 g of fibre.

high source of dietary fibre: a product that provides 4–6 g of fibre.

very high source of fibre: a product that provides 6 g (or more) of fibre.

sugar free: low in carbohydrates and kcalories; can be used as an extra food in the exchange system.

unsweetened or no sugar added: no sugar was added to the product; sugar may be found naturally in the food (for example, fruit canned in its own juice).

Canada's Food Guide

Canada's Food Guide to Healthy Eating, shown in Figure B-2, gives detailed information for selecting foods to meet the nutritional needs of all Canadians four years of age and older. Like the U.S. Daily Food Guide, Canada's Food Guide also takes a total diet approach, rather than emphasizing a single food, meal, or day's meals and snacks.

Figure B-2

Canada's Food Guide to Healthy Eating

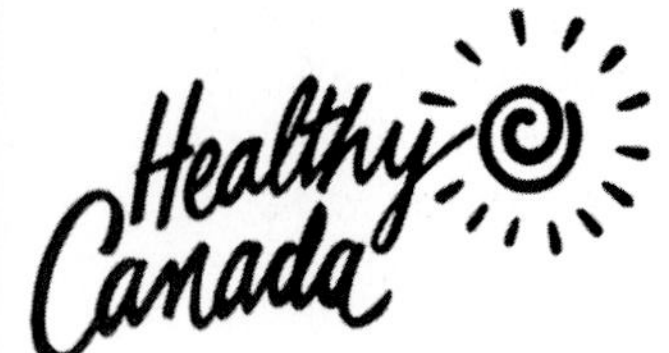

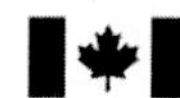

Healthy Canada
Health and Welfare Canada
Santé et Bien-être social Canada
CANADA'S
Food Guide
TO HEALTHY EATING
Enjoy a variety of foods from each group every day.
Choose lower-fat foods more often.
YOGOURT 2.2%
0.1%
SKIM MILK CHEESE
EVAP MILK
POWDER
2%
SKIM
TOFU
Grain Products
Choose whole grain and enriched products more often.
Vegetables & Fruit
Choose dark green and orange vegetables and orange fruit more often.
Milk Products
Choose lower-fat milk products more often.
Meat & Alternatives
Choose leaner meats, poultry and fish, as well as dried peas, beans and lentils more often.

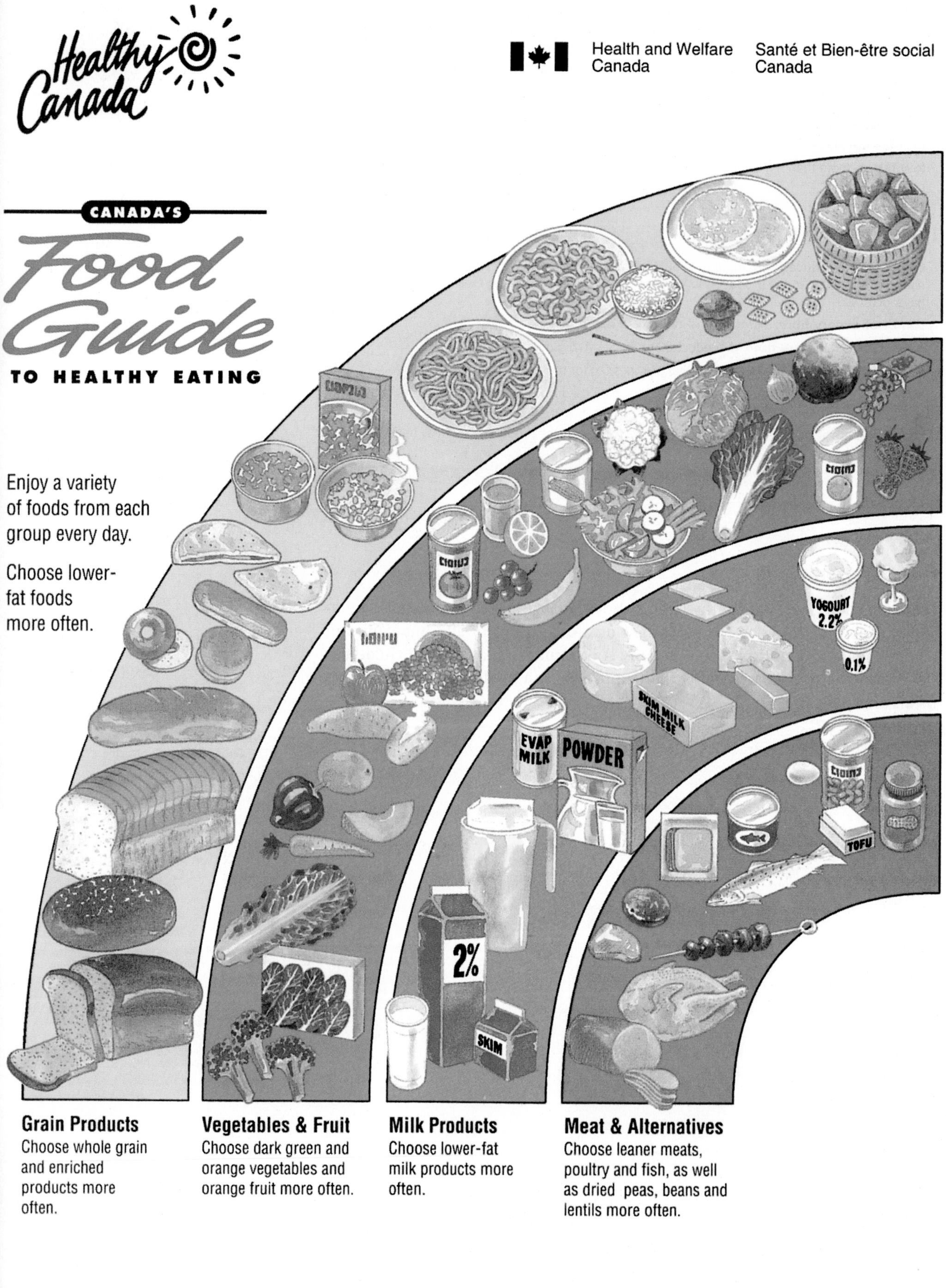

CANADA'S
Food Guide
TO HEALTHY EATING
FOR PEOPLE FOUR YEARS AND OVER

Different People Need Different Amounts of Food

The amount of food you need every day from the 4 food groups and other foods depends on your age, body size, activity level, whether you are male or female and if you are pregnant or breast-feeding. That's why the Food Guide gives a lower and higher number of servings for each food group. For example, young children can choose the lower number of servings, while male teenagers can go to the higher number. Most other people can choose servings somewhere in between.

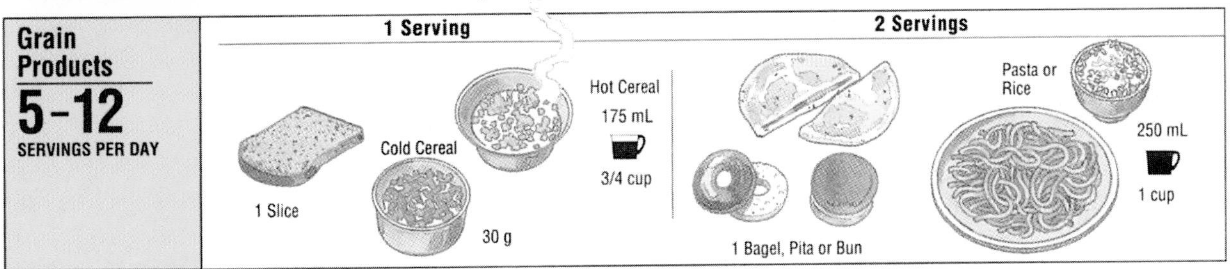

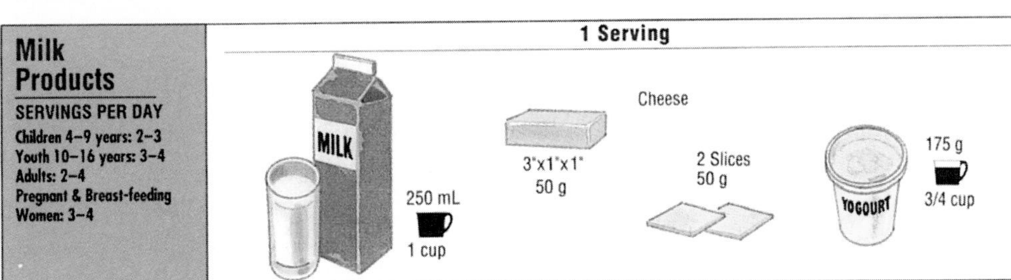

Other Foods

Taste and enjoyment can also come from other foods and beverages that are not part of the 4 food groups. Some of these foods are higher in fat or Calories, so use these foods in moderation.

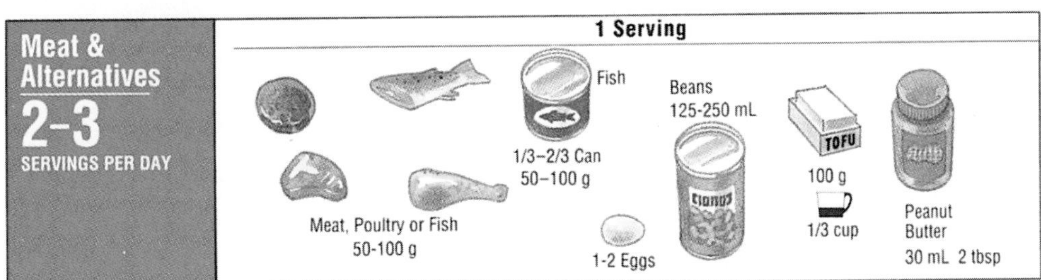

Enjoy eating well, being active and feeling good about yourself. That's VITALIT

CONTENTS

C

UNITED STATES: RECOMMENDATIONS AND EXCHANGES

WORLD HEALTH ORGANIZATION: RECOMMENDATIONS

Chapter 1 introduced the 1989 Recommended Dietary Allowances (RDA), and Chapter 25 introduced exchange systems; this appendix provides additional details. (See Appendix B for Canada's nutrition recommendations and exchange system.) Nutrition recommendations from the World Health Organization are also included.

RDA

As Chapter 1 mentioned, the Dietary Reference Intakes (DRI) reports are replacing the 1989 RDA in the United States and the 1991 RNI in Canada. Recommendations from the DRI reports are presented on the inside front covers. For nutrients that do not yet have new values, the RDA will continue to serve health professionals in the United States (Tables C-1, C-2, and C-3).

Table C-1

Estimated Safe and Adequate Daily Dietary Intakes of Additional Selected Minerals (United States)[a]

Age (yr)	Chromium (μg)	Molybdenum (μg)	Copper (mg)	Manganese (mg)
Infants				
0.0–0.5	10–40	15–30	0.4–0.6	0.3–0.6
0.5–1	20–60	20–40	0.6–0.7	0.6–1.0
Children				
1–3	20–80	25–50	0.7–1.0	1.0–1.5
4–6	30–120	30–75	1.0–1.5	1.5–2.0
7–10	50–200	50–150	1.0–2.0	2.0–3.0
11+	50–200	75–250	1.5–2.5	2.0–5.0
Adults	50–200	75–250	1.5–3.0	2.0–5.0

[a]Less information is available on which to base allowances for these nutrients. Therefore, they are not included in the main table of the RDA, and the figures provided here are in the form of ranges of recommended intakes; the toxic levels for many trace elements may be only several times usual intakes, so the upper levels for the trace elements given in this table should not be habitually exceeded.

Source: *Recommended Dietary Allowances,* © 1989 by the National Academy of Sciences, National Academy Press, Washington, D.C.

C

Table C-2
Estimated Minimum Requirements of Sodium, Chloride, and Potassium

Age (yr)	Weight (kg)	Sodium[a] (mg)	Chloride (mg)	Potassium[b] (mg)
Infants				
0.0–0.5	4.5	120	180	500
0.5–1.0	8.9	200	300	700
Children				
1	11.0	225	350	1000
2–5	16.0	300	500	1400
6–9	25.0	400	600	1600
Adolescents	50.0	500	750	2000
Adults	70.0	500	750	2000

[a]Sodium requirements are based on estimates of needs for growth and for replacement of obligatory losses. They cover a wide variation of physical activity patterns and climatic exposure but do not provide for large, prolonged losses from the skin through sweat.
[b]Dietary potassium may benefit the prevention and treatment of hypertension, and recommendations to include many servings of fruits and vegetables would raise potassium intakes to about 3500 milligrams per day.
Source: *Recommended Dietary Allowances,* © 1989 by the National Academy of Sciences, National Academy Press, Washington, D.C.

Table C-3
Median Heights and Weights and Recommended Energy Intakes (United States)

	Weight		Height		Average Energy Allowance			
Age (yr)	kg	lb	cm	in	REE[a] (kcal/day)	Multiples of REE[b]	kcal/kg	kcal/day[c]
Infants								
0.0–0.5	6	13	60	24	320		108	650
0.5–1.0	9	20	71	28	500		98	850
Children								
1–3	13	29	90	35	740		102	1300
4–6	20	44	112	44	950		90	1800
7–10	28	62	132	52	1130		70	2000
Males								
11–14	45	99	157	62	1440	1.70	55	2500
15–18	66	145	176	69	1760	1.67	45	3000
19–24	72	160	177	70	1780	1.67	40	2900
25–50	79	174	176	70	1800	1.60	37	2900
51+	77	170	173	68	1530	1.50	30	2300
Females								
11–14	46	101	157	62	1310	1.67	47	2200
15–18	55	120	163	64	1370	1.60	40	2200
19–24	58	128	164	65	1350	1.60	38	2200
25–50	63	138	163	64	1380	1.55	36	2200
51+	65	143	160	63	1280	1.50	30	1900
Pregnant (2nd and 3rd trimesters)								+300
Lactating								+500

[a]REE (resting energy expenditure) represents the energy expended by a person at rest under normal conditions.
[b]Recommended energy allowances assume light-to-moderate activity and were calculated by multiplying the REE by an activity factor.
[c]Average energy allowances have been rounded.
Source: *Recommended Dietary Allowances,* © 1989 by the National Academy of Sciences, National Academy Press, Washington, D.C.

C

Exchange Lists for Meal Planning

The U.S. exchange system groups together foods that have about the same amount of carbohydrate, protein, fat, and kcalories. Then any food on a list can be "exchanged" for any other food on that same list. Chapter 25 introduced the exchange lists and Tables C-4 through C-12 present the lists in detail.

Table C-4

U.S. Exchange System: Starch List

1 starch exchange = 15 g carbohydrate, 3 g protein, 0–1 g fat, and 80 kcal

NOTE: In general, a starch serving is ½ c cereal, grain, pasta, or starchy vegetable; 1 oz of bread; ¾ to 1 oz snack food.

Serving Size	Food
Bread	
½ (1 oz)	Bagels
2 slices (1½ oz)	Bread, reduced-kcalorie
1 slice (1 oz)	Bread, white (including French and Italian), whole-wheat, pumpernickel, rye
2 (⅔ oz)	Bread sticks, crisp, 40 x ½"
½	English muffins
½ (1 oz)	Hot dog or hamburger buns
½	Pita, 6" across
1 (1 oz)	Plain rolls, small
1 slice (1 oz)	Raisin bread, unfrosted
1	Tortillas, corn, 6" across
1	Tortillas, flour, 7–8" across
1	Waffles, 4½" square, reduced-fat
Cereals and Grains	
½ c	Bran cereals
½ c	Bulgur, cooked
½ c	Cereals, cooked
¾ c	Cereals, unsweetened, ready-to-eat
3 tbs	Cornmeal (dry)
⅓c	Couscous
3 tbs	Flour (dry)
¼ c	Granola, low-fat
¼ c	Grape nuts
½ c	Grits, cooked
½ c	Kasha
¼ c	Millet
¼ c	Muesli
½ c	Oats
½ c	Pasta, cooked
1½ c	Puffed cereals
½ c	Rice milk
⅓ c	Rice, white or brown, cooked
½ c	Shredded wheat
½ c	Sugar-frosted cereal
3 tbs	Wheat germ
Starchy Vegetables	
1/3 c	Baked beans
½ c	Corn
1 (5 oz)	Corn on cob, medium
1 c	Mixed vegetables with corn, peas, or pasta
½ c	Peas, green
½ c	Plantains
1 small (3 oz)	Potatoes, baked or boiled
½ c	Potatoes, mashed
1 c	Squash, winter (acorn, butternut)
½ c	Yams, sweet potatoes, plain
Crackers and Snacks	
8	Animal crackers
3	Graham crackers, 2½" square
¾ oz	Matzoh
4 slices	Melba toast
24	Oyster crackers
3 c	Popcorn (popped, no fat added or low-fat microwave)
¾ oz	Pretzels
2	Rice cakes, 4" across
6	Saltine-type crackers
15–2" (¾ oz)	Snack chips, fat-free (tortilla, potato)
2–5 (¾ oz)	Whole-wheat crackers, no fat added
Dried Beans, Peas, and Lentils	
½ c	Beans and peas, cooked (garbanzo, lentils, pinto, kidney, white, split, black-eyed)
⅔ c	Lima beans
3 tbs	Miso
Starchy Foods Prepared with Fat	
Count as 1 starch + 1 fat exchange.	
1	Biscuit, 2½" across
½ c	Chow mein noodles
1 (2 oz)	Corn bread, 2" cube
6	Crackers, round butter type
1 c	Croutons
16–25 (3 oz)	French-fried potatoes
¼ c	Granola
1 (1½ oz)	Muffin, small
2	Pancake, 4" across
3 c	Popcorn, microwave
3	Sandwich crackers, cheese or peanut butter filling
⅓ c	Stuffing, bread (prepared)
2	Taco shell, 6" across
1	Waffle, 4½" square
4–6 (1 oz)	Whole-wheat crackers, fat added

= 400 mg or more sodium per exchange.

Table C-5
U.S. Exchange System: Fruit List

1 fruit exchange = 15 g carbohydrate and 60 kcal

NOTE: In general, a fruit serving is 1 small to medium fresh fruit; ½ c canned or fresh fruit or fruit juice; ¼ c dried fruit.

Serving Size	Food
1 (4 oz)	Apples, unpeeled, small
½ c	Applesauce, unsweetened
4 rings	Apples, dried
4 whole (5½ oz)	Apricots, fresh
8 halves	Apricots, dried
½ c	Apricots, canned
1 (4 oz)	Bananas, small
¾ c	Blackberries
¾ c	Blueberries
⅓ melon (11 oz) or 1 c cubes	Cantaloupe, small
12 (3 oz)	Cherries, sweet, fresh
½ c	Cherries, sweet, canned
3	Dates
1½ large or 2 medium (3½ oz)	Figs, fresh
1½	Figs, dried
½ c	Fruit cocktail
½ (11 oz)	Grapefruit, large
¾ c	Grapefruit sections, canned
17 (3 oz)	Grapes, small
1 slice (10 oz) or 1 c cubes	Honeydew melon
1 (3½ oz)	Kiwi
¾ c	Mandarin oranges, canned
½ (5½ oz) or ½ c	Mangoes, small
1 (5 oz)	Nectarines, small
1 (6½ oz)	Oranges, small
½ (8 oz) or 1 c cubes	Papayas
1 (6 oz)	Peaches, medium, fresh
½ c	Peaches, canned
½ (4 oz)	Pears, large, fresh
½ c	Pears, canned
¾ c	Pineapple, fresh
½ c	Pineapple, canned
2 (5 oz)	Plums, small
½ c	Plums, canned
3	Prunes, dried
2 tbs	Raisins
1 c	Raspberries
1¼ c whole berries	Strawberries
2 (8 oz)	Tangerines, small
1 slice (13½ oz) or 1¼ c cubes	Watermelon
Fruit Juice	
½ c	Apple juice/cider
⅓ c	Cranberry juice cocktail
1 c	Cranberry juice cocktail, reduced-kcalorie
⅓ c	Fruit juice blends, 100% juice
⅓ c	Grape juice
½ c	Grapefruit juice
½ c	Orange juice
½ c	Pineapple juice
⅓ c	Prune juice

Table C-6
U.S. Exchange System: Milk List

Serving Size	Food
Nonfat and Very Low-Fat Milk	
1 nonfat/low-fat milk exchange = 12 g carbohydrate, 8 g protein, 0–3 g fat, 90 kcal	
1 c	Nonfat milk
1 c	½% milk
1 c	1% milk
1 c	Nonfat or low-fat buttermilk
½ c	Evaporated nonfat milk
⅓ c dry	Dry nonfat milk
¾ c	Plain nonfat yogurt
1 c	Nonfat or low-fat fruit-flavored yogurt sweetened with aspartame or with a nonnutritive sweetener
Low-Fat Milk	
1 reduced-fat milk exchange = 12 g carbohydrate, 8 g protein, 5 g fat, 120 kcal	
1 c	2% milk
¾ c	Plain low-fat yogurt
1 c	Sweet acidophilus milk
Whole Milk	
1 whole milk exchange = 12 g carbohydrate, 8 g protein, 8 g fat, 150 kcal	
1 c	Whole milk
½ c	Evaporated whole milk
1 c	Goat's milk
1 c	Kefir

Table C-7

U.S. Exchange System: Other Carbohydrates List

1 other carbohydrate exchange = 15 g carbohydrate, or 1 starch, or 1 fruit, or 1 milk exchange

Food	Serving Size	Exchanges per Serving
Angel food cake, unfrosted	1⁄12 cake	2 carbohydrates
Brownies, small, unfrosted	20 square	1 carbohydrate, 1 fat
Cake, unfrosted	20 square	1 carbohydrate, 1 fat
Cake, frosted	20 square	2 carbohydrates, 1 fat
Cookie, fat-free	2 small	1 carbohydrate
Cookies or sandwich cookies	2 small	1 carbohydrate, 1 fat
Cupcakes, frosted	1 small	2 carbohydrates, 1 fat
Cranberry sauce, jellied	¼ c	2 carbohydrates
Doughnuts, plain cake	1 medium, (1½ oz)	1½ carbohydrates, 2 fats
Doughnuts, glazed	33⁄40 across (2 oz)	2 carbohydrates, 2 fats
Fruit juice bars, frozen, 100% juice	1 bar (3 oz)	1 carbohydrate
Fruit snacks, chewy (pureed fruit concentrate)	1 roll (¾ oz)	1 carbohydrate
Fruit spreads, 100% fruit	1 tbs	1 carbohydrate
Gelatin, regular	½ c	1 carbohydrate
Gingersnaps	3	1 carbohydrate
Granola bars	1 bar	1 carbohydrate, 1 fat
Granola bars, fat-free	1 bar	2 carbohydrates
Hummus	⅓ c	1 carbohydrate, 1 fat
Ice cream	½ c	1 carbohydrate, 2 fats
Ice cream, light	½ c	1 carbohydrate, 1 fat
Ice cream, fat-free, no sugar added	½ c	1 carbohydrate
Jam or jelly, regular	1 tbs	1 carbohydrate
Milk, chocolate, whole	1 c	2 carbohydrates, 1 fat
Pie, fruit, 2 crusts	⅙ pie	3 carbohydrates, 2 fats
Pie, pumpkin or custard	⅛ pie	1 carbohydrate, 2 fats
Potato chips	12–18 (1 oz)	1 carbohydrate, 2 fats
Pudding, regular (made with low-fat milk)	½ c	2 carbohydrates
Pudding, sugar-free (made with low-fat milk)	½ c	1 carbohydrate
Salad dressing, fat-free (sodium symbol)	¼ c	1 carbohydrate
Sherbet, sorbet	½ c	2 carbohydrates
Spaghetti or pasta sauce, canned (sodium symbol)	½ c	1 carbohydrate, 1 fat
Sweet roll or danish	1 (2½ oz)	2½ carbohydrates, 2 fats
Syrup, light	2 tbs	1 carbohydrate
Syrup, regular	1 tbs	1 carbohydrate
Syrup, regular	¼ c	4 carbohydrates
Tortilla chips	6–12 (1 oz)	1 carbohydrate, 2 fats
Yogurt, frozen, low-fat, fat-free	⅓ c	1 carbohydrate, 0–1 fat
Yogurt, frozen, fat-free, no sugar added	½ c	1 carbohydrate
Yogurt, low-fat with fruit	1 c	3 carbohydrates, 0–1 fat
Vanilla wafers	5	1 carbohydrate, 1 fat

(sodium symbol) = 400 mg or more sodium per exchange.

C

Table C-8

U.S. Exchange System: Vegetable List

1 vegetable exchange = 5 g carbohydrate, 2 g protein, 25 kcal

NOTE: In general, a vegetable serving is ½ c cooked vegetables or vegetable juice; 1 c raw vegetables. Starchy vegetables such as corn, peas, and potatoes are on the starch list.

Artichokes	Mushrooms
Artichoke hearts	Okra
Asparagus	Onions
Beans (green, wax, Italian)	Pea pods
Bean sprouts	Peppers (all varieties)
Beets	Radishes
Broccoli	Salad greens (endive, escarole, lettuce, romaine, spinach)
Brussels sprouts	Sauerkraut 🧂
Cabbage	Spinach
Carrots	Summer squash (crookneck)
Cauliflower	Tomatoes
Celery	Tomatoes, canned
Cucumbers	Tomato sauce 🧂
Eggplant	Tomato/vegetable juice 🧂
Green onions or scallions	Turnips
Greens (collard, kale, mustard, turnip)	Water chestnuts
Kohlrabi	Watercress
Leeks	Zucchini
Mixed vegetables (without corn, peas, or pasta)	

🧂 = 400 mg or more sodium per exchange.

Table C-9
U.S. Exchange System: Meat and Meat Substitutes List

NOTE: In general, a meat serving is 1 oz meat, poultry, or cheese; ½ c dried beans (weigh meat and poultry and measure beans after cooking).

Serving Size	Food
Very Lean Meat and Substitutes	
1 very lean meat exchange = 7 g protein, 0–1 g fat, 35 kcal	
1 oz	Poultry: Chicken or turkey (white meat, no skin), Cornish hen (no skin)
1 oz	Fish: Fresh or frozen cod, flounder, haddock, halibut, trout; tuna, fresh or canned in water
1 oz	Shellfish: Clams, crab, lobster, scallops, shrimp, imitation shellfish
1 oz	Game: Duck or pheasant (no skin), venison, buffalo, ostrich
	Cheese with ≤ 1 g fat/oz:
¼ c	Nonfat or low-fat cottage cheese
1 oz	Fat-free cheese
	Other:
1 oz	Processed sandwich meats with ≤ 1 g fat/oz (such as deli thin, shaved meats, chipped beef, 🧂 turkey ham)
2	Egg whites
¼ c	Egg substitutes, plain
1 oz	Hot dogs with ≤ 1 g fat/oz
1 oz	Kidney (high in cholesterol)
1 oz	Sausage with ≤ 1 g fat/oz 🧂
Count as one very lean meat and one starch exchange:	
½ c	Dried beans, peas, lentils (cooked)
Lean Meat and Substitutes	
1 lean meat exchange = 7 g protein, 3 g fat, 55 kcal	
1 oz	Beef: USDA Select or Choice grades of lean beef trimmed of fat (round, sirloin, and flank steak); tenderloin; roast (rib, chuck, rump); steak (T-bone, porterhouse, cubed), ground round
1 oz	Pork: Lean pork (fresh ham); canned, cured, or boiled ham; Canadian bacon; 🧂 tenderloin, center loin chop
1 oz	Lamb: Roast, chop, leg
1 oz	Veal: Lean chop, roast
1 oz	Poultry: Chicken, turkey (dark meat, no skin), chicken white meat (with skin), domestic duck or goose (well-drained of fat, no skin)
	Fish:
1 oz	Herring (uncreamed or smoked)
6 medium	Oysters
1 oz	Salmon (fresh or canned), catfish
2 medium	Sardines (canned)
1 oz	Tuna (canned in oil, drained)
1 oz	Game: Goose (no skin), rabbit
	Cheese:
¼ c	4.5%-fat cottage cheese
2 tbs	Grated Parmesan
1 oz	Cheeses with ≤ 3 g fat/oz
	Other:
1½ oz	Hot dogs with ≤ 3 g fat/oz
1 oz	Processed sandwich meat with ≤ 3 g fat/oz (turkey pastrami or kielbasa)
1 oz	Liver, heart (high in cholesterol)
Medium-Fat Meat and Substitutes	
1 medium-fat meat exchange = 7 g protein, 5 g fat, and 75 kcal	
1 oz	Beef: Most beef products (ground beef, meatloaf, corned beef, short ribs, Prime grades of meat trimmed of fat, such as prime rib)
1 oz	Pork: Top loin, chop, Boston butt, cutlet
1 oz	Lamb: Rib roast, ground
1 oz	Veal: Cutlet (ground or cubed, unbreaded)
1 oz	Poultry: Chicken dark meat (with skin), ground turkey or ground chicken, fried chicken (with skin)
1 oz	Fish: Any fried fish product
	Cheese with ≤ 5 g fat/oz:
1 oz	Feta
1 oz	Mozzarella
¼ c (2 oz)	Ricotta
	Other:
1	Egg (high in cholesterol, limit to 3/week)
1 oz	Sausage with ≤ 5 g fat/oz
1 c	Soy milk
¼ c	Tempeh
4 oz or ½ c	Tofu
High-Fat Meat and Substitutes	
1 high-fat meat exchange = 7 g protein, 8 g fat, 100 kcal	
1 oz	Pork: Spareribs, ground pork, pork sausage
1 oz	Cheese: All regular cheeses (American 🧂, cheddar, Monterey Jack, swiss)
	Other:
1 oz	Processed sandwich meats with ≤ 8 g fat/oz (bologna, pimento loaf, salami)
1 oz	Sausage (bratwurst, Italian, knockwurst, Polish, smoked)
1 (10/lb)	Hot dog (turkey or chicken) 🧂
3 slices (20 slices/lb)	Bacon
Count as one high-fat meat plus one fat exchange:	
1 (10/lb)	Hot dog (beef, pork, or combination) 🧂
2 tbs	Peanut butter (contains unsaturated fat)

🧂 = 400 mg or more sodium per exchange.

C

Table C-10

U.S. Exchange System: Fat List

1 fat exchange = 5 g fat, 45 kcal

NOTE: In general, a fat serving is 1 tsp regular butter, margarine, or vegetable oil; 1 tbs regular salad dressing. Many fat-free and reduced fat foods are on the Free Foods List.

Serving Size	Food
Monounsaturated Fats	
⅛ medium (1 oz)	Avocadoes
1 tsp	Oil (canola, olive, peanut)
8 large	Olives, ripe (black)
10 large	Olives, green, stuffed 🖉
6 nuts	Almonds, cashews
6 nuts	Mixed nuts (50% peanuts)
10 nuts	Peanuts
4 halves	Pecans
2 tsp	Peanut butter, smooth or crunchy
1 tbs	Sesame seeds
2 tsp	Tahini paste
Polyunsaturated Fats	
1 tsp	Margarine, stick, tub, or squeeze
1 tbs	Margarine, lower-fat (30% to 50% vegetable oil)
1 tsp	Mayonnaise, regular
1 tbs	Mayonnaise, reduced-fat
4 halves	Nuts, walnuts, English
1 tsp	Oil (corn, safflower, soybean)
1 tbs	Salad dressing, regular
2 tbs	Salad dressing, reduced-fat
2 tsp	Mayonnaise type salad dressing, regular 🖉
1 tbs	Mayonnaise type salad dressing, reduced-fat
1 tbs	Seeds: pumpkin, sunflower
Saturated Fats*	
1 slice (20 slices/lb)	Bacon, cooked
1 tsp	Bacon, grease
1 tsp	Butter, stick
2 tsp	Butter, whipped
1 tbs	Butter, reduced-fat
2 tbs (½ oz)	Chitterlings, boiled
2 tbs	Coconut, sweetened, shredded
2 tbs	Cream, half and half
1 tbs (½ oz)	Cream cheese, regular
2 tbs (1 oz)	Cream cheese, reduced-fat
	Fatback or salt pork†
1 tsp	Shortening or lard
2 tbs	Sour cream, regular
3 tbs	Sour cream, reduced-fat

🖉 = 400 mg or more sodium per exchange

*Saturated fats can raise blood cholesterol levels.

† Use a piece 1″ × 10 × ¼″ if you plan to eat the fatback cooked with vegetables. Use a piece 2″ × 1″ × ½″ when eating only the vegetables with the fatback removed.

Table C-11

U.S. Exchange System: Combination Foods List

Food	Serving Size	Exchanges per Serving
Entrees		
Tuna noodle casserole, lasagna, spaghetti with meatballs, chili with beans, macaroni and cheese 🧂	1 c (8 oz)	2 carbohydrates, 2 medium-fat meats
Chow mein (without noodles or rice)	2 c (16 oz)	1 carbohydrate, 2 lean meats
Pizza, cheese, thin crust 🧂	¼ of 100 (5 oz)	2 carbohydrates, 2 medium-fat meats, 1 fat
Pizza, meat topping, thin crust 🧂	¼ of 10″ (5 oz)	2 carbohydrates, 2 medium-fat meats, 2 fats
Pot pie 🧂	1 (7 oz)	2 carbohydrates, 1 medium-fat meat, 4 fats
Frozen entrees		
Salisbury steak with gravy, mashed potato	1 (11 oz)	2 carbohydrates, 3 medium-fat meats, 3–4 fats
Turkey with gravy, mashed potato, dressing 🧂	1 (11 oz)	2 carbohydrates, 2 medium-fat meats, 2 fats
Entree with less than 300 kcalories 🧂	1 (8 oz)	2 carbohydrates, 3 lean meats
Soups		
Bean 🧂	1 c	1 carbohydrate, 1 very lean meat
Cream (made with water) 🧂	1 c (8 oz)	1 carbohydrate, 1 fat
Split pea (made with water) 🧂	½ c (4 oz)	1 carbohydrate
Tomato (make with water) 🧂	1 c (8 oz)	1 carbohydrate
Vegetable beef, chicken noodle, or other broth-type 🧂	1 c (8 oz)	1 carbohydrate
Fast Foods		
Burritos with beef 🧂	2	4 carbohydrates, 2 medium-fat meats, 2 fats
Chicken nuggets 🧂	6	1 carbohydrate, 2 medium-fat meats, 1 fat
Chicken breast and wing, breaded and fried 🧂	1	1 carbohydrate, 4 medium-fat meats, 2 fats
Fish sandwich/tartar sauce 🧂	1	3 carbohydrates, 1 medium-fat meat, 3 fats
French fries, thin	20–25	2 carbohydrates, 2 fats
Hamburger, regular	1	2 carbohydrates, 2 medium-fat meats
Hamburger, large 🧂	1	2 carbohydrates, 3 medium-fat meats, 1 fat
Hot dog with bun 🧂	1	1 carbohydrate, 1 high-fat meat, 1 fat
Individual pan pizza 🧂	1	5 carbohydrates, 3 medium-fat meats, 3 fats
Soft serve cone	1 medium	2 carbohydrates, 1 fat
Submarine sandwich 🧂	1 (60)	3 carbohydrates, 1 vegetable, 2 medium-fat meats, 1 fat
Taco, hard shell 🧂	1 (6 oz)	2 carbohydrates, 2 medium-fat meats, 2 fats
Taco, soft shell 🧂	1 (3 oz)	1 carbohydrate, 1 medium-fat meat, 1 fat

🧂 = 400 mg or more sodium per exchange.

Table C-12
U.S. Exchange System: Free Foods List

NOTE: A serving of free food contains less than 20 kcalories; those with serving sizes should be limited to 3 servings a day whereas those without serving sizes can be eaten freely.

Serving Size	Food
Fat-Free or Reduced-Fat Foods	
1 tbs	Cream cheese, fat-free
1 tbs	Creamers, nondairy, liquid
2 tsp	Creamers, nondairy, powdered
1 tbs	Mayonnaise, fat-free
1 tsp	Mayonnaise, reduced-fat
4 tbs	Margarine, fat-free
1 tsp	Margarine, reduced-fat
1 tbs	Mayonnaise type salad dressing, nonfat
1 tsp	Mayonnaise type salad dressing, reduced-fat
	Nonstick cooking spray
1 tbs	Salad dressing, fat-free
2 tbs	Salad dressing, fat-free, Italian
¼ c	Salsa
1 tbs	Sour cream, fat-free, reduced-fat
2 tbs	Whipped topping, regular or light
Sugar-Free or Low-Sugar Foods	
1 piece	Candy, hard, sugar-free
	Gelatin dessert, sugar-free
	Gelatin, unflavored
	Gum, sugar-free
2 tsp	Jam or jelly, low-sugar or light
	Sugar substitutes
2 tbs	Syrup, sugar-free
Drinks	
	Bouillon, broth, consommé
	Bouillon or broth, low-sodium
	Carbonated or mineral water
1 tbs	Cocoa powder, unsweetened
	Coffee
	Club soda
	Diet soft drinks, sugar-free
	Drink mixes, sugar-free
	Tea
	Tonic water, sugar-free
Condiments	
1 tbs	Catsup
	Horseradish
	Lemon juice
	Lime juice
	Mustard
1½ large	Pickles, dill
	Soy sauce, regular or light
1 tbs	Taco sauce
	Vinegar
Seasonings	
Flavoring extracts	
Garlic	
Herbs, fresh or dried	
Pimento	
Spices	
Hot pepper sauces	
Wine, used in cooking	
Worcestershire sauce	

= 400 mg or more sodium per exchange.

Nutrition Recommendations from WHO

Like the Committee on Diet and Health in the United States, the World Health Organization (WHO) has also assessed the relationships between diet and the development of chronic diseases.[1] Their recommendations are expressed in average daily ranges that represent the lower and upper limits:

- Total energy: sufficient to support normal growth, physical activity, and body weight (body mass index = 20 to 22).
- Total fat: 15 to 30 percent of total energy.
  - Saturated fatty acids: 0 to 10 percent total energy.
  - Polyunsaturated fatty acids: 3 to 7 percent total energy.
  - Dietary cholesterol: 0 to 300 milligrams per day.
- Total carbohydrate: 55 to 75 percent total energy.
  - Complex carbohydrates: 50 to 75 percent total energy.
  - Dietary fiber: 27 to 40 grams per day.
  - Refined sugars: 0 to 10 percent total energy.
- Protein: 10 to 15 percent total energy.
- Salt: upper limit of 6 grams/day (no lower limit set).

Note

1. Diet, nutrition and the prevention of chronic diseases: A report of the WHO Study Group on Diet, Nutrition and Prevention of Noncommunicable Diseases, *Nutrition Reviews* 49 (1991): 291–301.

CONTENTS

D

NUTRITION RESOURCES

People interested in nutrition often want to know where they can find reliable nutrition information. Wherever you live, there are several sources you can turn to:

- The Department of Health may have a nutrition expert.
- The local extension agent is often an expert.
- The food editor of your local paper may be well informed.
- The dietitian at the local hospital had to fulfill a set of qualifications before he or she became an RD (see Highlight 1).
- There may be knowledgeable professors of nutrition or biochemistry at a nearby college or university.

In addition, you may be interested in building a nutrition library of your own. Books you can buy, journals you can subscribe to, and addresses you can contact for general information are given below.

Books

For students seeking to establish a personal library of nutrition references, the authors of this text recommend the following books:

- *Present Knowledge in Nutrition,* 7th ed. (Washington, D.C.: International Life Sciences Institute—Nutrition Foundation, 1996).

This 646-page paperback has a chapter on each of 64 topics, including energy, obesity, each of the nutrients, several diseases, malnutrition, growth and its assessment, immunity, alcohol, fiber, exercise, drugs, and toxins. Watch for an update; new editions come out every few years.

- M. E. Shils, J. A. Olson, and M. Shike, eds., *Modern Nutrition in Health and Disease,* 8th ed. (Philadelphia: Lea & Febiger, 1994).

This two-volume set is a major technical reference book on nutrition topics. It contains encyclopedic articles on the nutrients, foods, the diet, metabolism, malnutrition, age-related needs, and nutrition in disease.

- Committee on Dietary Reference Intakes, *Dietary Reference Intakes for Calcium, Phosphorus, Magnesium, Vitamin D, and Fluoride* (Washington, D.C.: National Academy Press, 1997).
- Committee on Dietary Reference Intakes, *Dietary Reference Intakes for Thiamin, Riboflavin, Niacin, Vitamin B_6, Folate, Vitamin B_{12}, Pantothenic Acid, Biotin, and Choline* (Washington, D.C.: National Academy Press, 1998).

These two reports review the function of each nutrient, dietary sources, and deficiency and toxicity symptoms as well as provide recommendations for intakes. Watch for additional reports on the Dietary Reference Intakes for the remaining nutrients. Until they are published, you may need the following:

- Committee on Dietary Allowances, *Recommended Dietary Allowances,* 10th ed. (Washington, D.C.: National Academy Press, 1989).

The Canadian equivalent is *Nutrition Recommendations,* available by mail from the Canadian Government Publishing Centre, Supply and Services Canada, Ottawa, Ontario K1A OS9, Canada.

- Committee on Diet and Health, *Diet and Health Implications for Reducing Chronic Disease Risk* (Washington, D.C.: National Academy Press, 1989).

This 749-page book presents the integral relationship between diet and chronic disease prevention. Its nutrient chapters provide evidence on how diet influences disease development, and its disease chapters review the dietary patterns implicated in each chronic disease.

- E. M. N. Hamilton and S. A. S. Gropper, *The Biochemistry of Human Nutrition: A Desk Reference* (St. Paul, Minn.: West, 1987).

This 324-page paperback presents the biochemical concepts necessary for an understanding of nutrition. It is a handy reference book for those who have forgotten the basics of biochemistry or for those who are learning biochemistry for the first time.

We also recommend three of our own books that explore current topics in nutrition, health, and the life span:

- S. R. Rolfes, L. K. DeBruyne, and E. N. Whitney, *Life Span Nutrition: Conception through Life* (Belmont, Calif.: West/Wadsworth, 1998).
- F. S. Sizer and E. N. Whitney, *Nutrition: Concepts and Controversies,* 7th ed. (Belmont, Calif.: West/Wadsworth, 1997).
- E. N. Whitney, C. B. Cataldo, L. K. DeBruyne, and S. R. Rolfes, *Nutrition for Health and Health Care* (St. Paul, Minn.: West, 1995).

Journals

Nutrition Today is an excellent magazine for the interested layperson. It makes a point of raising controversial issues and providing a forum for conflicting opinions. Six issues per year are published. Order from Williams and Wilkins, 351 West Camden Street, Baltimore, MD 21201-2436.

The *Journal of the American Dietetic Association,* the official publication of the ADA, contains articles of interest to dietitians and nutritionists, news of legislative action on food and nutrition, and a very useful section of abstracts of articles from many other journals of nutrition and related areas. There are 12 issues per year, available from the American Dietetic Association (see "Addresses," later).

Nutrition Reviews, a publication of the International Life Sciences Institute, does much of the work for the library researcher, compiling recent evidence on current topics and presenting extensive bibliographies. Twelve issues per year are available from Nutrition Reviews, P.O. Box 1897, Lawrence, KS 66044-8897.

Nutrition and the M.D. is a monthly newsletter that provides up-to-date, easy-to-read, practical information on nutrition for health care providers. It is available from Lippincott-Raven Publishers, 12107 Insurance Way, Hagerstown, MD 21740.

Other journals that deserve mention here are *Food Technology, Journal of Nutrition, American Journal of Clinical Nutrition, Nutrition Research,* and *Journal of Nutrition Education. FDA Consumer,* a government publication with many articles of interest to the consumer, is available from the Food and Drug Administration (see "Addresses," below). Many other journals of value are referred to throughout this book.

Addresses

Many of the organizations listed below will provide publication lists free on request. Government and international agencies and professional nutrition organizations are listed first, followed by organizations in the following areas: aging, alcohol and drug abuse, consumer organizations, fitness, food safety, health and disease, infancy and childhood, pregnancy and lactation, trade and industry organizations, weight control and eating disorders, and world hunger.

U. S. Government

- Federal Trade Commission (FTC)
  Public Reference Branch
  (202) 326-2222
  www.ftc.gov
- Food and Drug Administration (FDA)
  Office of Consumer Affairs, HFE 1
  Room 16-85
  5600 Fishers Lane
  Rockville, MD 20857
  (301) 443-1544
  www.fda.gov
- FDA Consumer Information Line
  (301) 827-4420
- FDA Office of Food Labeling, HFS 150
  200 C Street SW
  Washington, DC 20204
  (202) 205-4561; fax (202) 205-4564
  www.cfsan.fda.gov
- FDA Office of Plant and Dairy Foods and Beverages
  HFS 300
  200 C Street SW
  Washington, DC 20204
  (202) 205-4064; fax (202) 205-4422
- FDA Office of Special Nutritionals,
  HFS 450
  200 C Street SW
  Washington, DC 20204
  (202) 205-4168; fax (202) 205-5295
- Food and Nutrition Information Center
  National Agricultural Library, Room 304
  10301 Baltimore Avenue
  Beltsville, MD 20705-2351
  (301) 504-5719; fax (301) 504-6409
  www.nal.usda.gov/fnic
- Food Research Action Center (FRAC)
  1875 Connecticut Avenue NW, Suite 540
  Washington, DC 20009
  (202) 986-2200; fax (202) 986-2525
- Superintendent of Documents
  U.S. Government Printing Office
  Washington, DC 20402
  (202) 512-1071
  www.access.gpo.gov/su_docs
- U.S. Department of Agriculture (USDA)
  14th Street SW and Independence Avenue
  Washington, DC 20250
  (202) 720-2791
  www.usda.gov/fcs
- USDA Center for Nutrition Policy and Promotion
  1120 20th Street NW, Suite 200
  North Lobby
  Washington, DC 20036
  (202) 208-2417
  www.usda.gov/fcs/cnpp.htm
- USDA Food Safety and Inspection Service
  Food Safety Education Office,
  Room 1180-S
  Washington, DC 20250
  (202) 690-0351
  www.usda.gov/fsis
- U.S. Department of Education (DOE)
  Accreditation Agency Evaluation Branch
  7th and D Street SW
  ROB 3, Room 3915
  Washington, DC 20202-5244
  (202) 708-7417
- U.S. Department of Health and Human Services
  200 Independence Avenue SW
  Washington, DC 20201
  (202) 619-0257
  www.os.dhhs.gov
- U.S. Environmental Protection Agency (EPA)
  401 Main Street SW
  Washington, DC 20460
  (202) 260-2090
  www.epa.gov
- U.S. Public Health Service
  Assistant Secretary of Health
  Humphrey Building, Room 725-H
  200 Independence Avenue SW
  Washington, DC 20201
  (202) 690-7694

Canadian Government

Federal

- Bureau of Nutritional Sciences
  Food Directorate
  Health Protection Branch
  3-West
  Sir Frederick Banting Research Centre, 2203A
  Tunney's Pasture
  Ottawa, Ontario K1A 0L2
  www.hc-sc.gc.ca
- Canadian Food Inspection Agency
  Agriculture and Agri-Food Canada
  59 Camelot Drive
  Nepean, Ontario K1A 0Y9
  (613) 225-CFIA or (613) 225-2342
  www.agr.ca
- Nutrition & Healthy Eating Unit
  Strategies and Systems for Health Directorate, 1917C
  17th Floor—Jeanne Mance Bldg.
  Tunney's Pasture
  Ottawa, Ontario K1A 1B4
  www.hc-sc.gc.ca
- Nutrition Specialist
  Health Support Services
  Indian and Northern Health Services Directorate
  Medical Services Branch
  20th Floor—Jeanne Mance Bldg., 1920B
  Tunney's Pasture
  Ottawa, Ontario K1A 0L3
  www.hc-sc.gc.ca

Provincial and Territorial

- Population Health Strategies Branch
  Alberta Health
  23rd Floor, TELUS Plaza, North Tower
  10025 Jasper Avenue
  Edmonton, AB T5J 2N3
- Nutritionist
  Preventive Services Branch
  Ministry of Health
  1520 Blanshard Street
  Victoria BC V8W 3C8

- Executive Director
  Health Programs
  2nd Floor 800 Portage Avenue
  Winnipeg, MB R3G 0P4
- Project Manager
  Public Health Management Services
  Health and Community Services
  P.O. Box 5100
  520 King Street
  Fredericton, NB E3B 5G8
- Director, Health Promotion
  Department of Health
  Government of Newfoundland and Labrador
  P.O. Box 8700
  Confederation Building, West Block
  St. John's, NF A1B 4J6
- Consultant, Nutrition
  Health & Wellness Promotion
  Population Health
  Department of Health and Social Services
  Government of the Northwest Territories
  Centre Square Tower, 6th Floor
  P.O. Box 1320
  Yellowknife, NT X1A 2L9
- Public Health Nutritionist
  Central Health Region
  201 Brownlow Avenue, Unit 4
  Dartmouth, NS B3B 1W2
- Senior Consultant, Nutrition
  Public Health Branch
  Ministry of Health, 8th Floor
  5700 Yonge St.
  North York, ON M2M 4K5
- Coordinator, Health Information Resource Centre
  Department of Health and Social Services
  1 Rochford Street, Box 2000
  Charlottetown, PEI C1A 7N8
- Responsables de la santé cardio-vasculaire et de la nutrition
  Ministère de la Santé et des Services sociaux, Service de la Prévention en Santé
  3e étage
  1075, chemin Sainte-Foy
  Quèbec (Quèbec) G1S 2M1
- Health Promotion Unit
  Population Health Branch
  Saskatchewan Health
  3475 Albert Street
  Regina, SK S4S 6X6
- Director, Nutrition Services
  Yukon Hospital Corporation
  #5 Hospital Road
  Whitehorse, YT Y1A 3H7

International Agencies

- Food and Agriculture Organization of the United Nations (FAO)
  Liaison Office for North America
  2175 K Street, Suite 300
  Washington, DC 20437
  (202) 653-2400
  www.fao.org
- International Food Information Council Foundation
  1100 Connecticut Avenue NW, Suite 430
  Washington, DC 20036
  (202) 296-6540
  ificinfo.health.org
- UNICEF
  3 United Nations Plaza
  New York, NY 10017
  (212) 326-7000
  www.unicef.com
- World Health Organization (WHO)
  Regional Office
  525 23rd Street NW
  Washington, DC 20037
  (202) 974-3000
  www.who.org

Professional Nutrition Organizations

- American Academy of Nutritional Sciences
  9650 Rockville Pike
  Bethesda, MD 20814
  (303) 530-7050; fax (301) 571-1892
  www.nutrition.org
- American Dietetic Association (ADA)
  216 West Jackson Boulevard, Suite 800
  Chicago, IL 60606-6995
  (800) 877-1600; (312) 899-0040
  www.eatright.org
- ADA, The Nutrition Hotline
  (800) 366-1655
- American Society for Clinical Nutrition
  9650 Rockville Pike
  Bethesda, MD 20814-3998
  (301) 530-7110; fax (301) 571-1863
  www.faseb.org/ascn
- Dietitians of Canada
  480 University Avenue, Suite 604
  Toronto, Ontario M5G 1V2, Canada
  (416) 596-0857; fax (416) 596-0603
  www.dietitians.ca
- Human Nutrition Institute (INACG)
  1126 Sixteenth Street NW
  Washington, DC 20036
  (202) 659-0789
  www.ilsi.org
- National Academy of Sciences/National Research Council (NAS/NRC)
  2101 Constitution Avenue, NW
  Washington, DC 20418
  (202) 334-2000
  www.nas.edu
- National Institute of Nutrition
  265 Carling Avenue, Suite 302
  Ottawa, Ontario K1S 2E1
  (613) 235-3355; fax (613) 235-7032
  www.nin.ca
- Society for Nutrition Education
  7101 Wisconsin Avenue, Suite 901
  Bethesda, MD 20814-4805
  (301) 656-4938

Aging

- Administration on Aging
  330 Independence Avenue SW
  Washington, DC 20201
  (202) 619-0724
  www.aoa.dhhs.gov
- American Association of Retired Persons (AARP)
  601 E Street NW
  Washington, DC 20049
  (202) 434-2277
  www.aarp.org
- National Aging Information Center
  330 Independence Avenue SW
  Washington, DC 20201
  (202) 619-7501
  www.aoa.dhhs.gov/naic
- National Institute on Aging
  Public Information Office
  31 Center Drive, MSC 2292
  Bethesda, MD 20892
  (301) 496-1752
  www.nih.gov/nia

Alcohol and Drug Abuse

- Al-Anon Family Group Headquarters, Inc.
  1600 Corporate Landing Parkway
  Virginia Beach, VA 23454-5617
  (800) 356-9996
  www.al-anon.alateen.org
- Alateen
  1600 Corporate Landing Parkway
  Virginia Beach, VA 23454-5617
  (800) 356-9996
  www.al-anon.alateen.org
- Alcohol & Drug Abuse Information Line
  Adcare Hospital
  (800) 252-6465

D

- Alcoholics Anonymous (AA)
General Service Office
475 Riverside Drive
New York, NY 10115
(212) 870-3400
www.aa.org
- Narcotics Anonymous (NA)
P.O. Box 9999
Van Nuys, CA 91409
(818) 773-9999; fax (818) 700-0700
www.wsoinc.com
- National Clearinghouse for Alcohol and Drug Information (NCADI)
P.O. Box 2345
Rockville, MD 20847-2345
(800) 729-6686
www.health.org
- National Council on Alcoholism and Drug Dependence (NCADD)
12 West 21st Street
New York, NY 10010
(800) NCA-CALL or (800) 622-2255
(212) 206-6770; fax (212) 645-1690
www.ncadd.org
- U.S. Center for Substance Abuse Prevention
1010 Wayne Avenue, Suite 850
Silver Spring, MD 20910
(301) 459-1591 ext. 244; fax (301) 495-2919
www.covesoft.com/csap.html

Consumer Organizations

- Center for Science in the Public Interest (CSPI)
1875 Connecticut Avenue NW, Suite 300
Washington, DC 20009-5728
(202) 332-9110; fax (202) 265-4954
www.cspinet.org
- Choice in Dying, Inc.
1035 30th Street NW
Washington, DC 20007
(202) 338-9790; fax (202) 338-0242
www.choices.org
- Consumer Information Center
Pueblo, CO 81009
(888) 8 PUEBLO or (888) 878-3256
www.pueblo.gsa.gov
- Consumers Union of US Inc.
101 Truman Avenue
Yonkers, NY 10703-1057
(914) 378-2000
www.consunion.org
- National Council Against Health Fraud, Inc. (NCAHF)
P.O. Box 1276
Loma Linda, CA 92354
(909) 824-4690
www.ncahf.org

Fitness

- American College of Sports Medicine
P.O. Box 1440
Indianapolis, IN 46206-1440
(317) 637-9200
www.acsm.org/sportsmed
- American Council on Exercise (ACE)
5820 Oberlin Drive, Suite 102
San Diego, CA 92121
(800) 529-8227
www.acefitness.org
- President's Council on Physical Fitness and Sports
Humphrey Building, Room 738
200 Independence Avenue SW
Washington, DC 20201
(202) 690-9000; fax (202) 690-5211
www.indiana.edu/~preschal
- Shape Up America!
6707 Democracy Boulevard, Suite 306
Bethesda, MD 20817
(301) 493-5368
www.shapeup.org
- Sport Medicine and Science Council of Canada
1600 James Naismith Drive, Suite 314
Gloucester, Ontario K1B 5N4, Canada
(613) 748-5671; fax (613) 748-5729
www.smscc.ca

Food Safety

- Alliance for Food & Fiber
Food Safety Hotline
(800) 266-0200
- FDA Center for Food Safety and Applied Nutrition
200 C Street SW
Washington, DC 20204
(800) FDA-4010 or (800) 332-4010
vm.cfsan.fda.gov
- National Lead Information Center
(800) LEAD-FYI or (800) 532-3394
(800) 424-LEAD or (800) 424-5323
- National Pesticide Telecommunications Network (NPTN)
Oregon State University
333 Weniger Hall
Corvallis, OR 97331-6502
(541) 737-6091
www.ace.orst.edu/info/nptn
- USDA Meat and Poultry Hotline
(800) 535-4555
- U.S. EPA Safe Drinking Water Hotline
(800) 426-4791

Health and Disease

- Alzheimer's Disease Education and Referral Center
P. O. Box 8250
Silver Spring, MD 20907-8250
(800) 438-4380
www.alzheimers.org
- Alzheimer's Disease Information and Referral Service
919 North Michigan Avenue, Suite 1000
Chicago, IL 60611
(800) 272-3900
www.alz.org
- American Academy of Allergy, Asthma, and Immunology
611 East Wells Street
Milwaukee, WI 53202
(414) 272-6071; fax (414) 276-3349
www.aaaai.org
- American Cancer Society
National Home Office
1599 Clifton Road NE
Atlanta, GA 30329-4251
(800) ACS-2345 or (800) 227-2345
www.cancer.org
- American Council on Science and Health
1995 Broadway, 2nd Floor
New York, NY 10023-5860
(212) 362-7044; fax (212) 362-4919
www.acsh.org
- American Dental Association
211 East Chicago Avenue
Chicago, IL 60611
(312) 440-2800
www.ada.org
- American Diabetes Association
1660 Duke Street
Alexandria, VA 22314
(800) 232-3472 or (703) 549-1500
www.diabetes.org
- American Heart Association
Box BHG, National Center
7320 Greenville Avenue
Dallas, TX 75231
(800) 275-0448 or (214) 373-6300
www.amhrt.org
- American Institute for Cancer Research
1759 R Street NW
Washington, DC 20009
(800) 843-8114 or (202) 328-7744; fax (202) 328-7226
www.aicr.org
- American Medical Association
515 North State Street
Chicago, IL 60610
(312) 464-5000
www.ama-assn.org

- American Public Health Association (APHA)
  1015 Fifteenth Street NW, Suite 300
  Washington, DC 20005
  (202) 789-5600
  www.apha.org
- American Red Cross
  National Headquarters
  8111 Gatehouse Road
  Falls Church, VA 22042
  (703) 206-7180
  www.redcross.org
- Canadian Diabetes Association
  15 Toronto Street, Suite 800
  Toronto, ON M5C 2E3
  (800) BANTING or (800) 226-8464
  (416) 363-3373
  www.diabetes.ca
- Canadian Public Health Association
  400-1565 Carling Avenue
  Ottawa, Ontario K1Z 8R1
  (613) 725-3769; fax (613) 725-9826
  www.cpha.ca
- Centers for Disease Control and Prevention (CDC)
  1600 Clifton Road NE
  Atlanta, GA 30333
  (404) 639-3311
  www.cdc.gov
- The Food Allergy Network
  10400 Eaton Place, Suite 107
  Fairfax, VA 22030-2208
  (800) 929-4040 or (703) 691-3179
  www.foodallergy.org
- Internet Health Resources
  www.ihr.com
- National AIDS Hotline (CDC)
  (800) 342-AIDS (English)
  (800) 344-SIDA (Spanish)
  (800) 2437-TTY (Deaf)
  (900) 820-2437
- National Cancer Institute
  Office of Cancer Communications
  Building 31, Room 10824
  Bethesda, MD 20892
  (800) 4-CANCER or (800) 422-6237
  www.nci.nih.gov
- National Diabetes Information Clearinghouse
  1 Information Way
  Bethesda, MD 20892-3560
  (301) 654-3327
  www.niddk.nih.gov
- National Digestive Disease Information Clearinghouse (NDDIC)
  2 Information Way
  Bethesda, MD 20892-3570
  (301) 654-3810
  www.niddk.nih.gov
- National Health Information Center (NHIC)
  Office of Disease Prevention and Health Promotion
  (800) 336-4797
  nhic-nt.health.org
- National Heart, Lung, and Blood Institute Information Center
  P.O. Box 30105
  Bethesda, MD 20824-0105
  (301) 251-1222
  www.nhlbi.nih.gov/nhlbi/nhlbi.htm
- National Institute of Allergy and Infectious Diseases
  Office of Communications
  Building 31, Room 7A50
  31 Center Drive, MSC2520
  Bethesda, MD 20892-2520
  (301) 496-5717
  www.niaid.nih.gov
- National Institute of Dental Research (NIDR)
  National Institute of Health
  Bethesda, MD 20892-2190
  (301) 496-4261
  www.nidr.nih.gov
- National Institutes of Health (NIH)
  9000 Rockville Pike
  Bethesda, MD 20892
  (301) 496-2433
  www.nih.gov
- National Osteoporosis Foundation
  1150 17th Street NW, Suite 500
  Washington, DC 20036
  (202) 223-2226
  www.nof.org
- Office of Disease Prevention and Health Promotion
  odphp.osophs.dhhs.gov
- Office on Smoking and Health (OSH)
  www.americanheart.org/heart.org/Heart_and_stroke_A_Z_Guide/osh.html

Infancy and Childhood

- American Academy of Pediatrics
  141 Northwest Point Boulevard
  Elk Grove Village, IL 60007-1098
  (847) 228-5005
  www.aap.org
- Association of Birth Defect Children, Inc.
  930 Woodcock Road, Suite 225
  Orlando, FL 32803
  (407) 245-7035
  www.birthdefects.org
- Canadian Paediatric Society
  100-2204 Walkley Road
  Ottawa, ON K1G 4G8
  (613) 526-9397; fax (613) 526-3332
  www.cps.ca
- National Center for Education in Maternal & Child Health
  2000 15th Street North, Suite 701
  Arlington, VA 22201-2617
  (703) 524-7802
  www.ncemch.org

Pregnancy and Lactation

- American College of Obstetricians and Gynecologists Resource Center
  409 12th Street SW
  Washington, DC 20024-2188
  (202) 638-5577
  www.acog.org
- La Leche International, Inc.
  1400 N. Meacham Road
  Schaumburg, IL 60173
  (847) 519-7730
  www.lalecheleague.org
- March of Dimes Birth Defects Foundation
  1275 Mamaroneck Avenue
  White Plains, NY 10605
  (914) 428-7100
  www.modimes.org

Trade and Industry Organizations

- B. Braun/McGraw, Inc.
  824 Twelfth Avenue
  Bethlehem, PA 18018
  www.bbraunusa.com
- Beech-Nut Nutrition Corporation
  P.O. 618
  St. Louis, MO 63188-0618
  (800) 523-6633
  www.beechnut.com
- Borden Inc.
  180 East Broad Street
  Columbus, OH 43215
  (800) 426-7336
- Campbell Soup Company
  Consumer Response Center
  Campbell Place, Box 26B
  Camden, NJ 08103-1701
  (800) 257-8443
  www.campbellssoup.com
- Elan Pharma/Hi Chem Diagnostics
  2 Thurber Boulevard
  Smithfield, RI 02917
  (401) 233-3526
  www.hi-chem.com
- General Mills, Inc.
  Number One General Mills Boulevard
  Minneapolis, MN 55426
  (800) 328-6787
  www.generalmills.com

- Hoffmann-LaRoche, Inc.
  340 Kingsland Street
  Nutley, NJ 07110
  (973) 235-5000
- Kellogg Company
  P.O. Box 3599
  Battle Creek, MI 49016-3599
  (616) 961-2000
  www.kelloggs.com
- Kraft Foods
  Consumer Response and Information Center
  One Kraft Court
  Glenview, IL 60025
  (800) 323-0768
  www.kraftfoods.com
- Mead Johnson Nutritionals
  2400 West Lloyd Expressway
  Evansville, IN 47721
  (800) 247-7893
  www.meadjohnson.com
- Nabisco Consumer Affairs
  100 DeForest Avenue
  East Hanover, NJ 07936
  (800) NABISCO or (800) 932-7800
  www.nabisco.com
- National Dairy Council
  10255 West Higgins Road, Suite 900
  Rosemond, IL 60018-5616
  (847) 803-2000
  www.dairyinfo.com
- NutraSweet/KELCO
  P.O. Box 2986
  Chicago, IL 60654-0986
  www.equal.com
- Pillsbury Company
  Consumer Relations
  P.O. Box 550
  Minneapolis, MN 55440
  (800) 767-4466
  www.pillsbury.com
- Procter and Gamble Company
  One Procter and Gamble Plaza
  Cincinnati, OH 45202
  (513) 983-1100
  www.pg.com/info
- Ross Laboratories, Abbot Laboratory
  625 Cleveland Avenue
  Columbus, OH 43215
  (800) 227-5767
  www.abbot.com
- Sherwood Medical
  1915 Olive Street
  St. Louis, MO 63103
  (800) 428-4400
- Sunkist Growers
  Consumer Affairs
  Fresh Fruit Division
  14130 Riverside Drive
  Sherman Oaks, CA 91423
  (800) CITRUS-5 or (800) 248-7875
  www.sunkist.com
- United Fresh Fruit and Vegetable Association
  727 North Washington Street
  Alexandria, VA 22314
  (703) 836-3410
- USA Rice Federation
  4301 North Fairfax Drive, Suite 305
  Arlington, VA 22203
  Phone: (703) 351-8161
  www.usarice.com
- Weight Watchers International, Inc.
  Consumer Affairs Department/IN
  175 Crossways Park West
  Woodbury, NY 11797
  (516) 390-1400; fax (516) 390-1632.
  www.weightwatchers.com

Weight Control and Eating Disorders

- American Anorexia & Bulimia Association, Inc.
  165 West 46th Street #1108
  New York, NY 10036
  (212) 575-6200
  members.aol.com/amanbu
- Anorexia Nervosa and Related Eating Disorders (ANRED)
  P.O. Box 5102
  Eugene, OR 97405
  (541) 344-1144
  www.anred.com
- National Association of Anorexia Nervosa and Associated Disorders, Inc. (ANAD)
  P.O. Box 7
  Highland Park, IL 60035
  (847) 831-3438
  members.aol.com/anad20/index.html
- National Eating Disorder Information Centre
  200 Elizabeth Street, College Wing 1-304
  Toronto, Ontario M5G 2C4
  (519) 253-7421; fax (519) 253-7545
- Overeaters Anonymous (OA)
  World Service Office
  6075 Zenith Court NE
  Rio Rancho, NM 87124
  (505) 891-2664; fax (505) 891-4320
  www.overeatersanonymous.org
- TOPS (Take Off Pounds Sensibly)
  4575 South Fifth Street
  P.O. Box 07360
  Milwaukee, WI 53207-0360
  (800) 932-8677 or (414) 482-4620
  www.tops.org

World Hunger

- Bread for the World
  1100 Wayne Avenue, Suite 1000
  Silver Spring, MD 20910
  (301) 608-2400
  www.bread.org
- Center on Hunger, Poverty and Nutrition Policy
  Tufts University School of Nutrition
  11 Curtis Avenue
  Medford, MA 02155
  (617) 627-3956
- Freedom from Hunger
  P.O. Box 2000
  1644 DaVinci Court
  Davis, CA 95617
  (530) 758-6200
  www.freefromhunger.org
- Oxfam America
  26 West Street
  Boston, MA 02111
  (617) 482-1211
  www.oxfamamerica.org
- SEEDS Magazine
  P.O. Box 6170
  Waco, TX 76706
  (254) 755-7745
  www.helwys.com/seedhome.htm
- Worldwatch Institute
  1776 Massachusetts Avenue NW, Suite 800
  Washington, DC 20036
  (202) 452-1999
  www.worldwatch.org

Contents

E

NUTRITION ASSESSMENT: SUPPLEMENTAL INFORMATION

♦

Chapters 15 and 16 described many nutrition assessment techniques that health care professionals can use to determine clients' nutrition status. From this assessment, they identify clients' nutrition needs and develop care plans for meeting those needs. This appendix provides additional details about nutrition assessment.

DRUG HISTORY: NUTRIENT INTERACTIONS

♦

Chapter 15 described nutrient-medication interactions and Chapters 18 through 28 provided a series of "prescription pads," listing medications used in the treatment of the specific diseases being discussed. Table E-1 provides examples of selected drugs, describes nutrition-related factors that affect the administration of medications, and lists the most common nutrition-related side effects.

E

Table E–1
Administration and Common Nutrition-Related Side Effects of Selected Medications

Medication Classification and Examples	Administration	Common Nutrition-Related Side Effects[a]
Analgesics		
Narcotic: codeine, merperidine, morphine sulfate	Give with food to reduce GI distress.	N/V, GI distress, reduced GI motility, constipation, lethargy.
Nonnarcotic (also act as nonsteroidal anti-inflammatory agents): aspirin, ibuprofen, naproxen		N/V, GI distress, GI bleeding, constipation. Aspirin may lower blood folate and vitamin C.
Antacids	Give with fluids between meals or at bedtime.	
Al-containing (also act as phosphate binders): Al carbonate, Al hydroxide, Al phosphate	Give with meals when used as phosphate binder.	Constipation, phosphorus deficiency. Long-term use in renal failure may cause Al toxicity.
Ca-containing (also act as phosphate binders and Ca supplements): Ca carbonate and Ca gluconate	When used as a phosphate binder or supplement, give with meals and separately from foods high in fiber, oxalate, or phytate and iron or fluoride supplements.	Constipation, chalky taste. Concurrent use with vitamin D supplements may lead to elevated blood Ca.
Mg-containing (also act as laxatives): Mg hydroxide, Mg oxide, and Mg citrate	Give separately from iron and folate supplements.	Diarrhea, chalky taste. Long-term use in renal failure may lead to Mg toxicity.
Antianginals		
Amyl nitrate, isosorbide dinitrate, nitroglycerin (see also *Antihypertensives*)	Give oral forms on an empty stomach. Limit alcohol.	Nutrition-related side effects are uncommon.
Antianxiety Agents	Give with food to reduce GI distress.	
Alprazolam, chlordiazepoxide	Avoid alcohol.	Increased appetite, weight gain, nausea, drowsiness.
Diazepam, lorazepam, oxazepam	Limit caffeine and avoid alcohol.	Constipation, diarrhea, dry mouth, drowsiness.
Meprobamate	Avoid alcohol.	N/V, diarrhea, drowsiness.
Anticoagulants, oral		
Ticlopidine	Give with food to improve drug absorption and reduce GI distress.	N/V, GI pain, diarrhea.
Warfarin	Maintain consistent vitamin K intake, avoid high doses of vitamins A and E, which can reduce the anticoagulant effect. Avoid high doses of vitamin C, which can reduce drug absorption.	Nausea.
Anticonvulsants	Give with meals to reduce GI distress.	
Phenytoin	Tube feedings may interfere with drug absorption (see Chapter 23).	N/V, swollen gums. May cause folate-deficiency anemia. Increases metabolism of vitamins D and K.

[a]Note that many other medications not listed in this table also have nutrition-related side effects. In addition, nutrition-related side effects other than those listed may occur. For example, almost all medications cause nausea in some people. In this table, nausea is only listed as a side effect if it occurs with relative frequency or does not resolve with time. More detailed texts should be consulted for the medications you routinely encounter in clinical practice.

Abbreviations: N/V = nausea/vomiting; Al = aluminum; Ca = calcium; Mg = magnesium; K = potassium; GERD = gastroesophageal reflux disease; ACE = angiotensin-converting enzyme; H2 = histamine$_2$.

E

Table E–1
Administration and Common Nutrition-Related Side Effects of Selected Medications (continued)

Medication Classification and Examples	Administration	Common Nutrition-Related Side Effects[a]
Anticonvulsants		
Primidone	Avoid alcohol.	N/V. May cause folate-deficiency anemia.
Valproic acid	Do not take tablets with milk or liquid form with carbonated beverages.	N/V, GI pain.
Antidepressants		
MAO inhibitors: phenelzine, tranylcypromine	Give with food to reduce GI distress. Avoid foods high in tyramine (see Chapter 15), alcohol, and tryptophan supplements. Limit caffeine.	Weight changes, dry mouth, constipation.
Tricyclic: amitriptyline, clomipramine, doxepin, imipramine, protriptyline	Give with food to reduce GI distress. Avoid alcohol and limit caffeine and high-fiber foods.	Dry mouth, constipation. Stimulates appetite, especially for sweets.
Other:		
bupropion	Give with food to reduce GI distress. Avoid alcohol.	Dry mouth, constipation.
fluoxetine	Give in morning without regard to food. Avoid tryptophan supplements.	Anorexia, weight loss, dry mouth, N/V, diarrhea.
nefazone	Food reduces drug absorption and bioavailability.	Dry mouth, N/V, constipation.
sertraline	Give at same time each day without regard to food. Avoid alcohol.	Dry mouth, N/V, constipation.
Antidiabetics		
Acarbose	Give at the start of each meal.	GI pain, flatulence, diarrhea, hypoglycemia.
Glipizide	Give 30 min before breakfast. Limit alcohol.	Hypoglycemia. GI side effects uncommon.
Glyburide	Give with breakfast. Limit alcohol.	Hypoglycemia. GI side effects uncommon.
Metformin	Give with meals to reduce GI distress. Limit alcohol.	N/V, bloating, flatulence, diarrhea. Lowers blood glucose, cholesterol, LDL, and triglycerides and raises HDL.
Troglitazone[c]	Give with meals.	Risk of hypoglycemia increases when used in combination with other antidiabetic agents. GI side effects uncommon.
Antidiarrheals		
Loperamide	Give without regard to food.	Nutrition-related side effects are uncommon.
Opium and paregoric	Give without regard to food.	N/V, constipation, sedation.

[a]Note that many other medications not listed in this table also have nutrition-related side effects. In addition, nutrition-related side effects other than those listed may occur. For example, almost all medications cause nausea in some people. In this table, nausea is only listed as a side effect if it occurs with relative frequency or does not resolve with time. More detailed texts should be consulted for the medications you routinely encounter in clinical practice.

Abbreviations: N/V = nausea/vomiting; Al = aluminum; Ca = calcium; Mg = magnesium; K = potassium; GERD = gastroesophageal reflux disease; ACE = angiotensin-converting enzyme; H2 = histamine$_2$.

[c]*Source:* Product advertisement in *Diabetes Care,* August 1977.

E

Table E–1
Administration and Common Nutrition-Related Side Effects of Selected Medications (continued)

Medication Classification and Examples	Administration	Common Nutrition-Related Side Effects[a]
Antihypertensives	Avoid natural licorice.	
ACE inhibitors:		
benazepril, enalapril, lisopril, ramipril	Limit alcohol and avoid salt substitutes. Monitor use of K supplements.	May elevate blood K.
captopril	Give 1 hr before or 2 hr after meals. Limit alcohol and avoid salt substitutes. Monitor use of K supplements.	Mouth ulcers. May elevate blood K.
fosinopril	Give separately from Ca or Mg supplements. Limit alcohol and avoid salt substitutes. Monitor use of K supplements.	May elevate blood K.
Alpha-adrenergic blockers: doxazosin, prazosin, terazosin	Limit alcohol.	Weight gain, fatigue.
Beta-blockers (also act as antiarrythmics and antianginals):		May mask signs of hypoglycemia.
atenolol	Give with food to reduce GI distress. Give separately from Ca supplements or antacids.	Nausea, dizziness.
metoprolol	Give with food to enhance bioavailability.	Diarrhea, confusion, dizziness.
nadolol	Limit alcohol.	Nutrition-related side effects are uncommon.
propranolol	Give with food to enhance bioavailability. Avoid alcohol and give separately from Ca supplements or antacids.	Dizziness, drowsiness, weakness.
Ca-channel blockers (also act as antianginals):		
amlopidine, isradipine	Give without regard to food.	Edema, headache.
dilitiazem	Give tablets or extended release capsules before meals.	Edema, dizziness.
felopidine	Do not give with grapefruit juice.	Edema, headache.
nicarpidine, nisoldipine	Do not give with high-fat foods or grapefruit juice.	Edema, headache, dizziness.
Other:		
clonidine	Avoid alcohol.	Dry mouth, constipation, edema, drowsiness, dizziness.

[a]Note that many other medications not listed in this table also have nutrition-related side effects. In addition, nutrition-related side effects other than those listed may occur. For example, almost all medications cause nausea in some people. In this table, nausea is only listed as a side effect if it occurs with relative frequency or does not resolve with time. More detailed texts should be consulted for the medications you routinely encounter in clinical practice.

Abbreviations: N/V = nausea/vomiting; Al = aluminum; Ca = calcium; Mg = magnesium; K = potassium; GERD = gastroesophageal reflux disease; ACE = angiotensin-converting enzyme; H2 = histamine$_2$.

E

Table E–1
Administration and Common Nutrition-Related Side Effects of Selected Medications (continued)

Medication Classification and Examples	Administration	Common Nutrition-Related Side Effects[a]
Antihypertensives (continued)		
guanfacine	Limit alcohol.	Dry mouth, constipation, drowsiness.
hydralazine	Give with food to enhance bioavailability. Limit alcohol.	Anorexia, N/V, edema, headache. Pyridoxine supplements correct drug-induced peripheral neuropathy.
methyldopa	Avoid alcohol. Do not give drug within 2 hr of giving iron supplements.	Dry mouth, headache, drowsiness, edema. High doses increase vitamin B_{12} and folate needs.
minoxidil		Bloating, edema.
Anti-Infectives		
Antibiotics:		
amoxicillin	Give without regard to food.	Diarrhea.
ampicillin	Give with 8 oz of water 1 hr before or 2 hr after meals.	Diarrhea.
cefaclor, cefoperazone, cefotaxime	Give parenterally. Avoid alcohol while using and for 3 days afterward. Give foods high in vitamin K or a vitamin K supplement with long-term use.	May interfere with bacterial vitamin K synthesis in intestine.
chloramphenicol	Give with 8 oz water 1 hr before or 2 hr after meals. Avoid alcohol. Limit use of iron supplements.	Increases risk of iron overload. Delays response to iron, folate, or vitamin B_{12}.
erythromycin	Give with 8 oz water 1 hr before or 2 hr after meals. Food decreases absorption of some forms (base and stearate), but may be given to reduce GI distress.	Epigastric pain, abdominal cramps.
ethambutol	Give with food to reduce GI distress.	Nutrition-related side effects are uncommon.
penicillin	Give penicillin K without regard to meals; penicillin G 1 hr before or 2 hr after meals. Use K supplements cautiously with penicillin K.	N/V, epigastric distress, mouth sores, diarrhea.
tetracycline	Give with water 1 hr before or 2 hr after meals. Separate the administration of Ca, iron, Mg, zinc, Al, or vitamin-mineral supplements or Al-, Ca-, or Mg-containing antacids by 3 hr.	N/V, diarrhea, cramps, dizziness.

[a]Note that many other medications not listed in this table also have nutrition-related side effects. In addition, nutrition-related side effects other than those listed may occur. For example, almost all medications cause nausea in some people. In this table, nausea is only listed as a side effect if it occurs with relative frequency or does not resolve with time. More detailed texts should be consulted for the medications you routinely encounter in clinical practice.

Abbreviations: N/V = nausea/vomiting; Al = aluminum; Ca = calcium; Mg = magnesium; K = potassium; GERD = gastroesophageal reflux disease; ACE = angiotensin-converting enzyme; H2 = histamine$_2$.

E

Table E–1
Administration and Common Nutrition-Related Side Effects of Selected Medications (continued)

Medication Classification and Examples	Administration	Common Nutrition-Related Side Effects[a]
Anti-Infectives (continued)		
Antifungals:		
amphotericin	Give parenterally and encourage fluids.	Anorexia, N/V, weight loss, stomach pain, fever, headache, soreness.
clotrimazole	Dissolve lozenge slowly in mouth over 15–30 min.	N/V.
flucytosine	Give capsules slowly over 15 min to reduce GI distress.	N/V, diarrhea, lethargy, dizziness.
ketoconazole	Give with foods to enhance absorption. Give Ca or Mg separately by 2 hr.	N/V.
Antivirals:		
acyclovir	Encourage fluids.	Headache.
famciclovir	Give without regard to food.	Headache.
foscarnet	Give parenterally and encourage fluids.	Anorexia, N/V, abdominal pain, diarrhea, fever, headache, weakness, dizziness.
ganciclovir	Give with food and encourage fluids.	N/V, abdominal pain, headache, fever, weakness.
lamivudine (3TC)	Give without regard to food.	Nausea, GI pain, diarrhea, headache, fever.
saquinavir	Give within 2 hr of a full meal.	Nausea, GI pain, diarrhea, headache.
stavudine (d4T)	Give without regard to food.	N/V, diarrhea, fever.
zalcitabine (DDC)	Give on empty stomach, if possible, to enhance absorption.	Anorexia, weight loss, N/V, mouth ulcers.
zidovudine (AZT)	Give without regard to food.	Anorexia, N/V, headache, anemia.
Anti-Inflammatory Agents		
Corticosteroids (also act as immunosuppressants): cortisone, dexamethasone, hydrocortisone, prednisone	Give with food to reduce GI distress. Encourage high-protein, low-sodium diet. Avoid alcohol. May need supplements of or diets high in K, Ca, and phosphorus; vitamins A, C, and D; and pyridoxine and folate.	Edema, osteoporosis. Increase appetite and weight. Induce negative nitrogen and Ca balances.
Diclofenac	Give with food, milk, or water to reduce GI distress. Avoid alcohol.	Nausea, GI pain, constipation, diarrhea, headache, edema.
Diflunisal	Give with food or milk to reduce GI distress. Avoid alcohol.	Nausea, GI pain, diarrhea, headache.
Mesalamine, olsalazin	Give with food and 8 oz water.	Nutrition-related side effects are uncommon.
Salsalate	Give with food, milk or water. Limit alcohol and caffeine.	Nausea, dizziness.

[a]Note that many other medications not listed in this table also have nutrition-related side effects. In addition, nutrition-related side effects other than those listed may occur. For example, almost all medications cause nausea in some people. In this table, nausea is only listed as a side effect if it occurs with relative frequency or does not resolve with time. More detailed texts should be consulted for the medications you routinely encounter in clinical practice.

Abbreviations: N/V = nausea/vomiting; Al = aluminum; Ca = calcium; Mg = magnesium; K = potassium; GERD = gastroesophageal reflux disease; ACE = angiotensin-converting enzyme; H2 = histamine$_2$.

E

Table E-1
Administration and Common Nutrition-Related Side Effects of Selected Medications (continued)

Medication Classification and Examples	Administration	Common Nutrition-Related Side Effects[a]
Anti-Inflammatory Agents (continued)		
Sulfasalazine (See also *Analgesics, nonnarcartic*)	Give with 8 oz water or food to reduce GI distress. Give folate supplement and encourage fluids.	Anorexia, N/V, GI pain, diarrhea, headache, dizziness. Lowers blood folate.
Antilipemics		
Cholestyramine	Give before meals. Mix powder form with water or fluids; never give dry or with carbonated beverages.	Nausea, belching, dyspepsia, constipation. May decrease absorption of fat, fat-soluble vitamins, folate, Ca, iron, zinc, and Mg.
Clofibrate	Give with food or milk to reduce GI distress.	Nausea, anemia.
Fluvastatin	Give without regard to food.	Nutrition-related side effects are uncommon.
Gemfibrozil	Give ½ hr before meals.	Taste alterations, dyspepsia, abdominal pain.
Lovastatin	Give with meals to enhance absorption. Give fiber, pectin, or oat bran separately by several hr. Avoid high doses of niacin and limit alcohol.	Constipation, headache.
Pravastatin, simvastatin	Avoid high doses of niacin and limit alcohol.	Nutrition-related side effects are uncommon.
Antinauseants, Antiemetics		
Dronabinol (marijuana derivative)	Give before lunch and dinner.	Stimulates appetite and causes weight gain. Euphoria.
Granisetron	Give without regard to food.	Constipation, headache, weakness.
Meclizine	Give without regard to food	Nutrition-related side effects are uncommon.
Metoclopramide	Give ½ hr before meals and at bedtime.	Nausea, diarrhea, lethargy.
Ondansetron	Give without regard to food.	Abdominal pain, constipation, headache, weakness.
Prochlorperazine	Give with food or milk. Avoid alcohol and limit caffeine.	Dry mouth, constipation. Stimulates appetite and causes weight gain. Increases urinary excretion of riboflavin.
Antineoplastics		
Aldesleukin	Give parenterally.	Anorexia, N/V, mouth inflammation, diarrhea, mental changes, dizziness, anemia, impaired renal function, fever, edema, infection.

[a]Note that many other medications not listed in this table also have nutrition-related side effects. In addition, nutrition-related side effects other than those listed may occur. For example, almost all medications cause nausea in some people. In this table, nausea is only listed as a side effect if it occurs with relative frequency or does not resolve with time. More detailed texts should be consulted for the medications you routinely encounter in clinical practice.

Abbreviations: N/V = nausea/vomiting; Al = aluminum; Ca = calcium; Mg = magnesium; K = potassium; GERD = gastroesophageal reflux disease; ACE = angiotensin-converting enzyme; H2 = histamine$_2$.

Table E–1
Administration and Common Nutrition-Related Side Effects of Selected Medications (continued)

Medication Classification and Examples	Administration	Common Nutrition-Related Side Effects[a]
Antineoplastics (continued)		
Bleomyin	Give parenterally.	Anorexia, N/V, mouth inflammation, weight loss, fever, respiratory impairment.
Carboplatin	Give parenterally with adequate fluids to maintain hydration.	Anorexia, N/V, mouth inflammation, GI pain, diarrhea, constipation, weakness, infections, anemia. Lowers blood levels of sodium, K, Ca, and Mg.
Carmustine	Give parenterally with adequate fluids to maintain hydration.	N/V, mild liver impairment.
Cisplatin	Give parenterally with adequate fluids to maintain hydration.	Severe N/V, taste alterations, diarrhea, infections, impaired renal function, anemia. Lowers blood levels of sodium, K, Ca, phosphorus, Mg, and zinc.
Cyclophosphamide	Give oral forms with food only if GI distress occurs. Encourage fluids.	Anoerxia, N/V, mouth inflammation, delayed wound healing.
Cytarbine	Give parenterally and provide adequate liquids.	Anorexia, N/V, mouth inflammation, diarrhea, anal ulcers, weight loss, infection. Lowers blood levels of K and Ca.
Dactinomycin	Give parenterally and provide adequate fluids. Encourage high-kcalorie foods. Raises vitamin B_{12} needs.	Anorexia, severe N/V, dry mouth, mouth and tongue inflammation, taste alterations, severe esophagitis, dysphagia, GI pain, diarrhea, weight loss, fatigue, anemia. Reduces absorption of fat, Ca, and iron.
Daunorubicin, doxorubicin, idarubicin	Give parenterally and provide adequate fluids.	Anorexia, N/V, weight loss, dry mouth, mouth and esophageal inflammation, anemia.
Estramustine	Store capsules in refrigerator.	Anorexia, N/V, diarrhea, edema, lethargy, Impairs glucose tolerance.
Etoposide, teniposide	Give parenterally or orally (etoposide).	Anorexia, N/V, mouth inflammation, diarrhea.
Floxuridine	Give parenterally.	Anorexia, N/V, diarrhea, weakness, lethargy.
Fluorouracil	Give parenterally.	Anorexia, severe N/V, mouth and esophageal inflammation, taste alterations, intestinal inflammation, diarrhea, weakness, anemia. May increase pyridoxine needs.
Interferon alfa 2a and 2b	Give parenterally and provide adequate liquids.	Anorexia, N/V, dry mouth, taste alterations, mouth inflammation, abdominal pain, diarrhea, weight loss, dizziness, headache, fatigue.

[a]Note that many other medications not listed in this table also have nutrition-related side effects. In addition, nutrition-related side effects other than those listed may occur. For example, almost all medications cause nausea in some people. In this table, nausea is only listed as a side effect if it occurs with relative frequency or does not resolve with time. More detailed texts should be consulted for the medications you routinely encounter in clinical practice.

Abbreviations: N/V = nausea/vomiting; Al = aluminum; Ca = calcium; Mg = magnesium; K = potassium; GERD = gastroesophageal reflux disease; ACE = angiotensin-converting enzyme; H2 = histamine$_2$.

E

Table E-1
Administration and Common Nutrition-Related Side Effects of Selected Medications (continued)

Medication Classification and Examples	Administration	Common Nutrition-Related Side Effects[a]
Antineoplastics (continued)		
Leuprolide	Give parenterally.	Anorexia, N/V, constipation, edema, headache, weakness, anemia.
Levamisole	Give without regard to food.	Nausea, taste alterations, diarrhea, fatigue.
Lomustine	Give on empty stomach to reduce GI distress. Provide adequate fluids.	N/V, anemia.
Mechlorethamine	Give adequate fluids.	Anorexia, severe N/V, anemia.
Megestrol acetate (see *Appetite Stimulants*)		
Melphalan	Give adequate fluids. Divide single daily dose if GI distress occurs (food reduces drug bioavailability).	Nutrition-related side effects are uncommon, but may include mild N/V and diarrhea.
Mercaptopurine	Give adequate fluids.	Anorexia, mild N/V, anemia.
Methotrexate	Give adequate fluids.	Anorexia, N/V, mouth and gum inflammation, weight loss, infection.
Mithramycin, plicamycin	Give parenterally.	Anorexia, N/V, mouth inflammation, diarrhea.
Mitomycin	Give parenterally.	Anorexia, N/V, weight loss, fever, weakness.
Mitoxantrone	Give parenterally.	N/V, mouth inflammation, GI bleeding, abdominal pain, diarrhea, fever.
Paclitaxel	Give parenterally.	N/V, mouth and esophageal inflammation, diarrhea, edema, fatigue, anemia.
Pegasparagase	Give parenterally.	N/V, fever, weakness.
Porfimer	Give parenterally.	N/V, dysphagia, abdominal pain, constipation, fever.
Streptozocin	Give parenterally.	N/V, renal and liver impairment.
Tamoxifen citrate	Give Ca and Mg supplements separately from enteric-coated tablet by 2 hr.	N/V.
Tretinoin	Give with meals. Do not give vitamin A supplements. Limit alcohol and fat.	N/V, abdominal pain, GI bleeding, respiratory impairment, elevated blood lipids, fever.
Antipsychotics		
Chlorpromazine (also acts as an antinauseant)	Give with food, milk, or water to reduce GI distress. Dilute oral concentrate in a 4 oz drink or soft food. Give Mg separately by 2 hr. Avoid alcohol.	Dry mouth, constipation, drowsiness, dizziness. May reduce vitamin B_{12} absorption.

[a]Note that many other medications not listed in this table also have nutrition-related side effects. In addition, nutrition-related side effects other than those listed may occur. For example, almost all medications cause nausea in some people. In this table, nausea is only listed as a side effect if it occurs with relative frequency or does not resolve with time. More detailed texts should be consulted for the medications you routinely encounter in clinical practice.

Abbreviations: N/V = nausea/vomiting; Al = aluminum; Ca = calcium; Mg = magnesium; K = potassium; GERD = gastroesophageal reflux disease; ACE = angiotensin-converting enzyme; H2 = histamine$_2$.

E

Table E–1
Administration and Common Nutrition-Related Side Effects of Selected Medications (continued)

Medication Classification and Examples	Administration	Common Nutrition-Related Side Effects[a]
Antipsychotics (continued)		
Fluphenazine, perphenazine, trifluoperazine	Give with food to reduce GI distress. Do not mix concentrates with caffeinated drinks, tea, or apple juice. Avoid alcohol.	Stimulate appetite. Weight gain, dry mouth, constipation, drowsiness. May increase riboflavin needs.
Haloperidol	Give with food or milk to reduce GI distress. Do not mix concentrates with coffee, tea, or fruit juice. Avoid alcohol.	Stimulates appetite. Weight gain, dry mouth, constipation, drowsiness.
Loxapine	Give with food, milk, or water to reduce GI distress. Dilute concentrate with orange or grapefruit juice.	Dry mouth, drowsiness.
Prochlorperazine (see *Antinauseants*)		
Resperidone	Avoid alcohol.	Stimulates appetite. Weight gain, drowsiness.
Thioridazine	Give with food, milk, or water. Dilute concentrate in orange or grapefruit juice.	Stimulates appetite. Weight gain, dry mouth, constipation, drowsiness, dizziness. May increase riboflavin needs.
Antisecretory Agents (see *Antiulcer Agents*)		
Antiulcer Agents		
Antisecretory agents (also act as anti-GERD): lanisoprazole, omeprazole	Give before a meal. Swallow capsules whole.	May reduce absorption of iron and vitamin B_{12}.
H2 blockers (also act as anti-GERD): cimetidine	Give iron supplements 1 hr before giving drug. Give Ca or Mg supplements separately by 2 hr. Limit caffeine, xanthine, and alcohol. Liquid form not compatible with tube feedings.	May reduce vitamin B_{12} absorption.
famotidine, nizatidine, rantidine	Limit caffeine, xanthine, and alcohol.	May reduce vitamin B_{12} absorption.
Other: sucralfate	Give with water on an empty stomach 1 hr before meals or at bedtime. Give Ca or Mg supplement separately by 30 min. Limit alcohol.	Nutrition-related side effects are uncommon. May cause Al-toxicity in people with end-stage renal failure.
Appetite Stimulants		
Dronabinol (see *Antinauseants*)		
Megestrol acetate, medroxy progesterone acetate	May take with food to reduce GI distress.	Increases appetite. Weight gain, edema. May increase blood sodium.
Appetite Suppressants		
Dexfenfluramine, fenfluramine, mazindol, phendimetrazine, phentermine	Give on empty stomach 1 hr before meals. Instruct client to follow a low-kcalorie diet.	Nausea, taste alterations, dry mouth, dizziness, drowsiness.

[a]Note that many other medications not listed in this table also have nutrition-related side effects. In addition, nutrition-related side effects other than those listed may occur. For example, almost all medications cause nausea in some people. In this table, nausea is only listed as a side effect if it occurs with relative frequency or does not resolve with time. More detailed texts should be consulted for the medications you routinely encounter in clinical practice.

Abbreviations: N/V = nausea/vomiting; Al = aluminum; Ca = calcium; Mg = magnesium; K = potassium; GERD = gastroesophageal reflux disease; ACE = angiotensin-converting enzyme; H2 = histamine$_2$.

Table E-1
Administration and Common Nutrition-Related Side Effects of Selected Medications (continued)

Medication Classification and Examples	Administration	Common Nutrition-Related Side Effects[a]
Cardiac Glycosides		
Digitoxin, digoxin, digitalis	Give separately from high-fiber, high-pectin foods. Encourage low-salt, high-K diet. Mg supplements may decrease drug absorption. Avoid natural licorice and give Ca or vitamin D supplements cautiously. Hypokalemia, hypomagnesemia, hypoalbuminemia, and hypercalcemia increase drug effects.	Anorexia, N/V, weight loss, diarrhea.
Diuretics		
K–losing (thiazide, thiazide-related, and loop diuretics):		
bendroflumethiazide, chlorothalidone, chlorothiazide, hydrochlorothiazide, methyclothiazide, metolazone, quinethazone	Give with food or milk to reduce GI distress. Avoid natural licorice and limit alcohol. May need K or Mg supplements. Carefully monitor use of Ca and vitamin D supplements.	Lowers blood sodium, K, chloride, Mg. Raises blood Ca and glucose.
indapamide	Same as above.	Headache, dizziness, fatigue. Lowers blood K, sodium, chloride, Mg, and phosphorus. Raises blood Ca and glucose.
ethacrynic acid, furosemide	Give with food or milk to reduce GI distress. Avoid natural licorice and limit alcohol. May need K or Mg supplements.	Lowers blood K, sodium, chloride, Ca, Mg, and zinc. Raises blood glucose.
K-sparing:		
amiloride	Give with food or milk to reduce GI distress. Avoid natural licorice and alcohol. Avoid foods high in K, K supplements, and salt substitutes.	Anorexia, N/V. Lowers blood sodium and chloride. Raises blood K.
spironolactone	Same as above.	N/V, cramps, diarrhea. Lowers blood sodium and chloride. Raises blood K and Mg.
triamterene	Same as above.	Decreases the metabolism of folate. Lowers blood sodium, Mg, and folate. Raises blood K.
Immunosuppressants		
Azathioprine	Give with food to reduce GI distress.	N/V.

[a]Note that many other medications not listed in this table also have nutrition-related side effects. In addition, nutrition-related side effects other than those listed may occur. For example, almost all medications cause nausea in some people. In this table, nausea is only listed as a side effect if it occurs with relative frequency or does not resolve with time. More detailed texts should be consulted for the medications you routinely encounter in clinical practice.

Abbreviations: N/V = nausea/vomiting; Al = aluminum; Ca = calcium; Mg = magnesium; K = potassium; GERD = gastroesophageal reflux disease; ACE = angiotensin-converting enzyme; H2 = histamine$_2$.

E

Table E–1
Administration and Common Nutrition-Related Side Effects of Selected Medications (continued)

Medication Classification and Examples	Administration	Common Nutrition-Related Side Effects[a]
Immunosuppressants (continued)		
Corticosteroids (see *Anti-Inflammatory Agents*)		
Cyclosporine	Mix liquid with milk or orange juice at room temperature; no grapefruit juice. Avoid K supplements or salt substitutes.	N/V, diarrhea, swollen gums, headache, impaired renal function, elevated blood pressure, elevated blood glucose. May raise blood K and lower Mg.
Muromonab-cd3	Give parenterally.	N/V, diarrhea, fever, infection.
Mycophenolate	Give with food to reduce GI distress.	N/V, diarrhea, constipation, impaired liver function, impaired kidney function, anemia, infection.
Tacrolimus	Give with food to reduce GI distress.	N/V, diarrhea, constipation, impaired liver function, impaired kidney function, headache.
Laxatives		
Bisacodyl	Give on empty stomach with water or juice. Encourage use of high-fiber foods and fluids.	Laxative dependency and low blood K and Ca with long-term use.
Docusate	Give syrup with milk or juice to mask taste. Encourage use of high-fiber foods and fluids.	Nausea, throat irritation. Long-term use may raise blood glucose and lower K.
Lactulose	Give with juice, milk, or water, or sweet food to improve palatability. Encourage use of high-fiber foods and fluids. Do not give to clients on lactose- or galactose-restricted diets. Dilute before using with tube feeding.	Belching, cramps, flatulence.
Magnesium salts (see *Antacids, Mg-containing*)		
Methycellulose	Give with fluids. Encourage use of high-fiber foods and fluids.	Increased peristalsis.
Mineral oil	Give 2 hr before or after eating. Encourage use of high-fiber foods and fluids.	May reduce absorption of fat-soluble vitamins.
Psyllium	Mix powder in water, juice, or milk. Encourage use of high-fiber foods and fluids.	Reduces blood cholesterol and LDL.
Senna	Give with water or juice. Encourage use of high-fiber foods and fluids.	Nausea, cramps.

[a]Note that many other medications not listed in this table also have nutrition-related side effects. In addition, nutrition-related side effects other than those listed may occur. For example, almost all medications cause nausea in some people. In this table, nausea is only listed as a side effect if it occurs with relative frequency or does not resolve with time. More detailed texts should be consulted for the medications you routinely encounter in clinical practice.

Abbreviations: N/V = nausea/vomiting; Al = aluminum; Ca = calcium; Mg = magnesium; K = potassium; GERD = gastroesophageal reflux disease; ACE = angiotensin-converting enzyme; H2 = histamine$_2$.

E

Table E-1
Administration and Common Nutrition-Related Side Effects of Selected Medications (continued)

Medication Classification and Examples	Administration	Common Nutrition-Related Side Effects[a]
Phosphate Binders		
Al carbonate, Al hydroxide, Ca acetate, Ca carbonate, Ca citrate (see *Antacids, Al- and Ca-containing*)		
Miscellaneous		
Calcitonin (Ca regulator)	Give at bedtime to reduce GI distress.	N/V. Lowers blood Ca and phosphorus.
Calcitriol (Ca regulator)	Do not give with vitamin D, Mg supplements, or Mg–containing antacids. Avoid high-phosphorus foods.	Raises blood Ca levels.
Colchicine (antigout)	Give with low-purine diet during acute attacks of gout.	N/V, GI pain, diarrhea.
Dextroamphetamine (stimulant)	Limit caffeine.	Anorexia, weight loss, dizziness, headache, confusion.
Epoetin alfa (erythropoietin)	May need iron, folate, or vitamin B_{12} supplement.	Raises hemoglobin and hematocrit.
Lithium carbonate (antimanic)	Give with meals to reduce GI distress. Consistent sodium intake helps maintain drug levels. Give adequate fluids. Do not use syrup form with tube feedings.	N/V, thirst, weight gain, dry mouth, fatigue, weakness, edema, dizziness. Raises blood Ca, phosphorus, and Mg.
Oral contraceptives (hormone)	Give with food. Give foods high in folate, pyridoxine, and vitamin B_{12} and vitamin C supplements. Limit caffeine.	N/V, weight changes, appetite changes, bone loss, edema. Lowers blood folate, pyridoxine, and vitamin B_{12}.
Sodium polystyrene sulfonate (K exchange resin)	Mix powder with cool water or sorbitol. Mix with sorbitol-containing syrup to combat constipation. Avoid K supplements. Do not give Ca-containing supplements or antacids for at least several hours.	Anorexia, constipation. Lowers blood K, Ca, and Mg.

[a]Note that many other medications not listed in this table also have nutrition-related side effects. In addition, nutrition-related side effects other than those listed may occur. For example, almost all medications cause nausea in some people. In this table, nausea is only listed as a side effect if it occurs with relative frequency or does not resolve with time. More detailed texts should be consulted for the medications you routinely encounter in clinical practice.

Abbreviations: N/V = nausea/vomiting; Al = aluminum; Ca = calcium; Mg = magnesium; K = potassium; GERD = gastroesophageal reflux disease; ACE = angiotensin-converting enzyme; H2 = $histamine_2$.

Sources: Z. M. Pronsky, *Food Medication Interactions*, 9th ed. (Pottstown, Pa.: Food Medication Interactions, 1993); *Nursing Drug Guide* (Philadelphia: Lippincott-Raven Publishers, 1997).

GROWTH CHARTS AND ANTHROPOMETRIC DATA

◆

Growth charts, shown in Figures E–1 through E–6, allow health care professionals to evaluate the growth and development of children from birth to 18 years of age. The assessor follows these steps to plot a weight measurement on a percentile graph:

- Select the appropriate chart based on age and gender. (When length is measured, use the chart for birth to 36 months; when height is measured, use the chart for 2 to 18 years.)
- Locate the child's age along the horizontal axis on the bottom or top of the chart.
- Locate the child's weight in pounds or kilograms along the vertical axis on the lower left or right side of the chart.
- Mark the chart where the age and weight lines intersect. Read the percentile.

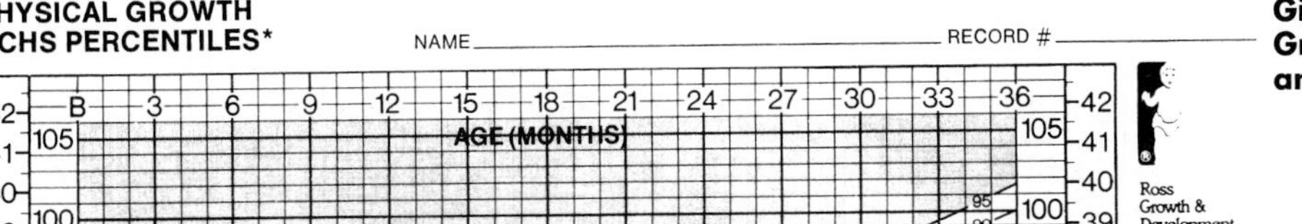

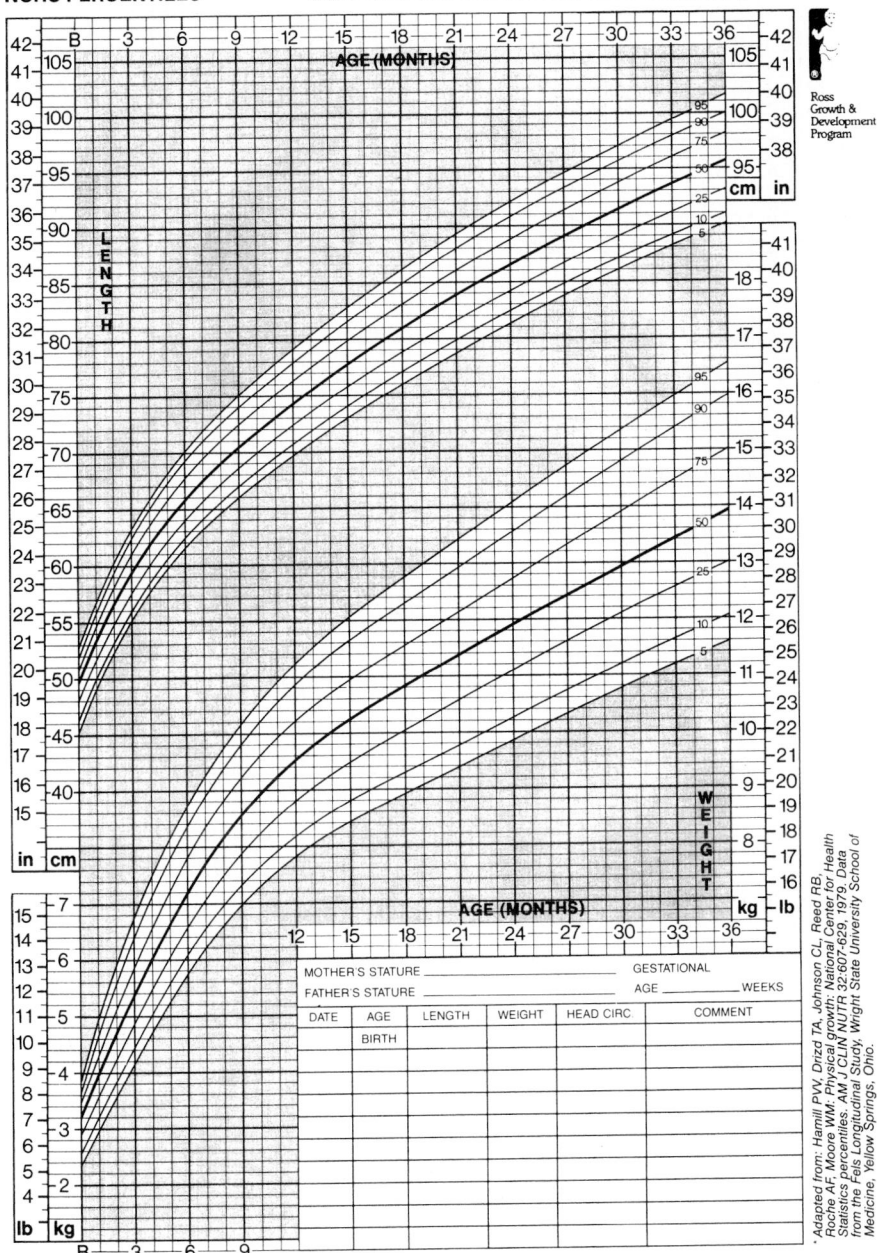

Figure E-1A
Girls: Birth to 36 Months Physical Growth NCHS Percentiles—Length and Weight for Age

*Adapted from: Hamill PVV, Drizd TA, Johnson CL, Reed RB, Roche AF, Moore WM: Physical growth: National Center for Health Statistics percentiles. AM J CLIN NUTR 32:607-629, 1979. Data from the Fels Longitudinal Study, Wright State University School of Medicine, Yellow Springs, Ohio.

To assess length, height, or head circumference, the assessor follows the same procedure, using the appropriate chart. Head circumference percentile should be similar to the child's height and weight percentiles.

With height, weight, and head circumference measures plotted on growth percentile charts, a skilled clinician can begin to interpret the data. Percentile charts divide the measures of a population into 100 equal divisions. Thus half of the population falls above the 50th percentile, and half falls below. The use of percentile measures allows for comparisons among people of the same age and gender. For example, a six-month-old female infant whose weight is at the 75 percentile weighs more than 75 percent of the female infants her age.

Figure E-1B
Girls: Birth to 36 Months Physical Growth NCHS Percentiles—Head Circumference for Age and Weight for Length

NAME ________________ RECORD # ________

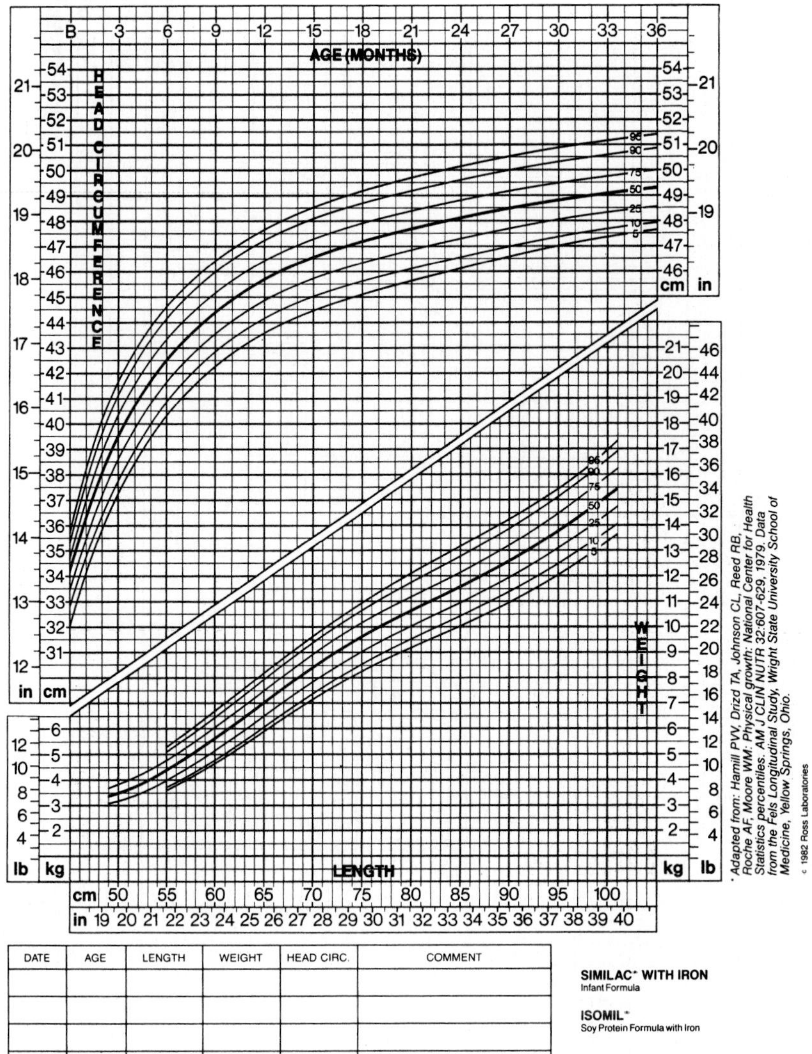

DATE	AGE	LENGTH	WEIGHT	HEAD CIRC.	COMMENT

SIMILAC® WITH IRON
Infant Formula

ISOMIL®
Soy Protein Formula with Iron

Reprinted with permission of Ross Laboratories

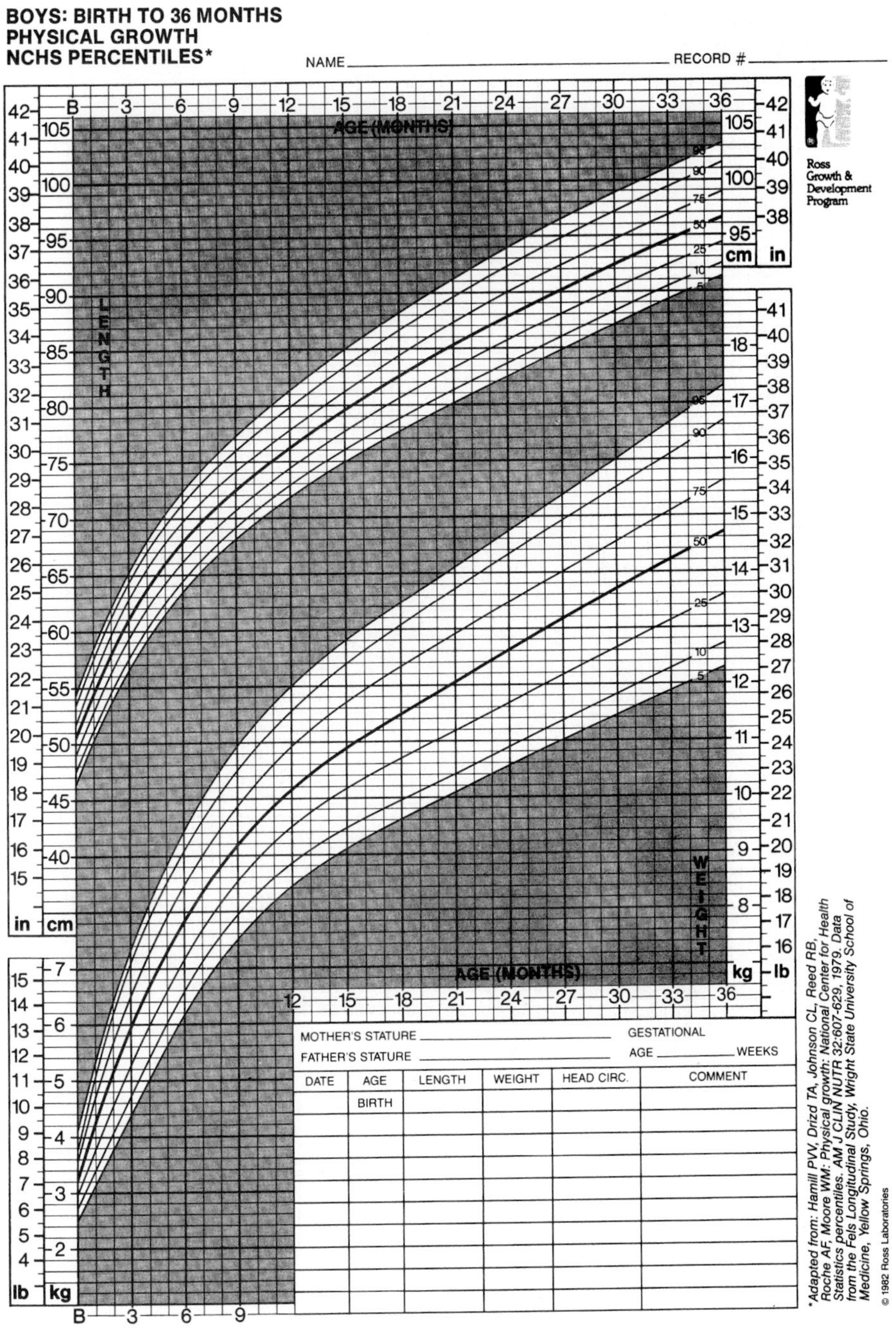

Figure E–2A
Boys: Birth to 36 Months Physical Growth NCHS Percentiles—Length and Weight for Age

BOYS: BIRTH TO 36 MONTHS
PHYSICAL GROWTH
NCHS PERCENTILES*

NAME ______________________ RECORD # ______________

Ross Growth & Development Program

AGE (MONTHS)

LENGTH

WEIGHT

MOTHER'S STATURE ______________ GESTATIONAL
FATHER'S STATURE ______________ AGE ________ WEEKS

DATE	AGE	LENGTH	WEIGHT	HEAD CIRC.	COMMENT
	BIRTH				

*Adapted from: Hamill PVV, Drizd TA, Johnson CL, Reed RB, Roche AF, Moore WM: Physical growth: National Center for Health Statistics percentiles. AM J CLIN NUTR 32:607-629, 1979. Data from the Fels Longitudinal Study, Wright State University School of Medicine, Yellow Springs, Ohio.

Figure E-2B
Boys: Birth to 36 Months Physical Growth NCHS Percentiles—Head Circumference for Age and Weight for Length

NAME ______ RECORD # ______

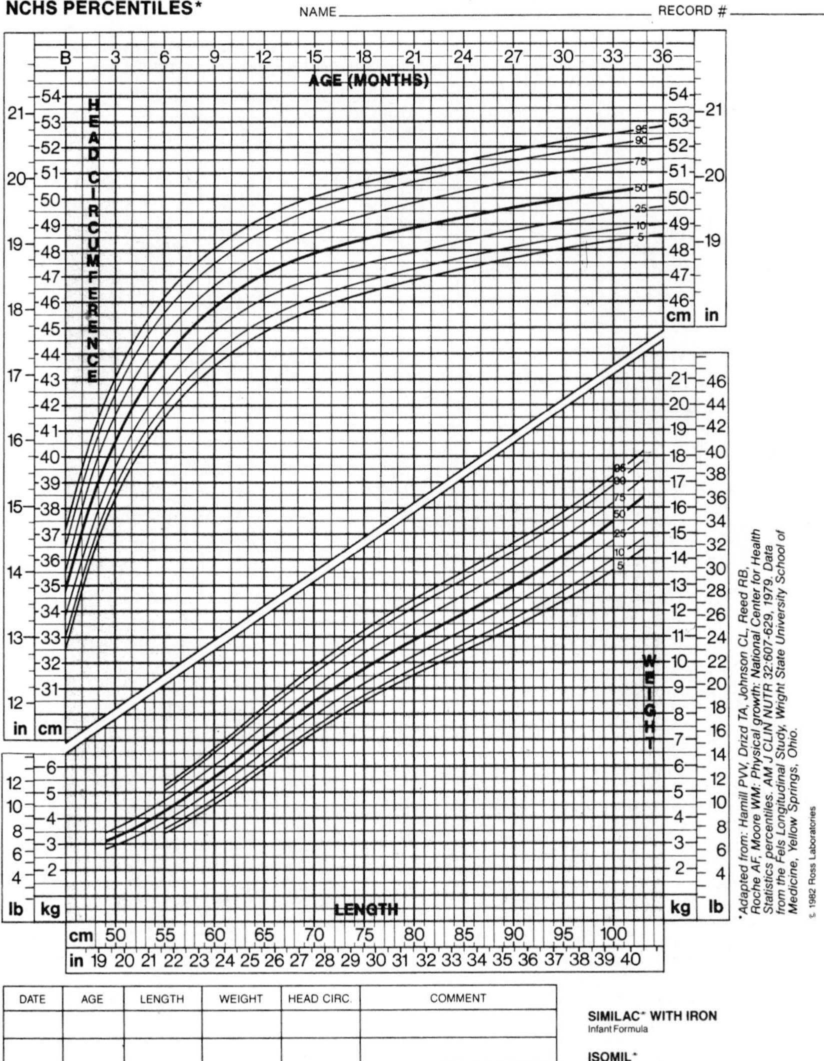

DATE	AGE	LENGTH	WEIGHT	HEAD CIRC.	COMMENT

SIMILAC® WITH IRON
Infant Formula

ISOMIL®
Soy Protein Formula with Iron

Figure E–3
Girls: 2 to 18 Years Physical Growth NCHS Percentiles—Height and Weight for Age

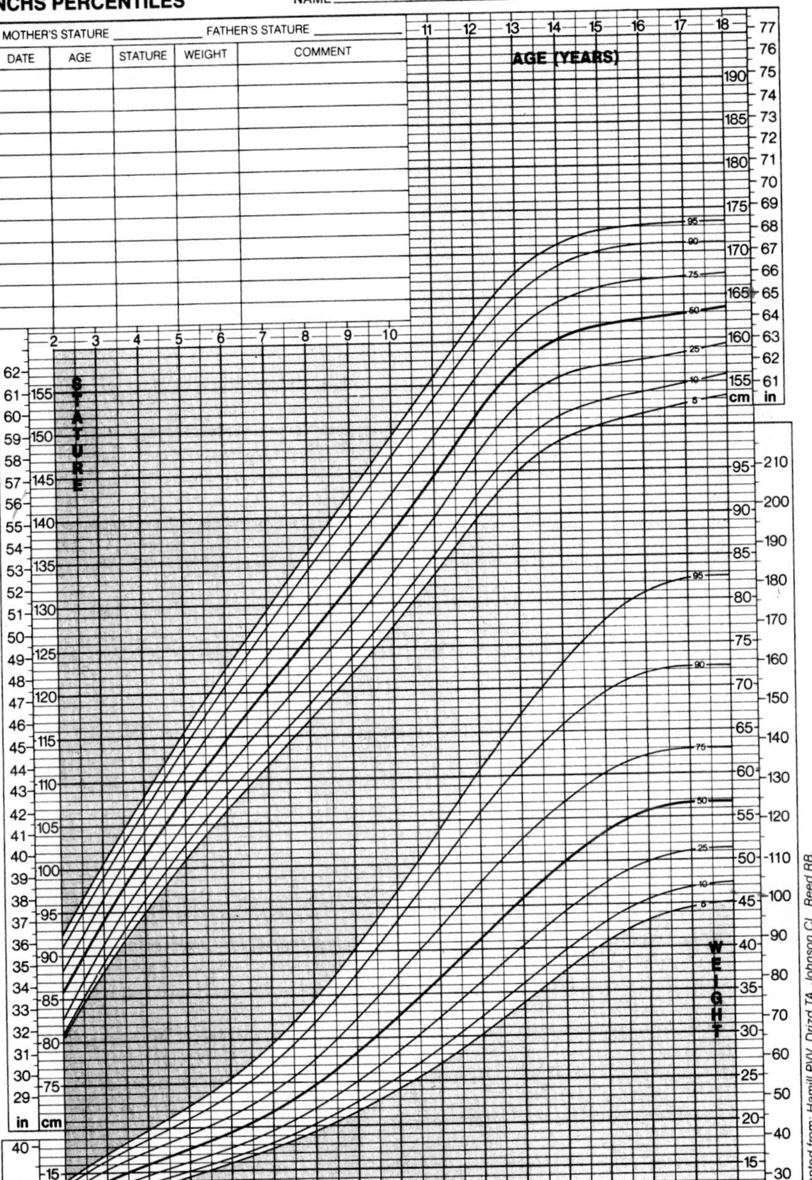

*Adapted from: Hamill PVV, Drizd TA, Johnson CL, Reed RB, Roche AF, Moore WM: Physical growth: National Center for Health Statistics percentiles. AM J CLIN NUTR 32:607-629, 1979. Data from the National Center for Health Statistics (NCHS), Hyattsville, Maryland.

E

E

Figure E–4
Boys: 2 to 18 Years Physical Growth NCHS Percentiles—Height and Weight for Age

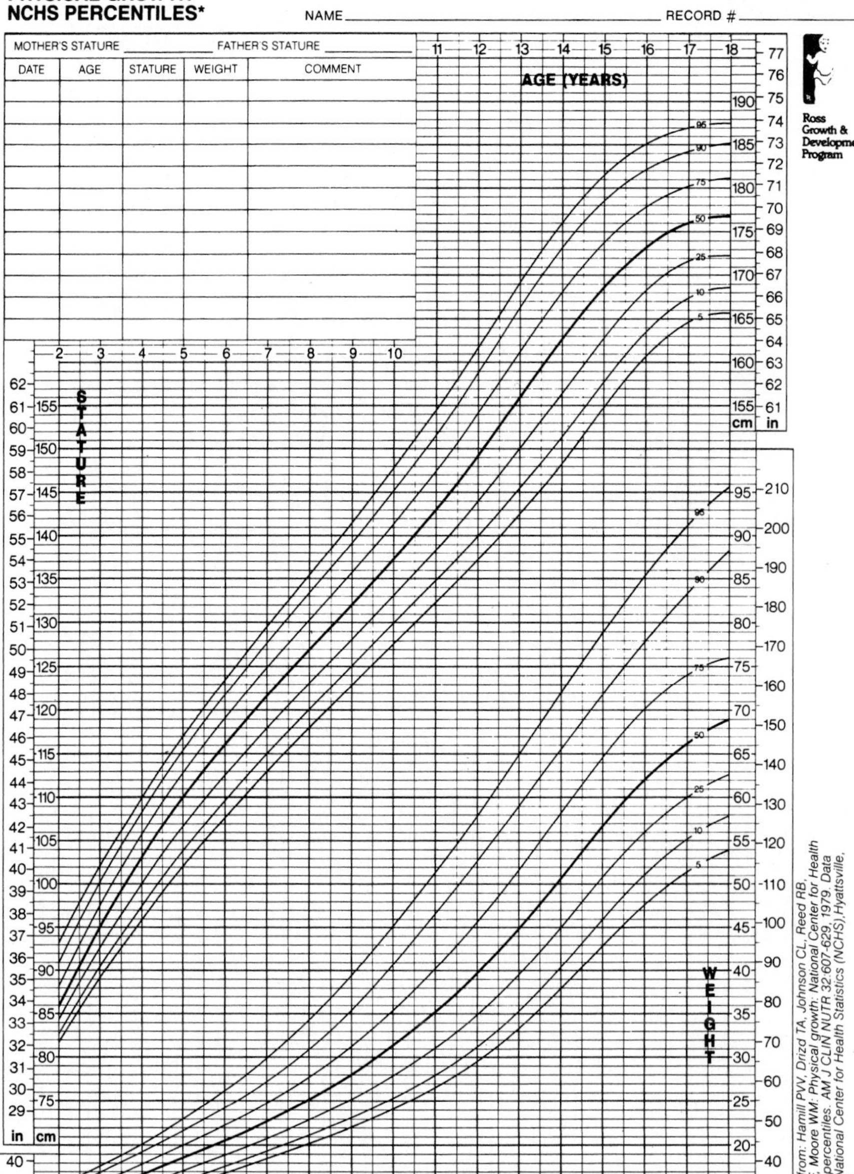

Figure E–5
Girls: Prepubescent Physical Growth NCHS Percentiles—Weight for Height

GIRLS: PREPUBESCENT
PHYSICAL GROWTH
NCHS PERCENTILES*

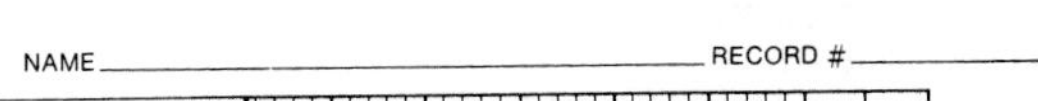

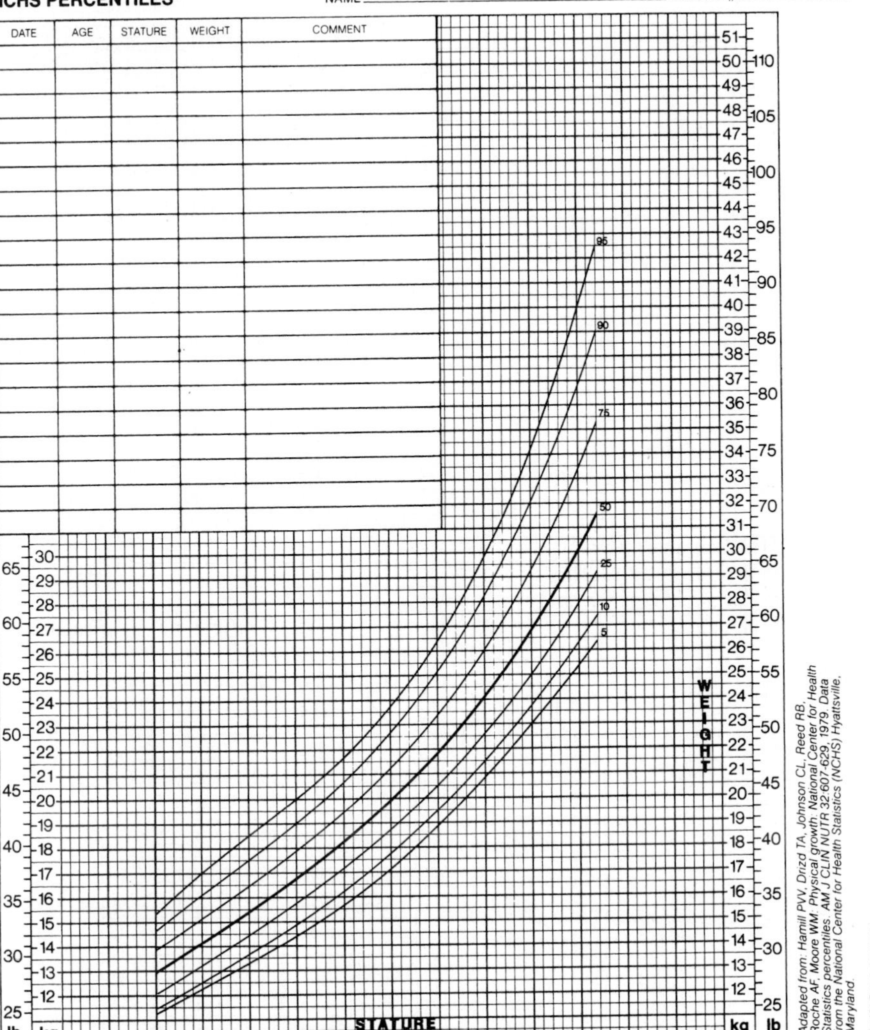

cm 85 90 95 100 105 110 115 120 125 130 135 140 145

in 34 35 36 37 38 39 40 41 42 43 44 45 46 47 48 49 50 51 52 53 54 55 56 57 58

SIMILAC® WITH IRON
Infant Formula

ISOMIL®
Soy Protein Formula with Iron

Reprinted with permission of Ross Laboratories

Figure E–6
Boys: Prepubescent Physical Growth NCHS Percentiles—Weight for Height

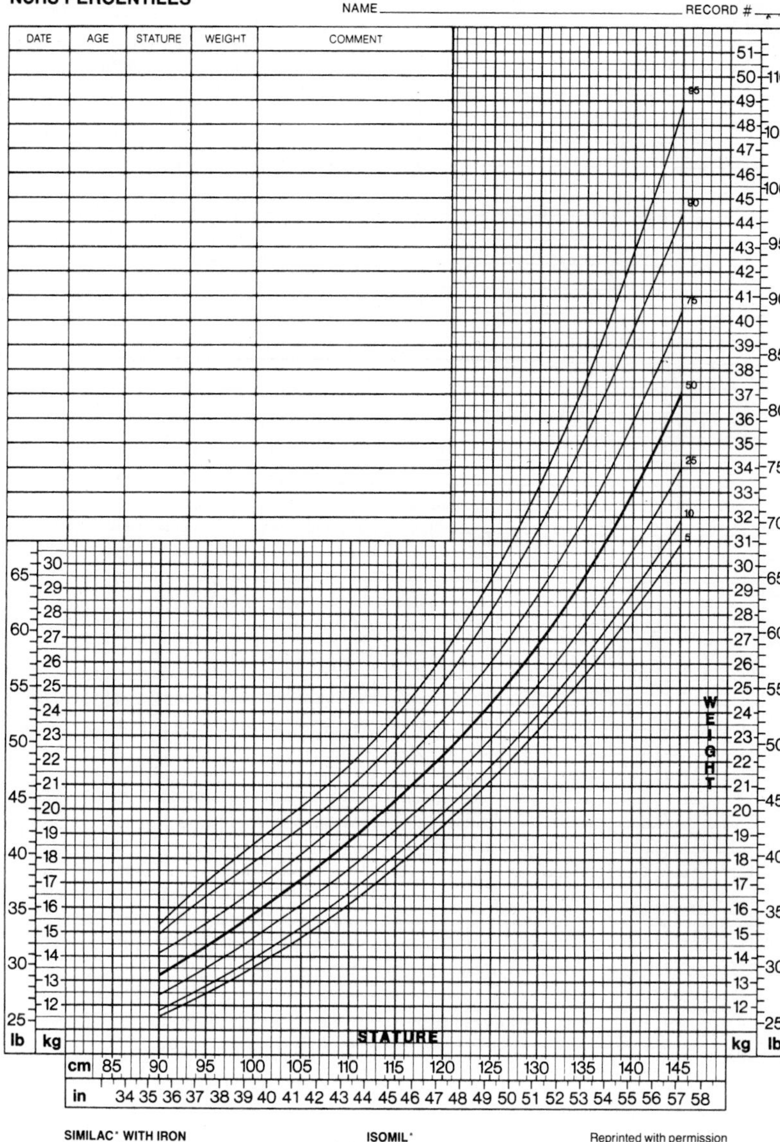

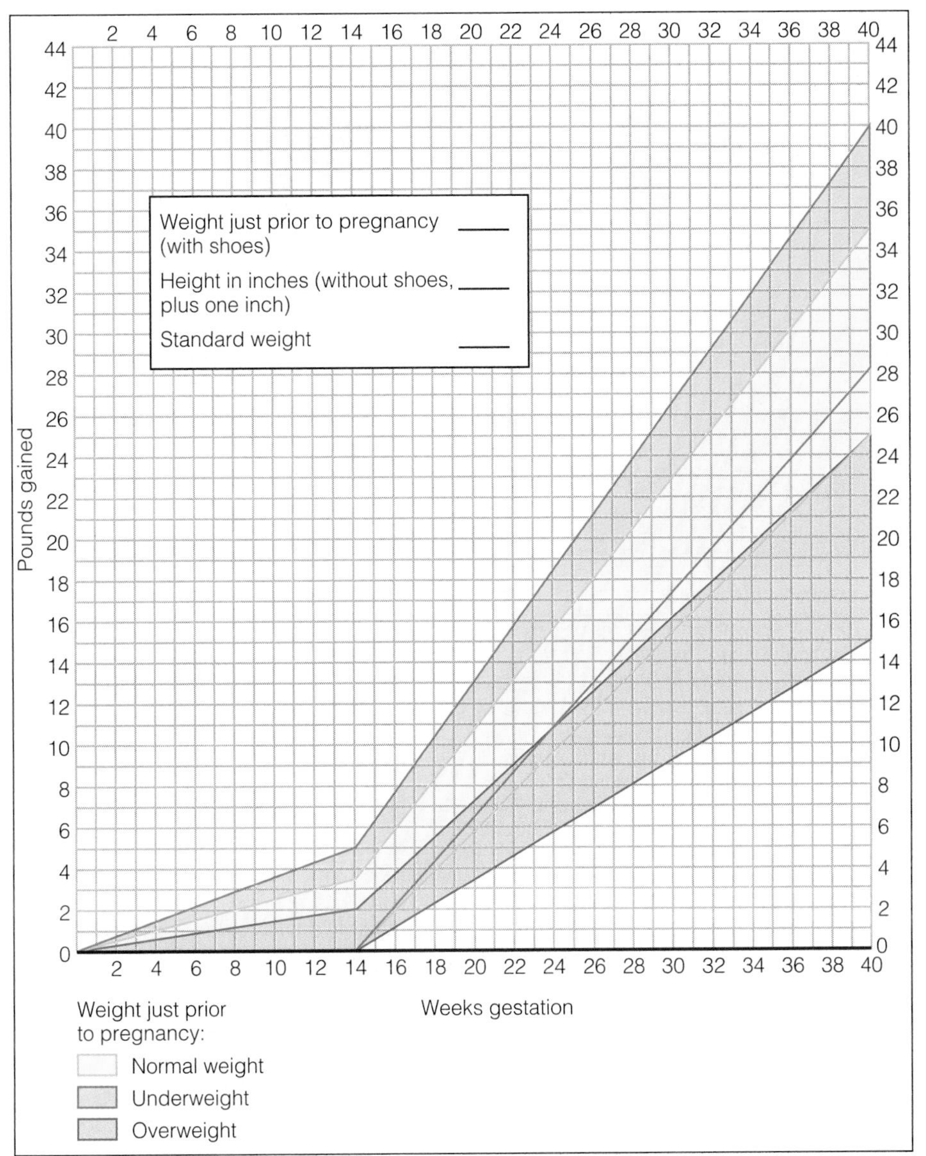

Figure E-7
Prenatal Weight-Gain Grid

Weight Gain During Pregnancy Chapter 12 described desirable weight gain patterns during pregnancy. Figure E-7 shows a prenatal weight-gain grid, used to plot the rate of weight gain during pregnancy. Normal-weight women should gain about 3½ pounds in the first trimester and just under 1 pound per week thereafter, achieving a total gain of 25 to 35 pounds by term; underweight women should gain about 5 pounds in the first trimester and just over 1 pound per week thereafter, achieving a total gain of 28 to 40 pounds by term; and overweight women should gain about 2 pounds in the first trimester and ⅔ pound per week thereafter, achieving a total gain of 15 to 25 pounds.

Figure E-8
How to Measure the Triceps Fatfold and Midarm Circumference

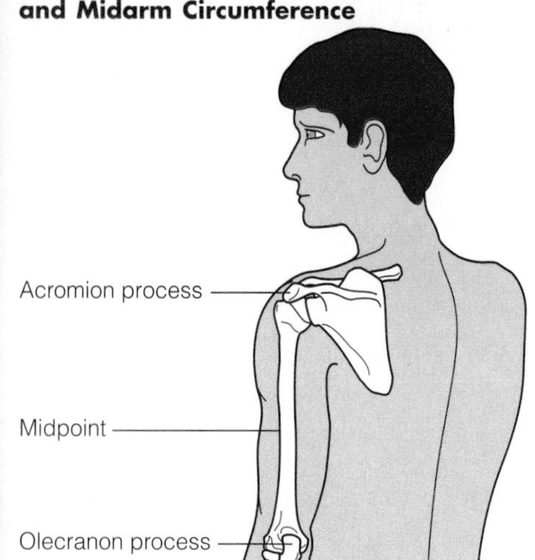

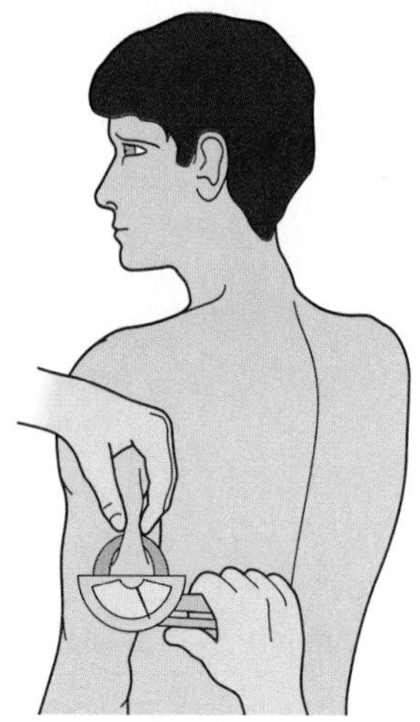

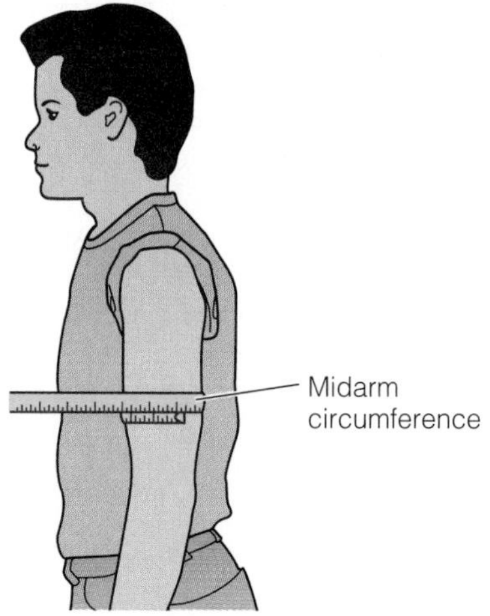

A. Find the midpoint of the arm:

1. Ask the subject to bend his or her arm at the elbow and lay the hand across the stomach. (If he or she is right-handed, measure the left arm, and vice versa.)

2. Feel the shoulder to locate the acromion process. It helps to slide your fingers along the clavicle to find the acromion process. The olecranon process is the tip of the elbow.

3. Place a measuring tape from the acromion process to the tip of the elbow. Divide this measurement by 2, and mark the midpoint of the arm with a pen.

B. Measure the fatfold:

1. Ask the subject to let his or her arm hang loosely to the side.

2. Grasp the fold of skin and subcutaneous fat between the thumb and forefinger slightly above the midpoint mark. Gently pull the skin away from the underlying muscle. (This step takes a lot of practice. If you want to be sure you don't have muscle as well as fat, ask the subject to contract and relax the muscle. You should be able to feel if you are pinching muscle.)

3. Place the calipers over the fatfold at the midpoint mark, and read the measurement to the nearest 1.0 millimeter in two to three seconds. (If using plastic calipers, align pressure lines, and read the measurement to the nearest 1.0 millimeter in two to three seconds.)

4. Repeat steps 2 and 3 twice more. Add the three readings, and then divide by 3 to find the average.

5. Ask the subject to let his or her arm hang loosely to the side. Place the measuring tape horizontally around the arm at the midpoint mark. This measurement is the midarm circumference. Use the following equation to calculate the midarm muscle circumference: Midarm muscle circumference (cm) = midarm circumference (cm) − [0.314[a] × triceps fatfold (mm)].

[a]This factor converts the fat fold measurement to a circumference measurement and millimeters to centimeters.

Arm Anthropometrics Triceps fatfold measurements (see Figure E-8) assist health care professionals in evaluating body composition, specifically body fat. As Chapter 16 explained, the triceps fatfold measurement together with the midarm circumference measurement (Figure E-8) can be used to calculate the midarm muscle circumference, a measure of lean body mass. Table E-2 and Table E-3 on p. E–23 shows percentile standards for the triceps fatfold and midam muscle circumference, respectively.

Table E-2
Triceps Fatfold Percentiles (Millimeters)

Age	Male					Female				
	5TH	25TH	50TH	75TH	95TH	5TH	25TH	50TH	75TH	95TH
1–1.9	6	8	10	12	16	6	8	10	12	16
2–2.9	6	8	10	12	15	6	9	10	12	16
3–3.9	6	8	10	11	15	7	9	11	12	15
4–4.9	6	8	9	11	14	7	8	10	12	16
5–5.9	6	8	9	11	15	6	8	10	12	18
6–6.9	5	7	8	10	16	6	8	10	12	16
7–7.9	5	7	9	12	17	6	9	11	13	18
8–8.9	5	7	8	10	16	6	9	12	15	24
9–9.9	6	7	10	13	18	8	10	13	16	22
10–10.9	6	8	10	14	21	7	10	12	17	27
11–11.9	6	8	11	16	24	7	10	13	18	28
12–12.9	6	8	11	14	28	8	11	14	18	27
13–13.9	5	7	10	14	26	8	12	15	21	30
14–14.9	4	7	9	14	24	9	13	16	21	28
15–15.9	4	6	8	11	24	8	12	17	21	32
16–16.9	4	6	8	12	22	10	15	18	22	31
17–17.9	5	6	8	12	19	10	13	19	24	37
18–18.9	4	6	9	13	24	10	15	18	22	30
19–24.9	4	7	10	15	22	10	14	18	24	34
25–34.9	5	8	12	16	24	10	16	21	27	37
35–44.9	5	8	12	16	23	12	18	23	29	38
45–54.9	6	8	12	15	25	12	20	25	30	40
55–64.9	5	8	11	14	22	12	20	25	31	38
65–74.9	4	8	11	15	22	12	18	24	29	36

Note: If measurements fall between the percentiles shown here, the percentile can be estimated from the information in this table. For example, a measurement of 7 millimeters for a 27-year-old male would be about the 20th percentile.

Source: Adapted from A. R. Frisancho, New norms of upper limb fat and muscle areas for assessment of nutritional status. *American Journal of Clinical Nutrition* 34 (1981): 2540–2545.

Table E-3
Midarm Muscle Circumference Percentiles (Centimeters)

Age	Male					Female				
	5TH	25TH	50TH	75TH	95TH	5TH	25TH	50TH	75TH	95TH
1–1.9	11.0	11.9	12.7	13.5	14.7	10.5	11.7	12.4	13.9	14.3
2–2.9	11.1	12.2	13.0	14.0	15.0	11.1	11.9	12.6	13.3	14.7
3–3.9	11.7	13.1	13.7	14.3	15.3	11.3	12.4	13.2	14.0	15.2
4–4.9	12.3	13.3	14.1	14.8	15.9	11.5	12.8	13.6	14.4	15.7
5–5.9	12.8	14.0	14.7	15.4	16.9	12.5	13.4	14.2	15.1	16.5
6–6.9	13.1	14.2	15.1	16.1	17.7	13.0	13.8	14.5	15.4	17.1
7–7.9	13.7	15.1	16.0	16.8	19.0	12.9	14.2	15.1	16.0	17.6
8–8.9	14.0	15.4	16.2	17.0	18.7	13.8	15.1	16.0	17.1	19.4
9–9.9	15.1	16.1	17.0	18.3	20.2	14.7	15.8	16.7	18.0	19.8
10–10.9	15.6	16.6	18.0	19.1	22.1	14.8	15.9	17.0	18.0	19.7
11–11.9	15.9	17.3	18.3	19.5	23.0	15.0	17.1	18.1	19.6	22.3
12–12.9	16.7	18.2	19.5	21.0	24.1	16.2	18.0	19.1	20.1	22.0
13–13.9	17.2	19.6	21.1	22.6	24.5	16.9	18.3	19.8	21.1	24.0
14–14.9	18.9	21.2	22.3	24.0	26.4	17.4	19.0	20.1	21.6	24.7
15–15.9	19.9	21.8	23.7	25.4	27.2	17.5	18.9	20.2	21.5	24.4
16–16.9	21.3	23.4	24.9	26.9	29.6	17.0	19.0	20.2	21.6	24.9
17–17.9	22.4	24.5	25.8	27.3	31.2	17.5	19.4	20.5	22.1	25.7
18–18.9	22.6	25.2	26.4	28.3	32.4	17.4	19.1	20.2	21.5	24.5
19–24.9	23.8	25.7	27.3	28.9	32.1	17.9	19.5	20.7	22.1	24.9
25–34.9	24.3	26.4	27.9	29.8	32.6	18.3	19.9	21.2	22.8	26.4
35–44.9	24.7	26.9	28.6	30.2	32.7	18.6	20.5	21.8	23.6	27.2
45–54.9	23.9	26.5	28.1	30.0	32.6	18.7	20.6	22.0	23.8	27.4
55–64.9	23.6	26.0	27.8	29.5	32.0	18.7	20.9	22.5	24.4	28.0
65–74.9	22.3	25.1	26.8	28.4	30.6	18.5	20.8	22.5	24.4	27.9

Source: Adapted from A. R. Frisancho, New norms of upper limb fat and muscle areas for assessment of nutritional status, *American Journal of Clinical Nutrition* 34 (1981): 2540–2545.

E

LABORATORY TESTS OF NUTRITION STATUS

♦

As Chapter 16 pointed out, urine and blood tests provide valuable information in a nutrition assessment. Two urine tests—creatinine excretion and urine urea nitrogen (UNN)—require a 24-hour urine collection and, therefore, are not routinely used.

Collecting a 24-hour urine sample presents many problems. Much effort is wasted if everyone involved does not conscientiously follow proper techniques. The collection of urine and recording of food intake data require client cooperation. The client must receive thorough instructions on how to collect and save all urine samples and must be advised to call for help if needed. Each urine sample is added to the collection container and refrigerated until the collection is complete. If even one urine sample is spilled or discarded, the test is invalid.

Urinary Creatinine Excretion Urinary creatinine provides an indirect measure of skeletal muscle mass. Assessors measure creatinine from a 24-hour urine collection and use it to calculate the creatinine-height index (CHI) as follows:

$$\text{CHI} = \frac{\text{measured urinary creatinine (24-hour sample)}}{\text{standard creatinine excretion for height and sex}} \times 100.$$

For example, to calculate the CHI in a man of medium frame, 5 feet 8 inches tall, who excretes 1090 milligrams of creatinine in 24 hours, follow these steps:

- ♦ Look up the standard creatinine excretion (Table E-4).

In this example, the standard creatinine excretion is 1.56 grams or 1560 milligrams.

- ♦ Complete the CHI equation: $\text{CHI} = \frac{1090 \text{ mg}}{1560 \text{ mg}} \times 100 = 70\%.$

Based on the CHI, the man in this example has a skeletal muscle mass 70 percent of that considered typical for a man of his height. Table E-5 shows the standard creatinine excretions for women.

Nitrogen Balance For nitrogen balance studies, food intake must be recorded during the same time period that urine is collected. In addition to saving all urine, the nurses and assistants caring for the client on each shift must record, or ensure that the client records, food intake data carefully. Protein intake is then calculated from the food intake record, and UUN is measured from the urine sample. Nitrogen balance equals the nitrogen intake minus the nitrogen output.

In most cases, protein contains about 16 percent nitrogen. Thus assessors multiply the protein intake by 16 percent or divide by 6.25—the mathematical equivalent—to calculate nitrogen intake. To determine nitrogen output, assessors must consider that in addition to nitrogen lost as UUN, the person also loses nitrogen through the skin, feces, and other sources of urinary nitrogen. To account for these losses, the assessor adds 4 grams of nitrogen to the UUN. Thus nitrogen balance is calculated as follows:

$$\text{Nitrogen balance} = \frac{\text{24-hour protein intake (g)}}{6.25} - [\text{24-hour UUN(g)} + 4\text{g}].$$

For example, if a woman's 24-hour protein intake is 90 grams and her 24-hour UUN is 17 grams, nitrogen balance can be calculated as follows:

$$\text{Nitrogen balance} = \frac{\text{90g protein}}{6.25} - (17 + 4)$$

$$\text{Nitrogen balance} = 14.4 \text{ g} - 21 \text{ g} = -6.6 \text{ g}.$$

The woman in this example is in negative nitrogen balance; that is, she is losing about 6.5 more grams of nitrogen than she is consuming.

E

Table E-4
Creatinine-Height Index Standards for Men

Height		Small Frame			Medium Frame			Large Frame		
		Ideal Weight	Creatinine		Ideal Weight	Creatinine		Ideal Weight	Creatinine	
in	cm	(kg)	(g/24 h)	(mmol/d)	(kg)	(g/24 h)	(mmol/d)	(kg)	(g/24 h)	(mmol/d)
61	154.9	52.7	1.21	10.7	56.1	1.29	11.4	60.7	1.40	12.4
62	157.5	54.1	1.24	11.0	57.7	1.33	11.8	62.0	1.43	12.6
63	160.0	55.4	1.27	11.2	59.1	1.36	12.0	63.6	1.46	12.9
64	162.5	56.8	1.31	11.6	60.4	1.39	12.3	65.2	1.50	13.3
65	165.1	58.4	1.34	11.8	62.0	1.43	12.6	66.8	1.54	13.6
66	167.6	60.2	1.39	12.3	63.9	1.47	13.0	68.9	1.59	14.1
67	170.2	62.0	1.43	12.6	65.9	1.52	13.4	71.1	1.64	14.5
68	172.7	63.9	1.47	13.0	67.7	1.56	13.8	72.9	1.68	14.9
69	175.3	65.9	1.52	13.4	69.5	1.60	14.1	74.8	1.72	15.2
70	177.8	67.7	1.56	13.8	71.6	1.65	14.6	76.8	1.77	15.6
71	180.3	69.5	1.60	14.1	73.6	1.69	14.9	79.1	1.82	16.1
72	182.9	71.4	1.64	14.5	75.7	1.74	15.4	81.1	1.87	16.5
73	185.4	73.4	1.69	14.9	77.7	1.79	15.8	83.4	1.92	17.0
74	187.9	75.2	1.73	15.3	80.0	1.85	16.4	85.7	1.97	17.4
75	190.5	77.0	1.77	15.6	82.3	1.89	16.7	87.7	2.02	17.9

Note: To convert urinary creatinine measures (g/24 h) to standard international units (mmol/d) multiply by 8.840.

Source: A. Grant and S. DeHoog, *Nutritional Assessment and Support*, 3rd ed., 1985.

Table E-5
Creatinine-Height Index Standards for Women

Height		Small Frame			Medium Frame			Large Frame		
		Ideal Weight	Creatinine		Ideal Weight	Creatinine		Ideal Weight	Creatinine	
in	cm	(kg)	(g/24 h)	(mmol/d)	(kg)	(g/24 h)	(mmol/d)	(kg)	(g/24 h)	(mmol/d)
56	142.2	43.2	0.79	7.0	46.1	0.83	7.3	50.7	0.91	8.0
57	144.8	44.3	0.80	7.1	47.3	0.85	7.5	51.8	0.93	8.2
58	147.3	45.4	0.82	7.2	48.6	0.88	7.8	53.2	0.96	8.5
59	149.8	46.8	0.84	7.4	50.0	0.90	8.0	54.5	0.98	8.7
60	152.4	48.2	0.87	7.7	51.4	0.93	8.2	55.9	1.01	8.9
61	154.9	49.5	0.89	7.9	52.7	0.95	8.4	57.3	1.03	9.1
62	157.5	50.9	0.92	8.1	54.3	0.98	8.7	58.9	1.06	9.4
63	160.0	52.3	0.94	8.3	55.9	1.01	8.9	60.6	1.09	9.6
64	162.5	53.9	0.97	8.6	57.9	1.04	9.2	62.5	1.13	10.0
65	165.1	55.7	1.00	8.8	59.8	1.08	9.5	64.3	1.16	10.3
66	167.6	57.5	1.04	9.2	61.6	1.11	9.8	66.1	1.19	10.5
67	170.2	59.3	1.07	9.5	63.4	1.14	10.1	67.9	1.22	10.8
68	172.7	61.4	1.11	9.8	65.2	1.17	10.3	70.0	1.26	11.1
69	175.2	63.2	1.14	10.1	67.0	1.21	10.7	72.0	1.30	11.5
70	177.8	65.0	1.17	10.3	68.9	1.24	11.0	74.1	1.33	11.8

Note: To convert urinary creatinine measures (g/24 h) to standard international units (mmol/d) multiply by 8.840.

Source: A. Grant and S. DeHoog, *Nutritional Assessment and Support*, 3rd ed., 1985.

Vitamin and Mineral Status Table E-6 shows laboratory tests that help assess vitamin and mineral status. Some vitamin and mineral deficiencies, notably folate, vitamin B_{12}, and iron, can lead to anemia. Anemia, defined as a reduced number of red blood cells, can also result from many medical conditions unrelated to nutrition. Table E-7 on p. E–27 shows the tests that help to define anemia and to distinguish among its major nutrition-related causes. Tables E-8 through E-13 provide standards for hemoglobin, hematocrit, serum ferritin, serum iron, percent transferrin saturation, and folate concentrations.

E

Table E–6
Biochemical Tests that may be Used to Assess Vitamin and Mineral Status

Nutrient	Assessment Tests	Nutrient	Assessment Tests
Vitamins		**Minerals[b]**	
Vitamin A	Retinol-binding protein, serum carotene	Iron	Hemoglobin, hematocrit, serum ferritin, total iron-binding capacity (TIBC), protoporphyrin, mean corpuscular volume (MCV), serum iron
Thiamin	Erythrocyte (red blood cell) transketolase activity, urinary thiamin	Iodine	Serum protein-bound iodine, radioiodine uptake
Riboflavin	Erythrocyte glutathione reductase activity, urinary riboflavin	Zinc	Plasma zinc, hair zinc
Vitamin B_6	Urinary xanthurenic acid excretion after tryptophan load test, urinary vitamin B_6, erythrocyte transaminase activity		
Niacin	Urinary metabolites NMN (N-methyl nicotinamide) or 2-pyridone, or preferably both expressed as a ratio		
Folate	Free folate in the blood, erythrocyte folate (reflects liver stores), urinary formiminoglutamic acid (FIGLU), vitamin B_{12} status (folate assessment tests alone do not distinguish between the two deficiencies)		
Vitamin B_{12}	Serum vitamin B_{12}, erythrocyte vitamin B_{12}, urinary methyl-malonic acid synthesis or DUMP test (from the abbreviation for the chemical name of DNA's raw material, deoxyuridine monophosphate), Schilling test		
Biotin	Serum biotin, urinary biotin		
Vitamin C	Serum or plasma vitamin C[a], leukocyte vitamin C, urinary vitamin C		
Vitamin D	Serum alkaline phosphatase		
Vitamin E	Serum tocopherol, erythrocyte hemolysis		
Vitamin K	Blood clotting time (prothrombin time)		

[a]Vitamin C shifts unpredictably between the plasma and the while blood cells known as leukocytes; thus a plasma or serum determination may not accurately reflect the body's pool. The appropriate clinical test may be a measurement of leukocyte vitamin C. A combination of both tests may be more reliable than either one alone.
[b]Serum levels of the major minerals including sodium, potassium, phosphorus, magnesium, and chloride are routinely evaluated in hospitalized clients.
Source: Adapted from A. Grant and S. DeHoog. *Nutritional Assessment and Support,* 3rd ed., 1985.

Table E–7
Laboratory Tests Useful in Evaluating Nutrition-Related Anemias

Test or Test Result	What It Reflects
General Tests for Anemia	
Hemoglobin (Hg)	Total amount of hemoglobin in the red blood cells (RBC)
Hematocrit (Hct)	Percentage of RBC in the total blood volume
Red blood cell (RBC) count	Number of RBC
Mean corpuscular volume (MCV)	RBC size; helps to determine if anemia is microcytic (iron deficiency) or macrocytic (folate or vitamin B_{12} deficiency)
Mean corpuscular hemoglobin concentration (MCHC)	Hemoglobin concentration within the average RBC; helps to determine if anemia is hypochromic (iron deficiency) or normochromic (folate or vitamin B_{12} deficiency)
Bone marrow aspiration	The manufacture of blood cells in different developmental states
Early Stages of Iron Deficiency	
↓ Serum ferritin	Early deficiency state with depleted iron stores
↓ Transferrin saturation	Progressing deficiency state with diminished transport iron
↑ Erythrocyte protoporphyrin	Later deficiency state with limited hemoglobin production
Folate-Deficiency Anemia	
↓ Serum Folate	Progressing deficiency state
↓ RBC Folate	Later deficiency state
Vitamin B_{12}-Deficiency Anemia	
↓ Serum vitamin B_{12}	Progressing deficiency state
Schilling test	Whether or not vitamin B_{12} is being absorbed

Table E–8
Standards for Hemoglobin Test Results

Age (yr)	Sex	Deficient (g/100 ml)	Acceptable (g/100 ml)
<2	M–F	<9.0	10.0 or >
2–5	M–F	<10.0	11.0 or >
6–12	M–F	<10.0	11.5 or >
13–16	M	<12.0	13.0 or >
	F	<10.0	11.5 or >
>16	M	<12.0	14.0 or >
	F	<10.0	12.0 or >
Pregnancy, 2nd trimester	F	<9.5	11.0 or >
Pregnancy, 3rd trimester	F	<9.0	10.5 or >

Note: To convert hemoglobin values (g/100 ml) to international standard units (g/L), multiply by 10.

Table E–9
Standards for Hematocrit Test Results

Age (yr)	Sex	Deficient (%)	Acceptable (%)
<2	M–F	<28	31 or >
2–5	M–F	<30	34 or >
6–12	M–F	<30	36 or >
13–16	M	<37	40 or >
	F	<31	46 or >
>16	M	<37	44 or >
	F	<31	38 or >
Pregnancy, 2nd trimester	F	<30	35 or >
Pregnancy, 3rd trimester	F	<30	33 or >

Note: To convert hematocrit values (%) to standard units, multiply by 0.01.

Table E–10
Standards for Serum Ferritin

Group	Deficient (ng/ml)
Children (3–14 years of age)	<10
Adolescents and adults	<12
Pregnant women	<10

Table E–11
Standards for Serum Iron

Age (yr)	Sex	Deficient (µg/100 ml)	Deficient (µmol/L)	Acceptable (µg/100 ml)	Acceptable (µmol/L)
<2	M–F	<30	<5.3	30 or >	5.3 or >
2–5	M–F	<40	<7.1	40 or >	7.1 or >
6–12	M–F	<50	<8.9	50 or >	8.9 or >
<12	M	<60	<10.7	60 or >	10.7 or >
	F	<40	<7.1	40 or >	7.1 or >

Note: To convert mg/100 ml to international units, multiply by 0.1791.

Table E–12
Standards for Percent Transferrin Saturation

Age (yr)	Sex	Deficient %	Acceptable %
<2	M–F	<15	15 or >
2–12	M–F	<20	20 or >
≥13	M	<20	20 or >
	F	<15	15 or >

Table E–13
Standards for Folate Concentrations

Measurement	Deficient (ng/ml)	Borderline (ng/ml)	Acceptable (ng/ml)
Serum folate	<3.0	3.0–6.0	>6.0
Erythrocyte folate	<140	140–160	>160

Note: To convert folate values (ng/ml) to international standard units (nmol/L), multiply by 2.266.

F AIDS TO CALCULATION

CONTENTS

Many mathematical problems have been worked out as examples at appropriate places in the text. This appendix aims to help with the use of the metric system and with problems not fully explained elsewhere.

Conversion Factors

Conversion factors are useful mathematical tools in everyday calculations, including those encountered in the study of nutrition. A conversion factor is a fraction in which the numerator (top) and the denominator (bottom) express the same quantity in different units. One example used many times throughout this text is that 2.2 pounds (lb) and 1 kilogram (kg) are equivalent; they express the same weight. The conversion factor used to change pounds to kilograms or vice versa is:

$$\frac{2.2 \text{ lb}}{1 \text{ kg}} \text{ or } \frac{1 \text{ kg}}{2.2 \text{ lb}}.$$

Because both factors equal 1, measurements can be multiplied by the factor without changing the value of the measurement. Thus the units can be changed.

To perform a conversion, use the factor with the unit you are seeking in the numerator (top) of the fraction. The following example illustrates the usefulness of conversion factors in the study of nutrition.

1. If you know that a 4-ounce hamburger contains 7 grams of saturated fat and you would like to determine how many grams (g) of saturated fat are contained in a 3-ounce (oz) hamburger, the conversion factor is:
$$\frac{7 \text{ g saturated fat}}{4 \text{ oz hamburger}}.$$
2. Multiply 3 ounces of hamburger by the conversion factor:
$$3 \text{ oz hamburger} \times \frac{7 \text{ g saturated fat}}{4 \text{ oz hamburger}} =$$
$$\frac{3 \times 7}{4} = \frac{21}{4} = \text{5 g saturated fat (rounded off to the nearest whole number).}$$

Percentages

A percentage is a comparison between a number of items (perhaps your intake of energy) and a standard number (perhaps the number of kcalories recommended for your age and sex—the energy RDA). The standard number is the number you divide by. The answer you get after the division must be multiplied by 100 to be stated as a percentage (*percent* means "per 100").

For example, if you wish to determine what percentage of the RDA your energy intake is, first find your energy RDA (see Appendix C). We'll use 2200 kcalories to demonstrate.

1. Total your energy intake for a day—for example, 1500 kcalories.
2. Divide your kcalorie intake by the RDA kcalories:
$$1500 \text{ kcal (your intake)} \div 2200 \text{ kcal (RDA)} = 0.68.$$
3. Multiply your answer by 100 to state it as a percentage:
$$0.68 \times 100 = 68 = 68\%.$$

In some problems in nutrition, the percentage may be more than 100. For example, suppose your daily intake of vitamin A is 3200 RE and your RDA (male) is 1000 RE. The following calculations show your vitamin A intake as a percentage of the RDA:

$$3200 \div 1000 = 3.2.$$

$$3.2 \times 100 = 320\% \text{ of RDA.}$$

Sometimes the comparison is between a part of a whole (for example, your kcalories from protein) and the total amount (your total kcalories). For example, suppose you wish to know what percentage of your total kcalories for the day come from protein, fat, and carbohydrate.

1. Find the total grams of protein, fat, and carbohydrate you consumed—for example, 60 grams protein, 80 grams fat, and 310 grams carbohydrate.
2. Multiply the number of grams by the number of kcalories from 1 gram of each energy nutrient (conversion factors):

$$60 \text{ g protein} \times \frac{4 \text{ kcal}}{1 \text{ g protein}} = 240 \text{ kcal.}$$

$$80 \text{ g fat} \times \frac{9 \text{ kcal}}{1 \text{ g fat}} = 720 \text{ kcal.}$$

$$310 \text{ g carbohydrate} \times \frac{4 \text{ kcal}}{1 \text{ g carbohydrate}} = 1240 \text{ kcal.}$$

$$240 + 720 + 1240 = 2200 \text{ kcal.}$$

3. Find the percentage of total kcalories from each energy nutrient (see Example 3):

- Protein: 240 ÷ 2200 = 0.109 × 100 = 10.9 = 11% of kcal.
- Fat: 720 ÷ 2200 = 0.327 × 100 = 32.7 = 33% of kcal.
- Carbohydrate: 1240 ÷ 2200 = 0.563 × 100 = 56.3 = 56% of kcal.
- 11% + 33% + 56% = 100% of kcal (total).

The percentages total 100 percent, but sometimes they total 99 or 101 because of rounding off. This is a reasonable error.

Ratios

A ratio is a comparison of two or three values in which one of the values is reduced to 1. A ratio compares identical units and so is expressed without units. For example, to find the potassium-to-sodium ratio of your diet, you first need to know how many milligrams of potassium and sodium you consumed, say, 3000 milligrams potassium and 2500 milligrams sodium.

1. Divide the potassium milligrams by the sodium milligrams:

   3000 mg potassium ÷ 2500 mg sodium = 1.2.

2. The potassium-to-sodium ratio is usually expressed as correct to one decimal point: 1.2.

The potassium-to-sodium ratio of your diet is 1.2:1 (read as "one point two to one" or simply "one point two"). A ratio greater than 1 means that the first value (in this case, milligrams of potassium) is greater than the second (sodium). When the second value is larger, the ratio is less than 1.

Weights and Measures

Length
1 inch (in) = 2.54 centimeters.
1 foot (ft) = 30.48 centimeters.
1 meter (m) = 39.37 inches.

To find degrees Fahrenheit (t_F) when you know degrees Celsius (t_C)*, multiply by 9/5 and then add 32:

$$9/5\ t_C + 32 = t_F.$$

To find degrees Celsius (t_C) when you know degrees Fahrenheit (t_F), multiply by 5/9 after subtracting 32:

$$5/9\ (t_F - 32) = t_C.$$

Volume
1 liter (L) = 1.06 quarts (qt) or 0.85 imperial quart.
1 liter = 1000 milliliters (ml).
1 milliliter = 0.03 fluid ounces.
1 gallon = 3.79 liters.
1 quart = 0.95 liter or 32 fluid ounces.
1 cup (c) = 8 fluid ounces.
1 tablespoon (tbs) = 15 milliliters.
3 teaspoons (tsp) = 1 tablespoon.
1 teaspoon = about 5 grams or 5 milliliters.
16 tablespoons = 1 cup.
4 cups = 1 quart.

Weight
1 ounce (oz) = approximately 28 grams (g).
16 ounces = 1 pound (lb).
1 pound = 454 grams.
1 kilogram (kg) = 1000 grams or 2.2 pounds.
1 gram = 1000 milligrams (mg).
1 milligram = 1000 micrograms (μg).

Energy units
1 kcalorie (kcal) = 4.2 kilojoules (kJ).
1 millijoule (mJ) = 240 kcalories.
1 kilojoule = 0.24 kcalories.
1 g carbohydrate = 4 kcal = 17 kJ.
1 g fat = 9 kcal = 37 kJ.
1 g protein = 4 kcal = 17 kJ.
1 g alcohol = 7 kcal = 29 kJ.

*Also known as *centigrade*.

G

ENTERAL FORMULAS

♦

The staggering number of enteral formulas available allows health care professionals to meet a variety of their client's medical needs, but also complicates the process of selecting an appropriate formula. The first step in narrowing the choice of formulas is to determine the client's ability to digest and absorb nutrients. Table G–1 lists examples of intact protein formulas for clients with the ability to digest and absorb nutrients and Table G–2 provides examples of hydrolyzed formulas for clients with limited ability to digest and absorb nutrients. Each formula is listed only once, although the formula may have more than one use. A high-protein formula, for example, may also be a fiber-containing formula. Tables G-3 through G-5 list modular formulas. Although this appendix provides many examples, the list of formulas is not complete. The information reflects the manufacturers literature and does not suggest endorsement by the authors. Be aware that formula composition changes periodically. Consult manufacturer's literature for updates. The following products are listed in this appendix:

- B. Braun-McGaw, Inc.[a]
  - Hepatic-Aid® II
  - Immun-Aid®
- Mead Johnson Nutritionals[b]
  - Casec®
  - Choice dm®
  - Comply®
  - Criticare HN®
  - Deliver® 2.0
  - Isocal®
  - Isocal® HN
  - Kindercal®
  - Lipisorb®
  - MCT Oil®
  - Microlipid®
  - Moducal®
  - Magnacal® Renal
  - Protain XL®
  - Respalor®
  - Sustacal®
  - Sustacal® Basic
  - Sustacal® with Fiber
  - Sustacal® Plus
  - TraumaCal®
  - Ultracal®
- Nestlé Clinical Nutrition[c]
  - Crucial®
  - Entrition® HN
  - Glytrol®
  - Nutren ®1.0
  - Nutren® 1.0 with Fiber
  - Nutren® 1.5
  - Nutren® 2.0
  - Nutren® Junior
  - Nutren® Junior with Fiber
  - NutriHep®
  - NutriVent®
  - Peptamen®
  - Peptamen® Junior
  - Peptamen® VHP
  - ProBalance®
  - Reabilan®
  - Reabilan® HN
  - Replete®
  - Replete® with Fiber
- Novartis Nutrition Corporation[d]
  - Compleat® Modified
  - Compleat® Regular
  - Compleat® Pediatric
  - DiabetiSource®
  - FiberSource®
  - FiberSource® HN
  - Impact®
  - Impact® 1.5
  - Impact® with Fiber
  - IsoSource® Standard
  - IsoSource® HN
  - IsoSource® VHN
  - IsoSource® 1.5
  - Resource® Diabetic
  - Resource® Standard
  - Resource® Plus
  - SandoSource® Peptide
  - Tolerex®
  - Vivonex® Pediatric
  - Vivonex® Plus
  - Vivonex® T.E.N.
- Nutrition Medical[e]
  - Fiberlan®
  - Gluco-PRO®
  - Isolan®
  - L-Emental®
  - L-Emental® Hepatic
  - L-Emental® Pediatric
  - L-Emental® Plus
  - Nitrolan®
  - PRO-Peptide®
  - PRO-Peptide® for Kids
  - PRO-Peptide® VHN
  - Ultralan®
- Ross Laboratories[f]
  - Advera®
  - Alitraq®
  - Ensure®
  - Ensure® Plus
  - Ensure® Plus HN
  - Ensure® with Fiber
  - Glucerna®
  - Jevity®
  - Jevity® Plus
  - Nepro®
  - Osmolite®
  - Osmolite® HN
  - Osmolite® HN Plus
  - PediaSure®
  - PediaSure® with Fiber
  - Perative®
  - Polycose®
  - ProMod®
  - Promote®
  - Promote® with Fiber
  - Pulmocare®
  - Suplena®
  - TwoCal® HN
  - Vital® HN

---

[a] *Enteral Nutrition Products Ready Reference* (provided June, 1997), *Hepatic-Aid II Instant Drink* (1992), and *Immune-Aid* (1992), McGaw Inc., Irvine, CA 92714.

[b] *Enteral Product Handbook* (1994), *Choice* (1995), *Comply* (1996), *Magnacal Renal* (1996), and *Protain XL* (1996), Mead-Johnson Nutritionals, Evansville, IN 47721.

[c] *Enteral Product Reference Guide* (1997), Nestlé Clinical Nutrition, Deerfield, IL 60015.

[d] *Enteral Products Guide* (1995) and *Compleat Pediatric* (1997), Sandoz Nutrition, Minneapolis, MN 55416.

[e] *Enteral Formulary, Gluco-PRO* (1997), *L-Emental* (1994), *L-Emental Hepatic* (1996), *L-Emental Pediatric* (1996), *L-Emental Plus* (1997), *PRO-peptide* (1997), *PRO-Peptide for Kids* (1997), and *PRO-Peptide VHN* (1997), Nutrition Medical, Inc., Minneapolis, MN 55442.

[f] *Ross Medical Nutritional System* (1996), Ross Laboratories, Columbus, OH 43215.

Table G–1
Intact Protein Formulas

Product	Form	Volume to Meet 100% RDI (ml)	Energy (kcal/ml)	Protein or Amino Acids (g/L)	Carbohydrate (g/L)	Fat (g/L)	Osmolality (mOsm/kg)	Notes
Lactose-Free, Isotonic or Near-Isotonic Formulas								
Compleat® Modified	liquid	1500	1.07	43	140	37	300	4.4 g fiber/L
Isocal®	liquid	1890	1.06	34	135	44	270	Low-residue, 20% fat from MCT
Isolan®	liquid	1250	1.06	40	144	36	300	Low-residue
IsoSource Standard®	liquid	1500	1.20	43	170	41	360	Low-residue, 50% fat from MCT
Nutren® 1.0	liquid	1500	1.00	40	127	38	300	Low-residue, 24% fat from MCT
Osmolite®	liquid	1887	1.06	37	151	35	300	Low-residue, 20% fat from MCT
Low- to Moderate-Residue Formulas[a]								
Compleat® Regular	liquid	1500	1.07	43	130	43	450	Blenderized formula, contains lactose, 4.4 g fiber/1000 ml
Ensure®	liquid	948	1.06	37	145	37	555	Lactose-free
Resource® Standard	liquid	1890	1.10	37	140	37	430	Lactose-free
Sustacal Basic®	liquid	1080	1.01	61	140	23	650	Lactose-free
Lactose-Free, Fiber-Containing Formulas								
Ensure® with Fiber	liquid	1391	1.10	40	162	37	480	14 g fiber/L
Fiberlan®	liquid	1250	1.20	50	160	40	310	Near-isotonic 14 g fiber/L
FiberSource®	liquid	1500	1.20	43	170	41	390	10 g fiber/L
Impact® with Fiber	liquid	1500	1.00	56	140	28	375	10 g fiber/L, enriched with arginine, nucleic acids, and omega-3 fatty acids
Jevity®	liquid	1321	1.06	44	154	35	300	Isotonic, 14 g fiber/L
Nutren® 1.0 with Fiber	liquid	1500	1.00	40	127	38	303	Near-isotonic, 14 g fiber/L
ProBalance®	liquid	1000	1.20	45	130	34	350	10 g fiber/L
Promote® with Fiber	liquid	1000	1.00	63	139	28	370	14.4 g fiber/L
Replete® with Fiber	liquid	1000	1.00	63	113	34	300	Isotonic, 14 g fiber/L
Sustacal® with Fiber	liquid	1500	1.06	46	139	35	480	11 g fiber/L
Ultracal®	liquid	1250	1.06	44	123	45	310	Near isotonic, 14.4 g fiber/L

[a]These formulas are frequently used as oral supplements. All formulas listed under "Intact Formulas: Isotonic or Near-Isotonic Formulas" can be used as a low- to moderate-residue formula.

Table G–1
Intact Protein Formulas (continued)

Product	Form	Volume to Meet 100% RDI (ml)	Energy (kcal/ml)	Protein or Amino Acids (g/L)	Carbohydrate (g/L)	Fat (g/L)	Osmolality (mOsm/kg)	Notes
Lactose-Free, High-kCalorie Formulas								
Comply®	liquid	1250	1.50	60	180	61	60	
Ensure® Plus	liquid	1420	1.50	55	200	53	690	
Ensure® Plus HN	liquid	947	1.50	63	200	50	650	
Deliver® 2	liquid	1000	2.00	75	200	102	640	
IsoSource® 1.5	liquid	933	1.50	68	170	65	650	8.0 g fiber/L
Nutren® 1.5	liquid	1000	1.50	40	113	45	430	50% fat from MCT
Nutren® 2.0	liquid	750	2.00	40	98	53	720	75% fat from MCT
Osmolite® HN Plus	liquid	1000	1.20	56	158	39	360	
Resource Plus®	liquid	1400	1.50	55	200	53	600	
Sustacal® Plus	liquid	1180	1.52	61	190	57	670	
TwoCal HN®	liquid	947	2.00	84	217	91	690	20% fat from MCT
Ultralan®	liquid	1000	1.50	60	202	50	540	50% fat from MCT
Lactose-Free, High-Protein Formulas								
Entrition® HN	liquid	1300	1.00	44	114	41	300	Isotonic, low-residue
FiberSource® HN	liquid	1500	1.20	53	160	41	390	7 g fiber/L
Isocal® HN	liquid	1250	1.06	44	123	46	270	Near-isotonic, low residue, 20% fat from MCT
IsoSource® HN	liquid	1500	1.20	53	160	41	330	Near-isotonic, low residue
IsoSource® VHN	liquid	1250	1.00	62	130	29	300	Isotonic, 10g fiber/L
Jevity® Plus	liquid	1000	1.20	56	175	39	450	12.0 g fiber/L
Libisorb®	liquid	1180	1.35	57	161	57	630	85% fat from MCT
Nitrolan®	liquid	1250	1.24	60	160	40	310	Near-isotonic, 50% fat from MCT
Osmolite® HN	liquid	1321	1.06	44	144	35	300	Lactose-free, low-residue, 20% fat from MCT
Promote®	liquid	1000	1.00	63	130	26	340	Fat primarily from polyunsaturated and monounsaturated sources, 20% fat from MCT
Sustacal®	liquid	1080	1.01	61	139	23	650	
Special Use Formulas: Pediatric (1 to 10 years)								
Compleat® Pediatric	liquid	900	1.00	38	126	39	N/A	Blenderized formula, 4.4 g fiber/L
Kindercal®	liquid	946	1.06	32	127	42	N/A	
Nutren Junior®	liquid	1000	1.00	30	128	42	350	
Nutren Junior® with Fiber	liquid	1000	1.00	30	128	42	350	6.0 g fiber/L
PediaSure®	liquid	1000	1.00	30	110	50	345	
PediaSure® with Fiber	liquid	1000	1.00	30	114	50	345	5.0 g fiber/L

Table G–1
Intact Protein Formulas (continued)

Product	Form	Volume to Meet 100% RDI (ml)	Energy (kcal/ ml)	Protein or Amino Acids (g/L)	Carbohydrate (g/L)	Fat (g/L)	Osmolality (mOsm/ kg)	Notes
Special-Use Formulas: Glucose Intolerance								
Choice dm®	liquid	948	1.06	45	106	51	440	13.6 g fiber/L, 25% kcalories from monounsaturated fat
DiabetiSource®	liquid	1500	1.00	50	90	49	360	4.4 g fiber/L
Glucerna®	liquid	1422	1.00	42	96	54	355	14.4 g fiber/L, high in monounsaturated fat
Gluco-PRO®	liquid	1000	1.06	45	106	51	300	Isotonic, 14.4 g fiber/L
Glytrol®	liquid	1400	1.00	45	100	48	380	15 g fiber/L
Resource® Diabetic	liquid	1890	1.06	63	99	47	450	12 g fiber/L
Special Use Formulas: Immune System Support								
Immun-Aid®	powder	2000	1.00	80	120	22	460	Enriched with arginine, glutamine, branched-chain amino acids, nucleic acids, omega-3 fatty acids, vitamins A, C, E, and B_6 and trace elements
Impact®	liquid	1500	1.00	56	130	28	375	Enriched with arginine, nucleic acids, and omega-3 fatty acids
Impact® 1.5	liquid	1250	1.50	80	140	69	550	Same as above
Impact® with Fiber (see Fiber-Containing Formulas)								
Special Use Formulas: Renal Insufficiency								
Magnacal® Renal	liquid	N/A	2.00	75	200	101	570	High in monounsaturated fat, 20% fat from MCT; intended for use once hemodialysis has been instituted
Nepro®	liquid	947	2.00	70	215	96	635	Lactose-free, high-calcium, low phosphorus; intended for use once dialysis has been instituted
Suplena®	liquid	947	2.00	30	255	96	600	Lactose-free, low in electrolytes; intended for use before dialysis is instituted

G

Table G–1
Intact Protein Formulas (continued)

Product	Form	Volume to Meet 100% RDI (ml)	Energy (kcal/ ml)	Protein or Amino Acids (g/L)	Carbohydrate (g/L)	Fat (g/L)	Osmolality (mOsm/ kg)	Notes
Special Use Formulas: Respiratory Insufficiency								
NutriVent®	liquid	1000	1.50	68	100	94	450	Lactose-free, 55% kcal from fat, 40% fat from MCT
Pulmocare®	liquid	947	1.50	63	106	93	475	Lactose-free, 55% kcal from fat, 20% fat from MCT, enriched with antioxidant nutrients
Respalor®	liquid	1420	1.52	76	148	71	580	Lactose-free; 41% kcal from fat; 30% fat from MCT; enriched with vitamins C and E, B-complex vitamins, zinc and trace elements
Special Use Formulas: Wound Healing								
Protain XL®	liquid	1200	1.00	57	129	30	340	14 g fiber/L, 20% fat from MCT, enriched with vitamins A and C and zinc
Replete®	liquid	1000	1.00	63	113	34	325	Enriched with vitamins A and C and zinc
TraumaCal®	liquid	2000	1.50	83	145	69	560	Enriched with vitamins C, B-complex,

Table G–2
Hydrolzed Protein Formulas

Product	Form	Volume to Meet 100% RDI (ml)	Energy (kcal/ ml)	Protein or Amino Acids (g/L)	Carbohydrate (g/L)	Fat (g/L)	Osmolality (mOsm/ kg)	Notes
Special Use Hydrolyzed Formulas: Heptatic Insufficiency								
Hepatic Aid® II	powder	—	1.20	44	169	36	560	Free amino acids, high in branched-chain amino acids, low in aromatic amino acids, no added vitamins or electrolytes.
L-Emental® Hepatic	powder	N/A	1.20	44	169	36	560	Free amino acids, high in branched-chain amino acids, low in aromatic amino acids, contains vitamins and minimal electrolytes
NutriHep®	liquid	1000	1.50	40	290	21	690	Free amino acids, high in branched-chain amino acids, low in aromatic amino acids, contains vitamins and minimal electrolytes
Special Use Hydrolyzed Formulas: HIV Infection or AIDS								
Advera®	liquid	1184	1.28	60	216	23	680	78% hydrolyzed and 22% intact protein, low fat, fiber added,enriched with vitamins E, C, B_6, B_{12}, and folate
Special Use Hydrolyzed Formulas: Immune System Support								
Alitraq®	powder	1500	1.00	53	165	16	575	47% free amino acids, 42% small peptides, enriched with glutamine and arginine
Crucial®	liquid	1000	1.50	63	90	45	490	Enriched with arginine
Perative®	liquid	1155	1.30	67	177	37	385	Enriched with arginine and β-carotene
Vivonex® Plus	powder	1800	1.00	45	190	7	650	100% free amino acids, enriched with and E and copper and zinc

Table G–2
Hydrolzed Protein Formulas (continued)

Product	Form	Volume to Meet 100% RDI (ml)	Energy (kcal/ml)	Protein or Amino Acids (g/L)	Carbohydrate (g/L)	Fat (g/L)	Osmolality (mOsm/kg)	Notes
Special Use Hydrolyzed Formulas: Malabsorption								
Criticare HN®	liquid	1890	1.06	38	220	5	650	50% free amino acids, 50% small peptides
L-Emental®	powder	N/A	1.00	38	205	3	630	100% free amino acids, enriched with glutamine
L-Emental™ Plus	powder	1800	1.00	45	190	7	650	100% free amino acids, enriched with glutamine
Peptamen®	liquid	1500	1.00	40	127	39	325	Near-isotonic
Peptamen VHP®	liquid	1500	1.00	63	105	39	365	70% fat from MCT
PRO-Peptide®	liquid	1500	1.00	40	127	39	270	Isotonic, 70% fat from MCT, enriched with glutamine
PRO-Peptide™ VHN	liquid	1500	1.00	62	104	39	300	Isotonic, 70% fat from MCT, enriched with glutamine
Reabilan®	liquid	2000	1.00	32	132	53	350	50% fat from MCT
Reabilan® HN	liquid	1500	1.33	44	119	41	490	50% fat from MCT
SandoSource® Peptide	liquid	1750	1.00	50	160	17	490	60% free amino acids and small peptides
Tolerex®	powder	3160	1.00	21	230	1.5	550	100% free amino acids
Vital® HN	powder	1500	1.00	42	185	11	500	87% hydrolyzed proteins, 13% essential amino acids
Vivonex® T.E.N.	powder	2000	1.00	38	210	3.0	630	100% free amino acids, enriched with branched-chain amino acids
Special Use Hydrolyzed Formulas: Pediatric (1 to 10 years)								
L-Emental® Pediatric	powder	1000* 1170**	0.8	24	130	24	360	100% free amino acids, contains glutamine
Peptamin Junior®	liquid	1000	1.0	30	138	39	310	60% fat from MCT
PRO-Peptide® for Kids	liquid	1000**	1.0	30	138	39	360	40% fat from MCT, enriched with trace elements, taurine, and carnitine
Vivonex® Pediatric	powder	1000* 1170**	0.8	24	130	24	360	100% free amino acids

*Children 1 to 6 years old
**Children 7 to 10 years old

Table G–3
Protein Modules

Product	Form	Major Protein Source	Energy (kcal/g)	Protein (g/100 g)
Casec®	powder	Calcium caseinate	3.7	88
Pro Mod®	powder	Whey protein	4.2	75.8

Table G–4
Carbohydrate Modules

Product	Form	Major Carbohydrate Source	Energy (kcal/ml or g)
Moducal®	powder	Hydrolyzed corn starch	3.8 kcal/g
Polycose Liquid®	liquid	Hydrolyzed corn starch	2.0 kcal/ml
Polycose Powder®	powder	Hydrolyzed corn starch	3.8 kcal/g

Table G–5
Fat Modules

Product	Form	Major Fat Source	Energy (kcal/ml)	Fat (g/100 ml)
MCT Oil®	liquid	Derived from coconut oil	7.7	87
Microlipid®	liquid	Safflower oil	4.5	50

APPENDIX H

Answers to Self Check Questions

Chapter 1

1. d 5. d 8. b
2. b 6. a 9. b
3. a 7. c 10. a
4. b

Chapter 2

1. b 5. a 8. a
2. b 6. b 9. d
3. c 7. d 10. c
4. a

Chapter 3

1. d 5. a 8. a
2. b 6. b 9. d
3. d 7. a 10. c
4. d

Chapter 4

1. b 5. b 8. b
2. b 6. d 9. b
3. d 7. b 10. b
4. a

Chapter 5

1. c 5. a 8. a
2. a 6. d 9. a
3. c 7. b 10. c
4a. 3, b. 4, c. 2, d. 5, e. 1

Chapter 6

1. a 5. a 8. a
2. a 6. b 9. a
3. a 7. c 10. d
4. a

Chapter 7

1. d 5. d 8. c
2. d 6. d 9. c
3. c 7. b 10. a
4. a

Chapter 8

1. d 5. c 8. d
2. b 6. a 9. d
3. a 7. d 10. b
4. d

Chapter 9

1. b 5. d 8. d
2. d 6. c 9. d
3. d 7. d 10. d
4. a

Chapter 10

1. b 5. d 8. b
2. a 6. d 9. a
3. b 7. c 10. d
4. c

Chapter 11

1. d 5. d 8. a
2. d 6. b 9. a
3. d 7. d 10. d
4. d

Chapter 12

1. a 5. b 8. d
2. d 6. b 9. c
3. c 7. c 10. c
4. b

Chapter 13

1. a 5. a 8. c
2. c 6. d 9. b
3. d 7. c 10. b
4. b

Chapter 14

1. d 5. d 8. b
2. d 6. d 9. a
3. a 7. c 10. b
4. c

Chapter 15

1. b 7. c 13. a
2. a 8. c 14. a
3. d 9. a 15. d
4. b 10. b 16. b
5. c 11. d 17. c
6. d 12. a

Chapter 16

1. c 5. b 8. c
2. c 6. b 9. b
3. d 7. c 10. a
4. a

Chapter 17

1. b 5. b
2. c 6. d
3. a 7. c
4. a

Chapter 18

1. c 5. b 8. d
2. c 6. d 9. a
3. a 7. a 10. b
4. b 11. c

Chapter 19

1. b 5. d 8. a
2. d 6. a 9. a
3. a 7. c 10. b
4. c

Chapter 20

1. b 7. b 13. d
2. a 8. a 14. c
3. c 9. b 15. b
4. c 10. d 16. a
5. a 11. c 17. c
6. d 12. b

Chapter 21

1. c 5. a 8. b
2. a 6. d 9. a
3. b 7. b
4. b

Chapter 22

1. a 5. d 9. b
2. c 6. b 10. a
3. b 7. a 11. c
4. d 8. c 12. d

Chapter 23

1. b 5. c 8. c
2. d 6. b 9. b
3. a 7. c 10. a
4. c

Chapter 24

1. b 6. a 10. b
2. c 7. d 11. b
3. b 8. c 12. c
4. a 9. c 13. d
5. d

Chapter 25

1. c 7. b 12. b
2. a 8. d 13. a
3. d 9. d 14. b
4. a 10. b 15. c
5. b 11. c 16. a
6. b

Chapter 26

1. a 5. c 9. a
2. c 6. b 10. b
3. d 7. c 11. c
4. b 8. b

Chapter 27

1. b 6. c 11. b
2. c 7. a 12. d
3. b 8. b 13. d
4. d 9. d 14. b
5. a 10. c 15. a

Chapter 28

1. c 5. a 9. c
2. a 6. c 10. d
3. c 7. b 11. b
4. d 8. d

Glossary

abscess: an accumulation of pus caused by local infection, that contains live microorganisms and immune system cells; treatment involves draining the abscess.

Acceptable Daily Intake (ADI): the amount of a sweetener that individuals can safely consume each day over the course of a lifetime without adverse effect. It includes a 100-fold safety factor.

acesulfame potassium: a low-kcalorie sweetener approved by the FDA and Health Canada; also known as acesulfame-K because K is the chemical symbol for potassium.

acid-base balance: the balance maintained between acid and base concentrations in the blood and body fluids.

acidosis: too much acid in the blood and body fluids.

acids: compounds that release hydrogen ions in a solution.

acquired immune deficiency syndrome (AIDS): the end stage of HIV infection, in which severe complications are manifested.

active vitamin D: the 1,25-dihydroxy form of vitamin D that promotes calcium balance and bone mineralization.

acupuncture (AK-you-PUNK-cher): a technique that involves piercing the skin with long thin needles at specific anatomical points to relieve pain or illness. Acupuncture sometimes uses heat, pressure, friction, suction, or electromagnetic energy to stimulate the points.

Adequate Intake (AI): a value used as a guide for nutrient intake when an RDA cannot be established.

additives: substances that are not normally consumed as foods by themselves, but are added to foods.

ADH (antidiuretic hormone): a hormone released by the pituitary gland in response to high salt concentrations in the blood. The kidneys respond by reabsorbing water.

adipose tissue: the body's fat, which consists of masses of fat-storing cells called *adipose cells.*

adolescence: the period of growth from the beginning of puberty until full maturity. Timing of adolescence varies from person to person.

advance directive: the means by which competent adults record their preferences for future medical interventions. The living will and durable power of attorney are types of advance directives.

aerobic (air-ROE-bic): energy-producing processes involving the immediate use of oxygen.

AI: see *Adequate Intake.*

aldosterone (al-DOS-ter-own): a hormone secreted by the adrenal glands that stimulates the reabsorption of sodium by the kidneys; aldosterone also regulates chloride and potassium concentrations.

alitame: a compound of two amino acids (alanine and aspartic acid) that is 2000 times sweeter than sucrose; FDA approval pending.

alkalosis: too much base in the blood and body fluids.

alpha-lactalbumin (lackt-AL-byoo-min): the chief protein in human breast milk, as **casein** (CAY-seen) is the chief protein in cow's milk.

alternative therapies: approaches to medical diagnosis and treatment that are not fully accepted by the established medical community; as such, they are not widely taught at U.S. medical schools or practiced in U.S. hospitals; also called adjunctive, unconventional, or unorthodox therapies.

Alzheimer's (ALTZ-high-mers) **disease:** a degenerative disease of the brain involving memory loss and major structural changes in the brain's nerve cells.

amino (a-MEEN-oh) **acids:** building blocks of protein; each has an amino group and an acid group attached to a central carbon, which also carries a distinctive side chain

amniotic (am-nee-OTT-ic) **sac:** the "bag of waters" in the uterus, in which the fetus floats.

amylase (AM-uh-lace): an enzyme that splits amylose (a form of starch). Amylase is a carbohydrase. The ending -ase indicates an enzyme; the root tells what it digests. Other examples: protease, lipase.

anabolism (an-ABB-o-lism): reactions in which small molecules are put together to build larger ones. Anabolic reactions consume energy.

anaerobic: energy-producing processes that do not involve the immediate use of oxygen.

angina: a painful feeling of tightness or pressure, felt in the area in and around the heart, often radiating to the back, neck, and arms, caused by a lack of oxygen to an area of heart muscle.

anorexia nervosa: an eating disorder characterized by a refusal to maintain a minimally normal body weight, self-starvation to the extreme, and a disturbed perception of body weight and shape, seen (usually) in teenage girls and young women.

antacids: acid-buffering agents used to counter excess acidity in the stomach.

anthropometric (an-throw-poe-MEH-trick): relating to measurement of the physical characteristics of the body, such as height and weight.

antibodies: large proteins of the blood and body fluids, produced in response to invasion of the body by unfamiliar molecules (mostly proteins); antibodies inactivate the invaders and so protect the body. The invaders are called **antigens.**

antidiuretic hormone: see *ADH.*

antimicrobial agents: substances used as food additives that prevent growth of illness-causing microorganisms in foods.

antioxidant: a compound that protects other compounds from oxygen by itself reacting with oxygen.

antisecretory agents: drugs that suppress or inhibit gastric acid secretion; also called **anti-GERD (GastroEsophageal Reflux Disease).**

anus (AY-nus): the terminal sphincter muscle of the GI tract.

appendix: a narrow blind sac extending from the beginning of the colon; the appendix stores lymphocytes.

appetite: the psychological desire to eat; a learned motivation that is experienced as a pleasant sensation that accompanies the sight, smell, or thought of appealing foods.

aroma therapy: a technique that uses oil extracts from plants and flowers (usually applied by massage or baths) to enhance physical, psychological, and spiritual health.

artery: a vessel that carries blood away from the heart.

arthritis: inflammation of a joint, usually accompanied by pain, swelling, and structural changes.

artificial colors: certified food colors, added to enhance appearance (*certified* means approved by the FDA).

artificial feeding: parenteral and enteral nutrition; feeding by a route other than the normal ingestion of food.

artificial flavors, flavor enhancers: chemicals that mimic natural flavors and those that enhance flavor.

artificial sweeteners: noncarbohydrate, nonkcaloric synthetic sweetening agents; sometimes called **nonnutritive sweeteners.**

ascites (ah-SIGH-teez): a type of edema characterized by the accumulation of fluid in the abdominal cavity.

ascorbic acid: one of the two active forms of vitamin C.

aspartame: a compound of two amino acids (phenylalanine and aspartic acid) that tastes like the sugar sucrose but is much sweeter. It provides about 4 kcalories per gram, as does protein, but because so little is used, it is virtually kcalorie-free. In powdered form it is sometimes mixed with lactose, however, so a 1-gram packet may contain 4 kcalories. It is used in both the United States and Canada.

aspiration: the flux of food or liquid into the lungs.

atherosclerosis: (ath-er-oh-scler-OH-sis): a type of artery disease characterized by accumulations of lipid-containing material on the inner walls of the arteries.

atrophic gastritis: a condition characterized by chronic inflammation of the stomach accompanied by a diminished size and functioning of the mucosa and glands.

atrophy (AT-ro-fee): of muscles, becoming smaller; a decrease in size because of disuse, undernutrition, or wasting diseases.

avian influenza: a newly discovered virus transmitted from birds to humans in Hong Kong; also known as "bird flu" or influenza A H5N1.

ayurveda (EYE-your-VAY-dah): a traditional Hindu system of improving health by using herbs, diet, meditation, massage, and yoga to stimulate the body to make its own natural drugs.

basal metabolism: the energy needed to maintain life when a person is at complete rest after a 12-hour fast. Basal metabolism, sometimes called *basal metabolic rate (BMR),* is normally the largest part of a person's daily energy expenditure.

bases: compounds that accept hydrogen ions in a solution.

bee pollen: a product consisting of bee saliva, plant nectar, and pollen that confers no benefit on athletes and may cause an allergic reaction in individuals sensitive to it.

behavior modification: the changing of behavior by the manipulation of *antecedents* (cues or environmental factors that trigger behavior), the behavior itself, and *consequences* (the penalties or rewards attached to behavior).

beriberi: the thiamin-deficiency disease; characterized by loss of sensation in the hands and feet, muscular weakness, advancing paralysis, and abnormal heart action.

beta-carotene: a vitamin A precursor made by plants and stored in human fat tissue; an orange pigment.

bicarbonate: an alkaline secretion of the pancreas; part of the pancreatic juice. (Bicarbonate also occurs widely in all cell fluids.)

bifidus (BIFF-id-us, by-FEED-us) **factors:** factors in colostrum and breast milk that favor the growth of the "friendly" bacteria *Lactobacillus* (lack-toe-ba-SILL-us) *bifidus* in the infant's intestinal tract; these bacteria prevent other, less desirable intestinal inhabitants from flourishing.

bile: a compound made by the liver from cholesterol, stored in the gallbladder, and released into the small intestine when needed. Bile is an emulsifier that prepares fat for digestion.

bilirubin: a pigment in the bile whose concentration in the blood may rise as a result of some disorders.

bioaccumulation: the accumulation of toxins in living tissues at concentrations that increase at higher levels of the food chain.

bioelectromagnetic medical applications: the use of electrical energy, magnetic energy, or both to stimulate bone repair, wound healing, and tissue regeneration.

biofeedback: the use of special devices to convey information about heart rate, blood pressure, skin temperature, muscle relaxation, and the like to enable a person to learn how to consciously control these medically important functions.

biofield therapeutics: a manual healing method that directs a healing force from an outside source (commonly God or another supernatural being) through the practitioner and into the client's body; commonly known as "laying on of hands."

biotechnology: the use of biological systems or organisms to create or modify products; also called **biogenetic engineering.**

bland diet: a diet that eliminates foods known to stimulate gastric acid secretion, irritate the gastric mucosa, or cause indigestion.

blind loop syndrome: the problems of fat malabsorption and vitamin B_{12} and folate deficiencies that result from the overgrowth of bacteria in a bypassed segment of the intestine.

blood doping: the process of injecting red blood cells to enhance the blood's oxygen-carrying ability. Risks include dangerous blood clotting, especially in athletes who become dehydrated; infections from nonsterile equipment; transfusion reactions; and dangers of improperly transferred blood. Blood doping is banned in Olympic competitions.

BMR: see *basal metabolism.*

body composition: the proportions of muscle, bone, fat, and other tissue that make up a person's total body weight.

body mass index (BMI): an index of a person's weight in relation to height, determined by dividing the weight (in kilograms) by the square of the height (in meters).

bolus (BOH-lus): the portion of food swallowed at one time.

bone marrow transplant: the replacement of diseased bone marrow in a recipient with healthy bone marrow from a donor; used as a treatment for breast cancer, leukemia, lymphomas, and certain blood disorders.

botulism: an often-fatal food poisoning caused by botulin toxin, a toxin produced by bacteria that grow without oxygen.

branched-chain amino acids: the amino acids leucine, isoleucine, and valine, which are present in large amounts in skeletal muscle tissue; packaged and promoted for athletes as ergogenic aids.

bronchitis (bron-KYE-tis): inflammation of the lungs' air passages.

brown sugar: refined white sugar crystals to which manufacturers have added molasses syrup with natural flavor and color; 91 to 96% pure sucrose.

buffers: compounds that can reversibly combine with hydrogen ions to help keep a solution's acidity or alkalinity constant.

bulimia (byoo-LEEM-ee-uh) **nervosa:** recurring episodes of binge eating combined with a morbid fear of becoming fat, usually followed by self-induced vomiting or purging.

caffeine: a stimulant that in small amounts may produce alertness and reduced reaction time in some people, but that also creates fluid losses. Overdoses cause headaches, trembling, an abnormally fast heart rate, and other undesirable effects.

calcium pangamate: a compound once thought to enhance aerobic metabolism, now known to have no such effect.

calcium rigor: hardness or stiffness of the muscles caused by high blood calcium.

calcium tetany: intermittent spasms of the extremities due to nervous and muscular excitability, which is caused by low blood calcium.

cancer: a disease that arises from the unchecked growth of malignant tumors.

cancer cachexia (ka-KEKS-ee-ah) **syndrome:** a syndrome that frequently accompanies many types of cancer; characterized by anorexia, inadequate intake of food, malnutrition, accelerated metabolism and wasting, and general ill health.

capillary (CAP-ill-ary): a small vessel that branches from an artery. Capillaries connect arteries to veins. Oxygen, nutrients, and waste materials are exchanged across capillary walls.

capitation: prepayment of a set fee per client in exchange for medical services.

carbohydrates: energy nutrients composed of monosaccharides.

cardiac output: the volume of blood discharged by the heart each minute. The amount of oxygenated blood the heart ejects toward the tissues at each beat is called the **stroke volume.**

cardiac sphincter (CARD-ee-ack SFINK-ter): the sphincter muscle at the junction between the esophagus and the stomach.

cardiorespiratory conditioning: improvements in heart and lung function and increased blood volume, brought about by aerobic training.

cardiorespiratory endurance: the ability to perform large-muscle, dynamic exercise of moderate-to-high intensity for prolonged periods.

carnitine: a nonprotein amino acid found in most body cells that serves as a carrier of unoxidized fatty acids. It is advertised as a "fat burner" to bodybuilders and as a means of sparing glycogen to endurance athletes, but studies show it does neither.

carrier: an individual who possesses one dominant and one recessive gene for a gentetic trait, such as an inborn error of metabolism. When a trait is recessive, a carrier may show no signs of the trait but can pass it on.

cartilage therapy: the use of cleaned and powdered connective tissue, such as collagen, to improve health.

catabolism (ca-TAB-o-lism): reactions in which large molecules are broken down to smaller ones. Catabolic reactions usually release energy.

cataracts: thickenings of the eye lenses that impair vision and can lead to blindness.

cathartic (ca-THART-ic): a strong laxative.

CD4+ lymphocyte: a type of circulating white blood cell that has the CD4+ protein on its surface and is a necessary component of the immune system.

celiac (SEE-lee-ack) **disease:** a sensitivity to a part of the protein gluten that causes flattening of the intestinal villi and generalized malabsorption; also called **gluten-sensitive enteropathy** (EN-ter-OP-ah-thee) or **celiac sprue.**

cell salts: a mineral preparation supposedly prepared from living cells.

central obesity: excess fat on the abdomen and around the trunk of the body.

central total parenteral nutrition (TPN): a method for meeting all nutrient needs by infusing formula either directly into a large-diameter central vein or threading a catheter through a peripheral vein and into a central vein.

central veins: the large-diameter veins located close to the heart.

cerebral cortex: the outer surface of the cerebrum.

cesarean section: surgical childbirth, in which the infant is delivered through an incision in the woman's abdomen.

chelation therapy: the use of ethylene diamine tetraacetic acid (EDTA) to bind with metallic ions, thus healing the body by removing toxic metals.

chemotherapy: the use of drugs to arrest or destroy cancer cells. Drugs used for chemotherapy are called **chemotherapeutic** or **antineoplastic agents.**

chiropractic (KYE-roe-PRAK-tik): a manual healing method of manipulating vertebrae to relieve musculoskeletal pain suspected of causing problems with internal organs.

choline: a nonessential nutrient that can be made in the body from an amino acid.

chronic obstructive pulmonary disease (COPD): one of several disorders, including emphysema and bronchitis, that interfere with respiration.

chronic or **congestive heart failure (CHF):** a syndrome in which the heart can no longer pump blood through the circulatory system.

chronological age: a person's age in years from his or her date of birth.

chylomicrons (kye-lo-MY-crons): the lipoproteins that transport lipids from the intestinal cells into the body.

chyme (KIME): the semiliquid mass of partly digested food expelled by the stomach into the duodenum (the top portion of the small intestine).

cirrhosis (sih-ROW-sis): an advanced form of liver disease in which scar tissue replaces liver cells that have permanently lost their function. The type of cirrhosis associated with alcohol abuse is called **Laennec's** (LAY-eh-necks) **cirrhosis. Postnecrotic cirrhosis** develops as a complication of chronic hepatitis. **Biliary cirrhosis** can develop from a prolonged inflammation or obstruction of the bile ducts. **Cardiac cirrhosis** is associated with congestive heart failure. In still other cases, cirrhosis may be **idiopathic,** having no identifiable cause.

clear liquids: foods that are liquid and transparent at body temperature.

clinical pathways, critical pathways, or **care maps:** charts or tables that outline a plan of care for a specific diagnosis, procedure, or treatment, with a goal of providing the best outcome at the lowest cost. The plan, developed by the health care team after a careful study of each facility's unique client population, is regularly reassessed and improved.

clinically severe obesity: a BMI of 40 or greater or 100 lbs or more overweight for an average adult. A less preferred term used to describe the same condition is *morbid obesity.*

CoA (coh-AY): a nickname for a small molecule (coenzyme A) that participates in metabolism. As pyruvate breaks down to the smaller compound acetic acid, a molecule of CoA is attached to it, making acetyl CoA (ASS-uh-teel or uh-SEET-ul co-AY).

coenzyme (co-EN-zime): a small molecule that works with an enzyme to promote the enzyme's activity. Many coenzymes have B vitamins as part of their structure.

coenzyme Q10: a lipid found in cells (mitochondria) shown to improve exercise performance in heart disease patients, but not effective in improving performance of healthy athletes.

cofactor: a mineral element that, like a coenzyme, works with an enzyme to facilitate a chemical reaction.

collagen: the characteristic protein of connective tissue.

collaterals: small blood vessels that develop to divert blood flow away from an obstructed organ; also called **shunts.**

colon or **large intestine:** the last portion of the intestine, which absorbs water. Its main segments are the ascending colon, the transverse colon, the descending colon, and the sigmoid colon.

colostrum (co-LAHS-trum): a milklike secretion from the breast that is rich in protective factors; it is present during the first day or so after delivery, before milk appears.

comatose: in a state of deep unconsciousness from which the person cannot be aroused.

competent: having sufficient mental ability to understand a treatment, weigh its risks and benefits, and comprehend the consequences of refusing or accepting the treatment.

complementary proteins: two or more proteins whose amino acid assortments complement each other in such a way that the essential amino acids missing from one are supplied by the other.

complementary therapy: therapy that incorporates both traditional medical and alternative treatments.

complete formulas: enteral formulas designed to supply all needed nutrients when provided in sufficient volume.

complete protein: a protein containing all the amino acids essential in human nutrition in amounts adequate for human use.

complex carbohydrates: long chains of sugars arranged as starch or fiber; also called **polysaccharides.**

concentrated fruit juice sweetener: a concentrated sugar syrup made from dehydrated, deflavored fruit juice, commonly grape juice; used to sweeten products that can then claim to be "all fruit."

conditionally essential amino acid: an amino acid that is normally nonessential but must be supplied by the diet in special circumstances when the need for it becomes greater than the body's ability to produce it.

conditioning: the physical effect of training; improved flexibility, strength, and endurance.

confectioners' sugar: finely powdered sucrose; 99.9% pure.

continuous renal replacement therapy (CRRT): a method for removing wastes from the blood that uses a semipermeable membrane to slowly remove all blood constituents that will cross the membrane; selected blood components are then returned to the client intravenously.

cool-down: five to ten minutes of light activity following a vigorous workout to gradually cool the body's core to near-normal temperature.

corn sweeteners: corn syrup and sugars derived from corn.

corn syrup: a syrup produced by the action of enzymes on cornstarch, containing mostly glucose. See also *high-fructose corn syrup (HFCS).*

cornea (KOR-nee-uh): the hard, transparent membrane covering the outside of the eye.

coronary heart disease (CHD): damage to the heart muscle that results from obstruction of blood flow to the heart muscle.

counterregulatory hormones: hormones such as glucagon, cortisol, and catecholamines that oppose insulin's actions and promote catabolism.

creatine: a nitrogen-containing compound that combines with phosphate to form the high-energy compound creatine phosphate (or phosphocreatine) in muscles.

cretinism (CREE-tin-ism): an iodine-deficiency disease characterized by mental and physical retardation.

Crohn's disease: inflammation and ulceration along the length of the GI tract, often with granulomas.

cyclamate: a 0-kcalorie sweetener; FDA approval pending in the United States; available in Canada on grocery store shelves but only as a tabletop sweetener, not as an additive.

cyclic parenteral nutrition: the continuous administration of TPN solutions for 8 to 12 hours with time periods when no nutrients are infused.

cystic fibrosis (CF): a hereditary disorder characterized by the production of thick mucus that affects many organs, including the lungs, pancreas, liver, heart, gallbladder, and small intestine.

cystinuria (SIS-tih-NEW-ree-ah): the presence of cystine in the urine; a symptom of an inherited metabolic disorder in which large amounts of the amino acids cystine, lysine, arginine, and ornithine are excreted in the urine.

cytokines (SIGH-toe-kynes): immune system factors that help regulate inflammatory responses. Some cause anorexia, fever, and discomfort.

Daily Food Guide: a food group plan for ensuring dietary adequacy that offers five categories of foods to choose from. (It treats vegetables and fruits as two separate groups.)

Daily Values: reference values developed by the FDA specifically for food labels.

dawn phenomenon: early morning hyperglycemia that develops in response to increased levels of counterregulatory hormones that act to raise blood glucose during an overnight fast. Without adequate insulin, the glucose cannot enter the cells and remains in the blood.

deamination: removal of the amino (NH_2) group from a compound such as an amino acid.

dehydration: the loss of water from the body that occurs when water output exceeds water input. The symptoms progress rapidly from thirst, to weakness, to exhaustion and delirium and end in death if not corrected.

delusions: inappropriate beliefs not consistent with the individual's own knowledge and experience.

dementia (dee-MEN-she-ah): irreversible loss of mental function.

denaturation (dee-nay-cher-AY-shun): the change in a protein's shape brought about by heat, acid, or other agents. Past a certain point, denaturation is irreversible.

dental caries: the gradual decay and disintegration of a tooth.

dextrose: an older name for glucose and the name used for glucose in intravenous solutions.

dextrose monohydrate: a form of glucose that contains some water and is stable in IV solutions. Dextrose solutions provide 3.4 kcal/g, whereas glucose provides 4 kcal/g.

DHEA (dehydroepiandrosterone): a hormone secreted by the adrenal glands that serves as a precursor to the male hormone testosterone. DHEA is available without prescription and is sold as an anti-aging remedy to improve energy, strength, and immunity. Side effects include acne, aggressiveness, and liver enlargement.

diabetes (DYE-uh-BEET-eez) **mellitus** (MELL-ih-tus or mell-EYE-tus): a group of metabolic disorders of glucose regulation and utilization.

diagnosis: the disease or condition a person has or is thought to have.

dialysate (dye-AL-ih-sate): a solution used during dialysis to draw wastes and fluids from the blood.

dialysis (dye-AL-ih-sis): removal of waste from the blood through a semipermeable membrane using the principles of simple diffusion and osmosis. The two main types of dialysis are hemodialysis and peritoneal dialysis.

dialyzer (dye-ah-LYES-er): the machine used for hemodialysis; also called an **artificial kidney.**

diet history: a record of eating behaviors and the foods a person eats.

diet manual: a book that describes the foods allowed and restricted on a diet, outlines the rationale and indications for use of each diet, and provides sample menus.

diet order: a physician's written statement in the medical record of what diet a client should receive.

diet pills: pills that depress the appetite temporarily; often, physician-prescribed amphetamines (speed). It is generally agreed that these drugs are of little value for weight loss and that their use can cause a dangerous dependency.

dietary adequacy: the characteristic of a diet that provides all the essential nutrients, fiber, and energy necessary to maintain health and body weight.

dietary balance: providing foods of a number of types in balance with one another such that foods rich in one nutrient do not crowd out foods that are rich in another nutrient.

Dietary Reference Intakes (DRI): a set of values for the dietary nutrient intakes of healthy people in the United States and Canada. These values are used for planning and assessing diets.

dietetic technician registered (D.T.R.): a professional who has earned an associate degree or higher; has completed a dietetic technician program approved by the American Dietetic Association (ADA); has passed a national registration exam; and assists in planning, implementing, and evaluating nutritional care.

dietetics: the practical application of nutrition, including the assessment of nutrition status, recommendation of appropriate diets, nutrition education, and the planning and serving of meals.

differentiation: development of specific functions different from those of the original.

diffusion: movement of solutes from an area of high concentration to one of low concentration.

digestion: the process by which complex food particles are broken down to smaller absorbable particles.

dioxins: toxic organic compounds containing chlorine, arising in industry as (among other things) by-products of the bleaching process.

dipeptide: two amino acids bonded together.

disaccharide: a pair of sugar units bonded together.

diuretic (dye-yoo-RET-ic): a medication that promotes the excretion of water through the kidneys. Only some diuretics increase the urinary loss of potassium. Others, called potassium-sparing diuretics, are less likely to result in a potassium deficiency (see Chapter 26).

diuretic abuse: use of diuretics to promote water excretion by dieters who believe their weight excesses are due to water accumulation. Diuretics promote water loss, not fat loss, and their use can cause dehydration and mineral imbalances.

DNA and **RNA (deoxyribonucleic acid** and **ribonucleic acid):** the genetic materials of cells necessary in protein synthesis; falsely promoted as ergogenic aids.

dominant gene: a gene that has an observable effect on an organism. If an altered gene is dominant, it has an observable effect even when it is paired with a normal gene; see also *recessive gene.*

drug history: a record of all the medications (over-the-counter and prescribed), illicit drugs, and nutrient supplements that a person takes routinely.

duodenum (doo-oh-DEEN-um or doo-ODD-num): the top portion of the small intestine (about "12 fingers' breadth" long, in ancient terminology).

durable power of attorney: a legal document in which one competent adult authorizes another competent adult to make decisions for her or him in the event of incapacitation. The phrase "durable power" means that the agent's authority survives the client's incompetence; "attorney" refers to an attorney-in-fact (not an attorney-at-law).

duration: length of time (for example, the length of time spent in an exercise session).

dysentery (DIS-en-terry): an infection of the gastrointestinal tract caused by an amoeba or bacterium that gives rise to severe diarrhea.

dysphagia (dis-FAY-gee-ah): difficulty in swallowing.

dysuria (dis-YOU-ree-ah): painful or difficult urination.

eating disorder: a disturbance in eating behavior that jeopardizes a person's physical and psychological health.

eclampsia: a severe stage that follows preeclampsia in which convulsions occur.

edema (eh-DEEM-uh): the swelling of body tissue caused by leakage of fluid from the blood vessels and accumulation of fluid in the interstitial spaces.

erythropoietin (e-RITH-row-POY-eh-tin): a hormone secreted by the kidneys in response to oxygen depletion or anemia that stimulates the bone marrow to produce red blood cells.

electrolyte solutions: solutions that can conduct electricity.

electrolytes: salts that dissolve in water and dissociate into charged particles called **ions.**

embolism: the obstruction of a blood vessel by an **embolus** (EM-boh-luss), or traveling clot, causing sudden tissue death.

emetic (em-ETT-ic): an agent that causes vomiting.

emphysema (EM-fe-SEE-ma): a type of COPD in which the lungs lose their elasticity and the victim has difficulty breathing; often occurs along with bronchitis.

emulsifiers: substances that mix with both fat and water and that permanently disperse the fat in the water, forming an emulsion.

end-stage renal disease: the severe stage of renal failure in which dialysis or a renal transplant is necessary to sustain life.

energy metabolism: all the reactions by which the body obtains and spends the energy from food or body stores.

energy-yielding nutrients: the fuel nutrients, those that yield energy the body can use.

engorgement: overfilling of the breasts with milk.

enrichment: the addition of nutrients to a food to meet a specified standard. In the case of refined bread or cereal, five nutrients have been added: thiamin, riboflavin, niacin, and folate in amounts approximately equivalent to, or higher than, those originally present and iron in amounts to alleviate the prevalence of iron-deficiency anemia.

enteral formulas: liquid diets intended for oral use or for tube feedings.

enterostomy (EN-ter-OSS-toe-mee): a gastric or jejunal opening made surgically or under local anesthesia through which a feeding tube can be passed.

enzyme replacements: extracts of pork or beef pancreatic enzymes that are taken as supplements to aid digestion.

enzymes: protein catalysts. A catalyst is a compound that facilitates chemical reactions without itself being changed in the process.

EPA, DHA: omega-3 fatty acids made from linolenic acid. The full name for EPA is eicosapentaenoic (EYE-cosa-PENTA-ee-NO-ick) acid. The full name for DHA is docosahexaenoic (DOE-cosa-HEXA-ee-NO-ick) acid.

epigastric: the region of the body just above the stomach.

epiglottis (epp-ee-GLOT-tiss): a cartilage structure in the throat that prevents fluid or food from entering the trachea when a person swallows.

epithelial (ep-i-THEE-lee-ul) **cells:** cells on the surface of the skin and mucous membranes.

epithelial tissues: the layers of the body that serve as selective barriers between the body's interior and the environment (examples are the cornea, the skin, the respiratory lining, and the lining of the digestive tract).

epoetin (eh-poy-EE-tin): a drug derived from the human hormone erythropoietin and marketed under the trade name Epogen; illegally used to increase oxygen capacity.

erythrocyte (er-REETH-ro-cite) **hemolysis** (he-MOLL-uh-sis): rupture of the red blood cells, caused by vitamin E deficiency.

esophageal hiatus (high-AY-tus): the opening in the diaphragm through which the esophagus passes.

esophageal ulcers: lesions or sores in the lining of the esophagus.

esophagus (e-SOFF-uh-gus): the food pipe; the conduit from the mouth to the stomach.

essential amino acids: amino acids that the body cannot synthesize in amounts sufficient to meet physiological need. Also called *indispensable amino acids.* Nine amino acids are known to be essential for human adults:

essential fatty acids: fatty acids that the body requires but cannot make in amounts sufficient to meet its physiological needs.

essential nutrients: nutrients a person must obtain from food because the body cannot make them for itself in sufficient quantity to meet physiological needs.

ethical: in accordance with moral principles or professional standards. Socrates described *ethics* as "how we ought to live."

ethnic diets: foodways and cuisines typical of national origins, races, cultural heritages, or geographic locations.

fad diets: diets based on exaggerated or false theories of weight loss; such diets are usually inadequate in energy and nutrients.

faith healing: healing by invoking divine intervention without the use of medical, surgical, or other traditional therapy.

fasting hypoglycemia: hypoglycemia that develops gradually and primarily affects the brain and central nervous system.

fatty acids: organic compounds composed of a chain of carbon atoms with hydrogens attached and an acid group at one end.

fatty liver: an accumulation of fat in the liver. In PEM, fat accumulates in the liver because no protein is available to form the lipoproteins that normally escort fat molecules in the blood (see Chapter 28).

fatty liver: an accumulation of triglycerides in the liver resulting from many disorders, including malnutrition and alcoholic liver disease; also called **hepatic steatosis.**

fatty streaks: an accumulation of cholesterol and other lipids along the walls of the arteries.

female athlete triad: a potentially fatal triad of medical problems: disordered eating, amenorrhea, and osteoporosis.

fetal alcohol syndrome (FAS): the cluster of symptoms seen in a person whose mother consumed excess alcohol during her pregnancy; includes mental and physical retardation with facial and other body deformities.

fetus (FEET-us): the developing infant from 8 weeks after conception until its birth.

fibers: a general term denoting in plant foods the polysaccharides cellulose, hemicellulose, pectins, gums, and mucilages, as well as the nonpolysaccharide lignins, that are not attacked by human digestive enzymes.

fibrocystic breast disease: a harmless condition in which the breasts develop lumps, sometimes associated with caffeine consumption. In some, it responds to treatment by abstinence from caffeine; in others, it can be treated with vitamin E.

fibrous plaques: mounds of lipid material, mixed with smooth muscle cells and calcium, which develop in the artery walls in atherosclerosis.

fistula: an abnormal opening between two organs or from an organ to the skin.

fitness: the characteristics of the body that enable it to perform physical activity; more broadly, the ability to meet routine physical demands with enough reserve energy to rise to a sudden challenge; or the body's ability to withstand stress of all kinds.

flapping tremor: uncontrolled movement of the muscle group that causes the outstretched arm and hand to flap like a wing; occurs in hepatic coma and other diseases that cause encephalopathy; also called **asterixis** (AS-ter-ICK-sis).

flexibility: the capacity of the joints to move through a full range of motion; the ability to bend and recover without injury.

fluid and electrolyte balance: maintenance of the necessary amounts and types of fluid and minerals in each compartment of the body fluids.

fluorapatite (floor-APP-uh-tite): the stabilized form of bone and tooth crystal, in which fluoride has replaced the hydroxy portion of hydroxyapatite.

fluorosis (floor-OH-sis): mottling of the tooth enamel from ingestion of too much fluoride during tooth development.

follicle (FOLL-i-cul): a group of cells in the skin from which a hair grows.

food allergies: adverse reactions to foods that involve an immune response; also called *food-hypersensitivity reactions.*

food aversion: a strong desire to avoid a particular food.

food consumption survey: a survey that measures the amounts and kinds of food people consume, estimates the nutrient intakes, and compares them with a standard such as the RDA.

food craving: a deep longing for a particular food.

food group plans: diet-planning tools that sort foods of similar origin and nutrient content into groups and then specify that people should eat certain numbers of servings from each group.

food intolerance: an adverse response to a food or food additive that does not involve the immune system.

food poisoning: illness transmitted to human beings through food and water, caused by infectious agents or by toxins that they produce.

food record: a log of all foods eaten over a period of time that may also include records of behaviors associated with eating, physical activity, medications, and symptoms associated with eating.

fortification: the addition to a food of nutrients that were either not originally present or present in insignificant amounts. Fortification can be used to correct or prevent a widespread nutrient deficiency, to balance the total nutrient profile of a food, or to restore nutrients lost in processing.

fossil fuels: coal, oil, and natural gas, which all come from the fossilized remains of plant life of earlier times.

free radical: a highly reactive chemical form that can cause destructive changes in nearby compounds, sometimes setting up a chain reaction.

frequency: the number of occurrences per unit of time (for example, the number of exercise sessions per week).

fructose: a monosaccharide; sometimes known as fruit sugar. It is abundant in fruits, honey, and saps.

full liquids: foods that are liquid or semiliquid at body temperature.

galactose: a monosaccharide; part of the disaccharide lactose.

galactosemia (ga-LAK-toe-SEE-me-ah): an inborn error of metabolism in which enzymes that normally metabolize galactose to compounds the body can handle are missing and an alternative metabolite accumulates in the tissues, causing damage.

gallbladder: the organ that stores and concentrates bile. When it receives the signal that fat is present in the duodenum, the gallbladder contracts and squirts bile through the bile duct into the duodenum.

garlic oil: extract of garlic.

gastrectomy (gas-TREK-tah-me): surgery that removes all (total gastrectomy) or part (partial or subtotal gastrectomy) of the stomach.

gastric glands: exocrine glands in the stomach wall that secrete gastric juice into the stomach.

gastric juice: the digestive secretion of the gastric glands containing a mixture of water, hydrochloric acid, and enzymes. The principal enzymes are pepsin (acts on proteins) and lipase (acts on emulsified fats).

gastric motility: spontaneous motion in the digestive tract accomplished by involuntary muscular contractions.

gastric partitioning: a surgical procedure used to treat clinically severe obesity. The operation limits food intake by effectively reducing the size of the stomach and delays gastric emptying by effectively restricting the outlet.

gastric residual: the volume of formula that remains in the stomach from a previous feeding. It is measured by gently withdrawing the gastric contents through the feeding tube using a syringe. If the measured gastric residual is acceptable, the residual is returned to the client through the feeding tube.

gastritis: inflammation of the stomach lining.

gastrostomy (gas-TROSS-toe-mee): an opening in the stomach made surgically or under local anesthesia through which a feeding tube can be passed. The technique for creating a gastrostomy under local anesthesia is called **percutaneous endoscopic gastrostomy,** or **PEG** for short. When the feeding tube is guided from such an opening into the jejunum, the procedure is called **percutaneous endoscopic jejunostomy (PEJ),** a misnomer because the enterostomy site is in the stomach.

gatekeeper: with respect to nutrition, a key person who controls other people's access to foods and thereby exerts a profound impact on their nutrition. Examples are the spouse who buys and cooks the food, the parent who feeds the children, and the caregiver in a day-care center.

genes: the basic units of hereditary information, made of DNA, that are passed from parent to offspring in the chromosomes. A pair of genes codes for each genetic trait.

gestational diabetes: the detection of abnormal glucose tolerance during pregnancy, which usually returns to normal postpartum.

GI tract: the gastrointestinal tract or digestive tract; the principal organs are the stomach and intestines.

ginseng: a plant whose extract supposedly boosts energy; side effects to chronic use include nervousness, confusion, and depression.

gland: a cell or group of cells that secretes materials for special uses in the body. Glands may be *exocrine glands,* secreting their materials "out" (into the digestive tract or onto the surface of the skin), or *endocrine glands,* secreting their materials "in" (into the blood).

glomerular filtration rate (GFR): the rate at which the kidneys form filtrate. Normally, the GFR is about 120 ml/min. Falling GFR signifies deteriorating kidney function. When the GFR falls drastically, to about 10 ml/min, dialysis is usually initiated.

glomerulus (glow-MARE-you-lus): a cup-shaped membrane enclosing a tuft of capillaries within a nephron. (The plural is *glomeruli.*)

glucose: a monosaccharide, the sugar common to all disaccharides and polysaccharides; also called blood sugar or dextrose.

glucose polymers: compounds that supply glucose, not as single molecules, but linked in chains somewhat like starch. The objective is to attract less water from the body into the digestive tract (osmotic attraction depends on the number, not the size of particles).

gluten (GLUE-ten): a vegetable protein found in wheat; **gliadin** is the protein fraction of gluten that causes the toxic effects in celiac disease.

glycemic effect: a measure of the extent to which a food raises the blood glucose concentration and elicits an insulin response, as compared with pure glucose.

glycerol (GLISS-er-ol): an organic compound, three carbons long, that can form the backbone of triglycerides and phospholipids.

glycine: a nonessential amino acid; promoted as an ergogenic aid because it is a precursor of the high-energy compound creatine phosphate.

glycogen (GLY-co-gen): a polysaccharide composed of glucose, made and stored by liver and muscle tissues of human beings and animals as a storage form of glucose. Glycogen is not a significant food source of carbohydrate and is not counted as one of the complex carbohydrates in foods.

glycolysis (gligh-COLL-uh-sis): the metabolic breakdown of glucose to pyruvate.

goiter (GOY-ter): an enlargement of the thyroid gland due to an iodine deficiency, malfunction of the gland, or overconsumption of a thyroid antagonist. Goiter caused by iodine deficiency is *simple goiter.*

gout (GOWT): a metabolic disease in which crystals of uric acid precipitate in the joints.

granulated sugar: common table sugar, crystalline sucrose; 99.9% pure.

granuloma (gran-you-LOH-mah): a tumor or growth that contains live microorganisms surrounded by immune system cells and covered with a fibrous coat.

GRAS (generally recognized as safe) list: a list of food additives, established by the FDA in 1958, that had long been in use and were believed safe.

growth hormone releasers: herbs or pills falsely promoted as enhancing athletic performance.

guarana: a reddish berry found in Brazil's Amazon Valley that is used as an ingredient in carbonated sodas and taken in powder or tablet form. Guarana is marketed as an ergogenic aid to enhance speed and endurance, an aphrodisiac, a "cardiac tonic," an "intestinal disinfectant," and a smart drug that supposedly improves memory and concentration and wards off senility. Because guarana contains seven times as much caffeine as its relative the coffee bean, there are concerns that high doses can stress the heart and cause panic attacks.

hazard: the ability of a substance to produce injury under the conditions of its use.

Hazard Analysis Critical Control Points (HACCP): a systematic plan to identify and correct potential microbial hazards in the manufacturing, distribution, and commercial use of food products.

HDL (high-density lipoprotein): the type of lipoprotein that transports cholesterol back to the liver from peripheral cells; composed primarily of protein.

health care team: a group of professionals representing different disciplines who work together to promote their clients' health.

health history: an account of the client's current and past health status and risk factors for disease. Traditionally, the health history has been called the medical history. The term *health history* now seems more appropriate, however, because the contents describe the client's health status, and current trends in the medical profession emphasize health promotion and disease prevention.

health maintenance organization (HMO): a form of managed care that limits the subscriber's choice of health care professionals to those affiliated with the organization and controls access to services by directing care through a primary care physician.

heart attack: sudden tissue death caused by blockages of vessels that feed the heart muscle; also called **myocardial infarction (MI)** or **cardiac arrest.**

heartburn: a burning pain felt behind the sternum caused by stomach acids irritating the esophagus.

heat stroke: the dangerous accumulation of body heat with accompanying loss of body fluid.

hematuria (HE-mah-TOO-ree-ah): blood in the urine.

hemlock: a poisonous herb having finely divided leaves and small white flowers.

hemochromatosis (heem-oh-crome-a-TOE-sis): iron overload characterized by deposits of iron-containing pigment in many tissues, with tissue damage. Hemochromatosis is a hereditary defect in iron metabolism.

hemodialysis: removal of wastes and fluids from the blood by passing it through a dialyzer.

hemoglobin: the oxygen-carrying protein of the red blood cells.

hemorrhagic (hem-o-RAJ-ik) **disease:** the vitamin K–deficiency disease in which blood fails to clot.

hemosiderosis (heem-oh-sid-er-OH-sis): a condition characterized by the deposition of the iron-storage protein hemosiderin in the liver and other tissues.

hepatic coma: a state of unconsciousness that results from severe liver disease.

hepatic encephalopathy: mental changes associated with liver disease that may include irritability, short-term memory loss, and an inability to concentrate.

hepatorenal syndrome: the symptoms of renal failure that occur as a consequence of liver failure.

herbal laxatives: laxatives containing senna, aloe, rhubarb root, castor oil, or buckthorn are commonly sold as "dieter's tea." Such products cause nausea, vomiting, diarrhea, fainting, and, in some users, possibly death.

herbal medicine: the use of plants to treat disease or improve health; also known as botanical medicine or phytotherapy.

herbal products: substances extracted from plants to mimic drugs that suppress appetite. Such substances may have dangerous side effects. St. John's wort, for example, is often prepared in combination with the herbal stimulant ephedrine, extracted from the Chinese plant ma huang. Ephedrine has been implicated in several cases of heart attacks and seizures.

herbal steroids or **plant sterols:** lipid extracts of plants, called *ferulic acid, oryzanol, phytosterols,* or *adaptogens;* marketed with false claims that they contain hormones or balance hormonal activity.

herpes virus: a virus that can lead to mouth lesions and may also affect the lower GI tract, causing diarrhea.

hiatal hernia: a protrusion of a portion of the stomach through the esophageal hiatus of the diaphragm. There are several types of hiatal hernias, but the *sliding hiatal hernia* is the most common.

high-density lipoprotein: see *HDL.*

high-fructose corn syrup (HFCS): the predominant sweetener used in processed foods today. HFCS is mostly fructose; glucose makes up the balance.

high-quality protein: an easily digestible, complete protein.

homeopathic (home-ee-OP-ah-thick) **medicine:** a practice based on the theory that "like cures like," that is, that substances that cause symptoms in healthy people can cure those symptoms when given in very dilute amounts.

homeostasis: the maintenance of constant internal conditions (such as chemistry, temperature, and blood pressure) by the body's control systems.

honey: sugar (mostly sucrose) formed from nectar gathered by bees. An enzyme splits the sucrose into glucose and fructose. Composition and flavor vary, but honey always contains a mixture of sucrose, fructose, and glucose.

hormones: chemical messengers. Hormones are secreted by a variety of glands in the body in response to altered conditions. Each travels to one or more target tissues or organs and elicits specific responses to restore normal conditions.

human immunodeficiency virus (HIV): the virus that causes AIDS. The infection progresses until it seriously hampers the function of the immune system and leaves its victims defenseless against numerous infections and cancer.

hunger: the physiological need to eat, experienced as a drive for obtaining food; an unpleasant sensation that demands relief.

hydrochloric acid (HCl): an acid composed of hydrogen and chloride atoms. The gastric glands normally produce this acid.

hydrogenation: the process of adding hydrogen to unsaturated fat to make it more solid and resistant to chemical change.

hydrolyzed formula: a liquid diet that contains broken-down molecules of protein, such as amino acids and short peptide chains; also called **monomeric formula.**

hydrophilic colloids: laxatives composed of fibers that work like fibers from food; they attract water in the intestine to form a bulky stool, which then stimulates peristalsis. Metamucil and Fiberall are examples of hydrophilic colloids.

hyperactivity: inattentive and impulsive behavior that is more frequent and severe than is typical of others a similar age; professionally called **attention deficit hyperactivity disorder (ADHD).**

hypercalciuria (HIGH-per-kal-see-YOU-ree-ah): excessive urinary excretion of calcium.

hyperoxaluria (HIGH-per-OX-all-YOU-ree-ah): excessive urinary excretion of oxalate.

hyperthermia: an above-normal body temperature.

hypertonic formula: a formula with an osmolality greater than that of blood serum.

hypertrophy (high-PER-tro-fee): of muscles, growing larger; an increase in size in response to use.

hypnotherapy: a technique that uses hypnosis and the power of suggestion to improve health behaviors, relieve pain, and heal.

hypothalamus (high-poh-THALL-uh-mus): a part of the brain that helps regulate many body balances, including fluid balance.

hypothermia: a below-normal body temperature.

ileocecal (ill-ee-oh-SEEK-ul) **valve:** the sphincter muscle separating the small and large intestines.

ileum (ILL-ee-um): the last segment of the small intestine.

imagery: a technique that guides clients to achieve a desired physical, emotional, or spiritual state by visualizing themselves in that state.

impaired glucose tolerance: inability of the body to adequately regulate its blood glucose concentration in response to either the intake of dietary carbohydrate or the release of glucose from cells during fasting or metabolic stress; sometimes called **borderline diabetes.**

inborn error of metabolism: an inherited flaw evident as a metabolic disorder or disease present from birth.

incidental food additives: substances that can get into food not through intentional introduction but as a result of contact with the food during growing, processing, packaging, storing, or some other stage before the food is consumed. The terms *accidental additives* and *indirect additives* mean the same thing.

incomplete protein: a protein lacking or low in one or more of the essential amino acids.

indemnity insurance: traditional fee-for-service insurance.

indirect calorimetry: an indirect estimate of resting energy needs made by measuring the ratio of carbon dioxide expired to the amount of oxygen inspired.

induration: a raised, hardened area of skin.

inflammatory bowel diseases (IBD): diseases characterized by inflammation of the intestine.

inflammatory response: the changes orchestrated by the immune system when tissues are injured by such forces as blows, wounds, foreign bodies (chemicals, microorganisms), heat, cold, electricity, or radiation.

inosine: an organic chemical that is falsely said to "activate cells, produce energy, and facilitate exercise," but has been shown actually to reduce the endurance of runners.

insoluble fibers: the tough, fibrous structures of fruits, vegetables, and grains; indigestible food components that do not dissolve in water.

insulin: a hormone secreted by the pancreas in response to high blood glucose; it promotes cellular glucose uptake and use or storage.

insulin resistance: the condition in which the cells fail to respond to insulin as they do in healthy people.

intensity: the degree of exertion while exercising (for example, the amount of weight lifted or the speed of running).

intermittent claudication: severe calf pain caused by inadequate blood supply; it occurs when walking and subsides during rest.

intestinal flora: the bacterial inhabitants of the GI tract.

intestinal juice: the secretion of the intestinal glands; contains enzymes for the digestion of carbohydrate and protein and a minor enzyme for fat digestion.

intestinal lumen: the inner open space within the intestine.

intra-abdominal fat: fat stored within the abdominal cavity in association with the internal abdominal organs, as opposed to the fat stored directly under the abdominal skin (subcutaneous fat).

intravenous (IV): through a vein.

invert sugar: a mixture of glucose and fructose formed by splitting sucrose in a chemical process; sold only in liquid form, sweeter than sucrose. Invert sugar is used as an additive to help preserve food freshness and prevent shrinkage.

iridology: the study of changes in the iris of the eye and their relationships to disease.

iron deficiency: having depleted iron stores.

iron-deficiency anemia: a blood iron deficiency characterized by small, pale red blood cells; also called **microcytic hypochromic anemia.**

iron overload: toxicity from excess iron.

irradiation: sterilizing a food by exposure to energy waves, similar to ultraviolet light and microwaves.

isotonic formula: a formula with an osmolality similar to that of blood serum (300 mOsm/kg).

IV catheter: a thin tube inserted into a vein through which nutrient solutions or medications can be given directly.

jaundice: yellowing of the skin due to spillover of bile pigments from the liver into the general circulation.

jejunostomy (JEE-ju-NOSS-toe-mee): an opening in the jejunum made surgically or under local anesthesia through which a feeding tube can be passed. The technique for creating a jejunostomy under local anesthesia is called a **direct endoscopic jejunostomy (DEJ).**

jejunum (je-JOON-um): the first two-fifths of the small intestine beyond the duodenum.

Kaposi's (cap-OH-seez) **sarcoma:** a type of cancer rare in the general population but common in people with HIV infection.

kcalorie control: management of food energy intake.

kcalories: units by which energy is measured.

keratin (KERR-uh-tin): a water-insoluble protein; the normal protein of hair and nails. Keratin-producing cells may replace mucus-producing cells in vitamin A deficiency.

keratinization: accumulation of keratin. The progression of this condition to the extreme is **hyperkeratosis.**

ketones (KEY-tones): acidic, fat-related compounds formed from the incomplete breakdown of fat when carbohydrate is not available; technically known as *ketone bodies.*

kwashiorkor (kwash-ee-OR-core or kwash-ee-or-CORE): a disease related to PEM.

lactic acid: a compound produced in muscles when they break down glucose anaerobically.

lactoferrin (lack-toe-FERR-in): a factor in breast milk that binds iron and keeps it from supporting the growth of the infant's intestinal bacteria.

lacto-ovo vegetarian diets: diets that include milk, cheese, and eggs (animal products) but exclude meat, fish, and poultry (animal flesh).

lacto-ovo vegetarians: people who include milk or milk products and eggs, but omit meat, fish, shellfish, and poultry from their diets.

lactose: a disaccharide composed of glucose and galactose; commonly known as milk sugar.

lacto-vegetarians: people who include milk or milk products, but exclude meat, poultry, fish, shellfish, and eggs from their diets.

latent: the period in the course of a disease when the conditions are present but the symptoms have not begun to appear.

lecithins: one type of phospholipids.

legal: established by law.

leptin: a protein produced by fat cells under the direction of the obesity gene that increases satiety and energy expenditure.

letdown reflex: the reflex that forces milk to the front of the breast when the infant begins to nurse.

levulose: an older name for fructose.

LDL (low-density lipoprotein): the type of lipoprotein derived from VLDL as cells remove triglycerides from them. LDL carry cholesterol and triglycerides from the liver to the cells of the body and are composed primarily of cholesterol.

life expectancy: the average number of years lived by people in a given society.

life span: the maximum number of years of life attainable by a member of a species.

limiting amino acid: the essential amino acid found in the shortest supply relative to the amounts needed for protein synthesis in the body. It limits the amount of protein the body can make.

linoleic acid, linolenic acid: polyunsaturated fatty acids, essential for human beings.

lipids: a family of compounds that includes triglycerides (fats and oils), phospholipids, and sterols.

lipoprotein lipase (LPL): an enzyme mounted on the surface of fat cells (and other cells). It hydrolyzes triglycerides in the blood into fatty acids and glycerol for absorption into the cells. There they are metabolized or reassembled for storage.

lipoproteins: clusters of lipids associated with proteins that serve as transport vehicles for lipids in the lymph and blood.

living will: a document signed by a competent adult that specifically states whether the person wishes aggressive treatment in the event of terminal illness or irreversible coma from which the person is not expected to recover.

longevity: long duration of life.

low birthweight (LBW): a birthweight less than 5½ lb (2500 g); indicates probable poor health in the newborn and poor nutrition status of the mother during pregnancy. Normal birthweight for a full-term baby is 6½ to 8¾ lb (about 3000 to 4000 g).

low-carbohydrate diets: diets designed to bring about metabolic responses similar to those of fasting. Without sufficient carbohydrate, the body cannot use its fat in the normal way, and ketosis results. Many physiological hazards accompany low-carbohydrate diets: high blood cholesterol, mineral imbalances, hypoglycemia, and more.

low-density lipoprotein: see *LDL.*

lymph (LIMF): the body fluid found in lymphatic vessels; lymph consists of all the constituents of blood except red blood cells.

lymphatic system: a loosely organized system of vessels and ducts that conveys the products of digestion toward the heart.

macrobiotic diet: a diet consisting of brown rice, miso soup, sea vegetables, and other traditional Japanese foods.

mad cow disease: a fatal virus that affects the nervous system of cattle. Formally known as bovine spongiform encephalopathy (BSE), this progressive neurological disorder resulted in the killing of 155,600 cattle in the United Kingdom as a precautionary measure to prevent transmission. Creutzfeldt-Jakob disease (CJD) is the human strain seen in the United Kingdom.

malnutrition: any condition caused by deficient or excess energy or nutrient intake or by an imbalance of nutrients.

maltose: a disaccharide composed of two glucose units; sometimes known as malt sugar.

managed care: health care delivery system that aims to provide cost-effective health care by coordinating services and limiting access to services.

maple sugar: a sugar (mostly sucrose) purified from concentrated sap of the sugar maple tree.

marasmus (ma-RAZZ-mus): a disease related to PEM; marasmus results from severe deprivation, or impaired absorption, of protein, energy, vitamins, and minerals.

margin of safety: as used when speaking of food additives, a zone between the concentration normally used and that at which a hazard exists. For common table salt, for example, the margin of safety is ⅕ (five times the concentration normally used would be hazardous).

massage therapy: a healing method in which the therapist manually kneads muscles to reduce tension, increase blood circulation, improve joint mobility, and promote healing of injuries.

mastitis: infection of a breast.

mechanical soft diet: a diet that excludes all foods that are difficult to chew or swallow; also called a **dental soft diet.**

medical nutrition therapy: a term introduced by the American Dietetic Association in 1994 to emphasize the role of nutrition in medical care.

medical record: a continuous written account of a client's health history, diagnosis, prognosis, and therapy.

meditation: a self-directed technique of relaxing the body and calming the mind.

melatonin: a hormone secreted by the pineal gland believed to help regulate the body's daily rhythms and promote sleep. Proof of safety or effectiveness is lacking.

metabolism: the sum total of all the chemical reactions that go on in living cells.

microvilli (MY-cro-VILL-ee or MY-cro-VILL-eye): tiny, hairlike projections on each cell of every villus that can trap nutrient particles and transport them into the cells. The singular form is **microvillus.**

moderate exercise: exercise that can be sustained comfortably for 60 minutes or so.

moderation: providing enough, but not too much of a dietary constituent.

modified atmosphere packaging (MAP): preservation of a perishable food by packaging it in a gas-impermeable container to which a gas mixture other than air has been added.

modified or **therapeutic diet:** a regular diet that is adjusted to meet special nutrient needs. Such diets can be adjusted in consistency, level of energy and nutrients, amount of fluid, or number of meals, or by the inclusion or elimination of certain foods.

modules: formulas or foods that provide primarily a single nutrient and are designed to be added to other formulas or foods to alter nutrient composition; they can also be combined together to create a highly individualized formula.

molasses: a thick brown syrup, left over from sugarcane juice during sugar refining. Blackstrap molasses contains iron, which comes from the machinery used to process it. This iron is not as well absorbed as the iron in meats and other foods.

molybdenum (mo-LIB-duh-num): a trace element.

monosaccharide : a single sugar unit.

monounsaturated fatty acid: a fatty acid that has one point of unsaturation; for example, oleic acid.

mood disorders: mental illnesses characterized by episodes of severe depression or excessive excitement (mania) or both.

mouth ulcers: lesions or sores in the lining of the mouth. Certain drugs, radiation therapy, and some disorders, such as oral herpes virus infections, can cause mouth ulcers.

MSG symptom complex: an acute and temporary intolerance reaction that may occur after eating MSG (monosodium glutamate). Symptoms include burning sensations, chest and facial flushing and pain, and throbbing headaches.

mucous membrane: the membranes composed of mucus-secreting cells that line the surfaces of body tissues.

mucus (MYOO-cuss): a mucopolysaccharide (a relative of carbohydrate) secreted by cells of the stomach wall that protects the cells from exposure to digestive juices (and other destructive agents). The cellular lining of the stomach wall with its coat of mucus is known as the mucous membrane. (The noun is mucus; the adjective is mucous.)

muscle endurance: the ability of a muscle to contract repeatedly within a given time without becoming exhausted.

muscle strength: the ability of muscles to work against resistance.

muscular dystrophy (DIS-tro-fee): a hereditary disease in which the muscles gradually weaken; its most debilitating effects arise in the lungs. This disease should not be confused with *nutritional* muscular dystrophy, a vitamin E–deficiency disease of animals characterized by gradual paralysis of the muscles.

mutation: an alteration in a gene such that an altered protein is produced.

mutual supplementation: the strategy of combining two protein foods in a meal so that each food provides the essential amino acid(s) lacking in the other.

myoglobin: the oxygen-carrying protein of the muscle cells.

nasoduodenal (ND): from the nose to the duodenum.

nasoenteric: from the nose to the stomach or intestine. *Nasoenteric feedings* include nasogastric, nasoduodenal, and nasojejunal feedings. Most clinicians use nasoenteric to refer to nasoduodenal and nasojejunal feedings only.

nasogastric (NG): from the nose to the stomach.

nasojejunal (NJ): from the nose to the jejunum.

naturopathic medicine: a system that integrates traditional medicine with botanical medicine, clinical nutrition, homeopathy, acupuncture, East Asian medicine, hydrotherapy, and manipulative therapy.

nephrons (NEF-rons): the working units of the kidneys; each nephron consists of a glomerulus and a tubule.

nephrotic syndrome: the complex of symptoms that occur when glomerular function falters; it includes proteinuria, low serum proteins, edema, and elevated blood lipids.

neural tube defects: malformations of the brain, spinal cord, or both during embryonic development.

neuron: a nerve cell; the structural and functional unit of the nervous system. Neurons initiate and conduct nerve transmissions.

niacin equivalents: the amount of niacin present in food, including the niacin that can theoretically be made from its precursor tryptophan present in the food.

night blindness: the slow recovery of vision after exposure to flashes of bright light at night; an early symptom of vitamin A deficiency.

nitrites: salts added to food to prevent botulism. An example is sodium nitrite.

nitrogen balance: the amount of nitrogen consumed (N in) as compared with the amount of nitrogen excreted (N out) in a given period of time. The laboratory scientist can estimate the protein in a sample of food, body tissue, or excreta by measuring the nitrogen in it.

nitrosamines (nigh-TROHS-uh-meens): derivatives of nitrites that may form when nitrites combine with amines.

nocturnal hypoglycemia: hypoglycemia that occurs while a person is sleeping.

nucleotides: nitrogen-containing components of RNA and DNA. Nucleotides can be synthesized in the bodies of healthy people, but the demands for nucleotides during severe stress may make them conditionally essential.

nursing bottle tooth decay: extensive tooth decay due to prolonged tooth contact with formula, milk, fruit juice, or other carbohydrate-rich liquid offered to an infant in a bottle.

nutrient density: a measure of the nutrients a food provides relative to the energy it provides. The more nutrients and the fewer kcalories, the higher the nutrient density.

nutrients: substances obtained from food and used in the body to provide energy and structural materials and to promote growth, maintenance, and repair.

nutrition assessment: the evaluation of many factors that influence or reflect nutritional health; the tools used for nutrition assessment include historical information, anthropometric measurements, physical findings, and laboratory tests.

nutrition care plan: a plan that translates nutrition assessment data into a strategy for meeting a client's nutrient and nutrition education needs.

nutrition care process: an organized approach to nutrition intervention that consists of assessing, planning, implementing, and evaluating. The nutrition care process parallels the nursing care process except that it focuses on nutrition concerns.

nutrition screening: a tool for quickly identifying clients who need complete nutrition assessments.

nutrition status: the state of the body's nutritional health.

nutrition status survey: a survey that evaluates people's nutrition status using food intake data, anthropometric measures, physical examinations, and laboratory tests.

nutritionist: a person who specializes in the study of nutrition. Some nutritionists are registered dietitians, but others are self-described experts whose training may be minimal or nonexistent. Some states make the term meaningful by allowing it to apply only to people who have master's (M.S.) or doctoral (Ph.D.) degrees from institutions accredited to offer such degrees in nutrition or related fields.

nutritive sweeteners: sweeteners that yield energy, including both the sugars and the sugar alcohols.

obesity: a chronic disease characterized by excessively high body fat in relation to lean body tissue.

octacosanol: an alcohol extracted from wheat germ; often falsely promoted to enhance athletic performance.

omega: the last letter of the Greek alphabet (ω, sometimes replaced by the letter *m* or *n*), used by chemists to refer to the position of the last double bond in a fatty acid.

omega-3 fatty acid: a polyunsaturated fatty acid with its endmost double bond three carbons back from the end of its carbon chain; relatively newly recognized as important in nutrition. Linolenic acid is an example.

omega-6 fatty acid: a polyunsaturated fatty acid with its endmost double bond six carbons back from the end of its carbon chain; long recognized as important in nutrition. Linoleic acid is an example.

oral antidiabetic agents: medications taken by mouth to lower blood glucose levels in people with type 2 diabetes.

organic: carbon containing. The four organic nutrients are carbohydrate, fat, protein, and vitamins.

organically grown crops: crops grown and processed according to USDA regulations defining the use of fertilizers, herbicides, insecticides, fungicides, preservatives, and other chemical ingredients.

orogastric: from the mouth to the stomach. This method is often used to feed infants because they breathe through their noses, and tubes inserted through the nose can hinder the infant's breathing. The tube is inserted before, and removed after, each feeding.

orthomolecular medicine: the use of large doses of vitamins to treat chronic disease.

osmolality (OZ-mow-LAL-eh-tee): a measure of the concentration of particles in a solution, expressed as the number of milliosmoles (mOsm) per kilogram.

osmosis: movement of water from an area of low solute concentration to one of high solute concentration.

osteoarthritis: a painful, chronic disease of the joints caused when the cushioning cartilage in a joint breaks down; joint structure is usually altered, with loss of function; also called *degenerative arthritis.*

osteomalacia (os-tee-o-mal-AY-shuh): a bone disease characterized by softening of the bones; symptoms include bending of the spine and bowing of the legs. The disease occurs most often in adult women.

osteoporosis (oss-tee-oh-pore-OH-sis): literally, porous bones; reduced density of the bones, also known as *adult bone loss.*

overload: an extra physical demand placed on the body; an increase in the *frequency, duration,* or *intensity* of exercise.

overnutrition: overconsumption of food energy or nutrients sufficient to cause disease or increased susceptibility to disease; a form of malnutrition.

overt: out in the open, full-blown.

overweight: body weight above some standard of acceptable weight that is usually defined in relation to height (such as the weight-for-height tables).

oxalate (OX-a-late): a nonnutrient found in significant amounts in rhubarb, spinach, beets, nuts, chocolate, tea, wheat bran, and strawberries.

ozone therapy: the use of ozone gas to enhance the body's immune system.

pancreas: a gland that secretes enzymes and digestive juices into the duodenum. (This is its exocrine function; it also has the endocrine function of secreting insulin and other hormones into the blood.)

pancreatic (pank-ree-AT-ic) **juice:** the exocrine secretion of the pancreas, containing enzymes for the digestion of carbohydrate, fat, and protein. Juice flows from the pancreas into the small intestine through the pancreatic duct. The pancreas also has an endocrine function, the secretion of insulin and other hormones.

pancreatitis: inflammation of the pancreas.

paranoia (PARA-NOY-ah): mental illness characterized by delusions of persecution.

parenteral nutrition: delivery of nutrient solutions directly into a vein, bypassing the intestine.

pasteurization: the treatment of food with heat sufficient to kill certain pathogens (disease-causing microbes) commonly transmitted through milk, juices, or foods. It is not a sterilization process; bacteria that cause spoilage are still present.

pathogens: disease-causing microorganisms.

pellagra (pell-AY-gra): the niacin-deficiency disease. Symptoms include the "4 Ds": diarrhea, dermatitis, dementia, and ultimately, death.

pepsin: a protein-digesting enzyme (gastric protease) in the stomach. It circulates as a precursor, pepsinogen, and is converted to pepsin by the action of stomach acid.

peripheral parenteral nutrition (PPN): the use of the peripheral veins to provide a solution that meets nutrient needs.

peripheral resistance: resistance to the flow of blood caused by the reduced diameter of the vessels at the periphery of the body—the smallest arteries and capillaries.

peripheral veins: the small-diameter veins that carry blood from the extremities (arms and legs).

peripherally inserted central catheter (PICC): a catheter inserted into a peripheral vein and advanced into a central vein.

peristalsis (peri-STALL-sis): successive waves of involuntary muscular contractions passing along the walls of the GI tract that push the contents along.

peritoneal dialysis: removal of wastes and fluids from the body using the peritoneal membrane as a semipermeable membrane.

peritonitis (pear-ih-toe-NIGH-tus): infection and inflammation of the membrane lining the abdominal cavity caused by leakage of infectious organisms through a perforation in an abdominal organ.

pernicious anemia: anemia caused by a lack of intrinsic factor and the consequent malabsorption of vitamin B_{12}.

persistent: of a stubborn or enduring nature; with respect to food contaminants, the quality of persisting, rather than breaking down, in the bodies of animals and human beings.

persistent vegetative state: exhibiting motor reflexes but without the ability to regain cognitive behavior, communicate, or interact purposefully with the environment.

pH: the concentration of hydrogen ions. The lower the pH, the stronger the acid. Thus pH 2 is a strong acid; pH 6 is a weak acid; pH 7 is neutral; and a pH above 7 is alkaline.

phosphate salt: a salt that has been demonstrated to raise the concentration of a metabolically important compound (diphosphoglycerate) in red blood cells and enhance the cells' potential to deliver oxygen to muscle cells; the salts may cause calcium losses from the bones if taken in excess.

phospholipids: one of the three main classes of lipids; these compounds are similar to triglycerides but have choline (or another compound) and a phosphorus-containing acid in place of one of the fatty acids.

physiological age: a person's age as estimated from her or his body's health and probable life expectancy.

phytates: nonnutrient components of grains, legumes, and seeds; phytates can bind minerals such as zinc, iron, calcium, and magnesium in insoluble complexes in the intestine, which the body excretes unused.

phytobezoar (FIGH-toh-BEE-zor): a mass of plant matter (fibers, leaves, and skins) that forms a ball and may block the outlet from the stomach to the intestine. **Trichobezoars** contain hair and nails. People with some psychiatric disorders may swallow their hair and nails.

phytochemicals: nonnutrient compounds found in plant-derived foods that have biological activity in the body.

pica (PIE-ka): a craving for nonfood substances; also known as *geophagia* (jee-oh-FAY-jee-uh) when referring to clay-eating behavior.

pituitary (pit-TOO-ih-tary) **gland:** in the brain, the "king gland" that regulates the operation of many other glands.

PKU, phenylketonuria (FEN-el-KEY-toe-NEW-ree-ah): an inborn error of metabolism in which phenylalanine, an essential amino acid, cannot be converted to tyrosine. Alternative metabolites of phenylalanine (phenylketones) accumulate in the tissues, causing damage, and overflow into the urine.

placenta (pla-SEN-tuh): an organ that develops inside the uterus early in pregnancy, in which maternal and fetal blood circulate in close proximity and exchange materials. The fetus receives nutrients and oxygen across the placenta; the mother's blood picks up carbon dioxide and other waste materials to be excreted via her lungs and kidneys.

plaques (PLACKS): mounds of lipid material that form within certain immune system cells (macrophages) that are embedded in artery walls. If atherosclerosis progresses to an advanced stage, the plaques may become hardened with fibrous connective tissue and calcium deposits.

platelets: tiny, disc-shaped bodies in the blood that are important in clot formation.

polypeptide: many amino acids bonded together. *Many* refers to ten or more. An intermediate strand of between four and ten amino acids is an **oligopeptide.**

polyunsaturated fatty acid (PUFA): a fatty acid with two or more points of unsaturation. For example, linoleic acid has two such points, and linolenic acid has three. Thus a polyunsaturated *fat* is composed of triglycerides containing a high percentage of PUFA.

portal hypertension: elevated blood pressure in the portal vein caused by obstructed blood flow through the liver.

precursor: a compound that can be converted into another compound; with regard to vitamins, compounds that can be converted into active vitamins; also known as **provitamins.**

preeclampsia: a condition characterized by hypertension, fluid retention, and protein in the urine.

preferred provider organization: a form of managed care that encourages subscribers to select health care providers from a group that has contracted with the organization to provide services at lower costs.

preformed vitamin A: vitamin A in its active form.

preservatives: antimicrobial agents, antioxidants, chelating agents, radiation, and other additives that retard spoilage or preserve desired qualities, such as softness in baked goods.

pressure sores: the breakdown of skin and underlying tissues due to constant pressure and lack of oxygen to the affected area; also called **decubitus** (dee-CUE-bih-tis) **ulcers** or **bedsores.**

prognosis: the predicted course and outcome of a disease or condition.

progressive diet: a diet that progresses as a client's tolerances permit; such diets often progress from clear liquids to full liquids to soft foods to regular foods.

progressive overload principle: the training principle that a body system, in order to improve, must be worked at frequencies, durations, or intensities that gradually increase physical demand.

protein isolate: a protein that has been separated from a food. Examples include casein from milk and albumin from egg.

protein-energy malnutrition (PEM): a deficiency of protein and food energy; the world's most widespread malnutrition problem, including both marasmus and kwashiorkor.

proteins: compounds composed of carbon, hydrogen, oxygen, and nitrogen atoms arranged into strands of amino acids. Some amino acids also contain sulfur atoms.

protein-sparing effect: the effect of carbohydrate in providing energy that allows protein to be used for other purposes.

puberty: the period in life in which a person becomes physically capable of reproduction.

pyloric (pie-LORE-ic) **sphincter:** the sphincter muscle separating the stomach from the small intestine (also called *pylorus* or *pyloric valve*).

pyloroplasty (pie-LOOR-o-PLAS-tee): surgery that enlarges the pyloric sphincter.

pyruvate (PIE-roo-vate): pyruvic acid, a 3-carbon compound derived from glucose, glycerol, and certain amino acids in metabolism.

radiation enteritis: inflammation and scarring of the intestinal cells caused by exposure to radiation.

radiation therapy: the use of radiation to arrest or destroy cancer cells.

rancid: the term used to describe fats when they have deteriorated, usually by oxidation; rancid fats often have an "off" odor.

raw sugar: the first crop of crystals harvested during sugar processing. Raw sugar cannot be sold in the United States because it contains too much filth (dirt, insect fragments, and the like). Sugar sold domestically as raw sugar has actually gone through about half of the refining steps.

reactive hypoglycemia: hypoglycemia experienced simultaneously with epinephrine-release symptoms one to three hours after a meal; also called **postprandial hypoglycemia.**

rebound hyperglycemia: hyperglycemia resulting from excessive secretion of counterregulatory hormones in response to excessive insulin and consequent low blood glucose levels; also called the **Somogyi** (so-MOHG-yee) **effect.**

recessive gene: a gene that has no observable effect on an organism. If an altered gene is recessive, it has no observable effect as long as it is paired with a normal gene that can produce a normal product. In this case, the normal gene is said to be dominant.

RDA (Recommended Dietary Allowances): the average daily amounts of nutrients considered adequate to meet the known nutrient needs of practically all healthy people.

rectum: the muscular terminal part of the GI tract extending from the sigmoid colon to the anus; the rectum stores waste prior to elimination.

refeeding syndrome: a set of physiologic and metabolic complications associated with reintroducing adequate nutrition too rapidly. These complications can include malabsorption, cardiac insufficiency, respiratory distress, congestive heart failure, convulsions, coma, and possibly death.

refined grain: a product from which the bran, germ, and husk have been removed, leaving only the endosperm.

reflux esophagitis (eh-sof-ah-JYE-tis): the backflow or regurgitation of gastric contents from the stomach into the esophagus, causing inflammation of the esophagus; also called **gastroesophageal reflux,** or **acid indigestion.**

registered dietitian (R.D.): a dietitian who has graduated from a university or college after completing a program of dietetics that has been accredited by the American Dietetic Association (or Dietitians of Canada). The dietitian must serve in an approved internship or coordinated program to practice the necessary skills, pass the association registration examination, and maintain competency through continuing education. Many states require licensing for practicing dietitians. Licensed dietitians (L.D.'s) have met all *state* requirements to offer nutrition advice.

renal: pertaining to the kidneys.

renal colic: the severe pain that accompanies the movement of a kidney stone through the ureter to the bladder.

renal failure: failure of the kidneys to maintain normal function to a degree that requires dialysis or kidney transplantation for survival.

renal filtrate: the fluid that passes from the blood through the capillary walls of the glomeruli, eventually forming urine.

renal insufficiency: reduced renal function but not to the degree that requires dialysis or kidney transplantation.

renal osteodystrophy (OS-tee-oh-DIS-tro-fee): a bone disorder resulting from calcium and phosphorus imbalances in renal disease.

renal threshold: the point at which a blood constituent that is normally reabsorbed by the kidneys reaches a level so high the kidneys cannot reabsorb it. The renal threshold for glucose generally is reached when blood glucose rises above 180 mg/100 ml.

renin (REN-in): an enzyme secreted by the kidneys in response to a reduced blood flow that triggers the release of the hormone aldosterone from the adrenal glands. Aldosterone, in turn, signals the kidneys to retain sodium and fluid.

requirement: the lowest continuing intake of a nutrient that will maintain a specified criterion of adequacy.

residue: the total amount of material in the colon; it includes dietary fiber and undigested food, intestinal secretions, bacterial cell bodies, and cells shed from the intestinal mucosa.

respiratory acidosis: a condition of too much acid in the blood caused by failure of the lungs to expel carbon dioxide properly. Excess carbon dioxide is normally released from the lungs during exhalation; failing lungs, however, are unable to perform this function rapidly enough.

retina (RET-in-uh): the layer of light-sensitive nerve cells lining the back of the inside of the eye; consists of rods and cones.

retinol-binding protein (RBP): the specific protein responsible for transporting retinol.

rheumatoid arthritis: a disease of the immune system involving painful inflammation of the joints and related structures.

rickets: the vitamin D–deficiency disease in children.

rooting reflex: a reflex that causes an infant to turn toward whichever cheek is touched, in search of a nipple.

royal jelly: a substance produced by worker bees and fed to the queen bees; often falsely promoted as enhancing athletic performance.

saccharin: a 0-kcalorie sweetener used in the United States, but available in Canada only in pharmacies and only as a sweetener, not as an additive.

saliva: the secretion of the salivary glands; the principal enzyme is salivary amylase.

salivary glands: exocrine glands that secrete saliva into the mouth.

salts: compounds composed of charged particles (ions). An example is potassium chloride (K^+Cl^-).

saturated fatty acid: a fatty acid carrying the maximum possible number of hydrogen atoms (having no points of unsaturation). A saturated fat is a triglyceride that contains three saturated fatty acids.

schizophrenia (SKITS-oh-FREN-ee-ah): mental illness characterized by an altered concept of reality and, in some cases, delusions and hallucinations.

science of nutrition: the study of nutrients in foods and of their ingestion, digestion, absorption, transport, metabolism, interaction, storage, and excretion. A broader definition includes the study of the environment and of human behavior as it relates to these processes.

scurvy: the vitamin C–deficiency disease.

sedentary: physically inactive (literally, "sitting down a lot").

segmentation: a periodic squeezing or partitioning of the intestine by its circular muscles that both mixes and slowly pushes the contents along.

selenium (se-LEEN-ee-um): a trace element.

semivegetarians: people who include some, but not all, groups of animal-derived foods in their diets; they usually exclude meat and may occasionally include poultry, fish, and shellfish; also called *partial vegetarians.*

senile dementia of the Alzheimer's type (SDAT): a degenerative disease of the brain involving memory loss and major structural changes in neuron networks.

senile dementia: the loss of brain function beyond the normal loss of physical adeptness and memory that occurs with aging.

sepsis: the presence of infectious microorganisms in the bloodstream.

septicemia (sep-tih-SEE-me-ah): the spread of an infection from a local area into the blood.

set-point theory: the theory that proposes that the body tends to maintain a certain weight by means of its own internal controls.

shock: a sudden drop in blood volume that disrupts the supply of oxygen to the tissues and organs and the return of blood to the heart.

short-bowel or **short-gut syndrome:** severe malabsorption that may occur when the absorptive surface of the small bowel is reduced, resulting in diarrhea, weight loss, bone disease, hypocalcemia, hypomagnesemia, and anemia.

simple carbohydrates (sugars): the monosaccharides (glucose, fructose, and galactose) and the disaccharides (sucrose, lactose, and maltose).

small intestine: a 10-foot length of small-diameter (1-inch) intestine that is the major site of digestion of food and absorption of nutrients.

soaps: chemical compounds formed between a basic mineral (such as calcium) and unabsorbed fatty acids. Soaps give steatorrhea its foamy appearance.

socioeconomic history: a record of a person's social and economic background, including such factors as age, education, income, and ethnic identity.

sodium bicarbonate: baking soda; an alkaline salt believed to neutralize blood lactic acid and thereby reduce pain and enhance possible workload. Some studies show that sodium bicarbonate in recommended doses can enhance performance of high-intensity exercise (in 1- to 5-minute exercise sessions), but its effects on endurance exercise are unknown; "soda loading" may cause intestinal bloating and diarrhea.

soluble fibers: indigestible food components that readily dissolve in water and often impart gummy or gel-like characteristics to foods. An example is pectin from fruit, which is used to thicken jellies.

sphincter (SFINK-ter): a circular muscle surrounding, and able to close, a body opening.

standard formula: a liquid diet that contains complete molecules of proteins; also called **intact** or **polymeric formula.**

standard or **regular diet:** a diet that includes all foods and meets the nutrient needs of a healthy person.

starch: a plant polysaccharide composed of glucose and digestible by human beings.

steatorrhea (stee-ah-toe-REE-ah): fatty diarrhea characterized by loose, foamy, foul smelling stools.

sterile: free of microorganisms, such as bacteria.

steroid (STARE-oid): a medication used to reduce tissue inflammation, to suppress the immune response, or to replace certain steroid hormones in people who cannot synthesize them.

sterols: one of the three main classes of lipids; sterols include cholesterol, vitamin D, and the sex hormones (such as testosterone).

stoma (STOH-ma): a surgically formed opening. After a colostomy or ileostomy, a stoma is formed by bringing the cut end of the intestine through the abdominal wall, rerouting the excretion of wastes.

stress response: an elaborate series of cardiovascular and metabolic events orchestrated by the body in response to *stressors* such as infection, surgery, burns, fractures, and deep, penetrating wounds (gunshot wounds, fistulas, or surgical incisions). These stressors may lead to tissue injury or necrosis (death), inflammation, and shock.

stroke: an event in which the blood flow to a part of the brain is cut off; also called a **cerebral vascular accident (CVA).**

sucralose: a 0-kcalorie sweetener that is 600 times sweeter than sucrose; FDA approval pending in the United States; approved in Canada.

sucrose: a disaccharide composed of glucose and fructose; commonly known as table sugar, beet sugar, or cane sugar.

sugar alcohols: sugarlike compounds; like sugars, they are sweet to taste but yield 2 to 3 kcal per gram, slightly less than sucrose. Examples are maltitol, mannitol, sorbitol, and xylitol.

sulfites: salts containing sulfur that are added to fresh and frozen fruits and vegetables to prevent changes in color and texture due to oxidation.

sustainable: a term used to describe the use of resources at such a rate that the earth can keep on replacing them—for example, a rate of cutting trees no faster than new ones grow.

syndrome X: the combination of insulin resistance, hyperinsulinemia, obesity, hypertension, elevated LDL and triglycerides, and lowered HDL that is frequently associated with diabetes and cardiovascular disease.

systemic inflammatory response syndrome (SIRS): the complex of symptoms that occur as a result of immune and inflammatory factors in response to tissue damage. In severe cases SIRS may progress to multiple organ failure.

terminal illness: a progressive, irreversible disease that will lead to death in the near future.

thermic effect of food: an estimation of the energy required to process food (digest, absorb, transport, metabolize, and store ingested nutrients).

"This Product Contains Olestra. Olestra may cause abdominal cramping and loose stools. Olestra inhibits the absorption of some vitamins and other nutrients. Vitamins A, D, E, and K have been added."

thrombosis: the formation of a **thrombus,** a blood clot that may obstruct a blood vessel, causing gradual tissue death. A **coronary thrombosis** blocks blood flow through an artery that feeds the heart. A **cerebral thrombosis** blocks blood flow through an artery that feeds the brain.

thrush: a fungal infection of the mouth and esophagus caused by *Candida albicans;* the technical term for this infection is **candidiasis.** Thrush is characterized by a thick white coating of the tongue that alters taste sensations and causes pain on chewing and swallowing.

tissue rejection: destruction of healthy donor cells by the recipient's immune system, which recognizes the donor cells as foreign; also called **graft-versus-host disease (GVHD).**

toxicity: the ability of a substance to harm living organisms. All substances are toxic if used in high enough concentrations.

toxins: poisons. Toxins produced by bacteria come in two varieties: *enterotoxins,* which act in the GI tract, and *neurotoxins,* which act on the nervous system.

trachea (TRAKE-ee-uh): the windpipe; the passageway from the mouth and nose to the lungs.

training: practicing an activity, which leads to conditioning. Training is what you do; conditioning is what you get.

transferrin (trans-FERR-in): the body's iron-carrying protein.

transient ischemic attack (TIA): a temporary reduction in blood flow to the brain that causes temporary symptoms that depend on the part of the brain that is affected. Some common symptoms include light-headedness, visual disturbances, paralysis, staggering, numbness, or dysphagia.

transnasal: through the nose. A **transnasal feeding tube** is one that is inserted through the nose.

traveler's diarrhea: nausea, vomiting, and diarrhea caused by consuming food or water contaminated by any of several organisms, most commonly, *E. coli, Shigella, Campylobacter jejuni,* and *Salmonella.*

triglycerides: the chief form of fat in the diet and the major storage form of fat in the body; composed of glycerol with three fatty acids attached.

tripeptide: three amino acids bonded together.

tubule: a tubelike structure that surrounds the glomerulus and descends through the nephron. A pressure gradient between the glomerular capillaries and the tubule returns needed materials to the blood and moves wastes into the tubule to be excreted in the urine.

tumor: a new growth of tissue forming an abnormal mass with no function; also called a **neoplasm** (NEE-oh-plazm). Tumors that multiply out of control, threaten health, and require treatment are **malignant** (ma-LIG-nant). Tumors that stop growing without intervention or can be removed surgically and pose no threat to health are **benign** (bee-NINE).

turbinado (ter-bih-NOD-oh) **sugar:** raw (brown) sugar from which the filth has been washed; legal to sell in the United States.

24-hour recall: a record of foods eaten by a person during one 24-hour period.

type 1 diabetes mellitus: the less common type of diabetes in which the person produces no insulin at all.

type 2 diabetes mellitus: the more common type of diabetes that develops gradually and is associated with insulin resistance.

ulcer: an open sore or lesion. A **peptic ulcer** is an erosion of the top layer of cells from the mucosa of the stomach **(gastric ulcer)** or duodenum **(duodenal ulcer).** Ulcers may also develop in the mouth, esophagus, and other parts of the intestine, or on the skin.

ulcerative colitis (ko-LYE-tis): inflammation and ulceration of the colon.

ultrahigh temperature (UHT) treatment: sterilizing a food by short-time exposure to temperatures above those normally used in processing.

umbilical (um-BIL-ih-cul) **cord:** the ropelike structure through which the fetus's veins and arteries reach the placenta; the route of nourishment and oxygen into the fetus and the route of waste disposal from the fetus.

undernutrition: underconsumption of food energy or nutrients severe enough to cause disease or increased susceptibility to disease; a form of malnutrition.

unsaturated fatty acid: a fatty acid with one or more points of unsaturation where hydrogens are missing (includes monounsaturated and polyunsaturated fatty acids).

unspecified eating disorders: eating disorders that do not meet the criteria for specific eating disorders previously defined.

uremia (you-REE-me-ah): abnormal accumulation of nitrogen-containing substances, especially urea, in the blood; also called **azotemia** (AZE-oh-TEE-me-ah).

uremic (you-REE-mic) **syndrome:** the many symptoms that accompany the buildup of toxic waste products in the blood.

uremic frost: the appearance of urea crystals on the skin.

uterus (YOO-ter-us): the womb, the muscular organ within which the infant develops before birth.

vacuum packaging (MAP): preservation of a perishable food by packaging it in a gas-impermeable container from which air has been removed.

vagotomy (vay-GOT-o-mee): surgery that severs the nerves that stimulate gastric acid secretion.

varices (VAIR-ih-seez): blood vessels that have become distended and twisted.

variety (dietary): using different foods to obtain the same nutrients on different occasions.

vegan diets: diets that exclude all animal products and include only plant foods; also known as **strict vegetarian diets.**

vegans: people who exclude all animal-derived foods (including meat, poultry, fish, shellfish, eggs, cheese, and milk) from their diets; also called *strict vegetarians* or *total vegetarians.*

vegetarian diets: a general term used to describe diets that exclude meat, poultry, fish, or other animal-derived foods. The two main categories are *lacto-ovo vegetarian* and *vegan* diets.

vein: a vessel that carries blood back to the heart.

villi (VILL-ee or VILL-eye): fingerlike projections from the folds of the small intestine. The singular form is **villus.**

vitamin A: a fat-soluble vitamin. Its three chemical forms are *retinol* (the alcohol form), *retinal* (the aldehyde form, which is active in the pigments of the eye), and *retinoic acid* (the acid form).

vitamins: essential, noncaloric, organic nutrients needed in tiny amounts in the diet.

VLDL (very-low-density lipoprotein): the type of lipoprotein made primarily by liver cells to transport lipids to various tissues in the body; composed primarily of triglycerides.

voluntary activities: the component of a person's daily energy expenditure that involves conscious and deliberate muscular work—walking, lifting, climbing, and other physical activities. Voluntary activities normally require less energy in a day than basal metabolism does.

waist-to-hip ratio: a valuable, commonly used indicator of fat distribution; waist-to-hip ratio = waist circumference ÷ hip circumference. A ratio of 0.8 or greater for a woman or 0.95 or greater for a man suggests a risk to health.

warm-up: five to ten minutes of light activity, such as easy jogging or cycling, to warm up the body in preparation for vigorous activity.

water balance: the balance between water intake and water excretion, which keeps the body's water content constant.

water intoxication: the rare condition in which body water contents are too high. The symptoms may include confusion, convulsion, coma, and even death in extreme cases.

weight cycling: repeated cycles of weight loss and subsequent regain that affect body composition and metabolism. With intermittent dieting, a person rebounds to a higher weight (and a higher body fat content) after each round. The weight-cycling pattern is popularly called the *ratchet effect* or *yo-yo effect* of dieting.

weight training (also called **resistance training**): the use of free weights or weight machines to provide resistance for developing muscle strength and endurance. A person's own body weight may also be used to provide resistance as when a person does push-ups, pull-ups, or abdominal crunches.

white sugar: pure sucrose, produced by dissolving, concentrating, and recrystallizing raw sugar.

xerophthalmia (zer-off-THAL-mee-uh): progressive blindness caused by vitamin A deficiency.

Zollinger-Ellison syndrome: marked hypersecretion of gastric acid and consequent peptic ulcers caused by a tumor of the pancreas, which releases gastrin.

Index

Numbers in bold face (such as **123**) indicate pages on which definitions appear. Numbers in italics *(123)* indicate figures; numbers followed by the letter *t* (123*t*) refer to tables. Numbers followed by the letter *n* (123*n*) refer to footnotes on the page. Letters and numbers (such as D-1 to D-3) refer to appendix pages.

Photo and Art Credits

1 Jennie Oppenheimer/Studio Zocolo; 2 © Michael Newman/PhotoEdit; 3 © Myrleen Ferguson Cate/PhotoEdit; 12, 13, 14 (all), 15 (both) © Polara Studios Inc.; 20 Courtesy of the FDA; 35 Jennie Oppenheimer/Studio Zocolo; 36 Ray Stanyard; 37, 38 © Thomas Harm and Tom Peterson/Quest Photographic Inc.; 40 © Ray Stanyard; 43 © Felicia Martinez/PhotoEdit; 45 © Tony Freeman/PhotoEdit; 47 (all) © Polara Studios Inc.; 50 (all) © Thomas Harm and Tom Peterson/Quest Photographic Inc.; 57 Jennie Oppenheimer/Studio Zocolo; 64, 68, 70 (all), 71 © Polara Studios Inc.; 79 Jennie Oppenheimer/Studio Zocolo; 88 (both) © Photo Library, UN Food and Agriculture Organization, Rome; 97 (both) © Polara Studios Inc.; 99 Jennie Oppenheimer/Studio Zocolo; 110 From D. W. Fawcett, The Cell, 2nd ed. (Philadelphia: Saunders, 1981), color by Kidd & Company; 119, 137 Jennie Oppenheimer/Studio Zocolo; 143 © Thomas Harm and Tom Peterson/Quest Photographic Inc.; 144 © Chip Henderson/Tony Stone Images; 147 (top) © Thomas Harm and Tom Peterson/Quest Photographic Inc.; 147 (bottom) © Polara Studios Inc.; 152 (both), 153, 154 © Polara Studios Inc.; 156, 160 © Thomas Harm and Tom Peterson/Quest Photographic Inc.; 173 Jennie Oppenheimer/Studio Zocolo; 174 © Lawrence Migdale/Stock, Boston; 179 © Polara Studios Inc.; 180 Courtesy of Gjon Mili; 182, 184 © Thomas Harm and Tom Peterson/Quest Photographic Inc.; 187 (both) © Martin Rotker; 188 © Michael Newman/PhotoEdit; 189, 191 © Thomas Harm and Tom Peterson/Quest Photographic Inc.; 205 Jennie Oppenheimer/Studio Zocolo; 210 Courtesy Amgen, Inc., photo © John Sholtis/The Rockefeller University; 211 © Frank Whitney/The Image Bank; 212 © Lori Adamski Peek/Tony Stone Images; 213 © Rick Schaff; 217 © Nathan Benn/Stock, Boston; 219 © David Lissy/The Picture Cube; 230 (left) © Tony Freeman/PhotoEdit; 230 (right) © Michael Newman/PhotoEdit; 235 Jennie Oppenheimer/Studio Zocolo; 236 © Jurgen Reisch/Tony Stone Images; 244, 247 © 1998 Photo Disc, Inc.; 250 © Michael Medford/Image Bank; 251 © 1998 Photo Disc, Inc.; 252 © Amy G. Etra/PhotoEdit; 253 © Greg Vaughn/Pacific Stock; 255 (all) © Polara Studios Inc.; 267 Jennie Oppenheimer/Studio Zocolo; 268 © Charles Gupton/Tony Stone Images; 270 (top) © Anne Dowie; 270 (bottom) © Thomas Harm and Tom Peterson/Quest Photographic Inc.; 271 © Michael Grecco/Stock, Boston; 273 (all) © Thomas Harm and Tom Peterson/Quest Photographic Inc.; 276 © Amanda Merullo/Stock, Boston; 283 (top) © Michael Newman/PhotoEdit; 283 (bottom) © Robert Brenner/PhotoEdit; 284 © Robert Finken/The Picture Cube; 287 © Thomas Harm and Tom Peterson/Quest Photographic Inc.; 288 © Susan Van Etten/PhotoEdit; 289 Smithsonian photo by Antonio Montaner; 290 © Tony Freeman/PhotoEdit; 294 (both) © David Young-Wolff/PhotoEdit; 296 © Elizabeth Zuckerman/PhotoEdit; 299 Jennie Oppenheimer/Studio Zocolo; 305 © David J. Sams/Stock, Boston; 310 © Streissguth, A. P., Clarren, S. K., & Jones, K. L. (1985, July). Natural History of the Fetal Alcohol Syndrome: A ten-year followup of eleven patients. Laurey, Il, 89-92; 311 © Myrleen Ferguson Cate/PhotoEdit; 313 © 1998 Photo Disc, Inc.; 315 © J. Gerard Smith/Photo Researchers, Inc.; 316 © K. L. Boyd, DDS/Custom Medical Stock Photos; 320 © Felicia Martinez/PhotoEdit; 329 Jennie Oppenheimer/Studio Zocolo; 330 (both) © Anthony Vannelli; 335, 338, 344 © Tony Freeman/PhotoEdit; 357 Jennie Oppenheimer/Studio Zocolo; 360, 364 © Bill Bachmann/PhotoEdit; 370 © David Young-Wolff/PhotoEdit; 371 © Elena Rooraid/PhotoEdit; 379 Jennie Oppenheimer/Studio Zocolo; 387 © 1998 Photo Disc, Inc.; 394 © Steve Goldberg/Monkmeyer Press; 398 © 1998 Photo Disc, Inc.; 401 Jennie Oppenheimer/Studio Zocolo; 402 Courtesy Seca Corporation; 403 (top) © Tony Freeman/PhotoEdit; 403 (bottom) © G. DeGrazia/Custom Medical Stock; 404 © Tom McCarthy/PhotoEdit; 408 © David Young Wolff/PhotoEdit; 410 © W & D McIntyre/Photo Researchers, Inc.; 413 © Kate S. Kinkade/Medichrome/The Stock Shop; 418 © Dan Reynolds/Medical Images Inc.; 421 Jennie Oppenheimer/Studio Zocolo; 426 © Michael Newman/PhotoEdit; 427 © David Young-Wolff/PhotoEdit; 432 (left) © Laurent Sazy/The Gamma Liaison Network; 432 (right) © Bob Daemmrich/Tony Stone Images; 437 Jennie Oppenheimer/Studio Zocolo; 445 Courtesy Diamond Crystal Specialty Foods, Inc.; 454 © Charles Gupton/Stock, Boston; 461 Jennie Oppenheimer/Studio Zocolo; 478 © L. O'Shaughnessy/Medichrome/The Stock Shop; 481 Jennie Oppenheimer/Studio Zocolo; 494 © Simon Fraser/RVI, Newcastle-Upon-Tyne/Science Photo Library/Photo Researchers, Inc.; 495 (both) Courtesy of the Crohn's & Colitis Foundation of America, Inc.; 506 © David Young-Wolff/PhotoEdit; 507 Jennie Oppenheimer/Studio Zocolo; 509 © Charles Gupton/Stock, Boston; 512 (top) © Michael Newman/PhotoEdit; 512 (bottom) Flexiflo® Over-The-Guidewire Nasojeunal Feeding Tube, Courtesy of Ross Products Division, Abbott Laboratories, Columbus, Ohio; 528 © Charles Gupton/Stock, Boston; 531 Jennie Oppenheimer/Studio Zocolo; 533 © David Young-Wolff/PhotoEdit; 541 Courtesy Zevex, Inc.; 547 © Grace Moore/Medichrome/The Stock Shop; 551 Jennie Oppenheimer/Studio Zocolo; 552 © Bruce Ayres/Tony Stone Images; 555 © David York/Medichrome/The Stock Shop; 569 Jennie Oppenheimer/Studio Zocolo; 572 © Michael Newman/PhotoEdit; 574 © Tom Tracey/Medichrome/The Stock Shop; 578 © Lauren Goodsmith/The Image Works; 580 © Gary Conner/PhotoEdit; 590 © Darrell Gulin/Tony Stone Images; 593 Jennie Oppenheimer/Studio Zocolo; 599 © Myrleen Ferguson/PhotoEdit; 602 (all), 603 (all) © Polara Studios Inc.; 604 © 1998 Photo Disc, Inc.; 609 (top) © John Griffin/Medichrome/The Stock Shop; 609 (bottom) © Will & Deni McIntyre/Photo Researchers, Inc.; 610 (both), 612 © Tony Freeman/PhotoEdit; 623 © Anne Dowie; 627 Jennie Oppenheimer/Studio Zocolo; 629 (both) Reproduced by permission of ICI Pharmaceuticals Division, Cheshire, England; 630 © John Greim/The Stock Shop; 632 © Melanie Brown/PhotoEdit; 636 © Dennis McDonald/PhotoEdit; 645 © Polara Studios Inc.; 647 Jennie Oppenheimer/Studio Zocolo; 658 Anne Dowie; 667 © Hank Morgan/Rainbow; 669 Jennie Oppenheimer/Studio Zocolo; 671 © Niccolini/Medical Images, Inc.; 677 © Polara Studios Inc.; 683 © David Young-Wolff/PhotoEdit.

Daily Values for Food Labels

The Daily Values are standard values developed by the Food and Drug Administration (FDA) for use on food labels. Daily Values for protein, vitamins, and minerals reflect average allowances based on the RDA. Daily Values for nutrients and food components, such as fat and fiber, that do not have an established RDA but do have important relationships with health are based on recommended calculation factors as noted (Chapter 2 provides more details).

Nutrient	Amount
Protein[a]	50 g
Thiamin	1.5 mg
Riboflavin	1.7 mg
Niacin	20 mg NE
Biotin	300 μg
Pantothenic Acid	10 mg
Vitamin B_6	2 mg
Folate	400 μg
Vitamin B_{12}	6 μg
Vitamin C	60 mg
Vitamin A[b]	5000 IU
Vitamin D[b]	400 IU
Vitamin E[b]	30 IU
Vitamin K	80 μg
Calcium	1000 mg
Iron	18 mg
Zinc	15 mg
Iodine	150 μg
Copper	2 mg
Chromium	120 μg
Selenium	70 μg
Molybdenum	75 μg
Manganese	2 mg
Chloride	3400 mg
Magnesium	400 mg
Phosphorus	1000 mg

[a]The Daily Values for protein vary for different groups of people: pregnant women, 60 g; nursing mothers, 65 g; infants under 1 year, 14 g; children 1 to 4 years, 16 g.

[b]The Daily Values for fat-soluble vitamins are expressed in International Units (IU), an old system of measurement. The current RDA and tables of food composition use a more accurate system of measurement. Equivalent values are as follows: for vitamin A, 1500 μg RE; for vitamin D, 10 μg; for vitamin E, 9 mg α-TE.

Food Component	Amount	Calculation Factors
Fat	65 g	30% of kcalories
Saturated fat	20 g	10% of kcalories
Cholesterol	300 mg	Same regardless of kcalories
Carbohydrate (total)	300 g	60% of kcalories
Fiber	25 g	11.5 g per 1000 kcalories
Protein	50 g	10% of kcalories
Sodium	2400 mg	Same regardless of kcalories
Potassium	3500 mg	Same regardless of kcalories

NOTE: Daily Values were established for adults and children over 4 years old. The values for energy-yielding nutrients are based on 2000 kcalories a day.

Glossary of Nutrient Measures

kcal: kcalories; a unit by which energy is measured (see Chapter 1).

g: grams; a unit of weight equivalent to about 0.03 ounces.

mg: milligrams; one-thousandth of a gram.

μg: micrograms; one-millionth of a gram.

mg NE: milligrams niacin equivalents; a measure of niacin activity (see Chapter 10).

- 1 mg NE = 1 mg niacin or 60 mg tryptophan

mg α-TE: milligrams alpha-tocopherol equivalents; a measure of vitamin E activity (see Chapter 11).

- 1 mg α-TE = 1 mg d-α tocopherol

μg RE: micrograms retinol equivalents; a measure of vitamin A activity (see Chapter 11).

- 1 μg RE = 1 μg retinol or 6 μg beta-carotene

IU: international units; an old measure of vitamin activity determined by biological methods (as opposed to new measures that are determined by direct chemical analysis).